HUMAN EVOLUTIONARY GENETICS

second edition

HUMAN EVOLUTIONARY GENETICS

second edition

Mark Jobling

Edward Hollox

Matthew Hurles

Toomas Kivisild

Chris Tyler-Smith

Garland Science
Taylor & Francis Group
NEW YORK AND LONDON

Garland Science
Vice President: Denise Schanck
Senior Editor: Elizabeth Owen
Assistant Editor: Dave Borrowdale
Production Editor: Georgina Lucas
Illustrator: Matthew McClements, Blink Studio Ltd.
Cover Design: Matthew McClements, Blink Studio Ltd.
Copyeditor: Sally Huish
Typesetting: EJ Publishing Services
Proofreader: Jo Clayton
Indexer: Medical Indexing Ltd.

ISBN: 978-0-8153-4148-2

Library of Congress Cataloging-in-Publication Data

Human evolutionary genetics / Mark Jobling ... [et al.]. -- 2nd ed.
p. ; cm.
Rev. ed. of: Human evolutionary genetics / Mark A. Jobling, Matthew Hurles, Chris Tyler-Smith. c2004.
Includes bibliographical references and index.
ISBN 978-0-8153-4148-2
I. Jobling, Mark A. II. Jobling, Mark A. b Human evolutionary genetics.
[DNLM: 1. Evolution, Molecular. 2. Adaptation, Biological. 3. Anthropology, Physical. 4. Genetic Variation. 5. Genome, Human. QU 475]
QH431
599.93'5--dc23
2013008586

Published by Garland Science, Taylor & Francis Group, LLC, an informa business, 711 Third Avenue, New York, NY 10017, USA, and 3 Park Square, Milton Park, Abingdon, OX14 4RN, UK.

Printed in the United States of America

15 14 13 12 11 10 9 8 7 6 5 4 3 2 1

Visit our web site at http://www.garlandscience.com

Front Cover
Antony Gormley
FEELING MATERIAL XIV, 2005
4 mm square section mild steel bar
225 x 218 x 170 cm (unextended size). 70 Kg
Photograph by Stephen White, London
© the artist

The authors
Professor Mark A. Jobling
Department of Genetics
University of Leicester, UK

Dr. Ed Hollox
Department of Genetics
University of Leicester, UK

Dr. Matthew Hurles
Wellcome Trust Sanger Institute
Hinxton, UK

Dr.Toomas Kivisild
Division of Biological Anthropology
University of Cambridge, UK

Dr. Chris Tyler-Smith
Wellcome Trust Sanger Institute
Hinxton, UK

PREFACE

This book is a completely revised edition of *Human Evolutionary Genetics*, first published in 2004. We decided to write the first edition because there were no textbooks available covering the areas that interested us. Once we had embarked upon the Herculean task of producing it, we realized why nobody had attempted to summarize this forbiddingly broad and contentious field before. But luckily the reception was positive, one eager person (our ideal reader and not, we point out, one of the authors in disguise) writing on Amazon that "I bought a copy for myself, and another one for my advisor. I have read it twice in a week!" A revised version seemed like a pretty good idea.

We cheerfully imagined that the second edition would be easier to write than the first. How wrong we were. First, all three original authors (MJ, MH, and CTS) had accumulated additional responsibilities that reduced the available time for writing. Second, the field obstinately continued to grow, and scarcely a week went by without some interesting and important development—the genomes of new species, genomewide surveys of human variation, next-generation sequencing and its data tsunami, spectacular ancient DNA discoveries, large-scale population studies, novel statistical methods, archaeological and paleontological revelations—the list goes on. We sometimes wished everyone would just stop working for a bit, so we could catch up. So, our deadline for the second edition passed, was revised, and passed again. HEG1 was becoming more and more out of date. We needed help.

The cavalry duly arrived in the form of two sterling new recruits to the authorial team—EH and TK. They brought their own areas of interest and expertise, but also a more efficient and energetic approach to the writing process, which revitalized the whole project. So, after a lengthy and difficult gestation, here is HEG2.

Following an initial introductory chapter, the book is divided into five sections, allowing it to be read by interested students and researchers from a broad range of backgrounds. "How do we study genome diversity?" (Chapters 2–4) and "How do we interpret genetic variation?" (Chapters 5–6) together provide the necessary tools to understand the rest of the book. The first of these sections surveys the structure of the genome, different sources of genomic variation, and the methods for assaying diversity experimentally. The second introduces the evolutionary concepts and analytical tools that are used to interpret this diversity. The subsequent two sections take an approximately chronological course through the aspects of our current state of knowledge about human origins that we consider most important. The section "Where and when did humans originate?" (Chapters 7–9) first considers our links to our closest living nonhuman relatives, the other great apes, then investigates the genetic changes that have made us human, and finally details the more recent African origin of our own species. "How did humans colonize the world?" (Chapters 10–14) describes how human genetic diversity is currently distributed globally and then discusses the evidence for early human movements out of Africa, and the subsequent processes of expansion, migration, and mixing that have shaped patterns of diversity in our genomes. Finally, "How is an evolutionary perspective useful?" (Chapters 15–18) demonstrates the wider applications of an evolutionary approach for our understanding of phenotypic variation, the genetics of diseases both simple and complex, and the identification of individuals. Extensive cross-referencing between these sections facilitates different routes through the book for readers with divergent interests and varying amounts of background knowledge.

An important feature is the use of "Opinion Boxes"—short contributions by guest authors who are experts in different aspects of this diverse subject area. These help to give a flavor of scientific enquiry as an ongoing process, rather than a linear accumulation of facts, and encourage the reader to regard the published literature with a more critical eye. Opinions about how data should be interpreted change, and often an objective way to choose between different interpretations is not obvious. This is particularly true of genetic data on human diversity. Many of the debates represented in the Opinion Boxes scattered through this book derive from methodological differences.

Additional resources have been incorporated to permit interested readers to explore topics in greater depth. Each chapter is followed by a detailed bibliography, within which the sources that should be turned to first for more detail are highlighted in purple text. Electronic references to internet sites are given throughout the book, both for additional information and for useful software and databases. We explain specialist terms where they are first used, and include an extensive glossary at the back of the book that defines all terms in the text that are in bold type. At the end of each chapter is a list of questions (some short-answer, and some prose) that allow the reader to test their knowledge as they proceed. Teachers may be interested to know that most of the figures are freely available from the Garland Science Website (www.garlandscience.com) for use in teaching materials.

An obvious difference from the first edition is the presence of two extra chapters, reflecting developments in understanding the human genome in the context of other hominid genomes, and in complex disease. A very welcome development is the availability of full-color printing, which makes complex figures much easier to understand.

ACKNOWLEDGMENTS

We have many people to thank for contributions to this book. Twenty-five researchers found time to write Opinion Boxes; we are very grateful to: Mark Achtman, Elizabeth Cirulli & David Goldstein, Graham Coop & Molly Przeworski, Dorian Fuller, Tom Gilbert, Boon-Peng Hoh & Maude Phipps, Doron Lancet & Tsviya Olender, Daniel MacArthur, Andrea Manica & Anders Eriksson, Linda Marchant & Bill McGrew, John Novembre, Bert Roberts, Aylwyn Scally & Richard Durbin, Sarah Tishkoff, George van Driem, Bernard Wood, and Richard Wrangham & Rachel Carmody. We must stress that any opinions outside these Boxes are our own, and not necessarily endorsed by the Opinion Box contributors.

We are very grateful to the external specialists who helped improve the manuscript by providing data and advice, commenting on figures and questions, and reviewing draft chapters. Of course, despite the efforts of reviewers, errors and omissions will remain, and we take full responsibility for these. Thanks to: John Armour (University of Nottingham), Chiara Batini (University of Leicester), Stefano Benazzi (University of Vienna), Antonio Brehm (University of Madeira), John Brookfield (University of Nottingham), Terry Brown (University of Manchester), Anne Buchanan (Pennsylvania State University), John M. Butler (The National Institute of Standards and Technology), Lucia Carbone (Oregon Health and Science University), Susana Carvalho (University of Oxford), Vincenza Colonna (Wellcome Trust Sanger Institute), Murray Cox (Massey University), Todd R. Disotell (New York University), Michael Dunn (Max Planck Institute for Psycholinguistics), Wolfgang Enard (Max Planck Institute for Evolutionary Anthropology), Greg Gibson (Georgia Institute of Technology), Tom Gilbert (Natural History Museum of Denmark), Bernard Grandchamp (University of Paris Diderot), Ryan Gutenkunst (University of Arizona), Wolfgang Haak (University of Adelaide), Phillip Habgood (University of Queensland), Pille Hallast (University of Leicester), Terry Harrison (New York University), Simon Hay (University of Oxford), Andy I.R. Herries (La Trobe University), Sarah Hill (University of Cambridge), Hirohisa Hirai (Kyoto University), Rosalind Howes (University of Oxford), Arati Iyengar (University of Central Lancashire), Peter de Knijff (Leiden University Medical Center), Vincent Macaulay (University of Glasgow), Ripan Malhi (University of Illinois Urbana-Champaign), Nicola Man (University of New South Wales), Tomas Marques-Bonet (University Pompeu Fabra), Celia May (University of Leicester), Patrick McGrath (Georgia Tech), Bill McGrew (University of Cambridge), Pierpaolo Maisano Delser (University of Leicester), Mait Metspalu (University of Tartu), Darren Monckton (University of Glasgow), Maru Mormina (University of East Anglia), Connie Mulligan (University of Florida), David Nelson (Baylor College of Medicine), Barbara Ottolini (University of Leicester), Svante Pääbo (Max

Planck Institute for Evolutionary Anthropology), Luca Pagani (Wellcome Trust Sanger Institute), Luisa Pereira (University of Porto), Fred Piel (University of Oxford), Sohini Ramachandran (Brown University), Christine Rees (Illumina), Tim Reynolds (Birkbeck, University of London), Jorge Macedo Rocha (University of Porto), Rebecca Rogers Ackermann (University of Cape Town), Alexandra Rosa (University of Madeira), Mark Seielstad (University of California San Francisco), Giorgio Sirugo (Tor Vergata University Medical School), Roscoe Stanyon (University of Florence), Jay Stock (University of Cambridge), Peter Sudbery (University of Sheffield), Dallas Swallow (University College London), Martin Tobin (University of Leicester), Richard Villems (University of Tartu), Cynthia Vigueira (Washington University St. Louis), Bence Viola (Max Planck Institute for Evolutionary Anthropology), Tim White (University of California, Berkeley), Mike Whitlock (University of British Columbia), Jinchuan Xing (Rutgers University), Yali Xue (Wellcome Trust Sanger Institute), and Bryndis Yngvadottir (Wellcome Trust Sanger Institute).

We thank Liz Owen at Garland for her persistence, pragmatism, and practical help in bringing HEG2 to fruition. Dave Borrowdale's efficient administration of the review process of text and figures has also been greatly appreciated. We thank the artist, Matt McClements, for the professional job he has done in redrawing our figures. We also thank Georgina Lucas for helping us through the proof stage.

We are also grateful for the funding bodies and institutions that have allowed us to maintain our research interests in human evolutionary genetics, and to produce this book. In particular, MJ thanks the Wellcome Trust for funding through a Senior Fellowship, and the University of Leicester for providing a collegial working environment. EH thanks the same institution for giving him gainful employment. TK thanks colleagues from the Division of Biological Anthropology, University of Cambridge, for their support; and MH and CTS thank the Wellcome Trust both for its direct support and for its creation of the stimulating environment of the Sanger Institute.

Writing a textbook is undoubtedly interesting and educational, but also onerous. Much of this burden has fallen on our families and colleagues, as well as ourselves. We all thank our current and former group members for putting up with the distractions of book preparation, and for interesting delicacies from around the world. In addition, MJ thanks Nicky, Bill, and Isobel, EH thanks Gill and Kirsten, MH thanks Liz, Edward, Jenny, and Audrey, TK thanks Dagne and Uku, and CTS thanks Yali and Jack. Perhaps this book can now provide an explanation for our preoccupations and absences.

CONTENTS

DETAILED CONTENTS

AN INTRODUCTION TO HUMAN EVOLUTIONARY GENETICS

CHAPTER ONE

In this chapter, we introduce human evolutionary genetics and describe the recent dramatic advances in knowledge and understanding that have led to its central role in all human genetic studies. We explore the diverse sources of information about the human past that are available to us. These complementary records tell us about different aspects of the past, and are informative over different time-scales. We also identify the fundamental human evolutionary questions that can now be addressed using genetics.

1.1 WHAT IS HUMAN EVOLUTIONARY GENETICS?

Evolutionary genetics is founded on the principle that the genetic record of life is contained in the **genomes** of living species and it reveals evolutionary processes and relationships all the way back to the last universal common ancestor of all species. To find out about this ancestral organism we have to compare and contrast the most distantly related branches on the tree of life. Comparisons among much more closely related individuals, such as those from the same species, provide evidence on much more recent evolutionary processes. Our ability to read this genetic record has developed enormously in the last few years, although our confidence that information on our past exists within our heritable material is somewhat older. Genetic evidence comes from two main sources:

- The genomes of living individuals that must have been passed down from ancestors
- Ancient DNA from well-preserved organic remains, which may or may not be represented now in living descendants

Human evolutionary genetics involves the study of how different copies of the human genome differ from one another, and from those of our closest relatives, other **primates**. Differences between genomes also form the basis of anthropological, medical, and **forensic** genetics. All of these fields are experiencing massive advances as a result of two developments. First, the public availability of human and nonhuman genome sequences, annotated with important functional elements such as **genes**, and with sites of genetic variation. Second, technology allowing analysis of most of this genetic variation across the genome, initially with the development of hybridization **microarrays** (or **chips**), followed by the development of methods to sequence whole genomes rapidly and cheaply (so-called **next-generation sequencing**). This explosion of information is driving an unprecedented period of innovation that gives us the tools to analyze huge datasets of unparalleled quality and quantity. This wealth of data is itself catalyzing the development of new interpretative methods. In addition, with the publication of genome sequences from other extant primates, as well as data from the genomes of extinct members of the genus *Homo*, we can now comprehensively catalog the genetic differences between humans and our closest relatives. As we shall see in this book, these impressive developments are allowing us to answer some of the most fundamental questions regarding human origins. These are indeed exciting times for human evolutionary genetics.

In times past, when writing materials were precious commodities, a scribe would often reuse an existing manuscript rather than obtain a new parchment. The manuscript would be turned through 90°, and overwritten. These overwritten manuscripts bearing the imprint of more than one text are known as **palimpsests**. The genetic record is similarly a complex palimpsest. Variation among modern individuals is shaped by cumulative past processes, which allows us to investigate different times in human prehistory with the same data.

Different layers of the past are accessible through the analysis of genetic diversity. Moving from the most ancient to the most recent, we encounter:

- Our phylogenetic relationship to other species (Chapter 7)
- The origin of our species (Chapters 8, 9)
- Prehistorical migrations (Chapters 10, 11, 12, and 13)
- Historical migrations (Chapter 14)
- Genealogical studies (Chapter 18)
- Paternity testing (Chapter 18)
- Individual identification (Chapter 18)

Extracting information on any one past period or event requires careful interpretation to isolate it from previous and subsequent processes. This information can tell us not only about the **demographic history** and origins of populations but also something about the environmental challenges faced by those populations, through the influence of **natural selection** on genetic variation.

It is said that "the past is the source of the present" and this is true in the academic field of human evolutionary genetics as much as elsewhere. This exciting subject owes its current status to developments and debates over the last 150 years in genetics, paleontology, archaeology, anthropology, and linguistics. In this book we have avoided cataloging this history, instead taking a twenty-first-century perspective, but we discuss key developments where they are relevant, and provide a time line in **Figure 1.1**.

1.2 INSIGHTS INTO PHENOTYPES AND DISEASES

What use can an evolutionary perspective on human genetic variation have beyond the reconstruction of the past for its own sake?

A shared evolutionary history underpins our understanding of biology

The great twentieth-century evolutionary biologist Theodosius Dobzhansky wrote that "Nothing in biology makes sense except in the light of evolution." All the sizes, shapes, chemistries, and genes of organisms alive today derive from ancestors that can be traced back over billions of years. All of these features have been shaped by the environmental challenges faced by these organisms and their ancestors. If it were not the case that humans share an ancestor with every other species on the planet, there would be no value in performing any form of comparative analysis. There would be nothing that the *Escherichia coli* bacterium, brewer's yeast, fruit fly, nematode worm, zebrafish, mouse, or **chimpanzee** could tell us about ourselves. None of these species is our ancestor: they are our cousins, equally distant in time from our common ancestor (**Figure 1.2**). It is our shared evolutionary heritage with these species that makes them such powerful **model organisms**.

1786	Recognition of language families
1856	Discovery of Neanderthal type specimen
1859	Publication of Darwin's "The Origin of Species"
1866	Publication of Mendel's "Experiments in Plant Hybrids"
1871	Publication of Darwin's "The Descent of Man"
1900	Discovery of first genetic polymorphism—ABO blood group (Landsteiner)
1908	Hardy–Weinberg principle formulated
1918	Fisher reconciles Darwin's natural selection and Mendel's mechanism of inheritance
1925	*Australopithecus* fossil described from South Africa
1930–32	Fisher, Haldane & Wright publish the foundations of modern population genetics
1944	DNA shown to be heritable material
1949	Radiocarbon dating introduced
1953	Double-helical structure of DNA described
1956	Human chromosome number described
1957	Hemoglobin amino acid sequences determined
1959	Y chromosome shown to be sex-determining
1966	Genetic code deciphered
1968	Neutral theory of molecular evolution (Kimura)
1969	Internet first successfully tested
1977	Publication of DNA sequencing methods
1978	First human restriction fragment length polymorphisms (RFLPs) described
1978	First human *in vitro* fertilization
1980	First genome (φX174 bacteriophage) sequenced
1981	Human mitochondrial DNA (mtDNA) genome sequenced
1984	DNA fingerprinting (minisatellites) discovered
1984	DNA-DNA hybridization shows human–chimpanzee common ancestry
1985	Invention of polymerase chain reaction (PCR)
1985	First human ancient DNA results published
1985	First Y-chromosomal polymorphism described
1987	Development of laser-induced fluorescent detection of DNA
1987	African origin of human mtDNA identified
1988	Launch of Human Genome Project
1989	Development of capillary electrophoresis for sequencing
1990	First human microsatellites described
1991	Human Genome Diversity Project proposed
1994	Publication of "The History and Geography of Human Genes" (Cavalli-Sforza et al.)
1996	First mammal cloned from adult cell (Dolly)
1997	First Neanderthal mtDNA sequence
1999	First human chromosome sequenced (Chr 22)
2001	Release of draft human genome sequence
2002	Release of draft mouse and *Plasmodium* genome sequences
2002	Human Genome Diversity Project (HGDP) Cell Line Panel released
2004	First maps of copy-number variation published
2005	First-generation human Haplotype Map (HapMap) published
2005	Release of draft chimpanzee genome sequence
2005	First development of next-generation sequencing methods
2006	1 Mb of Neanderthal genomic sequence published
2007	First large-scale genomewide association studies
2007	First personal human genome resequenced (Venter)
2007	Second-generation human Haplotype Map (HapMap) published
2009	Exome capture and sequencing methods published
2010	Denisovan mtDNA and genome sequences published
2010	1000 Genomes Project pilot study published
2012	All great ape genomes now sequenced

Figure 1.1: Time line of important developments in the field of human evolutionary genetics.

To take just one example, sequencing the mouse genome allows us to identify more genes in the human genome than does sequencing the human genome alone. By identifying segments of DNA that are more similar between the two species than could be expected by chance, we can identify regions whose evolution has been constrained by the need to perform a specific function. Some of these regions are genes. In other words, we can identify a gene not because it

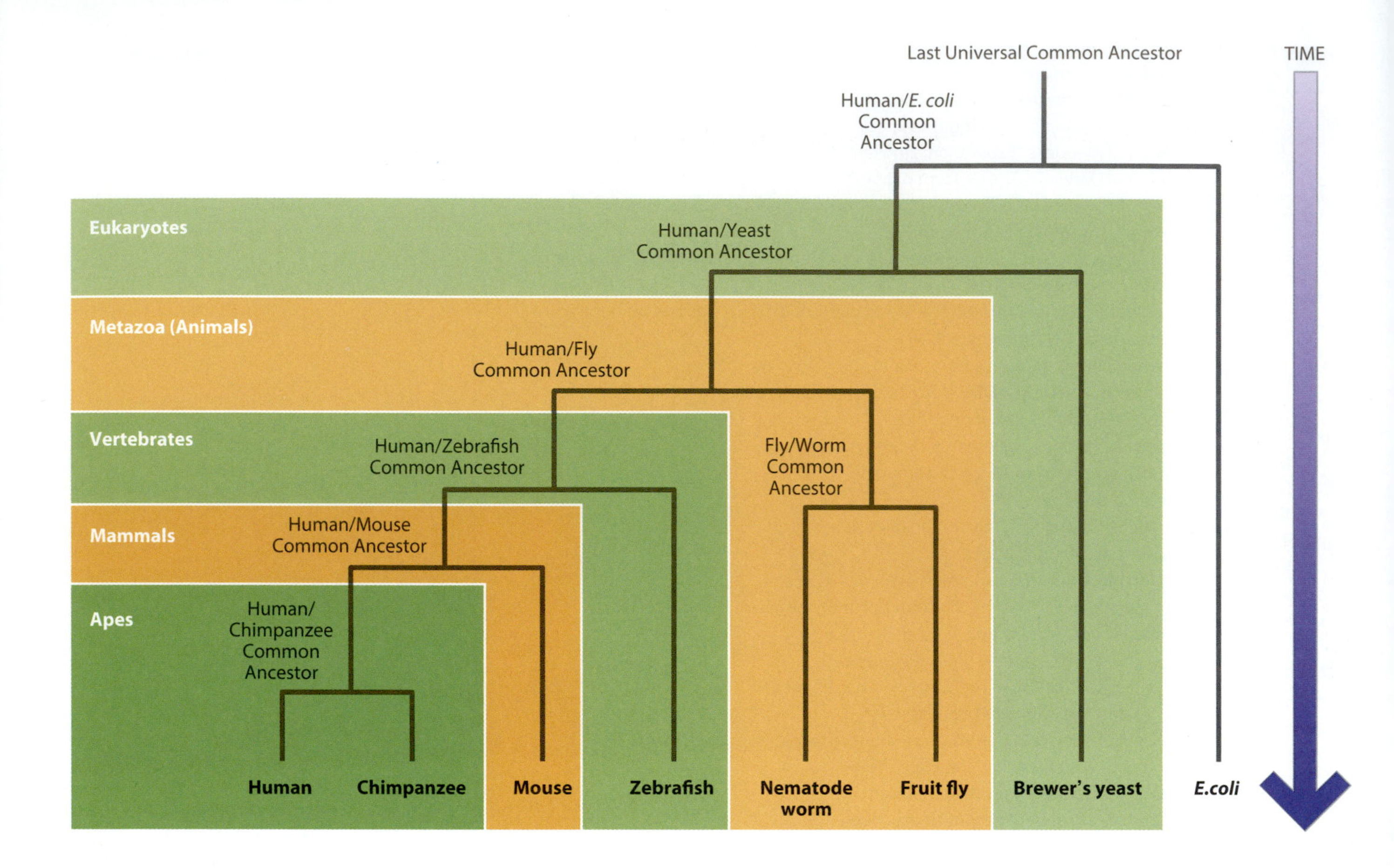

Figure 1.2: Cousins, not ancestors. A **phylogenetic tree** relates different branches of modern species, showing that they are all equally derived from their common ancestors in terms of time. Here branch lengths reflect evolutionary time, rather than evolutionary change.

looks like a gene, nor because an organism treats it like a gene (that is, makes a product from it), but because it *evolves* like a gene.

Understanding evolutionary history is essential to understanding human biology today

If we were to take a perspective to the biology of modern humans that neglected evolutionary history, what might we predict about the genetic diversity of our species, the significance of our phenotypic differences, and the prevalence of disease-causing **alleles**?

First, we would be struck by the huge numbers of humans, especially when compared with other animals of similar size. We might reasonably think that this should be mirrored by a correspondingly greater genetic diversity. Second, we might note the clustered distribution of phenotypic diversity (for example, skin color) among modern human groups and might expect this to be matched by a similar structuring of genetic diversity. Third, we might suppose that disease-causing alleles would be specific to different continental groups, in a similar manner to some of their easily observable "normal" phenotypes. As we shall discover in this book, all of these conclusions would be wrong.

To understand why this is so, we must comprehend that the past is not simply something that happened, and is studied for its own sake, but is more properly considered as the source of the present. If we are to improve our present circumstances, we must take account of how that present has come to be. An evolutionary perspective does not just address the question, What happened in the past? but also the question, Why is the present like it is?

Once we understand that the obvious differences between peoples' appearances can be unreliable indicators of biological origins, we start to appreciate the other factors that have shaped and continue to shape human biology. The interaction of humans and their surroundings comes to the fore, as does an understanding of human adaptability in the face of huge variability in inhabited environments.

Understanding evolutionary history shapes our expectations about the future

An evolutionary perspective on human genetic variation also allows us to make predictions, both about biological research, and about the future of our species. Today, we can pose many more biological questions than we are able to answer. An evolutionary perspective tells us how we might go about answering these questions, and about what kinds of answers we might expect.

Phenotypic traits of humans, be they skin color, height, or diseases such as **diabetes**, are controlled by a combination of inherited and environmental factors, and stochastic developmental and molecular processes. The easiest traits to dissect genetically are those determined in large part by single genes—so-called **Mendelian** traits. However, almost all of the phenotypic traits of most interest to both anthropologists and physicians are not so simple. Such complex traits are governed by interactions between multiple genes and the environment, and disentangling these interactions will help to relieve the considerable burden of complex diseases on individuals and economics.

Knowledge of our past helps us to predict the numbers and frequencies of genetic variants that influence a given trait and to choose the best strategy for identifying them—how best to define a human population (**Box 1.1**), which populations to choose, and which segments of the genome to concentrate on.

Box 1.1: Caucasians, Caucasoids, European-Americans, Whites? The confusing classification of human social groups

Population geneticists, forensic geneticists, anthropologists, and archaeologists need labels to refer to social groups of human beings. These labels differ between fields, and even within fields. In papers describing DNA diversity in a group of people living in the USA whose ancestors came from Europe, for example, you may find them referred to as:

- A ***US population***: because of population admixture in the last 500 years, the people of the USA are an extremely heterogeneous group, of which those of European descent form only a part.

- ***Caucasians***, or ***Caucasoids***: this is not meant to imply an origin in the Caucasus mountains, but refers to "beautiful people" in a racial classification scheme of the German anatomist Blumenbach (1752–1840). The skull that was claimed to best represent the characteristics of this group came from the Caucasus. The other classifications in this scheme were Mongoloid, Malay, Ethiopian, and American (referring to Native Americans).

- ***European-Americans***: usually in contradistinction to African-, Hispanic-, Japanese-, Native-, and other Americans. There is heterogeneity within this grouping—people who would classify themselves as European-American have ancestors from many different parts of Europe, such as Ireland, Italy, Poland, Russia, and Turkey.

- ***Whites***: this classification is favored by some scientific journals over "Caucasians," which might quite reasonably be reserved for people who really *do* come from the Caucasus. However, it seems odd to use "Whites," when authors of a paper may have no idea what skin color the donors of their DNA samples had.

Often these racial or ethnic labels disguise a great deal of biological heterogeneity, and the identification of DNA donors as members of social groups is not self-evident. Much confusion is possible when a paper has the title: "Strong Amerind/White sex bias and a possible Sephardic contribution among the founders of a population in northwest Colombia"[2]. It includes labels based on indigenous continental affiliation, skin color, membership of a group defined on religious-historical grounds, and current small-scale geography. A DNA donor may belong to a large number of categories simultaneously. In addition, when populations are compared in genetic studies, the level of classification in different samples may be unequal. The Hadza of Tanzania, with a population size of only 1000, have in some studies occupied the same analytical status as the South Chinese, whose population size is 600,000 times greater.

In general, the most suitable default method for classification is to use geographical information, rather than national, cultural, or phenotypic labels.

Not only that, but an evolutionary perspective also helps us to understand and predict which individuals will respond best to each therapy, and how to focus limited screening resources. Finally, genes of medical relevance are often sites of past **selection**, and evolutionary analyses of human diversity can offer a shortcut to identifying these important regions of the genome.

1.3 COMPLEMENTARY RECORDS OF THE HUMAN PAST

Any scientific investigation of the past should start with a consideration of the different types of evidence available, since this provides the basis for testing hypotheses and refining models. We are lucky in that while there was only one past, it is revealed to us in many ways (see **Figure 1.3**).

The historical record comprises written texts, the oldest of which are from Mesopotamia and date from as far back as four thousand years ago (KYA; "The Epic of Gilgamesh"). Writing itself goes back a further millennium but appears to have been mostly associated with accounting practices. These early texts were written in cuneiform (wedge-shaped) symbols indented into clay tablets using reeds, and it is only later that papyrus and modern alphabets were invented. Very few ancient languages were written down, and some of those that were recorded are still indecipherable (for example, the Cretan script known as Linear A). There are also oral histories, and folklore handed down through the centuries. It is difficult to judge the factual content of these inter-generational messages, let alone their time-depth. The boundaries between oral and written histories are often blurred; some histories that were initially oral have achieved special prominence and influence through being written down in antiquity.

Spoken languages retain evidence of their origins over thousands of years. The discipline of historical linguistics seeks to trace the ancestry of all ~6900 languages currently spoken in the world. Many of these languages can be traced to a number of ancestral languages known as **proto-languages**. For example, English, French, German, Russian, and Sanskrit all belong to the Indo-European **language family** (**Figure 1.4**) and share a common ancestral language known as proto-Indo-European. Although the origins of spoken languages can be traced back further than historical records, a common ancestry for all human

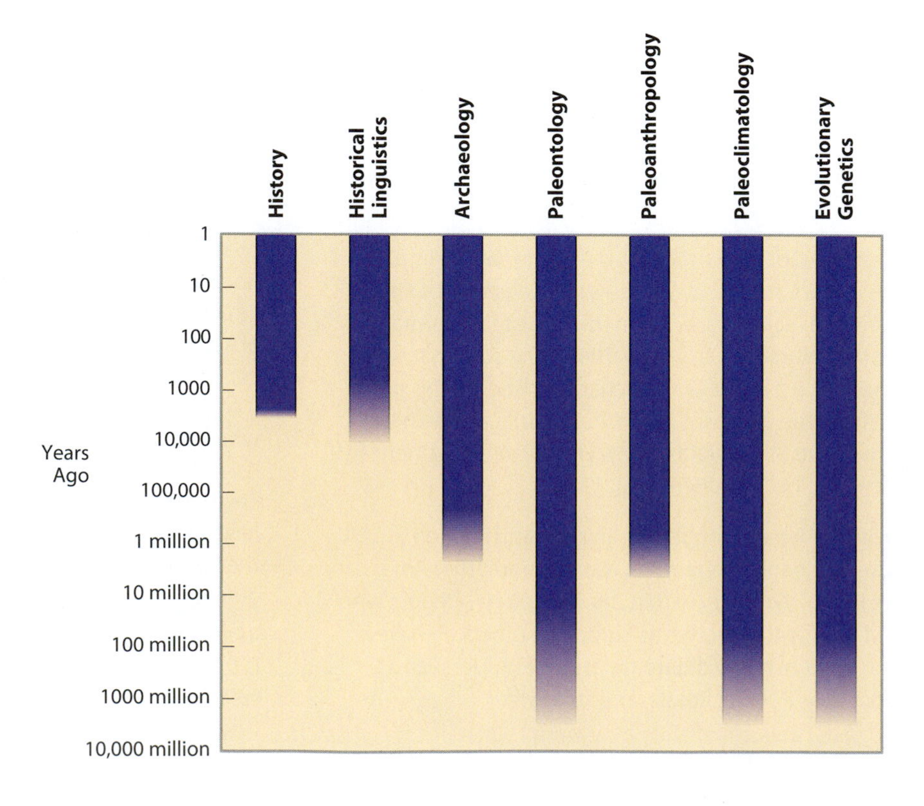

Figure 1.3: The informative time-depth of different records of the past.

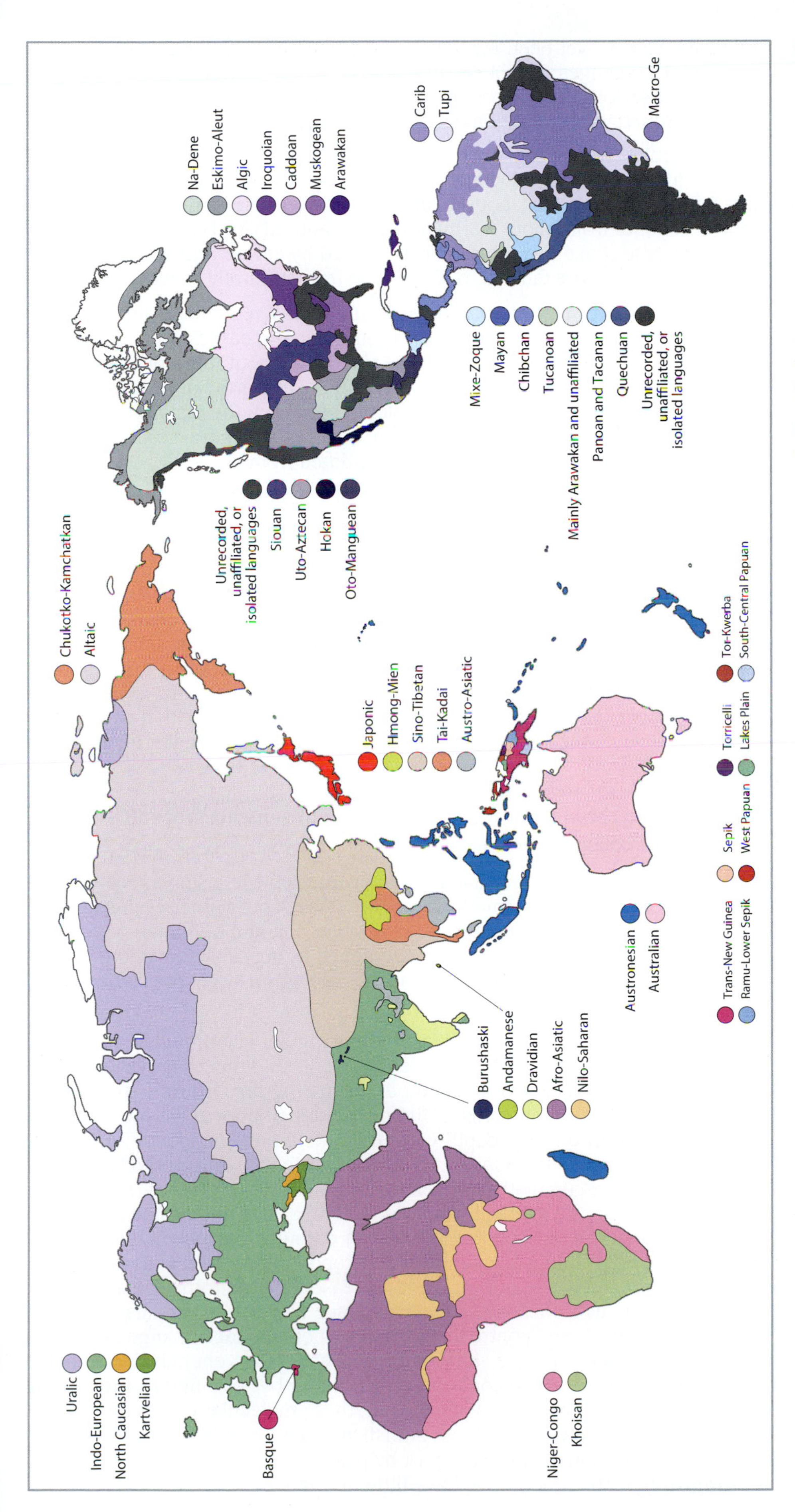

Figure 1.4: Distribution of some of the world's language families. The distributions of the major families are shown, together with selected others, and two languages that exist in isolation (Basque and Burushaski). Note that all the language families of the Americas, other than Na-Dene and Eskimo-Aleut, are lumped together by some linguists into one larger family, known as Amerind (Chapter 13). [Adapted from Diamond J & Bellwood P (2003) *Science* 300, 597. With permission from AAAS. Additional data and nomenclature from **www.ethnologue.com** (which lists 128 families), plus geographical information from The World Atlas of Language Structures (WALS; **http://wals.info/**), and **www.muturzikin.com/cartesoceanie/oceanie2.htm** for New Guinea.]

languages has not yet been identified, and some historical linguists have suggested that languages do not retain evidence of their origins over more than ~10,000 years. Nonetheless, comparisons of existing languages are a rich if controversial source of information about the past (**Opinion Box 1**).

The archaeological record consists of physical objects that have been shaped by human contact. These include not only tools, ornaments, and pottery, but also soils, waste deposits, houses, and landscapes. The earliest recognizable stone tools date from about 2.5 million years ago (MYA). Humans are not the only animals to make tools, but tools produced by earlier human ancestors or nonhuman species are difficult to distinguish from naturally occurring objects.

The paleontological record comprises the fossilized remains of living organisms or their traces, such as preserved footprints. The earliest microfossils are suggested to date from 3500 MYA. These would be about three-quarters of the age of the planet itself, suggesting that life on Earth started almost as soon as conditions were suitable. Paleoanthropologists focus on the remains of humans and their ancestors. The earliest fossils that appear to be more closely related to humans than to any other living species are dated from 5–7 MYA.

Paleoclimatology seeks information on past climates and helps us to reconstruct paleoenvironments. Data come from physical remains, both organic and geological in origin. An example of geological evidence comes from the cores of ice taken from ice sheets, which at increasing depths exhibit varying **isotope ratios** of elements such as oxygen that can be used to infer temperature changes progressively further back over the past million years. Similarly, cores of lake sediments reveal the predominance of pollen from different plants at different times, providing evidence about the biotic environment over shorter time-scales (typically up to 100 KYA). Geological records in rocks provide information on older climates as far back as 3800 MYA. We will see that climate has varied greatly over the last few million years, and has had a major influence on human populations.

Understanding chronology allows comparison of evidence from different scientific approaches

If we are to provide a wider context for an event visible in a single record of the past, we need to have some method to relate this event to other events in other records. For example, we might wonder whether a period of human population growth evident in the genetic record is matched by specific innovations apparent in the archaeological record that may indicate why a larger population could be supported.

The natural way to achieve this cross-referencing between multiple records is to relate them all chronologically, by dating events and processes visible in the different records. Consequently, methods for dating are of prime importance in the analysis of all the records of the past outlined above. We are fortunate in that information on time-depth can be extracted independently from each record, although not with equal accuracy or precision. Dating methods are typically dependent on cumulative processes (often known as "clocks" for obvious reasons) that occur, on average, at known uniform rates and can be quantified. The physicochemical methodologies used for dating in paleoclimatology, paleontology, and archaeology frequently depend upon the decay of **radioisotopes**. This process is stochastic but on average reasonably uniform and accurately quantifiable, and as a consequence can produce very accurate dates. Dating in linguistics is difficult and controversial, as rates of language change are highly variable. As a consequence, linguistic processes are generally not reliably dated. Dating the genetic record can utilize a number of different **molecular clocks**, but can also be controversial. First, these molecular clocks are difficult to calibrate accurately, and second, statistical confidence in the resulting date estimates is determined at least in part by past population sizes and structures (**demography**), about which we have little information.

OPINION BOX 1: How close is one language to another?

Just how closely is English related to Danish, or to Gujarati? The idea of putting a number on the distance between languages was first explored in 1831 by Samuel Rafinesque, who invented **lexicostatistics** in order to win a competition in Paris of the Société de Géographie, to determine the origin of Asiatic **negritos**. Since there were no other contenders, he was awarded a *médaille d'encouragement* worth only 100 francs rather than the advertised gold medal of 1000 francs. Rafinesque's negative finding showed the languages of disparate negrito peoples to be unrelated. It was never published, but his method was popularized by Jules Dumont d'Urville (**Figure 1**), who headed the jury.

Dumont d'Urville describes Rafinesque's technique thus: Between two terms expressing the same idea in two different languages we assign six degrees of relationship: 0 for completely unrelated terms, 1/5, 2/5, 3/5, or 4/5 for partially related terms, and 5/5 (or 1) for terms that are identical or nearly so. The total number of correspondences is then divided by the number of words compared. A score of 135/5 or 27 in a list of 45 words gives us a relationship of 0.60, whereas a score of 35/5 or 7, divided by 45, would give 0.15.

In the later nineteenth century, the idea that lexicostatistics could yield a separation date for branches of a language family gave rise to **glottochronology**. Following extensive criticism, lexicostatistics and glottochronology are anathema to most historical linguists (apart from a maverick subset) today.

Here are the criticisms in a nutshell. The distinction between vocabulary that is basic and words that are not is vague. In fact replacement of so-called basic vocabulary happens quite frequently. Rendering terms from different languages into English often masks substantive differences of meaning. Lexicostatistics often fails to distinguish between borrowed words (**identity by state**), and inherited words (**identity by descent**).

Figure 1: Jules Dumont d'Urville (1790–1842)

Recognizing **cognates** (words sharing a common origin) through systematic sound correspondences and clear similarities in form and meaning presumes detailed knowledge of language history. Many apparent cognates represent mere chance look-alikes. Languages change at highly variable rates. Back in 1850, Schleicher observed that languages spoken in tranquil backwaters (for example, Lithuanian, Georgian) change slowly, whereas languages in a constant maelstrom of social upheaval (for example, English, Mandarin) change quickly. Finally, laws about how speech sounds (**phonemes**) change regularly through time are vital, and correspondences between morphological systems and **grammatical** markers have far greater weight than mere lexical correspondences.

The math has undergone refinement, but Gray and Atkinson's[4] revolutionary use of **Bayesian** glottochronology to assess hypotheses for the origins of Indo-European (see **Section 12.5**) was greeted with indignation in conservative linguistic circles. In addition to objections against the misrepresentation and misinterpretation of language data, the methodological criticisms outlined above were reiterated, especially the issue of borrowings, chance resemblances, and false cognates.

Bayesian analysis of word correspondences is less problematic in Austronesian, where dispersal of the language family largely involved the colonization of previously uninhabited islands and therefore less contact. The claim[6] that the misidentification of borrowed vocabulary versus inherited word-forms does not compromise the validity of lexicostatistical findings is generally rejected by linguists. The other criticisms listed above also remain to be addressed.

In 1848, August Schleicher introduced the family tree model for language phyla by analogy to the **phylogeny** of biological species. In conventional models, whole integer values could be assigned to the nodes in the phylogeny, based on the intuitions of knowledgeable historical linguists. Yet linguists generally doubt that numbers derived from readily manipulable though complex mathematical models, based on simple and sometimes false lexical comparisons, can be very meaningful.

Nonetheless, the mathematization of linguistic phylogeny seems inevitable. The way forward is to accommodate the criticisms identified by linguists and to tweak the model, as Dunn et al. have done.[3] Mathematical approaches to linguistic phenomena are producing ever more intriguing results. A Bayesian analysis of lexical change may have shed some new light on the shape and rate of language evolution.[5] However, sometimes a headline-grabbing finding is just a foregone conclusion that could have been foretold by anyone familiar with, say, the phoneme inventories of **Khoisan** languages.[1] Numbers can be useful, but they actually ought to reflect realities accurately if they are to tell us something new and meaningful.

George van Driem, Department of Linguistics,
University of Berne

Having used chronology to identify a correlation between events and processes observed in different records of the past, it is tempting to ascribe causal relationships. However, it may not be possible to prove causality with the same degree of certainty available to other branches of science. The study of evolution is in many ways a historical discipline; being limited to a single past prevents us from demonstrating causal relationships.

It is important to synthesize different records of the past

No single record of the past is more important than any other, but each contains different features of the past. Therefore the record chosen for investigation in a given study is largely determined by the specific questions under consideration. If we want to address a cultural question, such as, What was life like for humans 10 KYA? archaeology will give us more insight than will linguistics or genetics. In contrast, if we want to answer a biological question, such as, Which nonhuman species is most closely related to us? archaeology and linguistics will be of no use.

To paint a fuller picture of the past we often seek to combine information from multiple records into a single synthesis. This requires us to consider more closely the similarities and unique characteristics of different records. For example, while we can be sure that the speakers of modern languages have ancestors, we cannot be sure that the makers of a certain style of pottery left any descendants. When similar information is being sought from several records of the past, it becomes particularly critical to appreciate the differences between them: if we are interested in migrations we must appreciate that artifacts can move through trade or technology transfer, without the concomitant movement of genes; similarly, not all **gene flow** need be accompanied by the movement of languages.

The different records are independent reflections of a single past—but they need not all tell us the same thing. Rather, conflicting signals allow us to reconstruct subtle and nuanced views of a past that we should expect to be just as complex as the present (**Figure 1.5**). Each individual discipline that interprets a record of the past contains competing hypotheses and many issues upon which there is no consensus. This could enable researchers in one field to cherry-pick hypotheses from other disciplines that agree best with their own thinking. Stringing together an initial contentious hypothesis from field A with equally contentious theories from fields B and C may make for a more exciting and interesting narrative, but does not make the original hypothesis any less contentious. As a result, human evolutionary studies are prone to heated debates, and the reader should realize that many ideas presented in this book are likely to be contested by someone. We recognize the importance of these ongoing debates and try to give readers a flavor of the diversity of opinions by incorporating Opinions written by active researchers who have new, interesting, or challenging theories, or who are particularly well placed to comment on an area of controversy.

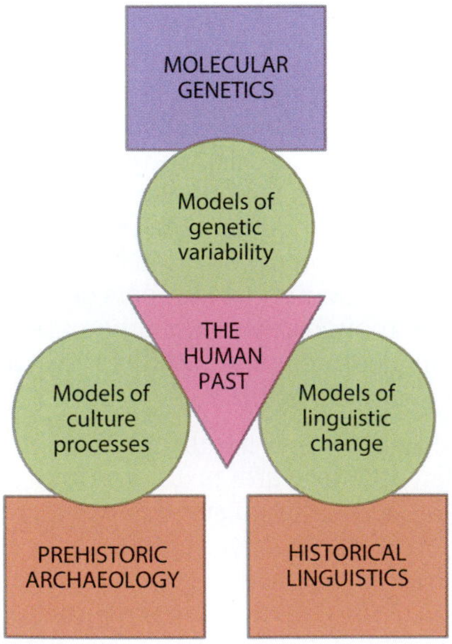

Figure 1.5: Reasonable optimism? Renfrew's "New Synthesis" showing how independent sources of information tell us about a single human past.
[From Renfrew C (1999) Reflections on the archaeology of linguistic diversity. In The Human Inheritance: Genes, Language, and Evolution (B Sykes ed.), pp. 1–32. Oxford University Prress. With permission from Oxford University Press.]

None of the different records represents an unbiased picture of the past

We do not have a time machine, and therefore must rely upon evidence that has survived to the present. This survival is selective. In the archaeological record we find many stone tools but few wooden ones: arrowheads but not shafts. In the paleontological record we find plenty of skeletal fossils, but soft tissues leave traces only very rarely. In the historical record we may not encounter those texts that displeased contemporaneous or subsequent heads of state, either because they were destroyed, or not written in the first place. Similarly, in the genetic record, survival is quite literally selective. Natural selection and other processes have shaped, and continue to shape, our genome in different ways. Even **ancient DNA** evidence, although not influenced by subsequent natural selection, is biased: for technical reasons it has so far told us more about the genetic diversity of our female ancestors than it has about our male forebears. Since the survival of ancient DNA is influenced by physical and chemical conditions, samples will be more plentiful from some regions of the world than from others.

Each of the disciplines studying these complementary records of the past is continually improving methods of collecting and interpreting data, both in terms of better coping with their selective nature, and in gaining access to entirely new forms of information. We can expect continued revolutions in understanding analogous to the introduction of radiometric dating in archaeology and ancient DNA studies in genetics.

1.4 WHAT CAN WE KNOW ABOUT THE PAST?

From the wealth of data on genetic variation among humans and between humans and other apes now at our fingertips, we can address many of the fundamental questions about human origins. For example, we can ask how genetically diverse humans are, when humans split from **Neanderthals**, or when and where humans originated. We can also address questions about individuals and groups, such as what proportion of genes in African-Americans came from Africa, or how genetically similar two humans are. More recently we have started to address questions about human phenotypes, such as asking which genetic changes make us human, or have allowed us to adapt to different environments.

All these questions, and more, are addressed in this book. Some of these new and exciting inferences can be drawn by analyzing the patterns of genetic variation throughout the genome. Others require us to study specific genomic regions in detail. With an entire human genome sequence a few mouse-clicks away, we can select the most appropriate regions to investigate to address the particular question in hand.

There are some fundamental questions about human origins that we cannot currently address, and some that we will never be able to address. What we *would like* to know about the past and what we *are able* to know about the past are two separate things. Some questions are unanswerable because of fundamental technical or ethical limitations. We cannot ask, for example, whether two contemporaneous **Australopithecine** species were **interfertile**, or what combination of mutations would be sufficient to give chimpanzees language skills akin to those of humans. It is important to note that some of the limits to our understanding of the past are not static: unforeseen technological innovations continue to surprise us. For example, prior to the advent of methods for sequencing ancient DNA, the genetic divergence between Neanderthals and modern humans was unknowable. Now we have a measure of this divergence, not just for one gene, but across the entire genome.

Other intuitively attractive questions fail because of mistaken implicit assumptions about the nature of genetic inheritance, such as, What was the ancestral biological homeland of population X? or, In which country did my ancestors live, a thousand years ago? Given the frequency with which these two questions about individual and population ancestry are asked, it is worth considering why they are unanswerable. To think of a single origin for an entire population is to misconceive how genetic diversity accumulates in a sexually reproducing, outbreeding species such as humans. The processes of **recombination** and independent chromosomal **assortment** divide the genome into segments that have independent genealogical histories: while the ancestry of a single segment of the genome converges on a single common ancestor, each segment has a separate such ancestor. These common ancestors are almost always far older than the population itself, and it is highly unlikely that they were ever present in the same place at the same time.

As we shall see in this book, human populations are fluid entities, giving and receiving genes from neighboring populations all the time. Alleles enter the population at different times and from different places. Four of the five authors

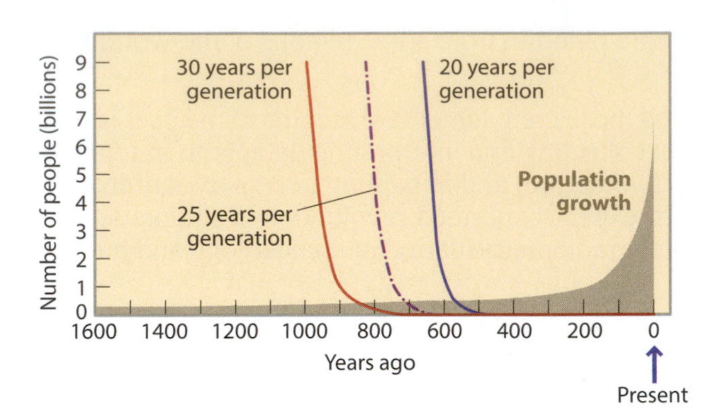

Figure 1.6: Past population growth compared with the accumulation of potential ancestors. The filled segment shows global population growth over the past 1600 years. The lines correspond to the potential numbers of ancestors of a single individual expected at a given point in time. These numbers depend on the length of **generation time** that is assumed, and three different possible generation times are illustrated.

of this book are British, each with more than 20,000 genes in their genomes. Some of these genes journeyed to Britain when it was still joined to the European continent as the ice sheets of the last ice age receded. Others were introduced by migrants as the sea levels rose, making what is now the bed of the North Sea uninhabitable. Still more genes may have been contributed by the first farmers, the Romans, the Angles, the Saxons, the Vikings, and the Normans, as well as other travelers less well documented in historical or archaeological records. What hope is there for a single biological homeland for the British?

The assumption of a simple ancestry for a single individual is similarly flawed. As a result of sexual reproduction, as we look back in time, individuals have ever-increasing numbers of ancestors (2, 4, 8, 16, 32, 64, 128, and so on). It is worth considering a few consequences of this seemingly trivial calculation. At 30 generations into the past (<900 years), each of us has more than a billion potential ancestors, and at 40 generations (<1200 years) more than a thousand billion (**Figure 1.6**). Since the current world population is only ~7 billion, and past populations were much smaller, such large numbers of ancestors never existed. This is because these potential ancestors were not all different individuals: the same person appears in different places in the family tree. There is another consequence of this reasoning that is worth spelling out. In Figure 1.6, the curves representing potential ancestors of a single individual and world population size intersect within the past 900 years. Before this time, everyone in the world was potentially an ancestor of every living person. Population substructure, the nonrandom breeding of individuals from different places, complicates this oversimplified calculation, but the essential point remains: the best answer to the question, Where did my ancestors live? is "everywhere."

1.5 THE ETHICS OF STUDYING HUMAN POPULATIONS

It is human nature to use knowledge of the past (see timeline in **Figure 1.7**) to guide actions in the present. This places an ethical responsibility upon those who seek to explore the past to do all they can to effectively and accurately communicate their findings, and to ensure that their work is not misused. This is not just a theoretical possibility: history is rife with the misuse of anthropological research to justify regimes that have cost the lives and livelihoods of many people. So much damage has been caused that some have even questioned whether the endeavor to reconstruct our evolutionary past should be undertaken at all. We believe that the potential intellectual and medical benefits of this work outweigh the potential dangers, but only when researchers take responsibility for the accurate popularization and public dissemination of this research, including active opposition to misinterpretation. Having said that, we must acknowledge that much of this work (including this book) is published by an unrepresentative subset of our species, namely men in developed countries, and that this cultural framework undoubtedly has an influence on interpretation.

As scientists, we must recognize that our work depends upon the support of wider society; indeed, most financial support for evolutionary studies around the world comes from the public purse. Irrespective of the source of funding,

Period	Date	Event
MYA	3500	Origin of life on Earth
	80–50	Origin of Primates
	>7	Human–Chimpanzee divergence
	7–6	First Hominid (*Sahelanthropus tchadensis*)
	6–4	Bipedalism (*Orrorin, Ardipithecus*)
	4.2–1	Gracile and robust *Australopithecus*
	2.5	First recognizable stone tools, Oldowan culture
	1.9	Origin of *Homo, H. erectus*
	1.8	*Homo erectus* outside Africa, in East Asia
	1.6	Acheulean bifaces
	1.0	*Homo heidelbergensis*
	1.2–0.3	Most recent common ancestors of extant diversity at autosomal loci
KYA	300	Levallois flaking technique
	250	First appearance of *Homo neanderthalensis*
	220–120	Most recent common ancestor of extant mtDNAs
	200	First fossils of anatomically modern humans
	140–40	Most recent common ancestor of extant Y chromosomes
	130–110	Warm interglacial period, cooling after 110 KYA
	100	Anatomically modern humans in the Levant
	50–100	Denisovan fossils
	80–75	Modern human behavior: art
	70–55	Glacial maximum
	50	Humans in Australia
	50	Later Stone Age
	47	Modern humans in West Asia: Upper Paleolithic (Aurignacian)
	46	Australian megafauna extinction
	42	Modern humans in Europe, southern Siberia
	28	Extinction of *Homo neanderthalensis*
	21–18	Last glacial maximum
	>15	Humans in the Americas
	14	American megafauna extinction
	13	Most recent date of *Homo floresiensis* fossils in Indonesia
	10	Holocene begins: warm, stable climate
		Neolithic transition begins: origins of agriculture, demographic expansion
	4	Written records in Mesopotamia
	3	Humans in Polynesia—last major expansion into new territory
AD	1492	Columbus reaches America: large-scale inter-continental travel begins
	c1750	Industrial revolution
	c1800	Human population exceeds one billion
	2001	Human germ-line modification

Figure 1.7: Time line of important events in human evolution. All times are approximate, and many are disputed. *Red* shading indicates events believed to have occurred in Africa, and illustrates the importance of Africa in our ancestry. Some crucial events, like the origin of language, cannot be dated.

public concerns about the implications of our work must be addressed. Although the most notorious historical misuse of anthropology has been the justification of genocide, recent public anxiety has been focused on issues of ownership, commercialization, and privacy. When work is being conducted in the public interest, the perception of misuse can be as important as the reality of misuse. Steps must be actively taken to ensure both that such misuse does not occur, and is seen not to occur. It is for these reasons that research projects on human subjects, whether they are medical patients or volunteers contributing a swab of cheek cells, should be scrutinized and approved by ethical committees before they begin. However, some individuals who undergo commercial genomic testing are now choosing to make their own personal genetic data fully and freely available, sometimes even including medical details and identification. Ideas about genetic privacy and ethics seem to be in flux and it remains to be seen how they will develop.

Human evolutionary genetics seems likely to undergo a revolution over the next few years, as a tsunami of genetic diversity data flows from new sequencing and genotyping platforms. These data will primarily come not from academic labs, but from the clinic and from consumer-driven testing. Similarly, the interpretation of such data seems likely to shift away from academia, toward people who are not professional scientists analyzing publicly available genetic datasets, often in more detail than the professionals can. "Citizen science" is on the rise.

SUMMARY

- Evolutionary genetics uses patterns of genetic differences between individuals and between species to make inferences about the past.
- New technologies are providing vast amounts of data, and new analytical methods are being developed to infer evolutionary processes from these data.
- Human evolutionary genetics can help to explain the distribution and frequencies of different phenotypes and diseases in humans today.
- Information on human evolution from genetic studies placed in the context of information from other scientific disciplines, including archaeology and linguistics, provides a synthetic view on our past.
- Asking clear, answerable questions is an important aspect of studies in human evolutionary genetics.

REFERENCES

1. **Atkinson QD** (2011) Phonemic diversity supports a serial founder effect model of language expansion from Africa. *Science* **332**, 346–349.
2. **Carvajal-Carmona LG, Soto ID, Pineda N et al.** (2000) Strong Amerind/white sex bias and a possible Sephardic contribution among the founders of a population in northwest Colombia. *Am. J. Hum. Genet.* **67**, 1287–1295.
3. **Dunn M, Greenhill SJ, Levinson SC & Gray RD** (2011) Evolved structure of language shows lineage-specific trends in word-order universals. *Nature* **473**, 79–82.
4. **Gray RD & Atkinson QD** (2003) Language-tree divergence times support the Anatolian theory of Indo-European origin. *Nature* **426**, 435–439.
5. **Gray RD, Bryant D & Greenhill SJ** (2010) On the shape and fabric of human history. *Philos. Trans. R. Soc. Lond. B Biol. Sci.* **365**, 3923–3933.
6. **Gray RD, Drummond AJ & Greenhill SJ** (2009) Language phylogenies reveal expansion pulses and pauses in Pacific settlement. *Science* **323**, 479–483.

SECTION 1

HOW DO WE STUDY GENOME DIVERSITY?

The success of the Human Genome Project gave us an unprecedented understanding of the structure and organization of our genome. In turn this has led to profound insights into its function and evolution, and the nature and dynamics of genome variation. This variation is the raw material of human evolutionary genetics.

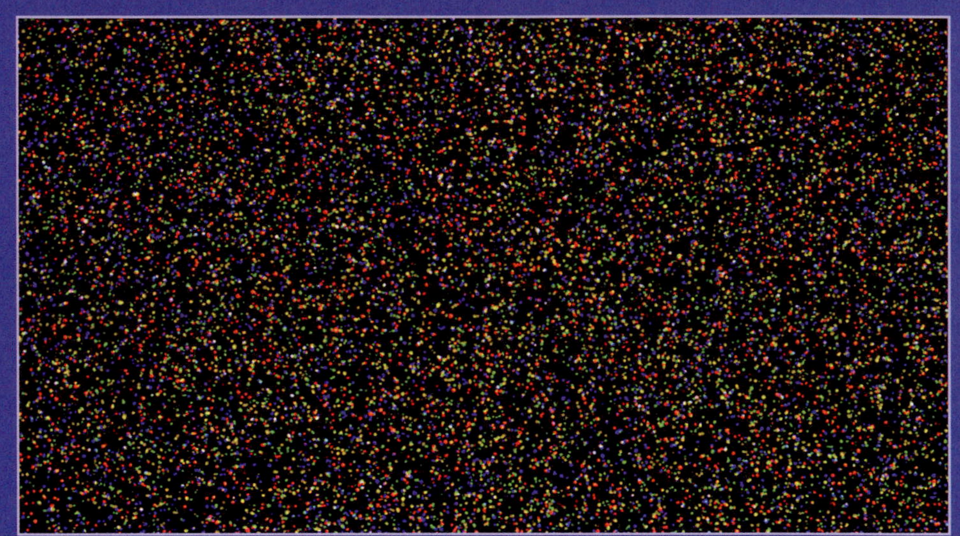

Courtesy of Illumina Cambridge Ltd.

CHAPTER 2
ORGANIZATION AND INHERITANCE OF THE HUMAN GENOME

This chapter introduces the structure and function of DNA, genes, and the genome, describing the packaging of DNA into chromosomes, and the means by which information encoded in genes is expressed as proteins. The inheritance patterns of different segments of the genome, and the process of genetic recombination, both key ideas in understanding the patterns of genome variation among human populations, are explored.

CHAPTER 3
HUMAN GENOME VARIATION

In this chapter we ask how the sequence and structure of the genome varies between individuals. This occurs over a wide range of scales, from single nucleotide polymorphisms, through differences in sequence copy numbers, to large-scale changes in the structures of chromosomes, each with characteristic mutation rates and processes. We explore the structures of haplotypes—combinations of different variants on a single DNA molecule—which are powerful tools for investigating our evolutionary past.

CHAPTER 4
FINDING AND ASSAYING GENOME DIVERSITY

The methods for discovering and assaying genome variation are described in this chapter. Technical advances—the polymerase chain reaction (PCR), together with the availability of the human genome sequence and novel DNA sequencing and genotyping technologies—have revolutionized the discovery and exploitation of variation. We discuss the methods available for typing variants in population samples, and the challenging field of ancient DNA studies.

ORGANIZATION AND INHERITANCE OF THE HUMAN GENOME

CHAPTER TWO

In this chapter we introduce the basic biological knowledge that is needed to understand the subject of human evolutionary genetics. We first step back and present the big picture, giving an overview of the scale, structure, organization, and inheritance of the human **genome**. Then, we focus on the essential details, taking a bottom-up approach. We start with the structure of **DNA** (**deoxyribonucleic acid**), moving on to how information in DNA is used to make **RNA** (**ribonucleic acid**) and **proteins** (**gene expression**), the details of genome organization, and the structure and function of **chromosomes**. We conclude by explaining how information in DNA is passed on from one cell to daughter cells, and from one generation to the next—the key issue of inheritance. Different parts of our genome are inherited in different ways, and understanding this is fundamental to appreciating how DNA can be used to study the evolutionary genetics of our species.

Some readers will approach this chapter having already studied human genetics, and will be familiar with most or all of its content. To them, it will serve as a source that they can refer back to from other parts of the book. Others may have come from different disciplines in which genetics is not central, and the chapter is written with these readers particularly in mind. Readers requiring more detail on any of these topics will find them excellently covered in Strachan and Read's *Human Molecular Genetics*.[17]

2.1 THE BIG PICTURE: AN OVERVIEW OF THE HUMAN GENOME

In February 2001, a milestone was reached in the attempt to sequence the entire human genome: substantially complete draft versions of the sequence were announced.[10, 19] Sequencing the genome was the grand big-science project in biology at the end of the twentieth century. It was a triumph not for the science of **genetics**, which is concerned with inheritance and variation, but for the newer science of **genomics**—the mapping, sequencing, and analysis of genomes. This enormous achievement provides us with a rich resource for identifying **genes**, understanding their evolution and that of our genome as a whole, and discovering sequence variation that can be analyzed in evolutionary and disease studies. It also facilitates comparative analysis of other animal genomes, and an understanding of the structure and function of chromosomes, the packages into which the genome is divided. Humans, like almost all animals, are **diploid**—that is, we have two copies of the genome in each of our **somatic** cells, the cells that make up our tissues. The human **haploid** genome (that is, a single copy) is composed of about 3.2 billion **nucleotides** (nt), the fundamental building blocks of DNA within which information is encoded.

Since the original announcement of the draft human genome sequence, efforts have focused on improving coverage and sequence quality, and ongoing refinements are reflected in the periodic release of new **builds** of the sequence assembly, known as the **reference sequence**. An up-to-date view of any feature of the genome is best obtained from an online tool such as the free-to-use UCSC or Ensembl Genome Browsers (**Figure 2.1**). These interfaces represent

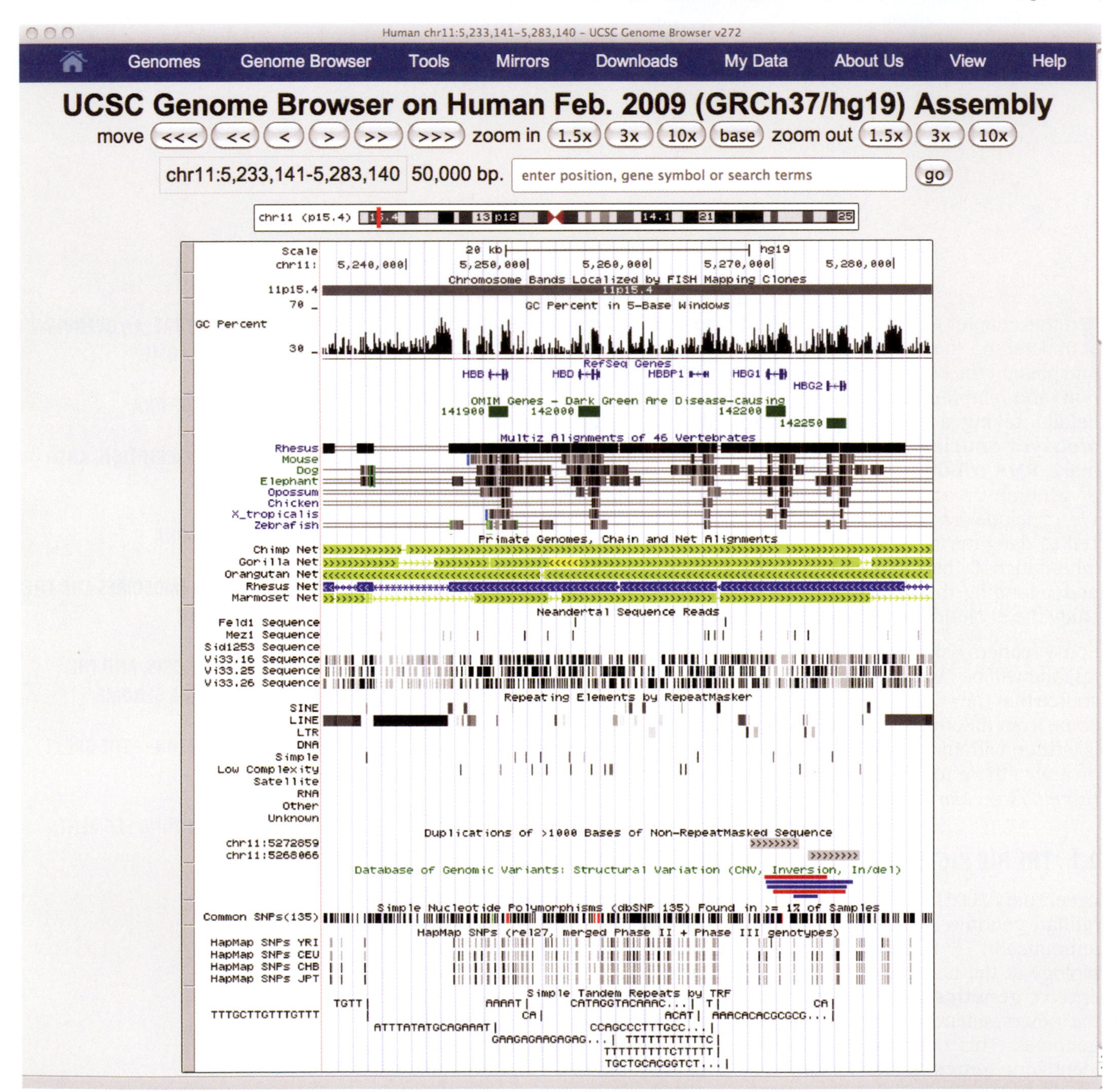

Figure 2.1: Browsing the human genome.
A 50-kb segment of chromosome 11, around the β-globin gene (*HBB*), displayed in the UCSC Genome Browser (**http://genome.ucsc.edu/**). The chromosome is arranged horizontally, reading from left to right, p arm to q arm, and features of the chromosomal DNA sequence, such as genes, are drawn below the sequence—these features are collectively called **annotations**, and each type of feature has its own horizontal track. There are four globin genes in the region (*HBB*, *HBD*, *HBG1*, and *HBG2*, annotated on the RefSeq Genes track), and the sequence conservation in other species, as well as the predicted regulatory elements, can be clearly seen. Other tracks include disease variants, **SNPs** (**single nucleotide polymorphisms**), the recombination map, and the distributions of repeated sequences of various kinds. There is also an alignment to chimpanzee sequence across the region, and the indication of a duplicated segment, which is due to the two related γ-globin genes (*HBG1* and *HBG2*). In the Web interface, all features shown in the browser are also links to detailed information, and many other possible annotation tracks can be selected.

Box 2.1: What is "the human genome sequence"?

Wonderful though the achievement of human genome sequencing is, the result is not perfect:

- The reference sequence contains sequencing errors, though remarkably rarely—about 1 per 100,000 bases.
- It contains gaps, though only ~250. Some of these will be closed by teams of people known as finishers, while 33 gaps represent large regions of complex and repetitive **heterochromatin** (including **centromeres**) and have not been seriously targeted in sequencing efforts.
- Since the initial announcement of the sequence, it has become clear that at least 5% of the genome is represented by very highly similar duplicated sequences, which can cause uncertainty and error in sequence assembly.
- Copy-number variation at different scales is now recognized to be common, and contributes to uncertainty of the assembly in some regions.
- The reference sequence is not derived from the DNA of one individual, but many. This means that it is an artificial composite, and does not reflect the true sequence of any real genome. A notable exception to this is the Y chromosome, the sequence of which was largely derived from the genomic DNA of one individual.[17] This allowed accurate assembly of the chromosome sequence, despite its very high (35–45%) proportion of highly similar duplicated sequences—an achievement that would have been impossible if sequences from different individuals were assembled.
- The reference sequence is continually being improved. Genomic coordinates generally differ between versions. Consequently, it is essential to know which version was used in any study. See http://www.ncbi.nlm.nih.gov/projects/genome/assembly/grc/human/index.shtml.

superb unions of biology and computing, allowing easy navigation over many orders of magnitude (from an entire chromosome to a specific nucleotide), and including annotations of features such as genes, the elements that regulate them, sequence variants, variation in sequence copy number, the types and distributions of repeated sequences, and sequence conservation with the genomes of other species.

It is important to understand that the reference genome sequence is not yet quite complete (and may never be so), and also that the idea that there is one single, archetypal human genome sequence is very misleading (**Box 2.1**); nonetheless, the current version covers 99% of the part that contains genes, and forms an invaluable standard with which to compare any other human DNA sequence.

DNA contains within it the instructions to produce the proteins that make up our cells. These instructions are contained within genes. Genes have undergone natural selection during human evolution, and variants within and around them contribute to the great variety of normal observable human variation, and can cause disease. Despite their importance, however, protein-coding genes comprise a mere 2% of the genome; some of the remaining 98% is essential for producing RNA molecules that are never used as intermediates in protein production, but for gene regulation and for the function of chromosomes; but much of it may have no specific function. As we shall see later in this chapter, comparisons between genome sequences of vertebrate species can help to identify these non-genic functional sequences. Using a variety of approaches, the large-scale research consortium **ENCODE** (the ENCyclopedia Of DNA Elements; genome.ucsc.edu/ENCODE/) set out to find all the functional elements in the human genome.[7]

The number of genes in the human genome has been a hotly debated issue; early assumptions were that vertebrates would possess substantially greater numbers of genes than other organisms, reflecting their apparently greater biological complexity. However, the initial draft sequence estimated that there were only ~32,000 protein-coding genes, which was regarded as surprisingly few. The number has been steadily falling as the sequence and gene predictions

Nuclear genome

- ~3,200,000,000 bp (haploid genome size)
- 2 copies per diploid cell
- 46 linear molecules (chromosomes) per diploid cell

diploid, somatic cell

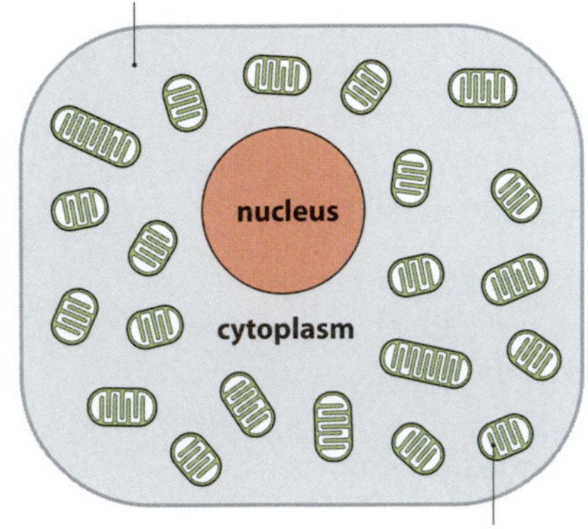

mitochondrion – 100s to 1000s per cell

mtDNA

- 16,569 bp
- 2–10 copies per mitochondrion, 1000s of copies per cell
- circular molecule

Figure 2.2: Compartmentalization of the human genome. bp, base pairs.

are refined, and currently (ensembl.org/Homo_sapiens) stands at only 23,532—only some 3000 greater than the nematode worm, *Caenorhabditis elegans*. Complexity in vertebrates is likely to be conferred by a greater repertoire of variation in how genes are expressed and regulated.

DNA is found in two subcellular compartments in human cells (**Figure 2.2**). In terms of DNA sequence, the vast majority is chromosomal DNA contained within the **nucleus**—sometimes called the **nuclear genome**. However, DNA is also found in specialized **organelles** within the **cytoplasm** of the cell, the **mitochondria**. At ~16,500 nucleotides, each mitochondrial DNA (**mtDNA**) molecule is only 0.0005% of the size of the nuclear genome, but is present in many copies in each cell, and is essential for life. While the nuclear genome is tightly packaged around proteins in a complex known as **chromatin**, mtDNA is not bound by proteins.

The diploid nuclear genome is divided into 46 chromosomes, made up of 23 pairs. Twenty-two pairs are common to all humans, but the remaining pair differs between the sexes, with females carrying two X chromosomes, and males one X and one Y chromosome. Male-specificity of the Y means that it is passed down only from father to son, unlike the other chromosomes, which are inherited from both parents. Mitochondrial DNA is borne by both sexes, but is passed on only from mothers to children, and so forms a maternally inherited counterpart to the Y chromosome. The unusual inheritance patterns of these segments of the genome have profound consequences for their genetic diversity and geographical differentiation among populations, and have made them particularly informative tools in human evolutionary genetic studies.

We now turn to the fundamental details of our heritable material—the structure of DNA itself.

2.2 STRUCTURE OF DNA

All organisms, with the exception of a few kinds of viruses, use deoxyribonucleic acid (DNA) as their genetic material. DNA is an extraordinary macromolecule that plays two central biological roles:

- It carries the instructions for making the components of a cell (mostly proteins, which themselves can manufacture further components)
- It provides a means for this set of instructions to be passed to the daughter cells when a cell divides

DNA is a polymer, and its monomeric subunits are called nucleotides. There are four varieties of nucleotide, which differ in portions known as **bases**. The bases are **adenine**, **guanine**, **cytosine**, and **thymine**, abbreviated as A, G, C, and T, and it is the order or sequence of these parts of each nucleotide that carries the genetic information. Adenine and guanine, double-ringed molecules, are **purines**, and cytosine and thymine, single-ringed, are **pyrimidines**. Each base is joined to a sugar molecule, **deoxyribose**, and each deoxyribose has a phosphate group attached to it; the sugar and phosphate play only a structural role in DNA, and themselves carry no information. The structures of these nucleotides are shown in **Figure 2.3**. Some nucleotides can be chemically modified after their incorporation into DNA: an important example is the **methylation** of cytosine, to give 5-methylcytosine. Methylation is an example of an **epigenetic** modification, which can be heritable, but involves no change to the DNA sequence itself. It plays a role in gene regulation, and has consequences for mutation that are discussed in the next chapter (**Section 3.2**).

DNA is not the only nucleic acid found within cells; there are many different species of RNA playing key roles, especially in protein production. RNA differs from DNA not only in the kind of sugar molecule it contains (ribose, rather than deoxyribose), but also in one of its bases. RNA contains the pyrimidine base **uracil** (U) in place of thymine (T). Like thymine, uracil pairs with adenine (A).

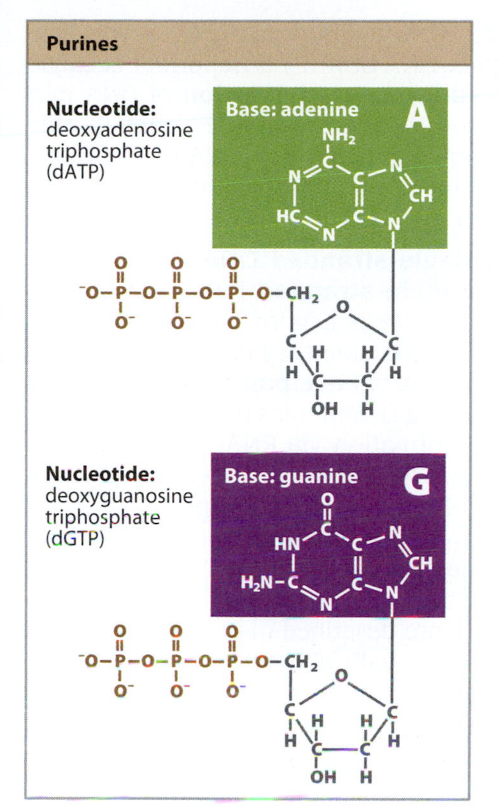

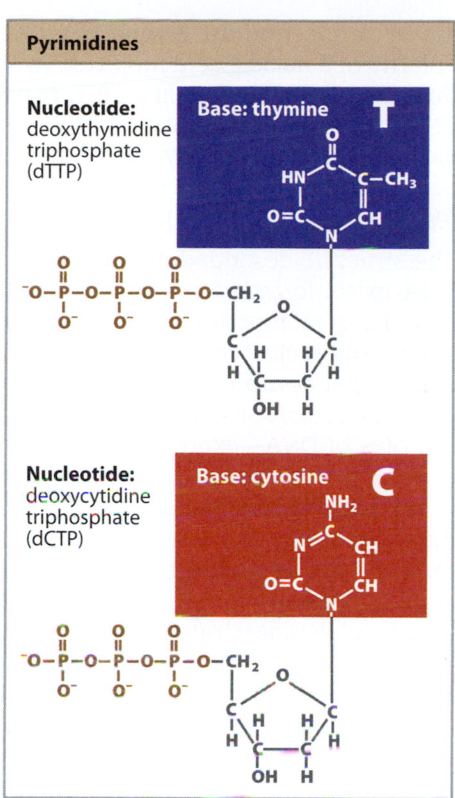

Figure 2.3: Chemical structures of the nucleotide components of DNA. These are the standard four nucleotides; some bases in DNA are modified by the addition of other chemical groups. When nucleotides are incorporated into DNA, two of the phosphate groups are lost.

In DNA, the phosphate group of one nucleotide is joined to the deoxyribose sugar of another (**Figure 2.4a**) and this forms the **sugar–phosphate backbone** of the molecule. Because the deoxyribose group itself is asymmetrical, it provides a **polarity** to the backbone. The five carbon atoms making up the deoxyribose molecule are given numbers from 1′ (pronounced "one prime") to 5′. Phosphate groups attach to the 3′ and 5′ carbon atoms of the ring, and this provides a convenient way to refer to the different ends of a DNA molecule: the **3′ end** has a free hydroxyl (-OH) group (unattached to another nucleotide) on

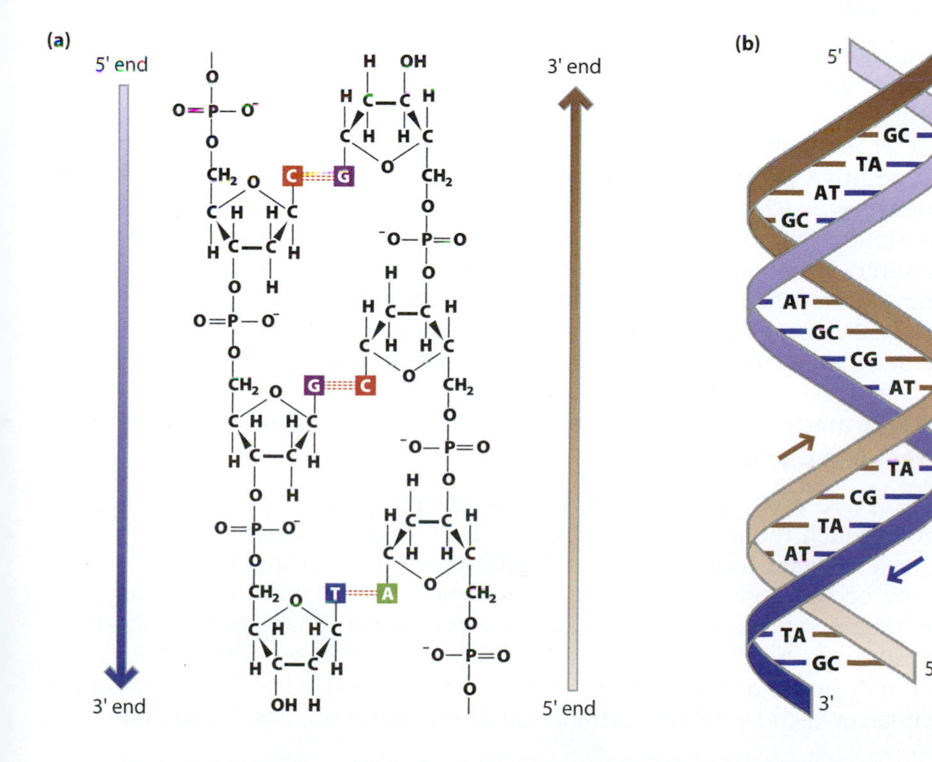

Figure 2.4: Double-stranded helical structure of DNA.
(a) The two strands of DNA are anti-parallel because the linking of the 3′ carbon atom of one base to the 5′ carbon of the next is in opposite directions on the two strands. Since sequences are written 5′ to 3′, the sequence of this three-base DNA molecule on the left-hand strand is CGT, while on the right-hand strand it is ACG.
(b) The two strands are wound round each other in a double helix. Arrows indicate the direction of each strand, 5′ to 3′.

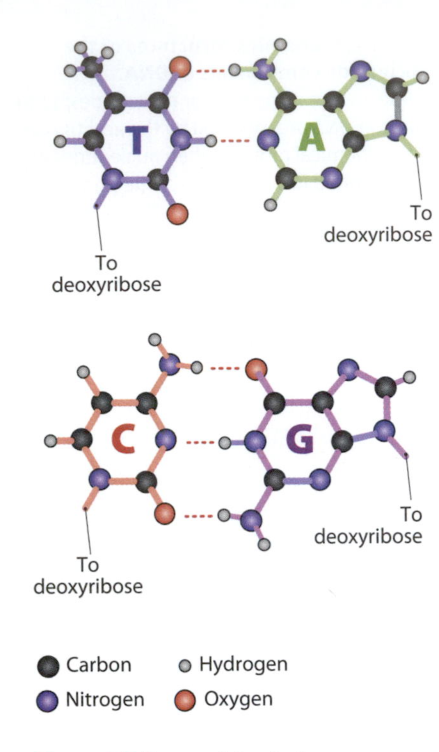

Figure 2.5: Base pairing between thymine and adenine, and between cytosine and guanine.
Hydrogen bonds are indicated as dashed lines.

the 3′ carbon, and the **5′ end** has a free hydroxyl group on the 5′ carbon. The polarity of a nucleic acid molecule (whether DNA or RNA) is important because fundamental processes like DNA **replication**, the **transcription** of DNA into RNA, and the **translation** of RNA into protein proceed in one direction only along its length. By convention, the sequence of bases in a DNA or RNA molecule is always written left to right in the direction 5′ to 3′, the direction in which the processes of transcription and translation proceed.

The structure described above is that of **single-stranded DNA**. Within a cell, DNA exists for most of the time in the **double-stranded** form, as two long strands, spiraling round each other in a double helix (Figure 2.4b). The bases of one strand project into the core of the helix, and here they pair with the bases of the other **complementary** strand. Specificity of **base pairing** (**Figure 2.5**), in which an A pairs strictly with a T, and a C with a G, underlies the two fundamental roles of DNA—expression of genetic information via RNA, and replication of genetic information prior to cell division. With the help of specific **enzymes** called **polymerases**, a single strand of DNA can act as a **template** either for the synthesis of a complementary strand of RNA (transcription), using **ribonucleotides** as building blocks (see **Section 2.3**), or of a complementary DNA strand (replication), when the building blocks are deoxyribonucleotides. The lengths of double-stranded DNA molecules are described in units of base pairs (bp), and for longer molecules kilobase pairs (usually abbreviated to **kilobases**, kb), megabase pairs (**megabases**, Mb), or even gigabase pairs (**gigabases**, Gb).

The chemical bonds that tie the atoms together within a single strand of DNA are strong and require considerable energy to break them—these are **covalent bonds**. The **hydrogen bonds** which exist within a double helix of DNA between the bases of one strand and another are much weaker, and can be broken ***in vitro*** by relatively gentle processes such as brief heating to 95°C or exposure to alkaline pH, or ***in vivo*** by active processes within cells. This separation of the strands of a double helix is called **denaturation**, or **melting**, and must occur prior to either transcription or replication to allow access to the genetic information. There are three hydrogen bonds between a G and a C, and two between an A and a T; more energy is therefore needed to denature GC-rich DNA than an equivalent length of AT-rich DNA.

2.3 GENES, TRANSCRIPTION, AND TRANSLATION

As we shall see, much of human DNA appears to have no specific function. However, some segments of DNA contain the instructions for the synthesis of proteins, or RNA molecules with specific functions. These segments are called genes. Protein production from a gene does not proceed directly from the DNA template. **RNA polymerase** first makes an intermediate RNA molecule, known as **messenger RNA** (mRNA) from the DNA template, in the process known as transcription. mRNA is then used as a template to convert the code of nucleotides into the code of **amino acids**, to produce a protein, via the process of translation. The many steps involved in making a protein product from a starting DNA template provide myriad opportunities for regulatory processes to influence the final amount and properties of active protein in the cell.

Genes are made up of introns and exons, and include elements to initiate and regulate transcription

The structure of a typical human protein-coding gene is shown in **Figure 2.6**. There is much more to a gene than the DNA segment that actually encodes the protein: gene transcription proceeds from the 5′ to the 3′ end, and outside the coding region at the 5′ end lie sequences necessary for transcription to be initiated and regulated, including the **promoter**. The gene may also have an **enhancer**, a short regulatory sequence that increases transcription, and which may lie distant from the gene in either direction. At the 3′ end there are sequences which signal the termination of transcription, and the addition of a

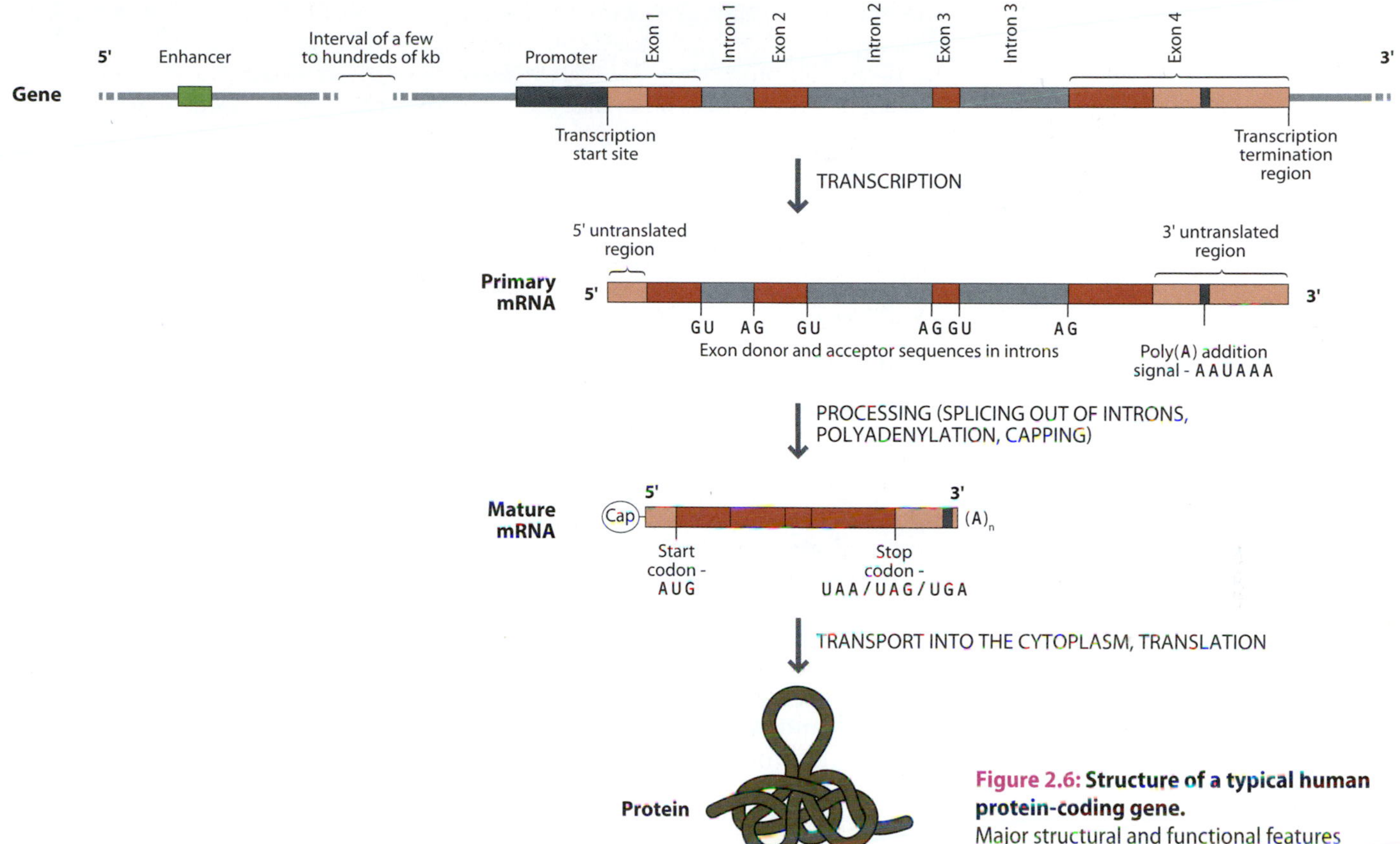

Figure 2.6: Structure of a typical human protein-coding gene.
Major structural and functional features of a typical gene are shown. Primary and mature mRNA products from the gene are also shown, together with the sequences of absolutely or very highly conserved elements involved in processing and translation; note that splicing can commence on incompletely transcribed mRNAs. The different elements of the gene are not to scale, and in particular the introns illustrated here are small relative to the exons (see Table 2.1 for typical sizes of these elements).

tail of multiple adenosine ribonucleotides [**poly(A) tail**] to the end of the mRNA molecule, which is important for mRNA stability.

Most human genes are not made up of a continuous block of sequence encoding the protein, but are divided into segments known as **exons**, with an average length of about 150 bp. Between these lie noncoding segments known as **introns**. Transcription produces a **pre-mRNA** that contains both the exons and the introns. A complex process known as **splicing** then removes the introns, before mature mRNA can be translated into protein. Because the introns are removed from the mRNA, their sequences do not directly affect the final protein product. However, the splicing process depends on some aspects of intron sequence, including the first two and last two nucleotides, which are usually GU and AG respectively. The mRNA molecule is stabilized by the addition of a cap consisting of a specialized nucleotide at the 5′ end, as well as the poly(A) tail at the 3′ end, signaled by the presence of the consensus sequence AAUAAA. As with every step in the process between DNA and protein, splicing is exploited as a powerful means to regulate the expression of genes and the composition of gene products. Many, perhaps most, human genes can be spliced in a variety of ways that include or exclude particular exons, so that one gene often codes for several related proteins.

Figure 2.6 shows a typical gene, but in reality human genes vary greatly in size and complexity (**Table 2.1**). The record-breaking dystrophin gene, mutations in which can cause the disease Duchenne muscular dystrophy (OMIM 310200), spans over 2.7 Mb of DNA, with 79 exons averaging only 180 bp in size marooned in an ocean of introns with an average size of 30 kb (and representing 99.4% of the gene). So large is this gene that the production of a single complete transcript is estimated to take as long as 16 hours,[18] compared to a few minutes for an average gene. The *SRY* gene, on the other hand, which is responsible for the initiation of testis development and thus male sex-determination, spans only a kilobase or so and has no introns at all.

TABLE 2.1:
PROPERTIES OF HUMAN GENES

	Mean values for 1804 genes	Dystrophin gene (*DMD*)	Sex-determining region, Y gene (*SRY*)
Exon length	145 bp	180 bp mean	612 bp
Exon number	8.8	79	1
Intron length	3365 bp	30,000 bp mean	–
5′ UTR length	300 bp	200 bp	140 bp
3′ UTR length	770 bp		133 bp
Coding sequence length	1340 bp (447 aa)	14,000 bp	612 bp (204 aa)
Genomic extent	27 kb	2700 kb	1 kb

[Data from International Human Genome Sequencing Consortium (2001) *Nature* 409, 860; Roberts RG et al. (1993) *Genomics* 16, 536; Koenig M et al. (1987) *Cell* 50, 509; Behlke MA et al. (1993) *Genomics* 17, 736.]
aa, amino acid; UTR, untranslated region.

Mature, spliced mRNA is transported out of the nucleus into the cytoplasm, where it acts as the template for protein production. The production of many copies of mRNA from a single gene amplifies the number of copies of the corresponding protein that can be made, as does the subsequent production of many protein molecules from each mRNA copy. Human mRNAs are eventually degraded with a median half-life of ~10 hours.[21]

The genetic code allows nucleotide sequences to be translated into amino acid sequences

While DNA and RNA are each composed of four kinds of nucleotides, proteins (sometimes called **polypeptides**) are polymers composed of 20 kinds of amino acids, and the complex process by which information encoded in one kind of polymer is converted into the other is called translation. The nucleotide code, known as the **genetic code**, is illustrated in Figure 2.7. In this near-universal system, a set of three adjacent nucleotides (a **codon**) specifies one of the amino acids, or alternatively represents an instruction to halt translation (**stop**

The genetic code

Codons	Amino acid			Codons	Amino acid			Codons	Amino acid			Codons	Amino acid		
AAA AAG	Lysine	**Lys**	K	CAA CAG	Glutamine	**Gln**	Q	GAA GAG	Glutamic acid	**Glu**	E	UAA UAG	STOP		
AAC AAU	Asparagine	**Asn**	N	CAC CAU	Histidine	**His**	H	GAC GAU	Aspartic acid	**Asp**	D	UAC UAU	Tyrosine	**Tyr**	Y
ACA ACG ACC ACU	Threonine	**Thr**	T	CCA CCG CCC CCU	Proline	**Pro**	P	GCA GCG GCC GCU	Alanine	**Ala**	A	UCA UCG UCC UCU	Serine	**Ser**	S
AGA AGG	Arginine	**Arg**	R	CGA CGG CGC CGU	Arginine	**Arg**	R	GGA GGG GGC GGU	Glycine	**Gly**	G	UGA	STOP		
AGC AGU	Serine	**Ser**	S									UGG	Tryptophan	**Trp**	W
												UGC UGU	Cysteine	**Cys**	C
AUA	Isoleucine	**Ile**	I	CUA CUG CUC CUU	Leucine	**Leu**	L	GUA GUG GUC GUU	Valine	**Val**	V	UUA UUG	Leucine	**Leu**	L
AUG	Methionine	**Met**	M									UUC UUU	Phenylalanine	**Phe**	F
AUC AUU	Isoleucine	**Ile**	I												

Figure 2.7: The genetic code.
Three-base codons and their corresponding amino acids (including three-letter and single-letter abbreviations) or translation STOP signals are shown. This is the code used in the human nuclear genome; the code in mtDNA differs (see Table 2.3).

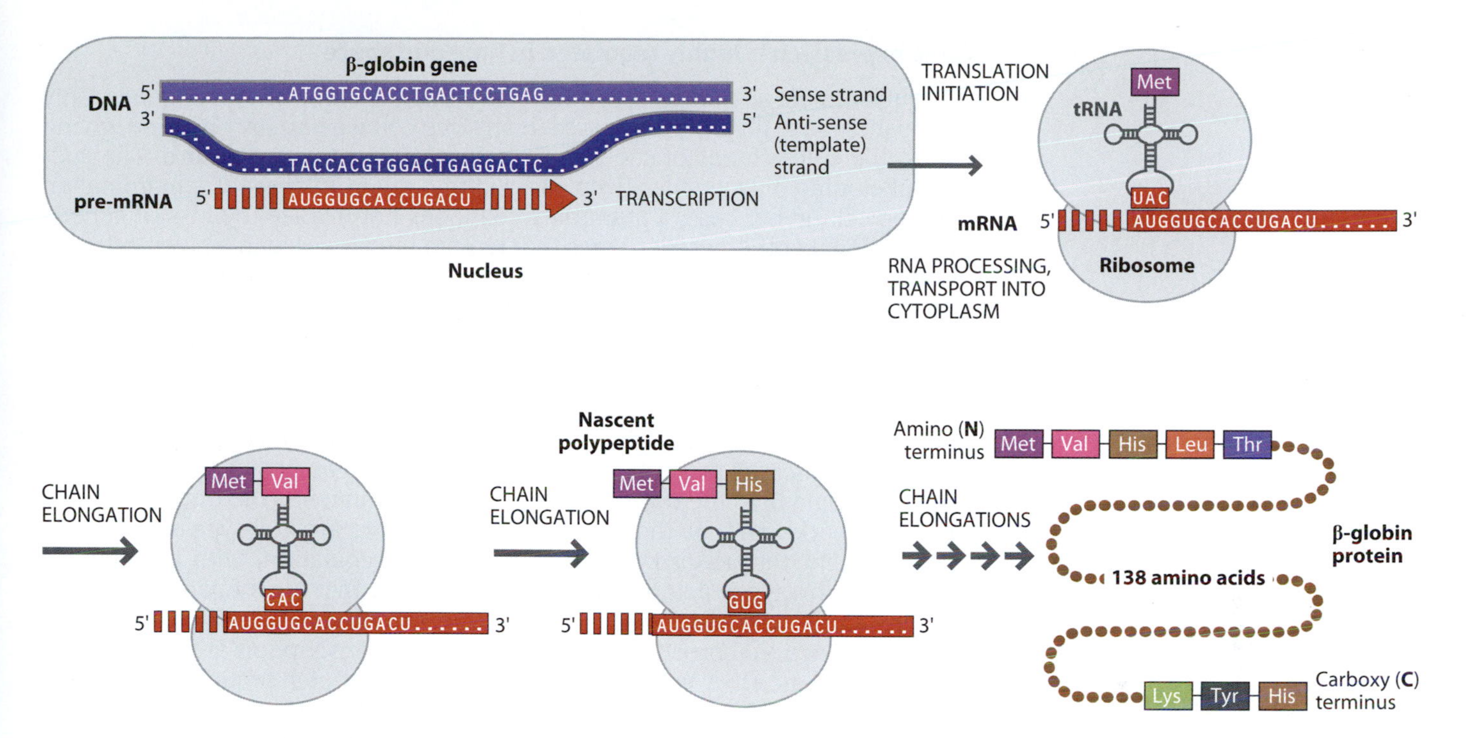

Figure 2.8: DNA makes RNA makes protein. Transcription of a protein-coding gene (here the β-globin gene) occurs in the nucleus, and after RNA processing and transport into the cytoplasm the mRNA is translated by the ribosome. Each codon in the mRNA is recognized by the complementary anticodon in a transfer RNA (tRNA) molecule that bears the appropriate amino acid. As the ribosome moves in a 5′ to 3′ direction along the mRNA, amino acids are added to the growing chain by formation of peptide bonds between them. Note that in this particular protein the initiating methionine is cleaved off after translation at a later stage (not shown); many proteins undergo other important post-translational modifications.

codon). The codons within a gene are arranged adjacent to each other, without intervening punctuation. The process of translating three bases into one amino acid involves one of a set of small intermediate RNA molecules (**transfer RNA**; **tRNA**), and is carried out in the cytoplasm by a large RNA–protein complex called the **ribosome** (**Figure 2.8**). Within its nucleotide sequence each type of tRNA molecule carries a specific set of three adjacent nucleotides, the **anticodon**, which binds to the relevant codon in the mRNA through base pairing. When primed to take part in protein synthesis, it also carries a specific amino acid, and thus represents the physical link between nucleotide sequence and amino acid sequence. Translation is initiated by the codon AUG, which specifies the amino acid methionine. A tRNA bearing this amino acid binds to the AUG codon via the anticodon CAU, and then, as the ribosome moves in a 5′ to 3′ direction along the mRNA, subsequent tRNAs bearing their specific amino acids enter the ribosome, and the chain of amino acids grows by the formation of **peptide bonds** between them. Translation ceases when the ribosome reaches one of the three possible stop codons (UAA, UAG, or UGA). The 5′-most translated region of the mRNA corresponds to the end of the protein bearing a free amino (-NH_2) group (the **N-terminus**) and the 3′-most translated region to that bearing a free carboxy (-COOH) group (the **C-terminus**).

Since there are 64 possible triplet codons, there is more than enough capacity in the genetic code to specify the 20 amino acids and a stop signal. The code is therefore **redundant**: at one extreme, each of the amino acids leucine, serine, and arginine has six different corresponding codons, while at the other, tryptophan and methionine each has only one. The pattern of redundancy in the code, and the relationship between the physicochemical properties of the amino acids and their codons, is nonrandom, and has important implications for the effects of mutations on protein function, which are discussed in Chapter 3 (**Section 3.2**).

Gene expression is highly regulated in time and space

Although almost all human cells contain a complete complement of genes, only a subset of these genes is expressed in any one cell type at any one time, giving rise to the cell's characteristic set of proteins, called the **proteome**. The successful cloning of Dolly the sheep showed that differentiation of a mammalian somatic cell (in her case, a mammary cell) does not involve the loss of genetic information. In other words, genes that remain unexpressed in a given cell type are not removed from the cell. The quantitative, spatial, and temporal regulation of the expression of genes is a complex process involving interactions of a very large number of factors within the cell, and evolution has exploited every stage of this process as an opportunity for regulation.

Different cell types often employ different transcription start sites, giving variability in the 5′ ends of transcripts, and thus the amino termini of proteins, which can confer different properties. Proteins functioning as positive and negative regulators bind 5′ to the transcription start site to influence the timing and rate of transcription, and, in cases where the regulators are cell-type specific, the cell specificity. Introns provide further scope for regulation, with **alternative splicing** pathways utilizing different exons from within a gene, and producing different forms of proteins with different functions, or even, under some conditions, nonfunctional proteins. Factors regulating transcription are classified as ***cis*** (on the same DNA molecule, for example a promoter sequence), or ***trans*** (on a different molecule, for example a protein transcription factor). Action of such transcriptional regulators leads to the production of some 160,000 distinct transcripts from the ~25,000 genes (www.ensembl.org/Homo_sapiens/Info/Index), emphasizing the importance of regulation in the diversity of gene products.

After transcription, different factors act to influence mRNA stability and half-life, and thus the amount of protein produced. The properties of the final protein product are also influenced by post-translational modifications such as specific cleavage, the addition of chemical groups such as phosphates or acetyl groups, cross-linking via cysteine side chains, and conjugation with sugar or lipid molecules.

Making sense of the tens of thousands of genes and gene products is a daunting task. This has been facilitated by a gene classification system called **gene ontology** (www.geneontology.org), which is a catalog of each gene product's cellular component of expression, molecular function, and the biological process in which it is involved.

2.4 NONCODING DNA

The end product of the genes described above (comprising only ~2% of the genome) is protein, with RNA playing the role of a temporary intermediate informational molecule. However, the process of protein production also requires noncoding RNAs (**ncRNAs**) which function in their own right, including molecules with key functions in translation, such as transfer RNAs and ribosomal RNAs. The human genome also encodes ~1500 **microRNAs** (abbreviated **miRNAs**, or **miRs**), short (22 nt on average) post-transcriptional regulatory molecules that bind to complementary sequences in target mRNAs, usually resulting in translational repression or mRNA degradation and hence gene silencing. It is now becoming clear that much more of the genome than previously thought is transcribed into ncRNA species which may have important and specific functions within cells. Indeed, the human **transcriptome** may result from the transcription of as much as 2400 Mb of the human genome (75%), with at least 25% being transcribed from both strands.[13] The functions of the great majority of these ncRNAs are unknown, and a proportion may represent transcriptional noise. However, evidence is accumulating that many ncRNAs play important roles in the control of chromosome dynamics, splicing, translational inhibition, and mRNA destruction, and our protein-centric view of gene expression and regulation is undergoing a change.

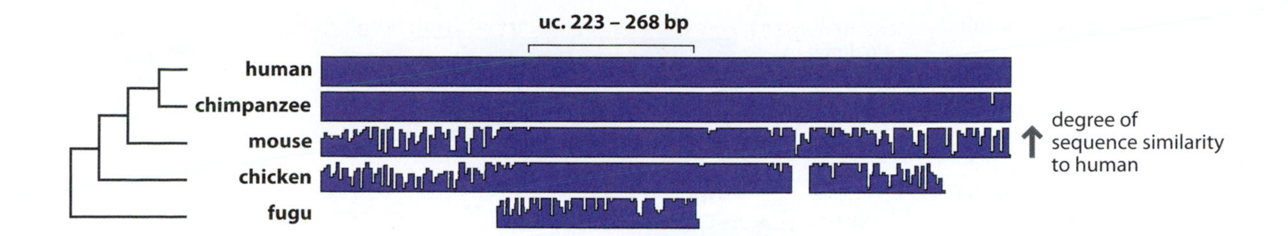

Figure 2.9: Recognizing functionally important elements by comparative genomics.
Sequence conservation around a 268-bp "ultraconserved element" (uc.223; Bejerano G et al. (2004) *Science* 304, 1321) within the *FOXP2* gene, showing sequence similarity as histograms compared to the human reference sequence for human, chimpanzee, mouse, chicken, and *Fugu rubripes* (puffer fish) genomes. Evolutionary relationships are shown in the tree to the left. Note that there is a very high degree of conservation between human and chimpanzee across the whole sequence, but that more distantly related species show restricted blocks of sequence conservation, thus pinpointing functionally important sequences. Conservation was identified using the UCSC Genome Browser.

A key method to identify human sequences that are functionally important is to compare them with their counterparts in other species, known as **orthologs**. This can be done on a genomewide scale, when the enterprise is called **comparative genomics**. DNA sequences change through time by mutation, but functionally important sequences change more slowly because changes are more likely to have deleterious effects (purifying selection; Chapter 5). Natural selection acts to remove the mutated versions, and so to conserve these sequences. In species that are very closely related to humans, such as chimpanzees (Chapter 7), there is a very high degree of conservation of almost all sequences. However, in more distantly related species such as the mouse, the chicken, or the puffer fish, overall sequence conservation is lower, and so specific highly conserved sequences stand out more clearly (Figure 2.9). Many coding sequences and important regulatory elements can readily be identified in this way. However, the approach also reveals several hundred mysterious **ultraconserved elements**[6] of at least 200 bp length, which are perfectly conserved between human, rat, and mouse, and mostly conserved in dog (99% identical) and chicken (95% identical). Despite this very strong evidence for functional importance, the functions of ultraconserved elements remain unclear; only about one-fifth are known to overlap exons, but general association with genes involved in RNA binding and splicing, and DNA binding and transcriptional regulation, suggests that they may play some unknown but important regulatory role in vertebrate development. This is supported by the finding that many elements possess enhancer activity *in vivo*.[15]

Some DNA sequences in the genome are repeated in multiple copies

The distinction is often made between single-copy and repetitive DNA, and genes tend to be placed in the former category. However, this is an oversimplification. Many genes have more than one copy, some have related copies that have arisen in the past through gene duplication events, and some lie in recent duplications of large chromosomal segments (**segmental duplications**) that cover many megabases and have their origins in recent primate evolution.[4] **Copy-number variation** exists,[8] so that some genes are single-copy in some genomes, but multi-copy in others (**Section 3.6**). At the deepest level, some human genes show sequence similarities to others that reflect ancient duplications of an entire ancestral genome that occurred early in the evolution of the vertebrate **lineage**, several hundred million years ago.[20]

Aside from the relatively low–copy-number repetition described above, the human genome contains a vast number of very highly repetitive sequences (Table 2.2). This highly repetitive component comprises dispersed repeat sequences with copy numbers in the hundreds to hundreds of thousands, comprising about 45% of the genome.[10] The extravagant production of this kind of DNA seems to be the result of "ignorant" or "selfish" processes within the genome that amplify sequences which then cannot readily be removed. Provided the presence of these sequences is not strongly deleterious, they will be tolerated and persist, although in some cases they may have evolved **genic** or regulatory functions. The structures of an **L1 element** and an ***Alu* element**, members of the two commonest classes of dispersed repeats, respectively the **LINEs** (long interspersed nuclear elements) and **SINEs** (short interspersed nuclear elements), are shown in Figure 2.10.

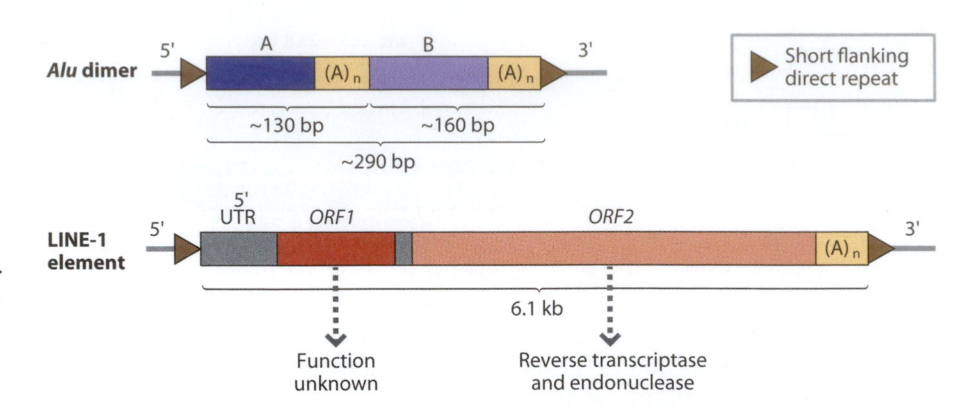

Figure 2.10: Structural features of full-length *Alu* and LINE-1 elements. The two monomers within the *Alu* dimer differ in length because of a 32-bp insertion in the B monomer. Lengths are approximate because of variation in the length of the poly(A) tails [$(A)_n$]. Note that the internal poly(A) tract is significantly more diverged than the tract at the 3′ end. Full-length LINE-1 elements are rare: most are truncated at the 5′ end. ORF, open reading frame; UTR, untranslated region. Diagrams are not to scale.

TABLE 2.2: CLASSES OF DISPERSED REPEATS IN THE HUMAN GENOME

Class	Copy no. per haploid genome	Fraction of genome	Autonomous transposition or retrotransposition?	Length of complete copies
LINEs	850,000	21%	yes	up to 6–8 kb
SINEs	1,500,000	13%	no	up to 100–300 bp
Retrovirus-like elements	450,000	8%	complete copies, yes	6–11 kb
DNA transposon copies	300,000	3%	complete copies, yes	2–3 kb

Incomplete elements, incapable of autonomous transposition, are common (see **Section 3.5**). [Data from International Human Genome Sequencing Consortium (2001) *Nature* 409, 860.]

2.5 HUMAN CHROMOSOMES AND THE HUMAN KARYOTYPE

The linear nature of DNA means that large genomes correspond to extremely long molecules. So great is the size of the haploid human genome that, if it were a single DNA duplex and stretched out, it would be about 1 m long. Clearly, since each of our nucleated somatic cells contains two copies of this genome in its nucleus, a mere 10 μm or so in diameter, it must be very tightly packaged, but at the same time it must be able to be faithfully replicated and segregated at cell division, and genes must be accessible for expression. Packaging of DNA occurs on many levels (**Figure 2.11**): the DNA molecule is first coiled round **nucleosomes**—octamers (eight-member complexes) of proteins called **histones**. This nucleosomal fiber is then itself coiled, and several successive levels of packaging give the required degree of condensation. Complex and highly regulated enzymatic modification of these histone proteins has evolved as an additional

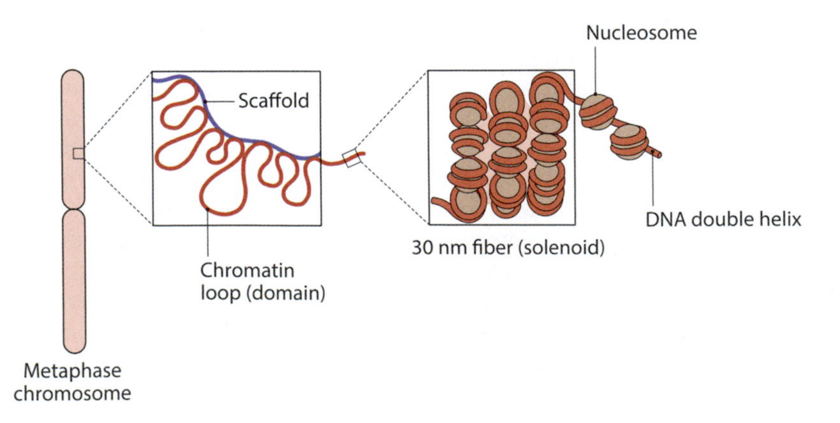

Figure 2.11: The packaging of DNA into chromatin and the chromosome.

mechanism for regulating gene expression by influencing DNA packaging. Like DNA methylation, this is an example of an epigenetic effect, which alters the behavior of the genome without involving any actual sequence change.

The human genome is divided into 46 chromosomes

The problem of managing such a large genome is also alleviated by subdivision: instead of existing as a single DNA molecule, the nuclear genetic material of the somatic cells of normal human individuals is divided into 46 separate molecules—the chromosomes. These in turn can be divided into 23 pairs (**Figures 2.12** and **2.13**): one of each pair is inherited from each parent. Twenty-two of

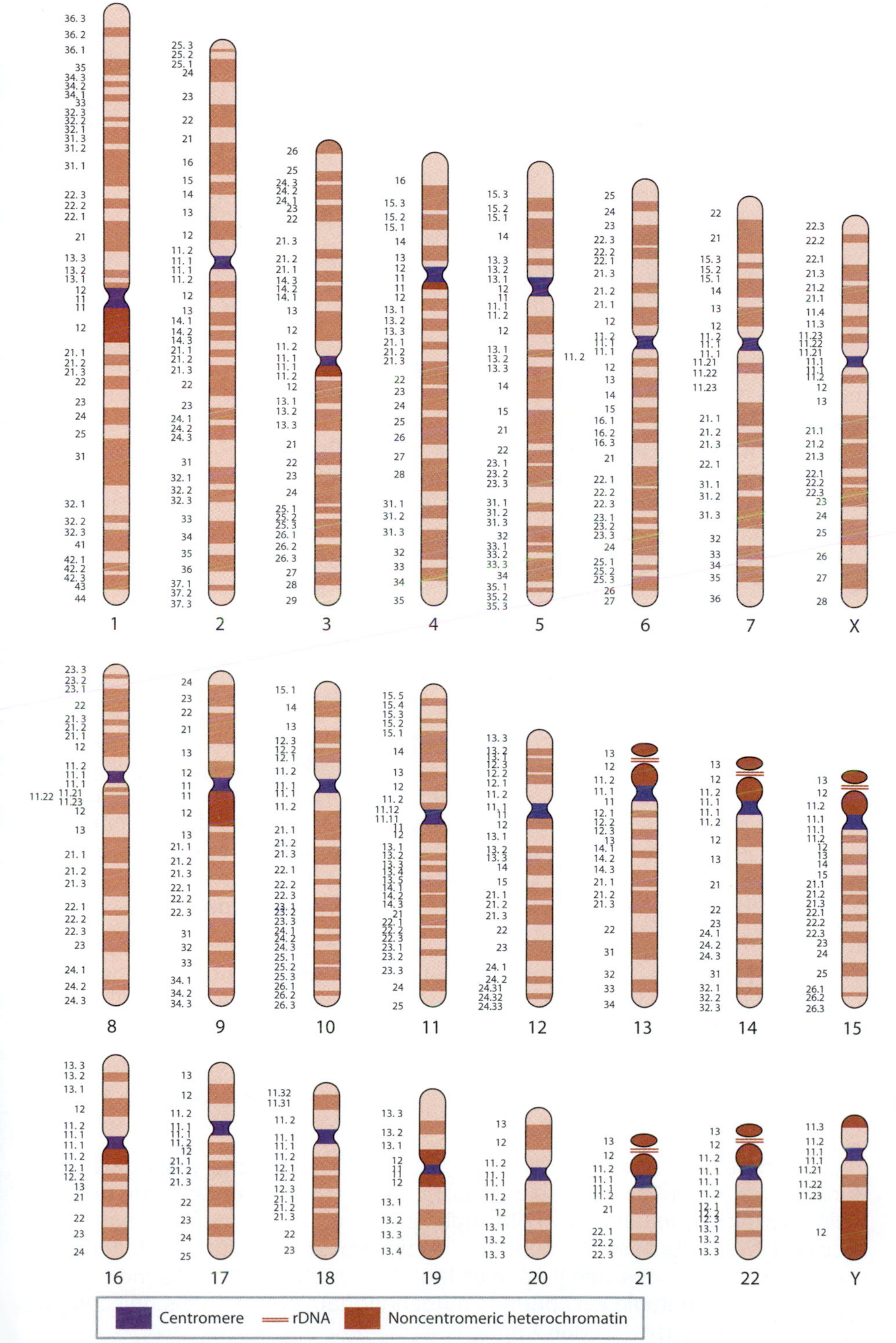

Figure 2.12: Human G-banded karyogram.
Knob structures interrupted by lines on the short arms of chromosomes 13, 14, 15, 21, and 22 contain arrays of ribosomal DNA (**rDNA**) repeats, encoding noncoding RNA molecules that form part of the ribosome. [From Strachan T & Read AP (2010) Human Molecular Genetics, 4th ed. Garland Science.]

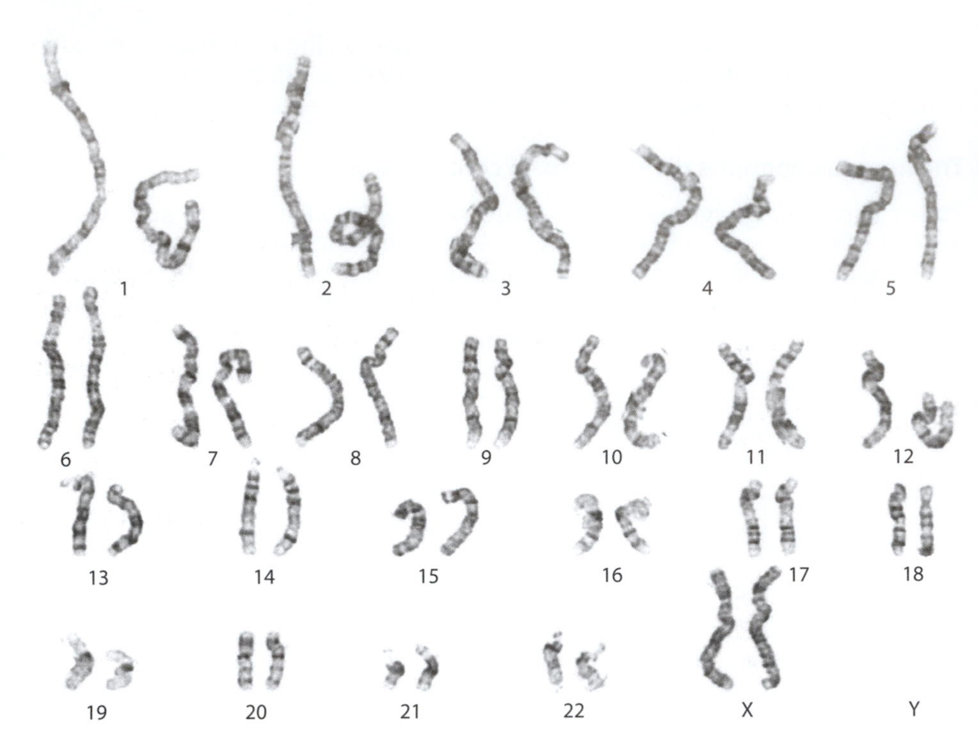

Figure 2.13: G-banded prometaphase chromosome karyogram of mitotic chromosomes from lymphocytes of a normal female.
Compare with idealized banding patterns in Figure 2.12. [Courtesy of Dr Sue Hamilton, Manchester University Hospital.]

the pairs, the **autosomes**, are identical between the sexes, and are numbered from 1 (the largest) to 22. The remaining two chromosomes are known as the **sex chromosomes**, because they differ between the sexes. Females have two copies of the **X chromosome**, which, at ~155 Mb, is about the same size as chromosome 7. Males have one X chromosome, but in addition they have one **Y chromosome** which is one of the smallest (~60 Mb) of all chromosomes. The Y is sex-determining in mammals through the action, in early development, of a gene, ***SRY*** (sex-determining region, Y), which causes the gonad precursor to become a testis rather than an ovary. Subsequent steps in sex determination are mediated by hormones released by the gonads. The chromosomal constitution of an individual is known as their **karyotype**, and the karyotypes of normal females and males are denoted 46,XX and 46,XY respectively.

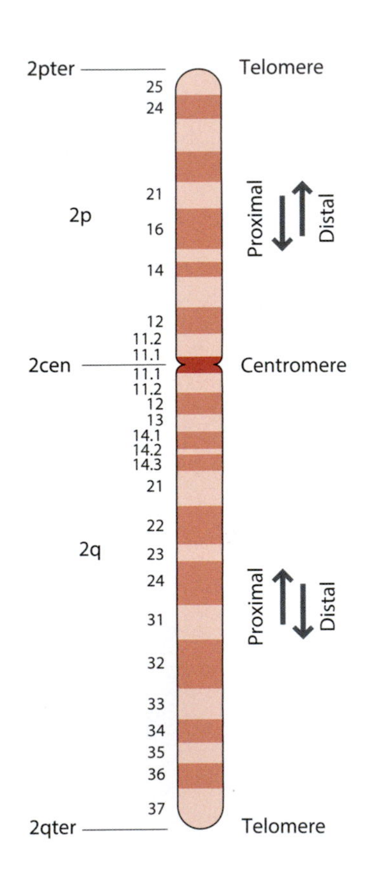

Figure 2.14: Structural features of a typical human chromosome.
This representation, of a G-banded chromosome 2, is known as an idiogram. The numbers of the dark and light bands are given to the left. The telomeres are given the suffix "ter" (for "terminus"), and the centromere "cen." Note that a two-digit band such as q23 is pronounced "two-three," not "twenty-three." Arrows to the right indicate the conventional arm-specific directions of proximal (toward the centromere) and distal (toward the telomere).

Figure 2.14 shows the structural features of a typical chromosome. It is important to stress that this kind of representation, showing a condensed, compact structure with easily recognizable chromosomal arms, reflects the situation for only a small part of the cell cycle, and that the characteristic banding pattern is produced artificially by staining methods. Except for a period during mitosis or meiosis, when cells are dividing, chromosomes are diffuse, and discrete structural features are difficult or impossible to recognize. Each chromosome has two essential structural elements to ensure its correct **segregation** and stability:

- The centromere, seen as a constriction in the chromosome at some stages of the cell cycle, is the point at which the **kinetochore** and **spindle fibers** attach when chromosomes are segregated into daughter cells at cell division.
- The **telomeres**, at each end of a chromosome, are essential to prevent chromosomes fusing with each other, and protect the chromosome against loss of sequences from its ends.

The chromosomes of some organisms, such as bacteria and yeast, possess a third feature essential for stable inheritance: these are **replication origins**, which permit the initiation of DNA replication, and which are associated with specific DNA sequences. In human chromosomes, replication is initiated at many hundreds of sites, and it appears that there is little or no sequence specificity involved. A notable exception is mitochondrial DNA, perhaps reflecting its bacterial ancestry (see **Section 2.8**).

Both centromeres and telomeres are complexes of DNA and specific structural proteins, and in each case the DNA component includes a tandem array of repeated sequences. In the case of centromeres, the **alphoid repeat** unit, of 170 bp, is arranged into a higher order repeat unit of up to a few kilobases, varying between chromosomes. This is then itself repeated many times to produce an array which is variable in length between chromosomes and individuals, but can be as large as 5 Mb. Telomeres have a 6-bp repeat unit, TTAGGG, arranged into tandem arrays of tens of kilobases at the very ends of chromosomes. Just **proximal** to these arrays are other more complex repeated sequences (pericentric or subtelomeric repeats) that tend to vary between centromeres or telomeres.

Size, centromere position, and staining methods allow chromosomes to be distinguished

Without the use of staining methods, differences between chromosomes that are visible through a microscope are their lengths, and the position of the centromere (seen as a constriction in most cases), which allow them to be divided into groups. While some human chromosomes are **metacentric**, with the centromere dividing the chromosome into two clearly recognizable arms, others are **acrocentric**, with the centromere lying very close to one end. The smaller of a chromosome's arms is designated the **p arm** (from the French: *petit*—small), and the larger, conventionally drawn at the bottom, is the **q arm** (for *queue*—tail). The use of particular staining methods reveals specific banding patterns, allowing all 24 different human chromosomes to be distinguished from each other. One widely used procedure is treatment with the protease trypsin, followed by Giemsa staining, which produces a pattern known as **G-banding**. Current methods allow 850 or more G-bands to be identified in a human **karyogram** (Figures 2.12 and 2.13), the **cytogenetic** representation of a chromosome set. Some chromosomes contain regions which behave abnormally in chromosomal banding, and which may vary in size between individuals as a result of their high repeat content. These regions, termed heterochromatin, are often associated with centromeres and represent highly condensed, transcriptionally inert segments of the genome. The remainder of the genome is termed **euchromatin**.

The underlying chemical basis of G-banding is unclear, but the availability of the genome sequence has revealed that differently staining bands reflect differences in nucleotide composition, usually described as **GC-content**, which expresses the percentage of base pairs that are G–C, as opposed to A–T. This percentage is 41% for the genome as a whole, but there are multi-megabase sections that depart substantially from this figure, ranging from 36% to 50% average GC-content. Dark-staining G-bands correlate with regions of the genome that have relatively low GC-content, and are also particularly poor in genes and SINEs such as *Alu* elements; light G-bands, in contrast, have high GC-content and are gene- and SINE-rich.

2.6 MITOSIS, MEIOSIS, AND THE INHERITANCE OF THE GENOME

When a cell divides, it passes on its genetic material to its daughter cells. There are two kinds of cell division, known as **mitosis** and **meiosis**. In mitosis, the process seen in somatic tissues, each daughter cell contains the same genetic material as the parental cell. The specialized division process of meiosis, however, is needed to produce a **gamete** (egg or sperm), which contains only half the diploid complement of genetic material (that is, is haploid). The coming together of egg and sperm at fertilization restores the diploid complement in the **zygote**. Meiosis is thus the critical process in the passage of the human genome from one generation to the next. As we will see below, different segments of the genome are inherited from the mother or father in different ways. **Figure 2.15** summarizes the features of the inheritance of the genome from one generation to the next.

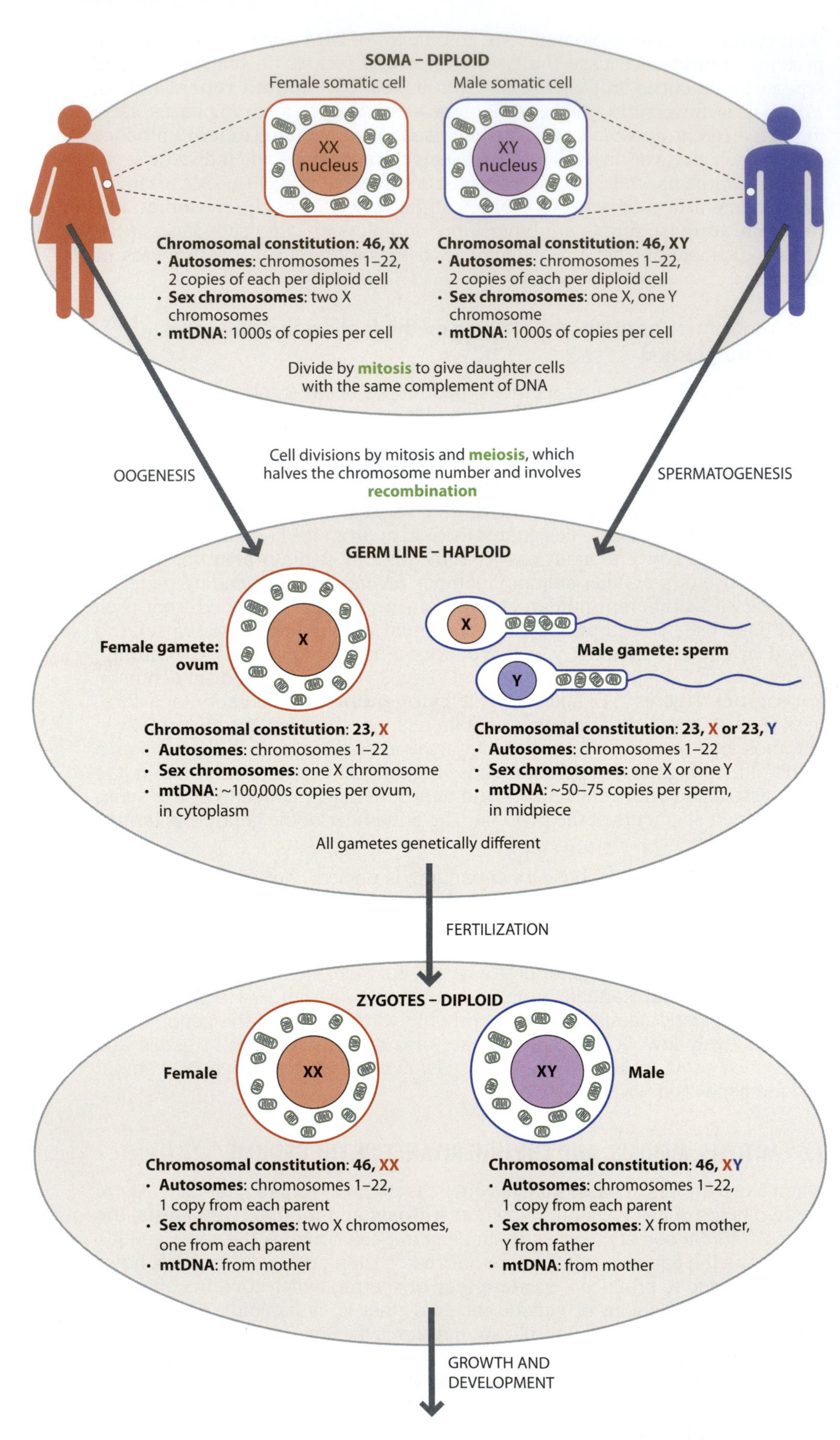

Figure 2.15: Overview of inheritance patterns of different segments of the human genome.

During mitosis a somatic cell's DNA replicates, so that it is temporarily tetraploid (contains four copies of the genome). The chromosomes condense, and the two copies of each chromosome are associated at their centromeres—these identical associated copies are known as **sister chromatids**. After the nuclear envelope has dissolved, the chromosomes align at the **metaphase plate**, a region in the center of the cell. The associated centromeres of sister chromatids then separate, and the two chromatids of each chromosome move to opposite poles of the cell. After this, the nuclear envelope re-forms around each set of segregated chromosomes as they de-condense, the cytoplasm divides, and cell division is complete, resulting in diploid daughter cells which are genetically identical to the diploid parent.

Mitosis is a critical process; each of us starts life as a single cell, the fertilized egg, and develops and survives as a result of the enormous number of mitotic cell divisions (estimated to be about 10^{17}) necessary for growth, development, and maintenance. Mitosis has great importance in disease, since errors in mitotic divisions can contribute to cancer development. However, in evolutionary terms the most important class of cell divisions is that which gives rise to the gametes, and it is these that enable the passage of genetic information to the next generation. Mitotic mutations will exist only as long as the individual who carries them, whereas any mutations that occur in meiosis can be passed on in perpetuity.

A gamete (egg or sperm; also known as a germ cell) is haploid; it contains only one copy of the genome, as opposed to the usual two copies in our somatic, diploid cells. The production of gametes in the **germ line** proceeds first by a series of mitotic cell divisions, leading in females from oogonia to primary **oocytes**, and in males from spermatogonia to primary spermatocytes (**Figure 2.16**). Following this, however, the cells enter meiosis, a cell division process

Figure 2.16: Gametogenesis in males and females.
Gametogenesis proceeds in males by mitotic divisions of diploid spermatogonia to give diploid primary spermatocytes. In females diploid oogonia divide to give primary oocytes. These cells can undergo meiosis. In females meiosis I occurs in a single primary oocyte at ovulation, and the fate of the secondary oocyte depends on whether or not it is fertilized. Unfertilized secondary oocytes are lost during menstruation, but contact of a sperm with the cell membrane triggers a rapid second meiosis. This entire process can take as much as 50 years. In contrast, spermatogenesis is rapid (48 days) and occurs throughout post-pubertal life.

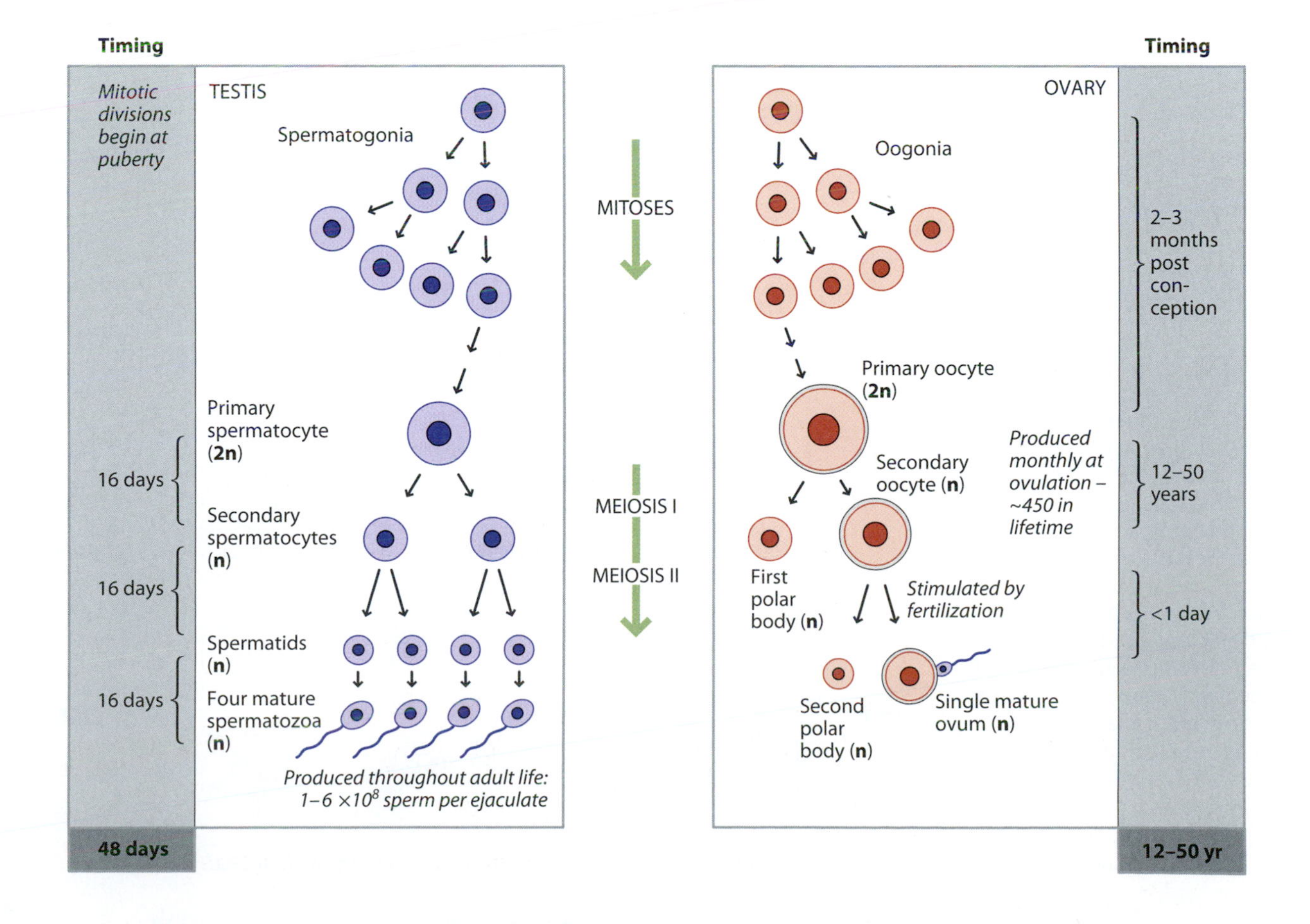

specific to the germ line. There are two fundamental distinctions between meiosis and mitosis:

- Meiosis includes a single round of DNA replication, but unlike mitosis it involves two subsequent cell divisions, reducing the genetic material from two copies to one.
- Cells produced by mitosis are genetically identical to each other (and to the parental cell), but the haploid cells (gametes) produced by meiosis are all genetically different.

The phases of meiosis are summarized in **Figure 2.17**. The genetic differences between gametes arise from two distinct processes: clearly, the reduction from diploidy to haploidy necessitates a choice of either the grand-paternal or the grand-maternal copy of each chromosome to pass into the gamete. This **independent assortment** of chromosomes alone leads to differences between gametes—provided the choice is random, the possible number of different combinations of haploid chromosome subsets is very large: 2^{23} (8,388,608). However, there is a second, important level of modification in the passage of genetic material to the gamete: recombination. During meiosis paternal and maternal chromosomal homologs align and exchange segments through recombination, also known as **crossing over**. This process is reciprocal, and there is no net loss of genetic information. Assortment and recombination, neither of which occurs during a normal mitotic cell division, ensure that any one gamete produced by a man or woman is genetically different from any other.

2.7 RECOMBINATION—THE GREAT RESHUFFLER

A key tool in human genetics is the pedigree: a collection of related individuals which allows us to observe how segments of DNA are inherited by children from their parents. These segments of DNA can be observed directly, as differences

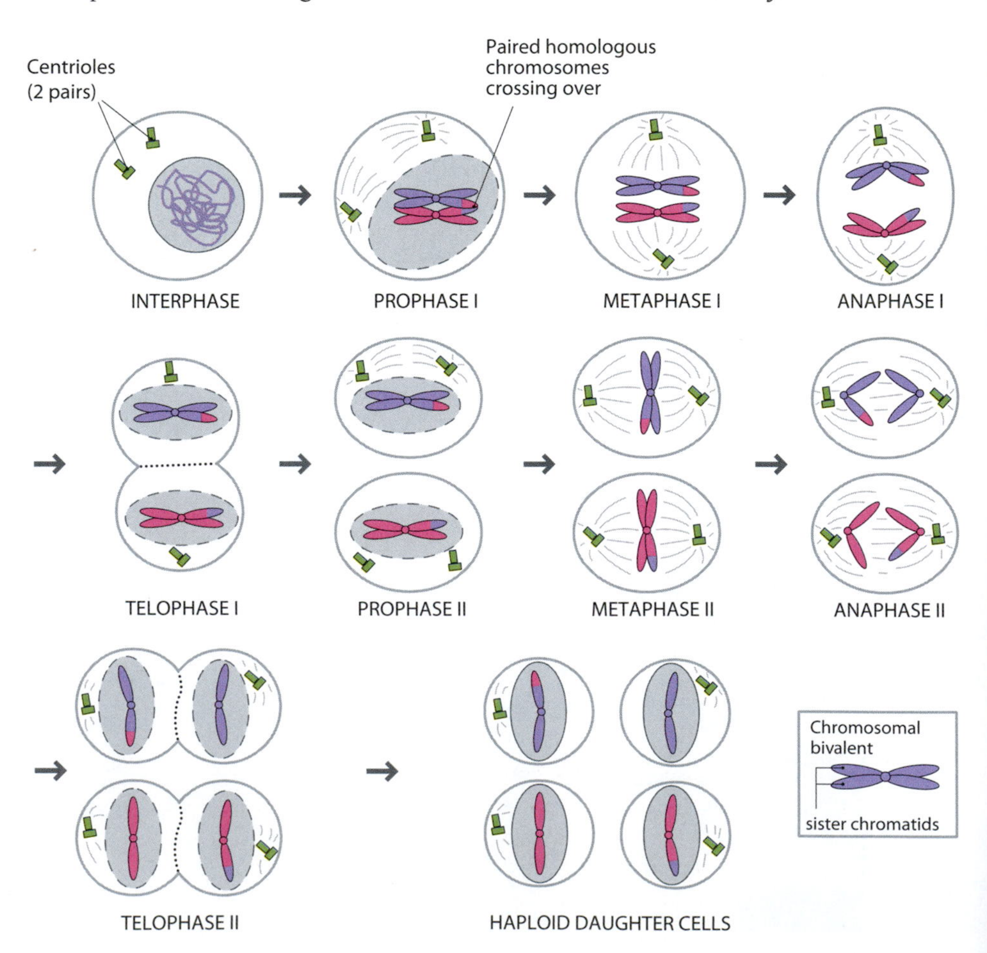

Figure 2.17: The stages of meiosis. DNA is replicated in interphase, and chromosomes condense and thicken in prophase I—a single chromosome is shown. At this stage crossing over also occurs and the nuclear envelope breaks down. During metaphase I and anaphase I the chromosomal bivalents line up and are then pulled apart, and diploid daughter cells form. After prophase II, the chromosomal bivalents again line up at metaphase II, and at anaphase II the sister chromatids are pulled apart by the spindle apparatus. Cell division is complete after telophase II, and each gamete contains one haploid copy of the chromosome. Two are recombinant, and two are nonrecombinant.

between homologous DNA sequences (**alleles**—see Chapter 3), or indirectly, in the form of phenotypes (such as diseases) that result from a specific underlying difference in sequence. Sequence variants (sometimes called **markers**) that lie on different chromosomes are independently inherited, and are said to be **unlinked**. However, two variants on the same chromosome, particularly ones that lie close together, may often be co-inherited within pedigrees; they are said to be **linked**. The co-inheritance of a pair of linked genetic variants can be disrupted by recombination events, when crossing over occurs between them, and a child inherits one variant, but not the other. The recognition of such recombination events allows us to determine the order of genetic variants along a DNA molecule, and the counting of recombination events allows us to estimate the **genetic distance** between variants; this is **genetic mapping**. Genetic distance is expressed in units of recombination, which are **centiMorgans** (cM): 1 cM of recombination between a pair of variants is a genetic distance corresponding to a 1% recombination frequency between them. With the advanced knowledge we now have of the physical structure of the human genome (measured in Mb, rather than cM), we can readily compare the **genetic map** with the **physical map** (**Figure 2.18**). As a genomewide average, the correspondence between genetic and physical distance is ~1 cM/Mb.

The frequency of recombination events varies between the sexes, because the production of eggs involves more recombination events than the production of sperm. Males have roughly 50 recombination events per meiosis, whereas females experience about 80, so the genetic map of females is expanded on average about 1.65-fold with respect to that of males (Figure 2.18). Also, the recombination rate across the genome within each sex is far from uniform. Recombination is more frequent toward the telomeres of chromosomes (up to 3 cM/Mb), and less frequent toward their centromeres (<0.1 cM/Mb). Regional variation is present on many scales: recombination rates are on average about twice as high (per Mb) on the smallest chromosomes compared with the largest because successful meiosis requires recombination, but both recombination "desert" and "jungle" segments can be identified within chromosome arms.

These average values over large chromosomal regions obscure substantial recombination rate heterogeneity at the sequence level. Direct molecular analysis of recombination indicates that, for some regions of the genome at least, events are concentrated in small (1–2-kb) segments of DNA known as **recombination hotspots**, separated by comparatively large regions of low recombinational activity.[11] There could be as much as 1000-fold difference in recombination rates between hotspots and cold domains. This non-uniformity of recombination takes on great importance when we come to consider the

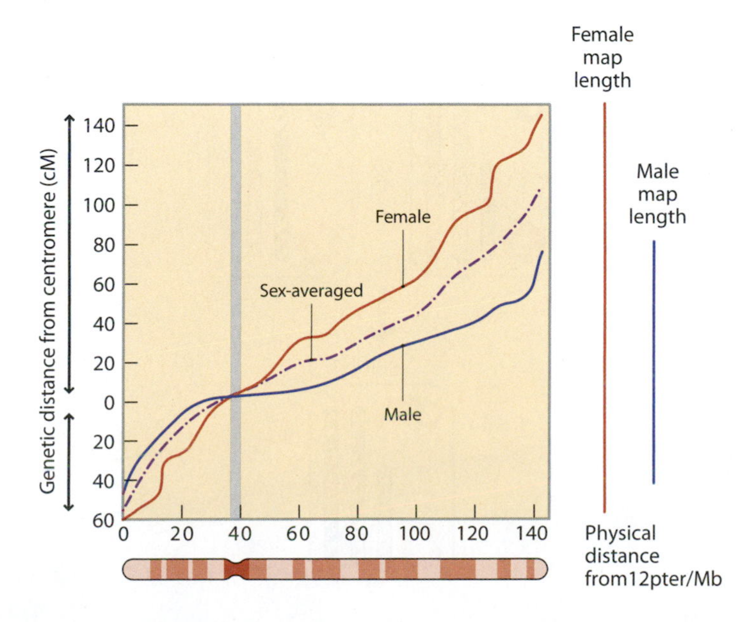

Figure 2.18: Genetic and physical distance compared.
Female, male, and sex-averaged genetic distance in cM plotted against physical distance in Mb for chromosome 12. Female recombination is elevated compared to male recombination. Recombination is generally elevated toward the telomeres, and depressed around the centromere. [Adapted from International Human Genome Sequencing Consortium (2001). *Nature* 409, 860. With permission from Macmillan Publishers Ltd.]

issue of the distribution of sequence variants along chromosomes in different human populations, in Chapter 17, and is discussed further in Chapter 3 (**Section 3.8**). Finally, there is individual heterogeneity: significant variation in recombination rates in pedigree data can be detected between individual women, although not between individual men,[12] and direct analysis of recombination in sperm DNA shows a high degree of variability in the recombination activity of specific hotspots between different men.[14] A specific protein, PR domain–containing 9 (PRDM9), has recently been shown to act as a master regulator of recombination hotspots, and is discussed further in Chapter 3.

Recombination undoubtedly plays an important role in generating new combinations of DNA sequences, and thus genetic and phenotypic diversity, from one generation to the next, but it also has importance for the behaviors of chromosomes in meiosis. Most chromosomes undergo on average just over one recombination event per chromosomal arm, and it seems that these events are necessary for the faithful segregation of chromosomes into daughter cells at division. Indeed, some chromosomal **aneuploidies** (departure from the correct number of chromosomes, due to **nondisjunction**) are accompanied by reduced recombination, or a failure to recombine.[9]

2.8 NONRECOMBINING SEGMENTS OF THE GENOME

The great majority of the human genome is inherited from both parents, and undergoes reshuffling each generation through recombination. However, there are two segments of our DNA that are atypical, being inherited from one parent only, and escaping recombination. These are the majority of the Y chromosome, and mitochondrial DNA (**Figure 2.19**). Uniparental inheritance and escape from recombination simplify interpretation of the historical record written in the DNA of these segments of the genome, as we shall see in later chapters. Furthermore, the sex-specificity of inheritance means that they contain records of male- and female-specific population processes and behaviors.

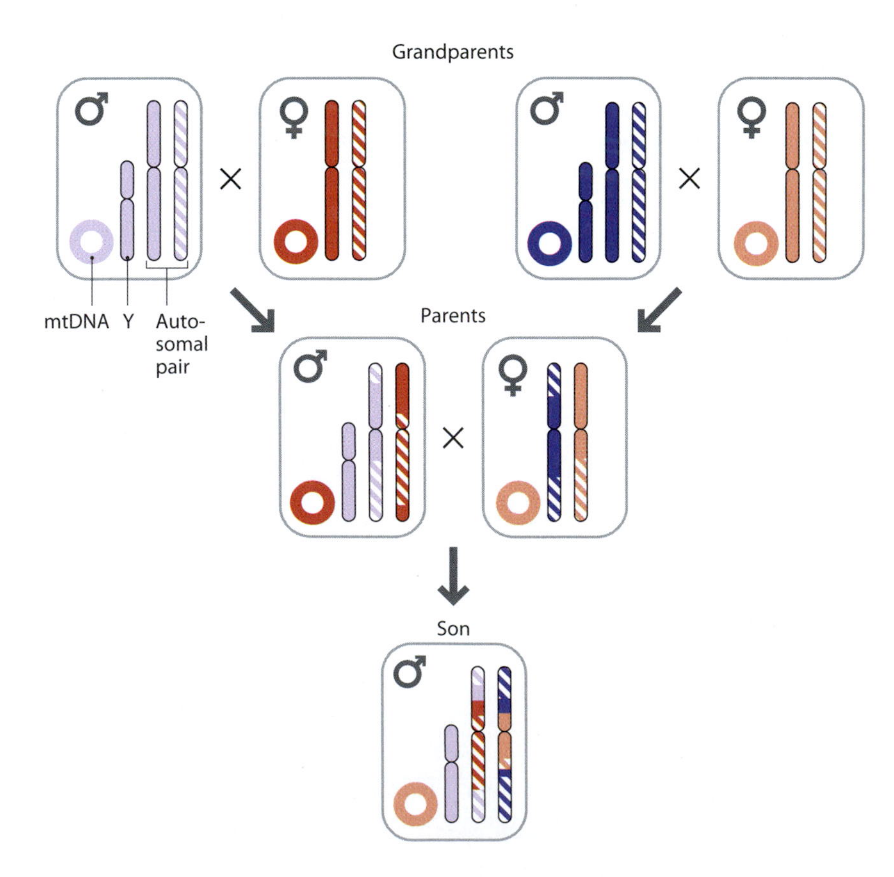

Figure 2.19: Inheritance of recombining and nonrecombining segments of the genome.
Schematic illustration of a three-generation pedigree. The son's Y chromosome and mtDNA each descends from a single grandparental ancestor, his paternal grandfather and maternal grandmother, respectively. In contrast, his autosomes descend from all of his grandparents, because of reshuffling of chromosomal segments in every generation through recombination.

The male-specific Y chromosome escapes crossing over for most of its length

Because of the Y chromosome's defining role in sex determination, it is male-specific, haploid, and passed from father to son. The Appendix describes this locus in greater detail. Since it has no homolog with which to recombine, we expect the Y chromosome to avoid recombination. This is true for more than 90% of its length—a region known as the **nonrecombining portion of the Y** (sometimes **NRPY**, or **NRY**), and sometimes the **male-specific region of the Y** (**MSY**). However, given the importance of recombination in ensuring correct segregation of chromosomes (**Section 2.7**), it is not surprising to find that the Y does actually recombine with the X chromosome in male meiosis, in specialized regions where sequence similarity with the X is preserved (**Figure 2.20**). Sequences within these regions can be inherited from either parent, like sequences on autosomes, and the regions are therefore referred to as **pseudoautosomal**. One of these regions, pseudoautosomal region 2 (PAR2), lying at the tips of the long arms of the X and Y chromosomes is a recent evolutionary acquisition specific to humans, and of little importance in **chromosomal segregation**. However, PAR1, a 2.6-Mb region at the tips of the short arms, derives from the ancient origin of the mammalian sex chromosomes as a pair of homologous autosomes some 300 MYA, and is the site of a recombination event in every male meiosis. The unusual behavior of the Y chromosome has consequences for its partner, the X (**Box 2.2**).

Maternally inherited mtDNA escapes from recombination

The other constitutively nonrecombining region of our genetic material is mitochondrial DNA (mtDNA), a circular double-stranded DNA molecule about 16.5 kb in size (**Figure 2.21**; and see the Appendix for a general description) whose entire sequence is known.[1,2] This is contained not within the nucleus, but within mitochondria—cytoplasmic organelles in which the energy-generating

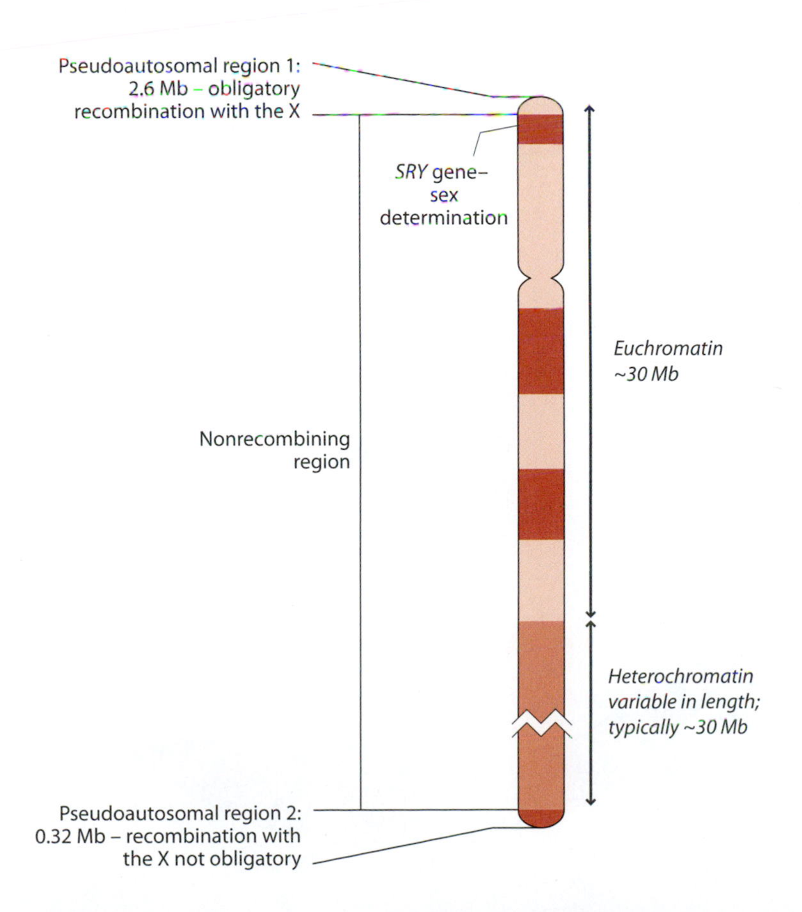

Figure 2.20: The Y chromosome. Idiogram of a G-banded Y chromosome, showing positions of intervals associated with specific phenotypes, and the sex-determining gene *SRY*. Correspondence of intervals with G-bands is only approximate.

Box 2.2: The unusual features of the X chromosome

The different sex chromosome constitutions of males (46,XY) and females (46,XX) have consequences not only for the Y, but also for the X chromosome:

- It is inherited twice as frequently from females as from males.
- In male meiosis the X chromosome behaves like the Y chromosome, recombining only in the pseudoautosomal regions, but in female meiosis it behaves more like an autosome, with recombination possible along its entire length.
- To balance the doses of X-chromosomal genes, one of the two X chromosomes in the somatic tissues of a female is largely transcriptionally silenced by **X-inactivation**. The initial choice of which X chromosome is inactivated is random and made early in development, but maintained clonally thereafter in daughter cells.
- The X has a low gene density compared with the rest of the genome, but is relatively rich in genes involved in reproductive and cognitive functions.
- The haploid nature of the X chromosome in males exposes them to increased risk of **X-linked** diseases, because in females the second copy can mask the effect of mutation. For this reason, there has been strong ascertainment of severe diseases linked to the X—it contains only 4% of human genes, but ~10% of Mendelian diseases have been assigned to it.
- Haploidy in males also has practical implications: the single copy makes X-chromosomal DNA variation relatively easy to analyze in male DNA samples.

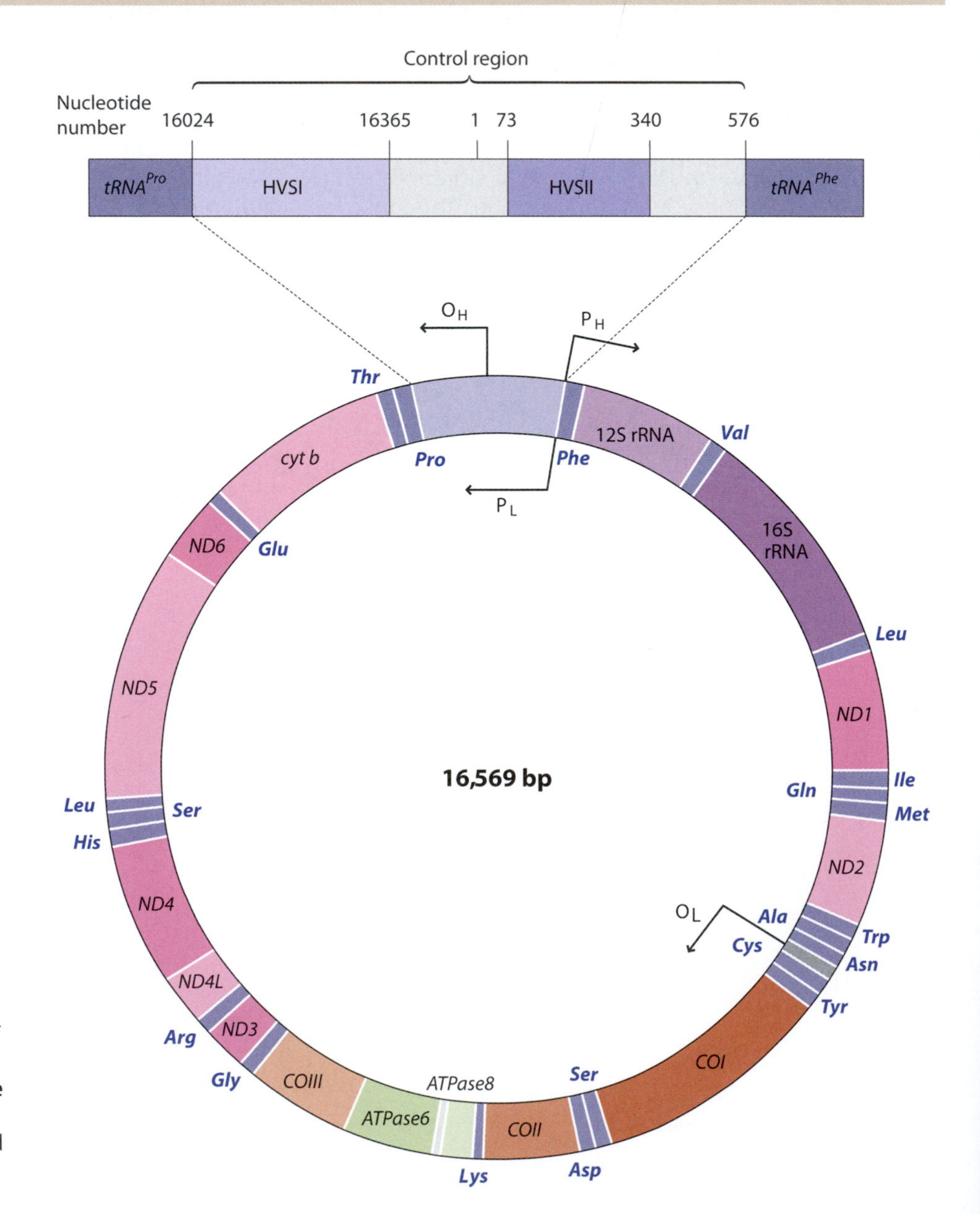

Figure 2.21: Mitochondrial DNA. Human mtDNA is a circular double-stranded molecule with one strand (the heavy strand) relatively rich in G bases, and the other (the light strand) rich in C. tRNA genes are indicated by the three-letter name of the corresponding amino acid (see Figure 2.7); remaining genes are protein coding, except for the two ribosomal RNA (rRNA) genes. Origins of replication of the light (O_L) and heavy (O_H) strands, and promoters for transcription of these two strands (P_L, P_H) are shown. The control region contains two **hypervariable segments** (also known as **hypervariable regions**, HVRs) that are commonly assayed for variability (HVSI and HVSII).

process of oxidative phosphorylation takes place. Many features of mitochondria and mtDNA suggest that these organelles have a bacterial origin.

This controversial theory was originally proposed by Lynn Margulis in the 1960s and is now accepted—mitochondria originated as **endosymbiotic** bacteria, taken up into proto-**eukaryotic** cells about 1.5 billion years ago, and providing energy generation in return for a safe environment. This **prokaryotic** past has left its traces in the many remarkable features of mtDNA that resemble those of modern bacterial genomes (**Table 2.3**). Perhaps the most striking evidence of the exogenous origin of mtDNA is in the genetic code: in mammals, five codons have different specificities in mtDNA compared with the nuclear genome.

Since its origin, mtDNA has lost most of its genes, and hence its autonomy. The 37 genes it now carries all play roles in either the oxidative phosphorylation pathway, or mitochondrial protein synthesis. The other genes essential for mitochondrial function, including those encoding mitochondrial-specific DNA- and RNA polymerases and many structural and transport proteins, have been transferred to the nuclear genome.

As well as this ancient transfer of genes, fragments of mtDNA have been inserted into the nuclear genome at various points in our more recent evolutionary history. Indeed, a survey of the draft sequence of the nuclear genome[7] indicates that it contains over 400 kb of copies of sequences currently residing on mtDNA—25 times as much as the mtDNA molecule itself! Some of these nuclear mtDNA insertions (**numts**) are ancient, and shared among all primates, while others are more recent, with some even showing variation in presence or absence among human populations. As we shall see in **Section 4.10**, numts are a potential source of problems in mtDNA studies in ancient samples, but also have their uses in the rooting of phylogenetic trees constructed using mtDNA.

The number of mitochondria in a cell varies with cell type: those requiring a lot of energy, such as nerve and muscle cells, contain thousands of mitochondria, each containing 2–10 copies of mtDNA, while other cell types may contain only a few hundred. Oocytes contain around 100,000 mitochondria, each containing a single mtDNA molecule, while sperm contain only about 50–75. Clearly, even if fertilization involved a complete mixing of paternal and maternal mtDNA molecules, the contribution of the father to the zygote's pool of mtDNA would

TABLE 2.3:
DIFFERENCES BETWEEN THE NUCLEAR AND mtDNA GENOMES

	Nuclear genome	mtDNA
Genome	linear chromosomes, ~50–250 Mb in size	circular molecule, ~16.5 kb
Packaging	packaged with histones as chromatin	not packaged
DNA replication origins	1000s of origins, sometimes diffuse	two discrete origins, one on each strand
Genes	~25,000 genes, mostly containing introns; much intergenic DNA	37 genes with no introns; almost no intergenic DNA
Transcripts	each gene has its own transcript(s)	only one from each strand; each mRNA covers many genes
Genetic code differences:		
UGA	STOP	Trp
AGA, AGG	Arg	STOP
AUA, AUU	Ile	Met

be relatively small. However, most evidence suggests that his contribution is actually zero, and that, under normal circumstances, mtDNA is exclusively maternally inherited. Sperm mitochondria, required for motility, are localized in the midpiece, between the sperm head and tail. After fertilization the midpiece and its paternal mitochondria can be seen within the zygote, and observed through several subsequent cell divisions. Note that some textbooks wrongly state that the midpiece of the sperm does not enter the egg.[3] The apparent absence of paternal mtDNA inheritance suggests that there may be an active system to eliminate paternal mitochondria; indeed, the one documented case of paternal inheritance, involving a **pathogenic** mtDNA mutation,[16] may reflect a defect in this elimination process.

If paternal mtDNA inheritance did occur, recombination between mtDNAs from different lineages could in principle happen—the mitochondrion contains the necessary molecular machinery to allow this. This fact, and some features of the pattern of polymorphic sites within the mtDNA molecule, have led to much debate over whether human mtDNA does indeed recombine, which is explored in Chapter 3.

SUMMARY

- DNA is a polymeric molecule composed of four nucleotide monomers, containing the bases A, G, C, and T. The order of bases contains instructions for the production of proteins, and for the transmission of identical genetic information to daughter cells. DNA is double-stranded, and its length is measured in base pairs (bp).
- Most human somatic cells are diploid (contain two genome copies), while germ cells (sperm and eggs) are haploid (contain one copy). The haploid genome comprises about 3200 Mb of DNA, and its sequence has been determined almost completely.
- The genome contains ~25,000 genes that contain instructions for producing proteins which build cells. DNA is transcribed into RNA in the nucleus, and this nucleotide sequence is then translated in the cytoplasm into amino acid sequences to give proteins, via the genetic code. As well as coding sequences, genes contain regulatory sequences, and are usually broken up into coding segments (exons) separated by introns. Exons are stitched together after transcription by splicing.
- Most of our DNA is noncoding, though as well as protein-coding genes, there are genes whose products are untranslated RNAs, and much of the genome is actually transcribed into RNA of unknown function. About 45% of the genome is made up of highly repetitive dispersed sequences including *Alu* and L1 elements.
- The diploid genome is divided into 46 chromosomes: 22 pairs of autosomes are shared between men and women, and women have an additional two X chromosomes while men have an X and a Y chromosome, which determines sex. Chromosomes have structural features essential for stability and segregation at cell division, the telomeres and centromere.
- Somatic cell division (mitosis) ensures orderly segregation of chromosomes into genetically identical daughter cells after DNA replication. Meiosis is a specialized cell division in the germ line that reduces the number of chromosomes by half to give haploid and genetically unique gametes (eggs and sperm), and includes recombination—exchange of segments between pairs of homologous chromosomes. Women show more recombination than men during meiosis. Fertilization re-creates the diploid state, and creates a unique combination of DNA sequences in a new individual.
- Recombination affects how DNA sequences are inherited. Autosomes recombine in both men and women, X chromosomes only in women, and

the paternally inherited Y chromosome doesn't recombine at all—with the exceptions of the pseudoautosomal regions shared between the sex chromosomes.

- Mitochondria contain their own circular genomes of 16.5 kb, which are maternally inherited and escape recombination because paternal mitochondria from sperm do not persist after fertilization.

QUESTIONS

Question 2–1: Identify the two chemical differences between the nucleic acids DNA and RNA.

Question 2–2: If one DNA strand in a duplex is composed of 50% pyrimidine bases, and one in eight bases on the complementary strand is a G, what is the T content of the first strand?

Question 2–3: For the DNA sequence:
5′-ATGGGACCACAAAAAGCGTAA-3′, give:
(a) The reverse complementary DNA sequence, that would form a base-paired double helix with it
(b) The mRNA sequence that would be derived from it, if it were the sense strand of a gene
(c) The resulting amino acid sequence if the mRNA were translated into protein

Question 2–4: Place in order, from the tip of the short arm to the tip of the long arm, the following chromosomal locations: 11cen, 11p15.5, 11q22.1, 11qter, 11p15.3, 11q24, 11p12, 11pter. Refer to Figure 2.14.

Question 2–5: Give two key distinctions between mitosis and meiosis.

Question 2–6: Describe the differences in DNA content between a mature egg and a mature sperm cell.

Question 2–7: What is the probability that a man inherits the following portions of the genome from his mother?
(a) His mtDNA
(b) His Y chromosome
(c) His X chromosome
(d) A random 1-kb segment of one of his copies of chromosome 11

Question 2–8: A 2-Mb segment of the genome has 49% GC-content and an average recombination frequency of 2.8 cM/Mb. Describe its likely content (high, or low) of genes and *Alu* elements, its G-banding characteristics, and its likely position on a chromosome arm.

Question 2–9: How is the structure of DNA adapted to its role in transmitting genetic information from one generation to the next?

Question 2–10: Describe how the amino acid sequence of a protein is related to the DNA sequence of a gene that encodes it.

Question 2–11: What is heterochromatin, where is it located, what is its function, and what problems does it cause in genome sequencing?

Question 2–12: Visit http://genome.ucsc.edu/cgi-bin/hgGateway and enter "chrX:31,000,000-33,150,000" into the "position/search" box, then click "Submit" to enter a region containing an extraordinary gene. Use the "zoom" and "move" buttons to navigate around the region, and explore the displayed features and available tracks.

REFERENCES

The references highlighted in purple are considered to be important (for this chapter) by the authors.

1. **Anderson S, Bankier AT, Barrell GB et al.** (1981) Sequence and organisation of the human mitochondrial genome. *Nature* **290**, 457–465.
2. **Andrews RM, Kubacka I, Chinnery PF et al.** (1999) Reanalysis and revision of the Cambridge reference sequence for human mitochondrial DNA. *Nat. Genet.* **23**, 147.
3. **Ankel-Simons F & Cummins JM** (1996) Misconceptions about mitochondria and mammalian fertilization: implications for theories on human evolution. *Proc. Natl Acad. Sci. USA* **93**, 13859–13863.
4. **Bailey JA & Eichler EE** (2006) Primate segmental duplications: crucibles of evolution, diversity and disease. *Nat. Rev. Genet.* **7**, 552–564.
5. **Bejerano G, Pheasant M, Makunin I et al.** (2004) Ultraconserved elements in the human genome. *Science* **304**, 1321–1325.
6. **Bensasson D, Zhang D-X, Hartl DL & Hewitt GM** (2001) Mitochondrial pseudogenes: evolution's misplaced witnesses. *Trends Ecol. Evol.* **16**, 314–321.
7. **ENCODE Project Consortium** (2012) An integrated encyclopedia of DNA elements in the human genome. *Nature* **489**, 57–74.
8. **Freeman JL, Perry GH, Feuk L et al.** (2006) Copy number variation: new insights in genome diversity. *Genome Res.* **16**, 949–961.
9. **Hassold T & Hunt P** (2001) To err (meiotically) is human: The genesis of human aneuploidy. *Nat. Rev. Genet.* **2**, 280–291.

10. **International Human Genome Sequencing Consortium** (2001) Initial sequencing and analysis of the human genome. *Nature* **409**, 860–921.
11. **Jeffreys AJ, Kauppi L & Neumann R** (2001) Intensely punctate meiotic recombination in the class II region of the major histocompatibility complex. *Nat. Genet.* **29**, 217–222.
12. **Kong A, Gudbjartsson DF, Sainz J et al.** (2002) A high-resolution recombination map of the human genome. *Nat. Genet.* **31**, 241–247.
13. **Mattick JS & Makunin IV** (2006) Non-coding RNA. *Hum. Mol. Genet.* **15** Spec No 1, R17–29.
14. **Neumann R & Jeffreys AJ** (2006) Polymorphism in the activity of human crossover hotspots independent of local DNA sequence variation. *Hum. Mol. Genet.* **15**, 1401–1411.
15. **Pennacchio LA, Ahituv N, Moses AM et al.** (2006) *In vivo* enhancer analysis of human conserved non-coding sequences. *Nature* **444**, 499–502.
16. **Schwartz M & Vissing J** (2002) Paternal inheritance of mitochondrial DNA. *New Engl. J. Med.* **347**, 576–580.
17. **Strachan T & Read AP** (2010) Human Molecular Genetics 4. 4th ed. Garland Science.
18. **Tennyson CN, Klamut HJ & Worton RG** (1995) The human dystrophin gene requires 16 hours to be transcribed and is cotranscriptionally spliced. *Nat. Genet.* **9**, 184–190.
19. **Venter JC, Adams MD, Myers EW et al.** (2001) The sequence of the human genome. *Science* **291**, 1304–1351.
20. **Wolfe KH** (2001) Yesterday's polyploids and the mystery of diploidization. *Nat. Rev. Genet.* **2**, 333–341.
21. **Yang E, van Nimwegen E, Zavolan M et al.** (2003) Decay rates of human mRNAs: correlation with functional characteristics and sequence attributes. *Genome Res.* **13**, 1863–1872.

HUMAN GENOME VARIATION

CHAPTER THREE

In the last chapter, we described the structure, organization, and inheritance of the human genome. This chapter is about the raw material of human evolutionary genetics: genome variation. The ~7 billion people living today carry ~14 billion genome copies, and, with the rare exceptions carried by identical twins, triplets, and so on, all of them are different. The differences between genomes encompass a wide range of scales: from **single nucleotide polymorphisms** (**SNPs**) through to structural variation involving millions of base pairs of DNA sequence. Most differences have no discernible effects on the people who carry them, but some result in phenotypic diversity with no obvious deleterious effect, and some cause, predispose to, or protect against, serious diseases.

In this chapter we give a description of human genome variation. To make useful inferences from this variation it is also necessary to understand how, and at what rates, it arises, so we include an account of the mechanisms of change (mutation) that generate variation. Our approach is to start with the simplest and smallest scale differences between genomes, involving just a single nucleotide, and then to move up in complexity, taking in variation involving tandemly repeated DNA sequences, transposable elements that can jump from one location to another, and structural differences that can encompass many kilobases of DNA. We include a discussion of copy-number variation, now recognized to be a major source of genetic differences between individuals. At the upper end of the scale, some variation is on such a large scale that it can even be observed when viewing chromosomes down the microscope. Along this journey we pay special attention to variation within mitochondrial DNA, which, befitting its unusual evolutionary history (**Section 2.8**), has its own unique rules for mutation and variability. Finally, we reach the key issue of the effects of genetic recombination on human genome diversity, and summarize the findings and implications of the **HapMap** and **1000 Genomes** projects, large-scale collaborative international research programs that have mapped in detail the patterns of variation across the entire genome in DNA samples originating from different continents.

3.1 GENETIC VARIATION AND THE PHENOTYPE

All humans are different from each other, and much of this difference has a genetic basis: differences in **phenotype** caused by differences in **genotype**. Some of these differences are easily observable, and we know from our own experience that they run in families—hair, eye, and skin color, stature, and some morphological features. Others feature in public outreach events on human genetics, and personal genomics companies' studies—color vision deficiencies, and the ability to roll the tongue (OMIM 189300), or to perceive the chemical **phenylthiocarbamide** as bitter-tasting (OMIM 171200; see **Section 15.5**), or to smell freesia flowers (OMIM 229250) or asparagus metabolites in urine (OMIM 108390). Other differences are less obvious but more important, and affect us medically—our blood group, should we require a transfusion, our HLA type, if we need an organ or bone-marrow transplant, and other factors which affect how we will respond to certain drugs and diets, or how likely we are to contract infectious diseases such as malaria or AIDS, or disorders such as diabetes, cancer, asthma, schizophrenia, or coronary heart disease. Some genetic differences are directly responsible for heritable single-gene disorders: an example is **cystic fibrosis** (**CF**; OMIM 219700; Table 16.1; Box 16.3), in which the production of thick, sticky mucus leads to serious problems with the lungs and digestive system. Others make more subtle, but still significant, contributions.

Despite the phenotypic impact of some genetic differences, it is important to appreciate that the vast majority of the many differences between human genome copies have little or no phenotypic effect. Most of the genetic differences between individuals and populations used in human evolutionary genetic studies and discussed in this book are of this kind. They are examples of **neutral** alleles, since they are thought not to affect the evolutionary fitness of someone who carries them, and hence their frequency is not directly affected by natural selection. Note that some alleles might change phenotypes but not fitness—these can also be considered as neutral. The neutral theory of **molecular evolution** is discussed in **Section 5.7**. If we consider a severely deleterious allele, it is easy to think of a "normal," or **wild-type** allele in contrast to it; however, in the context of the many alleles that are neutral or of unknown effect, it is difficult to conceive of a normal, or wild-type human genome (see Box 2.1). In Chapters 15 to 17 we turn specifically to some alleles that are known to have important effects upon our phenotypes. Below, we explain the terminology and general principles behind the influence of genetic variation on disease.

Some DNA sequence variation causes Mendelian genetic disease

We are diploid organisms, with two copies of most genes. When a potentially disease-causing mutation arises in one copy of a gene (one allele), the status of the other copy (allele) of the gene is usually important in determining whether or not the mutation affects the phenotype. Alleles that have a phenotypic consequence when only one copy is mutant can be recognized by the inheritance of a disorder from affected parent to affected child (**Figure 3.1a**), and are referred to as **dominant**. In some dominant phenotypes, such as achondroplasia (OMIM 100800), all people carrying the mutant alleles manifest the disorder fully; the dominant allele is fully **penetrant**. Many dominant alleles, however, do not manifest fully in all people who carry them: this reduced penetrance can complicate interpretation in family studies.

Many disease alleles, including those associated with CF, must be present in *both* copies of the gene for the disease to be expressed, and these are referred to as **recessive**—mutation in one copy can be masked by the presence of a normal, wild-type allele on the other homologous chromosome (Figure 3.1b). The mutant alleles of a patient manifesting a recessive disorder need not be identical at the DNA sequence level, but need only have similar effects on the function of the gene. When the two alleles are identical, the patient is **homozygous**; when they are different, the patient is **heterozygous**.

Most mutant alleles in recessive disorders result in either severe reduction or absence of gene product from that allele, or production of an abnormal, nonfunctional product. A CF patient carries two mutant alleles, and so no functional gene product (the cystic fibrosis transmembrane conductance regulator, a protein which functions as a chloride channel) is produced, leading to the disease phenotype. The parents of a CF patient are phenotypically normal heterozygous **carriers** (Figure 3.1b), who each carry one mutant and one normal allele. Often in these carriers, half as much gene product as normal is being produced, but this is sufficient for normal or near-normal function, and this situation is referred

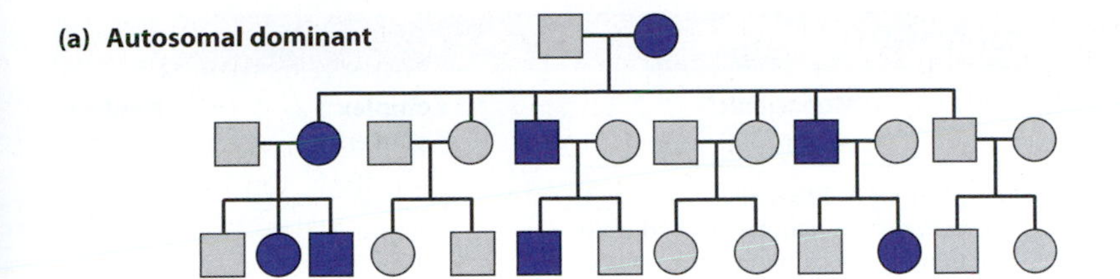

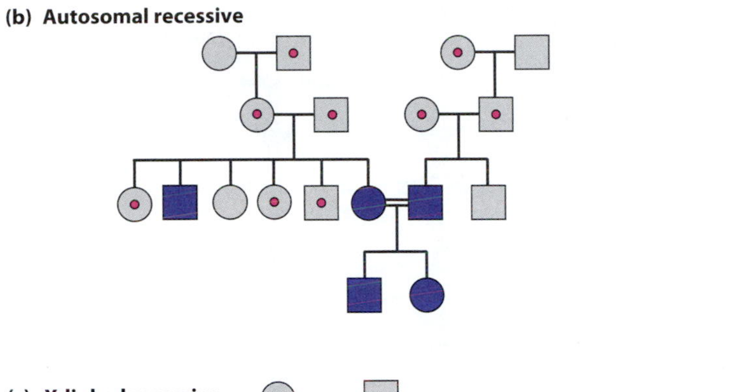

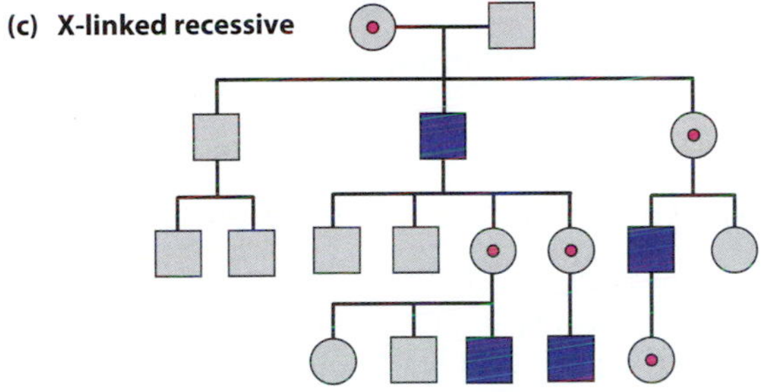

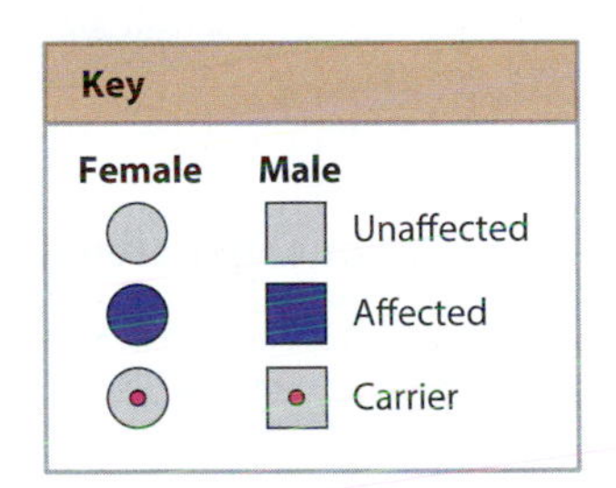

Figure 3.1: Dominant, recessive, and X-linked inheritance.
(a) An autosomal dominant mutant allele segregating in a three-generation pedigree. (b) Segregation of an autosomal recessive mutant allele. (c) Segregation of an X-linked recessive mutant allele.

to as **haplosufficiency** (literally, *a haploid dose is enough*). Many proteins are enzymes, and as catalysts of reactions it is not surprising that they often function adequately in a reduced, haploid dose.

The molecular basis of dominance is more complex than that of recessiveness. It can be **haploinsufficiency**—again, half as much product as normal is produced, but this half-dosage is *not* enough to ensure normal function. Dominance can also result from overproduction of protein from a mutant allele, or expression in an inappropriate place or time. Alleles that give a protein a modified or new function are often dominant. Such **gain-of-function** mutations cause achondroplasia: mutations in the fibroblast growth factor receptor 3 (*FGFR3*) gene lead to activation of the receptor at all times (constitutive activation), instead of in response to specific signals, disrupting normal bone maturation. The basis of dominance in many disorders remains to be elucidated.

Disease-causing alleles on the X chromosome can behave differently in males and females, because males have only one X, while females have two. An X-linked recessive mutant allele (for example, Duchenne muscular dystrophy; OMIM 310200; clinical features summarized in Table 16.1) will manifest in males (who are **hemizygous**), but not in females, who are heterozygous carriers (Figure 3.1c). X-linked dominant mutant alleles will manifest in both sexes; an example is the disease hypophosphatemia (OMIM 307800), with a phenotype of vitamin-D-resistant rickets and short stature.

Autosomal dominant and X-linked disorders are over-represented in disease databases, because they are easier to ascertain than autosomal recessive disorders.

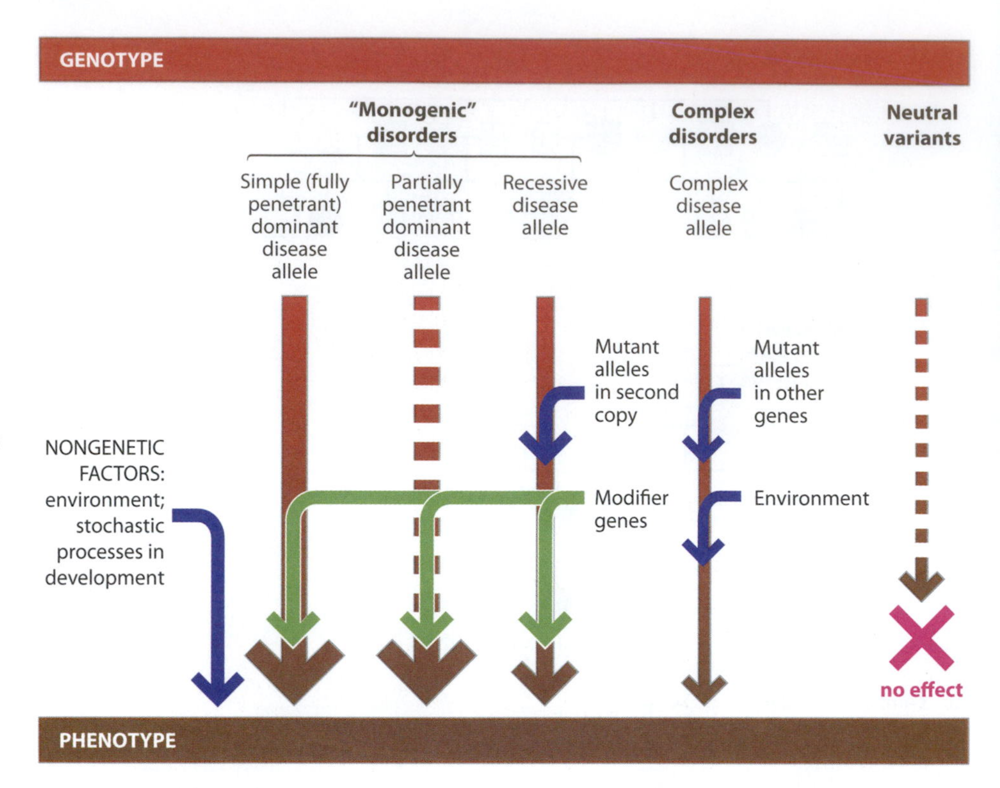

Figure 3.2: The roles of genetic and nongenetic factors in generating the phenotype. Dominant disease alleles have a phenotypic effect even though a normal copy of the allele is present. This can be a full effect in all cases (full penetrance), or in only a subset of cases (reduced penetrance). Recessive disease alleles manifest a phenotypic effect only when no normal copy is present. Alleles involved in complex disorders require mutant alleles in other genes and/or environmental factors for the phenotype to be manifested. Many sequence variants have no known effect on phenotype, and are therefore neutral. Note that some alleles that change phenotype, but not fitness, can also be considered as selectively neutral.

The relationship between genotype and phenotype is usually complex

The genetic basis of some of these phenotypic differences, such as whether or not a person suffers from CF, might seem straightforward and is quite well understood; however, even these so-called simple disorders can manifest variation in the severity of the phenotype, the cause of which is not well understood. The basis of other, often common, phenotypes (such as pigmentation, or schizophrenia predisposition) involves nongenetic factors and the relationship between genotype and phenotype is weaker. It is important to be aware that the path from genotype to phenotype is rarely simple (**Figure 3.2**).

For example, there are many observable differences between people that are not genetic, but due completely or in part to stochastic processes during development, or to environmental influences. **Monozygotic** twins demonstrate this: they share all of their genes and usually many aspects of their environment, but have phenotypes that, though very similar, are certainly *not* identical.

Furthermore, although **monogenic** (also known as simple, or Mendelian) disorders show relatively simple patterns of inheritance in families (Figure 3.1), there is often a more complex relationship between genotype and phenotype than the name monogenic suggests: different mutant alleles in the same gene can have different effects, and even the same allele can manifest differently, because other **modifier genes** can influence phenotype. Such factors affect, for example, the CF phenotype—whether or not the pancreas secretes enzymes normally and whether the *vas deferens* (the duct along which sperm pass) of affected males is occluded or open. There is therefore no sharp distinction between monogenic disorders and **complex**, or **multifactorial** disorders, such as asthma and diabetes, which are commonly said to involve the influence of more than one gene, as well as the environment (see Chapters 16 and 17).

Mutations are diverse and have different rates and mechanisms

This chapter is largely concerned with mutation—the ultimate source of all genetic diversity—and with the resulting kinds of differences at the DNA level that exist between people. Confusingly, the words used to describe such differences vary between contexts and even between professions. A **polymorphism** has been defined as existing when at least two different alleles are present in a population, and both are present at ≥1% frequency. Although all differences among sequences have mutation as their root cause, the word "mutation" itself is sometimes reserved by medical geneticists for variation causing disease, and is then contrasted with the term polymorphism, describing a sequence difference with no apparent effect on function. In some populations disease-associated "mutations" are present at frequencies >1%, and could possibly qualify as polymorphisms. Examples are a specific allele in Europeans responsible for CF, or β-globin gene alleles in some African populations responsible for **sickle-cell anemia** (OMIM 603903; clinical features summarized in Table 16.1). An allele below the 1% frequency used to define a polymorphism is sometimes called a **variant**, but this term is used by others to describe any form of DNA variation, irrespective of frequency or phenotypic effect. The literature also contains the frequency-independent term **single nucleotide variant** (**SNV**), which some have argued should replace single nucleotide polymorphism (SNP). Our practice is to use "polymorphism" (including SNP) to indicate ≥1% **allele frequency**, "variant" for any less frequent difference between genome copies, and "mutation" to refer to a **de novo** change.

A mutation is any change in DNA sequence, and covers a very broad spectrum of events with different rates and different molecular mechanisms. It can range from the substitution of a single base in the genome, and small insertions and deletions of a few bases, through expansions or contractions in the number of tandemly repeated DNA motifs, insertions of transposable elements, **insertion**, **deletion**, **duplication**, and **inversion** of megabase segments of DNA, to translocation of chromosomal segments and even changes in chromosomal number. **Figure 3.3** gives an overview of the scale and scope of these different mutational events, **Figure 3.4** shows the chromosomal distributions of the different classes of mutational event, and **Figure 3.5** summarizes the mutation rate ranges of the different classes.

Not all mutations are passed on to the next generation and contribute to evolutionary change; to do this they must fulfill two requirements. Firstly, they must occur in the germ line—the cell lineage culminating in the gametes. Because of this, germ-line mutation is what concerns us in evolutionary genetics; mutations in other cells of the body, the **somatic cells**, or **soma**, can have serious consequences such as cancers, but in evolutionary terms are not of direct importance.

Second, such mutations must be survivable, and compatible with fertility. Some mutations could be lethal before birth or before reproductive age. These will not be passed on to the next generation. Similarly, mutations which cause infertility are not passed on. Finally, changes in chromosomal number (aneuploidies) are not heritable, reflecting a failure to segregate chromosomes correctly in a parental meiosis.

Most modern analysis of human genetic variation is at the level of DNA, and this book is concerned with DNA sequence variation. However, before the 1980s analysis was at the level of proteins, and protein data have been of considerable importance in shaping hypotheses about population relationships; **Box 3.1** summarizes these classes of variation, sometimes known as **classical polymorphisms**, and the methods that were used to analyze them.

3.2 SINGLE NUCLEOTIDE POLYMORPHISMS (SNPS) IN THE NUCLEAR GENOME

The simplest and smallest-scale difference between two homologous DNA sequences is a **base substitution**, in which one base is exchanged for another (**Figure 3.6**). When a pyrimidine base is exchanged for another pyrimidine (for example, C for T), or a purine for another purine (for example, A for G), the change is called a **transition**. When a pyrimidine is exchanged for a purine, or vice versa, this is a **transversion**. These differences are examples of single nucleotide polymorphisms, or SNPs (pronounced "snips"). The insertion or deletion (**indel**) of a single base is also included in the category of SNPs—perhaps unadvisedly, because the mechanisms which underlie these indels, and the ways they are analyzed, differ from those for base substitutions.

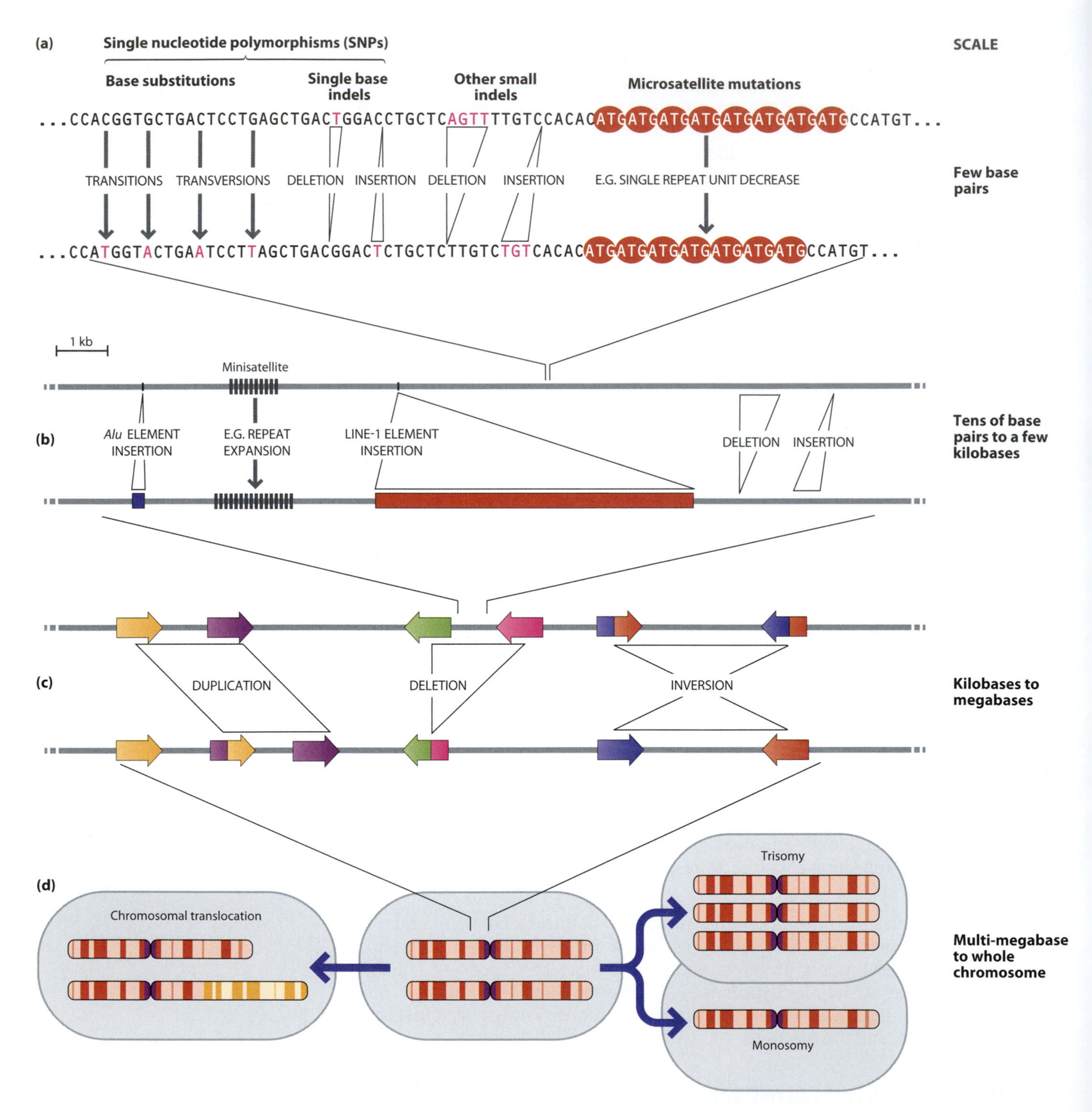

Figure 3.3: Overview of different classes and scales of mutation. (a) Examples of mutations affecting a few base pairs, including examples of transition and transversion substitutions (see Figure 3.6), and a change in the number of ATG repeats at a trinucleotide microsatellite. (b) Examples of mutations affecting tens of base pairs to a few kilobases, including insertion of a ~300-bp *Alu* element and a full length 6.1-kb LINE-1 element, and the expansion by several repeat units of a minisatellite. (c) Examples of segmental mutations on the kilobase to megabase scale. Arrows indicate the order and orientation of repeated segments of DNA. (d) Mutations on the multi-megabase or whole chromosome scale. Most translocations are deleterious, but are important in interspecies differences. Most aneuploidies (departure from the 46-chromosome complement, shown here as monosomy or trisomy) are early lethals; some, for example trisomy 21 and aneuploidies of the sex chromosomes, are tolerated. No aneuploidy can be passed on to offspring.

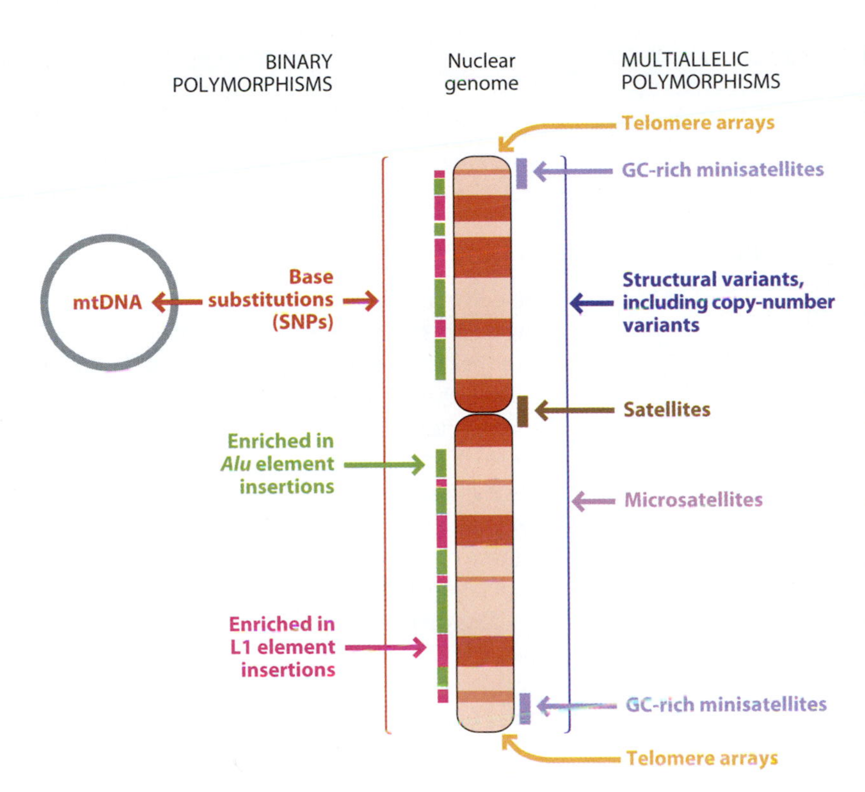

Figure 3.4: Overview of genomic distribution of different classes of polymorphisms.
SNPs are widely distributed and frequent, and microsatellites and non-GC-rich minisatellites are also widespread. Telomere arrays are at the termini of chromosomes, and most GC-rich minisatellites are concentrated toward chromosome ends. All chromosomes carry alpha satellite sequences in their centromeric regions, although there are other satellites on specific chromosomes at other positions. Although *Alu* elements are concentrated in GC-rich regions (light G-bands) and L1 elements are concentrated in AT-rich regions (dark G-bands), recent insertions of both elements are concentrated in the latter regions. See text for more details on distributions of all classes of loci.

Two fundamental processes lead to base substitutions: misincorporation of nucleotides during replication; and mutagenesis caused by the chemical modification of bases, or physical damage due, for example, to ultraviolet or ionizing radiation. These two processes are summarized in **Figure 3.7**.

Base substitutions can occur through base misincorporation during DNA replication

When a diploid cell divides, all ~6,400,000,000 base pairs of its DNA must be copied so that each daughter cell also contains exactly one diploid genome. DNA replication, the process that accomplishes this remarkable feat, proceeds with extremely high fidelity. A new base is incorporated if it pairs, through correct hydrogen bonding, with the existing base in the single-stranded template following strand separation. In addition, the **DNA polymerase**, the enzyme responsible for DNA synthesis, requires the correct overall geometry of the base

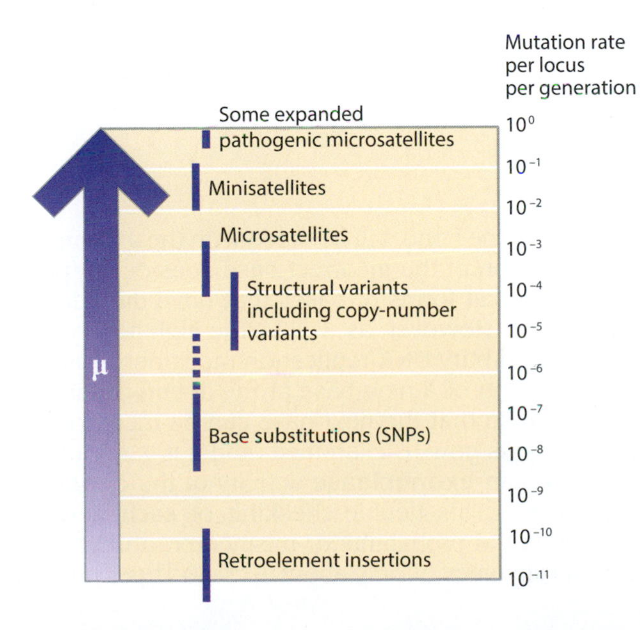

Figure 3.5: Overview of mutation rates (μ) of different classes of polymorphisms.
The figure summarizes information given throughout this chapter. Mutation rates of micro- and minisatellites are very variable between loci. The rates given here reflect ascertainment bias, in that they are those of the mostly widely used and polymorphic loci. The average rates are probably much lower.

Box 3.1: Classical polymorphisms

The ABO blood group system, discovered in 1900,[34] was the first human genetic polymorphism to be defined. It was detected serologically: red blood cells carry **antigens** on their surfaces that can react with specific **antibodies** carried in the serum (the fluid component of the blood). When this reaction occurs, the cells clump together (agglutinate), which is easily visible and provides a simple assay. Two antigens, A and B, underlie the ABO system, and individuals carry in their serum antibodies to the antigen(s) that they do not possess. Mixing blood and serum samples reveals four classes of individuals: those carrying only the A antigen, those carrying only the B, those carrying both (AB), and those carrying neither (O; see **Figure 1**). Early studies showed differences in the frequencies of these blood groups in different populations, and later more blood groups (such as MN and Rh) were defined and used in population studies. Each of these new blood group systems discovered was shown to behave independently of the ABO blood groups.

These and similar immunological methods were later used to detect specific variants of the immunoglobulins, proteins of the immune system found in serum or plasma, and also the extremely polymorphic **human leukocyte antigen** (**HLA**) system (see Box 5.3), analyzed since the late 1950s. HLA proteins are expressed on the surfaces of white blood cells, and are of great importance in transplant rejection or tolerance (**histocompatibility**).

More polymorphisms became available with the introduction of **protein electrophoresis** to separate proteins in an electric field on the basis of their size and charge. This method was used to separate different hemoglobins, in particular that responsible for sickle-cell anemia (Hb^S; OMIM 603903; Chapter 16), and thus to demonstrate the first molecular disease. Many different blood proteins were then shown by electrophoresis to contain genetic differences, and these could be used in population studies. The amount of population data (in particular, sample sizes) accumulated using classical polymorphisms is very large, and a review is included in *The History and Geography of Human Genes*.[10] Conclusions obtained with classical polymorphisms tend to agree with those obtained with later DNA data when they can be compared—a good example is the excellent agreement on the apportionment of genetic diversity among different populations (see Section 10.2).

The molecular basis of many of these classical polymorphisms at the DNA level is now understood, and DNA-based methods for their detection are available. For example, the gene underlying the ABO blood group system encodes a glycosyltransferase enzyme that adds a sugar molecule to a carbohydrate structure known as the H antigen on the surface of red blood cells. The A allele codes for an enzyme which adds the sugar *N*-acetylgalactosamine, whereas the enzyme coded for by the B allele has two amino acid differences which alter its specificity so that it instead adds D-galactose, forming the A and B antigens respectively. The O allele has an inactivating mutation in the gene and so the H antigen remains unmodified.[69]

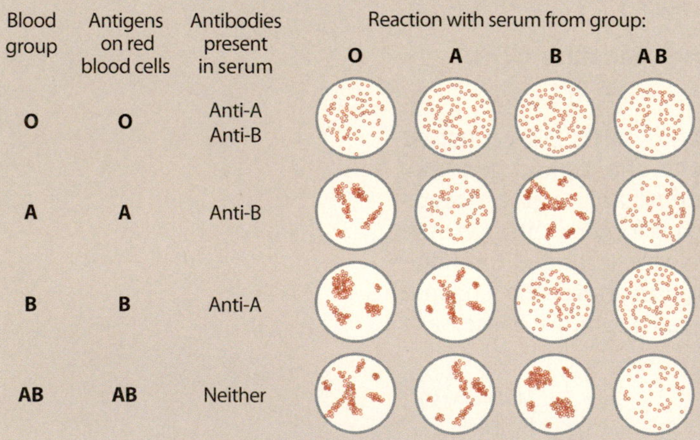

Blood group	Antigens on red blood cells	Antibodies present in serum	Reaction with serum from group: O	A	B	AB
O	O	Anti-A Anti-B				
A	A	Anti-B				
B	B	Anti-A				
AB	AB	Neither				

Figure 1: Defining the ABO blood group system by agglutination patterns.

pair before the bond will be made with the growing daughter strand. Occasional incorporation of the incorrect base does occur, possibly because of rare, transient chemical forms of bases that have different base-pairing properties and geometries. However, as well as the high fidelity involved in this incorporation step, the primary DNA replication machinery has a proofreading function: after incorporation of a new base at the 3′ end of the daughter strand, the enzyme moves on so that the next base can be incorporated. The previously incorporated base is now re-examined, and if it is not observed to be part of a correct base pair, an **exonuclease** activity of the enzyme excises it, and the enzyme tries again. This double-checking of each newly incorporated base greatly decreases the probability of misincorporating a base: replication errors occur with a frequency of only about 10^{-9}–10^{-11} per nucleotide per replication event.

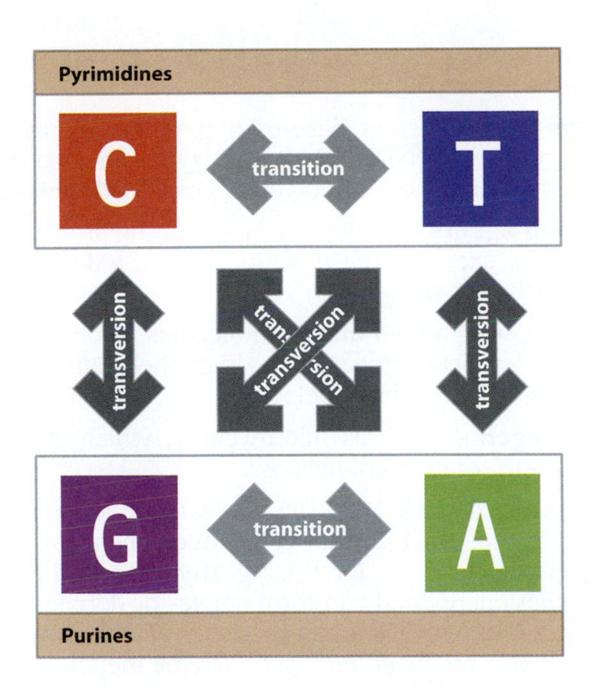

Figure 3.6: Transition and transversion mutations.
For each base there are two possible transversions, but only one transition. Nonetheless the mechanisms of mutation lead to transitions being more common than transversions (see text for details).

Base substitutions can be caused by chemical and physical mutagens

The integrity of the genetic material is under constant assault by chemical and physical processes that alter bases or damage the physical structure of DNA (Figure 3.7). There are spontaneous, endogenous chemical processes going on in all cells that lead to base modification or loss: one example is the **deamination** (loss of an amine group, $-NH_2$) of cytosine to produce uracil, which, unlike cytosine, pairs with adenine. About 400 cytosines are deaminated daily in each human cell, and between 10,000 and 1,000,000 bases per cell per day are damaged as a result of normal chemical reactions involving deamination, oxidation, methylation, and depurination (loss of a purine base).

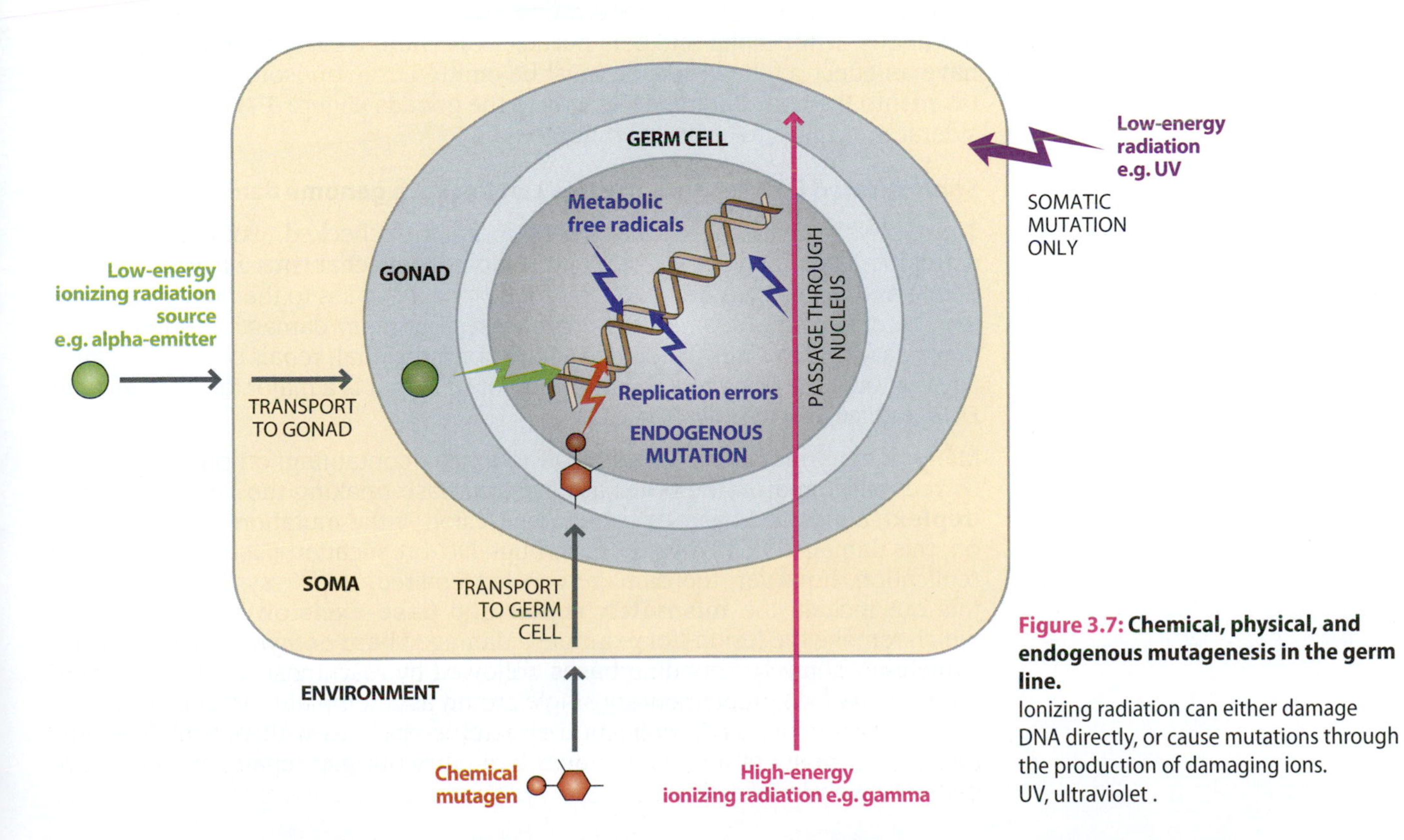

Figure 3.7: Chemical, physical, and endogenous mutagenesis in the germ line.
Ionizing radiation can either damage DNA directly, or cause mutations through the production of damaging ions.
UV, ultraviolet .

As well as this endogenous chemical damage, DNA damage can be caused by chemical **mutagens**. Examples are:

- **Base analogs**, with different base-pairing properties to natural bases, which can be incorporated and subsequently lead to base substitution. An example is 5-bromouracil, a T analog that occasionally pairs with G.
- **Base-modifying agents** that alter the base-pairing properties within DNA. Hydroxylamine, for example, reacts with C to give a derivative which pairs with A, rather than with G.
- **Intercalating agents**, which are flat molecules, such as **ethidium bromide**, that can insert between base pairs in the helix, distorting its structure and often leading to insertions or deletions.
- **Cross-linking agents**, such as mitomycin C, which cross-link different parts of a DNA helix, or different helices.

Physical agents can also damage DNA. Ultraviolet (UV) radiation is a low-energy form of electromagnetic radiation that can induce chemical linking between adjacent thymidine nucleotides (producing **thymidine dimers**). **Ionizing radiation** has higher energy and can form single- or double-strand breaks in the DNA backbone. Alternatively, it can produce reactive ions, known as free radicals. These are also produced endogenously by metabolic processes, and can modify DNA, leading to base substitutions, for example. Ionizing radiation includes gamma radiation, which is a form of high-energy electromagnetic radiation emitted by some radioisotopes, and X-rays, which are essentially the same as gamma rays, but generated by electron bombardment of heavy metal atoms. Forms of particulate radiation are also ionizing: neutrons, alpha particles, and beta particles are emitted by radioisotopes, while cosmic radiation consists of a range of subatomic particles with varying energies, and originates outside the Earth.

Chemical and physical mutagens are important contributors to many cancers, but their effects on germ-line mutation can be very different from their somatic effects. A good example is UV radiation, which is non-penetrating and therefore has only external mutagenic effects, causing skin cancers (see **Section 15.3**). Gamma radiation is highly penetrating and can cause germ-line and somatic mutations, while alpha and beta particles are more weakly penetrating, and to have an effect on the germ line must be emitted from an isotope that has been taken into the body and made its way to the gonads (Figure 3.7). The same consideration applies to chemical mutagens.

Sophisticated DNA repair processes can fix much genome damage

These physicochemical assaults on DNA, if left unchecked, would lead to catastrophic levels of genome damage. Elaborate mechanisms have evolved to detect and repair DNA damage. The mutations that pass to the next generation are therefore the outcome of two processes—primary damage and failures of secondary repair. After the death of a cell or individual, repair ceases but damage continues, leading to a deterioration of DNA that is a major factor in ancient DNA studies, discussed in Chapter 4.

Many primary mutagenic events lead to a helix containing, on one strand, the correct base, and on the other a mispaired base (making the helix a **heteroduplex**), a modified base, or some other lesion. If the mutation is to be passed on, this damaged helix must pass through DNA replication (**Figure 3.8**). Prior to replication, however, the damage can be repaired. Repair systems that carry this out include the **mismatch repair** and **base-excision repair** systems, which remove the lesion (for example a damaged base or thymidine dimer) and sometimes some surrounding bases, followed by resynthesis of the gap using the undamaged complementary single strand as a template. As was described above, the primary DNA replication apparatus operates with very high fidelity. In contrast, many of the polymerases that carry out gap repair synthesis have comparatively low fidelity.

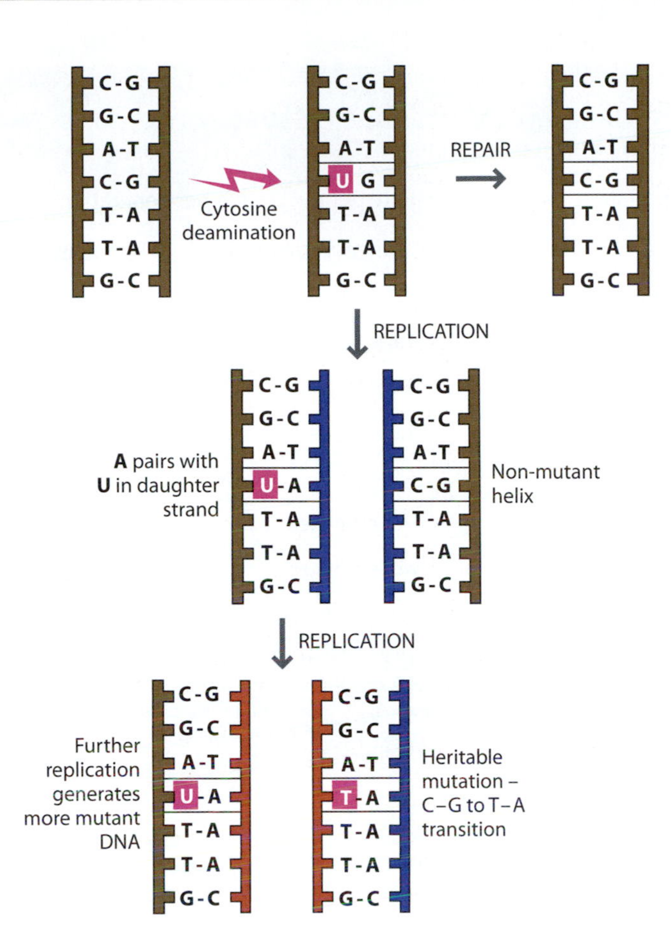

Figure 3.8: The fates of damaged DNA. Damage to one strand of DNA, in this case cytosine deamination to yield uracil, can be repaired. Generation of a stable and heritable base-pair mutation requires two rounds of DNA replication.

Double-strand breaks (**DSBs**), which can be created by ionizing radiation, free radical action, and events during DNA replication, are particularly damaging lesions that can result in chromosomal aberrations, cell malfunction, and cell death. Complementary strand information cannot be used to repair such lesions, and instead the DSB can be repaired through one of two pathways. In homologous recombination (HR), the undamaged sister chromatid acts as template. This ensures accurate repair, but because it relies on the presence of the sister chromatid, is dependent on the stage of the cell cycle at which damage occurs. In **nonhomologous end-joining**, a protein complex recognizes the DNA ends, juxtaposes and joins them. This sometimes involves removal or addition of bases at the DSB, and so is not as accurate as HR, and can itself be mutagenic. In general, cell-cycle progression is tightly coordinated with DNA repair and cell survival or death through a network of genome surveillance pathways.

The rate of base substitution can be estimated indirectly or directly

Knowledge of the rates of the different classes of base substitution mutation is extremely important in evolutionary genetics. These rates can be incorporated into models of sequence evolution that underlie our attempts to interpret patterns of diversity within and between species (see **Section 5.2**). The mutation rate in nuclear DNA is so low that, until recently, it was impossible to measure directly in pedigrees, and indirect approaches were therefore used. The first methods for estimating mutation rates were based on observing the phenotypic outcomes of specific kinds of mutations, namely genetic diseases, but these have disadvantages, and methods based on a direct comparison of DNA sequences are preferable.

For neutral mutations, the rate of mutation is expected to be equal to the evolutionary rate of change (see **Section 5.7**). Therefore a direct comparison of nonfunctional DNA sequences between species whose divergence times are known can give an estimate of mutation rate. One example of this approach

TABLE 3.1:
ESTIMATES FOR MUTATION RATES OF BASE SUBSTITUTIONS AND SMALL INDELS

Type of mutation	Rate per nucleotide per generation	
	Indirect estimation	Direct estimation
All base substitutions	2.2×10^{-8}	1.20×10^{-8}
Transitions	1.6×10^{-8}	8.15×10^{-9}
Transition at non-CpG	1.0×10^{-8}	6.18×10^{-9}
Transition at CpG	1.8×10^{-7}	1.12×10^{-7}
Transversions	5.8×10^{-9}	3.87×10^{-9}
Transversion at non-CpG	Not estimated	3.76×10^{-9}
Transversion at CpG	Not estimated	9.59×10^{-9}
Indel mutations (1–65 bp in length)	2.3×10^{-9}	Not estimated

Indirect data are from Ebersberger I et al. (2002) *Am. J. Hum. Genet.* 70, 1490, based on comparison of 1,944,162 bp of noncontiguous human and chimpanzee orthologous sequence, analyzed using the method of Nachman MW & Crowell SL (2000), *Genetics* 156, 297 assuming species divergence time of 5 MYA, ancestral effective population size of 10,000, and generation time of 20 years. Direct data are from Kong A et al. (2012) *Nature* 488, 471.

compared 1,944,162 bp of noncontiguous chimpanzee sequences with human orthologs:[14] mutation rates deduced from this study are given in **Table 3.1**.

The development of high-throughput DNA sequencing technologies has recently allowed entire genomes to be sequenced from members of human pedigrees. This approach can give direct estimates of the base substitution rate, but is complicated by the need to distinguish the large numbers of sequencing errors (false positives) from true mutations, and the possibility of missing some substitutions (false negatives). For example, sequencing of the genomes of 78 parent–child trios[30] yielded a mutation rate estimate of 1.2×10^{-8} per base per generation, in agreement with other small-scale genome sequencing studies. Such estimates are lower than the rates determined from human–chimpanzee comparisons. This may be accounted for by uncertainty about the date of species divergence, and the generation time (see Table 6.6), or even mutation rate variation between individual humans. The implications of different mutation estimates for the timing of events in human evolution are discussed in Opinion Box 11.2.

A number of important features of small-scale mutation in humans emerge from these and similar data (Table 3.1):

- Base substitutions, occurring at an average rate of the order of 10^{-8} per base per generation, are ~10 times more frequent than indels, although this relative frequency varies substantially between loci, because it depends on the sequence context.
- Transitions are almost three times more frequent than transversions, despite the fact that for a given base there are two possible transversions to only one transition (Figure 3.6)—if all mutations were equally probable we would expect a 2:1 ratio of transversions to transitions. This deviation from expectation might be explained by influences of efficiency of error detection and repair, sequence context, or perhaps by differences in misincorporation rates.
- Rates of both transitions and transversions at a particular dinucleotide, C followed 3′ by G (known as a **CpG**, where the "p" indicates a phosphate group) are an order of magnitude higher than those at other dinucleotides. The reasons for this are discussed in the next section.

Because of their low mutation rate, SNPs usually show identity by descent

It is important to note that mutation rates of base substitutions are in general so low that a given mutation at a particular position is unlikely to have recurred or reverted in the small populations of early humans; independent occurrences are therefore not usually found today at appreciable frequencies. Examples do, however, exist: for example, the set of DNA polymorphisms constituting the Y chromosome phylogeny (see Appendix Figure 3) includes 521 base substitutions, of which 15 (~3%) are recurrent,[26] and some have spread widely in human populations. Note that even though the base-substitutional mutation rate is low (Table 3.1) the current population size (~7×10^9) is so large that the recurrence or reversion of any SNP in some individual(s) is expected to occur each generation. The population frequency of such new mutations will of course be extremely low, so in practice they are not generally encountered.

The low mutation rate means that this class of mutation generally shows **identity by descent**, rather than **identity by state**: the presence of the same base at a SNP in two independent genome copies usually implies that the base has been inherited from a common ancestor (**Figure 3.9a**). This stability also means that the direction of evolutionary change can generally be established (Figure 3.9b) by examining the orthologous DNA sequence in the great apes, our closest living nonhuman relatives (see Chapter 7). The allele resembling the great ape allele is known as the **ancestral state** of the SNP (coded as 0), and the allele that differs is the **derived state** (coded as 1). The low mutation rate also means that base substitutions generally belong to the class of **binary polymorphisms**, also known as **biallelic**, or **diallelic polymorphisms**, or **unique event polymorphisms** (**UEPs**).

The CpG dinucleotide is a hotspot for mutation

The rate of base substitutions is not uniform throughout the genome. In particular, where a C nucleotide is followed by a G (denoted a CpG dinucleotide), mutation rate is elevated by more than tenfold (Table 3.1). This is because of a chemical modification that occurs at most CpGs.

About 75% of CpG dinucleotides in the human genome are targets of **DNA methylation**: a specific methyltransferase enzyme adds a methyl ($-CH_3$) group to the

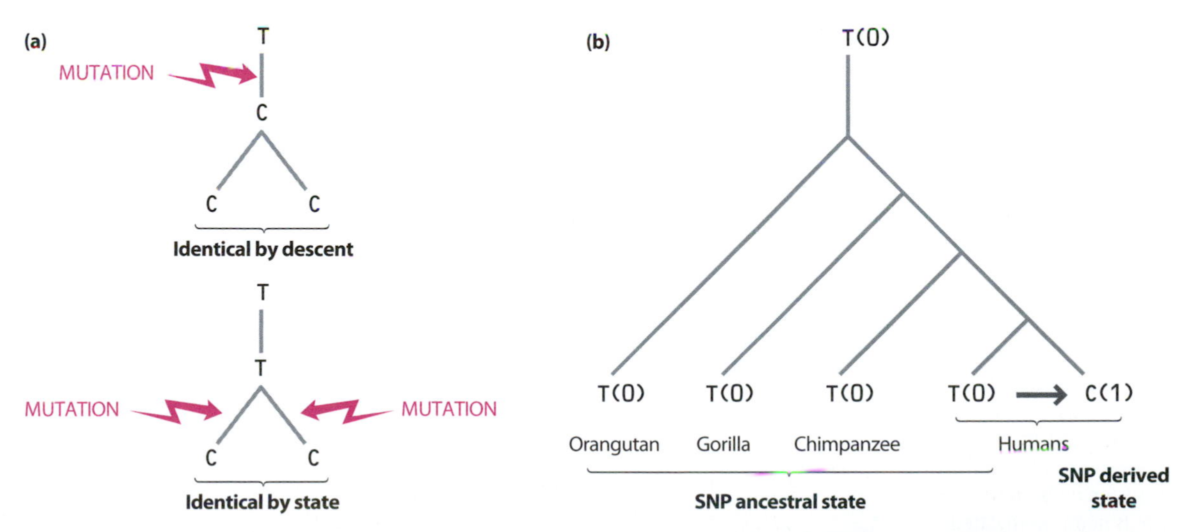

Figure 3.9: Identity by descent, and ancestral state determination of a SNP.
(a) When two alleles and their common ancestor share the same allelic state (here, a C allele at a SNP), they show identity by descent (top panel); when the common ancestor has a different allelic state (T allele), they show identity by state (bottom panel). (b) Given the known phylogeny of our great ape relatives (see Chapter 7), the presence of a T at a SNP site in great ape orthologs of a human sequence showing a T/C polymorphism defines T as the ancestral state (0). The mutation underlying the polymorphism is a T(0) to C(1) transition. Note that polymorphism at the same base is on rare occasions expected to occur in the other primates. Analysis of more than one member of each species is useful.

5-carbon of the cytosine ring to give 5-methylcytosine (**Figure 3.10a**). On the other strand, the complementary dinucleotide to CpG is also CpG, and so methylation affects C bases on both strands at adjacent base pairs (Figure 3.10b). The functions of this epigenetic modification are not entirely understood: it plays a role in the control of gene expression, the control of chromosome condensation, and even as a host defense mechanism protecting against the damaging activities of transposable elements (see **Section 3.5**).

When cytosine undergoes spontaneous (or mutagen-induced) deamination, it yields uracil, which is not a legitimate base in DNA, and is efficiently excised and replaced with cytosine by a repair system. In contrast, deamination of 5-methylcytosine yields thymine, a legitimate DNA base, resulting in a mispair of the new thymine with the guanosine on the other strand. To fix this T–G mispair, either the T can be changed to a C, or the G to an A, yielding either a T–A or C–G pair: however, there is nothing to show that the original base pair was C–G, and so the repair machinery will often make the incorrect choice. As a result, methylated cytosine nucleotides, and hence CpG dinucleotides, are hotspots for mutation. The tendency is for CpG to mutate to TpG or CpA (Figure 3.10b);

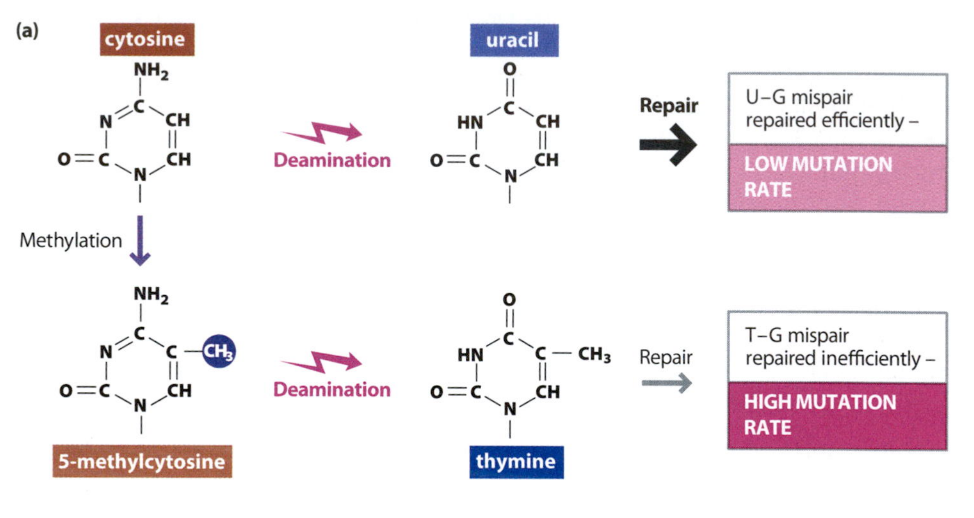

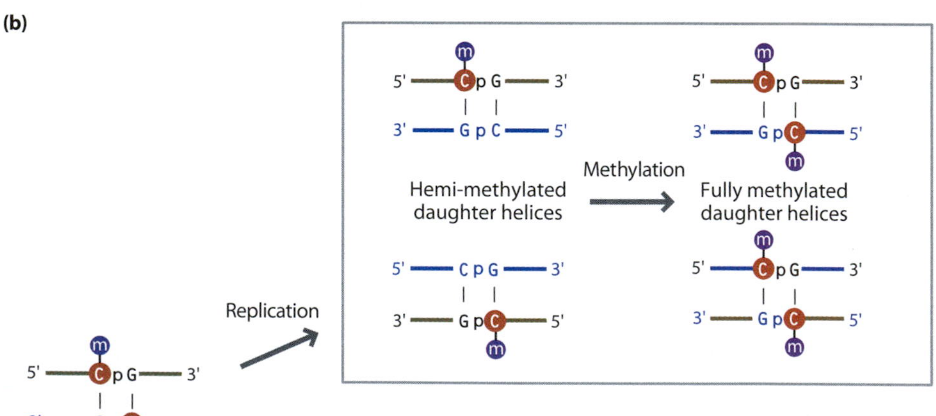

Figure 3.10: The CpG dinucleotide is a site for methylation and a hotspot for mutation. (a) Spontaneous or mutagen-induced deamination of cytosine yields uracil, which is efficiently recognized and removed by uracil glycosidase. Deamination of 5-methylcytosine yields thymine, which is not efficiently repaired. (b) Most CpG dinucleotides are methylated at the 5-carbon of cytosine on both strands. Replication yields hemi-methylated daughter helices that become fully methylated through the action of a specific methyltransferase. Unrepaired deamination followed by replication leads to a CpG to TpG mutation, or, if deamination affects the complementary strand, a CpG to CpA mutation.

in other words, the methylation of cytosine should lead to an increase of transition mutations. This is indeed observed (see Table 3.1); the accompanying greater than fivefold increase of transversion mutations at CpG dinucleotides is not understood.

The highly mutable nature of methylated CpG dinucleotides means that they tend to be lost by mutation. The average fraction of C and G in the genome is 21% each, so we might expect to find about 4% (0.21 × 0.21) of dinucleotides to be CpG, but the actual overall frequency of CpG in the genome is only about one-fifth of this value. There are, however, ~45,000 **CpG islands**, each extending over hundreds of base pairs, and typically associated with the 5′ ends of genes, where CpG dinucleotides exist at their expected frequency. In these islands CpGs remain unmethylated, and so do not represent mutation hotspots. Absence of methylation reflects a role for the islands in the regulation of transcription.

Evidence for hypermutability of CpG comes not only from interspecies comparisons (Table 3.1) but also from data on mutations in humans causing genetic disorders. For example, in hemophilia A (OMIM 306700), an X-linked disorder of blood coagulation resulting from a deficiency of clotting factor VIII, 42% of all base substitutions changing amino acids or introducing stop codons are either CpG to TpG or CpG to CpA mutations, even though CpG represents only ~1% of all dinucleotides in the gene's coding region. Data based on an analysis of 7271 disease-causing base substitutions in 547 different genes[32] show that 37% of transitions occur at CpG dinucleotides, which reflects a transition rate five times the base mutation rate [note that this figure for genes is not in agreement with that for largely nongenic sequences (Table 3.1), which may reflect differences in sequence composition].

Base substitutions and indels can affect the functions of genes

Evolutionary genetic studies often focus on alleles that are assumed to be neutral—having no effect on fitness, and usually no effect on phenotype. Many of these lie outside genes; however, because of the nature of the biological mechanisms that interpret information in DNA and express it as proteins, many sequence changes *within* genes are also expected to have little or no effect on function. On the other hand, as alluded to above, base substitutions within genes can be the cause of genetic disease, and it is therefore important to understand the potential phenotypic effects of such changes in DNA sequences. Most alleles affecting phenotype do so adversely—there are many more ways to impair the function of a gene than to enhance it. However, there is certainly an **ascertainment bias** here, since variants that cause diseases are noted in clinics and find their way into gene mutation databases. Variants improving the performance of a protein or changing its function advantageously are not necessarily noted, but are nevertheless the stuff of positive selection and evolution (see **Section 5.4**). This discussion focuses on deleterious variants.

Figures 3.11 and **3.12** show some of the potential effects of base substitutions and small indels on a protein-coding gene; **Table 3.2** gives the relative frequencies of the different types of variant found in the 1000 Genomes Project (**Box 3.2**) that are predicted to damage proteins.

Synonymous base substitutions

In the **open reading frame** (**ORF**) itself (Figure 3.11), base substitutions can have a wide range of possible outcomes, from no effect to a total abolition of protein production. The redundancy of the genetic code (see Figure 2.7) provides a buffer against deleterious effects of base substitution: many amino acids are encoded by more than one codon (six different codons encode serine, for example), and in general third-position changes cause relatively little effect. Substitutions that do not alter an amino acid are known as **silent-site**, or **synonymous** substitutions, and are often assumed to be selectively neutral. However, such substitutions can occasionally alter splicing, mRNA stability, or

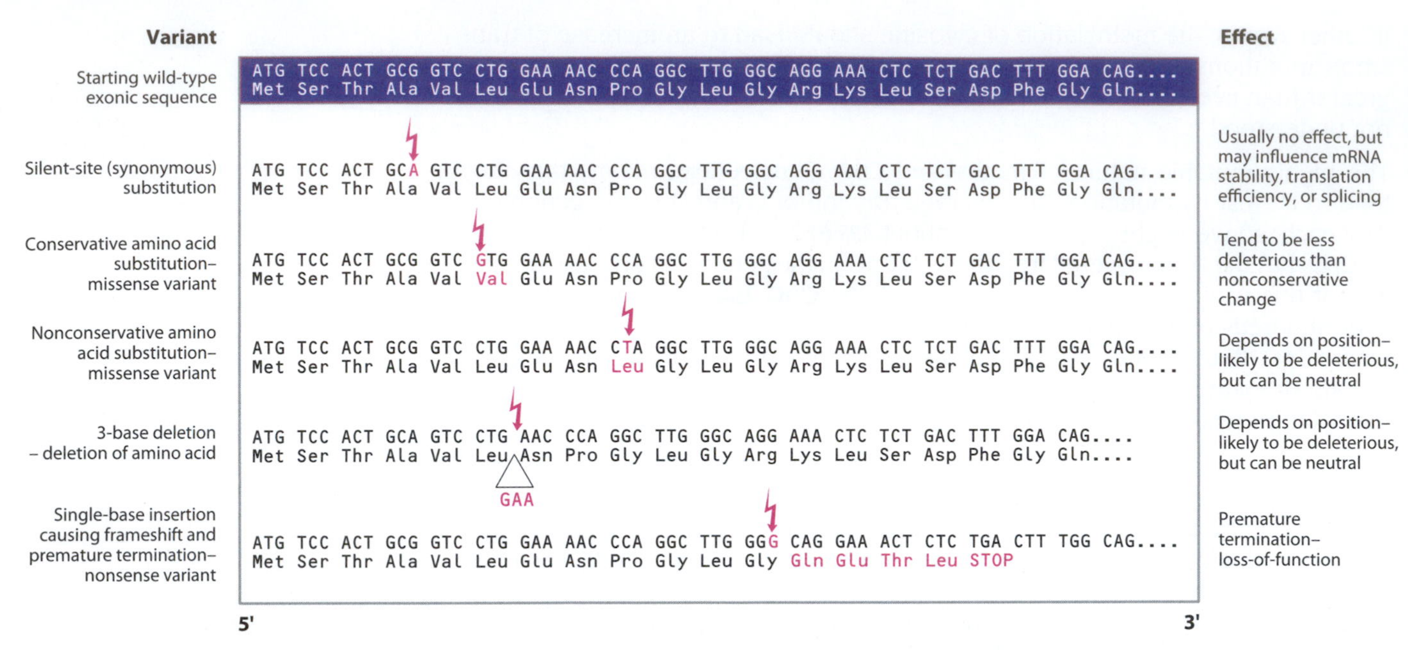

Figure 3.11: Small-scale variants within the open reading frame of a gene and their potential effects.
See Figure 2.7 for explanations of the three-letter amino acid abbreviations.

the efficiency of translation, and hence affect function. An increasing proportion of cases are now recognized to have deleterious effects through altered splicing[64]—for example, about 25% of synonymous variants within exons 9 and 12 of the *CFTR* gene result in altered splicing. The *IRGM* (immunity-related GTPase family, M) gene encodes a protein involved in the intracellular removal of bacteria, and contains a synonymous variant associated with the gut inflammatory disorder Crohn's disease; the mechanism is the disruption of a microRNA-binding site in the mRNA, so translation cannot be down-regulated.[8]

Nonsynonymous base substitutions

Base substitutions which lead to a change of amino acid are termed **nonsynonymous**, or **missense** mutations, and are denoted, for example, as "Arg702Trp" or "R702W"—a substitution of tryptophan for arginine at amino acid 702 of a protein. Beyond simple redundancy, the organization of the genetic code minimizes the deleterious effects of base substitution: the distribution of amino acids among the 64 possible codons means that many missense mutations will tend to replace one amino acid with another that shares similar chemical properties (a **conservative** substitution). The effects of these substitutions depend on their position within the protein: conservative substitutions outside the substrate-binding site of an enzyme, for example, might have little effect, while those within it might be deleterious. Some base substitutions, however, change one amino acid for another that has different chemical properties; such **nonconservative** substitutions are more likely to be deleterious than are conservative changes. An example is the sickle-cell anemia (Hb^S) allele, which changes a hydrophilic (water-loving) glutamic acid on the surface of hemoglobin to a hydrophobic (water-hating) valine: the consequence is that the hemoglobin molecules associate together, rather than remaining in solution, leading to damaging changes in cellular behavior. A mutation changing an amino acid codon into a **termination codon** is termed a **nonsense** or **stop** mutation, and is likely to be deleterious unless very close to the natural termination codon. Analysis of the 1000 Genomes Project data (Box 3.2) reveals ~10,000 nonsynonymous and ~30 nonsense SNPs per genome.[1, 38]

Large-scale sequencing of human genomes has revealed the existence of **multinucleotide polymorphisms** (MNPs), where two or more adjacent base pairs are altered compared with the reference sequence.[45] These, in effect, are adjacent SNPs, but occur at a much greater frequency than expected if the constituent SNPs were occurring independently; this suggests that MNPs may arise in a single mutation event. Nearly all MNPs within genes will result in amino acid changes, and are therefore likely to have marked effects on protein function.

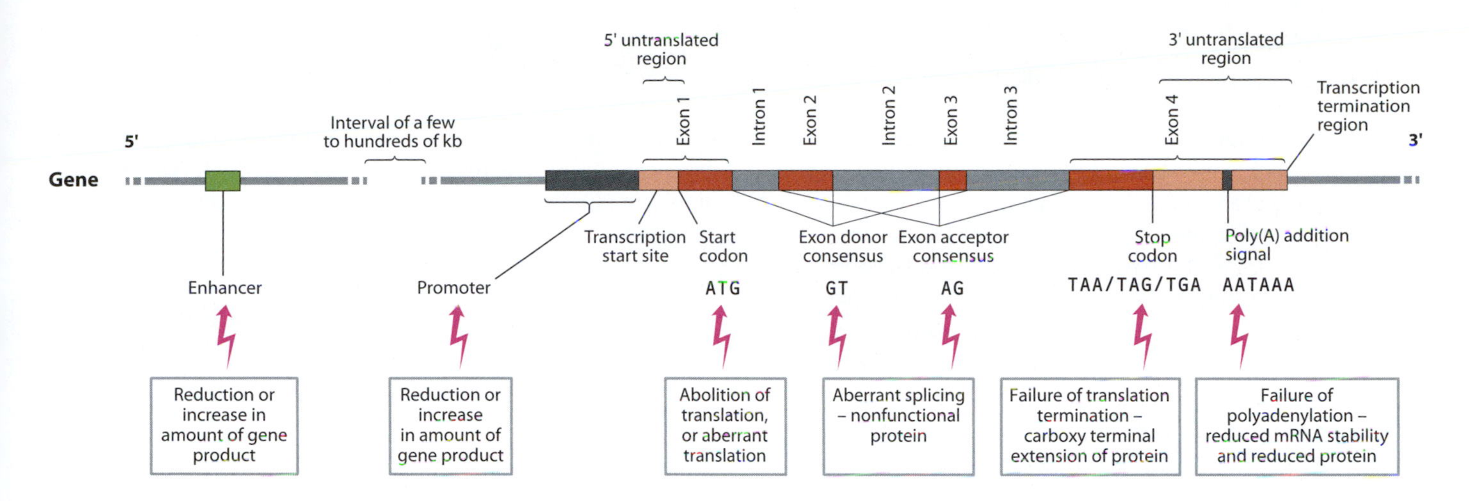

Figure 3.12: Small-scale gene mutations outside an open reading frame and their potential effects.
The different elements of the gene are not to scale, and in particular the introns illustrated here are small relative to the exons (see Table 2.1 for typical sizes of these elements).

Indels within genes

A small indel within an ORF is more likely to be deleterious than a base substitution. Because the genetic code is based on triplets, any indel of a multiple of three base pairs will lead to an insertion or deletion of amino acid(s). Again, the effects of such changes will depend on their position in the protein. The classic example of a single amino acid deletion is the commonest cystic fibrosis allele, ΔF508 (Box 16.3), which removes a phenylalanine from the CFTR protein, leading to defective intracellular processing. Another example (see **Section 17.4**) is a 6-bp deletion in the *APOL1* gene that removes two adjacent amino acids from apolipoprotein L-1 and confers resistance to sleeping sickness together with susceptibility to kidney disease.[15] If an indel is not a multiple of three bases, however, it changes the reading frame of translation. Such **frameshift** variants are often catastrophic, since the amino acid sequence 3′ of the mutation is completely different from that of the wild type. Translation can proceed until a termination codon is reached, leading to prematurely terminated protein, or to abnormally extended protein, particularly if the variant is close to the 3′ end of the coding region. In practice, a protein quality-control mechanism known as **nonsense-mediated decay** often intervenes, leading to degradation of the defective transcript before it can be translated. Alternatively, the change of reading frame may lead to unstable mRNA or aberrant splicing, which abolishes

TABLE 3.2:
ESTIMATED NUMBERS OF POTENTIAL CODING AND LOSS-OF-FUNCTION VARIANTS WITHIN PROTEIN-CODING GENES

Type of variant	Average number per genome
Synonymous	10,572–12,126[a]
Nonsynonymous (missense)	9966–10,819[a]
Generation of stop codon (nonsense)	26.2 (5.2)[b]
Splice site variant	11.2 (1.9)[b]
Small indel causing frameshift	38.2 (9.2)[b]
Large deletion	28.3 (6.2)[b]
Total number of LoF variants	103.9 (22.5)[b]

Data from the low-coverage dataset of 1000 Genomes Project Consortium (2010) *Nature* 467, 1061.
[a] Interquartile range of the number of variants per individual across the CEU, CHB, JPT, and YRI HapMap samples (see Box 3.6 for the three-letter abbreviations of the populations).
[b] Average number of variants in the CEU sample, with average number in homozygous state in parentheses; from MacArthur DG et al. (2012) *Science* 335, 823.
LoF, loss of function.

Box 3.2: The 1000 Genomes Project

The **1000 Genomes Project** is something of a misnomer, because neither in its pilot phase, published in 2010,[1] nor in its ongoing production phase, has it targeted one thousand genomes; in fact, the name "The MultiGen Project" was considered as an alternative, but did not stick. The primary aim of the project, coordinated by a large international consortium, was to use **next-generation sequencing** technologies to discover and characterize the genetic variation shared between individuals, specifically over 95% of the variants having allele frequencies of ≥1% in chosen populations from Africa, Europe, South Asia, East Asia, and the Americas. About 15% of the reference sequence is difficult to analyze because of complex and duplicated regions, and parts of the genome are not even included in the reference sequence because sequence assembly is poor, so the focus was on the **accessible genome**. Since functional variants are often found in coding regions, lower-frequency alleles (≥0.1%) were also sought here. Variants were identified by alignment of the sequence reads to the reference genome.

The project's pilot phase used population samples collected for the HapMap project (see Box 3.6 for the three-letter abbreviations of all the populations), and comprised three sub-projects:

- *Trio pilot*: whole-genome sequencing at high **coverage** (each base covered on average 42 times) of two families of African and European origins (YRI and **CEU**), each including two parents and one daughter.
- *Low-coverage pilot*: whole-genome sequencing at low coverage (2–6×) of 59 unrelated African individuals (YRI), 60 unrelated European individuals (CEU), and 60 unrelated East Asian individuals (30 CHB and 30 JPT).
- *Exon pilot*: targeted high-coverage (average >50×) sequencing of 8140 exons from 906 randomly selected genes (total of 1.4 Mb), in 697 individuals from seven populations of African (YRI, LWK), European (CEU, TSI), and East Asian (CHB, JPT, CHD) ancestry.

Some of the findings of the pilot project are discussed in the text of this chapter.

The second (production) phase of the Project is combining low-coverage (4×) whole-genome sequencing, genotyping, and deep targeted sequencing of all **CCDS** (www.ncbi.nlm.nih.gov/CCDS) coding regions in 2500 individuals from five major regions of the world. All samples analyzed in the pilot project were derived from **lymphoblastoid cell lines**, which carry somatic mutations that arise during cell culture. The production phase plans to focus on blood-derived DNA where possible, to minimize these artifacts. The sampling approach was based on choosing non-vulnerable populations relevant to medical genetic studies to maximize the utility of the information produced. In coding regions, the full project should have 95% power to detect variants at a frequency of 0.3% and ~60% power to detect variants at a frequency of 0.1%. Technological improvements should also allow access to >90% of the reference genome.

protein production. Among disease-causing mutations, at least, small deletions predominate over small insertions, and the great majority of events involve one to five bases. According to the 1000 Genomes Project (Box 3.2; Table 3.2) data, however, each individual carries ~40 in-frame indels contributing to the number of loss-of-function variants which together with nonsense mutations and splice site variants (see below) affect ~100 genes.

Base substitutions outside ORFs

Outside the ORF itself, mutations can affect expression of the gene by altering promoters, enhancers, miRNA-binding sites, or **polyadenylation signals**. In addition, mutation of the consensus splicing signals at the ends of introns can affect splicing, leading, for example, to the retention of an intron in the transcript or the **exon skipping**, and consequent loss of function. While the GT/AG dinucleotides that mark the ends of introns (Figure 3.12) are highly conserved and necessary for correct splicing, other adjacent and even distant nucleotides can influence splicing efficiency or accuracy.[64] Alternatively, a new splice site can be created within an intron. To complicate matters even further, an intronic variant can even affect the regulation of an adjacent gene; this is the case for the human **lactase** gene, down-regulation of which in adulthood can be prevented by a number of mutations 14 kb upstream, within an intron of an unrelated gene (see **Section 15.6**).

Whole-genome resequencing provides an unbiased picture of SNP diversity

Interest in the identification of SNPs has been driven largely by their potential use as **molecular markers** in **disease association studies** (see Chapters 16 and 17). Initially, SNP discovery studies were based on analyzing a few loci in several individuals, and as a result many of the widely used SNPs have diverse and poorly documented origins. Now, whole-genome resequencing studies (**Table 3.3**) are revealing relatively unbiased pictures of the numbers and distributions of variants between genomes.

SNPs discovered through many different methods are described by numbers prefixed with the letters **rs**, for **RefSNP**, and are deposited in the database **dbSNP** (www.ncbi.nlm.nih.gov/projects/SNP/). At the time of writing dbSNP (build 137) contains 53.6 million human SNPs. Many have been identified by the computational analysis of various sequence data held in other databases, and a proportion are artifacts. Validated SNPs are those that have been ascertained with a noncomputational method, or have been detected and genotyped in a population sample; these have associated allele frequency data. For a given SNP, an important property is its **minor allele frequency** (**MAF**), the frequency at which the less common allele is found within a given population. SNP-based studies focused on so-called common variation normally consider SNPs with a MAF of at least 5%.

When a diploid genome is sequenced, SNPs can be identified directly as heterozygous base positions; SNPs that are homozygous in the sequenced genome can be identified by comparison with a reference sequence. Each of the genomes sequenced to date contains ~3–4 million SNPs.[4, 35, 65, 67] Initial studies of sequences largely of European descent[25, 61] showed the average **nucleotide**

TABLE 3.3: SOME OF THE FIRST HUMAN GENOMES TO BE SEQUENCED

Sample	Population	Number	Method	Coverage	Source
J. Craig Venter	European	1	Sanger dideoxy	7.5×	Levy S et al. (2007) *PLoS Biol.* 5, e254
J.D. Watson	European	1	454	7.4×	Wheeler DA et al. (2008) *Nature* 452, 872
HapMap NA12878 trio	CEU	3	Illumina, SOLiD, 454	24.7–29.7×	1000 Genomes Project trio pilot
HapMap NA19239 trio	YRI	3	Illumina, SOLiD, 454	13.6–28.5×	1000 Genomes Project trio pilot
HapMap NA18507 trio	YRI	3	Illumina	37.1–43.0×	Bentley DR et al. (2008) *Nature* 456, 53
Family quartet	European	4	DNA nanoballs	51–88×	Roach JC et al. (2010) *Science* 328, 636
JF	European	1	Illumina	12.3×	Illumina
HapMap NA10851	CEU	1	Illumina	28.4×	Park H et al. (2010) *Nat. Genet.* 42, 400
AK1	Korean	1	Illumina	28×	Kim JI et al. (2009) *Nature* 460, 1011
SJK	Korean	1	Illumina	28×	Ahn SM et al. (2009) *Genome Res.* 19, 1622
YH-1	Han Chinese	1	Illumina	36×	Wang J et al. (2008) *Nature* 456, 60
S. Africans	Bantu (Desmond Tutu), Kalahari Bushman	2	454, SOLiD	7.0–23.4×	Schuster SC et al. (2010) *Nature* 463, 943
HGDP	Han Chinese, Papuan, San, Yoruba	4	Illumina	4.4–7.1×	Green RE et al. (2010) *Science* 328, 710

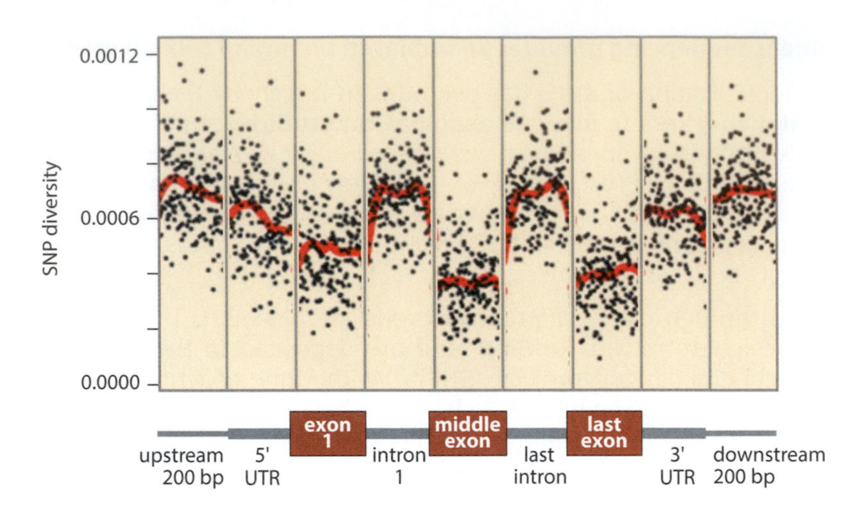

Figure 3.13: Local SNP diversity is affected by natural selection. Exons contain less diversity than other regions of genes. Here, SNP diversity is measured as average heterozygosity in the samples of European origin (CEU). [From 1000 Genomes Project Consortium (2010) *Nature* 467, 1061. With permission from Macmillan Publishers Ltd.]

diversity (π, representing the likelihood that a given nucleotide position differs across two randomly sampled sequences; see **Section 6.2**) in both genomewide and locus-specific studies to be ~7.6×10^{-4}. This means that, on average, we expect to find one SNP about every 1250 bp. More recent genome sequencing of an African individual[4] gives a higher value of 9.9×10^{-4}, equivalent to one SNP per 1006 bp; this increased diversity of African genomes is an important general finding that we will return to in Chapter 9.

SNP diversity varies significantly between different parts of the genome. In principle this could reflect variation in mutation rate, because of variable GC-content or variable efficiency of DNA mismatch repair, for example. However, as is discussed in Chapters 5 and 6, the sequence diversity of a **locus** is influenced not only by its mutation rate, but also by natural selection and population history. For instance, strong positive selection on a locus will cause it to show lower sequence diversity than loci not under positive selection, while **balancing selection** will favor the retention of more than one allele in the population and therefore enhance diversity. Such effects can be seen on a small scale, such as the reduced average diversity in the functional elements of genes (**Figure 3.13**), or on a large scale, such as the striking peak of SNP diversity around the **HLA** locus (Box 5.3) on chromosome 6p evident in genomewide sequencing data[1] (**Figure 3.14**).

3.3 SEQUENCE VARIATION IN MITOCHONDRIAL DNA

Variation in mtDNA warrants separate discussion because it has two distinctive aspects—its high mutation rate compared with nuclear DNA, and the unusual manner in which variation passes from one generation to the next.

mtDNA has a high mutation rate

Early comparisons of human and other primate DNAs indicated that the base-substitutional mutation rate in mtDNA is about 10 times higher than the average rate in nuclear DNA; indeed, the rate in the control region is even higher, rapid enough for mutations to be readily detectable in pedigree studies. This high rate leads to a large number of different sequences in human populations—one of the reasons for the popularity of mtDNA as a tool for evolutionary studies. Current rates for mtDNA mutation have been estimated by comparison of large numbers of whole human mtDNA sequences with the chimpanzee mtDNA sequence, and corroborated by considering diversity within some human populations whose times of origin are known from archaeological data.[51] The observed mutation rate (the outcome of both mutation and selection) throughout the 16.5-kb molecule is not uniform. The lowest rate (2×10^{-7} per bp per generation, assuming a 25-year generation time) can be observed at RNA genes

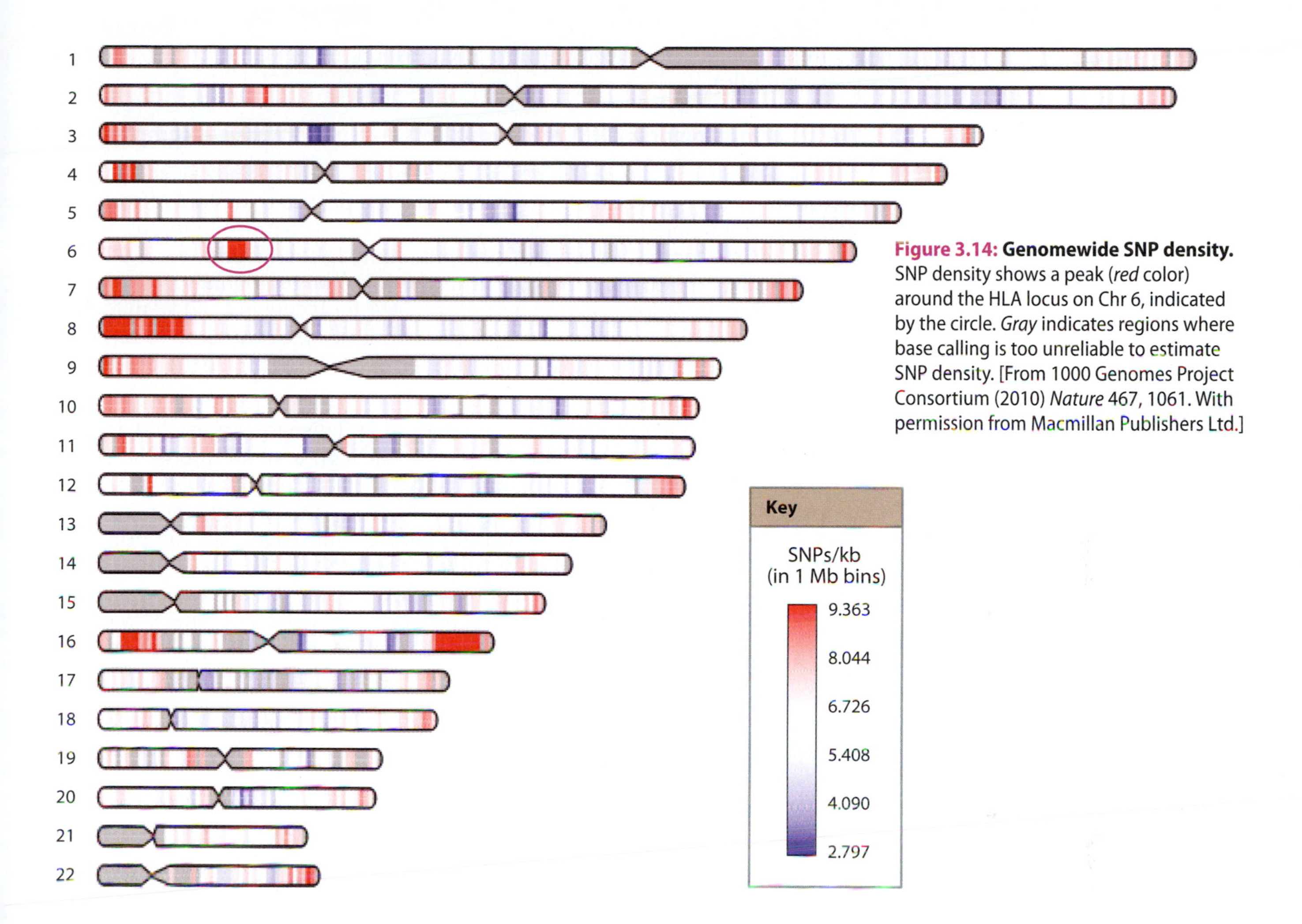

Figure 3.14: Genomewide SNP density. SNP density shows a peak (*red* color) around the HLA locus on Chr 6, indicated by the circle. *Gray* indicates regions where base calling is too unreliable to estimate SNP density. [From 1000 Genomes Project Consortium (2010) *Nature* 467, 1061. With permission from Macmillan Publishers Ltd.]

and the first and second codon positions of the protein-coding genes, and an intermediate rate (5×10^{-7}) at the third codon positions and conserved non-coding regions of the **control region**.[51] A more than tenfold higher mutation rate (5×10^{-6}) in **hypervariable segments** (**HVSI** and **HVSII**) of the control region is due to the presence of a number of evolutionarily unstable hotspot positions that make the use of control region data problematic in interspecies comparisons. Even the mutation rate at the third codon positions is so rapid that more than one-third of the nucleotide states at these positions in humans can be shared with the chimpanzee sequence by state and not by descent.[29] Mitochondrial DNA is also characterized by a particularly high ratio of transitions to transversions—at >30, this ratio is more than tenfold higher than that for nuclear DNA (see **Section 3.2**).

Systems for the repair of mtDNA are active within the mitochondrion, but several factors may account for its higher rate of mutation compared with nuclear DNA:

- Because of its function in energy generation through oxidative phosphorylation, the mitochondrion contains a high concentration of mutagenic oxygen free radicals.
- mtDNA may have a higher turnover rate than nuclear DNA, requiring more replications per unit of time.
- Because of the way it is replicated, mtDNA spends more of its time in the vulnerable single-stranded form than does nuclear DNA; indeed, the control region is sometimes called the **D-loop** to reflect its unusual structure.

- mtDNA is not packaged with histones, which may make it more vulnerable to mutation.

The transmission of mtDNA mutations between generations is complex

Mutation in the nuclear genome has been described above and is easy to understand—mutations occur in the germ line and are passed to diploid offspring via gametes in a single (haploid) dose. Mutations in mtDNA are different: as was explained in Chapter 2, each cell contains many mitochondria, each of which contains several mtDNA molecules; thus there is a large population of mtDNAs in any cell. A mutation arises in a *single* mtDNA molecule in a *single* mitochondrion within this population of many mitochondria, yet can come to represent all the mtDNA in the soma of an individual in a subsequent generation. How does this occur?

Observations of mtDNA variants, both neutral and disease-associated, indicate that the population of mitochondria passes through a small intergenerational bottleneck (**Figure 3.15**), and experiments in mice have been used to dissect this process.[63] After fertilization the zygote contains >100,000 mitochondria, each containing one or two mtDNAs. During female embryogenesis the primordial germ cells (**oogonia**) develop, and early in this process there is a physical bottleneck of a few hundred mtDNAs that allows intracellular selection against severely deleterious mutations. Subsequently, a genetic bottleneck occurs in the early postnatal period, involving the replication of only a subset of mtDNAs;

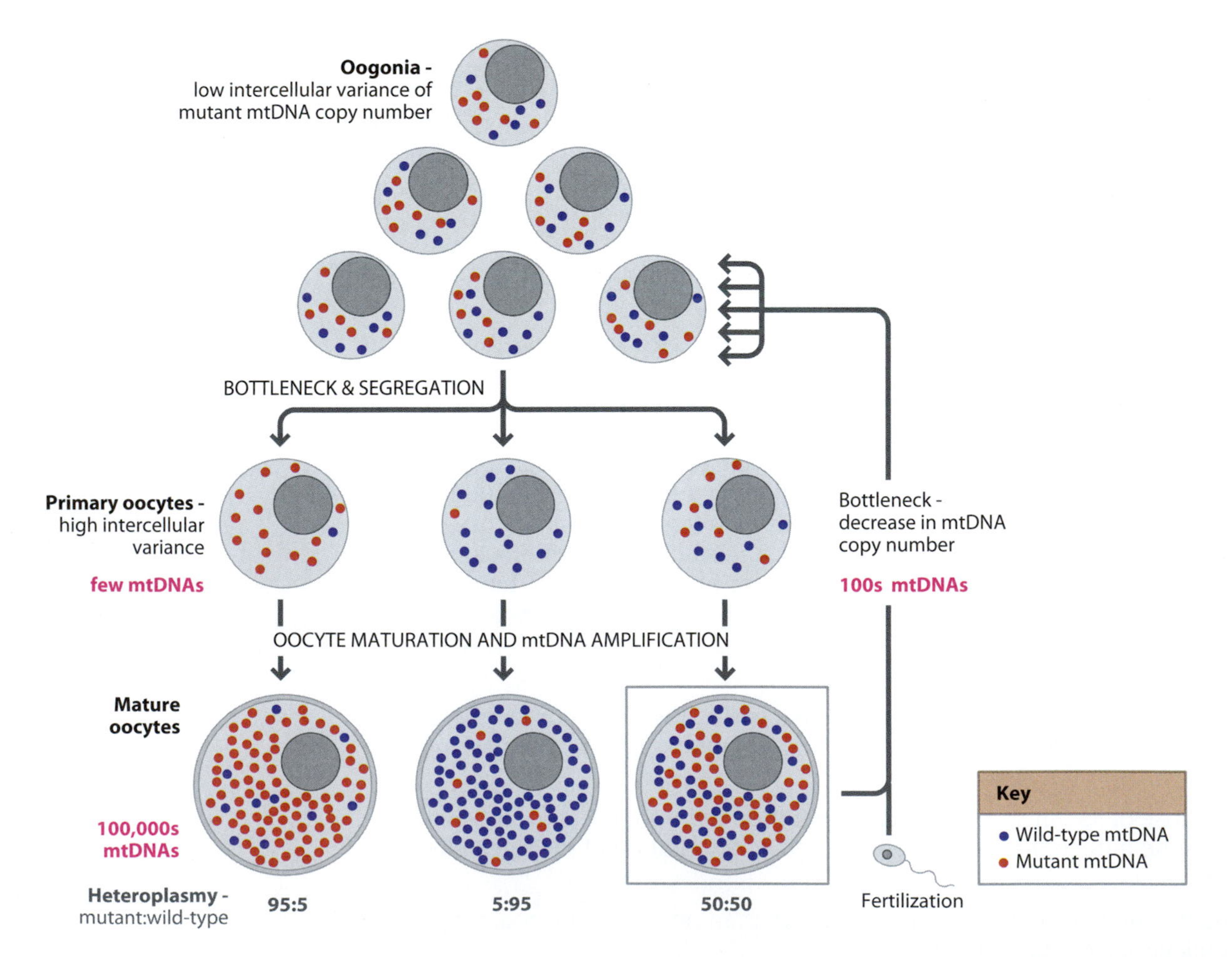

Figure 3.15: Intracellular population genetics: the fate of mutant mtDNAs.
A female developing from a fertilized oocyte that is 50% heteroplasmic for a mutant mtDNA type produces primordial germ cells (oogonia) which show similar levels of heteroplasmy (low variance). During the production of primary oocytes from these cells cytoplasmic segregation of mitochondria leads to a high variance in copy number of the mutant mtDNA type.

this allows segregation of neutral and mildly deleterious mutations. The result of these bottlenecks is that a mutant mtDNA representing a small minority of molecules in the soma of a mother can come to represent a range of proportions from zero to a large majority in her children. This phenomenon, known as **cytoplasmic segregation**, accounts for the wide variation in severity of mtDNA-associated diseases from one generation to the next, and also leads to interpretative problems when mtDNA mutation rates are considered from pedigree data where mutations that will never be fixed in the descendant mitochondrial lineage can nonetheless be observed.

The term **heteroplasmy** is used to refer to a situation where more than one mtDNA type occurs in a cell, and **homoplasmy** when all mtDNAs are identical. Clearly, it is difficult to apply these terms rigorously, since in a large population of mtDNA molecules within a cell at any one time there may be several mutant varieties present at undetectably low concentrations. In practice, heteroplasmy can only be recognized when a certain proportion of a particular class of mutant molecules has accumulated—this is usually >10% with Sanger sequencing methods. New sequencing technologies have great potential to provide an unbiased picture of heteroplasmy,[37] but, as with nuclear DNA analysis, the problem of false positives needs to be dealt with carefully. Note that the sensitivity of PCR (**Section 4.2**) does allow the targeted detection of a *specific* class of mutant molecule even at extremely low copy numbers.

3.4 VARIATION IN TANDEMLY REPEATED DNA SEQUENCES

Another class of genetic variation, on a generally larger scale and often mutating much more rapidly than that described in the previous sections, is common in eukaryotic genomes. This variation involves changes in the numbers of repeated DNA sequences arranged adjacently in **tandem** arrays, and the highly heterogeneous classes of loci undergoing these changes are collectively known as **variable number of tandem repeats** loci, or **VNTRs**. While the high variability of these **multiallelic** loci is a useful property in many respects, the underlying high mutation rates mean that, in contrast to SNPs:

- Alleles with the same size and sequence may not reflect identity by descent, but identity by state (coincidental resemblance, sometimes called **convergent evolution**).
- Ancestral states cannot be determined by reference to great ape DNAs.

VNTRs are classified according to the size of their repeat units, the typical number of units in repeat arrays, and sometimes their level of variability. **Figure 3.16** provides an overview of the scales of the different common classes of VNTRs, and their generally used names. Nomenclature is not systematic, and can be mystifying (see **Box 3.3**).

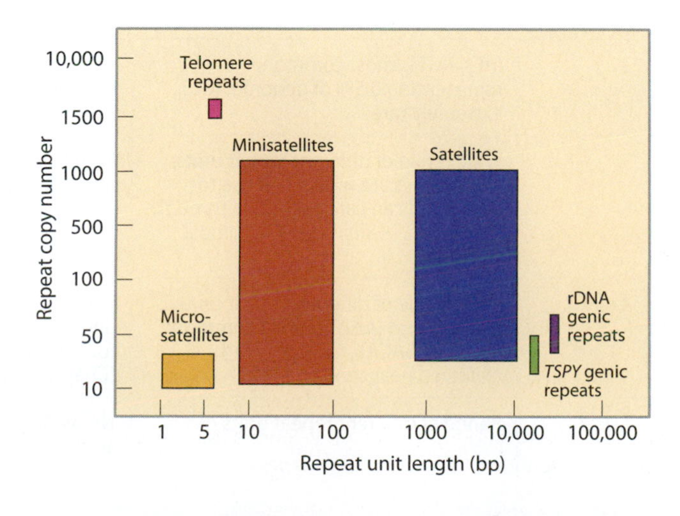

Figure 3.16: Overview of repeat unit length and repeat array length at VNTR loci.
Approximate repeat unit length and repeat copy number are shown on non-uniform scales for microsatellites, minisatellites, satellites, and telomere repeat arrays, which contain about 1500–2500 copies of a hexanucleotide repeat. Microsatellites and minisatellites with fewer repeat units are very abundant but not usually analyzed because of their low mutability and level of polymorphism. For satellites the repeat unit length represents the higher-order repeat; many satellites have a lower-order repeat unit length of a few to a few hundred base pairs. Ribosomal DNA (rDNA) arrays are on the short arms of the five acrocentric chromosomes (see Section 2.5); the array of the 20-kb unit containing the *TSPY* gene is on Yp.

Box 3.3: Why "satellite"?

One term that often crops up in VNTR nomenclature is **satellite**; since this is non-intuitive, it deserves some historical explanation. Early attempts to fractionate physically sheared DNA used buoyant-density gradient ultracentrifugation, in which separation was on the basis of density in cesium chloride density gradients, often in the presence of binding agents that accentuated differences in density between sequences of different compositions. In these experiments, first reported in 1961 when satellites like Sputnik were in the news, certain minor fractions formed separate bands from the bulk of the DNA, and were therefore called satellites of the major fraction. It later became clear that these fractions represented very high copy number tandemly repeated sequences, including those present at the centromeres of chromosomes in some species (see Section 2.5). The term satellite later became adopted as a term for a wide range of tandemly repeated sequences including micro- and minisatellites, even though these kinds of sequences (unlike true satellites—see main text) are not separable in an ultracentrifuge.

VNTR loci cover a vast range of different scales from microsatellites of a few tens of base pairs to multi-megabase satellites that are cytogenetically visible. It is important to note that, despite the shared property of varying tandem repeat number, these different loci have little in common in the mechanisms that generate them and maintain their variability. Their properties are described in the following sections, starting with the smallest-scale loci, and moving on to the largest.

Microsatellites have short repeat units and repeat arrays, and mutate through replication slippage

Microsatellites, also widely known as **short tandem repeats** (**STRs**), and sometimes as simple sequence repeats (SSRs), are tandem arrays of repeat units 1–7 bp in length, and those that have a useful degree of polymorphism have a typical copy number of 10–30. **Table 3.4** lists some general properties of microsatellites by repeat unit size, and **Figure 3.17** shows some examples. Some compound microsatellite loci contain repeat units of more than one length, such as adjacent di- and tetranucleotide arrays. Microsatellites composed of some specific repeat units show clustering, but most are distributed throughout the genome (**Figure 3.18**), which has been a useful property in **linkage analysis** (see Box 16.1).

While variation at most microsatellites is considered to have no influence on the well-being of the person carrying them, some have phenotypic effects. These include subtle influences on transcription, but the best-known examples are a small subset of trinucleotide microsatellites in which **dynamic mutations** cause dramatic expansions in repeat number, thus causing a number of genetic diseases:[43]

- A CAG repeat within the coding region of a gene, encoding a **polyglutamine tract**, expands from typically 10–30 repeats to 40–200 repeats. The resulting polyglutamine expansions cause proteins to aggregate within certain cells and kill them. These CAG expansion disorders, for example Huntington's

TABLE 3.4: PROPERTIES OF MICROSATELLITES BY REPEAT UNIT SIZE

Repeated unit/bp	Properties and distribution	Utility
1	Mostly poly(A)/poly(T), associated with *Alu*, LINE, and other retroelements	Not used, due to small differences in allele size and problem of allele-calling due to **PCR stutter**, resulting from slippage synthesis errors by the PCR polymerase
2	$(AC)_n/(GT)_n$ most common, representing 0.5% of genome; $(GC)_n$ extremely rare	Widely used in early studies because of ease of discovery; stutter a problem
3	Wide range of different repeat units; some arrays are within or close to genes and can cause diseases through expansion. $(AAT)_n$ and $(AAC)_n$ most common	Widely used. Alleles easily discriminated, and little stutter
4	Wide range of different repeat units. $(AAAC)_n$ and $(AAAT)_n$ most common; $(GATA)_n/(GACA)_n$ frequent, and clustered near centromeres	Widely used. Alleles easily discriminated, and little stutter; form basis of most forensic microsatellite profiling (Chapter 18)
5, 6, 7	Range of different repeat units	Not widely used because of relative scarcity

Figure 3.17: Examples of microsatellite structures.
Normal and pathogenic size ranges for the CAG repeat in the huntingtin open reading frame (expanded in Huntington's disease) are given.

disease (OMIM 143100; clinical features summarized in Table 16.1) and spinocerebellar ataxia type I (OMIM 164400), are mostly dominant.

- A trinucleotide repeat (for example, CGG, CCG, CTG, GAA) in the promoter, untranslated region, or an intron of a gene expands from typically 5–50 repeats to 50–4000 repeats. This can abolish transcription, as in the case of Fragile X syndrome (OMIM 309550) or Friedreich's ataxia (OMIM 229300).
- Expansion of a trinucleotide repeat in the 3′ UTR of a gene leads to accumulation of RNAs that interfere with the normal balance of RNA-binding proteins and disrupts splicing; this occurs in the disease myotonic dystrophy (OMIM 160900), with expansions from 5–37 CTG repeats, to 50–thousands.

Microsatellite mutation rates and processes

The mutation rates of microsatellites have been estimated by direct pedigree analysis,[9, 57] and general characteristics of mutability have also been studied by large-scale comparisons of human and chimpanzee microsatellites.[28] Typical figures for mutation rate estimates from direct approaches are around 10^{-3}–10^{-4} per polymorphic microsatellite per generation. Application of such rates to the dating of past events such as migrations and population expansions is somewhat controversial, since when populations with well-documented short-term histories are examined the **effective mutation rate** that explains the accumulated diversity seems to be considerably lower. For example, the average directly determined mutation rate for a widely used set of Y-chromosome-specific tetranucleotide microsatellites is 2.0×10^{-3} per microsatellite per generation, but the effective rate inferred from population data[71] is 6.9×10^{-4}, assuming a 25-year generation time (note that population data suggest a longer generation time is more appropriate for the Y—see Table 6.6). The discrepancy may be partly explained by back-and-forth changes in repeat unit number being completely observed in pedigree studies, but not in population studies, where the considerable **homoplasy** of alleles can be hidden.

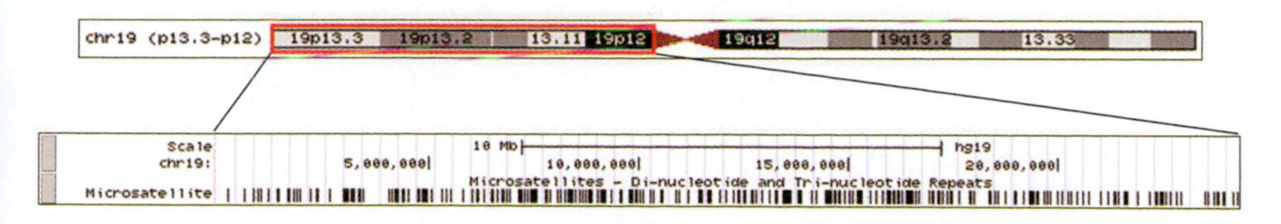

Figure 3.18: Microsatellites are distributed throughout human chromosomes.
Microsatellites representing a minimum of 15 perfect uninterrupted repeats of any di- or trinucleotide motif (288 and 14 examples respectively) are indicated by vertical bars along the ~25 Mb of the short arm of chromosome 19, with a schematic idiogram given above. Such loci are very likely to be polymorphic. Produced in the UCSC Genome Browser (**http://genome.ucsc.edu/cgi-bin/hgGateway**), based on Tandem Repeats Finder output (**http://tandem.bu.edu/trf/trf.html**).

A number of properties of the microsatellite mutation process emerge from direct mutation studies[9, 57] and from research which models mutation to explain diversity data:

- Most mutations (68% for dinucleotides, and >95% for tetranucleotides) involve an increase or decrease of a single repeat unit (**Figure 3.19a**).

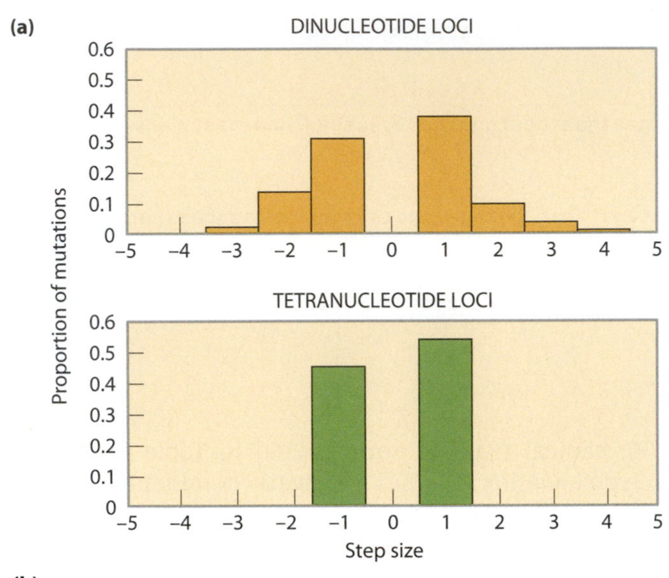

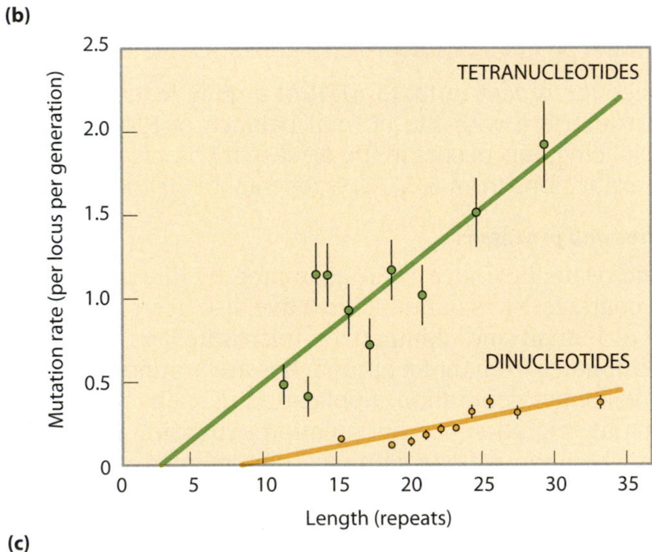

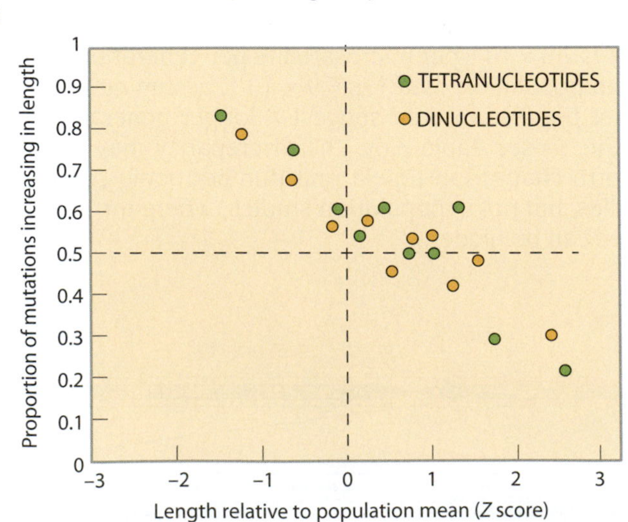

Figure 3.19: Properties of microsatellite mutation processes.
(a) Mutation events involve predominantly (dinucleotides) and near-exclusively (tetranucleotides) single-step changes. (b) Mutation rates increase with repeat array length. (c) Shorter alleles are biased toward repeat gains in mutation, while longer alleles are biased toward repeat losses. [From Sun JX et al. (2012) *Nat. Genet.* 44, 1161. With permission from Macmillan Publishers Ltd.]

- Homogeneous (pure) repeat arrays mutate faster than equivalent-sized interrupted arrays containing variant repeats.
- Overall mutation rate increases as array length increases (Figure 3.19b).
- Shorter alleles tend to increase in size, while longer alleles tend to decrease, leading to an equilibrium distribution of allele lengths (Figure 3.19c). An additional factor may be that larger microsatellites also represent larger targets for point mutation, which through interrupting the repeat array leads to reduced mutation rates.

The **stepwise mutation model** or **single-step mutation model** (**SMM**; see **Section 5.2**) is often used to model microsatellite evolution. According to this model, the length of a microsatellite varies at a fixed rate independent of repeat length and with the same probability of expansion and contraction. As is clear from the above list of properties, this model is oversimplistic.

It is widely accepted that microsatellite mutation occurs as a result of DNA **replication slippage** (**Figure 3.20**), even though direct evidence on the mechanism remains elusive. Slippage occurs *in vitro*, and the similar mutational properties of haploid, Y-chromosomal microsatellites to their diploid autosomal counterparts[27] suggests that interchromosomal recombination does not play a major role. Also, there is no excess of microsatellites found in autosomal recombination hotspots.[7]

Minisatellites have longer repeat units and arrays, and mutate through recombination mechanisms

Minisatellites consist of repeat units from about 8 to 100 bp in length, with copy numbers from as low as 5 to over 1000 (**Table 3.5**). They are not simply microsatellites on a larger scale, but are qualitatively different in their

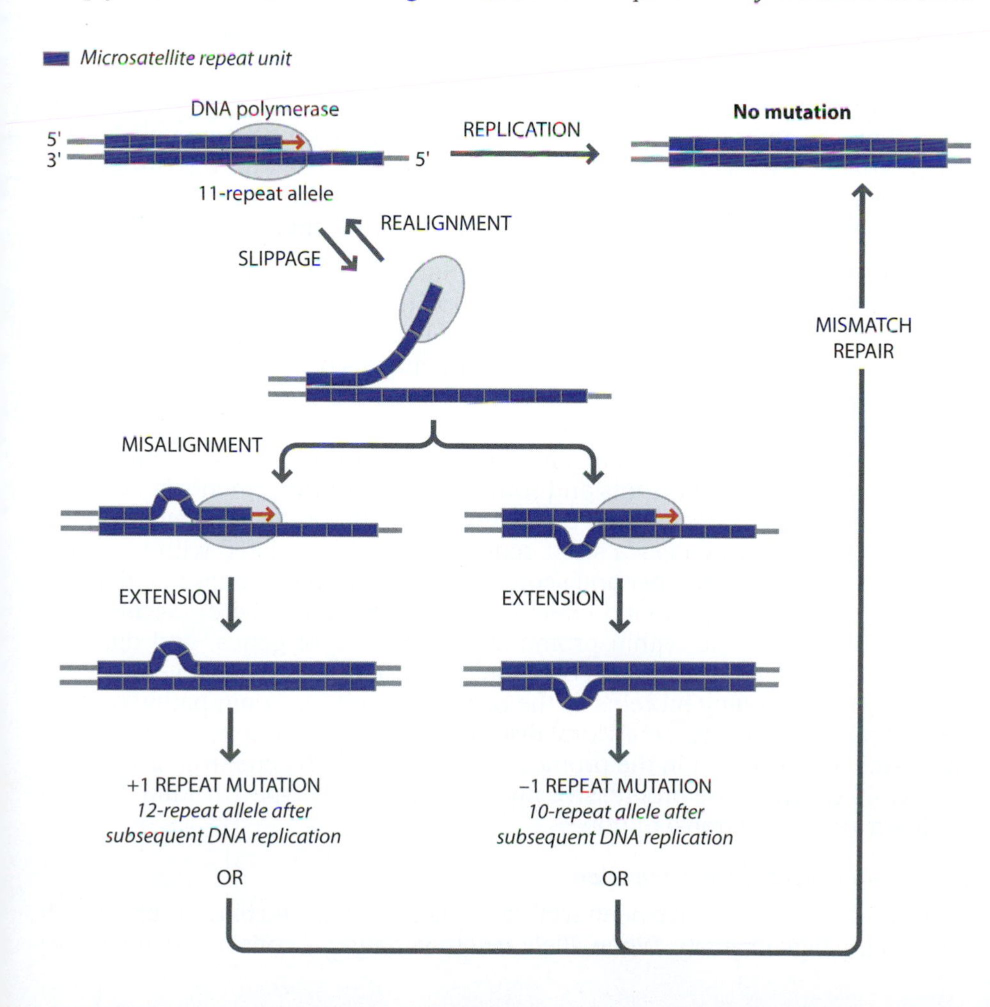

Figure 3.20: Mutation at microsatellites. Model of the effects of slippage and misalignment in microsatellite mutation. Defects in mismatch repair, such as those in some human cancers, lead to elevated mutation rates in microsatellites. Mutations shown here are a single-step increase or decrease; mutations involving greater repeat numbers occur, but are about tenfold rarer, and may involve other, non-slippage-based processes.

TABLE 3.5: GENERAL PROPERTIES OF SIX MINISATELLITES

Mini-satellite	Chromo-somal location	Repeat unit length (bp)	GC-content	Repeat copy number	Repeat sequences	Sex-averaged mutation rate/ generation
MS32	1q42–1q43	29	62%	12 to >800	`tgactcagaatggagcaggtggccagggg` `...................c.a.......`	0.8%

MS205	16pter	45–54	75%	8–87	`tgcatgccgacccgtctactcgcccgcccc`
					`c.---------...................`
					`ccacgtaccccgcccccggccc-ta`
					`.....c................c..`
						0.4%

CEB1	2q37	31–43	70%	5 to >300	`tcagcccagggacctcgcaggccaccctccc`
					`...a......a.....t....-...----..`
					`tccctcccccc`
					`t........t.-`
						7.6%

B6.7	20qter	34	56%	6–540	`gtggacagtgagggggtctctacaggccatg`
					`....................-.t...a....`
					`agg`
					`..a`
						5.4%

MSY1	Yp11	25	25%	22–114	`cataatatacatgatgtatattata` `..c..c....t.c.......a....`	2.7%

PRDM9	5p14	84 (coding)	60%	8–18	`tgtggacaaggtttcagtgttaaatcagatgt`
					`tattacacaccaaaggacacatacaggggaga`
					`agctctacgtctgcagggag` (plus many variants)
						~0.3–20 × 10^{-5}

The sequence of one repeat unit is given, with positions of variation underneath ("." same base as above; "-" deleted base). Variant positions exist in combination to give up to 20 different repeat types (CEB1, B6.7, PRDM9). [From Armour JA et al. (1993) *Hum. Mol. Genet.* 2, 1137; Berg IL et al. (2010) *Nat. Genet.* 42, 859; Buard J & Vergnaud G (1994) *EMBO J.* 13, 3203; Jeffreys AJ et al. (1991) *Nature* 354, 204; Jeffreys AJ et al. (2013) *Proc. Natl Acad. Sci. USA* 110, 600; Jobling MA et al. (1998) *Hum. Mol. Genet.* 7, 643; May CA et al. (1996) *Hum. Mol. Genet.* 5, 1823; Tamaki K et al. (1999) *Hum. Mol. Genet.* 8, 879.]

variability, mutation rates, mutation processes, and chromosomal distribution. They are among the most dynamic loci in our genome, some displaying hypervariability, with vast numbers of alleles of different lengths and structures, locus-specific mutation rates as high as 14% per generation, and complex mutation processes involving both inter- and intra-allelic events.

As is the case with microsatellites, minisatellites tend to be treated as neutral loci. Again, however, there are exceptions. Some lie in coding regions, resulting in repetitive and highly variable proteins. One case is the *MUC1* gene, which encodes the highly polymorphic glycoprotein mucin, containing 20-amino-acid repeat units varying in copy number from about 20 to over 120 in Europeans. Another particularly interesting example is the gene encoding PR domain–containing 9 (*PRDM9*), which has recently been identified as a major regulator of meiotic recombination hotspots and some genome rearrangements in humans. The minisatellite encodes the DNA-binding motif of the protein, a **zinc finger** array that recognizes a short sequence motif associated with hotspots, and variation in copy number and sequence of the 84-bp (28-amino-acid) minisatellite repeat has a profound effect on recombination activity[5] (**Section 3.8**). Some minisatellites lie within promoters or 3′ UTRs of genes, and different repeat numbers can have effects on gene expression—there are several examples in genes encoding proteins of the dopamine and serotonin pathways that have apparent effects on behavioral disorders and traits. Expansions of a 12-bp minisatellite repeat unit in the promoter of the cystatin B gene from 2–3 copies to over 40 copies cause progressive myoclonus epilepsy type 1 (OMIM 254800) by down-regulating transcription.[33]

Minisatellite diversity and mutation

The minisatellites that have been well studied are a small and biased subset with particularly high diversity. When allele length variation is considered, these loci

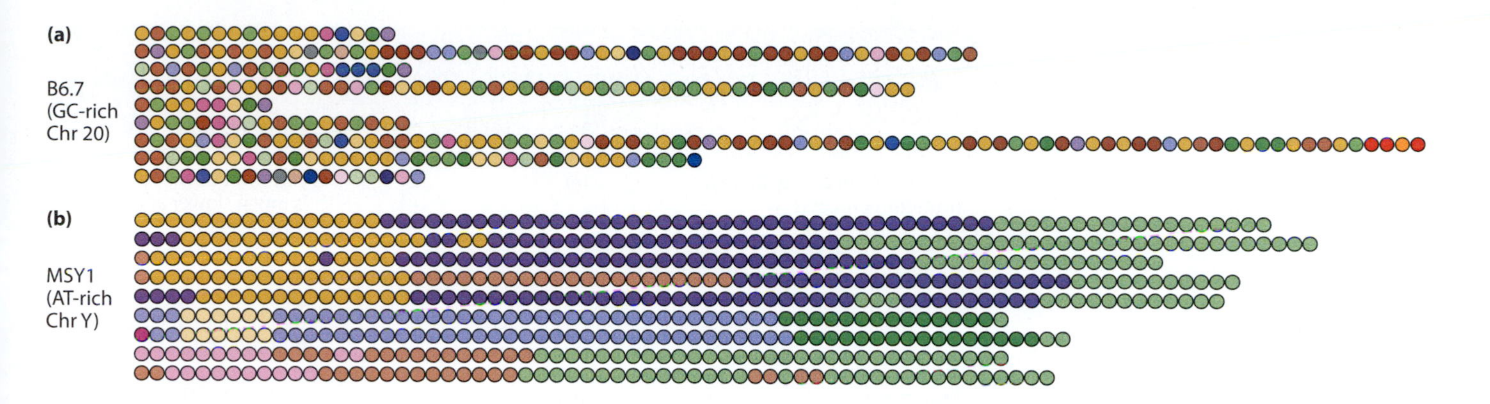

Figure 3.21: Hypervariability of minisatellite alleles.
A selection of alleles (each line representing a different allele) for (a) the autosomal minisatellite B6.7. Each different colored circle represents a sequence variant on a 34-bp repeat unit; (b) the Y-chromosomal minisatellite MSY1. Each different colored circle represents a sequence variant on a 25-bp repeat unit. The block-like structure of MSY1 alleles contrasts with the interspersed pattern for B6.7, and reflects the absence of crossing over on the Y chromosome. [a, data from Tamaki K et al. (1999) *Hum. Mol. Genet.* 8, 879. b, data from Jobling MA et al. (1998) *Hum. Mol. Genet.* 7, 643.]

show typical **heterozygosity** values (the probability that two alleles sampled at random from the population differ in length) of well over 90%. Sequencing reveals additional diversity—minisatellites contain variant repeats differing by base substitutions and small indels (**Figure 3.21**), though unlike microsatellites, this internal variation does not impair the mutation process. A method known as minisatellite variant repeat PCR (**MVR-PCR**) allows these variant repeats to be mapped within arrays conveniently, and allows access to the details of the mutation process in sperm DNA. This yields a far greater number of mutant molecules than does pedigree analysis, but of course is restricted to an analysis of male mutation—all information about female mutation remains pedigree based, and therefore limited to a relatively few observations. Mutation at many minisatellites is in fact highly elevated in males compared to females (see **Section 3.7**); as an extreme example, the minisatellite *CEB1* has a mutation rate of 15% per sperm, but only 0.2% per oocyte.[62]

Germ-line mutation events at minisatellites can be highly complex, involving interactions between alleles in which one allele acquires blocks of repeat units from the other in a nonreciprocal **gene conversion** process (see **Section 3.8**) and resulting in a general bias toward repeat unit gains over losses. For most minisatellites, mutations occur preferentially at one end of the repeat array.

GC-rich minisatellites tend to be clustered toward the ends of chromosomes where recombination is elevated, and this, together with some properties of their sequences, suggested that they might be associated with recombination hotspots—either as cause or consequence. This has now been confirmed by analysis of mutants in sperm DNA, and these minisatellites exist as a by-product of the meiotic recombination process. The primacy of interallelic processes is emphasized by the absence of variable, GC-rich minisatellites from the nonrecombining region of the Y chromosome, but their plentiful presence in the recombining pseudoautosomal region.

The main differences between mutation at minisatellites and microsatellites are summarized in **Table 3.6**.

Telomeres contain specialized and functionally important repeat arrays

Telomeres are essential structures at the ends of chromosomes, and include as their DNA component tandem arrays of the hexanucleotide repeat TTAGGG, typically 10–15 kb in length. They are sometimes included in the class of microsatellites, or even minisatellites, but their high repeat number, absence of **distal** flanking single-copy DNA, and specialized functional role put them in a class of their own. One end of the array is contiguous with sub-telomeric DNA, while the other is the true chromosomal terminus, and so has no flanking DNA. The most proximal parts of telomere arrays contain variant repeats (for example, TCAGGG), and this source of variation has been used to study the dynamics of telomeres. The distal majority of the telomere array appears to be homogeneous, and repeats are added to it by a specialized DNA polymerase, **telomerase**,

TABLE 3.6:
MUTATION AT MICRO- AND MINISATELLITES COMPARED

	Microsatellites	Minisatellites
Germ-line specificity of major mutation process?	no	yes, somatic mutation is much slower and intra-allelic
Lower threshold effect of array size on mutation rate?	yes	no, for example a 5-repeat allele at the CEB1 minisatellite maintains a mutation rate of 0.4%
Mutation rate increases with array homogeneity?	yes	no, some of the most variable loci have the most heterogeneous repeat units
Interallelic processes?	probably not	yes
Bias to repeat unit gains over losses?	for small alleles	yes
Effects on mutation from flanking DNA variation?	no evidence	yes

which is active in the germ line and regrows the telomere, which loses repeat units from its end at every DNA replication.

Satellites are large, sometimes functionally important, repeat arrays

Satellites, sometimes called macrosatellites, are large tandem arrays spanning hundreds of kilobases to megabases, and composed of repeat units of a wide range of sizes that can display a higher-order structure (**Figure 3.22**).

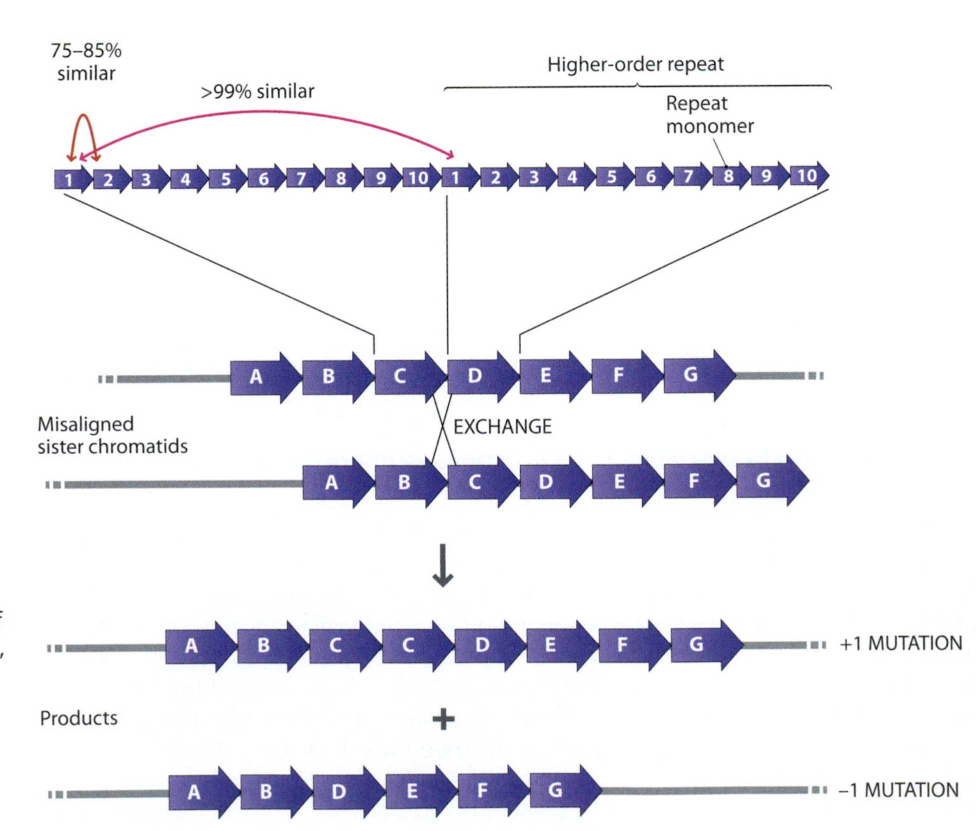

Figure 3.22: Structures of satellites, and mutation by unequal sister chromatid exchange.
This schematic diagram (*upper part*) shows an array of repeat monomers (for alphoid DNA these are 170 bp in length) and their higher-order structure: individual subunits are about 75–85% similar to each other, but higher-order repeats, which, for alphoid DNA, can be as large as 6 kb depending on the chromosome, are over 99% similar. This indicates that the mode of evolution has been as higher-order repeats, probably in an unequal exchange between sister chromatids of a bivalent in meiosis (see Figure 2.17), shown in the lower part of the figure. Mutations may involve changes of more than one higher-order repeat unit.

A good example is alpha satellite, or **alphoid** DNA, which forms an important component of centromeres. The alphoid repeat monomer is ~170 bp in length, but in many chromosomes it is arranged in higher-order repeat units of a few kilobases. This higher-order structure can be repeated thousands of times to form an array several megabases in size. The detailed structures of satellite arrays are very difficult to investigate because of their size and repetitive nature. The human reference sequence, often thought of as a complete sequence, actually contains gaps where major blocks of satellite lie, because the sequences of these regions cannot be assembled from individual sequence reads using standard methods. Some large VNTRs contain entire genes within their repeat units, and arrays of different numbers of repeats may affect levels of gene expression; a good example is the tandem arrays of ribosomal DNA (rDNA) genes, encoding **rRNA**.

The mutation processes at these loci cannot be studied directly, but probably involve **unequal crossing over** between homologous chromosomes or sister chromatids misaligned by integral numbers of higher-order units (Figure 3.22). Gene conversion (**Section 3.8**) between units is also prevalent, making such loci within species more similar to each other than to orthologous sequences found in other species (so-called concerted evolution). Historically, some satellite polymorphisms have been used in human evolutionary studies, but nowadays they have been superseded by loci that are easier to type, analyze, and understand.

3.5 TRANSPOSABLE ELEMENT INSERTIONS

Forty-five percent of the human genome is composed of dispersed repeat elements[24] with copy numbers ranging from a few hundred to several hundred thousand, including >850,000 LINEs (long interspersed nuclear elements) and >1,500,000 SINEs (short interspersed nuclear elements). The best-studied examples of these types are respectively the L1 and *Alu* **retrotransposons** (see **Section 2.4**). **Figure 3.23** shows the mechanisms of insertion of L1 and *Alu* elements and **Table 3.7** summarizes their properties. The **reverse transcriptase** enzyme encoded in full-length L1 elements is related to that used by **retroviruses** (for example, HIV) to integrate a DNA copy of their RNA genome into the host genome. Other dispersed repeat elements are more closely related to retroviruses. Human endogenous retroviruses (**HERV**s) have retained the ability to replicate themselves throughout a genome but do not have the ability to construct the protein coat that allows them to leave the cell. If these HERVs enter the germ line and replicate, integrated DNA copies of their genomes are inherited by future generations. Over time, HERVs accumulate mutations that prevent them from replicating themselves; almost all HERVs have suffered this fate.

Some transposable elements are human-specific—for example, ~5500 human *Alu* insertions and ~1200 human L1 insertions in the human reference sequence are not found in chimpanzee DNA.[40] These human-specific elements belong to particular subfamilies (defined by shared sequence variants) and a subset is still active in transposition. This is demonstrated by their ability to occasionally disrupt genes by insertional mutagenesis, thus causing genetic diseases[3] such as hemophilia A (OMIM 306700), Duchenne muscular dystrophy (OMIM 310200), and cystic fibrosis (OMIM 219700). As well as this direct mechanism for disease causation, *Alu* elements in particular can act as substrates for illegitimate recombination, leading to duplication or deletion of segments between a pair of elements, or to translocations when they lie on different chromosomes (see **Section 3.6**). This mechanism plays at least as significant a role in genetic disease as retrotransposition, and, as comparisons between human and chimpanzee genomes show,[19, 49] has also been important in genome evolution. *Alu* elements have also been an important source of new exons in primate evolution (**Section 8.2**), particularly affecting human genes, such as zinc finger genes, involved in regulation of transcription.[50]

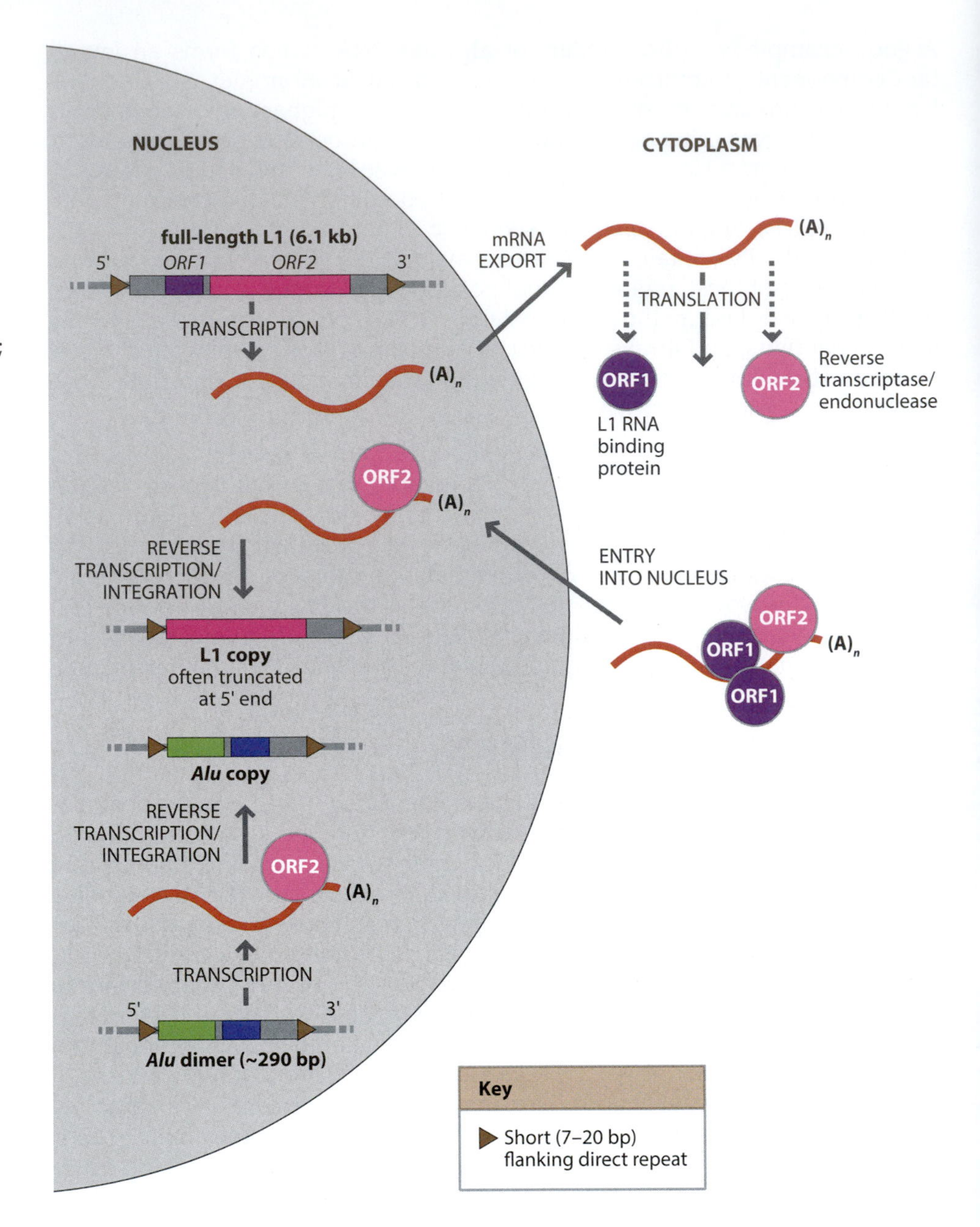

Figure 3.23: Steps in L1 and *Alu* retrotransposition. A full-length L1 element is transcribed and the transcript exits the nucleus. In the cytoplasm it is translated to yield ORF (open reading frame) 1 and 2 proteins; proteins and mRNA may assemble into a ribonucleoprotein particle for re-entry to the nucleus. There, the L1 RNA is reverse-transcribed and integrated into genomic DNA by a process called target-primed **reverse transcription**, catalyzed by ORF2; this is often accompanied by truncation. *Alu* elements are transcribed in the nucleus and their reverse transcription and integration is thought to require the L1 ORF2 protein. [Adapted from Ostertag EM & Kazazian HH (2001) *Ann. Rev. Genet.* 35, 501. With permission from Annual Reviews.]

Analysis of the 1000 Genomes Project pilot data discovered 7380 polymorphic elements in 185 individuals from the YRI, CEU, CHB, and JPT HapMap samples,[54] and some such elements have been used as binary molecular polymorphisms;[68] a much smaller number of polymorphic HERVs have also been found. All these elements have the advantage that they can insert, but having once inserted, are almost never removed—only 0.5% of *Alu* element differences between humans, chimpanzees, and macaques could be explained by precise excisions in one lineage, rather than by new insertions.[60] It is therefore straightforward to decide upon the ancestral state of the polymorphism—it is almost invariably absence of the element. Also, since there is little or no sequence specificity for insertion sites, when two chromosomes share an inserted element at exactly the same position, it is almost always an example of identity by descent, rather than identity by state, and therefore an example of a unique event polymorphism. The abundance and simplicity of typing of SNPs nevertheless makes them by far the most widely exploited class of binary polymorphism.

TABLE 3.7:
TRANSPOSITIONAL PROPERTIES OF L1 AND *ALU* ELEMENTS

	L1	*Alu*
Phylogenetic distribution	mammals	higher primates
Copy number in human genome	516,000	1,090,000
Distribution in human genome	enriched on X chromosome and in AT-rich DNA	enriched in GC-rich DNA
Transcriptionally competent copies	40–50	~2000
Autonomous transposition?	yes	no
Number of known transpositions causing disease[a]	18	43
Active subfamilies	Ta	Y, Ya5, Ya8, Yb8
Number of polymorphic elements	115 known	>1000 estimated
Approximate transposition rate (number of transpositions per diploid genome per generation)	0.004–0.1 (1 per 10–250 births)	~0.05 (~1 per 20 births)

[a] Data from Belancio VP et al. (2008) *Genome Res.* 18, 343, which also lists four cases of disease caused by insertion of the nonautonomous SINE element SVA. Other data from Cordaux R et al. (2006) *Gene* 373, 134.

3.6 STRUCTURAL VARIATION IN THE GENOME

Structural variation is a broad term covering changes in chromosome structure. Some of these changes are **balanced**, involving no alteration of sequence copy number—examples are the inversion of a chromosomal segment or a reciprocal **translocation** between a pair of chromosomes. Others involve differences in the numbers of particular sequences between alleles—**copy-number variation**. Structural variation exists over a very wide range of scales: at the upper end it includes changes involving entire chromosomes; at the lower end some researchers have argued that even a single-base indel should be counted as a structural variant. In practice an arbitrary threshold of >1 kb has generally been used to define structural variation, so it excludes small indels, including *Alu* insertions. In principle, insertions of L1 retroelements could be classed as structural variants, but they are generally not included because they form a coherent class with a well-understood basis.

Appreciation of the impact of structural variation on the genome has been growing rapidly in recent years, largely because of the introduction of technologies such as microarrays and novel sequencing methods (explained in Chapter 4) that allow it to be assayed more comprehensively and efficiently. Its prevalence means that the genome is far from the essentially invariant structure suggested by the idea of a reference sequence, onto which the polymorphisms described above (SNPs, VNTRs, transposable element insertions, and so on) can be mapped. Instead it is very dynamic—structurally polymorphic in different individuals, with many megabases of DNA present in one genome copy but absent in another, and prone to recurrent rearrangements. As with the other classes of genetic variation described in this chapter, much structural variation has little or no effect on carriers, but some clearly does,[53] and assessing its general functional and evolutionary importance is an area of intense research activity.

Some genomic disorders arise from recombination between segmental duplications

Analysis of the human reference sequence revealed the remarkable fact that ~5% of the genome exists as long, duplicated sequences (**paralogs**) that have arisen within the last 40 million years. These segmental duplications (also known as **low-copy repeats**, or **LCRs**) have been defined operationally as being between one and hundreds of kilobases in size and >90% similar in sequence. Their existence has profound implications for the evolution of our genome and for disease because such paralogous repeats have the potential to promote **non-allelic homologous recombination** (**NAHR**). NAHR typically occurs in repeats that are >10 kb in length, >97% similar in sequence, and contain a segment of sequence identity at least a few hundred base pairs in length. When NAHR occurs between LCRs in the same orientation (**direct repeats**) the result is the deletion or duplication of material lying between them; when the LCRs are in opposite orientations (inverted repeats), the result is inversion of the intervening sequence (**Figure 3.24**). NAHR between repeats lying on different chromosomes results in translocations.

There are many instances of human diseases that are caused by NAHR between paralogous repeats, often mediated by the deletion or duplication of dosage-sensitive genes. One example is rearrangement sponsored by NAHR between the *CMT1A* repeats on chromosome 17p: these 24-kb direct repeats flank a 1.5-Mb region containing the dosage-sensitive *PMP22* gene. Duplication of this region on one chromosome (that is, a total of three doses instead of the normal diploid two) leads to Charcot–Marie–Tooth disease (Table 16.1; OMIM 118220), while the reciprocal deletion (one dose rather than the diploid

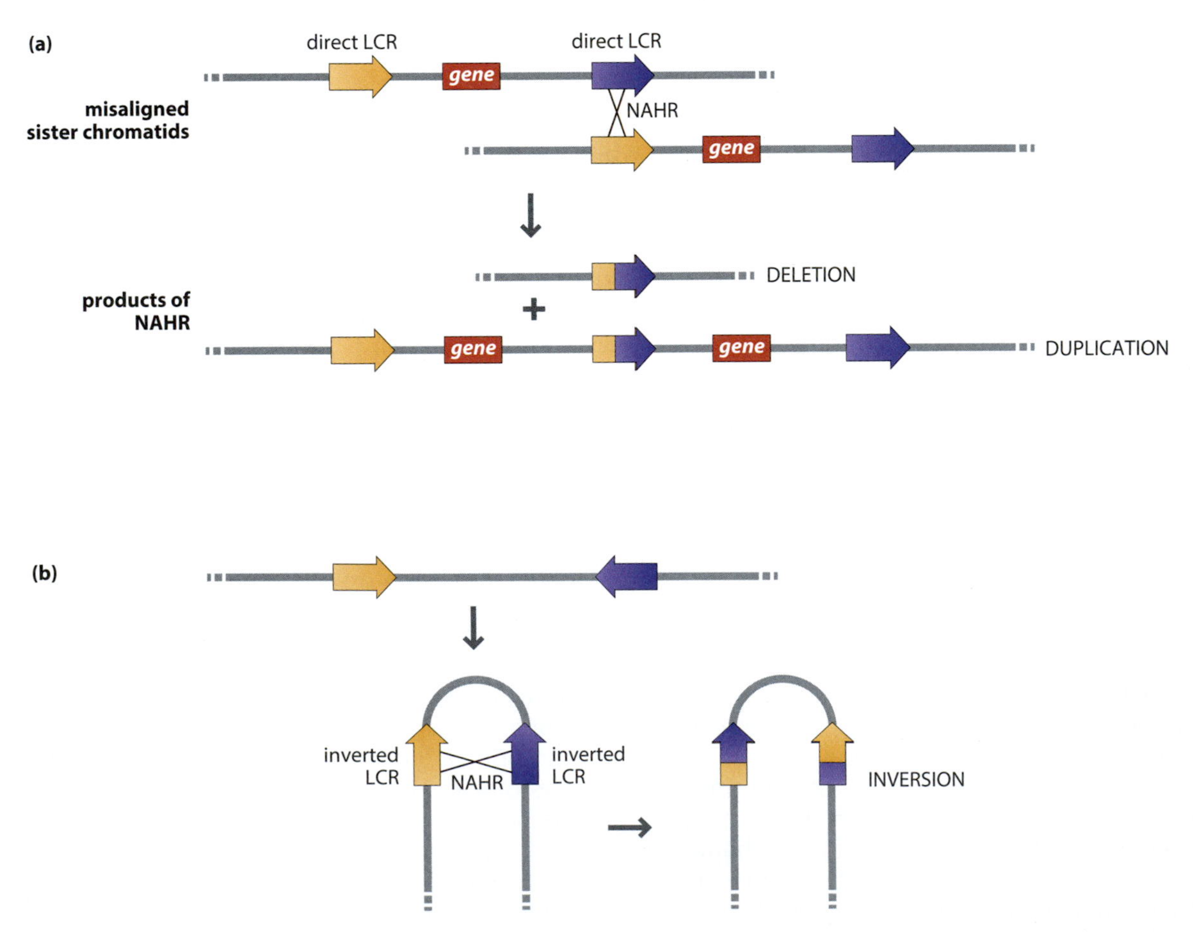

Figure 3.24: Products of NAHR between direct and inverted repeats.
(a) NAHR between direct repeats generates deletions or duplications. If a dosage-sensitive gene lies between them, a phenotype can result (a genomic disorder). (b) NAHR between inverted repeats generates inversions.

two) leads to hereditary neuropathy with pressure palsies (OMIM 162500). Diseases caused by rearrangements due to underlying structural features of the genome, rather than to relatively random events such as base substitutions, are called **genomic disorders** (**Section 16.3**). Rates of NAHR causing four different genomic disorders have been determined by PCR analysis in sperm DNA,[59] and are typically about 10^{-6}–10^{-5} events per generation.

Copy-number variation is widespread in the human genome

The term copy-number variant (CNV) traditionally refers to a DNA segment at least 1 kb in size, for which differences in copy number have been observed when comparing two or more genomes. The term **copy-number polymorphism** (CNP) is also sometimes used, and describes a CNV that exists at >1% frequency in a population.

Unlike small-scale variants that can be ascertained effectively by DNA sequencing, the detection of CNVs is highly dependent on the method used, which can make it difficult to compare studies. One systematic survey[44] (**Figure 3.25**) used two complementary methods to seek CNVs in the genomes of 270 individuals from four different populations (the HapMap samples). From this and other studies, a conservative estimate of the extent of copy-number variation between any two genomes is that it involves about 4 Mb of DNA; this is equivalent to about 1 in 800 bp on average, considerably greater than the figure of 1 in 1000–1250 bp given above for SNPs.

Although many CNVs are likely to have no phenotypic effect, some certainly cause genetic disease; indeed, genomic disorders are examples of CNVs. Others contribute to the susceptibility to complex disease, for instance higher copy number of β-defensin genes increases the risk of the common inflammatory skin disease psoriasis.[20] Others still may be targets of natural selection—an interesting example is the salivary amylase gene, where copy number varies between populations that have different dietary starch intakes (**Section 15.6**). Although many phenotypes associated with CNVs are mediated by gene-dosage effects, other causes are possible, such as disruption of a coding sequence, or alteration of transcriptional regulation. A general picture of the functional importance of CNVs is given by the observation that they explain ~18% of the variation in the expression of ~15,000 transcripts in HapMap **lymphoblastoid** cell lines, compared with ~83% for a genomewide set of SNPs.[55]

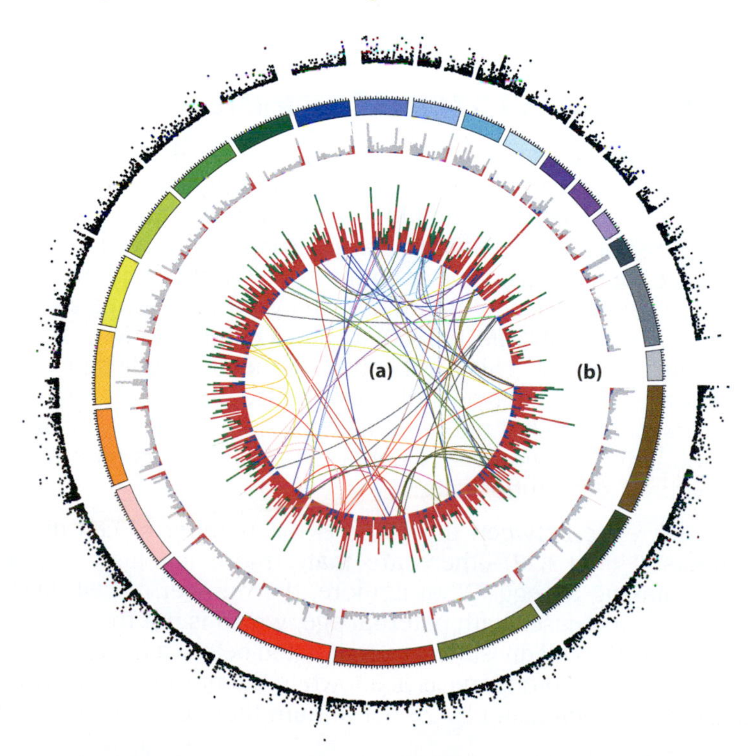

Figure 3.25: Distribution of CNVs in the human genome.
Circular map showing the genomic distribution of different classes of CNVs. Chromosomes are color-coded in the outer circle. (a) Curved lines connect the origins and new locations of 58 putative interchromosomal duplications, colored according to chromosome of origin.
(b) Stacked histogram representing the number of deletions (*red*), duplications (*green*), and multiallelic (*blue*) loci in 5-Mb bins. [Adapted from Conrad DF et al. (2006) *Nat. Genet.* 38, 75. With permission from Macmillan Publishers Ltd.]

The mechanisms of mutation of CNVs include NAHR, described above. However, sequencing of CNV regions has shown that other mechanisms must exist, since not all CNV breakpoints are marked by LCRs.[70] Some simple duplication/deletion CNVs can be explained by nonhomologous end-joining (NHEJ), a mechanism used to repair double-strand breaks. Breakpoints in these cases can exhibit no evidence of sequence homology, or **microhomology** involving just a few base pairs. Some CNVs suggest the existence of DNA replication-based mechanisms that can lead to deletions, duplications, or inversions, and more complex rearrangements arising in a single mutation event. Rates of CNV mutations are difficult to generalize, but when LCRs are involved typically lie in the range of 10^{-6} to 10^{-4} per generation—several orders of magnitude faster than SNPs.

Cytogenetic examination of chromosomes can reveal large-scale structural variants

Cytogenetic examination of chromosomes in individuals with genetic disorders has been a useful way to pinpoint disease genes—for example, cytogenetically visible deletions and translocations helped to localize the Duchenne muscular dystrophy gene (one of the first disease genes to be isolated by **positional cloning**; OMIM 310200) to Xp21. At this level there are also polymorphisms in normal individuals that can be recognized by cytogenetic analysis of banded chromosomes. These include inversions, deletions, duplications, length polymorphisms, and length variations in heterochromatin (as large as several tens of megabases, in the case of the Y chromosome). There are also some asymptomatic translocations—for example, those generating **satellited chromosomes** by the translocation of material from the short arms of acrocentric chromosomes such as chromosome 15. Because they are difficult to detect by molecular, PCR-based approaches, these chromosomal polymorphisms are probably under-ascertained, and have not been exploited as a source of useful variation in human evolutionary genetic studies.

Many of these large-scale rearrangements are probably caused through NAHR. For example, recombination between inverted ~400-kb clusters of olfactory receptor genes on chromosome 8p is the cause of a common benign inversion polymorphism. Heterozygous mothers can then undergo aberrant recombination in the inverted region, leading to deletions and **supernumerary** (that is, in excess of the normal diploid number of 46 chromosomes) inverted duplication chromosomes in offspring.[16]

3.7 THE EFFECTS OF AGE AND SEX ON MUTATION RATE

Mutations at many of the classes of loci described above show increasing rates with parental age. Furthermore, when the parent of origin of a de novo mutant chromosome is determined, for example by typing polymorphisms linked to the mutation, a strong bias toward paternal origin is often observed (see Box 5.4). The first evidence for sex bias came from the analysis of mutations associated with genetic disease:[13] 154 new mutations analyzed in a set of six disorders including achondroplasia (OMIM 100800) all had a paternal origin, and showed a strong paternal age effect. This has been confirmed by studies in which parent–child trios are analyzed by whole-genome sequencing (**Figure 3.26a,b**), showing a strong paternal, but not maternal effect on the rate of occurrence of base substitutions in the offspring. What could underlie this paternal-specific effect upon the base substitution rate?

One striking difference between the sexes is the number of cell divisions in **gametogenesis** (**Figure 3.27**)—there are many more involved in making a sperm than in making an egg. What is more, the number of cell divisions in **spermatogenesis** increases with paternal age, which is not the case for **oogenesis**, in which all the cell divisions are completed before a female is born. The number of cell divisions in oogenesis is 23, while the number in spermatogenesis at puberty is 36, increasing by 23 per annum thereafter: the octogenarian

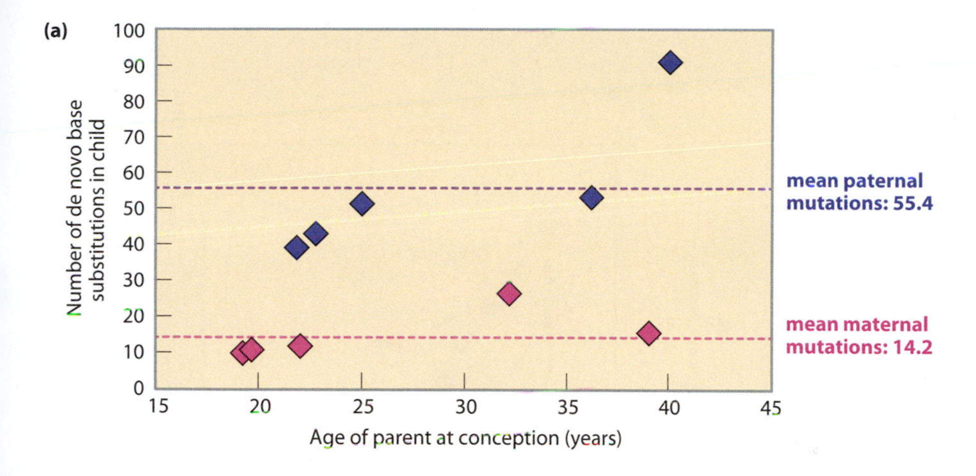

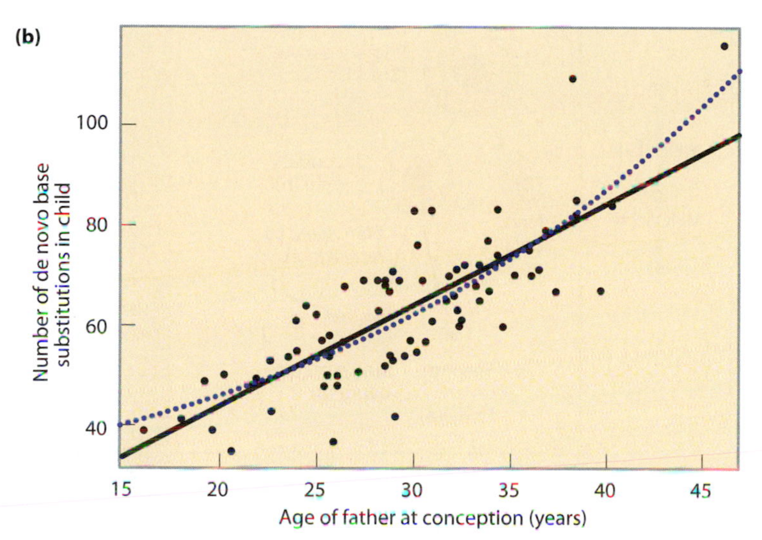

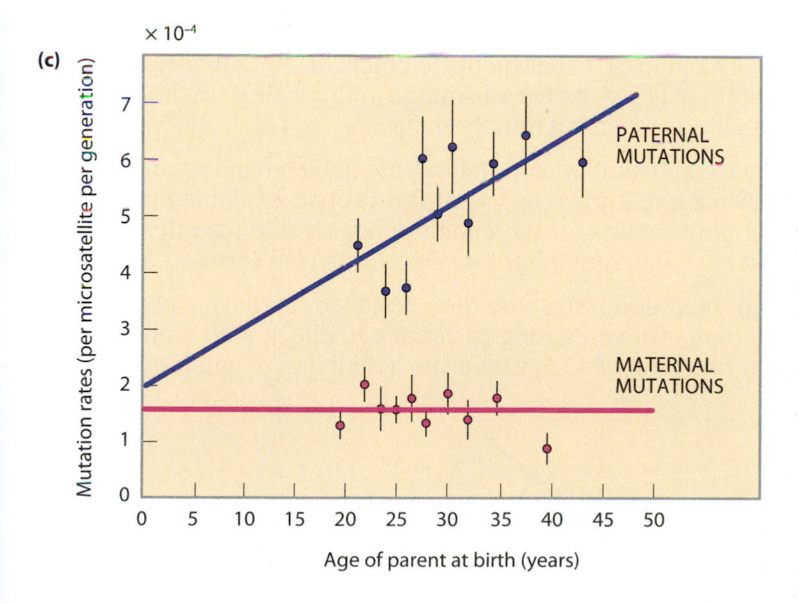

Figure 3.26: Sex and age effects on mutation for base substitutions and microsatellites.
(a) Parental age- and sex-effects on numbers of de novo base substitutions in whole-genome sequencing data from five Icelandic trios in which parent-of-origin could be determined. (b) Number of de novo base substitutions in 78 sequenced Icelandic trios as a function of paternal age. Here, the parent-of-origin is undetermined. The *black* line is the linear fit, while the dashed *blue* line represents a model where maternal mutations are assumed to occur at a constant rate of 14.2 per generation (see panel a) and paternal mutations increase exponentially with father's age. (c) Mutation rates for dinucleotide and tetranucleotide microsatellites as a function of parental age at childbirth, measured in 24,832 Icelandic trios for 2477 microsatellites. The paternal rate shows a positive correlation with age ($P = 9.3 \times 10^{-5}$), with a doubling of the rate from age 20 to 58. The maternal rate shows no evidence of increase with age ($P = 0.47$). [a and b, from Kong A et al. (2012) *Nature* 488, 471. c, from Sun JX et al. (2012) *Nat. Genet.* 44, 1161. With permission from Macmillan Publishers Ltd.]

father's sperm have undergone more than 40 times as many cell divisions as the 15-year-old's (**Figure 3.28**). Since base substitutions are associated with DNA replication through base misincorporation, this seems a plausible explanation for much of the excess of mutations in men compared with women, and for the paternal age effect. However, for genes such as *FGFR2*, *FGFR3*, and *RET* the extreme male-specific mutation effect seems too large to be accounted for

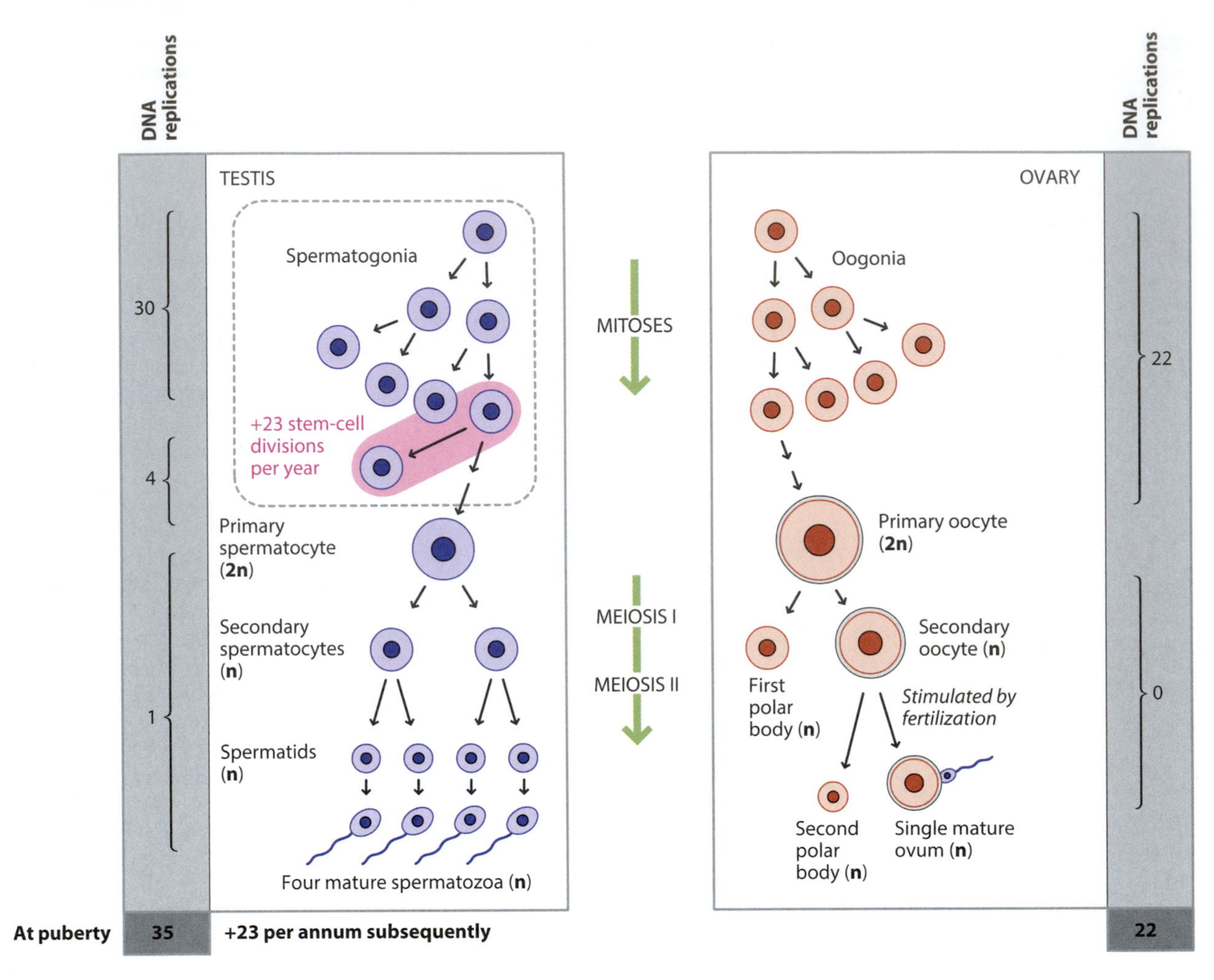

Figure 3.27: Gametogenesis involves more DNA replications in males than in females.
Gametogenesis proceeds in males by mitotic divisions of each diploid spermatogonium (stem cell) to give a diploid primary spermatocyte that proceeds into meiosis, and a new spermatogonium that can divide further in the same way. The total number of cell divisions in males at puberty is 36 (equating to 35 DNA replications because there is only a single replication during the two meiotic divisions). Following puberty there is one stem-cell division every 16 days, or about 23 in a year. In females, diploid oogonia divide to give primary oocytes, which can undergo meiosis. The whole process involves 23 cell divisions, or 22 DNA replications.

by male–female differences in cell division number, while for other genes the male–female differences in mutation rate are too small: additional factors must be involved. For *FGFR2* at least, experiments to detect and quantify mutations in sperm suggest that there is a selective advantage in the male germ line for stem cells that carry the disease-causing mutation.[18]

The difference in base-substitutional mutation rate between males and females is approximated by a figure known as the **alpha factor**, explained in Box 5.4, and has important implications for the evolutionary divergence rates between chromosomes that have different inheritance patterns (see **Section 7.3**).

Sex differences in mutation rates are not confined to base substitutions. Microsatellite mutations show a strong paternal age effect, with a paternal-to-maternal mutation rate ratio of ~3.3, as well as a doubling of paternal mutation

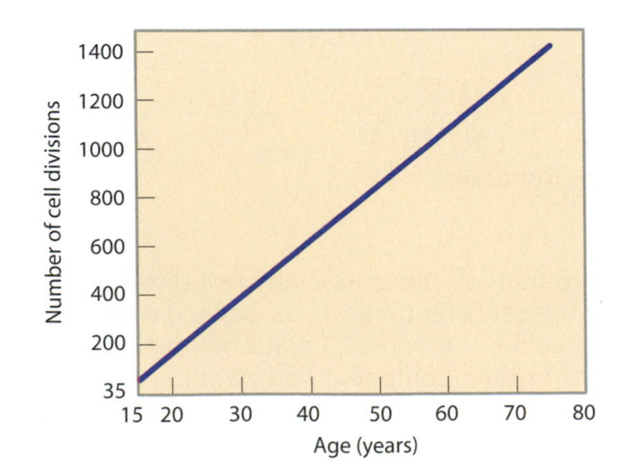

Figure 3.28: Increase of cell division number in sperm with paternal age.
[Data from Crow JF (2000) *Nat. Rev. Genet.* 1, 40.]

rate between the ages of 20 and 58 (Figure 3.26c). Many minisatellites also show a large excess of male mutation (see **Section 3.4**), but in most cases this is likely to be a reflection of sex differences in the behavior of meiotic recombination (see **Section 2.7**), rather than of differences in cell division number during gametogenesis. While most genomic disorders (see **Section 3.6**) show no parent-of-origin bias in their mutations, a few do: for example, about 87% of *CMTIA*-mediated duplications, causing Charcot–Marie–Tooth disease, arise during spermatogenesis, while around 80% of deletions causing neurofibromatosis type 1 (OMIM 162200) are maternal in origin.[52] Again, differences between the sexes in these cases could reflect differences in meiotic recombination behaviors. Perhaps the best-known sex difference is observed in Down syndrome (trisomy 21; OMIM 190685), where 95% of trisomies arise from chromosomal nondisjunction in maternal meiosis, and there is a strong age effect that is not well understood.

3.8 THE EFFECTS OF RECOMBINATION ON GENOME VARIATION

A **haplotype** refers to the combination of allelic states of polymorphisms along the same DNA molecule, in other words on the same chromosome or on mtDNA (**Figure 3.29**). These sites can include any class of DNA variant, from a base substitution to a CNV. If we type Y-chromosomal or mtDNA polymorphisms we automatically derive a haplotype, since these molecules are themselves haploid. Typing X-chromosomal polymorphisms in a male also yields a haplotype directly, since a male carries only one X chromosome. However, if we type X-chromosomal polymorphisms in a female, or autosomal polymorphisms in

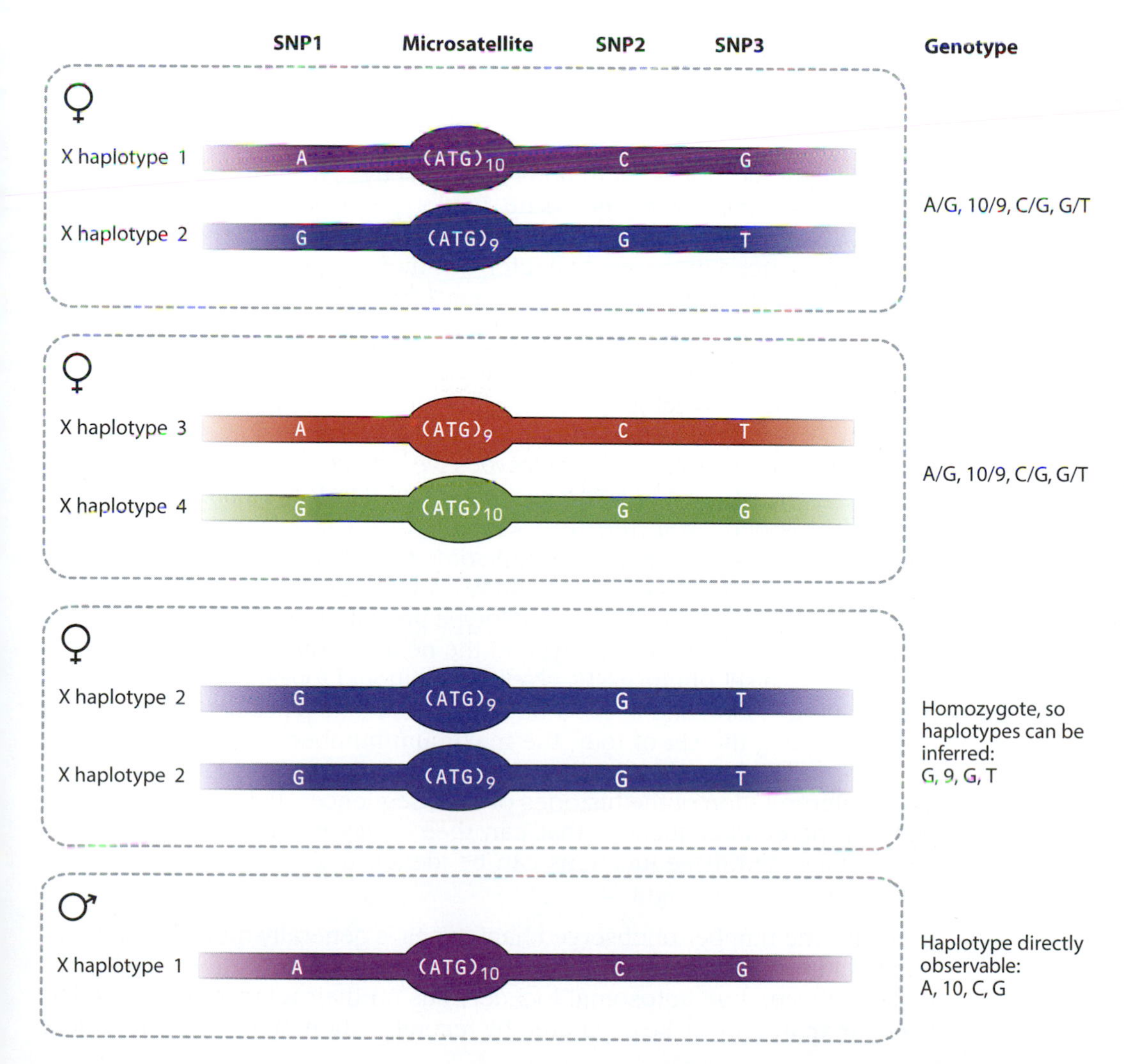

Figure 3.29: Genotypes and haplotypes. Hypothetical haplotypes involving three SNPs and a microsatellite linked on the X chromosome in females and a male. Direct analysis of these polymorphisms in females will yield genotypes, from which the haplotypes cannot be directly inferred except in the case of a homozygote. Note that the top two females have identical genotypes, but different underlying haplotypes. Males are hemizygous for the X chromosome, and therefore yield haplotypes directly; autosomal loci will yield genotypes in both sexes.

Box 3.4: Do mtDNA or the Y chromosome recombine?

The usefulness of mtDNA and the Y chromosome in evolutionary studies is based upon their escape from recombination. But are there any circumstances where recombination could occur on these molecules? And is there any evidence that it does?

The Y chromosome

Ectopic crossing over between the Y and the X or autosomes is well known but mostly leads to infertility because genes necessary for spermatogenesis are lost.

However, there are large segments of similar (>90%) sequences outside the pseudoautosomal regions on the X and the Y chromosomes that reflect their shared ancestry, or recent transpositions of material from X to Y (**gametologous** regions). In principle it is possible for gene conversion (see main text) to occur between blocks of such sequence, transferring segments of X DNA into a Y chromosome (or vice versa). Evidence for such events has been found by resequencing segments of the sex chromosomes,[46, 58] though they appear to be rare and small scale. Some of the recurrent mutations in the Y phylogeny may be due to this process.

Mitochondrial DNA

The molecular machinery for recombination exists within mitochondria, but for any recombination to have an evolutionarily important effect it must occur between mtDNAs that carry diverged haplotypes, deriving from two different individuals. In other words, it requires paternal inheritance of mtDNA, in addition to the usual maternal inheritance. Paternal inheritance has been demonstrated in humans,[48] but only in rare and special circumstances where pedigrees are segregating mitochondrial diseases that apparently allow some selective process to come into play. Under normal conditions, it appears that paternal mitochondria are actively eliminated after fertilization.

Despite these considerations, a number of controversial reports presented apparent evidence in the past that recombination has indeed occurred in human mtDNA. It appears that most of this evidence was based on errors in sequencing or databases, and the impact of any past rare recombination events is probably negligible.

either sex, we do not obtain haplotypes directly, except in the case where both chromosomes have the same haplotype (**homozygosity**). Instead, a diploid genotype—two haplotypes combined—is obtained (Figure 3.29). Methods described in the next chapter can then be used to deduce a pair of haplotypes from the genotype (see **Section 4.9**).

mtDNA and the major part of the Y chromosome are nonrecombining, so haplotypes on these molecules accumulate diversity only through mutation (**Box 3.4** considers whether there are, in fact, circumstances under which this is not true). If we consider four SNPs on a Y chromosome, for example, then the maximum number of haplotypes we expect to observe is five (**Figure 3.30a**): the ancestral state is the absence of all the derived alleles, and then each allele is derived in turn to yield an extra haplotype. We may in fact observe fewer than five haplotypes, if some have been lost, or are not sampled. In the rest of the genome, in addition to mutation, haplotype diversity is increased by recombination: a haplotype present in one generation can be broken up to yield a new haplotype in the next. Recombination occurring in the middle of a set of four SNPs gives an additional four recombinant haplotypes (Figure 3.30b), and if recombination could take place between any pair of SNPs among the set of four, the maximum number of haplotypes would be 2^4, or 16 (Figure 3.30c). Although recombination events cause problems in the interpretation of the histories of DNA sequences, they create junctions between parental sequences that can themselves be passed to successive generations, and these junctions can be identified and exploited as genetic markers in their own right.[39]

In reality, the number of observed haplotypes is generally much less than its maximum possible value. The likelihood of recombination (the genetic distance) between two autosomal loci depends on their relative positions: loci lying far apart tend to be separated by recombination often, while loci close

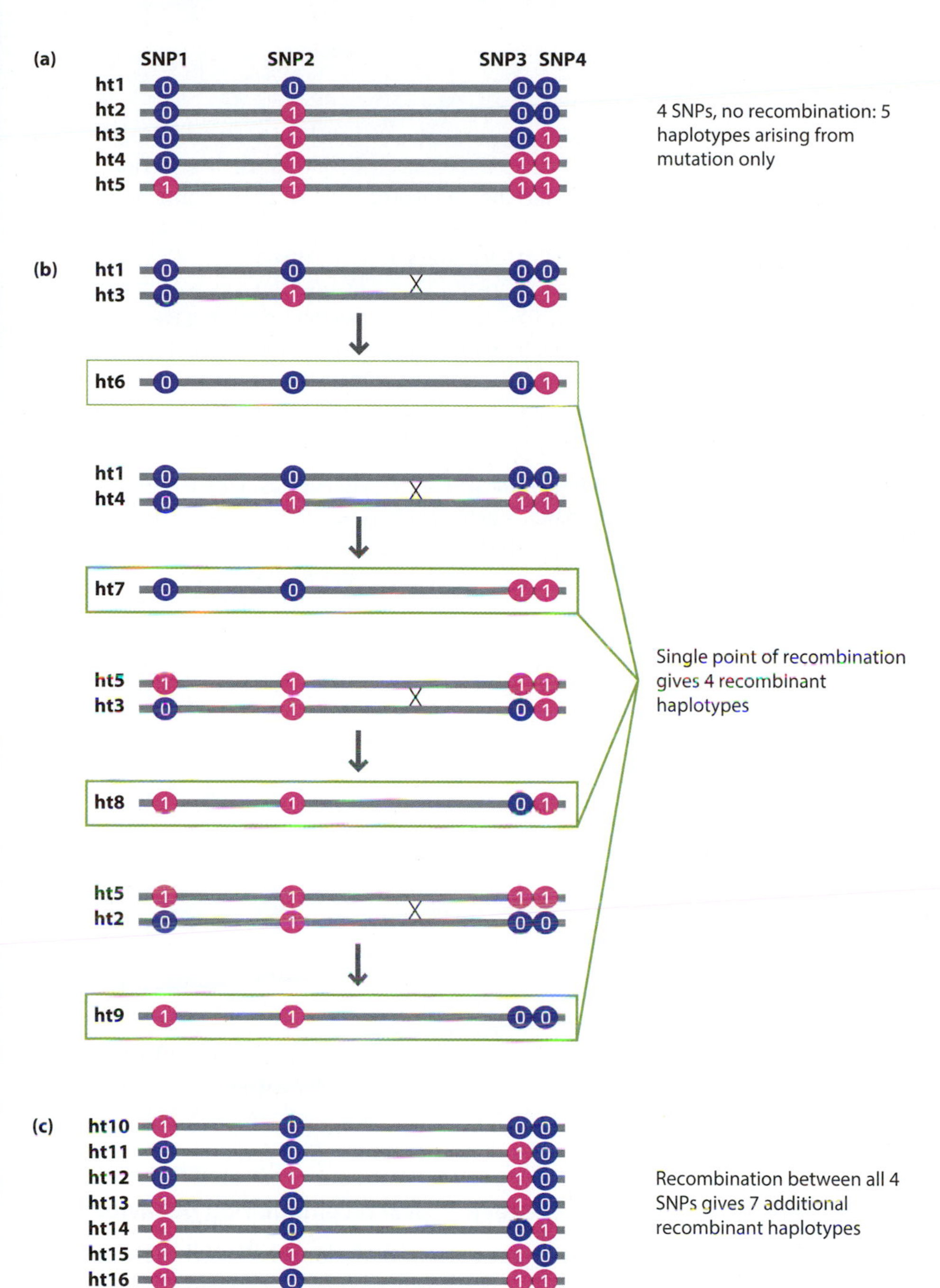

Figure 3.30: Influence of recombination on haplotype diversity.
(a) In the absence of recombination, four SNPs yield a maximum of five haplotypes.
(b) A single central recombination point yields four additional haplotypes.
(c) Free recombination yields a total of 16 haplotypes.

together are rarely separated. But it is not only simple physical distance that governs genetic distance, since the two are not linearly related (see **Section 2.7**). The tendency of particular alleles at neighboring loci to be co-inherited because of reduced recombination between them can lead to associations (correlations) between alleles in a population. This property is known as **linkage disequilibrium** (**LD**). LD is said to occur when two alleles are found together on the same chromosome more often than expected if the alleles were segregating at random. There are various different measures for assessing this that differ in their properties and usefulness (**Box 3.5**).

Haplotype structure and LD have been extensively investigated by two different methods: the genomewide analysis of SNP haplotypes in different populations, and the more direct investigation of recombination events in pedigrees and in sperm DNA. These are described in the next two sections.

Box 3.5: Measures of linkage disequilibrium

The oldest[36] and simplest measure of LD, D, defines it as the difference between the observed frequency of a two-locus haplotype and the expected frequency (based on observed allele frequencies) if the alleles were randomly segregating. For two loci, **A** (alleles A and a) and **B** (alleles B and b), the observed frequency of the haplotype AB (allele A in combination with allele B) is given by P_{AB}. Under random segregation, the expected frequency is the product of the allele frequencies, or $P_A \times P_B$, where P_A is the observed frequency of allele A, and P_B the frequency of allele B. Lewontin's measure of LD is therefore:

$$D = P_{AB} - P_A P_B$$

If D is significantly different from zero, as assessed using a statistical test (Fisher exact test), then LD is said to exist; whether it is positive or negative depends on the arbitrary labeling of alleles.

Though this simple measure is intuitively appealing, its dependence on allele frequencies means that comparisons of different values of D are of limited use. The measure D', which is the absolute value of D divided by its maximum possible value given the allele frequencies at the two loci, is better. It has the useful property that $D' = 1$ if, and only if, two alleles have not been separated by recombination during the history of the sample being analyzed. This case is known as complete LD, and at most three out of the four possible two-locus haplotypes are observed (see Figure 1). It is not clear how values of $D' < 1$ should be interpreted, though they indicate disruption of LD. D' values can be inflated in small samples, and when minor allele frequencies are low, they can give artifactual indications of LD.

Another measure, r^2 (sometimes given as Δ^2), the square of the correlation coefficient between the two loci, is derived by dividing D^2 by the product of the four allele frequencies at the two loci. It has some useful properties that have made it popular in comparing LD in disease association studies. $r^2 = 1$ (known as perfect disequilibrium) if, and only if, the alleles have not been separated by recombination, and have the same allele frequency. r^2 is less inflated by small sample sizes than is D'.

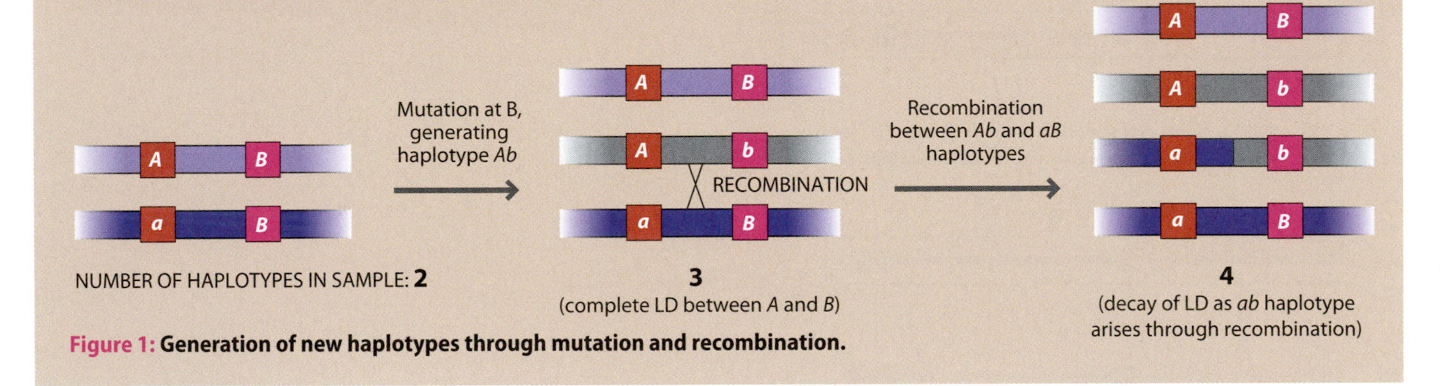

Figure 1: Generation of new haplotypes through mutation and recombination.

Genomewide haplotype structure reveals past recombination behavior

Small-scale early studies of particular regions had demonstrated the existence of LD in the human genome, but the availability of the reference sequence, the expanding SNP database, and the development of high-throughput genotyping methods have now allowed the genomewide patterns of LD to be appreciated. The aim of the **International HapMap Project** (HapMap for short; Box 3.6) was to characterize patterns of genetic variation and LD in a sample of 269 individuals from four geographically defined populations with ancestry in Africa, Europe, and Asia, thus facilitating the design of efficient **genomewide association studies** for common disease. The study began in 2002, and by 2007 had published a haplotype map based on over 3.1 million SNPs. Haplotypes were deduced from genotypes, using information from inheritance in parent–child trios as well as statistical methods (described in Section 4.9). The implications for population structure, natural selection, and disease-association studies have been profound, and will be discussed elsewhere; here, we focus on the findings for haplotype structure and LD.

The results of HapMap demonstrate an extremely non-uniform distribution of recombination in the genome, reflected in a discontinuous block-like structure of LD (Figure 3.31). These **haplotype blocks** can contain many SNPs that are

Box 3.6: The nuts and bolts of HapMap

HapMap proceeded in three phases:

- ***Phase I*** aimed to genotype at least one common (MAF ≥0.05) SNP per 5 kb in each of 269 samples (see **Table 1** and **Figure 1**), and published data on 1 million SNPs;[22] genotyping was carried out by nine genome centers using six different technologies.
- ***Phase II*** increased the typed SNPs on the same samples to 3.1 million,[23] including SNPs with MAF <0.05. It largely employed one genotyping technology, and estimated per-genotype accuracy at ≥99.5%.
- ***Phase III*** included additional samples from a more diverse set of populations (see Table 1 and Figure 1), and has typed 1.3–1.5 million SNPs using two standardized methods.

All the HapMap samples are available for purchase as DNA aliquots and as immortalized lymphoblastoid cell lines (**Section 4.1**). This has made them an invaluable standardized set of samples: the Phase I/II samples have been studied not only for SNP variation, but also for CNV and gene expression, and have undergone genome sequencing using next-generation methods as part of the 1000 Genomes Project (see Box 3.2). Note that because none of the samples was collected as a representative of a particular larger population (for example YRI to represent Yorubans, or Nigerians, or Africans), HapMap recommends that specific local identifiers be used to describe them.

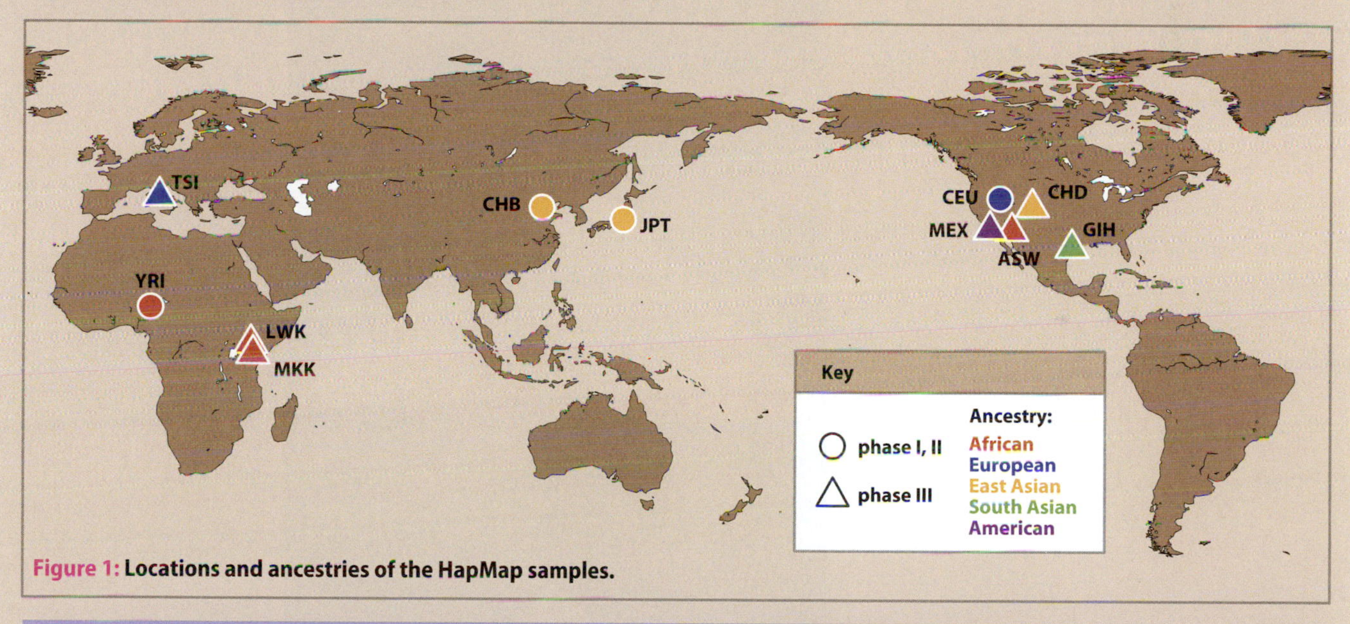

Figure 1: Locations and ancestries of the HapMap samples.

TABLE 1: DETAILS OF POPULATION SAMPLES USED IN HAPMAP

Standard abbreviation	Origin	Sample size and composition
Phase I and II		
YRI	Yorubans from Ibadan, Nigeria	90 (30 parent–child trios)
CEU	Utah (US) residents of N and W European ancestry	90 (30 parent–child trios)
CHB	Han Chinese from Beijing, China	45 unrelated individuals
JPT	Japanese from Tokyo, Japan	44 unrelated individuals
Phase III		
ASW	individuals of African ancestry from Southwest USA	90 (11 parent–child trios, 24 parent–child duos + 9 unrelated individuals)
CHD	Chinese from Metropolitan Denver, Colorado, USA	100 unrelated individuals
GIH	Gujarati Indians from Houston, Texas, USA	100 unrelated individuals
LWK	Luhya from Webuye, Kenya	100 unrelated individuals
MEX	individuals of Mexican ancestry from Los Angeles, California, USA	90 (30 parent–child trios)
MKK	Maasai from Kinyawa, Kenya	180 (30 parent–child trios + 90 unrelated individuals
TSI	Tuscans from Italy	100 unrelated individuals

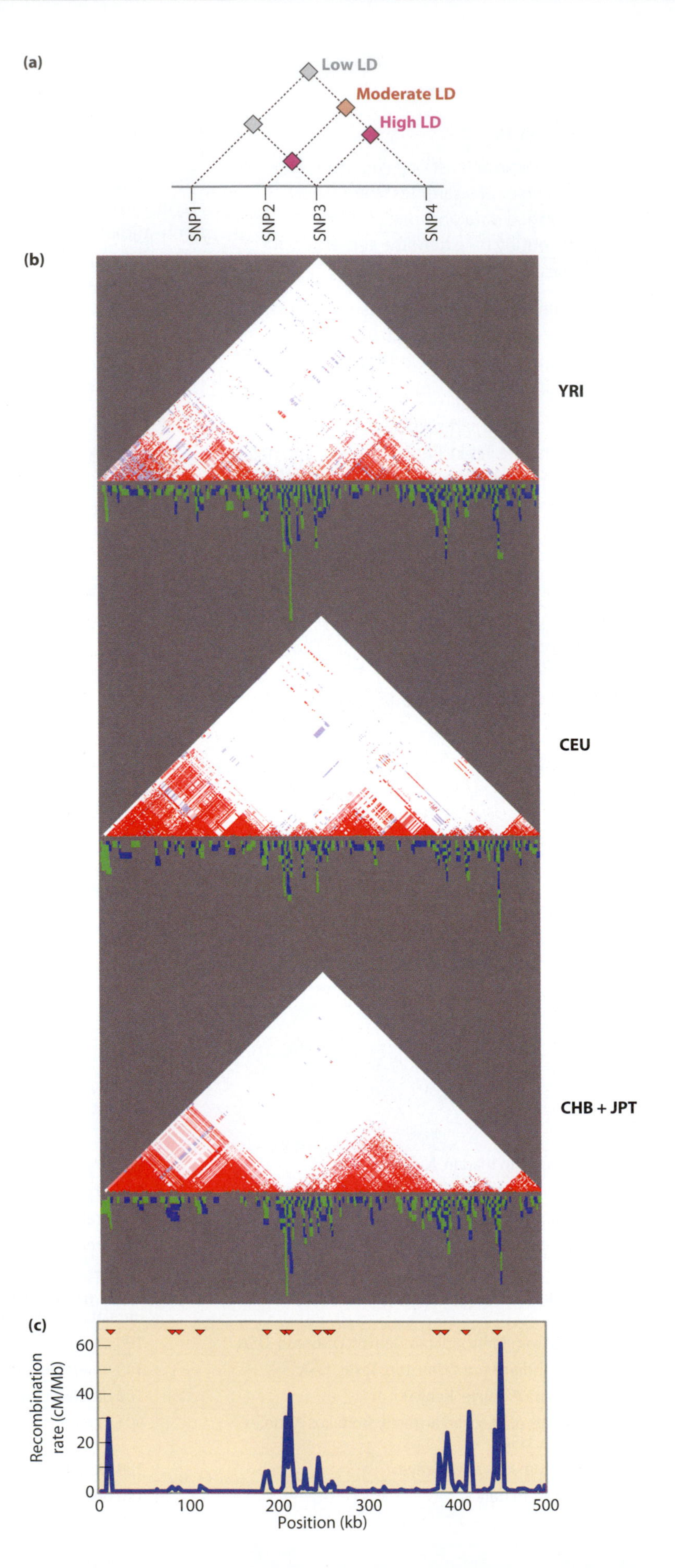

Figure 3.31: Haplotype structure and recombination rate in a 500-kb region. (a) Constructing triangular LD plots. Every pair of SNPs is connected along lines at 45° to the horizontal baseline. The color of the diamond at the position where two SNPs intersect indicates the amount of LD; more intense colors indicate higher LD. (b) LD in a 500-kb region on Chr 2q37.1. The triangular plots represent pairwise comparisons of *D'* for all SNPs within the region, for the YRI, CEU, and CHB + JPT population samples. Below each plot are shown the intervals where distinct recombination events must have occurred (*blue* and *green* indicate adjacent intervals). Stacked intervals represent regions where there are multiple recombination events. (c) Graph showing estimated recombination rates, with putative hotspots indicated by *red* triangles. [b and c, from International HapMap Consortium (2005) *Nature* 437, 1299. With permission from Macmillan Publishers Ltd.]

highly correlated with one another (Box 3.5), and display the limited haplotype diversity that is expected from relatively few historical recombination events. The breakdown of LD between blocks is often abrupt, reflecting the occurrence of most human recombination events within **hotspots**. Computational methods allow the local recombination rate to be calculated from the population-scale genotypic data across the whole genome: rates vary greatly, and from HapMap Phase II data ~33,000 hotspots showing greatly elevated recombination could be identified, with a mean width of 5.5 kb (constrained by SNP density—most empirically determined hotspots are considerably smaller). About 60% of recombination occurred within these hotspots, yet they comprised only ~6% of sequence.

The localization of thousands of hotspots across the genome allowed them to be searched for particular DNA sequence motifs that might be involved in the initiation of recombination at these sites. Such motifs would be examples of ***cis***-acting factors, which exert their influence upon the DNA molecule on which they lie. These searches led to the identification[42] of a 13-bp degenerate **Myers motif**, or **hotspot motif** (CCNCCNTNNCCNC), which is over-represented in human hotspots, and estimated to be associated with the activity of about 40% of them. This motif is thought to be a binding site for a meiosis-specific protein, PR domain–containing 9 (PRDM9), which acts to modify histone H3, thus possibly triggering hotspot activity epigenetically via chromatin remodeling. This protein is thus an example of a ***trans***-acting factor, influencing hotspots on DNA molecules other than the one that encodes it. Further support for the involvement of PRDM9 in hotspot activity comes from direct analysis, described below.

Recombination behavior can be revealed by direct studies in pedigrees and sperm DNA

The HapMap project gave information about recombination behavior by assessing the long-term cumulative effect of large numbers of recombination events upon haplotypes in populations. The approach has been powerful, but has the disadvantage that it reveals nothing about inter-individual or sex-specific differences in recombination behavior. An alternative and more direct approach is to study recombination events in families. The problem here is that human family sizes tend to be small, so correspondingly small numbers of potential recombination events will be available, leading to relatively low-resolution genetic maps. However, some recent studies have used large numbers of individuals and many genomewide polymorphisms, thus improving resolution. One study used ~300,000 genomewide SNPs to analyze 15,257 Icelandic parent–child pairs, deducing haplotypes based on inheritance patterns (**Section 4.9**). Directly observed recombination events gave a map with a resolution of ~10 kb,[31] and comparing maps of male and female recombination showed that ~15% of hotspots were sex-specific in their activity.

A further approach to the analysis of recombination is to analyze the products of meiosis directly. Access to large numbers of female meiotic events is impractical because sufficient numbers of oocytes cannot be harvested, but access to large numbers of male meiotic events is simple—a single semen donation contains tens of millions of sperm. The advantage of the approach is that it gives access to the fine detail of recombination and can detect inter-individual differences, but its technical difficulty means that it cannot be used on a genomewide scale, and to date only ~0.01% of the genome has been surveyed. Typically, information on allelic association is used to identify a region of LD breakdown (a putative hotspot), and then a PCR assay is designed to specifically recover rare recombinant molecules in sperm. These can then be analyzed by SNP typing to determine the exact location of the recombination event, and it is also possible to quantify the recombinants to give an estimate of the recombination rate. Using this method ~30 hotspots have been analyzed in detail.[66] These hotspots have rather uniform width of ~1.5 kb, but vary enormously in intensity (**Figure 3.32**).

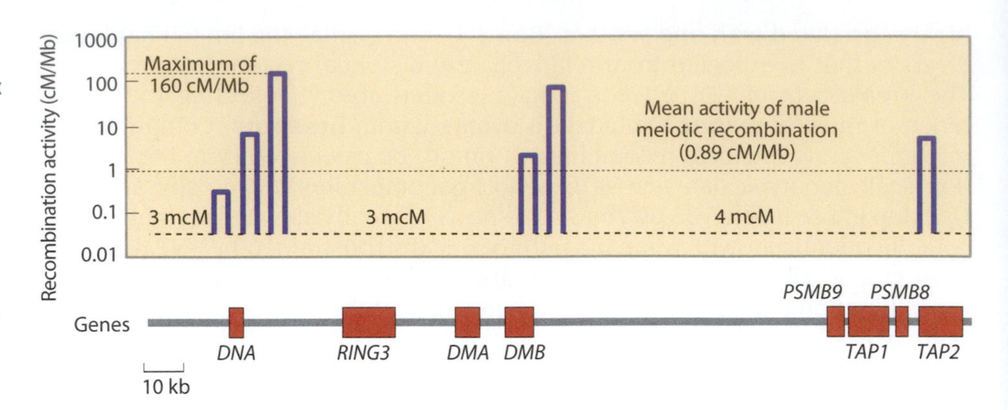

Figure 3.32: Recombination hotspots in male meiosis in the class II region of the major histocompatibility complex (MHC).
Sperm crossover activity is shown over a 216-kb segment lying in the MHC class II region on chromosome 6. The *black* line (*top*) shows the mean male recombination activity for between two and six sperm donors. The background recombination activity of 0.3–0.4 mcM/Mb is approximate. mcM, millicentiMorgans. [Adapted from Jeffreys AJ et al. (2001) *Nat. Genet.* 29, 217. With permission from Macmillan Publishers Ltd.]

Reduction of the intensity of specific hotspots in particular sperm donors can sometimes be attributed to individual SNPs, and in at least one case to a SNP that disrupts a Myers motif,[5] supporting its role as a *cis*-acting factor in recombination. Furthermore, sperm-based analysis of recombination behavior in men carrying different alleles of the *trans*-acting factor PRDM9 showed that subtle changes in the *PRDM9* minisatellite (see **Section 3.4**), encoding the zinc finger that binds the Myers motif, can profoundly affect hotspot activity.

Many different approaches are showing that recombination behavior, despite its fundamental importance to chromosomal segregation and haplotype diversity, is highly evolutionarily labile. Recombination hotspot locations are not conserved between human and chimpanzee, and some differ between human populations, partly due to differences in population frequencies of *PRDM9* alleles.[5] Some human hotspots have evolved very recently, while the human hotspot motif is inactive in chimpanzees, and the *PRDM9* gene itself is very rapidly evolving.[41]

The process of gene conversion results in nonreciprocal exchange between DNA sequences

Mutation and recombination are not the only processes affecting haplotype diversity. Information transfer can also occur between haplotypes through a process called gene conversion (despite its name, this process acts upon all DNA sequences, and not only genes[11]). Unlike classical crossing over, this is a *nonreciprocal* transfer of genetic information between homologous sequences, in which one sequence remains unchanged, but the other is converted to the sequence state of the first (**Figure 3.33a**). Gene conversion has been best studied in yeast, and is relatively poorly characterized in humans, where it cannot be formally distinguished from a double crossover. Conversion tract lengths are typically in the range of 10–1000 bp. Rather than being regarded as a specialized process, gene conversion should be seen as one of the consequences of homologous recombination via the four-stranded intermediate, the **Holliday junction** (**Figure 3.34**).

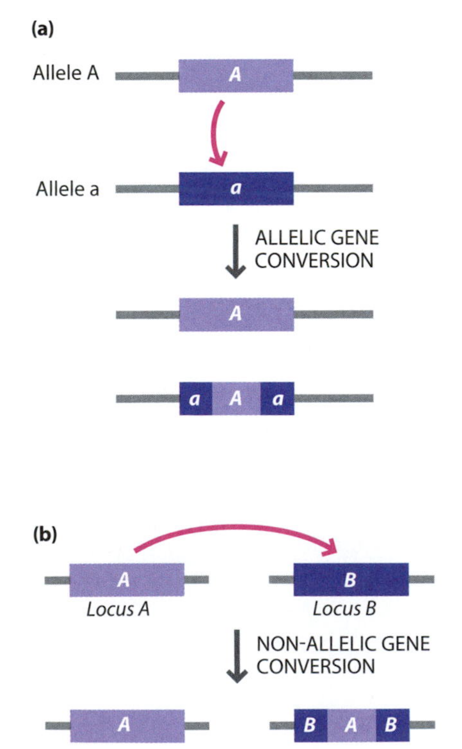

Figure 3.33: Allelic and non-allelic gene conversion.
Nonreciprocal transfer of genetic information between: (a) alleles, and (b) different loci (paralogs). In the latter case a segment of sequence identity of at least about 200 bp is required for conversion to initiate.

Gene conversion has a number of consequences for genome evolution and disease. First, when a short stretch of one chromosome is transferred into its homolog by gene conversion this is equivalent to two closely spaced recombination events, and can break down LD in a similar way to recombination. In an LD analysis of randomly chosen sequences having an average distance between SNPs of only 124 bp, a significant fraction of SNP pairs showed incomplete LD.[2] This unexpected finding is best explained by gene conversion, occurring at a rate around three- to tenfold higher than that of recombination. Gene conversion is very active within meiotic recombination hotspots.

Second, gene conversion acts not only between allelic sequences (homologs), but also between paralogs (Figure 3.33b)—highly similar sequences that are

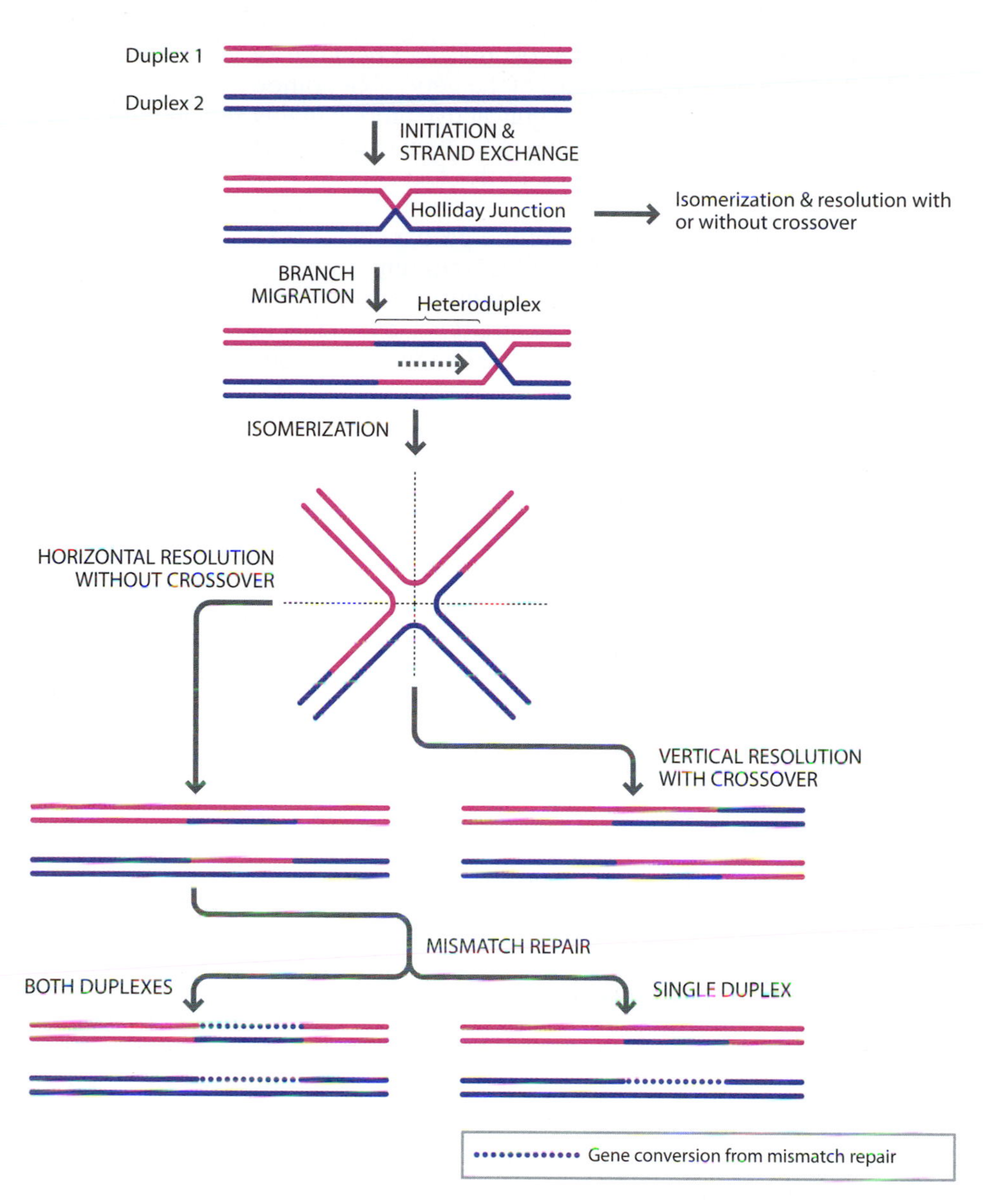

Figure 3.34: Association of homologous recombination and gene conversion through the Holliday junction.

non-allelic (**Section 3.6**). In doing so it can erase **paralogous sequence variants** (**PSV**), increasing the length of sequence identity, and thus provide a more readily utilized substrate for NAHR. The *CMT1A* repeats (see **Section 3.6**) have been shown to have been homogenized by gene conversion.[21]

Third, conversion between paralogous genes can act to maintain sequence homology (concerted evolution). An example is the extremely high sequence similarity (>99.97%) between megabase-scale inverted repeats on the Y chromosome that contain genes required for spermatogenesis, which is driven by rapid gene conversion.[47] On the other hand, gene conversion can have deleterious effects. It may be responsible for the spread of some β-thalassemia (OMIM 141900) mutations into different chromosomal haplotypes, and is the cause of 75% of mutations in the *CYP21* (steroid 21-hydroxylase) gene, causing congenital adrenal hyperplasia (OMIM 201910), by transferring sequence from a *CYP21* pseudogene into the nearby active copy.[12]

As well as generating haplotype diversity, gene conversion can generate excess sequence diversity in paralogous sequences. One example is the promoter of the growth hormone gene *GH1*, which shows approximately twentyfold greater nucleotide diversity than other autosomal loci because of gene conversions with neighboring paralogous copies of the gene.[17] Genomewide analysis shows that gene conversion is generally highly active among highly similar paralogous sequences.[56] The process also appears to display a bias, favoring some alleles over others. This is a consequence of the fact that A–C and G–T mismatches in heteroduplex recombination intermediates tend to be repaired to G–C base pairs during meiosis. This increases the probability of the fixation of G and C alleles and leads to an enrichment of GC-content. Genes showing accelerated evolutionary rates on the human lineage during recent evolution are candidates for involvement in human-specific adaptations, but in fact may be evolving rapidly through selectively neutral but biased gene conversion associated with the evolutionarily labile recombination process.[6]

SUMMARY

- Genetic differences between people arise by germ-line mutation. While any DNA sequence change is a mutation, underlying rates and processes differ greatly.
- Base substitutions (SNPs) are binary mutations with a low average nuclear rate of the order of 10^{-8} per base per generation. There is a bias toward transitions over transversions, and a tenfold elevated rate at CpG dinucleotides. Most base substitutions show identity by descent, and ancestral state can be determined from great ape sequences.
- Most base substitutions, and other variants, are selectively neutral. However, base substitutions in genes can adversely affect gene function through amino acid substitution, introduction of stop codons, and regulatory or splicing changes. Mutations outside genes can have functional consequences in less well understood ways.
- Mitochondrial DNA base substitutions occur at a higher rate than nuclear: approximately tenfold outside the control region, and approximately one hundredfold within it. Variant segregation from one generation to the next is complex, due to stochastic segregation of mutant mtDNA in the development of the oocyte.
- Multiallelic VNTR mutations vary in scale and process, but have relatively high rates—there is little identity by descent, and ancestral state determination is not possible.
- Microsatellites have repeats of 1–7 bp and variable examples have typical repeat numbers of 10–30. Mutation probably occurs through replication slippage. Minisatellites have repeat units of 8–100 bp and repeat numbers of 5 to >1000, often with complex interallelic mutation processes, and sex-averaged germ-line rates up to 7%.
- Retrotransposons such as *Alu* and L1 elements are very abundant, and some are polymorphic, with very low transposition rates. They show identity by descent, and the ancestral state is always absence.
- Structural polymorphisms are widespread and diverse. Over 5% of the genome comprises recent segmental duplications, and these very similar paralogs can sponsor rearrangements causing disease or neutral polymorphism.
- Many mutations show elevated rates in paternal meiosis, and increase with paternal age. This is partly explained by increased numbers of germ-line DNA replications in males compared to females.
- A haplotype is the combination of different allelic states along the same DNA molecule, and its diversity is increased by recombination. Linkage disequilibrium (LD) is the association of alleles because of reduced recombination between them, which can reflect either physical distance, or variable recombination activity.

- Haplotype diversity has been extensively investigated in genomewide-scale population studies, and reveals a block-like structure in which regions of low recombination are punctuated by highly active recombination hotspots.
- Population studies and direct studies in sperm DNA and families are revealing the *cis*- and *trans*-acting factors that regulate recombination, in particular the PRDM9 protein, which acts as a key regulator.
- Gene conversion, the nonreciprocal transfer of sequence between alleles or between paralogs, can reduce LD, homogenize repeats, and increase nucleotide diversity.

QUESTIONS

Question 3–1: Place in order of increasing mutation rate the following changes to the nuclear genome: (a) insertion of an *Alu* element; (b) change in repeat copy number at a minisatellite; (c) transition at a CpG dinucleotide; (d) transversion at a non-CpG sequence; (e) change in repeat copy number at a polymorphic microsatellite; (f) deletion of a single base; (g) transition at a non-CpG dinucleotide.

Question 3–2: Haplotypes of two SNPs (rs16911905 and rs1003586) have been determined in a Nigerian population sample. The haplotype counts are: G-T 16, C-T 23, G-C 64, C-C 17. Calculate the linkage disequilibrium value, *D*, between these two SNPs.

Question 3–3: You analyze a short region in a population and find the following five sequences (uppercase letters indicate SNPs, numbered 1–3):

```
         1                         2       3
(a)  ctgtGaccgatgtccatatcaaatgccaAagatacaCagtggc
(b)  ctgtGaccgatgtccatatcaaatgccaCagatacaCagtggc
(c)  ctgtAaccgatgtccatatcaaatgccaCagatacaCagtggc
(d)  ctgtAaccgatgtccatatcaaatgccaCagatacaTagtggc
(e)  ctgtGaccgatgtccatatcaaatgccaAagatacaTagtggc
```

Why might you conclude that recombination had taken place in this region? Between which SNPs is the recombination event? Suppose that the base immediately to the right of SNP 3 was a G, instead of an A: how might this affect your interpretation of the recombination history of the sequences?

Question 3–4: Correctly assign the terms paralogs, homologs, and orthologs to the following three descriptions:

(a) The human β-globin gene and the gorilla β-globin gene

(b) The human β-globin gene on one copy of chromosome 11 and the same gene on the other copy of the same chromosome

(c) The human β-globin gene and the human δ-globin gene

Question 3–5: Explain why and how estimates of mutation rates from population data depend on assumptions about generation time. Refer for Chapter 6 for additional information.

Question 3–6: What mechanisms generate human genetic variation in different types of tandemly repeated DNA?

Question 3–7: Describe some of the medical and evolutionary consequences of copy-number variation in the human genome. Refer to other chapters for examples—the index may help.

Question 3–8: Use online resources to investigate the following variants, including their ancestral states, genes and functional/phenotypic consequences (if any), and their population distributions: rs113993960, rs11209026, rs5030858, rs76171189.

REFERENCES

The references highlighted in purple are considered to be important (for this chapter) by the authors.

1. **1000 Genomes Project Consortium, Durbin RM, Abecasis GR et al.** (2010) A map of human genome variation from population-scale sequencing. *Nature* **467**, 1061–1073.
2. **Ardlie K, Liu-Cordero SN, Eberle MA et al.** (2001) Lower-than-expected linkage disequilibrium between tightly linked markers in humans suggests a role for gene conversion. *Am. J. Hum. Genet.* **69**, 582–589.
3. **Belancio VP, Hedges DJ & Deininger P** (2008) Mammalian non-LTR retrotransposons: for better or worse, in sickness and in health. *Genome Res.* **18**, 343–358.
4. **Bentley DR, Balasubramanian S, Swerdlow HP et al.** (2008) Accurate whole human genome sequencing using reversible terminator chemistry. *Nature* **456**, 53–59.
5. **Berg IL, Neumann R, Lam KW et al.** (2010) PRDM9 variation strongly influences recombination hot-spot activity and meiotic instability in humans. *Nat. Genet.* **42**, 859–863.
6. **Berglund J, Pollard KS & Webster MT** (2009) Hotspots of biased nucleotide substitutions in human genes. *PLoS Biol.* **7**, e26.
7. **Brandstrom M, Bagshaw AT, Gemmell NJ & Ellegren H** (2008) The relationship between microsatellite polymorphism and recombination hot spots in the human genome. *Mol. Biol. Evol.* **25**, 2579–2587.
8. **Brest P, Lapaquette P, Souidi M et al.** (2011) A synonymous variant in IRGM alters a binding site for miR-196 and causes deregulation of IRGM-dependent xenophagy in Crohn's disease. *Nat. Genet.* **43**, 242–245.
9. **Brinkmann B, Klintschar M, Neuhuber F et al.** (1998) Mutation rate in human microsatellites: influence of the structure and length of the tandem repeat. *Am. J. Hum. Genet.* **62**, 1408–1415.
10. **Cavalli-Sforza LL, Menozzi P & Piazza A** (1994) The History and Geography of Human Genes. Princeton University Press.
11. **Chen JM, Cooper DN, Chuzhanova N et al.** (2007) Gene conversion: mechanisms, evolution and human disease. *Nat. Rev. Genet.* **8**, 762–775.
12. **Collier S, Tassabehji M & Strachan T** (1993) A de novo pathological point mutation at the 21-hydroxylase locus: implications for gene conversion in the human genome. *Nat. Genet.* **3**, 260–264.
13. **Crow JF** (2000) The origins, patterns and implications of human spontaneous mutation. *Nat. Rev. Genet.* **1**, 40–47.
14. **Ebersberger I, Metzler D, Schwarz C & Pääbo S** (2002) Genomewide comparison of DNA sequences between humans and chimpanzees. *Am. J. Hum. Genet.* **70**, 1490–1497.
15. **Genovese G, Friedman DJ, Ross MD et al.** (2010) Association of trypanolytic ApoL1 variants with kidney disease in African Americans. *Science* **329**, 841–845.
16. **Giglio S, Broman KW, Matsumoto N et al.** (2001) Olfactory receptor-gene clusters, genomic-inversion polymorphisms, and common chromosome rearrangements. *Am. J. Hum. Genet.* **68**, 874–883.
17. **Giordano M, Marchetti C, Chiorboli E et al.** (1997) Evidence for gene conversion in the generation of extensive polymorphism in the promoter of the growth hormone gene. *Hum. Genet.* **100**, 249–255.
18. **Goriely A, McVean GA, Rojmyr M et al.** (2003) Evidence for selective advantage of pathogenic FGFR2 mutations in the male germ line. *Science* **301**, 643–646.
19. **Han K, Lee J, Meyer TJ et al.** (2008) L1 recombination-associated deletions generate human genomic variation. *Proc. Natl Acad. Sci. USA* **105**, 19366–19371.
20. **Hollox EJ, Huffmeier U, Zeeuwen PL et al.** (2008) Psoriasis is associated with increased beta-defensin genomic copy number. *Nat. Genet.* **40**, 23–25.
21. **Hurles ME** (2001) Gene conversion homogenizes the CMT1A paralogous repeats. *BMC Genomics* **2**, 11.
22. **International HapMap Consortium** (2005) A haplotype map of the human genome. *Nature* **437**, 1299–1320.
23. **International HapMap Consortium** (2007) A second generation human haplotype map of over 3.1 million SNPs. *Nature* **449**, 851–861.
24. **International Human Genome Sequencing Consortium** (2001) Initial sequencing and analysis of the human genome. *Nature* **409**, 860–921.
25. **International SNP Map Working Group** (2001) A map of human genome sequence variation containing 1.42 million single nucleotide polymorphisms. *Nature* **409**, 928–933.
26. **Karafet TM, Mendez FL, Meilerman M et al.** (2008) New binary polymorphisms reshape and increase resolution of the human Y-chromosomal haplogroup tree. *Genome Res.* **18**, 830–838.
27. **Kayser M, Roewer L, Hedman M et al.** (2000) Characteristics and frequency of germline mutations at microsatellite loci from the human Y chromosome, as revealed by direct observation in father/son pairs. *Am. J. Hum. Genet.* **66**, 1580–1588.
28. **Kelkar YD, Tyekucheva S, Chiaromonte F & Makova KD** (2008) The genome-wide determinants of human and chimpanzee microsatellite evolution. *Genome Res.* **18**, 30–38.
29. **Kivisild T, Shen P, Wall DP et al.** (2006) The role of selection in the evolution of human mitochondrial genomes. *Genetics* **172**, 373–387.
30. **Kong A, Frigge ML, Masson G et al.** (2012) Rate of de novo mutations and the importance of father's age to disease risk. *Nature* **488**, 471–475.
31. **Kong A, Thorleifsson G, Gudbjartsson DF et al.** (2010) Fine-scale recombination rate differences between sexes, populations and individuals. *Nature* **467**, 1099–1103.
32. **Krawczak M, Ball EV & Cooper DN** (1998) Neighboring-nucleotide effects on the rates of germ-line single-base-pair substitution in human genes. *Am. J. Hum. Genet.* **63**, 474–488.
33. **Lalioti MD, Scott HS, Buresi C et al.** (1997) Dodecamer repeat expansion in cystatin B gene in progressive myoclonus epilepsy. *Nature* **386**, 847–851.
34. **Landsteiner K** (1900) Zur Kenntnis der antifermentativen, lytisichen und agglutinierenden Wirkungen des Blutserums und der Lymphe. *Zbl. Bakt. I. Abt.* **27**, 357–362.
35. **Levy S, Sutton G, Ng PC et al.** (2007) The diploid genome sequence of an individual human. *PLoS Biol.* **5**, e254.
36. **Lewontin RC** (1964) The interaction of selection and linkage. I. General considerations; heterotic models. *Genetics* **49**, 49–67.
37. **Li M, Schonberg A, Schaefer M et al.** (2010) Detecting heteroplasmy from high-throughput sequencing of complete human mitochondrial DNA genomes. *Am. J. Hum. Genet.* **87**, 237–249.

38. **MacArthur DG, Balasubramanian S, Frankish A et al.** (2012) A systematic survey of loss-of-function variants in human protein-coding genes. *Science* **335**, 823–828.

39. **Melé M, Javed A, Pybus M et al.** (2010) A new method to reconstruct recombination events at a genomic scale. *PLoS Comput. Biol.* **6**, e1001010.

40. **Mills RE, Bennett EA, Iskow RC et al.** (2006) Recently mobilized transposons in the human and chimpanzee genomes. *Am. J. Hum. Genet.* **78**, 671–679.

41. **Myers S, Bowden R, Tumian A et al.** (2010) Drive against hotspot motifs in primates implicates the PRDM9 gene in meiotic recombination. *Science* **327**, 876–879.

42. **Myers S, Freeman C, Auton A et al.** (2008) A common sequence motif associated with recombination hot spots and genome instability in humans. *Nat. Genet.* **40**, 1124–1129.

43. **Orr HT & Zoghbi HY** (2007) Trinucleotide repeat disorders. *Annu. Rev. Neurosci.* **30**, 575–621.

44. **Redon R, Ishikawa S, Fitch KR et al.** (2006) Global variation in copy number in the human genome. *Nature* **444**, 444–454.

45. **Rosenfeld JA, Malhotra AK & Lencz T** (2010) Novel multi-nucleotide polymorphisms in the human genome characterized by whole genome and exome sequencing. *Nucleic Acids Res.* **38**, 6102–6111.

46. **Rosser ZH, Balaresque P & Jobling MA** (2009) Gene conversion between the X chromosome and the male-specific region of the Y chromosome at a translocation hotspot. *Am. J. Hum. Genet.* **85**, 130–134.

47. **Rozen S, Skaletsky H, Marszalek JD et al.** (2003) Abundant gene conversion between arms of massive palindromes in human and ape Y chromosomes. *Nature* **423**, 873–876.

48. **Schwartz M & Vissing J** (2002) Paternal inheritance of mitochondrial DNA. *New Engl. J. Med.* **347**, 576–580.

49. **Sen SK, Han K, Wang J et al.** (2006) Human genomic deletions mediated by recombination between Alu elements. *Am. J. Hum. Genet.* **79**, 41–53.

50. **Shen S, Lin L, Cai JJ et al.** (2011) Widespread establishment and regulatory impact of Alu exons in human genes. *Proc. Natl Acad. Sci. USA* **108**, 2837–2842.

51. **Soares P, Ermini L, Thomson N et al.** (2009) Correcting for purifying selection: an improved human mitochondrial molecular clock. *Am. J. Hum. Genet.* **84**, 740–759.

52. **Stankiewicz P & Lupski JR** (2002) Molecular-evolutionary mechanisms for genomic disorders. *Curr. Opin. Genet. Dev.* **12**, 312–319.

53. **Stankiewicz P & Lupski JR** (2010) Structural variation in the human genome and its role in disease. *Annu. Rev. Med.* **61**, 437–455.

54. **Stewart C, Kural D, Stromberg MP et al.** (2011) A comprehensive map of mobile element insertion polymorphisms in humans. *PLoS Genet.* **7**, e1002236.

55. **Stranger BE, Forrest MS, Dunning M et al.** (2007) Relative impact of nucleotide and copy number variation on gene expression phenotypes. *Science* **315**, 848–853.

56. **Sudmant PH, Kitzman JO, Antonacci F et al.** (2010) Diversity of human copy number variation and multicopy genes. *Science* **330**, 641–646.

57. **Sun JX, Helgason A, Masson G et al.** (2012) A direct characterization of human mutation based on microsatellites. *Nat. Genet.* **44**, 1161–1165.

58. **Trombetta B, Cruciani F, Underhill PA et al.** (2009) Footprints of X-to-Y gene conversion in recent human evolution. *Mol. Biol. Evol.* **27**, 714–725.

59. **Turner DJ, Miretti M, Rajan D et al.** (2008) Germline rates of de novo meiotic deletions and duplications causing several genomic disorders. *Nat. Genet.* **40**, 90–95.

60. **van de Lagemaat LN, Gagnier L, Medstrand P & Mager DL** (2005) Genomic deletions and precise removal of transposable elements mediated by short identical DNA segments in primates. *Genome Res.* **15**, 1243–1249.

61. **Venter JC, Adams MD, Myers EW et al.** (2001) The sequence of the human genome. *Science* **291**, 1304–1351.

62. **Vergnaud G, Mariat D, Apiou F et al.** (1991) The use of synthetic tandem repeats to isolate new VNTR loci—cloning of a human hypermutable sequence. *Genomics* **11**, 135–144.

63. **Wai T, Teoli D & Shoubridge EA** (2008) The mitochondrial DNA genetic bottleneck results from replication of a subpopulation of genomes. *Nat. Genet.* **40**, 1484–1488.

64. **Wang GS & Cooper TA** (2007) Splicing in disease: disruption of the splicing code and the decoding machinery. *Nat. Rev. Genet.* **8**, 749–761.

65. **Wang J, Wang W, Li R et al.** (2008) The diploid genome sequence of an Asian individual. *Nature* **456**, 60–65.

66. **Webb AJ, Berg IL & Jeffreys A** (2008) Sperm cross-over activity in regions of the human genome showing extreme breakdown of marker association. *Proc. Natl Acad. Sci. USA* **105**, 10471–10476.

67. **Wheeler DA, Srinivasan M, Egholm M et al.** (2008) The complete genome of an individual by massively parallel DNA sequencing. *Nature* **452**, 872–876.

68. **Witherspoon DJ, Marchani EE, Watkins WS et al.** (2006) Human population genetic structure and diversity inferred from polymorphic L1(LINE-1) and Alu insertions. *Hum. Hered.* **62**, 30–46.

69. **Yamamoto F, Clausen H, White T et al.** (1990) Molecular genetic basis of the histo-blood group ABO system. *Nature* **345**, 229–233.

70. **Zhang F, Carvalho CM & Lupski JR** (2009) Complex human chromosomal and genomic rearrangements. *Trends Genet.* **25**, 298–307.

71. **Zhivotovsky LA, Underhill PA, Cinnioglu C et al.** (2004) The effective mutation rate at Y chromosome short tandem repeats, with application to human population-divergence time. *Am. J. Hum. Genet.* **74**, 50–61.

FINDING AND ASSAYING GENOME DIVERSITY

CHAPTER FOUR

The previous chapter described the nature of variation in the human genome, and how it arises. In this chapter, we outline the methods that are used to find and assay this variation. Human genetic diversity has been studied since blood groups were discovered over a century ago.[31] Early methods for studying genetic differences were indirect, based on immunological reactions (see Box 3.1), and later, on **electrophoretic** analysis of gene products. Methods evolved toward the direct detection of DNA variation in the 1970s and 1980s, including laborious **Southern blotting** analysis (see **Box 4.1**, and Figure 18.1), but it was with the invention of the truly revolutionary technique, the **polymerase chain reaction** (**PCR**),[35, 46] that the analysis of human genetic diversity took its first great leap forward.

Five subsequent technological advances have further increased the scope and pace of diversity studies:

- The development of **capillary DNA sequencing** using fluorescent dyes.
- The subsequent determination of the human genome **reference sequence**, and the associated development of computational tools to display and share information about the genome sequence and its variation.
- Cheap and accurate methods for high-throughput genomewide SNP typing.
- Reliable methods for the measurement of **structural** and **copy-number variation**.
- The development of so-called **next-generation** and **third-generation sequencing** methods, which have massively reduced the cost of sequencing large genomes but introduced their own complications.

If we are interested in the variation in a sample of genomes, the ideal way to proceed is clearly to analyze all the variation in all the samples—in other words to carry out comprehensive **resequencing**. This is an unbiased way to assess diversity, and will detect DNA variation of all kinds. However, most studies thus far have involved a compromise between the number of samples and number of variants to be analyzed, driven by cost considerations, and so variants discovered in a small set of samples have often been typed in a larger set. It is important to realize that this is suboptimal because it introduces **ascertainment bias**. The recent technological developments mentioned above mean that unbiased sequencing-based approaches to diversity are now becoming possible.

As was set out in Chapter 3, variation between copies of the human genome exists over a wide range of physical scales, from single nucleotides to whole chromosomes. The methods that can be employed to detect and assay such

Box 4.1: Restriction enzymes as tools for finding and assaying DNA variation

As discussed in Section 4.2, before PCR there was a problem of sensitivity in analyzing single-copy sequences in the large and complex human genome. Specific sequence detection was accomplished by the use of DNA **probes**, cloned sequences labeled with the **radioisotope** phosphorus-32. In this method (see Figure 18.1), genomic DNA is cut by **restriction endonucleases**—enzymes that recognize and cleave DNA at specific short (4–8 bp; for example, GAATTC) sequences. The DNA fragments are then resolved on agarose gels, transferred to a special membrane, and hybridized with the denatured, radiolabeled probe. Washing to remove excess probe, followed by exposure of the membrane to X-ray film (**autoradiography**), reveals specific hybridizing restriction fragments as bands on the film. This process is known as Southern blotting, after its inventor.[47]

Mutations of any base in a restriction enzyme recognition sequence will lead to a failure to cleave, and conversely, mutation can sometimes create a new cleavage site; in addition, length mutations can occur at tandem repeat loci that lie between cleavage sites. Such changes are recognized by alterations in the lengths of restriction fragments detected by DNA probes: **restriction fragment length polymorphisms** (**RFLPs**). RFLPs were key markers in constructing genetic maps and locating disease genes through linkage analysis. However, they have limitations: many positions of variation are unrecognized by known restriction enzymes, disruption of a restriction sequence can be due to several different possible mutations, and cleavage by some restriction enzymes is inhibited by DNA methylation, so failure to cleave might be misinterpreted.

Today, a restriction enzyme can be used to type an individual SNP that creates or destroys a cleavage site in a PCR product (see Figure 4.10 for an example). Online tools can be used to ask if a suitable enzyme is available (for example, NEBcutter: tools.neb.com/NEBcutter2/). Restriction enzymes can also be used in a PCR-based approach to assay diversity in nonhuman species for which genome sequences are unavailable. **AFLPs** (**amplified fragment length polymorphisms**) are generated by digesting genomic DNA with restriction enzymes which produce 5′ single-stranded overhangs, then attaching a short adapter sequence. Primers targeted to the adapters give PCR products whose length is determined by the position of the genomic restriction sites, and polymorphism in these sites can thus be identified.

variation must also vary to suit the scale of variation to be detected, and some scales are more easily analyzed than others (**Figure 4.1**). The armory of available methods today is impressive, but it is important to recognize that no method, and no published study, can ever be completely free from error (**Box 4.2**).

In the sections below, we describe PCR and the other key advances, and we also consider the problem of deducing haplotypes from diploid genotypes, and the special case of analyzing genetic variation in ancient samples. Since any study of human genetic diversity requires DNA samples from a number of different individuals, we begin by asking what sources of DNA are available.

4.1 FIRST, FIND YOUR DNA

Here, we set aside the ethical considerations that affect sampling in particular populations (discussed in **Section 10.1**), and address only the practical questions regarding sources of human DNA. There are several possible tissue sources differing in the amount of DNA that they yield, and in other important aspects. Since the advent of PCR the amount of DNA needed for many analyses has dropped dramatically: previously, a genotyping experiment used 5–10 μg of each DNA sample to test a single polymorphism, while a typical PCR makes do with 1000-fold less.

Blood is a widely used source for human DNA, and the preferred source for next-generation sequencing. It can be taken easily from a vein in the arm, and 5 ml yields 50–200 μg of DNA. Disadvantages of blood are that taking it requires appropriately qualified personnel and involves some risk of blood-borne diseases. There are also practical and administrative difficulties in transporting fresh blood from the field, or between labs. Blood DNA is usually analyzed in aqueous solution, but alternatively a blood drop (often produced by stabbing the thumb with a lancet) dried on a specialized paper (FTA paper) allows PCR to be done using DNA immobilized *in situ*.

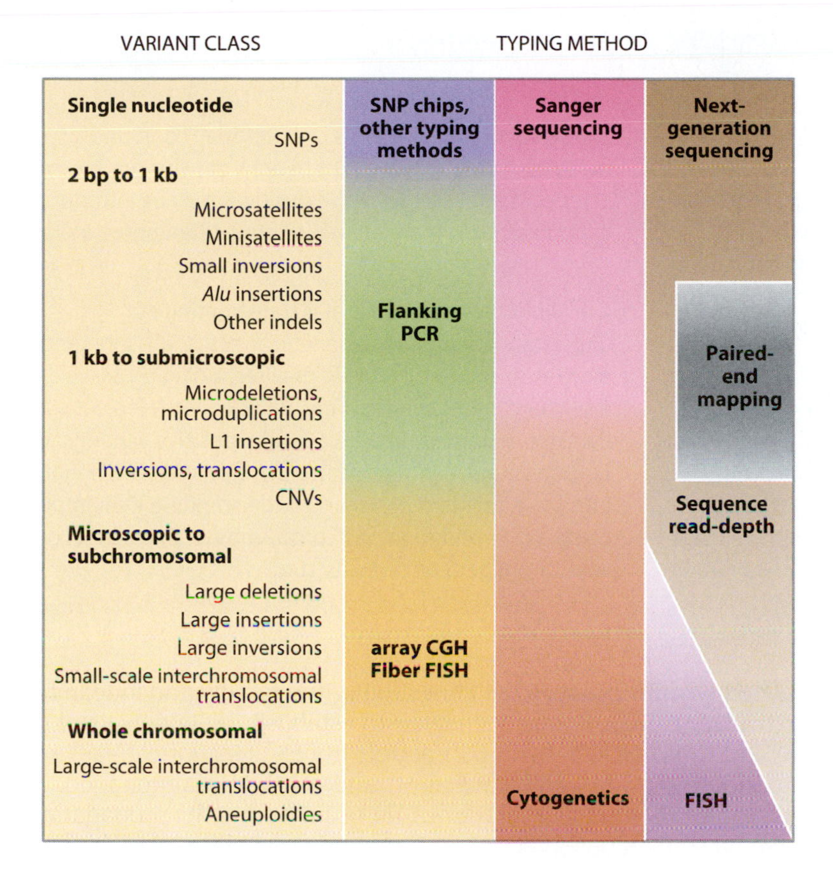

Figure 4.1: Different scales of human genetic variation, and the methods used for variant detection and typing. CGH, comparative genomic hybridization; FISH, fluorescence *in situ* hybridization.

Saliva or buccal (cheek) cells offer a less invasive alternative to blood. Qualified personnel are unnecessary, and self-sampling is possible—this method is favored by police when sampling individuals for genetic profiling, and by personal genomics companies (see Chapter 18). A brush or swab can be simply rubbed in the inside of the cheek and placed in preservative buffer, though this yields a small amount of DNA (<1 μg). Alternatively a "spit-kit" can be used in which several milliliters of saliva are mixed with a preservative solution prior to extraction. This method, as implemented in the Oragene™ kit, for example, can yield up to hundreds of micrograms of DNA, suitable for high-throughput SNP analysis or sequencing.

Semen is a good source of DNA (albeit available from only half the population) and is sampled noninvasively by donation. However, recruitment of semen donors can be sensitive, and semen is not widely used as a starting material in genetic diversity studies. In studies of mutation and meiotic recombination, however, semen samples have been of great importance (see **Sections 3.4** and **3.8**).

Hair has been used in some studies of human genetic diversity. Traditionally the major issue was whether the hair was cut or plucked, since the root contains much cellular material for DNA extraction. The hair shaft contains small amounts of very fragmented DNA, but recently has been shown to be an eminently suitable source for ancient DNA studies (see **Sections 4.10** and **11.4**), since it can be effectively decontaminated, and analyzed by next-generation sequencing methods.

Cell lines are established from primary sampled tissues, and offer the advantage that they can be cultured, sometimes indefinitely, yielding unlimited amounts of DNA. Furthermore, functional studies are also possible, including assays of gene expression. Cell lines can be established from fibroblasts taken from skin biopsies, but the primary sampling procedure is unpleasant, and the cell lines are slow to establish and relatively short-lived. A better source is lymphocytes, which can be isolated from blood and treated with Epstein–Barr virus, leading to

Box 4.2: Living with error in genetic diversity studies

Any experimental analysis of a large number of samples contains mistakes. The most simple errors are in labeling samples, and evidence from well-managed clinical contexts indicates that such errors are in the order of 0.1–0.25%. Efficient laboratory procedures, automation, and bar coding can reduce such errors to a minimum. Other errors occur in the transcription of data, or even in the formatting of data tables during publication.

Errors also occur in data generation. The finished portion of the draft human genome sequence[25] had an error rate of <0.01%, and in most SNP genotyping using chip technologies the rate is likely to be much higher than this. These random errors are a nuisance, and add an unwelcome aspect of noise to studies. What is more, the different typing systems and different kinds of errors may confound **meta-analyses** that attempt to draw conclusions by analysis across different studies. False positives may be introduced if cases and controls are analyzed using different technologies.

Although next-generation sequencing (NGS) is a revolutionary method that is transforming our understanding of human genetic diversity, it is inherently more error-prone than traditional technologies, as we discuss elsewhere in this chapter. The 1000 Genomes Project pilot data, for example, contain ~5% false positive variant calls. Recognizing the extent of false-positive and false-negative errors in any NGS dataset, and understanding their impact on data interpretation and conclusions, is an essential part of any current genetic diversity study.

immortalized **lymphoblastoid** cell lines that grow rapidly and indefinitely. This method (**transformation**) has been used to establish the widely used HapMap (Box 3.6) and CEPH-HGDP (Box 10.2) panels of cell lines. One disadvantage of this method is the cost of transformation and cell-line maintenance. Another potential problem is somatic mutations that occur during the propagation of the cells; these have been observed at mini- and microsatellite loci,[5, 45] in the form of chromosomal aneuploidies and deletions,[42] and as base substitutions identified by next-generation sequencing.[14]

Even though one PCR consumes very little DNA, a sample obtained from one individual is finite. **Whole-genome amplification** (**WGA**) methods have been developed to bypass this problem by amplifying, in as representative a way as possible, sequences from the entire genome. The best established of these is **multiple displacement amplification** (**MDA**).[16] This non-PCR-based method employs short **primers** with random sequences that bind to many different sites throughout the genome, using a high-fidelity polymerase from a bacterial virus (ϕ29) to synthesize long (>10 kb) molecules from the template. Each new strand is displaced from its 5′ end by a following strand, and then new strands themselves become templates for further synthesis (**Figure 4.2**). This method can produce 20–30 μg of representative, high-fidelity, high-molecular-weight product from as little as a few picograms of starting DNA.

4.2 THE POLYMERASE CHAIN REACTION (PCR)

Like the internet, PCR is now so universal that it is difficult for those of us who were around before it arrived to remember what life was like without it. Most readers of this book will not have known the pre-PCR era at all, nor appreciate how extraordinary a revolution PCR represented. Why is it so special?

Suppose we are interested in studying the sequence and variation of the β-globin gene (*HBB*; OMIM 141900). This small gene spans about 1.6 kb of DNA, and is therefore of a manageable size for conventional DNA sequencing in a number of different individuals. However, we have a problem: the β-globin gene represents only 0.00005% of the 3200-Mb genome. This is such a low concentration that direct analysis is impossible—its "signal" is lost in the overwhelming "noise" of the rest of the genome.

What PCR provides is a way to amplify individual gene copies directly, cheaply, and quickly (**Figure 4.3**): short (typically 18–24 nucleotides) **oligonucleotide primers** can be designed which flank the β-globin gene, and synthesized by a

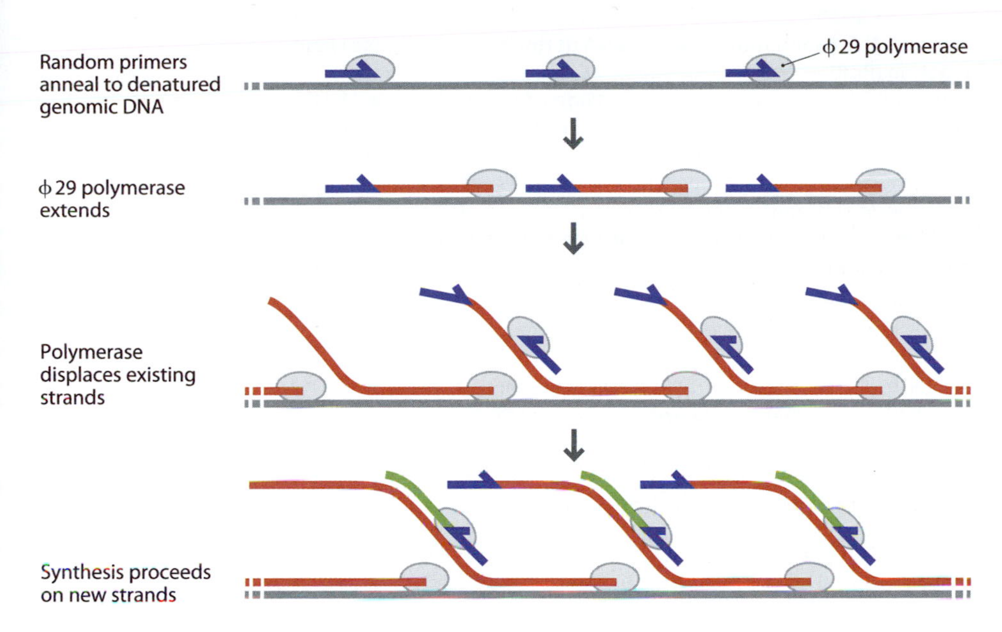

Figure 4.2: Whole-genome amplification.
The principle of multiple displacement amplification is illustrated; random primers are six nucleotides in length, and the reaction proceeds at a temperature of 30°C for 2–3 hours.

commercial supplier. In the PCR (**Table 4.1**), which is carried out in a programmable heating block called a **thermal cycler**, the genomic template DNA is **denatured** by heating, then the reaction is cooled to a specific temperature to allow the primers to **anneal** to their specific target sequence. The temperature is then raised in the **extension** phase, and a thermostable DNA polymerase, isolated from a thermophilic ("heat-loving") bacterium such as *Thermus aquaticus* (***Taq***), carries out DNA synthesis from the primers, using the genomic DNA as template. This cycle of denaturation, annealing, and extension is repeated, typically 25–30 times, and in each cycle previously synthesized product, as well as the original genomic DNA, acts as template: if the reaction were 100% efficient, each cycle would double the amount of target sequence (the **amplicon**). This would mean that, starting with 100 ng (nanograms) of genomic DNA (about 30,000 genome copies), the β-globin gene would be amplified to a copy number of over a billion, with a mass of over 50 μg. In practice, the amplification is not perfectly efficient, but can yield typically 100 ng of the specific target sequence, which is more than enough DNA for sequencing or other molecular analyses.

PCR primers must be carefully designed to give specific products from the large and complex human genome. Primers should not anneal strongly to either other primers in the PCR, or to themselves, as this can inhibit annealing with template DNA. Both primers in any pair should have similar lengths and sequence compositions, so that they anneal specifically to the template at the same temperature. Primer sequences must be selected to anneal only to a particular region of DNA, avoiding cross-amplification of similar sequences elsewhere. To aid this, **BLAST** searches are commonly used—allowing the detection of any alternative primer-binding sites—and can be done together with primer design in the **NCBI** tool Primer-BLAST. Computer simulations (**electronic PCR**; **ePCR**) can also aid primer design.

The specificity and sensitivity of PCR, combined with the ingenuity of researchers, has led to a wide range of applications, including:

- Direct determination of the DNA sequence of PCR products
- Selective amplification of a specific allele from a diploid genotype (**Section 4.9**), and typing of polymorphisms within amplicons (see **Sections 4.5, 4.7, and 4.8**)
- Detection of very rare mutant or recombinant DNA molecules within cell populations
- Quantification of template molecules in genomic DNA, allowing measurement of copy number of genomic sequences

TABLE 4.1: INGREDIENTS OF A TYPICAL PCR

Ingredient	Purpose
1–100 ng genomic DNA	template for synthesis
1 μM oligonucleotide primers	to prime synthesis of specific target DNA
4 × dNTPs @ 0.2 mM each	building-blocks of synthesized DNA
Taq DNA polymerase	thermostable enzyme for DNA synthesis
Mg^{2+} @ 1.0–5.0 mM	required for enzyme activity
Buffer: pH 7.5–9.0; salt (e.g. ammonium sulfate); mild non-ionic detergent (e.g. Tween-20)	required for stable, enzyme-friendly conditions

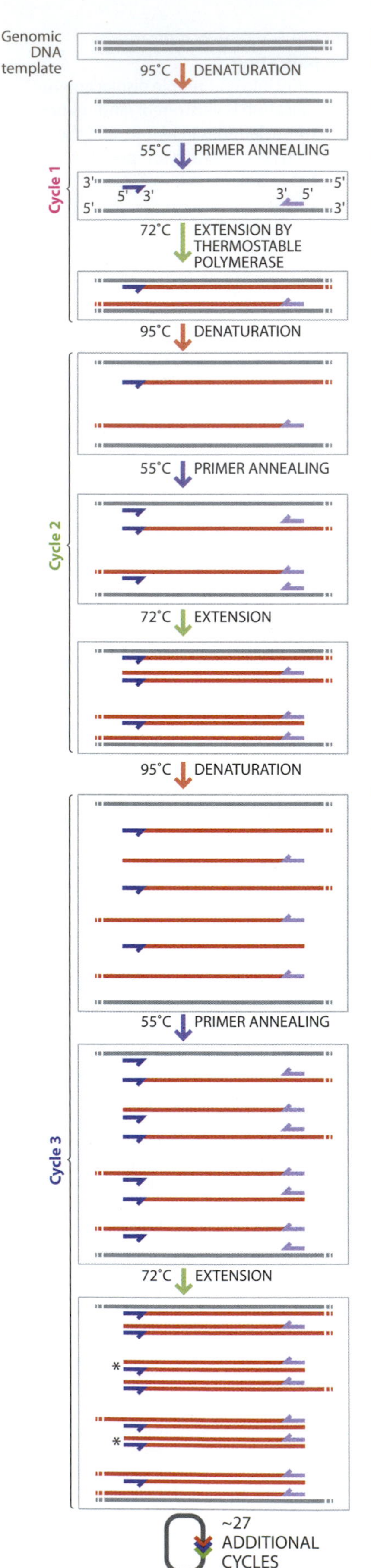

Figure 4.3: Accumulation of target DNA during the polymerase chain reaction (PCR).
In principle the number of copies of the targeted region, delineated by the primers (*dark blue and light blue arrows*), doubles each cycle. A 30-cycle PCR would therefore yield 2^{30} (over 1 billion) copies of the target, but in practice the reaction is not 100% efficient and the yield is less. Note that products delimited at both ends by the primers (*marked* *) do not start to accumulate until cycle 3. Denaturation and extension temperatures are those used typically; annealing temperature depends on the primer sequences and reagent concentrations.

- Detection of specific transcripts in cell extracts, after reverse-transcription of RNA into **cDNA** (**complementary DNA**).
- Quantification of transcripts and thus monitoring of gene expression
- Detection of methylated sites within DNA templates
- Introduction of specific mutations into DNA molecules (*in vitro* mutagenesis)
- Amplification of minute amounts of DNA from fossil material (ancient DNA; **Section 4.10**)
- Amplification of trace amounts of DNA present in crime-scene samples in forensic analysis (see Chapter 18)

Some of these methods will be discussed more fully in this chapter—in particular, the use of PCR in typing polymorphisms in DNA, and in retrieving DNA sequences from ancient samples. We first concentrate on analyzing diversity in modern samples, and ancient DNA studies are discussed in **Section 4.10**.

4.3 SANGER SEQUENCING, THE HUMAN REFERENCE SEQUENCE, AND SNP DISCOVERY

Because PCR relies on specific oligonucleotide primers, it requires prior knowledge of the sequence to be amplified. Therefore, aside from the invention of PCR itself, the other key advance in human genetic diversity studies has been the determination of the human reference genome sequence. This was made possible by the automation of **Sanger sequencing** (named after its inventor Frederick Sanger, and often known as **chain-termination**, **dideoxy**, or capillary sequencing). The method relies upon the incorporation into the growing daughter strand of labeled **dideoxynucleotides** (**Figure 4.4**), which allow the detection of DNA fragments of different lengths, each one terminated at one of the four bases. Initially, detection was by autoradiography of radiolabeled fragments separated by gel electrophoresis. The development of fluorescent labels that could be detected by laser excitation, and the replacement of gels by **capillary electrophoresis** of samples in 96- or 384-well microtiter plates, permitted the necessary automation and increased throughput for the sequencing of large genomes. In one day a single capillary sequencer (for example, the ABI3730xl) can process more than 1000 samples, generating ≥850-bp reads for each, with >99% accuracy. While this may seem impressive, it pales into insignificance compared with the throughput of next-generation methods, described in **Section 4.4**.

The drive to discover SNPs came from a desire to find the causative variants underlying human monogenic disorders, an interest in the evolutionary history of individual loci (including searches for the impact of natural selection), and the idea that a high-density SNP map of the genome would be a useful tool for the discovery of genes involved in complex disorders, via association studies (see Chapter 17). The strategy for sequencing the human genome was itself designed so that SNPs could be identified: large-insert clone libraries were constructed from the DNAs of different individuals, and since the amount of sequence obtained covered several genome equivalents, many putative SNPs

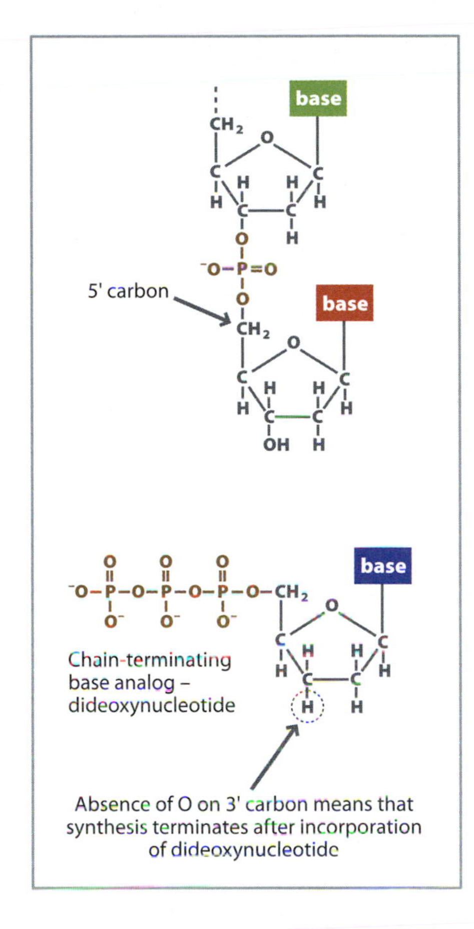

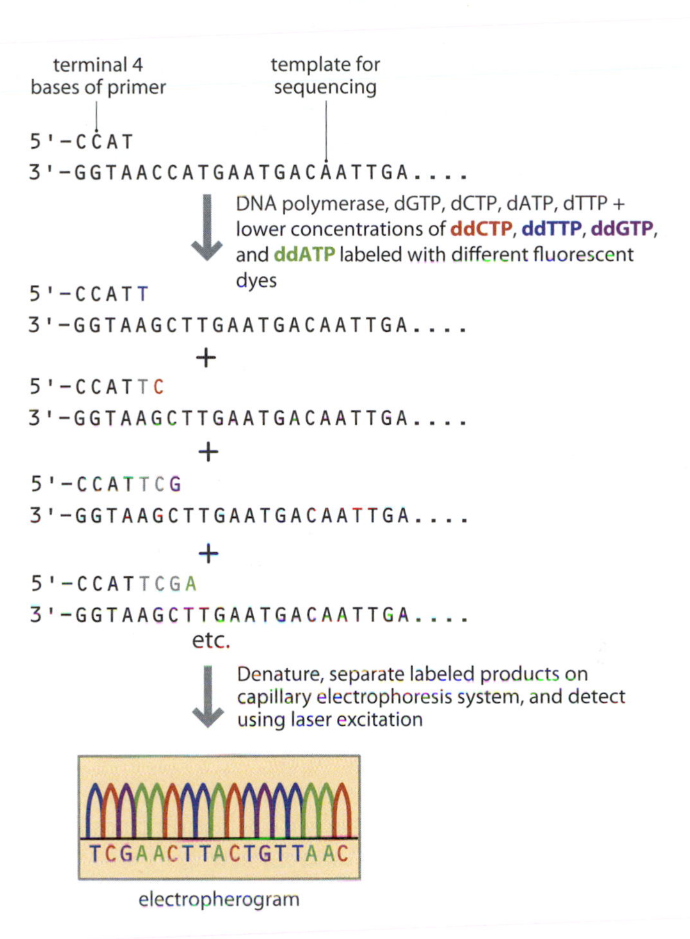

Figure 4.4: Principle of Sanger sequencing of DNA. When a 2′, 3′-dideoxynucleotide analog is incorporated into the growing DNA chain, it terminates synthesis because the 3′ carbon atom does not carry the hydroxyl (–OH) group necessary for the next phosphodiester bond to be made. Inclusion of the appropriate concentration of a fluorescent-dye-labeled dideoxy analog (for example, ddCTP) in a reaction will allow extension to proceed, but also give rise to a set of molecules terminated at C bases in the synthesized strand. Separation of these molecules by capillary electrophoresis maps the positions of C bases; the other three bases, each labeled with a different dye, are included in the same reaction, allowing the sequence to be read from an **electropherogram** after dye excitation using a laser. Before fluorescence detection became available, radioactivity detection methods were used.

were discovered.[26] Other SNPs were identified through the more directed resequencing efforts of the SNP Consortium, and of researchers focusing on specific loci. In principle, Sanger sequencing is the "gold standard" for SNP discovery, often being used to validate SNPs discovered by next-generation methods, and should find all the variation within a sequence. In practice this is almost true, although there are some regions which, because of high GC content or secondary structure, prove difficult to sequence reliably.

The automation of Sanger sequencing technology reduced the cost of resequencing human genomes: the personal genome of J. Craig Venter was sequenced using this method, costing \$70 million compared with the \$300 million estimated cost of producing the reference sequence. However, this is still prohibitively high for population-scale genome sequencing.

4.4 A QUANTUM LEAP IN VARIATION STUDIES: NEXT-GENERATION SEQUENCING

The high cost of existing sequencing methods led to the development (beginning in 2005) of a diverse set of novel technologies that aimed to massively increase sequencing throughput and to reduce cost. With Sanger sequencing being regarded as a first-generation technology, these methods are usually collectively known as **next-generation sequencing** (**NGS**) (sometimes

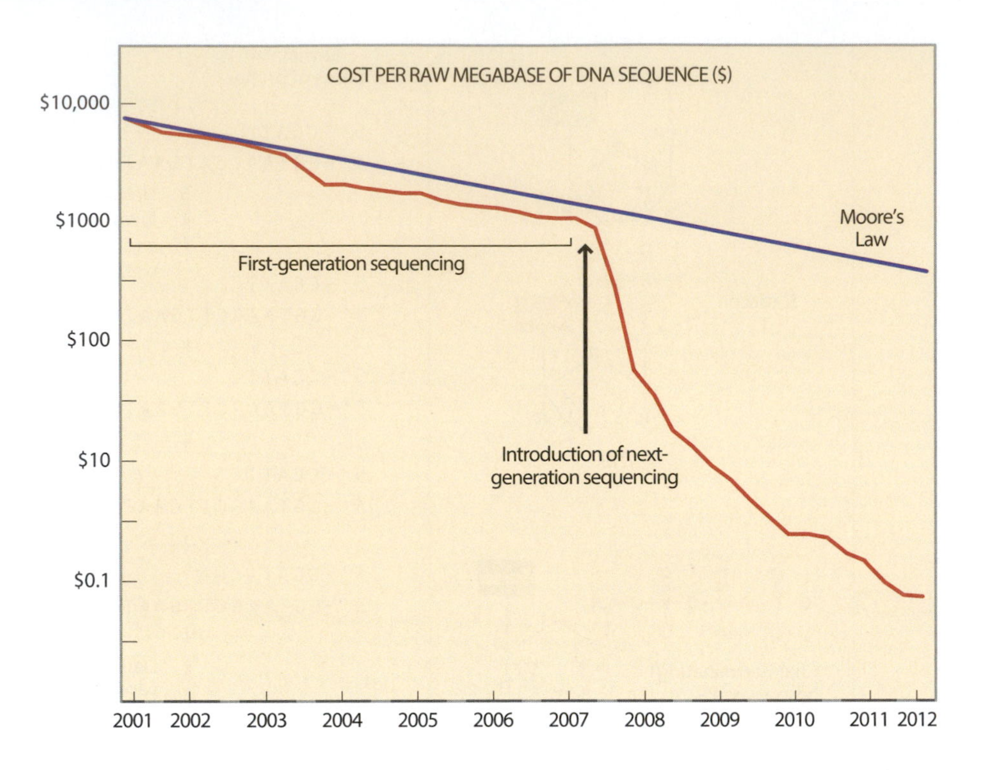

Figure 4.5: The rapidly declining cost of DNA sequencing. The line labeled "Moore's Law" describes the long-term trend whereby the cost of computing halves every two years. Taken from **www.genome.gov/sequencingcosts/**.

second-generation); a set of emerging methods involving sequencing from single DNA molecules are currently referred to as **third-generation**, and are briefly discussed below.

Two general features of the well-established NGS methods are: (1) the fragmentation of genomic DNA by, for example, **nebulization** (use of compressed air to shear DNA into ~200–500-bp fragments), and (2) the subsequent immobilization of spatially separated template DNA fragments on a solid surface or support prior to sequencing. These features allow the simultaneous **shotgun sequencing** of millions or billions of fragments, and hence NGS methods are sometimes referred to as **massively parallel**.

So rapid has been the decline of the cost of DNA sequencing since the development of NGS that a comparison is often made to **Moore's law**, which describes a long-term trend whereby computing power doubles every two years. **Figure 4.5** illustrates the fact that sequencing costs per megabase are declining at an even faster rate; a notional target of the $1000 genome is often discussed, and although the interpretation of this phrase depends upon definitions of quality and completeness, the target will effectively be achieved within a few years—at the time of writing the cost is already less than $10,000. The outstripping of computer power has important implications, since the rate-determining step in NGS is not the generation of the sequence itself, but its computational storage, processing, and analysis.

Illumina sequencing is a widely used NGS method

Rather than detailing the diverse methods developed by competing companies, we illustrate the NGS approach by focusing here on the technology that has become most widely established. A comparison of some of the most popular methods is in **Table 4.2**, and further information can be found in the literature.[33] Each NGS platform has its own methods for DNA template preparation, sequencing, and data processing, and the peculiarities of each of these steps leads to characteristic rates and types of errors. Combining data from different platforms within one project (as was done for the 1000 Genomes Project, for example) therefore presents a daunting logistical challenge.

Figure 4.6 illustrates the principle of **Illumina™ sequencing**, also sometimes known as **Solexa sequencing** after the original company that developed it.

TABLE 4.2:
SEQUENCING PLATFORMS COMPARED

Platform	Principle	Max. read length/bp	Run time/ days	Gb/run/ machine	Pros	Cons
Capillary Sanger sequencing	ddNTP termination and fluorescent detection	850	1	≤0.001	accurate; useful for validation of NGS data	low throughput, expensive
Illumina™ HiSeq™	polymerase-based sequence-by-synthesis	2 × 160 (paired end)	11	600	massive throughput	massive throughput and high costs can be a disadvantage for some projects; short reads
Life/APG SOLiD™ 4	ligation-based sequencing	2 × 50 (paired end)	12–16	100	high throughput	lengthy run times
Roche GS FLX Titanium™ XL+ (**454 sequencing**)	sensing of pyrophosphate release on base incorporation	1000	1	0.7	long reads aid mapping/assembly	inaccurate sequencing of *homopolymeric* tracts; expensive
IonTorrent™	semiconductor sensing of H^+ release on base incorporation	2 × 300 (paired end)	0.1	1	rapid runs, cheap instrument; useful for SNP validation	inaccurate sequencing of homopolymeric tracts

Note that run times are for sequencing only, and exclude prior preparation steps, which can be time-consuming.

Following DNA fragmentation, ligation of adapters, and immobilization on the surface of a **flow-cell**, bridging amplification generates spatially separated **clusters** each containing ~1000 identical molecules. This amplification process is necessary because the final detection method is not sufficiently sensitive to detect single-molecule fluorescence. Sequencing primers are annealed to each cluster template, and in each cluster a specialized DNA polymerase incorporates the fluorescently labeled terminating nucleotide that is the complement of the template base. The remaining unincorporated nucleotides are then washed away, and fluorescence imaging is performed to determine the identity of the incorporated nucleotide. This is followed by a cleavage step, which removes the terminating group and the fluorescent dye, and an additional wash prior to the next incorporation step. This cyclical process continues, giving typical read lengths of 100 bp (originally, 35 bp) before sequence quality becomes too compromised to be useful.

The above account describes sequencing from one end of each DNA fragment. However, there are considerable advantages to so-called **paired-end** approaches. Here, after the initial sequencing of a library of fragments from one end, clusters can be regenerated so that sequencing can then be undertaken from the other end of each fragment. This not only yields more sequence data, but also produces pairs of sequences that are known, a priori, to be separated by the average length of each fragment. Typically, this length is about 200 bp, but can be much greater. Here, DNA is fragmented and then size-selected (by gel electrophoresis) to a specific length between 2 and 10 kb. Such enriched fragment libraries can then be sequenced as described above. For an initial library of 5-kb fragments, for example, the derived reads yield short sequences that must lie ~5 kb apart in the sequenced genome. This information is particularly important in understanding structural variation compared to the reference sequence assembly, and is returned to in **Section 4.8**.

In some resequencing studies, particular regions of the genome must be sequenced in many individuals. In order to maximize the efficient use of NGS machines, which are usually designed to generate a vast amount of sequence data rather than a small amount, DNA **bar coding** or **indexing** can be carried

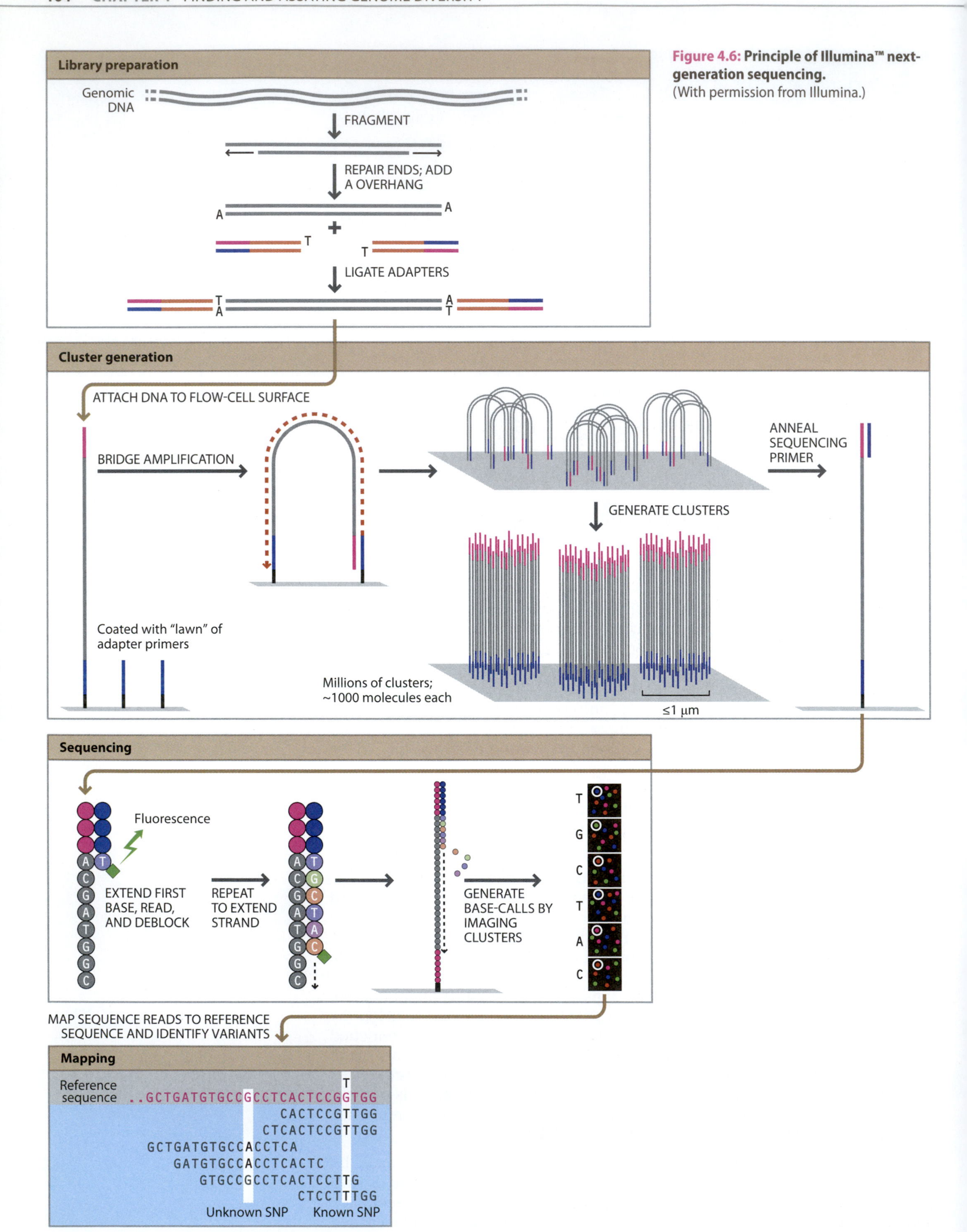

Figure 4.6: Principle of Illumina™ next-generation sequencing. (With permission from Illumina.)

out. Here, each **library** made from the DNA of a different individual has a specific short (typically 6 bp) sequence incorporated immediately after the adapter primers. Following sequence generation, the recognition of this motif allows a particular **sequence read** to be assigned to a particular individual within the sequencing run.

Sequencing can be targeted to regions of specific interest or the exome

Many published studies employing NGS methods in humans describe the analysis of whole-genome sequences. However, it is also possible to target specific regions of the genome for sequencing. One way to do this is by PCR-amplifying segments of interest prior to sequencing; an example is a study of a total of 480 kb of DNA per individual, amplified in 124 PCRs, including candidate genes involved in altitude adaptation.[38] This approach is laborious and time-consuming, and limited in the number of kilobases of DNA that can be targeted, though it does provide high and usually continuous sequence coverage for the examined regions.

An alternative approach is to use oligonucleotide **hybridization** to capture particular regions from genomic DNA (**Figure 4.7**). This method is known as **sequence capture**, or **target enrichment**; the latter name reflects the important fact that

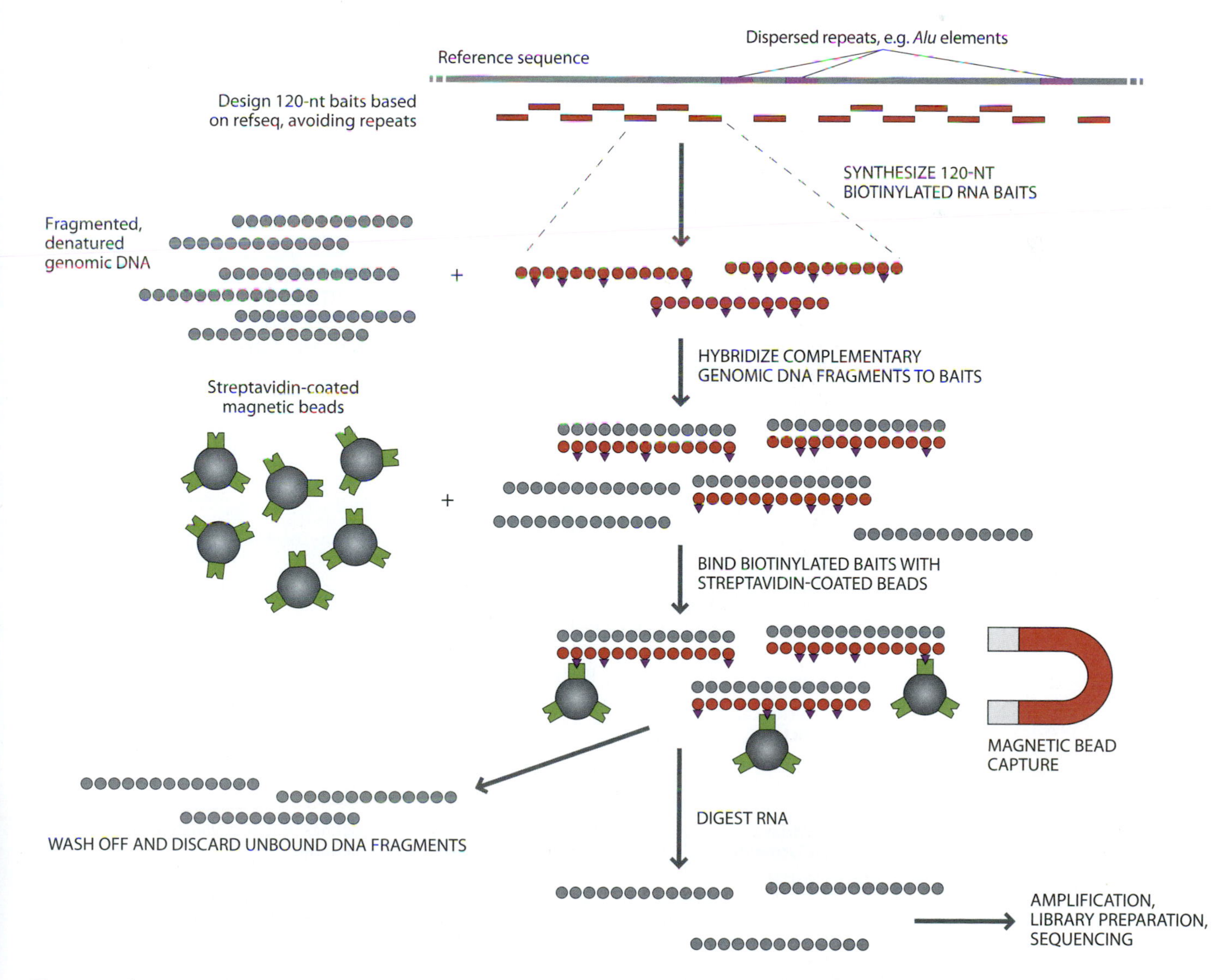

Figure 4.7: Sequence-capture approach for targeted next-generation sequencing. The method shown is that used in the Agilent SureSelect™ protocol.

although most of the sequence obtained originates from the targeted region, there is always some off-target sequence recovered and sequenced. Long (50–120 nt) oligonucleotides (sometimes known as **probes**, or **baits**) are designed based on the reference genome, and then used in solution to capture complementary single-stranded genomic DNA fragments which can then be used to construct a library for sequencing. The amount of sequence to be captured varies depending on the protocol, but can range from a few hundred kilobases up to tens of megabases. Inclusion of highly repeated sequences such as *Alu* or L1 elements among the baits would lead to over-representation of such sequences in the capture fragments. Therefore it is usual to **repeat-mask** during bait design. Although sequence reads from fragments captured by the baits can extend into the masked regions, this nonetheless leads to a discontinuous final sequence in which repeat regions are absent, which is a disadvantage of the approach.

The sequence-capture method can be carried out in a custom design, where researchers focus on their own particular regions of interest, but the commonest application of the method is in **exome** sequencing. This captures the ~1% of the genome (initially ~30 Mb, but increased to ~50 Mb in later designs[13]) comprising the exons of all annotated protein-coding genes (plus a small amount of flanking sequence including **splice sites**), in which the majority of highly pathogenic variants are likely to lie. Following its first application[36] in 2010 to discover the gene for a rare Mendelian disorder of unknown cause, Miller syndrome (OMIM 263750), exome sequencing has become a standard tool for analyzing families segregating Mendelian disorders, and is carried out using commercial kits. There have been debates about the virtues of the targeted and whole-genome approaches, and **Table 4.3** compares their advantages and disadvantages.

NGS data have to be processed and interpreted

Generating NGS data is becoming easier and cheaper, and the major task remains the computational storage and processing of these data to extract reliable and meaningful information.[37] The two main complications are the short read lengths and the high error rates; the presence of indels can also cause particular difficulties. **Figure 4.8** represents the steps needed in an intensive analysis of NGS data for a single genome, which are linked together in a bioinformatic **pipeline**. The first step is the mapping of the vast number of ~100-bp sequence

TABLE 4.3: COMPARISON OF WHOLE-GENOME AND EXOME SEQUENCING

	Whole-genome sequencing	Exome sequencing
Cost	costly for high-coverage sequencing, but rapidly decreasing, and likely to reach exome cost	approximately 10 × cheaper than whole-genome approach
Technical factors	no capture step	requires capture step; possible bias
	vast amount of data to store/process	less data for storage and processing
Variant discovery	defines all sequence and much structural variation	defines sequence variation in ~1.5% of genome
	uncovers functional coding and noncoding variation, including regulatory variants and noncoding RNAs	only coding and splice site variation in annotated protein-coding genes
	~3.5 million variants per genome	~20,000 variants per exome

read pairs to the reference genome (**Figure 4.9**). This inevitably introduces a bias, in that unmapped (and lost) reads might represent material present in the sequenced genome but absent from the reference. Methods are available for the de novo assembly of genomes, and have been used in nonhuman organisms such as the giant panda, but are not yet widely used for human genome sequencing because of issues with their performance, particularly for structurally complex regions. For example, de novo assemblies of the genomes of two humans were ~16% shorter than the reference sequence, and lacked over 99% of known duplicated sequences.[3]

There are a number of sources of error in NGS: first, the mapping procedure (which must be able to cope with sequencing mistakes as well as true variation including SNPs and indels) is not error-free, and some reads will be wrongly mapped. Second, the error rates associated with individual NGS reads are high compared with Sanger sequencing, although they are decreasing as the technology matures; for example, in Illumina™ sequencing an individual **read** has a mean error rate of >0.1%, and there is an increased probability of error toward the end of each read. A consequence of this is that the base call for a given nucleotide should be seen as probabilistic, unlike the more deterministic approach taken to data from traditional Sanger sequencing. Reducing and measuring the uncertainty requires algorithms that model the errors introduced in base calling, alignment, and assembly, together with prior information, such as allele frequencies. This results in a measure of uncertainty (a **quality score**) associated with each SNP and genotype call, which can be incorporated into downstream analyses. A widely used score is given by

$$Q = -10 \log_{10} P(\text{error})$$

such that Q scores of 20 and 30 correspond to 1% and 0.1% error rates, respectively, in base calling.

Because of the massive throughput of NGS methods, a particular base in a sequenced region or genome is often covered by many individual reads, and this **high coverage** can increase the degree of confidence about the nature of that base, primarily by increasing the probability that both alleles are sampled equally. For any described NGS dataset, the coverage is therefore a useful statistic (as a proxy for sensitivity and specificity of the assay), and in practice is determined by the cost and the study design. For example, in the population-scale sequencing undertaken by the 1000 Genomes Pilot Project, 179 DNA samples were sequenced at low coverage (2–6×), enabling the aim of efficient discovery of 95% of shared variants having a ≥1% frequency within 85% of the reference sequence (the accessible genome), but also meaning that many bases in individual genomes were not covered at all. If an individual genome is to be sequenced so that genotypes in 95% of the reference sequence can be determined accurately[1] (with two independent sequences of the same genome having >99.999% concordance), the average sequence read-depth for Illumina™ technologies must be at least 50×. Advances in technologies and algorithms will no doubt reduce this figure over the next few years. The company Complete Genomics offers a sequencing service using its proprietary "DNA nanoballs" technology, claiming 99.9998% accuracy, and providing public data at mean 80× coverage on a set of genomes (www.completegenomics.com/public-data/).

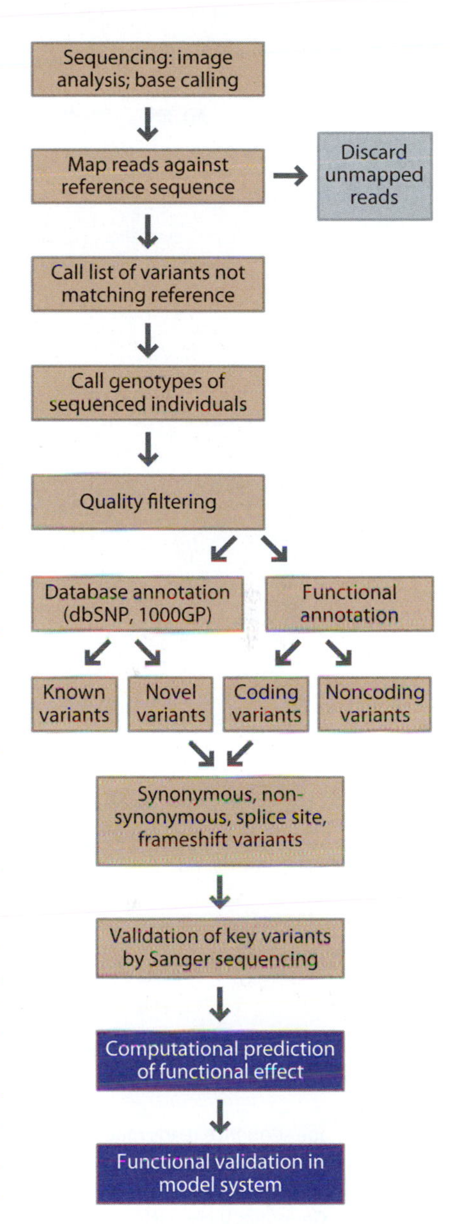

Figure 4.8: Steps in the generation and analysis of next-generation sequencing data.
1000GP, 1000 Genomes Project data.

Third-generation methods use original, unamplified DNA

All widely used NGS methods require a library-preparation step, in which DNA is fragmented, end-repaired, linked to adapters, and amplified. This can introduce biases in the recovery of sequence, and causes particular problems for ancient DNA analysis. By contrast, so-called third-generation sequencing methods produce sequence data from single, original template DNA molecules, avoiding the need for library preparation and amplification. The HeliScope™ sequencer has a throughput of 37 Gb per run and mean read length of 32 bp, and uses much less DNA than established NGS platforms. The recently released Pacific

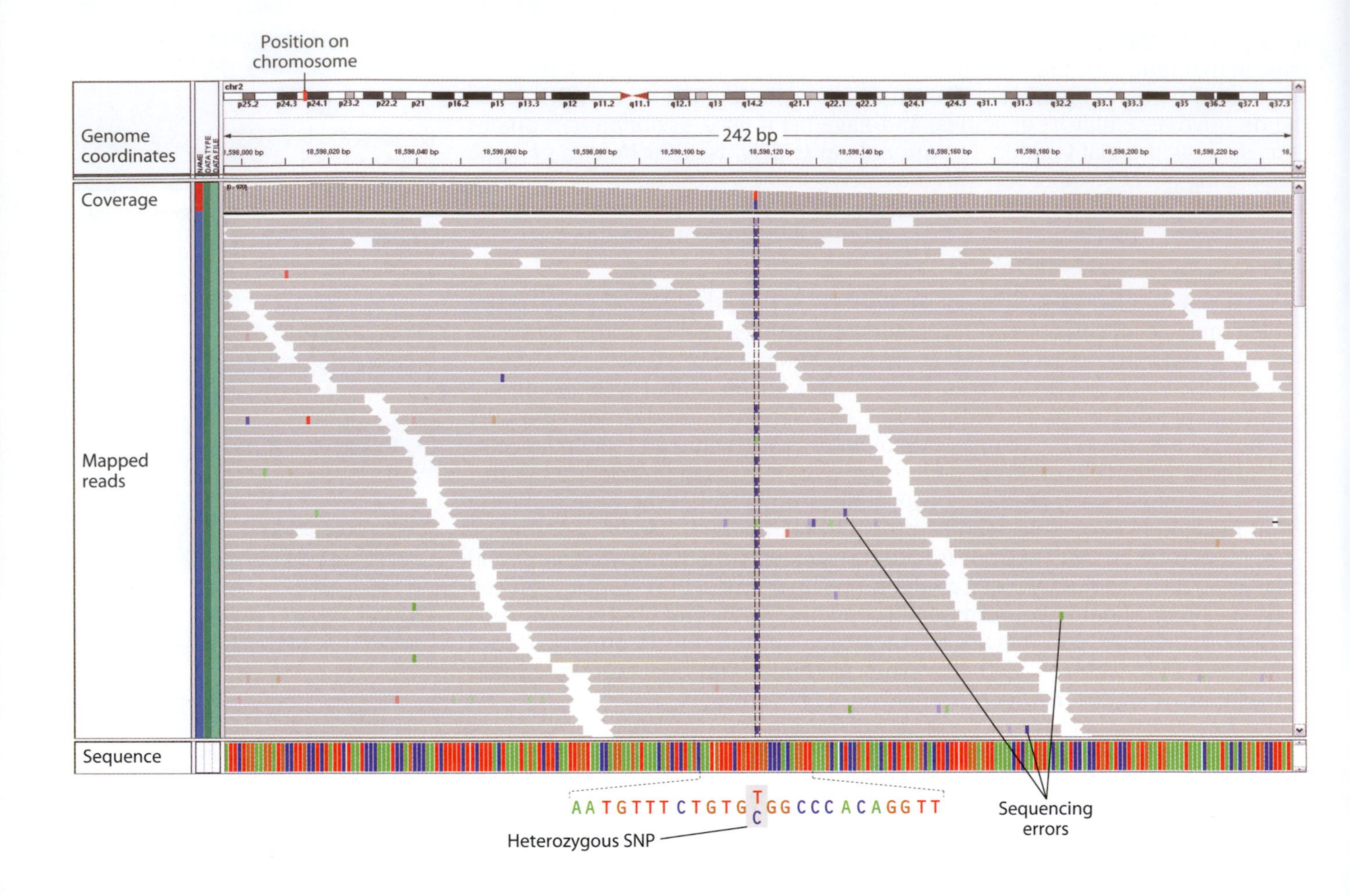

Figure 4.9: Mapping next-generation sequencing reads to the reference sequence.
High-coverage sequence reads mapped to a 242-bp window around a heterozygous SNP on Chromosome 2 are shown, in the Integrative Genomics Viewer [IGV: **www.broadinstitute.org/igv/**; Robinson JT et al. (2011) *Nat. Biotech.* 29, 24]. In the lower part of the figure, each horizontal gray box is a ~100-bp mapped sequence read. Some sequencing errors, appearing only sporadically in reads, are indicated.

Biosciences sequencer has lower throughput but the potential to achieve very long (mean of 3-kb) read lengths (albeit with >10% error rates) that could allow the direct determination of haplotypes. The company Oxford Nanopore promises hand-held devices for single-molecule sequencing. The area of sequencing technology continues to advance remarkably rapidly.

4.5 SNP TYPING: LOW-, MEDIUM-, AND HIGH-THROUGHPUT METHODS FOR ASSAYING VARIATION

Prior to the advent of next-generation sequencing, a few million SNPs had been discovered, and a wide variety of methods developed so that subsets of these could be typed in genomic DNA samples, differing in complexity, accuracy, throughput, and cost. Rather than cataloging these different methods, here we describe three currently used methods that exemplify the different scales of the technologies.

PCR-RFLP typing is a simple low-throughput method

If a SNP creates or disrupts a restriction enzyme recognition site, it can be typed by amplifying the flanking DNA by PCR, then digesting the PCR product with the restriction enzyme prior to separation and detection by agarose gel electrophoresis. This simple method is used for typing small numbers of SNPs in typically a few tens to a few hundred samples, and the simplicity of the methods and necessary equipment makes it suitable for use in student practical classes. PCR-RFLP typing has been quite widely used in the analysis of Y-chromosomal and mtDNA polymorphisms. The method has disadvantages: not all SNPs are suitable for PCR-RFLP analysis, because they may not affect a cleavage site for

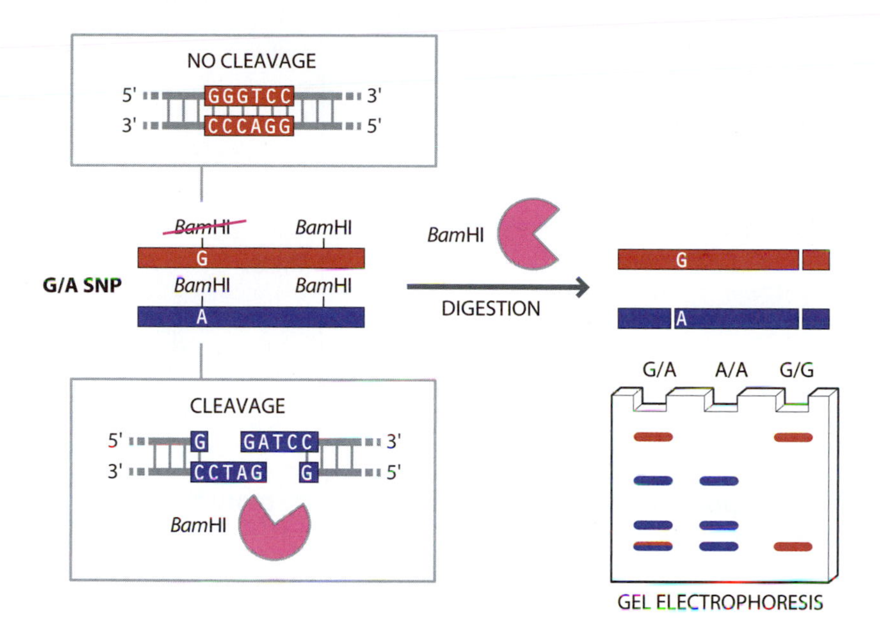

Figure 4.10: PCR-RFLP analysis: a simple, low-throughput method for typing a SNP.
In this hypothetical example, a G/A SNP creates or destroys a *Bam*HI restriction site. A second (constitutive) *Bam*HI site within the same PCR amplicon provides a positive control for the completeness of restriction enzyme digestion.

a known or available restriction enzyme; and partial digestion can occur, which can be difficult to distinguish from SNP heterozygosity. To alleviate the latter problem, design of the assay so that a constitutive restriction site is included in the PCR product helps, as illustrated in **Figure 4.10**. PCR-RFLP approaches are not readily adapted to a multiplex format, so are generally used to analyze one SNP at a time.

Primer extension and detection by mass spectrometry is a medium-throughput method

A simple and robust means to analyze a SNP is by primer extension, where a primer is designed to anneal with its 3′ end immediately adjacent to the SNP site, and then extended by one base, complementary to the SNP allele present in the sequence. This principle is simple and the reaction robust, so assays can be carried out under similar conditions, thus allowing multiplexing and minimizing optimization time. Various detection methods can be coupled to primer extension. For example, fluorescent dideoxynucleotides can be incorporated and terminated products separated by capillary electrophoresis (**mini-sequencing**, as implemented in the SNaPshot™ assay of Applied Biosystems), or the release of pyrophosphate upon base incorporation can be monitored (**pyrosequencing**).

One detection method that allows multiplexing of tens of SNPs, and the analysis of hundreds to tens of thousands of samples daily, is mass spectrometry (**MS**). The method measures the mass/charge (m/z) ratio of ions, from which their molecular weight can be inferred. The specific technique of **MALDI** (**matrix-assisted laser desorption-ionization**) allows the analysis of large, nonvolatile, thermolabile intact molecules by MS, and so is suited to examining changes in molecular weight of primers caused by base incorporation (**Figure 4.11**). The MALDI process uses a laser pulse to produce ions from the extended primer, which has been co-crystallized with a laser-absorbent chemical matrix. The time taken for these ions to traverse a **time-of-flight** (**TOF**) detector is inversely proportional to m/z, and allows mass to be calculated. The small mass difference represented by the incorporation of a specific dideoxynucleotide can be detected, or alternatively a larger mass difference can be exploited by generating products of different lengths for each allele (the DNA MassARRAY™ system), and can be used for high-throughput analysis.

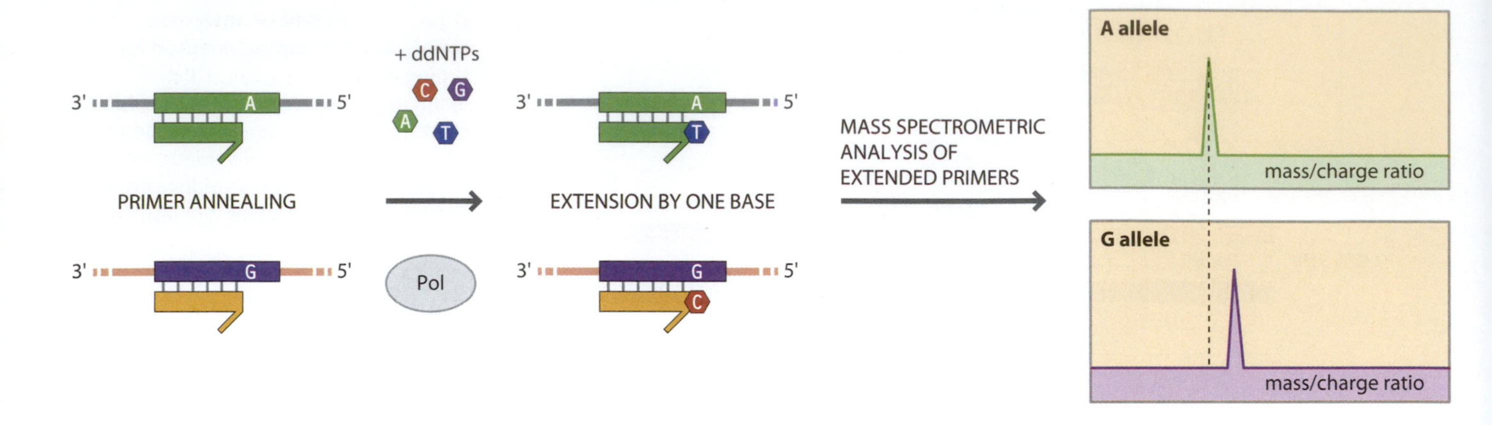

Figure 4.11: Primer extension and detection by mass spectrometry: a medium-throughput method for SNP typing.
The assay detects the incorporation of a specific base complementary to the SNP by measuring the mass/charge (*m/z*) ratio of the product by mass spectrometry (**MALDI-TOF**—see text). The method allows multiplexing of tens of SNPs.
Pol, polymerase.

High throughput SNP chips simultaneously analyze more than 1 million SNPs

The extreme technical challenge of simultaneously typing a sufficient number of SNPs to carry out genomewide association studies has been met by a number of microarray-based technologies, collectively and colloquially known as **SNP chips**. Current versions of such chips can accomplish the remarkable feat of assaying >1 million SNPs simultaneously in a DNA sample with >99% accuracy and reproducibility, for a few hundred dollars. An example of such a technology is the Infinium™ assay (**Figure 4.12**) used in Illumina™ bead chips such as the 1.2M-Duo, which assays ~1.2 million SNPs. The assay is based on primer extension, as outlined above. Following whole-genome amplification (using the method described in **Section 4.1**), DNA fragments are allowed to anneal to an array of silica beads, each of which carries many copies of a 50-nt oligonucleotide whose 3′ end lies adjacent to a particular SNP site. Primer extension incorporates the complementary modified dideoxynucleotide, which terminates synthesis, and also allows detection via fluorescence. Homozygous SNPs will produce single-color fluorescence, while heterozygous SNPs will produce approximately equal ratios of two colors, allowing the genotype at each bead to be called. Proprietary software is used to do this efficiently.

Whole-genome SNP chips are based on a tag SNP design

In order to make genomewide association studies as cost-effective as possible, it was necessary to select a subset of all known SNPs for analysis. The results of the International HapMap Project allowed this choice to be made rationally: in Phase I of the project ~1 million SNPs were typed in 269 DNA samples from four populations, and in Phase II the same samples were analyzed using 3.1 million SNPs. Deduction of haplotypes from the genotypic data (see **Section 4.9**) revealed blocks of strong linkage disequilibrium across the genome, allowing a subset of SNPs to be chosen as tags for nearby highly correlated SNPs, with minimal loss of information. The choice was made such that all SNPs with minor allele frequency (MAF) ≥5% were highly correlated ($r^2 \geq 0.8$) with a **tag SNP**.[24] In this way, reduction of the number of genotyped SNPs by 75–90% was possible: in Phase I of HapMap, 260,000 (**CHB** + **JPT**) to 474,000 (**YRI**) SNPs are required to capture variation at all common SNPs. In the much larger HapMap Phase II dataset, the corresponding figures were ~1.09 million and ~500,000 SNPs respectively. Not all SNPs are taggable: 0.5–1% of SNPs with MAF ≥ 20% could not be tagged, most likely because they lie in recombinationally active regions.

Based on these findings, and those of HapMap Phase III, a number of companies designed genomewide SNP-typing microarrays (SNP chips; Figure 4.12). These chips have increased in resolution, with widely used current versions such as the Affymetrix® SNP6.0 and Illumina™ 1.2M-Duo analyzing ~1 million SNPs, and Illumina's Omni5 chip exploiting data from the 1000 Genomes Project to

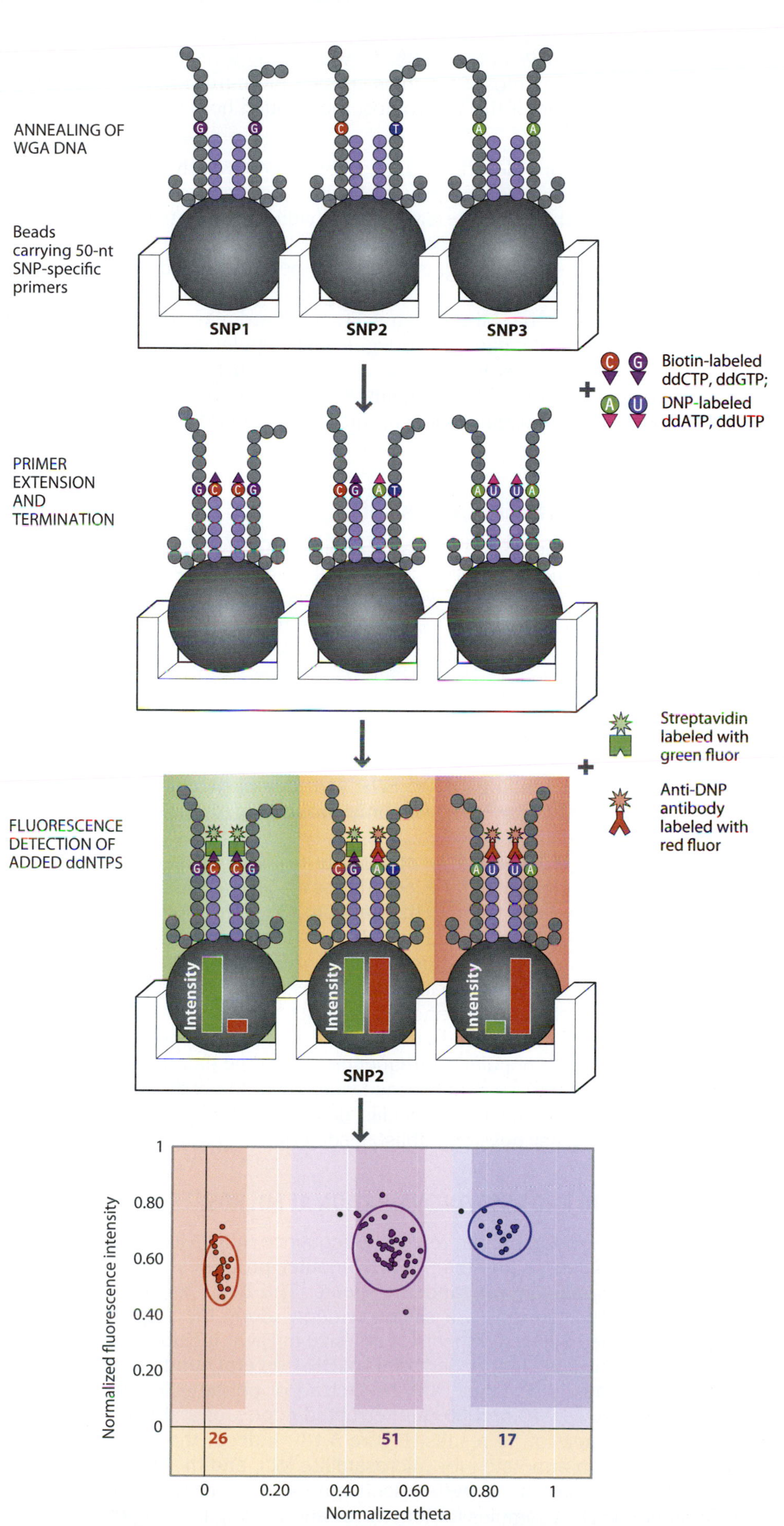

Figure 4.12: Typing principle for a commercial SNP chip: simultaneous analysis of over 1 million SNPs. DNP, 2,4-dinitrophenol. (With permission from Illumina.)

analyze ~4.3 million SNPs of MAF ≥1%. As well as tag SNPs on the autosomes and the X chromosome, most SNP chips contain SNPs from mtDNA and from the male-specific region of the Y chromosome, though how these are selected is not always clear.

Although the SNP chips described above have been widely used in human population genetics studies, this was not what they were designed for. The ascertainment and choice of the SNPs is not well documented, and the tagging design is not ideal for population studies.[28] To address this problem, the Affymetrix Axiom™ Human Origins Array has been designed. This analyzes ~630,000 SNPs ascertained by resequencing in 13 population-specific panels, including San, Yoruba, Mbuti Pygmy, French, Sardinian, Han, Cambodian, Mongolian, Karitiana, Papuan, and Bougainville populations from the CEPH-HGDP panel (Box 10.2), as well as variants identified from the Neanderthal Genome Project. The array also contains 87,000 SNPs that include sets from mtDNA and the Y chromosome, and variants from standard chips for data comparison purposes.

4.6 DATABASES OF SEQUENCE VARIATION

Prior to the publication of the human genome reference sequence in 2001, the number of known human SNPs was small, and any researcher interested in diversity within a particular genomic region had to resort to laborious resequencing, or to one of the diverse methods then used for SNP discovery. These methods are now essentially obsolete thanks to large-scale SNP discovery efforts and the advent of NGS, which have given rise to publicly available databases of diversity; at the time of writing dbSNP (build 135) contains over 52 million variant nucleotides. Each new genome that is sequenced contributes many more.

If we are interested in finding known SNPs within a particular region or gene, a good starting point is the UCSC Genome Browser (genome.ucsc.edu). Here, the region or gene can be specified, and the relevant tracks under "Variation and Repeats" activated. **Figure 4.13** shows the example of a ~24-kb region at the 5′ end of the huntingtin (*HTT*) gene. All known SNPs can be seen (based on dbSNP), as can those SNPs typed in the HapMap samples, and those present on various SNP chips. Data are also available on allele frequencies in the CEPH-HGDP panel of DNA samples, in this case for three SNPs in the region. Access to variation data from the 1000 Genomes Project for a particular segment of the genome can be gained from another browser (browser.1000genomes.org/index.html). An alternative, hand-curated, and more anthropologically based, route to SNPs and their population frequencies is via the database ALFRED (for ALlele FREquency Database;[40] alfred.med.yale.edu/). Many disease-associated SNPs are described in the locus-specific databases that focus on particular genes or diseases (www.hgvs.org/dblist/glsdb.html).

4.7 DISCOVERING AND ASSAYING VARIATION AT MICROSATELLITES

Today, availability of the human reference sequence makes finding microsatellites relatively trivial. This is thanks to computer programs that search DNA sequences systematically for tandem repeats. The best example is Tandem Repeats Finder,[6] which moves a sliding window along the sequence to seek candidate matched adjacent repeats of any size in DNA, including repeats containing mismatches and indels. Segments of DNA several hundreds of kilobases in size can be rapidly searched, and tri-, tetra-, or pentanucleotide loci identified having between 10 and 30 uninterrupted copies of a single repeat type (see, for an example of its results, Figure 3.18). Indeed, the entire reference sequence has been annotated for microsatellite loci, and the results are available via the UCSC Genome Browser (see Figure 4.13). Experimental analysis of several genomic DNA samples for length variation, using flanking PCR primers, will then demonstrate whether or not a locus is polymorphic in the sample. An example of this approach was the systematic isolation of 166 new microsatellites from the Y chromosome.[27]

Variation at microsatellites is analyzed by PCR, using flanking primers close enough to the array that single-repeat-unit size differences between alleles can be discerned readily. One primer for each microsatellite to be analyzed is labeled at its 5′ end with a fluorescent dye, allowing PCR products to be detected on capillary-based sequencing platforms using laser technology. The use of multiple dyes allows the co-amplification of several microsatellites (**multiplexing**)

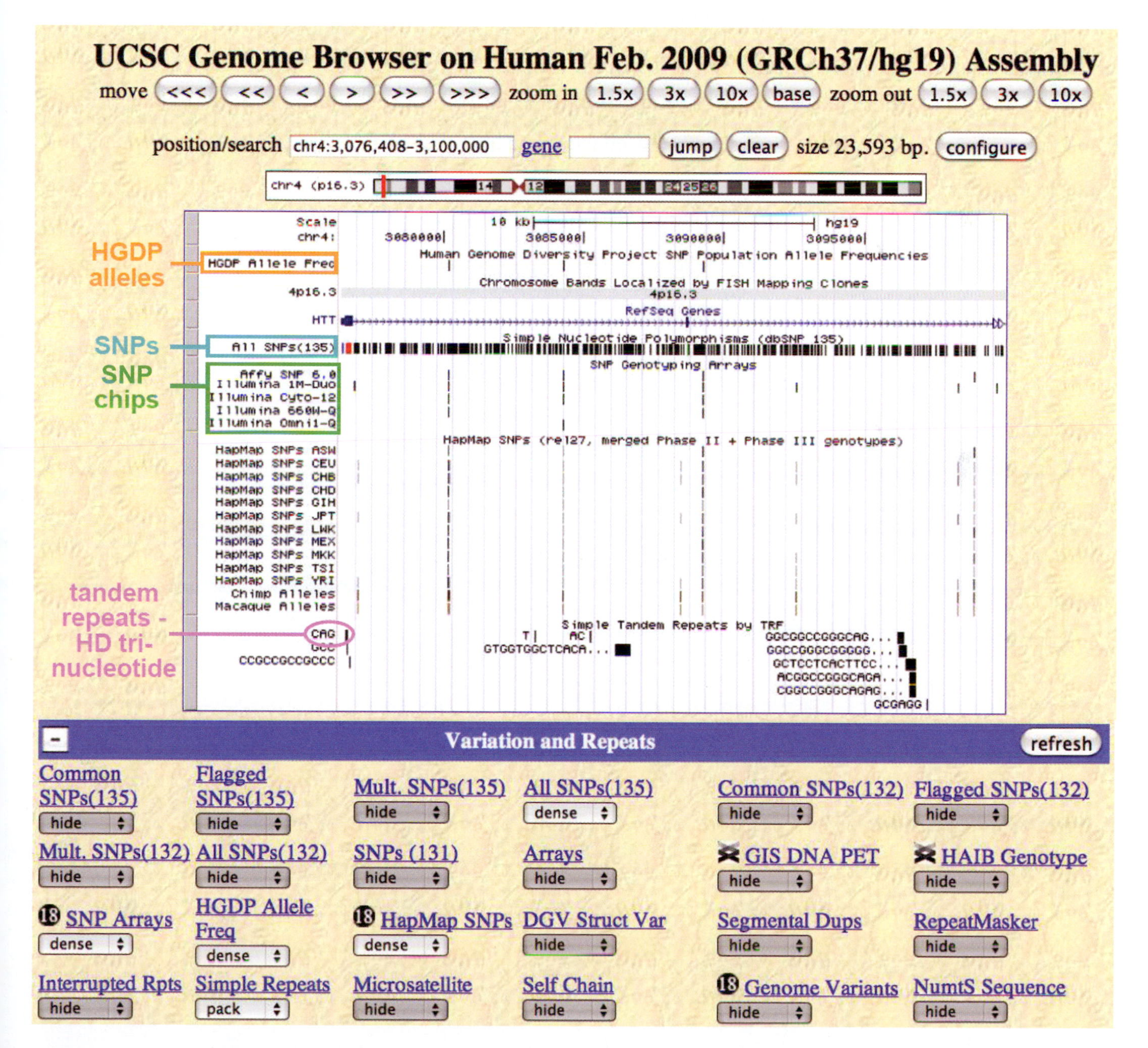

Figure 4.13: Finding human genetic variation via the UCSC Genome Browser.
This screenshot (**genome.ucsc.edu**) shows a ~24-kb region at the 5′ end of the huntingtin (*HTT*) gene: tracks highlighted include all SNPs; those SNPs included on a number of commercial SNP chips; alleles found in the Human Genome Diversity Project (HGDP) set of DNA samples; and simple tandem repeats predicted by the program Tandem Repeats Finder (TRF), including the trinucleotide repeat (CAG) responsible in expanded form for Huntington's disease. Many other possible annotation tracks that have been omitted here are available in the browser.

and their simultaneous separation and detection (**Figure 4.14**; see also Figure 18.3). Such methods were important in linkage analysis and association studies prior to the availability of SNP chips, and are now widely used in DNA profiling in forensic work (see Chapter 18) and in evolutionary studies of Y-chromosome diversity, for example. At the time of writing, the most widely used next-generation sequencing approaches are not well suited to recovering the repetitive sequences of microsatellites; however, as read lengths increase and sequence alignment algorithms mature, this is likely to improve.

4.8 DISCOVERING AND ASSAYING STRUCTURAL VARIATION ON DIFFERENT SCALES

Discovering and assaying variation at minisatellites

As with microsatellites, identification of new minisatellites is today made easy by the use of Tandem Repeats Finder. Length variation of minisatellites has traditionally been measured after Southern blotting: digestion of genomic DNA with a restriction enzyme which does not cut in the minisatellite repeat unit, followed by electrophoretic separation and hybridization with a radiolabeled repeat probe, will reveal an RFLP due not to sequence change in a restriction site, but to variation in the number of repeats lying between fixed restriction sites. PCR using flanking primers allows the size of the repeat array to be measured more easily, although some alleles are too large to amplify readily.

Discrimination of different allele lengths is limited in resolution. Longer alleles with short repeat units may appear identical by length (isoalleles), but actually differ by a small number of repeat units. Also, sequence analysis of minisatellites

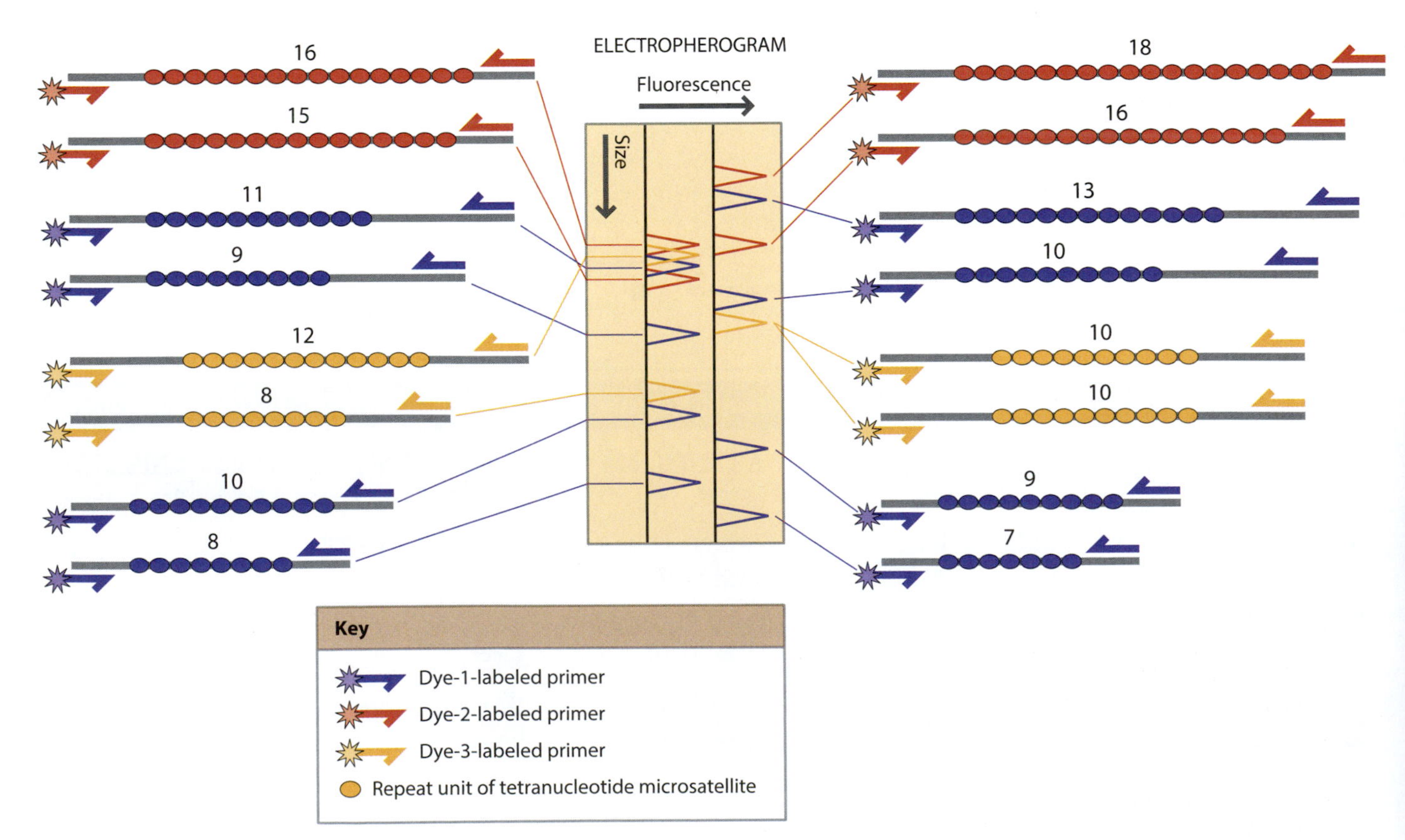

Figure 4.14: Typing microsatellites.
PCR primers labeled with fluorescent dyes allow multiplexing and simultaneous detection of several loci (here, four) on a capillary electrophoresis platform. Primers amplifying loci with non-overlapping size ranges (here, *blue*) can be labeled with the same dye. A fourth dye is used for a size standard (not shown) which is run in every lane. Five-dye systems are also available, which allow deeper multiplexing.

reveals an additional source of variation other than repeat number: repeat units themselves differ from each other, usually by base substitutions but sometimes by indels too. Using allele-specific PCR primers directed at particular repeat unit types, the positions of these variant repeats can be mapped along alleles, in the technique of **minisatellite variant repeat-PCR** (MVR-PCR; see Figure 3.21).

One special case in which such analysis has been useful is in the study of variation within telomere repeats at the termini of chromosomes. Toward their ends, telomere repeat arrays are pure stretches of the repeated hexanucleotide TTAGGG, but internally they include variant repeats whose positions can be mapped from proximal flanking DNA in a variant of MVR-PCR called TVR-PCR.[4] This has greatly facilitated studies of the mutational dynamics of telomeres.

Discovering and assaying variation at well-defined indels, including *Alu*/LINE polymorphisms

In principle, small indels involving up to a few tens of base pairs are readily detected by sequencing, and can either be typed by sequencing also, or by flanking PCR and electrophoretic measurement of size differences between alleles. The availability of whole-genome sequences generated by NGS should therefore allow the easy identification of large numbers of indel polymorphisms. However, short-read sequences arising from indel regions are difficult to map to the correct location in the genome, a problem that is particularly severe for larger insertions. Commonly, reads originating from indels align with multiple single-base mismatches to the reference sequence, rather than with a gap. To alleviate this problem, local realignment or de novo assembly can be used, and a Bayesian method[2] (Dindel) has been developed to identify indels from short-read NGS data in individuals, and has been applied to the 1000 Genomes Project data. Nevertheless, high-throughput identification and genotyping of short indels lags far behind SNP calling and genotyping.

Early in the study of polymorphic *Alu* elements, putative examples were analyzed one by one using flanking PCR primers, which yield a fragment ~300 bp larger from an *Alu*$^+$ allele than from an *Alu*$^-$ allele. The availability of whole-genome sequences now allows their near-comprehensive detection. Computational analysis of paired-end NGS data can rapidly identify clusters of paired-end reads in which one end of each cluster matches the reference genome, but the other matches an *Alu* element. Application of this approach[23] identified 4342 candidate polymorphic *Alu* insertions in the genomes of eight individuals. Once identified, PCR-based validation (successfully done for 63/64 tested cases in this study) is straightforward.

Polymorphisms of L1 elements have been less well studied; they may have lower rates of transposition (see Table 3.7), and the large size of full-length elements, together with their frequent truncation upon transposition, has made them a less tractable target for systematic surveys. Methods for surveying whole-genome sequences, similar to those described for *Alu* elements above, have been developed. A survey of pooled low-coverage (2–4×) 1000 Genomes Project data from 310 individuals identified 953 putative insertions, 24% of which appeared to represent full-length (>6 kb) L1 elements.[17]

Discovering and assaying structural polymorphisms and copy-number variants

Choice of methods for the discovery and analysis of larger-scale structural polymorphisms and copy-number variants depends on the amount of DNA sequence involved in the variants, and whether they are balanced (involving no change in sequence copy number) or unbalanced. As with all forms of variation, NGS methods are contributing greatly to our knowledge of structural polymorphism.

At the largest scale, variation in the copy number of whole chromosomes (aneuploidy) is readily identified by cytogenetic analysis of appropriately stained chromosomes under the microscope, as are large insertions and deletions.

Similarly, large-scale balanced rearrangements such as inversions and translocations can also be identified microscopically. The lower resolution limit for this method is 2–3 Mb, so detection of smaller-scale rearrangements must rely on different methods.

Development of **array-comparative genomic hybridization** (**array CGH**, or **aCGH**) was key in the discovery of the extent of copy-number variation (CNV) in the human genome (**Figure 4.15**). The method relies upon the labeling, with different fluorescent dyes, of genomic DNAs from a test and a reference sample, and then the co-hybridization of these labeled DNAs to an **array** of clone DNAs, or, more recently, **oligonucleotide probes**. If the copy number of a region within both DNAs is equal, then the fluorescence from each will also be equal, but if one is over-represented with respect to the other, its fluorescent signal will dominate. This allowed the genomewide surveying of CNVs. A key issue is the resolution of the array—the first-generation clone arrays[42] carried 26,574 large-insert clones covering ~94% of the **euchromatic** portion of the genome to discover 1447 CNV regions, but tended to overestimate the size of CNVs because of this coarse-grained resolution. A later-generation array based on 42 million oligonucleotides discovered 11,700 CNVs over ~440 bp in length.[15] As well as aCGH methods, genomewide SNP chips can also be used to identify CNVs, as departures from the expected 1:1 fluorescence intensity ratio for the alleles of physically consecutive heterozygous SNPs.

The genomewide methods described above provide clear indications of the presence of a CNV region at a particular location, but less clear information about the precise number of copies of a given sequence. Generally speaking, while it is comparatively easy to detect differences between one and two copies of a sequence, it is far harder to differentiate between, for example, four and five copies. Once a particular CNV has been identified, measurement of copy number can be attempted using a variety of methods. One widely used technique is **quantitative PCR** (**qPCR**), which monitors the accumulation of a specific PCR product from a candidate CNV region in real time, measured against an independent control sequence that is expected to be invariant in copy number. A more reliable method is the **paralog ratio test** (**PRT**), in which a primer pair is designed to simultaneously amplify both a sequence from a CNV region and a sequence of a slightly different size from a non-CNV region as a control (**Figure 4.16**). Software is available to assist the design of such assays (http://prtprimer.org/). Measurement of the fluorescence ratio of products from the two regions (one primer being fluorescently labeled) gives a measure of the ratio, and hence the copy number of the test locus.

The availability of short shotgun sequence reads from whole-genome NGS experiments has allowed genomewide surveys of CNVs, on the basis of three distinct signatures in such data: unusual **read-depth**, discrepant separation between read-pairs, and reads that span a breakpoint and thus map to two distinct locations in the reference sequence.[34] **Figure 4.17** shows an example of a region in which the comparison of two genomes indicates a copy-number difference. Use of read-depth data from low-coverage genomewide data was used to analyze 159 human genomes[50] and allowed accurate copy-number measurement for larger CNVs, as assessed by comparing read-depth to qPCR measurements, for example (**Figure 4.18**). Depth of coverage is a key determinant of the power to detect and genotype CNVs from NGS data using read-depth.

Most of the methods described above provide no information about the *per-chromosome* copy number of a variant: for instance, if they indicate that an autosomal CNV has a copy number of 5 in the genome of a particular individual, we do not know a priori if this reflects allele combinations of 5 + 0 copies, 4 + 1 copies, or 3 + 2 copies. Non-PCR-based methods that examine the long-range structure of each allele can be particularly useful for disentangling this. Cleavage by restriction enzymes cutting outside the CNV region, followed by **pulsed-field gel electrophoresis** (**PFGE**), a method for separating large (up to megabase-scale) molecules, and Southern blotting using a specific probe, can indicate both copy number, and the allele combinations that underlie it. An alternative and even more direct method is **fiber FISH** (**fluorescence *in situ* hybridization**), in which fluorescently labeled probes are hybridized to stretched chromatin fibers, allowing a resolution of a few kilobases. **Figure 4.19** shows an

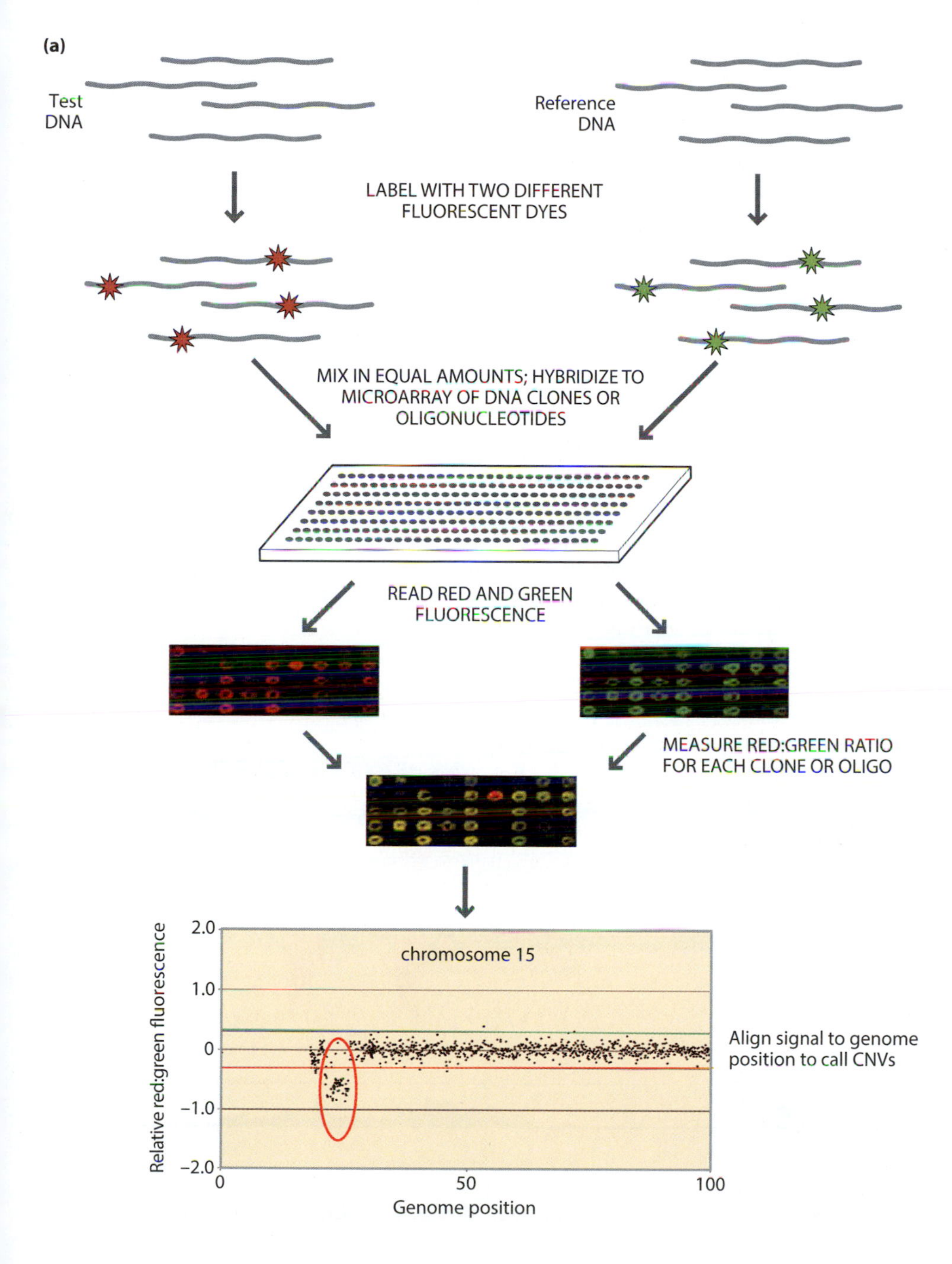

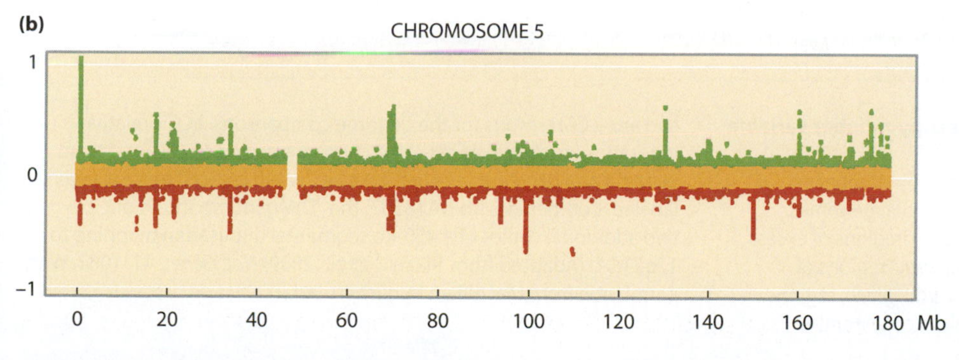

Figure 4.15: Array-comparative genomic hybridization (aCGH) for the detection of copy-number variation.
(a) Overview of method. (b) An example of the result of multiple aCGH experiments on the HapMap Phase I samples for chromosome 5. [a, Adapted from Read A & Donnai D (2006) *New Clinical Genetics*. With permission from Scion Publishing Ltd; b, from Redon R et al. (2006) *Nature* 444, 444. With permission from Macmillan Publishers Ltd.]

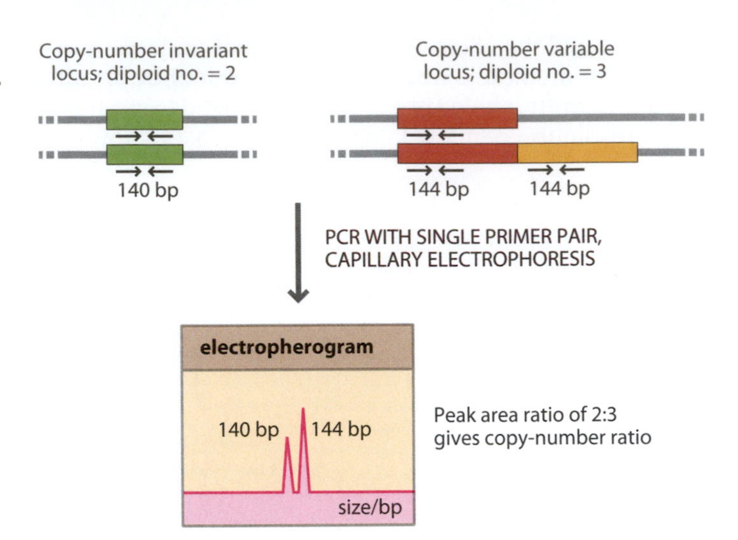

Figure 4.16: The paralog ratio test for the detection of copy-number variation. A single primer pair amplifies both an invariant and a copy-number variable locus, giving PCR products of slightly different sizes. Electrophoretic separation and quantitation (here, from peak area) gives the relative copy numbers. The use of the same primer pair for both reference and test loci ensures that the kinetics of amplification similar for each. For the original description of the paralog ratio test, see Armour JA et al. (2007) *Nucleic Acids Res.* 35, e19.

example in which the copy number of the salivary amylase gene (*AMY1*) can be measured directly in each chromosome of an analyzed individual.[39]

Balanced polymorphisms (translocations and inversions) cannot be detected using aCGH methods, SNP chips, or sequencing read-depth, but paired-end NGS approaches now allow them to be detected in an unbiased way.[18] Once identified, individual rearrangements can be assayed by the detection of novel junction sequences by PCR.

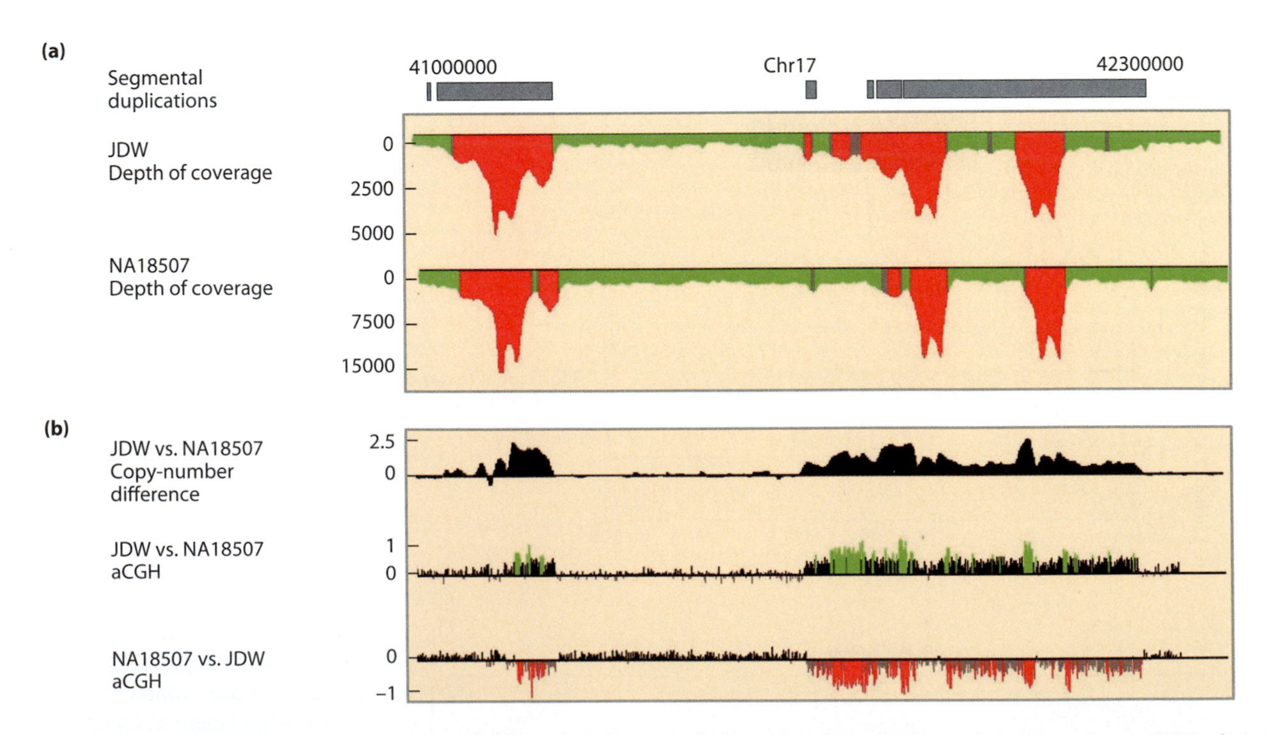

Figure 4.17: Sequence read-depth indicates copy-number variable regions.
(a) Read-depth for a region containing a known copy-number polymorphism on 17q21.31 among Europeans, for two genomes, James D. Watson (JDW) and NA18507. *Red* indicates regions of excess read-depth (mean + 3 standard deviations; s.d.); *gray*, regions of intermediate read-depth (mean + 2 s.d. and − 3 s.d.); *green*, regions of normal read-depth (mean ± 2 s.d.). (b) Absolute copy number and array CGH results for the genome comparisons. aCGH relative $\log_2$ ratios are shown as *red–green* histograms and correspond to an increase and decrease in signal intensity when test-reference is reverse-labeled. Compared with NA18507, the JDW genome shows one to two additional copies of a 459-kb segmental duplication mapping to 17q21.31. [Adapted from Alkan C et al. (2009) *Nat. Genet.* 41, 1061. With permission from Macmillan Publishers Ltd.]

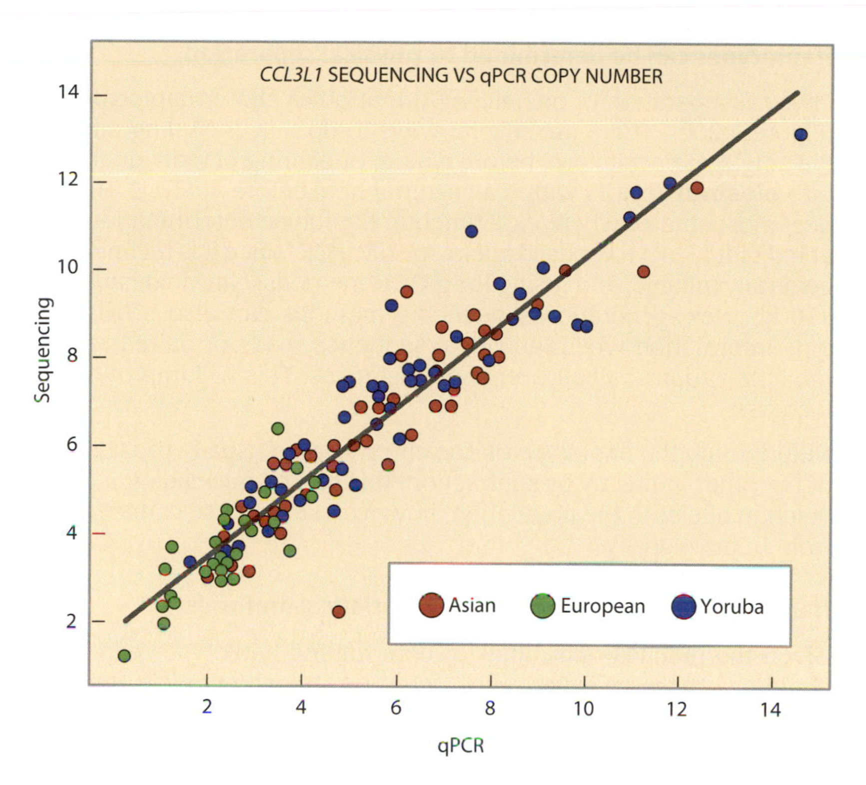

Figure 4.18: Performance of sequence read-depth in estimating copy number. qPCR-based copy-number genotyping is highly correlated with sequencing-based copy-number estimates ($r = 0.95$). [From Sudmant PH et al. (2010) *Science* 330, 641. With permission from AAAS.]

4.9 PHASING: FROM GENOTYPES TO HAPLOTYPES

Knowledge of haplotypes is essential for the construction of **gene trees**—phylogenies of individual loci—which, as we shall see throughout this book, are important in many areas of human evolutionary genetics. Haplotypes are often also the starting point for attempts to locate genes involved in disease, and for understanding the pattern of recombination and LD in the genome. Some loci—the Y chromosome, mtDNA, and the X chromosome in a male—are haploid, and typing polymorphisms on these molecules immediately yields haplotypes (**Figure 4.20a**; and see **Section 3.8**). For most of the genome, however, this is not so (Figure 4.20b; and see Figure 3.29). It may be easy enough to determine which alleles are present, but not how they are associated together on a chromosome—a property sometimes called **gametic phase**, or simply phase. The problem is how to obtain haplotypes from genotype data in diploid DNA. Essentially, there are three different approaches,[10] summarized below.

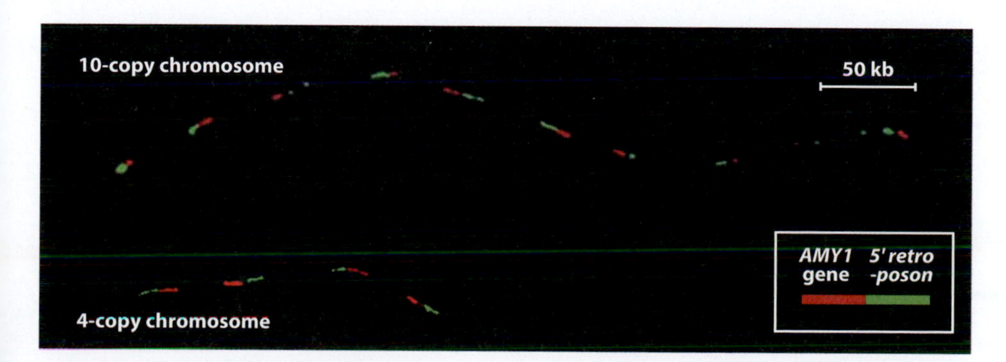

Figure 4.19: Fiber FISH demonstrates copy number of sequences on individual chromosomes.
High-resolution fiber FISH validation of copy-number estimates for the gene *AMY1*. The analyzed individual was estimated by qPCR to have 14 diploid *AMY1* gene copies, consistent with fiber FISH results showing one allele with 10 copies and the other with 4 copies. [From Perry GH et al. (2007) *Nat. Genet.* 39, 1256. With permission from Macmillan Publishers Ltd.]

Haplotypes can be determined by physical separation

Physical separation of one allele from another allows haplotype determination (Figure 4.20b). There are several ways to do this, including dilution of DNA to the single-molecule level before typing, or cloning of individual DNA molecules into **plasmid** vectors within a bacterial host before analysis. However, in practice, most studies where experimental haplotype determination has been done at individual loci have used allele-specific PCR, since it is technologically simple, generally reliable, and (using **long PCR** methods) can yield haplotypes covering >10 kb. Next-generation sequencing methods can also provide direct haplotype information when individual sequence reads or paired reads encompass multiple variants, albeit over short distances. This will improve as read lengths increase.

Determining the haplotype of one chromosome usually provides the haplotype of the other indirectly by subtraction; this can be misleading if there are **null** or deletion alleles in the population, however, since these cannot be distinguished from homozygosity.

Haplotypes can be determined by statistical methods

Given the practical difficulties of determining haplotypes experimentally, particularly on the massive scale required in genomewide studies of linkage disequilibrium and allele association involving many thousands of SNPs, statistical methods have been developed for deducing haplotypes from genotypic data. Two different varieties of method are described below.

Clark's algorithm[12] is a simple and intuitive method that first searches a sample of genotypes for homozygotes and single-site heterozygotes, thus identifying

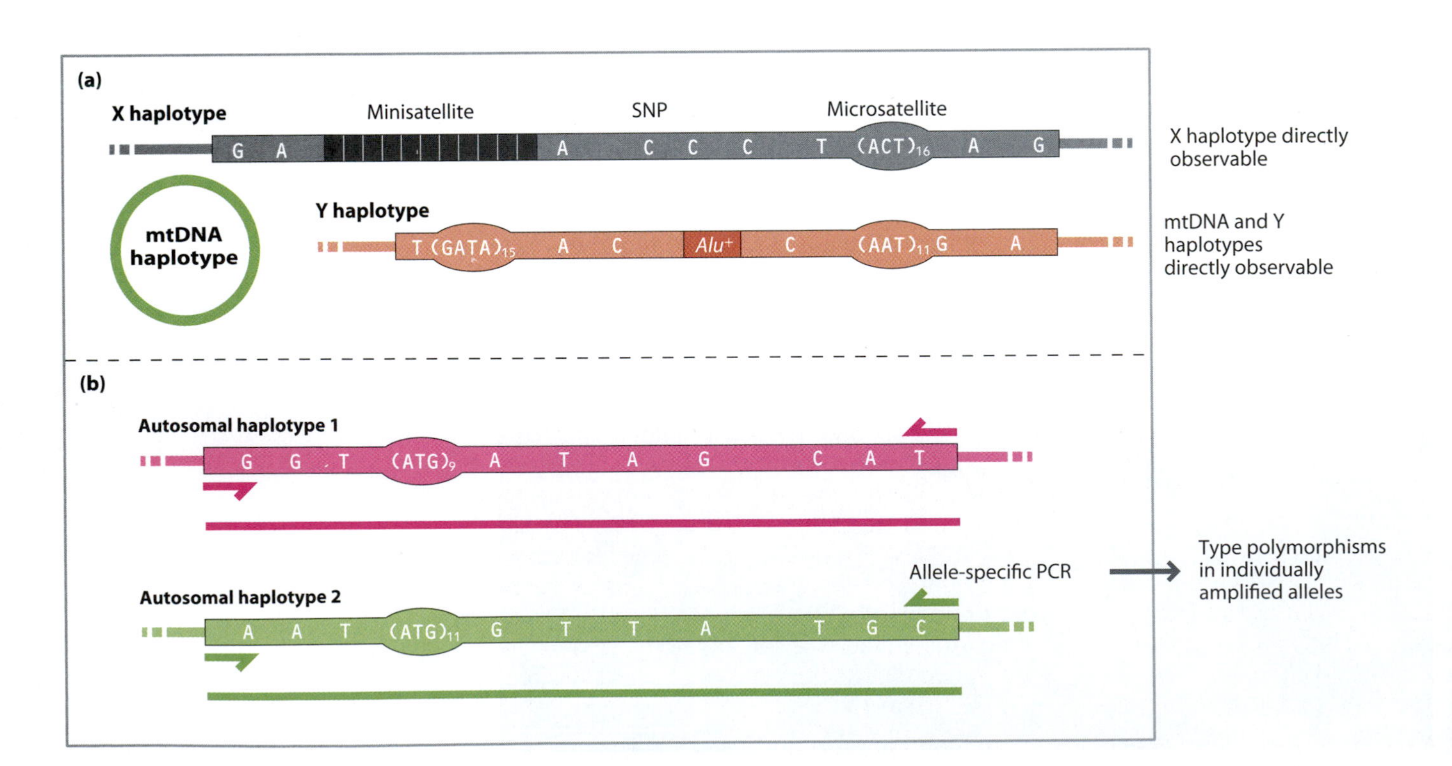

Figure 4.20: Experimental methods for deducing haplotypes from genotypes.
Hypothetical haplotypes composed of SNPs, microsatellites, a minisatellite, and an *Alu* insertion are shown in a male. (a) Haploid molecules (mtDNA, the Y chromosome, and the X chromosome in a male) yield haplotypes directly. (b) Autosomal haplotypes can be deduced by specific amplification of each allele (colored arrows indicate allele-specfic primers) followed by typing of polymorphisms. Typing of one allele should allow the haplotype of the other to be deduced by subtraction.

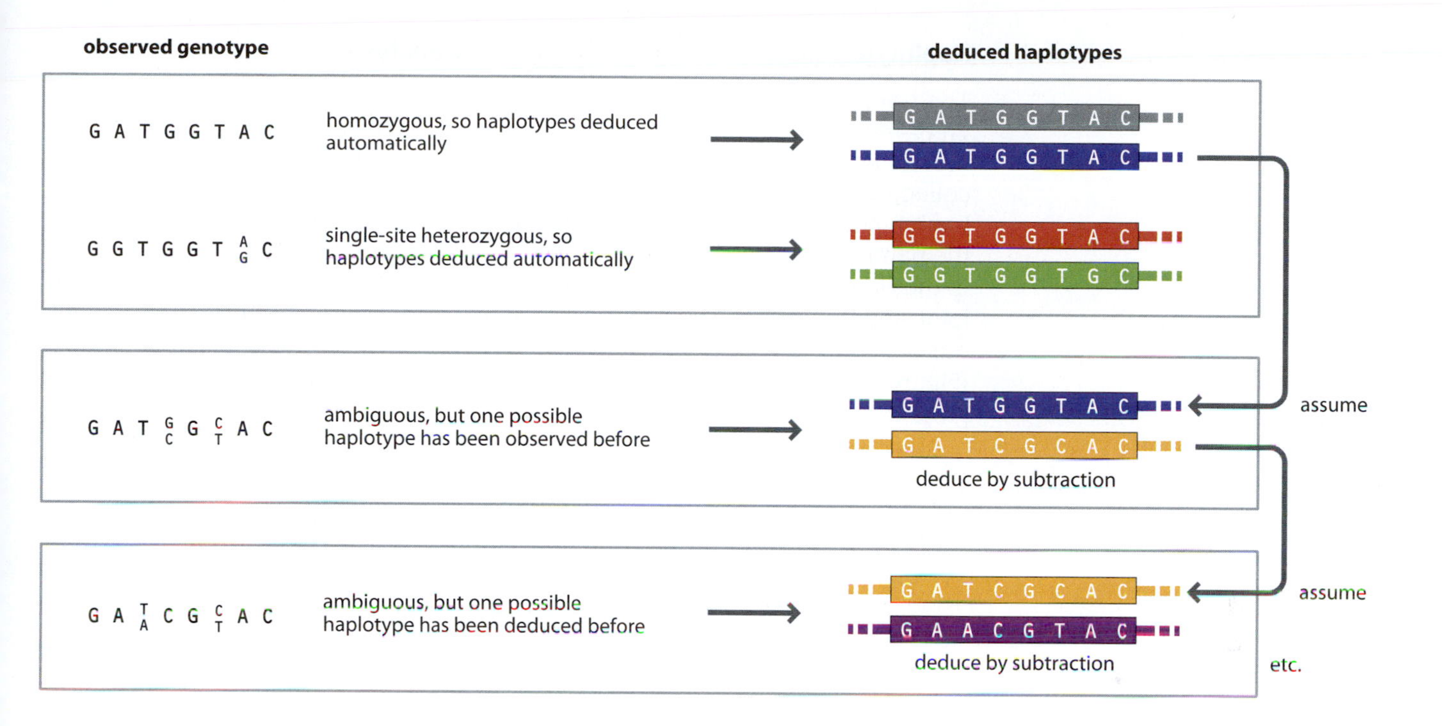

Figure 4.21: Deducing haplotypes from genotypes by Clark's parsimony algorithm.
A hypothetical diploid locus containing eight SNPs is shown. Sequencing reveals homozygous and heterozygous genotypes, and the underlying haplotypes are deducible using a parsimony principle.

a subset of underlying haplotypes unambiguously (**Figure 4.21**). The remaining unresolved genotypes are then examined to ask whether one of the possible underlying haplotypes is one that has already been found. If it has, this haplotype is assumed to be correct, and the other haplotype in the unresolved genotype is deduced by subtraction. In this way the pool of observed and deduced haplotypes increases in size, until, ideally, it contains all haplotypes. If the sample of genotypes is small, it may be that no homozygotes or single-site heterozygotes can be found, in which case the algorithm cannot begin; also, some "orphan" haplotypes may remain impossible to deduce. Nonetheless, this simple algorithm performs remarkably well, for both short segments of DNA and for genomewide patterns of LD. Clark's algorithm is sometimes called a "**parsimony**" approach, because it attempts to minimize the number of haplotypes in the sample.

Because new haplotypes are derived from existing haplotypes by the processes of mutation and recombination, **coalescent**-based models (**Section 6.5**) can be applied to phase reconstruction. Such models form the basis of many population-based statistical methods, including the widely used program PHASE.[48] As well as haplotype inference, PHASE can also draw inferences about the underlying evolutionary and demographic processes giving rise to the patterns of LD, including the estimation of recombination rates. Several programs have been developed in order to deal rapidly and accurately with very large datasets involving many samples and many SNPs. For example, BEAGLE[9] does not explicitly model recombination and mutation, but clusters haplotypes at each locus, adapting the clustering to the amount of available information so that the number of clusters increases with sample size. Statistical phasing methods have been applied to the large HapMap and 1000 Genomes Project reference datasets, and as a result phased versions are available.

As an extension of statistical haplotype inference, a number of programs have been developed to allow the **imputation** of SNPs that have not been typed in a sample, by exploiting the LD structure of the genome and the availability of reference population datasets:[32] this is a key element of genomewide association studies, and is particularly important when carrying out meta-analyses of several studies that have used different sets of SNPs.

Haplotypes can be determined by pedigree analysis

The traditional method for determining haplotypes is by examination of pedigrees—the principle is illustrated in **Figure 4.22a**. While some large pedigrees are assembled for studies of monogenic disease, many studies of complex disorders collect small pedigrees, such as father–mother–child trios. Determination of haplotypes in children (phase) in such cases depends on the informativeness of the parental haplotypes, and can often fail (Figure 4.22b). The inclusion of trios in the HapMap project was important to validate the statistical approaches, and joint approaches that consider both statistical estimation and pedigree methods are available. Next-generation sequencing of whole genomes from families (Figure 4.22c) can yield long-range genomewide haplotypes, and allows the pinpointing of recombination events.[43]

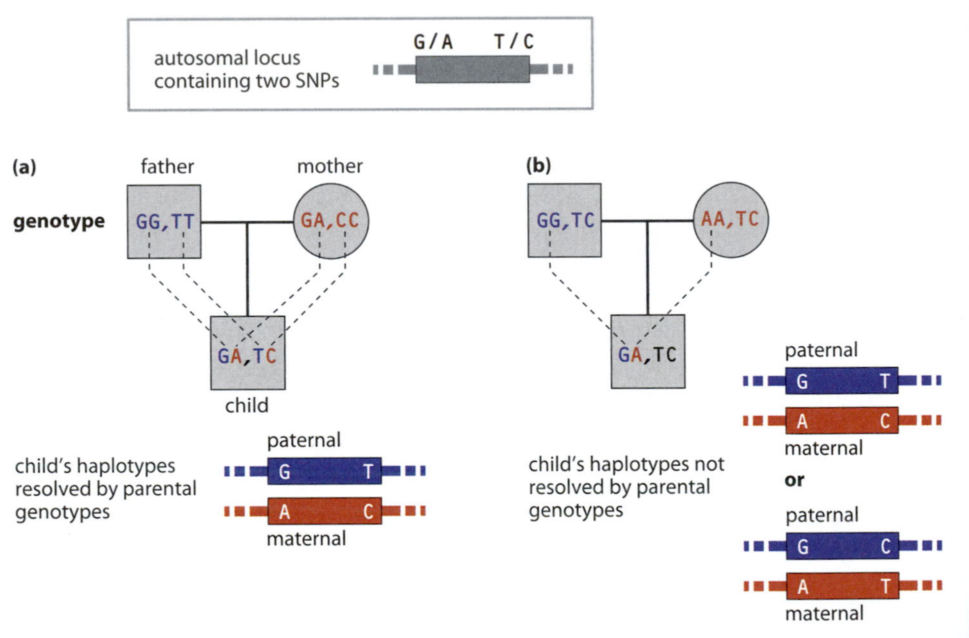

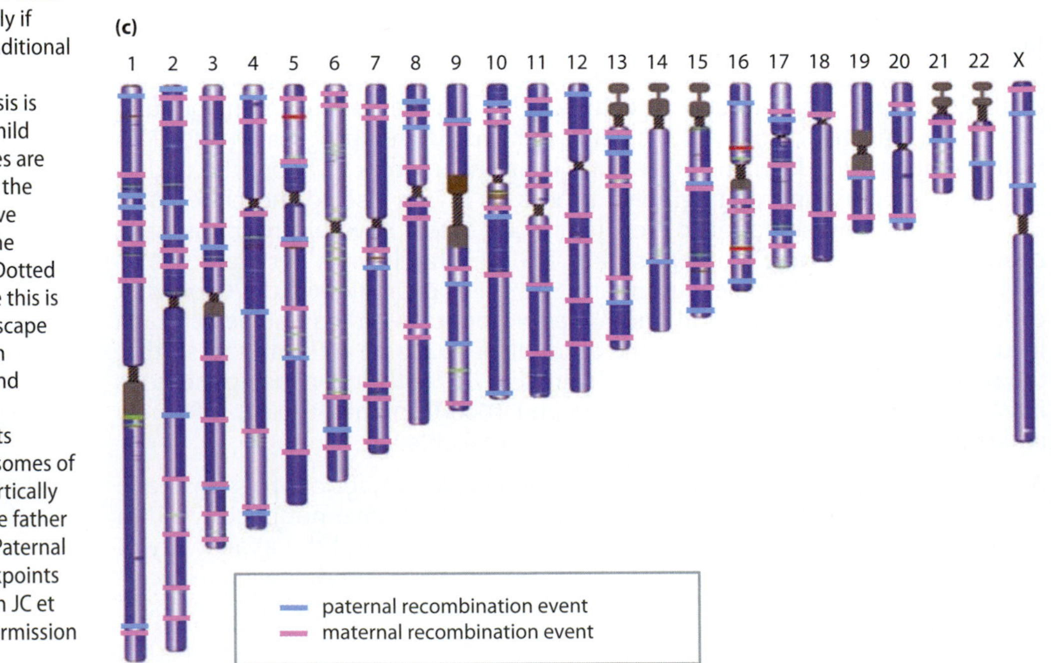

Figure 4.22: Deducing haplotypes from genotypes by pedigree analysis. A hypothetical locus containing two SNPs (G/A and T/C) is shown, and it is assumed that no recombination can occur between the two SNPs—this is likely if they are physically close. (a) The traditional method of determining haplotype (gametic phase) by pedigree analysis is successful in this father–mother–child trio because the parental genotypes are informative, allowing deduction of the child's haplotypes. (b) Uninformative parental genotypes do not allow the child's haplotypes to be deduced. Dotted lines show allele inheritance where this is deducible. (c) Recombination landscape from whole-genome sequencing in a pedigree of two parents, a son, and a daughter. Each chromosome in this schematic karyotype represents information from the four chromosomes of the two children, and is divided vertically to indicate the inheritance from the father (*left half*) and mother (*right half*). Paternal and maternal recombination breakpoints are indicated. [Adapted from Roach JC et al. (2010) *Science* 328, 636. With permission from AAAS.]

4.10 STUDYING GENETIC VARIATION IN ANCIENT SAMPLES

The length of this book is testimony to the difficulty of making inferences about the past from the genetic diversity of modern human populations. Imagine how much simpler our task would be if we could go back in time and sample the DNA of humans and extinct closely related species directly: this is the promise of ancient DNA (aDNA) studies. Provided human remains could be found from the appropriate population and era, and DNA from them isolated and analyzed, this could allow us to address directly many of the questions with which this book is concerned, for example:

- Determination of phylogenetic relationships of ancient hominins, and whether they interbred with modern humans, from DNA sequence data
- Analysis of the ancient sequences of candidate genes involved in important human-specific traits
- Direct understanding of mutational dynamics by sampling loci at different time-points
- Analysis of the diversities of ancient populations, and the ancestries of individuals
- Diagnosis of genetic disorders and measurements of allele frequencies in ancient populations
- Determination of past frequencies for alleles involved in adaptive traits such as pigmentation, dietary adaptations linked to agriculture, and responses to particular pathogens
- Determination of sex of remains, and deduction of kin relationships in group burials

Further issues could be investigated using DNA not from ancient humans themselves, but from their pathogens, their domesticated animals and plants, and their feces (preserved as **coprolites**), which would give information about diet.

Early aDNA studies were bedeviled by problems: rapid postmortem degradation of DNA frequently made it unamplifiable using the available PCR methods; old samples were often contaminated with ubiquitous modern DNA, which made it difficult or impossible to prove that putative ancient sequences were genuine; and there was a lack of reference genome sequences which made the authentication of novel sequences derived from ancient samples contentious. Reflecting the "Jurassic Park" craze of the early 1990s, a number of high-profile reports appeared describing putative ancient DNA sequences retrieved from 17–20 million years (MY)-old magnolia leaves, an 80-MY-old dinosaur bone, and a 120–135-MY-old amber-embedded weevil, amongst others. It is now believed that most or all of these exciting results were artifacts of modern DNA contamination (with bacterial, fungal, or human DNA), and that physicochemical processes set a probable upper limit of 100 thousand years (KY) to 1 MY on the survival of DNA.[22] In **Opinion Box 2**, Tom Gilbert, a practitioner in this difficult field, asks what the true limits to aDNA studies are.

DNA is degraded after death

After death, DNA is usually rapidly degraded by the action of **endonucleases**. Particular conditions can allow escape from these enzymes, but over time other spontaneous processes act to damage DNA. In the cells of a living person, as we saw in **Section 3.2**, DNA is continually being monitored and repaired. After death, the systems that accomplish this cease functioning, and the physicochemical assault can go on unimpeded. The result is that DNA recovered from bones or other tissues of long-dead humans is severely damaged by cleavage of the sugar–phosphate backbone, resulting in short DNA fragments; loss of bases (**abasic** sites); chemical modification of bases; and inter- or intramolecular cross-linking of sugar–phosphate backbones (**Figure 4.23**).

OPINION BOX 2: Reasonable expectations of ancient DNA

Following the series of unfortunate errors of the late 1980s and early 1990s, which led to unrealistic expectations of how long DNA will survive post mortem, and thus how far back it can contribute insights into the past, the view of what aDNA really has to offer stabilized. This was in light of the assessment of two limits: (1) What is the minimum size of DNA fragments needed in order to recover meaningful sequence information? (Answer: ~60–70 bp—the size of two PCR primers plus a minimum of 20–30 bp between their binding sites); and (2) For how long, and under what environmental conditions, might one expect fragments of at least this size to survive? (Answer: if cool and dry, maybe 100,000 years; if warmer and wetter, a lot shorter). Given this, if a research team had the luxury of plenty of money and time, they could dream of analyzing mitochondrial HVSI, or even complete mitochondrial genomes, from tens to hundreds of samples, with possibly the occasional autosomal or Y-chromosomal SNP thrown in for good measure. Then, in 2006, all this changed with the release of the next-generation sequencers, and specifically the elimination of PCR from the equation, which effectively reduced the minimum DNA size requirement to only 20–30 bp.

The power of such technologies is discussed elsewhere in this chapter, but what did this change actually mean for the scope of aDNA studies? Much larger parts of the genome could now be targeted—although, importantly, aDNA will never be able to give whole genome recovery. For example, despite containing a very pure fraction of human DNA, and being sequenced to 20× haploid coverage, the reported nuclear genome of the 4500-year-old **Saqqaq** Greenlander generated through shotgun sequencing[41] still represented only 79% of the reference sequence. Why? Because a proportion of the genome is comprised of repetitive regions that cannot be resolved without reads that are considerably longer than the usual length of aDNA templates. Furthermore, for the less well preserved samples that represent the bulk of old human material, shotgun sequencing to this level would be extremely expensive. For example, the generation of a 20×-covered genome (amount of sequence needed = 60 Gb) from a sample containing only 1% endogenous human DNA would require 6000 Gb of sequence—a huge amount, even taking into consideration the power of the latest sequencers. Fortunately, not many researchers require the complete genome to adequately answer their questions, and the use of sequence-capture methods[11] can greatly improve efficiency, for example by focusing solely on the exome. Thus, dreams of population studies at the exome level are becoming realistic, in particular as the cost of sequencing drops.

Nevertheless, we are still left with the challenge of aDNA survival—although shorter molecules can now be used, eventually the length of DNA drops below the critical 20–30 bp threshold (**Figure 1**). It is unlikely we will ever be able to use shorter molecules, simply due to the **bioinformatic** problem of mapping them uniquely to a reference genome. While a reduction of 40 bp from the lengths needed for PCR would not appear to make a major difference in maximum time of survival, due to the kinetics of DNA degradation it is reasonable to expect a much greater upper time limit—perhaps around several million years in natural frozen conditions, and several hundred thousand years at average temperate conditions. The ambient temperature and humidity remains a key factor, thus even with these improvements it will be surprising if DNA can be recovered from controversial specimens such as the tropical Flores hominid (*Homo floresiensis* to those that support its independent species status) or archaic hominids from Africa. But other key hominids may turn out to hold sufficient genetic information to provide insights into their evolutionary history and relationship to modern humans, in particular those that date to within the last million years in Europe and the colder regions of Asia, such as *H. heidelbergensis* and *H. erectus*.

M Thomas P Gilbert, Natural History Museum of Denmark, Copenhagen

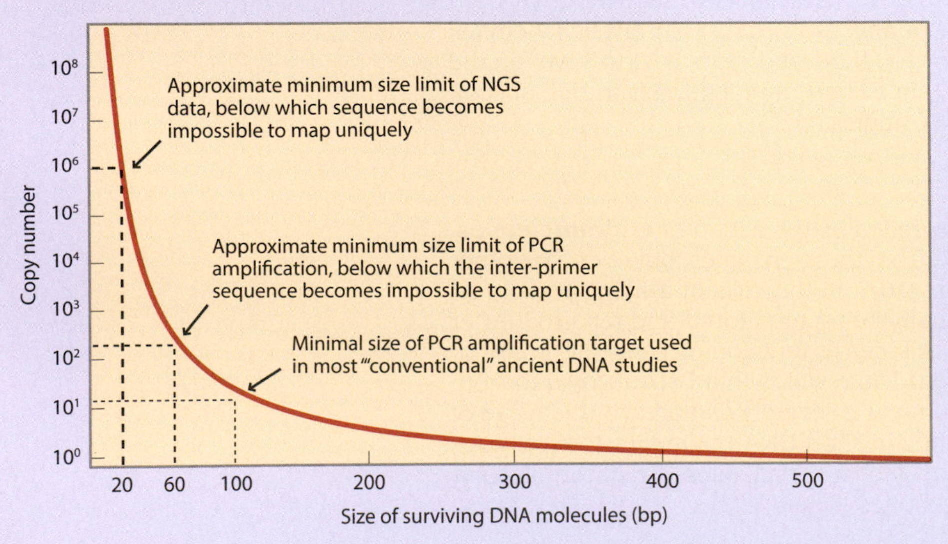

Figure 1: The relationship between size of surviving DNA fragments and copy number in ancient samples. As DNA degrades following death, it rapidly shortens. After sufficient time has passed, the size distribution of remaining DNA molecules can be represented as an exponential curve, with DNA copy number increasing exponentially with decreasing length (the actual values vary by sample age, conditions, and amount of material analyzed). This has important implications: (1) the shorter the size of DNA templates targeted in an aDNA study, the greater the chance of success; (2) because NGS approaches do not require primer binding to the DNA, the molecules used can be ~40 bp shorter than in PCR-based studies. This can lead to a significant increase in DNA molecules that can be analyzed, allowing older and more poorly preserved material to yield useful data.

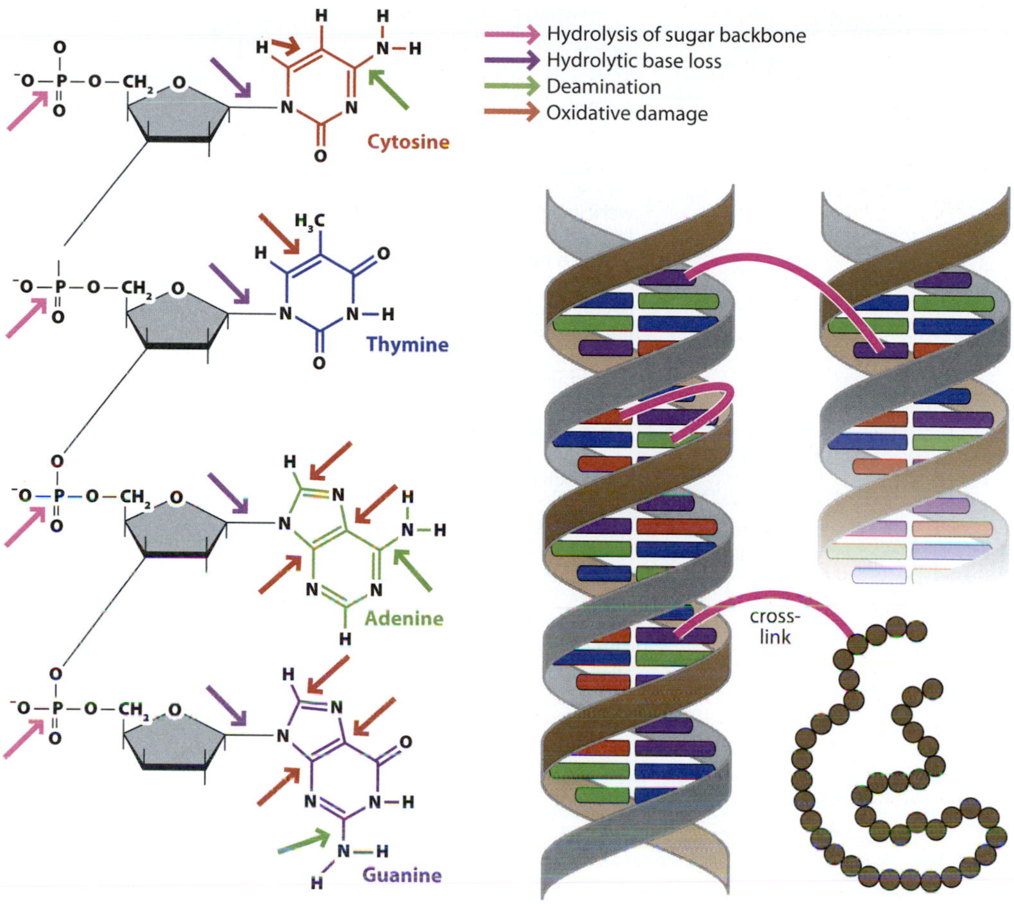

Figure 4.23: Postmortem DNA damage. DNA is prone to spontaneous damage and degradation, including hydrolytic and oxidative damage, and cross-linking between or within helices, as well as to proteins. [Adapted from Hebsgaard MB et al. (2005) *Trends Microbiol.* 13, 212. With permission from Elsevier.]

Cleavage leads to a very short average length of isolated DNA—only a few tens or (at best) hundreds of base pairs. Missing bases or base modification cause misincorporation by the polymerase during PCR, while cross-linked sites or oxidized derivatives of cytosine or thymine called **hydantoins** will block the polymerase and truncate synthesis. The longest fragments that can be amplified are only 100–200 bp, and yield of endogenous DNA is extremely low. The high copy number of mtDNA per cell means that it has a better chance of surviving than does nuclear DNA, and so it has traditionally been a source of more data than has nuclear DNA.

DNA survival in an ancient sample is influenced by the conditions under which it has existed since it was deposited: temperature, pH, humidity, and salt concentration affect the rates of the modifications that DNA undergoes post mortem. The **Denisova** cave, for example, which has yielded extensive ancient sequences, is in Siberia—dry, and with an annual average temperature around 0°C. Samples from Africa, which has been on average warmer and damper, and often has acidic forest soils, are less likely to yield useful data.

Contamination is a major problem

Provided that DNA can be analyzed at all in a sample, the most serious problem for aDNA researchers working on humans or their close evolutionary relatives is modern DNA contamination. As we have seen, ancient DNA is present in very small amounts, is seriously damaged, and is difficult to amplify; in contrast, modern DNA is plentiful, ubiquitous, in relatively good condition, and therefore very easy to amplify. Efforts to surmount the problem of contamination must be rigorous (**Table 4.4**). Contamination can occur in the laboratory, where it is typically due to the presence of the products of previous PCRs: a successful

TABLE 4.4:
PRODUCING AUTHENTIC ANCIENT DNA SEQUENCES

Aim	Procedure
Prevent modern DNA contamination	excavate and curate ancient specimens under 'clean' conditions
	work in a properly equipped and physically isolated clean laboratory facility for DNA extraction and NGS library preparation, in which no PCR work is ever done
	decontaminate surfaces, reagents, and tools frequently, using UV light and/or bleach
	decontaminate ancient specimens before processing
Detect and quantify modern DNA contamination	use multiple *blank* controls in extractions and PCRs
	clone PCR products and sequence multiple clones
Show that results are consistent with ancient DNA	quantify template DNA
	show that amplifiable PCR products are short (<500 bp)
	demonstrate inverse relationship between amplification length and efficiency, reflecting degradation and damage in the ancient DNA template
	ideally, observe age-dependent patterns in both sequence diversity and DNA damage
Show reproducibility	repeat findings within laboratory
	have results reproduced independently by another laboratory

reaction creates 10^{12}–10^{15} molecules, and invisible aerosols can be generated during laboratory manipulations that contaminate surfaces and equipment. Contamination can also occur in the sample itself: this can be genomic DNA from people who have handled a specimen during or after its excavation.

These problems apply to all aDNA studies, but some apply most acutely to studies of hominid samples. Imagine that you are working on samples from a woolly mammoth: you might inadvertently contribute your own DNA to the sample between the time that it is collected and the time that PCR is carried out, but choice of appropriate mammoth-specific PCR primers should mean that your DNA will not be amplified, while the mammoth's DNA will. Even if your DNA *were* amplified and sequenced, using universal 16S mtDNA primers, for example, it would be easily distinguishable from that of the ancient sample because of differences in sequence. In other words, there is a *phylogenetic signal* in the mammoth DNA that distinguishes it from likely modern contaminating DNAs, and thus provides support for its authenticity. Now consider an attempt to amplify DNA from a 5000-year-old anatomically modern human skeleton. The sequence you obtain might be similar to, or even identical to, many modern human sequences—this was the case with the mtDNA HVSI sequence of Ötzi, the Tyrolean Iceman[21] (**Box 4.3**). There is no phylogenetic signal here, and proof of authenticity becomes much more difficult. In the landmark study[30] of the Neanderthal **type specimen** (see **Section 9.5**) many appropriate controls were done, but the finding that the corresponding Neanderthal sequence lay outside the spectrum of variation of modern human mtDNA sequences was nonetheless of great importance in supporting authenticity. In the case of mtDNA studies, the existence of human numts (nuclear mitochondrial DNA insertions; see **Section 2.8**) makes life even more difficult, since they are diverged from human mtDNA itself and might easily be confused with ancient, divergent mtDNA sequences.

Bacterial cloning of PCR products into plasmid vectors has provided a defense both against contamination and misincorporation during PCR on damaged templates. Sequencing of multiple overlapping clones allows the reconstruction of the original sequence, and the identification of bases introduced by polymerase errors.

Application of next-generation sequencing to aDNA analysis

As described above, aDNA is highly fragmented—usually less than a few hundred base pairs in size. This presents a problem for traditional methods of analysis, but actually makes it well suited to analysis by next-generation sequencing, which normally requires DNA to be fragmented prior to library preparation. The application of NGS methods to ancient specimens has revolutionized this

Box 4.3: Ötzi, the Tyrolean Iceman

The Tyrolean Iceman, popularly known as Ötzi, was discovered in September 1991 (Figure 1a). He had emerged in a naturally mummified state from the ice of the Schnalstal glacier in the Ötztal Alps, ~5300 years after his death. Investigations of Ötzi, together with his beautifully preserved clothing, weapons, and equipment, offered a rare opportunity to learn about the life of a late Neolithic human.

Shortly after his discovery, ancient DNA studies were undertaken with the hope of investigating Ötzi's affiliations with modern European populations;[21] these studies were extremely difficult, since the DNA was badly degraded (mostly <150 bp) and there was demonstrable contamination with modern DNA. Short lengths of endogenous mtDNA could be amplified, and it was shown by painstaking analysis and replication of results in two different laboratories that a fragment of HVSI was closely related to the mtDNA reference sequence. In the 394-bp region analyzed, Ötzi's sequence differed only at two positions: T to C at 16224 and T to C at 16331. It was therefore typical of sequences found in many modern European populations, including those from the Mediterranean, Alpine, and Northern regions where it represented about 2% of the sample—an unsurprising, but hard-won result.

aDNA analysis was also carried out on the contents of Ötzi's colon and ileum.[44] PCR analysis was done using primers directed at species-informative segments of mammalian mtDNA, plant chloroplast DNA, and multi-copy plant or fungal nuclear genes. Phylogenetic analysis of the sequences, together with sequences of known species, suggested that Ötzi's last meals had included the meat of ibex and red deer, and possibly cereals.

Recent technical advances have allowed a complete genomic sequence to be determined.[29] Typically, the retrieved sequences (the **metagenome**) indicate a number of sources, but ~78% are human on the basis of alignment to the reference sequence (Figure 1b). Ancestry analysis indicates possible affiliation with the modern inhabitants of Sardinia, and analysis of functional variants suggests that Ötzi had brown eyes, belonged to blood group O, and was lactose intolerant. Recovery of ~60% of the genome of the bacterium *Borrelia burgdorferi* probably indicates that Ötzi suffered from Lyme disease.

(a)

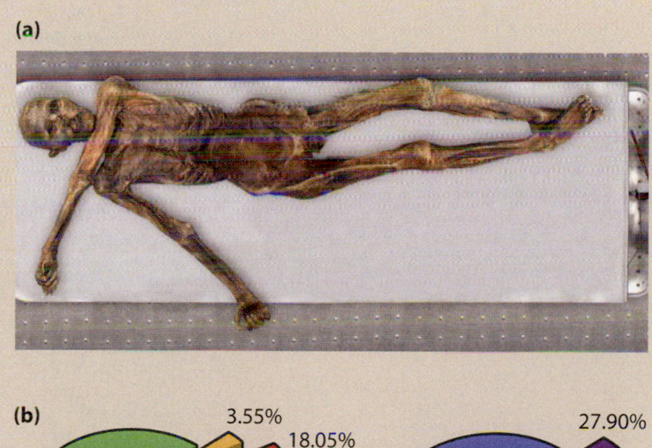

(b)

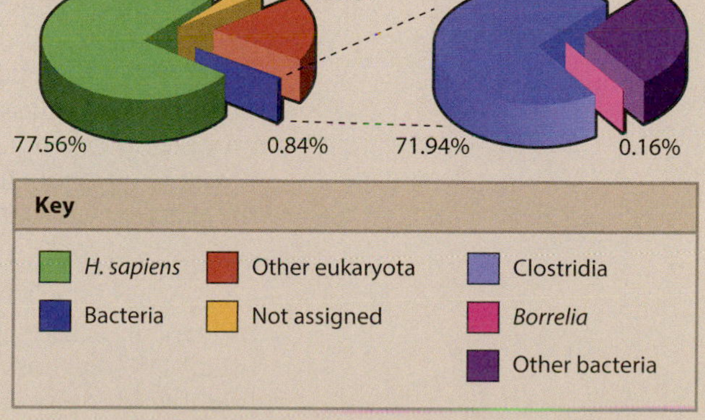

Figure 1: Ötzi and his metagenome.
(a) Ötzi in his purpose-built preservation facility in the Archaeological Museum of Bolzano. (b) Pie-charts showing uniquely assigned reads of BLAST searches of 8 million 50-bp reads against the NCBI nucleotide sequence collection: overall distribution of reads (*left*), and distribution of bacterial species (*right*). [From Keller A et al. (2012) *Nat. Commun.* 3, 698. With permission from Macmillan Publishers Ltd.]

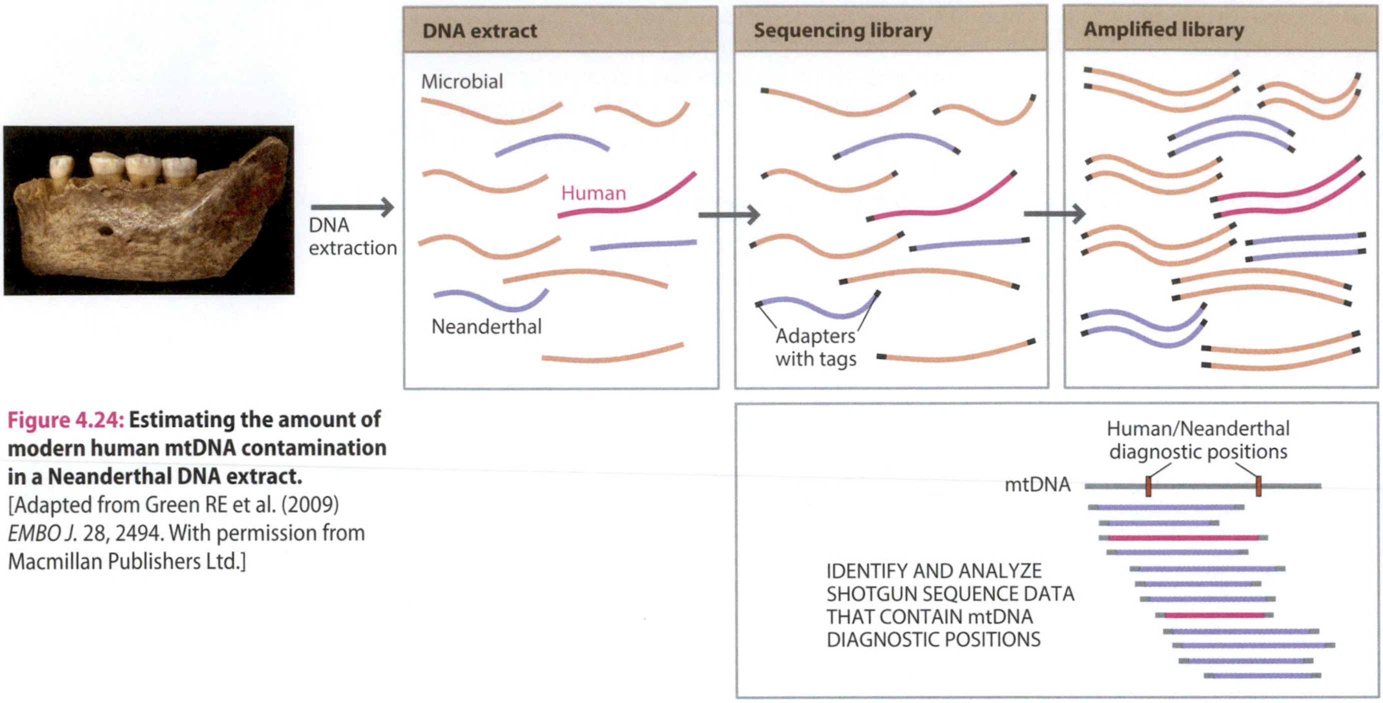

Figure 4.24: Estimating the amount of modern human mtDNA contamination in a Neanderthal DNA extract. [Adapted from Green RE et al. (2009) *EMBO J.* 28, 2494. With permission from Macmillan Publishers Ltd.]

field.[49] Efficient preparation of sequencing libraries from ancient DNA, followed by whole-library amplification can provide near-complete ancient genome sequences from >50 mg bone powder, compared with the ~1 g required for the PCR-based sequencing of the ~350 bp of Neanderthal mtDNA.

However, ancient DNA represents a complex mixture originating from many sources, dominated by bacterial, fungal, and environmental DNA introduced to the fossil since it was deposited, as well as likely DNA contamination introduced during or after sample collection. In Neanderthal specimens, endogenous DNA represents at most only a few percent of the total. This **metagenomic** nature of the DNA has led to innovations that improve the proportion of recovered hominid DNA, and other techniques have also been developed to improve sequence quality.

Restriction enzymes have been used to selectively cleave nonhuman DNA after library preparation.[20] Recognition sites rich in the dinucleotide CpG (such as that for *Bst*UI—CGCG) are rare in hominid DNA, but much more common in microbial DNA. In studies of Neanderthal samples, this treatment increased the proportion of endogenous sequences four- to sixfold. Although it biases the sequencing against GC-rich regions of hominid genomes, it greatly increases the efficiency of data production to provide an overview of the sequence.

Treatment with DNA repair enzymes can also improve sequence reliability from ancient samples.[8] Presence of uracil bases due to deamination is a major cause of error in ancient sequences, and methods including the use of the enzyme uracil-DNA-glycosylase remove these. The overall error rate can be reduced twentyfold for Neanderthal mtDNA compared with untreated DNA.

Modern human DNA contamination is almost inevitable, and methods to estimate its extent are important. Because Neanderthal mtDNA sequences fall outside the variation found among modern humans, there are base substitutions observed in most or all Neanderthal mtDNAs but not in modern human contaminants, and vice versa. Next-generation sequencing with sufficient read-depth allows the relative representation of these diagnostic positions to be estimated, giving the ratio of Neanderthal to modern human mtDNA in a DNA extract (**Figure 4.24**). Here, mtDNA is being used as a proxy for the genome as a whole, which is risky if the ratio of nuclear:mtDNA differs in different tissues and contaminant sources. Note that the approach here only works if the sample being analyzed is sufficiently diverged from modern sequences to carry a suite of diagnostic fixed mutations. An independent approach (requiring no such diagnostic sites) is the detection of modern Y-chromosomal DNA in ancient female specimens.[19] The addition of short project-specific bar codes to the ligated adapters in a clean environment before sequencing begins also helps in identifying contamination that might arise in the sequencing facility.

Finally, sequence-capture methods (Figure 4.7) can be used to improve efficiency of sequence generation for specific genomic regions. Array-based hybridization capture has been used to sequence complete mtDNAs,[7] and targeted autosomal regions[11] in Neanderthals.

SUMMARY

- Studies of human genetic variation have been revolutionized by technological developments, and the full range of variation at a genomewide scale in large population samples can now be analyzed.
- Blood, buccal cells, and immortalized cell lines are the most useful sources of DNA in population studies.
- The polymerase chain reaction (PCR) is an essential method that allows the easy analysis of specific loci within the human genome.
- The method of Sanger sequencing and its automation in fluorescence capillary sequencing was used to obtain the human genome reference sequence, a key starting point for any study of human genetic diversity.
- Next-generation sequencing (NGS) methods allow the parallel analysis of millions of DNA fragments, and have enormously reduced the cost of genome sequencing. Sequencing can be targeted to specific regions of the genome, including the exome, comprising the exons of all annotated protein-coding genes.
- Interpreting NGS results is challenging, due to the vast amounts of data, short read lengths, and high error rates. The number of sequence reads covering a particular base (coverage) is important in determining the confidence of the data.
- SNP typing systems vary in complexity, cost, throughput, and accuracy. Individual SNPs can be typed using restriction enzymes that cleave alleles differentially; high-throughput typing is done using chips based, for example, on hybridization and primer extension, and can genotype millions of SNPs simultaneously.
- Microsatellites and minisatellites are easily discovered from the reference sequence using bioinformatic methods, and microsatellites can be typed in multiplex PCRs using flanking primers and size separation by capillary electrophoresis.
- Structural polymorphisms and copy-number variants can be analyzed using methods such as array-comparative genomic hybridization, but the analysis of sequence read-depth in NGS allows less biased discovery and typing.
- Since most of human DNA is diploid, methods are needed to deduce haplotypes from genotypes. These include molecular separation of one allele from another (for example by allele-specific PCR), statistical methods, and pedigree analysis.
- Ancient DNA studies offer the prospect of studying past populations, their kinship relationships, and diets directly. The work is hampered by DNA degradation, and ubiquitous modern DNA contamination. Stringent criteria for authenticity must be applied to all studies. The application of NGS methods is allowing the recovery of megabase-scale sequences from ancient samples.

QUESTIONS

Question 4–1: You want to sequence a 5-kb gene in (a) one person; (b) 1000 people. What technologies would you choose and why? How would this change if the region was 5 Mb in size?

Question 4–2: You want to type a single SNP in (a) one person; (b) 100 people; (c) 10,000 people. What technologies would you choose and why? How would this change if the number of SNPs was 10, or 10 million?

Question 4–3: Consider the following four primer pairs (each written 5′ to 3′), and use BLAST and any other appropriate online tools to explain the PCR results described, using human genomic DNA as template.

	Primers	Result
(a)	CTAAAGATCAGAGTATCTCCCTTTG AAATTGTTTTCAATTTACCAG	379-bp single band, though only in half of the individuals you test
(b)	CTAAAGATCAGAGTATCTCCCTTTG CTGGTAAATTGAAAACAATTT	no product
(c)	TGCTTGAAGCTACGAGTTTG ATACATGCAGATACTCAGAG	a smear of products of diverse sizes
(d)	ACTAGCTCAAGTATAGAGTG AAATTGTTTTCAATTTACCAG	no product

Question 4–4: The PCR primers in Question 4-2 part (a) flank a SNP, rs9786153. Use dbSNP and NEBcutter to design a PCR-RFLP assay for this SNP.

Question 4–5: Two parents and their three children are genotyped for three linked microsatellites, A, B, and C. The genotypes (where the figures indicate numbers of repeat units) are:

Father	A(13,15)	B(17,19)	C(13,14)
Mother	A(12,15)	B(16,19)	C(13,14)
Child 1	A(13,15)	B(16,19)	C(13,14)
Child 2	A(12,15)	B(17,19)	C(13,14)
Child 3	A(12,13)	B(19)	C(14)

Assuming there is no recombination between A, B, and C, what are the mother's haplotypes?

Question 4–6: You have discovered a copy-number variant that influences a key trait in human evolution. Your colleagues dispute the existence of the variant. What technologies would you use to convince them that it existed? How would these depend on the nature of the copy-number variation (CNV)?

Question 4–7: Professor Pangloss has persuaded a museum curator to lend him the famous Heidelberg Jaw, and aims to improve his scientific reputation by carrying out ancient DNA analysis. What technical advice would you give him?

REFERENCES

The references highlighted in purple are considered to be important (for this chapter) by the authors.

1. **Ajay SS, Parker SC, Abaan HO et al.** (2011) Accurate and comprehensive sequencing of personal genomes. *Genome Res.* **21**, 1498–1505.

2. **Albers CA, Lunter G, MacArthur DG et al.** (2011) Dindel: accurate indel calls from short-read data. *Genome Res.* **21**, 961–973.

3. **Alkan C, Sajjadian S & Eichler EE** (2011) Limitations of next-generation genome sequence assembly. *Nat. Methods.* **8**, 61–65.

4. **Baird DM, Jeffreys AJ & Royle NJ** (1995) Mechanisms underlying telomere repeat turnover, revealed by hypervariable variant repeat distribution patterns in the human Xp/Yp telomere. *EMBO J.* **14**, 5433–5443.

5. **Banchs I, Bosch A, Guimera J et al.** (1994) New alleles at microsatellite loci in CEPH families mainly arise from somatic mutations in the lymphoblastoid cell-lines. *Hum. Mut.* **3**, 365–372.

6. **Benson G** (1999) Tandem repeats finder: a program to analyze DNA sequences. *Nucleic Acids Res.* **27**, 573–580.

7. **Briggs AW, Good JM, Green RE et al.** (2009) Targeted retrieval and analysis of five Neandertal mtDNA genomes. *Science* **325**, 318–321.

8. **Briggs AW, Stenzel U, Meyer M et al.** (2010) Removal of deaminated cytosines and detection of *in vivo* methylation in ancient DNA. *Nucleic Acids Res.* **38**, e87.

9. **Browning SR & Browning BL** (2007) Rapid and accurate haplotype phasing and missing-data inference for whole-genome association studies by use of localized haplotype clustering. *Am. J. Hum. Genet.* **81**, 1084–1097.

10. **Browning SR & Browning BL** (2011) Haplotype phasing: existing methods and new developments. *Nat. Rev. Genet.* **12**, 703–714.

11. **Burbano HA, Hodges E, Green RE et al.** (2010) Targeted investigation of the Neandertal genome by array-based sequence capture. *Science* **328**, 723–725.

12. **Clark AG** (1990) Inference of haplotypes from PCR-amplified samples of diploid populations. *Mol. Biol. Evol.* **7**, 111–122.

13. **Coffey AJ, Kokocinski F, Calafato MS et al.** (2011) The GENCODE exome: sequencing the complete human exome. *Eur. J. Hum. Genet.* **19**, 827–831.

14. **Conrad DF, Keebler JE, DePristo MA et al.** (2011) Variation in genome-wide mutation rates within and between human families. *Nat. Genet.* **43**, 712–714.

15. **Conrad DF, Pinto D, Redon R et al.** (2009) Origins and functional impact of copy number variation in the human genome. *Nature* **464**, 704–712.

16. **Dean FB, Hosono S, Fang LH et al.** (2002) Comprehensive human genome amplification using multiple displacement amplification. *Proc. Natl Acad. Sci. USA* **99**, 5261–5266.

17. **Ewing AD & Kazazian HH, Jr.** (2011) Whole-genome resequencing allows detection of many rare LINE-1 insertion alleles in humans. *Genome Res.* **21**, 985–990.

18. **Feuk L** (2010) Inversion variants in the human genome: role in disease and genome architecture. *Genome Med.* **2**, 11.

19. **Green RE, Briggs AW, Krause J et al.** (2009) The Neandertal genome and ancient DNA authenticity. *EMBO J.* **28**, 2494–2502.

20. **Green RE, Krause J, Briggs AW et al.** (2010) A draft sequence of the Neandertal genome. *Science* **328**, 710–722.

21. **Handt O, Richards M, Trommsdorf M et al.** (1994) Molecular genetic analyses of the Tyrolean Ice Man. *Science* **264**, 1775–1778.

22. **Hebsgaard MB, Phillips MJ & Willerslev E** (2005) Geologically ancient DNA: fact or artefact? *Trends Microbiol.* **13**, 212–220.

23. **Hormozdiari F, Alkan C, Ventura M et al.** (2011) *Alu* repeat discovery and characterization within human genomes. *Genome Res.* **21**, 840–849.

24. **International HapMap Consortium** (2005) A haplotype map of the human genome. *Nature* **437**, 1299–1320.

25. **International Human Genome Sequencing Consortium** (2001) Initial sequencing and analysis of the human genome. *Nature* **409**, 860–921.

26. **International SNP Map Working Group** (2001) A map of human genome sequence variation containing 1.42 million single nucleotide polymorphisms. *Nature* **409**, 928–933.

27. **Kayser M, Kittler R, Erler A et al.** (2004) A comprehensive survey of human Y-chromosomal microsatellites. *Am. J. Hum. Genet.* **74**, 1183–1197.

28. **Keinan A, Mullikin JC, Patterson N & Reich D** (2007) Measurement of the human allele frequency spectrum demonstrates greater genetic drift in East Asians than in Europeans. *Nat. Genet.* **39**, 1251–1255.

29. **Keller A, Graefen A, Ball M et al.** (2012) New insights into the Tyrolean Iceman's origin and phenotype as inferred by whole-genome sequencing. *Nat. Commun.* **3**, 698.

30. **Krings M, Stone A, Schmitz RW et al.** (1997) Neandertal DNA sequences and the origin of modern humans. *Cell* **90**, 19–30.

31. **Landsteiner K** (1900) Zur Kenntnis der antifermentativen, lytisichen und agglutinierenden Wirkungen des Blutserums und der Lymphe. *Zbl. Bakt. I. Abt.* **27**, 357–362.

32. **Marchini J & Howie B** (2010) Genotype imputation for genome-wide association studies. *Nat. Rev. Genet.* **11**, 499–511.

33. **Metzker ML** (2010) Sequencing technologies—the next generation. *Nat. Rev. Genet.* **11**, 31–46.

34. **Mills RE, Walter K, Stewart C et al.** (2011) Mapping copy number variation by population-scale genome sequencing. *Nature* **470**, 59–65.

35. **Mullis KB & Faloona FA** (1987) Specific synthesis of DNA *in vitro* via a polymerase-catalyzed chain-reaction. *Methods Enzymol.* **155**, 335–350.

36. **Ng SB, Buckingham KJ, Lee C et al.** (2010) Exome sequencing identifies the cause of a Mendelian disorder. *Nat. Genet.* **42**, 30–35.

37. **Nielsen R, Paul JS, Albrechtsen A & Song YS** (2011) Genotype and SNP calling from next-generation sequencing data. *Nat. Rev. Genet.* **12**, 443–451.

38. **Pagani L, Ayub Q, Macarthur DG et al.** (2012) High altitude adaptation in Daghestani populations from the Caucasus. *Hum. Genet.* **131**, 423–433.

39. **Perry GH, Dominy NJ, Claw KG, et al.** (2007) Diet and the evolution of human amylase gene copy number variation. *Nat. Genet.* **39**, 1256–1260.

40. **Rajeevan H, Soundararajan U, Kidd JR et al.** (2012) ALFRED: an allele frequency resource for research and teaching. *Nucleic Acids Res.* **40**, D1010–1015.

41. **Rasmussen M, Li Y, Lindgreen S et al.** (2010) Ancient human genome sequence of an extinct Palaeo-Eskimo. *Nature* **463**, 757–762.

42. **Redon R, Ishikawa S, Fitch KR et al.** (2006) Global variation in copy number in the human genome. *Nature* **444**, 444–454.

43. **Roach JC, Glusman G, Smit AF et al.** (2010) Analysis of genetic inheritance in a family quartet by whole-genome sequencing. *Science* **328**, 636–639.

44. **Rollo F, Ubaldi M, Ermini L & Marota I** (2002) Ötzi's last meals: DNA analysis of the intestinal content of the Neolithic glacier mummy from the Alps. *Proc. Natl Acad. Sci. USA* **99**, 12594–12599.

45. **Royle NJ, Armour JAL, Crosier M & Jeffreys AJ** (1993) Abnormal segregation of alleles in CEPH pedigree DNAs arising from allele loss in lymphoblastoid DNA. *Genomics* **15**, 119–122.

46. **Saiki RK, Scharf S, Faloona F et al.** (1985) Enzymatic amplification of β-globin genomic sequences and restriction site analysis for diagnosis of sickle-cell anemia. *Science* **230**, 1350–1354.

47. **Southern EM** (1975) Detection of specific sequences among DNA fragments separated by gel electrophoresis. *J. Mol. Biol.* **98**, 503–517.

48. **Stephens M, Smith NJ & Donnelly P** (2001) A new statistical method for haplotype reconstruction from population data. *Am. J. Hum. Genet.* **68**, 978–989.

49. **Stoneking M & Krause J** (2011) Learning about human population history from ancient and modern genomes. *Nat. Rev. Genet.* **12**, 603–614.

50. **Sudmant PH, Kitzman JO, Antonacci F, et al.** (2010) Diversity of human copy number variation and multicopy genes. *Science* **330**, 641–646.

SECTION 2

HOW DO WE INTERPRET GENETIC VARIATION?

Insights into the human past can be generated by analyzing modern genetic diversity because genetic variation has not arisen independently in each individual, but has a shared history that has been shaped by different processes, such as natural selection and population expansion. This section introduces the fundamental concepts of population genetics, molecular evolution, and phylogenetics that allow us to infer the history of genetic variation, and the processes responsible for the observed pattern of genetic variation. These concepts are central to the rest of this book.

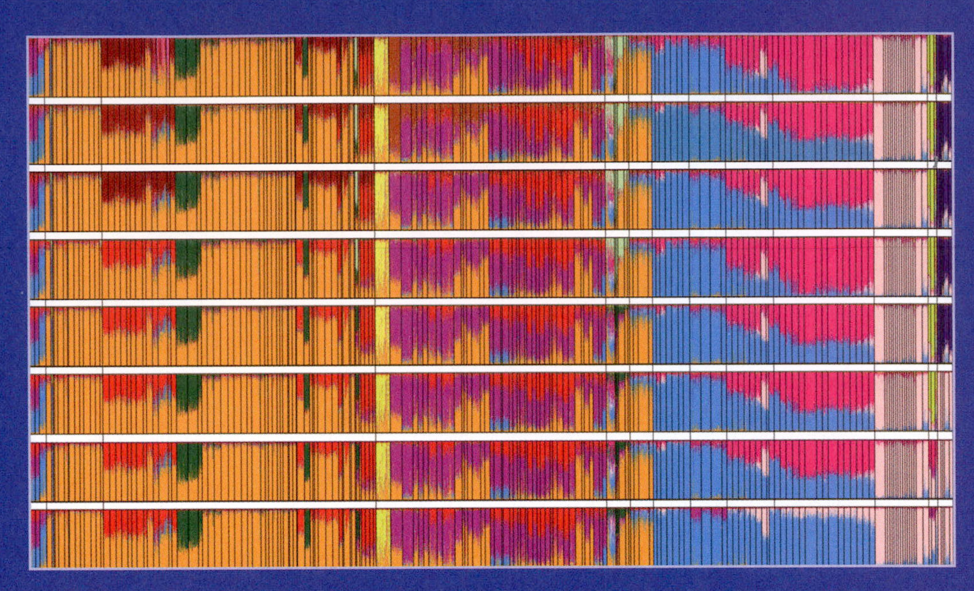

From Tishkoff et al. (2009) *Science* 324,1035. With permission from AAAS.

CHAPTER 5
PROCESSES SHAPING DIVERSITY

This chapter demonstrates that genetic diversity within a species is shaped by a number of different processes. These include mutation and recombination, which generate new genetic variation; genetic drift, which removes variation; and selection, which shapes preexisting variation. The interplay of these different forces is complex and requires us to develop mathematical models that describe simplified but fundamental features of how populations and molecules change over time.

CHAPTER 6
MAKING INFERENCES FROM DIVERSITY

The knowledge gained from modeling these different processes allows us to extract information about the past from genetic diversity data. In this chapter we describe methods of analyzing genetic diversity that allow us to investigate the role that each of these processes has played in shaping variation. Case studies to illustrate the application of these inferential methods are found in subsequent chapters and readers are referred forward to the relevant sections. Similarly, in the later chapters readers are directed back to this section for methodological detail. Some may wish to refer to the contents of Chapters 5 and 6 only as necessary—as if they were a guide-book for the journey through the complexities of human genetic diversity.

PROCESSES SHAPING DIVERSITY

CHAPTER FIVE

In *On the Origin of Species* Charles Darwin was primarily interested in the evolution of species over geological time. Others have been more concerned with processes operating on genetic diversity, within a single species, over a timescale of generations. These two scales of evolutionary change are often referred to as **macro**- and **microevolution**. While it is often assumed that species-level evolution is just an extrapolation of population-level evolution, the reconciliation of these two fundamental evolutionary levels is by no means complete. In this chapter, we will show that because microevolutionary processes shape genetic diversity, we can measure them by studying allele frequencies within populations. We will then have enough grounding in population genetic theory to understand its application to human evolutionary studies, particularly in Chapter 6, but also throughout the book. For further details on population genetic theory we recommend a specialist textbook, such as Hartl and Clark's *Principles of Population Genetics*.[16]

5.1 BASIC CONCEPTS IN POPULATION GENETICS

Why do we need evolutionary models?

We study evolutionary processes by considering how allele frequencies within a population change in time and space. By understanding the mechanisms through which evolutionary processes act, we can produce mathematical models that approximate reality. Such models are necessary to understand the subtle interplay between the processes, and allow us to infer past processes from modern diversity. Using mathematical models that represent simplified versions of reality we can estimate parameters from the data, such as population growth rate, the age of an allele, or the migration rate between two populations. Models also allow us to test different hypotheses about the past. Put simply, if the model does not fit the observed data well, at least one of the assumptions underlying the model must be wrong. Alternatively, we can make several models and test which one best fits the observed data: for example, does a prehistoric migration between two ancestral populations, or divergence during a period of isolation, better explain the current patterns of genetic diversity? There are a variety of methods for testing **goodness-of-fit**, some of which are explored in the next chapter (see Box 6.4 for more about likelihood-based methods), where we also give several examples of how real inferences about human evolution can be derived from analysis of data using mathematical models.

One of the strengths of many population genetic models is their generality: they can be applied to data from any species that share broad characteristics. For example, some models applied to humans might be equally applicable to all other species that reproduce sexually and do not self-fertilize. However, models require us to make assumptions that may not be true of all species. This problem drives mathematical models of evolution to become ever more sophisticated, abandoning simplifying assumptions one by one, and introducing new parameters that provide a better fit to biological reality. Nonetheless, even in the data-rich field of modern human genetics, no amount of data can compensate for an inappropriate model.

The concept of a **population** is central. We must define a population before we can measure the frequency of an allele within it. In addition, we are often interested in reconstructing past demographic events, and demography is a property of populations, not of individuals. It is for these reasons that this discipline is known as **population genetics**. Furthermore, many studies of human genetic diversity group individuals from a number of closely situated but distinct locations into a single population, often defined by political boundaries that may be only a few human generations old. An ecological approach to sampling, such as using regular grid squares, is rarely, if ever, adopted for humans (**Section 10.2**). This sampling of groups, rather than of individuals, leads to their being considered as a natural unit of investigation.

One type of model we will encounter is a mathematical approximation of populations, their interactions, and mating structures. When the term "population" is being used it is important to be clear how it was defined and whether it refers to individuals grouped together for the sake of analysis, or an idealized group, assumed to be adhering to the assumptions of a mathematical model (for example, randomly mating). In other words, does the term refer to a practical or theoretical entity?

The other types of mathematical model are those describing the molecular processes of mutation and recombination, which, as we saw in Chapter 3, differ between DNA sequences and genomic regions. These enable us to go beyond allelic definitions and allow us to make the connection between molecular diversity and population processes.

The Hardy–Weinberg equilibrium is a simple model in population genetics

The **Hardy–Weinberg equilibrium** (HWE) model describes the relationship between allele frequencies and genotype frequencies in a randomly mating population. In diploid organisms such as humans, two alleles, A_1 and A_2, at the same locus, with allele frequencies p and q respectively, can be sorted to make three possible genotypes: A_1A_1, A_1A_2, and A_2A_2. If we know the frequency of these two alleles in a population (p and q) we can predict the proportions of the genotypes in the succeeding generation by combining gametes (which contain single alleles) at random—a postulate known as the Hardy–Weinberg principle.[15] Thus the proportion of each genotype in the next generation is:

$$A_1A_1 = p^2,\ A_1A_2 = 2pq, \text{ and } A_2A_2 = q^2$$

If the genotype proportions in the next generation are calculated in this manner, and are found to be indistinguishable from those in the parental generation, then no evolution (defined as a change in allele frequencies) is occurring, and the population is at HWE. At the time of its discovery, the existence of this equilibrium was important as it showed that mating alone need not alter allele frequency.

For us to be able to estimate genotype proportions from one generation to the next in this way, the population must be made up of an infinite number of randomly mating, sexually reproducing diploid organisms. However, for HWE to be observed the idealized population must have certain additional properties, including:

- No selection
- No mutation

- No overlap between generations
- No migration
- No substructure

If the genotype proportions are not in HWE we might reasonably conclude that at least one of these assumptions has been broken.

How do we use the HWE model to test a hypothesis about human evolution? We can test the observed genotype frequency for an allele at a single nucleotide polymorphism (SNP, **Section 3.2**) against that expected from HWE given the allele frequencies deduced from the data. If the observed data do not fit the model well, one of the assumptions (for example, no selection) of the model does not apply to the SNP. Ability to digest **lactose** in milk as an adult is determined by a single SNP in Europeans (**Section 15.6**). Our hypothesis is that the trait, and therefore alleles at the responsible SNP, will be subject to natural selection.

Using genotype frequency data from over 3000 British people,[9] we can calculate allele frequencies for p and q as 0.747 and 0.253 respectively. **Table 5.1** shows that we can calculate the expected genotype frequencies using Hardy–Weinberg proportions, and compare them with the observed genotype frequencies using a goodness-of-fit test (in this case the χ-squared test). Following the calculation in Table 5.1, there is no significant difference between observed and expected genotype frequencies given the HWE model, showing that the assumptions of the HWE model have not been broken. So we would infer that this SNP is not subject to natural selection.

However, we would be wrong; indeed, as shown by other tests (**Section 15.6**), this SNP displays some of the strongest evidence of positive selection for any variant in the genome. So why doesn't testing for departure from HWE detect selection at this SNP? The answer is that this test is very weak, and is poor at rejecting the null hypothesis (no selection). There are two reasons for this:

- Very strong natural selection is required to distort genotype frequencies sufficiently to be detected by goodness-of-fit tests.
- One round of random mating in the absence of natural selection restores genotype frequencies to HWE. Therefore the selective events of the past are very effectively erased and the HWE is capable of detecting selection only in the current generation.

Departures from HWE are generally rare in humans, and would only be observed as a result of selection if extreme differential mortality occurred within a single generation, as in survivors of kuru (**Box 5.1**). More often, they can result from population structure (and hence departure from random mating) generated, for example, by regarding samples from different continents as a single population.

TABLE 5.1:
TESTING OBSERVED GENOTYPE COUNTS AGAINST HARDY–WEINBERG EQUILIBRIUM EXPECTATION

Genotypes	Observed genotype counts	Expected genotype frequencies	Expected genotype counts		$(O-E)^2/E$
TT	1881	0.567	1897.6		0.145
CT	1236	0.378	1264.4		0.639
CC	227	0.064	214.1		0.777
				Sum	1.561
				p (1 df)	0.21

Box 5.1: Kuru disease in the Fore of Papua New Guinea

Hardy–Weinberg equilibrium (HWE) of genotypes is expected in an outbreeding species such as humans, so well-established instances of deviations from HWE, Hardy–Weinberg disequilibrium, are very unusual and particularly interesting. A nonsynonymous SNP (rs1799990) at codon 129 of the human prion protein gene (*PRNP*) encodes either methionine or valine, and heterozygosity confers resistance to the acquired neurodegenerative disease **kuru**[29, 32] (OMIM 245300).

Kuru is caused and transmitted by a **prion** encoded by *PRNP*, and first came to the attention of Western medicine in the 1950s, when the Eastern Highlands of Papua New Guinea came under external administrative control. Inhabitants of this region included the Fore (Figure 1), who had a high incidence of kuru, with a peak mortality per year of around 2% in some villages. It was found that kuru is transmitted by consuming the brains of kuru-infected individuals, and that the Fore routinely ate deceased relatives at mortuary feasts. The men had the first choice of tissues, and left the less attractive brain, enriched for prions, to the women. Kuru was therefore more common among women than men. The practice subsequently stopped, so that young Fore do not engage in it.

Measuring the genotype frequencies of Fore women born before 1950, who had therefore been exposed to kuru-infected brains on multiple occasions yet were still surviving, showed a dramatic increase in frequency of heterozygotes. This increase is not seen in young modern Fore, nor in men born before 1950 who would have been less involved in brain consumption (Table 1). The departure from HWE is due to the selective mortality of *PRNP* homozygotes from kuru.

Figure 1: A group of Fore men.
In the 1950s and 1960s, the kuru epidemic killed a quarter of the female population in the South Fore, with few female survivors of marriageable age in some villages. [From Mathews JD (2008) *Philos. Trans. R. Soc. Lond. B Biol. Sci.* 363, 3679.]

TABLE 1:
***PRNP* CODON 129 GENOTYPES IN SUSCEPTIBILITY-STRATIFIED GROUPS FROM THE EASTERN HIGHLANDS OF PAPUA NEW GUINEA.**

Fore group	Methionine homozygotes	Heterozygotes	Valine homozygotes	Departure from HWE, p value
Women born before 1950	16	86	23	2.1×10^{-5}
Men born before 1960	34	111	60	0.15
Young modern Fore individuals	52	136	94	0.80

Indeed, departures from HWE as a result of biological effects are so rare and subtle that an apparent gross departure from HWE within a single population is routinely used to detect technical errors in genotyping and improve data quality prior to further analyses.[28]

The weakness of HWE as a test for events that alter allele frequency emphasizes the importance of more sophisticated population genetic models. These incorporate information about mutation rate, recombination rate, and population size: processes that are discussed in the rest of this chapter. How we use these improved models to test hypotheses in human evolutionary genetics is the subject of Chapter 6.

5.2 GENERATING DIVERSITY BY MUTATION AND RECOMBINATION

Mutation is the only process generating new alleles: indeed, by definition any change producing a new allele is called a mutation. It provides the raw material on which evolution can act. There are a broad variety of mutational changes, and these occur at widely varying rates (see Chapter 3). Each mutation is a single change occurring in a single cell. Evolutionary consequences follow only from those changes that occur in the germ line, and not those in somatic tissues, because somatic mutations are not heritable. The dynamics of many types of mutations vary between the soma and the germ line. Because of the high fidelity of DNA polymerases and the operation of DNA repair mechanisms, germ-line mutations occur at low rates for individual nucleotides, although (given the size of the human genome) they are inevitable in every generation. Estimates of the human nucleotide mutation rate from different studies are given in Table 6.4, and estimates of the mutation rate of different classes of substitution are given in Table 3.1.

Mutation changes allele frequencies

In the absence of other processes, a particular allele will decrease in frequency, because it will accumulate mutations changing it into different alleles. This phenomenon is known as mutation pressure. By knowing the mutation rate for the whole gene (μ) and the initial allele frequency (p_0), assuming no back mutation, and ignoring stochastic processes, we can calculate this allele's frequency (p_t) t generations later, by:

$$p_t = p_0 e^{-\mu t}$$

At low mutation rates, mutation pressure is a weak force that can only have appreciable impact over long time-scales. After 1000 generations, the wild-type sequence of a gene 1000 bp in size with a per-generation nucleotide mutation rate of 2×10^{-9} will only decrease in frequency from 1.0 to 0.998.

Mutation can be modeled in different ways

The example above introduced the model of a gene in which each new mutation creates a new allele; in other words we discounted the possibility of **back mutations** and **recurrent mutations**. This is known as the **infinite alleles model**. If we consider a gene 1000 bp in length then the number of possible SNP alleles is enormous: 4^{1000}. If the 1000-bp sequence has n mutational changes in n different nucleotides then the probability of a back mutation is small: $n/3000$.

However, if we consider the evolution of a polymorphic microsatellite, oscillating in size by whole numbers of repeats, we can see that the opportunity for back mutation and recurrent mutation is much greater than for SNPs. Thus the infinite alleles model does not always appear to be a close approximation of biological reality. We need different models for different types of mutation. The stepwise mutation model (SMM) provides a better fit to microsatellite evolution. According to this model, mutations increase and decrease allele length by one unit with equal probability (**Figure 5.1**).

Initially, the SMM considered single-step changes only, but there is good empirical evidence for a lower, but nevertheless appreciable, rate for multiple-step mutations and the model can be adapted to account for these.[10] There are, however, other known aspects of microsatellite evolution not incorporated within the SMM model (see also **Section 3.4**):

- A positive correlation between allele length and mutability
- A lower length threshold under which mutation rate becomes undetectable
- A possible small bias toward expansions of short alleles, resulting in an increase in size of the microsatellite

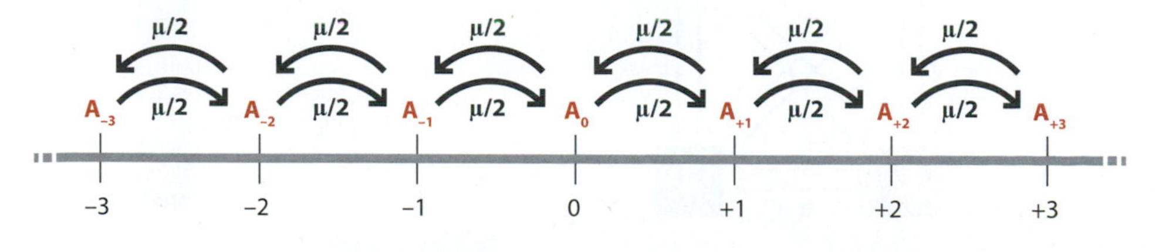

Figure 5.1: The stepwise mutation model. The model considers only single-step mutations, and regards an increase or decrease as equally probable and independent of allele length. The average mutation rate is μ, and any allele mutates to a smaller or larger allele with rate $\mu/2$.

- A possible preference for deletions rather than expansions in longer alleles; together with the previous point, this produces an equilibrium allele length distribution
- Very large expansions in triplet-repeat diseases, and consequent negative selection in these and other examples

Other types of mutations, such as genomic structural variation and GC-rich minisatellite mutations, fit neither of the above models.

If we are interested in aspects of sequence evolution involving the possibility of several changes occurring at the same site, then we need more complex models of mutation—for example, we may need to consider the probability that an A will mutate to a C and then subsequently back to an A again. These models come into play when considering sequence evolution over long time-scales, where back mutations result in the observed **sequence divergence** being an underestimate of the true number of mutational changes. We will come to applications of these models in Chapter 6.

In the simplest model all nucleotide substitutions occur at the same rate, while the most complex model allows a different rate for each nucleotide change. These models can be represented as a substitution scheme, and as a probability matrix, shown in **Figure 5.2**. The simplest example is known as the **Jukes–Cantor model** (JC), and one of the more complex models is the **general reversible model** (**REV**). There are a number of intermediate models that contain some, but not all, of the complexity of the REV model.

The frequency of each nucleotide clearly influences the probability of nucleotide changes averaged over an entire sequence. For example, an A to G transition may have the same rate as a C to T transition, but if there are twice as many As as Cs in a sequence then the probability of an A to G occurring within the sequence as a whole is not the same as that of a C to T. The JC model does not take potential bias in **base composition** into account, but the REV model does.

There are further aspects of sequence evolution known from empirical studies that are not accounted for in these models (**Section 3.2**). First, small (1–20 bp) insertion or deletion alleles (indels) occur on average once every 7.2 kb in the human genome.[34] Ignoring this kind of mutational change can have a large impact; for example, whether or not indels are removed prior

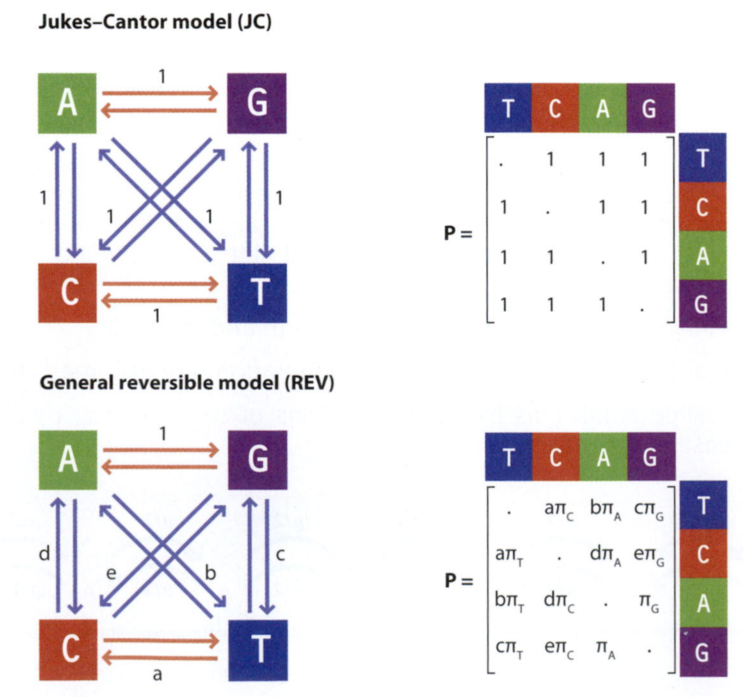

Figure 5.2: Models of sequence evolution.
The probability matrices of two different models of sequence evolution are shown. This matrix contains the relative rates of the different possible base substitutions, which are also shown on a substitution scheme that shows transitions in *red* and transversions in *blue*. The REV model includes the π_i parameter, which is the frequency of that base in the sequence.

to sequence analysis makes a fourfold difference to the apparent sequence divergence between humans and chimpanzees, according to one way of measuring it (**Section 7.3**). The probability of a small indel event occurring is largely determined by the repetitive nature of the surrounding sequence, in a manner that is poorly understood and therefore difficult to model. Such changes are rarely found as polymorphisms in coding regions because they often disrupt the **reading frame**. Second, the phenomenon of the increased mutability of CpG dinucleotides departs significantly from the REV model (**Section 3.2**). The mutability of a nucleotide depends on its neighbor, so that not all Cs and Gs have the same probability of mutating. Both transitions and transversions have increased probability at CpGs.

Models have been developed that can accommodate rate variation among sites within a sequence. These fit such variations in rate to a statistical distribution. Some, like the **gamma distribution**, have a single modal value, whereas other models allow multimodal rate distributions that may provide a better fit to the rate variation among sites, as suggested by the increased mutability of CpGs described above.

Meiotic recombination generates new combinations of alleles

Meiotic recombination occurs as a part of sexual reproduction, and enhances the ability of populations to adapt to their environments by combining advantageous alleles at different loci (**Figure 5.3**). By contrast, asexually reproducing species and nonrecombining portions of the human genome are prone to the operation of **Muller's ratchet**, the slow but inexorable accumulation of deleterious mutations. This process of degeneration may explain the low density of functional genes on the nonrecombining portion of the Y chromosome (**Appendix**).

Recombination generates new combinations of alleles on the same DNA molecule, known as haplotypes (**Section 3.8**), and in this way increases haplotype diversity. Consequently, recombination is capable of breaking up advantageous allelic combinations. This results in the theoretical possibility that outbreeding can result in a drop in fitness known as **outbreeding depression**.

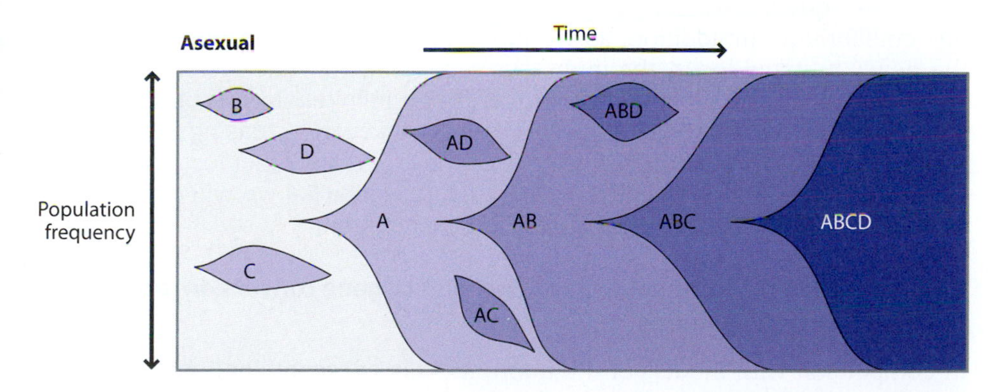

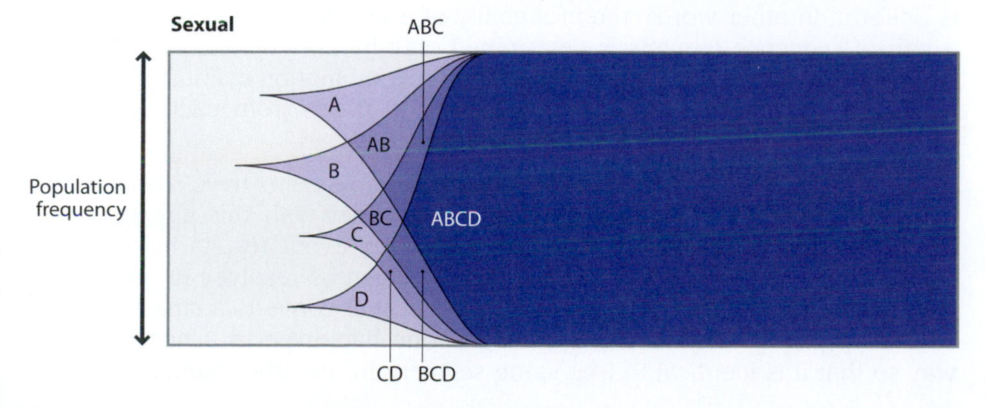

Figure 5.3: The advantage of sexual reproduction.
Four alleles (A–D) all increase the fitness of the organism, with the fittest having all four alleles. Only one allele at a time can prevail in an asexual organism, so they must be combined serially. By comparison, in a sexually reproducing organism these beneficial alleles can be combined in parallel. Thus it takes much less time to assemble the fittest genotype.

While alleles at loci on different chromosomes are randomly segregated during meiosis, alleles at loci closely linked on the same chromosome are not, as recombination between them occurs infrequently. Linked loci share a common evolutionary heritage: selection operating on one locus will affect diversity at the other. For example, an allele that rises to high frequency because of positive selection on a linked locus is said to be hitchhiking (**Section 6.7**). Conversely, negative selection at a locus also reduces diversity at linked loci, albeit at a slow rate, by a process known as **background selection**.[8]

Linkage disequilibrium is a measure of recombination at the population level

Recombination can be studied at the population level by investigating whether specific alleles at different loci are correlated with one another more or less often than would be expected by chance. This nonrandom correlation is known as linkage disequilibrium (LD; **Section 3.8**, Box 3.5).

In an analogous fashion to the reduction in frequency of an allele by mutation pressure, recombination can reduce the frequency of a haplotype. Rather than monitor this process through the decline in frequency of the haplotype itself, we can follow the decay of LD using the statistic D as follows. When a new mutation arises on a chromosome, it is linked to all other variant sites on the same chromosome forming a single haplotype. In other words, it will only be found associated with one allele at each of those other loci, and so is in complete LD with them (D is at its maximal possible value). However, over several generations the frequency of the new mutant allele may grow; if so, recombination events will introduce the new allele onto copies of the chromosome with different alleles at the other variant sites (see the figure in Box 3.5). As a consequence, LD starts to decay. If we know the recombination rate per generation (r) between the newly mutated locus and a given locus, after a certain number of generations (t) we can track the decay of LD over time, by relating the present value of D (D_t) to the initial value of D (D_0) using the equation:

$$D_t = (1 - r)^t \times D_0$$

From this equation we can see that as time increases, D_t tends to zero (linkage equilibrium). In addition, as we move along the chromosome away from the newly mutated locus, the interlocus recombination rate increases, meaning that D_t will tend to zero even sooner. In an infinitely large population, LD would continue to decay over time as a result of an ever-increasing frequency of recombination between the newly mutated locus and any other locus. However, real populations are not infinitely large, and in **Section 5.6** we will explore why an inexorable decay of LD is an unrealistic expectation.

Recombination results in either crossing over or gene conversion, and is not uniform across the genome

In comparison with models of mutation, models of recombination have traditionally been fairly simple. The simplest model is that the rate of recombination is uniform. In other words, the probability of a **crossover** occurring between a pair of sequence variants is determined only by the physical distance that separates them. The products of this type of recombination event are two new haplotypes containing contiguous stretches of alleles from each **ancestral haplotype** (**Figure 5.4a**).

Studies of recombination in humans and model organisms have revealed two biological properties of recombination that conflict with this simple model of recombination. First, not every recombination event results in a crossover (**Section 3.8**). A recombination intermediate can be resolved in one of two ways: a crossover, or a gene conversion event that converts a small segment of DNA (typically less than a kilobase) in one haplotype in a nonreciprocal way so that it is identical to that same segment in the other haplotype. Many

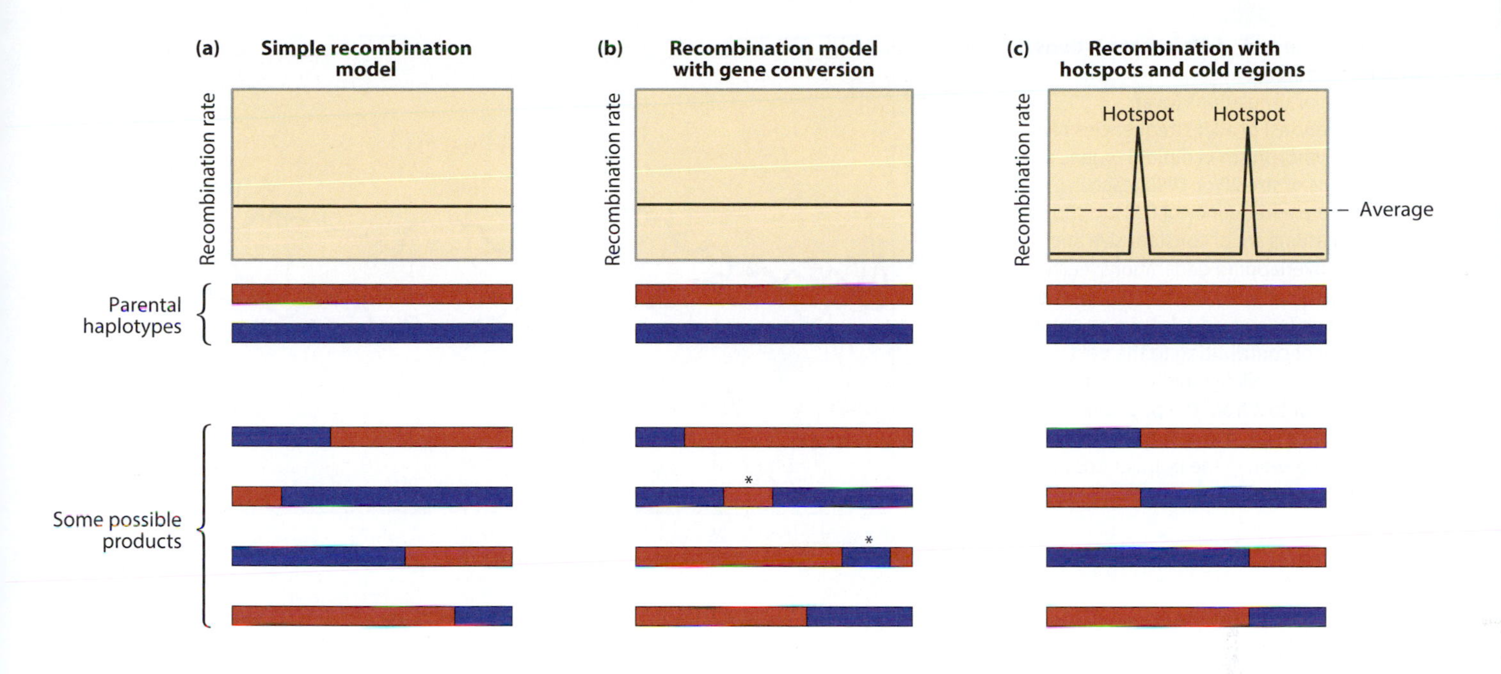

Figure 5.4: **Three models of recombination.**
The recombination rate along the haplotype is shown above the parental haplotype and four typical recombined haplotypes for three models of recombination: (a) simple recombination, (b) recombination with gene conversion, and (c) recombination with hotspots and cold regions. Sections resulting from gene conversion are highlighted by asterisks.

recombination models used on large datasets, for example in the initial HapMap study, use methods that do not distinguish the effects of crossovers and gene conversions. Second, recombination rates are not uniform along a segment of DNA (**Section 3.8**). Crossovers appear to be concentrated in hotspots between which lie recombinationally inert, "cold" regions, and, at larger scales, recombination rates vary along the chromosome, often being relatively low near centromeres and high near telomeres. Hotspot position is different between individuals, and between populations, because of genetic variation in the *PRDM9* gene (**Section 3.8**).

Some models of recombination have been proposed that incorporate either one of these two additional complexities, but few, if any, models have combined the two. Incorporating gene conversion into recombination models requires knowledge of the ratio of gene conversions to crossover events, and the length of the gene-converted segment[46] (Figure 5.4b). Recombination rate heterogeneity can be modeled by considering the size and spacing of recombination hotspots, and the ratio of the recombination rates in hotspots and in cold regions (Figure 5.4c).

5.3 ELIMINATING DIVERSITY BY GENETIC DRIFT

No population is infinitely large, as is assumed by the Hardy–Weinberg theorem. Each generation represents a finite sample from the previous one, and variation in allele frequency between generations occurs through the stochastic process of sampling. This source of variation is known as **random genetic drift**.[44]

Intuitively, we might expect that the magnitude of genetic drift relates to the size of the population being sampled, and this can be shown to be true. **Figure 5.5** illustrates the change in allele frequency over 100 generations in simulated populations, starting with an initial allele frequency of 0.5. The allele rapidly becomes either fixed (100% frequency) or lost from the populations of constant size 20, whereas both alleles persist in the populations of constant size 1000, with more subtle variations in frequency. As genetic drift is a random process, it is impossible to predict which allele survives. A model that describes genetic drift in a finite population in combination with the other assumptions of the HWE (**Section 5.1**) is known as the **Wright–Fisher model**. This model is fundamental to many aspects of population genetics.

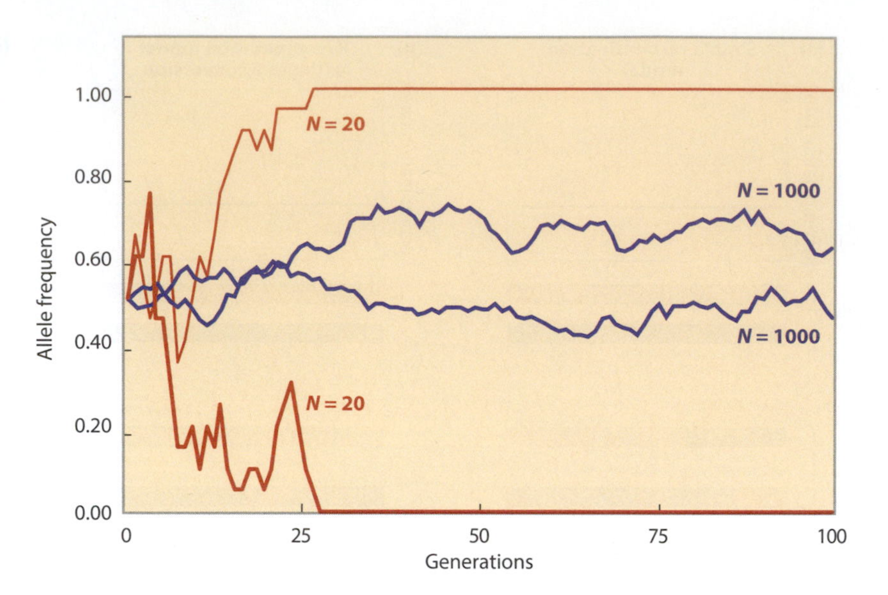

Figure 5.5: Genetic drift in populations of different sizes.
The results of simulations (over 100 generations) of allele frequencies of a binary polymorphism in diploid organism populations of size 20 or 1000 (each starting from a frequency of 0.5) are shown. The populations are of constant size and have non-overlapping generations; each generation is sampled randomly from the previous one (so individuals have an equal probability of contributing to the next generation). The allele rapidly becomes either fixed or lost from the populations of constant size 20, whereas more subtle variations are seen in the populations of constant size 1000.

The effective population size is a key concept in population genetics

The Wright–Fisher model, like the HWE model, contains many unrealistic assumptions when compared with real populations. First, generations overlap in real human populations; second, populations are rarely constant in size; and third, large populations do not exhibit random mating. These three factors differ in importance for any given population. Wright's concept of **effective population size** (N_e) allows us to compare the amount of genetic drift experienced by different populations.[7] N_e for any population represents the size of an idealized Wright–Fisher population that experiences the same amount of genetic drift as the one under study. It measures the magnitude of genetic drift: the smaller N_e, the greater the drift. We can understand the impact of different properties of real populations on genetic drift through the changes they cause in this value.

There are, in fact, two genetic ways of defining effective population sizes: one is based on the sampling variance of allele frequencies (that is, how an allele's frequency might vary from one generation to the next), and the other utilizes the concept of inbreeding (that is, the probability that the two alleles within an individual are identical by descent from a common ancestor). Both of these properties of a finite population depend on the size of that population. There also can be nongenetic definitions, such as the number of breeding individuals inferred from demographic studies. For the sake of simplicity in this chapter we treat these definitions interchangeably, but the reader should be aware that while under most simple population scenarios these definitions of effective population size give identical values for N_e, in more complex situations this is not always the case.

It is not easy to relate the effective population size (N_e) to the **census size** of a population (N), as there are many parameters that can affect this relationship, only some of which are relevant to humans. These are discussed later in this section. N_e is almost always substantially less than the actual population size. For example, the introduction of overlapping generations alone into the population model[11] reduces N_e to 25–75% of N. It is also important to remember that there is a distinction between **long-term effective population size** and recent effective population size. In descriptions of genetic diversity, in most genetic literature, and in this book, the value normally refers to long-term N_e. Recent effective population size, in humans, can be quite distinct from the pattern predicted from genetic diversity data, and reflects the very recent exponential expansion in human census size. We will discuss how human N_e can be estimated in **Section 6.6**, and some estimates of human N_e are given in Table 6.4.

TABLE 5.2:
RELATIVE EFFECTIVE POPULATION SIZES FOR DIFFERENT CHROMOSOMES

	Y chromosome	X chromosome	Autosome
Wright–Fisher population	1/4	3/4	1
Extreme male reproductive variance	1/8	9/8	1

Different parts of the genome have different effective population sizes

Up to this point, effective population size has been considered at the level of individuals; however, not all genomic loci are equally represented in all individuals. If we consider a single mating couple as a microcosm of a species with equal sex ratios, they have between them four copies of each autosome, three copies of the X chromosome, two copies of mitochondrial DNA (mtDNA), only one of which will be inherited by the succeeding generation, and a single Y chromosome. Thus, given a 1:1 sex ratio, the effective population size of the Y chromosome and mtDNA will be only a quarter that of the autosomes, and a third that of the X chromosome. This assumes that the reproductive variances of males and females are equal, as in the Wright–Fisher model.

If, however, we also take into account the differences in effective population sizes between the sexes the relationships for the different loci become more complex. As we saw in Chapter 2, the Y chromosome is inherited paternally and mtDNA is inherited maternally, while the X chromosome is inherited twice as often from females as it is from males. The higher reproductive variance (that is, variation in number of offspring) of males than females reduces the N_e of the Y chromosome relative to that of mtDNA, the X chromosome, and the autosomes, and increases the N_e of mtDNA and the X chromosome relative to that of the autosomes. In cases of extreme male reproductive variance it is possible that the N_e of the X chromosome may exceed that of the autosomes, up to a limit of 9/8 of autosomal N_e (**Table 5.2**). In such extreme cases the N_e of the Y chromosome approaches its lower limit of 1/8 that of the autosomes.[6] Such considerations may partially explain why the Y chromosome exhibits such a high degree of population differentiation (**Appendix**). However, discrepancies in generation times between the sexes also cause their effective population sizes to differ.[31] The sex with the shorter generation time will experience more genetic drift (all other factors being equal) as a result of more frequent episodes of sampling a new generation from the previous one. In humans, females appear to have the shorter generation time (**Section 6.6**), which should lower the N_e of mtDNA relative to biparentally and paternally inherited loci. The relative importance of these opposing factors may differ from population to population. For example, analysis of detailed Icelandic genealogies indicates that in the last few centuries generation-time discrepancies between the sexes have outweighed any differences in reproductive variance, with the consequence that the effective population size of Icelandic mtDNA is less than that of the Y chromosome.[18]

Genetic drift causes the fixation and elimination of new alleles

The concept of effective population size allows us to calculate the probability and rate of **fixation** (rise to 100% frequency) for a new allele in the absence of selection and mutation. Fixation itself is a rare event—a far more likely outcome for a new allele is that it will be lost. As intuition might suggest, with no favoring of either outcome, the fixation probability of an allele in the absence of selection is equal to its frequency in the population; a new allele would have a frequency of $1/2N$. Thus the smaller the population, the greater chance a new allele has

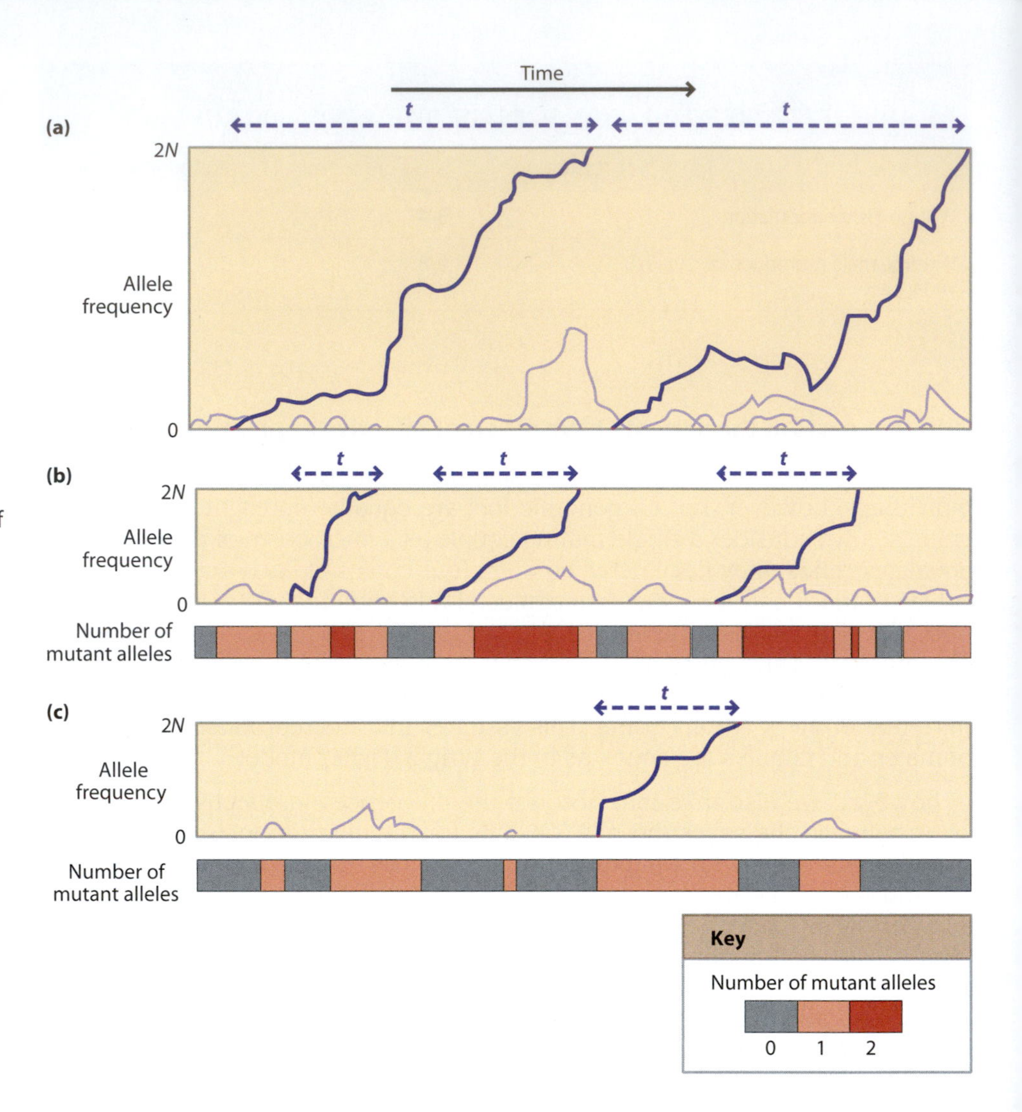

Figure 5.6: Schematic view of the fixation of new alleles in three different populations.
The change in allele frequency over time of new mutations in three different populations is shown. New alleles that arise and are then fixed are shown in *blue*, new alleles that are eliminated are shown in *gray*. The time taken for new alleles to fix (t) is longer in the larger population (a) than the smaller population (b). More new alleles are fixed in the smaller population than in the larger population. (c) A population of the same population size as (b), but with a lower mutation rate (μ). The time to fixation in (c) is no different from that in (b), but the time between fixation of new alleles is greater, as is the proportion of time spent with no polymorphism.

of becoming fixed (see **Figure 5.6**). The average time to fixation (t) in generations has been shown to be:

$$t = 4N_e$$

Therefore a new allele in a smaller population will not only have a higher probability of becoming fixed, but it will also be fixed more rapidly than it would in a larger population. Nonetheless, fixation under the influence of drift alone is substantially slower than if selection were acting (**Section 5.6**).

Variation in census population size and reproductive success influence effective population size

Few populations are constant in size for many generations, so what happens to the effective population size during these fluctuations? The long-term N_e is approximately equal to the harmonic mean rather than the arithmetic mean of the population sizes over time (**Figure 5.7**). The harmonic mean is the reciprocal of the mean of the reciprocals:

$$1/N_e = (1/t)\Sigma^{t}_{i=1}(1/N_i) \text{ for } t \text{ generations}$$

In practice, this means that N_e is disproportionately affected by the smaller population sizes. So in the recently expanded human population, the long-term effective population size (and hence the amount of neutral variation) is still largely determined by the smaller ancestral population sizes in our past. Table 6.4 gives estimates of N_e in humans, and Figure 12.6a shows estimates of the population growth over the past 100 KY.

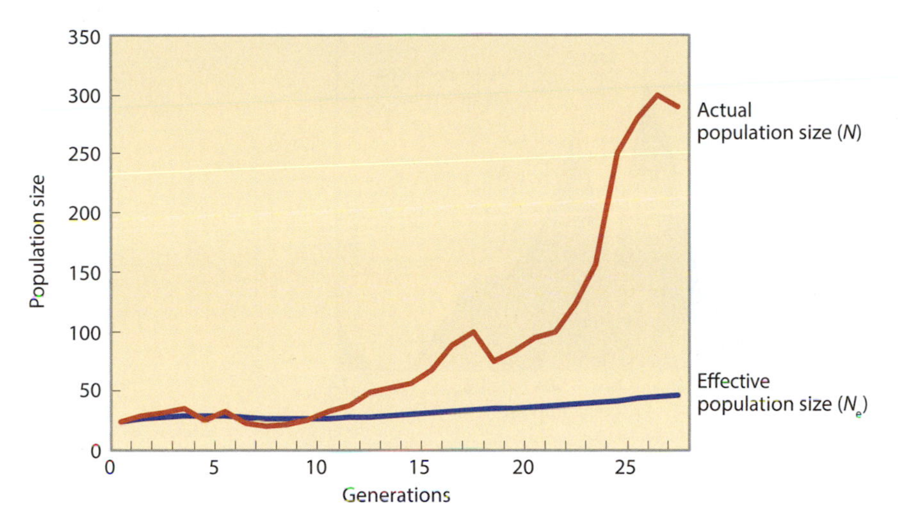

Figure 5.7: N_e in a population of variable size.
The harmonic mean of the census size barely changes despite a recent population expansion.

This dependence of present-day variation on past small population sizes brings us to two important population processes that shape the genetic diversity apparent in many human populations: size-reduction **bottlenecks** and **founder effects**. In many respects the two processes are similar because both involve reduced population size, but the difference between them can be seen in **Figure 5.8**. Founder effects relate to the process of colonization and the genetic separation of a subset of the diversity present within the source population. In contrast, bottlenecks refer to the reduction in size of a single, previously larger, population and a loss of prior diversity.

The Wright–Fisher model assumes that all parents have an equal chance of contributing to the next generation. This results in a **Poisson distribution** of numbers of offspring. However, in real human populations there is often substantial variation in the contribution of individuals to the succeeding generation. To put it another way, there is a higher variance in the number of offspring than that expected under a Poisson distribution (where the variance equals the mean). This can be due to social causes, and need not be attributed solely to

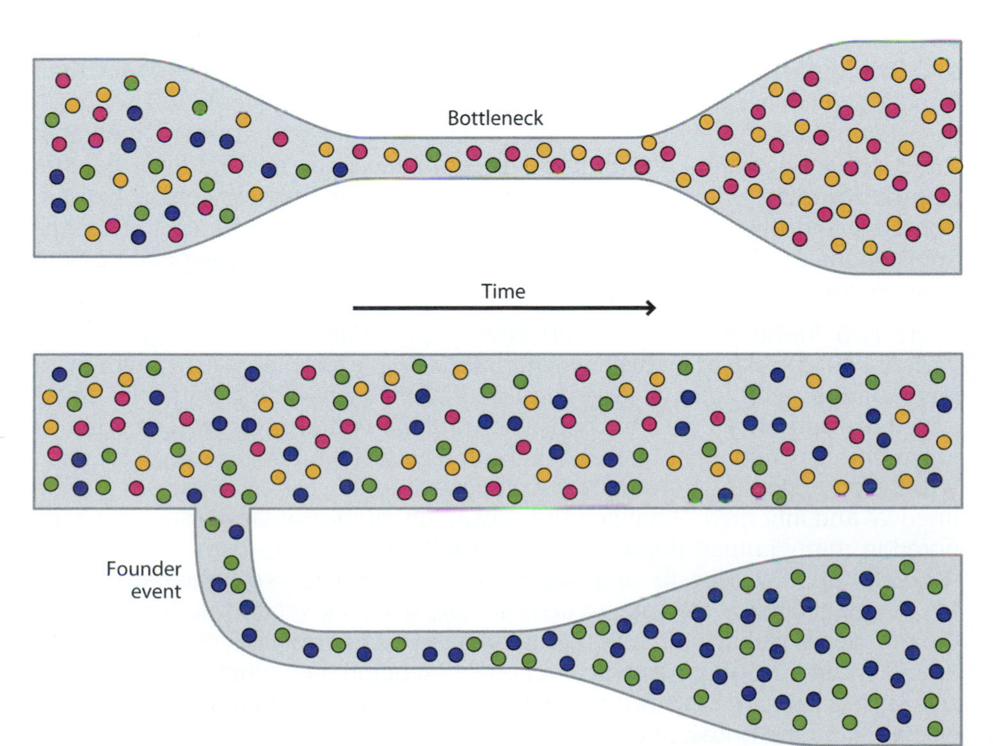

Figure 5.8: Bottlenecks and founder events.
Circles of different colors represent different alleles. Both bottlenecks and founder events result in a loss of allelic diversity.

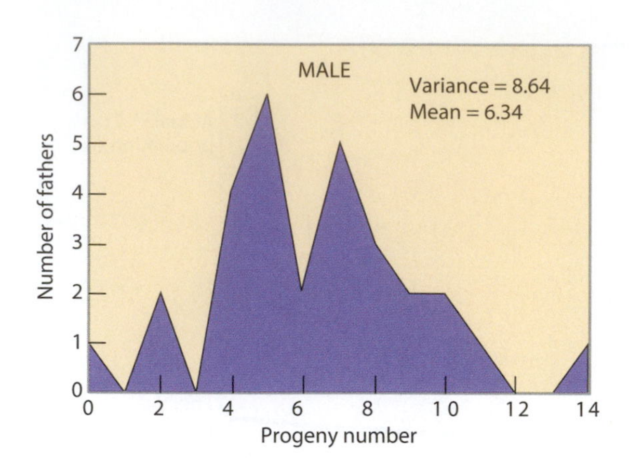

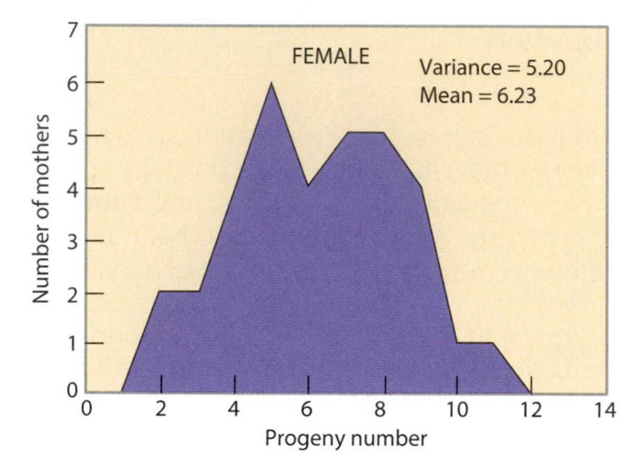

Figure 5.9: Difference in reproductive variance for male and female Aka pygmies.
The two sexes have almost the same mean value, but different variances. [Data from Hewlett BS (1988) in Human Reproductive Behaviour: A Darwinian Perspective. Betzig L, Borgerhoff Mulder M, Turke P (eds). Cambridge University Press.]

differences in **fertility**. The higher the **reproductive variance**, the lower the effective population size, because parental contributions become more and more unequal.

Because reproductive variance often differs between the sexes (see **Figure 5.9**), males and females may have different effective population sizes. Most anthropological studies show males to have higher reproductive variance than females, which is expected to result in a lower male effective population size. For example, using demographic data from Inuit hunter-gatherers of Greenland, the effective population size of males is estimated to be half the male **census population size**, whilst for females it is 70–90% of the female census population size, and this is due to a higher male reproductive variance.[31] This has implications for the effective population sizes of portions of the genome with different inheritance patterns.

There is a further reduction in effective population size when reproductive variance is correlated between generations, for example, when children of large families tend to have large families of their own. Biologically, this inheritance of **fecundity** could happen when a gene conferring greater fertility is polymorphic within a population. Alternatively, social mechanisms of inherited fertility may operate in structured societies where access to resources is both unequal and inherited. Whatever the cause, inheritance of family size has been noted in many human populations, from different types of demographic data (**Box 5.2**). In 1932, Huestis and Maxwell used completed questionnaires from University of Oregon students to demonstrate a significant correlation between the number of siblings of their parents and the number of children of their parents.[22] Alternatively, genealogical records can detail past inheritance of family size, as has been demonstrated with the Saguenay-Lac Saint Jean population in Quebec[2] and the British nobility.[39]

Population subdivision can influence effective population size

Previously we have considered only randomly mating populations; however, most human populations are not so homogeneous. In one respect, all human mating is nonrandom because it usually involves a conscious choice, but in our present context "random" means only that mating is random with respect to the genetic make-up of each individual. A population may be nonrandomly mating because it consists of smaller, partially isolated **subpopulations**, also known as **demes**. Alternatively, nonrandom mating may also occur because mate choice is not blind to genetic relatedness.

Population subdivision is often modeled in terms of a **metapopulation** that comprises partially isolated subpopulations. This isolation eventually leads to partial **genetic differentiation** as genetic drift operates independently within each subpopulation. Members of the same subpopulation are therefore more closely related, on average, than are members of different subpopulations. Depending on the nature of the **population structure**, the effective population size of the metapopulation can be increased or decreased relative to a randomly mating population of the same size. If there are substantial levels of extinction and recolonization of subpopulations then the effective population size of the metapopulation can be dramatically reduced relative to the census size.

If subpopulations are not completely isolated, then the migration of individuals between them results in gene flow, reducing differentiation. So to understand the impact of population subdivision on genetic drift we must model: (1) the number, size, and spatial arrangement of the subpopulations; and (2) gene flow by migration. These models are considered in greater depth in **Section 5.5**. One aspect shared by all these models is the specification of a measure of population structure that is used to estimate parameters such as the rate of gene flow,

Box 5.2: The F_{ST} statistic

F_{ST} is a statistic that was developed independently by Sewall Wright and Gustave Malécot in the 1940s and 1950s, and is possibly the most widely used statistic in population genetics. It is in fact one of a family of statistics called the **fixation indices** that measure the deviation of observed heterozygote frequencies from those expected under Hardy–Weinberg theorem.[20, 45]

F_{ST} measures the apportionment of genetic variation between subpopulations; in other words, it compares the genetic diversity found within subpopulations (the "S" of the subscript) to the genetic diversity of the total population (the "T" of the subscript). It can also be regarded as measuring the proportion of genetic diversity due to allele frequency differences among subpopulations.

F_{ST} varies between 0 and 1, can be defined in a number of different ways, and can be estimated from genetic diversity data by a variety of methods, most commonly:

$$F_{ST} = \frac{(H_T - H_S)}{H_T}$$

where H_T is the expected heterozygosity of the entire population and H_S is the mean expected heterozygosity across subpopulations.

For use as a genetic distance, F_{ST} can be formulated to compare two populations (known as pairwise F_{ST}) and can be defined as:

$$F_{ST} = \frac{V_p}{p(1-p)}$$

where p and V_p are the mean and variance of gene frequencies between the two populations respectively.

Wright suggested that qualitative guidelines shown in Table 1 could be used to interpret F_{ST} values. Using these guidelines, humans, with a genomewide average F_{ST} value of around 0.05, show little to moderate genetic variation.

TABLE 1: SEWALL WRIGHT'S QUALITATIVE GUIDELINES FOR INTERPRETING F_{ST}

F_{ST} values	Level of genetic differentiation
Less than 0.05	little
Between 0.05 and 0.15	moderate
Between 0.15 and 0.25	great
Greater than 0.25	very great

or the effective population size of the metapopulation. Some have argued that such estimates have little, if any, relevance to reality, because all current models are oversimplistic, and contain important assumptions that are violated by all human populations.

Perhaps the best-known measure of population structure is F_{ST} (Box 5.2). When gene flow is high and there is little differentiation between subpopulations F_{ST} is close to zero. When subpopulations are highly differentiated, then genetic diversity of the metapopulation is much greater than in any subpopulation and F_{ST} is close to 1. F_{ST} values between pairs of populations can also be considered to be a measure of genetic distance between them.

Subpopulation divergence results in an excess of **homozygotes** in the metapopulation and a corresponding deficiency of **heterozygotes**, a phenomenon known as the **Wahlund effect**.

Mate choice can influence effective population size

Nonrandom mating also results from individuals choosing their mates via some assessment of their mutual similarity. If individuals choose partners on the basis of shared phenotypic characteristics such as socioeconomic status, IQ, or skin color, this is known as **assortative mating**.

Assortative mating can be based on physical, **psychometric**, or cultural traits. Physical traits that are selected include similar attractiveness, age, and ethnicity. The last can be demonstrated by the statistical analysis of census data, whereas the first is trickier as it relies upon a subjective notion of beauty. Nevertheless, it has been argued that individuals do choose to mate with those of a similar level of attractiveness to themselves. Psychometric traits thought to have been selected during assortative mating include IQ, and the presence of a mental disorder. Other relevant traits include religion, deafness, and educational qualifications.

Disassortative mating (or negative assortative mating) results when partners are chosen on the basis of their phenotypic differences rather than similarities. Disassortative mating at a locus generates a greater heterozygote frequency, and assortative mating a lower heterozygote frequency, than that expected under random mating. Assortative mating augments genetic drift by decreasing the effective population size, whereas the opposite is true for disassortative mating. One of the best-known traits proposed as a candidate for disassortative mating is resistance to infectious diseases. Much of an individual's resistance is encoded in the MHC region of the genome (**Box 5.3**). This region contains several closely linked and highly polymorphic genes that are involved in immunological recognition and response. Disassortative mating is one of a number of plausible explanations for the surprisingly high degree of polymorphism in the MHC region.

Inbreeding and **outbreeding** occur when mating happens between individuals who are respectively more, or less, related than would be expected by random mating. The more closely related the two partners are, the higher the chance that they will pass on the same deleterious recessive allele to their offspring. Thus there is a fitness cost to inbreeding known as **inbreeding depression**. The degree of inbreeding, or **consanguinity** (from the Latin "of the same blood"), is measured by the **coefficient of kinship** (f), which is the probability that two alleles from two different individuals are identical by descent. An alternative measure is the **coefficient of relatedness**, which is simply equal to $2f$. Incest represents the extreme of inbreeding, and usually refers to sexual intercourse between close relatives. The definition of close relatives is usually regulated by religion or the state, with first-cousin marriage ($f = 0.0625$) allowed by many religions but marriage between closer relations generally proscribed. Incest taboos are nearly universal and may represent an adaptive behavioral strategy to minimize inbreeding depression. Nevertheless, there is evidence of institutional incest in certain dynasties: for example, sibling marriage was expected of the ruling Egyptian Pharaohs during the eighteenth and nineteenth dynasties (~1400–1700 years ago).[3]

Current surveys record a significant rate of consanguineous marriages in certain countries: it has been estimated that 10% of marriages in the global population are between partners related as second cousins or closer ($f \geq 0.0156$). Modern studies suggest that consanguineous marriages result in an increase in the female reproductive life span and the consequent higher average number of children, at least for first-cousin marriages, may outweigh the negative effects of inbreeding depression (estimated at about 4% more pre-reproductive deaths in offspring of first-cousin marriages).[4] Given the small size and extensive dispersal of prehistoric hunter-gatherer groups, it may be reasonable to suppose that there were similar levels of inbreeding throughout human evolution, although studies on present-day hunter-gatherer societies suggest social factors may promote breeding between, rather than within, tribes.[19]

Genetic drift influences the disease heritages of isolated populations

Some human populations exhibit high incidences of multiple genetic diseases that are rare in surrounding populations. They appear to have a distinct heritage of genetic disease. These populations also show high frequencies of usually rare, but neutral, alleles. Often these groups are known to have undergone demographic processes that have resulted in small effective population sizes: for example, founder effects [for example, Finns (**Section 16.2**), Afrikaners in South Africa] and **endogamy** (within-group marriage, for example, Roma).

However, in some cases, where there is good evidence that a disease allele has been imported into a population by a single founder, it appears that insufficient time has elapsed for genetic drift alone to account for the high frequency. An example is the increase in carrier frequency for the disorder of amino acid metabolism, tyrosinemia I (OMIM 276700), from 1/5000 to 1/22 within 12 generations in the Saguenay-Lac Saint Jean population of Quebec. In such cases it is tempting to invoke some form of selective process. However, a more sophisticated appreciation of the demographic factors underpinning genetic drift often provides an adequate explanation. In the Quebec case, inheritance of family size, which was well documented in the genealogical records for this population, increases genetic drift sufficiently to account for the observed carrier frequencies.

5.4 THE EFFECT OF SELECTION ON DIVERSITY

Natural selection, as defined by Darwin and elaborated by Fisher, is the differential reproduction of individuals of different genotypes in sequential generations. Genotypic variation produces individuals with varying capacities to survive and reproduce in different environments. Selection can occur at any stage on the long journey from the formation of a genotype at fertilization to the bearer of that genotype generating their own viable progeny, including:

- Survival into reproductive age—viability and mortality
- Success in attracting a mate—sexual selection
- Ability to fertilize—fertility and gamete selection (**meiotic drive**)
- Number of progeny—fecundity

The sum of these is the ability of an individual genotype to survive and reproduce, its **fitness**, which is partly dependent on the environment. The important factor is the relative fitness of a genotype compared with other genotypes competing for the same resources. Relative fitness is measured by a **selection coefficient** (s), which compares a genotype with the fittest genotype in the population. A selection coefficient of 0.1 represents a 10% decrease in fitness compared with the fittest genotype.

Box 5.3: The major histocompatibility complex

What is it?

When a tissue is transferred from one individual to another, it may be rejected or accepted by the host immune system; this is known as histocompatibility. Although a number of loci throughout the genome are involved in histocompatibility, in humans the major determinants are found in a large gene cluster on chromosome 6 known as the **major histocompatibility complex** (**MHC**). The different MHC-encoded proteins that can be recognized by the immune system are cell-surface proteins each known as a human leukocyte antigen (HLA). HLAs include proteins that are expressed on all nucleated cells.

How is the locus arranged?

Due to its medical importance, the gene-dense MHC locus was one of the first large regions to be sequenced during the Human Genome Project. The 3.6-Mb locus is divided into three regions, called classes (see Figure 1). Ancient gene duplication events have generated several expressed HLA genes and many pseudogenes within the class I and II regions.[21] These HLA genes are involved in the development of **adaptive immunity** through the presentation of bacterial and viral **antigens** to T lymphocytes. Different alleles at each individual gene vary in their ability to present antigens from different pathogens.

How diverse is it?

The most unusual feature of the MHC is the huge amount of variation contained within it. The HLA Sequence Database (http://www.ebi.ac.uk/imgt/hla/) currently contains over 8000 allele sequences from 35 different loci within the MHC (see Figure 1; and Figure 3.14).

As well as the sheer number of alleles, the differences between them are often many times greater than at other loci (many alleles at the same HLA locus differ by 5–17% of all nucleotides, whereas most alleles at other loci differ by less than 1%), indicating that their common ancestry is ancient. Many of these alleles are so old that they pre-date the human–chimpanzee split: that is, a human allele may be more closely related to a chimpanzee allele than to an alternative human allele, a characteristic known as **trans-species polymorphism**. In addition, there is also variation in the **copy number** of HLA-DRB genes.

Why is it so diverse?

High MHC diversity within modern humans could be explained by selection for diversity or by an elevated mutation rate. However, trans-species polymorphism can only be explained by the operation of selection in preventing the fixation of alleles over time. Three alternative selection pressures have been proposed, although they are not mutually exclusive:

- Heterozygote advantage—individuals with heterozygous MHC haplotypes are better able to resist infectious disease as a result of having a broader spectrum of antigen-binding specificities.
- Frequency-dependent selection—low-frequency alleles are favored if pathogens have evolved to evade immune detection in individuals carrying the higher-frequency alleles.
- Disassortative mating—mate preference for dissimilar MHC haplotypes would maintain a highly diverse MHC in a population.

A role for selection is supported by the concentration of variation in the exons coding for the antigen-binding groove of the protein,[23] presence of very old haplotypes,[41] and peaks of noncoding variation around variable genes, suggesting hitchhiking along with the balancing selection operating at these loci. Evidence for disassortative mating is contradictory; in support of this idea women show some evidence of preferring body odor of MHC-dissimilar men, and genetic evidence from a moderately inbred Anabaptist group, the Hutterites, suggests that married couples are more likely than by chance to be MHC-dissimilar. However, some studies have failed to support the evidence for a female body-odor preference, and studies of other, more outbred, populations find no evidence of disassortative mating at the MHC.[17]

What role does recombination play in generating diversity?

Recombination within heterozygous HLA genes creates new alleles, and interallelic and intergenic gene conversion generates additional variation. The MHC exhibits a high degree of linkage disequilibrium (LD) that most likely results from the localization of recombination events to certain hotspots within the locus, between which lie long regions of low recombination. As a consequence, linked MHC genes are frequently co-inherited in haplotype blocks (Section 3.8), making it easy to identify disease-related haplotypes, but difficult to locate disease-related alleles to a single gene. For example, the tightly linked class II loci, DRB1, DQA1, and DQB1, are often found to be in complete LD.

But what about selection?

Through linkage analysis and **association studies**, numerous MHC haplotypes have been associated with susceptibility to, and protection against, different diseases.

These include infectious diseases (for example, the protective effect of HLA-B*53:01 against malaria), autoimmune disorders [for example, susceptibility to multiple sclerosis (OMIM 126200) conferred by HLA-DRB1*15:01], and other diseases [for example, HLA-DQB1*06:02 predisposes to narcolepsy (OMIM 605841)]. Particularly interesting is the HLA-B*57 allele, which is significantly protective against HIV but is very strongly associated with the inflammatory disease ankylosing spondylitis (OMIM 142830). Because of the medical importance of these associations, the MHC haplotype project has sequenced eight full MHC haplotypes that are commonly associated with type 1 diabetes (OMIM 222100) and multiple sclerosis, to help disentangle these associations and determine their functional basis.

Can HLA variation be used to explore the human past?

The current geographical distribution of HLA alleles is shaped to some degree by events in the human past. High diversity at the protein level facilitated extensive study of these loci prior to the advent of DNA-based methods. However, the association of different MHC haplotypes with many different diseases raises the possibility that the spatial distribution of HLA alleles may be shaped not by population history, but by different selective environments. Selection can be expected to skew the frequencies not only of disease-related alleles, but also of alleles at any linked loci.

How are different alleles named?

The **nomenclature** for the different alleles has been complicated by the use of two different methods to define alleles. Initially, **serological** methods that detect some but not all variation at the protein level were used to identify alleles. More recently, direct analysis of DNA sequences at this locus has revealed that multiple alternative DNA sequences can encode the same serologically defined allele. Thus the nomenclature has evolved to include information on the gene at which the allele is found, the serological allele, and the underlying DNA sequence. For example, the serological allele HLA-A1 (the first allele at the A locus within the HLA) can be encoded by the DNA allele HLA-A***01**:01 or HLA-A***01**:02. The first two numbers (shown here in bold) define the serological allele, and the second two numbers define different nonsynonymous changes that yield different proteins with the same **immunoreactivity**. The colon between these numbers is a convention introduced in 2010 and many reports show allele names without it. A third level of numbers can be added to indicate any synonymous changes that might be present. So, for example, the two alleles that give the same HLA-A*01:01 protein sequence but differ by a mutation that does not cause an amino acid change, are defined as HLA-A*01:01:01 and HLA-A*01:01:02. A fourth level of numbers can be added to indicate any nucleotide differences in the noncoding regions. As an added complication, some of the serological alleles have been given new names for the purposes of the DNA naming system; so HLA-DR17 has become HLA-DRB1*03, of which there are 84 nonsynonymous alleles (HLA-DRB1*03:01 to HLA-DRB1*03:84), some of which have synonymous variants (for example, HLA-DRB1*03:05:01 and HLA-DRB1*03:05:02).

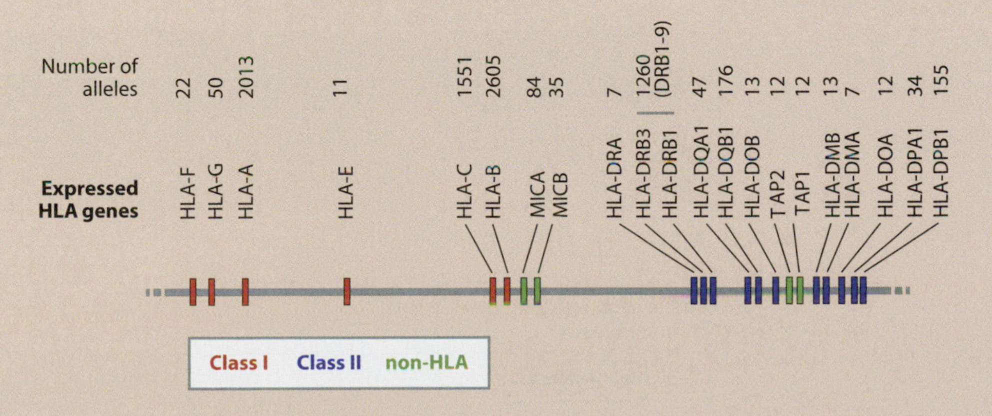

Figure 1: Structure of the MHC region on chromosome 6, showing the location of the genes and the number of alleles reported in the HLA Sequence Database.

Mutations that reduce the fitness of the carrier are subject to **negative selection**, also known as **purifying selection**, whereas mutations that increase fitness undergo positive, or **diversifying selection**. However, to understand the dynamics of selection at diploid loci we must consider the impact of mutants on the fitness of the genotypes, and not on the individual alleles. The two alleles within a diploid genotype can interact to determine the phenotypic fitness of an organism in different ways. This in turn affects the efficiency of natural selection in fixing or eliminating novel alleles. For example, a novel deleterious allele will be eliminated more rapidly from the population if it reduces the fitness of a heterozygote. Alternatively, a new allele may increase the fitness of a heterozygote relative to that of both homozygotes. The two homozygous genotypes may exhibit different reductions in fitness (s_1 and s_2). Such selection is known as **overdominant selection** (also known as **heterozygote advantage**) and creates a **balanced polymorphism**. By contrast, **underdominant selection** operates where new alleles reduce the fitness of the heterozygote alone. Several different selective regimes are summarized in Table 5.3.

Overdominant selection is not the only mechanism by which balanced polymorphisms can be generated, but is one of a number of processes described collectively as balancing selection (Box 6.6). An alternative mechanism is **frequency-dependent selection**, whereby the frequency of a genotype determines its fitness. If a genotype has higher fitness, relative to other genotypes, at low frequencies, but lower fitness at higher frequencies, an intermediate equilibrium value will be reached over time. Box 5.3 describes the MHC region, where genes have been suggested to be under both frequency-dependent and overdominant selection.

Other classic examples of balanced polymorphisms in humans are those that protect against **malaria** when heterozygous but have a reduced fitness compared with wild-type when homozygous, as a result of red blood cell disorders. A number of these types of balanced polymorphisms have arisen in different areas of malarial **endemicity**. The best known is the sickle-cell anemia allele of the β-globin gene, Hb^S (OMIM 603903), which dramatically reduces fitness when homozygous. Malarial endemicity is not spread equally across the world, and as a consequence these balanced polymorphisms exhibit a limited geographical range that closely parallels that of malaria (Figure 5.10, Section 16.4).

Even small selective forces are capable of causing appreciable changes in allele frequencies over many generations. The inter-generational change in allele frequencies can be calculated by incorporating the selection coefficients described in Table 5.3 into the Hardy–Weinberg theorem. Figure 5.11 compares the selection dynamics of a low-frequency advantageous allele under positive and **co-dominant** selection. It can be seen that selection achieves the most

TABLE 5.3: DIFFERENT TYPES OF SELECTION, AND THEIR EFFECTS ON GENOTYPE FITNESS

Type of selection	Genotype fitness		
	A_1A_1	A_1A_2	A_2A_2
Simple negative/positive selection (A_2 is recessive)	1	1	$1 - s$
Simple negative/positive selection (A_2 is dominant)	1	$1 - s$	$1 - s$
Co-dominant selection	1	$1 - s$	$1 - 2s$
Overdominant selection	$1 - s_2$	1	$1 - s_1$
Underdominant selection	1	$1 - s$	1

Note: s = selection coefficient.

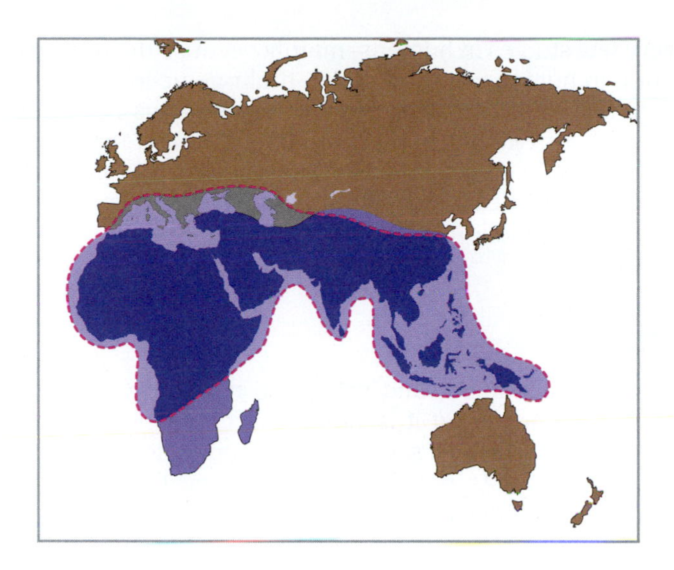

Figure 5.10: The overlapping geographical distributions of red blood cell disorders and malaria. *Blue* indicates regions of current malarial incidence and the red dashed line shows the distribution of red blood cell disorders.

rapid changes in allele frequencies when alleles are at intermediate frequencies. However, in **Section 5.6** we shall see that other processes acting on allele frequencies may outweigh small selective forces.

Mate choice can affect allele frequencies by sexual selection

In cases of limiting numbers of available partners, selection can operate at the level of **mate choice**. For humans, where the levels of investment of males and females in their offspring are unbalanced, the availability of females is the factor limiting male reproduction and therefore females can exercise mate choice. Desirable traits may be those that indicate health, access to resources, and ability or willingness to invest in offspring. An alternative mechanism of sexual selection in response to limited female mating resources is that of competition between males:

> We may conclude that the greater size, strength, courage, pugnacity, and energy in man, in comparison with woman, were acquired during primeval times, and have subsequently been augmented, chiefly through the contests of rival males for the possession of the females. (Charles Darwin, *Descent of Man*, 1871)

If these attractive or competitive traits are to some degree genetically determined, then the loci responsible are said to be under sexual selection. Mate choice can be differentiated from assortative mating on the basis that specific preferences are shared among all members of the same sex.

Darwin invoked sexual selection (Box 15.4) to explain the presence of secondary sexual characteristics among humans. Others have proposed that the human mind itself is largely a result of this selective process.[33] So far, there

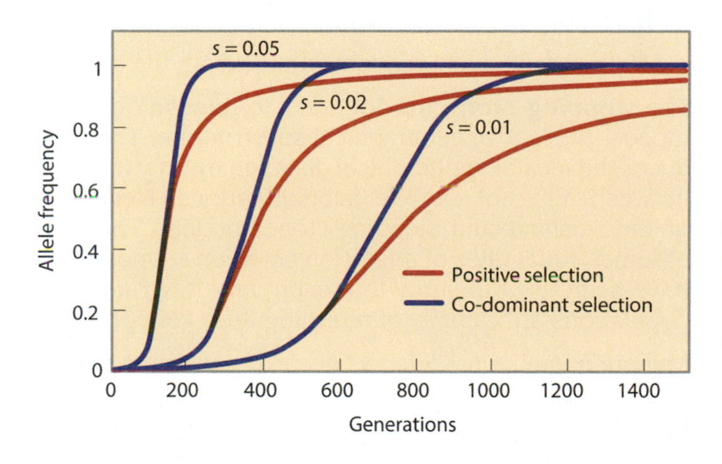

Figure 5.11: Positive and co-dominant selection for an advantageous allele. The selection dynamics of a low-frequency, advantageous allele are compared under positive and co-dominant selection. Selection achieves the most rapid changes in allele frequencies when alleles are at intermediate frequencies.

have been very few studies in humans, mainly because the traits possibly under sexual selection in humans show complex multi-genic inheritance that has yet to be elucidated. Studies of sexual selection in humans have been limited to observing indirect correlations between phenotypes and reproductive success, an example being the higher number of offspring born to taller men.[38] Sexual selection, whether genetically or culturally determined, could be expected to lower the effective population size of the selected sex by increasing the reproductive variance.

5.5 MIGRATION

Unlike genetic drift, mutation, and selection, migration cannot change specieswide allele frequencies, but it is capable of changing allele frequencies in populations. It thus belongs to a second tier of population processes that shape human genetic diversity. As noted previously, gene flow counteracts genetic differentiation and is modeled within the framework of a larger, subdivided, metapopulation.

First, we must be clear on some definitions, because they are often used interchangeably in the literature. Colonization is the process of movement into previously unoccupied land, thus entailing a founder effect. By contrast, **migration** is the movement from one occupied area to another. Gene flow is the *outcome* of a migrant contributing to the next generation in their new location. Thus, to observe gene flow directly we not only need to monitor the movement of migrants but also their reproductive success. Estimates of gene flow have, therefore, relied upon indirect methods that assess allele frequency differences among populations using simplified models.

There are several models of migration

Perhaps the simplest model of gene flow is the ***n*-island model** devised by Sewall Wright. A metapopulation is split into islands of equal size N, which exchange genes at the same rate per generation, m (where m represents a proportion of the population migrating rather than an actual number). Under the assumptions of this model the rate of migrant exchange can be related directly to F_{ST} (Box 5.2), by the equation:

$$F_{ST} = 1/(1 + 4Nm)$$

The assumptions of the n-island model include:

- No geographical substructure apart from the division into islands: all islands are equivalent
- Each population persists indefinitely
- No mutation
- No selection
- Each population has reached equilibrium between mutation and drift
- The migrants are a random sample from the source island

The **stepping-stone model** seeks to remove one obvious flaw of the n-island model—the lack of geographical substructure. The stepping-stone model introduces the idea of geographical distance by only allowing the exchange of genes between adjacent discrete subpopulations. Figure 5.12 shows a comparison of the n-island and stepping-stone models. The stepping-stone model also assumes equal rates of migration between subpopulations. Both kinds of model have been used to show that even very low rates of migration between subpopulations are capable of retarding their genetic differentiation.

Migration can be modeled as occurring within a continuous population, rather than discrete subpopulations, by considering that mating choices are limited by distance, and that these distances are typically less than the overall range

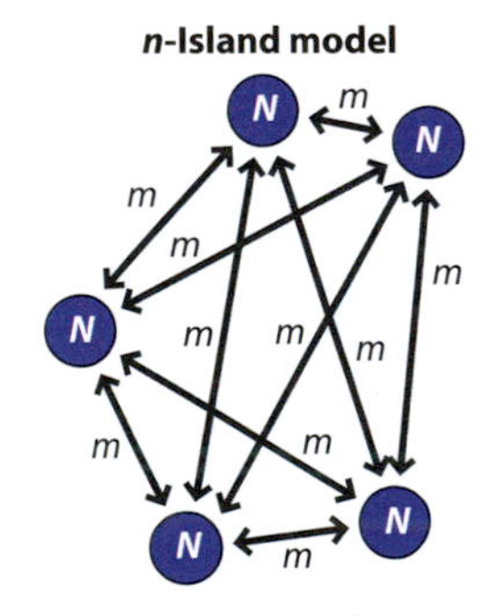

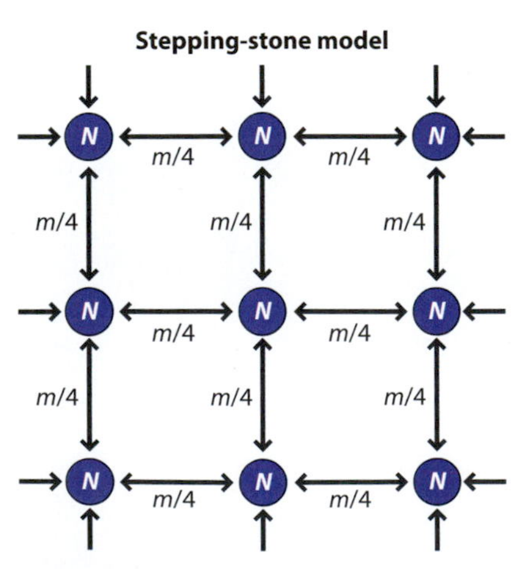

Figure 5.12: The *n*-island and stepping-stone models of gene flow. Each diagram represents one of a family of models: the N-island model, and the two-dimensional stepping-stone model, also known for obvious reasons as the lattice model. N, population size; m, rate of exchange of genes per generation.

of the population. This is the basis for **isolation by distance** (IBD) models. Within such models, genetic similarity develops in neighborhoods as a function of dispersal distances. These can be thought of either as the difference between birthplaces of parent and offspring, or marital distances. Different mathematical functions have been used to relate the decline in frequency of these dispersals to geographical distance. Once the system has reached equilibrium between gene flow and the differentiation caused by genetic drift, genetic similarity declines over distance in a predictable fashion. The stepping-stone model described above is a discontinuous example of IBD.

These migration models are mathematically tractable and can be generalized to many species. However, for many human populations (unlike those of other species) we often have detailed data on parameters such as migration rates, migration distances, and marital distances. The migration matrix model uses this detailed information and thus can incorporate different migration rates and asymmetric migration between subpopulations. In this way, a more complex and realistic relationship between distance and migration is obtained. Nevertheless, it seems unlikely that present-day migration rates have been constant for long enough to allow the system to reach a state of equilibrium, which is required by calculations using such models. The uneven pattern of most human habitation falls between the models of discrete subpopulations and uniform continuity assumed by the stepping-stone and isolation by distance models respectively. Furthermore, migration processes are far more complex than the current models allow. Migration processes often include long-distance movements as well as smaller-scale mating choices. The choice to migrate is taken by individuals on the basis of multiple "push" and "pull" factors, so that migration rates are rarely, if ever, symmetric between two populations. Migrants are seldom a random sample of their source population; they are often age-structured, sex-biased, and related to one another. The latter property of migrants is known as **kin-structured migration** and is well documented both ethnographically[12] and archaeologically.[1] In light of these complications, we should be cautious in attempting to estimate parameters of population structure and be skeptical of their relationship to reality.

There can be sex-specific differences in migration

If we consider possible differences between the sexes in their migration behavior, an intuitive hypothesis on observing the modern world might be that men tend to migrate over longer distances than women: intercontinental migrants tend to be male-biased and recent history documents explorers, traders, and soldiers as being almost exclusively male. Involuntary migration, particularly slavery, is often sex-biased: for example the Atlantic slave trade involved mostly males, while the Indian Ocean/Red Sea slave trade involved mostly females. However, when considering the impact of migration on genetic diversity, we must not only examine long-distance migration patterns but also small-scale local migrations.

Marital residence patterns are critical to investigating local migration patterns. **Patrilocality** describes the phenomenon by which a female from one village, when marrying a man from a different village, takes up residence in the man's village. In contrast, **matrilocality** describes the situation when the husband moves to the wife's village. It has been estimated that roughly 70% of modern societies are patrilocal.[5, 35] In other words, in the majority of societies, mtDNAs are moving between villages each generation, whereas Y chromosomes are staying put. Similarly, the X chromosome is more mobile than the autosomes, as it passes down the female line twice as often as it does down the male line.

How might we determine which of the above current phenomena—the apparent male bias of long-distance migration or the female bias of intergeneration marital movement—has had a greater role in shaping modern genetic diversity?

To resolve this issue, we can compare the geographical patterning of genetic diversity of chromosomes that have different inheritance patterns. Migration

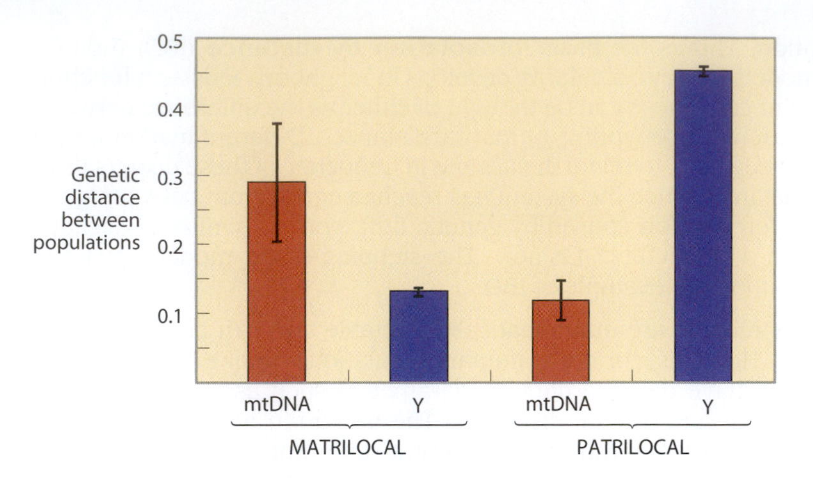

Figure 5.13: Influence of sex-specific migration on genetic diversity. [Data on three matrilocal and three patrilocal groups among the hill tribes of northern Thailand, from Oota H et al. (2001) *Nat. Genet.* 29, 20.]

reduces population differentiation, so by studying the relationship between geographical distances and genetic distances among the same set of populations, we can identify which loci appear to have been experiencing greater gene flow. These comparisons assume that migration rate is the only factor determining population differentiation; however, genetic drift also influences levels of population subdivision, and the effective population sizes of the Y chromosome and mtDNA need not be equal (Table 5.2). Consequently it has been argued that these differences between loci reflect differences in genetic drift as a result of sex differences in reproductive variance, and not migration rates. However, populations with a matrilocal pattern of marital residence show greater mtDNA than Y-chromosomal genetic differentiation (**Figure 5.13**). This suggests that a sex-specific difference in migration rate, and not drift, is responsible for the different patterns of genetic differentiation of mtDNA and the Y chromosome.

The sex-specific migration rate differs dramatically between populations and cultures, with a slight bias toward male migration overall (**Figure 5.14**). This suggests that long-range migration patterns (male-dominated) have contributed most to different patterns of diversity between mtDNA and Y chromosomes, but local effects can outweigh this overall pattern. Indeed, culture has an important effect: in sub-Saharan Africa, hunter-gatherer populations show matrilocality while pastoralist and agricultural populations show patrilocality.[43]

5.6 INTERPLAY AMONG THE DIFFERENT FORCES OF EVOLUTION

Thus far, we have examined individually some of the important factors influencing the level of variation in a population. Mutation, recombination, and

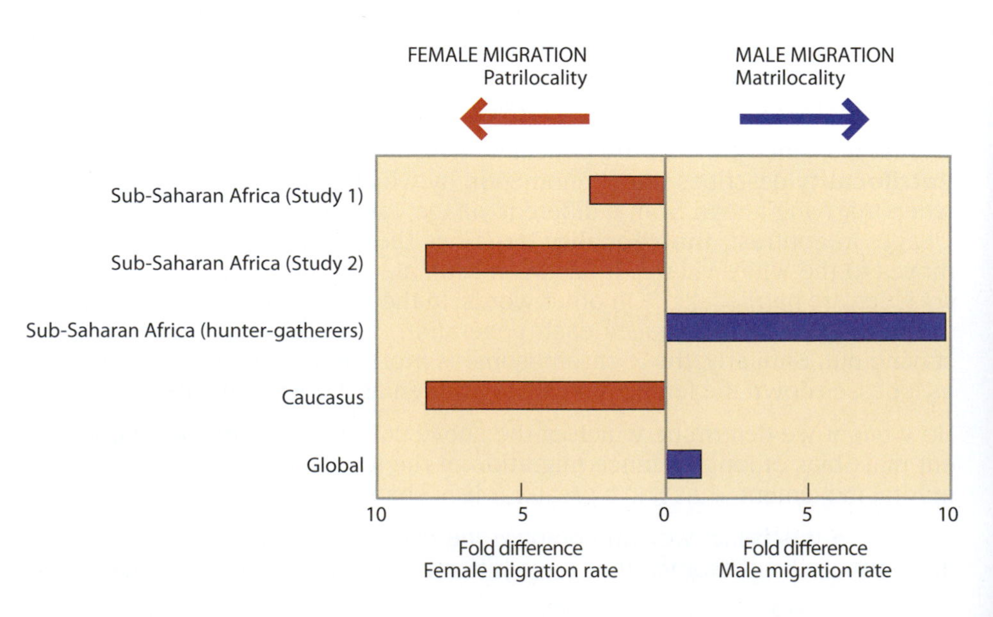

Figure 5.14: Sex-specific differences in migration in humans. Bars labeled Studies 1 and 2 reflect pattern in migration of sub-Saharan populations irrespective of lifestyle. [Study 1 is Destro-Bisol G et al. (2004) *Mol. Biol. Evol.* 21, 1673. Study 2 is Wilder JA et al. (2004) *Nat. Genet.* 36, 1122.]

migration increase diversity, random genetic drift decreases it, and selection can do either. In this section, we investigate how these opposing forces interact with one another.

In the previous section we saw that in a subdivided population the opposing forces of migration and drift can reach an equilibrium state whereby differentiation among subpopulations, as measured by F_{ST}, remains constant over time. It is only by assuming that this equilibrium has been attained that we can estimate migration rates from F_{ST} values in real populations.

There are important equilibria in population genetics

Mutation–drift balance

In the simplest model of a population with no selection or migration and the usual assumptions of constant size and random mating, diversity will reach an equilibrium value where the number of novel variants (generated by mutation) entering the population is balanced by the number lost by drift. This is known as **mutation–drift balance** or **mutation–drift equilibrium**. There is a simple analogy to illustrate this point: imagine a water tank fed by a dripping tap at the top, with another tap at the bottom to let water flow out (**Figure 5.15**). Water will accumulate in the tank until the amount entering from the tap at the top (= mutations) is balanced by the amount lost through the tap at the bottom (= drift), leading to a stable water level (= diversity). If the mutation rate increases (more water in) or decreases (less water in), diversity at equilibrium will increase or decrease correspondingly. Similarly, if drift increases (opening the bottom tap) or decreases (closing the bottom tap), diversity will decrease or increase. This equilibrium value of diversity is known as the **population mutation parameter** (**θ** or **theta**), and is discussed in more detail in Chapter 6.

Recombination–drift balance

Earlier in this chapter we saw that in an infinitely large population, linkage disequilibrium (LD, Chapter 3) decays over time as a result of recombination generating new haplotypes. However, random genetic drift is continually removing haplotypes from the population. As a consequence LD can reach an equilibrium value in finite populations. This equilibrium value of LD is determined by the **population recombination parameter** (**ρ** or **rho**). This parameter combines information on effective population size and recombination rate (c) using the equation:

$$\rho = 4N_e c$$

The precise relationship between different measures of LD (Box 3.5) and ρ is complex, but when ρ is large

$$r^2 \approx 1/\rho$$

Thus we can see that LD decreases as ρ increases, for example as a result of a larger effective population size (lower genetic drift) or higher recombination rate.

In real populations, it is apparent that LD is not simply an equilibrium between recombination and drift, but can be greatly affected by selection, mutation, gene conversion, and demography (Chapter 6). Because demography influences LD, analysis of LD within a population can allow inferences to be made on the prehistoric demography of that population (for an example of how patterns of LD can reveal ancient admixture events, see **Section 14.4**).

If we are examining LD over a large genomic region containing many polymorphisms, it is unclear how best to combine the information from measures of LD based on comparisons between individual pairs of variants (that is, D, D' or r^2, Box 3.5). Therefore attention has focused on estimating ρ itself for these kinds of data, as this gives a single measure of LD for the entire region. Estimating ρ requires the use of population models and is computationally intensive, but it allows the other forces that shape LD (for example, demography and mutation)

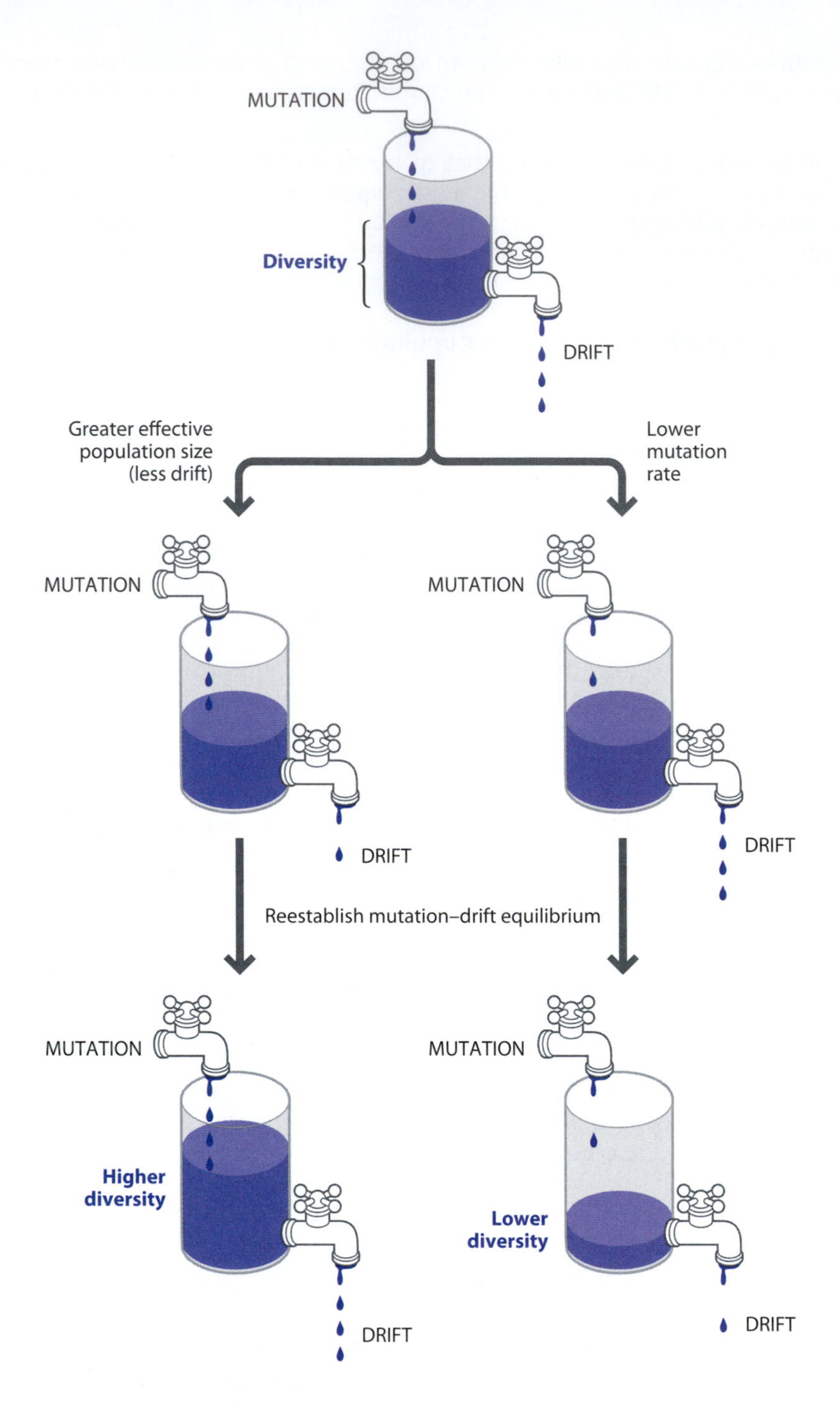

Figure 5.15: A metaphorical depiction of the relationship between mutation rate, drift, and diversity. A change in either the mutation rate or effective population size changes the diversity at mutation–drift equilibrium—see Section 5.6. (Original metaphor by John Relethford.)

to be taken into account.[40] An additional advantage of studying ρ is that it allows *c*—the recombination rate across the region—to be estimated, and compared between different regions of the genome.

Mutation–selection balance

Some deleterious mutational events have sufficiently high mutation rates that within a large population they occur several times within a single generation, and can be considered recurrent mutations. Mutation and selection are opposing forces determining the frequency of such mutant alleles in the population. The rate at which new alleles are generated by mutation can be balanced by the eventual elimination of each mutant allele by negative selection so that the average number of examples of a given mutant allele reaches an equilibrium value within the population. Lowering the selective cost of a mutation, or increasing its mutation rate, will increase this equilibrium value. Mutant alleles that cause monogenic diseases are often likely to be in mutation–selection balance and this is discussed in Chapter 16.

Rather than considering a single recurrent mutation in isolation, if we consider all deleterious alleles together, a balance between mutation and selection may operate over the genome as a whole, such that at equilibrium each genome contains a certain number of deleterious alleles. This has implications for the study of complex diseases where both genetic variation and environmental variation play a role (Chapter 17).

Does selection or drift determine the future of an allele?

So far in this section we have not considered the interplay between selection and drift, and their relative weight in influencing allele frequencies. For example, the selection dynamics described in **Section 5.4** assume an infinitely large population. What happens in finite populations where random genetic drift is also operating?

Because drift operates more effectively in smaller populations, stronger selection is required to influence fixation or elimination of alleles. Whether drift or selection predominates depends on a number of factors, which include:

- The effective population size
- The selection coefficient
- The type of selection
- The frequency of the allele under selection

Equations relating these parameters exist for different types of selection. They can be used to determine whether an allele is likely to be under the influence of selection. Relating these parameters together allows us to draw four important conclusions:

- Selection often substantially increases the probability that an **advantageous allele** becomes fixed compared to a **neutral allele**; in humans, most new advantageous alleles are still far more likely to be eliminated than fixed.
- If new alleles are almost exclusively deleterious, the optimal allele can persist unchanged over very long time-scales. This conforms to the hypothesis that functional constraint on important proteins such as histones underlies their extreme lack of variability among diverse species.
- The time taken to fix an advantageous allele is much shorter than that to fix a neutral allele.
- A general rule for diploid loci is that for selection to be effective then the following relationship should hold:

  $$s > 1/2N_e$$

  where s is the selection coefficient.

  For haploid loci that are transmitted by only one sex, with one-quarter the effective population size of diploid loci, the relevant rule is:

  $$s > 2/N_e$$

The use of this last rule can be seen in the following example. A polymorphic inversion on the human Y chromosome, present in roughly 70% of British males, protects against XY translocations during meiosis. The offspring resulting from these rearrangements (XX males and XY females) are infertile; infertility is evolutionary death for the individual. However, these translocations occur only at low rates, such that the selection coefficient (s) of this inversion has been calculated as 1/90,000.[24] Given that the Y chromosome is a haploid locus and the effective population size (N_e) of humans has been estimated at around 10,000 (Table 6.4), the selective advantage of this inversion is not sufficiently large to overcome the effects of drift: $1/90{,}000 < 2/10{,}000$).

5.7 THE NEUTRAL THEORY OF MOLECULAR EVOLUTION

Before information on molecular diversity became available it was hypothesized that **genetic load** (that is, the accumulation of deleterious alleles) would mean that only a limited amount of polymorphism would be compatible with a sustainable population, because most new mutations would be selected against. Muller famously predicted that only one in a thousand genes would be heterozygous. According to this view, polymorphisms were stable entities, maintained by balancing selection. In contrast, the substitution of one nucleotide for another in a DNA sequence of a species was the result of positive selection for new mutations spreading through the population to **fixation**. The process of DNA substitution during evolution and the processes affecting the frequency of polymorphisms within a species were therefore thought to be independent.

The first studies, in the 1960s, to measure genetic diversity directly used protein electrophoresis to measure the frequency of **allozymes** in populations. Their finding was startling at the time: the amount of polymorphism uncovered within human genomes, and those of other species, was many times greater than was expected. To explain this, Motoo Kimura developed the **neutral theory of molecular evolution**, often referred to as simply the **neutral theory**.[25, 36] Neutral theory states that the fate of mutations is largely determined by random genetic drift rather than selection. The theory holds that negative selection is the prevailing mode of selection that eliminates deleterious mutations whereas cases of positive and balancing selection are rare. The vast majority of polymorphisms observed in populations are transient, awaiting eventual fixation or elimination by genetic drift. The theory therefore predicts that most polymorphisms have little or no effect on fitness. Kimura showed that, for a polymorphism where $2\ N_e\ s \leq 1$ (a selective advantage of <0.00005 in humans) then genetic drift would determine the fate of that polymorphism.

The publication of the theory caused a polarized debate in population genetics. A consensus may now have been reached, suggesting that the neutral theory does not entirely explain the observed genetic variation and adaptation in humans and other species, but is nevertheless a useful conceptual framework for thinking about genetic variation and evolution. In particular, it provides a powerful null model against which empirical data can be tested for evidence of selection, and it is in that context that we show it being applied to human evolutionary genetics in Chapter 6, and throughout this book.

The molecular clock assumes a constant rate of mutation and can allow dating of speciation

Perhaps the most important result of the neutral theory is that it presented a unified model of allele frequencies in a population and molecular substitution rates in an evolutionary lineage. The neutral theory shows that the rate of sequence evolution is driven only by the rate of mutation. Assuming the rate of mutation is constant, the rate of evolution is approximately constant over all evolutionary lineages. Therefore measuring the number of differences between the DNA sequences of two species can, in principle, be used to date the divergence of those two species,[27, 47] if calibration points of dated lineage divergences are known. This requires accurate, independent dating of the lineages by other disciplines, most notably **paleontology**. We will address applications of the molecular clock hypothesis and **genetic dating** in the next chapter; however, here we need to consider whether this is a reasonable approximation of the evolutionary process.

The mutation rate and the substitution rate can be shown to be equal, independent of population size, by a simple mathematical proof. The rate of nucleotide substitution (k) is equal to the rate at which new mutations are generated ($2N\mu$), multiplied by their probability of fixation (u). In a population of size N, diploid loci have a population size of $2N$, and the rate of nucleotide substitution (k) is therefore:

$$k = 2N\mu u$$

Remember that for a neutral mutation the probability of fixation is its frequency, which for a new mutation is the reciprocal of the population size ($1/2N$).

$$k = 2N(1/2N)\ \mu$$

$$\text{therefore—}k = \mu$$

The regularity of mutation might not translate into a regularity of evolution if selection plays a dominant role in determining the survival of new mutations in some lineages. Indeed, the regularity of molecular evolution contrasts with the non-uniform change of morphological evolution.

Fossil dates used to calibrate estimates of species divergence dates may be unreliable, or have unacceptably broad confidence limits. The **relative rates test** does not require absolute divergence times, but simply the knowledge of the order in which a number of lineages diverged from one another. This test (**Figure 5.16**) compares the rate of evolution in two lineages by relating them to a third, which is known to be an **outgroup** (a lineage more distantly related to the other two lineages than they are to one another).

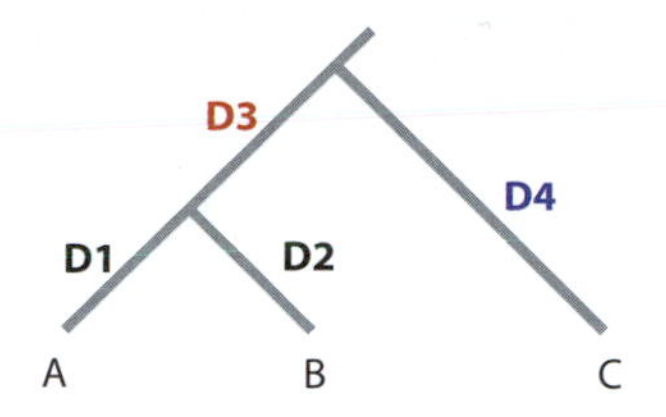

If molecular clock true (**D1** = **D2**) then:
$D_{AC} = D_{BC}$ (**D1**+**D3**+**D4** = **D2**+**D3**+**D4**)
Thus, test to see if $D_{AC} - D_{BC} = 0$

Figure 5.16: Testing the molecular clock with the relative rates test.
The rate of evolution in two lineages, leading to A and B, is compared with that in an outgroup lineage, C. D1 to D4 are mutational distances on different branches of the phylogeny relating the three species.

Although the lineages used in the relative rates test are often individual species, the test can also be used on nonrecombining regions of the genome within a species (where the phylogeny is known). The number of mutational events in each branch of the tree relating the three lineages is calculated. The significance of any differences between the mutational distances shown in Figure 5.16 can be assessed by a number of different methods, such as the **likelihood ratio test** (Box 6.3).

There are problems with the assumptions of the molecular clock

Comparisons of rates of evolution between species have often shown significantly varying rates of nucleotide substitution on different evolutionary lineages. Where the same rates are found across multiple loci it suggests that different mutation rates rather than selection are the cause of rate inequalities. There are several possible explanations, which relate the mutation rate to biochemical processes within the cell (**lineage effects**). The different processes that have been proposed to cause lineage effects need not be mutually exclusive, but could operate in concert (**Figure 5.17**).

The lineage effect that has received most attention is the **generation time hypothesis**. This assumes that most mutations occur during DNA replication in the germ line. As a consequence, the mutation rate is determined by the number of replications during a certain period of time. If species have the same number

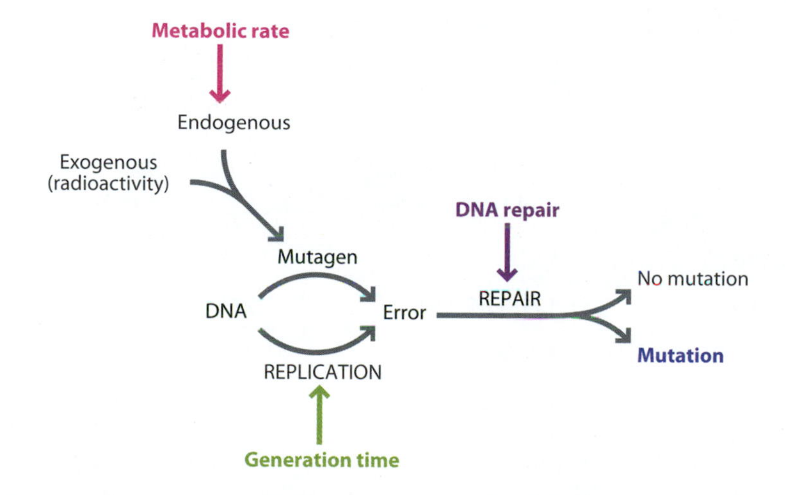

Figure 5.17: Different sources of lineage effects.
A number of processes are involved in generating a new mutation. Potential sources of lineage effects are shown in different colors.

of cell divisions per generation, then the mutation rate becomes dependent on the generation time. This hypothesis of replication errors causing mutations also underpins the hypothesis of **male-driven evolution** (see **Box 5.4**).

The generation time hypothesis has gained support from studies of mammalian species that show a better correlation of mutational distance with generation time than with calendar time. Many studies have demonstrated that while generation times within mammalian groups have the order rodents < monkeys < humans, evolutionary rates have the reverse order, both for indels and base substitutions. Similarly the mutation rate among higher primates seems to be negatively correlated with generation time. The mutation rate increase in rodents does not appear to be linear with respect to generation time as there is only a two- to fourfold increase in annual mutation rates relative to humans despite a fortyfold difference in generation times; however, the difference in replications per generation can account for some of this discrepancy. Human males have four times more replications per generation than do male mice, and this would be expected to lessen rate differences. Further problems for the generation time hypothesis appear in the difference in the substitution rate ratio of synonymous (silent) and nonsynonymous (amino-acid-changing) mutations (**Section 3.2**) between primates and rodents, suggesting a potential role for selection. In addition, there remain a number of cases where calendar time, rather than generation time, is better correlated with rates of evolution, even after correcting for the differing number of replications per generation among the different species.

Box 5.4: Male-driven evolution

J. B. S. Haldane first proposed that the greater number of genome replications in the male germ line should result in the male mutation rate being higher than that of the female.[14] His hypothesis is based on the same assumption—that mutation is caused by errors in replication—as the generation time hypothesis discussed in Section 5.7.

In humans (Figure 3.27), the number of replications required during oogenesis is constant, at about 22, whereas the number of replications in the male germ line increases with age. Thirty replication events are required to generate spermatogonial stem cells at puberty (~13 years old), which then go through ~23 replications per year, before the final five replications required to make mature spermatozoa. Thus a male reproducing at 25 will be using DNA that has gone through $30 + 23 \times (25 - 13) + 5 = 311$ replication cycles. This is about 14 times as many as for the oocyte DNA.

The ratio of male to female mutation rates (α—also known as the alpha factor) can be calculated by comparing the number of mutations that have accrued in autosomal, Y-chromosomal, and X-chromosomal sequences over the same time period. The ratios of these numbers can be related to α by the following equations:

$$\text{X/autosome} = 2/3(2 + \alpha)/(1 + \alpha)$$
$$\text{Y/autosome} = 2\alpha/(1 + \alpha)$$
$$\text{Y/X} = 3\alpha/(2 + \alpha)$$

Accurate estimates of α have been derived from analysis of large regions of genomic sequence. The estimates shown in **Table 1** incorporate a consideration of the diversity apparent in the human–chimpanzee common ancestor, which allows more accurate estimation of α than the equations given above.[42] Recent direct measurement of mutations from genome sequences of families[26] suggest a value of α of about 3.9.

TABLE 1: ESTIMATES OF α (95% CONFIDENCE INTERVALS) FROM HUMAN–CHIMPANZEE SEQUENCE COMPARISONS

Correction for ancient diversity	X/autosome	Y/autosome	Y/X
None	7.58 (7.04–8.20)	1.77 (1.64–1.94)	2.68 (2.50–2.89)
2× modern human diversity	6.92 (6.38–7.58)	3.04 (2.70–3.46)	4.03 (3.67–4.43)
4× modern human diversity	6.11 (5.58–6.78)	11.2 (8.04–17.6)	8.24 (7.01–9.84)

The **metabolic rate hypothesis** is based on the idea that most mutations result from the presence of **endogenous mutagens**.[30] Free radical by-products of aerobic respiration are the prime suspects, and, therefore, organisms with higher metabolic rates should produce more mutagenic free radicals. Differences in metabolic rates have been used to explain rate differences that were previously difficult to reconcile with generation time differences. In addition, considerations of metabolic mutagens may explain why rate differences are generally more pronounced among mitochondrial sequences than nuclear ones, as most oxidative free radicals are produced within mitochondria themselves.

Alternative sources of lineage effects could lie in the enzymatic mechanisms that act to repair the effects of mutagenic processes, rather than the processes themselves. While most attention has focused on varying efficiencies of **DNA repair**, another source could be differences in the many pathways that mop up mutagens of different kinds before they are able to damage DNA. At present, the relative efficiencies of these pathways in different lineages are too poorly characterized to allow these hypotheses to be tested.

In addition to rate variation amongst lineages there are often also differences between the rates of nonsynonymous and synonymous substitution. To explain this weakness of the molecular clock hypothesis, two alternatives have been proposed. The **episodic selection** model suggests that episodic selective pressures at nonsynonymous sites, created by environmental changes, distort the molecular clock.[13] The **nearly neutral** model suggests that nonsynonymous changes are slightly deleterious rather than neutral, so the interplay of drift and selection is sensitive to fluctuations in population sizes.[37] The idea is that negative correlation between population size and generation time allows the rate of nonsynonymous changes to operate largely independently of any generation-time effect. Small populations tend to have longer generation times and drift predominates, frequently fixing nonsynonymous mutations that occur at low rates. By contrast, negative selection predominates at nonsynonymous sites in large populations, so fixation occurs infrequently although mutations are generated more rapidly as a result of shorter generation times. These factors cancel out such that large and small populations have similar rates of evolution with respect to calendar time. This debate has not been resolved.

SUMMARY

- Mutation and recombination increase human diversity by generating new alleles and new haplotypes respectively. Genetic drift reduces diversity, and results from the random sampling of one generation from the preceding one, causing random change in allele frequencies. Selection can operate in a number of different ways to increase, decrease, or maintain diversity.
- Migration increases population-specific diversity by introducing new alleles.
- The different forces acting on allele frequencies could in principle balance one another out so that, with sufficient time, diversity within a population reaches an equilibrium value. Human populations, however, are not at equilibrium.
- The effective population size (N_e) can be very different from the census population size and is affected by past population size fluctuations, variance in reproductive success, and population structure. The effective population size varies between the Y chromosome, X chromosome, mtDNA, and autosomes, because of their different patterns of inheritance.
- The discovery of high levels of natural polymorphism led to the development of the neutral theory, which states that the fate of most mutations in the human genome is determined by genetic drift.
- The concept of a molecular clock is useful for dating splits in gene and species trees but its constant rate assumption is often violated by the finding of different mutation rates in different phylogenetic lineages, which can be explained by a number of factors, known as lineage effects.

QUESTIONS

Question 5–1: A SNP in the tensin gene (*TNS1*) was genotyped in 184 Maasai from Kinyawa in Kenya. The genotype counts were 4 AA, 35 AG, and 145 GG. Test whether or not this SNP is in Hardy–Weinberg equilibrium in this population.

Question 5–2: Using the genotype frequency data from the 1000 Genomes browser (browser.1000genomes.org), calculate whether the rs1799990 SNP in the *PRNP* gene is in HWE in the GBR, LWK, and CLM populations.

Question 5–3: Using 11,600 as an estimate of human long-term N_e (from Table 6.4), how strong does selection need to be to overcome the effect of genetic drift in changing the frequency of a particular allele on

(a) An autosome

(b) The Y chromosome

Comment on the value of *s* in both cases.

Question 5–4: Briefly compare and contrast these forms of selection: negative, balancing, purifying, and positive.

Question 5–5: What are the consequences of polygyny for the relative diversity of the sex chromosomes and mitochondrial DNA?

Question 5–6: In humans, why is long-term effective population size so different from real census population size?

Question 5–7: Giving examples, explain the consequences of a major reduction in population size followed by a rapid expansion for

(a) Neutral variation

(b) The ability of selection to drive an allele to fixation

Reading Chapter 13 will help with your answer.

Question 5–8: Discuss the consequences of kin-structured migration on population genetic parameters, giving examples in humans.

Question 5–9: Discuss the evidence for selection at HLA genes.

REFERENCES

The references highlighted in purple are considered to be important (for this chapter) by the authors.

1. **Anthony DW** (1990) Migration in archeology: the baby and the bathwater. *Am. Anthropol.* **92**, 895–914.
2. **Austerlitz F & Heyer E** (1998) Social transmission of reproductive behavior increases frequency of inherited disorders in a young-expanding population. *Proc. Natl Acad. Sci. USA* **95**, 15140–15144.
3. **Bittles A & Black M** (2010) Consanguineous marriage and human evolution. *Annu. Rev. Anthropol.* **39**, 193–207.
4. **Bittles AH & Neel JV** (1994) The costs of human inbreeding and their implications for variations at the DNA level. *Nat. Genet.* **8**, 117–121.
5. **Burton ML, Moore CC, Whiting JWM et al.** (1996) Regions based on social structure. *Curr. Anthropol.* **137**, 87–123.
6. **Caballero A** (1995) On the effective size of populations with separate sexes, with particular reference to sex-linked genes. *Genetics* **139**, 1007–1011.
7. **Charlesworth B** (2009) Effective population size and patterns of molecular evolution and variation. *Nat. Rev. Genet.* **10**, 195–205.
8. **Charlesworth B, Morgan M & Charlesworth D** (1993) The effect of deleterious mutations on neutral molecular variation. *Genetics* **134**, 1289–1303.
9. **Davey Smith G, Lawlor DA, Timpson NJ et al.** (2008) Lactase persistence-related genetic variant: population substructure and health outcomes. *Eur. J. Hum. Genet.* **17**, 357–367.
10. **Di Rienzo A, Peterson A, Garza J et al.** (1994) Mutational processes of simple-sequence repeat loci in human populations. *Proc. Natl Acad. Sci. USA* **91**, 3166–3170.
11. **Felsenstein J** (1971) Inbreeding and variance effective numbers in populations with overlapping generations. *Genetics* **68**, 581–597.
12. **Fix AG** (2004) Kin structured migration: causes and consequences. *Am. J. Hum. Biol.* **16**, 387–394.
13. **Gillespie JH** (1984) The molecular clock may be an episodic clock. *Proc. Natl Acad. Sci. USA* **81**, 8009–8013.
14. **Haldane J** (1947) The mutation rate of the gene for haemophilia, and its segregation ratios in males and females. *Ann. Eugen.* **13**, 262–271.
15. **Hardy GH** (1908) Mendelian proportions in a mixed population. *Science* **28**, 49–50.
16. **Hartl DL & Clark AG** (2007) Principles of Population Genetics, 4th ed. Sinauer Associates.
17. **Havlicek J & Roberts SC** (2009) MHC-correlated mate choice in humans: a review. *Psychoneuroendocrinology* **34**, 497–512.
18. **Helgason A, Hrafnkelsson B, Gulcher JR et al.** (2003) A populationwide coalescent analysis of Icelandic matrilineal and patrilineal genealogies: evidence for a faster evolutionary rate of mtDNA lineages than Y chromosomes. *Am. J. Hum. Genet.* **72**, 1370–1388.
19. **Hill KR, Walker RS, Božičević M et al.** (2011) Co-residence patterns in hunter-gatherer societies show unique human social structure. *Science* **331**, 1286–1289.
20. **Holsinger KE & Weir BS** (2009) Genetics in geographically structured populations: defining, estimating and interpreting F_{ST}. *Nat. Rev. Genet.* **10**, 639–650.
21. **Horton R, Wilming L, Rand V et al.** (2004) Gene map of the extended human MHC. *Nat. Rev. Genet.* **5**, 889–899.
22. **Huestis R & Maxwell A** (1932) Does family size run in families? *J. Hered.* **23**, 77–79.
23. **Hughes AL & Nei M** (1988) Pattern of nucleotide substitution at major histocompatibility complex class I loci reveals overdominant selection. *Nature* **335**, 167–170.

24. **Jobling MA, Williams G, Scheibel at al.** (1998) A selective difference between human Y-chromosomal DNA haplotypes. *Curr. Biol.* **8**, 1391–1394.

25. **Kimura M** (1968) Evolutionary rate at the molecular level. *Nature* **217**, 624–626.

26. **Kong A, Frigge ML, Masson G et al.** (2012) Rate of de novo mutations and the importance of father's age to disease risk. *Nature* **488**, 471–475.

27. **Kumar S** (2005) Molecular clocks: four decades of evolution. *Nat. Rev. Genet.* **6**, 654–662.

28. **Leal SM** (2005) Detection of genotyping errors and pseudo SNPs via deviations from Hardy–Weinberg equilibrium. *Genet. Epidemiol.* **29**, 204–214.

29. **Lindenbaum S** (2008) Understanding kuru: the contribution of anthropology and medicine. *Philos. Trans. R. Soc. Lond. B Biol. Sci.* **363**, 3715–3720.

30. **Martin AP & Palumbi SR** (1993) Body size, metabolic rate, generation time, and the molecular clock. *Proc. Natl Acad. Sci. USA* **90**, 4087–4091.

31. **Matsumura S & Forster P** (2008) Generation time and effective population size in polar Eskimos. *Proc. Biol. Sci.* **275**, 1501–1508.

32. **Mead S, Whitfield J, Poulter M, et al.** (2008) Genetic susceptibility, evolution and the kuru epidemic. *Philos. Trans. R. Soc. Lond. B Biol. Sci.* **363**, 3741–3746.

33. **Miller G** (2000) The Mating Mind: How Sexual Choice Shaped the Evolution of Human Nature. Heinemann.

34. **Mills RE, Luttig CT, Larkins CE et al.** (2006) An initial map of insertion and deletion (indel) variation in the human genome. *Genome Res.* **16**, 1182–1190.

35. **Murdock GP** (1967) Ethnographic Atlas. University of Pittsburgh Press.

36. **Nei M, Suzuki Y & Nozawa M** (2010) The neutral theory of molecular evolution in the genomic era. *Annu. Rev. Genomics Hum. Genet.* **11**, 265–289.

37. **Ohta T** (1992) The nearly neutral theory of molecular evolution. *Annu. Rev. Ecol. Syst.* **23**, 263–286.

38. **Pawlowski B, Dunbar R & Lipowicz A** (2000) Evolutionary fitness. Tall men have more reproductive success. *Nature* **403**, 156.

39. **Pearson K, Lee A & Bramley-Moore L** (1899) Mathematical contributions to the theory of evolution. VI. Genetic (reproductive) selection: Inheritance of fertility in man, and of fecundity in thoroughbred racehorses. *Philos. Trans. R. Soc. Lond. A* **192**, 257–330.

40. **Pritchard JK & Przeworski M** (2001) Linkage disequilibrium in humans: models and data. *Am. J. Hum. Genet.* **69**, 1–14.

41. **Raymond CK, Kas A, Paddock M et al.** (2005) Ancient haplotypes of the HLA class II region. *Genome Res.* **15**, 1250–1257.

42. **Taylor J, Tyekucheva S, Zody M et al.** (2006) Strong and weak male mutation bias at different sites in the primate genomes: insights from the human-chimpanzee comparison. *Mol. Biol. Evol.* **23**, 565–573.

43. **Wilkins JF** (2006) Unraveling male and female histories from human genetic data. *Curr. Opin. Genet. Dev.* **16**, 611–617.

44. **Wright S** (1931) Evolution in Mendelian populations. *Genetics* **16**, 97–159.

45. **Wright S** (1951) The genetical structure of populations. *Ann. Eugen.* **15**, 323–354.

46. **Yin J, Jordan MI & Song YS** (2009) Joint estimation of gene conversion rates and mean conversion tract lengths from population SNP data. *Bioinformatics* **25**, i231–i239.

47. **Zuckerkandl E & Pauling L** (1965) Evolutionary divergence and convergence in proteins. In Evolving Genes and Proteins (V Bryson & HJ Vogel eds), pp 97–166. Academic Press.

MAKING INFERENCES FROM DIVERSITY

CHAPTER SIX

Current human genetic diversity contains information about the sizes and movements of past populations, and on the history of human **adaptation** to changing environments. But how should we pose our questions about these topics in order to get useful answers? And what are the limits to the inferences we can make from modern genetic diversity? There are two related goals:

1. A *description* of the distribution of present diversity, which allows comparisons between species or between populations within a species. These include comparisons of genetic diversity, and its apportionment between subpopulations.
2. *Inferences* about how modern diversity evolved. Studies of human genetic diversity are usually limited to a single time-slice, such as analyzing diversity among modern populations, which constrains our ability to investigate the past. Prehistorical and historical processes must therefore be inferred. Such **inferential methods** require explicit or implicit models of the evolutionary processes, some of which were described in the previous chapter. Inferences of past processes are motivated by (i) an anthropological interest in the prehistory of populations, their origins, movements, and demographies; and (ii) an interest in the evolutionary history of specific segments of DNA, be they individual genes, chromosomes, or entire genomes.

These interests, although conceptually distinct, often lead to related questions, because population processes affect the molecular diversity of DNA. In this chapter, we will come across both descriptive and inferential methods. There is no formal distinction between the two; descriptions of present diversity almost inevitably lead to discussions of how it might have arisen. It is worth noting that there is often no simple and unique answer to questions such as, How could a particular pattern of genetic diversity have arisen within a gene? Therefore, a combination of several analytical approaches is advisable.

6.1 WHAT DATA CAN WE USE?

The first studies, over 40 years ago, used protein polymorphisms, characterized immunologically or by protein **electrophoresis**. Because the DNA sequences of these polymorphisms were unknown, evolutionary inference could only be based on the analysis of allele frequencies. These polymorphisms are typified by blood groups, and are commonly referred to today as **classical polymorphisms**, or **classical markers** (Box 3.1).

The advent of direct methods for investigating the sequence of protein and DNA molecules led to the development of **molecular polymorphisms**, or **molecular markers** (Chapter 4). By defining the underlying molecular differences between alleles at a locus, these methods allow the introduction of the concept of evolutionary distances between alleles—for example, the number of repeat units by which two microsatellite alleles differ, or the number of variant bases between two aligned sequences. In this way, evolutionary inference can be based both on the analysis of allele frequencies, *and* on the molecular comparison of different alleles. In addition, these methods allowed diversity in noncoding regions of the genome to be investigated.

Some analytical methods described here are applicable to both types of data, classical and molecular, whereas others are suitable only for one type. Several methods devised for classical data have been adapted to take account of the extra information within molecular data, and this may render a method suitable for only some kinds of molecular data. For example, a method devised for analyzing DNA sequences may not be applicable to microsatellite diversity or structural variation.

It is worth noting that a single locus inevitably contains less information on our evolutionary past than do many loci, no matter how informative that individual locus is. A single locus gives a single account of the evolutionary process. As any historian knows, the collation of several corroborative sources is vital to an accurate reconstruction of the past; any single account may be biased, whether by chance (drift) or by design (selection). The analysis of whole genomes provides a picture of the variation in a genome, and a huge amount of information for analysis. SNP chip technology has made collecting data genomewide, from millions of loci, routine (**Section 4.5**), and as we see in this chapter and later in the book, such data provide much information about our evolutionary history. Existing analytical methods have been scaled up to cope with genome-scale data, and novel methods specifically developed. We provide a reference table so that the reader can easily find in this chapter details on the methods that will be encountered in the rest of the book (**Table 6.1**). We also introduce a variety of computer programs that have been written to implement the methods we describe in this chapter (**Box 6.1**).

6.2 SUMMARIZING GENETIC VARIATION

Summary statistics that summarize the amount of variation do not encapsulate all information present in the data but allow comparisons between populations and between loci.

Heterozygosity is commonly used to measure genetic diversity

Perhaps the simplest way to describe the amount of diversity is to count the number of haplotypes present. This can be done either within many populations for a single locus, allowing a comparison of diversity between populations, or at many loci within a single population, allowing a comparison among loci. Clearly this measure does not account for molecular distances between haplotypes, and so is applicable to both classical alleles (for example HLA—see Box 5.3) and molecular haplotypes. It is, however, highly dependent on sample sizes, which is usually a serious disadvantage.

A commonly used measure of diversity that is also blind to **molecular distance** between alleles is Nei's **gene diversity** statistic. Despite its name, this measure is suitable for both coding and noncoding polymorphisms, and measures the probability that two alleles drawn at random from the population will be different from each other. Consequently, for diploid loci this statistic is referred to as a measure of heterozygosity, which becomes **virtual heterozygosity** at haploid loci. Gene diversity at a locus is defined as:

$$h = 1 - \sum_{i=1}^{q} x_i^2$$

where q is the number of alleles and x_i is the frequency of the ith allele. Often,

TABLE 6.1:
FREQUENTLY USED METHODS

Method	What it does	Type of data needed	Limitations	Section in this book
Calculation of θ	a measure of genetic diversity	DNA sequence data from the same region in different individuals	interpretation not intuitive	**6.2 Summarizing genetic variation:** Measuring nucleotide diversity
Principal component plots	shows genetic distance between individuals or populations	genetic variation data from many loci, typically from SNP chips	sometimes difficult to interpret results in terms of population history	**6.3 Measuring genetic distance:** Representing genetic distance and population structure using multivariate analyses
STRUCTURE-like plots	assigns an individual to a group based on genetic diversity	genetic variation data from many loci, typically from SNP chips	requires prior estimation of the number of groups	**6.3 Measuring genetic distance:** Population structure using individual genomic data
Phylogenetic trees and networks: maximum likelihood (ML) and Bayesian	compares the relationship between DNA sequences	DNA sequence data from different species or individuals	Bayesian and ML methods have different limitations; see text and Box 6.4	**6.4 Phylogenetics:** Character-based phylogenetic methods
Use of F_{ST}	a measure of population differentiation	genetic variation data from one or more loci	affected by both demography and selection	**6.3 Measuring genetic distance:** Distances between populations **6.7 Has selection been acting?** Selection tests based on interpopulation diversity and Box 5.2
Parameter estimation using coalescent methods	provides an estimate of various population parameters	DNA sequence data from different individuals	dependent on assumptions made in the coalescent model	**6.5 Coalescent approaches to reconstructing population history**
Extended haplotype tests	measures recent positive selection	SNP haplotype data from different individuals	only detects recent, strong, positive selection sweeps	**6.7 Has selection been acting?** Selection tests comparing allelic frequency and haplotype diversity
Allele frequency spectrum tests	measures departures from neutrality	DNA sequence data from the same region in different individuals	affected by both demography and selection	**6.7 Has selection been acting?** Selection tests can be based on the analysis of allele frequencies at variant sites

this value for h is corrected for sample size by multiplying by $n/(n-1)$, where n is the sample size, to give an **unbiased estimator** of population heterozygosity.

In situations where there is a high degree of polymorphism almost all alleles (or haplotypes) are different from one another, and thus gene diversity ceases to be useful as it is close to 1 in all populations. In such situations, measures of diversity are required that take account not of the *identity* of the allele, but the *distances* between alleles. Such measures must consider the molecular nature of allelic variation, and therefore are often specific to different types of data.

Nucleotide diversity can be measured using the population mutation parameter theta (θ)

The Wright–Fisher model (**Section 5.3**) shows that the level of diversity within a population can reach an equilibrium value whereby the generation of new alleles by mutation is balanced by the elimination of alleles by **genetic drift** (**Section 5.6**). It is therefore possible to relate the mutation rate (μ, per site per generation) to the genetic drift in a population using a theoretical measure of diversity, a parameter called θ—theta. In practice, since the rate of drift is inversely proportional to effective population size (**Section 5.3**), N_e is used and the equation relating these parameters for diploid loci is:

$$\theta = 4N_e\mu$$

Box 6.1: Software for making inferences about diversity

It is an inescapable fact that computer programs are needed to interpret the ever-increasing amounts of population genetic data. These programs have been written by many different academic scientists, and are often available free; however, this has fostered many different data formats and varying degrees of user-friendliness that can make an initial foray into this world quite daunting. On further investigation, the view may or may not become clearer, although there are several different data format conversion tools available, such as READSEQ. Also, many programs are available on the Web, or as user-friendly Windows® programs, although many still require a basic knowledge of MS-DOS or Unix command-line functions.

An encyclopedic list of programs, mainly concentrated on phylogenetics, is maintained by Felsenstein (http://evolution.genetics.washington.edu/phylip/software.html) and there are comprehensive reviews on population genetics software.[20,38] Any non-encyclopedic list that we provide will undoubtedly reflect our own biases, but Table 1 shows a few, selected for popularity, functionality, and user-friendliness, biased toward smaller datasets. At the moment, software for analyzing whole human genomes is lagging behind the generation of genome data, primarily because of the sheer computing power required to analyze such a large amount of data.

Some general advice on using software packages:

Read the manual. The quality of the manuals varies from program to program, but the programs in Table 1 all have well-written manuals, and sometimes extensive help pages. Read the manual thoroughly to understand what the program is doing, and have a go with the example data. Don't be afraid to play with the example data by changing parameters, conditions, and so on. If you have problems, some programs have internet forums or social networking pages where you can post questions to other users. If you still have problems, read the manual again, and only contact the author as a final resort—the authors will normally be friendly and will try to be helpful but often get a large number of questions from software users where the answers are already in the manual.

Be skeptical. No piece of software is perfect, so treat the results of an analysis with as much skepticism as you would treat the results of a scientific experiment.

Be careful with parameters. Understand what every parameter means, and don't always accept the default parameters.

TABLE 1:
SOME COMMONLY USED SOFTWARE IN EVOLUTIONARY GENETICS

Name	Platform	Type	Comments	Reference
ARLEQUIN	Windows®	population genetics, multifunctional	performs a wide variety of powerful analyses. The variety of data formats is complex	Excoffier L et al. (2005) *Evol. Bioinform. Online* 1, 47.
BEAST	Unix/Windows®/Mac	population genetics and phylogenetic analysis	powerful and flexible software for Bayesian analysis of sequence data	Drummond A & Rambaut A (2007) *BMC Evol. Biol.* 7, 214.
DNASP	Windows®	population genetics, multifunctional	performs a wide variety of analyses on DNA sequence data	Librado P & Rozas J (2009) *Bioinformatics* 25, 1451.
STRUCTURE	Unix/Windows®/Mac	population genetic analysis by clustering	can generate plots itself, or output files for the DISTRUCT software	Rosenberg NA et al. (2002) *Science* 298, 2381.
EIGENSOFT	Unix	population genetic analysis by principal component analysis	analysis of genomewide SNP and microsatellite data	Patterson N et al. (2006) *PLoS Genet.* 2, e190.
CLUSTAL OMEGA	Web/Unix/Windows®/Mac	multiple sequence alignment	latest version of a venerable sequence aligner	Sievers F et al. (2011) *Mol. Syst. Biol.* 7, 539.
NETWORK	Windows®	phylogenetic tree and network analysis	generates median-joining, maximum parsimony, and reduced median networks	www.fluxus-engineering.com
PHYLIP	Web/Windows® command line/Unix/Mac	phylogenetic analysis	over 30 years old, but still a very popular suite of programs using both parsimony and maximum-likelihood methods	Felsenstein J (1989) *Cladistics* 5, 164.
PAML	Windows® command line/Unix/Mac	phylogenetic analysis	a suite of programs using maximum-likelihood analysis. Particularly popular for inferring natural selection using interspecies data	Yang Z (2007) *Mol. Biol. Evol.* 24, 1586.
MEGA	Unix/Windows®/Mac	alignment and population genetics, multifunctional	a wide variety of tests with a user-friendly front-end	Tamura K et al. (2011) *Mol. Biol. Evol.* 28, 2731.

Thus by knowing the mutation rate and θ, and assuming an equilibrium state, we can estimate the effective population size of a diploid population from DNA sequence diversity. θ is a fundamental parameter of molecular evolution, sometimes referred to as the **neutral parameter** or the population mutation parameter and, in different forms, appears in many different analytical methods. The above equation is specific for diploid loci inherited from both parents; however, a more general form of the equation that considers loci with other inheritance patterns can be considered. Here, n represents the number of heritable copies of the locus per individual, which is 2 for diploid loci, 0.5 for the Y chromosome and mtDNA, and 1.5 for the X chromosome (the average of males and females):

$$\theta = 2nN_e\mu$$

We can see that, assuming the same mutation rate, the reduced number of heritable copies of the sex chromosomes should result in their having lower diversity at equilibrium than autosomal loci.

How can we summarize the observed diversity within a set of nucleotide sequences? We could count the number of nucleotide sites that vary within the entire set of aligned sequences (see **Box 6.2**), known as **segregating sites**. However, such an empirical measure is clearly dependent on the length of sequence analyzed: the longer the sequence, the greater the number of segregating sites. Measuring the proportion of all sites that are segregating would surmount this problem, but this itself depends on the number of sequences sampled—as more sequences are studied, more segregating sites are found. An empirical measure that takes these factors into account, nucleotide diversity, π, is analogous to Nei's gene diversity. Nucleotide diversity describes the probability that two copies of the same nucleotide drawn at random from a set of sequences will be different from one another. The **expected value** of π will be independent of sample size, unlike the value for the number of segregating sites. The definition of π is shown below:

$$\pi = \sum_{ij}^{q} x_i x_j d_{ij}$$

where q is the number of different allelic sequences, x_i and x_j are the frequencies of the ith and jth sequences respectively, and d_{ij} the proportion of different nucleotides between them. Like gene diversity, nucleotide diversity can be corrected for sample size by multiplying by $n/(n-1)$, where n is the sample size, to give the unbiased estimator.

As we have seen, the amount of variation expected at each nucleotide site under neutral evolution is given by θ. Thus, if selection is absent, and if each mutation occurs at a different nucleotide site (and, of course, if the other assumptions are met), the expected values of π and θ should be equal. This model is called the **infinite sites model**.

There are several different methods for estimating θ from sequence data. These methods use different parameters derived from the observed diversity, including:

- The number of alleles
- The number of segregating sites (S)
- The number of **singletons** (η)
- The observed homozygosity (F)
- The mean number of **pairwise** differences (π)

Each estimate of θ is represented as θ_S, θ_η, and so on, depending on the parameter used. In an ideal neutrally evolving population these different estimators of θ will have the same value. If the different estimators of θ give significantly different values for a single locus, we can infer that the population departs from the neutral model, for example because of a different demography or because

Box 6.2: Sequence alignment

Having obtained sequence data from a number of individuals, we want to identify the evolutionary changes between two homologous sequences (that is, sequences that share a common ancestor). To do this, we align the sequences to ensure that the bases that are compared derive from the same nucleotide position in the common ancestor. Sequences must be aligned before we carry out any comparative analysis, such as calculating sequence diversity, constructing phylogenetic trees, or testing for selection.

Several different methods produce **sequence alignments**. Some methods maximize the number of matching aligned bases (similarity methods), while others minimize the number of mismatched aligned bases (distance methods). Gaps within the alignment where a base in one sequence is not matched by a base in the other (perhaps as a result of an insertion or a deletion) are also kept to a minimum. Scores are calculated to compare alternative sequence alignments, which are based on arbitrary weighting of gaps, mismatches, and matches. For example, consider the two alternative alignments in Table 1; if we score 1 for a mismatch and 5 for a gap, then the top alignment has the lowest score (in bold) and is therefore preferable. However, if we score 1 for a mismatch and 0.5 for a gap then the bottom alignment has the lower score (again in bold). In this example we have not taken into account the size of the gap; most popular alignment methods give greater weighting to larger gaps than to smaller gaps.

The methods described above are for pairwise sequence alignments. Aligning multiple sequences is more complex, and there are a variety of different methods for comparing the quality of different alignments. Some human genome browsers (Figure 2.1) also include pre-computed multiple sequence alignments of human sequence with other species.

A related problem to sequence alignment is that of homology searching. For example, one might want to find a homologous sequence within a large database of genomic sequence, or find the position of a small sequence within a single much longer sequence. Homology search methods attempt to identify regions of *local* similarity, rather than optimize an alignment over the entire length of all sequences (known as *global* methods).

Many of the popular methods are based on an algorithm known as BLAST—Basic Local Alignment Search Tool. A BLAST search usually returns a number of homology matches between the query sequence and the database. Each match comes associated with a score and an E value that determines how much significance should be ascribed to each individual match. The score is a measure of the match identified, which is assessed in a manner similar to the scoring example for pairwise alignments given above, except that a similarity method is used such that the highest score represents the best alignment. The E (Expected) value describes the number of matches with that score that could be expected given the size of the database. As the E value gets closer to zero, the significance of the associated match increases.

For aligning DNA sequences from an organism to its reference genome, we only need to identify sequences that are very similar (typically >95% identity). **BLAT**—Blast-Like Alignment Tool—was developed for this particular purpose, and is much faster than BLAST because it keeps an index of the whole genome in the computer's memory. Mapping large numbers of very short sequences, such as those generated by high-throughput sequencing machines, is a computationally intensive task, requiring specific software and often a high-performance computer.

TABLE 1: SCORING SEQUENCE ALIGNMENTS

<table>
<tr><th></th><th>mismatch score = 1
gap score = 5</th><th>mismatch score = 1
gap score = 0.5</th></tr>
<tr><td>Alignment 1</td><td>TACTCTGATC
|| || |
TAGTC--GCC
1 +5+1+1 = 8</td><td>TACTCTGATC
|| || |
TAGTC--GCC
1+0.5+1+1 = 3.5</td></tr>
<tr><td>Alignment 2</td><td>TACTCTGATC
|| || | |
TAGTC-GC-C
1 +5+1+5 = 12</td><td>TACTCTGATC
|| || | |
TAGTC-GC-C
1+0.5+1+0.5 = 3</td></tr>
</table>

selection is acting. Such comparisons form the basis of several site frequency spectrum statistics used for detecting selection, which measure how different the various estimates for θ actually are (Section 6.7). Comparison of the value for these statistics between a single locus and the rest of the genome can help distinguish between selection occurring at the locus in question and demographic events that affect diversity across the genome.

The mismatch distribution can be used to represent genetic diversity

The **mismatch distribution**, also known as the distribution of pairwise differences, is appropriate for data where discrete differences between alleles can

Figure 6.1: Generation of a mismatch distribution from a matrix of pairwise distances.
A distance matrix is constructed using pairwise differences between a set of five sequences. Differences between the sequences are shown in *blue* and underlined to facilitate comparisons between sequences. The mismatch distribution is a histogram obtained by counting the number of pairwise comparisons that share the same number of differences between the two sequences.

Sequences	
I	AGTCTTACGTATC
II	AGTCTTGCGTATC
III	AGTTTTACGTATC
IV	AGTCTTGCGTGTC
V	AGTCTTACGTATC

Pairwise differences (mismatches)

	I	II	III	IV	V
I	-				
II	1	-			
III	1	2	-		
IV	2	1	3	-	
V	0	1	1	2	-

Frequency of pairwise differences

p.d.	Num.	Freq.
0	1	0.1
1	5	0.5
2	3	0.3
3	1	0.1

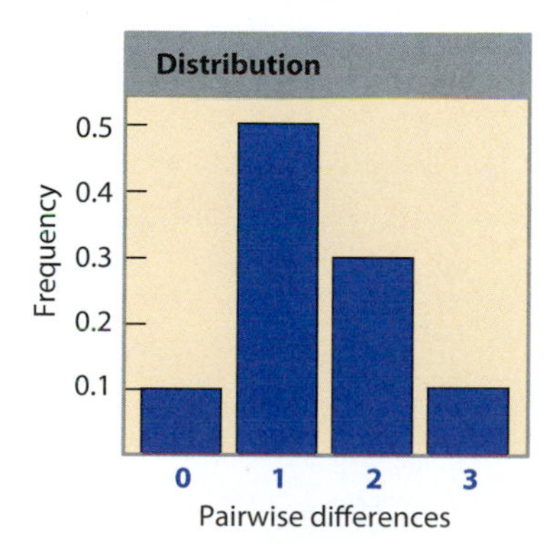

be counted; these differences can be nucleotide substitutions or microsatellite repeat units, for example. The distribution of the number of such differences between each allele and every other allele summarizes the discernible genetic diversity. **Figure 6.1** shows how the mismatch distribution is calculated from DNA sequences.

The mismatch distribution is a good example of a descriptive **summary statistic** that overlaps with inferential methods. As well as describing the diversity apparent within a sample, the shape of the distribution has been shown to be indicative of population history, in particular being influenced by episodes of population expansion. While the mean of the distribution provides a simple description of the overall diversity, the shape of the distribution is also informative. A smooth, bell-shaped mismatch distribution indicates a period of rapid population growth from a single haplotype, whereas a ragged, multimodal distribution indicates a different situation, for example a population whose size has been constant over a long period (**Figure 6.2**).

6.3 MEASURING GENETIC DISTANCE

Measures of **genetic distance** are statistics that allow us to compare the relatedness of populations or molecules. The greater the evolutionary distance between them, the greater the numerical value of the statistic. If a measure is greater between population A and B than between C and D, we can say that C and D are more closely related than are A and B.

Such measures allow us to explore population structure and molecular diversity in greater detail, by pairwise comparisons, rather than by averaging over all populations or molecules. As we shall see, by making certain assumptions, it becomes possible to convert distance measures to an evolutionary time-scale. This might allow us to say, for example, not only that C and D share a more recent common ancestor than do A and B, but that the common ancestor of A and B is twice as old as that of C and D.

Historically, genetic distances between populations were based solely on allele frequencies, but as with the measures of population subdivisions discussed above, we can also include molecular information, in the form of distances between alleles, haplotypes, and indeed genomes. Many of the methods used to determine genetic distance between two populations or between two molecules (known as a **pairwise distance**) can be expanded to examine more complex issues of population structure.

Genetic distances between populations can be measured using F_{ST} or Nei's *D* statistics

There are several ways to measure genetic distances between populations. Different measures are used depending on the type of data and different expectations about the underlying evolutionary processes. For example, diversity data from polymorphisms with a high mutation rate may be analyzed with a genetic distance measure that emphasizes the contribution of mutational processes to population divergence. Alternatively, genetic drift may be thought to be the predominant process causing population divergence at SNPs, and the genetic distance measures chosen to reflect that.

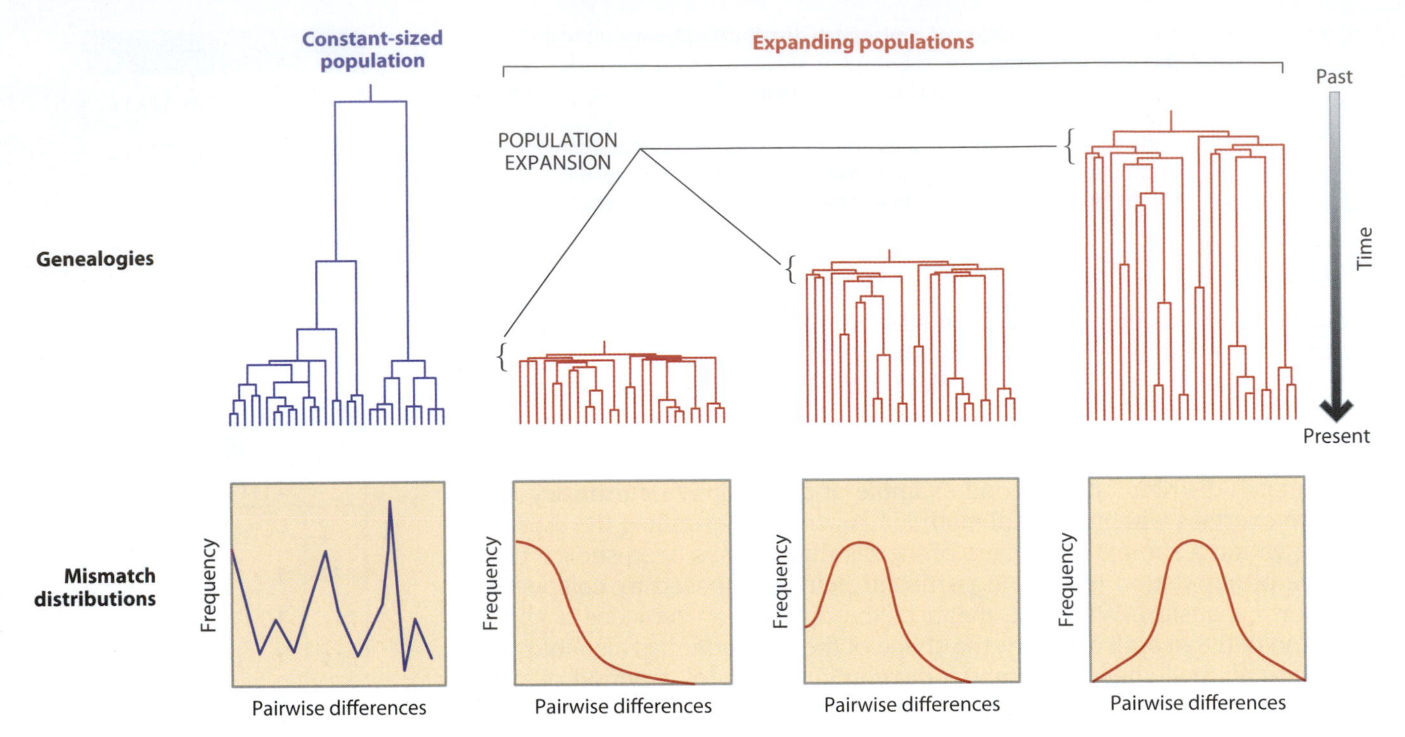

Figure 6.2: Genealogies and mismatch distributions for a constant-sized population and three populations that have undergone population expansions at different times.
The longer branches in the genealogy of the constant population are lineages ancestral to many individuals, whereas the longer branches in the expanding populations are often specific to individuals. Because branch length is indicative of time and therefore the accumulation of variants, more alleles are shared among individuals in the constant population than in the expanding population, and greater numbers of alleles differentiate individuals in populations with more ancient expansions. Thus it can be seen how the mismatch distributions summarize this information.

If genotype frequencies among our sampled individuals are not in Hardy–Weinberg equilibrium, but exhibit a deficiency of heterozygotes, we should question whether these individuals come from a single randomly mating population (**Section 5.1**). Indeed, if the population is comprised of several partially isolated subpopulations, this alone may cause deficiency in heterozygotes. This subdivision generates a **hierarchical population structure**. A single metapopulation composed of a number of subpopulations is the simplest, two-tiered, example of such a structure. Once data have been shown not to conform to randomly mating (**panmictic**) expectations, then it can be said that there is population structure, and we can measure the apportionment of diversity amongst these different tiers of the hierarchy.

If we consider two populations X and Y with the frequency of the *i*th allele being x_i and y_i respectively, the simplest measure of genetic distance between two populations sums the difference between the allele frequencies, $\Sigma(x_i - y_i)$. This needs to be squared to avoid differences in sign canceling each other out, $\Sigma(x_i - y_i)^2$. However, this quantity fails to give sufficient weight to alleles with frequencies close to 0% or 100%.

Two commonly used classical measures of genetic distance are F_{ST} (see **Section 5.3** and Box 5.2) and Nei's standard genetic distance, *D*. *D* varies between 0 and infinity, and relates the probability of drawing two identical alleles from the two different populations (which is $\Sigma x_i y_i$) to the probability of drawing identical alleles from the same population (Σx_i^2 and Σy_i^2) by the following equation:

$$D = -\ln\left(\frac{\Sigma x_i y_i}{\sqrt{\Sigma x_i^2 \Sigma y_i^2}}\right)$$

By making assumptions about the processes that are driving the divergence of populations, we can relate distance measures to time. This relationship can then be used to generate a "corrected" (or "transformed") version of the statistic that can be shown (under certain assumptions) to be linear with respect to evolutionary time.

For example, Wright showed that under the action of drift alone, F_{ST} varies with time (t) according to the equation

$$F_{ST} = 1 - e^{(-t/2N)}$$

Rearranging this gives:

$$t = -2N \ln(1 - F_{ST})$$

Thus the measure $-\ln(1 - F_{ST})$ is linear with respect to time.

Similarly, Nei's standard genetic distance, D, under certain assumptions, should also be linearly related to time by:

$$D = 2\alpha t$$

where α is the rate of fixation of nucleotide substitution. The implications of these equations for genetic dating are discussed in greater detail in **Section 6.6**. A clarification of the different D statistics is given in **Table 6.2**.

Linearity of the genetic distance measure with respect to time is a useful property especially when constructing phylogenies. However, some demographic factors can disrupt the linear relationship between a given genetic distance measure and time. These include population bottlenecks and even minimal amounts of migration between diverging populations. Most questions of anthropological interest involve processes occurring over relatively short time periods, during which few mutations accumulated, but genetic drift and migration may have been substantial. The other major property that affects the usefulness of a measure is its variance. How precisely does the estimate from an empirical sample reflect the true population value? If hundreds of individuals are needed from each population for an estimated measure to be accurate, this will decrease its usefulness. Whatever measure of genetic distance between populations is used, we must test its significance: that is, determine if the distance is significantly different from zero. This is especially important for human populations, which are closely related (Chapter 10). Our confidence in saying that two populations are genetically different depends on the variance of the statistic we use—the lower the variance of the statistic, the higher the confidence.

A number of measures of population genetic distance have been specifically developed for use with microsatellites. The mutational dynamics of these loci differ significantly from the standard infinite alleles model of molecular evolution (**Section 5.2**), or indeed the infinite sites model described above. The widely-used F_{ST} analog $\boldsymbol{R_{ST}}$ accounts for the molecular distances between alleles using the stepwise mutation model (SMM).[24]

Distances between alleles can be calculated using models of mutation

In principle, once the molecular basis of allelic variation has been defined, it is simple to estimate distances between alleles. Two homologous sequences will differ at a certain number of sites, and two microsatellite alleles will differ

TABLE 6.2: AN ABC OF *D* STATISTICS

D statistic	What does it measure?	Section in this book
Tajima's *D*	allele frequency spectrum	6.7
Lewontin's *D*	linkage disequilibrium	Box 3.5
Nei's *D*	genetic distance	6.3
Patterson's *D*	genomic distance	6.3

by a certain number of repeat units. By assuming that the differences between alleles accumulate one by one, the number of differences between alleles represents simple genetic distance.

As we saw previously, linearity with respect to evolutionary time is an important property of genetic distances. However, parallel mutations and reversions during the course of evolution disrupt the linearity of molecular genetic distances. Such events are much more likely over a given period of time for loci with a fast mutation rate, such as microsatellites, than for loci with a slower mutation rate, such as SNPs. They result in the number of differences observed between two molecules being an underestimate of the actual number of mutations that have occurred since their split from a common ancestor.

For genetic distances based on microsatellite data, we can use models of microsatellite mutation to correct for this effect. The model most frequently used is the stepwise mutation model (SMM) discussed in the previous section. This model has been used to derive a measure of microsatellite genetic distance, which is the average of the squared distance between alleles (known as **ASD**). This measure has been shown to be linear over longer time periods than other measures. Nevertheless, once more complex, and realistic, models of microsatellite evolution are considered, such as those that incorporate constraints on allele size, the linearities of all measures are substantially reduced.

Genetic distances between nucleotide sequences can be corrected using specific models of sequence evolution (discussed in greater depth in **Section 5.2**). However, the amount of sequence divergence observed between homologous human sequences is so low, typically less than 0.1%, that sites are not close to saturation, and so the observed sequence differences are a reasonable approximation of the actual number of mutational changes. Consequently, as long as mutation rates at all sites are similar, little if any correction is required. It becomes more important to correct the observed sequence divergences when comparing sequences from different species, especially if they are distantly related. Some Jukes–Cantor corrected sequence distances between hominoids are shown in Table 7.1.

Genomewide data allow calculation of genetic distances between individuals

Calculating a useful genetic distance between individuals becomes possible with large numbers of loci. When one SNP is genotyped, an individual can only be in one of three classes: homozygous for allele A_1, homozygous for allele A_2, or heterozygous. As the number of loci rises, the number of classes to which individuals can be assigned increases, particularly if the loci are independent and not in linkage disequilibrium. The ultimate aim is to have enough loci to identify each individual as genetically unique. This can be achieved (except for identical twins) by using a few loci with multiple alleles, such as microsatellites or minisatellites, which forms the basis for individual identification in forensics (**Section 18.1**), or a few tens of biallelic loci. Thus SNP chips that now routinely genotype over a million SNPs contain ample information for individual identification (**Section 4.5**).

How do we measure genetic distance between individuals? One way is to compare each individual against every other sampled individual, for all genotyped SNPs. This generates a covariance matrix that has as many rows and columns as sampled individuals. Each number in the matrix represents the degree of similarity between two individuals, as measured by all the genotyped SNPs. This is analogous to a distance matrix between populations, and can be visualized in two dimensions using multivariate analysis, such as **principal component analysis** (**PCA**, see below).

An alternative approach uses a co-ancestry matrix; that is, for N genomes a matrix of $N \times N$ where each element in the matrix x_{ij} is an estimate of the number of discrete segments of genome i that are most closely related to the

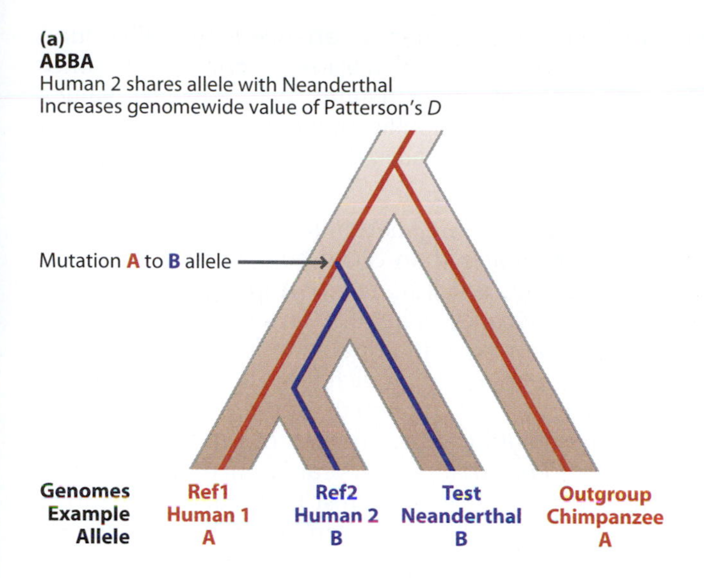

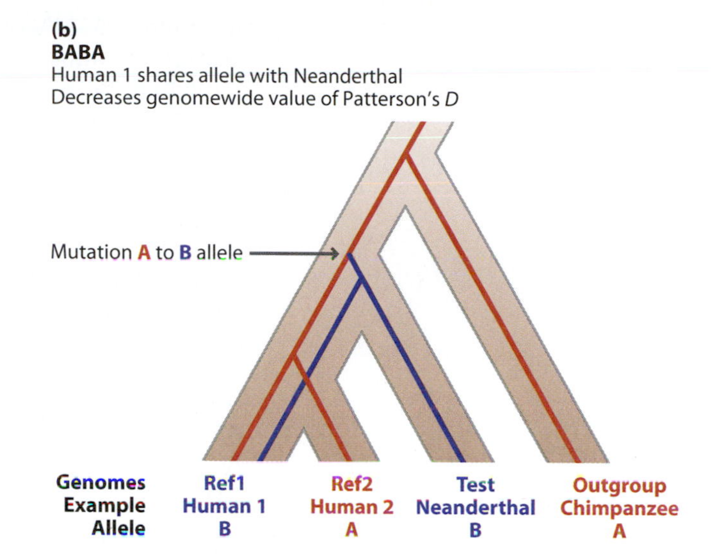

Figure 6.3: Measuring the distance between individual genomes by the ABBA/BABA approach. (a) A gene genealogy where one human genome (human 2) shares an allele at a SNP with a Neanderthal but not with another human (human 1). This "ABBA" configuration would increase the value of Patterson's *D* statistic, which is calculated for all SNPs across the genome. (b) A gene genealogy where human 1 shares an allele at a SNP with a Neanderthal but not with human 2. This "BABA" configuration would decrease the value of Patterson's *D* statistic.

corresponding region of genome *j*. This is a particularly powerful approach when the genomes have been genotyped at a lot of variants, by genome re-sequencing for example. This is because the co-ancestry matrix makes use of phased haplotype data, and as such can reveal population structure and relationships between individuals due not only to allele frequency differences, but also to differences in linkage disequilibrium.[48] It seems that this approach is able to detect finer population structure, and more subtle genetic relationships between individuals, than methods based on genetic distance. The co-ancestry matrix generated by this approach can then be analyzed by PCA, in a similar manner to the genetic distance matrices described above.

The distance between individuals' genomes can be calculated by counting the number of differences between the two genomes directly, or by counting the number of differences when compared with a reference genome. This can then be corrected using a specific model of sequence evolution as described in **Section 5.2**. An alternative is the statistic **Patterson's *D***, which was devised as a test statistic to analyze the relationship of ancient genomes to modern human genomes, but can be used for any four-way genome comparison. The format of the statistic is $D(\text{ref}_1, \text{ref}_2, \text{test}, \text{outgroup})$ where we want to compare the relative closeness of the test genome to the ref_1 and ref_2 genomes; the outgroup genome is usually chimpanzee.[31] If *D* is zero, then the derived alleles in the test sequence, for example in the Neanderthal genome, match alleles in the two human reference genomes equally often. If *D* is positive then the derived alleles in the Neanderthal match alleles in ref_2 more often than ref_1, and if *D* is negative then the derived alleles in the Neanderthal match alleles in ref_1 more often than ref_2. This approach is also known as the ABBA/BABA approach, referring to the ancestral (A) or derived (B) alleles of the four genomes (**Figure 6.3**).

Complex population structure can be analyzed statistically

Many of the methods used to determine genetic distance between two populations, or between two molecules, can be expanded to examine more complex population structure. For example, F_{ST} can be thought of as measuring the proportion of the total variance in allele frequencies that occurs between subpopulations (Box 5.2). If genetic drift results in the subpopulations being highly differentiated, this proportion (and F_{ST}) will be large, while if large amounts of gene flow between subpopulations maintain their similarity, this proportion will be much smaller ($F_{ST} \approx 0$). As noted, F_{ST} values vary between 0 and 1, and an F_{ST} of 0.3 means that 30% of the total allele frequency variance is found between subpopulations, with the corollary that 70% of allele frequency variance exists within the subpopulations themselves.

In humans 5–13% of common SNP allele frequency variance is typically found between continental groups (see Table 10.2), which is low compared with other large mammals with broad geographical ranges (Figure 10.4). However, values of population subdivision at some loci are significantly elevated or depressed when compared with this average. In some cases this is due to the impact of selection (**Section 6.7**).

A critical requirement is a significance test to determine whether or not any apparent population subdivision is greater than could be expected, based on chance differences in allele frequencies resulting from sampling effects alone. One method for determining the significance of population subdivision is to use a **permutation** test. Such tests are common in population genetics where the variation of a given measure cannot be easily predicted by standard statistical distributions (for example, normal, Poisson, binomial), and are also known as **Monte-Carlo methods**, a name resulting from the use of random numbers to emulate a casino-like situation. Permutation tests randomize the empirical data many times and calculate the measure of interest from each randomization. The real measure from the observed data is then compared against the permuted measures to see if it is significantly different.[50]

An alternative to allele frequency-based methods is to use the analysis of molecular variance (**AMOVA**), a method that considers, for example, the variance in the number of microsatellite repeat units at a given locus.[21] This method takes into account the molecular relationship of alleles, rather than just their frequency, when apportioning variance between tiers of the hierarchical population structure, and can calculate φ_{ST} (φ, the Greek "F," is pronounced in English "fie" to rhyme with sky), a molecular analog of F_{ST}. The AMOVA method can be applied to any data where genetic distances between alleles can be calculated.

If we assume that one of the simple models for population substructure outlined in the previous chapter (for example, an ***n*-island** or stepping-stone model) is a reasonable approximation of the metapopulation being studied, we can calculate certain parameters of that structure, for example the migration rate per generation between subpopulations. To do this, we have to assume that the metapopulation structure has reached **migration drift equilibrium** (**Section 5.6**); in other words the present rates of migration and genetic drift have remained unchanged for long enough that the level of population subdivision has reached a stable equilibrium value.

Population structure can be analyzed using genomic data

The development of SNP chips and genome resequencing has allowed population structure analysis to be performed at the individual level—asking the question, To which population does this individual belong? This has involved the application of **cluster analysis** to genetic data. The aim is to group different individuals into *K* different clusters based on genetic similarity, so that the individuals within a cluster are more similar to each other than to individuals outside the cluster. We can choose *K* based on our prior hypothesis—for example if we are analyzing a population affected by an admixture from one other population, we would choose $K = 2$ (**Section 14.3**, Figure 14.12). Alternatively, if we do not have a clear prior hypothesis for the number of clusters to expect, the clustering algorithm can be run several times with different values of *K*. Each run gives a different probability reflecting the fit of the clusters to the data, assuming Hardy–Weinberg equilibrium (**Section 5.1**). The *K* value that gives the best fit to the data is the best estimate of the true number of clusters, or populations. There are difficulties in interpreting the relationship of *K* to the population structure of the individuals analyzed. First, the probabilities reflecting the fit of the data for each value of *K* are computationally difficult to obtain, and are best regarded as approximate figures, particularly when several *K* values give similar probabilities. In these situations, it can be regarded as best to take the smallest value of *K* as the true value. Second, interpreting a statistically meaningful *K* (that is, one that maximizes the probability of the fit) as a biologically meaningful result can be difficult.

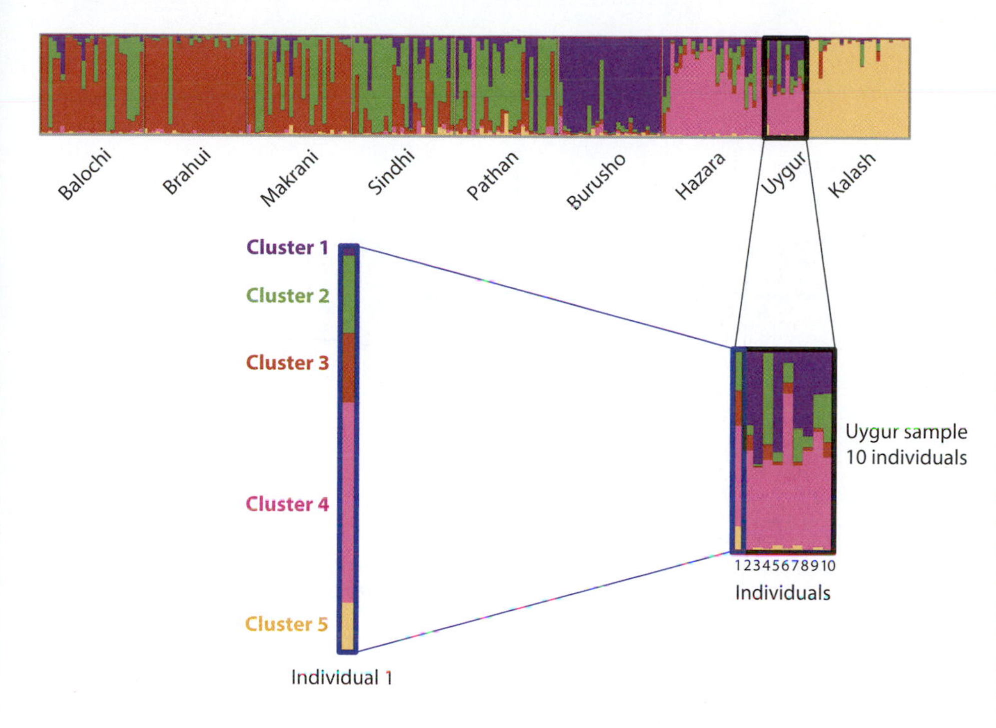

Figure 6.4: Detecting population structure from genomewide data by cluster analysis.
The top part shows a typical output from the program DISTRUCT, using data from the program STRUCTURE. In this case data from 377 microsatellites on 210 individuals from nine populations from Pakistan have been clustered. Five clusters ($K = 5$) have been specified, and each cluster is represented by a different color. Ten Uygur samples have been highlighted, of which one individual has been shown in detail. Each individual is shown as a vertical bar with the probability of that individual's genome being a member of each of five clusters shown as different colored segments of that bar. In this particular individual, there is about a 50% probability of the genome being a member of cluster 4, with smaller probability for clusters 3, 2, and 5, and a minimal probability of the genome belonging to cluster 1. [Adapted from Rosenberg NA et al. (2002) *Science* 298, 2381. With permission from AAAS.]

The result of this cluster analysis is a matrix where each individual is given a value or **membership coefficient** for every cluster, which reflects the probability of that individual belonging to that cluster. These coefficients can also be interpreted as the fractions of the genome having membership in those particular clusters, and can be shown graphically for each individual, visualizing the population structure across all individuals (**Figure 6.4**, Figure 14.8). The programs STRUCTURE and CLUMPP are used to cluster genetic data,[22, 44] with graphical output produced using the programs DISTRUCT or STRUCTURE itself (see below). More recently, the program ADMIXTURE has been widely used for clustering genetic data for the detection of recent admixture events (**Section 14.3**).

Genetic distance and population structure can be represented using multivariate analyses

How can we display a set of pairwise genetic distances in a comprehensible manner? If we have *n* populations, we require *n* dimensions to fully display their pairwise genetic distances on a graph (**Figure 6.5**). However, we often have more than four populations or molecules that we want to compare, yet we cannot conceive of, or represent, the four or more dimensions required to display these data. **Multivariate analyses** allow us to reduce these multiple dimensions to the two or three dimensions we can draw and comprehend, while minimizing the inevitable loss of information. **Multi-dimensional scaling** (MDS) is one of these methods, where the loss of information is represented as a stress statistic, with the lower the stress statistic, the better the MDS fit to the data and therefore the less information lost.

It is also possible to use multivariate analysis to generate two- or three-dimensional graphical representations of distances between populations using the raw data of allele frequencies, rather than genetic distances. Principal component analysis (PCA) is a commonly used example of this approach. Individual axes, known as principal components (PCs) or **eigenvectors**, are extracted sequentially, with each PC being independent and encapsulating as much of the remaining variation as possible. Using PCA it is possible to estimate the proportion of the total variance in the dataset that has been summarized within these reduced dimensions. PCA can reveal global relationships from a set of pairwise distances between populations (**Figure 6.6**). PCs have been used to construct

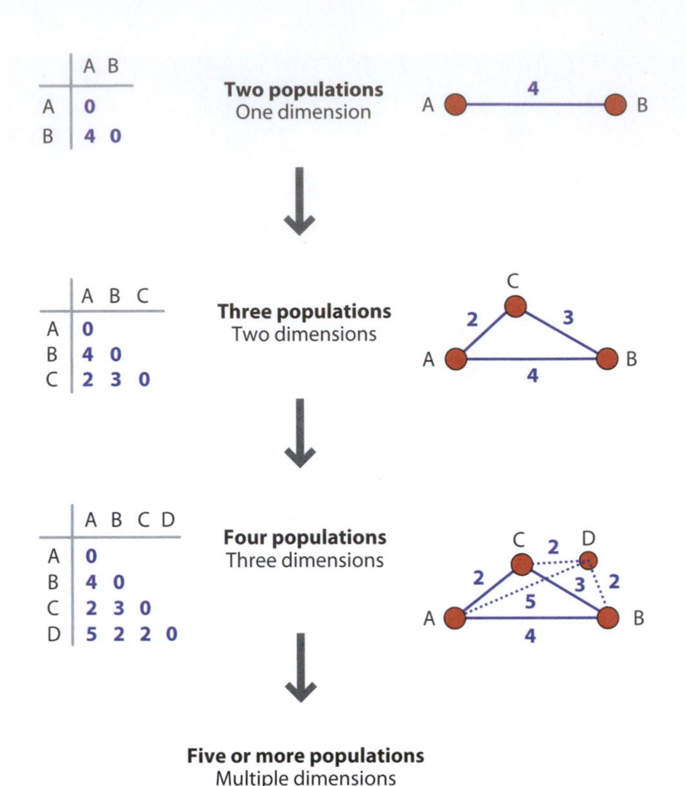

Figure 6.5: How many dimensions are needed to display population relationships? Distance matrices are shown relating genetic distances between increasing numbers of populations. As the number of populations increases, the number of dimensions needed to display these distances visually such that the lengths of the lines represent the genetic distances between them quantitatively also increases. A single line of a given length can relate two populations. Three populations can be related by a triangle, whose sides have lengths proportional to the genetic distance. Four populations can be represented by a pyramid. With five or more populations it is no longer possible to represent population relationships in three-dimensional space.

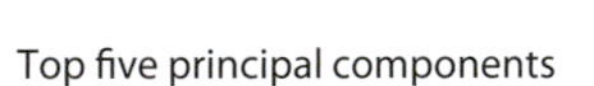

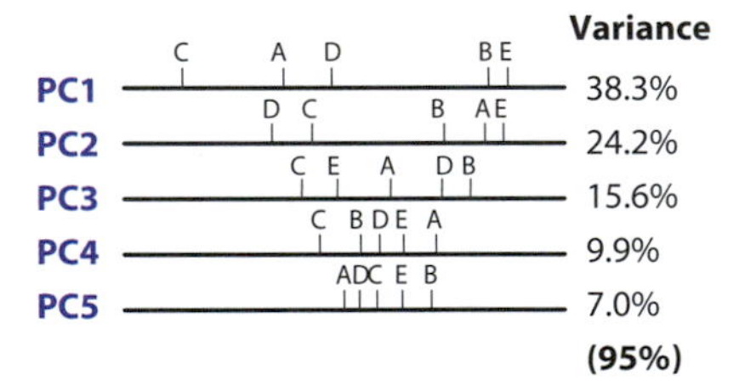

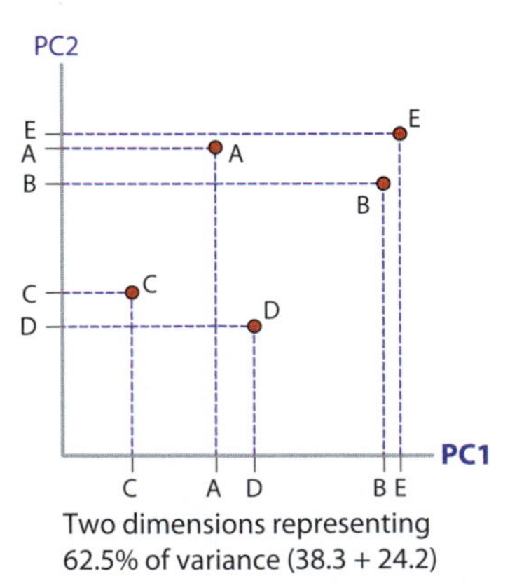

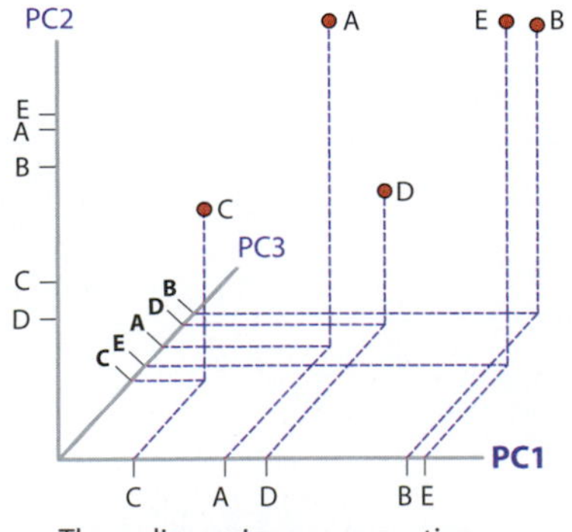

synthetic maps that summarize information from several alleles with similar geographic distributions (Figure 12.12). Interpretation of these maps is not necessarily straightforward (**Opinion Box 3**).

PCA is also used to display genetic distances between a set of individuals, rather than a set of populations. **Figure 6.7** shows an example from real genomewide SNP data. Each data point represents an individual person, plotted according to the **eigenvalues** of the first two PCs of the data calculated from the covariance matrix of a genomewide SNP dataset. Plotting values for two PCs for every individual allows the spatial separation of the different groups; the first PC (*x* axis) separates the Fulani individuals from the other Africans, and the second PC (*y* axis) separates three populations from Chad and northern Cameroon from the other populations closer to the coast. Examples of how PCA plots can be used in admixture analysis are in Chapter 14, and other examples of PCA plots in this book are Figure 10.6 and Figure 13.14.

For the plots of genomewide data, the first two PCs are normally shown, but any two PCs can be plotted to search for useful information. It should be remembered that, for genomewide analysis, although the first two PCs will represent the largest amount of variation of all PCs, they may still represent a small amount of the total variation in the dataset. Principal components can also be rotated, a statistical procedure that is sometimes used to emphasize the relationship between the genetic structure and geography (Figure 6.7b, Figure 18.8).

Figure 6.6: Graphical representations of principal component analysis of five populations in both two and three dimensions. Principal components (PCs) are extracted from multivariate data from five populations (A–E) such that each successive PC contains a smaller proportion of the overall variance. Each PC is displayed as a single axis. The first five PCs account for 95% of the variance within the dataset. These PCs can then be used as axes for graphs that maximize the amount of variation displayed in a fixed number of dimensions.

Principal component analysis (PCA) is a useful tool for population geneticists, but it does not come without challenges. While it often lays bare important patterns of variation, it can sometimes belie intuition: one colleague once moaned to me that parsing through the first few PCs of a complex dataset was like "reading tea leaves." Thankfully much progress has been made in understanding how PCA behaves when applied to genetic data, resulting in better awareness of its pitfalls and improved interpretation of results.

Much progress has come from careful study of the underlying mathematics and from making novel connections between PCA and other areas, such as random matrix theory[60] and expected pairwise coalescent times.[53] In my work with Matthew Stephens,[58] we found how the PCs of spatially structured genetic data are connected to the Discrete Cosine Transform. Beyond the math, simulations have also been useful.

When a sample is composed of individuals from K discrete, differentiated populations, the PCA results are fairly straightforward to interpret. If the dataset is above a critical size, visual inspection will show K distinct clusters of samples, stratified across $K - 1$ PCs associated with sizeable eigenvalues. Admixed individuals will appear intermediate between these clusters at distances that correspond to their relative ancestry in each source population. One complication is that the positions of the groups depend on the sample sizes, as well as the relative levels of differentiation. If these are highly variable, the spacing will produce a distorted view of the true genetic differentiation.

A common alternative to discrete population structure is the case where individuals that are distributed across a spatial continuum and ones that are closer geographically are more genetically similar. If such a spatial structure is homogeneous and sampling is uniform, then PCA will no longer show discrete clusters; instead the PC coordinates will form sinusoidal functions over geographic space. For example, in two dimensions, PC1 and PC2 will each typically be a gradient, one perpendicular to the other, and PC3 and PC4 will be saddle- and mound-like shapes (**Figure 1**). These sinusoidal functions might seem surprising, but they emerge mathematically, in part as a consequence of PCA's attempt to summarize the variation as efficiently as possible (there are connections here to image compression techniques such as JPEG).

One consequence of these sinusoidal functions is that one can use PC coordinates as proxies for spatial ancestry, for example reconstructing a geographic map from PC1 versus PC2 plots (see Figure 18.8). PC coordinates can therefore be used to address population stratification in genomewide association studies or, if training data are available, to infer the geographic ancestry of individuals of unknown origin (though admixed individuals are problematic).

The sinusoidal functions also suggest that gradient and wave-like shapes in PC maps should not be interpreted as evidence for anything more than spatial structure in populations. Previous researchers have sometimes linked them to specific processes such as population expansions (famously, a NW–SE gradient in PC1 as evidence of the Neolithic expansion in Europe—see Section 12.5). In a simulation study we found that gradients in PC1 emerge with and without expansions, and can even lie perpendicular to the expansion wave.[28]

In practice, many datasets show a blend of discrete and spatial population structure, and often include recently admixed individuals. The core ideas just discussed help in interpreting such complex scenarios. Also, PCA is sensitive to factors that are independent of population history, including batch effects due to how samples were grouped in the genotyping process. If variants in linkage disequilibrium (LD) are not filtered prior to analysis, some PCs can describe long LD blocks, such as occur in inversions or after selective sweeps. Despite these complications, if used carefully, PCA is a powerful tool for exploring data and gaining more insight into large genetic datasets.

John Novembre, Department of Ecology and Evolutionary Biology, University of California, Los Angeles, USA

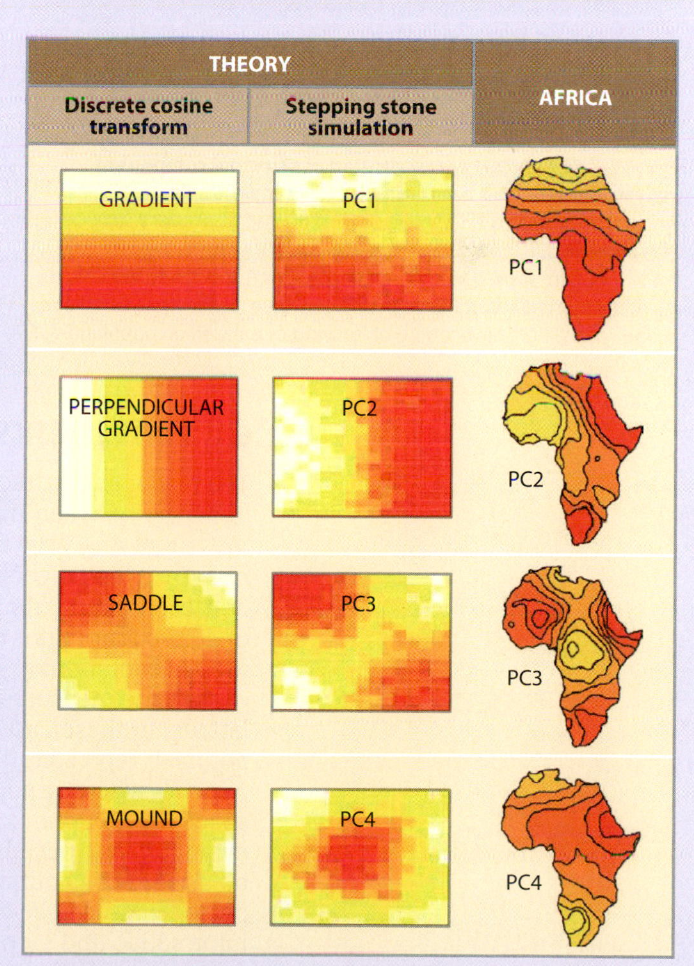

Figure 1 Theoretical, simulated and real PC maps. First column: Theoretical expected PC maps for models in which genetic similarity decays with geographic distance. Second column: PC maps for population genetic data simulated with no range expansions, but constant homogeneous migration and drift. Third column: PC maps from classical polymorphism data for Africa.

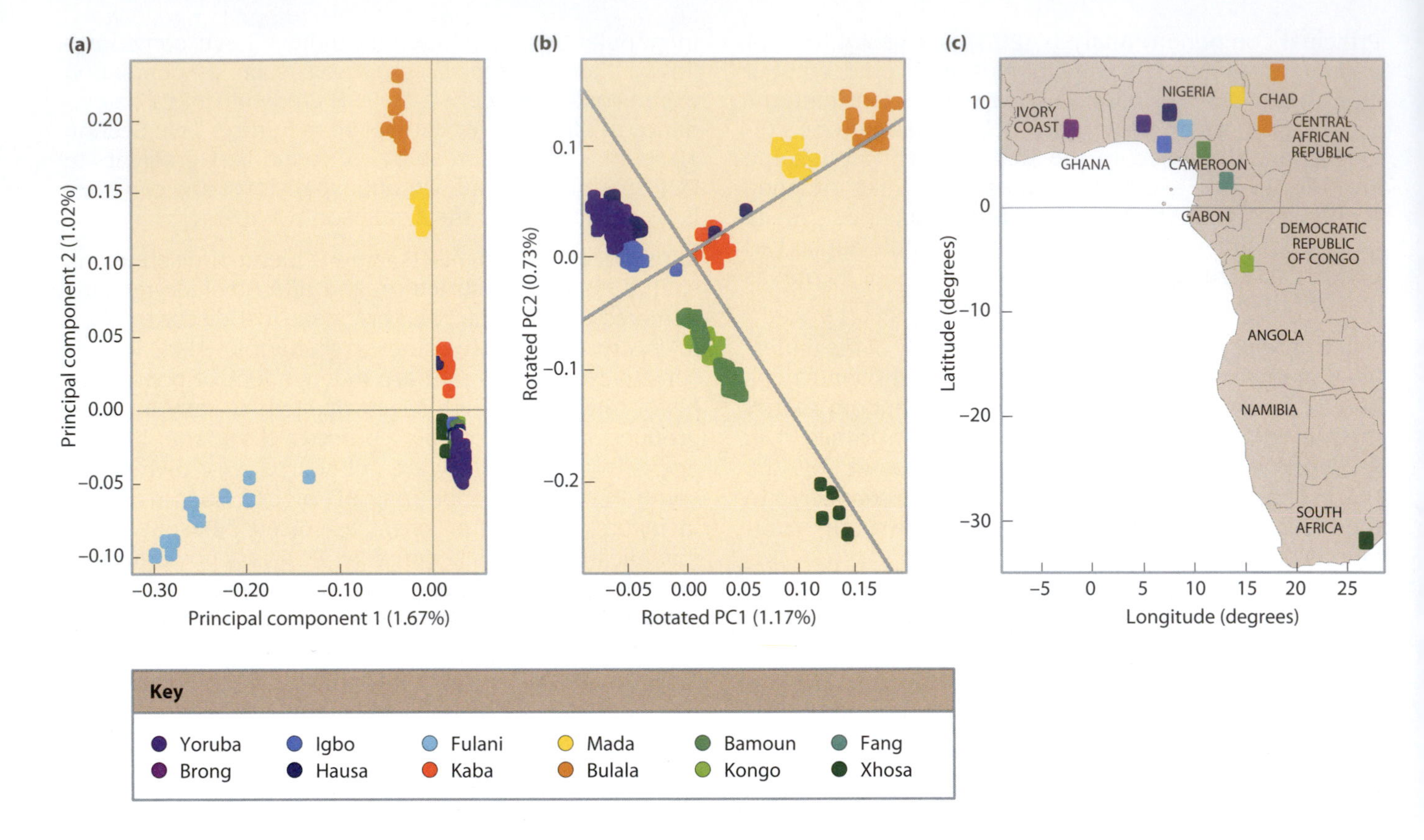

Figure 6.7: Graphical representation of principal component analysis (PCA) of SNP data from different individuals. Principal components (PCs) have been extracted from over 250,000 SNPs typed in 225 individuals from 11 populations in West Africa. (a) The first two PCs are shown on the *x* axis and *y* axis respectively, with each individual plotted according to their eigenvalue for the first (*x* axis) and second (*y* axis) PC. Color-coding of individuals is based on population of origin. Note the relatively small proportion of total variation described by PC1 (1.67%) and PC2 (1.02%), typical of PCA plots of genomewide data. (b) A rotated PC plot, without the Fulani population, is used to highlight the spatial similarity described by the first two PCs and the geographical distribution of samples shown in (c). [From Bryc K et al. (2010) *Proc. Natl Acad. Sci. USA* 107, 786. With permission from National Academy of Sciences.]

6.4 PHYLOGENETICS

The phylogenetic tree is an intuitively attractive method for displaying the relationships between many kinds of entities. Sometimes the tree itself describes the actual ancestral relationships of these entities; it encapsulates the *mechanism* by which diversity arose. Such is the case for trees of separate species or nonrecombining haplotypes of unique variants—for example, DNA sequences of different primates or human Y-chromosomal SNPs—and this is how trees are used in evolutionary genetics today. A tree is also a tool for the graphical display of distance measures of other kinds, such as the relationships of human populations. In these cases, it may imply a *model* for how diversity arose, but does not itself represent the *mechanism*. For example, a tree of populations implies a model whereby populations split (fission) from common ancestors, and subsequently do not mix. This certainly does *not* represent the reality of population evolution, and because of the potential for confusion, trees should not normally be used for this purpose. Further reading on this, and other matters relating to phylogenies, can be found in Felsenstein's book *Inferring Phylogenies*,[25] which is informative and surprisingly entertaining for a field which is generally thought to be rather dry.

Phylogenetic trees have their own distinctive terminology

Due to their inherent attractiveness as graphical tools, trees are used across a wide variety of disciplines, for a broad range of purposes. This has led to a

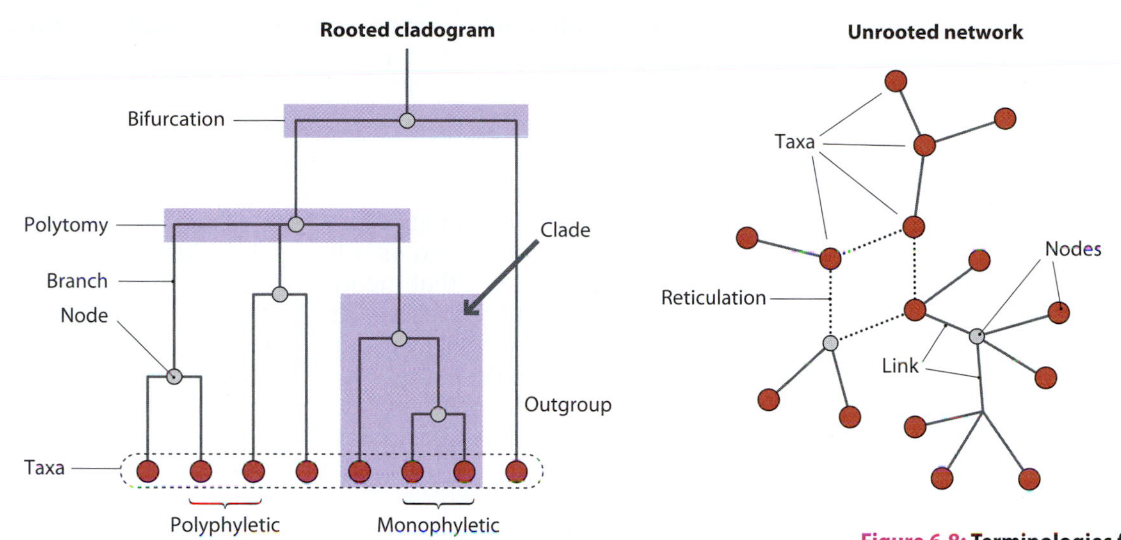

Figure 6.8: Terminologies for trees and networks. A number of different features of evolutionary trees and networks described in the text are highlighted.

set of partially redundant terminologies that can easily confuse the uninitiated (**Figure 6.8**).

A tree consists of **branches** (also known as **edges**) between **nodes**. The ultimate aim is to relate groups of populations or molecules, known as **taxa** (singular, **taxon**) or operational taxonomic units (OTUs). These taxa occupy special nodes. Nodes unoccupied by taxa represent hypothetical ancestors (or hypothetical taxonomic units—HTUs). In a **cladogram**, all terminal nodes (known as leaves) represent taxa, whereas all internal nodes are hypothetical ancestors. However, other types of evolutionary trees also allow taxa to occupy internal nodes.

Trees can be **rooted** or **unrooted**. Rooted trees are trees where a node that is ancestral to all taxa is known. The root defines the directionality in the tree with respect to evolutionary time, meaning that evolutionary changes are assumed to have occurred from ancestral to derived states. An unrooted tree can be rooted either by assuming that the root falls midway along the longest branch on the tree (mid-point rooting), or preferably by incorporating a taxon known to be an outgroup to all other taxa, and seeing where it joins the unrooted tree.

The proximity of an internal node to the root of the tree determines the relative antiquity of the divergence event it represents. Nodes that are closer to the root are more ancient than those further from the root. Unrooted trees are not able to relate the ancestry of different nodes in this way; it is not immediately obvious which nodes are descendants and which are ancestors. The number of possible trees for any given number of taxa increases rapidly and is different for rooted and unrooted trees. For nine taxa, there are 135,135 possible unrooted trees but over 2 million rooted ones.

The branching pattern of a tree is known as its **topology**, and the descendants of a single node form a **clade**. Some trees allow only two branches to descend from each node, a process known as **bifurcation**. Alternatively, more than two branches can descend from the same internal node, forming a **polytomy** (or **multifurcation**). A cladogram represents solely the relationships between taxa—branch lengths between nodes are irrelevant. By contrast, **additive** trees use branch lengths to reflect evolutionary distance quantitatively. Thus additive trees can vary not only topologically, but also quantitatively in the length of their branches.

The relationship between a group of taxa and clades allows taxon groupings to be classified on the basis of the tree. A grouping of taxa that fall into a single clade is **monophyletic**. However, if this grouping excludes other members of the same clade it is **paraphyletic**. Taxa that span multiple clades are **polyphyletic**. Only a monophyletic grouping of taxa is considered to be a coherent evolutionary lineage.

An important property of trees is that as evolutionary time progresses toward the present, branches diverge but never coalesce. However, some biological processes (for example, recombination) can cause lineages to merge, while others (for example, parallel mutation) cause them to *appear* to merge. In either case the result can be represented as a four-sided closed structure known as a **reticulation**, or cycle. Trees that incorporate such structures, in an attempt to include these biological processes, are known as **networks**. Some network methods represent taxa solely as terminal nodes, while others also allow taxa to occupy internal nodes. Lines connecting nodes are known as **links**. Alternatively, such reticulations can be removed by **pruning**, an approach that has been used to simplify trees of mtDNA sequences by removing the tips (Figure 7.19a).

There are several different ways to reconstruct phylogenies

There are many different methods for constructing trees and networks from genetic data. Not all are suitable for all the different types of data available, and no single method predominates. Tree construction methods are generally classified using two important criteria: first, the type of data used as input, and second, the means by which a tree is constructed.

Input data fall into two major classes: *distances* and *characters*. Genetic distances between populations or molecules must first be calculated from raw data, as described in **Section 6.3**, and are represented in the form of a **distance matrix**. Characters are discrete units of evolution, whether single-base changes in a nucleotide sequence, or changes in numbers of repeat units of a microsatellite. Character-based methods allow us to infer the character content of ancestors.

There are two main classes of phylogeny construction methods. The first, known as *clustering* methods, uses an iterative algorithm to combine taxa together in a hierarchical fashion (one by one). The second class, known as *searching* methods, considers the whole range of possible trees and chooses that which best fits the data according to some **optimality criteria**. In practice, the range of possible trees is often so large that it becomes computationally unfeasible to compare all trees. It has been shown that if a million trees could be compared every second, it would still take 10 million years to compare all possible trees relating just 20 taxa. Consequently these methods seek to sample a representative subset of all trees. In practice this is done by: (1) jumping between separated regions of the entire range of trees; (2) finding the best tree in each region; and (3) comparing the best trees from each region.

How can we choose one phylogenetic method over another? A good phylogenetic method should have five characteristics; it should be

- Efficient—a tree is constructed rapidly
- Consistent—the same tree is obtained as more data are added
- Robust—the tree is insensitive to violations of the method's assumptions
- Powerful—few data are required to get the correct tree
- **Falsifiable**—the validity of the method's assumptions can be tested

As we shall see below, no single method has all five characteristics—each emphasizes some desirable properties over the others. Fast methods are not always robust, and powerful methods are often slow.

Trees can be constructed from matrices of genetic distances

The unweighted pair-group method with arithmetic mean (UPGMA) is perhaps the simplest phylogenetic method. The tree is built by an iterative clustering process that combines the two taxa that have the smallest genetic distance between them. When taxa are combined they form a new taxon. The genetic distances between this composite taxon and other taxa are the average of the distances from the individual constituent taxa. The UPGMA produces a special form of additive tree, known as an **ultrametric** tree. This tree has the interesting property that the distance between each taxon and the root is the same. If we draw a scaled version of this tree, we find that all the terminal nodes are aligned. Therefore, the UPGMA method has the advantage of being a convenient representation of taxa living in the same moment in time. However, the UPGMA method assumes equal rates of evolution for all taxa, and if evolutionary rates are unequal among lineages, topological errors will result.

Cavalli-Sforza and Edwards suggested in 1967 that the best tree based on a distance matrix would be that which gave the shortest sum of branch lengths (S). This is known as the principle of "minimum evolution." Whilst this is a good example of an optimality criterion that can be used as the basis for a "searching" phylogenetic method, it also provides a rationale for designing alternative clustering methods to UPGMA. Clustering methods are much faster than search-based methods.

Neighbor-joining (NJ) is a clustering method[68] that attempts to find the tree with the minimal value for S. The iterative procedure used to reconstruct the phylogeny is very fast to compute and often produces trees that are very close to the minimum evolution tree.

Trees can be generated using character-based methods

The principle of **maximum parsimony** (MP) defines the best tree as the one that requires the smallest number of evolutionary changes to account for the data. The branch lengths of trees produced from character-based methods are the numbers of individual evolutionary changes along each branch. Thus MP is for character data what minimum evolution is for distance matrix methods. When two (or more) trees are equally parsimonious, there is no criterion for choosing between them, and no unique tree can be inferred.

Having defined the optimality criterion for MP, how is this ideal tree sought? For the number of taxa commonly used in studies of human genetic diversity, it is often not possible to examine all possible trees because this would be computationally too laborious. The method only considers the subset of nucleotide sites known as **informative sites**. These are polymorphic sites at which at least two alleles are present in two or more individuals. This reduces some of the computational load, but a search strategy is still required, both for jumping between different locales within the range of possible trees, and for identifying the most parsimonious tree in each locale. There are some examples where an initially published tree has been shown in subsequent analyses not to be the most parsimonious, because the initial search strategies missed more globally optimal trees.[61, 80]

MP methods can incorporate information about the relative rate of different mutations, for example if **transversions** are known to occur less frequently than **transitions**, or if certain sites are known to be hypermutable. These mutational events can be weighted accordingly, such that the rarer changes carry more influence. MP methods can be sensitive to unequal rates of evolution or undersampling of certain parts of the phylogeny, via a process known as **long branch attraction** (**Figure 6.9**).

A tree can be regarded as a hypothesis that attempts to explain the data. Thus alternative trees can be considered to be competing hypotheses which can be compared through a likelihood framework, as described in **Box 6.3**. The optimality criterion used is that the chosen tree should be the most likely one. Under a given evolutionary model, the best tree is that which has the **maximum likelihood** (ML) of producing the data. Moreover, alternative evolutionary models can be compared by applying the likelihood ratio test. This lends falsifiability,

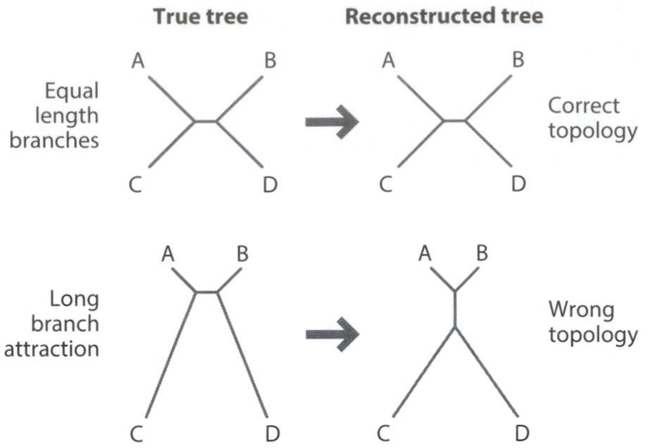

Figure 6.9: Long branch attraction. Two trees relating the evolutionary relationships of four taxa are shown: one in which the rate of evolution is equal for all lineages and the other for which taxa C and D have experienced much faster mutation rates, which gives them longer branch lengths. When MP is used to reconstruct the phylogeny relating these four taxa from their sequence divergences, it identifies the correct phylogeny in the first scenario but not in the second.

which was one of the desirable properties for a phylogenetic method considered earlier. Both **tree topology** and branch lengths should be inferred at the same time using a ML approach, as both affect the likelihood of the tree. As a result, ML methods are computationally intensive.

ML methods require models of sequence evolution (discussed in Chapter 5). The more complex of these models can incorporate the differential rates of different types of sequence changes, as well as variable mutation rates among different sites. However, these models incorporate parameters (for example, how much more likely transitions are than transversions) that are unknown for the dataset in question. This introduces a "chicken and egg" problem. If we knew the tree relating a set of sequences we could estimate the parameters of the model more accurately, but we need the parameters of the model to get the tree in the first place. This means that trees and model parameters must both be varied, which again increases the computational load enormously.

Box 6.3: Maximum likelihood and the likelihood ratio test

The likelihood of a hypothesis is the probability of obtaining the observed data given that that hypothesis is correct. Maximum likelihood asks, What is the most likely hypothesis given the outcome (data)? A hypothesis, such as a particular model of human population growth, can be varied by changing the parameters within a model, or by changing the model itself, and comparing the likelihoods of the different models to identify the model with the highest likelihood. Commonly, individual likelihoods are very small, even for the most likely model, and so are often expressed as log-likelihoods.

Evolutionary models tend to have many parameters. Often, the true values of the model parameters are unknown, but we would like to estimate them. Typical parameters within evolutionary models include: mutation rate, population size, population growth rate, and ages of alleles. Maximum likelihood can be used to estimate the parameters of these models. The criterion used to choose one value for a parameter over another is that a given value maximizes the likelihood of the chosen model.

In the above scenario, the investigator is at the mercy of the evolutionary model. An alternative model may very well give an alternative maximum likelihood estimate for the same parameter. How can we decide which model is the more appropriate? A likelihood ratio test compares the ability of alternative models to explain the data, by considering the significance of the test statistic below:

$$2 \log \frac{\text{maximum likelihood under alternative hypothesis}}{\text{maximum likelihood under null hypothesis}}$$

This test statistic often approximates to the χ^2 distribution with one degree of freedom, allowing easy assessment of the relative merits of different hypotheses. It is only suitable for comparing models with different numbers of parameters if they are **nested models**, whereby the simpler model can be obtained directly from the more complex model by constraining its parameters. Generally, specifying more parameters gives a better fit of the model to the data. The principle of the likelihood ratio test is to make sure that the increase in the model's complexity does significantly improve our ability to account for the observed data.

As evolutionary models become more complex, more parameters are required, and the information present in the data can be spread more thinly amongst them. Consequently, more data are often required to maintain similar levels of certainty when more complex models introduce new parameters.

Box 6.4: Likelihood, probability, and Bayes' theorem

Probability and **likelihood** are different. Probability is concerned with making predictions of outcomes (data) from a solid set of hypotheses (model plus parameters). For example, with a balanced coin (hypothesis), what is the probability of getting five "heads" from 10 spins (outcome)? This can be written as Prob(O|H)—the probability of the outcome given the hypothesis. This forms the basis for classical **frequentist** statistics based on the *p* value, so that a *p* value of 0.05 means that the probability of a particular outcome is 0.05, or 1 in 20, given the null hypothesis.

We can rephrase the value Prob(O|H)—the probability of the outcome given the hypothesis—as the *likelihood* of the hypothesis. Maximum likelihood is concerned with generating better hypotheses having observed the outcome (Box 6.3). For example, how likely is it that the coin is balanced (hypothesis) given that five "heads" were observed in 10 spins (outcome)?

Bayesian approaches to data analysis focus on the probability of the hypothesis given the outcome—Prob(H|O). The two probabilities are proportional:

$$\text{Prob}(H|O) \propto \text{Prob}(O|H)$$

This can be rewritten as an equation if an extra term is introduced: the probability of the hypothesis Prob(H), called the **prior probability distribution** of the hypothesis.

$$\text{Prob}(H|O) = \frac{\text{Prob}(O|H) \times \text{Prob}(H)}{\Sigma_H \text{Prob}(H) \times \text{Prob}(O|H)}$$

This is Bayes' theorem, named after the Reverend Thomas Bayes who died before his work on his eponymous theorem was published. Σ_H Prob(H) is the sum of probabilities of all possible hypotheses, and Prob(H|O) is the **posterior probability distribution** of the hypothesis.

It can be seen that Bayesian approaches are related to maximum-likelihood approaches, but the key difference is the determination of the prior probability of the hypothesis. Proponents of Bayesian approaches argue that it fits much better with the course of scientific research in that we have prior hypotheses (that is, models based on what we know already) and we place new data in the context of our prior hypotheses to test whether we should change our hypotheses about the world. Critics of Bayesian approaches suggest that selection of prior hypotheses is too subjective and, where a prior hypothesis is not obvious, the choice of prior hypothesis can determine the posterior probability of the hypothesis. Supporters of Bayesian approaches argue that it is more scientifically justified to frame our hypotheses in the context of previous knowledge rather than in a metaphorical vacuum.

Bayesian approaches are now used throughout human evolutionary genetics. They are well-developed in phylogenetics, but are also used for calling genotypes and phasing haplotypes for genotype data, quantitative trait mapping, and population modeling, as described in the text. For a discussion on Bayesian approaches in phylogeny see Huelsenbeck et al.[41] and for a comprehensive review on other applications see Beaumont and Rannala.[8]

Although maximum parsimony was originally proposed as an approximation to maximum likelihood methods, it has been shown that the shortest tree (the MP tree) is not always the most likely tree (the ML tree), although in practice they are often very similar. Some conditions under which MP and ML trees are prone to differ are known: when there has been a large amount of evolutionary change within the tree, or substitution rate variation among lineages. An example of a MP tree is shown in Figure 7.16b, and examples of ML trees are shown in Figures 7.16a and 7.22.

Bayesian methods (**Box 6.4**) are now commonly used in phylogenetic analysis, driven by the availability of increased computing power. This power is needed for the repeated **Markov chain Monte-Carlo** simulations involved in Bayesian calculations. If an ML approach determines the probability of the data (such as the sequence alignment) given the hypothesis (the tree), a Bayesian approach determines the probability of the hypothesis (the tree) given the data (the sequence alignment). So, given a prior distribution of parameters, and a likelihood of the tree, a posterior distribution of parameters, including the tree, can be generated. Bayesian approaches do not generate a single tree, but a posterior distribution of trees. Either the most probable tree, or the most frequently observed tree, amongst this distribution of trees is chosen to reflect the real phylogeny.

How confident can we be of a particular phylogenetic tree?

Given a dataset of unknown quality, every phylogenetic method will reconstruct at least one phylogeny. But having obtained a phylogeny for our data, how confident can we be that it has high *accuracy* (proximity to the true tree) and *precision* (the number of alternative trees that can be excluded)?

The accuracy of phylogenetic methods can be tested by generating data from a known phylogeny, and then asking which phylogenetic method reconstructs the tree closest to the known tree. Such datasets can be generated either by simulating sequence divergence ***in silico*** along the branches of a predetermined tree, or by manipulating experimental organisms or molecules in the laboratory in a controlled fashion, and subsequently analyzing polymorphisms within the different lineages.

The level of confidence we should have in a reconstructed phylogeny can be assessed statistically using the **bootstrap**. This method is based on the idea that if a dataset strongly supports a certain statistical result (in this case a tree), then randomly chosen subsets of the data should support the same result (tree). In practice, the phylogenetic bootstrap is performed by resampling sites from the sequence alignment *with replacement*. The same number of sites as were present in the original alignment is selected, but each time a site is selected from the alignment it remains able to be selected again. Therefore, some of the sites in the original alignment are present more than once in these synthetically replicated datasets, and others are absent. Each of a large number of synthetic datasets (typically 100–1000) is used to reconstruct a phylogeny, and each time an internal node in the original tree is precisely replicated in a bootstrapped dataset this is noted. Bootstrap values are usually displayed on the original phylogeny in the form of percentages next to the nodes to which they refer. Thus a value of 92 means that the same node was reconstructed from 92% of all synthetic datasets. There is no agreement over what constitutes a good bootstrap value, although bootstrapping is generally regarded as conservative and values over 70% are often considered reasonably reliable.

Bayesian methods generate a posterior distribution of trees that can be used to determine the statistical support for a particular tree. The proportion of trees in the posterior distribution that have the same structure as the most likely tree determines the support for that tree. The numbers at each node on the tree represent the posterior probability that the nodes are correct.

Networks are methods for displaying multiple equivalent trees

Some biological processes are not adequately represented by a phylogeny in which taxa are forever splitting but never joining. These processes include recombination, which merges two previously divergent haplotypes. Similarly, gene flow between populations can result in their sharing young alleles despite a more ancient fission from a common ancestor. These kinds of processes generate loops within phylogenies known as reticulations or cycles. Such phylogenies are called networks.

A single network contains within it several trees. Thus if two trees are similarly well supported by the data, we do not have to choose one over the other, but can summarize them in a network. This network represents more of the information present in the data than does either tree alone. As with trees, networks can be constructed from distance or character data, and there are a number of alternative methods of construction.

The method of **split decomposition** displays incompatible and ambiguous links, which favor different phylogenies within genetic distance data, as a **split network**. For example, this method can represent phylogenetic information from DNA sequence alignments where one section of the alignment supports a particular phylogeny and another part of the alignment supports a different phylogeny. This approach can be used to visualize and measure the degree of statistical support for a given phylogenetic tree.[43]

Figure 6.10: Constructing a minimum spanning network from microsatellite haplotypes.
The data consist of five haplotypes (A–E) comprising numbers of repeat units at seven linked microsatellites. Assuming that microsatellite mutations occur in single steps, a distance matrix can be calculated. For example, haplotypes A and C differ by one repeat unit at the second microsatellite, and two repeat units at the fourth microsatellite, making a total of three repeat-unit differences between these two haplotypes. The network is then constructed from this distance matrix using the procedure described in the text.

Seven-locus microsatellite haplotype

A	15	10	16	24	9	16	13
B	15	11	16	25	9	16	14
C	15	11	16	26	9	16	13
D	14	11	16	24	9	16	13
E	15	10	17	24	9	16	13

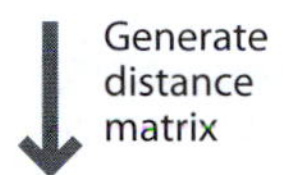

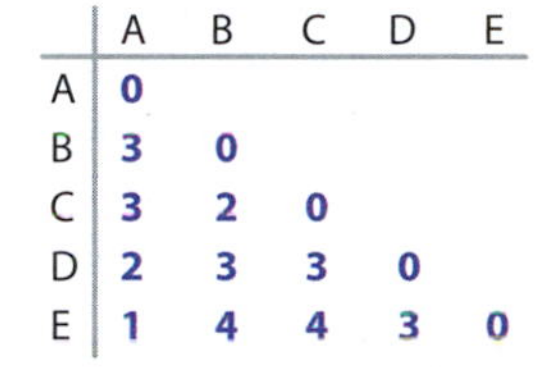

	A	B	C	D	E
A	0				
B	3	0			
C	3	2	0		
D	2	3	3	0	
E	1	4	4	3	0

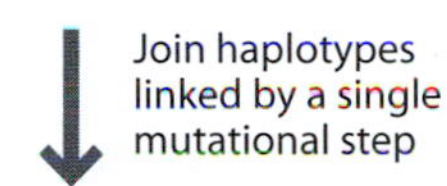

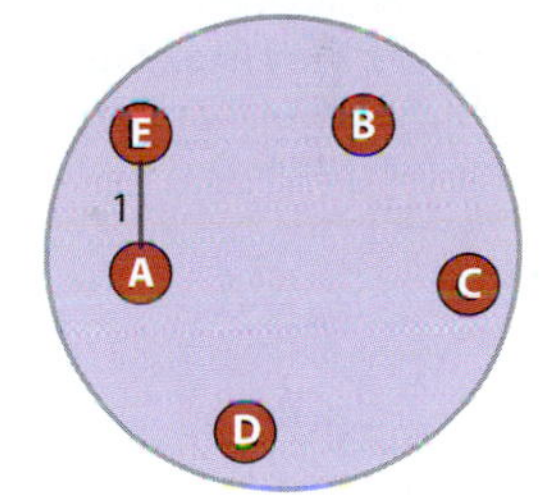

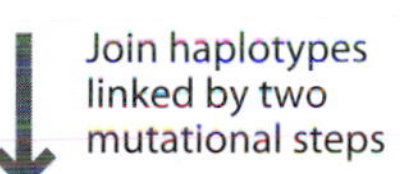

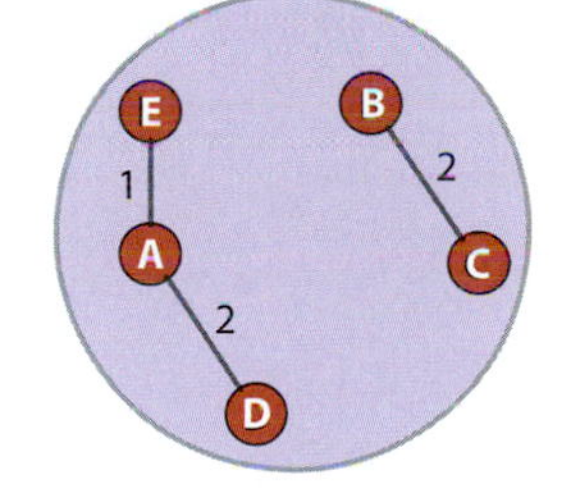

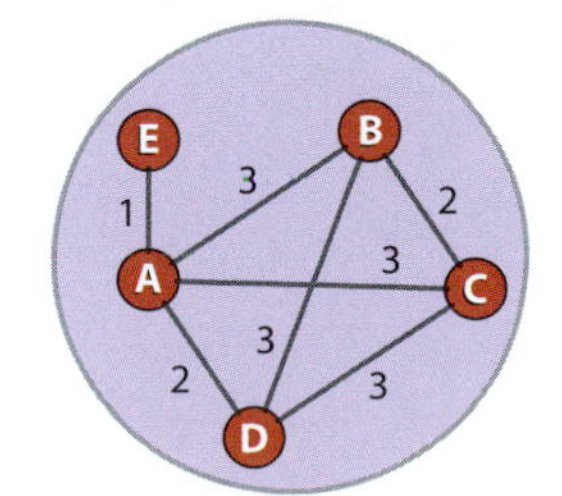

As an alternative to split decomposition, **minimum spanning networks** and **median networks** also reflect reticulation events in the data that could be caused by recombination or recurrent mutation.

A minimum spanning network can easily be constructed by hand (illustrated for microsatellite haplotypes in **Figure 6.10**), by following a simple procedure:

1. A character-based distance matrix between all taxa is computed.
2. Links are drawn between all taxa separated by single mutational steps.
3. Mutational steps of increasing size are considered until all taxa are linked into a single network and all most-parsimonious links of equal length from any given taxa have been reconstructed.

The minimum spanning network shown in Figure 6.10 is by no means the most parsimonious method for linking all taxa. The overall length of a network can often be shortened by adding hypothetical ancestral, unobserved nodes to the network (**Figure 6.11a**). Median networks provide a systematic method for reconstructing these ancestral nodes whilst guaranteeing that the resultant network contains all the most parsimonious trees.[6] Median networks are also constructed by a simple procedure, generally performed computationally:

1. Variant sites within sequences or microsatellite haplotypes are converted to binary characters.
2. Characters showing perfectly correlated variation across all haplotypes are combined into a single character, which is weighted for the number of sites incorporated within it.
3. For each triplet of haplotypes in turn, a median haplotype is calculated, which is the consensus of the three haplotypes, and, if novel, is added to the set of haplotypes.
4. The observed haplotypes plus the median haplotypes are then linked by a single-step network, and the lengths of branches defined by multiple characters are weighted accordingly.

Median networks are commonly used when **homoplasies** resulting from parallel mutations or reversions are frequent. This type of data includes mitochondrial DNA **control region** sequences (where the base substitution mutation rate is high) and microsatellite haplotypes. Median networks are prone to producing hyperdimensional cubes of reticulations when the number of taxa becomes large, which quickly make the network unintelligible. A coherent set of rules has therefore been developed to reduce the network's complexity by removing some reticulations through elimination of the least likely links.[6]

An alternative method for constructing networks with limited levels of reticulation, known as **median joining**, has also been developed.[5] The algorithm used to construct these median-joining networks is based on the limited introduction of likely ancestral sequences/haplotypes into a minimum spanning network of the observed sequences. Again, these likely ancestral sequences are identified through the calculation of median haplotypes. The median-joining algorithm is fast, has the advantage of being applicable to multiallelic polymorphisms, and is useful for large datasets.

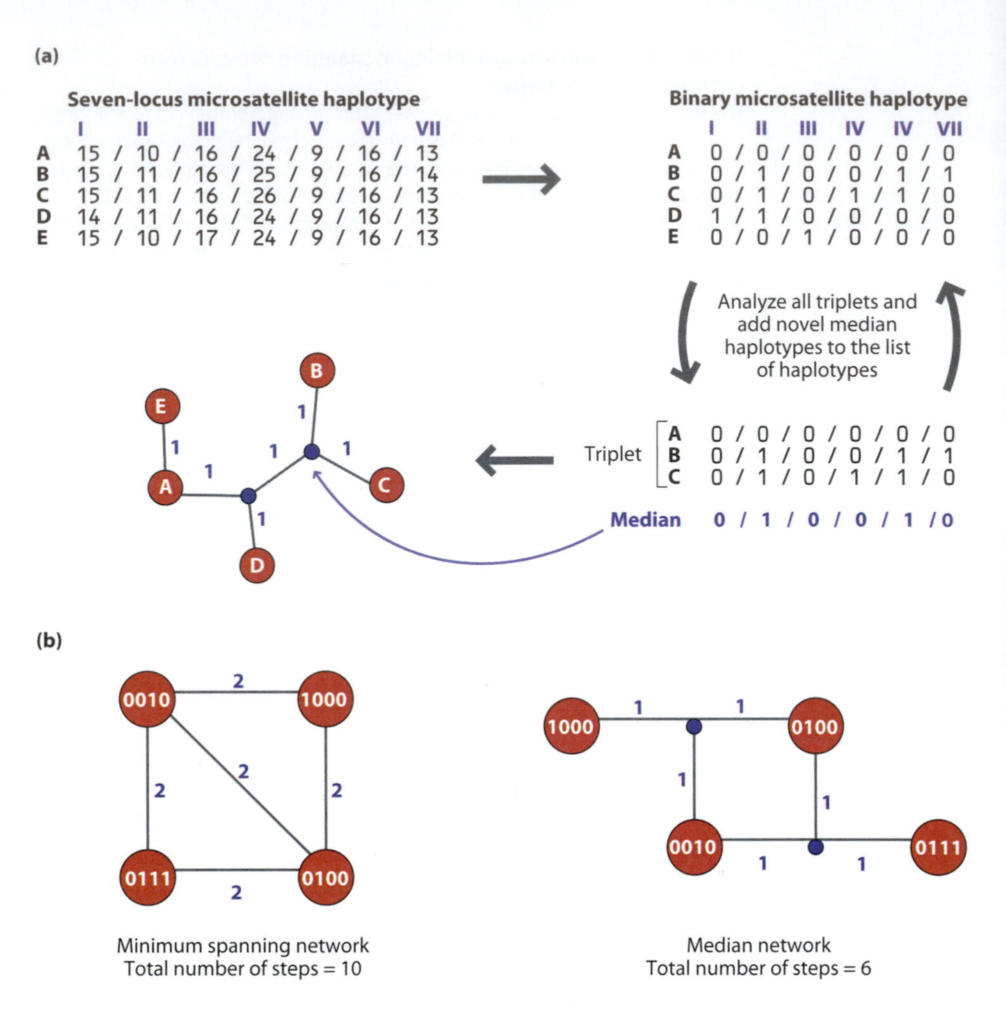

Figure 6.11: Constructing a median network from microsatellite haplotypes. (a) The haplotype data used are the same as for the minimum spanning network example in Figure 6.10. First, haplotypes of allele repeat numbers are converted to binary format. Invariant loci (V and VI) are ignored. The smallest allele at a microsatellite is designated 0. An allele one repeat larger is designated 1. For microsatellite IV, where three alleles are present, the smallest is 00, the allele one repeat unit longer is 01, and the allele two repeats longer than the shortest is 11. This takes account of a single-step mutational mechanism. Ancestral nodes not present in the sampled data are represented by small filled circles. In this example, a single most parsimonious tree is produced by the median network algorithm. (b) A minimum spanning network and median network for the same four binary haplotypes; both networks contain reticulations (cycles). Again, the median network is shorter.

Thus reduced median and median-joining networks represent alternative methods for obtaining intelligible networks with limited amounts of reticulation. The former method generates all possible ancestral sequences and then eliminates the least likely, whereas the latter method introduces limited numbers of the most likely ancestral sequences into a phylogeny of the observed sequences. Ancestral recombination graphs (see **Section 6.5**) are an alternative approach to representing recombination in a region, and an approach that has a mathematical framework rooted in **coalescent theory**.

6.5 COALESCENT APPROACHES TO RECONSTRUCTING POPULATION HISTORY

We have discussed in **Section 6.4** that phylogenetic trees are most frequently used to represent a model of the evolutionary process. Coalescent analysis is another way of modeling trees, but of individual alleles in a population. However, the intention of coalescent analysis is very different from that of phylogenetic analysis, because we are *not* seeking to make a representative tree of the history of a DNA sequence, but rather to use a distribution of many thousands of possible DNA sequence trees to infer various parameters concerning the population, such as effective population size or the **time to the most recent common ancestor (TMRCA)**. A coalescent approach is very powerful in making these inferences from within-species data, such as DNA sequence data from different human individuals in different populations. Because of this, it forms a crucial analytical method in human evolutionary genetics. Further details can be found in accessible reviews[57] and in the textbook *Coalescent Theory—An Introduction*, which is written for a biologist audience and introduces complex mathematical concepts clearly.[81] **Table 6.3** lists some applications of coalescent analysis described elsewhere in this book.

TABLE 6.3:
SOME EXAMPLES OF STUDIES THAT USE COALESCENT ANALYSIS

Application	Data used	Software used	Section of this book
Timing of speciation of great apes	genome sequences	COALHMM	7.4
Timing of speciation of Bornean and Sumatran orangutans	genome sequences	COALHMM	7.4
Developing demographic models of human evolution	6 Mb of autosomal DNA sequence	SIMCOAL 2	9.4
Ancient population structure	genome sequences	MS	9.5
Demography in European and African populations	mtDNA sequences	BEAST	12.5, 12.6
Dating of Y-chromosomal lineages in Europe	Y-chromosomal SNPs and microsatellites	BATWING	12.5
Comparing models of the settlement of Oceania	genomewide SNP genotypes	COSI	13.3
Dating origin of disease gene mutations	microsatellite genotypes linked to disease alleles	FOUNDER_TEST.C	16.2
Dating origin of G6PD deficiency alleles	5.2 kb of sequence within the *G6PD* gene	GENETREE	16.4
Dating origin of the major cystic fibrosis allele	data from linked microsatellites	Not stated	Box 16.3
	data from 23 linked SNPs	COLDMAP	

The genealogy of a DNA sequence can be described mathematically

Coalescent analysis is a genealogical description of the ancestry of a DNA sequence sampled from different individuals.[39, 81] This set of ancestral relationships is known as the **gene genealogy** (see **Figure 6.12**). As we move backward in time through the generations, we will start to encounter DNA sequences that

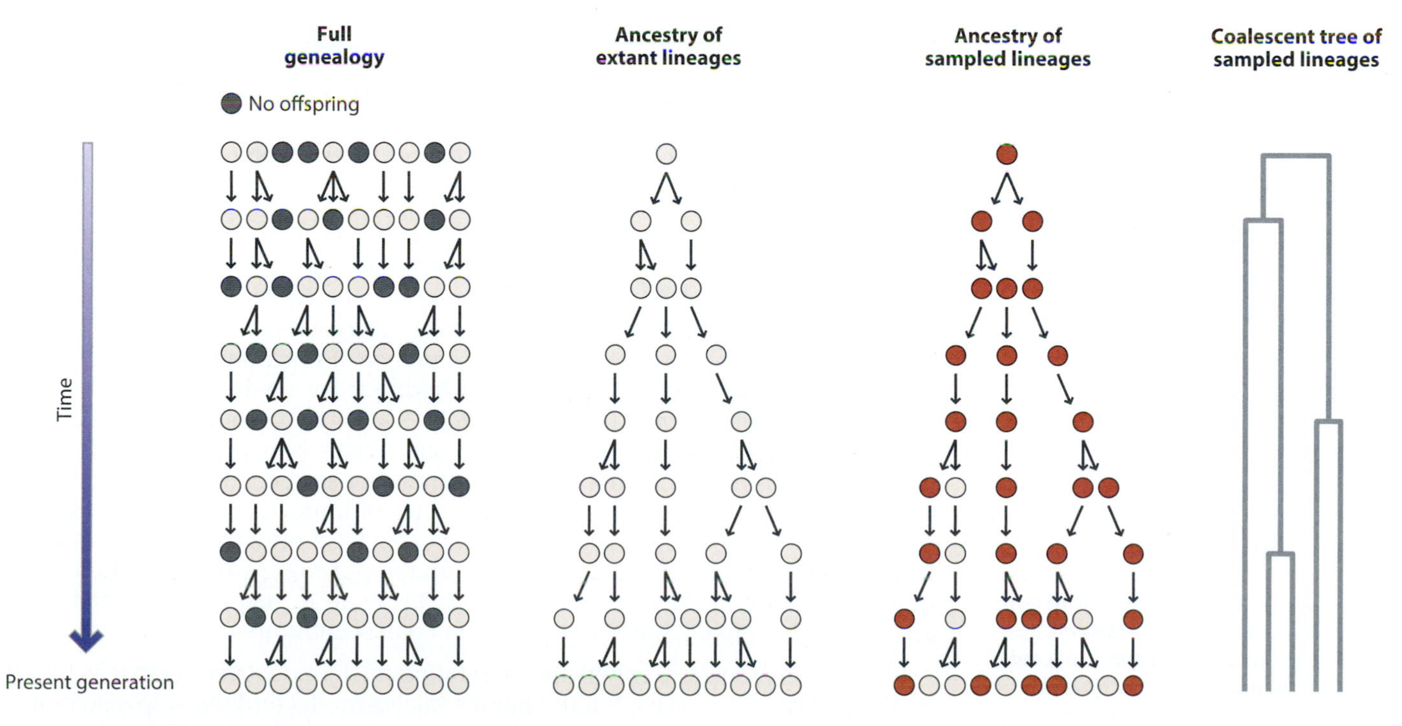

Figure 6.12: Genealogical relations among nine generations of a population with constant size 10.
Considering only those individuals who have contributed to the present generation reveals a most recent common ancestor of all lineages. If only a subset of lineages is sampled (sampled lineages are shaded in *red*) they may share the same common ancestor as that shared by all extant lineages.

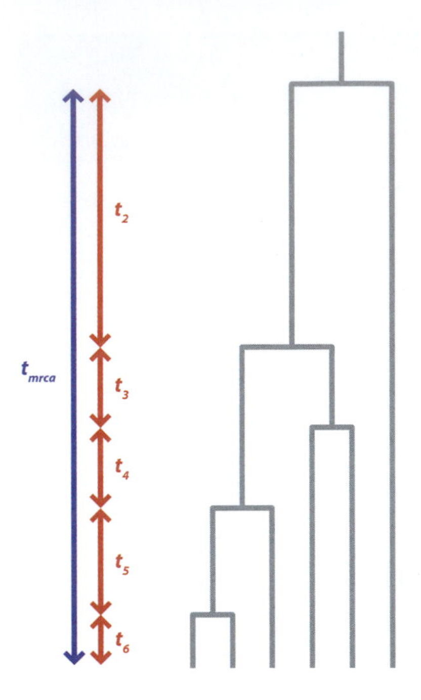

Figure 6.13: Times to coalescence in a gene genealogy.
Six sampled lineages coalesce to form one lineage. Each coalescence time from k lineages to $k - 1$ lineages is given as t_k. For example, the time from four to three lineages is t_4. The TMRCA is the sum of t_k between $k = 2$ and $k = n$, where n is the total number of lineages.

are ancestral to two existing DNA sequences. This process of the merging of lineages is known as coalescence. As we go further back these ancestral DNA sequences continue to coalesce until a single common ancestor of all modern DNA sequences is encountered. Note how in Figure 6.12, although only a subset of lineages is sampled, they have the same most recent common ancestor as the entire population (see **Section 6.7**).

The insight of coalescent analysis is to separate the gene genealogy (Figure 6.12) from the neutral mutation process. This is because, by definition, neutral variants do not affect reproductive success, and therefore cannot affect the genealogy. **Figure 6.13** shows a gene genealogy for 6 sequences, and, traveling backward in time, the time taken to go from 6 lineages to 5 lineages (a coalescence event) is t_6. We can generalize this so that the time taken to go from k to $k - 1$ lineages is t_k. The relationship between the value of t_k [strictly speaking, our **expectation** of t_k which is written $E(t_k)$] is given by:

$$E(t_k) = \frac{2}{k(k-1)}$$

For example, if there are 6 lineages then $E(t_k) = 2/(6 \times 5) = 0.067$. We can add up all the possible t_k values:

$$E(t_{mrca}) = \sum_{k=2}^{n} \frac{2}{k(k-1)}$$

For example, for a sample of 6 sequences then $E(t_{mrca}) = 2/2 + 2/6 + 2/12 + 2/20 + 2/30 = 1.67$. This value for the TMRCA is in **coalescent units**. An important point is evident by comparing the time to first coalescence (2/30) and the time to last coalescence from the last-but-one coalescence (2/2). This shows that most of the TMRCA is determined by most ancient coalescence, and increasing the sample size (increasing n) does not increase the estimate of TMRCA substantially. In fact, the probability that the genealogy of a sample of size n contains the most recent common ancestor (MRCA) of the population as a whole is $(n - 1)/(n + 1)$; for example 0.6 for a sample as small as 4.

An alternative way of representing the process above is by giving the rate of coalescence of lineages. Rate is the reciprocal of time—we think of time being in years or minutes and a rate being measured per year or per minute. So the rate of coalescence of k lineages is simply $k(k - 1)/2$.

Neutral mutations can be modeled on the gene genealogy using Poisson statistics

Following the modeling of the genealogy, neutral mutations can be placed on its branches, in a process that is known as **mutation dropping**. The Poisson distribution can be used to model this process, because mutations, particularly single nucleotide substitutions, are generally rare events. Also, in the infinite sites model, we know that the expected number of nucleotide differences in a sample of two sequences is the **nucleotide diversity** π, which under a Wright–Fisher model will be equal to θ (**Section 6.2**). So, for two sequences we can say that the expected number of nucleotide differences between them will equal θ, and therefore the expected number of mutations on each branch will equal $\theta/2$ (**Figure 6.14**). We need to add a parameter of time, because the deeper the genealogy, the more mutations we would expect. So now we have a model of the mutation process that we can place on the genealogy; the probability of a mutation follows a Poisson distribution with mean (or expectation) and variance both equal to $\theta t/2$, where t is time in coalescent units.

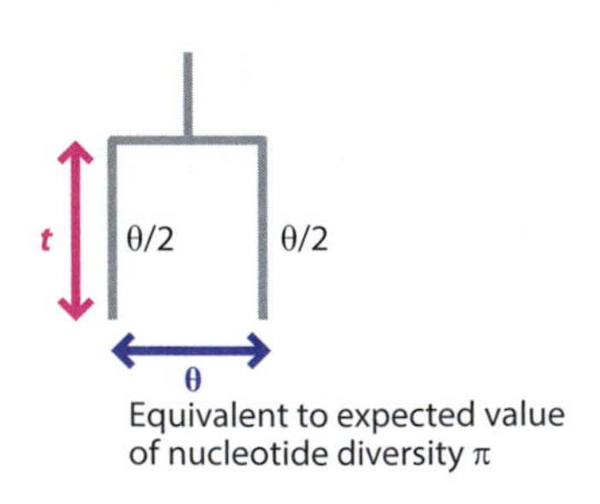

Figure 6.14: A gene genealogy of two sequences showing the relationship of nucleotide diversity and expected number of mutations on a branch.
The time to coalescence of the two sequences is represented by t and we expect θ to be equivalent to the nucleotide diversity π in a standard neutrally evolving sequence in a Wright–Fisher population.

The mutation model described above, relating nucleotide diversity to θ, is based on the infinite sites model, but the infinite alleles model and the stepwise mutation model (**Section 5.2**) can also be incorporated into a coalescent model. Similarly, the model described above is for a population of constant size, but models can be built allowing for changes in population size, which have characteristic effects on the gene genealogy (Figure 6.2).

Coalescent analysis can be a simulation tool for hypothesis testing

Once we are happy with our model we can test a hypothesis by asking, Are the observed genetic data compatible with a certain model of evolution? It is impossible to precisely reconstruct the true genealogy, as this would require us to know in which generation each mutation occurred, and in which generation any pair of alleles shared their most recent common ancestor. Consequently, the coalescence of lineages is simulated stochastically according to a specified evolutionary model and thousands of plausible gene genealogies are generated. Then the observed data are compared with these replicates to see if they are significantly different. Selection tests (Section 6.7) are one example of hypothesis testing using coalescent simulations. In these tests, a neutrally evolving population model is used to generate simulated data for comparison with observed genetic diversity. If the observed data are significantly different from the simulated data then the null hypothesis (neutral evolution) can be rejected.

For example, we could analyze 1 kb of diploid sequence from 10 different individuals and find no polymorphisms in the sequences. How unusual would this be? Perhaps purifying selection is acting on this sequence, removing new mutations from the population? Or is it just chance—that the 10 individuals happen to be closely related to each other? We can simulate the data under the standard neutral population model, and compare our observed data with the simulated data. This approach can account for the random nature (**stochasticity**) of sampling of sequences from a population, and the random nature of the coalescent processes that occurred in the past.

How can we compare the simulated data with the observed data? The most common way is to calculate certain population parameters from the simulated data, and compare them with the same parameters estimated from the observed data. In the example given above, we wish to see whether observing no polymorphisms (also termed segregating sites) in 20 sequences of 1 kb length is unusual. We would construct a simple coalescent model and simulate a genealogy many thousands of times, calculating the number of segregating sites (S, **Section 6.2**) for each simulated genealogy. This would give us a distribution of the number of segregating sites that we would expect under the neutral model, and the fraction of simulated genealogies that gave S of zero is an estimate of the probability of S being zero given the model. We can estimate that there is about a 30% chance of no polymorphism in 20 human sequences of 1 kb length under the standard neutral model, so this observation is not unusual and is most likely to be due to chance.

Coalescent analysis uses ancestral graphs to model selection and recombination

If we find that our data do not fit the null hypothesis, can we incorporate selection into our coalescent model? Standard coalescent theory assumes that each individual DNA sequence is equally likely to have been derived from the previous generation. This is not true when natural selection acts—a beneficial allele is more likely to be in the next generation than a less-beneficial one. This situation is tackled by using the **structured coalescent**, which splits the alleles into two classes: selected and non-selected, and treats each class as a separate gene genealogy. The pattern of the sampled proportion of the population in either of these two classes will change over time as the frequency of the selected allele grows, and this trajectory itself can be modeled.

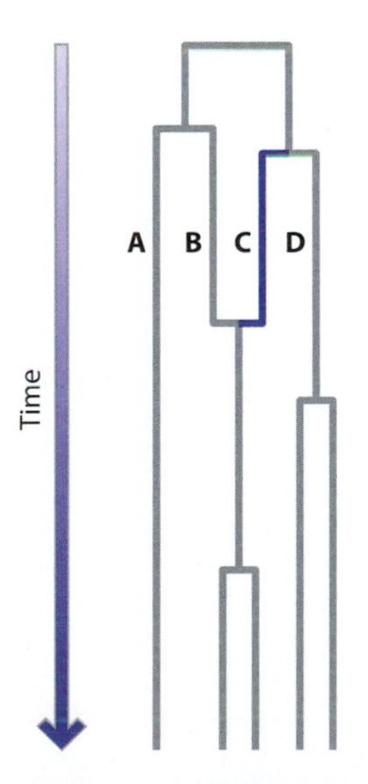

Figure 6.15: A gene genealogy represented by an ancestral graph. As well as lineages coalescing, ancestral graphs incorporate lineages splitting as we move back in time. These splits can represent recombination events or selection involving the sequence of interest.

Alternatively, an **ancestral selection graph** can be generated to model the genealogy of a locus under selection. Here a graph is a mathematical object, similar to a network, which includes lineages splitting as well as coalescing, as the genealogy moves backward in time (**Figure 6.15**). We have already seen that k lineages coalesce with a rate $k(k-1)/2$. We can introduce an extra transition, that of lineage splitting as we move back in time, which happens at a rate $k\lambda/2$. λ is the rate parameter that represents here the strength of selection.

So why can selection be represented as this complex branching structure? In Figure 6.15, traveling down the genealogy in forward time, we can imagine that lineage C has a selective advantage over lineage B, and this differential reproductive success causes lineage B to, in effect, be replaced by lineage C.

Standard coalescent theory also assumes that there has been no recombination, so that a single genealogy can be calculated for a segment of DNA. While this is true for **nonrecombining loci** such as the Y chromosome and mtDNA, this is clearly not the case for autosomal sequences. This issue can be approached genetically by choosing autosomal regions that have no evidence of ancestral recombination as measured by linkage disequilibrium. Alternatively, the problem can be resolved mathematically by incorporating a model of recombination into the coalescent model. Recombination means that a single locus will not have a simple, tree-like genealogy but will have a lineage that will both coalesce and split as we move backward in time. We can therefore use **ancestral recombination graphs** to model the genealogy of a locus that has undergone recombination. These are related to ancestral selection graphs, except the rate parameter is the population recombination parameter ρ (**Section 5.6**). In Figure 6.15, the recombination has happened between lineage C and lineage B, which would result in a B/C recombinant DNA sequence that continues down to the tip of the genealogy, with one further mutation. Once ancestral graphs have been generated, mutation events can be modeled on them in the normal way, using mutation dropping at a known rate.

Coalescent models of large datasets are approximate

Mathematical models describing the process of coalescence going backward in time are simpler and more efficient than models of the evolution of genetic variation in populations forward in time, although both approaches can be used to estimate population parameters. Coalescent methods need only consider those DNA sequences that have been sampled, not all of the DNA sequences present in the population as a whole. Huge computational savings are achieved through not having to consider the DNA sequences in previous generations that did not contribute to the current sample. Nevertheless, larger datasets generated by large-scale resequencing studies can potentially contain a large amount of information about the history of a region of the genome, and of the entire genome itself. Huge amounts of data require detailed models, including specifications regarding demography, recombination, and selection. Simulation of many possible genealogies using ancestral graphs is possible, but given the number of parameters of a model including demography, recombination, and selection, the parameter space is astonishingly large. This has led to the development of **approximate Bayesian computation** (**ABC**), which avoids the calculation of an exact likelihood of a particular evolutionary model, given certain parameters[16, 18] (see Box 6.4 for more about Bayesian statistics). Instead it uses summary statistics calculated from each one of a large number of coalescent simulations of the data, and these summary statistics are compared with the summary statistic of the observed data. If the difference is small enough (based on an arbitrary threshold), then the summary statistics from the simulated sample are accumulated to form the **posterior distribution** of the parameter estimates.

6.6 DATING EVOLUTIONARY EVENTS USING GENETIC DATA

A rooted phylogeny provides a *relative* chronology for genetic changes: changes close to the tips of a tree must have occurred after those closer to the root within the same clade. However, when integrating genetic data with those from other disciplines, it is often desirable to produce an *absolute* chronology, in other words, to provide actual time estimates of when such changes occurred. In this way, the timing of a change, be it in a population or a molecule, can be placed in a wider context, perhaps related to an archaeological culture or a paleoclimatological event.

We will consider dating methods that attempt to date population splits and molecular changes separately. For both of these classes of methods, it is usually assumed that selection has not been acting on the loci being studied, as this would complicate the relationship between variation and time (see **Section 5.7** on the neutral theory). A number of different methods exist which estimate the TMRCA of a set of sequences sharing a variant at a particular site. Note that this is different from dating the variant itself. There may be a substantial time lag between the origin of a new variant and the MRCA of the sampled chromosomes that carry it.

Dating population splits using F_{ST} and Nei's D statistics is possible, but requires a naive view of human evolution

Dating population splits revealed by population phylogenies usually requires the assumption that populations are like species: that no gene flow has occurred between them after they split. Classical population genetics allows us to relate the difference in gene frequencies between two populations to the time since they shared a common ancestor. In principle, the measures of genetic distance discussed above (for example, F_{ST}, Box 5.2) should give us some measure of the time, as they increase as the time passes since two populations split. However, many of these distances do not show linear relationships with time, or if they are linear, only exhibit linearity over short time spans. Accordingly, various transformations of these standard statistics have been proposed that improve their linearity.

Sometimes it is assumed that any differences in gene frequencies are due solely to genetic drift, with no mutation or selection. In such cases, the rate of population divergence depends only on the effective population sizes. Simple approaches, using F_{ST} to estimate population divergence times, for example, assume a constant effective population size. More sophisticated approaches use comparisons of genomewide allele frequency spectra between populations. In **Section 6.7**, we will examine examples of how comparisons between populations can identify genes under positive selection by detecting alleles at high frequency in one population but not in another (**Section 6.7**, Figure 15.9). Because the overall allele frequency spectrum is influenced by selection and demography, this principle can be extended to infer demographic parameters of human populations, including the date of a population split between, for example, Europeans and Africans, and changes in population size.[30]

Another approach is to measure genetic distance generated by mutation rather than allele frequency difference generated by drift. A commonly used method adopting this approach considers the genetic distance measure, Nei's D, which as we saw above is related linearly to time via the parameter α, the rate of fixation of nucleotide substitution. In Chapter 5, we saw that the rate of fixation of neutral variants is independent of the population size, and is equal to the neutral mutation rate. Although much initial attention focused on the dating of population splits, such putative events rarely if ever represent the reality of population evolution; an absence of post-fission gene flow seems unlikely for most human populations. Indeed, dating **admixture** events using genomewide data can provide useful information about population histories (see **Section 14.4**).

Evolutionary models can include the timing of evolutionary events as parameters

The timing of population splits is heavily dependent on assumptions made about the change, or constancy, of the effective population size,[10] and the role of subsequent migration. The most effective approach is to model the time of the population split jointly with other demographic parameters, such as the effective population size (**Figure 6.16**). Even the simple model shown in Figure 6.16 requires estimation of seven different parameters, and the more complex the model, the more parameters that must be estimated and therefore the more data that are required to accurately infer parameter values. Different models

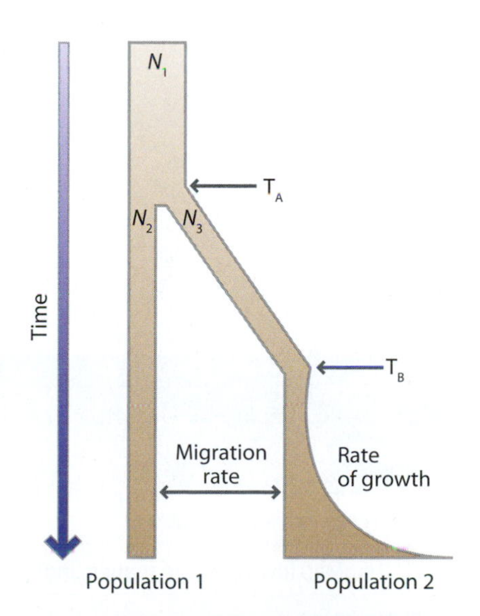

Figure 6.16: A simple demographic model for two populations.
This simple model has seven parameters: the three effective population sizes N_1, N_2, and N_3; two times, T_A and T_B; the migration rate between the two populations; and the growth rate of population 2.

can be compared using likelihood-based approaches (Box 6.3) or statistics such as **Akaike's information criterion**. There are currently two approaches to building models:

- **Coalescent approaches.** For these, data are simulated using several different coalescent models and the model that best fits the data is chosen. These analyses typically use a Bayesian approach, which requires prior estimates of the parameters, usually based upon existing diversity data. Care must be taken so that the choice of prior distribution of a parameter does not overly influence the likelihood of a particular model fitting the data, and therefore the final choice of model.[32]
- **Diffusion approximation.** An alternative to coalescent approaches are models that simulate population divergence forward in time, and are therefore more closely related to classical population genetic models. The diffusion approximation approach attempts to model a discrete number of individuals in a discrete number of generations as a continuous process described by an equation. One implementation of this approach, which is suited to large-scale datasets, uses the allele frequency spectrum as a summary of the genetic data from a given population generated by, for example, resequencing many genomes.[35]

Evolutionary models and effective population size

Large-scale datasets and population models allow effective population sizes to be estimated at particular times in human evolution. This approach contrasts with the classical view of effective population size (N_e, **Section 5.3**), which is, in effect, the effective population size over a long but unspecified time in human evolution. This is often known as **long-term effective population size**, usually calculated as a classical parameter from genetic diversity data. Indeed, genomewide estimates have generally supported the accepted value of human long-term N_e of 10,000, at least for non-African populations (**Table 6.4**). Analysis of genomewide long-range LD can also provide estimates for N_e,[76] but these estimates are lower than others. The reason for this is not known, but patterns of LD are strongly influenced by population bottlenecks, so the N_e estimates may be strongly influenced by the reduction of population sizes during these bottlenecks. Now whole-genome coalescent analyses are providing estimates of the effective population size throughout human evolutionary history, and this looks set to continue with more data and more complex models. One approach is called **pairwise sequentially Markovian coalescent analysis** (**PSMC**) which calculates a history of ancestral effective population sizes from the pattern of heterozygous polymorphisms in a single diploid genome. This analysis suggests that between 100,000 and 1 million years ago the N_e varied between 7000 and 15,000—similar to the range estimated previously.[49]

TABLE 6.4: ESTIMATES OF EFFECTIVE POPULATION SIZE IN HUMANS

Sample	N_e (× 10,000)	Data used	Reference
Various human populations	1.0	nucleotide diversity at different loci	Takahata N (1993) *Mol. Biol. Evol.* 10, 2.
CEU (HapMap low-coverage sequencing)	1.16	per-generation recombination rates and linkage disequilibrium	The 1000 Genomes Project Consortium (2010) *Nature* 467, 1061.
YRI (HapMap low-coverage sequencing)	2.01		
CHB/JPT (HapMap low-coverage sequencing)	1.30		
CEU HapMap (SNP data)	0.31	linkage disequilibrium across the genome	Tenesa A et al. (2007) *Genome Res.* 17, 520.
YRI HapMap (SNP data)	0.75		

An allele can be dated using diversity at linked loci

Summary statistics of haplotype diversity can be used to date an allele.[70] A new haplotype defined by a unique allele arises as a new mutation on a single chromosome, which, by definition, has no diversity at linked polymorphisms. As this haplotype increases in frequency, diversity accumulates through mutation and recombination at linked polymorphisms. The amount of diversity carried on a particular haplotype can be related to the age since that haplotype last shared a single common ancestor. The decay of the ancestral haplotype resulting from both recombination and mutation can be modeled, and the age of the allele calculated, assuming that the mutation rate of the linked polymorphism and the rate of recombination between it and the allele being dated are known.[29] Ideally, the recombination rate is estimated by studying linkage of the two loci through pedigrees. However, in the absence of such data, the recombination rate of the entire chromosomal region, as determined from LD maps, has been used.

For nonrecombining haplotypes, mutation drives diversification, and so it alone represents the molecular clock. The ρ ("rho") statistic (not to be confused with the population recombination parameter ρ, **Section 5.6**) represents the average number of nucleotide changes between the root haplotype and every individual in the sample. This requires a haplotype phylogeny, which must contain the root haplotype, and so median networks are often used, as they allow the reconstruction of ancestral haplotypes even if they are not observed within the sample. These mutational changes are counted from the network itself, rather than by estimation from the observed number of differences between two haplotypes; this takes account of possible reversions and **parallelisms** at sites with greater mutation rates (**Figure 6.17**). Confidence intervals for ρ can be calculated directly.[67] ρ is related to time by the equation:

$$\rho = \mu t$$

The ρ statistic does not perform well when trying to date the TMRCA using simulated data, suggesting that it may not always reliably determine the TMRCA on real data.[17] This is likely to be due to the randomness of mutation within the time-scales used to date TMRCAs of polymorphism data within a species, and the variation in population size over time, which determines the rate of fixation of neutral alleles. While phylogenetic methods assuming a constant molecular clock are fine for phylogenetics and for dating older events because this randomness is averaged out over time, within shorter time-scales the randomness of the mutation rate can seriously distort any TMRCA estimate based on a "clock-like" statistic.

An alternative approach is to use coalescent analysis to estimate the TMRCA of a given set of DNA sequences. We can see that a gene genealogy produced by a coalescent process shows the MRCA as the root of the tree, and therefore the estimate of the time to the root of the coalescent gene genealogy is the estimate of the TMRCA. The coalescent approach would generate a large number of trees

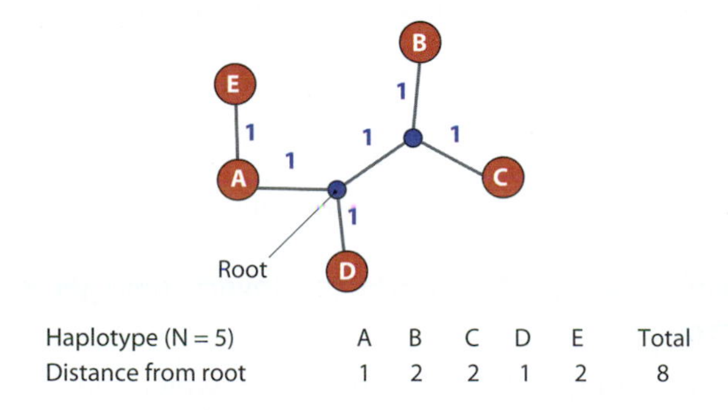

Haplotype (N = 5)	A	B	C	D	E	Total
Distance from root	1	2	2	1	2	8

Mean distance from root (ρ) = 8/5 = 1.6
Mutation rate (μ) = 5×10^{-5} (1 mutation every 20,000 years)
Time to MRCA (ρ/μ) = 32,000 years

Figure 6.17: Basis of dating using haplotype diversity (rho dating). The phylogeny used is identical to that constructed using the median network algorithm in Figure 6.11. The root is assigned to one of the unobserved internal nodes, and the distance of each individual taxon from the root is calculated from the phylogeny. Eight mutation changes are observed between the root and the five observed haplotypes, giving a ρ value of 1.6. The time to the MRCA of these haplotypes is ρ/μ, from rearranging the equation given in the text.

consistent with the data, which can be used to generate a mean value for the TMRCA and an estimate of that value's variance.[84]

Interpreting TMRCA

Having obtained a date for the MRCA of a set of extant haplotypes, what relevance does this have for particular population processes? We must remember that the date of a MRCA is a purely genetic estimate and need not have any obvious correlate in the other prehistorical records. This date does not represent the timing of a migration that spread the chromosomes carrying those haplotypes, although it does represent an upper bound to the age of the migration. The TMRCA of a set of alleles is typically much older than the age of the population split. Only when there is a strong bottleneck or founder effect is the TMRCA similar to the age of the population split.

Barbujani uses the following analogy to demonstrate this point:[7]

> ...suppose that some Europeans colonize Mars next year: If they successfully establish a population, the common mitochondrial ancestor of their descendants will be Paleolithic. But it would not be wise for a population geneticist of the future to infer from that a Paleolithic colonization of Mars.

It is also worth remembering that the TMRCA has in most cases broad confidence intervals, and coalescent theory shows that simply increasing the number of samples does not add much power to defining the TMRCA (**Section 6.5**). This has the consequence that a range of historical and prehistorical explanations may be consistent with a TMRCA estimate. So, the introduction of more powerful, statistically robust methods based on coalescent analysis has meant that the value of the TMRCA is not always as useful as we had hoped. Increased resolution may come from analysis of the larger number of variants discovered and genotyped by large-scale sequencing projects.

Estimations of mutation rate can be derived from direct measurements in families or indirect comparisons of species

Coalescent analysis is considerably strengthened by accurate estimates of the mutation rate. For intra-allelic methods, which assume a molecular clock, accurate mutation rates are essential. Given the importance of a value for mutation rate, how do we decide on a value to use? Mutation rates are likely to vary between individuals and over time, so an average mutation rate, relevant for the time period in question, is required for dating analyses.

Mutation rate calibration can either be performed directly on individual meioses (for example, in pedigrees), or indirectly through the observation of a certain amount of divergence across a known time span. The faster mutation rates of microsatellite loci can be measured using direct methods,[46] and whole-genome sequences of parent–child trios can be analyzed to estimate the number of mutations in a single generation.[78] Indirect approaches often use divergence between species whose divergence is well dated in the fossil record. Different studies often appear to use different mutation rates for the same locus; however, sometimes these rates are in fact identical, but are expressed in different ways (see **Box 6.5** for more details).

Different estimates of the human mutation rate are given in **Table 6.5**, and the estimation of mutation rates is discussed more fully in **Section 3.2**.

An estimate of generation time is required to convert some genetic date estimates into years

Another key value in dating is the generation time, which may be required to convert a date estimate from generations into years. Due to the likely increase in generation times over the past 7 million years, the generation time of choice depends to some degree on the time-depth being investigated. Generation

Box 6.5: Why are there so many ways of expressing mutation rates?

One of the most confusing aspects of the evolutionary genetics literature is that mutation rates quoted for some loci appear to be different between different publications. Given the importance of mutation rates for calibrating the molecular clock, these potential sources of conflict between interpretations must be understood. Some mutation rates are different because they genuinely reflect different rates of evolution, while other rates only appear to be different; they are in fact the same rate expressed in different ways.

A mutation rate is essentially the number of mutations that can be expected to occur within a segment of DNA over a certain period of time. Typically, the segment of DNA considered is often either the entire locus under study, or a nucleotide within that locus. Similarly, there are two time periods that are frequently considered, a year, and a generation. If we consider the simple case of a locus 1500 bp in size, and a generation time of 25 years, there are four ways of expressing the same mutation rate:

- 0.6 mutations per locus per year
- 15 mutations per locus per generation
- 0.0004 mutations per nucleotide per year
- 0.01 mutations per nucleotide per generation

Whether a mutation rate is expressed per year or per generation often depends on how that mutation rate is estimated. If the mutation rate is estimated from pedigrees (as for microsatellites), it is often given per generation, whereas if it is calculated using calibration points from the fossil record (more commonly used for sequence evolution), an annual rate is more frequently used. We can be more confident about the number of years to a dated common ancestor than we can be about the number of generations, which depends in part on the reproductive behavior of all intermediate ancestors.

The earliest mutation rate estimates were based on the incidence of dominant disorders that must have occurred de novo, rather than been inherited. These estimates first appeared prior to the advent of DNA sequencing technologies, and are only observable through the pathological outcome. As a consequence, the length of the gene could not be known, and these mutation rates could only be expressed per locus. There are other population genetics parameters which can similarly be expressed in per nucleotide or per locus form, most notably the population mutation parameter (also known as θ—Section 5.6).

time estimates used for modern humans tend to vary between 20 and 35 years, whereas, due to the expected shorter generation time of our common ancestor, 15 years has often been taken as the average generation time of hominins since the split from chimpanzees. For chimpanzees themselves, the average generation time is about 25 years, but for gorillas it is much shorter at around 19 years.[47] Given that variation of the generation time estimate can have a drastic effect on the dating estimate, it is important to consider generation time in more detail.

TABLE 6.5:
ESTIMATES OF NUCLEOTIDE SUBSTITUTION MUTATION RATE IN HUMANS

Sample	μ (× 10^{-8}) per nucleotide per generation (95% confidence intervals)	Method	Reference
European-American	1.1 (0.68–1.7)	direct analysis of pedigrees	Roach JC et al. (2010) *Science* 328, 636.
CEU HapMap (European) trios	1.2 (0.94–1.73)	direct analysis of pedigrees	The 1000 Genomes Project Consortium (2010) *Nature* 467, 1061.
YRI HapMap (African) trios	1.0 (0.72–1.44)	direct analysis of pedigrees	
Icelandic families	1.2	direct analysis of pedigrees	Kong A et al. (2012) *Nature* 488, 471.
Genome reference sequence	2.4	human–chimpanzee divergence[a]	Ebersberger I et al. (2002) *Am. J. Hum. Genet.* 70, 1490.
Mostly Northern European	1.8	de novo mutation rate of 20 Mendelian disease loci	Kondrashov AS (2003) *Hum. Mutat.* 21, 12.
Y chromosome	3.6	human–chimpanzee divergence[a]	Kuroki Y et al. (2006) *Nature* 38, 156.
Y chromosome	3.0 (0.89–7.0)	direct analysis of pedigrees	Xue Y et al. (2009) *Curr. Biol.* 19, 1453.

[a] Based on human–chimpanzee divergence time of 5 MYA, autosomal N_e of 10,000, and generation time of 20 years.

TABLE 6.6:
ESTIMATES FOR AVERAGE HUMAN GENERATION TIMES (YEARS)

Source	X chromosome	Y chromosome	Autosome	Mitochondrial DNA	Reference
French-Canadian parish records from 1850s–1990s	31	35	32	29	Tremblay M & Vézina H (2000) *Am. J. Hum. Genet.* 66, 651.
Census data of Polar Eskimos in Greenland 1820–1906	29	32	29	27	Matsumura S & Forster P (2008) *Proc. Biol. Sci.* 275, 1501.
Pedigree data of Icelanders 1742–2002	–	32	30	29	Helgason A et al. (2003) *Am. J. Hum. Genet.* 72, 1370.
Hunter-gatherers (review of anthropological literature)	–	32	26	29	Fenner JN (2005) *Am. J. Phys. Anthropol.* 128, 415.
Developed nations (review of anthropological literature)	–	31	27	29	

One striking finding from studies of genealogical records is that males tend to have significantly longer generation times than females. One large study of French-Canadians revealed an average female generation time of 29 years, but an average male generation time of 35 years.[79] This is due in part to menopause in women, with no analogous reproductive limit for men. Nevertheless, there are clearly other factors involved. For example, female fertility declines rapidly between the ages of 20 and 40 years, with no such dramatic change in male fertility. In addition, in many societies men tend to be older, on average, than their partners. Whatever the causal mechanisms in modern societies, the possibility that male and female generation times in prehistory may also have been significantly different cannot be excluded. This would mean that different generation times should be used for loci with different patterns of inheritance. **Table 6.6** shows estimates of generation times for different loci. In general, we assume that life expectancies, and therefore generation times, were shorter in prehistory,[83] although modern hunter-gatherers have very similar generation times to modern individuals from developed nations.[26]

6.7 HAS SELECTION BEEN ACTING?

As we have seen in the previous chapter, selection comes in many guises, each of which is expected to have a different effect on genetic diversity (see **Section 5.4**). Negative or purifying selection removes new, deleterious variants from the population and we would expect most human genes (and other functional elements) to be subject to such selection. This is indeed the case, and can be seen from the lower diversity observed in exons compared to introns (Figure 3.13). It can also be observed by comparing the allele frequency spectrum of nonsynonymous (amino acid altering) SNPs observed in gene exons with the allele frequency of synonymous SNPs (those that do not change an amino acid, and can mostly be regarded as neutral). **Figure 6.18** shows that nonsynonymous SNPs tend to be rarer than synonymous SNPs. This is because nonsynonymous SNPs are more likely to affect the protein function and are therefore preferentially removed by purifying selection before reaching high allele frequency. Figure 6.18 also shows the allele frequency spectrum of **nonsense** SNPs, which truncate the protein due to a premature stop codon. These are very likely to be deleterious, because the functional protein is lost, and purifying selection is even stronger for these variants, with none observed here with an allele frequency of greater than 0.1.

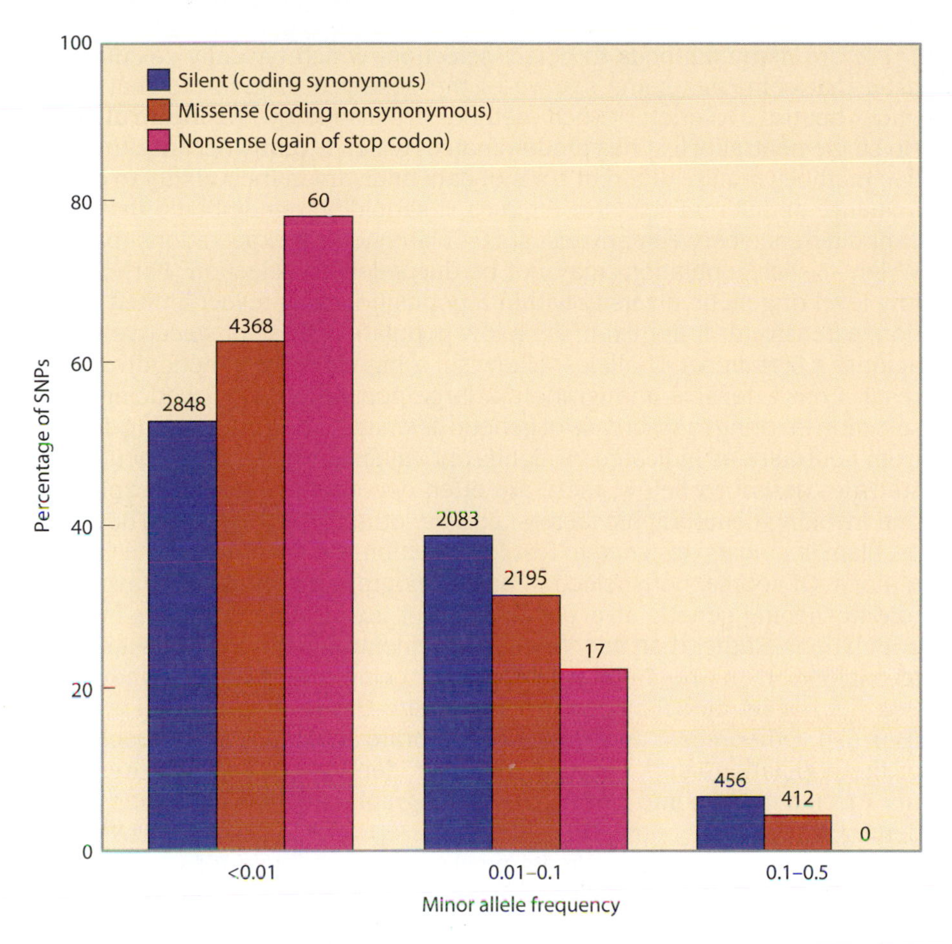

Figure 6.18: Enrichment of nonsynonymous SNPs at low allele frequencies.
Nonsynonymous SNPs are divided into missense SNPs and nonsense SNPs. The higher proportion of nonsynonymous SNPs at a frequency of less than 0.01 suggests that purifying selection is acting against these SNPs because they are likely to disrupt protein function. The trend is particularly noticeable for nonsense SNPs which are likely to result in a truncated, nonfunctional protein. Numbers refer to SNPs in the 1000 Exon Pilot Project. [Data from Marth G et al. (2011) *Gen. Biol.* 12, R84.]

There are a few examples of balancing selection in humans that are well established, for example at the HLA locus (Box 5.3), and the hemoglobin variant leading to sickle-cell anemia (**Section 16.4**)—approaches to detect this form of selection are discussed in **Box 6.6**. However, positive selection, which increases the probability that a new variant will become fixed, and reflects a new adaptation of a gene to the environment, is of particular interest and relevance to evolutionary genetics.

Box 6.6: Detecting balancing selection

There are a few examples of balancing selection (Section 5.4) acting on the human genome, including the MHC locus (Box 5.3) and hemoglobin variant Hb^S (Section 16.4). However, the extent of the role of balancing selection in shaping genetic diversity in the human genome, and human phenotypic diversity, is not clear and has been controversial.

Balancing selection maintains genetic variation in a population, so scanning genomes for higher than expected levels of variation should provide a list of candidate loci where balancing selection is operating. Whole-genome sequencing data confirm this increase in diversity across the MHC region (Figure 3.14), but in less clear-cut examples it can be difficult to distinguish whether an increase in diversity is due to balancing selection or neutral population processes. Recent balancing selection at a locus is also predicted to generate extended haplotypes, detectable by extended haplotype tests (see Section 6.7 and Table 6.7), but this signature may be indistinguishable from a signature of recent positive selection. One study used an allele frequency spectrum method to detect genes that are enriched for intermediate frequency alleles (that is, with frequencies around 50%).[3] Many of the genes, like those in the MHC region, have a role in defense against pathogens.

Another characteristic of balancing selection is the maintenance of **trans-species polymorphisms**. In the MHC region, there are a higher number of SNPs shared between humans and chimpanzees than would be expected given a neutral model, and several other genes including the antiviral gene *TRIM5* show trans-specific SNPs.[12] Deeper analyses require more complete polymorphism datasets of humans and chimpanzees based on whole-genome resequencing.

There are many methods to detect selection, which typically calculate a statistic that compares some feature of the observed diversity to that expected under neutral evolution.[56] Such methods are also known as **neutrality tests**. No single neutrality test has predominated, which is partly because the alternative methods require different sorts of data and vary in their ability to detect the influence of different selective regimes. It should be remembered that a significant difference between any test statistic and neutral expectations (based on a Wright–Fisher population) may not be due solely to selection. For example, a low level of genetic diversity within a population may reflect limited immigration, extensive drift (for example, a low population size), or selective pressures against a certain set of alleles. Likewise, a high level of genetic diversity may result from extensive immigration, a large population size (reducing drift), or selection favoring the increase of genetic diversity. When interpreting departures from neutrality, as indicated by significant values for many of the neutrality test statistics described below, there are often two equally plausible explanations: one invoking demographic factors, and the other selective ones. The recurring problem in studies designed to test for selection is disentangling the two explanations. Of course, both selection and demographic factors will have played a role in shaping genetic diversity at different loci at different times in different populations. Rather than adopting a simple neutral null hypothesis it is possible to compare many nonneutral hypotheses, using a likelihood framework (see Box 6.3), or by modeling using approximate Bayesian computation methods (Box 6.4). Other recent approaches incorporate an explicit demographic model of the populations under consideration to identify loci with features that are unexpected in the context of a given demography, such as Schaffner's "best fit" demographic model of the YRI, CEU, and CHB HapMap populations.[69] A limitation is that a demographic model needs to be developed for each population investigated, and such models are inevitably imperfect.

The availability of large genomewide datasets has led to an alternative approach to this problem which relies on the reasonable assumption that most of the genome is not positive selected, but has been influenced by the population-specific demographic processes in the past, thus providing an empirical null distribution of any statistic of interest. Instead of comparing a value of a test statistic for a candidate gene against the value of that test statistic for a theoretical equilibrium model of neutral evolution, we can compare it with this empirical distribution from the whole genome (**Figure 6.19**). If the observed value lies in the extremes of the distribution (typically the top or bottom 1% or 5%) it can be regarded as statistically significant, an approach known as **outlier analysis**. Such analyses, based on genomewide datasets, have the advantage of being "hypothesis-free" in the sense of not requiring a particular gene to be a prior candidate for selection and thus more likely to discover unexpected candidates. However, they have the disadvantage that many of the top 1% or 5% of genes or SNPs may be there by chance rather than selection. In addition, we have no idea how much of the genome has really experienced a particular form of selection. Is it 5%, or 0.5%, or 0.005%? With this approach we cannot compare different populations to ask whether there has been more selection within Africa or outside, for example. Another problem is that certain demographic histories, such as population subdivision, are likely to produce outlier loci even in the absence of selection. More recent approaches incorporate a demographic model of human population history to identify outlier loci in the context of a given demography, such as a recent population expansion. Thus it is best to consider outlier analysis as a method to enrich for candidates for natural selection, rather than evidence in itself, with further analyses required to clearly identify selection.[1, 77]

We have classified the tests described below on the basis of the different effects of selection each attempts to detect. The power of these tests to detect selection depends on, amongst other factors, the type and strength of selection operating, when selection operated, and whether the selection was acting on a new

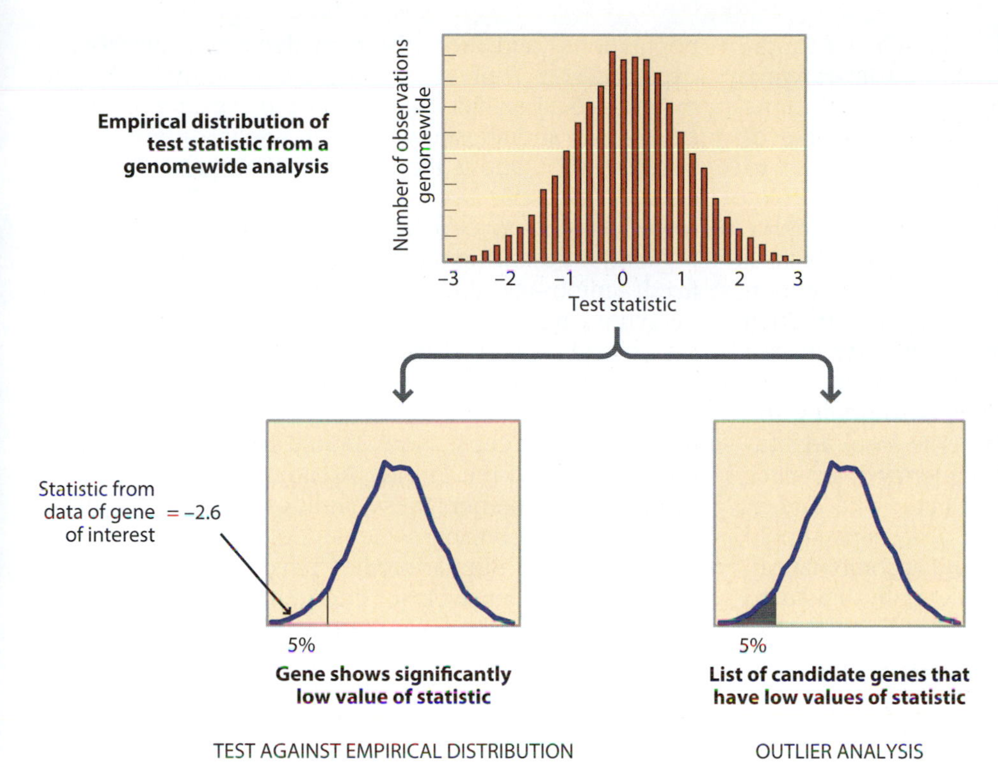

Figure 6.19: Candidate and outlier empirical distributions of genomewide data.
If we apply an imaginary test statistic to a set of genomewide data, we obtain a distribution with a mean of zero and standard deviation of 1. A candidate approach compares the observed value for this statistic for a candidate locus (−2.6) with the empirical genomewide distribution. An outlier approach takes all the loci from the empirical distribution below a threshold for further analysis. In this case, the threshold is 5%, meaning that 5% of the loci with the lowest value of the statistic are included (*shaded region*).

variant or on variation that has been present in the population for some time. There is much debate on the relative importance of recent strong positive selection on a new variant compared to other forms of positive selection (**Opinion Box 4**). This is centered around the concept of **hard sweeps**, where a single haplotype has increased in frequency due to recent positive selection on one variant (classic **selective sweep**), versus **soft sweeps**, where positive selection has acted on one or more variants that have existed in the population for some time (**standing variation**, **Figure 6.20**). If the multiple variants are within different genes, then this is termed **polygenic adaptation** (Figure 6.20). Hard sweeps have been tested extensively by modeling, and strong hard sweeps are reasonably straightforward to distinguish from no selection and soft sweeps using several of the tests we describe below. However, distinguishing soft sweeps from no selection is difficult and ideally needs the functional effect of each variant to be determined (**Figure 6.21**).

Differences in gene sequences between species can be used to detect selection

Variants within coding sequence can be classified nonsynonymous or synonymous, as described above (and **Section 3.2**). The proportion of synonymous sites that are variable within a set of sequences (dS or K_S) is assumed to be independent of selection, and therefore a product of neutral evolution. The ratio of nonsynonymous sites that are variable within the same set of sequences (dN or K_a) to dS will be greater than 1 under positive selection, and less than 1 under purifying selection. By testing if dN/dS (K_a/K_S) is significantly different from 1, we are in effect testing the dN statistic against the neutral expectation.

If a short **orthologous** sequence is taken from two closely related species, there are often too few variable sites to give us statistical power to detect differences between dN and dS. This means that, in studying selection in human genes, comparisons with the rhesus macaque orthologous sequence may be more powerful than comparisons with the chimpanzee orthologous sequence. However, if there are very large differences between the sequences (such as

Humans have evolved numerous phenotypic adaptations, most recently to changing diets and far-ranging environments. Understanding how these adaptations came about at the molecular level is a major goal of human evolutionary genetics. The main approach to date has been to look for the footprint of recent adaptations in patterns of genetic variation, using the model of a selective sweep. Under this model, positive selection acts on a new, beneficial mutation with a large effect on fitness, and brings it to fixation in the population. Such a fixation should lead to a decrease in diversity (hence the term "sweep") and a skew in the frequency spectrum toward new rare alleles at sites surrounding the beneficial substitution. If sweeps underlie local adaptation, they should also lead to large frequency differences between populations, both at the selected site and at linked sites. Moreover, these signals of sweeps should be most pronounced in regions of low recombination, where a beneficial substitution is physically linked to more sites.

Scans for selected sweeps have identified thousands of genes as potential targets of positive selection in the human genome. Unfortunately, while there are beautiful examples of sweeps (for example, *EDAR*, *SLC24A5*), many or most of these candidates may be spurious. Simulations can be used to estimate the false positive rate but, because our knowledge of human demographic history is necessarily incomplete, they provide only limited insight. Moreover, a sweep is only one of many ways in which adaptation can proceed, and many other forms of selection can lead to similar signatures in polymorphism data.

So how compelling is the evidence for widespread sweeps in humans? To address this question, researchers have turned to genomewide patterns of variability, rather than focusing on specific instances. They find that human diversity levels increase with recombination rates (more so than divergence between humans and other primates). Moreover, near genic regions, which are likely enriched for changes of functional (and hence potentially selective) importance, there is both a dip in diversity levels and increased allelic differentiation between human populations. Unfortunately, this evidence is consistent with frequent sweeps but also with many other forms of adaptation as well as with purifying selection against deleterious alleles in the background (background selection).

To home in on the role of sweeps, we (and others) have used the fact that each sweep involves a focal substitution, which should be fixed in a species or between populations, and around which decreased diversity is expected.[15, 37] Thus, sites that are a priori more likely to experience a beneficial substitution (for example, amino acid sites) should show greater allelic differentiation between populations than what is observed at "control" sites (for example, synonymous sites). Yet there appears to be little or no enrichment genomewide, with most populations differing only by a handful of fixed (or nearly fixed) amino acid differences, and by a similar number of synonymous differences. A related test considers amino acid substitutions on the human lineage and contrasts average diversity patterns around them to what is seen around synonymous substitutions. Under a model of prevalent sweeps, there should be more of a dip in diversity around the former. Again, this is not observed in humans. Together, these findings strongly constrain how frequently strong sweeps could have occurred in coding regions. Similar studies have not yet been conducted comprehensively for intergenic regions, but there are similarly few fixed differences between populations. Moreover, there is little or no evidence for an excess of rare alleles in regions of low recombination, again inconsistent with sweeps being the dominant mode of linked selection in the human genome.

These findings may seem implausible, as humans are clearly adapted in numerous respects. So how might adaptation have proceeded? One possibility is that selection often utilizes variation already present in the population.[36] If the allele was not initially rare in the population, then the difference in frequency between populations may not stand out compared with the genomic background. Moreover, the selected allele will have been present on multiple haplotypes, which distorts or weakens the footprints of the selective sweep. Similarly, if multiple, independent new alleles underlie adaptations, such that no one reaches fixation rapidly, there may be little recognizable signature. These scenarios are extremely plausible, all the more so as we begin to understand the complex genetic basis of traits through genomewide association studies. Selection on standing variation and polygenic selection have been the objects of investigation in evolutionary quantitative genetics for many decades; understanding their genetic underpinnings is now a central challenge for human evolutionary genetics.

Graham Coop, Department of Evolution and Ecology, University of California Davis

Molly Przeworski, Department of Human Genetics, University of Chicago

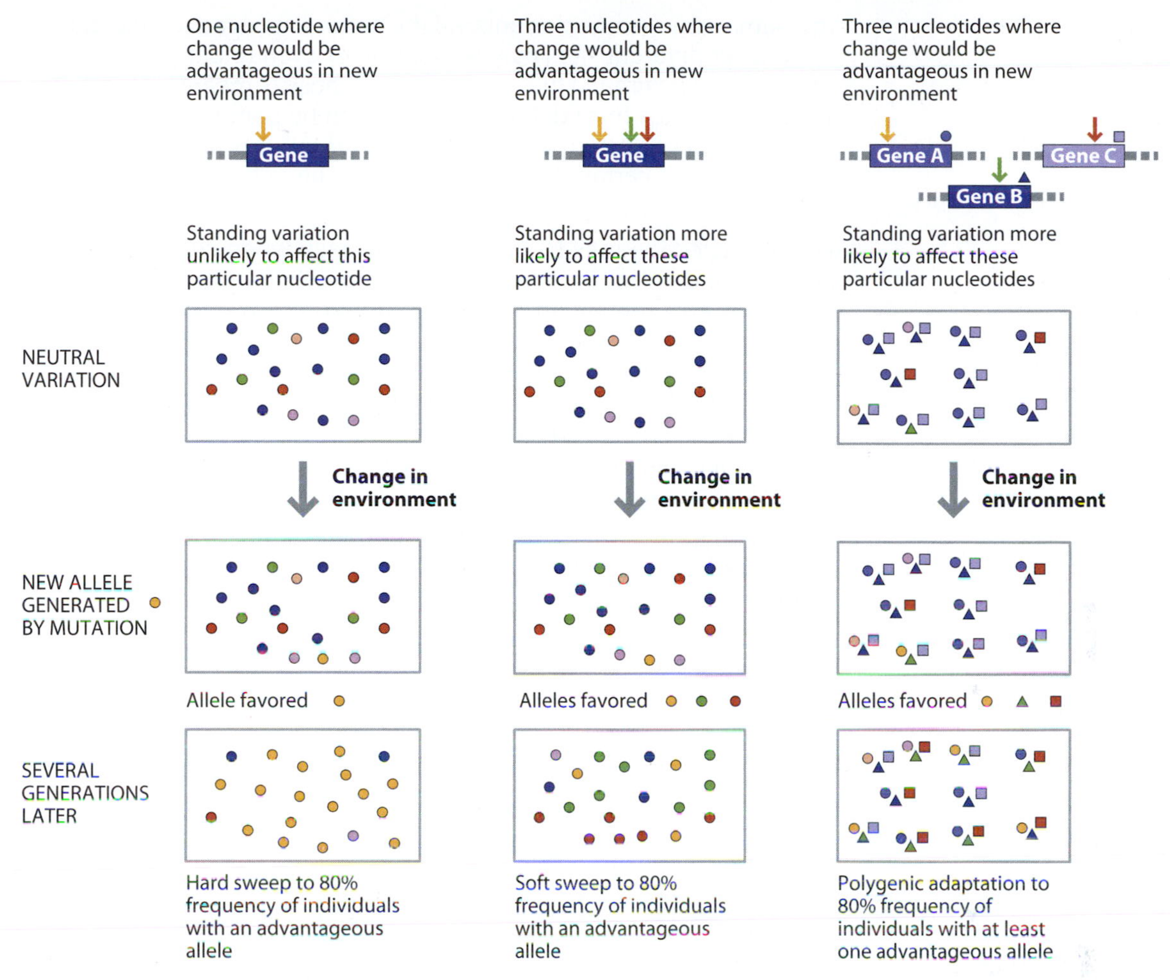

Figure 6.20: Hard sweeps, soft sweeps, and polygenic adaptation. Three different selective scenarios are shown. Each column of three boxes represents the same population at different times, with different colored shapes within the box representing different alleles of a gene. In the first example, following a change in the environment, only the *yellow* allele is advantageous, which leads to a hard sweep. In the second example, three alleles of one gene are advantageous following a change in the environment, and are present as standing variation in the population, resulting in a soft sweep. In the third column, three alleles of different genes are advantageous, resulting in polygenic adaptation from standing variation. In all three cases, the lowest box shows the population with 80% of individuals carrying advantageous alleles. Individuals are shown as haploid for simplicity, but the principle applies for diploid individuals.

between humans and non-mammals) then the possibility arises that there are parallel mutations or **reversions** between sequences. The observed number of differences between sequences becomes an underestimate of the true number of changes and this value must be corrected using a model of sequence evolution.[86]

Using the synonymous substitution rate as a proxy for neutral evolution avoids possible problems associated with differential mutation rates at different loci. Rates of neutral evolution vary from region to region throughout the genome, perhaps because of differences in sequence composition. However, a further complication arises from the fact that many organisms show biases in their codon usage resulting from both **mutational bias** and selection for the efficient translation of proteins. As a consequence, evolution at synonymous sites may not be truly neutral, and this may have to be taken into account using a more complex evolutionary model.

As described above, dN/dS represents an average over the whole gene, often several thousand nucleotides. Comparisons between human and chimpanzee have identified genes where dN/dS is greater than 1 for the whole gene, such as the protamine-1 gene *PRM1*, which is involved in packaging of DNA in the sperm. However, selective pressures are likely to be different among nucleotides within the same gene (some variants might be under negative selection while other variants at different sites may be neutral, or even under positive selection), and thus it could be argued that an average of these is meaningless. However, when many sequences from different species can be compared, it does become possible to detect selective pressures at individual sites using codon-based tests to detect whether certain functional modules have been under different selective regimes.[87]

Positive selection may be episodic, especially when considered over long time periods, and selection pressures may have been limited to certain lineages. We know, for example, that selection pressures on globin genes have not been equal among all human populations: resistance to malaria in regions of endemicity has resulted in localized, distorted patterns of diversity (**Section 16.4**). By reconstructing ancestral sequences within a **phylogeny** relating the different observed sequences, codon-based selection tests can indicate whether selection has been acting equally in all regions of the tree, or has been episodic, and so exhibits lineage-specific evolutionary pressures.[89]

The comparative analysis of whole genome sequences provides an opportunity to use dN/dS to detect selection patterns over a large number of genes. A key question is whether we can identify genes that have undergone accelerated evolution in the human lineage after human–chimpanzee divergence, as such genes may have undergone natural selection as part of human speciation (Chapter 8). While the high sequence similarity between chimpanzee and

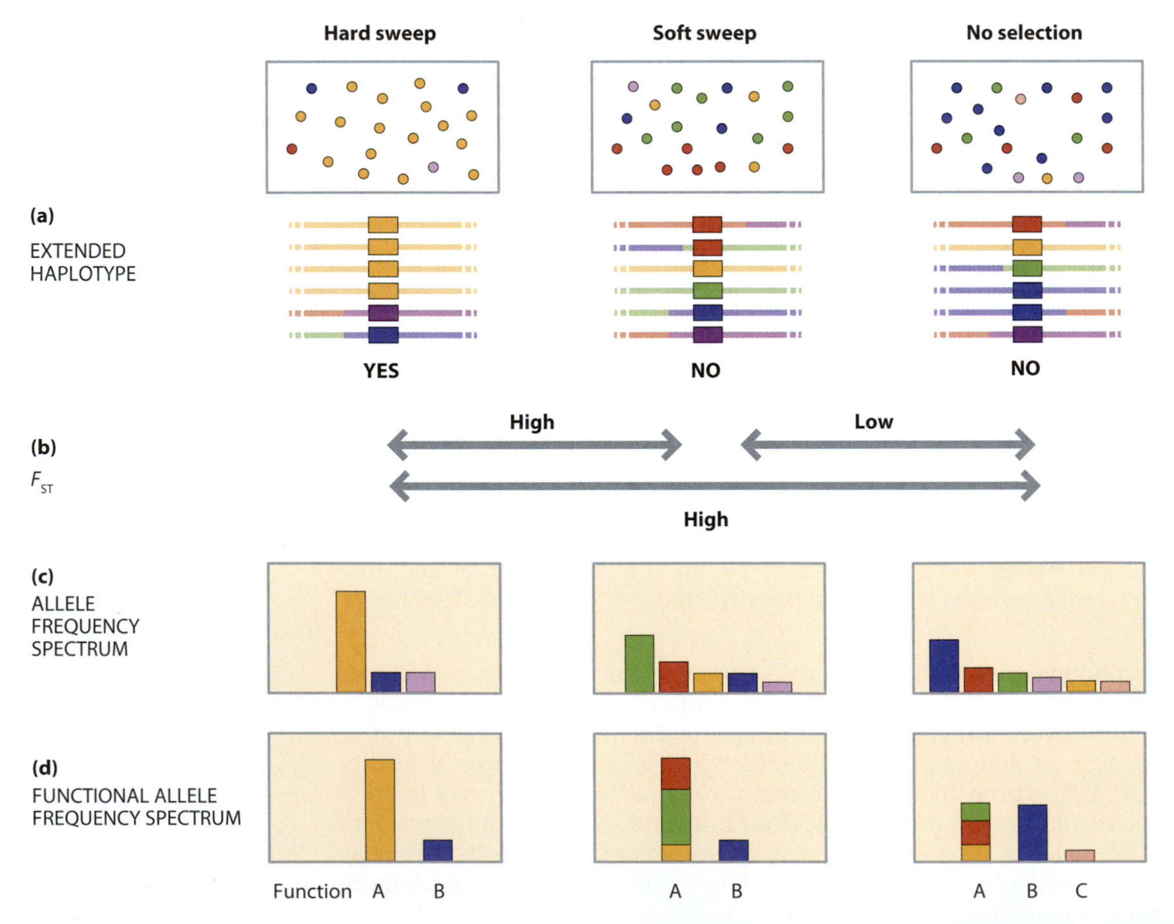

Figure 6.21: Effect of hard sweeps and soft sweeps on statistics designed to detect selection.
Illustrated are three populations, where one has undergone a hard sweep, one a soft sweep, and one no selection at all. Different colored circles represent different alleles of the same gene. The results on (a) haplotype diversity, as measured by extended haplotype statistics; (b) population differentiation at the selected gene, as measured by F_{ST}; (c) allele frequency spectrum; and (d) frequency spectrum of functionally distinguishable alleles—for example, in the soft sweep situation the *green*, *yellow*, and *red* alleles are different at the DNA sequence level but all have the same effect on a particular function.

human may limit the ability to detect selection at a single gene, genomewide analyses can identify specific functional classes of genes under greater selection. Including orthologous loci from a third, more distantly related species, in this case the mouse or macaque, identifies changes that have occurred in the human lineage rather than the chimpanzee lineage. Such outlier analyses have identified a list of selected genes, including ones involved in the sense of smell, skeletal development, neurogenesis, and pregnancy.[14] These analyses can be extended to include other mammalian species, providing more evidence for unusually rapid evolution in certain genes in the particular lineages[54] (Figure 7.25).

Selection can also be inferred from comparison between modern genomes and archaic genomes from ancient DNA. The small number of nucleotide sites that have changed between Neanderthal and modern human, and are now fixed in modern human, are candidates for natural selection, particularly if more than one nonsynonymous change has occurred in a gene, which is unlikely by chance. Table 8.2 shows genes that have more than one nonsynonymous difference between humans and Neanderthals.

Comparing variation *between* species with variation *within* a species can detect selection

Rather than comparing the amount of change at synonymous and nonsynonymous sites between two or more species, an alternative codon-based test compares diversity within a species with divergence between species. The **McDonald–Kreitman test** compares the amount of nonsynonymous and synonymous polymorphism within a species, with the amount of nonsynonymous and synonymous fixed differences between species.[52] Under neutral evolution, intraspecific polymorphism levels and interspecific substitutions are both determined by the rate of mutation, and thus the dN/dS ratios should be equal between and within species. Positive selection in the lineage leading to a species could be expected to increase *inter*specific nonsynonymous substitutions relative to *intra*specific nonsynonymous polymorphisms (**Figure 6.22**).

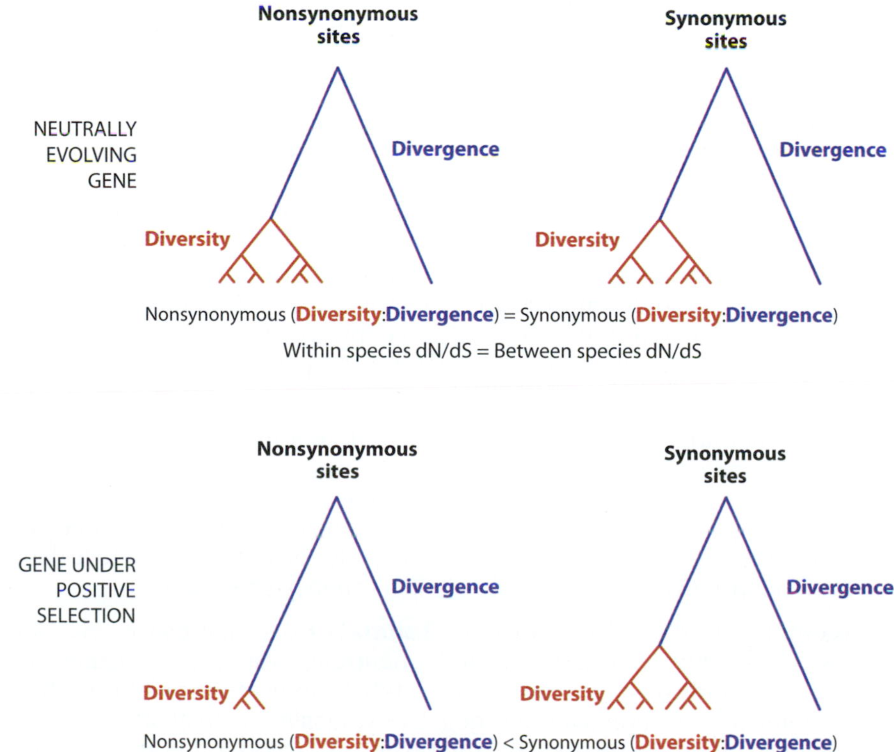

Figure 6.22: The basis of the McDonald–Kreitman neutrality test. Intraspecific diversity and interspecific divergence is shown for two genes, one that is evolving in a neutral fashion and the other that is under positive selection. Under neutral evolution the ratio of diversity to divergence should be the same at synonymous and nonsynonymous sites. If the gene has undergone adaptive evolution since divergence from the common ancestor, many more advantageous alleles will have become fixed than might otherwise have been expected. This positive selection can be detected as a reduction in the ratio of diversity to divergence at nonsynonymous (N), but not synonymous (S), sites.

The McDonald–Kreitman test has been used to identify genes and gene classes likely to be under positive selection in humans in a genomewide outlier analysis.[11] The data suggest that strong positive selection is rare, and most nonneutral evolution in the human genome is in fact weak negative selection, acting against mildly deleterious alleles of genes, revealed by an excess of nonsynonymous polymorphism compared with synonymous divergence at those genes.

The **Hudson–Kreitman–Aguadé** (**HKA**) test can be regarded as a more general form of the McDonald–Kreitman test.[40] It compares within-species polymorphism and between-species divergence at two (or more) loci. It does not use information on whether the variation is synonymous or nonsynonymous, and therefore can be used on both coding and noncoding sequences. Under neutrality, the level of within-species polymorphism should be correlated with between-species divergence. This ratio should be the same at both loci if they are evolving in a neutral fashion. Thus the null hypothesis of neutrality can be tested at a chosen locus by comparing it with sequence polymorphism and divergence at a neutral control locus in the same two species.

Under neutrality, three parameters are expected to relate observed diversity within and between the two species: θ, the time since divergence, and the ratio of effective population sizes. In essence, these are estimated from the observed diversity at all loci, and are then used to generate expected polymorphism and divergence for *each* locus. A statistical method known as a goodness-of-fit test is then used to compare these locus-specific measures of observed and expected diversity. If neutral evolution is operating at both loci, the expected and observed measures of diversity should agree reasonably well.

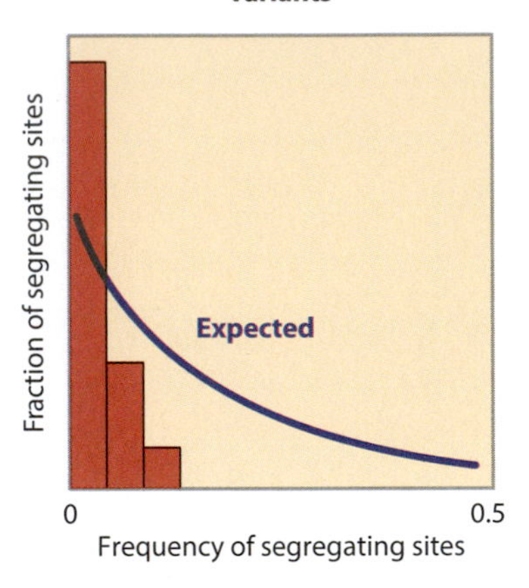

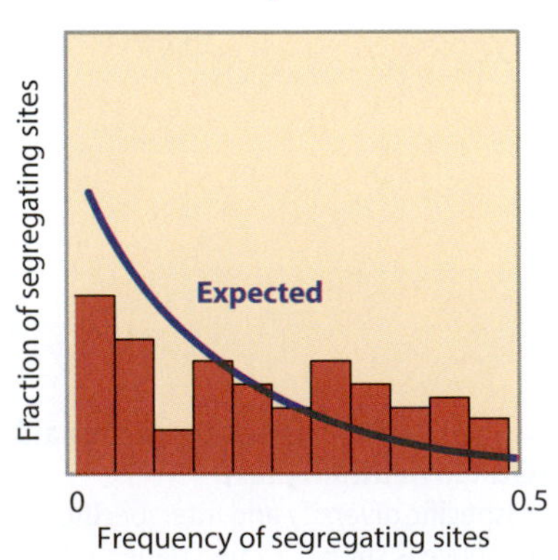

Figure 6.23: Frequency spectrum of segregating sites under different types of selection.
Two types of deviation from the site frequency spectrum expected in a constant-sized, neutrally evolving population (smooth line) are shown. One spectrum shows rare alleles (those at low population frequency) being more prevalent than expected, and the other shows intermediate alleles being over-represented. Both scenarios can be caused by either selection or demographic factors.

Selection tests can be based on the analysis of allele frequencies at variant sites

Variants within a set of sequences are present at different frequencies; some are found in only a single sequence (singletons), while others are in multiple sequences. Both selection and demography shape the spectrum of these frequencies (known as the **site frequency spectrum** or **allele frequency spectrum**). For example, in a population that has undergone a recent expansion many variant sites will have arisen relatively recently and so will be present only at low frequencies at all loci. Such a population can be considered to have an excess of rare alleles relative to a Wright–Fisher model. Similarly, if a specific haplotype at a locus is undergoing positive selection, it will be increasing in frequency relative to other haplotypes, and so a haplotype-specific excess of rare alleles may be observed at that locus, but not at others. By contrast, population subdivision and balancing (overdominant) selection maintain multiple haplotypes in the population for longer than would be expected under neutral evolution, and so produce an excess of intermediate frequency alleles (**Figure 6.23**). Population subdivision increases the number of intermediate frequency alleles at all loci, whereas balancing selection only influences the site frequency spectrum at or near the locus at which selection is acting.

Methods addressing the frequencies of variant sites are often based on the expectation that under neutral evolution the different estimates of θ should be equal. Only some estimates of θ incorporate information on allele frequencies; the number of segregating sites (S) is independent of frequencies, but nucleotide diversity (π) is not. Consequently, discrepancies between estimates of θ that incorporate frequency information differently (or not at all) detect departures from neutral expectations of the allele frequency spectrum.

One commonly used statistic, known as **Tajima's *D***, compares two estimates of θ, based, respectively, on S and π. Under neutrality, Tajima's D is expected to be zero. Significantly positive values of this statistic indicate population subdivision or balancing selection, whereas negative values indicate positive selection or population growth. A related statistic takes into account the ancestral state of the variants (to see how the ancestral state is identified, see **Section 3.2** and

Figure 3.9). Under neutrality, very few high-frequency derived alleles are present in a population. However, directional selection leads to a selective sweep that increases the proportion of linked high-frequency derived alleles as the sweep is nearing completion or in adjacent regions after fixation. This pattern can be measured by **Fay and Wu's *H*** statistic.[23]

Site frequency spectrum statistics have been used in candidate gene studies and are now starting to be used in genomewide outlier analyses.[78] These approaches are particularly useful because they can be applied genomewide, irrespective of whether the sequences being considered are coding or not.

Comparing haplotype frequency and haplotype diversity can reveal positive selection

A discrepancy between the frequency of a haplotype and its linked diversity expected under neutrality provides evidence of past selection.[70] High-frequency haplotypes are expected to be old, because time is required for the haplotype to increase in frequency by drift. They therefore have accumulated high levels of diversity. Low-frequency haplotypes can be either old or young, and may therefore have high or low diversity. Positive selection for one variant can rapidly increase the frequency of its surrounding haplotype, so it becomes common without achieving a high level of haplotype diversity. Haplotype diversity can be measured by examining the number of variants at a linked multiallelic polymorphism, such as a microsatellite. This approach has been used to identify positive selection on the lactase gene (*LCT*), involved in the digestion of fresh milk, where one particular haplotype has been selected in Europeans (**Section 15.6**).

The extent of homozygosity at SNPs spanning a genomic region, which reflects the length of a conserved haplotype, can also be used as a measure of haplotype diversity. Haplotypes carrying an allele that has undergone recent selection extend over significantly longer distances than other haplotypes at the same locus with similar frequencies. This effect is generated by a recent selective sweep, where a new variant is subject to positive selection and therefore increases in frequency (**Figure 6.24**). The increase in frequency of the selected allele is accompanied by an increase in frequency of alleles on the same haplotype (that is, in LD with the selected allele). With increasing time since the selective event, recombination breaks down this haplotype so that eventually the signature of selection disappears. Several closely related test statistics have been developed that detect this signature, collectively called **extended haplotype tests** (**Table 6.7**).

Extended haplotype tests have been used to provide evidence of recent selection at several genes, including some of those discussed in Chapter 15. They have also been used as a statistic for outlier analysis on genome diversity data from different populations. An example is given in **Figure 6.25a**, where the lactase gene (*LCT*) is a clear outlier showing an extended haplotype in Europeans. The breakdown of a haplotype, as demonstrated in Figure 6.24, is often plotted as a haplotype bifurcation diagram (Figure 6.25b). Extended haplotype tests can be modified to test for a selective sweep having occurred in one population but not in others (cross-population extended haplotype homozygosity, XP-EHH), and this has been applied in an outlier analysis on **HapMap** SNP data.[66]

Analysis of frequency differences between populations can indicate positive selection

We might expect that some selection has operated on one population but not on another, particularly if the selective agent is an environmental difference between the two populations. In these cases, selection will have had an impact on the diversity of one population, usually on specific loci, but not on the other population. Therefore measures of interpopulation differences may also measure the effect of selection at a gene in one population but not another.

Earlier (**Section 6.2**) we introduced several statistics that can be used to measure the differences between populations. A simple approach is to compare the nucleotide diversity π across a genomic region of one population with another. A striking example from a nonhuman genome, but showing the selective effect of domestication by humans, is the comparison of nucleotide diversity between

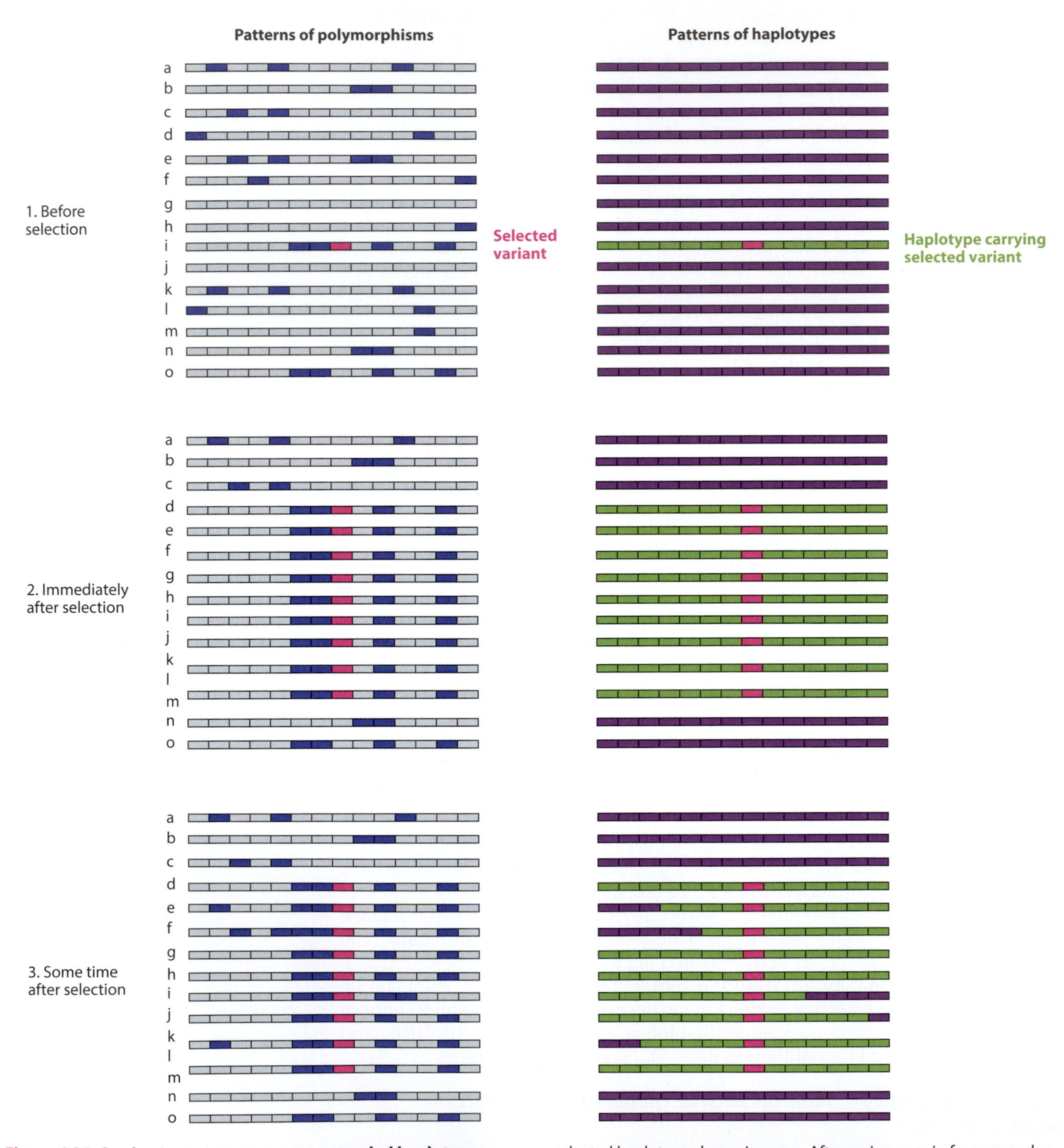

Figure 6.24: A selective sweep generates extended haplotypes. A genetic region is shown here as a series of polymorphisms (*boxes*), with different alleles shown as dark and light *blue* boxes. A sample of this genetic region from a population is shown as a–o. A *pink* mutation arises on a particular haplotype, which is selected (1) and increases in frequency (2). Recombination breaks down this selected haplotype (3). The right-hand side shows the effect on haplotype structure, with the selected haplotype shown in *green*. After an increase in frequency due to selection (2) we can see that an extended haplotype (*green*) has risen to high frequency through a selective sweep. This extended haplotype is subsequently broken down by recombination over time (3). [Adapted from Oleksyk TK et al. (2010) *Philos. Trans. R. Soc. Lond. B Biol. Sci.* 365, 185. With permission from Royal Society Publishing.]

TABLE 6.7: EXTENDED HAPLOTYPE TEST STATISTICS

Name	Abbreviation	Software	Reference
Extended haplotype homozygosity/ relative extended haplotype homozygosity/ cross-population extended haplotype homozygosity	EHH/REHH/XP-EHH	SWEEP	Sabeti PC et al. (2007) *Nature* 449, 913.
Integrated haplotype score	iHS	WHAMM	Voight BF et al. (2006) *PLoS Biol.* 4, e72.
Whole genome long-range haplotype test	WGLRH	WGLRH	Zhang C et al. (2006) *Bioinformatics* 22, 2122.
Log-ratio of integrated EHH scores between populations	Ln(Rsb)	–	Tang K et al. (2007) *PLoS Biol.* 5, e171.
Extended haplotype-based homozygosity score test	EHHST	available as C++ source code	Zhong M et al. (2010) *Eur. J. Hum. Genet.* 18, 1148.
Delta-integrated EHH	ΔiHH	*rehh* (on R statistical platform)	Grossman SR et al. (2010) *Science* 327, 883.

domesticated maize and its non-domesticated relative **teosinte** across the *tb1* locus (Figure 12.29).

If a SNP shows an extreme allele frequency difference between two populations, when compared with allele frequency differences of other SNPs, then

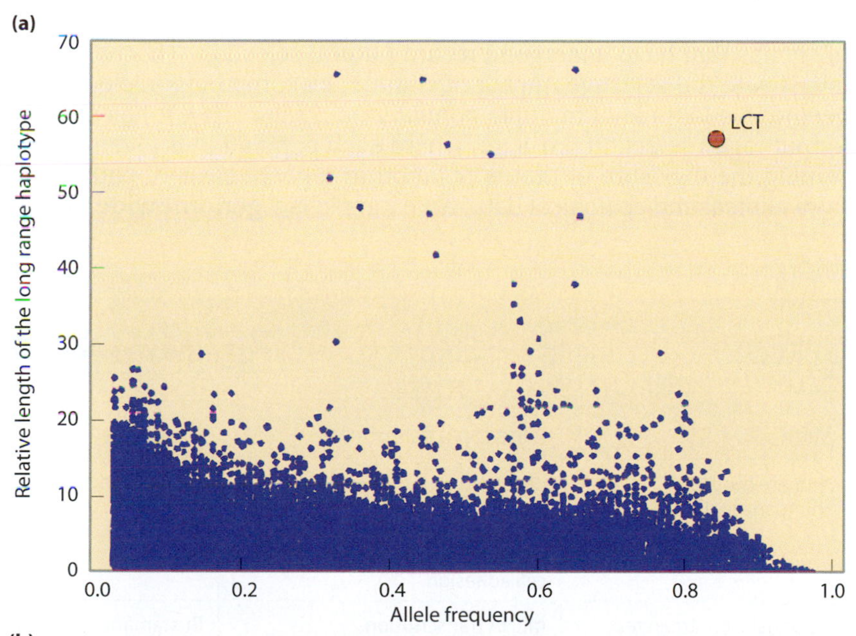

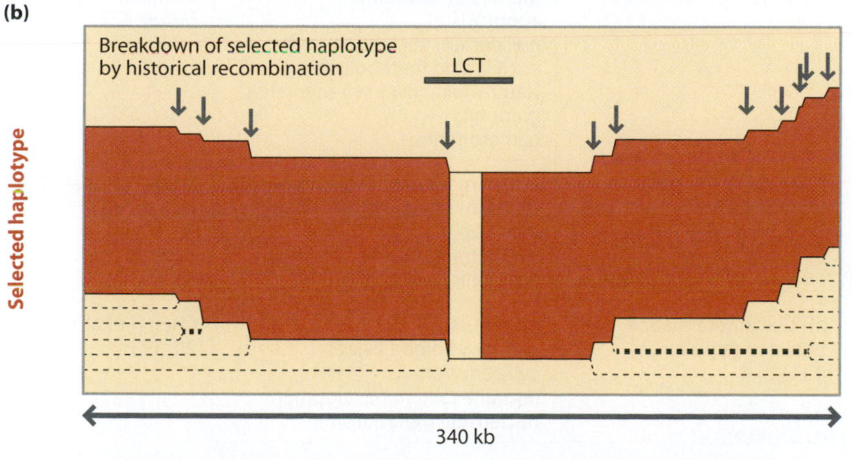

Figure 6.25: An extended haplotype at the lactase (*LCT*) gene.
(a) For all genotyped SNPs from the CEU HapMap sample (Box 3.6), the relative length of the haplotype (*y* axis) and allele frequency (*x* axis) have been plotted. A SNP at the lactase gene (*LCT*) is a clear outlier, indicating a high frequency of an extended haplotype. (b) The extended haplotype at *LCT* can be plotted as a haplotype bifurcation diagram, with each SNP on the haplotype indicated by an arrow, and a bifurcation at that point. [Adapted from Sabeti P et al. (2006) *Science* 312, 1614. With permission from AAAS.]

selection could be a cause of that difference. An example of this approach is the comparison of allele frequencies in Han Chinese and Tibetans of over 100,000 SNPs (Figure 15.9), where two SNPs in the *EPAS1* gene show very large allele frequency differences and are likely to have undergone selection (**Section 15.4**). This allele frequency statistic for each SNP can be formally written as $|\Delta DAF|$, which is the absolute difference (that is, always positive) in derived allele frequencies between two populations. The 1000 Genomes Project systematically identified SNPs with high $|\Delta DAF|$, including SNPs in *EDAR* (**Section 10.3**), *SLC24A5* (**Section 15.3**), the Duffy antigen receptor *DARC* (**Section 17.4**), and *TLR1*, a Toll-like receptor where the SNP is involved in the protection against leprosy.[45]

The statistic F_{ST} (Box 5.2) has been used in outlier analyses to identify genes that might have undergone selection. The initial study used a threshold of $F_{ST} > 0.45$ to identify SNPs within a total of 156 genes that may have undergone selection in European-Americans, African-Americans, or East Asians, including genes that have functional roles such as *EDAR* (**Section 10.4**).[2] F_{ST}, however, depends on the allele frequency of the SNP tested. Other tests of population differentiation have been developed, such as the across-population-composite likelihood ratio (XP-CLR),[13] which is more robust to SNP ascertainment bias on the basis of frequency, and has been applied to genomewide datasets (**Table 6.8**). Another example is the population branch statistic, which generates a tree based on the F_{ST} for three populations (**Section 6.3**) for many different loci, with an unusually long branch leading to one of the populations suggesting selection at that locus in that population.[88]

An alternative approach, where allele frequency data from different populations are compared with different environmental exposures, can also be used to identify genes that have undergone recent natural selection. This has been used to investigate the selective impact of climate, by comparing allele frequency and geographic latitude of multiple populations (**Figure 6.26**). It has also been used for pathogen selection pressure, where each population is given a value representing the exposure of that population to a given class of pathogen, derived from clinical and ecological data. An example is a **genomewide scan** for SNPs

TABLE 6.8:
FUNCTIONAL CATEGORIES SHOWING ENRICHMENT OF PUTATIVE POSITIVELY SELECTED GENES

Method	Functional category	Reference
Human–mouse–chimpanzee dN/dS	olfaction nuclear transport G-protein-mediated signaling signal transduction cell adhesion	Clark AG et al. (2003) *Science* 302, 1960.
McDonald-Kreitman test	mRNA transcription apoptosis nucleoside, nucleotide, and nucleic acid metabolism Natural killer (NK) cell-mediated immunity gametogenesis	Bustamante CD et al. (2005) *Nature* 437, 1153.
Extended haplotypes (iHS) (Europeans)	MHC-I-mediated immunity electron transport olfaction mRNA transcription fertilization	Voight BF et al. (2006) *PLoS Biol.* 4, e72.
XP-CLR differentiation between populations (CEU-YRI)	apoptosis pattern of blood vessels cell adhesion immune system development nucleoside metabolism	Chen H et al. (2010) *Gen. Res.* 20, 393.

(a)

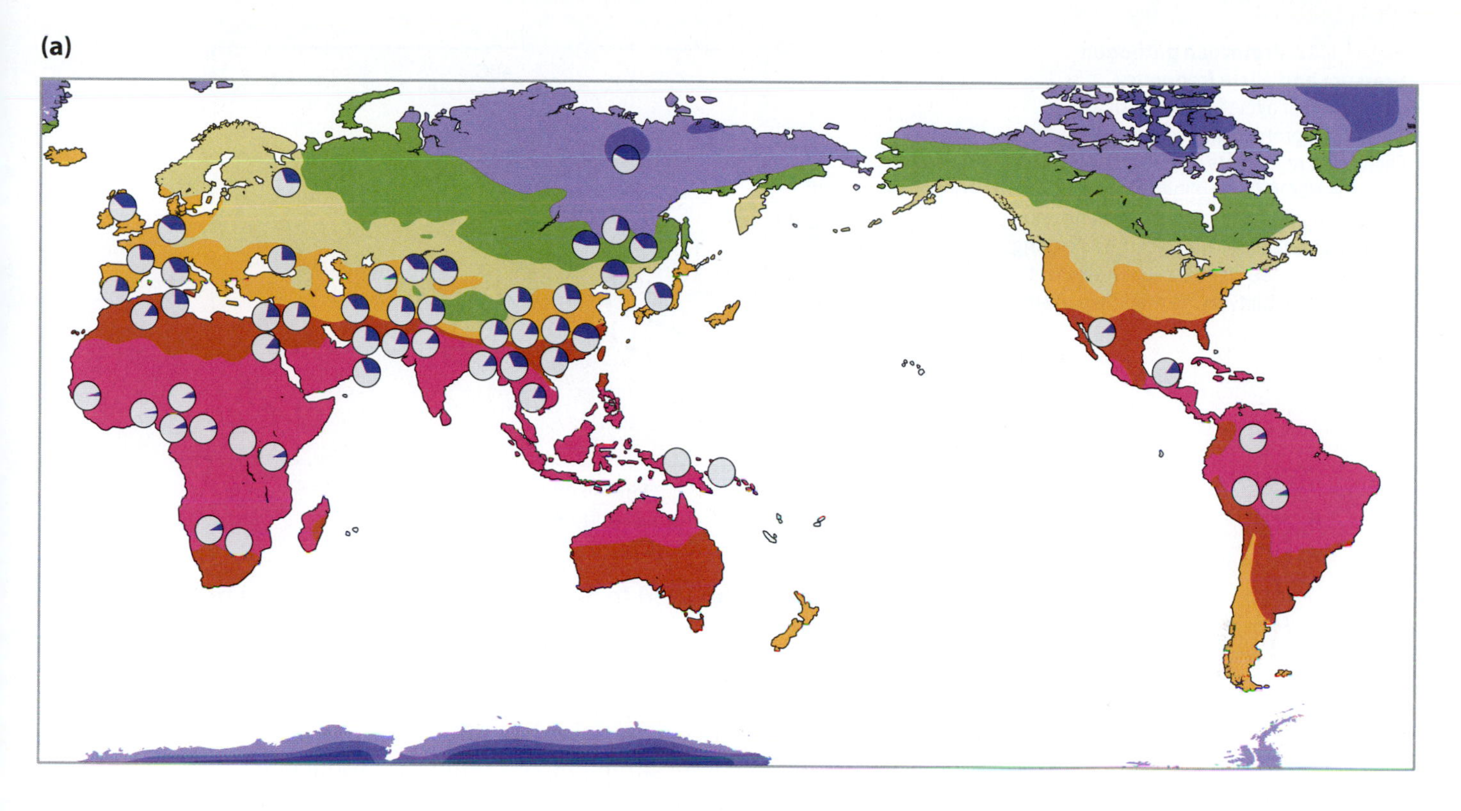

(b)

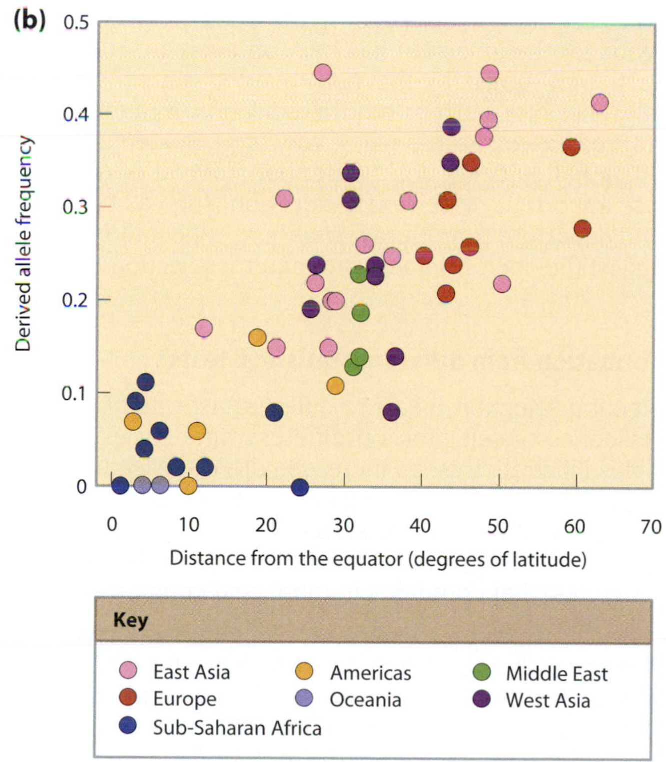

Figure 6.26: Global temperature and allele frequency. (a) The map shows allele frequencies in different populations of a SNP (rs12946049) in the regulatory-associated protein of MTOR (*RPTOR*) gene (pie charts), together with winter maximum temperature ranging from *blue* (cold) to *pink* (hot) (gradient plot). (b) The scatter plot shows the correlation between allele frequency at this SNP and latitude. [From Novembre J & Di Rienzo A (2009) *Nat. Rev. Genet.* 10, 745. With permission from Macmillan Publishers Ltd.]

showing a high correlation between allele frequency and protozoan diversity across the Human Genome Diversity Project populations.[62] Outliers included SNPs in known susceptibility genes for malaria (a parasitic protozoan), as well as previously unidentified genes that may have been selected by protozoan infection (**Figure 6.27**).

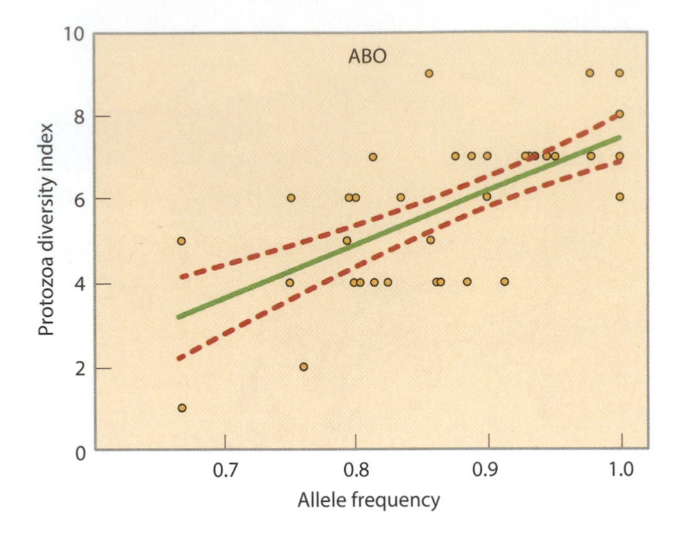

Figure 6.27: Protozoan pathogen pressure and allele frequency. Correlation of allele frequency of a SNP with a protozoan diversity index. Each point represents a population from the HGDP-CEPH panel (Box 10.2). Protozoan pathogens include malaria and *Trypanosoma brucei*, which causes sleeping sickness. The *ABO* gene has been shown to affect susceptibility to the malaria parasite.

Other methods can be used to detect ongoing or very recent positive selection

Natural selection is the differential reproduction of genotypes, so if there is a correlation between a particular genotype at a locus and reproductive success, we can infer that the locus is under selection. This approach requires very large datasets including a measurement of lifetime reproductive success, and has been used in one study based on a large database of Icelanders to demonstrate that an **inversion** on chromosome 17q has undergone positive selection. A woman who is heterozygous for the inversion has had, on average, 10% more children than individuals homozygous for the non-inverted state.[73] Similar approaches can potentially measure current natural selection by correlating phenotype with lifetime reproductive success using very large epidemiological studies (**Section 15.7**). Functional evidence for differential survival of different genotypes can also provide important evidence for selection, such as selection by malaria (Table 16.3), and departure from Hardy–Weinberg expectation can in certain circumstances suggest the operation of strong natural selection (**Section 5.1**, Box 5.1).

How can we combine information from different statistical tests?

Different methods for detecting selection measure different aspects of genetic diversity, and so detect evidence of selection on different time-scales (**Figure 6.28**). Two extreme examples illustrate this: an increased dN/dS reflecting multiple amino acid changes present in all humans will not detect a recent selective event 4 KYA. Conversely, a population-specific selective event in a promoter may be detected by an interpopulation diversity statistic such as F_{ST} but will not be reflected in a codon-based test that compares humans and chimpanzees.

If we are to combine statistics, we need to take into account the confounding effect of certain measures not being entirely independent—for example, a region where SNPs show high F_{ST} values is more likely than average to show a significant value for Tajima's D, because both statistics measure population structure revealed by a particular pattern of genetic diversity. Two approaches have been developed to combine different statistics that detect selection. These are likely to be useful when a selection pressure has operated over a long time-scale or if it has occurred when the informative time span of two statistics overlap. One rationale behind this is to pinpoint the exact SNP where selection is likely to have occurred, when only extended haplotypes (which can extend for several hundred kb) have provided evidence of selection. One approach, called the **composite of multiple signals** (**CMS**) test,[33] combines five statistical tests: F_{ST}, a measure of the change in derived allele frequency between populations |ΔDAF|, and three different extended haplotype tests (iHS, ΔiHH,

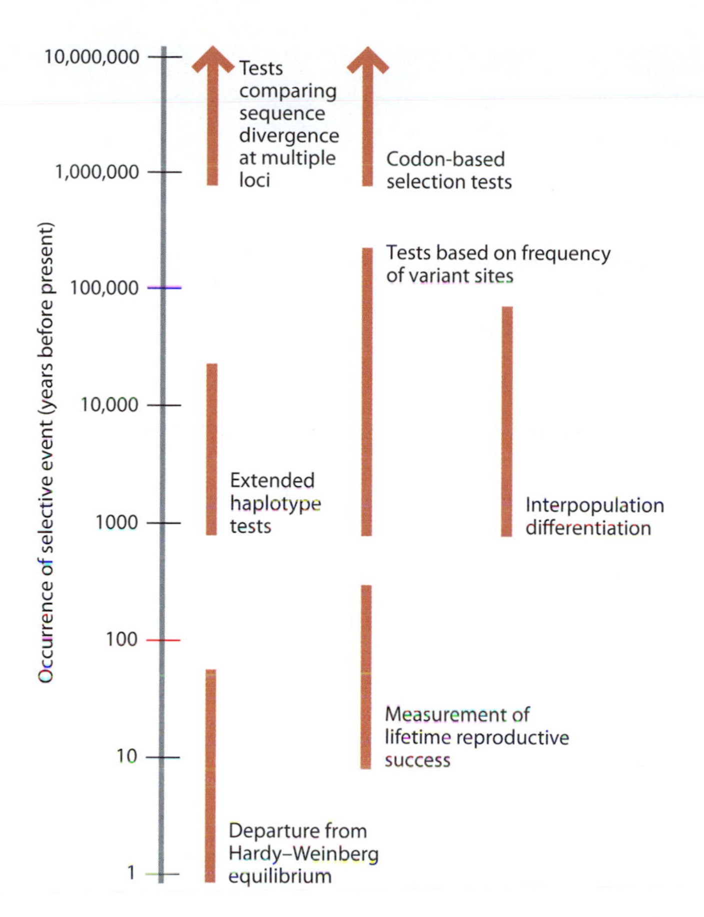

Figure 6.28: Detection of past selection events.
Different methods detect selective events on different time-scales (time-scales shown are approximate). [Adapted from Oleksyk TK et al. (2010) *Philos. Trans. R. Soc. B Biol. Sci.* 365, 185. With permission from Royal Society Publishing.]

and XP-EHH; Table 6.7). Another approach, the **composite likelihood ratio** (**CLR**) test, combines different aspects of the site frequency spectrum analyzed in tests such as Tajima's *D* and Fay and Wu's *H*. Both approaches have been used to provide evidence for natural selection at the *EDAR* gene (**Section 10.3**, Figure 10.9).

Tests for positive selection have severe limitations

A fundamental limitation of all the selection tests described here is the paucity of well-founded examples where we are certain, independent of the statistical tests themselves, that selection has occurred (**Table 6.9**). Such examples could come from functional studies involving humans, human cells, or model organisms. But without them, it is difficult to evaluate how well our selection tests are really performing. It is thus sobering, but perhaps not entirely surprising, to note that genomewide scans have identified gene lists and functional categories with limited overlap (Table 6.9), and often do not detect well-known cases of selection, like the Duffy antigen receptor gene *DARC*, where a null allele has been selected because it confers resistance to *Plasmodium vivax* malaria (**Section 17.4**). This is partly due to different time-scales of selection detected by the different methods (Figure 6.28), but it could also suggest a significant **false negative** rate, in addition to the **false positive** rate implicit in outlier analyses. It should also be noted that many of the methods discussed here have little power to detect selection when intragenic recombination is operating,[82] or when multiple rounds of selection have acted at the same locus. Population structure may also play an important role in reducing our ability to detect positive selection.[63]

It is likely that many adaptations in human evolution, such as those involving infectious disease, involved **fluctuating selection**. This would result in short bursts of strong selection in the past, followed by drift to loss or fixation of the allele in the absence of selective pressure.[42] There is extensive theoretical literature on this type of selection, but, as yet, little practical application to human

TABLE 6.9:
SOME GENES SHOWING EVIDENCE OF POSITIVE SELECTION IN HUMANS SUPPORTED BY BOTH POPULATION-GENETIC AND FUNCTIONAL EVIDENCE

Gene name	Protein name	Selected phenotype	Tests providing evidence of positive selection	Section in the book
EDAR	ectodysplasin A receptor	hair thickness, tooth shape, or sweating	population differentiation, extended haplotype, allele frequency spectrum	10.3
SLC24A5	solute carrier family 24 member 5	pale skin	population differentiation, extended haplotype	15.3
LCT	lactase	ability to digest fresh milk as an adult	extended haplotype, haplotype diversity, population differentiation	15.6
G6PD	glucose-6-phosphate dehydrogenase	resistance to severe malaria	extended haplotype	16.4
CASP12	caspase-12	resistance to severe sepsis	allele frequency spectrum	17.4
DARC	Duffy antigen receptor for chemokines	resistance to malaria caused by *Plasmodium vivax*	allele frequency spectrum, population differentiation	17.4
APOL1	apolipoprotein L1	improved lysis of *Trypanosoma brucei* parasites	extended haplotype	17.4

genetic diversity. Frequency-dependent selection, and other forms of balancing selection suggested to be responsible for the astonishing diversity at the major histocompatibility complex[74] (MHC, Box 5.3), are also not well measured using these approaches (Box 6.6). Frequency-dependent selection is particularly difficult to simulate, since the relationship between allele frequency and selection coefficient is variable and, depending on the locus, arbitrary. This area requires new advances in mathematical population genetics.[59]

The hope is that as studies of both simulated data and real data suggest different approaches in detecting different types of selection, the two approaches will inform evolutionary hypotheses as to how, when, where, and why a particular selective event occurred in human history. However, providing evidence of natural selection is problematic in that there is no definitive test for what is almost always a historical explanation for a particular pattern of diversity. This is the case both for statistics applied to a single gene, and for genes identified as outliers in a particular test. Evidence for selection often accumulates slowly, from different studies often in different scientific disciplines, using different statistics and different experimental methods. It is an area where multidisciplinary studies, though difficult in practice, may bear fruit. We discuss specific cases of natural selection for human phenotypes in Chapter 15, and the role of natural selection in disease is considered in Chapters 16 and 17.

6.8 ANALYZING GENETIC DATA IN A GEOGRAPHICAL CONTEXT

Until now our consideration of population subdivision has focused solely on a qualitative measure—the population affiliation of a given individual. This neglects some quantitative knowledge about each data point, namely its geographical location. Individuals 1, 2, and 3 may belong to populations A, B, and C respectively; however, if populations A and B are neighbors, whereas C is further away, some information has been overlooked. By incorporating this information we can start to integrate genetic data with information about the landscape from which they were sampled. For example, high levels of differentiation between two nearby populations may result from the presence of a mountain range separating them. By relating genetic data to specific geographical locations we can also attempt to integrate patterns of modern diversity with past geographical processes, for example, sea level changes that resulted in the formation of land bridges.

Geographical analyses of genetic data allow us to partially disentangle the relative contributions of history and geography to modern genetic diversity.[34] In the absence of external, nongenetic, information, it can be difficult to discern whether geographical constraints or historical episodes account for observed patterns of genetic diversity. In many cases, the first step is to test whether the genetic data show any form of geographic structure. There must be demonstrable geographical patterning before any explanation need be sought. This is analogous to the tests for population subdivision discussed earlier in this chapter. Once patterning has been detected, it is necessary to determine the nature of the pattern: Are certain alleles found in patches? Or do smooth gradients of allele frequencies span the sampled area? Such gradients of allele frequencies are known as **clines**, and they can be generated by a number of evolutionary processes.

For these analyses it is not enough to have a few sites that are sparsely sampled over a wide geographical region. Any landscape being investigated should be extensively sampled, with as regular as possible a distribution of sample sites. Some analyses require a population at each sample site, whereas others can cope with a separate geographical location for each sampled individual.

When the geographical region being investigated is large, geographical distance calculations between sites need to take account of the Earth's curvature. This is done using a trigonometric equation that calculates the **great circle distance**, which is the distance between any two points on the surface of a sphere. Other estimates of the distance between two sites take into account the topography of the Earth, rather than regarding it as a simple uniform sphere. The **walking distance** between two points can be calculated, for example by only allowing distances across land areas under 2000 m in altitude to be used in the calculation (**Figure 6.29**). This approach can be developed further, using models of human demography and geography to simulate genetic diversity at different stages of human colonization.[19]

Genetic data can be displayed on maps

The simplest approach is to plot graphical representations of allele frequencies within a population onto the geographical location of that population. Pie charts allow the frequencies of several (but not many) alleles to be shown in the same map (examples are shown in **Sections 12.5 and 12.6**).

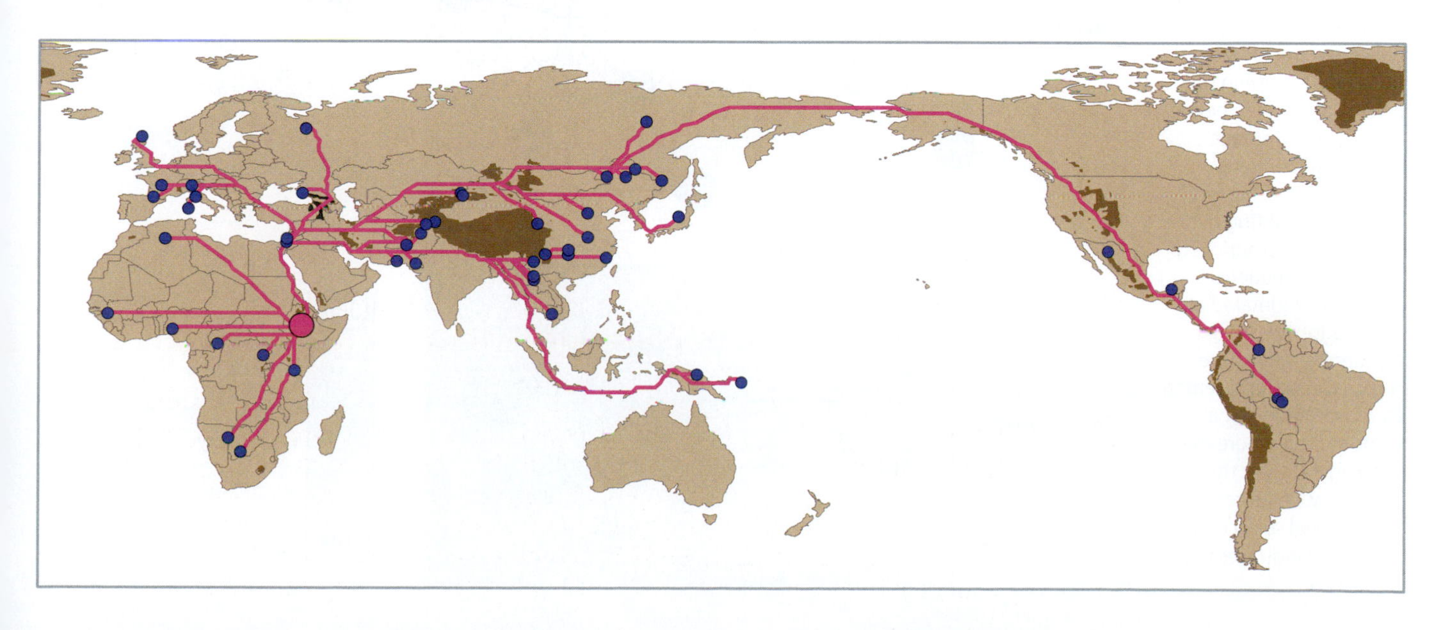

Figure 6.29: Geographical distances between Addis Ababa and different global locations, using the "walking distance" approach.
"Walking distances" between Addis Ababa (*pink circle*) and the 54 sampling locations of the HGDP-CEPH panel (Table 10.1) shown as *blue* circles are indicated by *pink* lines. Land with an altitude of more than 2000 m is shaded dark brown. [From Liu H et al. (2006) *Am. J. Hum. Genet.* 79, 230. With permission from Elsevier.]

Alternatively, a contour map of allele frequencies can be produced. The contour lines, known as isogenic lines, join points of equal allele frequency, like lines of altitude on a topological map, or lines of equal air pressure on a weather map, generated by a method called **interpolation** (**Figure 6.30**). This regular grid of estimated gene frequencies can then be used to generate contour maps more easily.

Interpolation is performed by laying a fine grid on top of a map and estimating the gene frequency at each point where the grid lines intersect (pseudo-sites). Several different methods are available, which rely on identifying the nearest sampled locations to the pseudo-site, and using the allele frequencies at these locations to estimate the value at the pseudo-site. The distance between the sampled locations and the pseudo-site is often taken into account by weighting the allele frequency at the observed site according to its proximity. Thus, the gene frequency at the closest observed site has more impact on the gene frequency at the pseudo-site than that at the furthest observed site (see Figure 6.30). Other interpolation methods can be far more statistically involved. One such example, known as **kriging**, estimates a mathematical function that describes the genetic surface, rather than estimating single pseudo-values on the grid. A problem with gene frequency maps is that the process of interpolation is itself

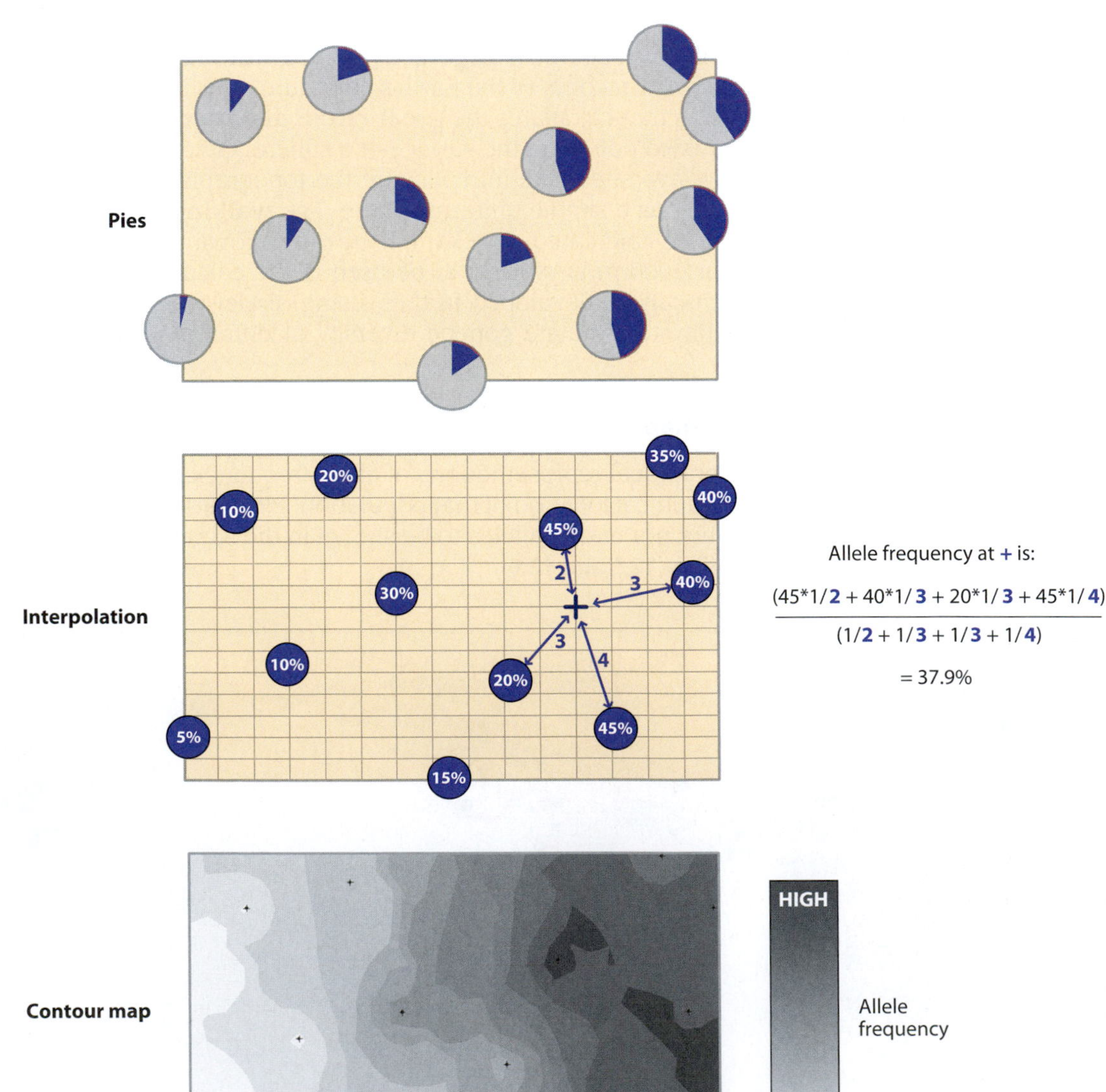

Figure 6.30: Pie charts and the method of interpolation.
The frequencies of an allele at 12 sampled locations distributed irregularly throughout a geographical region are shown as pie charts (*blue slice*). These locations can be used to estimate allele frequencies at regularly spaced pseudo-sites defined by intersecting gridlines by the process of interpolation. Allele frequencies at the four closest sampled locations to a pseudo-site are identified, and the gene frequency is determined as the average of the gene frequencies at these observed sites, weighted according to the inverse of distance between them. Other distance weighting functions could also be used, for example weighting by the inverse distance squared. Once allele frequencies have been estimated at all pseudo-sites, allele frequencies across the region can be represented by a two-dimensional contour map, where the shading indicates the frequency, or a three-dimensional surface, where the height indicates the frequency.

capable of generating clines. After interpolation it can be impossible to tell the difference between genuine clines present in the data and artificial clines that were generated during interpolation.[64, 72]

The use of densely sampled genetic data throughout a geographical region allows the reconstruction of a "genetic landscape," the features of which have been shaped by past processes, be they historical or geographical. **Phylogeography** refers to the study of the geographical distribution of the clades within a phylogeny, and provides a temporal dimension to be combined with the spatial dimension of geography. Many analyses rely on visual inspection of the allele frequency data overlaid onto maps, but geographic patterns are often open to alternative interpretations. More quantitative methods such as nested cladistic analysis[75] and founder analysis[65] have been developed and applied to questions about human evolution, yet these methods have little power to distinguish one evolutionary explanation of a geographical pattern over others.[55] There have been some bitter arguments over the value of phylogeographical approaches (Box 12.3).

Genetic boundary analysis identifies the zones of greatest allele frequency change within a genetic landscape

A zone of abrupt genetic change results from populations dwelling on either side of a **genetic boundary** being more differentiated than might otherwise be expected, and points to limited gene flow between them. That may mean that the populations are currently isolated because of some **genetic barrier** to mating and/or dispersal, physical or cultural, or that the populations evolved independently and came into contact only recently. The former explanation emphasizes the effects of geography, the latter the effects of history. Once identified, genetic barriers can be correlated with physical or cultural barriers to gene flow, to identify which of these has had the greatest impact on the distribution of modern diversity.

A method to detect genetic boundaries by combining data from all alleles was devised by Womble and has subsequently become known as Wombling.[4, 85] By calculating the gradient (first derivative) of allele frequency change at a number of locations for each allele separately, regions of greatest change (that is, steepest gradients) have high values (represented by peaks in three dimensions) and regions of least change (flattest gradients) have lower values. Consequently, these **surfaces** of allele frequency change can be summed from all alleles. Peaks present in the same location for multiple alleles reinforce one another. Only peaks over a certain height are deemed significant, and these represent the genetic boundaries within the genetic landscape. Perhaps inevitably, Bayesian Wombling has also been developed.[27]

Spatial autocorrelation quantifies the relationship of allele frequency with geography

Processes whereby populations exchange genes in a geographically restricted fashion (such as isolation by distance) leads to the non-independence of genetic variation throughout space. In other words, the frequency of an allele within one population is to some degree correlated with the frequency of the same allele in neighboring populations. This relationship of genetics and geography can be quantified using **spatial autocorrelation**. It has been shown that the nature and extent of this **autocorrelation** can be used to investigate the processes causing it. Different patterns of spatial autocorrelation are expected under:

- Random distribution of alleles in space (a rather unlikely condition for humans)
- Isolation by distance
- Clinal variation
- The presence of multiple non-overlapping clines in the region of interest

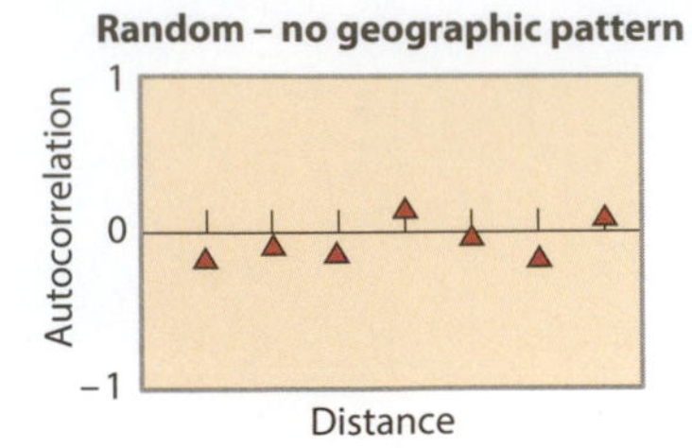

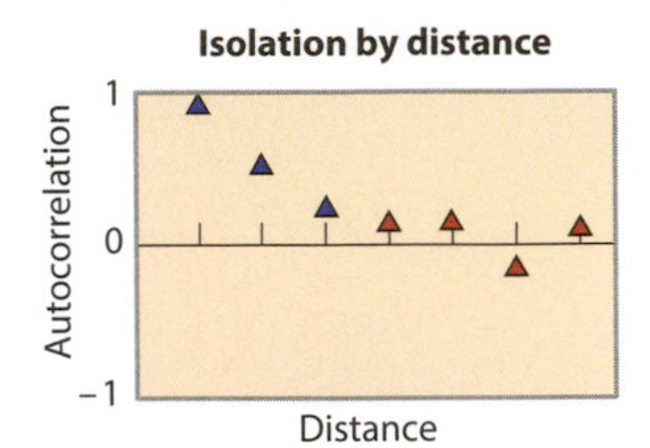

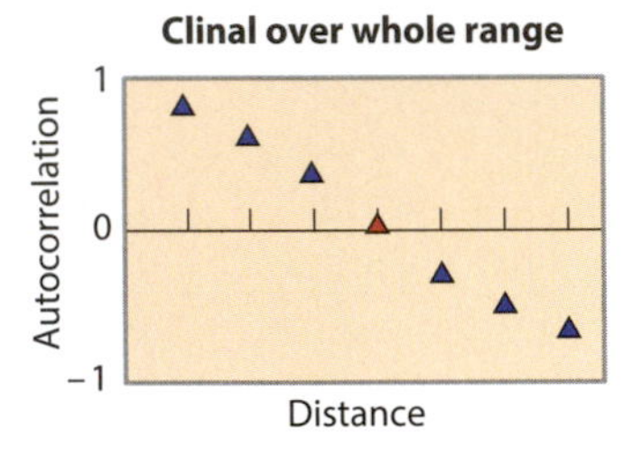

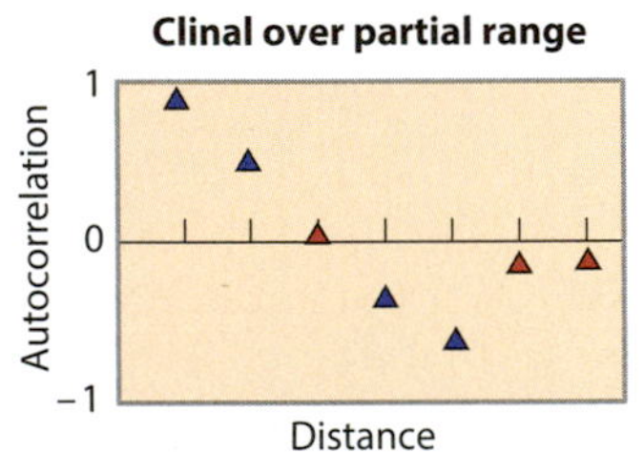

Figure 6.31: Spatial autocorrelation shows different patterns for clines, isolation by distance, and random mating.
Pairwise comparisons of autocorrelation between populations are combined into different classes on the basis of the distance between them. Triangles represent the average levels of autocorrelation observed within each distance class. *Blue* triangles indicate levels of autocorrelation that are significantly different from those obtained when haplotypes are distributed randomly across the geographical space.

While the latter three processes generate significant geographical structure, it can be difficult to determine which has been in operation from observation of this structure alone.

To quantify spatial autocorrelation, pairs of sampled localities are pooled into different arbitrary distance classes. For each distance class the level of spatial autocorrelation is calculated using a measure of genetic similarity that varies between 1 (strong positive correlation) and –1 (strong negative correlation). Plotting the level of spatial autocorrelation against the distance class generates a **correlogram**, which describes quantitatively the geographical pattern of genetic variation, and tests for its departure from spatial randomness. From its shape, inferences on the likely processes generating the geographical structure can often be drawn. There are different measures of genetic similarity that can be used for spatial autocorrelation analysis. Some are applicable to classical allele frequency data, whereas others have been specially formulated for molecular data, taking into account genetic distances between alleles.[9] **Figure 6.31** shows the patterns of spatial autocorrelation, in the form of correlograms, expected under random mating, isolation by distance, and clinal variation. Spatial autocorrelation analysis has been especially prominent in analyses of European genetic diversity, where different interpretations of clinal patterns of diversity have been hotly debated (**Section 12.5**).

Mantel testing is an alternative approach to examining a relationship between genetic distance and other distance measures

For a set of human populations, we often have a number of distance matrices that contain pairwise comparisons of different aspects of these populations. These distances can be genetic distances, linguistic distances, geographical distances—in fact any form of measurement. Often we want to detect correlations between these matrices, which might tell us something about their evolutionary past. However, we have a problem, because each pairwise distance within a matrix is not independent, and consequently, the assumptions of classical statistical methods for detecting correlations between variables (for example, bivariate regression methods) are violated. A **Mantel test** detects correspondence between matrices and provides an alternative method for disentangling history and geography.[34, 51]

Correspondence between any two distance matrices may result from a causal relationship between them, for example a migration that disperses both genes and languages. Alternatively, correspondence between them may result from both being correlated with a third matrix. In this case superficial correspondence between genetic and linguistic distances may result from both being correlated with geographical distances rather than any direct relationship between them. In such situations, a geographical process, for example isolation by distance, generates geographic structuring of both languages and genes. Thus both historical and geographical processes may underlie correspondences between any two matrices.

Extensions to the Mantel test allow correspondences between three distance matrices to be tested. In this way, a common historical process causing similarity between genetic and linguistic datasets need only be invoked once it has been demonstrated that geographical distances are not themselves the determining factor. In such a case, genetic and linguistic distances would be better correlated with each other, than either would be to geographical distances.[71]

SUMMARY

- Although describing present genetic diversity and inferring how it arose are two separate analytical objectives, the methods by which they are accomplished are often interrelated.

- Study of data from many genomewide loci is now the norm, giving a more complete picture of the demographic history of humans. New analytical methods are being developed for whole genome sequences of different individuals.
- Demography influences variation genomewide whereas selection operates in a locus-specific manner, thus genomewide analyses help to identify outlier loci that are likely to be under selection.
- Phylogenies inform us about the evolutionary relationship of different species, and gene genealogies or networks represent the evolutionary relationship of different DNA sequences. Principal component analysis or cluster analysis tells us about the evolutionary relationship of individuals or populations.
- Coalescent analysis is a mainstay of modern population genetic analysis, and can be applied to almost any question about human evolutionary history.
- Chronological information can be obtained from genetic data, and can be useful for integrating genetic evidence with that from other disciplines. However, relating genetic diversity to time requires many assumptions to be made and confidence limits around age estimates are often very broad.
- Genomewide studies have shown that purifying selection is the dominant evolutionary force on variation throughout the genome but the proportion of the genome that is influenced by balancing selection or positive selection is still not known.
- Integrating geographic information with genetic data is a powerful approach for making inferences from genetic diversity, and will only increase in the future as sampling density increases.

QUESTIONS

Question 6–1: Express the mutation rate of 1.1×10^{-8} per nucleotide per generation as

(a) Mutations per gene per generation, for a 2.5-kb gene

(b) Mutations per genome per generation for a 3000-Mb genome

(c) Mutations per genome per year, for a 3000-Mb genome assuming a 29-year average generation time

Question 6–2: State whether you would use a tree or a network to display the following types of genetic data, and why.

(a) SNP haplotypes from the nonrecombining part of the Y chromosome

(b) SNP haplotypes of part of chromosome 14

(c) Data from 20 microsatellites from the X chromosome

(d) Mitochondrial DNA control region sequence data

(e) Mitochondrial DNA genome sequences

Question 6–3: You suspect that a gene is a biological candidate for long-term balancing selection, and you have sequence data from 100 individuals.

(a) What test would you use to test your suspicion?

(b) Comparable data are provided from 100 chimpanzees—would this help you?

Question 6–4: Compare and contrast the different methods used to detect positive selection at the genes listed in Table 6.9.

Question 6–5: What determines whether selection occurs by a soft sweep, hard sweep, or polygenic adaptation?

Question 6–6: Discuss how different estimates of θ can be used in different tests of natural selection.

Question 6–7: (a) Download the following four complete mtDNA genome sequences from the Genbank database and save as FASTA files: Human X haplogroup (accession number JN828961.1), Human T haplogroup (JX415318.1), Neanderthal (FM865411.1), Denisovan (FN673705.1). Go to Clustal Omega online (type "EBI Clustal Omega" into a search engine) and paste your sequences one-by-one into the box, selecting DNA from the pull down box, and "PHYLIP" as the output alignment format. Click "Submit," wait, then check your sequence alignment. Can you see an indel, a transition, a transversion, and an STR? Save your alignment as a text file. (b) Go to an online server for PHYML (type "PHYML Interface" into a search engine), a program from the PHYLIP suite that calculates a maximum-likelihood tree. Paste your alignment into the box, click submit, and view the tree. Convince yourself that the two modern human sequences are the most closely related and go back and redraw the tree, investigating some of the parameters, including bootstrapping.

REFERENCES

The references highlighted in purple are considered to be important (for this chapter) by the authors.

1. **Akey JM** (2009) Constructing genomic maps of positive selection in humans: Where do we go from here? *Genome Res.* **19**, 711–722.
2. **Akey JM, Zhang G, Zhang K et al.** (2002) Interrogating a high-density SNP map for signatures of natural selection. *Genome Res.* **12**, 1805–1814.
3. **Andrés AM, Hubisz MJ, Indap A et al.** (2009) Targets of balancing selection in the human genome. *Mol. Biol. Evol.* **26**, 2755–2764.
4. **Balanovsky O, Dibirova K, Dybo A et al.** (2011) Parallel evolution of genes and languages in the Caucasus region. *Mol. Biol. Evol.* **28**, 2905–2920.
5. **Bandelt HJ, Forster P & Röhl A** (1999) Median-joining networks for inferring intraspecific phylogenies. *Mol. Biol. Evol.* **16**, 37–48.
6. **Bandelt HJ, Forster P, Sykes BC & Richards MB** (1995) Mitochondrial portraits of human populations using median networks. *Genetics* **141**, 743–753.
7. **Barbujani G, Bertorelle G & Chikhi L** (1998) Evidence for Paleolithic and Neolithic gene flow in Europe. *Am. J. Hum. Genet.* **62**, 488–492.
8. **Beaumont MA & Rannala B** (2004) The Bayesian revolution in genetics. *Nat. Rev. Genet.* **5**, 251–261.
9. **Bertorelle G & Barbujani G** (1995) Analysis of DNA diversity by spatial autocorrelation. *Genetics* **140**, 811–819.
10. **Brookfield J** (1997) Importance of ancestral DNA ages. *Nature* **388**, 134.
11. **Bustamante CD, Fledel-Alon A, Williamson S et al.** (2005) Natural selection on protein-coding genes in the human genome. *Nature* **437**, 1153–1157.
12. **Cagliani R, Fumagalli M, Biasin M et al.** (2010) Long-term balancing selection maintains trans-specific polymorphisms in the human *TRIM5* gene. *Hum. Genet.* **128**, 577–588.
13. **Chen H, Patterson N & Reich D** (2010) Population differentiation as a test for selective sweeps. *Genome Res.* **20**, 393–402.
14. **Clark AG, Glanowski S, Nielsen R et al.** (2003) Inferring nonneutral evolution from human–chimp–mouse orthologous gene trios. *Science* **302**, 1960–1963.
15. **Coop G, Pickrell JK, Novembre J et al.** (2009) The role of geography in human adaptation. *PLoS Genet.* **5**, e1000500.
16. **Cornuet JM, Santos F, Beaumont MA et al.** (2008) Inferring population history with DIY ABC: a user-friendly approach to approximate Bayesian computation. *Bioinformatics* **24**, 2713–2719.
17. **Cox MP** (2009) Accuracy of molecular dating with the rho statistic: deviations from coalescent expectations under a range of demographic models. *Hum. Biol.* **80**, 335–357.
18. **Csilléry K, Blum MGB, Gaggiotti OE & François O** (2010) Approximate Bayesian computation (ABC) in practice. *Trends Ecol. Evol.* **25**, 410–418.
19. **Currat M & Excoffier L** (2011) Strong reproductive isolation between humans and Neanderthals inferred from observed patterns of introgression. *Proc. Natl Acad. Sci. USA* **108**, 15129–15134.
20. **Excoffier L & Heckel G** (2006) Computer programs for population genetics data analysis: a survival guide. *Nat. Rev. Genet.* **7**, 745–758.
21. **Excoffier L, Smouse PE & Quattro JM** (1992) Analysis of molecular variance inferred from metric distances among DNA haplotypes: application to human mitochondrial DNA restriction data. *Genetics* **131**, 479–491.
22. **Falush D, Stephens M & Pritchard JK** (2003) Inference of population structure using multilocus genotype data: linked loci and correlated allele frequencies. *Genetics* **164**, 1567–1587.
23. **Fay JC & Wu CI** (2000) Hitchhiking under positive Darwinian selection. *Genetics* **155**, 1405–1413.
24. **Feldman MW, Krumm J & Pritchard JK** (1999) Mutation and migration in models of microsatellite evolution. In Microsatellites—Evolution and Applications (D Goldstein & C Schlotterer eds), pp 98–115. Oxford University Press.
25. **Felsenstein J** (2004) Inferring Phylogenies. Sinauer.
26. **Fenner JN** (2005) Cross-cultural estimation of the human generation interval for use in genetics-based population divergence studies. *Am. J. Phys. Anthropol.* **128**, 415–423.
27. **Fitzpatrick MC, Preisser EL, Porter A et al.** (2010) Ecological boundary detection using Bayesian areal Wombling. *Ecology* **91**, 3448–3455.
28. **François O, Currat M, Ray N et al.** (2010) Principal component analysis under population genetic models of range expansion and admixture. *Mol. Biol. Evol.* **27**, 1257–1268.
29. **Goldstein DB, Reich DE, Bradman N et al.** (1999) Age estimates of two common mutations causing factor XI deficiency: recent genetic drift is not necessary for elevated disease incidence among Ashkenazi Jews. *Am. J. Hum. Genet.* **64**, 1071–1075.
30. **Gravel S, Henn BM, Gutenkunst RN et al.** (2011) Demographic history and rare allele sharing among human populations. *Proc. Natl Acad. Sci. USA* **108**, 11983–11988.
31. **Green RE, Krause J, Briggs AW et al.** (2010) A draft sequence of the Neandertal genome. *Science* **328**, 710–722.
32. **Gronau I, Hubisz MJ, Gulko B et al.** (2011) Bayesian inference of ancient human demography from individual genome sequences. *Nat. Genet.* **43**, 1031–1034.
33. **Grossman SR, Shylakhter I, Karlsson EK et al.** (2010) A composite of multiple signals distinguishes causal variants in regions of positive selection. *Science* **327**, 883–886.
34. **Guillot G, Leblois R, Coulon A & Frantz AC** (2009) Statistical methods in spatial genetics. *Mol. Ecol.* **18**, 4734–4756.
35. **Gutenkunst RN, Hernandez RD, Williamson SH & Bustamante CD** (2009) Inferring the joint demographic history of multiple populations from multidimensional SNP frequency data. *PLoS Genet.* **5**, e1000695.
36. **Hermisson J & Pennings PS** (2005) Soft sweeps. *Genetics* **169**, 2335–2352.
37. **Hernandez RD, Kelley JL, Elyashiv E et al.** (2011) Classic selective sweeps were rare in recent human evolution. *Science* **331**, 920–924.
38. **Hoban S, Bertorelle G & Gaggiotti OE** (2012) Computer simulations: tools for population and evolutionary genetics. *Nat. Rev. Genet.* **13**, 110–122.
39. **Hudson RR** (1990) Gene genealogies and the coalescent process. In Oxford Surveys in Evolutionary Biology (D Futuyama & J Antonovics eds), Vol **7**, pp 1–44. Oxford University Press.
40. **Hudson RR, Kreitman M & Aguadé M** (1987) A test of neutral molecular evolution based on nucleotide data. *Genetics* **116**, 153–159.
41. **Huelsenbeck JP, Ronquist F, Nielsen R & Bollback JP** (2001) Bayesian inference of phylogeny and its impact on evolutionary biology. *Science* **294**, 2310–2314.
42. **Huerta-Sanchez E, Durrett R & Bustamante CD** (2008) Population genetics of polymorphism and divergence under fluctuating selection. *Genetics* **178**, 325–337.
43. **Huson DH & Bryant D** (2006) Application of phylogenetic networks in evolutionary studies. *Mol. Biol. Evol.* **23**, 254–267.
44. **Jakobsson M & Rosenberg NA** (2007) CLUMPP: a cluster matching and permutation program for dealing with label switching and

multimodality in analysis of population structure. *Bioinformatics* **23**, 1801–1806.

45. **Johnson CM, Lyle EA, Omueti KO et al.** (2007) Cutting edge: A common polymorphism impairs cell surface trafficking and functional responses of TLR1 but protects against leprosy. *J. Immunol.* **178**, 7520–7524.
46. **Kayser M, Roewer L, Hedman M et al.** (2000) Characteristics and frequency of germline mutations at microsatellite loci from the human Y chromosome, as revealed by direct observation in father/son pairs. *Am. J. Hum. Genet.* **66**, 1580–1588.
47. **Langergraber K, Prüfer K, Rowney C et al.** (2012) Generation times in wild chimpanzees and gorillas suggest earlier divergence times in great ape and human evolution. *Proc. Natl Acad. Sci. USA* **109**, 15716–15721.
48. **Lawson DJ, Hellenthal G, Myers S & Falush D** (2012) Inference of population structure using dense haplotype data. *PLoS Genet.* **8**, e1002453.
49. **Li H & Durbin R** (2011) Inference of human population history from individual whole-genome sequences. *Nature* **475**, 493–496.
50. **Manly BFJ** (2007) Randomization, Bootstrap and Monte Carlo Methods in Biology. Chapman & Hall/CRC.
51. **Mantel N** (1967) The detection of disease clustering and a generalized regression approach. *Cancer Res.* **27**, 209–220.
52. **McDonald JH & Kreitman M** (1991) Adaptive protein evolution at the Adh locus in *Drosophila*. *Nature* **351**, 652–654.
53. **McVean G** (2009) A genealogical interpretation of principal components analysis. *PLoS Genet.* **5**, e1000686.
54. **Moreno-Estrada A, Tang K, Sikora M et al.** (2009) Interrogating 11 fast-evolving genes for signatures of recent positive selection in worldwide human populations. *Mol. Biol. Evol.* **26**, 2285–2297.
55. **Nielsen R & Beaumont MA** (2009) Statistical inferences in phylogeography. *Mol. Ecol.* **18**, 1034–1047.
56. **Nielsen R, Hellmann I, Hubisz M et al.** (2007) Recent and ongoing selection in the human genome. *Nat. Rev. Genet.* **8**, 857–868.
57. **Nordborg M** (2007) Coalescent theory. In Handbook of Statistical Genetics (DJ Balding, M Bishop & C Cummings eds), pp. 843–872. Wiley.
58. **Novembre J & Stephens M** (2008) Interpreting principal component analyses of spatial population genetic variation. *Nat. Genet.* **40**, 646–649.
59. **Nowak MA & Sigmund K** (2004) Evolutionary dynamics of biological games. *Science* **303**, 793–798.
60. **Patterson N, Price AL & Reich D** (2006) Population structure and eigenanalysis. *PLoS Genet.* **2**, e190.
61. **Penny D, Steel M, Waddell PJ & Hendy MD** (1995) Improved analyses of human mtDNA sequences support a recent African origin for *Homo sapiens*. *Mol. Biol. Evol.* **12**, 863–882.
62. **Pozzoli U, Fumagalli M, Cagliani R et al.** (2010) The role of protozoa-driven selection in shaping human genetic variability. *Trends Genet.* **26**, 95–99.
63. **Przeworski M** (2002) The signature of positive selection at randomly chosen loci. *Genetics* **160**, 1179–1189.
64. **Rendine S, Piazza A, Menozzi P & Cavalli-Sforza LL** (1999) A problem with synthetic maps: reply to Sokal et al. *Hum. Biol.* **71**, 15–25.
65. **Richards M, Macaulay V, Hickey E et al.** (2000) Tracing European founder lineages in the Near Eastern mtDNA pool. *Am. J. Hum. Genet.* **67**, 1251–1276.
66. **Sabeti PC, Varilly P, Fry B et al.** (2007) Genome-wide detection and characterization of positive selection in human populations. *Nature* **449**, 913–918.
67. **Saillard J, Forster P, Lynnerup N et al.** (2000) mtDNA variation among Greenland Eskimos: the edge of the Beringian expansion. *Am. J. Hum. Genet.* **67**, 718–726.
68. **Saitou N & Nei M** (1987) The neighbor-joining method: a new method for reconstructing phylogenetic trees. *Mol. Biol. Evol.* **4**, 406–425.
69. **Schaffner SF, Foo C, Gabriel S et al.** (2005) Calibrating a coalescent simulation of human genome sequence variation. *Genome Res.* **15**, 1576–1583.
70. **Slatkin M & Bertorelle G** (2001) The use of intra-allelic variability for testing neutrality and estimating population growth rate. *Genetics* **158**, 865–874.
71. **Smouse PE, Long JC & Sokal RR** (1986) Multiple regression and correlation extensions of the mantel test of matrix correspondence. *Syst. Zool.* **35**, 627–632.
72. **Sokal RR, Oden NL & Thomson BA** (1999) A problem with synthetic maps. *Hum. Biol.* **71**, 1–13.
73. **Stefansson H, Helgason A, Thorleifsson G et al.** (2005) A common inversion under selection in Europeans. *Nat. Genet.* **37**, 129–137.
74. **Takahata N & Nei M** (1990) Allelic genealogy under overdominant and frequency-dependent selection and polymorphism of major histocompatibility complex loci. *Genetics* **124**, 967–978.
75. **Templeton AR** (1998) Nested clade analyses of phylogeographic data: testing hypotheses about gene flow and population history. *Mol. Ecol.* **7**, 381–397.
76. **Tenesa A, Navarro P, Hayes BJ, et al.** (2007) Recent human effective population size estimated from linkage disequilibrium. *Genome Res.* **17**, 520–526.
77. **Teshima KM, Coop G & Przeworski M** (2006) How reliable are empirical genomic scans for selective sweeps? *Genome Res.* **16**, 702–712.
78. **The 1000 Genomes Project Consortium** (2010) A map of human genome variation from population-scale sequencing. *Nature* **467**, 1061–1073.
79. **Tremblay M & Vézina H** (2000) New estimates of intergenerational time intervals for the calculation of age and origins of mutations. *Am. J. Hum. Genet.* **66**, 651–658.
80. **Vigilant L, Stoneking M, Harpending H et al.** (1991) African populations and the evolution of human mitochondrial DNA. *Science* **253**, 1503–1507.
81. **Wakeley J** (2008) Coalescent Theory. Roberts & Company.
82. **Wall JD** (1999) Recombination and the power of statistical tests of neutrality. *Genet. Res.* **74**, 65–79.
83. **Weiss KM & Wobst HM** (1973) Demographic models for anthropology. *Mem. Soc. Am. Archaeol.* **27**, 1–186.
84. **Wilson IJ, Weale ME & Balding DJ** (2003) Inferences from DNA data: population histories, evolutionary processes and forensic match probabilities. *J. R. Statist. Soc. A* **166**, 155–201.
85. **Womble WH** (1951) Differential systematics. *Science* **114**, 315–322.
86. **Yang Z & Bielawski JP** (2000) Statistical methods for detecting molecular adaptation. *Trends Ecol. Evol.* **15**, 496–503.
87. **Yang Z, Wong WSW & Nielsen R** (2005) Bayes empirical Bayes inference of amino acid sites under positive selection. *Mol. Biol. Evol.* **22**, 1107–1118.
88. **Yi X, Liang Y, Huerta-Sanchez E et al.** (2010) Sequencing of 50 human exomes reveals adaptation to high altitude. *Science* **329**, 75–78.
89. **Zhang J, Nielsen R & Yang Z** (2005) Evaluation of an improved branch-site likelihood method for detecting positive selection at the molecular level. *Mol. Biol. Evol.* **22**, 2472–2479.

SECTION 3

WHERE AND WHEN DID HUMANS ORIGINATE?

In this section we turn from considerations of what questions we might ask, what sources of genetic information are available, and what analytical tools we can use, to a discussion of the answers that can be provided. We will follow a loosely chronological course in the next two sections, and begin in this section by examining the path leading to the origin of modern humans.

Courtesy of Susana Carvalho, University of Oxford, UK.

CHAPTER 7
HUMANS AS APES

This chapter places humans in the context of other primate species: how similar are we to other apes, and which are our closest relatives? Evidence comes from our morphology, behavior, and chromosomes, and especially from our DNA. We will see that our closest living relatives are chimpanzees and bonobos, but that we differ from them substantially in our large population size and wide distribution coupled with low genetic diversity.

CHAPTER 8
WHAT GENETIC CHANGES HAVE MADE US HUMAN?

In this chapter we ask what changes in our evolutionary history have occurred to make us human. Although a number of human-specific traits can be recognized in our morphology and behavior, we are only just beginning to understand the genetic changes that have allowed these to develop, a search that is now also assisted by the sequences of two extinct hominins.

CHAPTER 9
ORIGINS OF MODERN HUMANS

This chapter turns to the human-specific or hominin line, and we follow its development from the chimpanzee–human split to the appearance of anatomically modern humans. Crucial evidence comes from fossils and archaeological remains, and again especially from our genetic diversity, with key contributions from ancient DNA. For most of this time, our ancestors were just one of several hominin species living in Africa, and our unique demographic characteristics appear only with fully modern humans.

HUMANS AS APES

CHAPTER SEVEN

We are apes, but a unique type of them. The similarities and differences between humans and other living apes are the subject of this chapter. There are two major themes.

Which nonhuman animals are the closest living relatives of humans?

This question has been considered within an evolutionary framework since Darwin wrote that "much light will be thrown on the origin of man and his history" in *On The Origin of Species*. Initial work used morphological methods (**Section 7.1**); subsequently, studies of chromosomes (**Section 7.2**) and molecules (**Section 7.3**) have come to dominate the field. These studies address two main questions: What was the branching order of the species? and What was the time-scale? A crucial point to appreciate is the distinction between the **species tree** and what is usually called a **gene tree** (even if it represents a noncoding DNA sequence rather than a gene). Populations contain many genes and many copies of each gene. As a result of recombination and independent assortment of chromosomes, each segment of DNA (defined as a region between recombination positions) has an independent history (**Figure 7.1**). DNA phylogenies differ because of many factors, including the stochastic nature of evolution and selection; for further discussion, see **Section 7.4**. In contrast, there is a single species tree.

Are humans typical apes?

Despite humans being very similar genetically to other apes, it is clear that humans are atypical in many respects: diversity, population size, geographical distribution, and population structure. These unusual features of human population genetics result from a distinct human evolutionary history and underlie the remainder of this book.

We will see that these comparisons lead us to ask some rather obvious evolutionary questions, such as How did humans come to be so unusually abundant and widespread? but also some that we might not otherwise ask, such as Why is there only one species of living human? The studies we discuss in the next three sections compare humans with other apes, mostly using the simple terms "chimpanzees," "gorillas," and "orangutans." However, these three groups can be subdivided into at least five different species and many subspecies—this is discussed in **Section 7.4.**

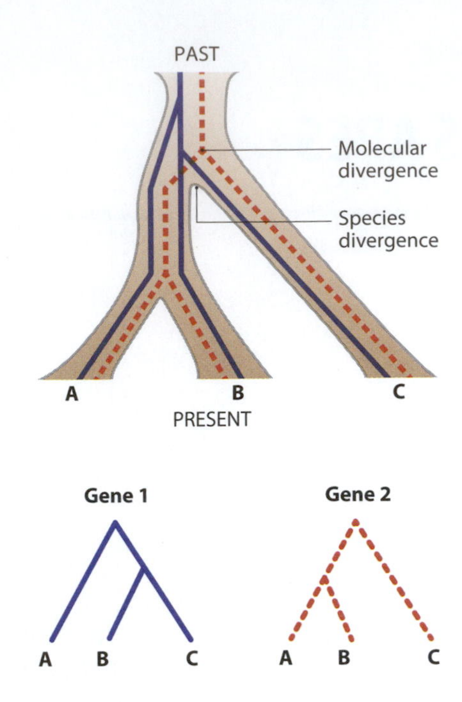

Figure 7.1: Gene trees and species trees. The species tree (*gray*) shows the evolutionary relationships of three species: species A and B are most closely related. One gene tree (*red line*) has the same branching order as the species tree, although the divergence times are earlier. A second gene tree (*blue line*) shows a different branching order and divergence times: for this DNA sequence, species B and C are the most closely related.

7.1 EVIDENCE FROM MORPHOLOGY

Before the development of molecular techniques during the past 50 years, morphological characters were the major source of evidence available to place humans in the tree of life (**Box 7.1**). Furthermore, given the probable limits of ancient DNA techniques to within the past 100 KY (**Section 4.10**), **morphology** is the only tool we have for reconciling most fossils with living species. As with other **cladistic** enterprises, the emphasis in paleontology—the study of prehistoric forms of life—is on identifying derived characters that are phylogenetically informative for revealing the underlying tree of ancestral relationships. If derived characters are shared among multiple taxa then they can be used to infer phylogenetic branching whereas the sharing of "primitive" characters between taxa does not inform us about their phylogenetic relationship. The distinction of unique changes from those that are **convergent** enables us to decide which morphological characters are **homologous** and which are **analogous**.

The work of Carl Linnaeus in the eighteenth century laid the foundation for all future attempts to classify organisms by producing a hierarchical system in which species could be lumped together into ever more inclusive groupings. A current classification of humans within this hierarchy is shown below:

- **Kingdom**: Animalia
  - **Phylum**: Chordata
    - **Class**: Mammalia
      - **Infraclass**: Eutheria
        - **Order**: Primates
          - **Superfamily**: Hominoidea (hominoids)
            - **Family**: Hominidae (hominids)
              - **Tribe**: Hominini (hominins)
                - **Genus**: *Homo*
                  - **Species**: *sapiens*

But how does this classification relate to those of other species, both extant and extinct?

Primates are an Order of mammals

Although some levels in the mammalian taxonomy are clearly resolved, the **phylogenetic** relationships of the groupings within these levels often remain hotly disputed. There are over 20 living Orders of mammals (of which Primates is one) which themselves fall into three accepted groupings on the basis of

Box 7.1: A history of the classification of humans

Whilst classifications of nature and our place within it were doubtless made earlier in human history, the first person known to have recognized the similarities between ourselves and other primates was the Roman gladiatorial physician Galen of Pergamum. He noted similar morphologies between the Barbary apes (Old World Monkeys, **OWM**) he dissected and the unfortunate gladiatorial recipients of his unanesthetized surgery. After his death (c. 200 AD) Galen's anatomical work remained largely unchallenged for over 1000 years.

The Swedish botanist Linnaeus (1707–1778) devised the binomial system of nomenclature (*Genus species*), recognized the grouping **primates**, and coined the name *Homo sapiens* in his *Systema Naturae* (1758). Interestingly, this is one of only two of Linnaeus' 44 primate binomial names that are retained today. Linnaeus boldly included chimpanzees in the same genus (as *Homo troglodytes*), but did not believe in evolution.

Thomas Henry Huxley, by contrast a fervent believer in Darwinian evolution by natural selection, published *Evidence as to Man's Place in Nature* in 1863, and in it dispelled the then popular idea that humans should be classified within their own order, stating:

> It is quite certain that the Ape which most nearly approaches man, in the totality of its organization, is either the Chimpanzee or the Gorilla.

He also noted that:

> Thus, whatever system of organs be studied, the comparison of their modifications in the ape series leads to one and the same result—that the structural differences which separate Man from the Gorilla and the Chimpanzee are not so great as those which separate the Gorilla from the lower apes.

Nonetheless, Huxley still classified *Homo sapiens* within its own family, Anthropini, and later even within its own suborder, Anthropidae. Charles Darwin, however, disagreed in *The Descent of Man* (1871):

> ...from a genealogical point of view it appears that this rank is too high, and that man ought to form merely a family, or possibly even only a sub-family.

Many former taxonomies distinguished humans as a distinct family. However, a recent consideration of species separation times, rather than the more abstract concept of morphological discontinuity, considers that all living great apes fall into the family Hominidae.

reproductive mechanisms: the egg-laying monotremes (Prototheria) found only in Australia and New Guinea (for example, platypus, echidna); the marsupials (Metatheria) found in Australia and the Americas; and the placental mammals (Eutheria) found over most of the globe. In contrast to these relative certainties, the phylogenetic tree relating the 20 extant Eutherian orders is difficult to resolve using morphological data. Phylogenetic analyses based on gene sequences, **indels**, and **retroposon** insertion data have agreed that all placental mammals can be grouped into four clades: Afrotheria, Xenarthra, Laurasiatheria, and Euarchontoglires.[44] These superordinal groups correspond to the biogeographic distribution of early mammals, with Afrotheria and Xenarthra being restricted to Africa and South America respectively, and Laurasiatheria and Euarchontoglires having origins in the northern hemisphere. The 414 living species of primates[55] belong to the latter clade, along with rodents, lagomorphs (for example, rabbit and pika), flying lemurs, and tree shrews (**Figure 7.2**). The grouping of rodents, primates, and tree shrews is supported by these being the only Eutherian orders known to have *Alu*-like **SINEs** (see **Section 3.5**) derived from 7SL RNA.[45]

The term Primate (from the Latin for "first") was coined by Linnaeus himself and remains well accepted, yet compared with many of the other Eutherian orders the Primates are not very clearly defined by a suite of derived morphological characters. For example, pentadactyly (five digits) is a primitive character of land vertebrates that has been frequently modified in other mammalian orders. However, shared derived characters include the binocular vision and a shortened muzzle or snout that indicates a greater reliance on sight than on smell. In addition most primates have grasping hands and feet and appreciably larger brains relative to their body size compared with other mammals.

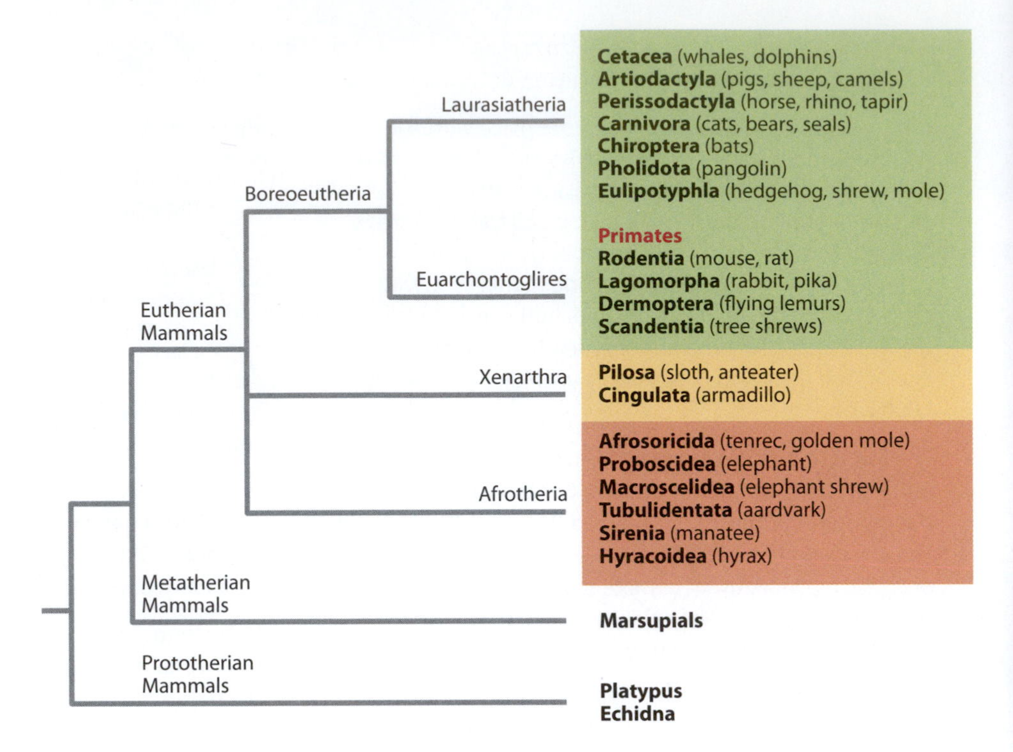

Figure 7.2: Phylogeny of the four major clades of Eutherian mammals. The higher order taxonomic groupings within each of the four superordinal mammalian clades are shown, together with representatives of that group. Marsupials are placed as an outgroup. The exact branching order among Eutherians is left unsolved according to the model, which is based on retroposon and geological data [data from Nishihara H et al. (2009) *Proc. Natl Acad. Sci. USA* 106, 5235]. This model proposes that the nearly simultaneous division of Africa, Laurasia (North America together with Eurasia), and South America ~120 MYA lead to the concomitant divergence of the placental ancestors into three extant groups that are highlighted by different colors.

Hominoids share a number of phenotypic features with other anthropoids

Traditional morphological classifications of primates identified two groupings: the "lower" primates (prosimians), which included the tarsiers, lemurs, and lorises; and the "higher" primates (Anthropoidea), which included the apes and monkeys. More recently it has been shown on both morphological and molecular grounds that the tarsiers share a common ancestor with the Anthropoidea and are now grouped together in the Suborder Haplorrhini, rather than deriving from an ancestor uniquely shared with lemurs and lorises, now Strepsirrhini (**Figure 7.3**). As their subordinal name indicates, tarsiers have a dry nose in common with Anthropoidea, as well as several other anatomical features such as sperm morphology and a hemochorial placenta (in which maternal blood is in direct contact with the chorion, as in humans). The features that have been used to support the now abandoned prosimian taxon, such as nocturnal lifestyle, small body size, two-horned uterus, and saltatory locomotion (jumping) have to be understood either in terms of ecological adaptations or retention of primitive characters.[22] Small body size and limb morphology adapted to arboreal habitat are generally characteristic of the earliest primate fossils from the Eocene period (**Figure 7.4**). Despite their nocturnal lifestyle, the eye anatomy of tarsiers is more similar to that of Anthropoidea, including the absence of a specific layer of tissue in the eye, the tapetum lucidum, which in lemurs, lorises, carnivores, and many other mammals improves vision in low light conditions. While tarsiers lack such night-vision adaptations, they have evolved their characteristically large eyes with many more photoreceptors to support their nocturnal lifestyle (**Figure 7.5**).

The extant Anthropoidea are characterized by complete bony eye sockets and lower molars with rounded puffy cusps and can be further resolved morphologically into three groups. The New World Monkeys (NWM, Platyrrhini) have twelve premolars and side-facing nostrils, while the Old World Monkeys (OWM, Cercopithecoidea) and the Hominoidea (which includes humans and other apes) have eight premolars and a nose with downward-facing nostrils. These distinctive morphological features are in their derived state in the hominoids and OWM and therefore define the Catarrhini clade; this is also supported by molecular evidence (Figure 7.3). On the other hand, most OWM and NWM species share numerous primitive traits, for example the tail and slightly shorter forelimbs (whereas hominoids primitively had longer arms), features which cannot, however, be taken to support the grouping of monkeys as a clade.

One of the characteristic innovations found in apes and OWM is their **trichromatic** vision. Their ability to distinguish green and red is due to the long-wavelength-sensitive (LWS) gene duplication on the X chromosome. The duplication event probably occurred at the start of the catarrhine radiation, some 30–40 MYA, because most extant catarrhines carry two functionally distinct copies (*OPN1LW* and *OPN1MW*) of the ancestral LWS gene while the NWM have only one copy. In contrast to catarrhines, almost all other mammals have two photopigments produced by the autosomal short-wavelength-sensitive SWS1 (blue) and the X-chromosomal long-wavelength-sensitive LWS (red–green) genes.[21] Notably, the three polymorphic amino acid substitutions that in combination distinguish the red- and green-sensitive copies of the single X-chromosomal color-vision gene in the NWM, are the same three mutations that have been fixed in the two descendant copies of the same gene in the ancestors of OWM and apes. Many NWM females that are polymorphic for these substitutions have a mosaic of red and green opsin cones, because of random X-chromosome

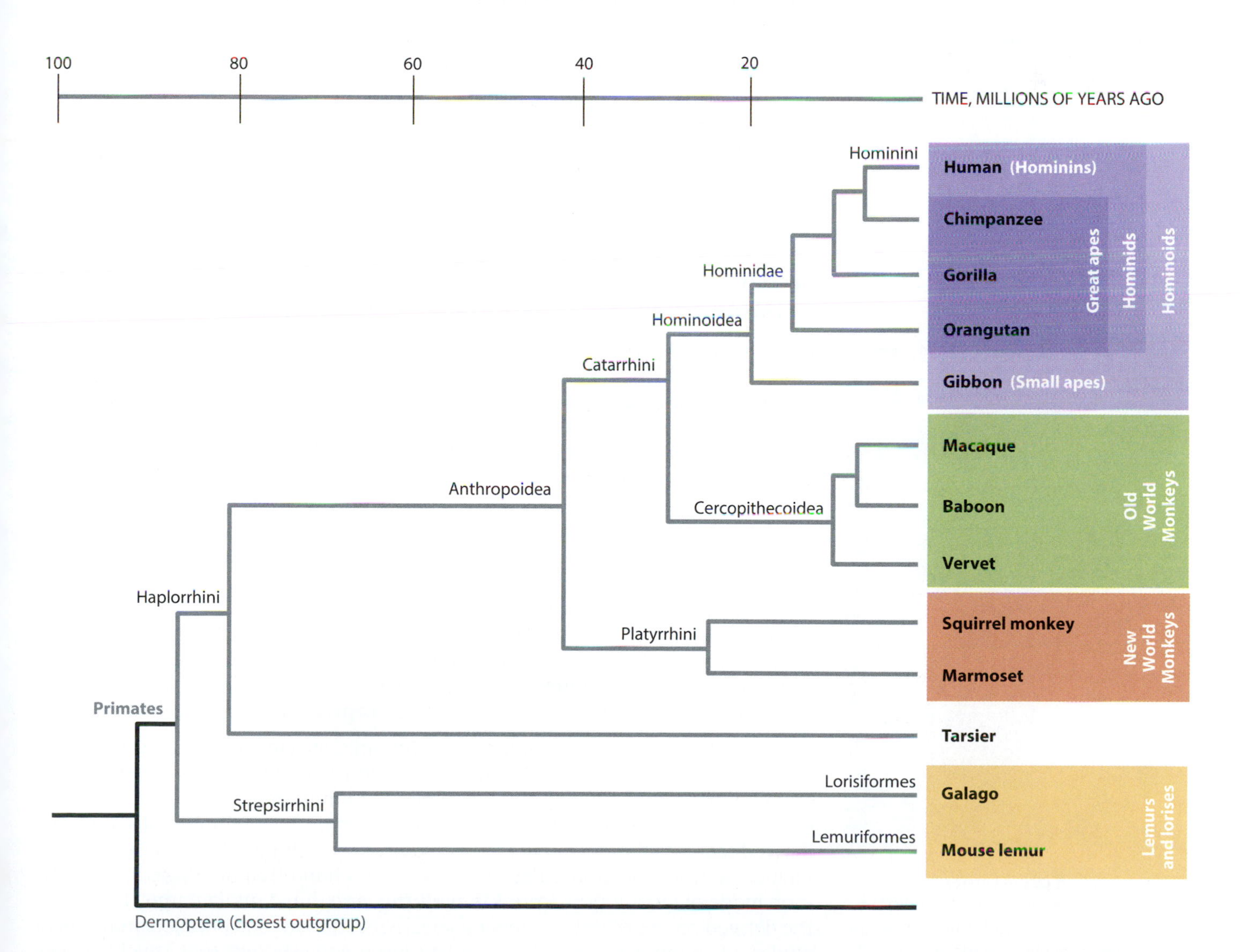

Figure 7.3: A phylogeny of extant primate groups. The phylogeny is shown for species that have undergone complete genome sequencing. The branching order is shown on a time-scale inferred from molecular data [data from Perelman P et al. (2011) *PLoS Genet.* 7, e10001342]. While the oldest known fossil primates date to the Eocene period, statistical modeling of species preservation [data from Tavaré S et al. (2002) *Nature* 416, 726] suggests the last common ancestor of primates may have lived more than 80 MYA, a date that agrees well with genetic data.

Figure 7.4: A fossil primate *Darwinius masillae*.
One of the best-preserved early fossil primates from the Middle Eocene (~47 MYA) sediments of Messel, Germany. The fossilized individual was a juvenile female belonging to an agile arboreal species and might have attained a body weight of 650–900 g had she lived to adulthood. [From Franzen JL et al. (2009) *PLoS One* 4, e5723. With permission from Public Library of Science.]

inactivation (Box 2.2), and can therefore see the world in three different colors whilst males are **dichromatic**. There is evidence of convergent evolution in howler monkeys, which have evolved trichromatic vision as a result of an independent duplication event of the same ancestral gene (**Figure 7.6**). While the advantages of trichromatic vision include, for example, the distinction of the ripeness of fruits, experiments on NWM have shown that dichromatism may also have some advantages: potentially in superior visual ability to discriminate color-camouflaged shapes and patterns.[52] Compared with other primates catarrhines are less able to rely on smells as cues and this may be because of their improved vision. More olfactory receptor genes have become pseudogenes in those species that have trichromatic vision than in dichromats[17] (Figure 7.6). Functional polymorphisms in the red-sensitive copy of the color-vision gene are more abundant in human populations than expected by chance. It has been suggested that some of these polymorphisms are associated with the shift in light detection from red toward orange spectrum. Their prevalence may be due to selective advantage in female carriers—due to random X-inactivation the carriers of such mutations may have improved ability to distinguish subtle differences of color of certain fruits, insects, and background foliage.[63]

Ancestral relationships of hominoids are difficult to resolve on morphological evidence

It is difficult to consider morphological classifications of the Hominoidea in isolation from molecular evidence, as this superfamily has undergone substantial revision over the past 40 years. However, two families are currently recognized: the Hylobatidae or small apes (gibbons and siamang) and the Hominidae (hominids) including great apes (orangutans, chimpanzees, bonobos, and gorillas) and humans. Among the hominids the Hominini tribe (hominins) combines humans with a number of extinct species that are more closely related to humans than to other apes (see Chapter 8). Compared with OWM, the hominoids share features in upper limb anatomy related to their specialized arboreal locomotion, which includes vertical climbing and hanging by the forelimbs. On gross anatomical grounds the hominids are clearly differentiated from the Hylobatidae, and all are differentiated from the OWM, by, amongst other characters, the absence of a tail. Note that if humans and only a subset of hominids share a most recent common ancestor then the former classification of great apes under the Pongidae family becomes **paraphyletic**.

Figure 7.5: Philippine tarsier.
Tarsiers represent one of our most distant relatives in the clade Haplorrhini. All extant tarsier species are now restricted to island Southeast Asia. They have developed large eyeballs to cope with low light conditions. (Courtesy of Jeroen Hellingman under the Creative Commons Attribution-Share Alike 3.0 Unported license.)

The ancestral relationships of the living hominid species are difficult to resolve on morphological evidence. **Bonobos** (sometimes called pygmy chimpanzees) and common chimpanzees clearly share a common ancestor to the exclusion of the others, and, on morphological grounds, there are a number of seemingly derived features that group these two species with the gorillas. These features include anatomical similarities of the wrist and hand that are linked to a common mode of locomotion, **knuckle walking**, and thin dental enamel. However, the derived nature of thin enamel has been questioned, and detailed anatomical studies of chimpanzee and gorilla wrist morphology suggest that knuckle walking may have evolved independently in the two genera of African apes.[30]

In light of these difficulties, could fossil apes shed light on the evolution of the important morphological characters? The global fossil record contains only a few representatives of hominoid forms over the past 25 MY that are considered ancestral to modern apes. However, confounded by the problems of small sample sizes and **sexual dimorphism** the fossil hominoids have undergone substantial taxonomic tinkering, which may also have been intensified by a tendency among researchers to assert their own discoveries as direct hominin ancestors. Analyses of the dentition of the fossil hominoid *Ramapithecus* in the 1960s and 1970s placed it on a lineage leading to humans, and the early dates suggested a split of humans from other hominoids >12 MYA. However, subsequent discoveries linked *Ramapithecus* to *Sivapithecus* (a close relative of the extant orangutan) through the identification of a number of derived facial characteristics shared solely with orangutans. Thus it appears that these species, and not humans, split first from the other great apes.[2]

Although many morphological characters are shared between humans and the African apes (chimpanzees and gorillas), most are also present in gibbons and are therefore primitive.[2] However, the presence of derived features such as the frontal sinuses and axillary glands in the armpits of African apes and humans indicates a shared ancestor. The hominoid fossil record in Africa between 10 and 5 MYA is notoriously sparse (**Section 9.1**), hindering resolution

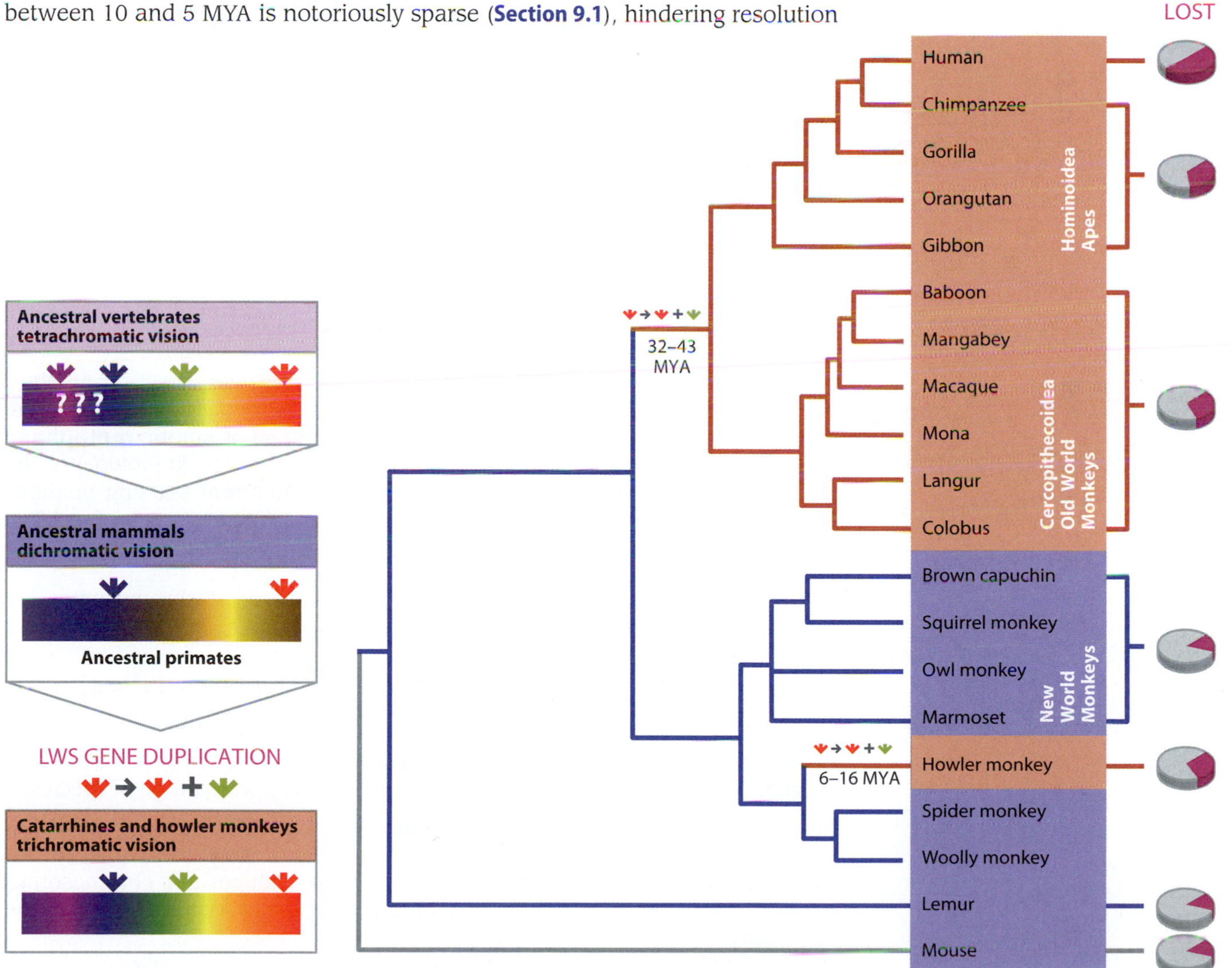

Figure 7.6: Origins of trichromatic vision in catarrhines. Ancestral vertebrates were able to distinguish four colors including one in the ultraviolet spectrum. The ancestors of modern mammals, who were probably nocturnal animals, lost two of the color-vision genes and retained only the ability to distinguish short (blue) and long (red and yellow) wavelength colors. The ability to distinguish red and green may have arisen firstly as a polymorphism in the X-chromosomal color-vision gene in the ancestors of New and Old World Monkeys. The polymorphism has been independently fixed via gene duplication in catarrhines and howler monkeys. Age estimates on the primate phylogeny are coalescent dates based on 8 Mb of genomic sequence data [data from Perelman P et al. (2011) *PLoS Genet.* 7, e1001342]. The proportions of **pseudogenes** among olfactory receptor (OR) genes in different primate groups are shown in *red* pie charts according to the corrected estimates of data from Gilad Y et al. [Gilad Y et al. (2004) *PLoS Biol.* 2, e5].

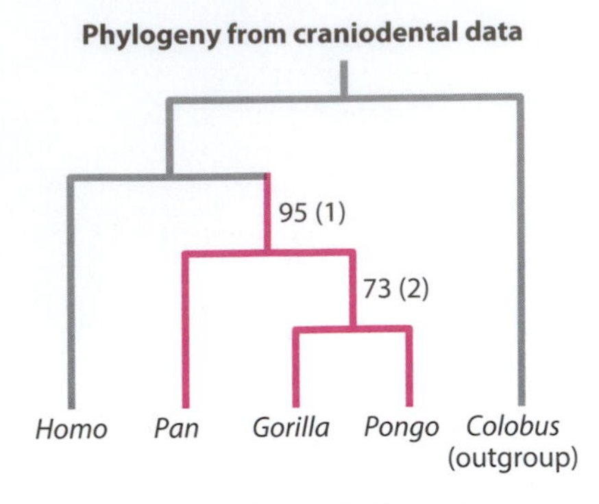

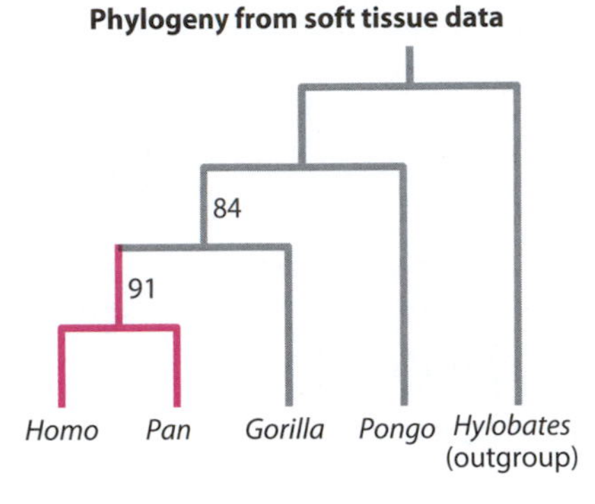

Figure 7.7: Comparisons of hominoid phylogenies from two sources of morphological characters.
Numbers next to branches indicate percentage bootstrap support. Numbers in parentheses indicate bootstrap support for that clade from the alternative dataset. The chimpanzee–gorilla–orangutan (*Pan–Gorilla–Pongo*) clade, supported by 95% of bootstraps of the craniodental data, is supported by only 1% of bootstraps from the soft tissue data.

of the gorilla–chimpanzee–human **trichotomy** so that any conclusions remain weakly supported by morphological data alone. Small sets of molar and premolar teeth from East Africa dating to 5–12.5 MYA have been interpreted as having morphological features shared with modern gorilla species.[33, 50, 61] Given the small number of morphological characters that can be collected from such dental evidence, it remains to be seen whether these shared features are indeed uniquely derived and characteristic of the clade of gorilla ancestors, or represent **homoplasic** diversity of the Miocene apes.[66]

Phylogenies of living hominoids constructed from either hard tissue (**craniodental**) or soft tissue characters are different. The soft tissue characters reflect the molecular phylogeny much better than the craniodental characters. The hominoid phylogeny produced from craniodental data not only supports an incorrect tree (**Figure 7.7**), but certain erroneous clades within the tree attain **bootstrap** support of 95%.[8, 16] The branching order in the tree based on craniodental data appears to be highly sensitive to the choice of the outgroup, selection of characters, and the method of tree construction.[5] This is unfortunate, because soft tissue characteristics are rarely preserved in the fossil record, and so their use is limited to existing species.

7.2 EVIDENCE FROM CHROMOSOMES

Like the morphological features of the animals themselves, the morphologies of their chromosomes provide information about the relationships of primate species. Study of the number and physical appearance of mitotic **metaphase** or late prophase chromosomes under the microscope (the karyotype) is the province of cytogeneticists, who use a number of different staining methods to reveal underlying features of chromosomal DNA structure (see **Section 2.5**).

Karyotypes are useful in phylogenetic comparisons for a number of reasons:

- Generally, they change relatively slowly, because most chromosomes are constrained by the requirement to pair with homologs, limiting their extent of gross rearrangement.
- They provide a global comparison of genomes, rather than of specific segments that may be under the influence of particular selective effects.
- Methods of study are technically straightforward, requiring universal reagents, and are therefore readily applicable across a wide range of species.

Because of their relative stability, comparisons of karyotypes among the higher primates can give quite robust information about branching order. However, no cytogenetic clock is available to calibrate the gross changes that chromosomes undergo, so karyotypic analysis gives no information about dates within the phylogeny.

Human and great ape karyotypes look similar, but not identical

The most obvious difference between the chromosomes of humans and the great apes is their number: while humans have only 46 chromosomes, chimpanzees, gorillas, and orangutans all have 48. Comparison of **G-banded** karyotypes, however,[68] demonstrates a very high degree of similarity between the four species, and shows that the difference in chromosome number results from an apparently simple end-to-end fusion of two small chromosomes seen in

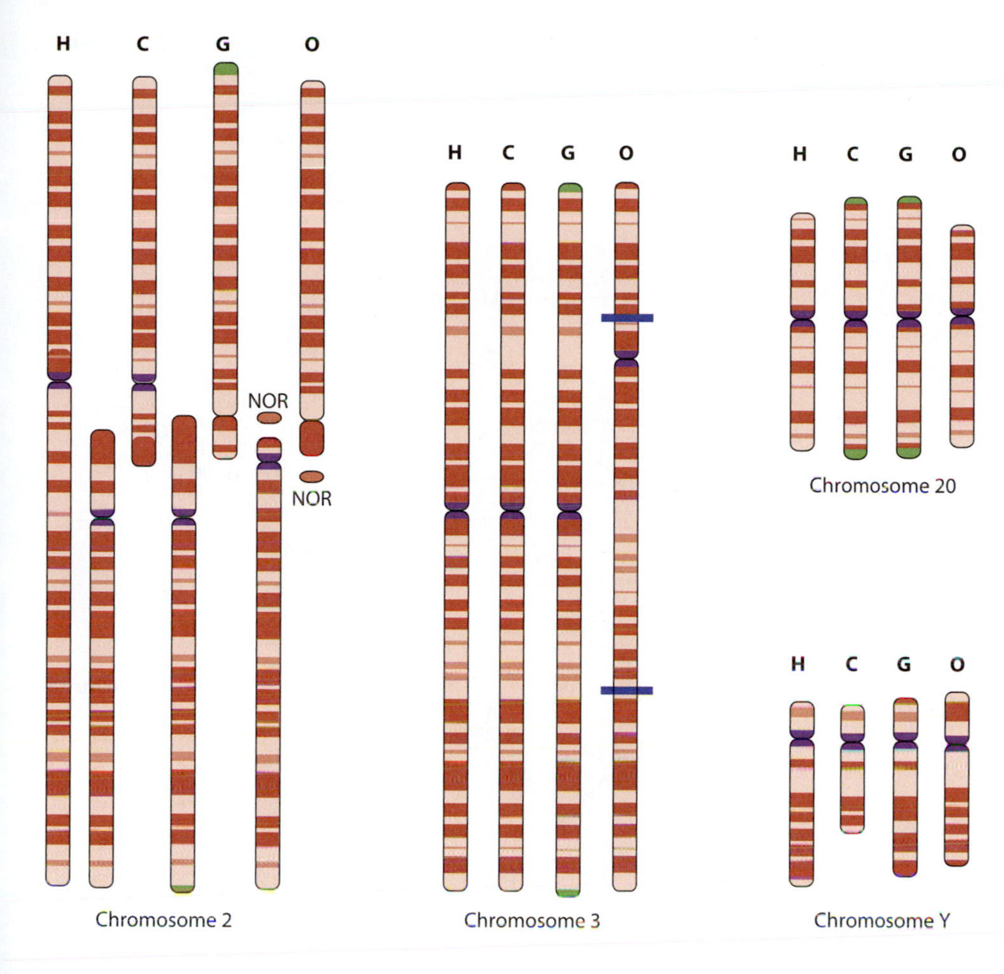

Figure 7.8: Examples of human and great ape chromosomes. Idiograms of G-banded chromosomes 2, 3, 20, and Y from humans (H), with their orthologs in chimpanzees (C), gorillas (G), and orangutans (O). Conservation of banding pattern is evident. Human chromosome 2 results from a terminal fusion of two chromosomes since the last common ancestor with chimpanzees. One centromere was inactivated. Orangutan orthologs have nucleolar organizer regions (**NOR**) at the tips of their short arms, where rRNA genes lie. The location of gorilla- and chimpanzee-specific terminal satellites are shown in *green*; *blue* horizontal lines mark the positions of complex inversion breakpoints on the orangutan ortholog to human chromosome 3. The Y chromosome is poorly resolved and highly variable between species. [From Yunis JJ & Prakash O (1982) *Science* 215, 1525. With permission from AAAS.]

the great apes to form the large **metacentric** chromosome 2 in humans (**Figure 7.8**). Alignment of G-banded chromosomes allows other less dramatic rearrangements to be seen. Most of these are inversions of chromosomal segments, variations in **heterochromatin** adjacent to centromeres, or the presence of G-bands at the ends of chimpanzee and gorilla chromosomes that are absent from those of humans and orangutans. Translocations seem to be limited to a reciprocal event between the equivalents of human chromosomes 5 and 17 in the gorilla with respect to the other species.

Careful consideration of these differences allowed possible ancestral karyotypes to be reconstructed, and a phylogeny to be deduced (**Figure 7.9**). This has chimpanzees as the closest relative to humans, chimpanzee–human as a sister-group to gorilla, and chimpanzee–human–gorilla as a sister-group to the orangutan. Although this branching order agrees with the more recent consensus from molecular data (see **Section 7.3**), this was not a truly objective attempt at tree building, because the status of orangutan as outgroup was assumed from the beginning.

Molecular cytogenetic analyses support the picture from karyotype comparisons

Molecular techniques allow a more fine-grained analysis of the organization of DNA sequences on chromosomes, and so can show whether or not the picture revealed by simple inspection of G-banded karyotypes is consistent with more detailed analyses.

In **chromosome painting**, DNA from a single human chromosome, for example from a chromosome-specific library or PCR products amplified from a specific flow-sorted chromosome, is labeled and hybridized back to a **metaphase**

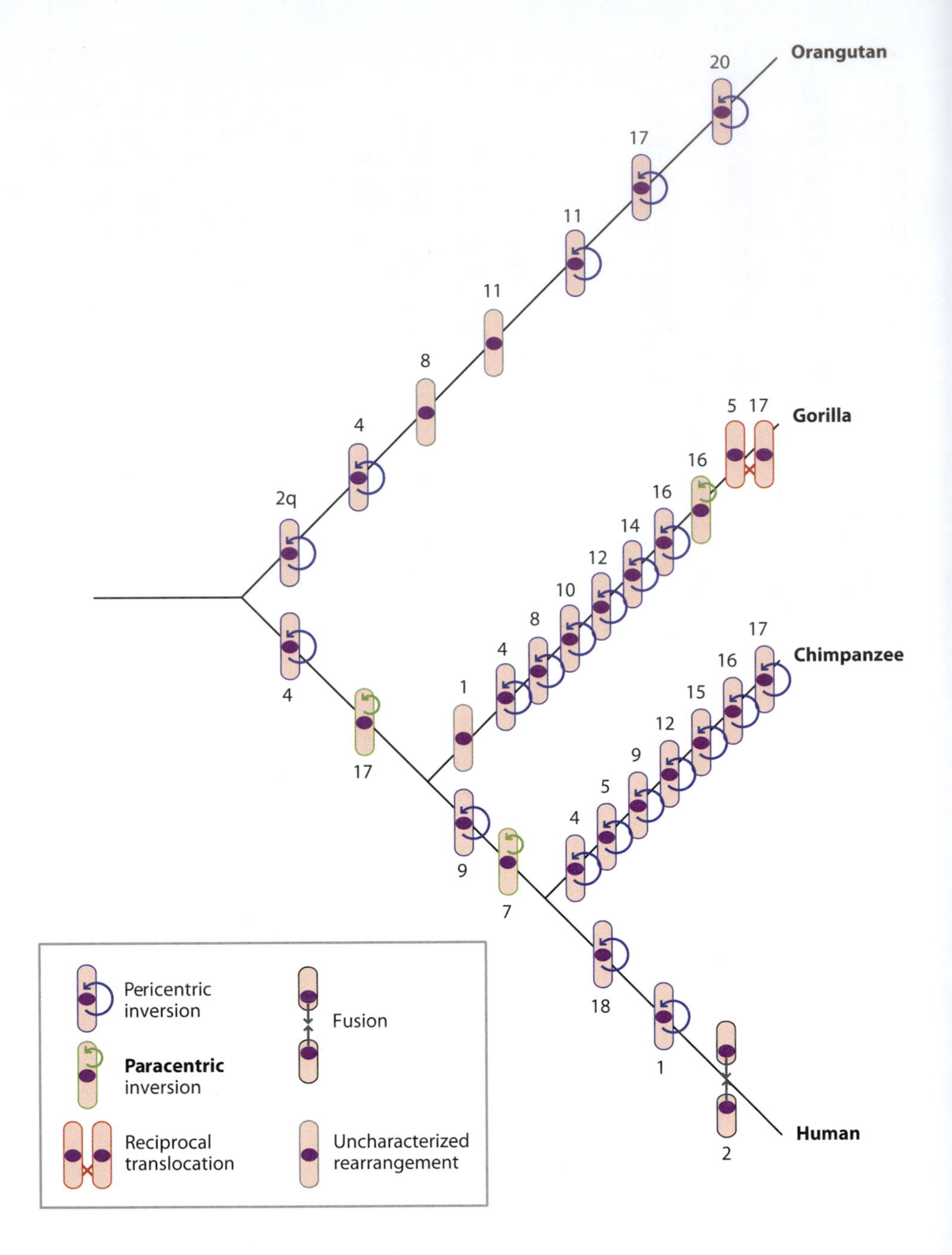

Figure 7.9: Phylogeny of humans and other great apes from karyotypic evidence.
Numbers above or below the symbols identify the chromosomes involved, using the human nomenclature. Rearrangements involving heterochromatin are not shown. [Data from Yunis JJ & Prakash O (1982) *Science* 215, 1525.]

spread under conditions that allow only unique sequences to generate a signal. This method reveals a picture consistent with that gained from inspection of G-banded chromosomes.[23] The origin of human chromosome 2 is confirmed, as is the 5;17 translocation in gorilla.

Chromosome painting methods cannot detect inversions or other intrachromosomal rearrangements. However, these can be revealed using a particularly high-resolution method known as **chromosomal bar-coding**, which uses chromosome paints derived by amplifying human-specific PCR products from somatic hybrid rodent cell lines containing fragments of a subset of human chromosomes. This process allows 160 molecular cytogenetic landmarks to be identified[43] and detailed ancestral karyotype reconstruction. Even more detailed analysis of particular regions can be achieved by **fluorescence *in situ* hybridization (FISH)** using specific clones rather than DNA from whole chromosomes. For example, human bacterial artificial chromosome (**BAC**) clones forming a **contig** over a region thought to be involved in an inversion are hybridized to metaphase chromosomes from several species: a clone which spans a breakpoint gives a single signal on the human chromosome, but a "split" signal on the ape chromosome. Such comparative analyses of the human and chimpanzee genomes have identified nine **pericentric** inversions most of which are well characterized by analysis of the breakpoint regions.[28] Using a comparative FISH approach, several large duplications have been mapped to the pericentromeric region of the Y chromosome of hominoids indicating a burst of duplications at the time of their divergence from OWM.[29] One interchromosomal rearrangement, a 100-kb segment originating from chromosome 1 that has inserted into the Y chromosome, is shared by humans, chimpanzees, and bonobos, but is absent from other great apes.[65] This chromosomal **synapomorphy** provides further support to the *Homo–Pan* clade.

Knowing the phylogenetic relationships among more than 50 different primate species for which karyotype data are available enables us to reconstruct the hypothetical karyotype of a common ancestor. According to such reconstructions, the ancestral primate karyotype had 24 pairs of autosomes—two pairs more than humans have today (**Figure 7.10**). Only two interchromosomal fusions and three chromosome fissions are required to derive this karyotype from that reconstructed for the ancestor of all eutherian mammals.[58]

Unsurprisingly, sharing of chromosomal banding patterns between primate chromosomes is reflected in a high degree of conserved **synteny**—location of corresponding genes on the same chromosome in different species. One striking illustration of this at the phenotypic level is the fact that the commonest human autosomal **aneuploidy**, **trisomy 21**, has also been observed in chimpanzees,[40] gorillas,[62] and orangutans,[3] and in each species confers a phenotype similar to human Down syndrome (OMIM 190685).

All mammalian X chromosomes are subject to **X-inactivation** to balance gene dosage in males and females.[36] Placental mammals have extensive, chromosomewide inactivation, whereas in marsupials it is incomplete and locus-specific. Although the mechanisms of inactivation vary across orders,[46]

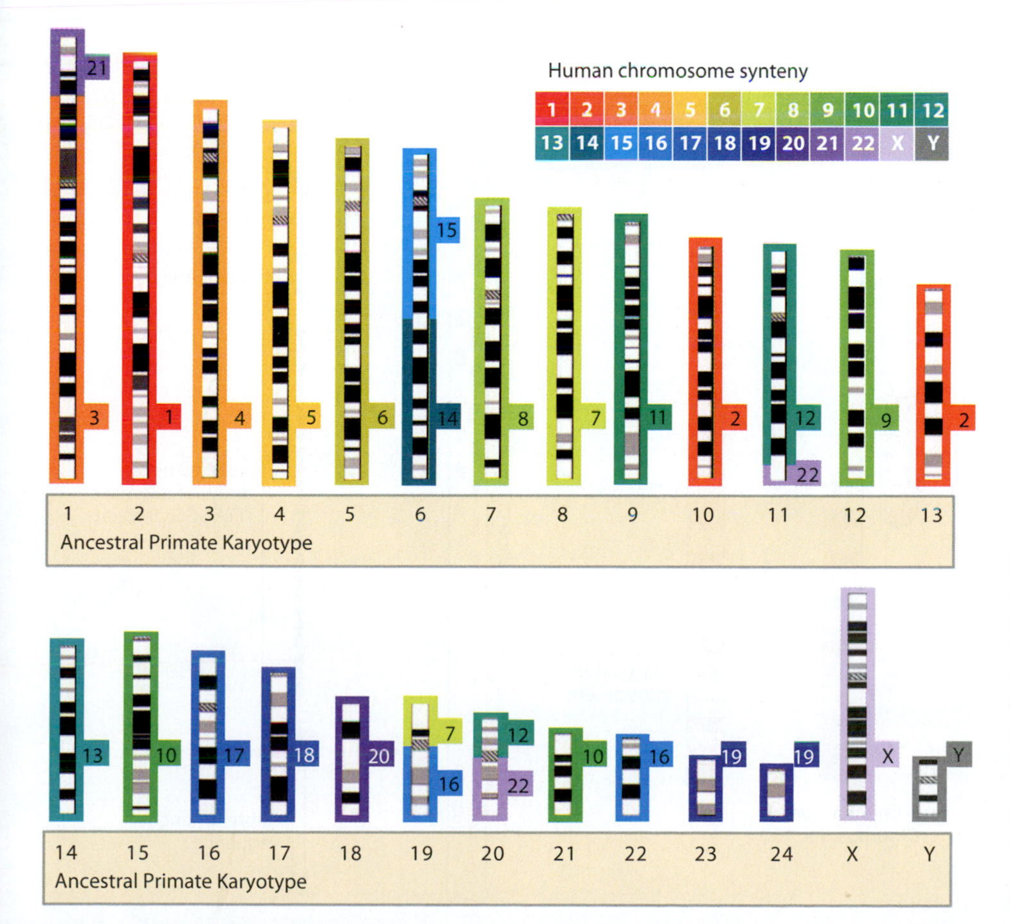

Figure 7.10: Hypothetical karyotype of the last common ancestor of living primates.
Chromosomes of the reconstructed karyotype of the primate ancestor are arranged according to size and numbered below. Syntenies to human chromosomes are shown in color and chromosome numbers on the side of each ancestral primate chromosome. [Adapted from Stanyon R et al. (2008) *Chromosome Res.* 16, 17. Courtesy of Springer Science and Business Media.]

the gene content as well as banding pattern of the X chromosome are highly conserved and thus very similar patterns are seen not only in great apes, but also in all apes and monkeys.[10] The Y chromosome is unique in being **constitutively** haploid; this absence of a homolog, and, therefore, of a requirement to pair along the whole of its length, has apparently freed it from the constraints borne by other chromosomes, and it is therefore cytogenetically more dynamic. FISH analysis using **yeast artificial chromosome** (**YAC**) clones[4] shows that inversions have been frequent, and confirms the presence of a block of sequence (about 4 Mb in size) on the short arm which has **transposed** from the long arm of the X chromosome since humans and chimpanzees last shared an ancestor.[47] This is the largest human-specific variant in the euchromatin. The short-arm XY-homologous **pseudoautosomal** region, in which recombination in male meiosis takes place, is highly conserved among humans and great apes. However, molecular analysis reveals another evolutionary novelty in humans, namely a second pseudoautosomal region (PAR2) on the tip of the long arms of the sex chromosomes,[13] which again is X-specific in great apes.

The origin of the most striking landmark in the human karyotype, chromosome 2, has been investigated in detail (**Figure 7.11**). *In situ* hybridization showed that telomeric DNA could be found at band 2q13, consistent with a terminal fusion event. **Cosmids** were isolated from this region and were found to contain $(TTAGGG)_n$ telomere repeats.[20] Sequencing identified arrays of the telomere repeat in a head-to-head arrangement, as would be expected for a simple ancestral telomere–telomere fusion, and flanking sequences that are typical of human subterminal repeats, and that cross-hybridize with other human telomeres. FISH analysis[26] has shown that the order of 38 cosmid clones that span the human chromosome 2 fusion point is conserved between human and chimpanzee, where they are divided between the tips of the short arms of chimpanzee chromosomes 12 and 13.

7.3 EVIDENCE FROM MOLECULES

Molecular analyses can provide information about both the order and timing of the branches in the great ape–human tree. In the last few decades there has

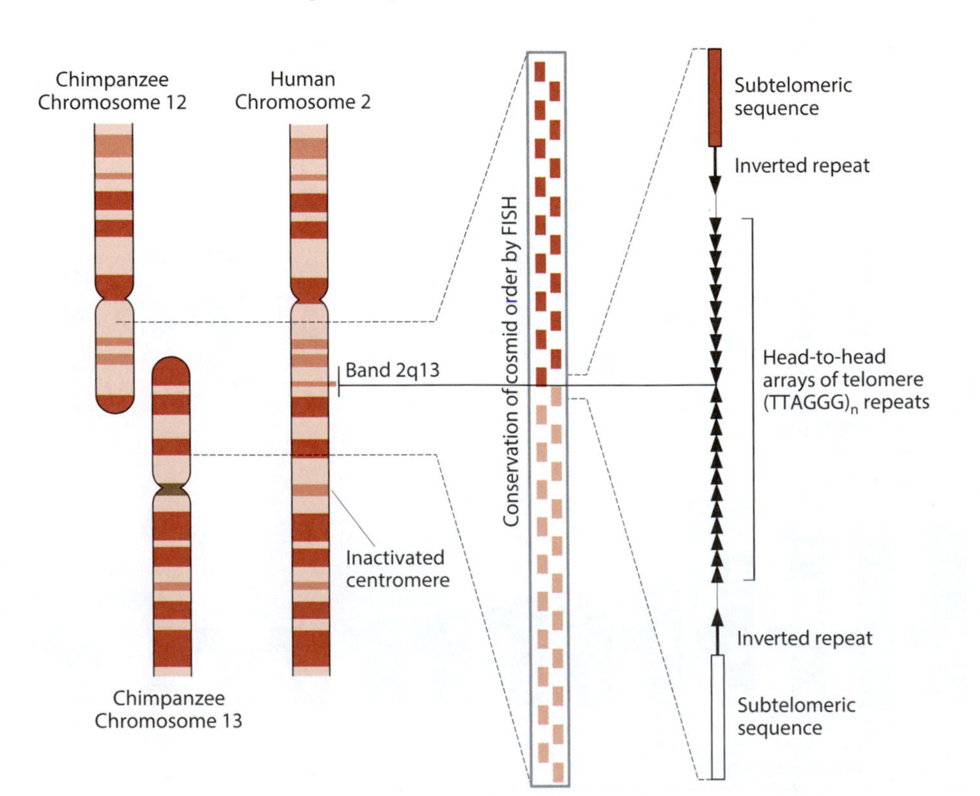

Figure 7.11: Structure of the ancestral fusion point on human chromosome 2. [Data from Ijdo JW et al. (1991) *Proc. Natl Acad. Sci. USA* 88, 9051; and Kasai F et al. (2000) *Chromosome Res.* 8, 727.]

been a marked improvement in the methods available, progressing from immunological distance measurements in the 1960s to DNA hybridization studies and now DNA sequence information on the scale of entire genomes. Interpretation of all of these datasets requires assumptions about whether and how selection is acting on the loci; most changes are assumed to be neutral, or nearly so, but it is important to remember that selection can give rise to quite different trees, for example at the **HLA** region (Box 5.3).

Molecular data support a recent date of the ape–human divergence

Molecular anthropology can perhaps be said to have begun in the 1960s with the pioneering work of Linus Pauling and Morris Goodman and the publication in 1967 of a key paper entitled "Immunological time-scale for Hominid evolution."[53] Sarich and Wilson presented the first use of molecular methods to estimate a date for the great ape–human split and, to the surprise of many, claimed that this date was as recent as 5 MYA; at that time *Ramapithecus* fossils dating to ~12–15 MYA were classified as hominins (**Section 7.1**). This work was based on quantitative measurement of the structural differences between **orthologous** proteins (serum albumins) in OWMs, great apes, and humans using an immunological method, microcomplement fixation; and on the calibration of these measurements using a date of 30 MYA for the split between OWMs and hominoids.

The results were expressed as an immunological distance, a measure which, by definition, is 1.0 for **antigen** and antiserum from the same species, and increases for more distantly related antigens as more antiserum is required to produce the same immunological response. Using an antiserum to human albumin, a value of 1.09 was obtained for the gorilla, 1.14 for both the chimpanzee and bonobo, and an average of 2.46 for six species of OWMs. Taking into account the immunological distances obtained when antisera to chimpanzee or gibbon albumin were used, and assuming that the log of the measure was proportional to time, a time of 5 MYA was calculated for the split between gorillas, chimpanzees, and humans (**Figure 7.12**).

While the results with antisera against human albumin (quoted above) and gibbon albumin placed humans closer to gorillas than chimpanzees, the measurements with antiserum to chimpanzee albumin identified humans as the closest relative of chimpanzees. Thus the method could not resolve the gorilla–chimpanzee–human split in a consistent way and this was presented as an unresolved trichotomy. Note that this is a failure to resolve a gene tree and is distinct from the gene tree–species tree difference. The resolution of this trichotomy formed the focus of much subsequent work. Other limitations were that only a single locus was assayed and that immunological distance is an indirect measure of evolutionary distance.

Genetic data have resolved the gorilla–chimpanzee–human trichotomy

An attractive approach to determining the branching order of the species tree in a single experiment is **DNA–DNA hybridization**, which effectively compares the entire **single-copy** components of two genomes with one another. If most regions of the genome are evolving neutrally, genetic distances estimated from such an approach are expected to correspond well to the species tree.

In DNA–DNA hybridization experiments, two denatured DNA samples (from either the same or different species) are re-associated and the thermal stability of the hybrid is measured. **Heteroduplexes** formed between molecules of different species are less stable than homoduplexes formed between molecules of the same species, and this reduction in stability provides a measure of the divergence between the species. The method is well suited to comparing species that have diverged for >10 MY, but for closely related species the small differences can be masked by random experimental error. For gorillas, chimpanzees, and humans the differences are small and the subject was for a time

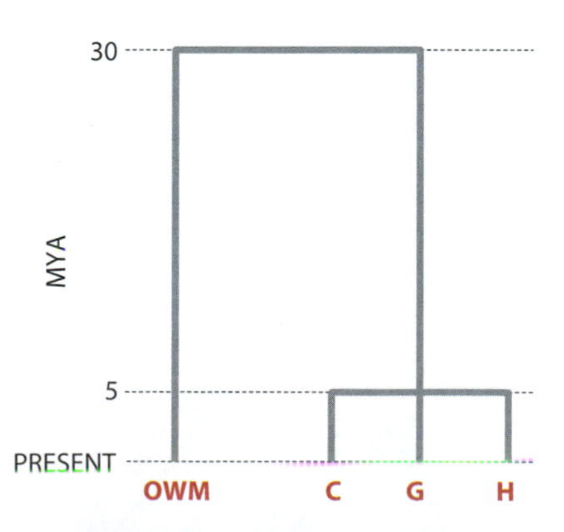

Figure 7.12: The gorilla–chimpanzee–human trichotomy.
DNA–DNA hybridization data support a closer relationship between chimpanzees (C) and humans (H) than between either species and gorillas (G). [DNA–DNA hybridization data from Sibley CG & Ahlquist JE (1984) *J. Mol. Evol.* 20, 2.]

unusually contentious, even by the standards of human evolutionary studies. Thus although the work[57] provided evidence for a closer relationship between chimpanzees and humans than between either species and gorillas, and thus appeared to resolve the trichotomy, concerns about its interpretation meant that Sibley and Ahlquist's conclusion was not universally accepted. Nevertheless, subsequent work has supported the view that chimpanzees and humans are the most closely related of these three ape species.[38]

Gene trees do not necessarily have the same topology as species trees (Figure 7.1). If polymorphism survives from one speciation event to the next then there is a possibility that alleles may be distributed through **lineage assortment** in such a way that the most similar sequences are not necessarily in the most closely related species (**Figure 7.13**). Unsurprisingly, for any pair of speciation events, the degree to which we might expect allele mis-assortment due to such incomplete lineage sorting (ILS) depends on the probability of alleles becoming fixed over a given length of time. For alleles not under selection this depends on the effective population size. Thus, on average, there is a higher probability that gene trees from haploid mitochondrial and Y-chromosomal data more accurately reflect the species tree, due to their lower effective population sizes (see **Sections 5.3** and **6.2**), than trees from autosomal data.

Some loci within the human genome carry polymorphisms that are maintained by balancing selection, over much longer time spans than would be expected of neutral alleles (**Section 5.4**, Box 6.5). This explains why alleles at some human HLA loci are more closely related to their chimpanzee counterparts than they are to other human alleles (Box 5.3). Such loci are to be avoided when reconstructing species trees. Besides selection, unequal rates of evolution amongst lineages and biased base composition can hamper phylogeny reconstruction (see **Section 6.4** for more details).

Consequently, one approach to resolving the gorilla–chimpanzee–human trichotomy has been to examine gene trees from many neutral, single-copy, orthologous loci in the three species.[7] These data can subsequently be analyzed in a number of ways. The sequences can be concatenated (combined into a single, long sequence) and bootstrap support (see **Section 6.4**) obtained for clades within the single resulting phylogeny. Alternatively, phylogenies can be reconstructed for individual loci and multi-locus tests applied to determine whether any topology is significantly predominant. **Figure 7.14** shows how data from 53 intergenic loci provided 100% bootstrap support for a human–chimpanzee species clade as the majority consensus whereas 42% of the individual locus-specific phylogenies support human–gorilla or chimpanzee–gorilla clades.

Similarly, comparisons of great ape genomes have provided robust support for the human–chimpanzee clade with only 30% of loci resolving to alternative clades in which gorilla would be closest to either the human or chimpanzee genome.[54] Notably, the effect of ILS is less pronounced in genomic regions that contain genes, and within exons the proportion drops to 22%, which is likely due to physical linkage of the variation with functionally important substitutions that

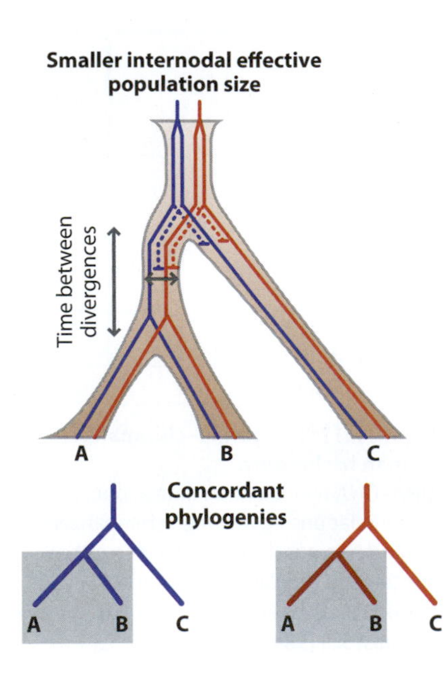

Figure 7.13: The effect of ancestral effective population size on the frequency of discordant gene trees.
The degree to which gene trees of neutral loci fail to reflect the same phylogeny as the species tree depends on the time between successive speciation events and the effective population size of the ancestral population. In these two examples, the time between speciation events is the same in the two scenarios. Two gene trees (*red* and *blue*) are shown within each species tree. Lineages being lost by drift are shown in dashed lines. In the first example, the greater width of the species ancestral to A and B represents a larger effective population size that allows a polymorphism in one of the gene trees to persist between the two speciation events. The subsequent lineage assortment of this polymorphism results in discordant phylogenies.

have been targeted by natural selection. Natural selection reduces the effective population size (N_e) of coding regions that are under functional constraints as compared with the neutrally evolving loci, which retain more ILS over time. Consistent with this view, the X chromosome has a lower effective population size than the autosomes and also appears to be affected by less (13%) ILS.[54]

Sequence divergence is different among great apes across genetic loci

Once a definitive phylogeny has been obtained, and the branch-points dated, the proportion of gene trees that do not support this phylogeny can be used to estimate the **effective population size** of the ancestral population that existed between the species' splits. In the above scenario, this ancestral population corresponds to the human–chimpanzee common ancestor. Although, as we shall see in the next section, dating branch-points is not without controversy, it appears that the effective population size of this common ancestor was 5–10 times greater than that of extant humans. The finding of ~1% of genome regions where ILS extends back to the ancestor of humans and orangutans supports the view that the ancestors of humans, chimpanzees, and gorillas had a large effective population size and did not pass through severe genetic bottlenecks over the course of several million years.[19] There are many possible explanations for the dramatic reduction of N_e in the human lineage, and some will be considered later in this book.

Different loci within the great ape genomes exhibit different degrees of sequence divergence. Regions under greater selective constraint will exhibit less sequence divergence; consequently, nonsynonymous changes within coding regions (dN) are less frequent than synonymous changes (dS). The reasons underlying other differences in sequence divergences shown in **Table 7.1** are less immediately apparent.

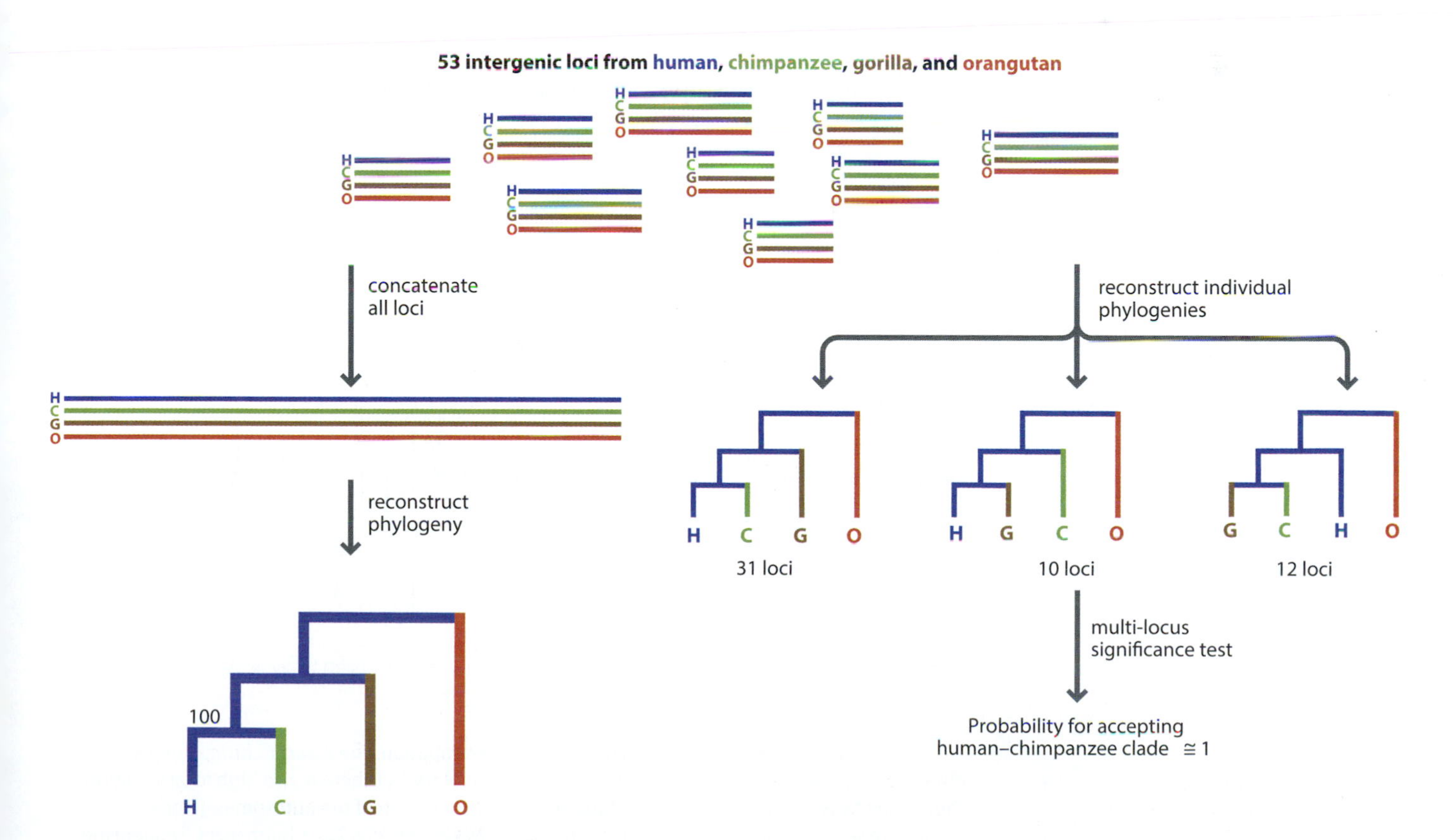

Figure 7.14: Combining data from multiple orthologous loci to resolve the gorilla–chimpanzee–human trichotomy. The number (100) next to the human–chimpanzee clade indicates percentage bootstrap support. H, human; C, chimpanzee; G, gorilla; O, orangutan. [Data from Chen FC & Li WH (2001) *Am. J. Hum. Genet.* 68, 444.]

TABLE 7.1:
PERCENTAGE SEQUENCE DIVERGENCES (JUKES–CANTOR DISTANCES) BETWEEN HOMINOIDS

Locus	H-C	H-G	C-G	H-O	C-O	G-O
Noncoding (Chr Y)	1.68 ± 0.19	2.33 ± 0.2	2.78 ± 0.25	5.63 ± 0.35	6.02 ± 0.37	6.17 ± 0.37
Pseudogenes (autosomal)	1.64 ± 0.10	1.87 ± 0.11	2.14 ± 0.11	–	–	–
Pseudogenes (Chr X)	1.47 ± 0.17	–	–	–	–	–
Noncoding (autosomal)	1.24 ± 0.07	1.62 ± 0.08	1.63 ± 0.08	3.08 ± 0.11	3.12 ± 0.11	3.09 ± 0.11
Genes (dS)	1.11	1.48	1.64	2.98	3.05	2.95
Introns	0.93 ± 0.08	1.23 ± 0.09	1.21 ± 0.09	–	–	–
Noncoding (Chr X)	0.92 ± 0.10	1.42 ± 0.12	1.41 ± 0.12	3.00 ± 0.18	2.99 ± 0.17	2.96 ± 0.17
Genes (dN)	0.80	0.93	0.90	1.96	1.93	1.77
Genomewide	1.37	1.75	1.81	3.40	3.44	3.50

H, human; C, chimpanzee; G, gorilla; O, orangutan. Data from Chen FC & Li WH (2001) *Am. J. Hum. Genet.* 68, 444; Scally A et al. (2012) *Nature* 483, 169; and references therein.

Inheritance pattern has a dramatic effect on sequence divergence. It has long been suspected that the larger number of mitotic cell divisions in the male germ line than in the female increases mutation rates in males relative to females, and that consequently evolution is male driven. The ratio of these sex-specific mutation rates is known as the **alpha factor**, estimates for which are given in Box 5.4. Y chromosomes are inherited solely through the male germ line, whereas X chromosomes pass through twice as many female meioses as male. Thus, male-driven evolution explains why sequence divergence amongst non-coding regions is greatest among Y-linked loci, and least among X-linked loci, as illustrated, for example, by the different levels of sequence divergence of human and chimpanzee chromosomes (**Figure 7.15**).

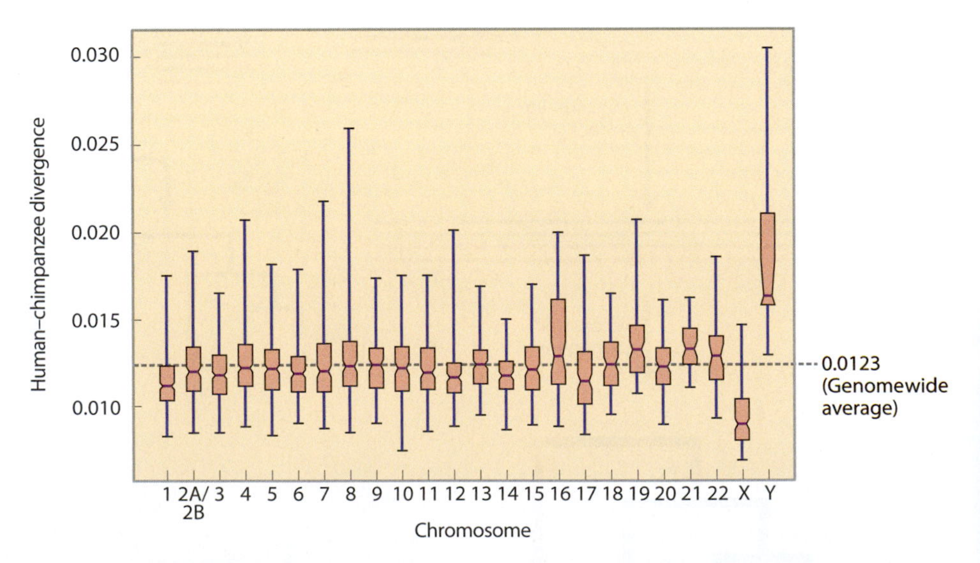

Figure 7.15: Distribution of human and chimpanzee divergence by chromosome. Differences between human and chimpanzee chromosomes are shown as a box plot where the edges of each box represent quartiles, the notches show the standard error of the median, and the vertical bars the range of variation. The X and Y chromosomes are outliers, but there is also high local variation within each of the autosomes. [From Mikkelsen TS & The Chimpanzee Sequencing and Analysis Consortium (2005) *Nature* 437, 69. With permission from Macmillan Publishers Ltd.]

Great apes differ by gains and losses of genetic material

It is worth noting that the commonly used measures of sequence divergence only take into account base-substitutional differences between orthologous loci. The comparison of ~2.4 gigabases of human and chimpanzee genomes yielded an overall estimate of 1.23% divergence (Figure 7.15) of which 0.17% is polymorphism within the species and only about 1.06% fixed divergence.[41] DNA sequences may also differ by insertions or deletions (indels) of small numbers of bases (see **Section 3.2**). Typically these indels are stripped from alignments before the calculation of sequence divergence measures such as the **Jukes–Cantor** distances (see **Section 6.3**) used in Table 7.1. A different measure of sequence divergence takes indels and base substitutions into account by considering what fraction of bases in one species are exactly matched in another species. This measure has been applied to 779 kb of aligned human and chimpanzee sequence in which ~106,000 base substitutions were supplemented by 1000 indels.[6] The base-substitutional sequence divergence in these alignments (1.4%) was similar to the genomewide estimate, yet, despite their tenfold lower frequency than nucleotide substitutions, indels contributed an additional 3.4% of sequence divergence, generating an overall divergence that approached 5%! This method for calculating sequence divergence is only accurate for comparisons of such closely related species; this is because the mutational mechanism of indel events is not well understood, which makes it impossible to correct for parallel and reversion mutations in comparisons of more distantly related species. In addition, it does not take account of the large number of nucleotides that may be involved in a single indel mutational event, so it does not reflect the number of mutations and its interpretation is not simple. Perhaps for these reasons, simple base-substitutional differences are generally still used.

Comparison of the chimpanzee and human genomes has identified 53 functional human genes that are either completely or partially deleted in the chimpanzee.[41] In total, the amount of lineage-specific gains or losses of functional sequence in humans, gorillas, and chimpanzees is 3–7 Mb per species.[54] The most common targets of loss and gain appear to be olfactory receptor genes and genes related to immunity or male fertility. Interestingly, gorillas, which rely more on a leafy diet than other hominoids, have a higher copy number of the *FLJ22004* gene; this gene plays a role in the regulation of cellulase and pectinase, enzymes involved in the digestion of cellulose and pectin from plants, and may thus enhance the utilization of the key staple foods of the species.[12] Similarly, several immunity-related genes were found to have undergone duplication in the chimpanzee lineage, including one encoding a sialic acid that humans cannot synthesize. Analysis of ancient retroviral elements in hominoid genomes suggests that, in contrast to human ancestors, the chimpanzee and gorilla lineages have been more subject to retroviral infections over the last 3–4 MY.[67] The human lineage, on the other hand, is characterized by the abundant expansion of the *Alu* family, characteristic of primates, while the orangutan genome shows a low number of lineage-specific *Alu* transpositions.[35]

The DNA sequence divergence rates differ in hominoid lineages

The global **molecular clock** is the simplest method for estimating species divergence dates from molecular diversity. It assumes that, for a given locus, mutation is occurring at a constant rate across all taxa (**Section 5.7**). Thus, species bifurcations that are well resolved in the fossil record can be used to date less well-characterized divergences. However, a number of factors may alter rates of sequence change over evolutionary time, including differences in generation times, in DNA repair mechanisms, and in metabolic rates. Empirical data indicate that rates do vary between mammalian lineages, and unsurprisingly the molecular clock becomes less reliable in the deeper branches of the phylogeny. For example, it has been demonstrated that mitochondrial evolution has proceeded faster in primates than in many other mammalian orders, and that within primates there has been an evolutionary slowdown in the hominoids, perhaps due to their longer generation times (see Figure 7.16). Analysis of 8 Mb of genomic sequence in 186 primates has revealed that the mutation rate slowdown is characteristic of all Catarrhini rather than being uniquely associated with hominoids.[48]

Because of these rate differences, a number of statistical methods have been devised to detect mutation rate heterogeneity (see Section 5.7). Most are based on the **relative rates test**, which compares molecular distances between pairs of species that are equally related: two species compared with a known **outgroup** (see Figure 5.16). Under a constant rate these distances should be equal. If rate heterogeneity is detected, various analytical techniques allow it to be taken into account when calculating divergence times. These techniques require accurate estimates of branch lengths, and in the case of faster-mutating loci or deeper divergences, this involves a correction of observed sequence distance measures for multiple mutations at the same site. As with other applications of **phylogenetic reconstruction** methods, changing the model of sequence evolution often gives different inferences.

There are three potential sources of error in estimation of molecular divergence dates: the chosen calibration point; the methods used to reconstruct the phylogeny and to take into account rate heterogeneity; and the choice of data. These three factors are not unrelated—for example, the calibration point needs to be one for which data are available.

One way to validate a calibration point is to have a number of paleontologically well-dated divergences incorporated within the data: this allows the reciprocal calibration of one well-dated divergence by another. If these reciprocal calibrations do not agree, then the data, the method, or the calibration must be changed. By using different variations of all three factors, reciprocal calibrations can reveal which combinations are least accurate. However, note that genetic estimates of clade age are expected to be older than estimates made using the fossil record. This is because genetic data identify the earliest stages of divergence, which pre-date the important morphological changes. Multiple calibration points derived from fossil evidence can thus be used as priors in Bayesian estimation of the branch lengths and divergence times[48] (Figure 7.3).

Studies of many individual loci have estimated the date of the human–chimpanzee split as 4–7 MYA, with the gorilla splitting off earlier, 6–9 MYA. This range has become the consensus, even though most of these estimates rely on the patchy primate fossil record for calibration. A calibration date of 12–16 MYA for the split of the orangutan from the other extant hominids has often been used, and a date of 25–30 MYA for the Cercopithecoidea–Hominoidea split. Genomewide comparisons of human, chimpanzee, and gorilla sequences, using a rate estimate of 10^{-9} substitutions per base pair per year, as derived from human–macaque divergence, would yield 4 and 6 MY for the human–chimpanzee (H-C) and human/chimpanzee–gorilla (HC-G) splits, respectively.[54] However, the empirical sequence data from human populations and familial trios suggest ~40% lower mutation rates which would push the H-C date to 6.5 MYA and HC-G date to 10 MYA (Figure 7.17). Projecting this low mutation rate further to earlier speciation events in the phylogenetic history of great apes would be inconsistent with the interpretations based on the Middle Miocene (12–16 MYA) fossil record. But this inconsistency could be resolved by assuming that great ape mutation rate has decreased over the past 10 MY as body size and generation times have increased.[54]

7.4 GENETIC DIVERSITY AMONG THE GREAT APES

Having arrived at a consensus about the phylogenetic relationship of humans with their closest nonhuman relatives, we now ask whether the pattern of genetic diversity in the great apes resembles that of our own species. Polymorphism maintained between species may indicate balancing selection,

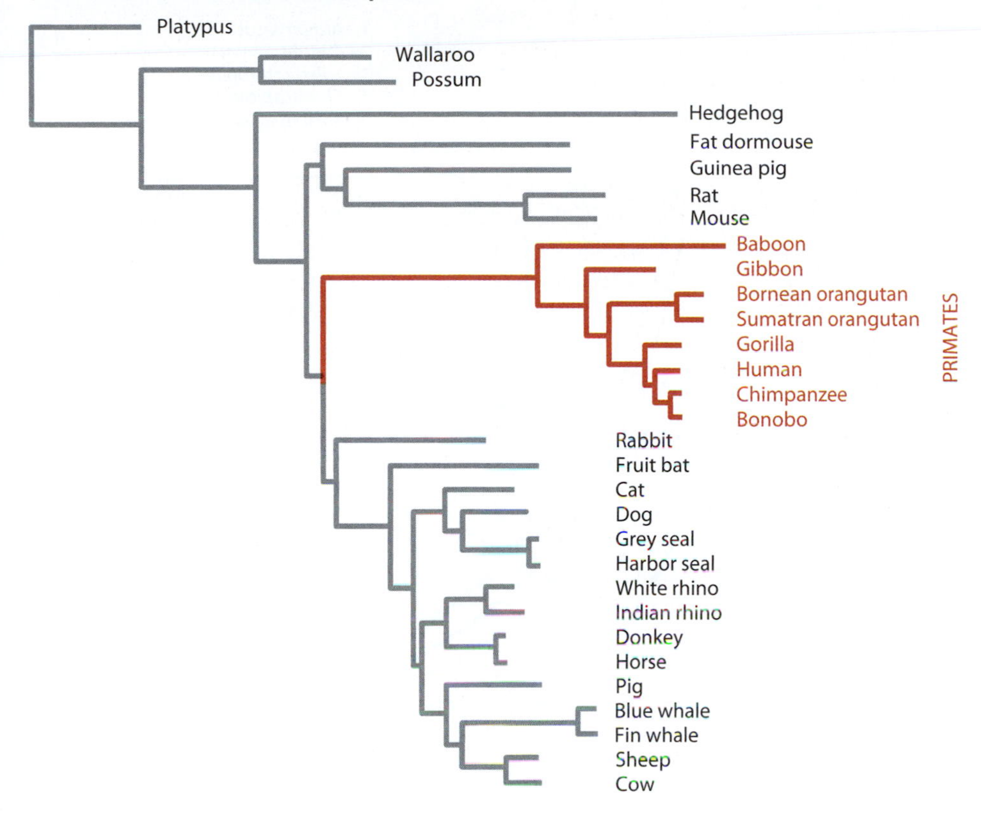

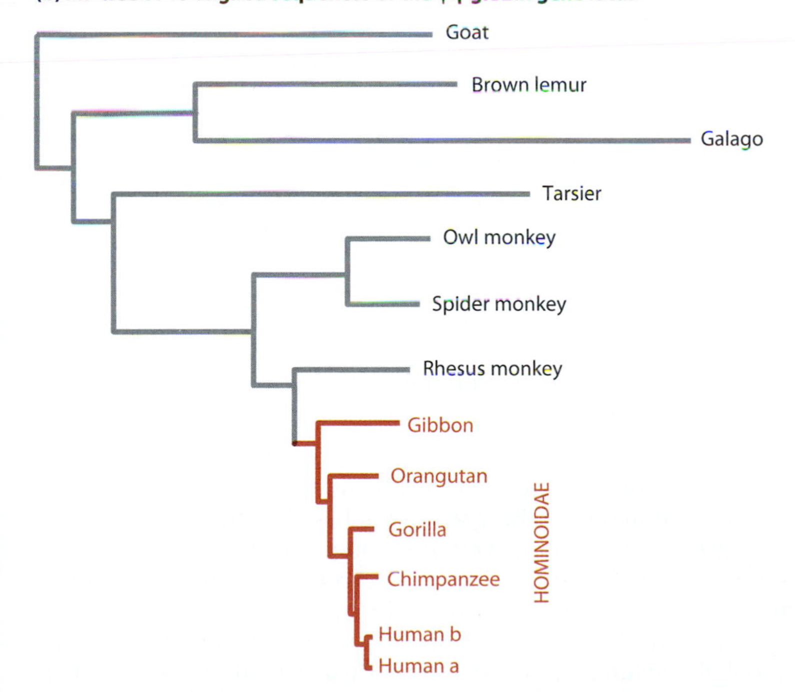

Figure 7.16: Phylogenies exhibiting rate heterogeneity among lineages.
If mutation rates are constant across lineages, the sum of branch lengths leading to each taxon in a rooted tree should be equal, and thus the tips should be level. Longer branch lengths indicate a faster rate and shorter branch lengths a slower rate. (a) The mtDNA tree shows a higher rate of mtDNA mutation in primates compared with other mammals. (b) The ψη-globin tree shows a decrease in mutation rate at the ψη-globin gene locus in hominoids compared with other primates. [a, from Yoder AD & Yang Z (2000) *Mol. Biol. Evol.* 17, 1081; b, from Bailey WJ et al. (1991) *Mol. Biol. Evol.* 8, 155. With permission from Oxford University Press.] ML, maximum likelihood; MP, maximum parsimony.

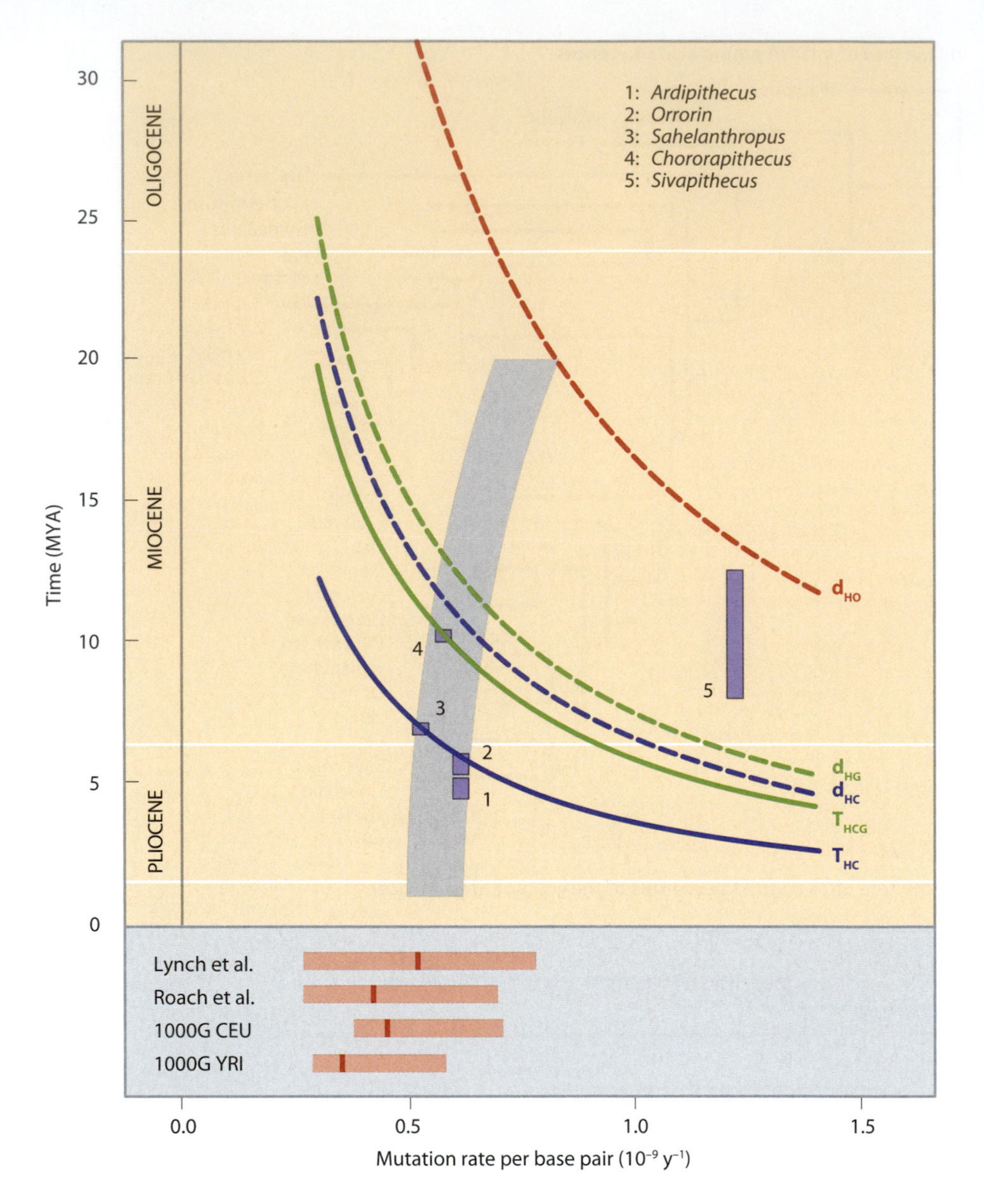

Figure 7.17: Modeling great ape speciation and divergence times. Solid *blue* (H-C) and *green* (HC-G) lines show speciation dates as a function of mutation rate, estimated using a coalescent model. Dashed lines show the corresponding average sequence divergence times (including *red* for HCG-O). *Blue* blocks represent hominid fossil species, each with a vertical extent spanning the range of dates proposed and a horizontal position at the maximum mutation rate consistent both with its proposed phylogenetic position and the coalescent model estimate. The bending of the *gray* shaded region shows that an increase in mutation rate going back in time can accommodate present-day estimates of mutation rate, hypotheses about the phylogenetic position of the fossils, and a mid-Miocene speciation for orangutan. The lower panel shows four estimates of the average mutation rate in humans with 95% confidence intervals [1000 Genomes Project Consortium (2010) *Nature* 467, 1061; Lynch M (2010) *Proc. Natl Acad. Sci. USA* 107, 16577; Roach JC et al. (2010) *Science* 328, 636] in present-day humans. [Data from Scally A et al. (2012) *Nature* 483, 169. With permission from Macmillan Publishers Ltd.]

and discrepancies of levels of diversity within and between species may signal the action of natural selection, or different population structures and histories, and different effective population sizes.

Certainly there are enormous differences in census population sizes today. These are difficult to determine for the great apes because they live in forest environments that are not easy to survey, but rough estimates are given in **Figure 7.18**. One way to appreciate the disparity is to consider that currently it takes the human population less than 2 days to increase by a number equal to the total world population of great apes. Further, while humans are now distributed over almost all the land surface of the Earth, distributions of great apes are highly restricted (Figure 7.18), and are becoming more so through human activities. The World Conservation Union lists all great ape species as either

Figure 7.18 (*right*): Current distributions and approximate population sizes of the great apes. Distribution maps and approximate census sizes of the subspecies of great apes are presented according to various sources, including **http://apesportal.eva.mpg.de/status/species/**; **http://www.iucnredlist.org**; **http://www.wcs.org**; **http://www.ellioti.org**, and **hww.unep.org/grasp**. Many of the given population sizes are the subject of debate and continuing revision

(a) Chimpanzees and bobonos

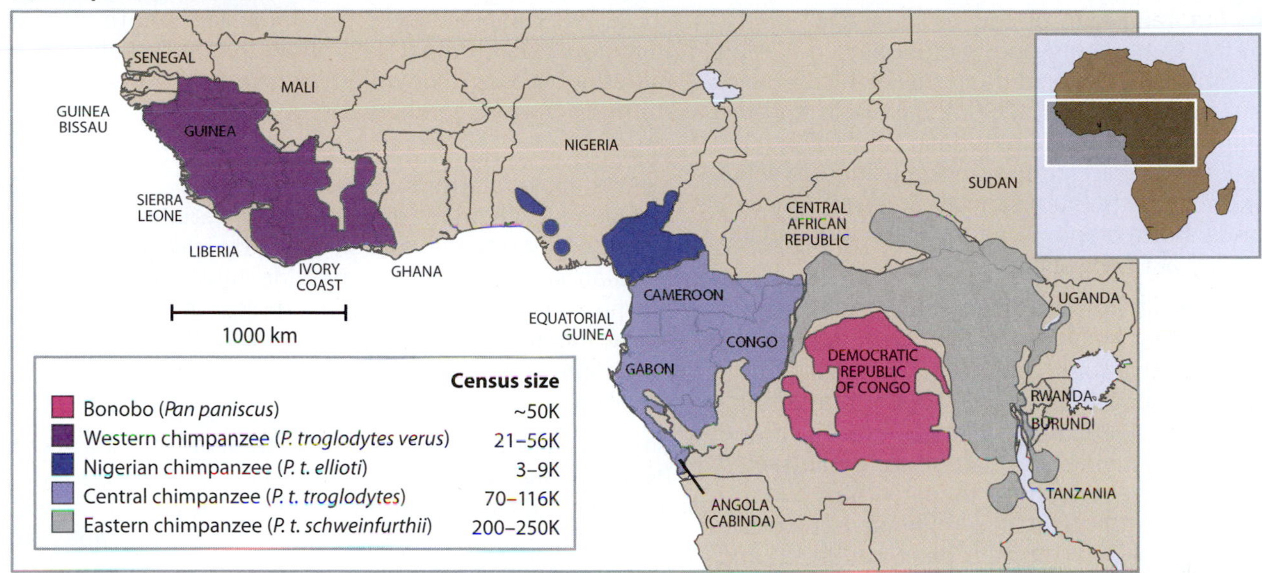

(b) Gorillas

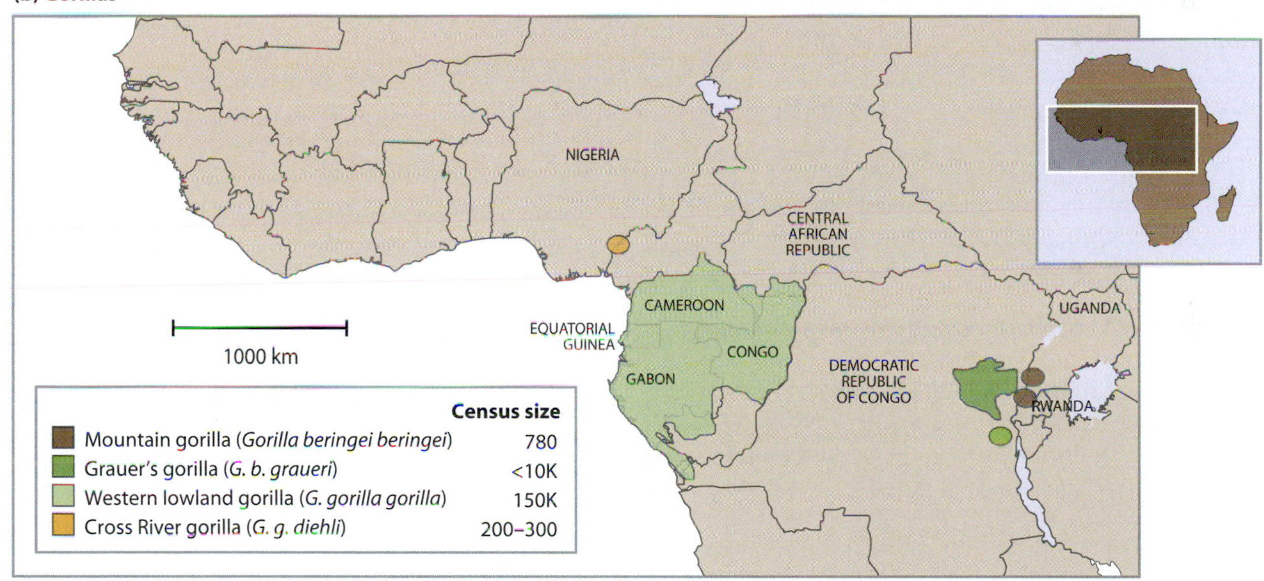

(c) Orangutans

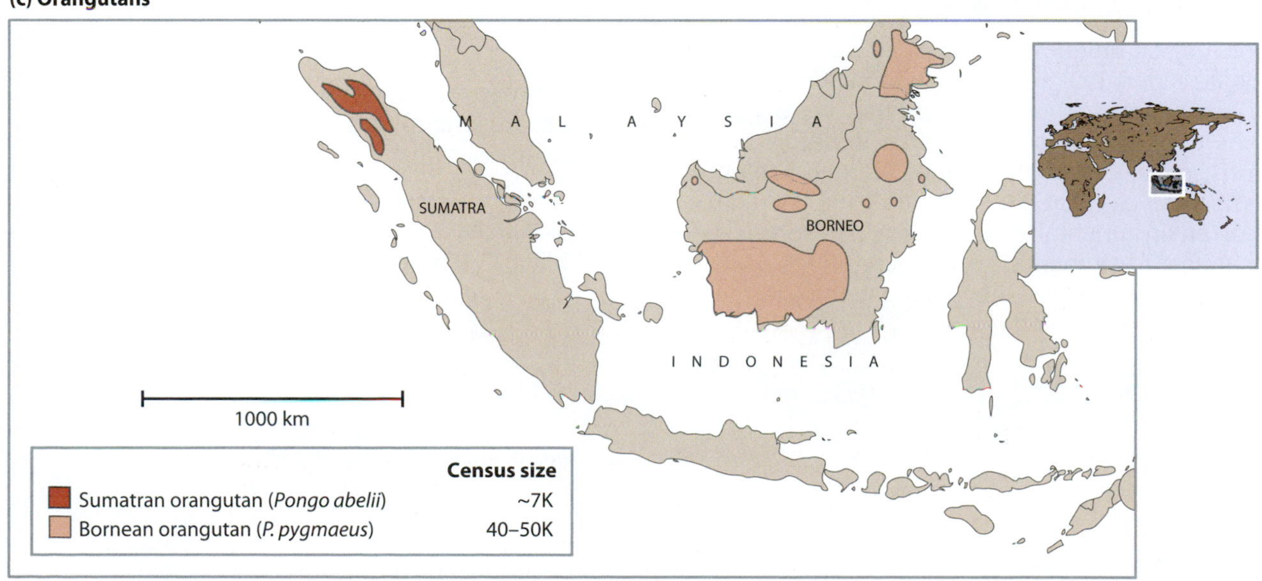

Of the living apes, only humans are secure. The other species are but remnants of the "golden age" of the apes, in the Miocene. Only one evolutionary radiation remains, the gibbons (Hylobatidae), the highly derived, small-bodied, arboreal apes of Southeast Asia and the Malay archipelago. The remaining taxa (bonobo, chimpanzee, gorilla, and orangutan) are increasingly isolated relicts, in shrinking populations in sub-Saharan Africa or on two islands, Borneo and Sumatra. If they are inevitably on their way out, why should we care?

We should strive to preserve them because the other apes are magnificent creatures in their own right: intelligent, complex, fascinating, and challenging. Moreover, they offer unique opportunities to study form and function in action, and genes and environment in process, while still living in their environments of evolutionary adaptedness. Short of inventing a time machine, we will never see our ancestors and their ilk behave; thus, we can only guess at fleshing out the paleontological and archaeological records of our forebears. Bones and artifacts are mute, so all they offer is inference; living organisms think, feel, and act, and we can record this.

Comparative studies inform us at various levels: we contrast apes as species, subspecies, populations, lineages, and individuals. Chimpanzees are xenophobic and "pongicidal," while bonobos are neophilic (that is, with strong affinity for novelty) and non-fatally aggressive. West African chimpanzees have stone tools and an archaeological record, while East African chimpanzees apparently have neither. Mountain gorillas are terrestrial herbivores, while lowland gorillas are more arboreal and frugivorous. Sumatran orangutans are sociable tool-users, while Bornean orangutans are neither. High-ranking chimpanzee matrilines sequester resources, resist emigration, and are reproductively successful, while subordinate ones are peripheralized. Silverback male gorillas and alpha male chimpanzees show individually distinctive styles of dominance, from despotism to laissez-faire, and these rank factors affect day-to-day social relations, such as food-sharing.

Human apes may be unique in some ways, such as language, obligate bipedality, or even-handedness, but we share with our cousins other attributes that were long thought to be derived human traits. Apes in nature are cultural, in the sense of having socially learned behavioral patterns that characterize a group, independent of genetic or environmental determinism. These traditions are manifest in elementary technology, such as in extractive foraging, as acquired by youngsters from the mother. Or, they occur as customs, such as group-typical social grooming patterns, that indicate conformity and imitation. Longitudinal data show that such behavioral patterns are passed down through vertical transmission, from generation to generation of apes.

Findings from cultural primatology show us that when groups or populations of apes go extinct, then it is not just genetic diversity or phenotypic variation or taxonomic continuity that is lost. Similarly, although captive populations may suffice to provide subjects for genetic, morphological, or physiological research, they will not allow the full expression of behavioral, cognitive, or emotional acts. Thus, preservation of apes in nature is essential.

So, what are the threats to the survival of wild apes? Paradoxically, some of the obvious ones are no longer a problem. Capture and trade in apes for zoos, circuses, or laboratories is virtually banned. Keeping apes as household pets is increasingly unacceptable or illegal, at least in industrialized countries. But three main threats remain: bush meat, disease, and habitat loss. Killing apes for food has a long history amongst tropical foraging peoples, and so long as it was done locally with traditional technology "for the cooking pot," it was no problem. However, hunting with firearms for an international commercial trade is unsustainable.

Epidemiologically, we now know that highly contagious diseases such as Ebola can annihilate wild populations of apes, although the extent to which anthropogenic influences are involved is unclear. We know from bitter experience that infectious disease can be accidentally introduced by tourists, or even by researchers, to wild apes who have been habituated to humans at close range. Finally, the largest threat to apes in nature is deforestation, or conversion of forest to horticulture. This not only deprives the apes of natural resources, but also places them in the invidious position of having to turn to crop-raiding to survive.

All of these threats are being addressed, but to widely varying extents, and current efforts are not enough to ensure the apes' survival, even to the midpoint of the current century. Fortunately, field primatologists and their allies are committed to the preservation of our nearest living relations.

Linda F. Marchant, Miami University, Oxford, Ohio, USA, and
William C. McGrew, Corpus Christi College, University of Cambridge, Cambridge, UK

endangered or critically endangered, and a survey of Western Equatorial Africa documented a catastrophic decline in numbers due to hunting and Ebola hemorrhagic fever.[64] This sad fact gives the understanding of great ape diversity a vital and urgent purpose beyond the academic activities of primate taxonomists (**Opinion Box 5**).

The practical reasons for studying great ape diversity include:

- Defining species and subspecies is important because these are the units of protection and captive breeding.
- Estimates of heterozygosity and genetic variation in a population reflect the level of inbreeding, which is considered to be one of the indicators of extinction risk.
- Understanding the diversity of wild gene pools allows the preservation of as much diversity as possible within captive populations. This is important for maintaining fitness in response to selective pressures.
- Defining population genetic subdivision is important for relocation and reintroduction programs, which need to consider the genetic distinctiveness of isolated populations.
- Defining population substructure aids in identifying the geographical origins of confiscated illegal pets.

How many genera, species, and subspecies are there?

Describing the genetic diversity within a species or subspecies first requires these categories to be defined. This is not straightforward,[37] and indeed there is some circularity when genetic data themselves are sometimes used as criteria for defining species or subspecies, as is the case for chimpanzees.[18, 42]

Since Darwin's time, a wide range of species definitions has been proposed. Traditional versions have been based on taxonomically useful characters such as morphology, while more recent species concepts have focused on shared ecology (and geographical distributions), or genetic similarity. A prominent model has been the **biological species concept**,[39] which defines a species in terms of an interbreeding natural population that is reproductively isolated from others by one or a number of possible mechanisms (**Box 7.2**). Models such as this regard species as real entities, and have been challenged by ideas suggesting that species have no objective reality, but are man-made categories.[37] One view suggests that it is the local population, united by gene flow within it, and not the species, which is the evolutionary unit. This ambiguity can have a profound effect on the number of species that we recognize.

Recent changes in great ape classification (**Table 7.2**) have introduced a new subspecies of chimpanzee (*P. t. ellioti*), and have divided gorillas into two species (*G. gorilla* and *G. beringei*). On the basis of the large genetic differences between Sumatran and Bornean orangutans, they are now also classified as distinct species.

Intraspecific diversity in great apes is greater than in humans

Phylogenies drawn from two loci, mtDNA HVSI[14] and a 10-kb segment of Xq13.3[24] (**Figure 7.19**), illustrate dramatically reduced diversity in humans compared with great apes. One explanation for lower diversity in Xq13.3 and mtDNA in humans might be selection, but this would have to act upon both loci similarly—which seems unlikely—and statistical tests (see **Section 6.7**) find no support for selection at Xq13.3. Alternatively, the difference may represent a difference in population history, such as a recent expansion from a founder population in humans, and tests based on multiple loci sequenced across the genome support this. Based on the genetic variation observed within and among species, the estimates of **long-term effective population size** show that despite the fact that humans currently outnumber all other hominoids by several orders

Box 7.2: Reproductive barriers between species

The reproductive isolation of species can be divided into two parts:

- Prezygotic—mechanisms preventing fusion of egg and sperm, so no zygote can form; due to species differences in traits such as
    - Sexual behavior
    - Geographical range or habitat preference
    - Seasonal breeding
    - Gamete compatibility (post-mating, but still prezygotic)
- Postzygotic—mechanisms preventing a zygote from developing into a fertile offspring; due to
    - Hybrid inviability
    - Hybrid sterility

TABLE 7. 2:
SPECIES AND SUBSPECIES DISTINCTIONS AMONG THE GREAT APES

	Species	Taxonomic issues
Humans	*Homo sapiens*	Should *Pan* and *Homo* be united as the same genus?
Common chimpanzees	*Pan troglodytes verus* (western); *P. t. troglodytes* (central); *P. t. schweinfurthii* (eastern); *P. t. ellioti* (eastern Nigerian–west Cameroon)*	Subspecies distinctions genetic Behavioral differences do not follow subspecies lines Subspecies are interfertile in captivity, but no data on hybrid inviability
Bonobo (aka pygmy chimpanzee)	*Pan paniscus*	Behavioral (including social), morphological, genetic, and geographical distinctions from *P. troglodytes* Interfertile with common chimpanzees in captivity
Eastern gorillas	*Gorilla beringei graueri* (Grauer's) *G. b. beringei* (mountain)	Species distinction morphological, geographic, and genetic Subspecies distinction morphological and geographic
Western gorillas	*Gorilla gorilla gorilla* (western lowland) *G. g. diehli* (Cross River)	Subspecies distinction craniodental and geographic
Orangutans	*Pongo abelii* (Sumatran) *P. pygmaeus* (Bornean)	Distinctions morphological, behavioral, cytogenetic (pericentric inversion of chromosome 2), genetic Genetic differences large enough to have warranted different species status; other differences relatively small Species interfertile in captivity; hybrids fertile

Taxonomy based on Wilson DE & Reeder DM (2005) Mammal Species of the World, 3rd ed. http://www.bucknell.edu/msw3/
* *P. t. vellerosus* has been renamed as *P. t. ellioti* [Oates JF et al. (2009) *Primates* 50, 78]

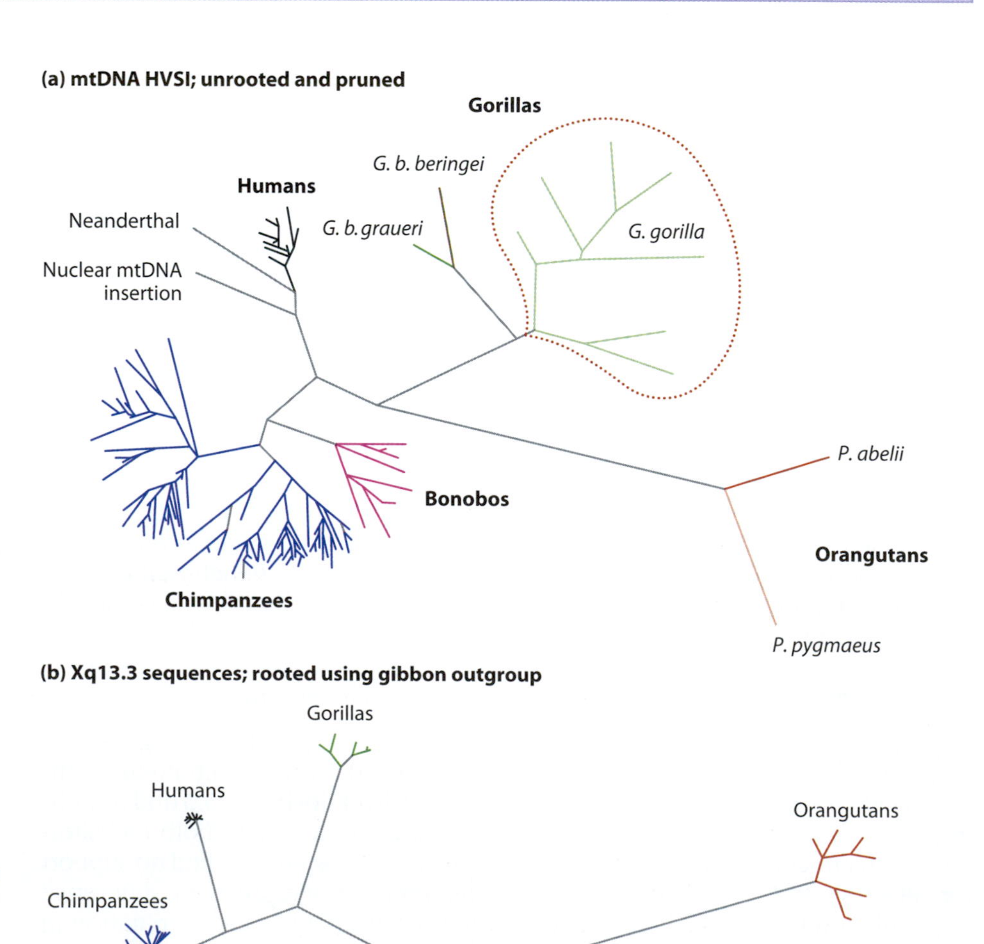

Figure 7.19: Phylogenies showing relative diversities among apes. (a) mtDNA phylogeny from HVSI sequences. The tree ("pruned" to remove **homoplasies**) includes sequences from a Neanderthal, and from a human nuclear mtDNA insertion. Gorilla species are differentiated; for more on chimpanzee subspecies, see Figures 7.21–22. (b) Xq13.3 sequence phylogeny. [a, from Gagneux P et al. (1999) *Proc. Natl Acad. Sci. USA* 96, 5077. With permission from the National Academy of Sciences. b, from Kaessmann H et al. (2001) *Nat. Genet*. 27, 155. With permission from Macmillan Publishers Ltd.)

of magnitude, the large human population is a relatively recent phenomenon (**Figure 7.20**). If effective population sizes were to be used as a measure of evolutionary success of a species then we could say that most other hominoids were outcompeting humans—up to the last 100 KYA, or possibly even up to the last 10–20 KYA. Because the TMRCA (time to most recent common ancestor) for any autosomal locus is expected to be four times the effective population size in generations, the coalescence times of chimpanzee loci are typically more than 2 MY, pre-dating the speciation split with bonobos,[11] whereas most human gene trees have coalescence times younger than 1 MY. Similarly to the chimpanzees, the orangutans and gorillas are characterized by high levels of genetic diversity, although their species separation, as estimated from the genetic data, is relatively recent—about 400 KYA for Bornean/Sumatran orangutans.[35]

As mentioned above, several revisions in great ape taxonomy have been undertaken based on morphology, geographic data, and genetic diversity estimates. Within the chimpanzee species, the clear separation of eastern Nigerian and west Cameroonian individuals from both the central and western chimpanzees (**Figure 7.21**) has been used as an argument to consider them a distinct subspecies, *P. t. ellioti*. Separation of other subspecies of chimpanzees is similarly clear in the phylogeny based on complete mtDNA sequences as well as in a tree of Y-chromosomal haplotypes based on sequence variation.[59] By contrast,

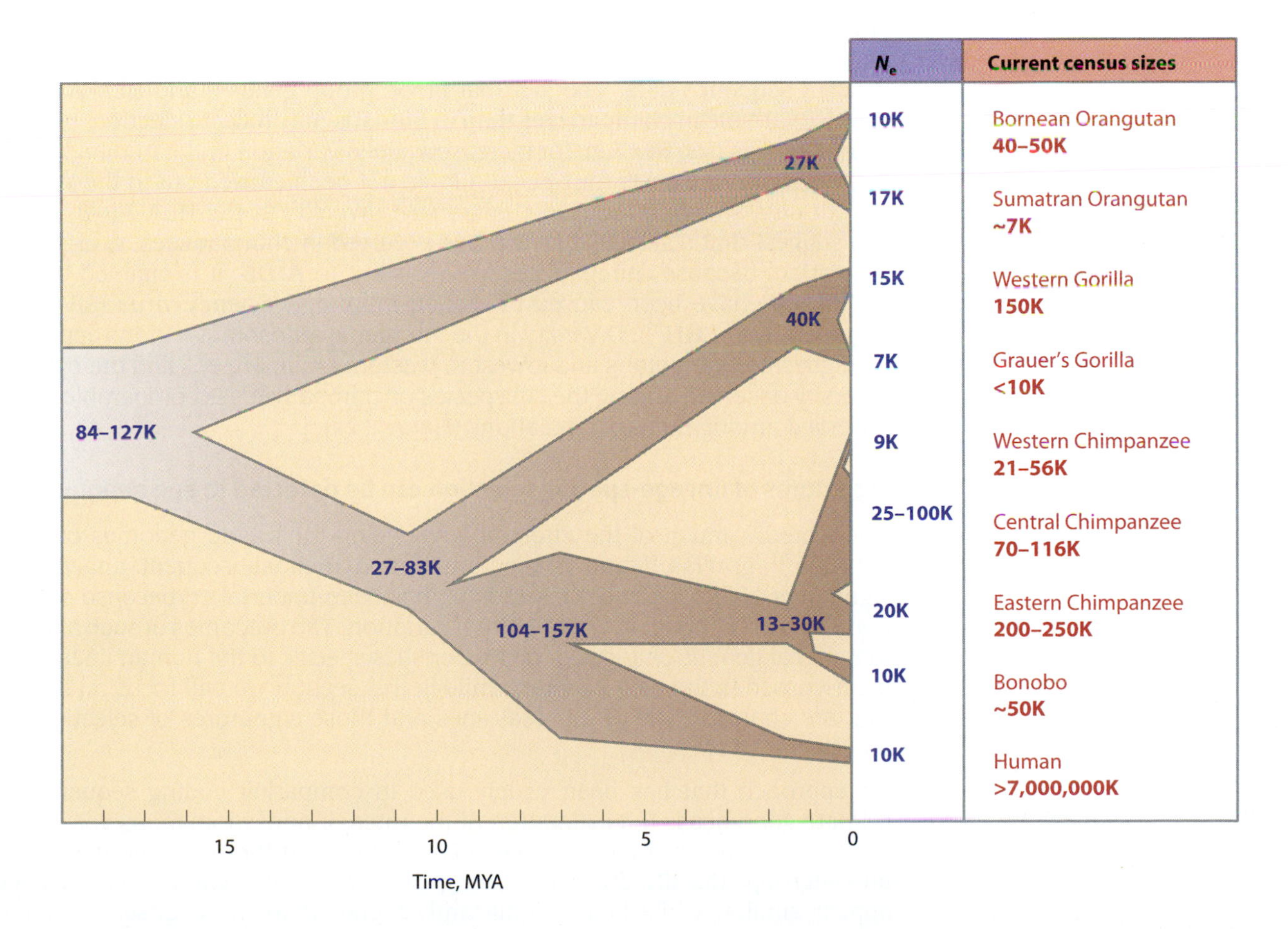

Figure 7.20: Estimated ancestral effective population size changes in the evolution of great apes.
Great ape census size estimates are as in Figure 7.18 based on IUCSN list of threatened species (**http://www.iucnredlist.org/**). Estimates of effective population sizes and ancestral population sizes (*blue*) are shown. The thickness of the branches is shown proportional to the effective population size estimates. [Adapted from Mailund T et al. (2011) *PLoS Genet.* 7, e1001319. With permission from Public Library of Science and from Marques-Bonet T et al. (2009) *Annu. Rev. Genomics Hum. Genet.* 10, 355. With permission from Annual Reviews.]

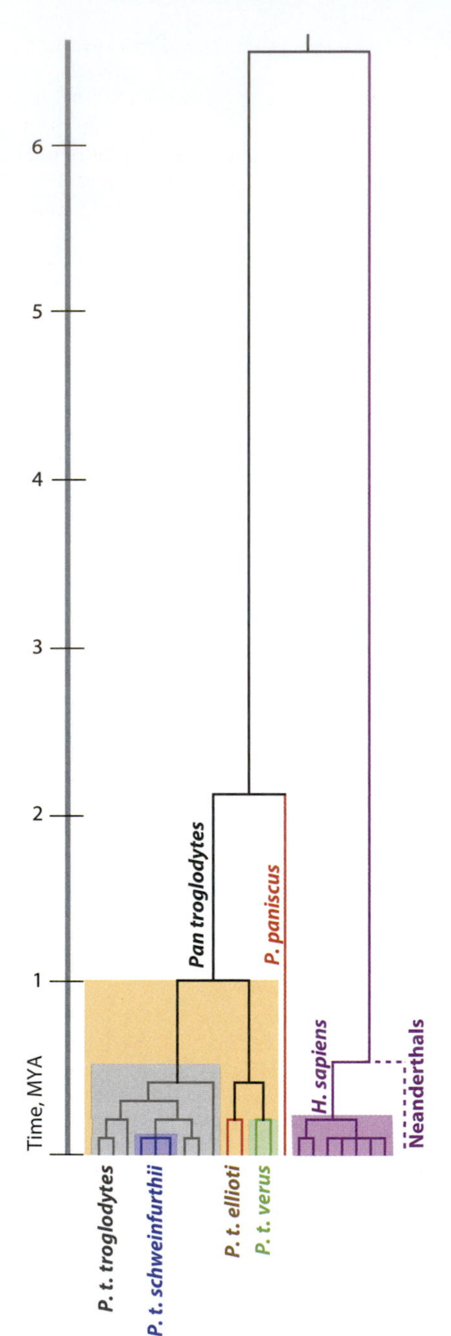

Figure 7.21: Phylogeny and genetic diversity of complete mtDNA sequences among chimpanzee subspecies.
Note the complete lineage sorting of the mtDNA clades of *P. t. ellioti* and *P. t. verus* and the paraphyletic relationship of *P. t. troglodytes* lineages with regards to those of *P. t. schweinfurthii*. [From Bjork A et al. (2011) *Mol. Biol. Evol.* 28, 615. With permission from Oxford University Press.]

in trees based on sequence variation in the autosomal genome, the subspecies differences are less clear (**Figure 7.22**). Although bonobos cluster together, chimpanzees have retained genetic variation that pre-dates their split with bonobos and as a result of extensive incomplete lineage sorting their subspecies are not well separated. This apparent discrepancy between the mtDNA and Y chromosome on the one hand and autosomal genes on the other can be explained by the fourfold greater effective population size of the autosomes, which results in a greater time-depth. If subspecies divergence occurs between these two coalescent times then the mtDNA gene tree will exhibit subspecies-specific clades, whereas the autosomal gene trees will not (**Figure 7.23**).

Even if we considered Neanderthals as part of human diversity, genetic variation observed in the mtDNA phylogeny of the chimpanzees (Figure 7.21) would be greater than among humans. The general picture for the nuclear genome is more complex. Early studies of **electrophoretic** variation among red blood cell enzyme and serum proteins indicated that chimpanzees were much less variable than humans, and this picture has been supported for a few nuclear loci by DNA sequence data. The fact that blood group genes[60] and HLA-A genes[1] are less variable in chimpanzees than in humans may reflect selection, by infectious diseases (see **Box 7.3**), for increased allelic variation in the human lineage; however, rigorous tests for selection have not been carried out in these cases. Recent data showing reduced chimpanzee diversity at the HLA-A, -B, and -C loci suggest that a selective sweep has occurred in chimpanzees, prior to subspeciation; because chimpanzees are resistant to **AIDS**, it is claimed that the sweep may have been caused by simian immunodeficiency virus (SIVcpz), a close relative of **HIV**[9]). Diversity in the intergenic autosomal regions appears to be highest in orangutans and lowest in western chimpanzees and the extent of diversity observed among the subspecies of chimpanzees is comparable to that observed among human populations (**Figure 7.24**).

Signatures of lineage-specific selection can be detected in ape genomes

Undoubtedly, much of the effort of sequencing the whole genomes of chimpanzees,[41] gorillas,[54] and orangutans,[35] and their subsequent analyses for signatures of lineage-specific selection, has been undertaken because of a primary interest in understanding human variation. The outcomes of such analyses that reveal genetic changes or characteristics specific to the human lineage will be discussed in the next chapter, while in this section we will focus on findings that are characteristic of all great apes and those signatures of selection that are specific to apes.

An approach that has been widely used in comparing coding sequences of primate genomes is the estimation of the rates of nonsynonymous versus synonymous changes (dN/dS, **Section 6.7**). When using the macaque genome as an outgroup, the dN/dS values along the human and chimpanzee lineages appear similar, while being significantly higher than those observed among macaques and non-primate mammals (**Figure 7.25**).[15, 32] These differences in dN/dS have been explained in terms of relaxed evolutionary constraints in the ancestry of great apes, due to a decreased intensity of purifying selection. On the other hand, an excess of amino-acid-changing mutations in certain classes of genes has been also interpreted as a signature of directional positive

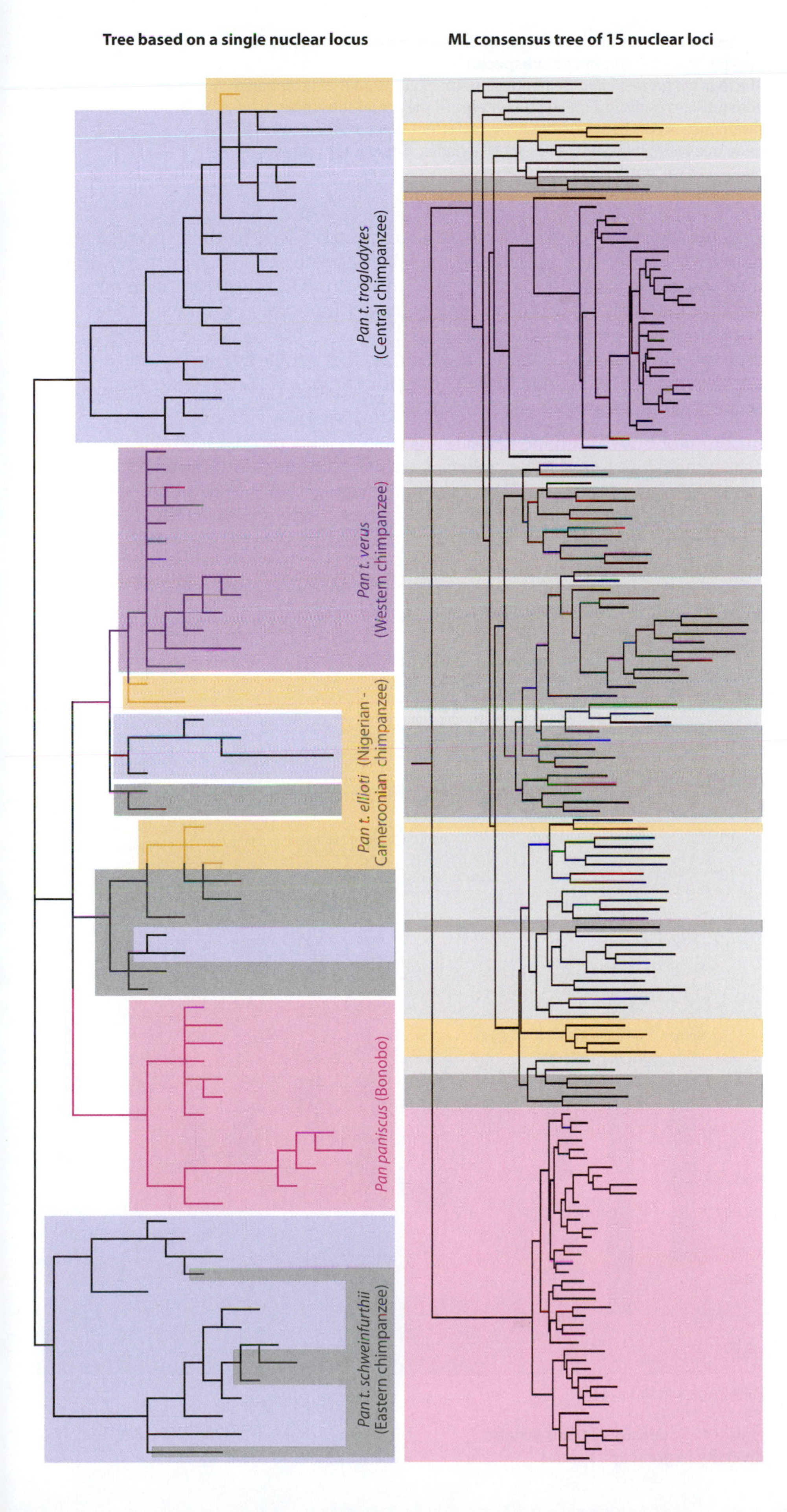

Figure 7.22: Genetic diversity of nuclear DNA among chimpanzee subspecies. A tree based on approximately 10 kb of noncoding sequence of a single nuclear DNA locus is compared against a consensus tree of 15 such loci. [From Fischer A et al. (2011) *PLoS One* 6, e21605. With permission from Public Library of Science.]

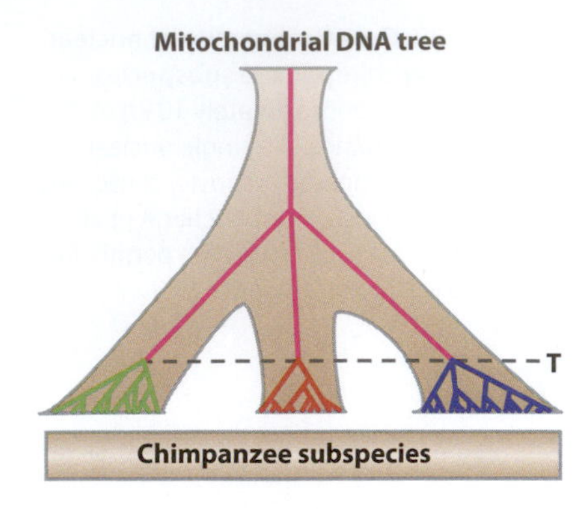

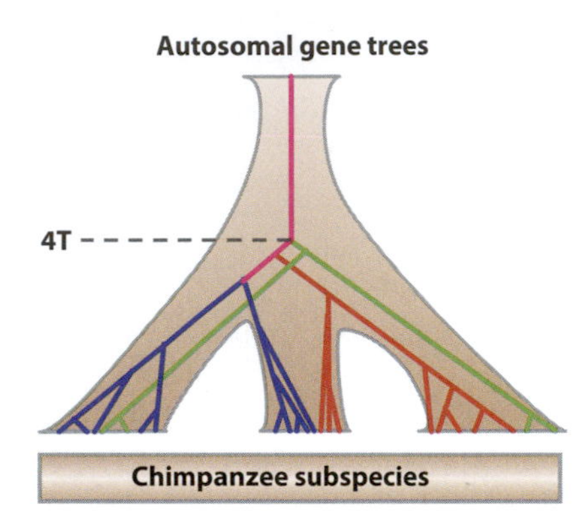

Figure 7.23: Schematic gene trees of different time-depth within a species tree of chimpanzee subspecies.
The fourfold greater effective population size of autosomal loci compared with mtDNA results in a greater time-depth. If the actual subspecies divergence occurs between these two coalescent times then the mtDNA gene tree will exhibit subspecies-specific clades, whereas the autosomal gene trees will not.

selection. More functional change than expected based on the genome average has been observed in great ape genes involved in sensory perception, immunity and defense, and reproduction.[38] In addition, analysis of the genomes of orangutans, which have lower energy usage than other primates, has revealed enrichment of positive selection signals in metabolic genes of the glycolipid and sphingolipid pathways.[35] However, in the gorilla lineage, genes showing accelerated dN/dS include those coding for ear, hair follicle, gonad, brain development, and sensory perception of sound.[54]

In comparisons of humans and other great apes certain characteristic evolutionary changes are found to be specific to some nonhuman primates. For example, the expansion of hundreds of copies of the retroviral PTERV1 element has been observed in chimpanzee and gorilla genomes, while no copies are found in the human or orangutan genomes.[67] Arguably the difference could be due to variation in the retrovirus restricting proteins, such as TRIM5α[25] or APOBEC3[49] in the cells of our ancestors. It is also possible that this viral epidemic in the ancestry of the chimpanzees and gorillas explains some of the estimates of lower genetic diversity in their immunity genes.

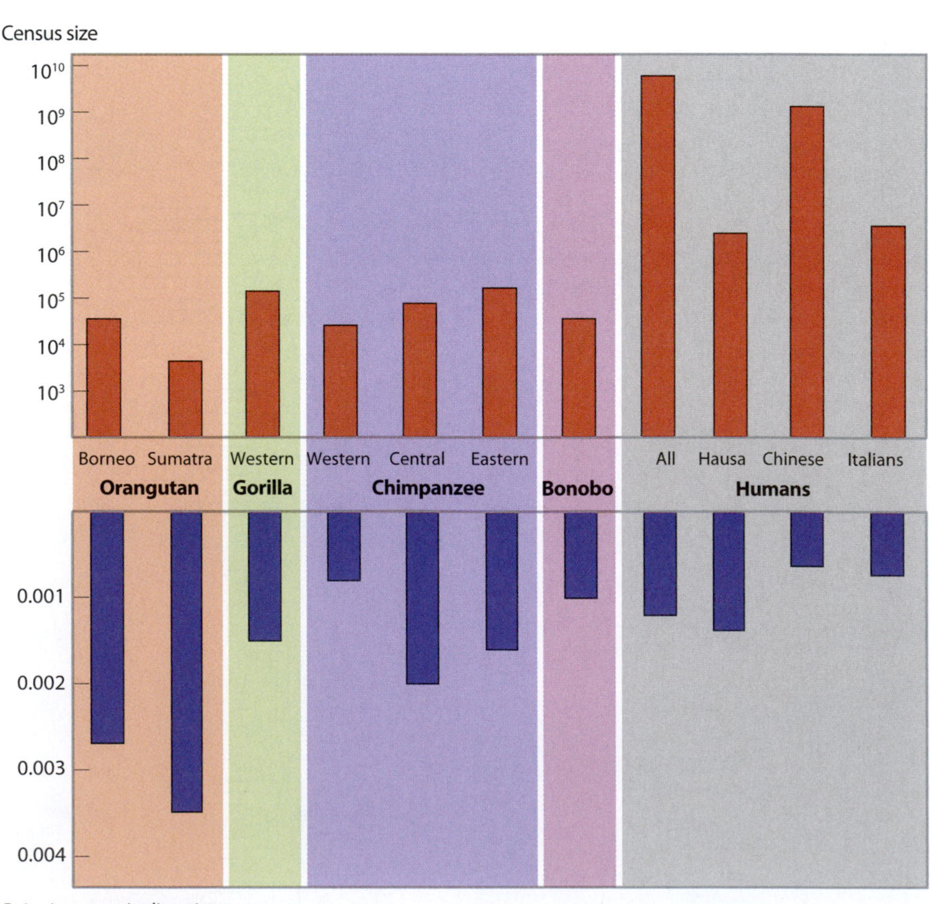

Figure 7.24: Census size and genetic diversity in great ape species.

[From Fischer A et al. (2006) *Curr. Biol.* 16, 1133. With permission from Elsevier.]

Box 7.3: Nonhuman primate origins of human infectious disease

Pathogens can jump between host species in a process known as **zoonosis**. This can be a repeated, occasional transfer from an animal reservoir to a human host and can be combated by limiting contact of humans with the donor animal. However, occasionally the pathogen can then transmit from human to human, and a true new human infectious disease arises. Nonhuman primate parasites represent a substantial reservoir of potential human pathogens, which may be able to make the jump from wild nonhuman primates to humans because of the close phylogenetic relationship. The two major human infectious diseases affecting Africa are both zoonoses from other great ape species: human immunodeficiency virus-1 (HIV-1, causing AIDS) from chimpanzees and *Plasmodium falciparum* protozoan (which causes severe malaria) from gorillas (see **Figure 1**).

The closest relatives of HIV-1, as determined by DNA sequence analysis, are simian immunodeficiency viruses (**SIV**s) that infect chimpanzees and gorillas in west central Africa, and in most cases do not cause disease in their hosts. The SIV that is most closely related to HIV-1, SIVcpz, is carried by chimpanzees and the HIV-1 strain M, which is responsible for the global human AIDS pandemic, is most likely to have been derived from a chimpanzee in south-east Cameroon early in the twentieth century.[27] HIV-2, which is responsible for some cases of AIDS in West Africa, is a distinct virus most closely related to SIVsm whose natural host is the sooty mangabey (*Cercocebus atys*), an Old World Monkey found in West Africa. The different strains of HIV-2 represent separate zoonotic acquisitions of SIVsm from these monkeys.[56]

There are many different species of *Plasmodium* that infect wild nonhuman apes, yet only two, *P. falciparum* and *P. vivax*, have successfully crossed over to infect and transmit between humans. *P. falciparum* is most closely related to *Plasmodium* that infects western gorillas in the wild, although it has also infected chimpanzees and bonobos in captivity.[34] The origin of *P. vivax*, which causes milder malaria in humans, is less clear. *P. vivax*-related parasites infect other ape species, yet the *Plasmodium* most closely related to *P. vivax* infects macaques in Southeast Asia, suggesting this may be the origin of the zoonosis.[51]

The transfer of pathogens is not one-way, because there are examples of human diseases transferring to other primates (**anthropozoonosis**), particularly in situations where there is close contact between the species. Human-derived *P. falciparum* can infect captive bonobos and chimpanzees. Chimpanzees can also suffer from respiratory disease caused by human respiratory syncytial virus (HRSV) and human metapneumovirus (HMPV), both of which are a major cause of respiratory disease in humans, and phylogenetic analysis of viral DNA sequences confirms that the viruses infecting chimpanzees originated from a human virus within the previous 10 years.[31] This, and other possible cases of anthropozoonosis, has occurred in habituated nonhuman ape populations, where free-living chimpanzees or gorillas have close contact with humans because of tourism or research.

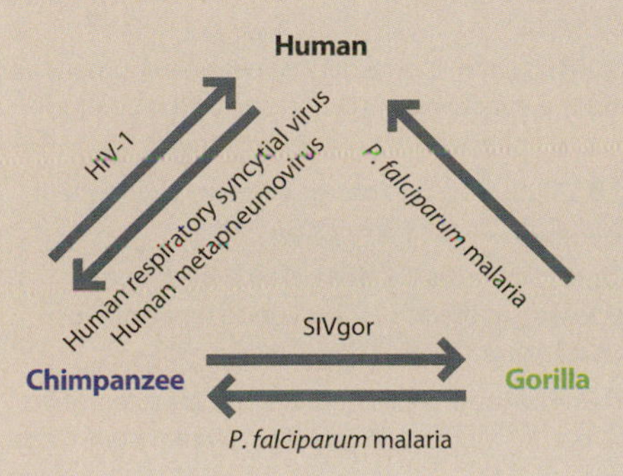

Figure 1: Some examples of zoonoses between great apes.

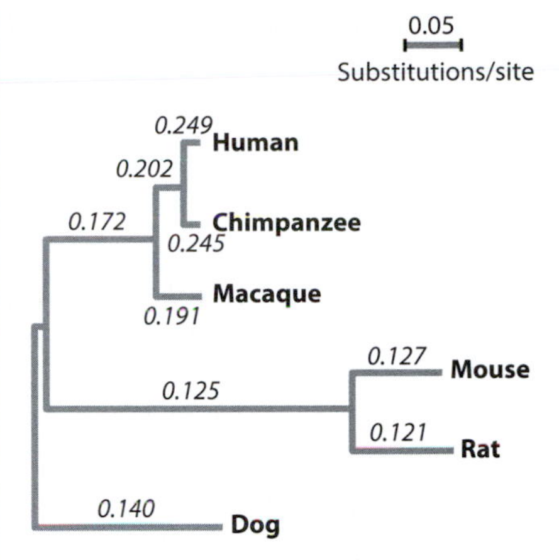

Figure 7.25: Branch-specific dN/dS rates in six mammalian species.
dN/dS ratios are shown in italic on respective branches. Branch lengths are proportional to divergence. [From Kosiol C et al. (2008) *PLoS Genet*. 4, e1000144. With permission from Public Library of Science.]

SUMMARY

- Morphological evidence from fossils and extant species places humans in the taxonomic order of primates and reveals humans to be most closely related to the chimpanzee, bonobo, gorilla, and orangutan.
- The karyotypes of humans and great apes are similar but differ in a number of respects, the most notable being the reduction in chromosomal number from 48 to 46 as a result of a chromosomal fusion on the hominin lineage.
- Molecular comparisons among great ape species reveal that in two-thirds of their DNA sequence humans and chimpanzees are more closely related to one another than either is to gorillas. This is compelling evidence to show that the species split between humans and chimpanzees occurred after the split with gorilla.
- The time since humans and great apes last shared their common ancestors can be estimated from molecular comparisons between the species. These calculations can be reconciled with fossil-derived estimates of the timing of the speciation events, although precision is hampered by difficulties in calibrating the molecular clock and also because of probable changes in generation lengths over time.
- Genetic diversity in great ape species appears to be significantly greater than in humans despite their now having much smaller, and decreasing, population sizes.

QUESTIONS

Question 7–1: Place in order of increasing divergence time from the human lineage, the following primate species: slender loris, Sumatran orangutan, lion tamarin, red-shanked douc, Madame Berthe's mouse lemur, siamang.

Question 7–2: The hominoids form a clade, but the monkeys, taken as a whole, do not. Explain why this is.

Question 7–3: What chromosomal differences exist between humans and orangutans, and what might these have contributed to speciation?

Question 7–4: Which of the following features are specific only to hominids?: lack of tail, axillary glands, knuckle walking, trichromatic vision, *Alu*-like SINE elements.

Question 7–5: How does genetic diversity vary between different great ape (including human) species and subspecies? What factors might account for these differences?

Question 7–6: How does human demographic history differ from that of chimpanzees, and how has this influenced genetic diversity in the two species?

Question 7–7: An individual of a chimpanzee subspecies can be accurately classified using mtDNA, but not using any single autosomal locus—why is this?

Question 7–8: Use information available from www.wcs.org to draw up a list of the main challenges in great ape and small primate conservation.

REFERENCES

The references highlighted in purple are considered to be important (for this chapter) by the authors.

1. **Adams EJ, Cooper S, Thomson G & Parham P** (2000) Common chimpanzees have greater diversity than humans at two of the three highly polymorphic MHC class I genes. *Immunogenetics* **51**, 410–424.
2. **Andrews P & Cronin JE** (1982) The relationships of *Sivapithecus* and *Ramapithecus* and the evolution of the orang-utan. *Nature* **297**, 541–546.
3. **Andrle M, Fiedler W, Rett A et al.** (1979) A case of trisomy 22 in *Pongo pygmaeus*. *Cytogenet. Cell Genet.* **24**, 1–6.
4. **Archidiacono N, Storlazzi CT, Spalluto C et al.** (1998) Evolution of chromosome Y in primates. *Chromosoma* **107**, 241–246.
5. **Bjarnason A, Chamberlain AT & Lockwood CA** (2011) A methodological investigation of hominoid craniodental morphology and phylogenetics. *J. Hum. Evol.* **60**, 47–57.
6. **Britten RJ** (2002) Divergence between samples of chimpanzee and human DNA sequences is 5%, counting indels. *Proc. Natl Acad. Sci. USA* **99**, 13633–13635.
7. **Chen FC & Li WH** (2001) Genomic divergences between humans and other hominoids and the effective population size of the common ancestor of humans and chimpanzees. *Am. J. Hum. Genet.* **68**, 444–456.
8. **Collard M & Wood B** (2000) How reliable are human phylogenetic hypotheses? *Proc. Natl Acad. Sci. USA* **97**, 5003–5006.
9. **de Groot NG, Otting N, Doxiadis GG et al.** (2002) Evidence for an ancient selective sweep in the MHC class I gene repertoire of chimpanzees. *Proc. Natl Acad. Sci. USA* **99**, 11748–11753.
10. **Dutrillaux B** (1979) Chromosomal evolution in primates: tentative phylogeny from *Microcebus murinus* (Prosimian) to man. *Hum. Genet.* **48**, 251–314.

11. **Fischer A, Prufer K, Good JM et al.** (2011) Bonobos fall within the genomic variation of chimpanzees. *PLoS One* **6**, e21605.

12. **Fortna A, Kim Y, MacLaren E et al.** (2004) Lineage-specific gene duplication and loss in human and great ape evolution. *PLoS Biol.* **2**, e207.

13. **Freije D, Helms C, Watson MS & Donis-Keller H** (1992) Identification of a second pseudoautosomal region near the Xq and Yq telomeres. *Science* **258**, 1784–1787.

14. **Gagneux P, Wills C, Gerloff U et al.** (1999) Mitochondrial sequences show diverse evolutionary histories of African hominoids. *Proc. Natl Acad. Sci. USA* **96**, 5077–5082.

15. **Gibbs RA, Rogers J, Katze MG et al.** (2007) Evolutionary and biomedical insights from the rhesus macaque genome. *Science* **316**, 222–234.

16. **Gibbs S, Collard M & Wood B** (2000) Soft-tissue characters in higher primate phylogenetics. *Proc. Natl Acad. Sci. USA* **97**, 11130–11132.

17. **Gilad Y, Przeworski M & Lancet D** (2004) Loss of olfactory receptor genes coincides with the acquisition of full trichromatic vision in primates. *PLoS Biol.* **2**, e5.

18. **Gonder MK, Oates JF, Disotell TR et al.** (1997) A new west African chimpanzee subspecies? *Nature* **388**, 337.

19. **Hobolth A, Dutheil JY, Hawks J et al.** (2011) Incomplete lineage sorting patterns among human, chimpanzee, and orangutan suggest recent orangutan speciation and widespread selection. *Genome Res.* **21**, 349–356.

20. **Ijdo JW, Baldini A, Ward DC et al.** (1991) Origin of human chromosome 2: an ancestral telomere-telomere fusion. *Proc. Natl Acad. Sci. USA* **88**, 9051–9055.

21. **Jacobs GH** (2009) Evolution of colour vision in mammals. *Philos. Trans. R. Soc. Lond. B Biol. Sci.* **364**, 2957–2967.

22. **Jameson NM, Hou ZC, Sterner KN et al.** (2011) Genomic data reject the hypothesis of a prosimian primate clade. *J. Hum. Evol.* **61**, 295–305.

23. **Jauch A, Wienberg J, Stanyon R et al.** (1992) Reconstruction of genomic rearrangements in great apes and gibbons by chromosome painting. *Proc. Natl Acad. Sci. USA* **89**, 8611–8615.

24. **Kaessmann H, Wiebe V, Weiss G & Pääbo S** (2001) Great ape DNA sequences reveal a reduced diversity and an expansion in humans. *Nat. Genet.* **27**, 155–156.

25. **Kaiser SM, Malik HS & Emerman M** (2007) Restriction of an extinct retrovirus by the human TRIM5α antiviral protein. *Science* **316**, 1756–1758.

26. **Kasai F, Takahashi E, Koyama K et al.** (2000) Comparative FISH mapping of the ancestral fusion point of human chromosome 2. *Chromosome Res.* **8**, 727–735.

27. **Keele BF, Van Heuverswyn F, Li Y et al.** (2006) Chimpanzee reservoirs of pandemic and nonpandemic HIV-1. *Science* **313**, 523–526.

28. **Kehrer-Sawatzki H & Cooper DN** (2007) Understanding the recent evolution of the human genome: insights from human–chimpanzee genome comparisons. *Hum. Mutat.* **28**, 99–130.

29. **Kirsch S, Munch C, Jiang Z et al.** (2008) Evolutionary dynamics of segmental duplications from human Y-chromosomal euchromatin/heterochromatin transition regions. *Genome Res.* **18**, 1030–1042.

30. **Kivell TL & Schmitt D** (2009) Independent evolution of knuckle-walking in African apes shows that humans did not evolve from a knuckle-walking ancestor. *Proc. Natl Acad. Sci. USA* **106**, 14241–14246.

31. **Kondgen S, Kuhl H, N'Goran PK et al.** (2008) Pandemic human viruses cause decline of endangered great apes. *Curr. Biol.* **18**, 260–264.

32. **Kosiol C, Vinar T, da Fonseca RR et al.** (2008) Patterns of positive selection in six Mammalian genomes. *PLoS Genet.* **4**, e1000144.

33. **Kunimatsu Y, Nakatsukasa M, Sawada Y et al.** (2007) A new Late Miocene great ape from Kenya and its implications for the origins of African great apes and humans. *Proc. Natl Acad. Sci. USA* **104**, 19220–19225.

34. **Liu W, Li Y, Learn GH et al.** (2010) Origin of the human malaria parasite *Plasmodium falciparum* in gorillas. *Nature* **467**, 420–425.

35. **Locke DP, Hillier LW, Warren WC et al.** (2011) Comparative and demographic analysis of orang-utan genomes. *Nature* **469**, 529–533.

36. **Lyon MF** (1961) Gene action in the X-chromosome of the mouse (*Mus musculus* L.). *Nature* **190**, 372–373.

37. **Mallet J** (2001) Species, concepts of. In Encyclopedia of Biodiversity, 3rd ed. (SA Levin et al., eds), vol 5, pp. 427–440. Academic Press.

38. **Marques-Bonet T, Ryder OA & Eichler EE** (2009) Sequencing primate genomes: what have we learned? *Annu. Rev. Genomics Hum. Genet.* **10**, 355–386.

39. **Mayr E** (1970) Populations, Species, and Evolution. Harvard University Press.

40. **McClure HM, Belden KH, Pieper WA & Jacobson CB** (1969) Autosomal trisomy in a chimpanzee: resemblance to Down's syndrome. *Science* **165**, 1010–1012.

41. **Mikkelsen TS & The Chimpanzee Sequencing and Analysis Consortium** (2005) Initial sequence of the chimpanzee genome and comparison with the human genome. *Nature* **437**, 69–87.

42. **Morin PA, Moore JJ, Chakraborty R et al.** (1994) Kin selection, social structure, gene flow, and the evolution of chimpanzees. *Science* **265**, 1193–1201.

43. **Müller S & Wienberg J** (2001) "Bar-coding" primate chromosomes: molecular cytogenetic screening for the ancestral hominoid karyotype. *Hum. Genet.* **109**, 85–94.

44. **Nishihara H, Maruyama S & Okada N** (2009) Retroposon analysis and recent geological data suggest near-simultaneous divergence of the three superorders of mammals. *Proc. Natl Acad. Sci. USA* **106**, 5235–5240.

45. **Nishihara H, Terai Y & Okada N** (2002) Characterization of novel Alu- and tRNA-related SINEs from the tree shrew and evolutionary implications of their origins. *Mol. Biol. Evol.* **19**, 1964–1972.

46. **Okamoto I, Patrat C, Thepot D et al.** (2011) Eutherian mammals use diverse strategies to initiate X-chromosome inactivation during development. *Nature* **472**, 370–374.

47. **Page DC, Harper ME, Love J & Botstein D** (1984) Occurrence of a transposition from the X-chromosome long arm to the Y-chromosome short arm during human evolution. *Nature* **311**, 119–123.

48. **Perelman P, Johnson WE, Roos C et al.** (2011) A molecular phylogeny of living primates. *PLoS Genet.* **7**, e1001342.

49. **Perez-Caballero D, Soll SJ & Bieniasz PD** (2008) Evidence for restriction of ancient primate gammaretroviruses by APOBEC3 but not TRIM5α proteins. *PLoS Pathog.* **4**, e1000181.

50. **Pickford M & Senut B** (2005) Hominoid teeth with chimpanzee- and gorilla-like features from the Miocene of Kenya: implications for the chronology of ape–human divergence and biogeography of Miocene hominoids. *Anthropol. Sci.* **113**, 95–102.

51. **Rayner JC, Liu W, Peeters M et al.** (2011) A plethora of *Plasmodium* species in wild apes: a source of human infection? *Trends Parasitol.* **27**, 222–229.

52. **Saito A, Mikami A, Kawamura S et al.** (2005) Advantage of dichromats over trichromats in discrimination of color-camouflaged stimuli in nonhuman primates. *Am. J. Primatol.* **67**, 425–436.

53. **Sarich VM & Wilson AC** (1967) Immunological time scale for hominid evolution. *Science* **158**, 1200–1203.

54. **Scally A, Dutheil JY, Hillier LW et al.** (2012) Insights into hominid evolution from the gorilla genome sequence. *Nature* **483**, 169–175.

55. **Schipper J, Chanson JS, Chiozza F et al.** (2008) The status of the world's land and marine mammals: diversity, threat, and knowledge. *Science* **322**, 225–230.

56. **Sharp PM & Hahn BH** (2010) The evolution of HIV-1 and the origin of AIDS. *Philos. Trans. R. Soc. Lond. B Biol. Sci.* **365**, 2487–2494.

57. **Sibley CG & Ahlquist JE** (1984) The phylogeny of the hominoid primates, as indicated by DNA–DNA hybridization. *J. Mol. Evol.* **20**, 2–15.

58. **Stanyon R, Rocchi M, Capozzi O et al.** (2008) Primate chromosome evolution: ancestral karyotypes, marker order and neocentromeres. *Chromosome Res.* **16**, 17–39.

59. **Stone AC, Griffiths RC, Zegura SL & Hammer MF** (2002) High levels of Y-chromosome nucleotide diversity in the genus Pan. *Proc. Natl Acad. Sci. USA* **99**, 43–48.

60. **Sumiyama K, Kitano T, Noda R et al.** (2000) Gene diversity of chimpanzee ABO blood group genes elucidated from exon 7 sequences. *Gene* **259**, 75–79.

61. **Suwa G, Kono RT, Katoh S et al.** (2007) A new species of great ape from the late Miocene epoch in Ethiopia. *Nature* **448**, 921–924.

62. **Turleau C, De Grouchy J & Klein M** (1972) Chromosomal phylogeny of man and the anthropomorphic primates. (*Pan troglodytes*, *Gorilla gorilla*, *Pongo pygmaeus*). Attempt at reconstitution of the karyotype of the common ancestor. *Ann. Genet.* **15**, 225–240.

63. **Verrelli BC & Tishkoff SA** (2004) Signatures of selection and gene conversion associated with human color vision variation. *Am. J. Hum. Genet.* **75**, 363–375.

64. **Walsh PD, Abernethy KA, Bermejo M et al.** (2003) Catastrophic ape decline in western equatorial Africa. *Nature* **422**, 611–614.

65. **Wimmer R, Kirsch S, Rappold GA & Schempp W** (2002) Direct evidence for the *Homo–Pan* clade. *Chromosome Res.* **10**, 55–61.

66. **Wood B & Harrison T** (2011) The evolutionary context of the first hominins. *Nature* **470**, 347–352.

67. **Yohn CT, Jiang Z, McGrath SD et al.** (2005) Lineage-specific expansions of retroviral insertions within the genomes of African great apes but not humans and orangutans. *PLoS Biol.* **3**, e110.

68. **Yunis JJ & Prakash O** (1982) The origin of man: a chromosomal pictorial legacy. *Science* **215**, 1525–1530.

WHAT GENETIC CHANGES HAVE MADE US HUMAN?

CHAPTER EIGHT

As we learned in the previous chapter, the vast majority of human genes are almost identical to those of chimpanzee and gorilla, yet there are differences in a number of morphological, physiological, biochemical, and behavioral traits. In this chapter, we address one of the central questions of human evolutionary genetics: what genetic changes have made us human, with all the characteristics that distinguish us from both other living apes and extinct **hominins**? In the broadest terms, we have two sources of relevant information. The first consists of the phenotypic characteristics, including morphological and behavioral ones, that are specific to humans and our hominin ancestors, examples of which are bipedalism and language; there are also human-specific changes among characteristics that evolve rapidly in most species, such as defense against pathogens, but we will be less interested in these here. The second consists of the genome sequences of the great apes and of two extinct hominins, which in principle allow us to identify almost all the genetic differences, from large-scale changes creating, duplicating, or deleting whole genes or gene sets, to single nucleotide changes that might influence the location, level, or timing of expression of genes. We assume that some of these genetic differences underlie the phenotypic differences, and so we aim to connect them. But we will see that this is a formidable task, and progress has been limited.

Human characteristics have accumulated over the 6–7 MY of evolution that separate us from our common ancestor with chimpanzees and bonobos. The fossil and archaeological records, considered in more detail in Chapter 9, help us to identify when some of these characteristics first appeared. We will see that some, such as **bipedalism**, developed millions of years ago while others, such as abstract art, are much more recent, appearing <100 KYA. Similarly, when we examine the genetic data, we can assign changes that lie on the human lineage to before or after the split between humans, **Neanderthals**, and **Denisovans** ~0.5 MYA, and sometimes make more precise estimates. These time-scales can help us to link phenotypic and genetic changes chronologically, but generally functional evidence is required and this is seldom available.

Throughout this chapter, we will concentrate on the phenotypic and genetic characteristics of humans as a species; that is, those that are shared by all humans. Later chapters will consider those that contribute to diversity among present-day humans and are found in some but not all humans. In many cases, our definition of a human-specific phenotype is based more on our lack of knowledge of the detailed phenotype of the other great apes rather than on a rigorous comparison across species. The fossil record of extinct hominins can help us, because we can use this to distinguish *Homo sapiens*-specific traits from those characteristic to all hominins (**Figure 8.1a**). However, by using the evidence from fossils we are largely restricted to comparison of phenotypes manifested in the bones.

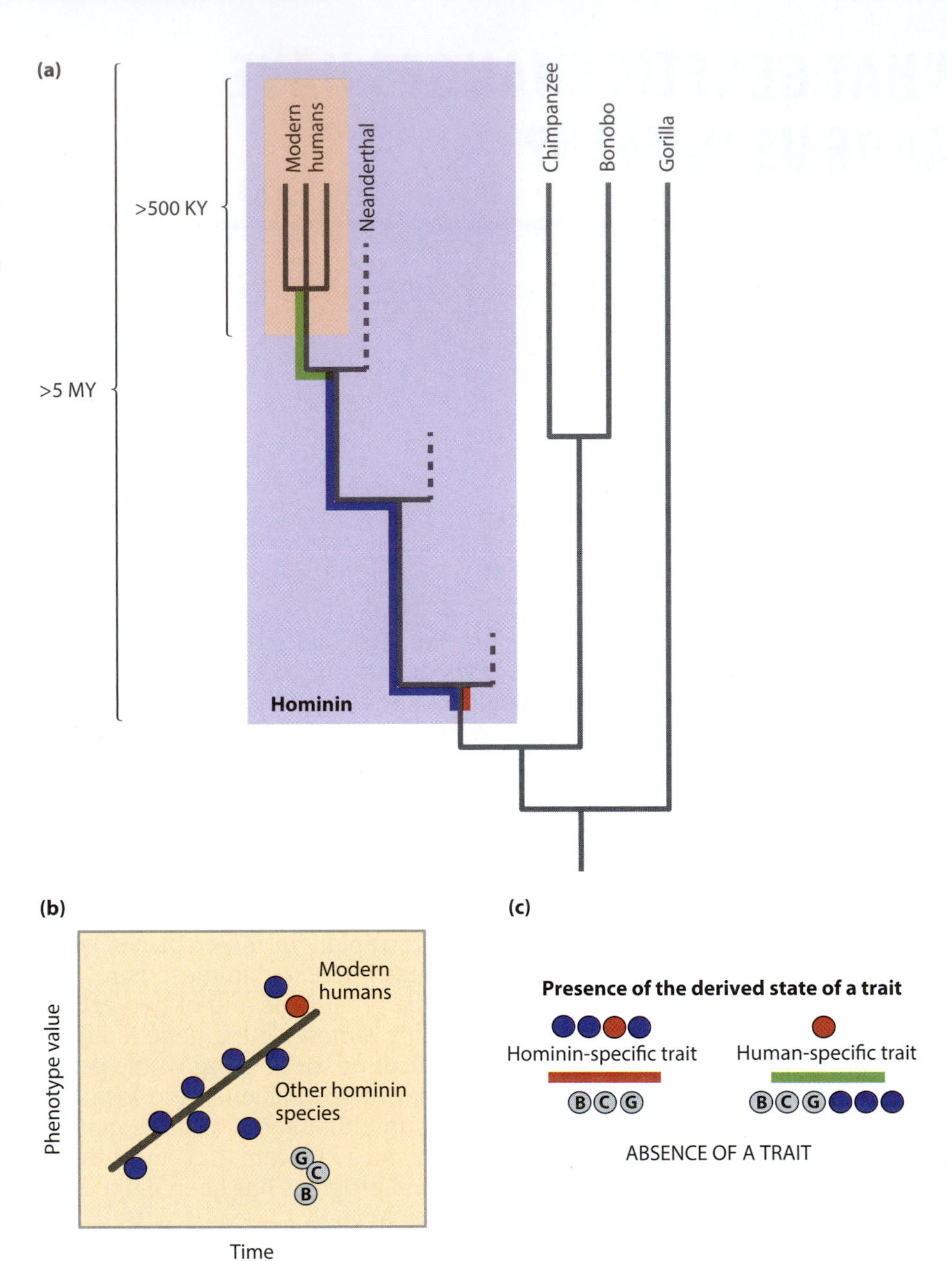

Figure 8.1: Evolution of human- and hominin-specific phenotypic traits. (a) Phylogeny of hominins and African apes. Broken lines indicate lineages to extinct species. *Red, blue*, and *green* lines indicate changes occurring on evolutionary lines specific either to modern humans or hominins in general, as further illustrated in parts b and c. (b) Some of the phenotypic variation between humans and great apes can be seen as series of gradual changes in the course of hominin evolution. (c) Other traits show discrete presence or absence. C, chimpanzee; B, bonobo; G, gorilla.

8.1 MORPHOLOGICAL AND BEHAVIORAL CHANGES EN ROUTE TO *HOMO SAPIENS*

What aspects of our morphology, development, and behavior separate us from our closest primate relatives? In answering this question we can compare ourselves with other living nonhuman primates, or attempt to draw upon the increasingly detailed fossil record of the hominins, to discern not only the strictly human versus general hominin-specific changes (Figure 8.1c) but also the tempo and pattern of change (Figure 8.1b). However, there is an element of circularity in this, as what determines a fossil's classification as a hominin (as opposed to lying on a lineage leading to one of the other great apes) is determined by the prevailing theory of what features define hominin status. Another potential problem is that the hominin phylogeny appears to be very bushy, with all but one lineage going extinct (Figure 8.1a). The ancestral line of modern humans through this tangled heritage remains open to interpretation.

More than 200 phenotypic traits have been listed where humans differ from other great apes.[71] Many such traits represent differences in scale, for example larger brain size, or reduced muscular strength and skeletal robustness, while others can be viewed as discrete innovations, such as cooking (see **Opinion Box 6**) or ice

The composition of human diets varies widely around the world but in every culture cooked food is the major component of the evening meal (see **Figure 1**). Cooking increases the net energy extracted from the diet, whether by gelatinizing starch, denaturing proteins, or reducing the structural integrity of food such that the costs of digestion are lower. Cooking offers other advantages too, including detoxification, reduced pathogen and parasite load, and flavor enhancement. Not surprisingly, therefore, most animals share humans' preference for cooked food.

Only among humans, however, does the consumption of cooked food appear to be obligatory. No cases are known of humans surviving on raw wild foods for long periods. Urban "raw-foodists" suffer energy shortage despite their food items being high-quality agricultural products, available year-round and often physically processed. This energy shortage leads many female urban raw-foodists of reproductive age to stop ovulating, impairing fertility. Such evidence indicates that a cooked diet may today be a necessary component of normal biological function.

Presumably the explanation for our unique dependence on cooked items is that, unlike other animals, fire-using human ancestors had such predictable access to cooked food that they could take evolutionary advantage of it. Candidate responses include increased activity levels, reductions of the digestive system, and larger brains. Compared with chimpanzees, humans in traditional societies expend substantially more energy on activity each day, even accounting for our larger body size. Humans have total gut volume that is small in relation to body mass, sufficient for cooked but not for raw wild food. Humans also have particularly small molars, poor for tough raw food but adequate for foods softened by heat.

The proposal that humans cannot survive in the wild without eating cooked food predicts critical differences in digestive physiology from great apes. Differences are expected in metabolic processes concerned with rates of digestion and absorption, in the rate of uptake of glucose (potentially higher in humans), in the regulation of the **microbiome** (fermentation less important than in nonhuman apes), in detoxification ability (generally less necessary with cooked food), and in specific detoxification pathways related to chemical by-products of the cooking process.

The last of these suggestions derives from the fact that the concentrations of particular compounds are higher in cooked food than in raw foods. Heterocyclic amines result from a high-heat reaction between the amino acids and creatinine present in meat; the most common of these, PhIP, has been associated with colorectal cancer risk in a dose-dependent manner in rats, but the association is less clear in humans. Similarly, Maillard compounds are a diverse group of molecules resulting from complexes between reducing sugars and amino acids. Many produce attractive flavors such as those associated with the browning of bread, but Maillard reactions can also lead to toxic compounds such as acrylamide. Acrylamide is toxic for rats and mice, but appears less so in humans. Possibly an evolutionary history of exposure to heterocyclic amines and Maillard compounds has favored metabolic pathways adapted to minimizing their potential damage in humans.

The time when cooked food became obligatory for human ancestors is contentious because biological and archaeological evidence do not agree. The fossil record indicates that in human evolutionary history major increases in energy budget and the largest reductions in gut size and molar size happened concurrently with the origin of *Homo erectus* around 1.9 MYA. Because there is no other time showing anatomical changes indicative of a soft, high-quality diet, the biological evidence suggests that human ancestors have been adapted to cooked food for almost 2 MY. By contrast archaeological records of the control of fire are limited to more recent times, with robust evidence for the control of fire being rare beyond 400 KYA and absent beyond 1 MYA. Signals of cooking are likewise only occasional beyond 250 KYA.

Whether humans became adapted to cooked food around 2 MYA or much later, ample time has elapsed for extensive genetic adaptations. Future genetic investigation into the biological processes and pathways affected by consumption of a cooked diet may ultimately enable us to detect a molecular signal of such adaptation.

Richard Wrangham
Department of Human Evolutionary Biology, Harvard University, Cambridge, USA

Rachel Carmody
Department of Human Evolutionary Biology, Harvard University, Cambridge, USA

Figure 1: Noodles dating to 4000 YA in China.
Shown here on top of an in-filled sediment cone; revealed after the inverted earthenware bowl containing them was removed. [From Lu H et al. (2005) *Nature* 437, 967. With permission from Macmillan Publishers Ltd.]

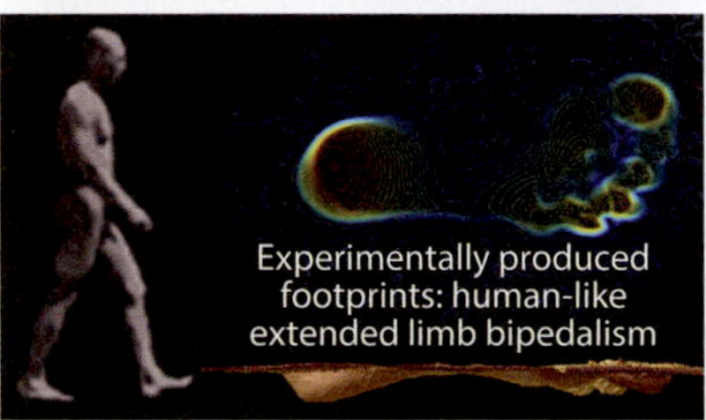

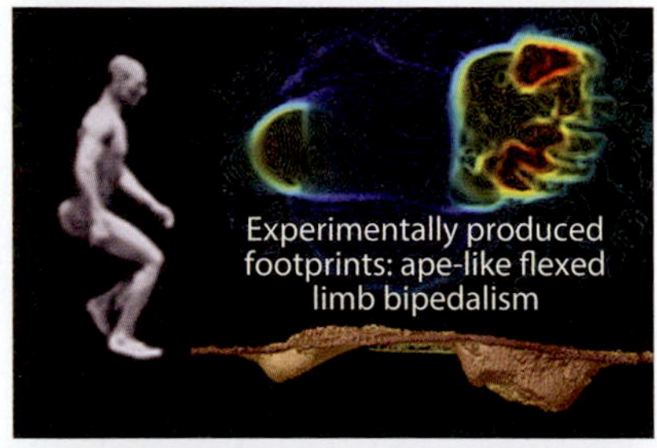

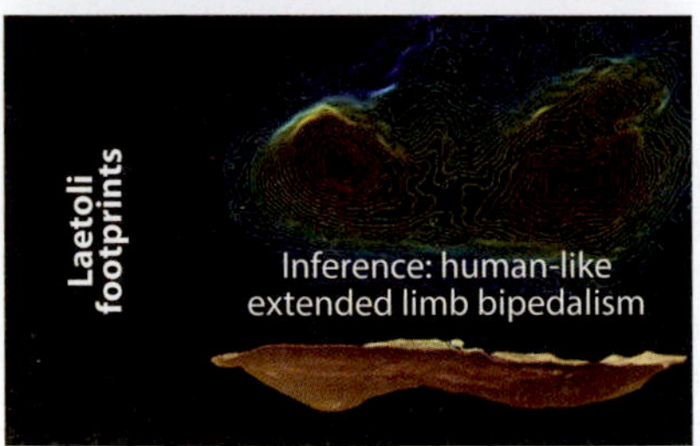

Figure 8.2: Inference of locomotion type of the Laetoli hominins from their footprints.
The footprints were uncovered in 1978 in Tanzania by Mary Leakey and co-workers, and were dated to 3.6 MYA. The locomotion type of the hominins has been inferred from the experimental biomechanical evidence to be more human- than ape-like. [From Raichlen DA et al. (2010) *PLoS One* 5, e9769. With permission from the Public Library of Science.]

hockey. Some uniquely human or hominin states can be defined by a loss or a lack of a specific character, such as the hairy skin common to other great apes. The uniqueness can be seen in specific aspects of our cultural evolution and our mind, for example meal times, somnambulism, drug use, and construction of different types of shelter. Some of these features have heritability due to cultural transmission and are not genetically encoded—an example is making fire—and such abilities can be, and have been, lost in certain human populations in the past. Diseases restricted to humans can be regarded as human-specific traits, particularly those where their diagnosis is based on the nature of the human mind, such as schizophrenia. As the time-scale of human evolution since the divergence from the other apes extends over millions of years it is not always apparent which of these traits are shared by hominins in general and which ones define us uniquely as human.

Some human traits evolved early in hominin history

One of the key features defining hominins in the fossil record is their method of locomotion. It is only on rare occasions that the evidence for locomotion is directly fossilized—as in the famous Laetoli footprints (**Figure 8.2**)—but it can be inferred indirectly from certain skeletal features such as the position of the connection point of the spinal cord to the skull, the shape of pelvis, or other characteristics of the bones of the lower limbs. Some evidence for a **bipedal** mode of locomotion appears amongst the earliest hominin fossils (see further discussion in Chapter 9), but it is unclear whether or not this early (perhaps **arboreal**) type of bipedalism was directly ancestral to the terrestrial bipedalism of later hominins, including those that left their footprints at Laetoli. The rapid expansion in **brain size** begins much later, at approximately the same time that recognizable stone tools appear in the archaeological record, although it is unclear whether or not there is any causal link. All great apes show a wide range of variation in their reproductive biology and behavior, and also socio-sexual behaviors not directly related to reproduction, many of which are species specific (**Table 8.1**). Early hominins, such as australopithecines, show extensive sexual dimorphism in body mass but not in canine morphology[55] whereas a sharp reduction in sexual dimorphism of body size, which may reflect a change in environment, mating practices, and/or diet, appears later in the hominin fossil record, around 1.8 MYA.

It is not only the differences in the adult morphology and behavior of humans and great apes that have changed over the past few millions of years, but also their pre- and postnatal development, or, more generally, **life-histories**. A small number of heterochronic differences—that is, the altered rate, start, and/or ending points of developmental processes—could potentially explain a wide range of phenotypic differences among the species. Humans have been viewed as **neotenic apes** because several aspects of our morphology and behavior can be seen as resulting from an incomplete ape developmental program. The cranial shape of adult humans, for example, is more similar to the infant than to the adult chimpanzee form (**Figure 8.3**). Human newborns are twice as large as those of chimpanzees and gorillas, relative to their respective maternal body weight. This is mainly due to fat deposits and, in fact, human babies are the fattest of all mammalian species, fatter even than infant seals.

Despite being relatively larger than other great ape neonates, human newborns are much more dependent on adult care and have a prolonged period of slow growth called **childhood** (**Figure 8.4**). These human deviations from the ape developmental program could be seen as evolutionary compromises for bigger and energetically costly brains. On the other hand, fat reserves accumulated in the fetus may also have helped human ancestors to cope with nutritional and immune stresses of high infant morbidity. Weaning in humans is also considerably earlier than in other great apes, with a consequent reduction of time spent by the mother on costly lactation. Together, these can be seen as part of an evolutionary strategy of having more babies at shorter time intervals. Rapid

TABLE 8.1:
SEXUAL DIMORPHISM IN HUMANS AND OTHER GREAT APES

Species	Mating system	Body size (male/ female) dimorphism	Canine dimorphism	Number of copulations per birth	Species-specific dimorphisms
Human	various	moderate (1.1)	slight	100–600	males with facial and body hair; females with enlarged breasts; different distribution of body fat
Chimpanzee	polygynous, multi-male	moderate (1.3)	moderate	100–700	females with exaggerated perianal swellings
Bonobo	polygynous, multi-male	moderate (1.2)	moderate	100–1300	exaggerated version of chimpanzees
Gorilla	polygynous, one male	high (1.5)	strong	25	older males develop silver back hair
Orangutan	polygynous, solitary	high (2)	strong	5	older males with throat pouches, cheek flanges
Gibbon	monogamous	slight (1.02)	slight	3	males have enlarged hyoid bones

Adapted from Plavcan JM (2001) *Am. J. Phys. Anthropol. Supp.* 33, 25. With permission from John Wiley & Sons, Inc; and Wrangham RW (1993) *Human Nature* 4, 47. With permission from Springer Science and Business Media.

reproduction and increased phenotype plasticity could have been key components of the "colonizing ape" adaptive strategies that allowed hominin dispersals out of Africa.[74] Indirect evidence from fossils suggests that early weaning is likely to pre-date human–Neanderthal divergence,[32] while the earliest evidence

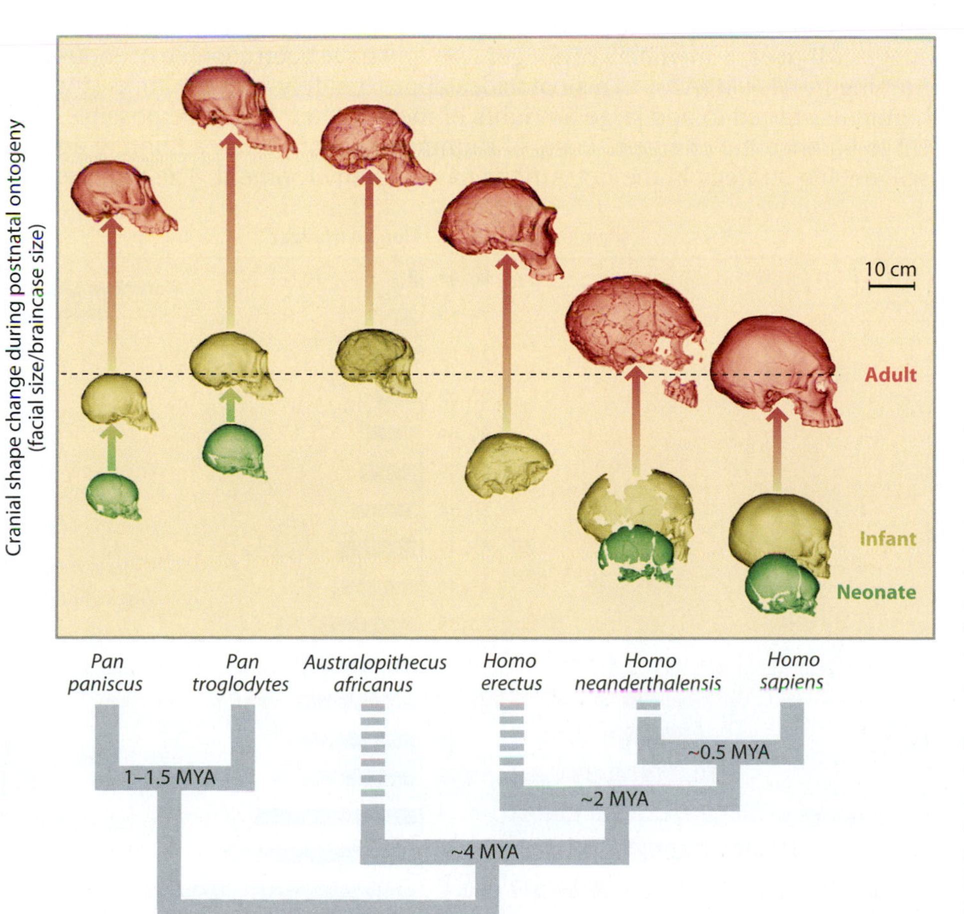

Figure 8.3: Comparison of cranial development in humans, fossil hominins, chimpanzees, and bonobos. The position of each skull along the vertical axis indicates the ratio of facial and braincase size. Extant species and Neanderthals are represented by neonates, infants (before eruption of the first permanent teeth), and adults; and earlier hominins are represented by those fossil specimens which best correspond to these ontogenetic stages. Human adult cranial shape resembles that of an infant chimpanzee. Bonobos (*P. paniscus*) show developmental change of cranial morphology similar to early hominins. Note the evolutionary trend toward short ontogenetic trajectories, especially of the early phase (neonate to infant) in fossil hominins. [Adapted from Zollikofer CP & Ponce de León MS (2010) *Semin. Cell Dev. Biol.* 21, 441. With permission from Elsevier.]

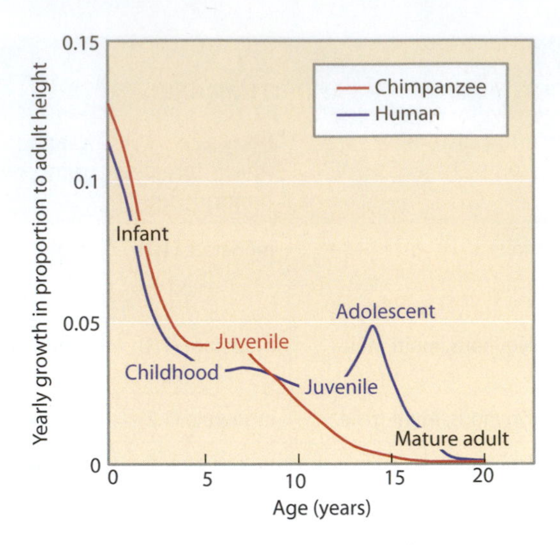

Figure 8.4: Stages of postnatal growth in human and chimpanzee. Infancy lasts from birth to the end of lactation and completion of deciduous tooth eruption. In great apes it is followed by a juvenile stage which lasts until first birth. Humans have a unique 4-year period of post-weaning childhood feeding dependency with slow growth, and an adolescent life history stage with a skeletal growth spurt. [Adapted from Bogin B (2009) *Am. J. Hum. Biol.* 21, 567. With permission from John Wiley & Sons, Inc.]

of prolonged childhood comes from the 160 KYA teeth of the Jebel Irhoud juvenile from Morocco.[63] Slower growth during childhood is compensated for by a growth spurt during adolescence, which is another human-specific stage of development not found in any other mammal.[6] In addition, humans have a characteristic postmenopausal longevity (grandmotherhood; **Figure 8.5**), which otherwise has only been described in some whale species.[35]

Humans are outperformed by many mammals in sprinting, but when it comes to running long distances, humans can do better than any other mammal. Notably, no other primate is capable of endurance running. But when and for what benefit did this unique behavior evolve in our ancestors? Arguably, human endurance running could not have evolved before relevant features of limb and head morphology and effective sweating-based thermoregulation had themselves evolved. Morphometric evidence from fossils assigned to *Homo erectus* suggests that early members of our genus might have been capable of endurance running by ~1.9 MYA.[8, 59] This coincides approximately with the time at which hominins started to add large amounts of meat to their diet. One possible benefit to balance the energetic costs of running may have been a hunting and/or scavenging strategy in the hot African savanna environment. The requirement

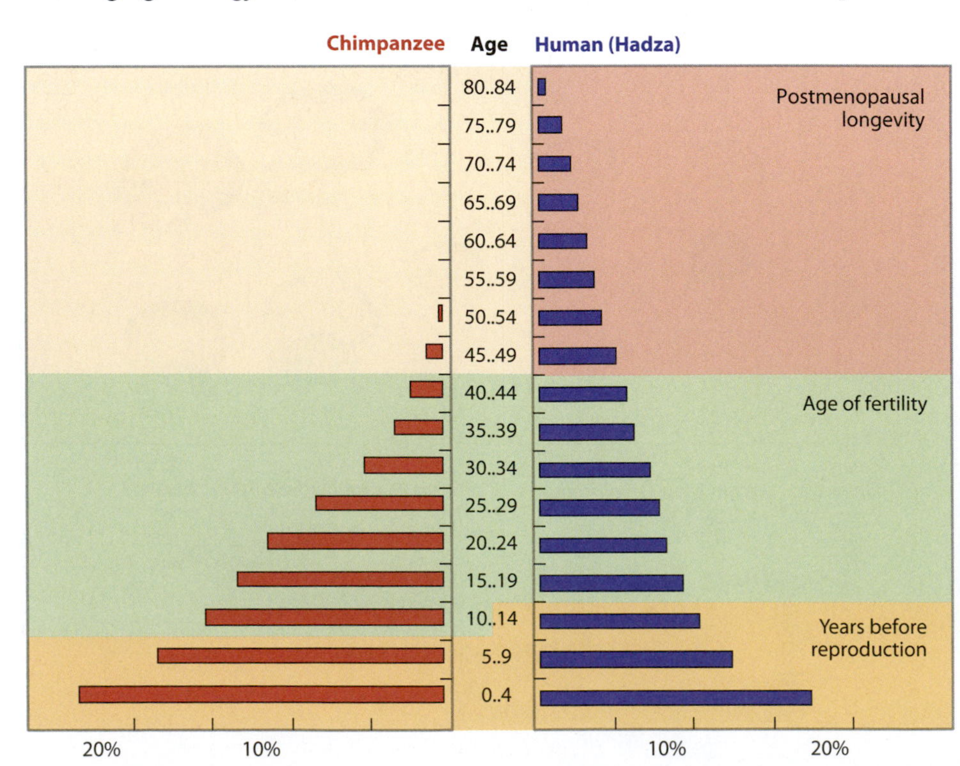

Figure 8.5: Female age structures in human and chimpanzee. Each bar shows the percentage of the population in the 5-year age class indicated in the vertical axis. Fertility ends by ~45 years in both species. Fewer than 3% of the adult chimpanzees live beyond that age. [Adapted from Hawkes K (2010) *Proc. Natl Acad. Sci. USA* 107 Suppl 2, 8977. With permission from Kristen Hawkes, University of Utah, USA.]

for effective thermoregulation to allow long-distance running means, however, suggests that by 1.9 MYA, *H. erectus* would likely have already lost its body hair (**Box 8.1**). Its naked skin would have had a human-like dense distribution of sweat glands producing watery rather than oily sweat.[34] By sweating, our bodies necessarily lose water and salt and, therefore, like other tropical animals, we are salt-craving. Salt helps to keep our blood pressure up during dehydration but high blood pressure, as the evolutionary cost for the benefits of effective thermoregulation, is also associated with a number of major health problems in later life, such as heart disease.

The human mind is unique

Ninety percent of humans, regardless of their culture or lifestyle, consistently show a preference to use their right hand over their left. The differences in muscle strength and bone morphology between left and right hands are the consequences, rather than the causes, of such preferences. The evidence from wild chimpanzees (and other primates) shows that they have no consistent preferences at the population or species level (**Figure 8.6**), even though individual preference of hand use can be observed.[47, 48] The **handedness** in humans is mirrored in the lateral asymmetry of the motor and sensory regions of the forebrain. It is possible that the lateralized brain eventually became the substrate for the evolution of language itself. When in human evolution did the shift toward right-handedness occur? Indications of asymmetry in tool use, Paleolithic art, bone morphology, and tooth scratches indicative of directionality in food processing suggest that Neanderthals, like modern humans, were predominantly right-handed on a similar scale, and that these lateral preferences therefore emerged earlier than 500 KYA.[24] Whether *Homo habilis* and *H. erectus* or even australopithecines were lateralized like modern humans is unknown because of the scarcity of fossils and relevant data.

(a)

(b)

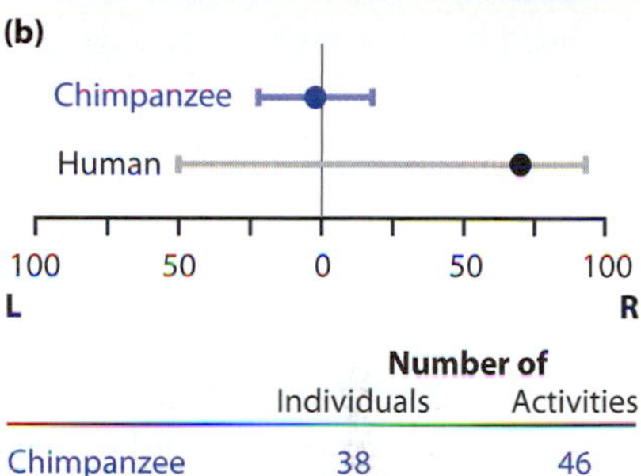

	Number of Individuals	Number of Activities
Chimpanzee	38	46
Human	1960	75

Figure 8.6: Hand preference in chimpanzees and humans compared. (a) The photograph shows wild chimpanzees in Bossou, Guinea, West Africa, using a pair of stones to crack open oil-palm nuts with different hands. (b) Data for chimpanzees refer to a community of wild chimpanzees (*P. t. schweinfurthii*) observed in the Gombe National Park. In each case the data relate to a wide range of everyday activities. Medians (*filled circles*) and boundary values (*horizontal bars*) for 95% of the population have been extracted from graphs in the original publications. [a, courtesy of Matsuzawa T, Humle T & Sugiyama Y (eds) (2011) The Chimpanzees of Bossou and Nimba. Springer. With permission from Springer Science and Business Media. b, data from Marchant L & McGrew WC (1996) *J. Hum. Evol.* 30, 427 and Provins KA et al. (1982) *Percept. Mot. Skills* 54, 179.]

The fossil record provides no direct evidence for when our ancestors started to speak, but paleontologists have still searched for indirect hints. Tool use and tool-making were long considered uniquely human-specific traits. Yet, chimpanzees make simple tools including some from stones, and their tool cultures, the subject of the emerging field of primate archaeology, differ by region and population.[29] It has been argued that greater levels of complexity in tool manufacturing and consistent styles over time would have required eloquent speech to transmit the details of tool production. The lower position of the larynx in adult humans is seen by many anthropologists as pivotal to a capability for modern language; however, given that this organ does not fossilize, whether archaic *Homo* had essentially modern vocal tracts, let alone the neurological ability to use them, remains open to debate. The finding that red deer, too, show lower laryngeal position demonstrates that this feature is not uniquely human, and suggests that it may exist as an adaptation allowing a species to exaggerate its perceived body size by producing lower-frequency resonances in the vocal tract.[22] The ape-like **hyoid** bone morphology of australopithecines suggests that they also had air sacs, while *H. heidelbergensis* and Neanderthals did not, and were potentially able to articulate vowels in the way that we do.[18]

In addition to using spoken language, humans also uniquely express themselves using a wide range of facial expressions and tears. **Emotional weeping** is often regarded to be a human-specific trait, but we have no clear understanding of how and why it evolved. Human tears contain a number of growth factors that induce healing and it has been proposed that tears may have evolved firstly as an ecological adaptation of some sort. Alternatively, it has been argued (including by Darwin[17]) that they are specifically linked to the evolution of human emotion expression (**Figure 8.7**)—tears appear as visual enhancers when transferring the message of sadness to our empathic fellows and may have therefore evolved to enhance parent–offspring bonding. In addition to the visual cues, human tears contain a chemosignal that, like the eye secretions of mice, may specifically affect the behavior of the opposite sex: male volunteers sniffing tears shed by women show a significant reduction in their testosterone levels.[25]

Box 8.1: Origins of the naked skin

Humans are unique among apes not only in brain-related phenotypes but also in several morphological and physiological traits including, perhaps most palpably, the lack of dense layer of hair covering our bodies. As we saw in Section 8.1 the loss (or substantial shortening) of body hair was essential for our thermoregulation, and enabled our ancestors to cope with hot conditions and to run long distances. On the other hand, humans have uniquely long head-hair, not seen in any other animals. Was the loss of body hair and becoming naked associated with loss-of-function mutations, consistent with the "degenerative ape" or "less is more" hypothesis[53] (see Section 8.2)? When did this change happen? Did it occur gradually via changes in many genes, or suddenly by a single genetic event? And when did our ancestors start to wear clothes? Approximately 30 keratin and 100 keratin-associated proteins form the structural basis of the human hair shaft and some of their genes are known to affect hair phenotype in mammals, including loss of hair. Yet, the evidence for a role of these genes in the loss of body hair in human ancestors has not been conclusive.[58]

Analyses of genetic variation of lice have been used to estimate the time when human ancestors lost body hair and when they started using clothing. The human head louse (*Pediculus humanus capitis*) and body louse (*P. h. humanus*) are phylogenetically related to the chimpanzee louse (*Pediculus schaeffi*) with which they share a common ancestor 6–7 MYA, consistent with the split time estimated for their hosts (**Figure 1**). Loss of body hair restricted *P. humanus* to the head. Human head lice are differentiated further into three genetically distinct strains and only one of these appears to have given rise to the body louse after humans started using clothing. According to Bayesian coalescent time estimates, the body or clothing louse may have emerged from a subset of head louse ancestors 80–170 KYA.[68] In contrast to body and hair lice, human pubic lice (*Pthirus pubis*) are genetically more closely related to the lice of gorillas (*P. gorillae*) than to *P. schaeffi* of chimpanzee. Notably, the coalescent time of *P. pubis* and *P. gorillae* is only 2–4 MY, suggesting that human pubic lice have originated separately, likely through a host shift. It is possible, considering this evidence, that the loss of human body hair occurred relatively early and that the first members of our genus were already walking naked and required fire and shelter to keep them warm at night. However, further evidence is still lacking to support this view and the genetic changes underlying the loss of human body hair have remained elusive.

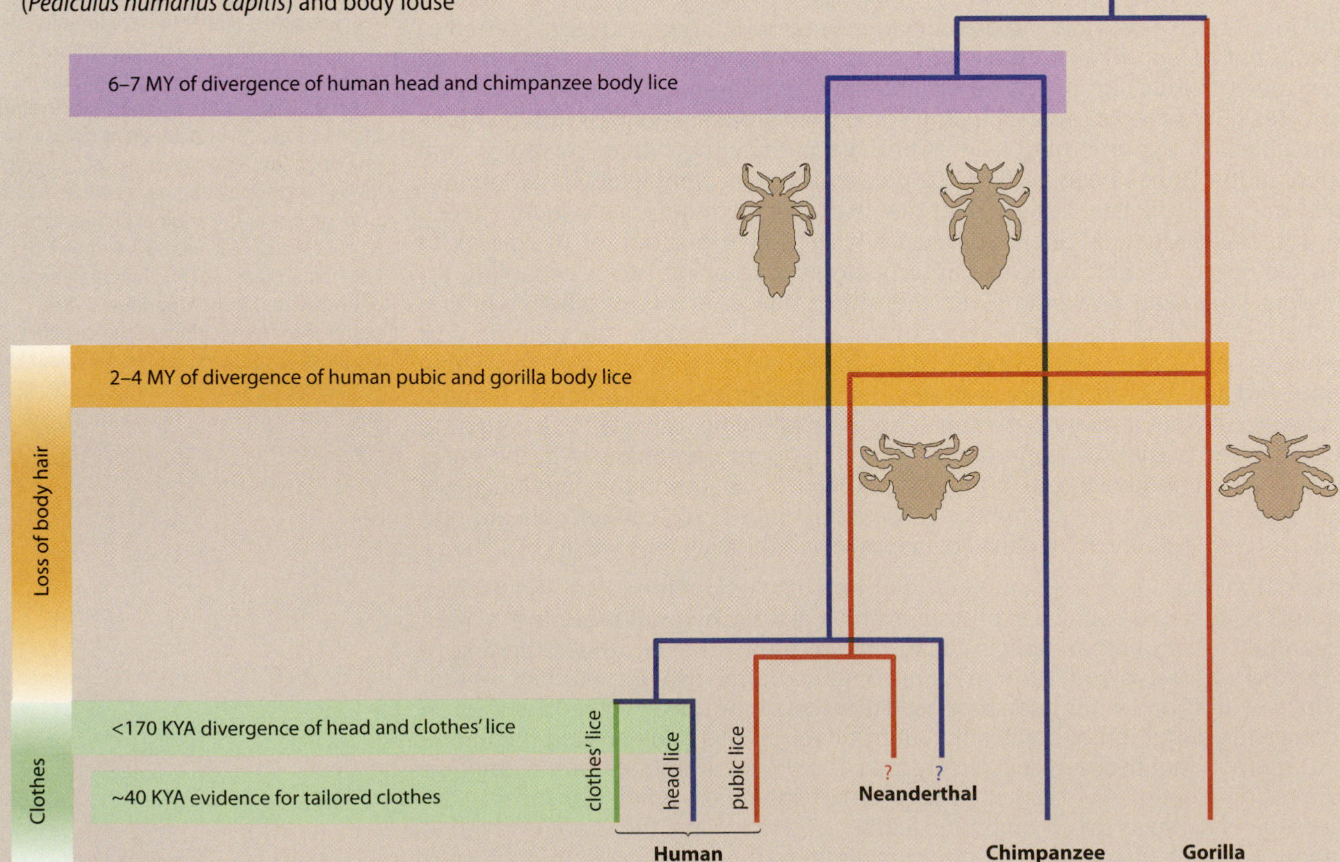

Figure 1: Loss of human body hair and origin of clothing. Divergence times of louse species are shown on the background of the phylogenetic tree of their hosts. [Adapted from Reed DL et al. (2007) *BMC Biol.* 5, 7. With permission from BioMed Central.]

Figure 8.7: Emotional weeping as one of the unique expressions of emotion in humans.
[From Darwin C (1872) The Expression of the Emotions in Man and Animals. London: John Murray.]

Only a few phenotypes are unique to modern humans

Our understanding of the phenotypic and behavioral features that are modern-human-specific is far more limited than the general list of features distinguishing us from other living apes, many of which were shared by extinct hominins. As will be discussed in more detail in Chapter 9, Omo I, the earliest known (195 KYA) anatomically modern human fossil shares with extant human populations several characteristics that make it distinct from the earlier forms of *Homo* and in particular from *H. heidelbergensis* and Neanderthals, our closest cousins. These features include a high and rounded forehead (**Figure 8.8**), presence of a chin, lack of brow ridges, gracile skeleton, and a distinct shape of the **clavicle**.[69, 73] It has been argued that many of the cranial features of modern humans are associated with their generally less robust skeleton and that the package of these gracile traits in the cranium may relate to a reduction in the length of a single element—the **sphenoid**, the central bone of the cranial base.[42] The differences in human and Neanderthal clavicles, on the other hand, suggest functionally different shoulder architectures that enable humans to raise their arms to the extent that allows them not only to climb, but also to throw, carry, and manipulate heavy objects.[73]

Compared with this consensus on the package of traits defining the **anatomical modernity** of humans, the package of traits defining **behavioral modernity** has been more debated. Material evidence in the archaeological record indicative of symbolic thinking, personal ornaments, art, and long-distance trade of raw materials was in the past seen as the hallmark of our species. But many components of this package have now been found in archaeological contexts of other hominins, such as Neanderthals. Instead of being diagnostic for "archaic" versus "modern" human behavior, such traits could instead be considered to reflect the behavioral variability of a range of hominin species[62] (see also Chapter 11).

8.2 GENETIC UNIQUENESS OF HUMANS AND HOMININS

This section focuses on the genetic changes that characterize us as humans. Although overall the human genome differs by only a few percent from those of other great apes, there are still many millions of structural and single nucleotide differences. Most of these are probably of no phenotypic relevance, but a few are likely to be associated with human-specific morphological and behavioral features regardless of whether their evolution has been affected by natural selection or not. Genomewide comparisons of humans and other apes became possible after the publication of the chimpanzee,[14] orangutan,[44] and gorilla[60] reference sequences. The publication of the Neanderthal and Denisovan genomes has opened up the possibility of identifying the genetic changes that differentiate us from them: we can now distinguish the changes that our ancestors shared with these extinct hominin species from those that are truly unique to modern humans.

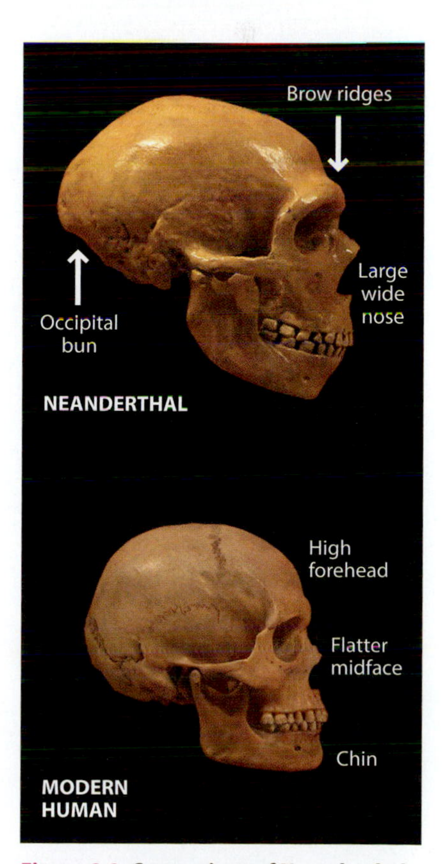

Figure 8.8: Comparison of Neanderthal and modern human skulls.

The sequence and structural differences between humans and other great apes can be cataloged

Human and chimpanzee genomes have approximately 40 million differences, of which 35 million are nucleotide substitutions and ~5 million are indels. How many of these have had significant effect on the phenotype? Simple reasoning from the number of phenotypic differences alone suggests that there must be many important changes. If there were just one or a few genetic changes that converted the ancestral ape's phenotypic package as a whole into a human one, we would expect to observe rare back-mutations; but among all of the mutant human phenotypes known, there are none that result in an ape-like phenotype. Nevertheless, this line of reasoning does not tell us whether there have been 100, 1000, or 10,000 functionally important changes. Many human-specific phenotypes—for example, opposable thumb and white covering to the eye-ball—probably have completely different genetic causes from one another and have evolved along independent paths over the course of evolution; but many other characteristics, such as brain development and culture-defining traits, may share some common genetic basis.

Considering the high sequence similarity of human and chimpanzee genomes, a priori expectations have been that regulatory mutations, that is, those that change the transcription pattern of individual genes, will be the most relevant in explaining human phenotype uniqueness.[38] If these mutations affect genes such as those encoding transcription factors, which themselves regulate other genes, a single mutation may indirectly affect a large number of genes. Hypotheses implying the existence of single mutations with major effects governing the evolution of key human traits, such as brain polarity or language, have sometimes been favored in the anthropological literature to explain the success of anatomically modern humans.[20] Although, as we will see in the following sections, most human-specific phenotypes are complex and polygenic, there might also have been genetic changes with large effects on the evolution of human- and hominin-specific traits. A well-known example from another species is body size in dog breeds, which is largely determined by a polymorphism in a single growth factor gene, *IGF1*.[67]

Phenotype differences can emerge also by structural changes of the genome that can lead either to the duplication or deletion of genes. The role of gene duplications in evolution has been emphasized by the "more is better" view[52] that has received wide support in vertebrate genomics by virtue of the fact that, after whole genome duplication events, many duplicated genes are retained because they are able to gain new functions. An alternative, "less is more," view is that we are in some respects "degenerate apes" and some of our specific characteristics—such as slow development, or loss of hair and muscle strength—could be due to loss-of-function mutations.[53]

All the classes of mutation discussed in Chapter 3 can potentially contribute to great ape–human differences. The largest scale, cytogenetically visible, changes among great apes were discussed in **Section 7.2**. Tandemly repeated DNA sequences evolve rapidly; indeed, the most rapidly changing regions of the genome are the blocks of heterochromatin near the centromeres and telomeres. These have low gene densities and are therefore poor candidates for functionally important changes. Differences in interspersed repeats such as **SINE**, **LINE**, and retrovirus-derived long terminal repeats (LTRs) may, however, occasionally have functional consequences. An example of differentiated LTRs includes the human-specific endogenous retrovirus family HERV-K cluster 9 and LTR13. Interspersed repeats, when inserted into genes, have affected human phenotype evolution by gene inactivation and also as modulators of gene expression.[10]

Humans have gained and lost a few genes compared with other great apes

Humans share all but a handful of their genes with apes and the birth of new genes from noncoding sequence is considered to be both extremely rare and difficult to identify because it requires accurate orthologous sequence data from nonhuman species. Nevertheless, the idea that human-specific genes might be responsible for human-specific phenotypes is attractive. Following the publication of human and great ape genome sequences, more than 60 de novo human-specific genes have been characterized.[39, 75] These are predominantly expressed in cerebral cortex and testes (**Figure 8.9**) but this would be true for many randomly chosen sets of genes. Moreover, the new genes tend to be short and their functional role in human phenotype uniqueness has yet to be determined.

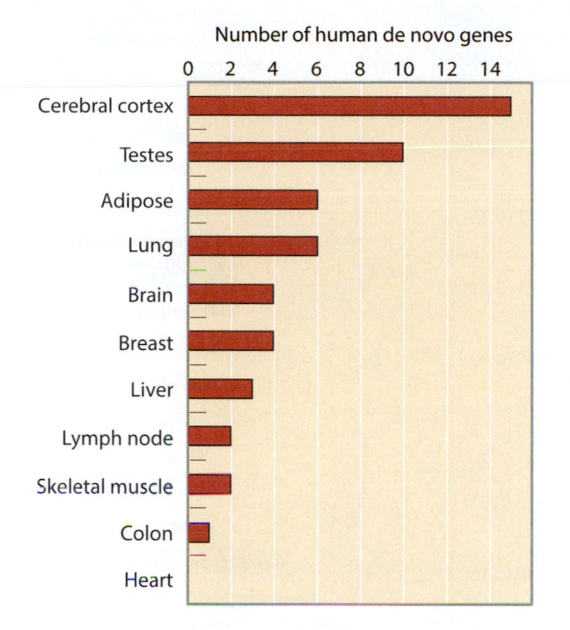

Figure 8.9: Expression of de novo genes by tissue type. The plot shows the distribution of 53 de novo genes by their expression in different tissues. Each gene has been allocated to the tissue that showed the highest normalized expression level among 11 tissue types. [Adapted from Wu DD et al. (2011) *PLoS Genet.* 7, e1002379. With permission from the Public Library of Science.]

Segmental deletions and duplications that have occurred on the human lineage affect more base pairs than do SNPs, and are enriched with genes associated with reproduction, immunity, and chemosensory activity. However, fine-scale comparisons with great ape genomes have been hampered by the low quality of their genomic sequences in duplicated regions. But even when considering reliably sequenced genomes, the rate of gene loss and gain in humans and great apes is more than twice that seen in other mammals.[27] And, among great apes, several gene families show an expansion of copy number that is unique to humans (**Figure 8.10**). One of these is *SRGAP2*, considered further below.

A few genes that are active in apes and other mammals are inactive in humans; both humans and chimpanzees are characterized by a twofold higher number of gene deletions than macaques.[72] A functionally characterized example of human gene loss is the mandible-expressed masticatory myosin heavy chain 16 gene (*MYH16*, OMIM 608580), which in humans has become inactivated due to a frameshift mutation dated to 2.4 MYA.[66] Another example of human-specific gene loss is the Alu-mediated inactivation of the cytidine monophospho-*N*-acetylneuraminic acid hydroxylase gene (*CMAH*, OMIM 603209) that determines the cell-surface distribution of sialic acid types, which are involved in cell–cell communication.[33] Consequently, human cells largely lack *N*-glycolylneuraminic acid (Neu5Gc), and the trace amounts detectable may be derived from the diet. The inactivity of the enzyme is due to a genomic deletion which removes a 92-bp exon and introduces a **frameshift** resulting in a truncated protein. It is possible that this change was fixed just by genetic drift, and therefore has no relevant consequence for the human phenotype. However, since the gene is intact in most other species and is involved in a variety of biological processes, it is also plausible that its loss was positively selected. Because many pathogens bind to these cell-surface sugars, the loss of Neu5Gc could have provided a way of resisting a pathogen affecting past populations. Sialic acids such as Neu5Gc survive better in fossils than DNA does, and it has been possible to show that the lack of Neu5Gc is shared by modern humans and Neanderthals, so the inactivation of the gene is likely to pre-date the common ancestor of these groups ~500 KYA.[13]

Gene inactivation may be linked to another difference between ape and human genomes. Apes have a SIGLEC12 protein which binds to Neu5Gc, and this binding requires a normally conserved arginine within the polypeptide.[1] Humans have SIGLEC12, but the human form lacks the arginine necessary for Neu5Gc binding. The role of SIGLEC12 in humans is unknown. Because it has lost or substantially changed its activity, there appears to be a set of functionally related genes that are co-evolving in a specific fashion on the human lineage.

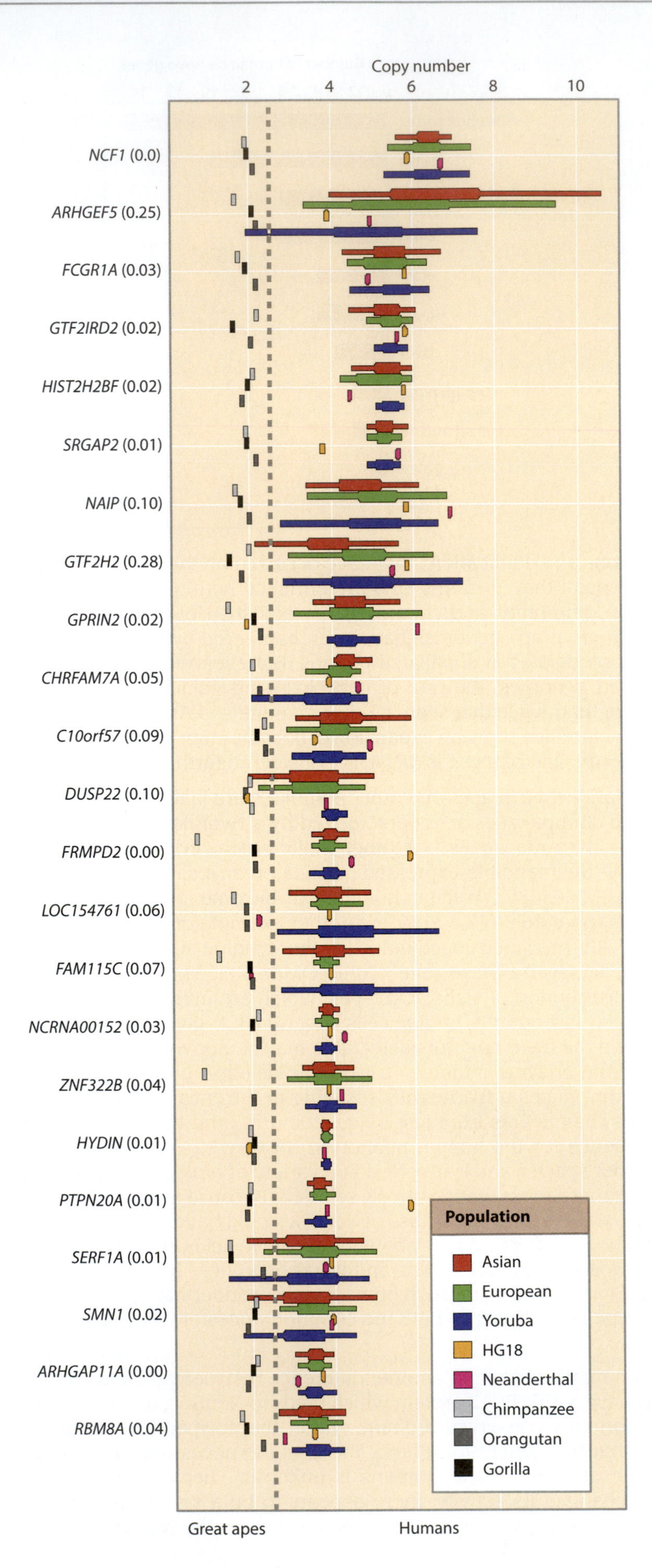

Figure 8.10: Human-specific gene family expansions.
HG18 refers to one version of the human reference genome. Numbers in parentheses reflect the values of the variation statistic, VST, measuring copy-number variation between human populations. [From Sudmant PH et al. (2010) *Science* 330, 641. With permission from AAAS.]

Altogether more than 10 uniquely human changes have been identified in siglec (sialic acid-binding immunoglobulin superfamily lectins) genes, including changes in binding efficiency of the enzyme, gene expression pattern, sequence change through gene conversion, and deletion or inactivation, suggesting that sialic acid biology has been a hotspot of human evolution.[70] Some of the deleted siglec genes that have been highly variable in human evolution have functional relevance to immunity and nervous tissue.

A systematic genomewide scan of human-specific deletions of regulatory sequences otherwise conserved across mammals identified 510 regions. One of these mapped near to a tumor suppressor gene *GADD45G*, loss of which is functionally correlated with the expansion of specific brain regions. A second case, a deletion of the regulatory region of the androgen receptor gene (*AR*, OMIM 313700), has been associated with the loss of facial vibrissae and penile spines in the human lineage.[49] Interestingly, some of the genes, like the immunity-related GTPase family, *M* gene (*IRGM*, OMIM 608212), have been lost in ancestral primates and been "resurrected" in the human lineage.[5]

Humans differ in the sequence of genes compared with other great apes

Of the 35 million single nucleotide positions at which humans differ from chimpanzees, 100,000 are in their **exomes**. Of these, in turn, approximately 40,000 are amino acid-changing substitutions.[14] Half of these mutations have occurred on the human lineage, and can be identified using gorilla or orangutan as an outgroup; most are not likely to affect phenotype and therefore evolve neutrally. For example, the analysis of the exome sequence of Venter's personal genome (see **Section 18.4**) concluded that ~80% of nonsynonymous mutations are unlikely to substantially change the properties of the protein.[50] Given that we have approximately 25,000 protein-coding genes (Chapter 2), this means that every gene has on average one human-specific amino acid-changing mutation that is fixed in all humans. However, only a small proportion of these mutations are likely to have made a detectable phenotypic difference in our evolution.

Nevertheless, one of the most common methods of assessing evolutionary changes on the human lineage with respect to great ape genomes focuses on the relative rate of nonsynonymous mutations (see Chapter 6). The **dN/dS ratio** is an informative summary statistic for assessing general species and genomewide patterns of protein-coding gene evolution. In genomewide surveys, several biological processes are significantly associated with high dN/dS ratios on the human lineage, including fatty acid metabolism and G-protein-mediated signaling.[4] Comparison of the polymorphic functional variation in the protein-coding genes among human populations with fixed variation of the same genes between humans and chimpanzees (McDonald–Kreitman test, **Section 6.7**) revealed that 13% of the genes have a paucity of amino acid differences between the two species due either to purifying or balancing selection,[11] while 10–20% of the amino acid differences have likely been fixed by positive selection.[7] Among these, genes involved in sensory perception, immunity, apoptosis, and reproduction have experienced the highest number of amino acid replacements whereas brain-expressed genes show the fewest changes between humans and chimpanzees.[51]

While the dN/dS ratio has been widely used, it is not very powerful for identifying particular genes that have been under positive selection in humans. A significant signal is only obtained if positive selection acts repeatedly on the same protein and thereby leads to a significant excess of nonsynonymous changes. Such patterns of protein evolution are rarely observed and if they are identified it is important to consider technical artifacts, such as wrong alignments, because these can also produce such a signal. Furthermore, because thousands of different statistical tests are performed in a genomewide screen, a large number of apparently statistically significant results are expected by chance, making it very difficult, or even impossible, to unambiguously assign human-specific positive selection to a particular gene based on dN/dS ratio tests alone. Finally,

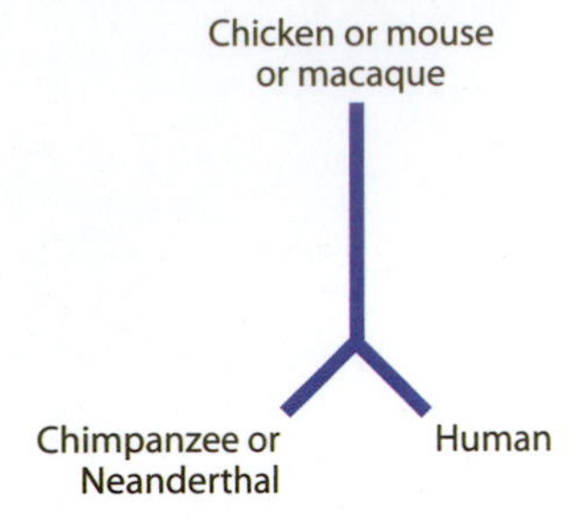

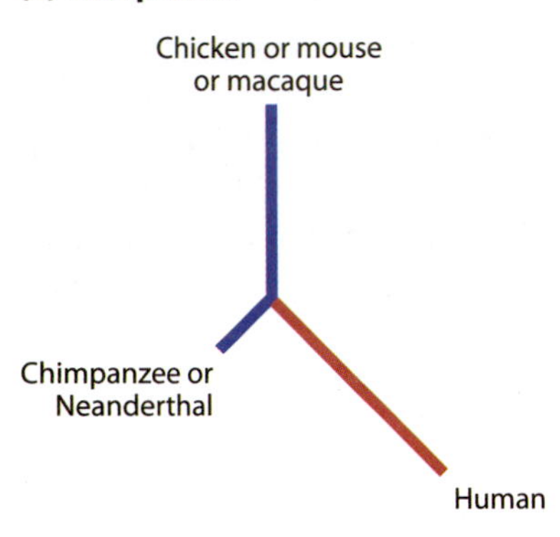

Figure 8.11: Detecting human accelerated regions (HARs). Each plot represents one locus. Branch lengths are proportional to genetic differences at orthologous sequences that are specific to the species shown at the tips of the branches. (a) The orthologous regions are identified, then (b) those showing a conserved pattern of low sequence divergence among the nonhuman species are selected. HARs are identified as those genomic regions where the human sequence shows statistically significant increase in substitution rate.

the dN/dS ratio can only test protein evolution involving amino acid changes. Although the same principle of comparing the rates of putative functional and neutral changes can be applied to regulatory regions, the assignment of the two categories is much more difficult than for coding regions.

A single substitution or indel can also lead to alternative splicing, which is a major mechanism for the enhancement of transcriptome and proteome diversity in mammals, affecting >90% of the genes.[36] For example, the kallikrein-related peptidase 8 (*KLK8*) gene shows fixed splicing variation between humans and chimpanzees while variations of this gene among humans have been associated with learning and memory. The human-specific splice variant of this gene is absent in the nonhuman primates studied so far.[46]

Many, probably most, **adaptive** changes that have occurred on the human lineage may have been due to mutations outside the protein-coding genes, for example in regulatory regions adjacent to genes or in distant noncoding regions. While less than 2% of the genome is protein-coding, at least 5.5% of the human genome is evolutionarily conserved, that is, can be regarded as functional.[43] The search for lineage-specific deviations (**Figure 8.11**) from otherwise conserved regions thus represents another method of genetically mapping the evolution of species-specific variation. Applied to humans, such searches for **human accelerated regions** (**HARs**) have revealed a number of new candidates for functional RNA genes.[56] Among the 49 HARs that were identified, only two mapped to known genes; and the topmost accelerated region, HAR1, turned out to be a novel RNA gene (OMIM 610556, *HAR1A*; the authors call it *HAR1F*) that is expressed in the Cajal–Retzius neurons in the human neocortex during fetal development. Scans of positive selection on the human lineage at *cis*-regulatory sequences have revealed enrichment of genes involved in neural function and nutrition, particularly in glucose metabolism.[30] The search for accelerated regions is thus a method that has potential for discovering new functionally important elements in the genome. Like the dN/dS ratio approach, however, the scans for HARs focus on detecting accumulated mutations and thus cases where single or few mutations at noncoding functional elements have a phenotypic impact will go undetected. Furthermore, distinguishing positive selection from relaxation of selection can be difficult.

Humans differ from other apes in the expression levels of genes

The amount, time, and place of expression of a transcript can be influenced by changes in the copy number of its gene, by changes in its regulatory sequences, and by changes in *trans*-acting proteins. For example, humans have between one and eight copies of the salivary amylase gene per haploid genome and corresponding higher levels of this enzyme in their saliva (Chapter 15 describes copy-number variation of this gene within human populations). Because chimpanzees and bonobos have just one copy per haploid genome, it has been suggested that higher copy numbers have been selected due to a shift in diet, to one containing starch-rich tubers, during hominin evolution.[54] Another example where changes in gene expression have been predicted from sequence changes is the human-derived allele of the microRNA miR-1304. In contrast to the ancestral allele in other apes and Neanderthals, the derived allele present in most humans does not repress its target genes, including enamelin and

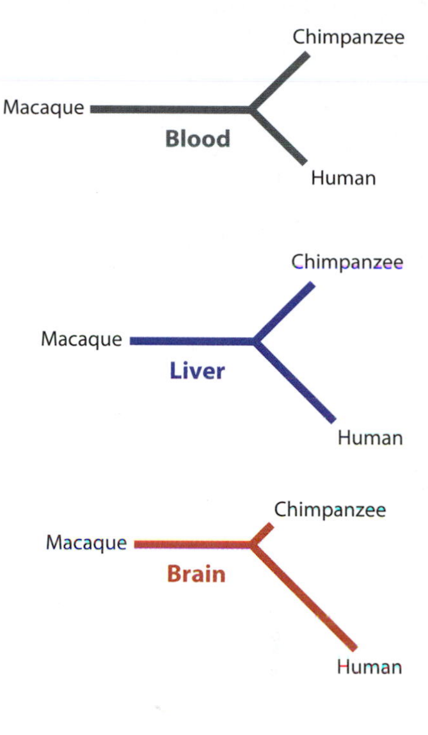

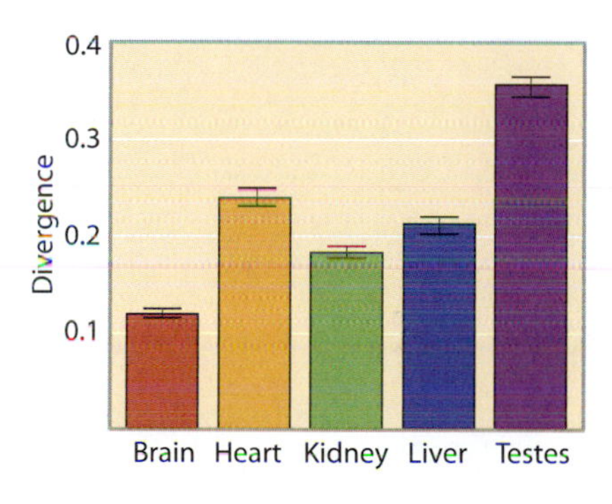

Figure 8.12: Comparison of gene expression patterns in humans, chimpanzees, and rhesus macaques.
Expression levels of ~18,000 human genes were measured using RNA from different tissues (blood, liver, or brain) from the three species, and a distance measure between the profiles was calculated. In blood and liver, human and chimpanzee expression patterns are the most similar, as would be expected from their evolutionary relationship. However, in brain, chimpanzee and rhesus macaque are the most similar, interpreted as indicating particularly pronounced changes in gene expression in the human brain. [Data from Enard W et al. (2002) *Science* 296, 340. Divergence of tissue-specific expression patterns is shown according to Khaitovich P et al. (2006) *Nat. Rev. Genet.* 7, 693.]

amelotin, which are important in tooth formation and hence could contribute to the known dental differences between modern humans and Neanderthals.[45]

Although there are exceptions as described above, it is generally not possible to reliably predict changes in gene expression from the DNA sequence, so it is important to measure these molecular phenotypes directly. With the development of high-throughput gene expression **microarrays** and RNA sequencing, it became possible to investigate systematically which genes show differences in expression between apes and humans. In one early set of experiments, arrays comprising ~18,000 human cDNAs were hybridized with probes generated from RNA of blood, liver, and brain samples, comparing pools of human, chimpanzee, and rhesus macaque individuals (**Figure 8.12**). Interestingly, while the amount of change on the human and chimpanzee lineage was about equal in blood and liver, it was considerably higher for humans in brain; even though, overall, brain expression levels were less differentiated than those of blood and liver. Further studies measuring more samples individually and using different platforms partly confirmed and extended these first findings, although the difficulty of obtaining samples from several species, tissues, and age-groups has so far hindered an exhaustive comparison of humans and their closest relatives. Overall, comparative expression studies[37] using microarray data have shown the following:

- Many genes (for example ~30% in testis or 10% in brain) show highly significant differences in expression level between humans and chimpanzees.
- Although environmental factors such as diet may have substantial temporary effects on gene expression levels among individuals within the species, their effect on among-species differences is minor.
- Brain-expressed genes show less variation in level than genes expressed in other tissues.
- Different cortical areas of the brain show the same extent of differentiation between humans and chimpanzees.
- The most divergent expression patterns between humans and chimpanzees are observed in testis-expressed genes.
- Those genes that are expressed more in the brain than other tissues show more rapid evolution on the human rather than the chimpanzee lineage.
- The expression differences in brain tend more often to lead to higher rather than lower expression of respective proteins in the human lineage.

One limitation of microarrays is that probes often are based on the human reference sequence and, although probes that show sequence differences between the human and chimpanzee reference sequences have been computationally removed from the analyses, this limits comparisons among more distantly related species. RNA sequencing methods do not have this limitation and a study comparing **polyadenylated** RNA sequence data collected from six organs of nine mammal species[9] revealed that the rates of evolution vary between evolutionary lineages and by genes. Interestingly, gorilla rather than chimpanzee or bonobo showed the highest similarity to human in gene expression patterns in testis, brain, and heart tissue. Humans were found to show a concerted pattern of gene expression differences in a module involving 259 genes enriched for gene ontology terms related to neuron insulation. A highly up-regulated gene in the human prefrontal cortex, limb expression 1 (*LIX1*, OMIM 610466), has a crucial role in motor neuron development and maintenance.

Genome sequencing has revealed a small number of fixed genetic differences between humans and both Neanderthals and Denisovans

Neanderthal genome studies have enabled us to explore in further detail whether the fixed differences that are observed between humans and other great apes appeared before or after the speciation event that led to the emergence of anatomically modern humans.[26] Because of methodological difficulties in distinguishing DNA-damage-induced changes from authentic variation of ancient DNA, the list of apparent Neanderthal-specific mutations necessarily contains a substantial proportion of variation that never existed in living individuals, and therefore the human–Neanderthal comparisons have focused only on differences that are inferred to have occurred on the human lineage.

The number of amino acid changes fixed in a sample of five human genome sequences but ancestral in the Neanderthal draft sequence was low: just 78, with only six genes (**Table 8.2**) carrying more than one change or in which a start or stop codon has been lost or gained. Three of these six are expressed in skin, leading the authors to suggest that selection on skin morphology or physiology may have been important on the human lineage after the split from Neanderthals. The fact that only a small proportion (0.8%) of the functional derived alleles are uniquely fixed in anatomically modern humans was further demonstrated by an array-based sequence-capture study successfully retrieving Neanderthal sequence data from ~11,000 nonsynonymous positions (**Figure 8.13**).

Positive selection was also investigated in the draft Neanderthal genome by searching for genomic regions where the common ancestor in humans arose after the split from Neanderthals.[26] The logic of this screen is that for most

TABLE 8.2:
GENES SHOWING MORE THAN ONE FIXED AMINO ACID DIFFERENCE BETWEEN HUMANS AND NEANDERTHALS

Gene	Description/function	Tissue of expression (UniGene, UniProt)
SPAG17	involved in structural integrity of the central apparatus of the sperm tail (axoneme)	highly expressed in testis but also in other organs containing cilia-bearing cells (brain, oviduct, lung, and uterus)
PCD16	calcium-dependent fibroblast-specific cell-adhesion protein	brain, skin, liver, muscle
TTF1	RNA polymerase I termination factor	widely expressed
CAN15	small optic lobes homolog, linked to visual system development	widely expressed with higher expression in brain
RPTN	multifunctional epidermal matrix protein	mouth, skin, mammary gland
TRPM1	member of melastatin subfamily encoding calcium-permeable cation channel with potential role in melanin synthesis	melanocytes; skin, eye, testis

Data from Green RE et al. (2010) *Science* 328, 710.

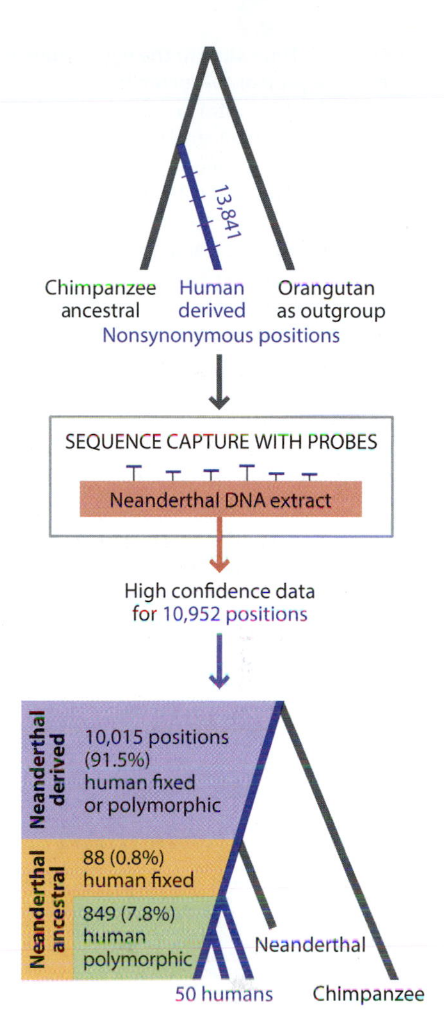

Figure 8.13: Distinguishing modern-human-specific nonsynonymous mutations from those shared with Neanderthal.
As an initial step, human-derived nonsynonymous mutations were defined by comparing human and chimpanzee genomes, and using orangutan as an outgroup. For each human-derived position a probe was designed to capture DNA from an extract of Neanderthal bone sample. The captured DNA was sequenced and high-quality reads were compared against similar sequence data from 50 human DNA samples. A total of 88 nonsynonymous changes shared by all 50 humans were identified. [From Burbano HA et al. (2010) *Science* 328, 723. With permission from AAAS.]

of the genome, a typical **coalescence time** is ~1 MY, so derived alleles will generally be shared between Neanderthals and humans, whose populations split <440 KYA. But if an advantageous variant in humans arose after this split, derived alleles will only be present in humans. This identified 212 regions where Neanderthals had fewer than expected derived alleles. Among the top 20, five were devoid of protein-coding genes, suggesting that selection must be acting on other functional elements. In the remaining 15, the genes present were implicated in processes including energy metabolism (*THADA*), cognitive function (*DYRK1A, NRG3, CADPS2*), and morphology (*RUNX2*). Interestingly, inactivation of one copy of *RUNX2* leads to the condition cleidocranial dysplasia (OMIM 119600), characterized by deformities of the collarbone (cleido-), cranium, and other parts of the skeleton, including bossing (bulging) of the forehead and a bell-shaped rib cage. Several of these skeletal features differ between humans and Neanderthals, and it has been speculated that the human-specific form of *RUNX2* has contributed to the morphology of the upper body and cranium characteristic of humans.[26]

Comparison of human and Denisovan genomes has identified 129 amino acid substitutions and 14 indels in the coding sequences of genes where the Denisovan individual carries the ancestral alleles while humans are fixed for the derived alleles.[57] Among these, 10 genes had two amino acid replacements, including two associated with skin disease. This finding is generally consistent with the results of similar analyses performed on the Neanderthal genome (Table 8.2), although multiple nonsynonymous substitutions were detected on different genes.

8.3 GENETIC BASIS OF PHENOTYPIC DIFFERENCES BETWEEN APES AND HUMANS

The genetic basis of phenotype variation is commonly determined by experimental or observational studies within species. These include the study of Mendelian disorders and linkage or association studies in humans, or functional analyses by gene manipulation in model organisms. In comparative genomic studies the information about gene function can then be related with observed interspecies differences. The existence of tens of thousands of human-specific amino acid-changing mutations makes the interpretation of their functional significance difficult. Genes that have changed more than others can be analyzed using tools that search for enrichment of functional annotations assigned to the genes or expression in particular cells or tissues. As mentioned in Chapter 7, reproduction- and immunity-related genes generally stand out among primates, and thus their enrichment in selection scans is not restricted to humans. The discovery of the genetic changes behind the evolution of uniquely human traits has been relatively slow compared with the pace of the generation of whole genome sequences, and of the elucidation of the genetic bases of human diseases.

Mutations causing neoteny have contributed to the evolution of the human brain

As we saw in **Section 8.1**, the evolutionary trend toward increasing brain volumes in hominins was paralleled by the reduction of their face sizes and changes in

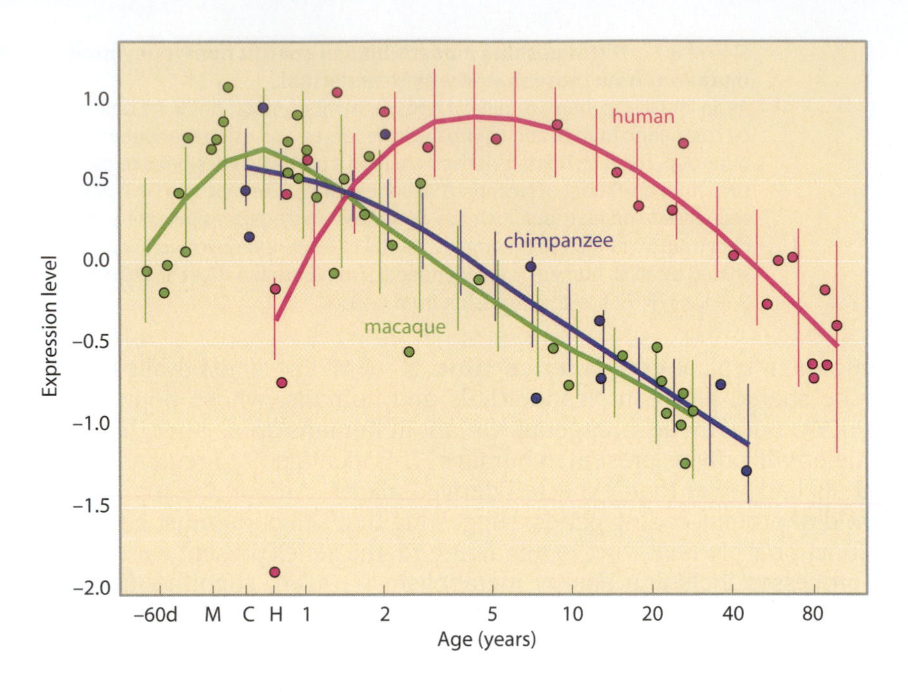

Figure 8.14: Time shift in the expression time of the prefrontal cortex synaptic genes in the postnatal development of human, chimpanzee, and macaque. Points indicate individuals and error bars show standard deviation across genes. M, C, and H indicate time of birth for macaque, chimpanzee, and human, respectively. [From Liu X et al. (2012) *Genome Res.* 22, 611. With permission from Cold Spring Harbor Laboratory Press.]

several life-history traits. How can such changes be understood at the molecular level? A small number of regulatory changes in developmental genes shared with other primates could lead to the human-specific development program. A subset of human genes that are expressed in the brain show age-specific mRNA expression levels in the prefrontal cortex when compared with chimpanzees and rhesus macaques.[64, 65] Among these three species, it is only humans who display three to five times faster divergence rates in age-specific patterns of some genes. Certain developmental changes in the human brain are delayed relative to other primates, consistent with the view of human **neoteny**. The timing of peak expression of synaptic genes in the prefrontal cortex, which in chimpanzee and macaque occurs before the first year of postnatal development, is shifted to the end of childhood, at year 5, in humans (**Figure 8.14**). A significant proportion of genes expressed in the human brain are switched on later than in chimpanzee and these genes appear neotenic—their employment in adult human brain corresponds to the infant or juvenile stage of chimpanzees. This human-specific pattern of gene expression is seen in the prefrontal cortex, but not in the cerebellum, and includes many neuron-related genes. A small number of *trans*-acting regulators, such as microRNA genes miR-92a, miR-454, and miR-320b, could arguably explain the rapid evolution of the human brain.[65]

As well as regulatory mutations in existing genes, the generation of new genes by duplication or their removal by deletion may provide mechanisms for different patterns of development in human evolution. One gene involved in cortical development, Slit-Robo Rho GTPase activating protein 2 (*SRGAP2*, OMIM 606524), has gone through three duplication events in hominin evolution (**Figure 8.15**). The second duplication event occurred 2.4 MYA, corresponding approximately to the time of emergence of the genus *Homo*, and is seen in the Neanderthal and Denisovan genomes as well as in modern humans. This event generated a novel human-specific paralogous gene *SRGAP2C*, which is expressed in both the developing and adult human brain and is likely to encode a functionally active but truncated protein. *SRGAP2C* is highly conserved in copy number among modern humans showing the least variation among human-specific duplicated genes, suggesting an important functional role. The *SRGAP2* mouse ortholog promotes spine maturation and limits spine density of neural dendrites in the mouse neocortex, and expression of the human-specific truncated *SRGAP2C* in mouse inhibits this function of *SRGAP2*. This leads to neoteny

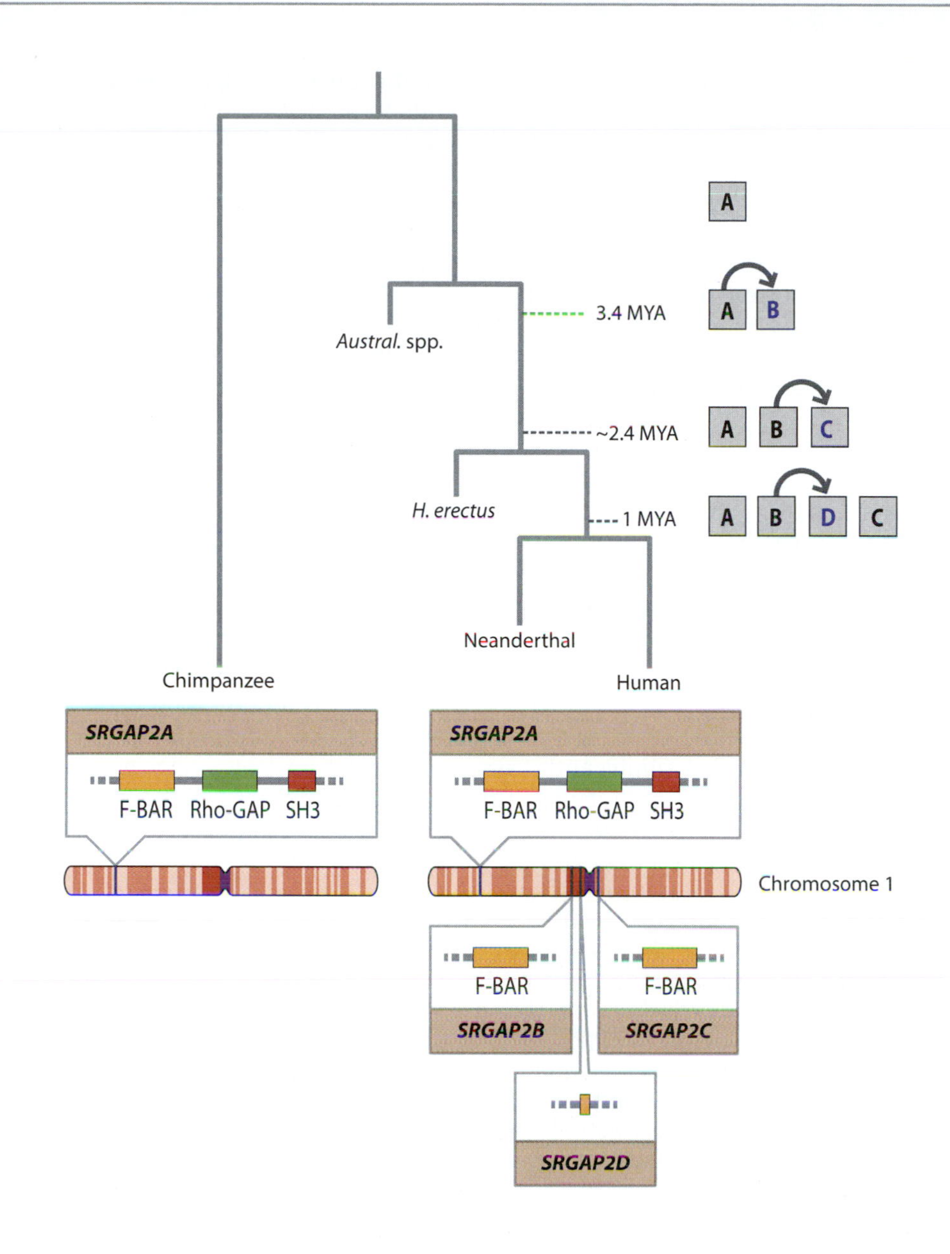

Figure 8.15: Duplication events of *SRGAP2* genes in hominin history. Location of *SRGAP2* paralogs on human chromosome 1 with putative protein products and their functional domains based on cDNA sequencing. Arrows show the reconstructed evolutionary history of *SRGAP2* duplication events. The first duplication involved only the first 9 of 22 exons of the *SRGAP2A* gene containing the F-BAR domain and all subsequent copies, therefore, contain only the F-BAR domain of the protein which is further truncated in paralog D. Copy numbers of the paralogs, and expression analyses, suggest that paralogs B and D are nonfunctional pseudogenes, whereas A and C are likely to encode functional proteins. [Adapted from Dennis MY et al. (2012) *Cell* 149, 912. With permission from Elsevier.]

during dendritic spine maturation in the neocortex.[12] The human-specific *SRGAP2C* gene therefore may have contributed to the emergence of human-specific features, including neoteny in the human neocortex.

Comparative studies of **microRNA** (miR) expression in humans, chimpanzees, and rhesus macaques have found enrichment of neural functions on the human evolutionary lineage and identified a signature of positive selection and human-specific expression of miR-34c-5p.[31] The change in expression pattern of this gene is associated with a high number of derived SNP alleles in humans that are not observed in Neanderthals. Because Neanderthals had a brain volume similar to or exceeding that of modern humans this evidence suggests that natural selection has continued to affect brain-expressed genes in our recent evolutionary history independently of changes in size.

The genetic basis for laterality and language remains unclear

As discussed in **Section 8.1**, humans differ from other great apes in laterality. It has been argued that an allelic change in one master gene may have been sufficient to determine the development of hemispheric asymmetry in humans.[2, 16] According to such a view, the derived allele of this hypothetical gene fosters a shift toward one hand preference, whereas the ancestral allele provides no

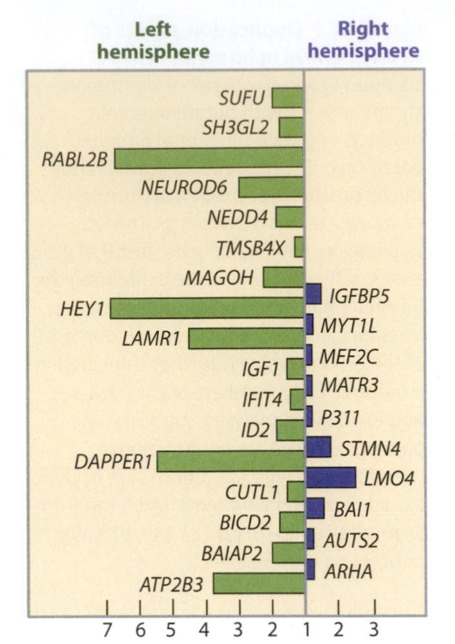

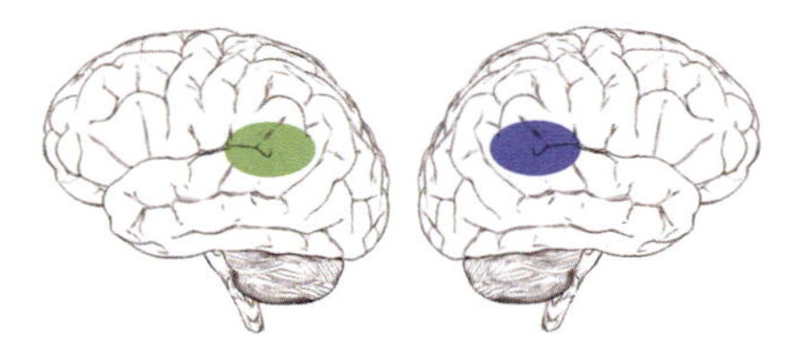

Figure 8.16: Asymmetric gene expression in 12-week-old human fetal brain.
The relative gene expression levels in the **perisylvian** region are average ratios of gene expression detected by real time-PCR between the left and right hemispheres of two brains. Only 27 genes that showed consistent hemispheric differences are shown. [Adapted from Sun T & Walsh CA (2006) *Nat. Rev. Neurosci.* 7, 655. With permission from Macmillan Publishers Ltd.]

directional tendency and homozygotes for the ancestral allele would show either ambidexterity or equally probable shifts to the left or right. Genomewide association studies have found several candidate loci for handedness including a gene on chromosome 2, *LRRTM1* (OMIM 610867). Yet this association holds only if left-handedness occurs in combination with schizophrenia and dyslexia.[23] Because the gene is paternally **imprinted** (maternally suppressed) the associated allele has to be located on the paternally inherited active chromosome to have an effect on this combined phenotype. Other regions in chromosomes 10, 12, 15, and 17 have been highlighted by both linkage and association studies suggesting that the genetic basis of handedness may be complex, and that more than one gene is involved in determining the lateral asymmetry of the brain.[15, 61] Analysis of gene expression in 12-week-old human embryos has detected 27 genes that consistently show hemispheric differences in expression (**Figure 8.16**). Early onset of lateral expression differences supports the observations made from ultrasound images that most fetuses prefer to suck their right thumbs, consistent with their postnatal handedness. Overall, the genetic evidence suggests that the basis of lateral asymmetry of human brain is complex and set early in the prenatal development.

Regions of human brain cortex associated with word formation (Broca's area) and recognition (Wernicke's area) also show significant lateral asymmetry. Analyses of neurons from these areas have identified hundreds of genes with expression differences between the hemispheres. These differentially expressed genes appear to be significantly associated with fast-evolving noncoding sequences on the human lineage. Among these is the *CNTNAP2* gene which has been associated with neuro-developmental disorders affecting language.[41] Several language disorders have high heritability and some of them segregate in families in Mendelian form. Linkage analysis of these families has revealed a number of genes, some of which have been further analyzed using comparative genomic approaches.[21] The best-studied gene is the forkhead box P2 (*FOXP2*, OMIM 605317), which was initially discovered in a pedigree of 15 affected individuals with developmental verbal dyspraxia characterized by impaired learning, poor mobility of the lower face, and deficits in language reception or grammar.[40] The amino acid sequence of *FOXP2* protein is highly conserved in vertebrates, and it is known to regulate the development of diverse regions of the brain, including those involved in speech production (for example, Broca's area). Humans and Neanderthals share the same sequence of the protein, which differs from the conserved ancestral copy found in chimpanzee, gorilla, and macaque by two nonsynonymous substitutions (**Figure 8.17**). When these two substitutions are inserted into the mouse *Foxp2* gene, the mice show specific changes in dopamine levels, neuronal morphology, synaptic plasticity in the striatum, and pup vocalizations. These findings together with the role for FOXP2 in vocal learning in birds, suggest that the two human-specific amino acid changes affect specifically cortico-striatal circuits, which are generally involved in the learning of motoric and cognitive skills. Since such learning is also relevant for speech and language, the amino acid changes in FOXP2 could have contributed to the evolution of language acquisition rather than adult language processing.[19] Interestingly, other genomic regions associated with language acquisition, in particular specific language impairment loci (*SLI1*, OMIM 606711 and *SLI2*, OMIM 606712) related to a child's ability to repeat pronounceable nonsense words, show specific change in copy number in the human lineage.[21]

Another so far less-studied example where a gene involved in human language disorder is also found to have a related function in other animals is *CYP19A1*. *CYP19A1* mutations in humans have been found in association with dyslexia, while in other vertebrates (fish and birds) orthologs are known to be involved in sexual differentiation of the brain and the regulation of vocalization.[3] Such examples show that although human language may have some qualitative differences from nonhuman primate communication, its genetic and biological basis is complex and includes components that show some degree of continuity with other species.

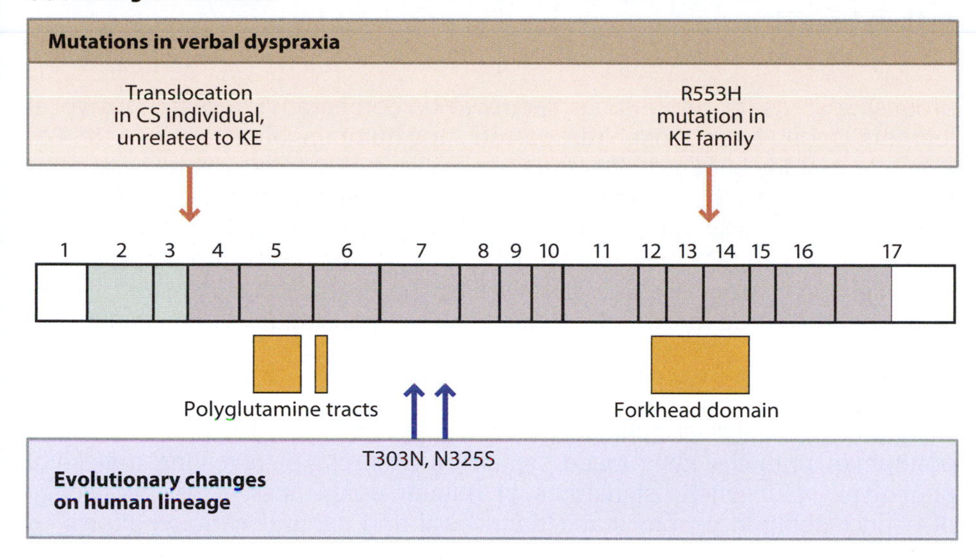

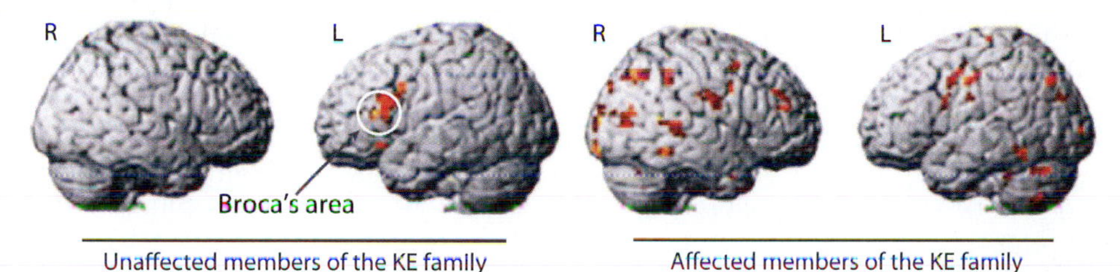

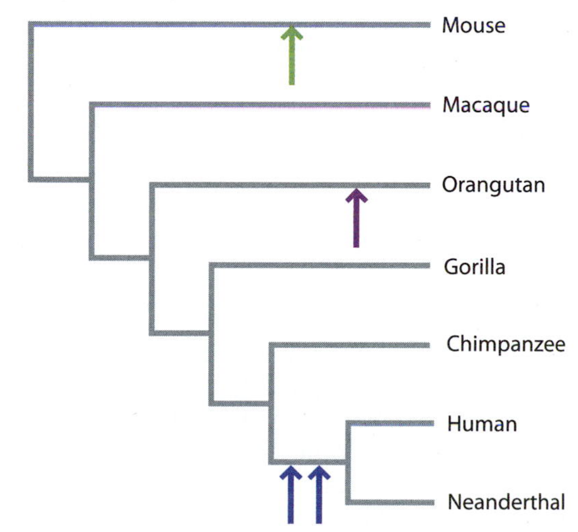

Figure 8.17: Structure and evolution of the *FOXP2* gene implicated in speech and language development.
(a) Structure of the *FOXP2* transcript showing 17 exons (*boxes*); note that several variant transcripts resulting from alternative splice patterns are known, and the exon numbering can be different. The open reading frame may extend from exon 2 to 17 (*light gray*) or from exon 4 to 17 (*dark gray*). The locations of structural features within the protein are shown with *yellow* boxes, inactivating mutations (translocation, R553H) that lead to speech disorder in a family, KE, and an unrelated individual, CS, are shown by *red* arrows, and evolutionary changes within the human lineage (Thr303Asn, Asn325Ser) by *blue* arrows according to Enard W et al. (2002) *Nature* 418, 869 and Lai CS et al. (2001) *Nature* 413, 519. (b) Neuroimaging: group-average activation in the unaffected and affected members of the KE family carrying out a language task. (c) Amino acid changes during *FOXP2* evolution within mammals are shown by arrows on the phylogeny, which represents the branching order but not the time-scale. Note that the protein is highly conserved: chimpanzee, gorilla, and macaque share the mammalian ancestral state of the protein. Considering this degree of conservation, it is unusual to observe two amino acid changes on the short branch leading to humans and Neanderthals [a, data from Enard W et al. (2002) *Nature* 418, 869; Krause J et al. (2007) *Curr. Biol.* 17, 1908. b, from Fisher SE & Marcus GF (2006) *Nat. Rev. Genet.* 7, 9. With permission from Macmillan Publishers Ltd.]

Human language abilities are certainly complex in terms of their biological basis and result from the interaction of many genes. With many candidate loci being discovered in association with specific language disorders, these can now be subjected to comparative genomic analyses involving great ape and extinct hominin genomes. Further progress in understanding the evolutionary changes that led to the human-specific eloquent verbal communication will then require functional studies in model organisms, following the example of diverse functional studies already performed on the *FOXP2* gene. One of the

Hummingbird (*Calypte anna*) with chicks

Red-and-green Macaw (*Ara chloropterus*)

Nightingale (*Luscinia megarhynchos*)

Figure 8.18: Vocal learners among birds—songbirds, parrots, and hummingbirds.

promising areas for future focus seems to be comparative research on vocal learners among birds—songbirds, parrots, and hummingbirds (**Figure 8.18**). For example, a study of genes differentially regulated in brain nuclei necessary for producing learned vocalizations in avian vocal learners versus non-learners revealed a calcium-binding protein, parvalbumin (PVALB, OMIM 168890) as the top candidate gene.[28] Consistent with the vocal learner versus non-learner pattern, higher parvalbumin expression was also observed in the brainstem tongue motor neurons used for speech production in humans relative to macaques.

What next?

Comparative studies of behavior, morphology, and genetics in humans and nonhuman primates have made significant progress in revealing individual phenotypic and genetic signatures of human uniqueness. Also, knowledge of extinct hominin genomes and their fossil and cultural heritage allows us to distinguish traits shared by several hominins from traits strictly specific to modern humans. However, we still do not know what genetic changes have made us human. Becoming human has certainly not been a singular moment but rather a journey across multiple adaptive and possibly neutral phenotype changes separated in time. In the near future a plethora of human and nonhuman genome sequences will be generated thanks to technological advances in sequencing, and ancient DNA studies will continue to contribute important data from extinct species. Great progress is also expected in our understanding of gene expression, whereas the key developments in bridging phenotypic and genetic information will rely upon advances in functional annotation of the genome.

SUMMARY

- The 6–7 million years of hominid evolution since humans and chimpanzees last shared a common ancestor was characterized firstly by a move toward bipedal locomotion, and subsequently by a dramatic increase in brain size.
- Humans differ from other apes in a number of phenotypic traits. Sometimes, particularly in behavioral aspects, the definition of human uniqueness depends mostly on our knowledge of nonhuman primates.
- Identifying the important genetic changes involved during hominid evolution can be approached in one of two ways: defining important human-specific phenotypes (for example, language) and researching their genetic bases, or identifying all genetic changes and asking which are of functional importance.
- Humans differ from chimpanzees by ~35 million nucleotide substitutions and another ~5 million structural differences, of which most are insertions and deletions of one or a few bases, but some comprise millions of base pairs. Only a small subset of these changes may have had biological significance, including some of the mutations that delete or insert genes or change gene expression patterns.
- The greatest consistent difference in gene expression patterns between humans and chimpanzees is seen in the testis. Expression patterns in brain are conserved, but show relatively higher rates of change during human compared with chimpanzee evolution. Changes at a few microRNA genes may explain concerted gene expression changes at many genes.
- The search for genes involved in human-specific characters, such as language, laterality, and life-histories, has identified a number of candidates that may contribute a small amount to these phenotypes.

QUESTIONS

Question 8–1: Give an example of a gene inactivation, a regulatory mutation, and a copy-number change that are believed to be important in the differentiation of human and great ape phenotypes.

Question 8–2: List five important phenotypic differences between humans and chimpanzees.

Question 8–3: Linnaeus placed humans and chimpanzees in the same genus. Was he right to do so?

Question 8–4: Are humans just apes with unusually delayed development?

Question 8–5: Discuss the time-scale and genetic basis of the evolution of the large human brain.

Question 8–6: The finding that Neanderthals carry the same FOXP2 amino acid sequence as modern humans prompted the headline: "Neanderthals could speak." Was this justified?

Question 8–7: Use the Genome Browser (http://genome.ucsc.edu/cgi-bin/hgGateway) Denisova Assembly to determine whether the Denisovan genome carries any of the common human red hair defining mutations: rs1805008(T), rs1805009(C), and rs1805007(T).

Question 8–8: Curiously, genetic studies of lice have illuminated human evolutionary history. Describe briefly how genetic analyses of other nonprimate species have also proved informative.
See Chapters 12, 13, and more...

REFERENCES

The references highlighted in purple are considered to be important (for this chapter) by the authors.

1. **Angata T, Varki NM & Varki A** (2001) A second uniquely human mutation affecting sialic acid biology. *J. Biol. Chem.* **276**, 40282–40287.
2. **Annett M** (2002) Handedness and Brain Asymmetry: The Right Shift Theory. Hove, UK: Psychology Press.
3. **Anthoni H, Sucheston LE, Lewis BA et al.** (2012) The aromatase gene CYP19A1: Several genetic and functional lines of evidence supporting a role in reading, speech and language. *Behav Genet.* 42, 509–527.
4. **Bakewell MA, Shi P & Zhang J** (2007) More genes underwent positive selection in chimpanzee evolution than in human evolution. *Proc. Natl Acad. Sci. USA* **104**, 7489–7494.
5. **Bekpen C, Marques-Bonet T, Alkan C et al.** (2009) Death and resurrection of the human IRGM gene. *PLoS Genet.* **5**, e1000403.
6. **Bogin B** (2009) Childhood, adolescence, and longevity: a multilevel model of the evolution of reserve capacity in human life history. *Am. J. Hum. Biol.* **21**, 567–577.
7. **Boyko AR, Williamson SH, Indap AR et al.** (2008) Assessing the evolutionary impact of amino acid mutations in the human genome. *PLoS Genet.* **4**, e1000083.
8. **Bramble DM & Lieberman DE** (2004) Endurance running and the evolution of *Homo*. *Nature* **432**, 345–352.
9. **Brawand D, Soumillon M, Necsulea A et al.** (2011) The evolution of gene expression levels in mammalian organs. *Nature* **478**, 343–348.
10. **Britten RJ** (2010) Transposable element insertions have strongly affected human evolution. *Proc. Natl Acad. Sci. USA* **107**, 19945–19948.
11. **Bustamante CD, Fledel-Alon A, Williamson S et al.** (2005) Natural selection on protein-coding genes in the human genome. *Nature* **437**, 1153–1157.
12. **Charrier C, Joshi K, Coutinho-Budd J et al.** (2012) Inhibition of SRGAP2 function by its human-specific paralogs induces neoteny during spine maturation. *Cell* 149, 923–935.
13. **Chou HH, Hayakawa T, Diaz S et al.** (2002) Inactivation of CMP-*N*-acetylneuraminic acid hydroxylase occurred prior to brain expansion during human evolution. *Proc. Natl Acad. Sci. USA* **99**, 11736–11741.
14. **Consortium CSaA** (2005) Initial sequence of the chimpanzee genome and comparison with the human genome. *Nature* **437**, 69–87.
15. **Corballis MC** (2009) The evolution and genetics of cerebral asymmetry. *Philos. Trans. R. Soc. Lond. B Biol. Sci.* **364**, 867–879.
16. **Crow T** (2002) Sexual selection, timing and an X-Y homologous gene: did *Homo sapiens* speciate on the Y chromosome? In The Speciation of Modern *Homo sapiens* (T Crow ed.), pp 197–216. Oxford, UK: Oxford University Press.
17. **Darwin C** (1872) The Expression of the Emotions in Man and Animals. London: John Murray.
18. **de Boer B** (2011) Loss of air sacs improved hominin speech abilities. *J. Hum. Evol.* **62**, 1–6.
19. **Enard W** (2011) FOXP2 and the role of cortico-basal ganglia circuits in speech and language evolution. *Curr. Opin. Neurobiol.* **21**, 415–424.
20. **Evans PD, Gilbert SL, Mekel-Bobrov N et al.** (2005) Microcephalin, a gene regulating brain size, continues to evolve adaptively in humans. *Science* **309**, 1717–1720.
21. **Fisher SE & Marcus GF** (2006) The eloquent ape: genes, brains and the evolution of language. *Nat. Rev. Genet.* **7**, 9–20.
22. **Fitch WT & Reby D** (2001) The descended larynx is not uniquely human. *Proc. Biol. Sci.* **268**, 1669–1675.
23. **Francks C, Maegawa S, Lauren J et al.** (2007) LRRTM1 on chromosome 2p12 is a maternally suppressed gene that is associated paternally with handedness and schizophrenia. *Mol. Psychiatry* **12**, 1129–1139.
24. **Frayer DW, Lozano M, Bermudez de Castro JM et al.** (2011) More than 500,000 years of right-handedness in Europe. *Laterality* 17, 51–69.
25. **Gelstein S, Yeshurun Y, Rozenkrantz L et al.** (2011) Human tears contain a chemosignal. *Science* **331**, 226–230.
26. **Green RE, Krause J, Briggs AW et al.** (2010) A draft sequence of the Neandertal genome. *Science* **328**, 710–722.
27. **Hahn MW, Demuth JP & Han SG** (2007) Accelerated rate of gene gain and loss in primates. *Genetics* **177**, 1941–1949.

28. **Hara E, Rivas MV, Ward JM et al.** (2012) Convergent differential regulation of parvalbumin in the brains of vocal learners. *PLoS One* **7**, e29457.

29. **Haslam M, Hernandez-Aguilar A, Ling V et al.** (2009) Primate archaeology. *Nature* **460**, 339–344.

30. **Haygood R, Fedrigo O, Hanson B et al.** (2007) Promoter regions of many neural- and nutrition-related genes have experienced positive selection during human evolution. *Nat. Genet.* **39**, 1140–1144.

31. **Hu HY, Guo S, Xi J et al.** (2011) MicroRNA expression and regulation in human, chimpanzee, and macaque brains. *PLoS Genet.* **7**, e1002327.

32. **Humphrey LT** (2010) Weaning behaviour in human evolution. *Semin. Cell Dev. Biol.* **21**, 453–461.

33. **Irie A, Koyama S, Kozutsumi Y et al.** (1998) The molecular basis for the absence of *N*-glycolylneuraminic acid in humans. *J. Biol. Chem.* **273**, 15866–15871.

34. **Jablonski N** (2010) The naked truth. *Sci. Am.* 302, 42–49.

35. **Johnstone RA & Cant MA** (2010) The evolution of menopause in cetaceans and humans: the role of demography. *Proc. Biol. Sci.* **277**, 3765–3771.

36. **Keren H, Lev-Maor G & Ast G** (2010) Alternative splicing and evolution: diversification, exon definition and function. *Nat. Rev. Genet.* **11**, 345–355.

37. **Khaitovich P, Enard W, Lachmann M & Pääbo S** (2006) Evolution of primate gene expression. *Nat. Rev. Genet.* **7**, 693–702.

38. **King MC & Wilson AC** (1975) Evolution at two levels in humans and chimpanzees. *Science* **188**, 107–116.

39. **Knowles DG & McLysaght A** (2009) Recent de novo origin of human protein-coding genes. *Genome Res.* **19**, 1752–1759.

40. **Lai CS, Fisher SE, Hurst JA et al.** (2001) A forkhead-domain gene is mutated in a severe speech and language disorder. *Nature* **413**, 519–523.

41. **Lambert N, Lambot MA, Bilheu A et al.** (2011) Genes expressed in specific areas of the human fetal cerebral cortex display distinct patterns of evolution. *PLoS One* **6**, e17753.

42. **Lieberman DE** (1998) Sphenoid shortening and the evolution of modern human cranial shape. *Nature* **393**, 158–162.

43. **Lindblad-Toh K, Garber M, Zuk O et al.** (2011) A high-resolution map of human evolutionary constraint using 29 mammals. *Nature* **478**, 476–482.

44. **Locke DP, Hillier LW, Warren WC et al.** (2011) Comparative and demographic analysis of orang-utan genomes. *Nature* **469**, 529–533.

45. **Lopez-Valenzuela M, Ramirez O, Rosas A et al.** (2012) An ancestral miR-1304 allele present in Neanderthals regulates genes involved in enamel formation and could explain dental differences with modern humans. *Mol. Biol. Evol.* 29, 1797–1806.

46. **Lu ZX, Peng J & Su B** (2007) A human-specific mutation leads to the origin of a novel splice form of neuropsin (KLK8), a gene involved in learning and memory. *Hum. Mutat.* **28**, 978–984.

47. **Marchant L & McGrew WC** (1996) Laterality of limb function in wild chimpanzees of Gombe National Park: comprehensive study of spontaneous activities. *J. Hum. Evol.* **30**, 427–443.

48. **McGrew WC, Marchant LF & Hunt KD** (2007) Etho-archaeology of manual laterality: well digging by wild chimpanzees. *Folia Primatol. (Basel)* **78**, 240–244.

49. **McLean CY, Reno PL, Pollen AA et al.** (2011) Human-specific loss of regulatory DNA and the evolution of human-specific traits. *Nature* **471**, 216–219.

50. **Ng PC, Levy S, Huang J et al.** (2008) Genetic variation in an individual human exome. *PLoS Genet.* **4**, e1000160.

51. **Nielsen R, Bustamante C, Clark AG et al.** (2005) A scan for positively selected genes in the genomes of humans and chimpanzees. *PLoS Biol.* **3**, e170.

52. **Ohno S, Wolf U & Atkin NB** (1968) Evolution from fish to mammals by gene duplication. *Hereditas* **59**, 169–187.

53. **Olson MV & Varki A** (2003) Sequencing the chimpanzee genome: insights into human evolution and disease. *Nat. Rev. Genet.* **4**, 20–28.

54. **Perry GH, Dominy NJ, Claw KG et al.** (2007) Diet and the evolution of human amylase gene copy number variation. *Nat. Genet.* **39**, 1256–1260.

55. **Plavcan JM, Lockwood CA, Kimbel WH et al.** (2005) Sexual dimorphism in *Australopithecus afarensis* revisited: how strong is the case for a human-like pattern of dimorphism? *J. Hum. Evol.* **48**, 313–320.

56. **Pollard KS, Salama SR, Lambert N et al.** (2006) An RNA gene expressed during cortical development evolved rapidly in humans. *Nature* **443**, 167–172.

57. **Reich D, Green RE, Kircher M et al.** (2011) Genetic history of an archaic hominin group from Denisova Cave in Siberia. *Nature* **468**, 1053–1060.

58. **Roca AL, Ishida Y, Nikolaidis N et al.** (2009) Genetic variation at hair length candidate genes in elephants and the extinct woolly mammoth. *BMC Evol. Biol.* **9**, 232.

59. **Ruxton GD & Wilkinson DM** (2011) Thermoregulation and endurance running in extinct hominins: Wheeler's models revisited. *J. Hum. Evol.* **61**, 169–175.

60. **Scally A, Dutheil JY, Hillier LW et al.** (2012) Insights into hominid evolution from the gorilla genome sequence. *Nature* **483**, 169–175.

61. **Scerri TS, Brandler WM, Paracchini S et al.** (2011) PCSK6 is associated with handedness in individuals with dyslexia. *Hum. Mol. Genet.* **20**, 608–614.

62. **Shea JJ** (2011) *Homo sapiens* is as *Homo sapiens* was. *Curr. Anthropol.* **52**, 1–35.

63. **Smith TM, Tafforeau P, Reid DJ et al.** (2007) Earliest evidence of modern human life history in North African early *Homo sapiens*. *Proc. Natl Acad. Sci. USA* **104**, 6128–6133.

64. **Somel M, Franz H, Yan Z et al.** (2009) Transcriptional neoteny in the human brain. *Proc. Natl Acad. Sci. USA* **106**, 5743–5748.

65. **Somel M, Liu X, Tang L et al.** (2011) MicroRNA-driven developmental remodeling in the brain distinguishes humans from other primates. *PLoS Biol.* **9**, e1001214.

66. **Stedman HH, Kozyak BW, Nelson A et al.** (2004) Myosin gene mutation correlates with anatomical changes in the human lineage. *Nature* **428**, 415–418.

67. **Sutter NB, Bustamante CD, Chase K et al.** (2007) A single IGF1 allele is a major determinant of small size in dogs. *Science* **316**, 112–115.

68. **Toups MA, Kitchen A, Light JE & Reed DL** (2011) Origin of clothing lice indicates early clothing use by anatomically modern humans in Africa. *Mol. Biol. Evol.* **28**, 29–32.

69. **Trinkaus E** (2005) Early modern humans. *Annu. Rev. Anthropol.* **34**, 207–230.

70. **Varki A** (2010) Colloquium paper: uniquely human evolution of sialic acid genetics and biology. *Proc. Natl Acad. Sci. USA* **107** Suppl 2, 8939–8946.

71. **Varki A & Altheide TK** (2005) Comparing the human and chimpanzee genomes: searching for needles in a haystack. *Genome Res.* **15**, 1746–1758.

72. **Varki A, Geschwind DH & Eichler EE** (2008) Explaining human uniqueness: genome interactions with environment, behaviour and culture. *Nat. Rev. Genet.* **9**, 749–763.

73. **Voisin JL** (2008) The Omo I hominin clavicle: archaic or modern? *J. Hum. Evol.* **55**, 438–443.

74. **Wells JC & Stock JT** (2007) The biology of the colonizing ape. *Am. J. Phys. Anthropol.* **134** Suppl 45, 191–222.

75. **Wu DD, Irwin DM & Zhang YP** (2011) De novo origin of human protein-coding genes. *PLoS Genet.* **7**, e1002379.

ORIGINS OF MODERN HUMANS

CHAPTER NINE

This chapter presents our current understanding of where and when modern humans arose. In the last two chapters, we saw that the split between humans and our closest living relatives, chimpanzees and **bonobos**, occurred approximately 6–7 MYA, and that some of the important genetic changes on the human lineage could be identified. Here, we will consider events after the chimpanzee–human split, leading to the origin of our species, modern humans, a much more recent event that occurred within the last 200 KY. We will consider evidence of many kinds: for much of the period, the only substantial source is the fossil record, and we will have to confront a plethora of names and opinions in order to extract the crucial points. After ~2.6 MYA, the exquisitely rare fossil finds start to be supplemented by a more abundant record, archaeology, and more detailed inferences about how our ancestors behaved become possible. And since the genetic diversity apparent among humans alive today has mostly accumulated over the past 1 MY or so, after this time we can begin to make use of the ever-expanding genetic information that forms the core of this book. We will see again and again the overwhelming importance of Africa for human evolution, as the place where both our genus *Homo* and species *H. sapiens* originated, and the continent that contains most of our genetic diversity and the roots of most of our genetic lineages. While it is clear that there were multiple dispersals of our genus out of Africa, we will have to consider the extent to which early expansions contributed to contemporary human gene pools, a question debated for decades and initially presented as a choice between a multiregional model, in which the early migrants contributed extensively, and an out-of-Africa model, in which they were completely replaced. A marriage of paleontology with genetics has given rise to the field of ancient DNA analysis, and has provided data supporting limited genetic contributions from both **Neanderthals** (*H. neanderthalensis*) and **Denisovans** to some but not all modern humans, giving us a new way of looking at old questions. We will thus set the stage for Section 4, in which we will see how modern humans were able to colonize the entire planet.

There can be ambiguity about the meaning of the term "human": we know that we are human, and that chimpanzees are not, but where in the continuum of evolutionary ancestors that link us should we draw the line between human and nonhuman? It makes no sense to ask which human baby had a nonhuman mother. Any decision is somewhat arbitrary; here, we will draw the line between the genus *Homo* and other genera, but will sometimes refer to "archaic" and "modern" humans when we need to refer to early and late members of our genus. We will also use the phrase the "human lineage," meaning the lineage that *led to* humans, and includes all species on the human line since the chimpanzee–human split. There is also ambiguity about the meaning of the term "ape": some would contrast apes with humans as mutually exclusive groups, but here we consider humans as one of the great apes.

Humans differ from our last common ancestor with other apes in several respects:

1. Morphology: the structure of our bodies, including our brains
2. Behavior: from the way we walk to our social organization, complex tool use, and language
3. Genetics: many neutral and some selected changes. Individual changes could have occurred independently, or in packages linked by selection or drift

Two kinds of evidence have been particularly important for understanding the likely times and places of the changes. Fossils and archaeology provide information about the environment, morphology, and, to some extent, the behavior of our ancestors and the related species present at different times. Genetics reveals the history of the lineages that have survived in living humans, and is also starting to include information from more ancient individuals who lived within the last ~100 KY.

The anthropologist Vincent Sarich once contrasted the two by remarking "I know my molecules had ancestors, the paleontologist can only hope that his fossils had descendants."

9.1 EVIDENCE FROM FOSSILS AND MORPHOLOGY

Both extinct and living species more closely related to humans than to chimpanzees are known as **hominins**; the term **hominid** has been used as an alternative to hominin in the past, but now hominid generally includes all fossils of humans, other great apes, and their immediate ancestors (**Section 7.1**). Candidate fossils can be examined for characteristics that differ between humans and other apes (**Box 9.1**) and their likely position in the phylogeny determined. This aim is hindered by the rarity of fossil hominins: our ancestors appear to have existed at low population densities and were seldom fossilized; furthermore, many of the early fossils are likely to have been predated by carnivores, but this became rarer as defenses against carnivores improved. In addition, the fossils that are found are very incomplete: teeth are the most frequent finds, then the relatively tough bones of the head, the **cranium** (skull excluding lower jaw) and

Box 9.1: Human characteristics that can be preserved in rock

- Brain cavity: absolutely and relatively larger in humans compared with other apes; attachment to spinal cord is more centrally located in the base of the skull.
- Teeth: enamel is thicker; canines and incisors are smaller in humans than other apes.
- Chest: more cylindrical in humans; that of other apes widens toward the base to accommodate larger gut.
- Legs longer, feet and pelvis adapted for upright walking in humans.
- Bipedalism can also be recognized from tracks of preserved footprints.
- Hands adapted for grasping with longer thumb in humans.
- Slow development and prolonged childhood in humans, identified from the "age at death" of fossils, measured, for example, from growth lines on teeth.
- Tool use more extensive in humans; complex tools and fire specific to humans.

mandible (lower jaw), but other bones (often collectively called **post-cranial**) are less often preserved, and traces of soft body tissues only in very exceptional circumstances, such as **endocasts** of the brain. Moreover, there are conceptual problems: there would undoubtedly have been variation within each species due to differences between individuals within the same population including **sexual dimorphism**, and also geographical differences, as well as changes over time. A decision has to be made about which fossils should be grouped under the same name (genus and species), and this inevitably has an arbitrary element. The person who discovers and names a new hominin species gains considerable credit and even celebrity, especially if it can be claimed as a direct human ancestor. The cynic might consider that there is thus a danger of excessive splitting and overemphasis of human similarities at the expense of features shared with other apes. The overabundance of names can be very confusing and the reader should be aware that the field is subject to continuous revision; opinions about the number of genera and species, and their relationships, differ considerably between experts, and change over time. Fortunately, images and information about many of the key fossils are readily available, from Websites, for example, http://www.archaeologyinfo.com or http://www.modernhuman-origins.com, or from books.[58]

Some fossils that may represent early hominins from 4–7 MYA are known from Africa

We saw in Chapter 7 that the human and chimpanzee lineages are generally assumed to have diverged about 6–7 MYA. The earliest hominin fossils should therefore originate from this period, but not before (see **Table 9.1** for a summary of physical dating methods). A few candidate fossils of this age are known

TABLE 9.1:
PHYSICAL DATING METHODS FOR FOSSIL SITES

		Useful time span		
Method	Basis	Type[a]	(KY)	Materials used
Isotopic:	radioactive decay	absolute		
Uranium–Thorium (U–Th)			>500	carbonates
Uranium–Lead (U–Pb)			1–500	carbonates/zircon
^{40}K-^{40}Ar, ^{40}Ar/^{39}Ar			>10	volcanic rocks
^{14}C			<50	organic
Cosmogenic nuclide burial	*in situ* formation of ^{26}Al, ^{10}Be and subsequent decay		<5000	quartz
Trapped electron dating:	electrons caught in defects in crystal lattices	absolute		
Thermoluminescence (TL)			1–200	quartz, feldspar
Optically Stimulated Luminescence (OSL)			1–500	quartz, feldspar
Electron Spin Resonance (ESR)			1–3000	carbonates, silicates
Amino acid racemization	chemical instability of L amino acids	relative	40–200	organic materials
Paleomagnetism	direction of magnetic field (N or S)	correlative	>780	iron-rich rocks
Paleontology	comparison of fossils with other sites	relative	all	fossils

[a] Absolute, provides a date in years; relative, allows comparisons with other sites or materials; correlative, provides findings which may allow refinement of an approximate date determined by other methods.

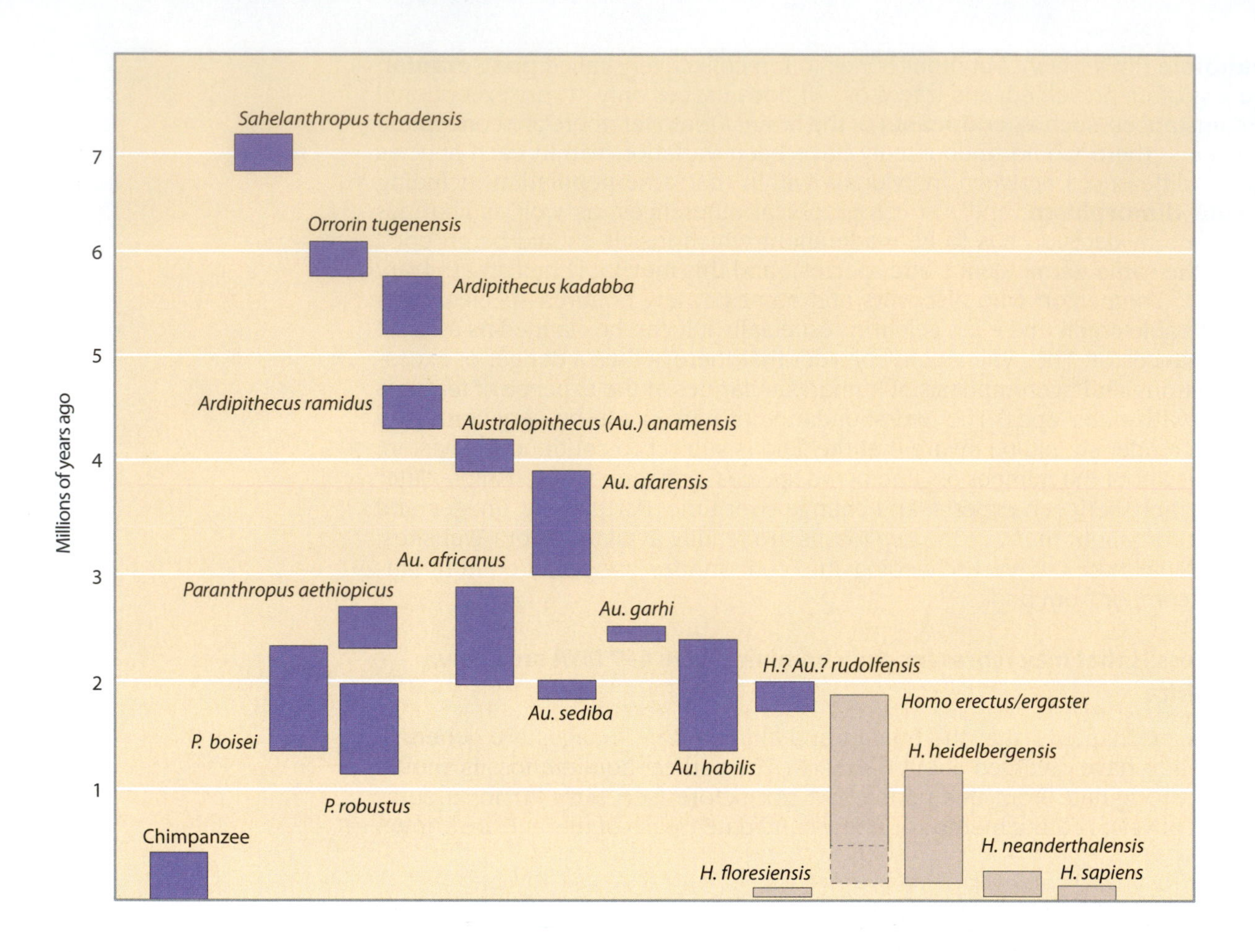

Figure 9.1: Fossil hominins. The time span of each species indicates either the uncertainty in dating or the times of the earliest and latest fossils, whichever is larger. Dotted lines indicate particular uncertainty about the later dates for *Homo erectus*. *Blue*: found only in Africa. *Gray*: found in Africa and elsewhere, or only outside Africa. Many aspects of the classification of these fossils are still debated and are likely to be revised.

(**Figures 9.1** and **9.2**). The oldest of these consists of a nearly complete cranium (Toumai, meaning "hope of life"), jaw fragment, and teeth from Chad, designated *Sahelanthropus tchadensis*,[12] dated to 6.8–7.2 MYA. This date would place it at the upper limit of the divergence time between humans and chimpanzees as estimated from genetic data (**Section 7.3**), and thus the date and relationship of *S. tchadensis* to the hominin lineage are key issues about which there is debate. Despite having a chimpanzee-sized brain, *S. tchadensis* has a number of features that link it to the hominin lineage, including a relatively flat face, attachment of the spinal cord at the bottom of the skull rather than at the back, and intermediate tooth enamel thickness. Unfortunately, an understanding of its mode of locomotion awaits the discovery of post-cranial remains: the position of attachment of the spinal cord hints at an upright stance, but is also consistent with other postures, and interpretation is hindered by the distorted and fragmentary nature of this part of the skull. The environment appears to have had a mosaic structure including forest, **savanna**, desert, and lakes.

Two genera dating to between 5 MYA and 6 MYA have been described: *Orrorin* and *Ardipithecus*. *Orrorin tugenensis*, also called Millennium Man, is represented by 13 fossils dating to about 5.8–6.1 MYA from the Tugen Hills in Kenya, East Africa.[56] These fossils include three fragmentary thigh bones which indicate upright walking, and small thick-enameled molars judged to link *Orrorin* to the human lineage. *Ardipithecus kadabba* is represented by 11 fossils, including a nearly complete foot, from the Middle Awash in Ethiopia, dated to between 5.2 and 5.8 MYA.[20] Despite the presence of thin enamel characteristic of chimpanzees, some researchers consider *Ardipithecus* to lie on the human lineage and also note the similarity of *Ardipithecus kadabba* teeth with those of *Orrorin* and *Sahelanthropus*, suggesting that "it is possible that all of these remains represent specific or subspecific variation within a single genus."[20] However, none of the key features of this group of fossils from 5–7 MYA—tooth morphology,

Figure 9.2: Sites of the earliest hominin and gracile australopithecine fossils. Sites of the earliest hominin fossils (7.0–4.5 MYA) in Chad and East Africa (*red*) and gracile australopithecine fossils (4.0–2.0 MYA), spread over more of Africa but absent from the rest of the world (*blue*). In the overlapping locations, the points representing the later sites have been offset to retain visibility.

attachment position of the spinal cord, or foot structure—provides unequivocal evidence for hominin status and after a decade of debate there is still no consensus about their relationships to one another or to chimpanzees and humans.[76]

The period between 5 MYA and 4 MYA is represented by an extensive but crumbling collection of fossils from 36 or more *Ardipithecus ramidus* individuals (4.7–4.3 MYA) from the Middle Awash in Ethiopia, including much of the skull, pelvis, lower arms, and feet from one female, Ardi[70] dated to ~4.4 MYA (see **Opinion Box 7**). Ardi would have weighed around 50 kg when alive and stood ~1.2 m tall: both the pelvis and foot structures provide strong evidence for bipedalism, although with a gait distinguishable from later hominins, and she may have spent substantial time in the surrounding trees. The earliest *Australopithecus* fossils are only slightly younger, and are considered in the next section.

Where the appropriate hominin features can be identified, all of these early fossils appear to represent chimpanzee-sized but apparently upright-walking species that lived in or near a wooded environment. The evidence for **bipedalism** at such an early date is an important finding and raises the question of how the human–chimpanzee common ancestor moved around. The most radical possibility is that bipedalism is the **primitive** trait, present in the common ancestor of all three species, and **knuckle walking** in chimpanzees and gorillas is **derived**, although this possibility requires the nonparsimonious assumption that knuckle walking arose independently in the two great ape lineages (**Section 7.1**). The difficulty of distinguishing between human and chimpanzee lineages at times close to their split should not be a surprise (and see Opinion Box 7), but is made worse by the lack of chimpanzee fossils dating to before 0.5 MYA.

Fossils of australopithecines and their contemporaries are known from Africa

Most hominin fossils dating after about 4.2 MYA and before the appearance of *Homo* are ascribed to the genus *Australopithecus* (*Au.*). The presence of a second genus, *Kenyanthropus,* has also been proposed for a series of 3.5–3.2 MYA fossils,[38] but many researchers consider these specimens to belong to *Au. afarensis*. The earliest known *Australopithecus* species is *Au. anamensis* from Lake Turkana and other sites from Kenya to Ethiopia dating between 3.9 and 4.2 MYA[37, 72] and it is generally considered to be the oldest indisputable hominin. This species is distinguished from the better-known and slightly later *Au. afarensis*[28] mainly by the larger size of the males (weight estimated as ~55 kg

The ongoing and sometimes contentious debates about whether this or that fossil taxon is the first or earliest hominin,[70] or the first or earliest member of the genus *Homo*,[8] have much in common. In both cases it boils down to the combination of a matter of principle plus arguments about the presence or absence of a particular morphology in a particular fossil. This opinion box focuses on the matter of principle.

Those who argue in favor of a fossil taxon being the "first" or "earliest" member of a clade do so on the basis that the presence of *any* of the distinctive morphological features seen in later members of that clade "clinches the deal." They assume there is a direct and simple relationship between morphological similarity and genetic relatedness; *all* shared morphology means shared recent ancestry, period.

So why are other researchers[76] skeptical about this principle, especially when it is applied to the hominin fossil record? The main reason is that when it has been "tested" in other mammal groups it has been found wanting. So, for example, when the relationships among the members of a contemporary group that are supported by morphology are compared with the relationships that are supported by molecular evidence, there are discrepancies. And when paleontologists have examined the fossil records of other mammal clades, they also see compelling evidence that similar morphology must have evolved more than once. This has been the case for investigations of bovids, equids, elephantids, carnivores, and Old World monkeys. There is no reason to assume that extinct higher primate lineages that lived at the same time in the same territory were immune from the tendency to adapt in similar morphological and phylogenetically confounding ways to similar ecological challenges. Long ago, the zoologist Ray Lankester suggested the term **homoplasy** should be used for morphology that is seen in sister taxa, but not in their most recent common ancestor. Because homoplasy can be mistaken for shared derived similarity, it complicates attempts to reconstruct relationships. Homoplasies give the impression that two taxa are more closely related than they really are.

One could cope with the confounding effects of homoplasy if the "noise" that it generates was trivial compared with the strength of the phylogenetic "signal." But in some attempts to infer relationships among extant higher primates using skeletal and dental (that is, hard-tissue) data in the form of either traditional non-metrical characters or characters generated from metrical data, the ratio between noise and signal was in the order of 1:2. The results of these analyses were not only frustratingly inconclusive, but when they were compared with the pattern of relationships generated using molecular data, some were found to be misleading. Other researchers suggested this dismal performance was due to the exclusion of character state data from fossil taxa, but this is arguable because soft-tissue characters (for which there are no fossil data) *are* capable of recovering a pattern of relationships among extant higher primates that is consistent with the molecular evidence.[15] It is not just the absence of fossils, it must be something about hard-tissue evidence. Thankfully not all hard-tissue evidence is problematic; it *can* produce results congruent with the relationships generated from molecular data as long as the anatomical regions targeted have a high enough signal to noise ratio and as long as the information about morphology is detailed enough. It is not good news for paleoanthropologists that the type of data the fossil record provides (that is, mostly craniodental hard-tissue morphology) seems to be particularly prone to homoplasy when used at this relatively fine taxonomic level.

The important point is that shared similarities can only take one so far in determining phylogenetic relationships, because homoplasy, as well as uncertainties in determining the polarity of character transformation, have the potential to generate substantial noise that serves to confound attempts to generate reliable hypotheses about relationships. These considerations have clear implications for generating hypotheses about the phylogenetic position of *Ardipithecus* (**Figure 1**) and *Australopithecus sediba*. Even if these taxa share *some* derived features with either later Pliocene hominins or with later *Homo*, it would be rash to simply presume those features are immune from homoplasy, especially when other aspects of their respective phenotypes suggest more distant relationships with, respectively, the hominin clade and later *Homo*.

Bernard Wood, The George Washington University, Department of Anthropology, Washington, DC, USA

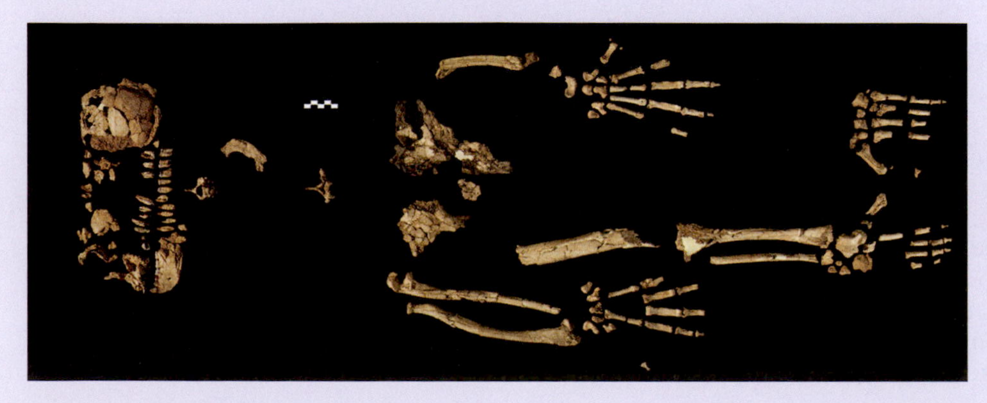

Figure 1: *Ardipithecus ramidus* partial skeleton.
Composite image including bones that may come from more than one individual from the same site. [From White TD et al. (2009) *Science* 326, 75. Reproduced with permission from AAAS.]

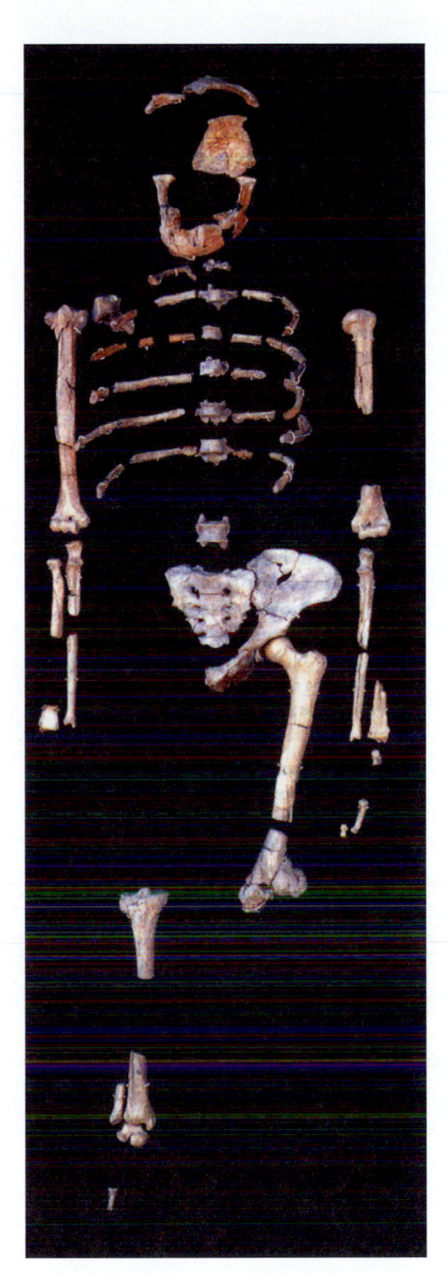

Figure 9.3: Skeleton of *Australopithecus afarensis* (A. L. 288–1 or Lucy). *Au. afarensis* lived in East Africa between 3 and 4 MYA and Lucy, dating to ~3.2 MYA and named after the Beatles' song "Lucy in the sky with diamonds," is its best-known representative. Lucy was a mature adult female when she died, but was only just over 1 m in height. (Photograph by Denis Finnin and Jackie Beckett, © American Museum of Natural History.)

and 45 kg, respectively); the females of the two species were similar (~30 kg). *Au. afarensis* (~3.0–3.9 MYA) is present at a number of sites in East Africa from Ethiopia to Tanzania, and includes the famous partial skeleton Lucy (3.2 MYA; **Figure 9.3**) and probably the Laetoli footprints (3.5 MYA; see Figure 8.2),[36] dramatically illustrating the existence of bipedal locomotion at this time.

The fossil material available from *Au. afarensis* is extensive enough to allow many of its characteristics to be deduced. The species is estimated to have been 1–1.5 m tall (and, as mentioned, bipedal); weight was between 25 and 50 kg, with considerable dimorphism between the sexes. Brain size was 400–500 cc [cubic centimeters, a non-SI unit (1 cc = 1 cm^3) still used in the field and adopted here]: similar, in proportion to body mass, to that of the chimpanzee. The habitat is thought to have been more open than that inhabited by the earlier hominins, perhaps with grassland as well as trees. *Au. africanus*, considered below, was probably a similar species, although with less sexual dimorphism and perhaps more human-like in some ways.

Australopithecus africanus, the first member of the genus to be discovered and named (the Taung Child),[14] is also known from a number of sites, all from the south of the continent, and most of these fossils date to between 2.0 and 2.9 MYA.[24] Later *Australopithecus* is represented in East Africa by fossils from the Middle Awash in Ethiopia designated *Au. garhi* (~2.5 MYA),[3] characterized by relatively large teeth; and in South Africa by *Au. sediba*, well dated at 1.977 MYA.[50] Thus **gracile** (lightly built) australopithecines were present in many areas of Africa from around 4.2 MYA to ~2.0 MYA (Figure 9.2). The relationships between the species mentioned here are unclear: the simplest scheme would consider *Au. africanus* and *Au. garhi* as geographical variants; *Au. afarensis* would be a descendant of *Au. anamensis*, and *Au. africanus/garhi* of *Au. afarensis*; *Au. sediba* is interpreted as a descendant of *Au. africanus*, although the Mrs Ples *Au. africanus* skull from Sterkfontein is virtually contemporary with *Au. sediba* at around 2 MYA. An as yet unclassified third species of *Australopithecus* has been suggested from South Africa in the form of the 2.6–2.2 MYA Little Foot skeleton from Sterkfontein, which has similarities to *Paranthropus* (see below).

Robust (heavily built) hominins with small brains and large jaws and chewing teeth were originally included in the genus *Australopithecus*, but are now commonly placed in a separate genus, *Paranthropus*. *P. aethiopicus* is represented by only a small number of specimens between 2.7 and 2.4 MYA, but these include the Black Skull, a fairly complete ~2.5-MY-old skull from Lake Turkana. *Paranthropus boisei* fossils, including the skull Zinj from Olduvai Gorge, Tanzania,[34] are found mostly in Ethiopia, Tanzania, and Kenya, and span the range 1.4–2.3 MYA. *Paranthropus robustus* remains are known from several sites in South Africa (Swartkrans, Gondolin, Drimolen, Coopers D, Sterkfontein, and Kromdraai B), and have been dated to between ~2.0 and ~1.2 MYA.[25] The robust morphology of these species is thought by some to represent an adaptation to a diet that required heavy chewing, such as low-quality fibrous vegetable food, for example, roots and nuts.

The question as to which of these hominin species is our direct ancestor has attracted considerable attention. It is widely agreed that the *Paranthropus* species form a separate lineage with no surviving descendants. *Au. anamensis* and *Au. afarensis* are good candidates for human ancestors before 3 MYA, but there seems to be no consensus about which fossils represent our ancestors between 3 MYA and the emergence of *Homo* (**Figure 9.4**). Recently, *Au. sediba* has been suggested as such an ancestor due to its mixture of australopithecine and *Homo*-like traits.[50] Its young age at 1.98 MYA might preclude this, but the

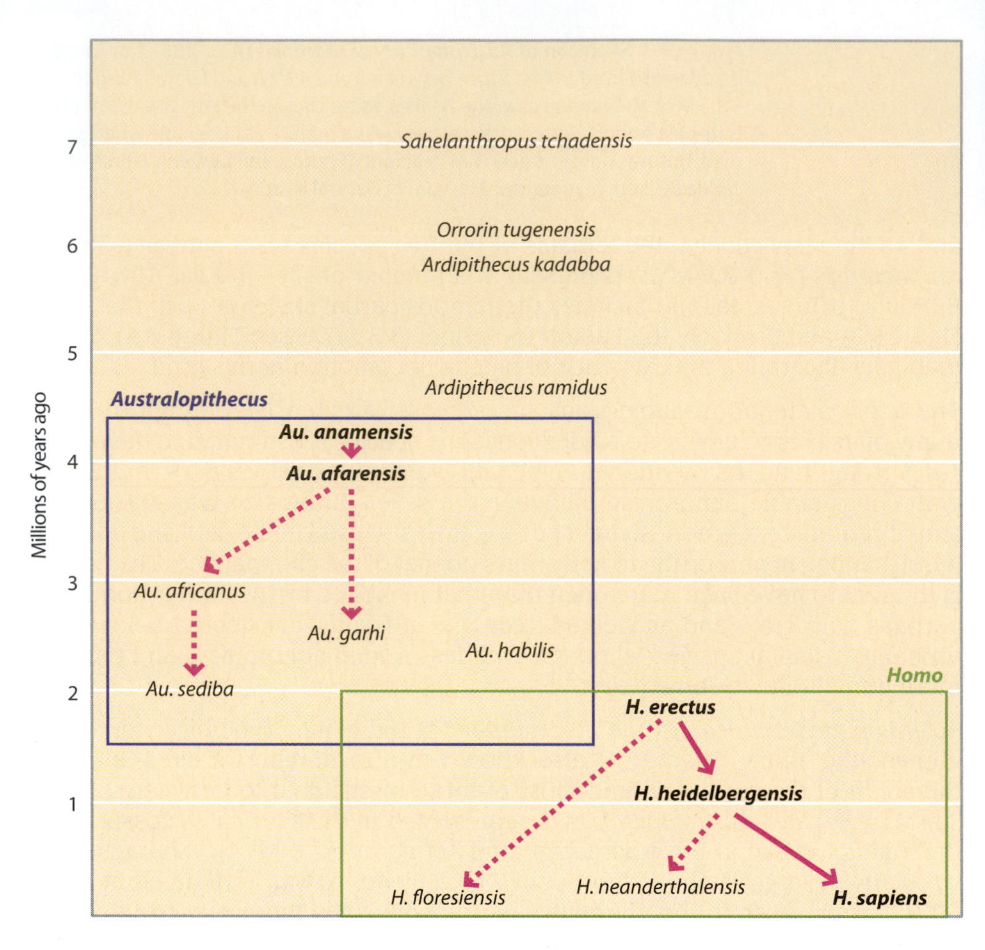

Figure 9.4: Relationships of fossil hominin species, indicating plausible human ancestors.
Species relationships are shown by *red* arrows, solid when on the human ancestral line; species in bold are likely human ancestors. Note the uncertainty about the relationships and human line affinities of the early hominins, and the uncertainty about which later australopithecine is ancestral to *Homo*. Compare with Figure 9.1.

type specimen from the site of Malapa may simply represent a late-occurring individual from a species that had existed for some time. Much debate has occurred over whether the large series of fossils (over 500 from Sterkfontein alone) attributed to *Au. africanus* represent two species or whether they simply represent a single highly variable species. This sample of fossils may include earlier specimens of *Au. sediba* that have yet to be identified, or the variability may be temporal with the Sterkfontein Member 4 deposit estimated to have formed over half a million years. If *Au. sediba* were confirmed as the ancestor of *Homo*, this would suggest a southern African, rather than eastern African, origin for our genus.

The genus *Homo* arose in Africa

The reader should not be surprised to learn that there are disagreements about which species should be included within our own genus, *Homo*, and thus about its origin. For many years, the earliest member of the genus was considered to be *H. habilis*, "handy man," named on the basis of a partial skull and jaw, OH 7, from Olduvai Gorge, Tanzania;[35] some *H. habilis* specimens may date back to about 2.3 MYA. However, *H. habilis* has been described as "a mishmash of traits and specimens, whose composition depends upon what researcher one asks" (Kreger, http://archaeologyinfo.com/homo-habilis/); in addition, *H. habilis* does not show the body size and shape, or small teeth, characteristic of humans, while later species do. These features appear shortly after 2 MYA in fossils described as *Homo ergaster* or *erectus*, and it therefore seems reasonable to draw the distinction between *Australopithecus* and *Homo* here;[74, 75] thus *habilis* would be assigned to *Australopithecus* and *erectus/ergaster* would be the first *Homo*. We will therefore refer to "*Au. habilis*" in the following sections.

A small group of fossils from Koobi Fora in northern Kenya, dating to 1.78–2.03 MYA, have been identified with flatter faces and shorter and more rectangular jaws than *Au. habilis*,[39] and assigned to the species *H. rudolfensis*. Debate about

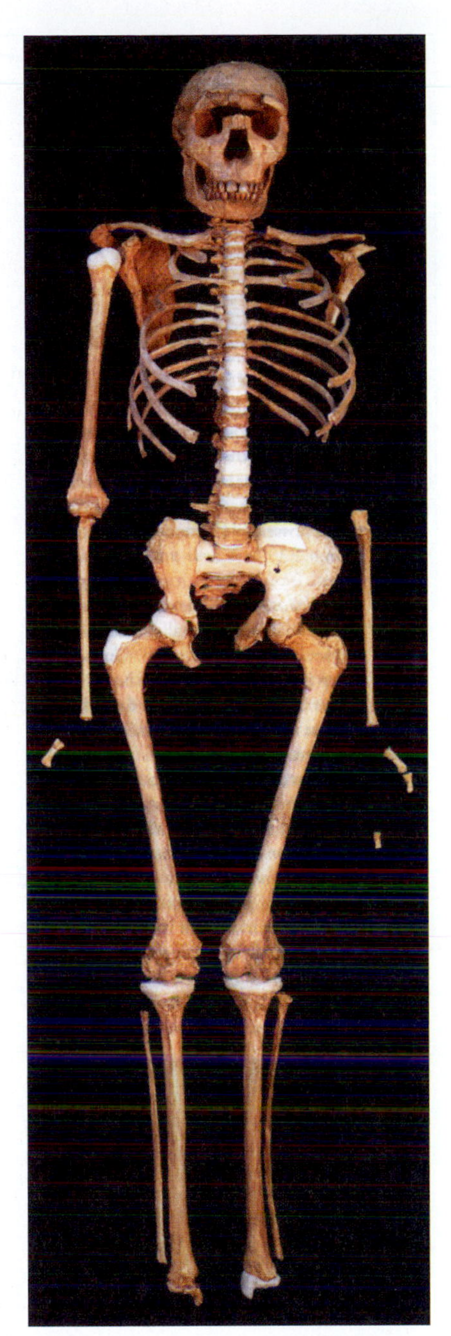

Figure 9.5: Skeleton of *Homo erectus* (WT 15000 or the Nariokotome Boy).
H. erectus (African specimens are sometimes called *H. ergaster*) is known from Africa at ~1.9 MYA and from Asia soon afterward (~1.8 MYA). This specimen, from Lake Turkana, Kenya, dates to about 1.6 MYA. The Nariokotome Boy was an adolescent male when he died, with the body size and shape of modern humans but a smaller brain. (Photograph by Denis Finnin and Jackie Beckett, © American Museum of Natural History.)

their affinities to contemporary hominins and taxonomic status, at the level of both genus and species, continues. In view of the lack of information about their body size and shape, they are assigned below to *Australopithecus*.

There is also debate about the distinctiveness of the two species *H. ergaster* and *H. erectus*: it is difficult to find morphological characteristics that separate them reliably. While one view considers *H. ergaster* to cover African individuals and reserves *H. erectus* for those found outside Africa, an alternative analysis would include all these specimens as a single widespread and variable species, *H. erectus*. The latter view is somewhat strengthened by the finding of a ~1.0-MY-old specimen resembling Asian *H. erectus* in Africa,[2] although this individual could alternatively have migrated back to Africa. Here, the name *H. erectus* will be used for this whole group of fossils. The first examples date to between 1.8 and 1.9 MYA[50] and, like all earlier hominins, are found in Africa, demonstrating an African origin for our genus. *H. erectus* fossils include the outstanding Nariokotome Boy (~1.6 MYA; **Figure 9.5**),[67] the most complete early hominin thus far found, which provides important insights into this species. He is thought to have been in early adolescence when he died, male, tall and thin at about 1.5 m high and weighing 47 kg. As an adult he would probably have reached 1.8 m and 68 kg, common figures for modern humans. His limb proportions and tooth size were also similar to those of modern humans, but his brain size (880 cc, corresponding to 909 cc at maturity) was significantly smaller than the modern human average (around 1450–1500 cc), although just within the modern human range (830–2300 cc).

H. erectus is the earliest hominin to be found outside Africa (**Figure 9.6**), and includes influential fossils discovered in the late nineteenth and early twentieth centuries, such as Java Man (Indonesia, the site of the type specimen Trinil 2 discovered in 1891) and Peking Man (China). The earliest *H. erectus* dates outside Africa, from Dmanisi (Georgia),[43, 59] are ~1.8 MYA,[17] a little younger than the earliest African *H. erectus* from Swartkrans Member 1 and Koobi Fora in Kenya at ~1.9 MYA. It has been suggested that the large body size providing tolerance to heat stress and dehydration, coupled with improved stone toolkits, may have allowed the species to live in a wide range of environments and thus expand out of Africa rapidly.[74] In Java, *H. erectus* may have survived until 135 KYA;[26] if so, they would have been contemporaries of fossils on the East Asian mainland that have in most cases been referred to as archaic *H. sapiens*. The analysis of aDNA from a >50-KY-old finger bone from Denisova Cave in Siberia has rekindled interest in Asian hominins, and we discuss this enigmatic taxon below (**Section 9.5**).

The report of fossils representing a tiny 1-m-tall hominin species, *H. floresiensis*, from the island of Flores in Indonesia in 2004[11] surprised paleontologists so much that the first reaction of some was to think that the story must be a hoax. But these "hobbits," named in tribute to the imaginary characters of J.R.R. Tolkien, were supported by the remains of multiple individuals and archaeological deposits spanning the period 17–74 KYA, including the fairly complete 380 cc cranium and skeleton of the type specimen, LB1. Despite the suggestion that *H. floresiensis* might represent modern humans suffering from **microcephaly** (a neurodevelopmental disorder in which the head is abnormally small), most paleontologists now accept them as distinct hominins, perhaps *H. erectus* descendants, surviving on an isolated island poor in resources since ~1 MYA, resulting in selection for small size.

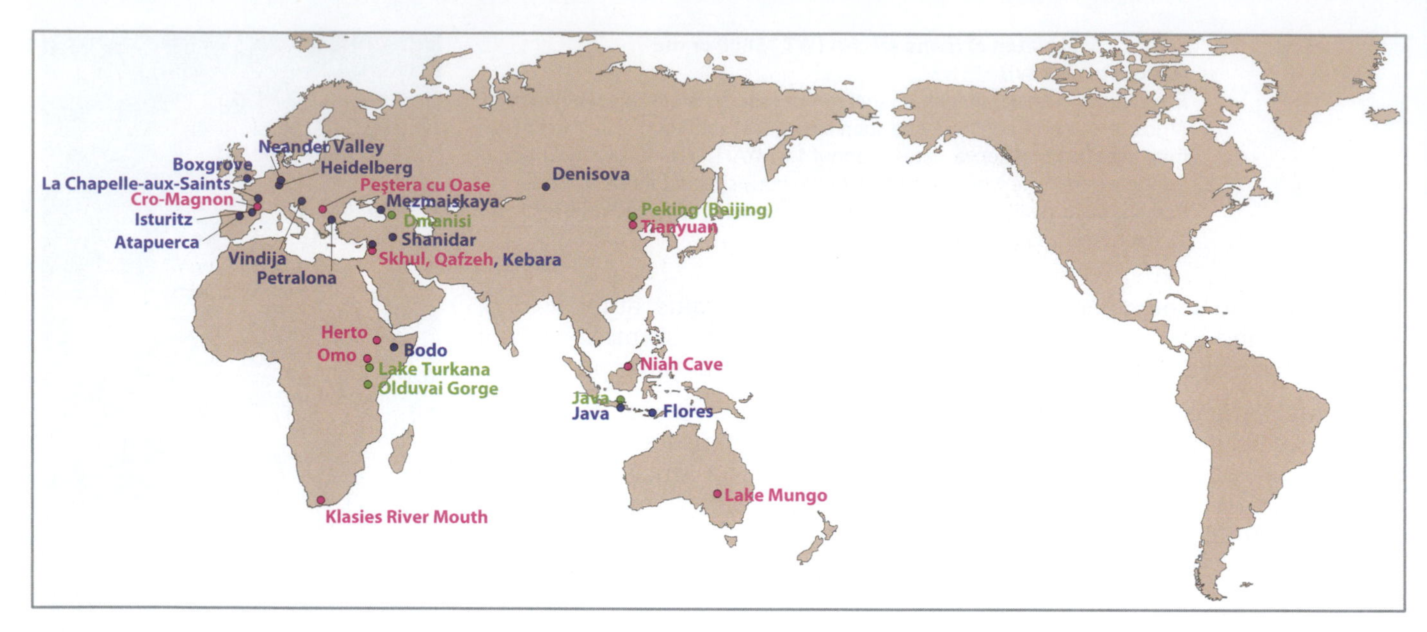

Figure 9.6: Sites of *Homo* fossils.
Early sites (1.9–1.6 MYA) are shown in *green*, later sites (800–12 KYA) for species other than *H. sapiens* in *blue*, and *H. sapiens* in *red*. These sites are spread throughout much of the world, illustrating the extensive spread of *Homo* compared with earlier hominins (Figure 9.2).

Later *Homo* from Africa and Europe (**Figure 9.7**) are less robust and have larger brains (~1200 cc instead of ~900 cc) than early *H. erectus* and are often designated *H. heidelbergensis*, the type specimen of which is the ~609 KYA Heidelberg Jaw from Germany[66] (**Figure 9.8**). Related specimens include the massive Bodo cranium (Ethiopia, ~600 KYA), the tibia (lower leg bone) from Boxgrove (England, ~500 KYA), and the Petralona 1 cranium (Greece, age uncertain but with estimates between 200 and 700 KYA). Many would also place the 1.2–0.80 MYA specimens from Gran Dolina and Sima del Elefante, Spain, designated *Homo antecessor* by their discoverers,[9] within *H. heidelbergensis*. According to this view, *H. heidelbergensis* would have been a widespread and somewhat variable species, perhaps originating from *erectus* in Africa some time prior to 1 MYA and giving rise to more recent *Homo* species, including *H. sapiens* and *H. neanderthalensis*. Other researchers, however, prefer to call the post ~1 MYA African specimens *H. rhodesiensis* after the ~300–125-KY-old Kabwe (or Broken Hill 1) skull from Zambia. In South Africa, other potential specimens assigned to *H. rhodesiensis* include the Saldanha Man skullcap from Elandsfontein dated between 1.1 and 0.6 MYA and the Cave of Hearths material which perhaps dates to between 800 and 400 KYA.[23] In this more complex scenario, not followed here, *H. heidelbergensis* is a European species giving rise to *H. neanderthalensis*, and *H. rhodesiensis* an African species giving rise to *H. sapiens*.

Neanderthals (also spelled "Neandertals"; **Figure 9.9**) form a morphologically distinct group of fossils from Europe and Western Asia between ~250 KYA and ~28 KYA, robust and with large brains (~1400 cc, larger than those of many modern humans) and well-developed brow ridges. Well-known examples include the type specimen Feldhofer 1 ~40 KYA from the Neander valley in Germany, The Old Man of La Chapelle-aux-Saints ~50 KYA from France (a 40–50-year-old individual showing evidence of arthritis, which was not recognized as pathological when the specimen was first described in the early twentieth century, leading to Neanderthals being wrongly stereotyped as "brutish" and "bent-kneed"), Kebara 2 (Israel, ~60 KYA), and Shanidar 4 from Iraq (~60 KYA), sometimes interpreted

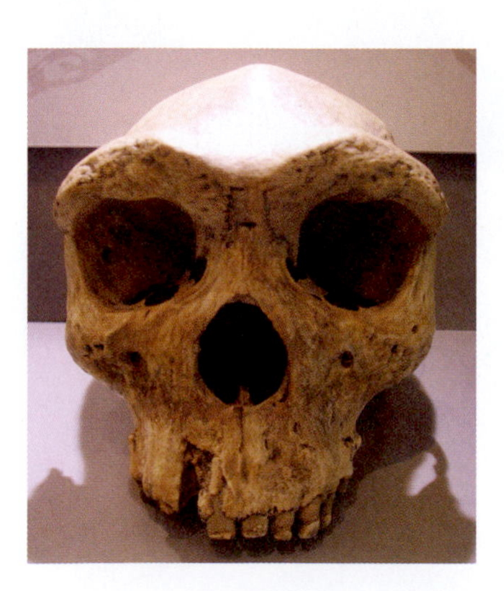

Figure 9.7: *Homo heidelbergensis* (Broken Hill 1 or the Kabwe Cranium).
This example was found in a lead and zinc mine in Zambia and its context is uncertain, but a date of 125–300 KYA has been suggested. [(Courtesy of Gerbil under Creative Commons Attribution-Share Alike 3.0 Unported license.]

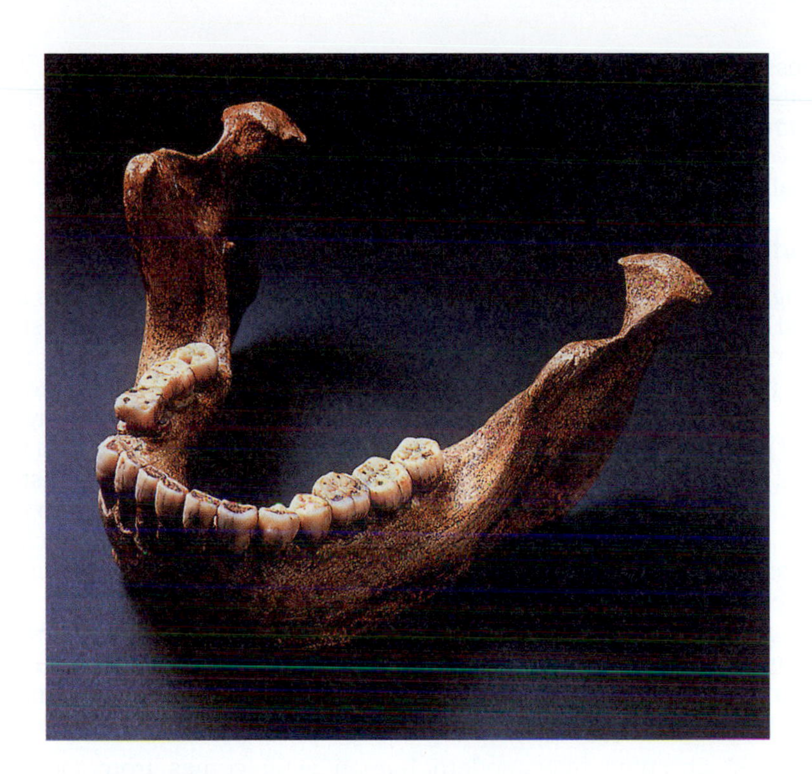

Figure 9.8: *Homo heidelbergensis* mandible (Mauer 1 or the Heidelberg Jaw).
Opinions differ about which African and European fossils dating from 1100–200 KYA should be ascribed to *H. heidelbergensis*, but this mandible is the type specimen and so must belong to *H. heidelbergensis*. It was found near Heidelberg, Germany, and dates to ~609 KYA. (Reproduced with permission of the Science Photo Library.)

as representing a deliberate burial. In the era of successful aDNA sequencing, bone fragments from Vindija, Croatia (see Figure 9.20), have become well known for the molecular, rather than archaeological, information they have provided. Neanderthals are thought to be descendants of *H. heidelbergensis* and are usually assigned to a distinct species, *Homo neanderthalensis*, but their relationship to modern humans has aroused intense debate, now informed by a **low-coverage** genome sequence, and is considered further below (**Section 9.5**).

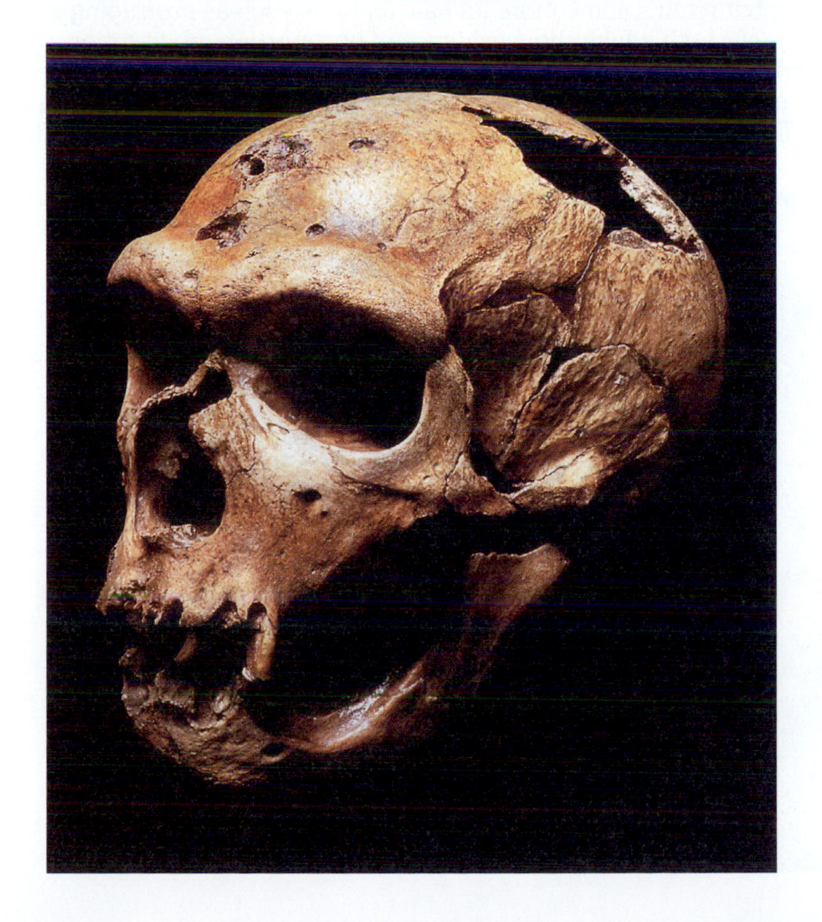

Figure 9.9: *Homo neanderthalensis* skull (The Old Man of La Chapelle-aux-Saints).
H. neanderthalensis lived in Europe and Western Asia from ~250 to 28 KYA. This specimen from France dates to ~50 KYA and was derived from a 40–50-year-old man. Note the large brow ridges and small chin. Several pathological features, including arthritis and resorption of the tooth sockets, are also present and contributed to the misinterpretation of Neanderthals as shuffling, brutish cavemen. (Reproduced with permission of the Science Photo Library.)

Needless to say, these classifications are not universally accepted. Indeed, all *Homo* species arising after *H. erectus* and before modern *H. sapiens* have sometimes been referred to collectively as "archaic *H. sapiens*," with specimens definitively assigned to *H. sapiens* (defined strictly) being referred to as **anatomically modern humans** (**AMH**).

The earliest anatomically modern human fossils are found in Africa

The origin of modern humans has probably been the most contentious issue in the field over the last 30 years. We will see that there is an important distinction between **morphology** and **behavior**, and will begin by considering modern human morphology and ask where and when this first appeared. Anatomically modern humans differ from earlier hominins ("archaic humans" or "archaic *H. sapiens*"), but these differences are not easy to define; indeed, it is often pointed out that there is no type specimen for *Homo sapiens*. Paleontologists have focused mainly on cranial features, which can be summarized by two characteristics, derived from a comparison of 100 recent humans, 10 fossils classified as anatomically modern *H. sapiens*, and 9 classified as *H. neanderthalensis* or *H. heidelbergensis*:[42] (1) extent of the globular shape of the skull; (2) degree of retraction of the face (Figure 8.8). This system allows a clear distinction between AMH and archaic humans, with zero overlap, but has the disadvantage that relatively complete specimens are needed; for fragmentary specimens it is necessary to use less reliable criteria.

The earliest accepted fully modern human skull comes from Omo-Kibish (Ethiopia, **Figure 9.10**) and dates to ~195 KYA.[45] Slightly later crania of one child and two adults from Herto (also Ethiopia) date to 154–160 KYA and show many of the features of modern human morphology,[71] yet the authors created a new subspecies *Homo sapiens idaltu* to accommodate them, emphasizing the morphological variation at this time. Despite this, few researchers use this subspecies level classification and they are most often defined as AMHs. The most complete cranium is large (1450 cc) and has the globular braincase of modern humans, but retains some more archaic features such as protruding eyebrows. Interestingly, both adults show evidence of postmortem modification, including cut marks, interpreted as resulting from mortuary practices. Other fragmentary specimens are known from Klasies River Mouth in South Africa at 90–120 KYA, and two sites from Israel dated to between 90 and 130 KYA: the cave at Qafzeh

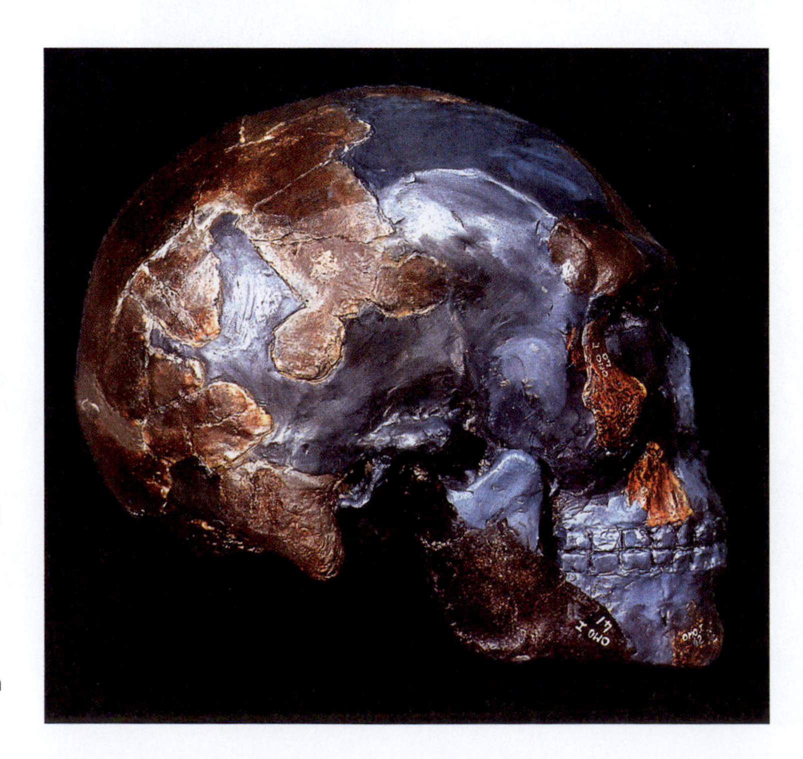

Figure 9.10: The earliest anatomically modern human cranium (Omo I). Modern features include the high forehead and developed chin; note that the gray portions are reconstructed. This specimen from Omo-Kibish in southern Ethiopia provides crucial evidence that modern human anatomy had developed in Africa by ~195 KYA. (Reproduced with permission from Michael Day.)

TABLE 9.2:
THE OLDEST ANATOMICALLY MODERN HUMAN FOSSILS FROM DIFFERENT REGIONS OF THE WORLD

Continent	Location	Remains	Date (KYA)	Reference
Africa	Omo-Kibish, Ethiopia	Omo I skull	195	McDougall I et al. (2005) *Nature* 433, 733.
Africa	Herto, Ethiopia	three crania	154–160	White TD et al. (2003) *Nature* 423, 742.
Africa	Klasies River Mouth, South Africa	multiple fragments	90–120	Royer D et al. (2009) *Am. J. Phys. Anthropol.* 140, 312.
Middle East	Qafzeh and Skhul, Israel	multiple, >30 individuals	90–130	Grün R et al. (2005) *J. Hum. Evol.* 49, 316.
East Asia	Niah Cave, Borneo	cranium and leg bones	34–46	Barker G et al. (2007) *J. Hum. Evol.* 52, 243.
East Asia	Tianyuan Cave near Beijing, China	partial skeleton including mandible	39–42	Shang H et al. (2007) *Proc. Natl Acad. Sci. USA* 104, 6573.
Australia	Lake Mungo	Lake Mungo 3	40 ± 2	Bowler J et al. (2003) *Nature* 421, 837.
Europe	Grotta del Cavallo	two molars	43–45	Benazzi S et al. (2011) *Nature* 479, 525.

with parts of more than 20 skeletons and the rock shelter at Skhul with at least 10 individuals, both including some likely burials. The **Levantine** nonhuman fauna at this time is interpreted as a temporary extension of the African fauna, and thus all of these early human remains, like the animals, can be considered African. Outside Africa (interpreted in this sense), the earliest accepted dates for modern fossils are all <45 KYA,[6] with fossils dating close to 40 KYA known from Europe, East Asia, and Australia (**Table 9.2**).

Evidence for the appearance of modern human behavior will be discussed in **Section 9.2**, and timing of the first modern human presence in different regions of the world will be considered in more detail in Chapters 11 and 13. Here, we note that, despite uncertainties in classification and dating, and the extremely incomplete nature of the fossil record, the earliest dates outside Africa are much more recent than dates inside Africa: it is clear from the fossil evidence that modern human morphology appeared considerably earlier in Africa than elsewhere.

The morphology of current populations suggests an origin in Africa

Morphological variation among present-day populations should also carry information about the origins of modern human anatomy: if this variation is predominantly neutral, the simple expectation is that it should be greatest close to the origin. However, morphology is shaped by selection as well as by neutral forces, so a combination of a large dataset (37 measurements each from 4666 male skulls belonging to 105 worldwide populations) and allowance for the correlations with climate was necessary to detect a signal of the origin. This showed that populations in sub-Saharan Africa were the most variable, and variance fell with distance away from this region, with distance from Africa accounting for 19–25% of the variation,[44] a striking parallel with the genetic pattern (**Section 9.4**). This study was not able to pinpoint a specific area within sub-Saharan Africa as the most likely origin, but interestingly found no evidence for a second origin, thus providing no support from cranial morphology for a **multiregional model** (see **Section 9.3**).

9.2 EVIDENCE FROM ARCHAEOLOGY AND LINGUISTICS

Archaeological evidence may be considered as the preserved signs (other than fossils) of hominin activity, although this definition could be extended to include the activity of nonhuman apes as well. While hominin fossils are very rare, archaeological remains, such as stone tools, are much more common.

Assemblies can be classified and associated with one or more hominin type through rare sites that contain both archaeological remains and fossils, and then allow the presence of these hominins to be inferred elsewhere, albeit with the limitation that there is no one-to-one correspondence between technology and species.

Chimpanzees use a range of tools, including sticks to extract termites and stones to break open nuts,[73] and even manufacture and use sticks for hunting,[51] while orangutans also use tools in a variety of ways, including for seed extraction and autoerotic purposes,[65] so a parsimonious assumption is that our common ancestor used tools as well. Most of these would not be preserved in the archaeological record, but a 4.3 KYA chimpanzee archaeological site containing modified stones carrying starch residues has been recognized in Ivory Coast in West Africa.[46] However, the identification of any tools used by the earliest hominins remains an area for future research, so known archaeology currently begins with the **Oldowan** culture (**Mode 1** technology) starting about 2.6–2.5 MYA (**Figures 9.11** and **9.12a**).

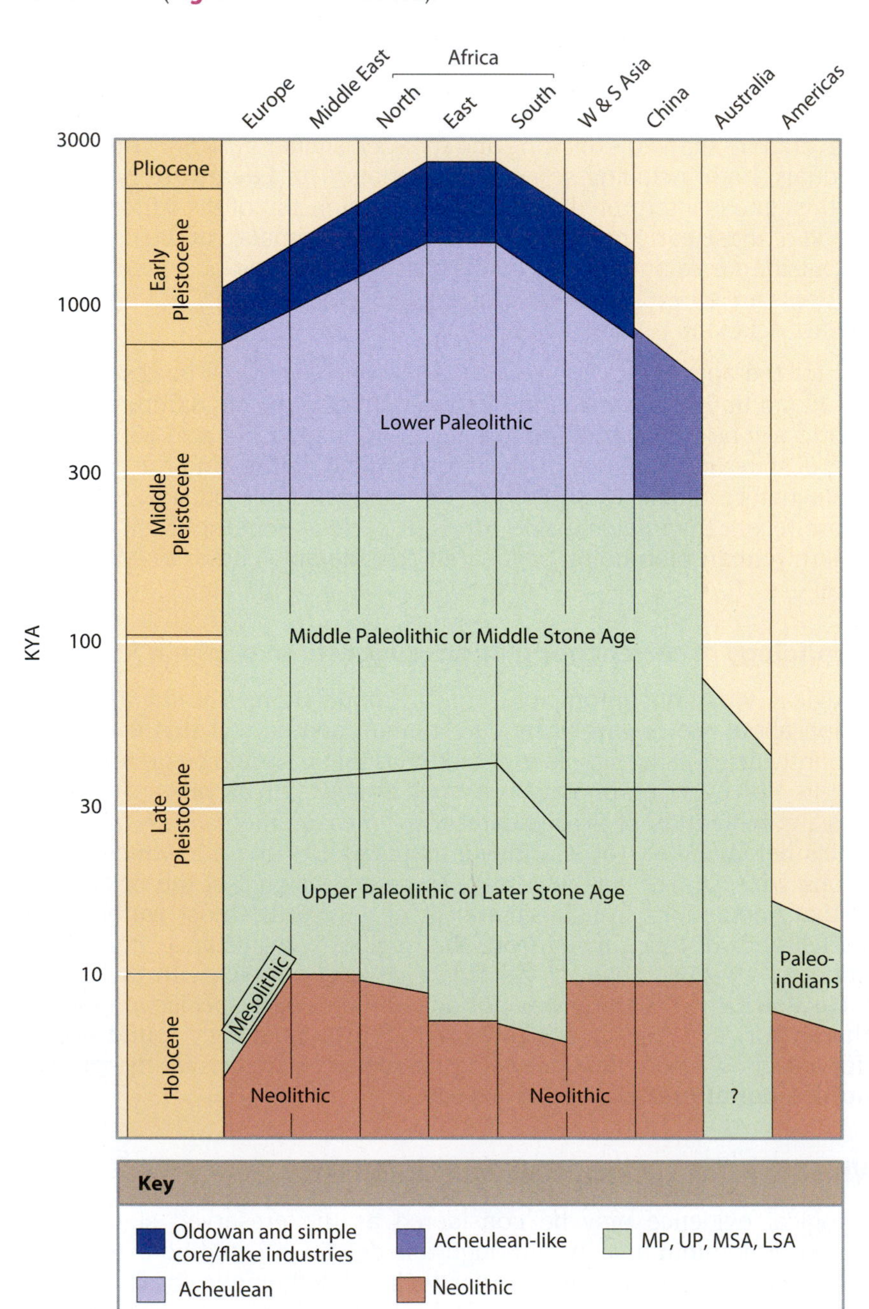

Figure 9.11: Chronology of archaeological stages in different regions of the world. The first column shows the time-scale (note that it is nonlinear), and the second column gives the geological epoch. Subsequent columns show the archaeological stage in selected world regions. Archaeological remains appear earlier in Africa than elsewhere. Opinions vary about whether or not there was a Neolithic period in Australia. MP, Middle Paleolithic; UP, Upper Paleolithic; MSA, Middle Stone Age; LSA, Later Stone Age.

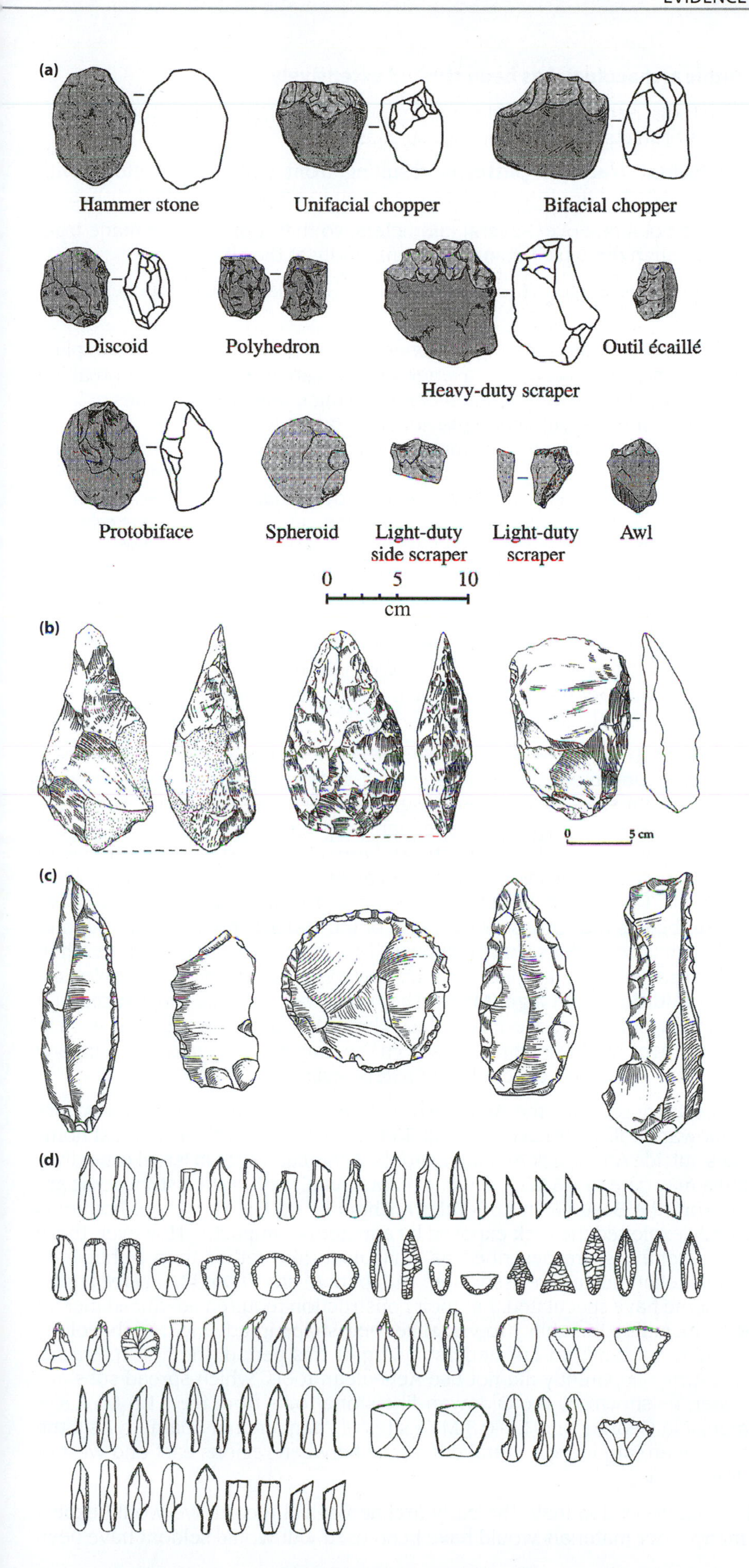

Figure 9.12: Stone tool technologies. (a) The oldest recognized stone tools are Oldowan, manufactured from pebbles and dating back to ~2.6 MYA. (b) After ~1.8 MYA, Acheulean tools are found, including bifaces. They continued to be used until ~150 KYA. (c) Mousterian tools were manufactured by the Levallois technique after ~300 KYA, associated with both Neanderthals and early anatomically modern humans. (d) Upper Paleolithic tools dating after 50 KYA showing the wide range of forms. [From Lewin R (1999) Human Evolution. With permission from John Wiley & Sons Inc.]

Paleolithic archaeology has been studied extensively

Oldowan tools, named after Olduvai Gorge, Tanzania, East Africa, could be recognized as artifacts by characteristics such as:

- **Conchoidal fracture patterns** resulting from striking one stone with another, which differ from the fractures seen in naturally cracked stones
- Transport of stone over several kilometers, so that tools may be made from lavas or quartzite which do not occur naturally at the site
- Concentrations of stone tools, sometimes in association with butchered animal bones

Oldowan tools consist of hammer stones, flakes, and cores, and, as implied above, were probably used to scavenge animal carcasses, including breaking open the bones. Even hyenas cannot crack the thick-walled limb bones of large animals, so tool use would have provided the hominins with a rich and novel food source—bone marrow. It is impossible to be certain of the identity of the toolmakers at Olduvai Gorge, but they are usually assumed to be *Au. habilis*. Other potential toolmakers include *Au. garhi* and *Au. rudolfensis*. In South Africa, the earliest stone tools are found in the same deposits as *Paranthropus robustus* remains, like the much younger Oldowan (<1.8 MYA) from Sterkfontein and Kromdraai.[25] The older Oldowan deposits from Swartkrans Member 1 are associated with both *P. robustus* and *H. erectus* and thus the latter may be responsible for the stone tool manufacture. At Drimolen and Swartkrans bone tools are associated with *P. robustus*.[5] As yet, no stone tools have been associated with the pre-2-MYA *Au. africanus*, but the ~2 MYA remains of *Au. sediba*, while not thus far associated with stone tools, have a hand morphology capable of making and using stone tools.[29]

By 1.76 MYA, strikingly different tools started to be made: symmetrical teardrop-shaped **handaxes**, worked around all or most of either one (unifacial working) or both sides (bifacial working, **bifaces**)[40] (Figure 9.12b). These are called **Acheulean** (also spelled "Acheulian" and referred to as **Mode 2** technology) after the French site St. Acheul, and are often found in association with larger flake tools than the Oldowan, as well as the same Oldowan tools, at least to begin with. The uses of these handaxes are poorly understood, but they have been described as the "Swiss Army knife" of the **Paleolithic** and continued to be used, with little obvious change in overall shape, until around 125 KYA when they are associated with stone tools characteristic of the next technological stage, the **Middle Stone Age** (**MSA**). They were, perhaps, the most successful of all human tools. They are found throughout Africa, in Europe, and in Asia south of the **Movius Line** which runs from the Caucasus mountains to the Bay of Bengal, but are largely absent from Eastern Asia.

The earliest evidence for the Acheulean is found in Africa, where it co-occurs with Oldowan tools near Lake Turkana, Kenya, at 1.76 MYA;[40] the earliest hominin sites outside Africa lack Acheulean tools. Acheulean sites in Israel (Ubeidiya) and India may date to 1.5 MYA,[49] while in East Asia, well-crafted stone tools are known from the Bose basin in Southern China by ~803 KYA, where hominins apparently exploited the rock exposed by a meteorite impact.[77] However, these eastern assemblies are described as Acheulean-like rather than Acheulean. Acheulean technologies are associated with both *H. erectus* and *H. heidelbergensis*, and some have speculated that their construction required advanced mental capacity, including the ability to visualize their shape in advance. Archaeology thus adds significantly to our understanding of this period: the first *H. erectus* to leave Africa apparently did not use Acheulean tools, which spread substantially later, as shown by the Oldowan-like stone tools from the ~1.8 MYA site of Dmanisi in Georgia and 1.2–0.8 MYA sites of Atapuerca in Spain; in Asia the tools reveal an important cultural difference between regions east and west of the Movius Line.

While stone tools dominate the early archaeological record, we would expect that many other materials would have been used, but would seldom have been preserved. A set of wooden throwing spears from Schöningen in Germany was found in association with butchered horses and is dated to ~400 KYA,[62] providing evidence for use of multiple materials and sophisticated hunting activity at this time.

Between ~800 KYA and 280 KYA a series of innovations occurred with the development of **prepared core technology**, followed by smaller flake-based stone tools that are characteristic of the MSA. The first prepared core technology was developed to create standardized blanks for the construction of handaxes and occurs between ~800 and ~300 KYA in Africa and Israel. By ~550–500 KYA the first blades are seen, and between 500 KYA and 280 KYA the first points occur, along with the potential early occurrence of the **Levallois technique** (**Mode 3**). The Levallois technique is a more complex form of prepared core technology that is generally associated with the MSA, in which the shaping of the tool was accomplished by removing flakes from a core, followed by removal of one final shaped flake which would form the tool itself (Figure 9.12c). The MSA first occurs by at least 280 KYA at Gademotta in Ethiopia and around 250–200 KYA in the Kapthurin Formation of central Kenya. In South Africa there is the suggestion, as at Kapthurin, that all the elements of the MSA may have been in place by 500–400 KYA, either marking the earlier beginnings of the MSA or the occurrence of a transitional industry often called the Sangoan or Fauresmith.[23] In Africa a variety of MSA industries have been defined which range from simplistic small flake-based industries to industries with refined bifacial points (Lupemban and Still Bay) to the early use of microliths (Howieson's Poort; ~65 KYA) that are normally characteristic of the **Later Stone Age** (**LSA**) of Africa or the **Upper Paleolithic** (**UP**) of Europe after 40 KYA.

The Levallois technique was being used in Europe by 200 KYA. Among other Mode 3 **Middle Paleolithic** industries in Europe is the **Mousterian**, characterized by flakes described as side scrapers and points. The human remains associated with Mousterian artifacts are usually Neanderthal, but at Qafzeh and Skhul they are early modern humans, and late Neanderthals may have used non-Mousterian tools, so again there is no simple correspondence between species and technology (**Section 11.2**). Mousterian-like toolkits are found in Asia as far to the east as Lake Baikal, and in southern Asia tools have been labeled as "Mousteroid" because of the high incidence of scrapers. Artifacts classified as MSA are also associated with the earliest modern human remains in Australia.

The Upper Paleolithic in Eurasia and Later Stone Age in Africa are defined by a greater diversity of stone tools and artifacts including microlithic technology and the use of bone and wood (Figure 9.12d). While the earliest potential art forms, such as the ~72 KYA Blombos Cave engraved and shell necklaces, occur within the MSA, the oldest unequivocal art occurs in the form of cave paintings and carved bone and ivory. In the Upper Paleolithic, the predominant (**Mode 4**) tools are described as **blades** instead of flakes; blades are long narrow flakes made from specialized cores and then reworked in a number of ways (Figure 9.12d). In particular, they may be retouched at the end rather than the side. Objects made from other materials, such as wood and bone, become much more abundant in the Upper Paleolithic and unequivocal art is found. The Upper Paleolithic is often associated with modern humans although, as we have seen above, there is no simple correspondence between toolkits and species. Discussion of subsequent developments will be continued in later chapters.

Evidence from linguistics suggests an origin of language in Africa

Languages change rapidly, even within the span of a human lifetime, so it may seem surprising that linguistics can be informative about ancient human origins. The relevant evidence comes not from the study of vocabulary, which turns over quickly, but from the basic units of sound: **phonemes**. A study of 504 diverse languages[4] found that phonemic diversity was highest in Africa and declined with distance from central/southern Africa; after correcting for population size, distance from Africa accounted for 19% of the variance in phonemic diversity. The parallels with the morphological and genetic patterns are striking (**Sections 9.1** and **9.4**).

9.3 HYPOTHESES TO EXPLAIN THE ORIGIN OF MODERN HUMANS

While many of the fossil discoveries described in the previous parts of this chapter have been made in the last few years, and dates have often been refined or revised, the basic pattern of an early exodus of *H. erectus* from Africa to occupy much of the Old World, followed by a much later appearance and expansion of modern *H. sapiens*, has been clear for decades, and has conditioned the debate that dominated the field during the second half of the twentieth century. This debate can be most easily appreciated by first considering two extreme views (**Figure 9.13**):

- The multiregional model proposed that the transition from *H. erectus* to *H. sapiens* took place in a number of areas of the Old World, with different modern human characteristics arising at different times in different places.
- In contrast, the **out-of-Africa model** proposed that the transition took place in Africa, and that these humans recently (<100 KYA) replaced the hominins already present on other continents.

One way of characterizing the difference is that, according to the multiregional model, our ancestors lived on several continents over the past 1 MY; in contrast, according to the out-of-Africa model, we descend entirely from the ancestors who lived less than a few hundred thousand years ago in Africa; their contemporaries from other continents did not contribute to our ancestry. These models were formulated before the classification of many of the species between the times of *H. erectus* and *H. sapiens* was adopted, and it is not entirely clear how all the additional species would fit into them.

Intermediate models are obviously possible, for example involving a recent origin of most human characteristics in Africa, but also interbreeding with archaic populations inside or outside Africa—a **leaky replacement** model. Fossil, archaeological, and genetic evidence provided little support for an extreme multiregional model, instead generally being interpreted to favor an out-of-Africa model, with or without a low level of admixture. Archaeological evidence, for example, suggested cultural contact between Neanderthals and modern humans in Eurasia before Neanderthals went extinct, and some paleontologists have identified intermediate morphology in some fossils including a ~25 KYA boy's skeleton from the Lapedo Valley in Portugal that was interpreted as a hybrid of Neanderthals and AMHs. We will see that aDNA data have provided new insights into this topic, transforming the debate in ways that had not been anticipated (**Section 9.5**).

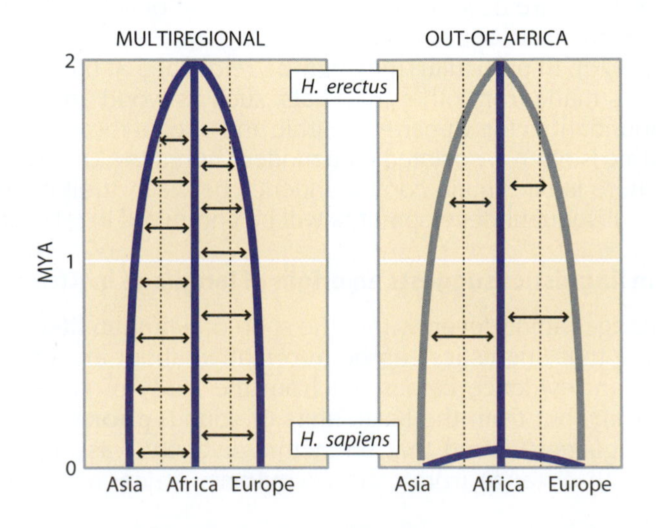

Figure 9.13: Two extreme models for the origins of modern humans.
Both models begin with *H. erectus* shortly after 2 MYA and lead to contemporary humans; many intermediate models could also be proposed. Horizontal arrows indicate gene flow between populations on different continents. In the multiregional model, extensive gene flow is required; the out-of-Africa model requires less. *Blue* lines: ancestors of modern humans. *Gray* lines: lineages that are not ancestors of modern humans.

With this background, we will now consider data from present-day samples: patterns of genetic variation in current populations should contain information about modern human origins.

9.4 EVIDENCE FROM THE GENETICS OF PRESENT-DAY POPULATIONS

The scenarios in Figure 9.13 make different predictions about the geographical distribution of genetic variation. According to the multiregional model, there is no strong reason for any one geographical region to show more diversity than another, or be the source of a majority of lineages; in contrast, the out-of-Africa model predicts both greater diversity in Africa, and that Africa would be the root of the majority of genetic lineages. These predictions would remain true for intermediate models close to one or other of these extremes. In order to evaluate such predictions, it is important to use datasets where the results are not significantly influenced by **ascertainment bias**. Suitable data can best be obtained by resequencing, but genotypes consisting of microsatellites (**Section 3.4**), which are variable in all populations, or haplotypes consisting of multiple SNPs, where different sets of common SNPs from a particular genomic region tend to identify the same set of haplotypes, are also suitable. We now have extensive genetic data that can be used to study the geographical distribution of genetic variation.

As explained in Chapter 6, the amount of genetic variation can be assessed in a number of ways, ranging from simple direct measures like the number of variants or nucleotide diversity, to more indirect statistics such as **effective population size** or the extent of **linkage disequilibrium**. We will see that all of these are informative, and identify a consistent pattern.

Genetic diversity is highest in Africa

Genetic diversity can best be evaluated using whole genome sequences, and such data are beginning to become available. The 1000 Genomes Pilot Project resequenced the genomes of population samples originating from Africa, Europe, and East Asia, providing a genomewide and reasonably unbiased view of the variation in each sample. The total numbers of SNPs discovered in the populations, the numbers per individual, and the corresponding numbers of **indels** were all highest in the YRI from Africa, intermediate in the CEU from Europe, and lowest in the CHB+JPT from East Asia[61] (**Table 9.3**).

In this study, the variant ascertainment (including sample size) was similar for the three areas, so it is meaningful to compare these raw numbers. Nucleotide diversity (**Section 6.2**) measured far from genes also shows its highest value in the YRI, an intermediate level in the CEU, and the lowest in the CHB+JPT: 1.3×10^{-4}, 1.0×10^{-4}, and 0.9×10^{-4}, respectively.[22]

Although the genome coverage in this study was close to the maximum possible—the entire **accessible genome**—the number of populations was small. Additional populations were examined in another component of the 1000 Genomes Pilot Project, which sequenced ~700 genes in seven populations.

TABLE 9.3:
GENETIC DIVERSITY, MEASURED AS SNP AND INDEL NUMBERS, FROM AFRICAN, EUROPEAN, AND EAST ASIAN SAMPLES IN THE 1000 GENOMES PILOT PROJECT

Continent	Sample	Total SNPs	SNPs per individual	Total indels	Indels per individual
Africa	YRI	10,938,130	3,335,795	941,567	383,200
Europe	CEU	7,943,827	2,918,623	728,075	354,767
East Asia	CHB+JPT	6,273,441	2,810,573	666,639	347,400

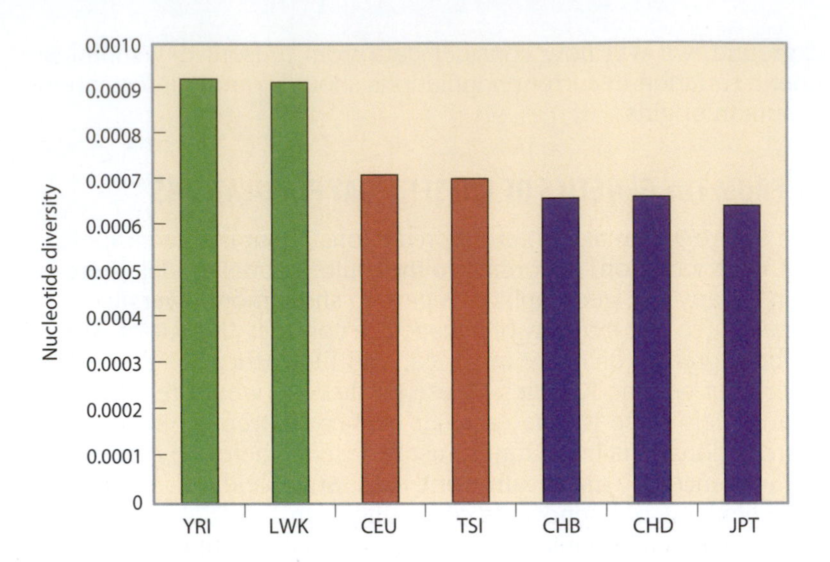

Figure 9.14: Nucleotide diversity of samples from Africa, Europe, and East Asia.
Data are from **fourfold degenerate** (approximately neutral) sites produced by the 1000 Genomes Project exon pilot. The abbreviations refer to population samples that are part of the HapMap Phase III project (see Box 3.6).

Again, there was a clear pattern, with the African samples showing the highest diversity, the European samples intermediate values, and the East Asian samples the lowest (**Figure 9.14**). This shows that such a pattern is a general feature of these geographical regions. Nevertheless, the geographical representation in these large-scale resequencing studies is still very limited (Figure 10.4). To compare levels of genetic variation in larger sets of populations, we need to turn to other datasets.

Microsatellite variation (**Section 3.4**) was typed at 377 autosomal loci in 51 populations from the **HGDP-CEPH panel** (Box 10.2) and the highest heterozygosity values were found in sub-Saharan Africa,[55] a finding confirmed by large-scale SNP genotyping in the same populations.[41] Strikingly, heterozygosity showed a strong linear decrease with distance from East Africa along plausible migration routes ($R^2 = 0.85$, $p < 10^{-4}$, **Figure 9.15**).[52, 53] Y-chromosomal variation in the same panel similarly showed decreases in TMRCA, expansion time, and N_e with distance from East Africa.[57] Such observations emphasize the importance of studying diversity within Africa (**Opinion Box 8**) and suggested a

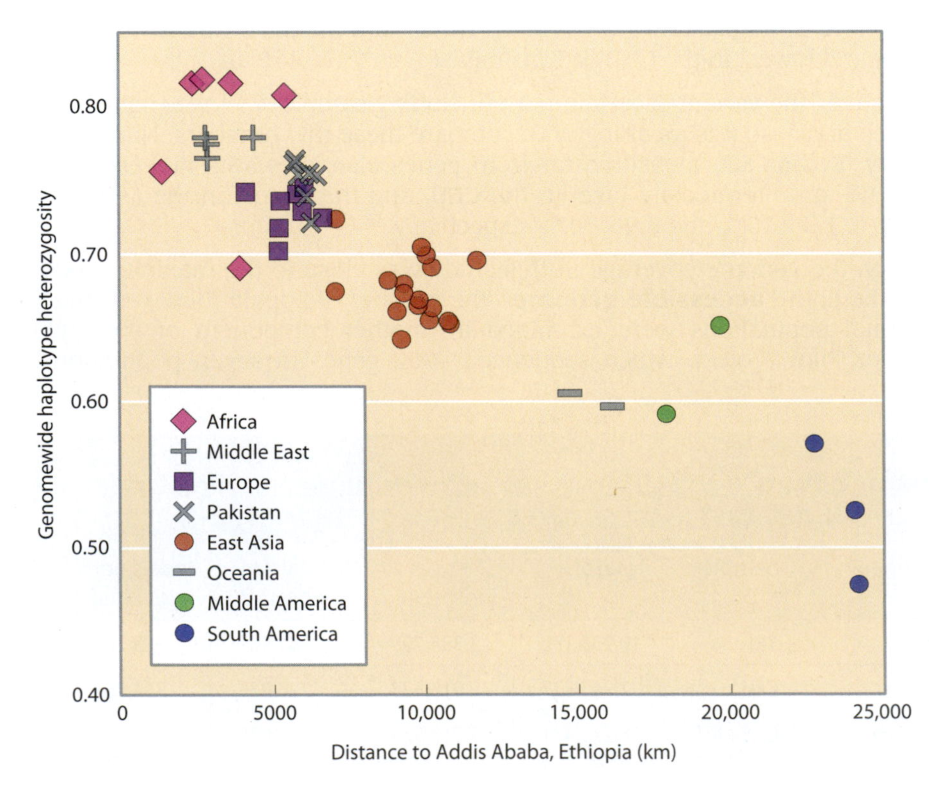

Figure 9.15: Decline of genomewide diversity with distance from East Africa.

Africa is a region of great genetic, linguistic, cultural, and phenotypic diversity. It contains more than 2000 distinct ethno-linguistic groups, speaking nearly a third of the world's languages, and practicing a wide range of subsistence patterns including agriculture, **pastoralism**, and hunting-gathering. Africans live in environments ranging from the world's largest desert and second-largest tropical rainforest to savanna, swamps, and mountain highlands, and these environments have undergone dramatic changes in the past. Differences in diet, climate, and exposure to pathogens among ethnically and geographically diverse African populations are likely to have produced distinct selection pressures, resulting in local genetic adaptations, some of which may play a role in disease susceptibility.

Given this great cultural and environmental variation, several important questions exist: (1) How much genetic structure exists among African populations? (2) How old is that genetic structure? (3) When and where did modern humans originate in Africa? (4) What are the source populations for migration(s) of modern humans out of Africa? and (5) Has introgression from archaic species shaped the African genomic landscape? A study of ~800 microsatellites and ~400 indel polymorphisms genotyped in >2500 Africans indicated high levels of population substructure.[64] Fourteen genetically divergent "ancestral population clusters" correlate with self-described ethnicity and shared cultural and/or linguistic properties. Most African populations have mixed ancestry from these different clusters, reflecting high levels of migration and admixture among ethnically diverse groups. Although some of the inferred ancestral populations likely reflect recent differentiation (for example, eastern and western Niger–Congo speakers which split within the last few thousand years), other structure is likely to be ancient. Indeed, there is evidence of shared common ancestry between several of the major hunter-gatherer populations in Africa who currently reside in Central, Southern, and Eastern Africa (Pygmies, San, Hadza, Sandawe). Analyses of mtDNA and Y-chromosome lineages that differentiated >50 KYA in these populations suggest that the common ancestry could have been quite ancient.

The oldest anatomically modern human fossil, dated to ~195 KYA, was found in southern Ethiopia.[45] However, the most divergent mtDNA, Y chromosome, and autosomal lineages are found in the San hunter-gatherer populations currently residing in southern Africa. Additionally, archaeological data suggest that the earliest modern behavior occurred in both southern and eastern Africa. However, there is a dearth of fossil and archaeological data for modern human origins, particularly from central and western Africa where material is poorly preserved due to the tropical climate. Heterozygosity based on microsatellite and SNP variation is highest in the **click-language**-speaking San, consistent with them being derived from a population ancestral to other populations. Given their current geographic location in southern Africa, Henn et al.[21] argued for a southern African origin of modern humans. However, linguistic data suggest that click languages may have originated in eastern Africa, as far north as Ethiopia. Therefore, it is possible that the San may have originated in eastern Africa and migrated south within the past 10–50 KY.

Furthermore, recent targeted and whole-genome[32] sequencing of African hunter-gatherers suggests that modern humans in Africa may have admixed with archaic populations that were as divergent from modern humans as Neanderthals were. The implication is that introgression from different archaic species occurred across the globe. Thus, although most of the modern human genome originated directly from Africa, some genomic regions have much older lineages that may have originated via non-African groups such as Neanderthals and Denisovans, and as-of-yet unknown archaic African populations. Thus, on a global level, a recent African origin model incorporating low levels of ancestry from local archaic populations is most appropriate. However, within Africa, many questions remain. For example, it is possible that modern human origins could involve multiple locations within the continent, given the closer geographic proximity of populations and opportunities for long-distance gene flow and admixture. Such a model would imply ancient substructure (and hence, ancient lineages) within African genomes (**Figure 1**). Additionally, the transition to modern human morphology could have been gradual, rather than abrupt.[69] Inference of modern human origins within Africa will ultimately require integration of novel paleobiological, archaeological, and whole-genome sequence data from diverse Africans, together with development of sophisticated computational modeling approaches.

Sarah A. Tishkoff, Departments of Genetics and Biology, University of Pennsylvania, USA

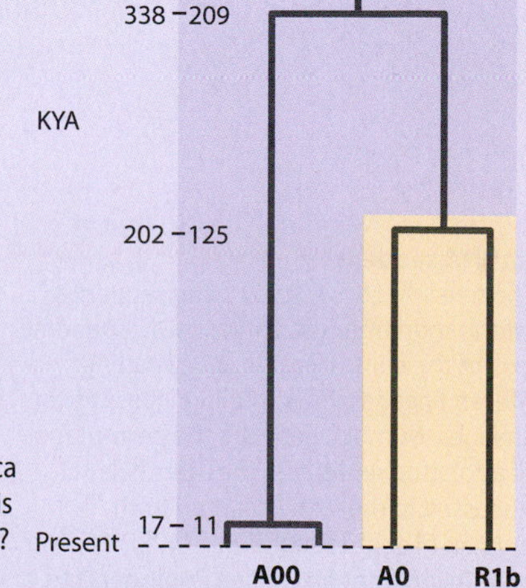

Figure 1: Outstanding questions about Africa Y-chromosomal history.
Until 2012, the first known branch in the Y-chromosomal phylogeny was between haplogroup A0 and the other haplogroups, represented here by R1b (*yellow*). In 2013, a much deeper-rooting haplogroup, A00, was reported as a very rare lineage in African Americans and the Mbo from Cameroon in West Africa (*blue*). Using different calibration scales, the new root could be placed at either 338 or 209 KYA. Does this indicate ancient population structure, archaic introgression or some other complexity of human origins? [Adapted from Mendez FL et al. (2013) *Am. J. Hum. Genet.* 92, 454. With permission from Elsevier.]

serial founder model for human expansion, described below. Although genotyping previously discovered SNPs provides a biased view of variation, haplotypes of ≥20 kb constructed from such genotypes escape this bias and show a similar pattern of highest haplotype diversity in sub-Saharan Africa, decreasing to lowest in South America.[13]

Having established that the highest levels of genetic variability are found in sub-Saharan Africa, it is of interest to ask how precisely the most likely place of origin for modern humans can be pinpointed. This is less straightforward than might be expected, because current genetic patterns in Africa are dominated by the spread of agriculturalists in the last few thousand years, which erased many earlier patterns (**Section 12.6**). Studies based on the HGDP-CEPH samples have pointed to East Africa,[52] while a combination of two studies that incorporated more samples from both hunter-gatherer populations (including the Hadza and Sandawe from Tanzania and ‡Khomani from South Africa) and Ethiopians favored a southern African origin, because the lowest LD values were found in this region.[21, 48] At present, considerable uncertainty remains, and it is possible that modern African populations do not retain sufficient genetic information to reach a clear conclusion about this topic.

Genetic phylogenies mostly root in Africa

The phylogeny of a locus can provide information about the time and place of its origin. This involves some assumptions: (1) that the phylogeny can be reconstructed accurately; and (2) that geographical movement has been limited, so that the modern distribution provides information about the ancient distribution. Molecular phylogenetic information was first applied to the question of human origins when mtDNA data became available. The features of this locus that make it particularly suitable for such studies, and aspects of the nomenclature of clades, are explained in the **Appendix**.

Mitochondrial DNA phylogeny

An mtDNA phylogeny based on the complete mtDNA sequences of 53 individuals of diverse geographical origins, rooted by comparison to a chimpanzee sequence, was constructed by Ingman et al.[27] All sequences were different and 657 variable positions were found, 516 of which were outside the hypervariable control region. Despite the elevated mutation rate of mtDNA compared with nuclear sequences, a robust phylogeny could be obtained from the complete sequence excluding the control region (**Figure 9.16**). This has some striking features:

- Complete separation of African and non-African lineages
- The first three branches lead exclusively to African lineages, while the fourth branch contains both African and non-African lineages

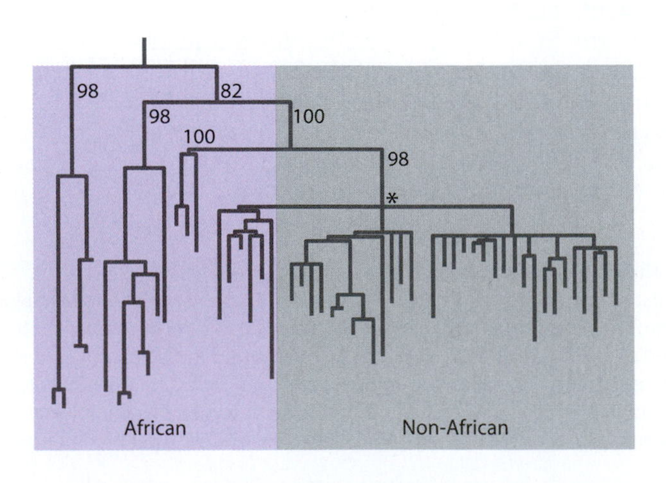

Figure 9.16: Neighbor-joining tree of mtDNA sequences.
The tree was constructed using sequence information from the entire mtDNA genome except the control region. *Blue* shading: African lineages. *Gray* shading: non-African lineages. Numbers indicate the percentage of bootstrap replicates. The asterisk is discussed in the text. [Adapted from Ingman M et al. (2000) *Nature* 408, 708. With permission from Macmillan Publishers Ltd.]

- Deep branches within African lineages, contrasting with **star-like** structure within non-African lineages
- **TMRCA** for the entire phylogeny: 172 ± 50 KY
- TMRCA for the branch containing African plus non-African lineages, marked with an asterisk in Figure 9.16: 52 ± 28 KY
- Expansion time for non-African lineages estimated at 1925 generations or 38.5 KY at 20 years/generation by the authors (48 KY at 25 years/generation)

A study focused specifically on the African lineages investigated 624 complete sequences. It found that the deepest phylogenetic split was between L0 lineages and the rest (L1′5), and proposed that this also corresponded to long-lasting population substructure originating before 90 KYA,[7] thus emphasizing an origin in sub-Saharan Africa and adding more detail to the model.

Y-chromosomal phylogeny

The Y chromosome is also a highly informative locus for such phylogenetic studies (**Appendix**). A Y-chromosomal phylogeny derived from **DHPLC**-based mutation detection in 64 kb of DNA from 43 individuals was, for many years, the largest ascertainment-bias-free global survey.[63] It revealed 56 variants which distinguished 32 lineages that fell into the parsimony tree shown in **Figure 9.17**, again rooted by comparison with other ape sequences. Although less detailed than the mtDNA phylogeny, its structure shows close parallels:

- Complete separation of African and non-African lineages
- The first two branches lead exclusively to African lineages, while the third branch contains both African and non-African lineages
- TMRCA for the entire phylogeny: 59 (40–140) KY, assuming 25 years/generation
- TMRCA for the branch containing African plus non-African lineages, marked with an asterisk in Figure 9.17: 40 (31–79) KY

This **point estimate** (best single estimate, but not taking account of the uncertainty) for the Y phylogeny TMRCA is very recent, and is discussed further in the **Appendix**.

Other phylogenies

Phylogenies from several autosomal and X-chromosomal loci are available. These are potentially complicated by recombination, but by analyzing very closely linked polymorphisms, usually within 10 kb or less, haplotypes showing little recombination can be identified and the effects of recombination minimized or excluded. An examination by Takahata and co-workers[60] in 2001 of

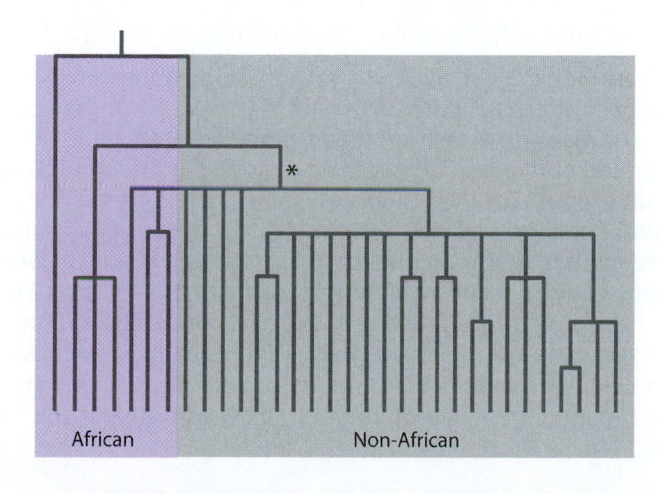

Figure 9.17: Y-chromosomal phylogeny. *Blue* shading: African lineages; *Gray* shading: non-African lineages. The asterisk is discussed in the text. [Adapted from Thomson R et al. (2000) *Proc. Natl Acad. Sci. USA* 97, 7360. With permission from the National Academy of Sciences.]

published data found that it was possible to infer the ancestral origin of a total of 10 loci. Nine of the 10 origins were in Africa, and one was in Asia. The latter, *Glycerol Kinase* (*GK*), was based on a sample size of 10 and a single variable position, so this conclusion might change if more data were available.

Unusual phylogenies including highly divergent non-African haplotypes have been identified in a number of genomic regions, including an inversion polymorphism on chromosome 17q, the microcephalin gene (*MCPH1*), and within the *HLA* region, leading to suggestions that they might represent examples of introgression from other hominins. Available aDNA evidence (see **Section 9.5**) has not supported this scenario for the first two examples, but has led to the remarkable suggestion (awaiting independent confirmation) that more than half of the *HLA-A* ancestry in Europe and Asia derives from archaic hominins.[1] It was proposed that this exceptional ancestry reflects a selective advantage of alleles already adapted to a Eurasian environment for humans migrating out of Africa.

We must remember that these analyses are of *loci*, not populations. If the population size were the same on each continent, the finding of an African origin for at least 9/10 loci would support the out-of-Africa model. In reality, population sizes have not been the same and it is likely that African populations were larger than those on other continents for much of human prehistory, and there is some evidence from phylogeny for ancient non-African contributions to our gene pool. Nevertheless, phylogenetic analyses overwhelmingly point to an African origin for most loci, and the simple conclusion is that most of our ancestors lived in Africa before 60 KYA.

Insights can be obtained from demographic models

It should be possible to formalize the insights from the variation observed in modern DNA into tests that seek to distinguish between explicit alternative models. The strength of such an approach is that it can quantify the likelihood of the alternatives, but the weakness is that available models are grossly oversimplified, and explore only a small proportion of potential models. Two types of model have, however, been used widely.

"Best-fit" demographic models have been sought to model characteristics of ancestral populations that would lead to the observed levels of genetic variation in current populations. **Parameters** have generally included effective population sizes at different times (including bottlenecks, expansions), the order and times at which populations split, and the migration rates between them. Such models of African, European, and East Asian populations have mostly supported an African origin, single exit involving a bottleneck, and large expansions of the European and East Asian populations. Some have suggested a divergence between African and Eurasian populations ~100 KYA[19] (**Figure 9.18a**) or a divergence between Europeans and East Asians as recently as 22.5 KYA[33] (Figure 9.18b). Some models have included archaic admixture as one alternative, and have found support for this, for example 14(2–20)% archaic contribution to Europeans and 1.5(0-5–2.5)% to East Asians[68] (Figure 9.18c).

Serial founder models have sought to capture and suggest explanations for the observed global patterns of genetic variation (**Figure 9.19**): they can incorporate many more populations than the best-fit models above, but use fewer parameters. The underlying observations are that diversity decreases while LD and population differentiation (F_{ST}) increase with distance from Africa,[52, 53] leading to two general conclusions. (1) These trends are linear with migrational (walking) distance, rather than with direct (great circle) distance (**Section 6.8**). (2) Sharp discontinuities, which would imply distinct types of human rather than a continuum, are not seen. A model that explains these observations in a simple and effective way starts from a single source population; a new population is formed by a subset of individuals (that is, founders), and after growth a subset of this subset founds the next population, and so on—this is the "serial" aspect (Figure 9.19).

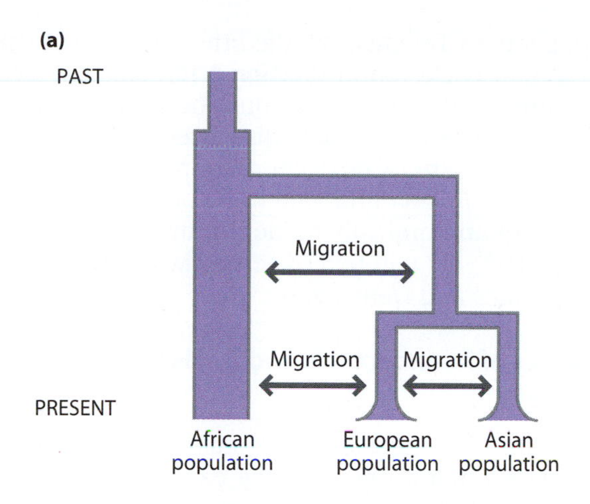

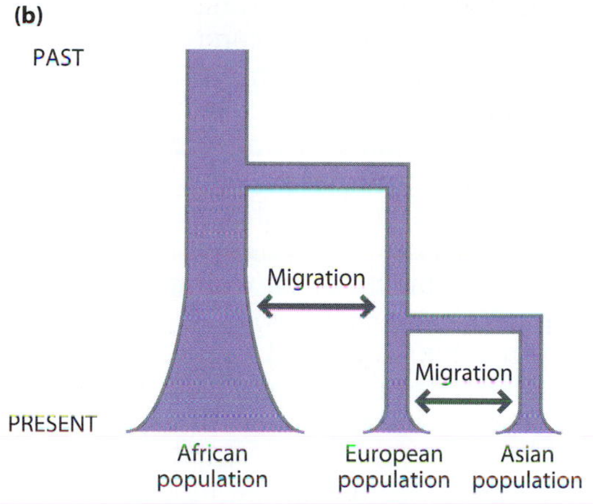

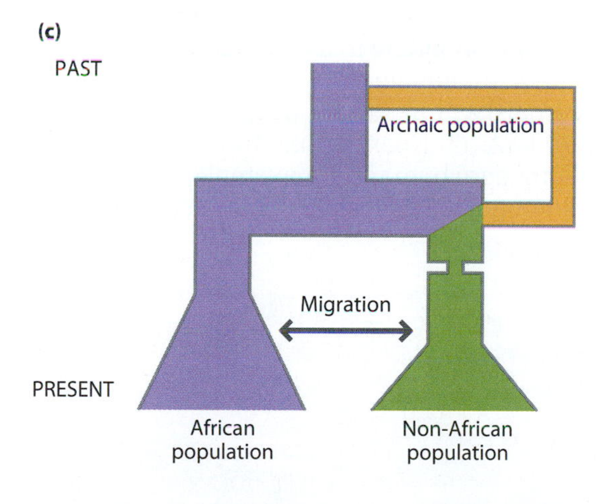

Figure 9.18: Best-fit demographic models.
Modern populations are represented at the bottom of each model, and the past population splits, size changes, and migration events considered by the model are shown. Model (c) includes admixture with archaic hominins (*yellow* and *green*). [a, adapted from Gutenkunst RN et al. (2009) *PLoS Genet.* 5, e1000695. With permission from Public Library of Science. b, adapted from Laval G et al. (2010) *PLoS One* 5, e10284. With permission from Public Library of Science. c, adapted from Wall JD et al. (2009) *Mol. Biol. Evol.* 26, 1823. With permission from Oxford University Press.]

Overall, demographic models support an origin of modern humans in sub-Saharan Africa and a stepwise expansion throughout the rest of the world involving multiple small bottlenecks. But specific details vary so much between models that conclusions drawn from current models need support from independent evidence for credibility.

9.5 EVIDENCE FROM ANCIENT DNA

The analysis of ancient DNA should be an ideal way to distinguish between different hypotheses about the origins of modern humans: by investigating a time series of fossils from any region, it should tell us directly whether there was regional continuity or replacement of early lineages by African ones. Unfortunately, it is impossible to obtain trustworthy DNA sequence data from most fossils: DNA does not survive well, and contaminating DNA, from the

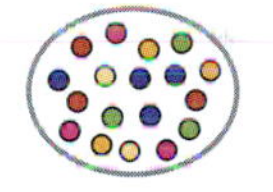

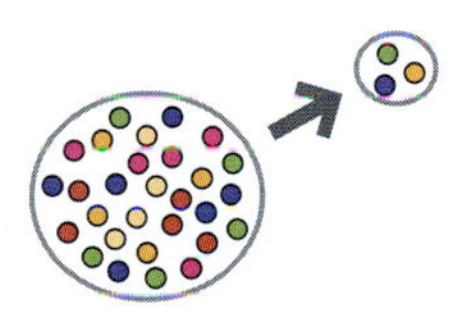

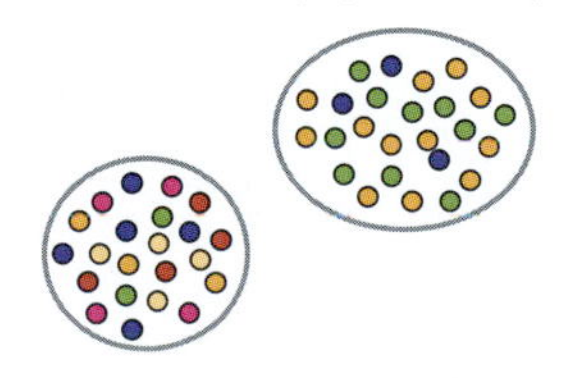

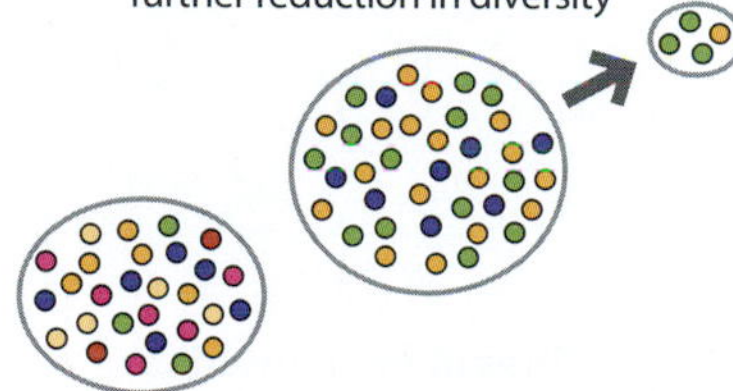

Figure 9.19: Serial founder model.

environment and from people who have handled the fossils or carried out the analysis, can provide a high background (**Section 4.10**). Ancient DNA work is technically demanding and stringent criteria must be met before results can be accepted as authentic. These criteria are discussed in the Opinion Box 2 in Chapter 4. Only with the advent of next-generation sequencing technology (**Section 4.4**) has aDNA fulfilled its potential and transformed our view of human evolution. aDNA analyses of anatomically modern human remains will be discussed in later sections. Here, we will be concerned with aDNA insights from archaic humans: Neanderthals and Denisovans.

Ancient mtDNA sequences of Neanderthals and Denisovans are distinct from modern human variation

Neanderthal mtDNA was an early target for aDNA studies: its generally high copy number in the cell suggested that it might be the easiest aDNA to derive data from, while its phylogenetic informativeness promised insights into the relationship between this extinct branch of hominin and the closely related current modern humans. The determination of a partial mtDNA hypervariable region sequence from the Neanderthal type specimen (called Feldhofer 1 after the cave of origin) in 1997 was thus a major step forward for aDNA workers.[31] Since then, complete mtDNA sequences from six individuals have been determined (**Figure 9.20**),[10] and partial sequences from 10 others. Some general conclusions about Neanderthal mtDNA sequences are possible: (1) Neanderthal mtDNAs are distinct from modern human mtDNAs: for example, a comparison of the six complete Neanderthal mtDNA sequences with 54 present-day humans and one ~30-KY-old early modern human identified an average of 202 substitutions (range 185–220) between Neanderthals and humans, compared with 60 (range 1–106) among this set of modern humans; this corresponds to a divergence time of 466 (321–618) KYA:[30] see **Figure 9.21**. (2) Neanderthal mtDNAs show low diversity, apparently lower even than that within modern humans, and much less than that within other apes: an average of 20 substitutions among the six Neanderthal sequences, corresponding to a coalescence time of ~100 KY.[10] This observation is particularly striking since the six sampling sites range from El Sidron in Spain to Mezmaiskaya in Russia (Figure 9.20), and the fossil dates from ~38 KYA to ~60–70 KYA. (3) Neanderthal mtDNAs are no more similar to Europeans than to other modern humans: for example,[31] Neanderthal–European hypervariable region differences = 28.2 ± 1.9, while Neanderthal–African differences = 27.1 ± 2.2.

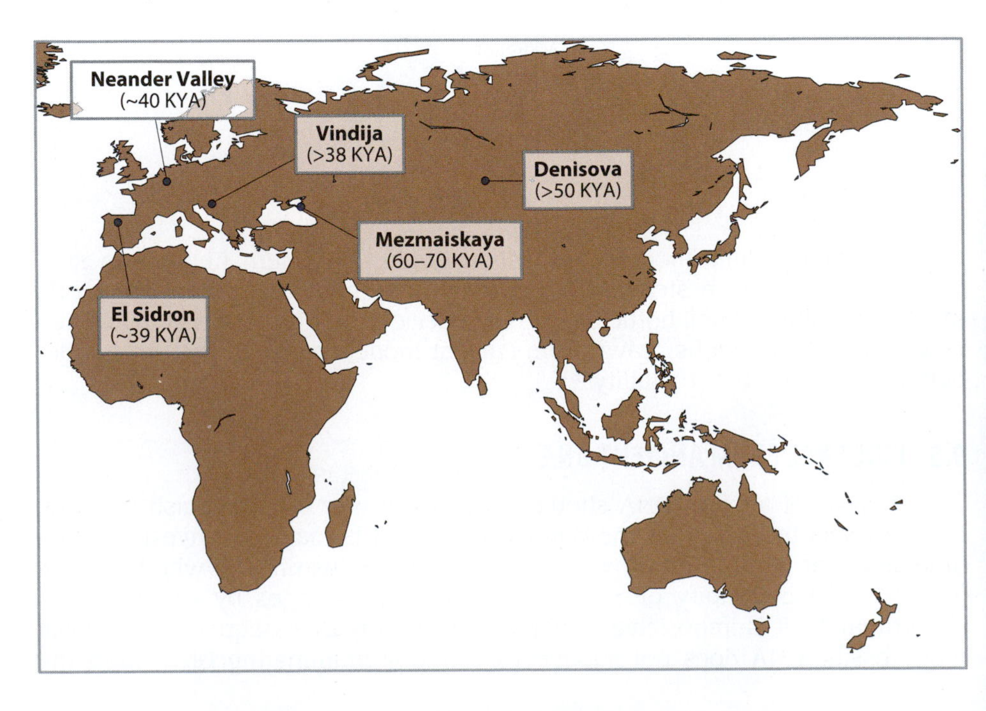

Figure 9.20: Sites of fossils used to generate aDNA data from Neanderthals and Denisovans.

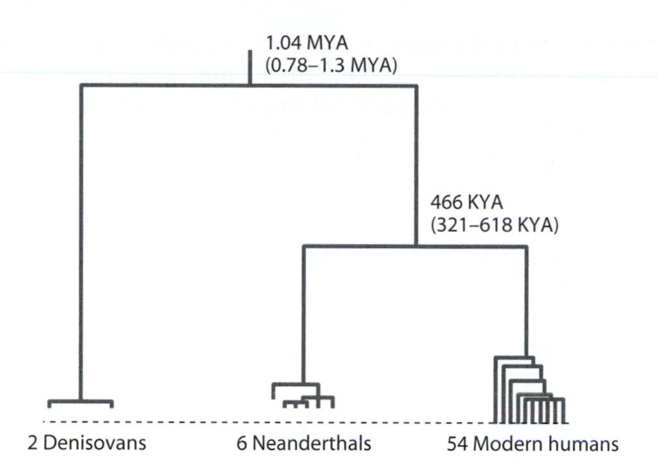

Figure 9.21: Phylogeny of Neanderthal, Denisovan, and modern human mtDNAs.
[Adapted from Krause J et al. (2010) *Nature* 464, 894. With permission from Macmillan Publishers Ltd.]

Indeed, by 2010 aDNA analysis had become such an effective way of characterizing fossils that it was being applied to hominin bones of uncertain affinity in order to identify them. A juvenile hominin "distal manual **phalanx** of the fifth digit" (tip of the little finger) too small to be dated directly but from deposits estimated to date to between 50 and 100 KYA, was excavated from Denisova Cave in the Altai Mountains of Siberia (Figure 9.20), and a 156×-coverage full mtDNA sequence was determined by Illumina GAII™ sequencing.[30] Surprisingly, this sequence did not match either Neanderthals or modern humans, showing on average 385 differences (range 372–396) from 55 modern humans, almost twice as many as from Neanderthals. The new hominins were subsequently designated Denisovans after their place of origin. An mtDNA sequence from a tooth from the same cave differed at just two positions.[54] Assuming a chimpanzee–human split 6 MYA, the Denisovan sequences diverged from the human–Neanderthal mtDNA clade 1.04 (0.78–1.3) MYA (Figure 9.21). The extraordinarily good DNA preservation in the phalanx, partly but not entirely explained by the low temperature in Siberia, allowed a sequence of the entire genome to be determined; this, and the affinity of this enigmatic hominin, is described below.

A Neanderthal draft genome sequence has been generated

The analysis of ancient nuclear DNA is more difficult than for ancient mtDNA because of its much larger size and lower concentration. Studies between 2007 and 2009 investigated individual genes of particular interest involved in skin pigmentation (*MC1R*, **Section 15.3**), speech and language (*FOXP2*, see **Section 8.3**), the ABO blood group (Box 3.1), and phenylthiocarbamide taste perception (*TAS2R38*, **Section 15.5**). The application of new sequencing technologies provided the potential to overcome these limitations, but was initially plagued by contamination, and improved procedures involving tagging of the aDNA library with a short oligonucleotide in the clean room used for extraction, before transfer to the contamination-prone sequencing environment, were developed as a response. Using these, a ~1.3× rough draft sequence of the Neanderthal genome was determined in 2010,[18] providing the basis for the rest of this section; insights into functional variants specific to modern humans are covered in Chapter 8.

The draft Neanderthal sequence was mainly derived from three bones from Vindija Cave in Croatia (Figure 9.20: Vi33.16, Vi33.25, and Vi33.26, yielding 1.2 Gb, 1.3 Gb, and 1.5 Gb, respectively). These came from three different females, although two carried the same mtDNA sequence, and dated to ~45 KYA. Small amounts of additional sequence were generated from additional Neanderthals from El Sidron, Feldhofer, and Mezmaiskaya. The Neanderthal sequence was shown to have <1% contamination with modern human DNA, and thus allowed an initial comparison of the modern human and Neanderthal genomes.

This comparison had to take into account the Neanderthal low coverage and DNA damage, which together resulted in a high error rate: the number of substitutions specific to the Neanderthal lineage was apparently 30× that on the human lineage. Consequently, Neanderthal-specific changes could not be identified with any confidence, but the Neanderthal sequence could be used to identify which changes on the lineage leading to humans arose before the human–Neanderthal split, and which after. Overall, this proportion was 12.7%, corresponding to an average split time ~825 KYA of individual segments of the genome assuming chimpanzee and human DNAs diverged 6.5 MYA. Lineage divergences are always earlier than population divergences, and the ancestral populations of humans and Neanderthals were estimated to have split 270–440 KYA.

The aspect of the Neanderthal genome study that attracted most attention was the comparison of allele sharing between Neanderthals and modern humans from different geographical regions, showing an excess of sharing with all humans outside Africa. SNPs were ascertained by resequencing modern humans and a statistic *D* (here designated Patterson's *D* to distinguish it from other *D* statistics; Table 6.2) was developed to quantify the sharing (see **Section 6.3**). A pair of humans was picked, and a random copy of each binary SNP chosen from each individual. When these differed, and Neanderthals carried the derived allele, the position contributed to *D*, which combined these differences over the genome. If the two humans were equally closely related to Neanderthals, *D* would be zero, but if one were more closely related, *D* would depart from zero. Application of this test led to the following conclusions:

- When the two humans were both from sub-Saharan Africa, or both from outside Africa, *D* was not significantly different from zero.
- However, when any human from inside Africa was compared with any from outside, *D* departed from zero, revealing excess sharing outside Africa.
- Two hypotheses could account for this excess sharing: (1) ancestral population structure, such that humans outside Africa were derived from a source population more closely related to Neanderthals; or (2) ~1–4% gene flow from Neanderthals to non-African humans.
- Green et al. favored hypothesis (2), suggesting mixing in the Middle East soon after the exit from Africa to account for the uniform *D* values in non-African populations. Note that the standard models for the migration out of Africa do not predict contact between humans and Neanderthals (Figure 11.7). An alternative interpretation is presented in **Opinion Box 9**.
- No excess of allele sharing specific to Europeans was found, excluding, at this level of sensitivity, admixture in Europe during the ~10 KY of possible coexistence and contact between ~40 KYA and ~30 KYA.

The possibility of admixture between humans and Neanderthals was particularly intriguing. Although it could not be distinguished from the less exciting explanation of ancestral population structure in this analysis (Opinion Box 9), debate about this issue had hardly begun by the time the Denisovan genome sequence was published later in 2010.

A Denisovan genome sequence has been generated

The extraordinarily good preservation and low contamination level of the Denisovan finger bone described above allowed a 1.9× draft genome sequence[54] to be determined in 2012. This was derived from a female and contained <1% contamination, and the sequence was higher in quality than the Neanderthal genome, in part because of the greater coverage, but also because improvements in aDNA technology allowed enzymatic removal of uracil residues, resulting from damage, from the starting DNA.

Analyses applied to the Neanderthal genome could also be used on the Denisovan sequence. A similar polarization of human lineage variants into ones that arose

The debate on the possible admixture between anatomically modern humans and other hominins has been revolutionized by the recovery of reliable genetic information from ancient specimens. The availability of these genomes opens the possibility of directly testing for localized admixture: in principle, if hybridization only occurred within part of the range of anatomically modern humans, we would expect the hominin genome to be genetically more similar to modern populations in that area than to modern populations in other areas. Indeed, a first analysis of the Neanderthal genome revealed Neanderthals to be genetically more similar to present-day Eurasians than to present-day Africans.[18] This asymmetry has been interpreted as evidence for admixture between Neanderthals and anatomically modern humans during the latter's exit out of Africa. Given that there is no significant difference between Europeans and Asians in their similarity to Neanderthal, it has been argued that such admixture would have had to happen at the very beginning of the out-of-Africa exodus, before the split between these two groups.

A possible complication in interpreting spatial patterns of similarity between any ancient hominin and modern human populations is that, while such differences might arise through recent hybridization, they could also, in principle, be the consequence of population structure in early humans and Neanderthals (that is, a case of **incomplete lineage sorting**—see **Section 7.3**). Because there is both archaeological and genetic evidence for ancient population structure in Africa, the effect of population structure in early humans has to be taken into account. To see how incomplete lineage sorting could generate the observed patterns, let us consider a simple hypothetical scenario. The common ancestor of both anatomically modern humans and Neanderthals would have inhabited the whole of Africa, Europe, and Central Asia at some point in the past, let say half a million years ago. The populations of this ancestor would have most certainly shown isolation by distance, such that populations in the northern part of the African continent would have been more similar to European ones than populations found further south. Approximately 300 KYA, the link between African and European populations was severed by a change in climate, with the European populations differentiating into Neanderthals and the African part of the range eventually becoming anatomically modern humans. It is likely that the population structure found in Africa would have persisted, at least to some extent, thus implying that the northern range of anatomically modern humans in that continent would have been more similar to Neanderthals than the southern range. When anatomically modern humans expanded out of Africa approximately 60–70 KYA, it is then quite likely that populations in the northern part of the range would have contributed most of the colonists due to their proximity to the exit points out of the continent (see **Figure 1**).[16] Thus, it would have been the populations that were more similar to Neanderthals who exited Africa and founded the European and Asian lineages of anatomically modern humans, generating exactly the pattern that we see of equal higher similarity of European and Asians to Neanderthals.

While the above logic cannot disprove admixture, it invalidates current tests for admixture, bringing us back to where we were before the Neanderthal genome was sequenced. It should be noted that the issues described above are a possible complication for any attempt to use geographic patterns in similarity between anatomically modern humans and ancient hominins, irrespective of the hominin in question, and which measure of similarity is used. Simple demographic models (for example, using two populations to represent African structure) are unlikely to capture the subtle patterns generated by the fine-grained structure that is likely to have existed across the whole continent. The only real solution to properly investigate admixture will be to obtain sequences of a number of genomes from ancient hominins (and ideally ancient anatomically modern humans), such that population structure in both sets of populations can be reconstructed. Only then will it be possible to show quantitatively whether differential similarity in certain populations is truly a sign of admixture, or whether population structure is a simpler and more parsimonious explanation.

Andrea Manica and Anders Eriksson,
Department of Zoology, University of Cambridge, UK

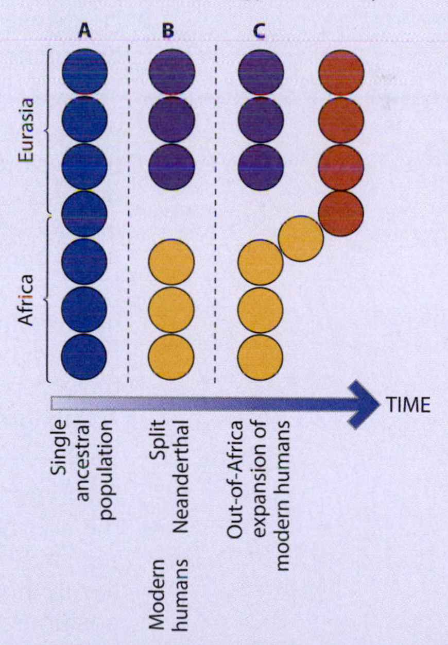

Figure 1: Schematic representation of the relationship between Neanderthal and anatomically modern humans.
A chain of connected populations spanning Africa and Eurasia represents the common ancestor (*blue*). Following the split (B), the northern part of the range became Neanderthals (*purple*), while the African part of the range eventually became modern humans (*orange*). From this African range, modern humans later expand to colonize Eurasia (*red*, C). It is likely that this expansion would have received a large contribution from the northern populations in Africa, which would be more similar to Neanderthals than other African populations due to their proximity to Eurasia before the split. [From Eriksson A & Manica A (2012) *Proc. Natl Acad. Sci. USA* 109, 13956. With permission from Andrea Manica, Cambridge University, UK.]

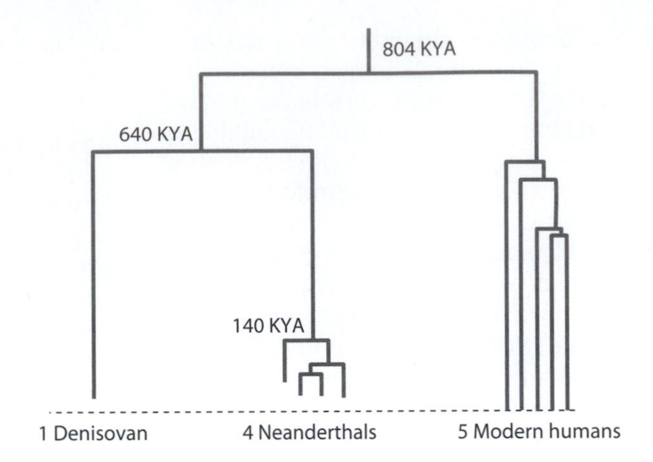

Figure 9.22: Average autosomal sequence divergence times of Neanderthals, Denisovans, and modern humans. Average autosomal divergences were calculated from low-coverage whole-genome sequences and converted into time estimates assuming that humans and chimpanzees diverged 6.5 MYA. [Data from Green RE et al. (2010) *Science* 328, 710, Reich D et al. (2010) *Nature* 468, 1053.]

before the split with Denisova and after indicated that the lineage split occurred ~760 KYA, not significantly different from the time of the Neanderthal split from the human lineages. Indeed, the same analysis applied to the Neanderthal and Denisova sequences suggested a split between these extinct hominins ~640 KYA. Thus the autosomal analysis indicated shared ancestry between the Neanderthal and Denisovan lineages after the split from humans (**Figure 9.22**), which contrasts with the mtDNA phylogeny (Figure 9.21). This difference could in principle be explained by drift in a large ancestral population, or by introduction of the Denisovan mtDNA lineage from another hominin; in view of the small effective population size inferred for Denisovans as described below, the former explanation appears unlikely. Further comparison of the Neanderthal and Denisovan sequences also suggested a Neanderthal-specific bottleneck, with a divergence date of the Vindija and Mezmaiskaya autosomal DNA sequences 140 ± 30 KYA (Figure 9.22).

Allele sharing was also examined using Patterson's *D* statistic. In contrast to the Neanderthal comparison, no excess sharing between Denisova and all non-Africans was detected. However, excess sharing was seen with Papuan and Melanesian samples, suggesting that 4.8 ± 0.5% of their genomes derive from Denisovans. Excess allele sharing between Denisovans and the ancestors of populations from Oceania cannot be explained by any simple model of ancestral population structure, so this conclusion is most readily accounted for by admixture, but does raise the question of where such admixture might have occurred, since Denisova Cave, in Siberia, lies far away from Oceania. Admixture between modern humans and Denisovans, in turn, makes the possibility of admixture between modern humans and Neanderthals more plausible.

It was not possible to identify, from the morphology of the Denisova finger bone, the source hominin taxon. A tooth (upper molar, either third or second) from the same cave, however, was potentially more informative. As described earlier in this section, this tooth yielded an mtDNA sequence differing at just two positions from the Denisovan reference, and thus is Denisovan. The tooth itself is large, outside the range of Neanderthal and early modern human variation, and, although within the size range of *H. erectus*, still distinguishable from the few known Chinese examples of *H. erectus* second and third molars. It therefore reinforces the distinction between Denisovans, Neanderthals, and modern humans, but leaves the morphological relationships between Denisovans and other hominin taxa in Asia 50–100 KYA unresolved. The discoverers have, with admirable restraint, declined to propose a Linnaean species name for Denisovans.

Subsequent improvements in the construction of libraries for sequencing, in particular the use of single-stranded instead of double-stranded aDNA as the starting material, allowed a higher coverage sequence to be generated from the

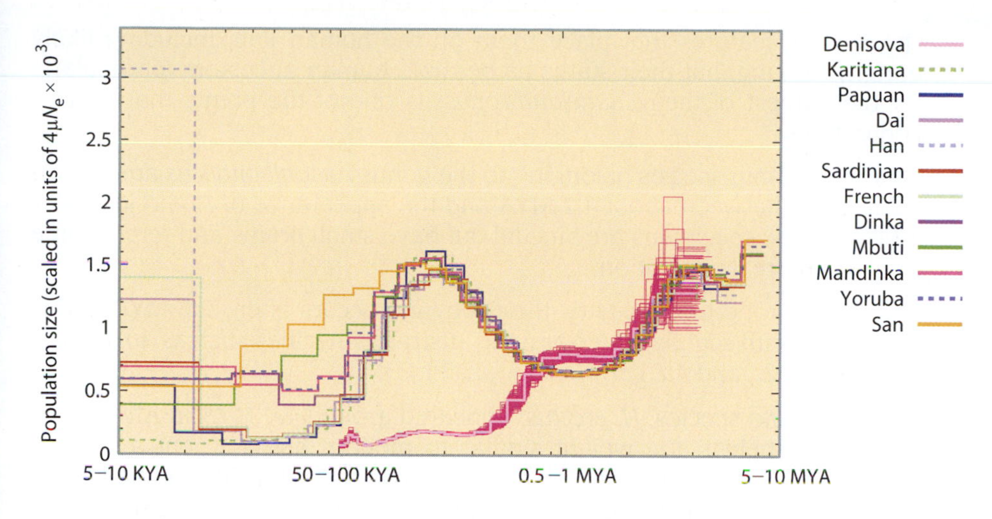

Figure 9.23: Changes in human and Denisovan inferred population size over time.
Past population sizes were estimated from high-coverage whole-genome sequences using the PSMC approach (Chapter 6). PSMC, pairwise sequential Markov coalescence. [Adapted from Meyer M et al. (2012) *Science* 338, 222. With permission from AAAS.]

same finger bone.[47] This more accurate sequence provided several additional insights:

- The Denisovan genome had accumulated 1.16% fewer inferred nucleotide substitutions since the chimpanzee–human common ancestor than 11 present-day humans, suggesting that if this ancestor had lived 6.5 MYA, the Denisovan individual died 74–82 KYA, consistent with the age estimates for the fossil.
- Denisovan heterozygosity was low (2.2×10^{-5}), about one-fifth of the level seen in a present-day African genome. Inference of the change in population size over time (Chapter 6, **Section 6.6**) suggested a demography shared with modern *H. sapiens* ancestors before 1–2 MYA and a decline in numbers after 400–800 KYA, coincidentally or not as modern human ancestors were increasing (**Figure 9.23**). As expected from a small effective population size, the Denisovan genome was enriched for slightly deleterious variants such as nonsynonymous changes.
- Comparison with the human genome led to the identification of 111,812 single-nucleotide changes and 9499 indels where the humans examined were fixed for the derived state and the Denisovan was ancestral; 260 of these coded for amino acid substitutions, including in *CNTNAP2*, a gene regulated by FOXP2 and implicated in language disorders. Further analyses of these fixed differences should yield rich insights into the genetic basis of human uniqueness (Chapter 8).

aDNA studies have thus increased the hominin taxa known from the period 50–150 KYA from four (*H. neanderthalensis, H. sapiens, H. erectus, H. floresiensis*) to five, and provided evidence for low levels of gene flow from both Neanderthals and Denisovans into modern humans. While any level of admixture excludes an extreme out-of-Africa model, the current data suggest a 2.5 ± 0.6% contribution to all non-African populations from Neanderthals and an additional 4.8 ± 0.5% to some Pacific populations from Denisovans, still supporting a predominantly African origin of all modern humans.

SUMMARY

- Information about modern human origins is provided by fossils, archaeological remains, and studies of both present-day human genetic variation and ancient DNA.
- Interpretations of almost all sources of evidence are hotly debated and it is difficult to identify a consensus view about many topics.
- Fossils that date to the approximate time of the chimpanzee–human split, about 5–7 MYA, have been described from three locations in Africa. They

show some features that place them on the human line, including likely upright walking, but their status as possible human ancestors (particularly for the earliest of them, *Sahelanthropus*) is one of the points that remain contentious.

- Several hominin species belonging to the genus *Australopithecus* are known from Africa between about 4.2 MYA and the appearance of *Homo* just after 2 MYA. These specimens are bipedal but have small brains and retain some form of arboreal adaptation.
- The oldest stone tool industry, the Oldowan, occurs after ~2.6 MYA and is associated with slightly larger-brained australopithecines such as *Au. habilis*, *Au. rudolfensis*, and *Au. garhi*.
- The first *Homo* species, *H. erectus*, appeared around 1.9 MYA in Africa, and exhibited a height and weight similar to modern humans, but a smaller brain. *H. erectus* was the first hominin to leave Africa and did so initially with simple Oldowan technology. *H. erectus* is known from Southeast Asia by 1.7 MYA, but hominins have not been found in Europe until after 1.2 MYA.
- After 1.8 MYA, *H. erectus* is associated with the more complex Acheulean technology that is found in Israel and India by 1.5 MYA but not in Europe until after 900 KYA. This technology is absent from Eastern Asia.
- Several later *Homo* taxa are known, including *H. heidelbergensis*, *H. neanderthalensis*, *H. floresiensis*, Denisovans, and *H. sapiens*, although their relationships are still debated. *H. heidelbergensis* was associated initially with Acheulean technology and later Middle Paleolithic or Middle Stone Age technology. While Mousterian technology is generally associated with Neanderthals, it is associated with modern humans and perhaps Denisovans at some sites.
- Modern human morphology is first found in Africa at about 200 KYA, but only much later (after 45 KYA) in other parts of the world.
- These observations led to the development of several hypotheses about the origins of modern humans, two extremes being the multiregional and out-of-Africa models.
- Genetic diversity is higher in Africa than on other continents, which is consistent with a longer period of evolution in Africa, and/or a larger population size.
- Most phylogenies of individual loci show a root in Africa and a subset of lineages in other parts of the world; this is seen particularly clearly in the well-resolved phylogenies of mtDNA and the Y chromosome. Such results imply an African origin for most of our ancestors.
- DNA cannot be extracted from most fossils, but ancient DNA analysis has been successful in generating a draft of Neanderthal genome sequence and a high-coverage Denisovan sequence. Analyses suggested that their genetic lineages diverged from those of modern humans about 800 KYA. However, there is likely to have been a small amount of subsequent mixing as modern humans expanded out of Africa and encountered these species.
- The current consensus view is therefore that an out-of-Africa model with minor archaic admixture explains the fossil, and modern morphological, linguistic, and genetic data most effectively.

QUESTIONS

Question 9–1: Human genetic diversity is generally highest in Africa and decreases with distance from Africa. What explanations could you suggest for finding the following exceptions to this pattern:

(a) An African population with low diversity?
(b) An American population with high diversity?

Question 9–2: To what extent is the assumption that *Sahelanthropus* was a member of the human lineage compatible with genetic data for human origins?

Question 9–3: A haplotype present in many modern humans is also found in Neanderthals. In the light of current interpretations of the Neanderthal genome sequence, what different evolutionary explanations could account for this observation and how might they be distinguished?

Question 9–4: You have managed to generate a draft "hobbit" genome sequence with low contamination, equivalent to the low-coverage Denisova sequence. Given the following genomewide comparisons and calculation of *D*, how would you interpret:

(a) *D* value of zero when comparing a Yoruban and a Japanese genome?
(b) A nonzero *D* value when comparing a Yoruban and a Melanesian genome?

Question 9–5: In a comparison of human, Neanderthal and Denisovan mtDNA sequences, the Denisovan is an outlier, while in an autosomal sequence comparison, the outlier is human (compare Figures 9.21 and 9.22). What explanations can you suggest for these patterns and what additional analyses and datasets could help to distinguish them?

REFERENCES

The references highlighted in purple are considered to be important (for this chapter) by the authors.

1. **Abi-Rached L, Jobin MJ, Kulkarni S et al.** (2011) The shaping of modern human immune systems by multiregional admixture with archaic humans. *Science* **334**, 89–94.
2. **Asfaw B, Gilbert WH, Beyene Y et al.** (2002) Remains of *Homo erectus* from Bouri, Middle Awash, Ethiopia. *Nature* **416**, 317–320.
3. **Asfaw B, White T, Lovejoy O et al.** (1999) *Australopithecus garhi*: a new species of early hominid from Ethiopia. *Science* **284**, 629–635.
4. **Atkinson QD** (2011) Phonemic diversity supports a serial founder effect model of language expansion from Africa. *Science* **332**, 346–349.
5. **Backwell L & d'Errico F** (2008) Early hominid bone tools from Drimolen, South Africa. *J. Arch. Sci.* **35**, 2880–2894.
6. **Barker G, Barton H, Bird M et al.** (2007) The 'human revolution' in lowland tropical Southeast Asia: the antiquity and behavior of anatomically modern humans at Niah Cave (Sarawak, Borneo). *J. Hum. Evol.* **52**, 243–261.
7. **Behar DM, Villems R, Soodyall H et al.** (2008) The dawn of human matrilineal diversity. *Am. J. Hum. Genet.* **82**, 1130–1140.
8. **Berger LR, de Ruiter DJ, Churchill SE et al.** (2010) *Australopithecus sediba*: a new species of *Homo*-like australopith from South Africa. *Science* **328**, 195–204.
9. **Bermudez de Castro JM, Arsuaga JL, Carbonell E et al.** (1997) A hominid from the lower Pleistocene of Atapuerca, Spain: possible ancestor to Neandertals and modern humans. *Science* **276**, 1392–1395.
10. **Briggs AW, Good JM, Green RE et al.** (2009) Targeted retrieval and analysis of five Neandertal mtDNA genomes. *Science* **325**, 318–321.
11. **Brown P, Sutikna T, Morwood MJ et al.** (2004) A new small-bodied hominin from the Late Pleistocene of Flores, Indonesia. *Nature* **431**, 1055–1061.
12. **Brunet M, Guy F, Pilbeam D et al.** (2002) A new hominid from the Upper Miocene of Chad, Central Africa. *Nature* **418**, 145–151.
13. **Conrad DF, Jakobsson M, Coop G et al.** (2006) A worldwide survey of haplotype variation and linkage disequilibrium in the human genome. *Nat. Genet.* **38**, 1251–1260.
14. **Dart R** (1925) *Australopithecus africanus*: the man-ape of South Africa. *Nature* **115**, 195–199.
15. **Diogo R & Wood B** (2011) Soft-tissue anatomy of the primates: phylogenetic analyses based on the muscles of the head, neck, pectoral region and upper limb, with notes on the evolution of these muscles. *J. Anat.* **219**, 273–359.
16. **Eriksson A & Manica A** (2012) Effect of ancient population structure on the degree of polymorphism shared between modern human populations and ancient hominins. *Proc. Natl Acad. Sci. USA* **109**, 13956–13960.
17. **Ferring R, Oms O, Agusti J et al.** (2011) Earliest human occupations at Dmanisi (Georgian Caucasus) dated to 1.85–1.78 Ma. *Proc. Natl Acad. Sci. USA* **108**, 10432–10436.
18. **Green RE, Krause J, Briggs AW et al.** (2010) A draft sequence of the Neandertal genome. *Science* **328**, 710–722.
19. **Gutenkunst RN, Hernandez RD, Williamson SH & Bustamante CD** (2009) Inferring the joint demographic history of multiple populations from multidimensional SNP frequency data. *PLoS Genet.* **5**, e1000695.
20. **Haile-Selassie Y, Suwa G & White TD** (2004) Late Miocene teeth from Middle Awash, Ethiopia, and early hominid dental evolution. *Science* **303**, 1503–1505.
21. **Henn BM, Gignoux CR, Jobin M et al.** (2011) Hunter-gatherer genomic diversity suggests a southern African origin for modern humans. *Proc. Natl Acad. Sci. USA* **108**, 5154–5162.
22. **Hernandez RD, Kelley JL, Elyashiv E et al.** (2011) Classic selective sweeps were rare in recent human evolution. *Science* **331**, 920–924.
23. **Herries AI** (2011) A chronological perspective on the acheulian and its transition to the Middle Stone Age in southern Africa: the question of the Fauresmith. *Int. J. Evol. Biol.* **2011**, 961401.
24. **Herries AI, Hopley PJ, Adams JW et al.** (2010) Letter to the editor: Geochronology and palaeoenvironments of Southern African hominin-bearing localities—A reply to Wrangham et al., 2009. "Shallow-water habitats as sources of fallback foods for hominins". *Am. J. Phys. Anthropol.* **143**, 640–646.
25. **Herries AIR, Curnoe D & Adams JW** (2009) A multi-disciplinary seriation of early *Homo* and *Paranthropus* bearing palaeocaves in southern Africa. *Quatern. Int.* **202**, 14–28.

26. **Indriati E, Swisher CC 3rd, Lepre C et al.** (2011) The age of the 20 meter Solo River terrace, Java, Indonesia and the survival of *Homo erectus* in Asia. *PLoS One* **6**, e21562.

27. **Ingman M, Kaessmann H, Pääbo S & Gyllensten U** (2000) Mitochondrial genome variation and the origin of modern humans. *Nature* **408**, 708–713.

28. **Johanson DC & White TD** (1979) A systematic assessment of early African hominids. *Science* **203**, 321–330.

29. **Kivell TL, Kibii JM, Churchill SE et al.** (2011) *Australopithecus sediba* hand demonstrates mosaic evolution of locomotor and manipulative abilities. *Science* **333**, 1411–1417.

30. **Krause J, Fu Q, Good JM et al.** (2010) The complete mitochondrial DNA genome of an unknown hominin from southern Siberia. *Nature* **464**, 894–897.

31. **Krings M, Stone A, Schmitz RW et al.** (1997) Neandertal DNA sequences and the origin of modern humans. *Cell* **90**, 19–30.

32. **Lachance J, Vernot B, Elbers CC et al.** (2012) Evolutionary history and adaptation from high-coverage whole-genome sequences of diverse African hunter-gatherers. *Cell* **150**, 457–469.

33. **Laval G, Patin E, Barreiro LB & Quintana-Murci L** (2010) Formulating a historical and demographic model of recent human evolution based on resequencing data from noncoding regions. *PLoS ONE* **5**, e10284.

34. **Leakey LSB** (1959) A new fossil skull from Olduvai. *Nature* **184**, 491–493.

35. **Leakey LSB, Tobias PV & Napier JR** (1964) A new species of genus *Homo* from Olduvai Gorge. *Nature* **202**, 7–9.

36. **Leakey MD & Hay RL** (1979) Pliocene footprints in the Laetolil Beds at Laetoli, northern Tanzania. *Nature* **278**, 317–323.

37. **Leakey MG, Feibel CS, McDougall I & Walker A** (1995) New four-million-year-old hominid species from Kanapoi and Allia Bay, Kenya. *Nature* **376**, 565–571.

38. **Leakey MG, Spoor F, Brown FH et al.** (2001) New hominin genus from eastern Africa shows diverse middle Pliocene lineages. *Nature* **410**, 433–440.

39. **Leakey MG, Spoor F, Dean MC et al.** (2012) New fossils from Koobi Fora in northern Kenya confirm taxonomic diversity in early *Homo*. *Nature* **488**, 201–204.

40. **Lepre CJ, Roche H, Kent DV et al.** (2011) An earlier origin for the Acheulian. *Nature* **477**, 82–85.

41. **Li JZ, Absher DM, Tang H et al.** (2008) Worldwide human relationships inferred from genome-wide patterns of variation. *Science* **319**, 1100–1104.

42. **Lieberman DE, McBratney BM & Krovitz G** (2002) The evolution and development of cranial form in *Homo sapiens*. *Proc. Natl Acad. Sci. USA* **99**, 1134–1139.

43. **Lordkipanidze D, Jashashvili T, Vekua A et al.** (2007) Postcranial evidence from early *Homo* from Dmanisi, Georgia. *Nature* **449**, 305–310.

44. **Manica A, Amos W, Balloux F & Hanihara T** (2007) The effect of ancient population bottlenecks on human phenotypic variation. *Nature* **448**, 346–348.

45. **McDougall I, Brown FH & Fleagle JG** (2005) Stratigraphic placement and age of modern humans from Kibish, Ethiopia. *Nature* **433**, 733–736.

46. **Mercader J, Barton H, Gillespie J et al.** (2007) 4300-year-old chimpanzee sites and the origins of percussive stone technology. *Proc. Natl Acad. Sci. USA* **104**, 3043–3048.

47. **Meyer M, Kircher M, Gansauge MT et al.** (2012) A high-coverage genome sequence from an archaic Denisovan individual. *Science* **338**, 222–226.

48. **Pagani L, Kivisild T, Tarekegn A et al.** (2012) Ethiopian genetic diversity reveals linguistic stratification and complex influences on the Ethiopian gene pool. *Am. J. Hum. Genet.* **91**, 83–96.

49. **Pappu S, Gunnell Y, Akhilesh K et al.** (2011) Early Pleistocene presence of Acheulian hominins in South India. *Science* **331**, 1596–1599.

50. **Pickering R, Dirks PH, Jinnah Z et al.** (2011) *Australopithecus sediba* at 1.977 Ma and implications for the origins of the genus *Homo*. *Science* **333**, 1421–1423.

51. **Pruetz JD & Bertolani P** (2007) Savanna chimpanzees, *Pan troglodytes verus*, hunt with tools. *Curr. Biol.* **17**, 412–417.

52. **Prugnolle F, Manica A & Balloux F** (2005) Geography predicts neutral genetic diversity of human populations. *Curr. Biol.* **15**, R159–160.

53. **Ramachandran S, Deshpande O, Roseman CC et al.** (2005) Support from the relationship of genetic and geographic distance in human populations for a serial founder effect originating in Africa. *Proc. Natl Acad. Sci. USA* **102**, 15942–15947.

54. **Reich D, Green RE, Kircher M et al.** (2010) Genetic history of an archaic hominin group from Denisova Cave in Siberia. *Nature* **468**, 1053–1060.

55. **Rosenberg NA, Pritchard JK, Weber JL et al.** (2002) Genetic structure of human populations. *Science* **298**, 2381–2385.

56. **Senut B, Pickford M, Gommery D et al.** (2001) First hominid from the Miocene (Lukeino Formation, Kenya). *C. R. Acad. Sci.* **332**, 137–144.

57. **Shi W, Ayub Q, Vermeulen M et al.** (2010) A worldwide survey of human male demographic history based on Y-SNP and Y-STR data from the HGDP-CEPH populations. *Mol. Biol. Evol.* **27**, 385–393.

58. **Stringer C & Andrews P** (2011) The Complete World of Human Evolution. Thames and Hudson.

59. **Swisher CC 3rd, Curtis GH, Jacob T et al.** (1994) Age of the earliest known hominids in Java, Indonesia. *Science* **263**, 1118–1121.

60. **Takahata N, Lee SH & Satta Y** (2001) Testing multiregionality of modern human origins. *Mol. Biol. Evol.* **18**, 172–183.

61. **The 1000 Genomes Project Consortium** (2010) A map of human genome variation from population-scale sequencing. *Nature* **467**, 1061–1073.

62. **Thieme H** (1997) Lower Palaeolithic hunting spears from Germany. *Nature* **385**, 807–810.

63. **Thomson R, Pritchard JK, Shen P et al.** (2000) Recent common ancestry of human Y chromosomes: evidence from DNA sequence data. *Proc. Natl Acad. Sci. USA* **97**, 7360–7365.

64. **Tishkoff SA, Reed FA, Friedlaender FR et al.** (2009) The genetic structure and history of Africans and African Americans. *Science* **324**, 1035–1044.

65. **van Schaik CP, Ancrenaz M, Borgen G et al.** (2003) Orangutan cultures and the evolution of material culture. *Science* **299**, 102–105.

66. **Wagner GA, Krbetschek M, Degering D et al.** (2010) Radiometric dating of the type-site for *Homo heidelbergensis* at Mauer, Germany. *Proc. Natl Acad. Sci. USA* **107**, 19726–19730.

67. **Walker A & Leakey REF** (eds) (1993) The Nariokotome *Homo erectus* Skeleton. Harvard University Press.

68. **Wall JD, Lohmueller KE & Plagnol V** (2009) Detecting ancient admixture and estimating demographic parameters in multiple human populations. *Mol. Biol. Evol.* **26**, 1823–1827.

69. **Weaver TD** (2012) Did a discrete event 200,000–100,000 years ago produce modern humans? *J. Hum. Evol.* **63**, 121–126.

70. **White TD, Asfaw B, Beyene Y et al.** (2009) *Ardipithecus ramidus* and the paleobiology of early hominids. *Science* **326**, 75–86.

71. **White TD, Asfaw B, DeGusta D et al.** (2003) Pleistocene *Homo sapiens* from Middle Awash, Ethiopia. *Nature* **423**, 742–747.

72. **White TD, WoldeGabriel G, Asfaw B et al.** (2006) Asa Issie, Aramis and the origin of Australopithecus. *Nature* **440**, 883–889.

73. **Whiten A, Goodall J, McGrew WC et al.** (1999) Cultures in chimpanzees. *Nature* **399**, 682–685.

74. **Wood B** (1996) Human evolution. *BioEssays* **18**, 945–954.

75. **Wood B & Collard M** (1999) The human genus. *Science* **284**, 65–71.

76. **Wood B & Harrison T** (2011) The evolutionary context of the first hominins. *Nature* **470**, 347–352.

77. **Yamei H, Potts R, Baoyin Y et al.** (2000) Mid-Pleistocene Acheulean-like stone technology of the Bose basin, South China. *Science* **287**, 1622–1626.

SECTION 4

HOW DID HUMANS COLONIZE THE WORLD?

The transformation of humans from a rare African species into a numerous one with a worldwide distribution is an unprecedented biological phenomenon, and is central to understanding why humans are genetically so similar to one another, and explaining the small, but appreciable, geographical differences that do exist among human populations. We continue along a path that is approximately chronological, discussing the early movements of modern humans out of Africa before considering the major effects that have followed the subsequent introduction of farming and the meeting of populations.

CHAPTER 10
THE DISTRIBUTION OF DIVERSITY

Studying human diversity raises important ethical and methodological questions, and we begin this chapter with these. Next, we see that most human genetic variation is found within any individual population, except for a few loci affected by natural selection.

CHAPTER 11
THE COLONIZATION OF THE OLD WORLD AND AUSTRALIA

A key event was the development, in Africa, of modern human *behavior* 100,000–60,000 years ago. A single expansion soon afterward peopled most of Asia, Australia, and Europe by 40,000 years ago, and included mixing with Neanderthals and Denisovans along the way.

CHAPTER 12
AGRICULTURAL EXPANSIONS

When the climate warmed and stabilized 10,000 years ago, agriculture appeared independently in several locations. Agriculture led to an enormous increase in the number of people, but did the farmers themselves expand, or did the neighbors learn farming and then expand themselves?

CHAPTER 13
INTO NEW-FOUND LANDS

In this chapter we consider the last great regions to be inhabited. In the Americas, most of the current indigenous gene pool may date back to one migration before 15,000 years ago. Most of the Pacific region was uninhabited until 3500 years ago, and the major migration was from the west.

CHAPTER 14
WHAT HAPPENS WHEN POPULATIONS MEET

This chapter considers the mixing of populations, or admixture. This is usually sex-biased, affecting the mtDNA, autosomes, and Y chromosome differentially, and establishing linkage disequilibrium: a legacy that persists for many generations.

THE DISTRIBUTION OF DIVERSITY

CHAPTER TEN

In the last chapter, we saw that our species *Homo sapiens* originated in Africa very recently on an evolutionary time-scale, <200 KYA. Now, humans have a worldwide distribution. How did this transformation from a rare tropical population to one with seven billion people inhabiting every continent come about? The key starting point is an understanding of patterns of genetic variation in different populations and that is the subject of this chapter. We have the molecular and statistical tools to generate and interpret such data. Yet, since we want to investigate the full range of human diversity and focus on the differences between populations, we often need the consent of some of the most disadvantaged people in the world to carry out these studies, and may produce findings that are not value-free, may conflict with the sample donors' beliefs, and are always open to misuse and misinterpretation. We therefore begin by considering the distressing history of this field, and the lessons we can learn about how best to carry out such studies in an appropriate way. We will discuss how to sample human genetic diversity, and encounter the leading international projects in this field. These will reveal both how neutral variants are distributed in the world, and how departures from these general patterns can inform us about natural selection and unusual demographic events.

10.1 STUDYING HUMAN DIVERSITY

Observations and inquiries in the area of human diversity are probably as old as humanity, and have in the past been linked to both blatant and subtle forms of **racism**. We will begin by considering some of the historical aspects of the field, ethical issues raised by such work, and the ways in which modern studies have learned from this early history.

The history and ethics of studying diversity are complex

Should we study human genetic diversity at all, or is this an area of work where the potential for misuse (see **Figure 10.1** and **Box 10.1**) outweighs the potential benefits to such an extent that it should not be pursued? Such studies already have a long history, and a pragmatic answer to this question is that information on human genetic diversity is needed, and is therefore generated, for medical and forensic applications, and thus is already available whatever evolutionary geneticists decide to do, so we must be ready to consider its implications and consequences. Furthermore, genetic information, in fact, refutes any scientific basis for racism as the existence of discrete human groups. It can therefore be used to argue that racism—the belief that discrimination between apparent groups is justifiable—is an entirely social construct. In addition, genetic information is of enormous intrinsic interest to many people, not just scientists.

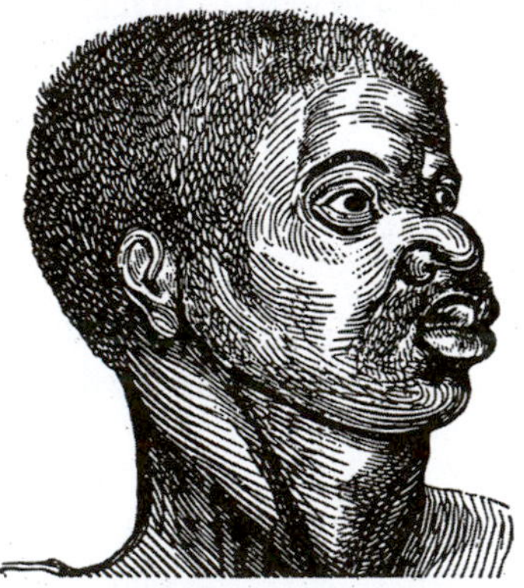

Figure 10.1: A racist view of humanity. Note not only the hierarchy, but also the falsification of Negro and chimpanzee skulls. [From Nott JC and Gliddon GR (1868) Indigenous Races of the Earth. Ayer Co Publishing Inc.]

Linnaeus' classification of human diversity

Our biological classification system originates with Linnaeus (1707–1778), who subdivided humans into two species (*diurnus* and *nocturnus*; see **Figure 10.2**) and a total of seven categories:

- *diurnus* (also referred to as *Homo sapiens* by Linnaeus)
  - *americanus*: red, with black hair and a scanty beard, obstinate, free, painted with fine red lines, regulated by customs
  - *europeus*: white, long flowing hair, blue eyes, sanguine, muscular, inventive, covered with tight clothing, governed by laws
  - *asiaticus*: yellow, melancholy, black hair and brown eyes, severe, haughty, stingy, wears loose clothing, governed by opinions
  - *afer* (that is, African): black, cunning, phlegmatic, black curly hair, women without shame and lactate profusely, anointed with grease, ruled by impulse
  - *monstrosus*: a miscellaneous collection including dwarfs and large, lazy Patagonians
- *nocturnus*
  - *troglodytes*: nocturnal, hunts only at night, lives underground

In some classifications he also included *ferus*: wild, hairy, runs about on all fours. Apart from the misleading notion that humans can be categorized into a small number of such groups and the language that now sounds offensive, this classification is notable for its inclusion of imaginary categories, and mixing of physical, intellectual, and cultural characteristics.

Galton's "Comparative worth of different races"

In *Hereditary Genius: an Inquiry Into its Laws and Consequences* published in 1869, Francis Galton (1822–1911) established a grading system, A, B, C, and so forth, for people within each race, and then compared the grades between races. At the top of the racial hierarchy were the ancient Greeks; Galton was slightly critical of his own race, the English, "the calibre of whose intellect is easily gauged by a glance at the contents of a railway book-stall," and placed them two grades below the Greeks, although "the average standard of the Lowland Scotch and the English North-country men is decidedly a fraction of a grade superior to that of the ordinary English." Inevitably, "the average intellectual standard of the negro race is some two grades below our own" and "the Australian type is at least one grade below the African negro." We see here and in other attempts to identify discrete categories of people how the numbers of categories and criteria used to define them differ substantially.[2, 7]

Modern attitudes to studying diversity

If genetic diversity studies of humans are an acceptable part of science, what are the prerequisites for such studies? There is a fundamental requirement, enshrined in the Declaration of Helsinki and recognized by international law, that research on humans can only be undertaken with **informed consent** from the subject. This means that, with some exceptions for forensic investigations and limited medical circumstances, individuals must not only freely agree to the research before it is undertaken, but must do this on the basis of an understanding of the nature and purpose of the research, its risks and benefits, and the potential outcomes/information produced (see **Opinion Box 10**). The risks associated with the physical procedures of donating saliva, cheek cells, hair, or blood to provide DNA are minimal; debate has focused on the risks associated with the use of the information obtained.[4] The results may have implications for:

- Health and, in some countries, health insurance: what if the donor is found to have a high chance of developing a particular disease later in life?

Box 10.1: Race and racism

> On May 7, 1876, Truganini, the last full-blood Black person in Tasmania, died at seventy-three years of age. Her mother had been stabbed to death by a European. Her sister was kidnapped by Europeans. Her intended husband was drowned by two Europeans in her presence, while his murderers raped her.
>
> It might be accurately said that Truganini's numerous personal sufferings typify the tragedy of the Black people of Tasmania as a whole. She was the very last. 'Don't let them cut me up,' she begged the doctor as she lay dying. After her burial, Truganini's body was exhumed, and her skeleton, strung upon wires and placed upright in a box, became for many years the most popular exhibit in the Tasmanian Museum and remained on display until 1947. Finally, in 1976—the centenary year of Truganini's death—despite the museum's objections, her skeleton was cremated and her ashes scattered at sea.

From *Black War: the Destruction of the Tasmanian Aborigines* by Runoko Rashidi, http://www.cwo.com/~lucumi/tasmania.html

The genocide of the Tasmanian Aborigines by the European settlers in the nineteenth century provides one of the worst of many examples of racism, and is notable for the "anthropological" justification of the public exhibition of Truganini's, and other, remains. Yet racism does not consist only of such crude episodes: racist thinking penetrates deeply into Western, and perhaps all, culture, evolutionary thought, and genetics. Consider these two quotations:

> I advance it, therefore, as a suspicion only, that the blacks, whether originally a distinct race, or made distinct by time and circumstance, are inferior to the whites in the endowment both of body and mind.
>
> We hold these truths to be self-evident: that all men are created equal.

Both are from Thomas Jefferson, and the next is Charles Darwin, writing about the gap between humans and apes after an anticipated future extinction of gorillas and "Hottentots":

> The break will then be rendered wider, for it will intervene between man in a more civilized state, as we may hope, than the Caucasian, and some ape as low as a baboon, instead of as at present between the negro or Australian and the gorilla.

Hierarchies of humans and apes, such as that illustrated in Figure 10.1, were common in anthropological and biological literature. They were usually based on a small number of visible characteristics such as skin color, hair color and morphology, and facial features.

These characteristics are influenced by both environmental and genetic factors, but even if allowance is made for the environment, the genes affecting these phenotypes have probably been subject to particular selection pressures and perhaps sexual selection (Section 15.3). Thus they are unrepresentative of the majority of the genome. Inevitably, the compilers put their own group at the top and those they wanted to exploit at the bottom (see Figure 10.1). Another notable feature of these schemes was that the number of "races" identified varied greatly between authors.

"Race," in addition to its everyday usages, is a biological term with a clear meaning: it refers to a group of individuals who can be cleanly distinguished from other groups of the same species. Of course, this requires that we specify what is meant by "cleanly," and an F_{ST} value ≥0.25 is commonly used: that is, 25% or more of the variation needs to be found between groups for these groups to be classified as "races." Some species are divided into races; the question of whether or not humans are such a species is an empirical one. We will see that the answer is "no."

- Stigmatization: what if the donor carries a trait judged to be undesirable, or is assigned to a group that experiences discrimination because of its identity? Related to this, what if a particular trait or disease becomes associated with a population, and the entire population is stigmatized as a result?
- Commercial applications: what if a **cell line** or DNA sequence leads to a patentable or saleable product?

Additional novel questions are raised by genetic research because we share DNA variants with our relatives, so study of one individual provides information about other members of their family and population. Therefore **group informed consent** is required in some situations and it would be unethical to sample consenting individuals from a group that had not given consent. The appropriate authority to provide such group consent, if there is one, can only be determined on a case-by-case basis for each population.

Benefits from genetic diversity studies are: (1) increased understanding of genetic history and relationships; (2) medical advances such as the identification

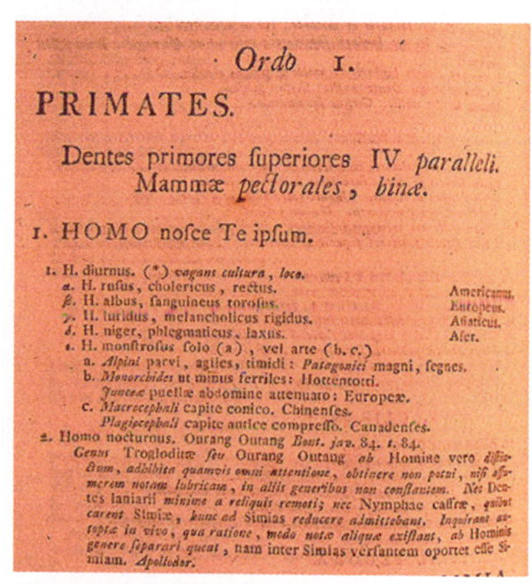

Ordo I.

PRIMATES.

Dentes primores ſuperiores IV *paralleli.* Mammæ *pectorales*, *binæ*.

1. HOMO noſce Te ipſum.

1. H. diurnus. (*) *varians cultura, loco.*
α. H. rufus, cholericus, rectus. — Americanus.
β. H. albus, ſanguineus toroſus. — Europeus.
γ. H. luridus, melancholicus rigidus. — Aſiaticus.
δ. H. niger, phlegmaticus, laxus. — Afer.
ε. H. monſtroſus ſolo (a), vel arte (b. c.)
a. *Alpini* parvi, agiles, timidi: *Patagonici* magni, ſegnes.
b. *Monorchides* ut minus fertiles: Hottentotti.
Junceæ puellæ abdomine attenuato: Europeæ.
c. *Macrocephali* capite conico. Chinenſes.
Plagiocephali capite antice compreſſo. Canadenſes.
2. Homo nocturnus. Ourang Outang *Bont. jav.* 84. *t.* 84.
Genus Trogloditæ *ſeu* Ourang Outang *ab* Homine vero *diſtinctum, adhibita quamvis omni attentione, obtinere non potui, niſi aſſumerem notam lubricam, in aliis generibus non conſtantem.* Nec Dentes laniarii *minime a reliquis remoti; nec* Nymphae caſſæ, *quibus carent* Simiæ, *hunc ad* Simias *reducere admittebant. Inquirant autoptæ in vivo, qua ratione, modo notæ aliquæ exiſtant, ab* Hominis *genere ſeparari queat*, nam inter Simias verſantem oportet eſſe Simiam. *Apollodor.*

Figure 10.2: Linnaeus' 1756 classification of humans.

The indigenous populations that occupy Peninsular Malaysia are locally known as *Orang Asli*, or the "original peoples." Collectively, they represent approximately 0.6% of Malaysians, and have been classified into three distinct groups—the Negrito, Senoi, and Proto-Malay—based on differences in languages, socio-cultural practices, physical appearance, and habitats. Each group can be further subdivided into six subgroups (**Table 1**, **Figure 1**). Based on the studies of their genetic history, the *Orang Asli* are thought to be the descendants of humans who arrived in Southeast Asia some 60 KYA.[9]

There are 869 recorded *Orang Asli* settlements in Malaysia. Only 1.4% of these are in or close to urban centers, whilst the majority are located in the rural and forest areas. The past decade has witnessed the relocation of some forest fringe communities into government resettlement schemes closer to urban areas. Whilst a significant proportion of Proto-Malays and Senoi work in orchards, plantations, and the fishing industry, the majority of Negrito communities still depend largely on foraging and collection of jungle produce for sale.

The health of most Malaysians has improved in the twentieth century, but communicable diseases abound in *Orang Asli* communities, especially in rural areas with poor access to health care. As a result of increased resettlement and adoption of a sedentary lifestyle, non-communicable diseases, including hypertension, obesity, and cardiovascular diseases, have also increased.

Fieldwork and genetic research amongst the *Orang Asli* require approval from the Malaysian Department of Indigenous Development for each project, community, and study time frame. It is important to include anthropologists in any research team so that the researchers appreciate the different cultural systems in different communities. For example, the Temuan people have firm hierarchies, and the headman "Tok Batin" and his elders must first be consulted during a customary courtesy visit, before any fieldwork can begin. Only with their consent can other members of the community be recruited. Translators are often necessary, as the elderly *Orang Asli* only speak their own languages. Many settlements are remote and only accessible by dirt tracks or by boats in good weather. These settlements may also lack electricity and running water. As a consequence of the history of resettlement and the loss of native lands, some *Orang Asli* can be hostile toward outsiders.

To build trust with the community, it is mandatory to pay courtesy calls before researching any community to explain to the elder members of the tribes the rationale of the study and how samples would be collected. Our sampling activities also include the sharing of direct benefits, addressing the communities' requests for sundry provisions and clothing. We conduct health screening and the physicians from our team conduct basic clinical examinations for any member of the community who requests it, regardless of whether or not they participate in the research project. In addition, we revisit the villages annually and it has been possible to provide health reports to the relevant individuals and health officers in charge for follow ups and treatment in the local government health centers.

TABLE 1:
MAJOR GROUPS AND SUBGROUPS OF *ORANG ASLI* AND THEIR POPULATION SIZES

Major groups	Subgroups	Population size
Negrito	Kensiu	240
	Kintak	132
	Lanoh	349
	Jahai	2072
	Mendriq	216
	Bateq	1542
Senoi	Temiar	25,233
	Semai	43,505
	Semoq Beri	3629
	Che Wong	665
	Jahut	5082
	Mah Meri	2858
Proto-Malay	Temuan	22,819
	Semelai	6584
	Jakun	29,263
	Kanaq	87
	Kuala	3716
	Seletar	1431

(From data collected in 2004 by the Jabatan Kemajuan Orang Asli, Malaysia.)

Boon-Peng Hoh, Institute of Medical Molecular Biotechnology, Universiti Teknologi MARA, Selangor, Malaysia
Maude E. Phipps, Jeffrey, Cheah School of Medicine and Health Sciences, Monash University Sunway Campus, Selangor, Malaysia

Figure 1: Locations of major groups and subgroups of *Orang Asli* in Malaysia.

of genes predisposing to disease (see Chapter 16); (3) accurate paternity testing, victim and assailant identification, and other forensic applications (see Chapter 18); (4) sometimes, immediate benefits to the population, such as medical advice or treatment. However, complications also arise, for example because the people who receive most of the long-term benefits may not be the donors.

Outstanding issues that have not been fully resolved include:

- Is informed consent from members of cultures that do not ascribe to Western scientific values truly "informed"? Indeed, can even leading geneticists such as Jim Watson and Craig Venter, who have volunteered to have their whole genomes sequenced and made public (Chapter 18), appreciate the full implications when these may only become apparent in the future as research reveals the medical implications of DNA variants?
- How much information about the donor should accompany a cell line or DNA sample, so that the privacy of the donor is not infringed?
- Can samples collected with no written consent many years ago, or perhaps decades ago, still be used?
- Can samples collected for one study be used in another?
- Can an individual give broad consent for all future studies, which may involve techniques that do not yet exist and have implications that are not currently understood (related to the first point, above)?

It is difficult to give general answers to many of the ethical questions that diversity studies raise; indeed, the possibility of ever more comprehensive genetic studies is one of the driving forces in the field of medical ethics. Answers may emerge more satisfactorily through the consideration of individual cases than through prior reasoning based on principles. We will encounter examples of such cases throughout this book.

Who should be studied?

The starting point for any study of human diversity is a set of humans, and this raises the question, Who should be studied? Sampling always creates problems: is the sample appropriate and representative? If not, conclusions drawn from the sample may not be applicable to the rest of the population that was sampled. Analyzing everyone would avoid the complications introduced by sampling, and some have argued that it is fairer, but at present it is impractical for DNA studies and even for DNA-free genetics based, for example, on phenotypic traits. This is likely to remain true for the foreseeable future, so the issues raised by sampling must be addressed.

Although human genetic diversity has been investigated for a long time, the early studies using pre-DNA polymorphisms aroused little controversy or public interest. Attitudes changed with the launch of the **Human Genome Diversity Project** (**HGDP**) in 1991, and all subsequent large-scale projects, including the **HapMap**, **Genographic**, and **1000 Genomes Projects**, have been influenced by this legacy.

A few large-scale studies of human genetic variation have made major contributions to human evolutionary genetics

The HGDP was announced in a paper published by Cavalli-Sforza, Wilson, and others[5] (http://hsblogs.stanford.edu/morrison/human-genome-diversity-project/). The authors called for the collection of "material to record human ethnic and geographic diversity," particularly from populations "that have been isolated for some time [and] are likely to be linguistically and culturally distinct." The planned scale was large: ~25 individuals each from around 500 populations: 12,500 in all. However, the project failed to raise major funding, instead attracting criticism, including from several of the indigenous peoples it aimed to involve. Although it did not achieve its aims, it did establish a panel of 1064 cell

Box 10.2: The HGDP-CEPH samples

The Human Genome Diversity Project (HGDP) did not achieve all of its aims (see text), but has left a legacy that has transformed the fields of human genetic variation and human evolutionary genetics: a panel of 1064 diverse DNA samples derived from **lymphoblastoid cell lines**, representing 51 populations.[3] The cells are held by the Centre d'Etude du Polymorphisme Humain (CEPH) in Paris, hence the name HGDP-CEPH, and DNA samples are distributed on a cost-recovery basis. The availability of cell lines makes the amount of DNA available inexhaustible. Before this panel was available, individual labs had to go to enormous effort to collect or, for the vast majority, assemble by collaboration, a smaller and less representative set of samples. Now, any lab can genotype their variant(s) of interest in this collection, and also benefit from the large amount of data already available on this common resource, resulting from the work of over 100 investigators (http://www.cephb.fr/en/hgdp/diversity.php/).

The samples (Figure 10.3) include representation of indigenous populations from Africa, Europe, Asia, and the Americas. There are, inevitably, limitations and drawbacks. Several important areas of the world are missing, such as Australia, North America, India, and most of Northern Asia. The cell lines themselves cannot be distributed, so cellular phenotypes cannot be investigated; in addition, DNA cannot be sent to commercial companies. There is no information about the phenotypes of the donors except the sex of the individual, and population and geographic origin, and donors cannot be re-contacted. The informed consent provided by the donors was appropriate for the twentieth century when the panel was established, but may not meet all the criteria expected in the twenty-first century.

Nevertheless, the importance of this collection has been immense. It provides the only shared resource for investigating indigenous human diversity, and genomewide short tandem repeat (STR)[24] and SNP[15] genotypes are available both for further analysis and as a reference panel for comparison with new data. Our understanding of the distribution of diversity,[15, 24] and developments such as the serial founder model[20, 21] for the spread of humans, derive directly from it. Human evolutionary geneticists owe a great debt to the large numbers of unnamed donors and sample collectors, as well as to the scientists who established this panel.

lines[3] that has become a standard resource in the field and provided the basis for many of the studies we will describe in later chapters (**Box 10.2**). Why did the HGDP fail to attract more widespread support and funding? The lack of direct benefits to the participating communities, the perceived risks of **biopiracy**, and the proposal to establish immortal cell lines from populations on the brink of extinction were all factors. The contrast with another project proceeding at the same time, the human genome sequencing project, may also be informative. Despite initial doubts about its implications for human genetics and the scientific community, the public sequencing project's policy of making data freely and immediately available to all, instead of just to the labs generating the data, seems to have been a key ingredient in its success. The medical relevance of the sequencing project was obviously another important factor, and its concentration in scientifically advanced countries simplified its organization, but perhaps a more "open" diversity project would have been more successful.

Subsequent human variation projects provide some contrasts to the HGDP, and indeed to one another (**Table 10.1**, **Figure 10.3**). The HapMap Project (Box 3.6; http://hapmap.ncbi.nlm.nih.gov/), initiated a decade after the HGDP, established an exemplary consent process including extensive community involvement and feedback, and has escaped most of the criticisms aimed at the HGDP. Its choice of mainly urban populations and explicit goal of benefiting human health have contributed to this outcome. Many of its accomplishments lie outside the scope of this book, but we will see how it has become central to our understanding of many aspects of human diversity and evolution. The 1000 Genomes Project (Box 3.2; http://www.1000genomes.org/) can be seen as its successor, beginning in 2008 as new sequencing technologies became available, but involving many of the same scientists and initially using HapMap samples. Its stated main objective was to develop a public resource of genetic variation to support the next generation of medical association studies, specifically by finding all accessible variants at a frequency of ≥1% across the genome and down to 0.1–0.5% in gene regions. However, secondary objectives included

TABLE 10.1:
LARGE-SCALE PROJECTS PROVIDING INFORMATION ON HUMAN GENETIC VARIATION

Project	Launch date	Primary aims	Sample size	Populations	Genetic analyses	Cell lines	Data release	Website (Key reference) Further information
HGDP	1991	collection of isolated population samples	1064	51, worldwide	chosen by investigator	yes	on publication	http://www.cephb.fr/en/hgdp/diversity.php/ [Cann HM et al. (2002) *Science* 296, 261.] Box 10.2
HapMap	2002	haplotype map for medical genetics	1184	11, Africa, Europe, South and East Asia	SNP genotyping	yes	full public release	http://hapmap.ncbi.nlm.nih.gov/index.html.en [The International HapMap Project (2003) *Nature* 426, 789.] Box 3.6
Genographic	2005	elucidate migration history	~500,000	many, worldwide; including public participation	Y-SNP and Y-STR genotyping, mtDNA HVSI sequencing	no	on publication	https://genographic.nationalgeographic.com/genographic/lan/en/index.html [Wells, Deep Ancestry: Inside the Genographic Project (2006) National Geographic Books]
1000 Genomes	2008	discover variants at ≥1% frequency for medical genetics	2500	27, Africa, Europe, South Asia, East Asia, Americas	whole-genome sequencing	yes	full public release	http://www.1000genomes.org/ [The 1000 Genomes Project Consortium (2010) *Nature* 467, 1061.] Box 3.2

investigating questions of evolutionary interest. The Genographic Project (https://genographic.nationalgeographic.com/genographic/index.html) illustrates an alternative way of exploring human diversity. Funded by National Geographic, IBM, and the Waite Family Foundation, as well as by public participants who each contributed $99, it focused on ancestry information from mtDNA and the Y chromosome (**Appendix**) and explicitly avoided studying variants of medical significance, or establishing cell lines. With this model, it achieved a sample size more than 100-fold greater than the other projects discussed here, and many of these samples can potentially be studied by further genotyping or sequencing for those participants who have consented appropriately. Along these lines, a Geno 2.0 project has begun (https://genographic.nationalgeographic.com/about/), which is genotyping large numbers of ancestry-informative, medically uninformative, SNPs.

Medical-genetic studies involving genomewide genotyping often involve >100,000 participants (Chapter 17) and so provide additional large datasets that can potentially be used for evolutionary studies as well, although ethical and bureaucratic factors usually limit the availability of datasets, even when these have been published in the scientific literature. In addition, the choice of populations sampled will restrict their relevance to many evolutionary questions.

Individuals may choose to make their own genotypes and genome sequences freely available, and as costs decrease, these are likely to become an increasingly important source of data (**Section 18.4**). An early project in this area has been the Personal Genome Project (PGP), founded by its self-described "guinea pig #1," George Church, and with >1800 participants by 2012[1]

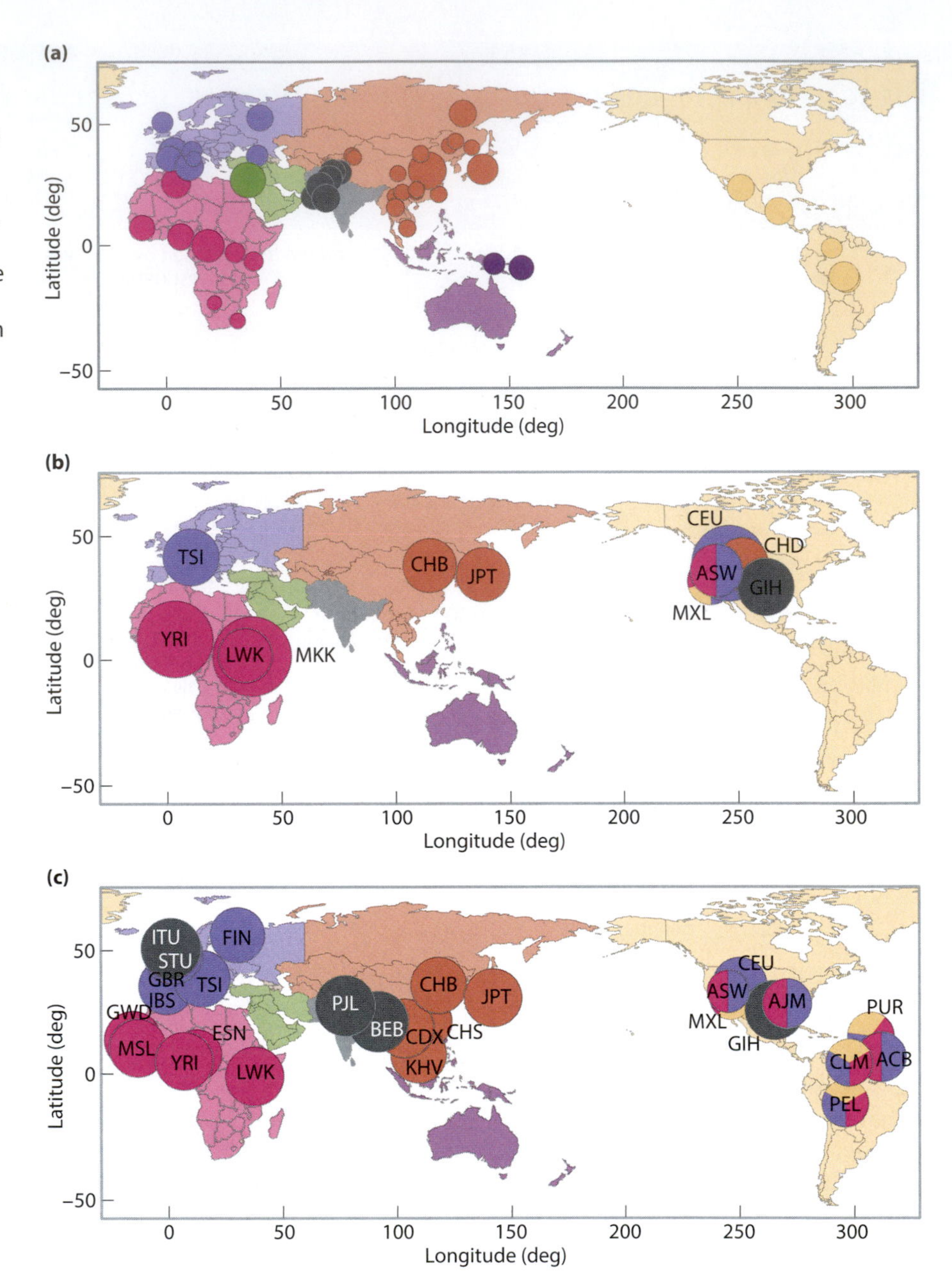

Figure 10.3: Samples included in three large-scale sample collections. (a) HGDP, (b) HapMap, (c) 1000 Genomes. Samples are represented by circles, with an area proportional to sample size (smaller circles in the second panel, for example TSI = 100), placed in the sampling location. Broad regions of the world are assigned colors, and these colors are used to indicate the geographical ancestry of the samples. While sampling location and ancestry often coincide, there are several discrepancies due to sampling migrant and mixed populations in the HapMap and 1000 Genomes collections. Note also the poor representation of indigenous populations from many regions of the world, such as North America or Australia. [Adapted from Colonna V et al. (2011) *Genome Biol.* 12, 234. With permission from BioMed Central Ltd.]

(http://www.personalgenomes.org/). Volunteers share their genotypes and genome sequence, and, crucially, other personal information such as health and medical data with the scientific community and the general public. They also donate tissue specimens which may be used in many ways, including for studies of gene expression and transformation into somatic cell-derived stem cells (that is, **induced pluripotent stem cells** or **iPS cells**). Thus far, to our knowledge, no harm to an individual from participating in genetic research of this kind has yet been documented.

What is a population?

The phrase "human population" is used widely, including in this book, but we now need to examine more carefully what is meant by the term "a population" (see also **Section 5.3**). When sets of individuals are phenotypically distinct and seldom interbreed, it is easy to distinguish populations: for example eastern mountain and eastern lowland gorillas (**Section 7.4**). However, although humans from anywhere in the world are potentially capable of interbreeding, they clearly do not form one worldwide randomly mating (**panmictic**) population, but exhibit structure. What criteria can be used to identify this structure?

Among the ways in which we can decide whether or not people belong to the same population are:

- Geographical proximity: individuals from the same population must be able to meet
- A common language: they must be able to communicate with each other
- Shared ethnicity, culture, religion: they are more likely to intermarry if they share history and values

None of these criteria is an absolute: we do not say that someone belongs to a different population if they move to a different country, if grandparents and grandchildren speak different languages, or if someone converts to a new religion. Nevertheless, after several generations, such changes could lead to the establishment of a new population. If these criteria were used alone, each individual might be considered to belong to many populations (geographical, linguistic, and so forth) defined in different ways, and these might change with time. The extent to which such memberships are correlated between individuals is unclear. If they were highly correlated, we could meaningfully identify distinct populations that would summarize the relationships between individuals, but if they were poorly correlated, it would be difficult to identify populations, reflecting the fact that, as we have seen earlier, human groups are not discrete, but are social constructs. The criteria used in the studies reported in this book are not always consistent. One common practice is to use self-determination: a person is a member of the group they identify with. The Czech writer Karl Deutsch described a nation as "a group of people united by a mistaken view of the past and a hatred of their neighbors"; although cynical, this encapsulates the ambiguity in defining "a population."

How many people should be analyzed?

A single individual can sometimes provide key evolutionary insights. We have seen the importance of the Neanderthal mtDNA and genome sequences, and the Denisovan genome (**Section 9.5**), and will encounter more examples in the next chapters, including Aboriginal Australian (**Section 11.4**) and Paleo-Eskimo genomes (**Section 13.2**). But these insights were possible because these sequences could be compared with many others: the analyses were not based entirely on a single genome. Another approach was based on the insight that a single individual's genome actually contains a "population" of regions with independent evolutionary histories because of recombination; these maternal and paternal copies of regions **coalesce** at different times and the distribution of these coalescence times provides insights into past demography[14] (**Section 9.5, Figure 9.23**).

Nevertheless, for most studies in population and evolutionary genetics, we need to analyze multiple individuals, and different projects have used very different sample sizes (Table 10.1). We thus need to consider questions such as, How many individuals need to be examined in order to address a particular question? How should these be distributed among different populations? There are no simple answers and in future sections we will see a range of strategies employed.

A related issue is what kind of weighting scheme should be used in choosing numbers of individuals and populations. It is generally agreed that, except in forensic investigations, weighting according to current population size is a poor option because it biases strongly toward recent expansions and these are not usually the main focus of interest in evolutionary studies. A geographical scheme is probably the most common, and aims to sample roughly equally from each geographical area; the projects listed in Table 10.1 do approximately this. A third possibility is to use linguistic criteria, on the grounds that the major linguistic divisions pre-date recent demographic expansions and thus sample according to the population structure that existed millennia ago (although such a scheme would complicate investigations of correlations between genetics and linguistics). For example, a study of Xq diversity[10] examined 69 individuals distributed among 16 out of 17 major language phyla.

In practice, sample availability is often the major criterion. Throughout this book, we will see that sample sizes of 20–50 per population are common in evolutionary DNA analyses, and 100, used in the 1000 Genomes Project, is quite large. Studies often examine a few hundred individuals in total. Advances in technology (Chapter 4) should allow much larger studies to be undertaken in the near future, and current medically motivated sequencing studies such as those in the UK (UK10K, http://www.uk10k.org/), The Netherlands (GoNL, http://www.nlgenome.nl/), and Sardinia (SardiNIA, http://genome.sph.umich.edu/wiki/SardiNIA) provide a foretaste of the datasets that may yield evolutionary insights as a by-product. Overall, the influence of sampling strategy on conclusions from genetic studies remains an under-appreciated and under-studied aspect of the field.

10.2 APPORTIONMENT OF HUMAN DIVERSITY

How distinct, genetically, are different populations? When information on the variation of a set of **classical polymorphisms** (Box 3.1) became available from different populations, it was possible to investigate how diversity was apportioned between human populations or groups of populations, and a pioneering study was presented by Lewontin in 1972.[13] Despite limitations in technology and dated terminology, this work identified the basic view that is still current, more than 40 years later, and so we present it in some detail.

The apportionment of diversity shows that most variation is found within populations

Lewontin[13] used 17 loci (blood groups, serum proteins, and red blood cell enzymes) for which variation had been detected by immunological or electrophoretic methods (see Box 3.1), and had allele frequency data available for several populations. The populations were classified into seven "races" termed **Caucasians**, Black Africans, Mongoloids, South Asian Aborigines, Amerinds, Oceanians, and Australian Aborigines, based on morphological, linguistic, historical, and cultural criteria. Diversity for each locus was measured by H, the Shannon information measure:

$$H = -\sum_{i=1}^{n} p_i \ln 2p_i$$

where p_i is the frequency of the ith allele; H is a somewhat similar measure to Nei's gene diversity (**Section 6.2**) and in this study ranged from 0 (no variation) to 1.9 (high variation). It was calculated at three levels for each locus:

- H_{pop}, the value for each individual population averaged over populations within a "race"
- H_{race}, the value for each "race," calculated from the average gene frequency over all populations within that "race," averaged over all "races"
- $H_{species}$, calculated from the frequency averaged over all populations within the species

These values were then used to apportion the diversity:

- Within populations = $H_{pop}/H_{species}$
- Between populations within races = $(H_{race} - H_{pop})/H_{species}$
- Between races = $(H_{species} - H_{race})/H_{species}$

The proportion of variation within populations ranged from 63.6% for the **Duffy blood group** (**Section 17.4**) to 99.7% for Xm, although only four populations were available for the latter marker; the mean proportion within populations was 85.4%. On average, 8.3% (range, 2.1–21.4%) corresponded to differences between populations within "races," and only 6.3% (range, 0.2–25.9%) was found between "races."

The overwhelming conclusion was that most variation lies within populations, and that "races" had no genetic reality, a conclusion reinforced by many subsequent analyses using independent population samples and DNA polymorphisms. Lewontin concluded:

> Human racial classification is of no social value and is positively destructive of social and human relations. Since such racial classification is now seen to be of virtually no genetic or taxonomic significance either, no justification can be offered for its continuance.

The apportionment of diversity can differ between segments of the genome

Subsequent studies differ in that they used DNA polymorphisms of several kinds, often analyzed data by F_{ST} or **AMOVA** (Box 5.2; **Section 6.3**), and referred to "continental groups" rather than "races," but the conclusions are strikingly similar: ~83–88% of autosomal variation is found within populations and ~9–13% between continental groups (**Table 10.2**). It is important to realize that such values depend on the frequency of the polymorphisms, and would be even lower if rarer variants were used.

Results from mtDNA and the Y chromosome are somewhat different, with less of the variation within populations and more between groups, as might be

TABLE 10.2: EARLY STUDIES OF THE APPORTIONMENT OF HUMAN DIVERSITY

	Variation (%)			
Locus	Within population	Between populations within groups	Between groups	Reference[a]
Autosomal				
17 classical polymorphisms	85.4	8.3	6.3	Lewontin RC (1972) *Evol. Biol.* 6, 381.
30 microsatellites	84.5	5.5	10.0	Barbujani G et al. (1997) *Proc. Natl Acad. Sci. USA* 94, 4516.
79 RFLPs	84.5	3.9	11.7	Barbujani G et al. (1997) *Proc. Natl Acad. Sci. USA* 94, 4516.
60 microsatellites	87.9	1.7	10.4	Jorde L et al. (2000) *Am. J. Hum. Genet.* 66, 979.
30 SNPs	85.5	1.3	13.2	Jorde L et al. (2000) *Am. J. Hum. Genet.* 66, 979.
21 *Alu* insertions	82.9	8.2	8.9	Romualdi C et al. (2002) *Genome Res.* 12, 602.
mtDNA				
RFLPs	75.4	3.5	21.1	Excoffier L et al. (1992) *Genetics* 131, 479.
RFLPs	81.4	6.1	12.5	Seielstad M et al. (1998) *Nat. Genet.* 20, 278.
HVSI	72.0	6.0	22.0	Jorde L et al. (2000) *Am. J. Hum. Genet.* 66, 979.
Y chromosome				
22 binary polymorphisms	35.5	11.8	52.7	Seielstad M et al. (1998) *Nat. Genet.* 20, 278.
30 polymorphisms, several types	59	25	16	Santos F et al. (1999) *Am. J. Hum. Genet.* 64, 619.
6 microsatellites	83.3	18.5	−1.8	Jorde L et al. (2000) *Am. J. Hum. Genet.* 66, 979.
14 binary polymorphisms	42.5	17.4	40.1	Romualdi C et al. (2002) *Genome Res.* 12, 602.

[a] Reference for apportionment analysis, which may use data first published elsewhere.

expected from their smaller effective population sizes and hence greater drift (**Appendix**). The latter is particularly marked for the Y chromosome, which may be partly explained by **patrilocality** (**Section 5.5**) although there were large differences between studies. One detected no variation between groups, probably because the polymorphisms used were a small set of rapidly mutating microsatellites, while another actually found that most of the variation (53%) was between groups (Table 10.2).

A comparison of the distribution of diversity in humans with that found in other species of large-bodied mammals provides a useful perspective.[27] Humans, despite their worldwide distribution, show a low F_{ST} value comparable to that seen in either waterbuck or impala, each from a limited geographical region, Kenya; this reflects our recent expansion out of Africa starting from a much smaller and more geographically restricted population (Chapter 9). Species with long-established wider ranges, such as coyotes from North America or gray wolves from Eurasia, have higher values (**Figure 10.4**).

Patterns of diversity generally change gradually from place to place

Having established that there *are* small but nonzero genetic differences between populations, we can ask what patterns are found, and how we can explain their origins and maintenance. The most important observation is that variant frequencies as a rule change gradually from place to place. Sharp changes in frequency over small distances are unusual, and generally have a special explanation, such as a barrier to migration. For example, populations separated by the Strait of Gibraltar [just 14.3 km (8.9 miles) of sea between Spain in Europe and Morocco in Africa] differ substantially in their frequencies of autosomal, mtDNA, and Y loci. This is because these populations lie at the western end of the Mediterranean Sea and both originated from the east, following parallel expansion paths through Europe to the north of the Mediterranean and Africa to the south, respectively, with little gene flow across the sea. They accumulated gradual but independent east-to-west differences, which are therefore maximal at the western extremity: Gibraltar.

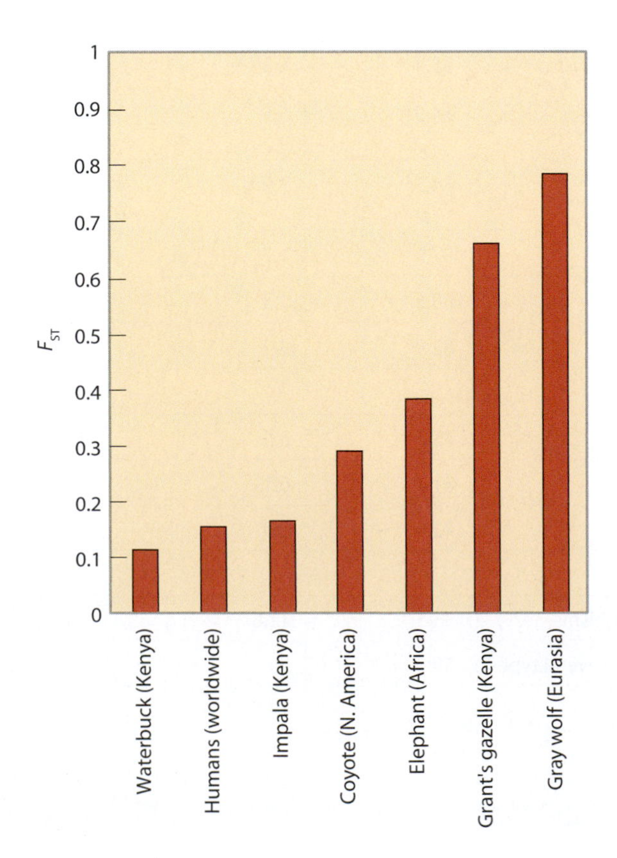

Figure 10.4: F_{ST} values for humans and other large-bodied mammals. Note the low human F_{ST}, typical of species with a restricted geographical distribution, despite the wide distribution of humans. [Data from Templeton AR (1999) *Am. J. Anthropol.* 100, 632.]

In **Section 9.4** and Figure 9.19, we saw that broad global patterns of diversity, highest in Africa and decreasing with distance from Africa, could be explained by the **serial founder model**. This model accounts for the origin of the gradual, **clinal** patterns of variation found throughout the world. An **isolation by distance model**, in which migrants tend to move short distances (**Section 6.8**), but large numbers of people seldom move long distances, explains how such patterns could persist for tens of thousands of years.

Thus we can interpret the pattern of human diversity as clinal. However, in the next section and other chapters, we will see that clustering methods, which assume that humans belong to discrete groups, also provide useful insights into human diversity, and are widely used.[24] So is human genetic variation better described as clinal or clustered?[26] One way of thinking about this question is to consider how the sampling scheme can influence the conclusion. Evenly spaced sampling from a continuous distribution produces a clinal pattern, while sampling groups of populations from locations separated by gaps, from the same distribution, produces clusters (**Figure 10.5**). Is this a good model for the distribution of human variation? Once sampling has been taken into account, some striking discontinuities can still be recognized in human genetic data, often associated with geographical barriers such as the Himalaya Mountains or the Strait of Gibraltar mentioned above, or social conventions such as those associated with the **caste system** in India. A follow-up study re-investigating worldwide patterns of variation concluded that while patterns within a continent might be largely clinal, there were robust clusters corresponding to different continents, arising from small discontinuous jumps in genetic distance for most population pairs on opposite sides of geographic barriers.[23]

Figure 10.5: Clusters or clines of human diversity?
The way in which samples are chosen from a continuous distribution (*central rectangle*) determines whether discrete clusters are seen (*circles at left*) or a more clinal pattern (*circles at right*).

The origin of an individual can be determined surprisingly precisely from their genotype

The small proportion of variation that lies between populations or continents can still be informative about their origin. A question is therefore how precise an origin can be specified. With genomewide datasets of hundreds or thousands of microsatellites or SNPs, analyzed using methods that summarize the information from multiple polymorphisms, or model ancestry using **STRUCTURE-like methods** (Table 6.1), the answer is that surprising precision can be obtained.

As we saw in the previous chapter, migration distance from East Africa explains ~85% of the genomewide heterozygosity level observed in an individual, meaning that heterozygosity alone can predict distance from East Africa with ~85% accuracy. In **Figure 10.6** and Figure 18.8, we see that even more detailed information can be obtained. PC plots based on genomewide SNP analyses show that world populations from the HGDP-CEPH collection cluster according to continent[15] (Figure 10.6a), European populations cluster approximately according to country[17] (Figure 18.8), and most remarkably of all, individuals from three Scottish Isles, or three Italian valleys, cluster according to the individual isle or valley[18] (Figure 10.6b). Thus these genomes contain detailed information about geographical ancestry, a finding with important implications for personal genomics and **forensic genetics**, discussed further in Chapter 18.

How can the finding that it is possible, with a sufficiently large set of polymorphisms, to deduce so much about the population of origin of an individual be reconciled with the earlier conclusion that most variation exists within populations, not between them? Particular alleles are usually not continent- or population-specific, although specific searches for these have had some success (**Section 14.2**, **Section 18.2**, and below); extraction of the geographical information generally requires analysis of large numbers of loci. If it is possible to identify genetic groups within humans, are these groups then "races"? No, because, as we have seen, the groups identified by current genetic techniques do not correspond to traditional races, and the differences between them are too small to justify being called races, which would require more than 25–30%

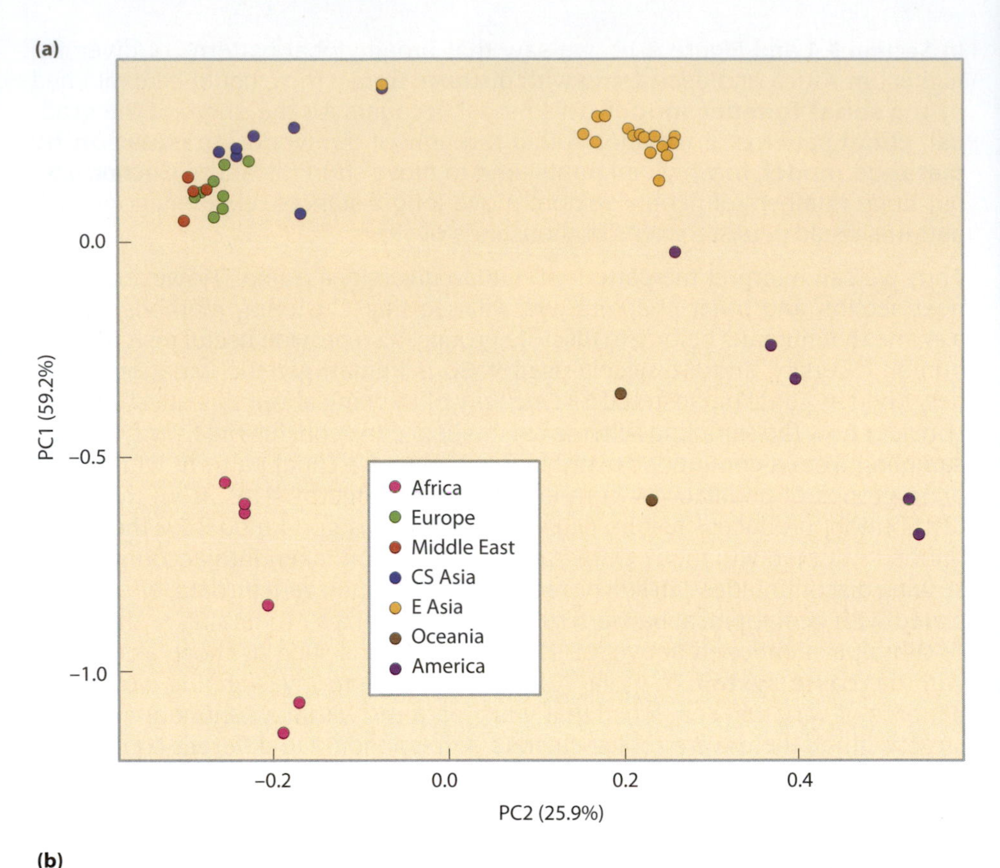

Figure 10.6: Prediction of geographical ancestry from genomewide SNP data. Each panel shows a PC analysis based on genomewide SNP genotypes. (a) Worldwide samples from the HGDP panel. Each dot represents a population and is colored according to the region of origin. (b) Samples from three Orkney Isles off the coast of Scotland (Sanday, *red*; Stronsay, *blue*; and Westray, *green*). [a, from Li JZ et al. (2008) *Science* 319, 1100. With permission from AAAS. b, from O'Dushlaine C et al. (2010) *Eur. J. Hum. Genet.* 18, 1269. With permission from Macmillan Publishers Ltd.]

genomewide difference between groups.[28] Irrespective of the genetic data, any idea that individuals should be treated by society according to their perceived ancestry or "race" is a social construct, not a biological one.

The distribution of rare variants differs from that of common variants

It may seem surprising to find a separate section focusing specifically on rare variants, but rare variants differ in some important ways from common ones. Common variants are invariably old, because it takes time for alleles to increase in frequency, even in the occasional cases where they are selectively advantageous (**Sections 6.6 and 6.7**). In contrast, rare variants *can* be old, but the vast majority of them are young. This has two major consequences: first, there has been less time for purifying selection to act on them, so they are enriched for deleterious variants (including medically relevant ones: Chapters 16 and 17) compared with common variants; second, there has been less time for them to spread geographically, so they tend to be restricted to a single population or a group of nearby populations. This is illustrated strikingly by the patterns of sharing of the variants that were found just twice in the 1092 individuals from the

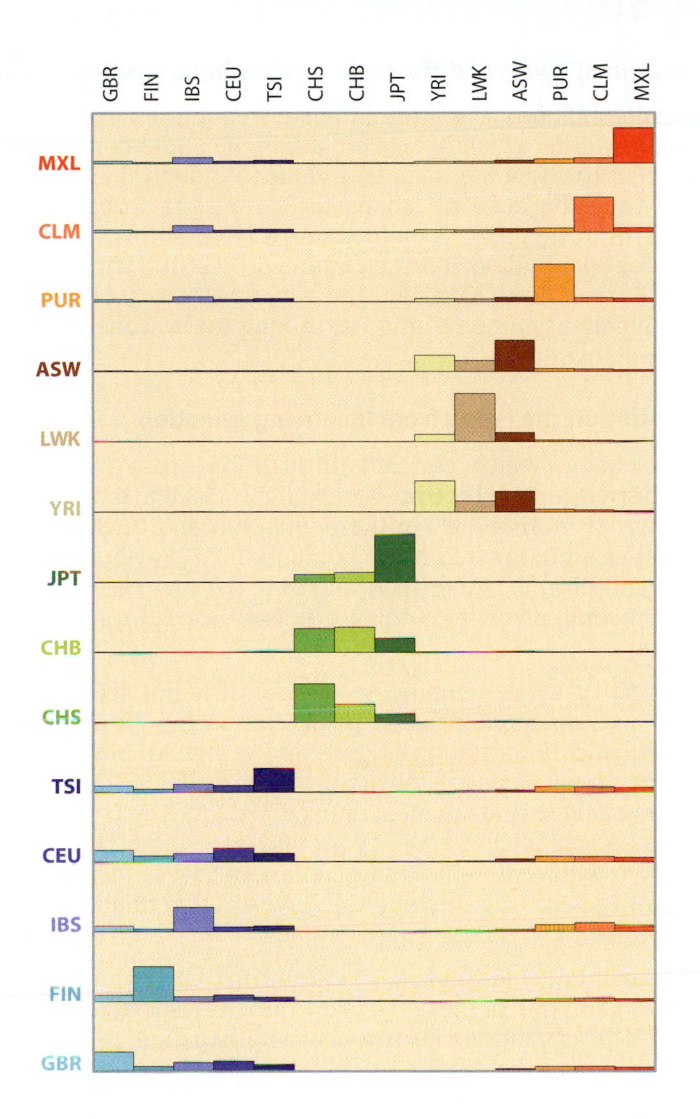

Figure 10.7: Distribution of rare variants within and between human populations.
Each row shows the sharing of variants detected twice in the 2184 chromosomes (1092 individuals) in the 1000 Genomes Project Phase 1 study between the population designated on the left, and all the other populations in the study (*top*). Note both the tendency for these rare variants to be shared within a population or with nearby populations, and the wider sharing in the American populations because of their admixture within historical times. [From The 1000 Genomes Project Consortium (2012) *Nature* 491, 56. With permission from Macmillan Publishers Ltd.]

14 populations included in Phase 1 of the 1000 Genomes Project[30] (**Figure 10.7**). These are the rarest shared variants in the study (with a frequency of <0.1%), and so should show any unusual pattern in its most extreme form. In every population, the second copy is most likely to be found within the same population. When it is found in a different population, the second population usually comes from the same continent. The main exception to this pattern is seen in the American populations, which reveal widespread sharing with both African and European populations, reflecting their history of admixture within historical times (Chapter 14). In contrast, common variants would generally be shared between all populations.

10.3 THE INFLUENCE OF SELECTION ON THE APPORTIONMENT OF DIVERSITY

In the previous sections of this chapter, we have considered studies conducted using variants that were often chosen for historical reasons (data were available when the study was carried out), or because they were considered neutral. We saw in **Section 10.2** that, although the mean between-population component of diversity in an early study was 14.6%, there was variation among the loci used: it ranged from as little as 0.3% for Xm to as much as 36.4% for the Duffy blood group. How much variation in these values is generally found? What effects does selection have? (see also **Section 6.7**). In the current section, we will encounter two measures of population differentiation: F_{ST} and ΔDAF (**Sections 6.3 and 6.7**). Both vary between 0 and 1, with 0 corresponding to no differentiation between populations, and 1 corresponding to complete differentiation; they are highly correlated.

The distribution of levels of differentiation has been studied empirically

Under neutral conditions, variation in allele frequencies between populations is determined by drift, which affects all loci in the same population equally, so all are expected to show the same F_{ST} value, although there will be a spread around this value because of stochastic factors. The observed distribution of F_{ST} values from the 1000 Genomes Project[30] is shown in **Figure 10.8**. The mean value for 35.6 million SNPs ascertained by sequencing 1092 individuals from Africa, Europe, East Asia, and the Americas was 0.05. The distribution, however, is highly asymmetric and has a long tail toward higher values, the maximum being over 0.8.

Low differentiation can result from balancing selection

Balancing selection, which can act through **heterozygote advantage** or **frequency-dependent selection**, for example (**Section 6.7**), favors the maintenance of two or more alleles in the population and thus high diversity, but if the same alleles are favored in all populations, F_{ST} will be low. Among the classical polymorphisms, some **HLA** alleles show low F_{ST} (**Table 10.3**) despite the very high overall diversity at this locus (see Box 5.3 for an introduction to the HLA locus).

With the benefit of whole-genome sequences, it is possible to identify genes with low F_{ST} in a comprehensive way, at least in the limited populations for which the sequence information is currently available. Since F_{ST} depends on allele frequency, low values are inevitable for rare alleles, and are only of interest for common alleles, and are most unexpected (and thus most interesting) for alleles with a derived allele frequency around 50%. Table 10.3 lists the 40–60% frequency SNPs within protein-coding genes that were identified by the 1000 Genomes Pilot Project[29] as showing the lowest differentiation in comparisons of African, European, and East Asian samples.

The HLA variant rs1129740 (Table 10.3) is found in DQA1*01 alleles (DQA1*01:01 to DQA1*01:07) and one of these (DQA1*01:03) is associated with both protection against some conditions such as chronic hepatitis C,[8] and susceptibility to others such as leprosy[22] and AIDS.[12] While there is no evidence that these diseases themselves have been the selective forces in human evolutionary

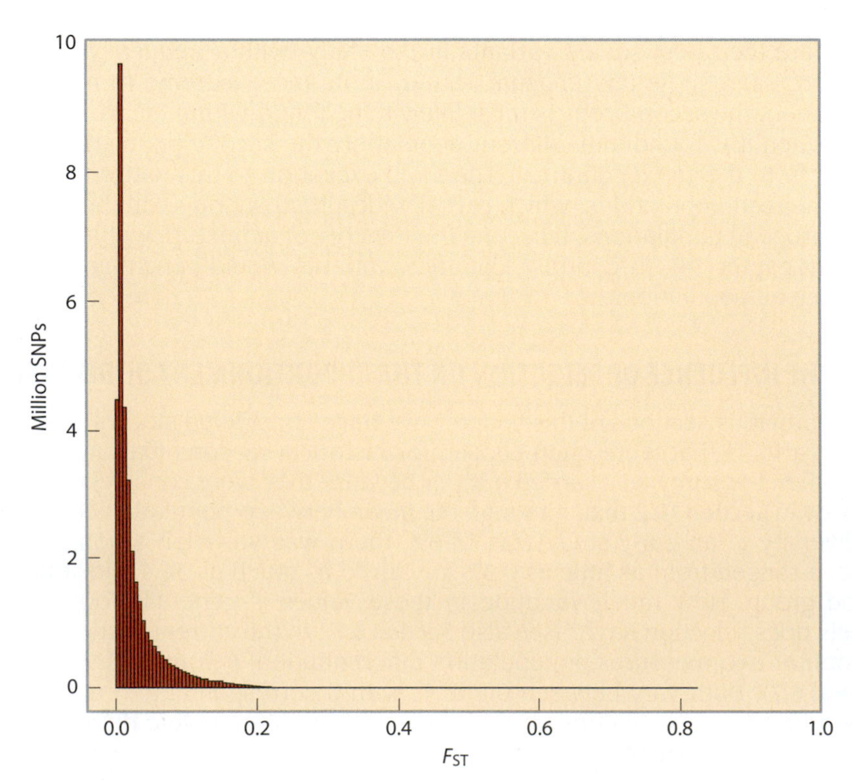

Figure 10.8: Distribution of human F_{ST} values.
Worldwide F_{ST} was calculated for 36.5 million SNPs discovered by Phase 1 of the 1000 Genomes Project. Note the low average value and skewed distribution with a few high values. (Unpublished analysis provided by Vincenza Colonna.)

TABLE 10.3:
THE LEAST-DIFFERENTIATED COMMON CODING SNPs IDENTIFIED BY THE 1000 GENOMES PILOT PROJECT

Chr	Position (GRCh37)	rs_id[a]	Gene	YRI_DAF[b]	CEU_DAF	CHB + JPT_DAF	DAF difference	Change to amino acid	Possible evolutionary consequence
11	32,874,926	rs11032025	*PRRG4*	0.40	0.40	0.40	0.000	synonymous	unknown
7	100,395,588	rs67377634	*MUC3A*	0.51	0.51	0.51	0.000	synonymous	unknown
14	68,042,574	rs11158685	*PLEKHH1*	0.46	0.47	0.46	0.007	nonsynonymous	unknown
20	18,445,963	rs6035051	*C20orf12*	0.47	0.47	0.46	0.007	synonymous	unknown
20	61,048,549	rs41305803	*GATA5*	0.42	0.41	0.42	0.007	synonymous	unknown
22	50,480,108	rs9628315	*TTLL8*	0.44	0.44	0.45	0.007	nonsynonymous	unknown
16	70,954,774	rs1798529	*HYDIN*	0.57	0.55	0.55	0.013	nonsynonymous	unknown
19	1,052,005	rs3764652	*ABCA7*	0.41	0.41	0.39	0.013	synonymous	unknown
22	50,658,424	rs11703226	*TUBGCP6*	0.41	0.39	0.41	0.013	nonsynonymous	unknown
6	32,609,105	rs1129740	*HLA-DQA1*	0.51	0.51	0.49	0.013	nonsynonymous	defense against infection
11	5,373,114	rs5006888	*HBG2/ HBE1*	0.44	0.45	0.43	0.013	nonsynonymous	unknown
11	5,373,646	rs5024041	*OR51B6*	0.44	0.45	0.43	0.013	synonymous	unknown
7	135,406,176	rs4596594	*SLC13A4*	0.53	0.55	0.55	0.013	synonymous	unknown
21	15,481,365	rs7278737	*LIPI*	0.49	0.49	0.47	0.013	nonsynonymous	unknown
6	146,755,140	rs2942	*GRM1*	0.50	0.52	0.52	0.013	synonymous	unknown
20	44,238,741	rs2245898	*WFDC9*	0.58	0.58	0.60	0.013	nonsynonymous	unknown
3	183,558,402	rs3732581	*PARL*	0.49	0.50	0.48	0.013	nonsynonymous	heart condition?

[a] rs_id is a unique identifying number given to each variant
[b] DAF, derived allele frequency

history that maintained both alleles of rs1129740 in the population, the link to several disease-related phenotypes suggests the more general possibility of such a cause. Biological explanations for the even frequencies of the other variants listed are lacking, and many may just be due to chance sampling, showing more variation in other populations. However, the association of the rs3732581 C-allele in the *PARL* gene with both a protective effect in reducing artery wall thickness and a susceptibility effect in increasing the risk of coronary artery disease[19] hints at possible balancing selection here as well.

High differentiation can result from directional selection

In contrast to balancing selection, directional selection acting in a subset of populations will lead to different alleles being at high frequencies in different populations (**Section 6.7**). Diversity within any one population may be low, but worldwide F_{ST} or ΔDAF will be high. The availability of whole-genome sequences from the 1000 Genomes Pilot Project[29] allows us to identify the regions of the genome that show the largest differences between populations. **Table 10.4** lists the five SNPs within protein-coding genes showing the largest ΔDAF values in pairwise analyses between the African (YRI), European (CEU), and East Asian (CHB+JPT) samples included in this project. We can draw

TABLE 10.4:
THE MOST HIGHLY DIFFERENTIATED CODING SNPs IDENTIFIED BY THE 1000 GENOMES PILOT PROJECT

Comparison	Chr	Position (GRCh37)	Frequency difference	Gene	Change to amino acid sequence	Possible evolutionary consequence
CEU vs CHB+JPT	15	46,213,776	0.992	*SLC24A5*	nonsynonymous	light skin color in Europeans
CEU vs CHB+JPT	5	33,987,450	0.968	*SLC45A2*	nonsynonymous	light skin color in Europeans
CEU vs CHB+JPT	4	38,475,043	0.939	*TLR1*	nonsynonymous	altered innate immunity (?)
CEU vs CHB+JPT	17	18,937,740	0.894	*AC007952.8*	nonsynonymous	unknown
CEU vs CHB+JPT	2	108,880,033	0.892	*EDAR*	nonsynonymous	hair and tooth morphology in East Asians
CEU vs YRI	15	46,213,776	0.979	*SLC24A5*	nonsynonymous	light skin color in Europeans
CEU vs YRI	5	33,987,450	0.979	*SLC45A2*	nonsynonymous	light skin color in Europeans
CEU vs YRI	8	145,610,489	0.972	*AF205589.2*	nonsynonymous	unknown
CEU vs YRI	9	126,302,623	0.966	*NR5A1*	nonsynonymous	sexual development, reproduction (?)
CEU vs YRI	4	38,475,043	0.937	*TLR1*	nonsynonymous	altered innate immunity (?)
CHB+JPT vs YRI	2	72,561,382	0.981	*EXOC6B*	synonymous	unknown
CHB+JPT vs YRI	15	39,936,798	0.978	*SPTBN5*	nonsynonymous	unknown
CHB+JPT vs YRI	17	60,251,208	0.976	*AC103810.1*	nonsynonymous	unknown
CHB+JPT vs YRI	8	145,610,489	0.972	*AF205589.2*	nonsynonymous	unknown
CHB+JPT vs YRI	16	46,815,699	0.968	*ABCC11*	nonsynonymous	dry earwax in East Asians

a number of conclusions from these findings. (1) These are the genes showing the most extreme geographical differentiation in our genomes, yet none of the frequency differences quite reach a value of one, which would correspond to complete fixation of one allele in one population and the other allele in the other population: even among these outliers, we cannot find fixed differences between populations. (2) Among the 11 genes identified, we know so little about the function of five that we cannot even speculate plausibly about the likely evolutionary implications (so these could be interesting avenues for future research). (3) For the other six genes, we can suggest either a general class of selective force (related to innate immunity, *TLR1*; or sexual development/reproduction, *NR5A1*), or more specific selection influencing earwax type (*ABCC11*), hair or tooth morphology (*EDAR*; Figure 15.2), or skin pigmentation (*SLC24A5*, *SLC45A2*). However, only for the last two genes can we be reasonably confident that the phenotype mentioned was the direct target of selection (see **Section 15.3**). (4) The skin pigmentation, hair/tooth, and earwax phenotypes were all noted and studied extensively before the era of genomic analyses, illustrating both the non-neutrality of some of the traits favored by anthropologists, and the effectiveness of these early scientists in identifying several of the most highly differentiated genes in humans.

Positive selection at EDAR

The *EDAR* (ectodysplasin A receptor) gene, carrying one of the most highly differentiated SNPs in the genome (Table 10.4), illustrates both the significant evolutionary insights that can be obtained by studying such unusual diversity patterns, and some of the complexities in interpretation that can arise. *EDAR* came to the attention of evolutionary geneticists via the HapMap1 Project, where it stood out because it carries a Val370Ala nonsynonymous variant (rs3827760) with a low (zero) frequency of the derived allele in the African (YRI) and European (CEU) samples, but a very high frequency in the East Asian (CHB) sample. Subsequent studies confirmed the high level of geographical differentiation in a larger sample[31] (**Figure 10.9a**), and identified a strong signal of

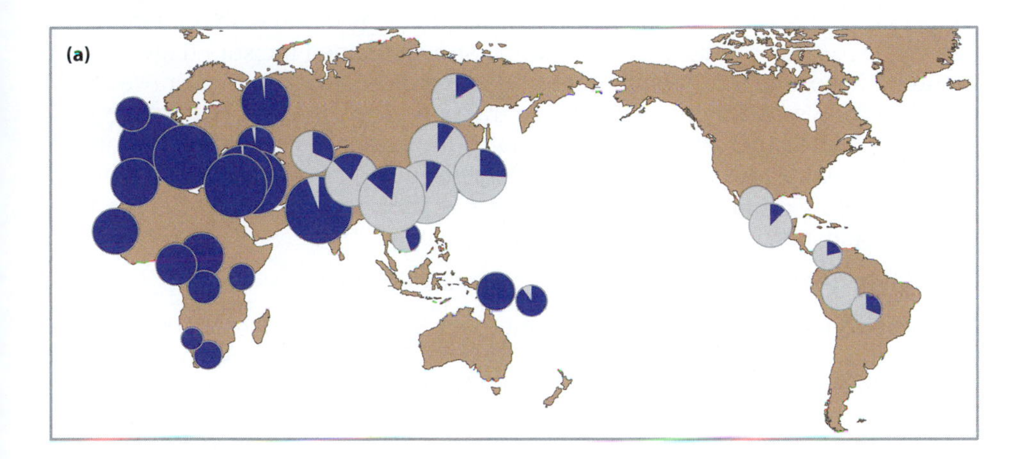

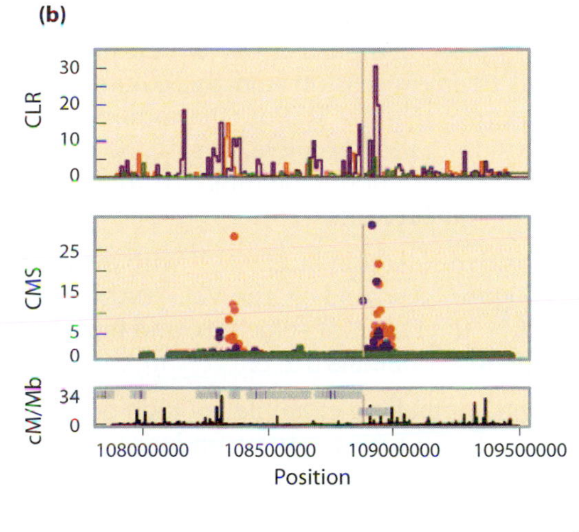

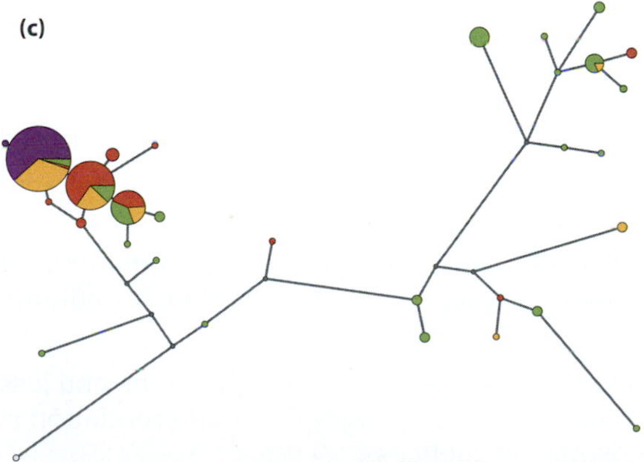

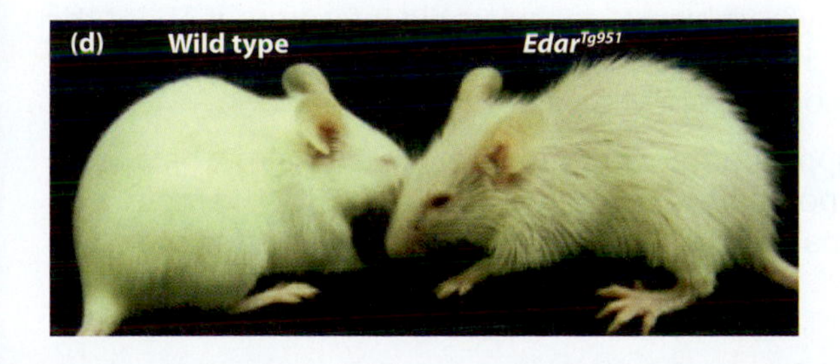

Figure 10.9: Positive selection at the *EDAR* gene.
(a) Geographical distribution of the ancestral (*blue*) and derived (*gray*) alleles of the Val370Ala variant of the *EDAR* gene. Note the 100% frequency of the ancestral (Val) allele in the west, contrasted with the high frequency of the derived (Ala) allele in the east. (b) Signals of positive selection in the vicinity of *EDAR*. CLR, composite likelihood ratio test for selection; CMS, composite of multiple signals; cM/Mb, centimorgans/megabase (recombination rate); *green*, selection in YRI (Africa, no signal); *orange*, selection in CEU (moderate signal, Europe); *purple*, selection in CHB+JPT (strong signal, East Asia). Note that the strongest signal is displaced from the Val370Ala substitution (*vertical gray line*). (c) Signals of positive selection in a 5-kb region surrounding the Val370Ala substitution. Selection is illustrated in these networks by the large circle size. The haplotype carrying the Ala-allele is selected in East Asians and Hispanics (*purple*, *yellow*) while a different haplotype carrying the Val-allele is selected in Europeans (*orange*). (d) Modeling the human thick hair phenotype in mice. In humans, the Ala allele is associated with thick hair, but in transgenic mice a similar phenotype was seen when the Val allele was overexpressed. [a and c, from Xue Y et al. (2009) *Genetics* 183, 1065. With permission from Genetics Society of America. b, from The 1000 Genomes Project Consortium (2010) *Nature* 467, 1061. With permission from Macmillan Publishers Ltd. d, from Mou C et al. (2008) *Hum. Mutat.* 29, 1405. With permission from John Wiley & Sons, Inc.]

positive selection (Figure 10.9b and c) using both haplotype-based tests (**Section 6.7**)[25] and allele frequency spectrum-based tests (**Section 6.7**).[31] Moreover, the gene had long been known to medical geneticists because its inactivation led to anomalies of skin, hair, teeth, and sweat glands, a combination known as hypohidrotic ectodermal dysplasia (OMIM 224900). It therefore seemed natural to suggest that the Val370Ala substitution might lead to a milder change in the same parts of the body, and indeed an association was found with both the thick hair characteristic of East Asians[6] and the shovel-shaped incisors (Figure 15.2) found in these and other populations.[11] It is therefore possible to propose models of positive selection based on these traits: perhaps the hair phenotype was considered attractive and selected sexually, for example, or a different pattern of sweating was advantageous. It is difficult to confirm any such model, and it remains possible that some trait that has not yet been identified was the true target of selection.

In addition, there are more substantial complexities:

- The strongest signal of positive selection in East Asians does not correspond to the Val370Ala substitution; instead it lies about 60 kb away within an intron of the same gene (Figure 10.9b). Modeling shows that a signal of selection *can* be located at such a distance from the target of selection, but the observation nevertheless raises the question of whether the Val370Ala substitution is really the main target of selection.
- Transgenic mice carrying multiple copies of mouse *Edar* (which has the ancestral Val-allele) show increased *Edar* expression and thick hair characteristic of human Ala-allele carriers[16] (Figure 10.9d). In humans, the Ala-allele is expressed at a higher level than the Val-allele, raising the possibility that the underlying cause of the human phenotype might be increased expression rather than the amino acid substitution.
- Europeans also show evidence of positive selection at *EDAR*,[31] with a signal in a similar location to East Asians (Figure 10.9b and c). The Val370Ala allele is absent from Europeans, so cannot be the target of selection, and both the target variant and selected phenotype in Europeans are entirely unknown.

Thus even for one of the best-studied and compelling examples of high population differentiation and positive selection, much remains to be understood.

SUMMARY

- Studies of human diversity raise important ethical questions which need to be addressed before each individual project is carried out.
- The sampling strategy will influence the conclusions, and strategies based on current population size and geographical or linguistic criteria have all been used in different studies.
- The definition of a "population" is not simple, but reflects a combination of geographical proximity, a common language, and shared ethnicity, culture, and religion.
- Most (at least 85%) autosomal variation is found within populations and less than 10% between different continents; more geographical differentiation is seen for mtDNA and Y-chromosomal sequences.
- With a large number of polymorphisms, considerable information about the population of origin of an individual can be obtained, sometimes down to the level of an individual country or even village.
- Selection influences the apportionment of diversity for a few loci: balancing selection can lead to low population differentiation levels (for example, *HLA* DQA1*01), while directional selection can lead to high levels (for example, *EDAR*, *SLC24A5*, *TLR1*).

QUESTIONS

Question 10–1: Professor Pangloss' reorganization of a lab freezer has led to the loss of labels on 150 tubes of DNA. However, on the box is written that it contains 50 DNA samples each from a Chinese, Nigerian and Irish population. Undaunted, and with his access to modern genotyping facilities and statistical methods, Pangloss says that he can assign each of the 150 samples to its population of origin in a short time, with moderate effort and expense. How?

Question 10–2: A population previously unknown to scientists has been contacted in the Amazon region. Would it be ethical to carry out genetic analyses on them? If not, why not? If so, of what kind and in what ways?

Question 10–3: It is proposed to sequence the genomes of the entire population of Erewhon and make the sequences publicly available. What benefits and risks would this pose for the Erewhonese and what advice would you offer to the Erewhon Genome Management Committee about how to maximize the benefits and minimize the risks?

Question 10–4: A company has developed an improved sequencing methodology and has offered to generate near-perfect sequences of 100,000 people. Who should be sequenced, why, with what accompanying details, and with what safeguards on the information?

Question 10–5: The geographical ancestry of a DNA sample can readily be deduced from its sequence. Does this prove that human races exist?

Question 10–6: What evidence suggests that the *EDAR* gene has experienced positive selection in humans? What additional analysis or experiment would you perform to identify the target(s) of selection and the selective force?

REFERENCES

The references highlighted in purple are considered to be important (for this chapter) by the authors.

1. **Ball MP, Thakuria JV, Zaranek AW et al.** (2012) A public resource facilitating clinical use of genomes. *Proc. Natl Acad. Sci. USA* **109**, 11920–11927.
2. **Barbujani G & Colonna V** (2010) Human genome diversity: frequently asked questions. *Trends Genet.* **26**, 285–295.
3. **Cann HM, de Toma C, Cazes L et al.** (2002) A human genome diversity cell line panel. *Science* **296**, 261–262.
4. **Caulfield T, Fullerton SM, Ali-Khan SE et al.** (2009) Race and ancestry in biomedical research: exploring the challenges. *Genome Med.* **1**, 8.
5. **Cavalli-Sforza LL, Wilson AC, Cantor CR et al.** (1991) Call for a worldwide survey of human genetic diversity: a vanishing opportunity for the Human Genome Project. *Genomics* **11**, 490–491.
6. **Fujimoto A, Kimura R, Ohashi J et al.** (2008) A scan for genetic determinants of human hair morphology: EDAR is associated with Asian hair thickness. *Hum. Mol. Genet.* **17**, 835–843.
7. **Gould SJ** (1981) The Mismeasure of Man. Penguin Books.
8. **Hohler T, Gerken G, Notghi A et al.** (1997) MHC class II genes influence the susceptibility to chronic active hepatitis C. *J. Hepatol.* **27**, 259–264.
9. **HUGO Pan-Asian SNP Consortium, Abdulla MA, Ahmed I et al.** (2009) Mapping human genetic diversity in Asia. *Science* **326**, 1541–1545.
10. **Kaessmann H, Heissig F, von Haeseler A & Pääbo S** (1999) DNA sequence variation in a non-coding region of low recombination on the human X chromosome. *Nat. Genet.* **22**, 78–81.
11. **Kimura R, Yamaguchi T, Takeda M et al.** (2009) A common variation in EDAR is a genetic determinant of shovel-shaped incisors. *Am. J. Hum. Genet.* **85**, 528–535.
12. **Kroner BL, Goedert JJ, Blattner WA et al.** (1995) Concordance of human leukocyte antigen haplotype-sharing, CD4 decline and AIDS in hemophilic siblings. Multicenter Hemophilia Cohort and Hemophilia Growth and Development Studies. *AIDS* **9**, 275–280.
13. **Lewontin RC** (1972) The apportionment of human diversity. *Evol. Biol.* **6**, 381–398.
14. **Li H & Durbin R** (2011) Inference of human population history from individual whole-genome sequences. *Nature* **475**, 493–496.
15. **Li JZ, Absher DM, Tang H et al.** (2008) Worldwide human relationships inferred from genome-wide patterns of variation. *Science* **319**, 1100–1104.
16. **Mou C, Thomason HA, Willan PM et al.** (2008) Enhanced ectodysplasin-A receptor (EDAR) signaling alters multiple fiber characteristics to produce the East Asian hair form. *Hum. Mutat.* **29**, 1405–1411.
17. **Novembre J, Johnson T, Bryc K et al.** (2008) Genes mirror geography within Europe. *Nature* **456**, 98–101.
18. **O'Dushlaine C, McQuillan R, Weale ME et al.** (2010) Genes predict village of origin in rural Europe. *Eur. J. Hum. Genet.* **18**, 1269–1270.
19. **Powell BL, Wiltshire S, Arscott G et al.** (2008) Association of PARL rs3732581 genetic variant with insulin levels, metabolic syndrome and coronary artery disease. *Hum. Genet.* **124**, 263–270.
20. **Prugnolle F, Manica A & Balloux F** (2005) Geography predicts neutral genetic diversity of human populations. *Curr. Biol.* **15**, R159–160.
21. **Ramachandran S, Deshpande O, Roseman CC et al.** (2005) Support from the relationship of genetic and geographic distance in human populations for a serial founder effect originating in Africa. *Proc. Natl Acad. Sci. USA* **102**, 15942–15947.

22. **Rani R, Fernandez-Vina MA, Zaheer SA et al.** (1993) Study of HLA class II alleles by PCR oligotyping in leprosy patients from north India. *Tissue Antigens* **42**, 133–137.

23. **Rosenberg NA, Mahajan S, Ramachandran S et al.** (2005) Clines, clusters, and the effect of study design on the inference of human population structure. *PLoS Genet.* **1**, e70.

24. **Rosenberg NA, Pritchard JK, Weber JL et al.** (2002) Genetic structure of human populations. *Science* **298**, 2381–2385.

25. **Sabeti PC, Varilly P, Fry B et al.** (2007) Genome-wide detection and characterization of positive selection in human populations. *Nature* **449**, 913–918.

26. **Serre D & Pääbo S** (2004) Evidence for gradients of human genetic diversity within and among continents. *Genome Res.* **14**, 1679–1685.

27. **Templeton AR** (1999) Human races: a genetic and evolutionary perspective. *Am. J. Anthropol.* **100**, 632–650.

28. **Templeton AR** (2003) Human races in the context of recent human evolution: A molecular genetic perspective. In Genetic Nature/Culture: Anthropology and Science Beyond the Two-Culture Divide (AH Goodman, D Heath & MS Lindree, eds), pp 234–257. University of California Press.

29. **The 1000 Genomes Project Consortium** (2010) A map of human genome variation from population-scale sequencing. *Nature* **467**, 1061–1073.

30. **The 1000 Genomes Project Consortium** (2012) An integrated map of genetic variation from 1,092 human genomes. *Nature* **491**, 56–65.

31. **Xue Y, Zhang X, Huang N et al.** (2009) Population differentiation as an indicator of recent positive selection in humans: an empirical evaluation. *Genetics* **183**, 1065–1077.

THE COLONIZATION OF THE OLD WORLD AND AUSTRALIA

CHAPTER ELEVEN

Humans have expanded out of Africa on several occasions, as we saw in Chapter 9, yet one of these expansions is of particular significance for understanding our evolutionary history. This is because the expansion ~50–70 KYA led ultimately to the extinction of all earlier hominins, albeit with some interbreeding, and subsequently to the occupation of the entire globe. It marked the start of a transition from an ape-like demography (subdivided populations, small numbers) to a modern human demography (little subdivision, enormous numbers). The basic geographical patterns of variation that were established at that time persist to this day and provide the background for much of human genetics, whether population, evolutionary, or medical in emphasis. In this chapter, we will explore the details of this expansion, and seek to understand some of the underlying issues: How was this expansion different from earlier expansions? Why did it occur ~50–70 KYA, rather than earlier or later? What routes did it take and why? Where and when did admixture with Neanderthals and Denisovans take place? What were the consequences for other species, nonhuman as well as human?

We will see that answering these questions requires an understanding of the environment and how it changed during the period, particularly the climate. During the last **interglacial**, ~120 KYA, the climate was warm, but between 100 and 15 KYA, it was generally much colder than at present—an **ice age**. Human fossils provide unequivocal evidence for the presence of anatomically modern humans at particular places and times, but only limited evidence about behavior. This limitation, and the rarity of human fossils, means that we also need to examine the complementary archaeological record. With this background, we then turn to the evidence from genetics. We will see that while mtDNA and the Y chromosome provide important insights into these events, autosomal evidence from both ancient and modern DNA samples is becoming available on a scale that is transforming our understanding. We will stop before considering the colonization of the Americas or the Pacific, or the changes in lifestyle that followed the climatic warming that began ~20 KYA, since these will be the subjects of later chapters.

11.1 A COLDER AND MORE VARIABLE ENVIRONMENT 15–100 KYA

Over the last two billion years, the average temperature of the surface of the Earth has varied by more than 10°C, oscillating between warm (sometimes entirely ice-free) and cold (**glacial** or ice age) conditions. These changes are thought to result from variations in the Earth's orbit, the positions of the continents, and the amount of carbon dioxide in the atmosphere. For the last 3 MY, the Earth has been in a cool period, but there has been considerable variation

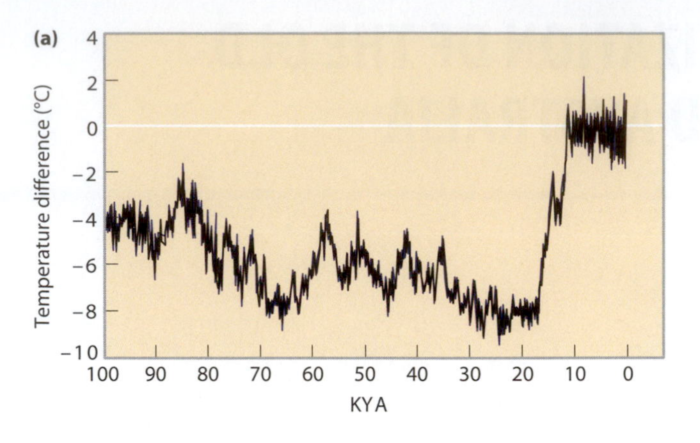

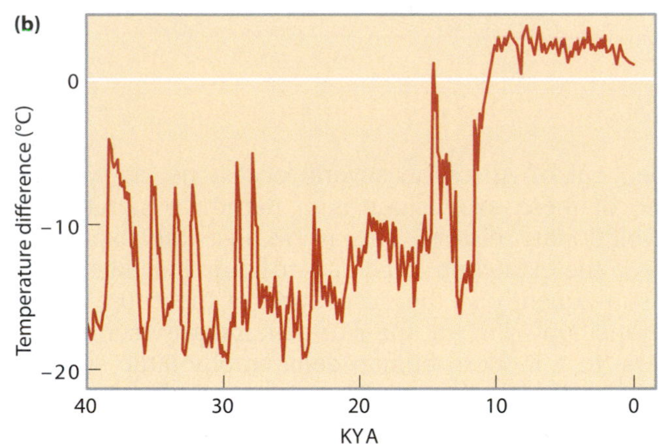

Figure 11.1: Temperature variation over the last 100 KY.
(a) Note that for most of this period the temperature has generally been colder than during the last 10 KY, particularly around 17–30 KYA. (b) Higher-resolution data from the Greenland GISP2 core for the last 40 KY showing a similar overall pattern but very rapid fluctuations of up to 10°C. [a, data from the Vostok ice core in Antarctica, Petit JR et al. (1999) *Nature* 399, 429, and http://cdiac.esd.ornl.gov/trends/temp/vostok/jouz_tem.htm. b, from Cuffey KM et al. (1995) *Science* 270, 455. With permission from AAAS.]

within this, including warm interglacials at around 250 KYA and 110–130 KYA with temperatures similar to the present; indeed, a ~100 KY warm–cold periodicity can be traced back for ~900 KY. For the period of particular interest here, the last 100 KY, detailed records of variables such as temperature, precipitation, and volcanic activity have been obtained from ice cores, and slightly less detailed records from lake and ocean sediments. The conditions during the last 12 KY we think of as "normal" are actually exceptional for both their high temperature and stability (**Figure 11.1a**). Before ~12 KYA the temperature was 2–9°C colder in the Antarctic, with a mean of 6.8°C colder than at present for the period 18–70 KYA. A higher-resolution record from the Arctic (Figure 11.1b) for the last 40 KY shows a similar overall profile where the temperature before ~12 KYA was 10–20°C colder than at present, but with very rapid fluctuations of up to 10°C. It has been suggested that such changes occurred in sudden jumps over periods of just a few decades, perhaps mediated by alterations to the circulation of ocean currents such as the present-day Gulf Stream in the North Atlantic. Twenty-three **interstadials**, short-lived warm events, have been identified within the last 100 KY, and at least six **Heinrich events**, short-lived periods of extreme cold. The most recent warming event at ~11.5 KYA (the **Younger Dryas** to **Holocene** transition) has been interpreted as "a series of warming steps, each taking less than 5 years. About half of the warming was concentrated into a single period of less than 15 years."[1] Thus, extreme climatic changes would have occurred within a human lifetime on many occasions during the spread of modern humans.

Low temperature had several consequences: precipitation was low, so that large areas of the world were cold deserts. Glaciers formed where there was sufficient snowfall at high latitudes, and combined to form huge ice caps that rendered large regions uninhabitable. These ice caps locked up a significant proportion of the world's water, so that sea levels were lower. Some environments with no modern parallels existed, such as **Mammoth Steppe**, cold and dry but productive with an extensive **megafauna**. While it is difficult to generalize about the climate over the entire Old World during this period, the following features have been suggested (http://www.esd.ornl.gov/projects/qen/nerc.html).

100–70 KYA: After the previous interglacial, when temperatures were similar to the present, had ended at around 110 KYA, the temperature subsequently fell, so it was generally slightly cooler than the present, and the period included some intense cold periods in Asia around 91, 83, 75, and 70 KYA. Nevertheless, the African fauna (including humans), which had expanded during the interglacial, remained in areas such as the Levant (Eastern Mediterranean) during this period.

Glacial maximum, **70–55 KYA:** One hypothesis suggests that this event was triggered by the eruption of Mount Toba in Sumatra, Indonesia, the largest explosive volcanic eruption in the last 450 MY (perhaps 3000 times larger than the 1980 Mount St. Helens, USA, eruption), which produced substantial deposits of volcanic ash as far away as India, and could have had a major environmental impact.[3] However, archaeological evidence for continuity before and after the eruption does not support the idea of a major influence on hominin activities,[32] and large-scale changes to flora and fauna are not seen. Fragmentation of the environment occurred within Africa separating sub-Saharan, Northeastern, and Northwestern regions; conditions elsewhere, and sea levels, may have been similar to those during the more recent glacial maximum (see below and **Figure 11.2**) with extensive ice caps and a hostile high-latitude environment. Australia was cooler and drier than at present.

55–25 KYA: Somewhat warmer than the glacial maximum, but still colder than the present and highly variable; sea level still as much as 70 m below the current level. There is evidence for wood or forest in Central Siberia in the later part, **Taiga** (cold coniferous forest) vegetation in the Altai mountains, and cold dry **steppe** in northern China and perhaps more widely across Asia and North America; some areas of Australia were wetter than at present. In Europe, there was an interstadial as warm as the present at 43–41 KYA which seems to have allowed the first modern human colonists to spread rapidly, followed by a cold Heinrich event 41–39 KYA and then relatively mild temperatures.

Last glacial maximum (LGM), **23–14 KYA:** The sea level fell to as much as 120 m below the present level. In Africa, it was dry (especially 16–17 KYA) and the Sahara extended southward (Figure 11.2). Northern Asia was a dry, treeless, polar desert or semi-desert since there was not enough snowfall for ice caps to build up; average temperatures in Siberia were 12–14°C colder than at present. Northern China was semi-desert, but there were more hospitable grasslands and cool temperate forest in the south. Australia was very dry with a large area of extreme desert in the center. Northern Europe was covered by ice sheets or polar desert; to the south there was semi-desert or grassland.

Figure 11.2: The Earth during the last glacial maximum at ~18 KYA.
Note the prominent ice caps and lower sea level leading to larger land areas, particularly in Southern Asia/Northern Australia and Beringia, the connection between Siberia and North America. Such a detailed reconstruction is not available for the earlier glacial maximum, but the appearance may have been similar.

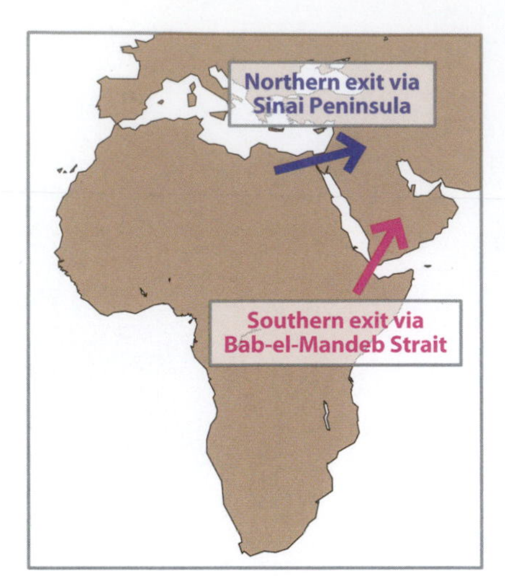

Figure 11.3: Proposed northern and southern exit routes from Africa.

Holocene, 12 KYA to present: During the Holocene (meaning wholly recent), temperatures rose rapidly to the current level in a very irregular fashion. For most of the period, they have been unusually stable.

In summary, the environment during the period when humans spread from Africa to the rest of the Old World was significantly different from our present one: mostly colder and drier with lower sea level, accompanied by marked short-term fluctuations in climate.

11.2 FOSSIL AND ARCHAEOLOGICAL EVIDENCE FOR TWO EXPANSIONS OF ANATOMICALLY MODERN HUMANS OUT OF AFRICA IN THE LAST ~130 KY

It seems likely that the hominins living in Africa expanded out of (and probably back to) Africa whenever conditions permitted. We have seen the early expansion of *H. erectus* soon after 2 MYA, and the suggestion that *H. heidelbergensis* evolved in Africa before 1 MYA and subsequently spread into Europe and perhaps Western Asia as well (Chapter 9). Simple geographical logic shows that if movement out of Africa was by land, or involved at most short journeys across the sea, the choice of routes was limited. One possibility was across the Mediterranean, either via the Strait of Gibraltar, or via some of the islands such as Sicily. However, there is no evidence for a Mediterranean route being used by Paleolithic people. Two other routes are possible: one from North Africa to the Levant via the Sinai Peninsula (Northern exit), and the other from East Africa to the Arabian Peninsula via the Bab-el-Mandeb Strait (Southern exit; **Figure 11.3**). One question will therefore be about which of these last two possibilities was used. We also saw that modern human anatomy appeared in Africa by ~200 KYA, but in contrast we will see in this chapter that there is clear archaeological evidence for art and symbolic behavior only after 100 KYA. Securely dated fossils of anatomically modern humans would provide the most direct evidence for the times at which modern humans reached different places. However, human fossils are rare and dating is often disputed, so the age of the earliest accepted human fossil from any region may be considerably later than the time of initial entry. Archaeological remains are more plentiful and *can* provide evidence for modern human behavior (**Box 11.1**), but it can still often

Box 11.1: Archaeological signs of modern human behavior

It might seem surprising that human behavior can be identified from the archaeological record, and indeed there are key aspects, such as the use of complex language, where archaeology has not provided any substantial insights. Nevertheless, archaeologists have identified several kinds of evidence that indicate **behavioral modernity**. Among them are:

- Finely made tools often involving the application of complex techniques and utilizing diverse materials such as bone, antler, or ivory as well as stone
- Long-distance transport, perhaps requiring exchange between different groups of humans
- Use of pigments such as ochre
- Production of objects for personal ornamentation such as beads and other jewelry
- Production of art such as cave or rock paintings and figures
- Music, indicated by the presence of musical instruments such as flutes
- Burial of the dead

One writer summarized these characteristics as the five behavioral Bs: blades, beads, burials, bone tool-making, and beauty.[10] Some of these are considered as signs of symbolic thinking, where one object represents another. Symbolic thinking would be strong evidence for behavioral modernity. There is, however, no sharp line between modern and pre-modern behavior: a tool may appear finely made to one person but crude to another. And it is not necessary for a population to demonstrate all of these characteristics in order to be judged behaviorally modern: cave painting, for example, is impossible in environments that lack caves. Thus there is still plenty of room for debate about whether the archaeology of a site does or does not support the behavioral modernity of its creators.

be difficult to decide whether particular finds were created by archaic or modern humans. However, because of their abundance, the earliest archaeological dates are likely to be closer to the time of entry than the earliest fossil dates. Humans share their environment and many aspects of their behavior with other animals, so the study of past animal and plant distributions is a rich source of comparative information.

Archaeological evidence is key to our understanding of the expansion of modern humans for two related reasons. First, it can demonstrate their presence at sites or times where no human fossils have been found, at least when we can distinguish between archaeological deposits left by archaic and modern humans. Second, and more crucially, it provides evidence for forms of behavior that are considered unequivocally "modern," such as art (Box 11.1). It is important to recognize that archaeological terms such as "Upper Paleolithic" refer to assemblages of remains such as stone or bone tools that represent *cultures*, not time periods. These cultures may have been present in different places at different times, so the date of an archaeological type needs to be related to a particular location. The Upper Paleolithic, for example, may have reached one place at 45 KYA but another only at 35 KYA; in other regions it may never have been present.

Anatomically modern, behaviorally pre-modern humans expanded transiently into the Middle East ~90–120 KYA

The African fauna expanded into nearby areas outside Africa during the warm climate of the last interglacial 110–130 KYA. Humans who were present in the Levant and Arabian Peninsula at this time can be seen as part of the same expansion. There is direct evidence for anatomically modern humans at Qafzeh and Skhul in Israel dating to 90–130 KYA (**Section 9.1**, Table 9.2). In addition, an archaeological site at Jebel Faya in the Arabian Peninsula (**Figure 11.4**) dated to 127 ± 16 KYA produced stone tools similar to finds of the same age from East Africa. Since the only known toolmakers in Africa from this time were anatomically modern humans, the authors suggest that the toolmakers in the Arabian Peninsula were anatomically modern as well.[4] Another site at Dhofar (Figure 11.4) dated to 106 ± 9 KYA yielded extensive tools produced using the **Levallois technique** (**Section 9.2**) and classified as belonging to the Nubian Complex.[37] Using similar reasoning, these authors also argued that the toolmakers were most likely modern humans. While the presence of anatomically modern but

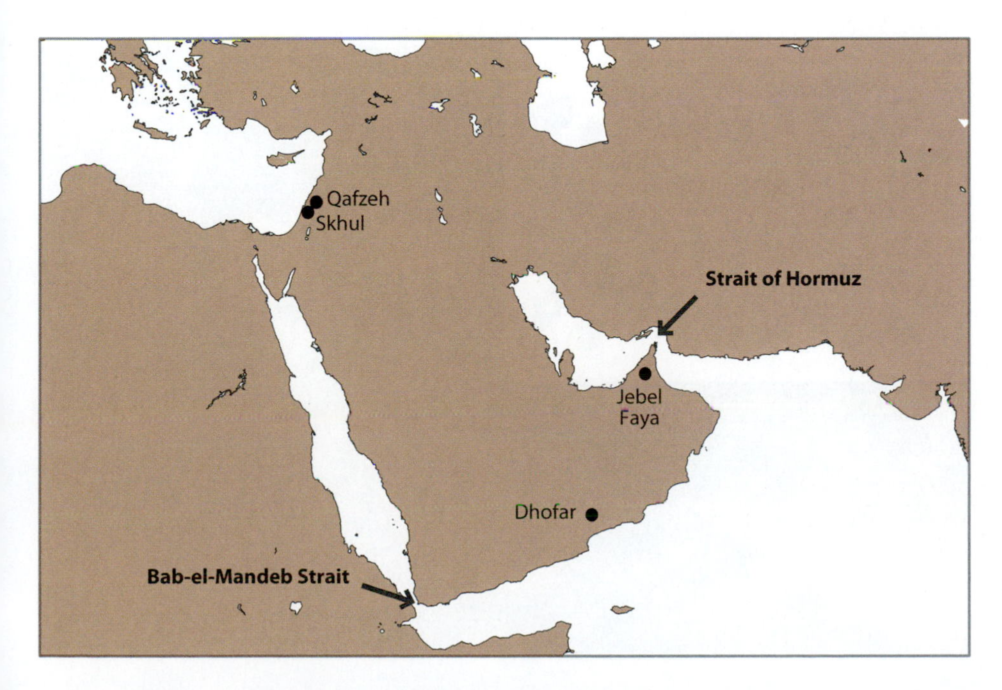

Figure 11.4: Fossil and archaeological sites in the Levant and Arabian Peninsula dating to 80–130 KYA.

behaviorally pre-modern humans outside Africa in the Levant before 100 KYA is well established, the extent of their distribution, and their links with Arabian Peninsula archaeological sites, are far from clear.

A plausible hypothesis is that this occupation of these regions outside Africa was not permanent, and that the early colonists were replaced by the cold-adapted Asian fauna, including Neanderthals in the Levant, when the climate deteriorated. Nevertheless, the full geographical extent of this expansion, and the way the colonists responded to the colder conditions, are unknown, and are the subject of continuing investigations.

Modern human behavior first appeared in Africa after 100 KYA

We saw in Chapter 9 that archaeologists recognize a transition between Mode 3 (Middle Stone Age/Middle Paleolithic) and Mode 4 (Later Stone Age/Upper Paleolithic) technologies. Developments in Mode 4 include the production of finer tools, the use of more diverse materials and their transport over longer distances, and the use of pigments. Here, we examine this transition in more detail. African archaeological deposits dating from around 250 KYA to 40 KYA belong to the **Middle Stone Age** (MSA); similar assemblies outside Africa are described by the alternative, but broadly equivalent, term Middle Paleolithic (MP). Archaic humans of this period, such as Neanderthals, as well as early anatomically modern humans, are thus both associated with MSA/MP remains. However, there is considerable diversity within the MSA industries.

A striking early development was the processing and use of ochre, a natural red or yellow pigment deriving its color from oxides or hydroxides of iron. For example, a pigment-processing toolkit consisting of an abalone (*Haliotis midae*) shell with a tight-fitting rounded stone inside and traces of red ochre and bone (a potential source of fat which would have provided a binder for the pigment) was found in a layer with an optically stimulated luminescence (OSL) date of 101 ± 4 KYA at Blombos Cave in South Africa.[17] The pigment produced would not have been an effective adhesive, but might have been used to protect or decorate a surface. Perhaps most striking of all, lumps of red ochre incised with abstract linear patterns dating to 75–80 KYA have been found at the same cave[18] and have even been interpreted as the first art: strong evidence for modern human behavior (**Figure 11.5**).

Figure 11.5: Middle Stone Age artifacts from Blombos Cave, South Africa. Middle Stone Age artifacts from Blombos Cave, South Africa, dating to ~60–80 KYA demonstrating (a) and (b) production of pierced shells; (c) pressure flaking; and (d) engraved ochre. Scale bar: 1 cm. [a and b, Henshilwood C et al. (2004) *Science* 304, 404. c, Mourre V et al. (2010) *Science* 330, 659. d, Henshilwood C et al. (2002) *Science* 295, 1278. All with permission from AAAS.]

Two MSA technologies notable for their sophistication are Still Bay and Howieson's Poort. Like the ochre fragments, they are again found in southern Africa and sites include Klasies River Mouth at the extreme south of the African continent (Figure 9.6), where anatomically modern human fossils have been found (Table 9.2), as well as the same Blombos Cave. Although distinct, both technologies are characterized by **hafted** weapons, use of bone, ostrich eggshell beads, and engraved ochre. A comprehensive study using optical dating at nine sites dated both technologies to short intervals of less than 5 KY between 60 and 80 KYA. Still Bay was the earlier (~71–72 KYA) and was separated by ~6 KY from Howieson's Poort (~60–65 KYA).[22] Production of some Still Bay stone tools involved heat-treatment followed by pressure-flaking, techniques that result in thinner, narrower, and sharper tips[31] (Figure 11.5). These features are otherwise recognized only in industries dating to ~20 KYA such as the Solutrean in Western Europe, or later. Perforated *Nassarius kraussianus* shell beads from the same level provide further evidence for modern behavior.[16] The Howieson's Poort layers at Diepkloof Rock Shelter, also in South Africa, provided evidence for extensive standardized engraving of linear and hatched patterns on ostrich eggshells, which were probably functional items used as containers.[43]

Evidence for signs of early modern behavior is not, however, confined to southern Africa. Perforated shells of the sea snail *Nassarius gibbosulus*, interpreted as beads used for personal ornamentation and thus diagnostic of behaviorally modern humans, dated to 82 (73–91) KYA have been found as far away as the Grotte des Pigeons (Morocco) in North Africa.[8] These must have been transported >40 km from the Mediterranean Sea and show traces of red ochre, interpreted as possibly transferred from decorated skin. The implications of these early northern findings for the location of the origins of modern behavior remain to be integrated with the more extensive data from further south.

The earliest evidence for the transition from the MSA to the **Later Stone Age** (LSA) has been reported from East Africa, where sites such as the Enkapune Ya Muto rock shelter in the Rift Valley of Kenya would place it at around 50 KYA,[2] although accurate dating is difficult because such dates are close to the limits of the radiocarbon method (**Box 11.2**). Flake-based industries were replaced by **blades**; ground bone tools and perforated ornaments became common, and long-distance trading was established.

From a present-day perspective, it can be tempting to view the exceptional MSA developments described above as forerunners of behavior that became common during the Late Stone Age/Upper Paleolithic. However, it is clear that there was no steady accumulation of such behaviors during the MSA. Instead, the evidence suggests that they appeared sporadically. This must to some extent reflect the incompleteness of the archaeological record, but may also indicate that these technologies and their use were appropriate at some times and places, and not in other circumstances, irrespective of the potential abilities of the makers.

In conclusion, evidence for modern behavior is found substantially earlier in Africa than on any other continent, and in Africa it appears to develop gradually between 100 and 60 KYA. Elsewhere, clear evidence for such behavior is only available after ~50 KYA and appears much more suddenly with the Upper Paleolithic. We infer that modern behavior evolved in Africa, but the source region within this continent remains unclear because there is early evidence from both northern and southern regions. Nevertheless, by 60 KYA, some African populations clearly had the potential for fully modern behavior.

Fully modern humans expanded into the Old World and Australia ~50–70 KYA

Modern human fossils in Asia, Australia, and Europe

We next examine the evidence for modern human fossils in different areas of the Old World and Australia, except for the Middle East, which was considered

Box 11.2: Radiocarbon (^{14}C) dating and its calibration

Radiocarbon dating is based on the principle that ^{14}C produced in the upper atmosphere by cosmic ray bombardment of nitrogen-14 (^{14}N) is incorporated into biological materials during their lifetime, along with normal ^{12}C. This ceases at death, and the ^{14}C then decays exponentially with a **half-life** of 5730 years. Measurement of the amount of ^{14}C remaining in a biological specimen therefore reveals the amount of time since death. With current techniques for detecting ^{14}C, samples as old as ~50 KY or slightly older can be dated, so it is central to understanding the timing of the spread of modern humans, but at its limits for some of the most interesting dates. If the amount of ^{14}C in the atmosphere had been constant, it would be simple to convert a measured proportion of ^{14}C into a date: half the present level would give a date of 5.7 KYA, one-quarter a date of 11.5 KYA, and so on. However, the amount of ^{14}C has varied considerably and there can be substantial differences between "^{14}C years" and "calendar years." Thus ^{14}C dates need to be calibrated against an independent standard. For dates ≤10 KYA, counting of annual growth rings in trees or timbers (**dendrochronology**) can be used, but this calibration cannot be extended far back into the glacial period because no usable record is available. Alternative calibration sources include annually layered sediments and corals, but the most extensive comparison, extending back to 45 KYA, is derived from a stalagmite that contained both thorium-230 (allowing measurement of the calendar age) and ^{14}C in the calcium carbonate (allowing measurement of the ^{14}C age).[5] The resulting comparison curve—it is not yet universally agreed for calibration (see Figure 1)—showed that ^{14}C levels were higher between 30 and 40 KYA, so that ^{14}C ages from this period translate into considerably older calendar ages. For example, 31 KYA (^{14}C) would correspond to ~38 KYA (calendar). Thus it is essential to know whether ^{14}C dates encountered in the literature are "raw" or "corrected." Note that dates from methods such as thermoluminescence (TL) and electron spin resonance (ESR) (Table 9.1), which are measured in calendar years, are sometimes "decalibrated" by converting to radiocarbon years.

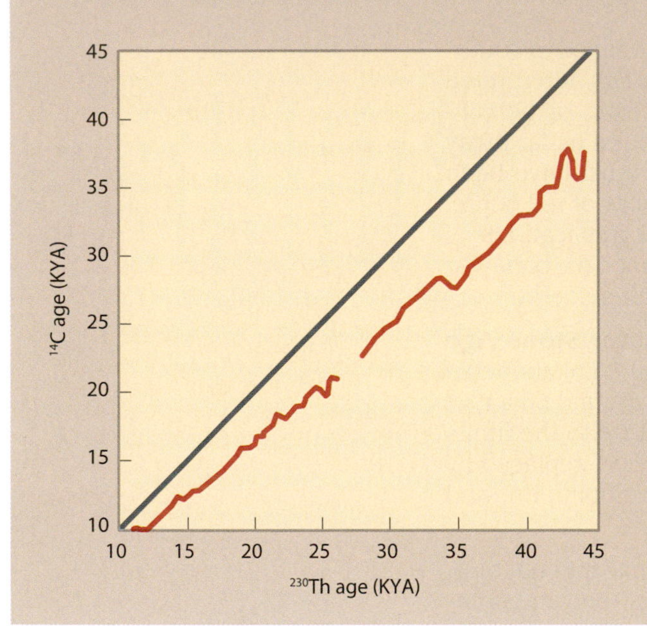

Figure 1: Comparison curve between radiocarbon and calendar dates for the period ~45–10 KYA.
Comparisons were made using sections from a stalagmite that could be dated by both methods. Perfect correlation (*blue*); actual correlation (*red*). (Data provided by David Richards.)

above. In East Asia, two molar teeth and a partial mandible from Zhirendong (= "*Homo sapiens* cave"), South China, dated to >100 KYA, have been interpreted as showing anatomically modern features,[23] raising the possibility that the early expansion into Arabia and the Levant might have reached China. But simpler and more compelling alternative interpretations would see this material as **gracile** *H. erectus*. The earliest widely accepted date for unequivocally modern human fossils in China is 39–42 KYA for a set of 34 bones and teeth from Tianyuan Cave, Zhoukoudian.[40] In Niah Cave, Borneo, a cranium and leg bones date from the same period (Table 9.2).

For most of the last 100 KY, the level of the sea has been lower than at present and Australia was part of a larger continent that included New Guinea and Tasmania, known as **Sahul**. Despite this, Sahul was always separated from Asia (**Sunda**) by a significant barrier: >90 km (56 miles) of water even when sea level was at its lowest (shown in Figure 11.14). The biological significance of this barrier is illustrated by the maintenance of very distinct faunas, with kangaroos, koalas, and platypus on the Australian side of the **Wallace line** and apes, tigers, and pigs on the Asian side. Entry into Australia was thus impossible for most animals, apparently including even the *H. erectus* present in Southeast Asia from ~1.8 MYA, so the crossing of this distance of water has itself been interpreted as a sign of modern human technology. LM3 (Lake Mungo 3), a red ochre-covered burial discovered in New South Wales in the southern part of Australia in 1974 (**Figure 11.6**), has been dated using optical methods to around 40 ± 2 KYA[9] (Table 9.2). The nearby LM1, a fragmentary female cremation, was assigned the same date by the same study. Thus there is plausible evidence for human fossils in Australia by 40 KYA, associated with varied rituals (and see **Opinion Box 11**). There is considerable morphological diversity among these and other Australian fossils. While this could arise from multiple colonization events, variation in a single population due to drift or selection is a more plausible explanation, but the uncertain chronology of many of the specimens makes it difficult to identify a temporal pattern or relationships.

Europe is the most intensively studied region, and as a result, the earliest accepted date of 43–45 KYA for two molars from the Grotta del Cavallo, southern Italy, based on associated shell beads[7], occurs around the time of entry of modern humans, as judged from the associated archaeological remains [**Uluzzian**, a pre-**Aurignacian** culture from Italy at the beginning of the **Upper Paleolithic** (UP)], and is among the oldest documented outside Africa. This is not an isolated finding: an upper jaw fragment from Kent's Cavern, southwest England, was dated from associated faunal material to 41–44 KYA,[19] implying that modern humans were widely spread in Europe by this time. The earliest directly dated modern human bone is from Peştera cu Oase, Romania, at 38–42 KYA.[45]

In conclusion, accepted dates for currently known fully modern human fossil material outside Africa are around 45 KYA or later, but the record is so incomplete in most regions that these dates are often unlikely to reflect the real time of appearance, and little significance should be attached to the apparent differences between continents. Nevertheless, these are much later than the dates in Africa and the Middle East.

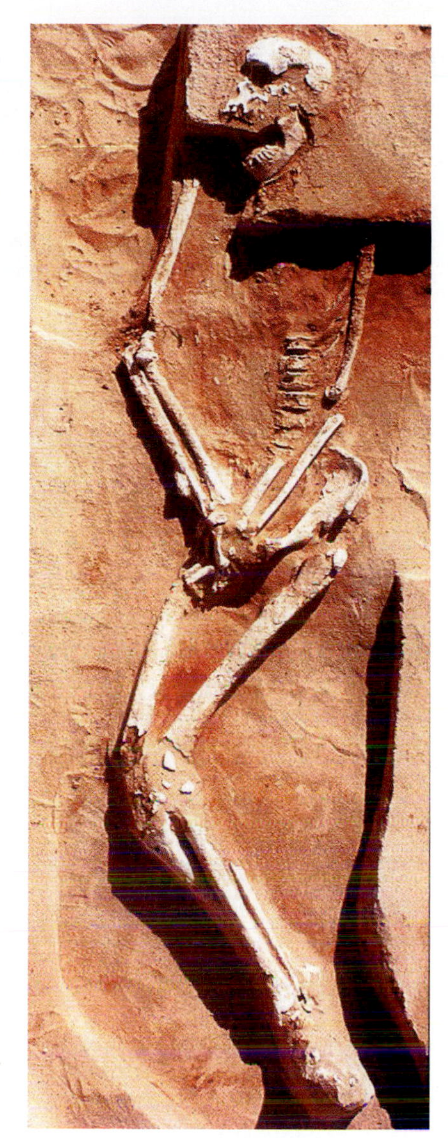

Figure 11.6: Lake Mungo 3, a well-preserved early Australian skeleton dated to ~42 KYA.

Initial colonization of Australia

In addition to the human fossils discussed in the last section, there have been substantial archaeological studies in Australia examining the initial colonization. Thermoluminescence (TL) dating of Malakunanja II in northern Australia—a 4-m-deep sand deposit containing artifacts such as flakes, red and yellow ochre, and a grindstone in the upper 2.6 m, and an absence of such artifacts lower down—provided a date of ~50 KYA for the lowest occupation level.[36] The study mentioned above which provided dates of ~42 KYA for the LM1 and LM3 fossils[9] suggested that the earliest occupation of the Lake Mungo site in southern Australia dated to 50–46 KYA. Older dates have been reported several times, but are controversial (see Opinion Box 11.1). By 35 KYA, however, sites are found in Tasmania, so most of the continent except the central desert seems to have been occupied by this time.

It is striking that, during the Pleistocene, Australia was populated by 24 genera of large marsupials, including 3-m-high kangaroos, reptiles and birds, and 1-tonne carnivorous lizards; but 23 of these megafauna genera (including 55 marsupial species weighing >10 kg) became extinct at around 46 KYA (95% confidence interval, 40–51 KYA).[35] Although the cause of this massive extinction continues to be debated (see Box 13.1), the balance of evidence attributes it to humans, either directly through hunting or indirectly through alterations to the environment such as burning. A detailed study at Lynch's Crater (Queensland) with a time resolution of ~100 years showed that megafauna loss (measured using *Sporormiella* fungus spores that are dependent on herbivore dung) preceded the appearance of charcoal (signaling environmental burning), indicating a direct role for hunting. The link between the extinctions and human activity suggests a substantial human presence by 46 KYA.

The stone tools used by the Australians from 45 KYA until as recently as 5–7 KYA are simple flake-based forms, quite unlike the finer blade-based UP technologies. If the people who migrated out of Africa were using finely made tools, why

OPINION BOX 11: When did modern humans first colonize Southeast Asia and Australia?

When modern humans first left Africa and entered Eurasia, they spread eastward, along the rim of the Indian Ocean. Australia lies at the end of this arc of dispersal, and our ancestors needed advanced planning capabilities and watercraft to safely island-hop through Southeast Asia and make landfall in northern Australia. Knowing when *Homo sapiens* first colonized this island continent has long been viewed, therefore, as providing a minimum date for the emergence of the cognitive skills and behaviors usually associated with our species.

Dating the time of human arrival has, however, not been straightforward. Radiocarbon dating has traditionally been the method of choice for the last 40 KY or so of human history. But radiocarbon dating cannot reliably extend back much further—at least not with the techniques available in the late 1980s, when I began working in northern Australia with Rhys Jones and Mike Smith. The 40 KY ceiling posed a conundrum for Australian archaeologists, because it was becoming clear that the continent had already been colonized by that time. So we took a different approach and used **thermoluminescence** (TL) to date the time of deposition of sediments at Malakunanja II rock shelter. The ages of 50 to 60 KY for the sediment layers containing the oldest stone tools, ochre, and ground hematite proved controversial. They extended the accepted time of human occupation of Australia by 10–20 KY, conflicting with the expectations of some archaeologists that modern human behavior emerged much later, along with the initial wave of migration of *Homo sapiens* out of Africa.

Optically stimulated luminescence (OSL) dating has since helped to largely resolve this controversy. OSL is better suited than TL for dating sediments exposed to sunlight before deposition, and analysis of individual grains of sand allows issues of post-depositional mixing and other site-formation processes to be investigated.[21] I first used this method to date the sediments surrounding the oldest stone tools and ground hematite at Nauwalabila I (**Figure 1 and 2**)—a rock shelter located close to Malakunanja II (**Figure 3**)—and obtained similar ages of 53–60 KY. Single-grain OSL dating of Malakunanja II subsequently confirmed that it was first occupied more than 50 KY ago, and that sediment mixing was not a significant problem. But the benefits of single-grain dating came to the fore at another site in northern Australia—Jinmium rock shelter—where claims had been made for human colonization by 120 KYA. By dating individual grains of quartz sand, I showed that the archaeological deposit had been contaminated by decomposed rubble and that the artifacts were an order-of-magnitude younger than proposed originally.

OSL dating studies at Devil's Lair and Lake Mungo in southern Australia have also added weight to the case for continental colonization before 50 KYA, and advances in methods of charcoal preparation for radiocarbon dating have resulted in ages of between 45 and 50 KY for several sites in Australia, neighboring Papua New Guinea, and at Niah Cave in Borneo. A date of 50 KYA is now widely accepted as the latest possible time for human arrival in Australia,[20] but some tantalizing questions remain. The oldest reliably dated skeletal remains of anatomically modern humans in Australia (Lake Mungo) and Southeast Asia (Niah Cave) are only 40 KY in age, and both sites also have stone tools at least 5 KY older. Is it safe to assume that *Homo sapiens* made these tools, given the evidence of genetic admixture between modern humans and the enigmatic Denisovans in Southeast Asia? Similarly, given the presence of Denisovan DNA in Aboriginal Australians, perhaps it is timely to ask which hominin group made the 50–60 KY-old artifacts at the northern Australian sites, where skeletal remains are absent and the earliest stone tools are several millennia older than any found further south? Also, the diminutive "hobbits" (*Homo floresiensis*) survived until at least 60 KYA on the Indonesian island of Flores, while the human metatarsal found at Callao Cave in the Philippines—and dated to 67 KY—raises further questions about the diversity and distribution of hominin populations in Southeast Asia during the period in which pioneering modern humans were dispersing through the region. There are several intriguing possibilities, therefore, of interactions between different hominin groups that are ripe for enquiry and that may radically revise our views of which group was the first to reach Australia's shores.

Richard "Bert" Roberts, Centre for Archaeological Science, University of Wollongong, Australia

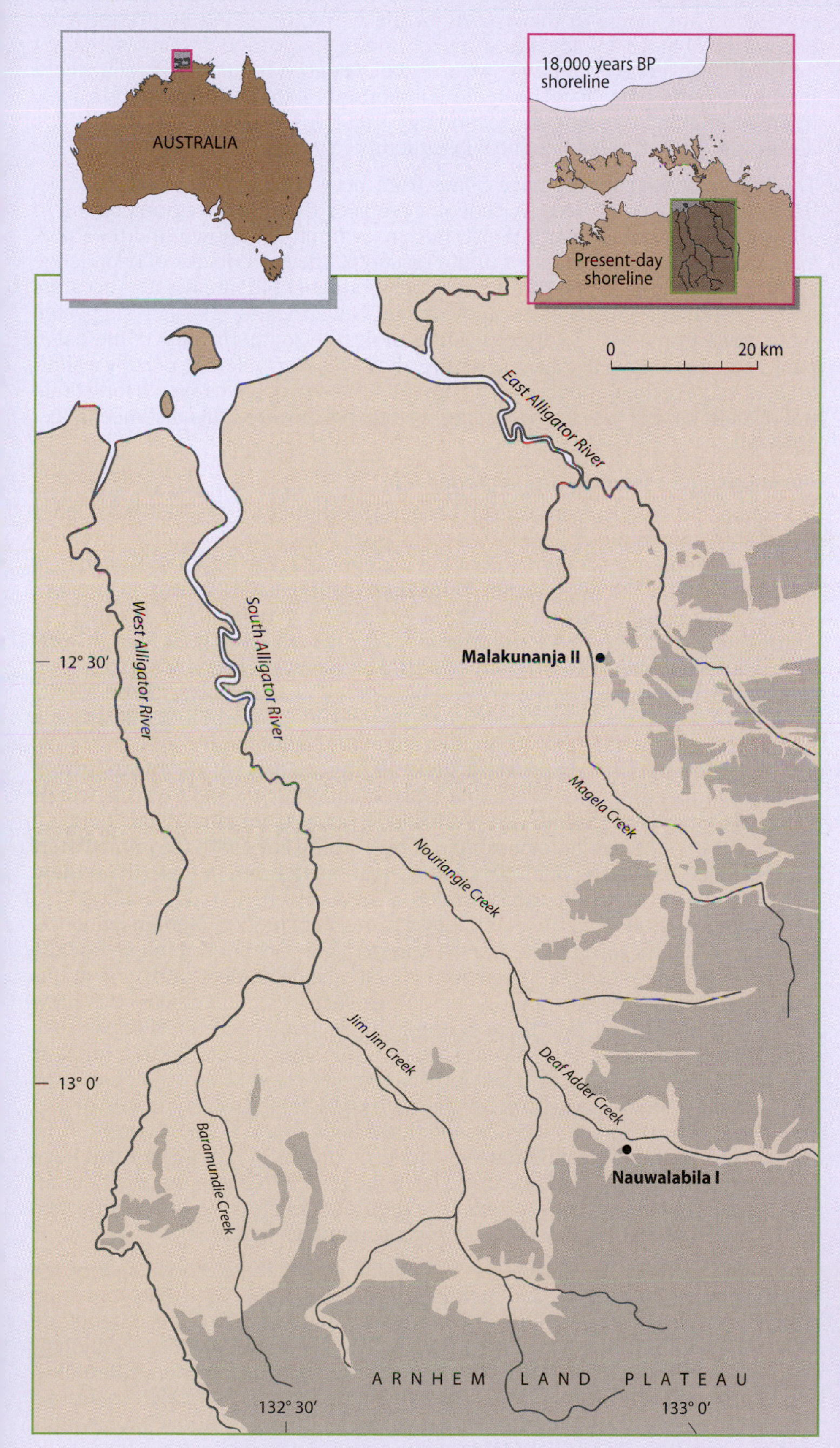

Figure 1: Map showing locations of some important sites in northern Australia.

Figure 2: The Nauwalabila I site.

Figure 3: The Malakunanja II site.

did their descendants in Australia use much cruder ones? Three relevant factors have been suggested. First, the fine-grained rocks such as flint needed for blade production are scarce in many parts of the world, including Southeast Asia. Second, fine blades are advantageous for hunting large land mammals, but not necessarily advantageous along the postulated coastal route to Australia where marine resources may have been more important. Third, **cultural drift** is likely in small migrating groups, so technology traditions could be lost and in the absence of contact would not be subsequently regained.[25]

The first Australians must have come from nearby Southeast Asia, and earlier probably via South Asia. Archaeological sites from these regions dating to ≥50 KYA would be of great interest, but are currently unknown and thus provide important opportunities for future research. The current lack of knowledge about such early sites makes it impossible to identify the immediate ancestors of the Australians and trace their movements. Nevertheless, it is often suggested that a route from East or Northeast Africa along the southern coast of the Asian continent was followed; this would have allowed the travelers to occupy a similar ecological niche throughout. The migration need not have taken a long time: travel at 16 km per year, for example, would have covered this distance in just 1000 years.

Upper Paleolithic transition in Europe and Asia

In Europe and nearby, there was an abrupt transition from the Middle to Upper Paleolithic, sometimes referred to as the "Upper Paleolithic revolution" because of the rapid and extensive nature of the changes in some regions. The first UP traditions include the Aurignacian, **Châtelperronian**, and Uluzzian. In Western Asia, the Aurignacian appears at about 47–49 KYA at sites in the Levant.[27] In Europe, the spread of the UP occurred a few thousand years later than in West Asia, with the earliest sites in southeastern Europe at ~46 KYA and rapid spread reaching Spain in ~5 KY, by 41 KYA,[26] a period that included a warm interstadial 41–43 KYA which could have facilitated the spread. It can be interpreted as a transition from Neanderthals to modern humans. Considerable archaeological and fossil data are available, and it is generally agreed that the MP (**Mousterian**) tradition at this time was associated with Neanderthals. Similarly, it is widely believed that UP Aurignacian was produced by modern humans. While there is a consensus that the UP Châtelperronian shows strong links with the immediately preceding Mousterian technologies and was associated with Neanderthals, there has been debate about whether Neanderthals underwent an independent MP to UP transition, or adopted UP technology as a result of modern human influence, in the way that present-day hunter-gatherers might use, but not independently invent, mobile phones. It has been argued that an independent transition at this time would represent an "impossible coincidence."[24] Nevertheless, this view is not universally held, and there is, for example, evidence from Aviones cave in southern Spain, dating to 45–50 KYA and thus well before modern humans reached Iberia, for signs of behavioral modernity (Box 11.1). Pierced marine shells and a *Spondylus gaederopus* (thorny oyster) shell showing traces of pigment containing charcoal and hematite have been found.[47] The reason for the rapid replacement of Neanderthals by modern humans in Europe has also been much debated; a key factor may have been the tenfold higher numbers of modern humans,[28] although this observation then just transfers the debate to which of their characteristics allowed these greater numbers to develop.

In Siberia, less detailed information is available, but there is evidence of MP sites dated to 70–40 KYA and Neanderthal fossils have been reported from Denisova Cave. Other archaic humans, Denisovans, were also present, but are currently identifiable only from their DNA. Their morphology and technology are poorly understood, so we cannot yet use archaeological evidence to understand how widespread they were. The transition to the UP is dated to around 42 KYA at Kara-Bom in the Altai region where UP layers lie above Mousterian, and to older than 39 KYA at Makarovo-4 near Lake Baikal, with a total of 20 sites older than 30 KYA,[12] so modern humans appear to have been widespread soon after 40 KYA. All these sites are south of 55° N latitude, but even the more northern areas with colder environments were colonized by mammoth hunters by 32 KYA, as shown by a Yana River site of this date (**Section 13.2**; Figure 13.4) which yielded sophisticated worked mammoth ivory and rhinoceros horn characteristic of modern humans.[13]

In this section, it will have become apparent that the fossil and archaeological records are very sparse in many parts of the world, and that our current conclusions are critically dependent on the chronology, which is often disputed. Thus, future discoveries or recalibrations may drastically change our view of human presence during this crucial period.

At present, our conclusions are that hominins probably expanded out of Africa whenever conditions permitted, including during the last interglacial, but that these pioneers did not necessarily persist through the subsequent cold periods. Fully modern humans expanded later, some time before 50 KYA. This last expansion populated most of the Old World and Australia by 40 KYA, except for the most northerly regions.

11.3 A SINGLE MAJOR MIGRATION OUT OF AFRICA 50–70 KYA

We now move from the consideration of the background information to an explicit examination of key questions about the expansion of fully modern humans out of Africa: (1) How many migrations were there? (2) What route(s) did they take? (3) When did they occur? (4) Where and when did admixture with archaic humans take place? The question about route(s) can be further subdivided into consideration of the route(s) out of Africa, and the route(s) into more distant parts of the world.

We turn to genetic information as the primary source of insights, although we will frequently refer back to other types of evidence. Genetic analyses can provide information about dates, and these are often calibrated using a particular mutation rate. Estimates of the autosomal mutation rate differ according to whether they are based on direct measurements in modern families, or inferred indirectly on the basis of sequence divergence between humans and other species, with consequences for dating that are explored in **Opinion Box 11.2**.

Populations outside Africa carry a shared subset of African genetic diversity with minor Neanderthal admixture

Genetic comparisons of African and non-African populations provide two simple and powerful conclusions: non-Africans carry a small subset of African diversity, and this subset is shared by all populations outside Africa (Chapter 9). These are, of course, overgeneralizations. They ignore recent migrations such as those back to North and East Africa over the last few millennia and those both into and out of Africa associated with the slave trade and colonial era. They also ignore the variation that has arisen by mutation and recombination after the initial migration, and the genomic regions added as a result of archaic admixture. Nevertheless, they provide a key basis for understanding the broad features of our genetic variation. Moreover, they are supported by inferences about past population sizes from single genomes using **PSMC** analysis (Figure 9.23). Before ~100 KYA, African and non-African individuals all show the same demographic history; after that time, their population sizes diverge, with a more pronounced bottleneck in non-Africans.

These simple observations imply that there was a single ancestral African source population for non-Africans. But if the source that carried this particular subset of African diversity remained isolated for a period of time, the observations are equally consistent with one or multiple migrations out of Africa starting from the same source.

OPINION BOX 12: Implications of a revised human mutation rate for dating based on sequence divergence

Estimating the times of demographic events from sequence data depends in essence on dividing the observed sequence divergence by the mutation rate. Most estimates for humans have used an autosomal mutation rate of ~10^{-9} per base per year, derived in part from calibrating the sequence divergence to macaques and other Old World primates against dates from fossil evidence (the phylogenetic rate). Recent advances in sequencing technology have made it possible to obtain an alternative value for the mutation rate based on identifying de novo mutations in human pedigrees. The observed mutation rate per base per generation from this approach is ~$1.2–1.6 \times 10^{-8}$. Assuming a generation time of 25–30 years during human evolution, this suggests a yearly de novo mutation rate of ~0.5×10^{-9} per base per year, half that used previously.[39] Given that our current picture of human evolution is based in part on genetic analyses scaled using the higher rate, the implications of this revision need to be understood.

Although shifting population divergence dates back in time by a factor of two seems a radical proposal, there are aspects of the paleoanthropological evidence which may actually accord better with older genetic dates and a revised evolutionary time-scale. For example, the first fossil and archaeological evidence for modern humans in Europe is dated as early as 40–45 KYA, and there is similar evidence for their presence in East and Southeast Asia from the same period. Nuclear genomic estimates of the split time between Europeans and Asians are in better agreement with this when scaled using the de novo mutation rate (giving a split 40–80 KYA) than with the higher phylogenetic rate (giving a split 20–40 KYA).

Further back in time (see **Figure 1**), estimates of the genetic separation between Africans and non-Africans would be rescaled to approximately 110 KYA, thereby pre-dating the appearance of modern humans in Europe and East Asia by up to 60 KY. However, there is fossil evidence for modern human occupation of the Levant as early as 100 KYA, and growing archaeological evidence for their subsequent presence elsewhere in the Middle East.[4, 30] There are also indications that modern human dispersal into Asia may have begun earlier than previously thought, from recent paleoanthropological finds in India and Southeast Asia[32, 41] and also from genetic data, specifically from sequencing of an Aboriginal Australian genome[33] and the finding of differing levels of archaic Denisovan admixture in present-day Asian populations.[29] In this context, a much earlier African/non-African split may reflect a scenario in which the modern human exodus from Africa occurred gradually, via an intermediate population in East Africa and the Middle East, over the period 100–60 KYA. Such a model is also more compatible with several other aspects of the genetic data. One example is the inference of continued gene flow between the ancestors of Africans and non-Africans for an extended period after divergence,[14] which would be difficult to sustain if divergence was followed immediately by dispersal across Eurasia. Another is the fact that the ancestors of all non-African populations seem to have experienced admixture with Neanderthals: an extended period of neighboring or overlapping presence in the Middle East provides a longer interval for this to have occurred.

Looking further back still, the revised chronology hints at significant changes in our interpretation of genetic diversity from the earliest phase of human origins. Studies of worldwide mtDNA and Y variation have estimated times to the most recent common human ancestor at those loci of around 150–200 KY. However, with the revised mutation rate, nuclear DNA comparisons between the **Khoe-San** people of the Kalahari Desert and other Africans (the deepest split within the human tree) suggest a divergence 250–300 KYA. Such discordance between single-locus and genomewide estimates could arise from a number of factors, such as differences in ancestry sampling, or introgression of haplotypes between ancient African populations. The sparse fossil evidence from that period suggests a transition from archaic to modern human forms but places few constraints on early demography. Given this limited amount of information, it seems likely that future discoveries, better sampling of African genetic diversity, and perhaps more data from ancient DNA will continue to suggest further significant changes in our picture of human evolution.

Aylwyn Scally and Richard Durbin,
Wellcome Trust Sanger Institute, UK

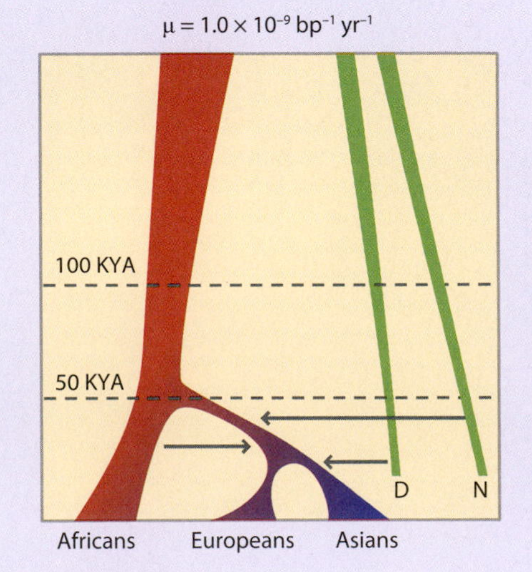

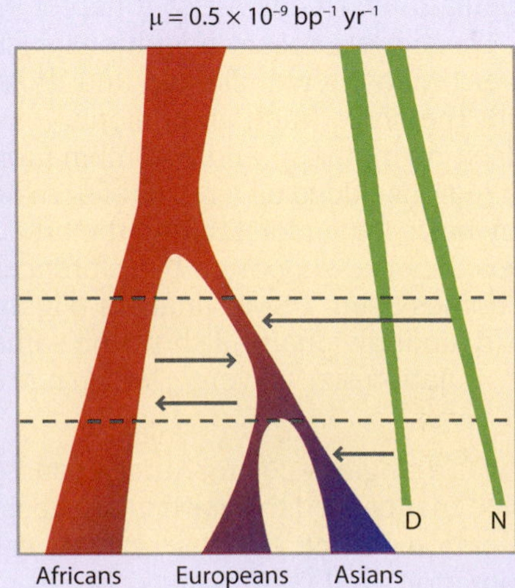

Figure 1: Implications of revising the mutation rate.
Implications of revising the mutation rate from the faster phylogenetic rate (*left*) to the slower pedigree rate (*right*). Dates of fossils (D and N) are unchanged, but inferred population divergence times are substantially older.

Archaic genetic material can be used to study ancient events "much as a medical imaging dye injected into a patient allows the tracing of blood vessels."[34] The archaic regions originating from an admixture event can be detected in descendants, and thus used to identify these descendants. We have seen that Neanderthal admixture has been detected at the same low level (~1.3%) in all non-African populations examined.[15] This is interpreted as the result of a single admixture event, and thus supports the idea of a single source population for all non-Africans. And, since evidence for Neanderthals inside Africa has never been found in the African fossil record, it implies that the common source population for all non-Africans was already living outside Africa at the time of admixture. It is, however, difficult to formulate a model that reconciles these genetic conclusions with all the additional evidence. We will see below that many geneticists favor the hypothesis of a southern route out of Africa (Figure 11.3) for fully modern humans, but according to the available fossil evidence for the distribution of Neanderthals, people migrating along a southern route would not have encountered Neanderthals. Possible explanations (**Figure 11.7**) for this conundrum are:

- The migration used the southern route 50–70 KYA, but the distribution of Neanderthals extended further to the south than currently documented (Figure 11.7, top).
- The migration occurred 50–70 KYA, but used the northern route, where Neanderthals could have been encountered in the Levant (Figure 11.7, middle).
- The admixture and migration were separate events, with admixture occurring in the Levant during the first expansion of anatomically modern but behaviorally pre-modern humans ~100 KYA. The admixed population retreated to Africa, remained isolated from other African populations, and migrated out of Africa using the southern route 50–70 KYA (Figure 11.7, bottom).

The extent of LD between SNPs shared between modern humans and Neanderthals suggested admixture 37–86 KYA,[38] making the third explanation unlikely. Further work is needed to distinguish between the first two explanations: for example, more fossil/archaeological surveys of this under-studied region.

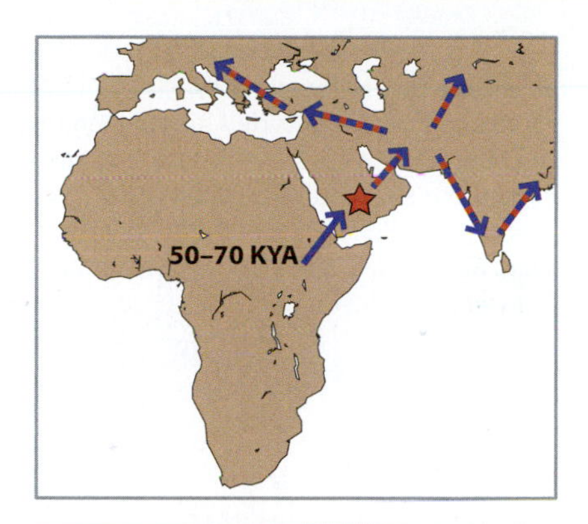

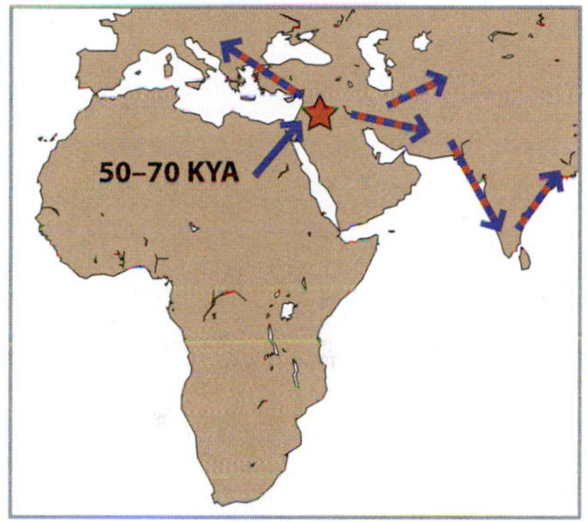

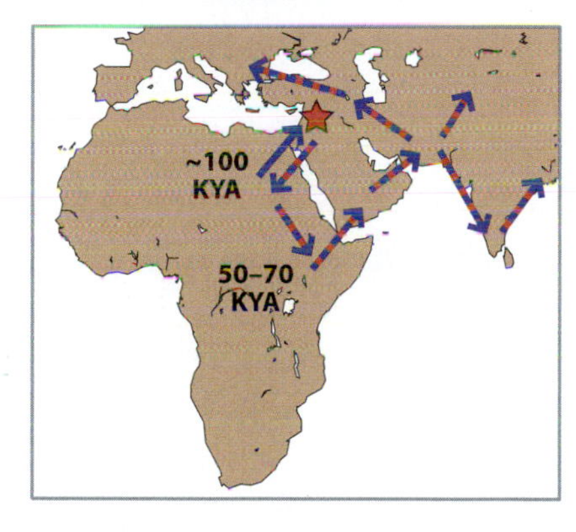

Figure 11.7: Models for the movement of modern humans out of Africa and their admixture with Neanderthals that reconcile genetic, archaeological, and fossil data.
Migration of African population shown by *blue* arrow, *red* star indicates the possible site of admixture, and *red/blue* arrows indicate migration of admixed population.

mtDNA and Y-chromosomal studies show the descent of all non-African lineages from a single ancestor for each who lived 55–75 KYA

The two haploid lineages (mtDNA and the Y chromosome, see **Appendix**) have been particularly informative. Both provide highly resolved haplotype trees in which the branches show high levels of geographical resolution, so that African lineages can generally be distinguished from non-African ones. In addition, the branch lengths of trees based on sequence information provide information about time-depth, so that estimates of the times of nodes in the tree, and thus of events related to them, can be made. For both loci, as we have seen, the root and most deep-rooting branches lie in Africa. Here, we examine more detailed phylogenies of these loci, and in particular consider what they can tell us about the timing of the out-of-Africa migration.

An mtDNA phylogeny based on 18,843 complete mtDNA sequences[6] represents world mtDNA diversity effectively (**Figure 11.8**). Almost all African mtDNA lineages belong to L clades, and non-African lineages to three clades M, N, and R that all descend from L3. This again supports a single source for all non-Africans. Other L3 descendant lineages are confined to Africa, so it is likely that L3 arose in Africa before the out-of-Africa movement. M, N, and R, however, most likely originated and diversified outside Africa. According to this logic, the migration occurred between the times of the origin of the L3 node and the oldest of the M, N, and R nodes, node N (Figure 11.8). The dates of these nodes have therefore been examined in some detail. A study focusing on the emergence of L3 used 328 complete sequences representing the known sublineages

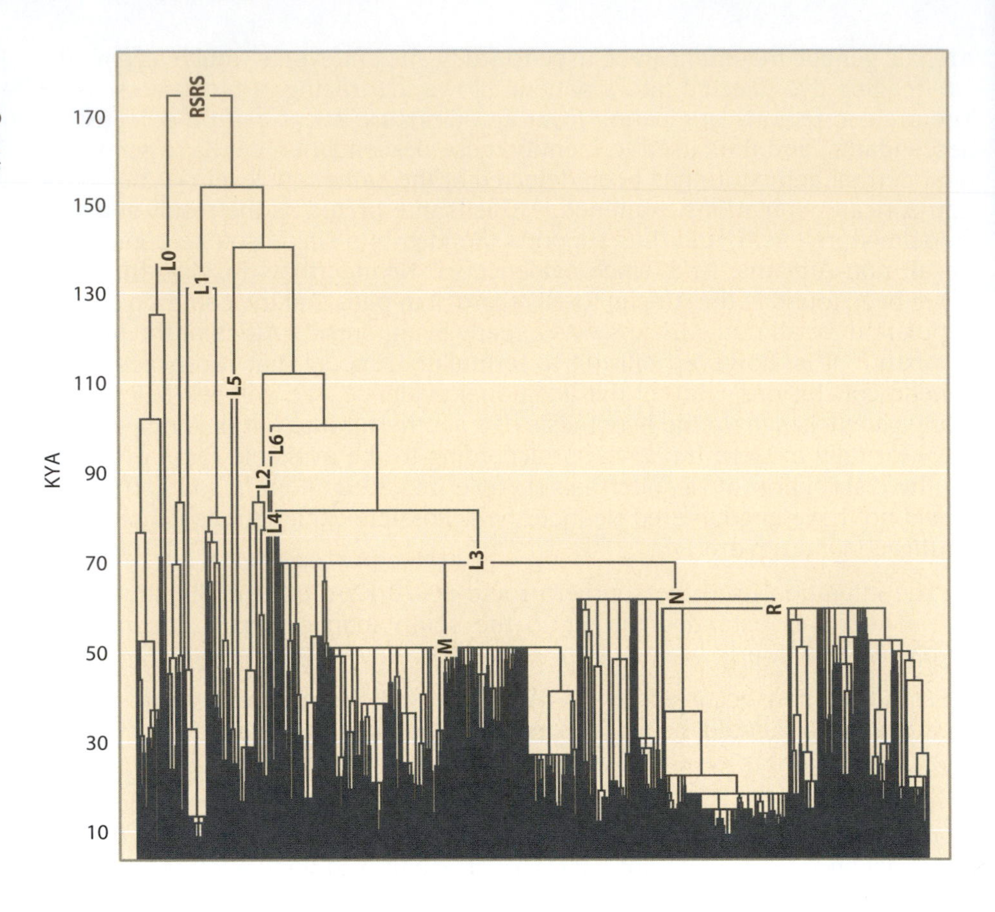

Figure 11.8: Phylogeny of 18,843 complete mtDNA sequences. Almost all lineages outside Africa belong to the M, N, and R clades, which show marked expansions soon after the expansion out of Africa. [From Behar DM et al. (2012) *Am. J. Hum. Genet.* 90, 675. With permission from Elsevier.]

and calculated a TMRCA in a number of ways using the **ρ statistic** or a maximum likelihood method (**Section 6.4**), assuming one mutation per 3624 years for the complete sequence.[42] Estimates ranged from 59 to 70 KYA, with a time of ~65 KYA considered most plausible, and an origin in East Africa. Applying the same approach to 385 complete N sequences suggested a time of ~61 KYA (range 57–65 KYA).[11] These two values together constrain the most likely time for the movement out of Africa to the interval of 61–65 KYA. While each date is associated with known uncertainty of several thousand years, and there may be systematic error in some assumptions such as the mutation rate, this interval provides the most carefully reasoned estimate for the time of the exit from Africa currently available.

The Y chromosome phylogeny provides, in principle, similar information. Analyses of a phylogeny based on complete sequences of 29 Y chromosomes identified a node that gave rise to all lineages outside Africa (and some inside Africa; **Figure 11.9**). Application of the same ρ statistic provided an estimate of ~57 KYA, although alternative dating methods applied to the same data (GENETREE and BEAST—Table 6.3) suggested slightly earlier dates of ~60–75 KYA.[46] The mutation rate used to calibrate these Y estimates has wide confidence intervals, but they support the mtDNA estimates. Y-based analyses are benefitting from an explosion of data from next-generation sequencing, and are expected to improve considerably in the near future.

In addition to providing dates to constrain the time of the out-of-Africa migration, the structures of both phylogenies are informative about demographic events. In the mtDNA tree, extensive **multifurcations** in the M, N, and R clades (Figure 11.8) indicate expansions in the numbers of lineages within each clade that were so close together in time that few intermediate mutations could accumulate. These may have begun in an **Arabian Cradle** where environmental conditions were more favorable 60–65 KYA,[11] or may have occurred slightly later. The Y-chromosomal phylogeny also shows evidence of a major and rapid

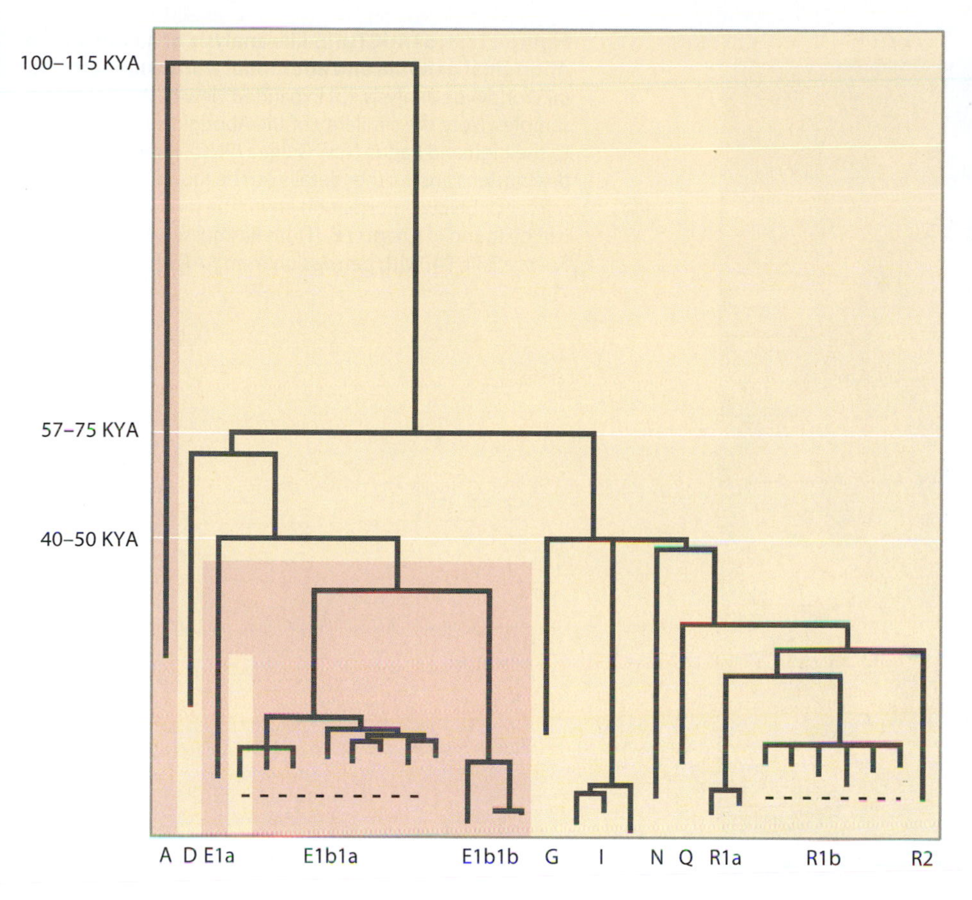

Figure 11.9: Phylogeny of 29 Y chromosomes based on full sequence data.
Pink: Africa. *Yellow*: outside Africa. Haplogroup names are given at the bottom. African lineages are poorly represented in the sample, and the most deep-rooting A sublineages are not included, so the TMRCA of the sample (100–115 KYA) does not represent the TMRCA of all extant Y chromosomes. In contrast, the TMRCAs of the branch that includes both African and non-African chromosomes (57–75 KYA) and the later expansion (40–50 KYA) are representative. [Data from Wei W et al. (2013) *Genome Res.* 23, 388.]

expansion of male lineages postdating the out-of-Africa movement, with this later expansion estimated to have occurred at ~40–50 KYA.[46] The authors speculated that this might reflect successful adaptation to inland, compared with coastal, environments.

11.4 EARLY POPULATION DIVERGENCE BETWEEN AUSTRALIANS AND EURASIANS

Given the archaeological evidence for modern human presence in Australia by ~50 KYA, analysis of Australian Aborigine genomes and the history of Australian populations is of great interest. Such studies are potentially complicated by the factors discussed in Chapter 10 (**Sections 10.1 and 10.2**) and by the mixed ancestry of many present-day Australian Aborigines, which generally includes European admixture. However, in the 1920s, a young man from Golden Ridge, near Kalgoorlie, Western Australia, donated a lock of hair to a visiting anthropologist, which was then preserved as part of the Duckworth Laboratory collections in Cambridge, UK (**Figure 11.10**). In 2011, genetic analysis of this sample was approved by Aboriginal representatives from the area where the sample was collected. Short DNA fragments with a mean length of 69 bp were successfully extracted and a **low-coverage** (6.4×) genome sequence was generated.[33] Levels of contamination were low, and both the mtDNA (haplogroup O1a) and Y (haplogroup K*) lineages fell within known Australian Aboriginal branches. It was therefore possible to investigate questions about population history in an unbiased way, starting from genomewide variants ascertained by sequencing.

PC and STRUCTURE-like analyses (see **Section 6.3**) demonstrated that the Aboriginal genome was most closely related to Papua New Guinea highlanders, among the populations used for comparison (**Figure 11.11**). This result fits expectations based on our understanding of the environment 50 KYA, since Australia and Papua New Guinea were connected, as part of the same larger landmass,

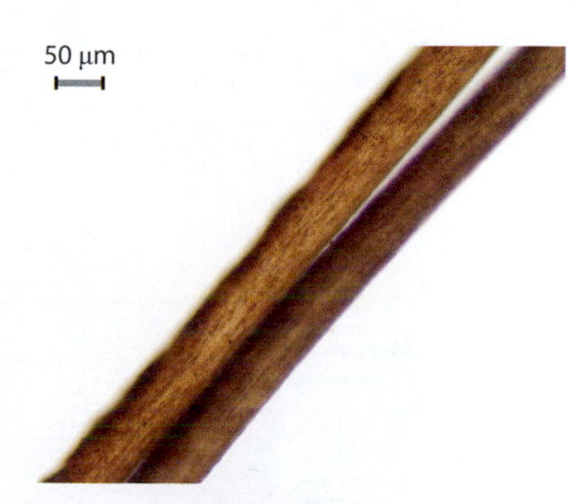

Figure 11.10: Australian Aboriginal hair used to generate a genome sequence. [From Rasmussen M et al. (2011) *Science* 334, 94. With permission from AAAS.]

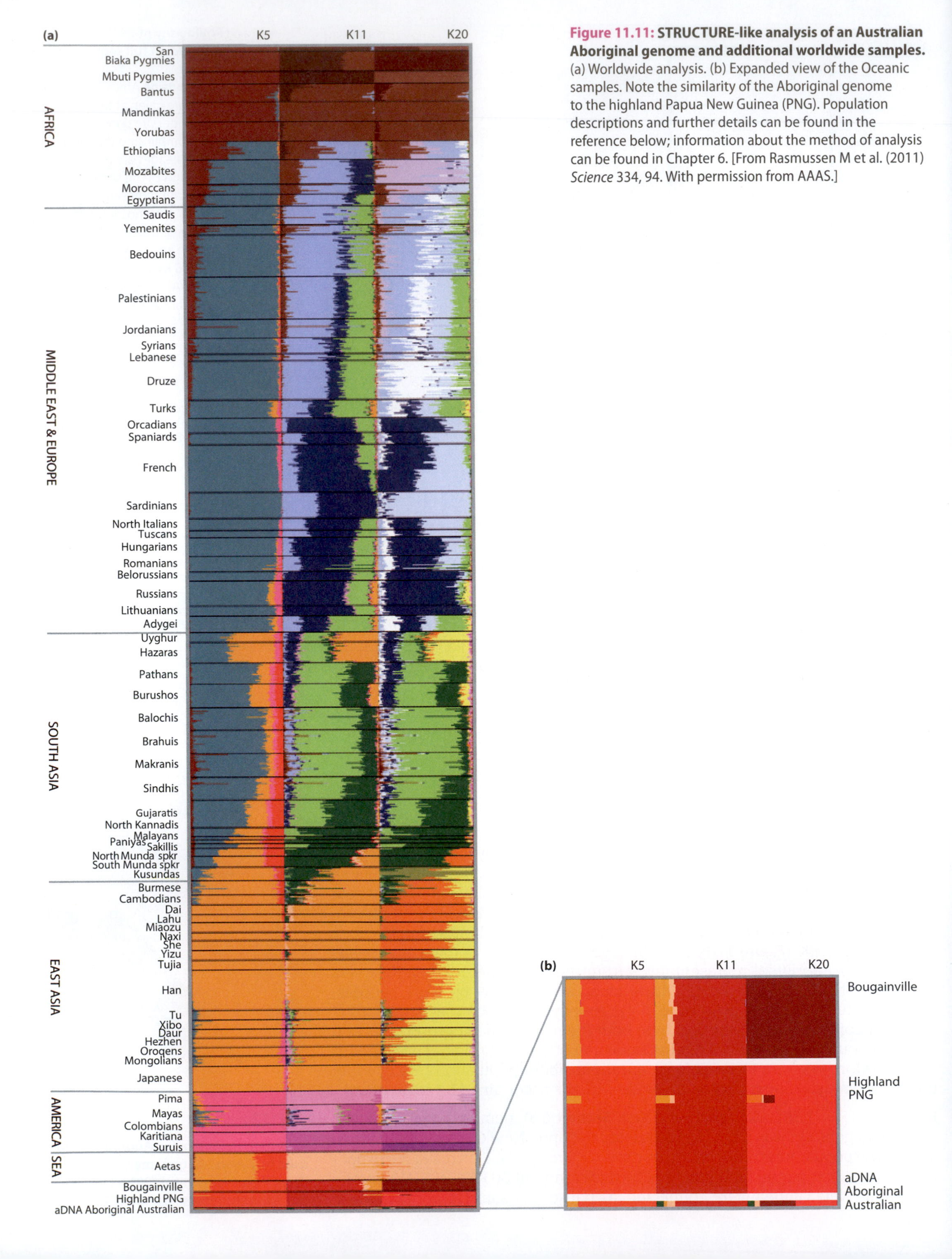

Figure 11.11: STRUCTURE-like analysis of an Australian Aboriginal genome and additional worldwide samples. (a) Worldwide analysis. (b) Expanded view of the Oceanic samples. Note the similarity of the Aboriginal genome to the highland Papua New Guinea (PNG). Population descriptions and further details can be found in the reference below; information about the method of analysis can be found in Chapter 6. [From Rasmussen M et al. (2011) *Science* 334, 94. With permission from AAAS.]

Sahul, at that time. It also provides additional confidence in the authenticity of the aDNA sequence and the lack of admixture of the donor. The authors were then able to use individual genome sequences to compare simple demographic models of the expansion outside Africa. Was the initial split between Australians on the one hand, and Asians plus Europeans on the other (**Figure 11.12**, Model 1), or between Australians plus Asians compared with Europeans (Figure 11.12, Model 2)? A version of **Patterson's *D* test** (**Section 6.3**) was used to compare the Australian Aboriginal genome with genomes from Africa (Yoruba, YRI), Asia (China, CHS), and Europe (Western Europe, CEU). The test favored Model 1, supporting the idea of an initial split between Australians (probably including Papuans) and the rest of the populations of Europe and Asia.

Denisovan admixture has also been highly informative for tracing some of the earliest population splits. A study of 33 populations from Asia and Oceania estimated levels of Denisovan admixture at up to ~7%.[34] The highest levels were detected in Papuans and Australians, and a lower level (due to dilution by recent gene flow from populations without Denisovan admixture) in Mamanwa from the Philippines (**Figure 11.13**); there were also detectable levels in nearby populations with ancestry from these sources, identifiable because the level of Denisovan ancestry was proportional to the level of Near Oceanian ancestry. A significant negative finding was the lack of Denisovan admixture in Andaman Islanders,[34] seen as direct descendants of the 50–70 KYA migration out of Africa.[44]

These findings imply that the early branch of humans ancestral to Australian Aborigines encountered Denisovans. If the southern coastal route was used, Denisovans must have been present far to the south of the one currently known fossil site, Denisova Cave in Siberia. The location of the encounter is entirely unknown, but on geographical grounds, Southeast Asia/Near Oceania north/west of the Wallace line would be obvious candidates.

Our current view of the most likely timings and routes of the initial expansions of fully modern humans into the Old World and Australia are summarized in

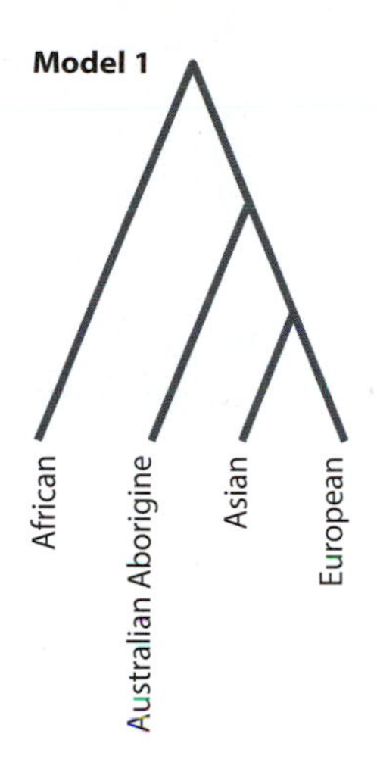

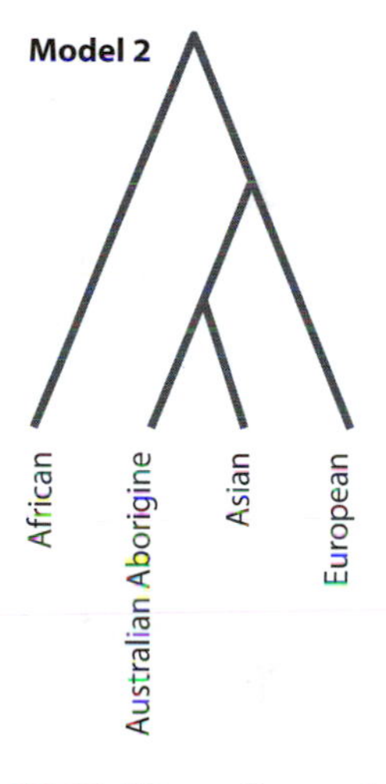

Figure 11.12: Alternative models for the initial population split outside Africa. Model 1: between Australian Aborigines and (Asians + Europeans). Model 2: between (Australian Aborigines + Asians) and Europeans. Genetic evidence favors Model 1.

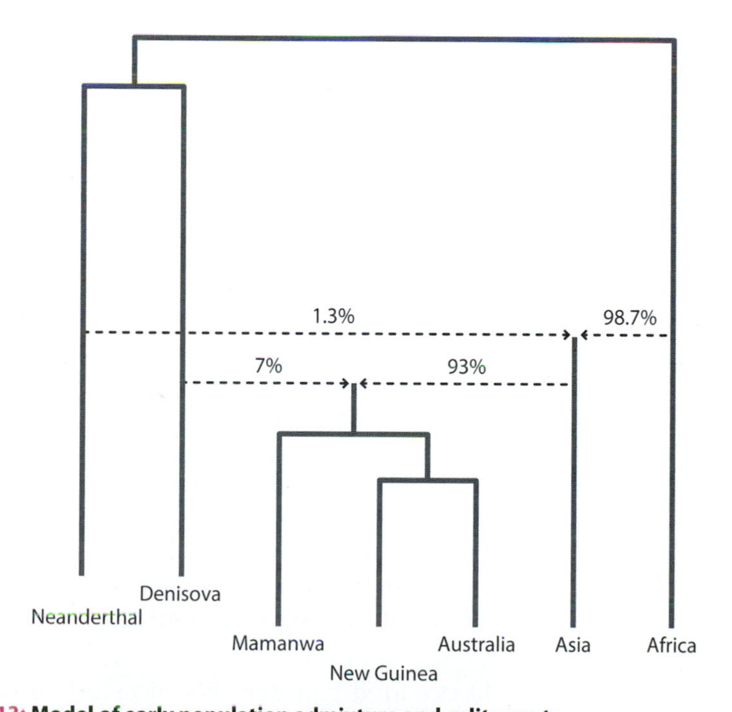

Figure 11.13: Model of early population admixture and split events. Model of early population admixture (*dotted lines*) and split (*solid lines*) following the migration out of Africa. Note the 1.3% Neanderthal contribution to all non-African populations and 7% Denisovan contribution to the populations from the Philippines (Mamanwa), New Guinea, and Australia. Early events are toward the top, later events toward the bottom, but are not drawn to scale. [Adapted from Reich D et al. (2011) *Am. J. Hum. Genet.* 89, 516. With permission from Elsevier.]

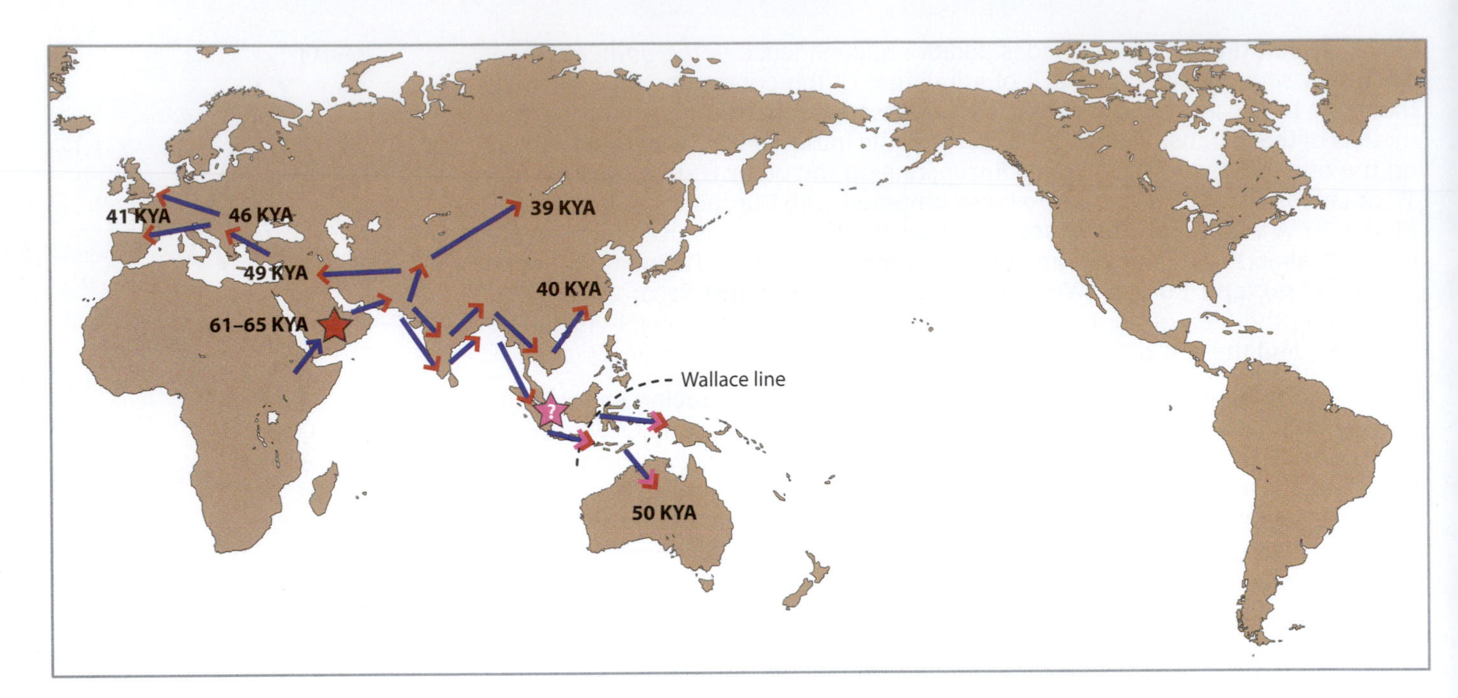

Figure 11.14: Initial expansion of fully modern humans into the Old World and Australia.
Note the single exit from Africa, and the early split of this non-African lineage into one lineage leading to Australia and New Guinea and a second lineage leading to both East Asia and Europe. There may have been gene flow between the Australia- and East Asia-bound lineages. *Red* star: admixture with Neanderthals. *Pink* star: admixture with Denisovans.

Figure 11.14. We will not describe subsequent demographic events in a comprehensive way, but in the remaining chapters of this section we will focus on a limited number of events and processes of particular importance for understanding human evolutionary genetics: the development of agriculture, the peopling of the last major areas of the globe, and the consequences of distinct populations meeting.

SUMMARY

- Climate has had a major influence on the spread of humans: after the last warm interglacial period (130–110 KYA), the climate cooled and was very unstable, with glacial maxima at 70–55 and 23–14 KYA. Only after 12 KYA did the climate become warm and stable in the way that we now consider normal.
- Fossil and archaeological evidence suggest that modern humans expanded out of Africa whenever the climate allowed. Anatomically modern but behaviorally pre-modern humans expanded into the Levant and Arabian Peninsula during the last interglacial 90–120 KYA. However, this early expansion appears not to have persisted.
- Archaeological evidence for modern human behavior, including the use of complex technologies, pigments, and art, is found in Africa 60–100 KYA, but outside Africa only after this period. Thus behavioral modernity probably developed gradually in Africa between 100 KYA and 60 KYA.
- Fully modern humans expanded out of Africa 50–70 KYA and had reached Australia and most of the Old World, except for the most northerly regions, by 40 KYA.
- Genetic evidence suggests that there was a single expansion out of Africa at this time, involving admixture with Neanderthals immediately after the exit, so that ~1.3% of the genome of all people outside Africa is of Neanderthal origin.
- The initial population split occurred between the ancestors of Australian Aborigines, some Papuans, and nearby populations on the one hand, and the ancestors of other Eurasians on the other.
- A second admixture event took place with Denisovans, probably when humans reached Southeast Asia, so that ~7% of the genome of Australian Aborigines, Papuans, and related populations is of Denisovan origin.

QUESTIONS

Question 11–1: There is archaeological evidence for the presence of modern humans outside Africa (in the Middle East and Arabian Peninsula) during the period 90–120 KYA. How does this fit into our understanding of the origins and spread of modern humans?

Question 11–2: What is behavioral modernity, and what do we know about where and when it evolved? How could its genetic basis (if any) be investigated?

Question 11–3: When did fully modern humans expand out of Africa and what routes did they take? What future evidence from archaeology or genetics could be most important in resolving current uncertainties?

Question 11–4: 'Human expansions have been the main cause of the megafauna extinctions that occurred soon after their arrival'. Discuss the evidence for and against this claim.

Question 11–5: Denisovan fossils are currently known only from Siberia, while present-day populations with Denisovan admixture occur much further south (see Figure 11.14). How can these observations be reconciled? What additional information would be most useful for this?

REFERENCES

The references highlighted in purple are considered to be important (for this chapter) by the authors.

1. **Adams J, Maslin M & Thomas E** (1999) Sudden climate transitions during the Quaternary. *Prog. Physical Geog.* **23**, 1–36.
2. **Ambrose SH** (1998) Chronology of the Later Stone Age and food production in East Africa. *J. Archaeol. Sci.* **25**, 377–392.
3. **Ambrose SH** (1998) Late Pleistocene human population bottlenecks, volcanic winter, and differentiation of modern humans. *J. Hum. Evol.* **34**, 623–651.
4. **Armitage SJ, Jasim SA, Marks AE et al.** (2011) The southern route "out of Africa": evidence for an early expansion of modern humans into Arabia. *Science* **331**, 453–456.
5. **Beck JW, Richards DA, Edwards RL et al.** (2001) Extremely large variations of atmospheric 14C concentration during the last glacial period. *Science* **292**, 2453–2458.
6. **Behar DM, van Oven M, Rosset S et al.** (2012) A "Copernican" reassessment of the human mitochondrial DNA tree from its root. *Am. J. Hum. Genet.* **90**, 675–684.
7. **Benazzi S, Douka K, Fornai C et al.** (2011) Early dispersal of modern humans in Europe and implications for Neanderthal behaviour. *Nature* **479**, 525–528.
8. **Bouzouggar A, Barton N, Vanhaeren M et al.** (2007) 82,000-year-old shell beads from North Africa and implications for the origins of modern human behavior. *Proc. Natl Acad. Sci. USA* **104**, 9964–9969.
9. **Bowler JM, Johnston H, Olley JM et al.** (2003) New ages for human occupation and climatic change at Lake Mungo, Australia. *Nature* **421**, 837–840.
10. **Calvin WH** (2004) A Brief History of the Mind. Oxford University Press.
11. **Fernandes V, Alshamali F, Alves M et al.** (2012) The Arabian cradle: mitochondrial relicts of the first steps along the southern route out of Africa. *Am. J. Hum. Genet.* **90**, 347–355.
12. **Goebel T** (1999) Pleistocene human colonization of Siberia and peopling of the Americas: an ecological approach. *Evol. Anthropol.* **8**, 208–227.
13. **Goebel T, Waters MR & O'Rourke DH** (2008) The late Pleistocene dispersal of modern humans in the Americas. *Science* **319**, 1497–1502.
14. **Gravel S, Henn BM, Gutenkunst RN et al.** (2011) Demographic history and rare allele sharing among human populations. *Proc. Natl Acad. Sci. USA* **108**, 11983–11988.
15. **Green RE, Krause J, Briggs AW et al.** (2010) A draft sequence of the Neandertal genome. *Science* **328**, 710–722.
16. **Henshilwood C, d'Errico F, Vanhaeren M et al.** (2004) Middle Stone Age shell beads from South Africa. *Science* **304**, 404.
17. **Henshilwood CS, d'Errico F, van Niekerk KL et al.** (2011) A 100,000-year-old ochre-processing workshop at Blombos Cave, South Africa. *Science* **334**, 219–222.
18. **Henshilwood CS, d'Errico F, Yates R et al.** (2002) Emergence of modern human behavior: Middle Stone Age engravings from South Africa. *Science* **295**, 1278–1280.
19. **Higham T, Compton T, Stringer C et al.** (2011) The earliest evidence for anatomically modern humans in northwestern Europe. *Nature* **479**, 521–524.
20. **Hiscock P** (2008) Archaeology of Ancient Australia. Routledge.
21. **Jacobs Z & Roberts R** (2007) Advances in optically stimulated luminescence dating of individual grains of quartz from archeological deposits. *Evol. Anthropol.* **16**, 210–223.
22. **Jacobs Z, Roberts RG, Galbraith RF et al.** (2008) Ages for the Middle Stone Age of southern Africa: implications for human behavior and dispersal. *Science* **322**, 733–735.
23. **Liu W, Jin CZ, Zhang YQ et al.** (2010) Human remains from Zhirendong, South China, and modern human emergence in East Asia. *Proc. Natl Acad. Sci. USA* **107**, 19201–19206.
24. **Mellars P** (2005) The impossible coincidence. A single-species model for the origins of modern human behavior in Europe. *Evol. Anthropol.* **14**, 12–27.
25. **Mellars P** (2006) Going east: new genetic and archaeological perspectives on the modern human colonization of Eurasia. *Science* **313**, 796–800.
26. **Mellars P** (2006) A new radiocarbon revolution and the dispersal of modern humans in Eurasia. *Nature* **439**, 931–935.
27. **Mellars P** (2011) Palaeoanthropology: the earliest modern humans in Europe. *Nature* **479**, 483–485.
28. **Mellars P & French JC** (2011) Tenfold population increase in Western Europe at the Neandertal-to-modern human transition. *Science* **333**, 623–627.

29. **Meyer M, Kircher M, Gansauge MT et al.** (2012) A high-coverage genome sequence from an archaic Denisovan individual. *Science* **338**, 222–226.

30. **Millard AR** (2008) A critique of the chronometric evidence for hominid fossils: I. Africa and the Near East 500–50 ka. *J. Hum. Evol.* **54**, 848–874.

31. **Mourre V, Villa P & Henshilwood CS** (2010) Early use of pressure flaking on lithic artifacts at Blombos Cave, South Africa. *Science* **330**, 659–662.

32. **Petraglia M, Korisettar R, Boivin N et al.** (2007) Middle Paleolithic assemblages from the Indian subcontinent before and after the Toba super-eruption. *Science* **317**, 114–116.

33. **Rasmussen M, Guo X, Wang Y et al.** (2011) An Aboriginal Australian genome reveals separate human dispersals into Asia. *Science* **334**, 94–98.

34. **Reich D, Patterson N, Kircher M et al.** (2011) Denisova admixture and the first modern human dispersals into Southeast Asia and Oceania. *Am. J. Hum. Genet.* **89**, 516–528.

35. **Roberts RG, Flannery TF, Ayliffe LK et al.** (2001) New ages for the last Australian megafauna: continent-wide extinction about 46,000 years ago. *Science* **292**, 1888–1892.

36. **Roberts RG, Jones R & Smith MA** (1990) Thermoluminescence dating of a 50,000-year-old human occupation site in northern Australia. *Nature* **345**, 153–156.

37. **Rose JI, Usik VI, Marks AE et al.** (2011) The Nubian Complex of Dhofar, Oman: an African middle stone age industry in Southern Arabia. *PLoS One* **6**, e28239.

38. **Sankararaman S, Patterson N, Li H et al.** (2012) The date of interbreeding between Neandertals and modern humans. *PLoS Genet.* **8**, e1002947.

39. **Scally A & Durbin R** (2012) Revising the human mutation rate: implications for understanding human evolution. *Nat. Rev. Genet.* **13**, 745–753.

40. **Shang H, Tong H, Zhang S et al.** (2007) An early modern human from Tianyuan Cave, Zhoukoudian, China. *Proc. Natl Acad. Sci. USA* **104**, 6573–6578.

41. **Shen G, Wang W, Cheng H & Edwards RL** (2007) Mass spectrometric U-series dating of Laibin hominid site in Guangxi, southern China. *J. Arch. Sci.* **34**, 2109–2114.

42. **Soares P, Alshamali F, Pereira JB et al.** (2012) The Expansion of mtDNA Haplogroup L3 within and out of Africa. *Mol. Biol. Evol.* **29**, 915–927.

43. **Texier PJ, Porraz G, Parkington J et al.** (2010) A Howiesons Poort tradition of engraving ostrich eggshell containers dated to 60,000 years ago at Diepkloof Rock Shelter, South Africa. *Proc. Natl Acad. Sci. USA* **107**, 6180–6185.

44. **Thangaraj K, Chaubey G, Kivisild T et al.** (2005) Reconstructing the origin of Andaman Islanders. *Science* **308**, 996.

45. **Trinkaus E, Moldovan O, Milota S et al.** (2003) An early modern human from the Peştera cu Oase, Romania. *Proc. Natl Acad. Sci. USA* **100**, 11231–11236.

46. **Wei W, Ayub Q, Chen Y et al.** (2013) A calibrated human Y-chromosomal phylogeny based on resequencing. *Genome Res.* **23**, 388–395.

47. **Zilhao J, Angelucci DE, Badal-Garcia E et al.** (2010) Symbolic use of marine shells and mineral pigments by Iberian Neandertals. *Proc. Natl Acad. Sci. USA* **107**, 1023–1028.

AGRICULTURAL EXPANSIONS

CHAPTER TWELVE

In the previous chapter we saw how modern humans subsisting as hunter-gatherers moved out of Africa and colonized Eurasia and Australasia within the last 60–65 KY. Since then, these regional populations have undergone substantial demographic change and additional migrations. In this chapter we explore the dramatic impact that the shift to agricultural food production has had on population size and genetic diversity over the past 10 KY.

It is hard to overestimate the impact of the advent of agriculture on human genetic diversity, and more generally on global biodiversity. Agriculture was a prerequisite for the development of the modern world as we know it, driving the rampant expansion (approaching 1000-fold) of our species, the development of complex societies, and our very conception of the world around us. A comprehensive study of the impact of agriculture would be nothing less than the history of the world over the past 10 millennia. Here we confine ourselves to the direct effect of agricultural innovation and its spread on genetic diversity in both humans and the species our ancestors domesticated.

Farming has been a very successful cultural innovation that spread rapidly over most of the globe. A major concern of this chapter is whether this flow of culture was accompanied by a flow of genes. Was the spread of agriculture mediated by indigenous peoples learning the techniques of their neighbors (**acculturation**) or by migrations of the farmers themselves (gene flow; **Figure 12.1**)? A number of different simple models for gene flow have been developed, the most popular of which has been the **demic diffusion** model of Ammerman and Cavalli-Sforza, which envisages a demographically driven **wave of advance** of "farmer" genes.[2]

In this chapter we first consider the background evidence from other disciplines on the origins, impact, and spread of agriculture. We then address two case studies of human genetic diversity within well-characterized zones of agricultural spread: first, out of the Near East into Europe; and second, out of tropical West Africa into sub-Equatorial Africa. Finally, we examine the genetic evidence from domesticated animals and plants themselves for clues as to the nature of the domestication process.

12.1 DEFINING AGRICULTURE

A general dictionary definition of **agriculture** is "the science, art, and business of cultivating soil, producing crops, and raising livestock." However, the word derives from *ager* (field) and *cultura* (cultivation) and can also refer more specifically to the intensive growing of crops in fields, with the more general term "farming" including the tending of animals. Agricultural populations

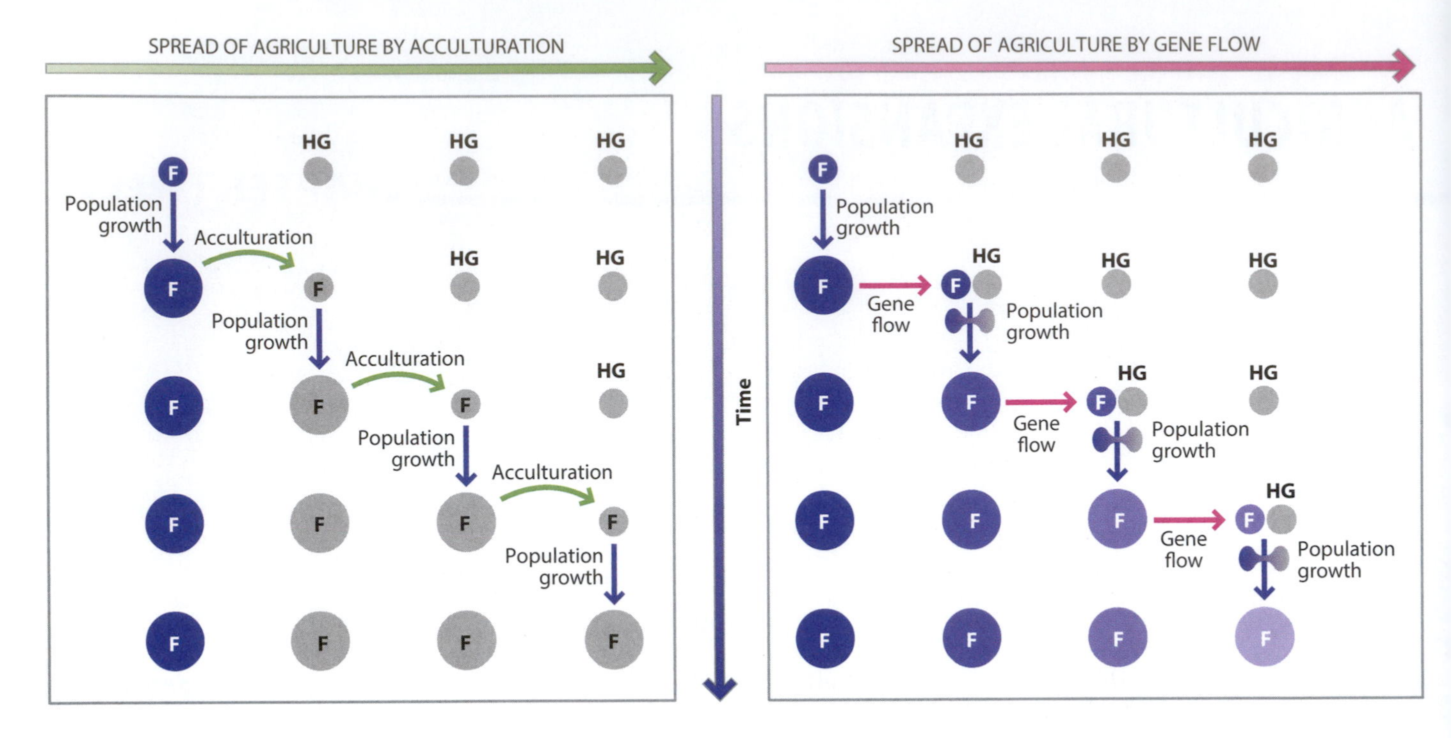

Figure 12.1: Acculturation and gene flow models for the spread of agriculture.
Shading indicates the genetic ancestry, and the labels F and HG refer to farming and hunter-gatherer economies.

typically live in settlements neighboring their cultivated land, and this increased **sedentism** contrasts with the relatively mobile lives of most hunter-gatherers whose movements are governed more by the availability of seasonal resources.

Although the development of agriculture has often been presented as a sudden burst of innovation, it is actually an assemblage of features that developed much more gradually—it was a process, not a single event. Many hunter-gatherer groups were adopting more settled lifestyles before the advent of agriculture, and were actively managing the plant and animal resources at their disposal.

For crop cultivation, the process of agricultural change can be appreciated by dissecting it into its component processes: soil preparation; selective breeding; propagation (planting seed or taking cuttings); crop protection: weeding, use of pesticides, and protection from the weather; harvesting; and storage. While some of these processes were introduced rapidly at the dawn of agriculture, others have a more ancient origin among hunter-gatherer societies and a few (such as food storage) are even practiced by some nonhuman species. Although the adoption of individual practices leads to improved food production, it is the domestication of plants and animals that represents the dominant innovation. **Domestication** is defined as the selective management and breeding of a species to make it more useful to humans, as distinct from the general planting and harvesting of plants (**cultivation**), or the taming of wild animals whose breeding is not manipulated. Many of the modern domesticated animals and plants have undergone great morphological divergence from their ancestral wild counterparts (**Section 12.7**). Note that it was only much later that the additional applications of domesticated animals were appreciated. This **secondary product revolution** includes the use of hair and hides for clothing, and large mammals for traction and transport.

The advent of agriculture is associated with such a dramatic change in tool usage that it is assigned its own cultural period—the **Neolithic** (New Stone Age), and the term "Neolithic revolution" was coined by the archaeologist Vere Gordon Childe (**Figure 12.2a**). Although originally defined by the use of ground or polished stone implements and weapons (Figure 12.2b), other characteristic features of the Neolithic include the manufacture of pottery (Figure 12.2c) and the appearance of human settlements. The Neolithic appears at different times in different regions, and the reasons for this are explored in the next section. Note that when terms such as "**Mesolithic**" and "Neolithic" were first proposed,

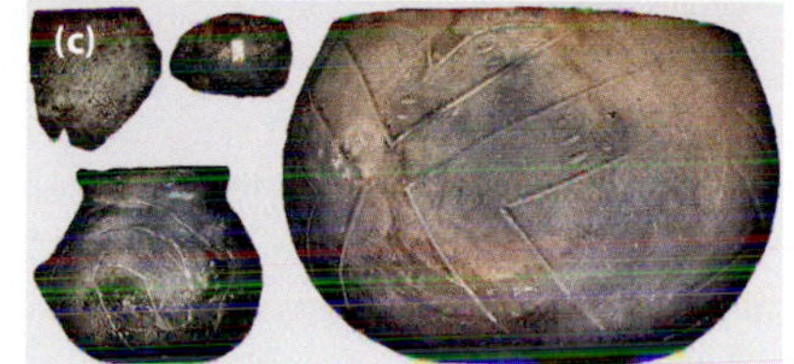

Figure 12.2: Vere Gordon Childe and some of the elements of his "Neolithic revolution."
(a) Vere Gordon Childe. (b) Neolithic polished stone tools. (c) Typical Linear Band Pottery (LBK). (a, courtesy of Department of Archaeology, University of Edinburgh. b, © Michael Greenhalgh. Source: http://rubens.anu.edu.au/raider5/greece/thessaloniki/museums/archaeological/neolithic/. c, collection University of Jena. Courtesy of Roman Grabolle.)

the definitions were simple: Mesolithic people were early Holocene hunter-gatherers with flaked stone tools and no pottery, while Neolithic people were later farmers with polished stone tools and pottery. However, evidence of Mesolithic societies with pottery, and of people with characteristically Neolithic pottery but who practiced hunting and gathering, has blurred these terms. Confusion is caused when they are used to refer to periods of time, to ways of life, or to types of material culture, but these labels nevertheless persist.

In the modern era, agricultural means of food production vastly predominate over hunting and gathering lifestyles. Hunter-gatherers today tend to reside in marginal areas in which farming is not ecologically feasible (**Figure 12.3**), for example the Inuit north of the Arctic Circle, the Khoe-San of the Kalahari Desert, and the Aborigines of central Australia. The environmental marginality of many of these populations makes them poor models for prehistoric hunter-gatherers, who presided over the richer lands subsequently occupied by farmers.

12.2 THE WHERE, WHEN, AND WHY OF AGRICULTURE

Where and when did agriculture develop?

From archaeological evidence (**Box 12.1**), the dates of the first appearance of agriculture vary widely in different regions. In addition, the domesticated plants and animals differ. These two observations suggest multiple independent origins of farming practices. Dating the prehistoric ranges of these individual farming cultures reveals a common feature: an ancient zone of agricultural innovation from which farming subsequently expanded. The difficulty in distinguishing between a truly independent origin and a secondary zone in which

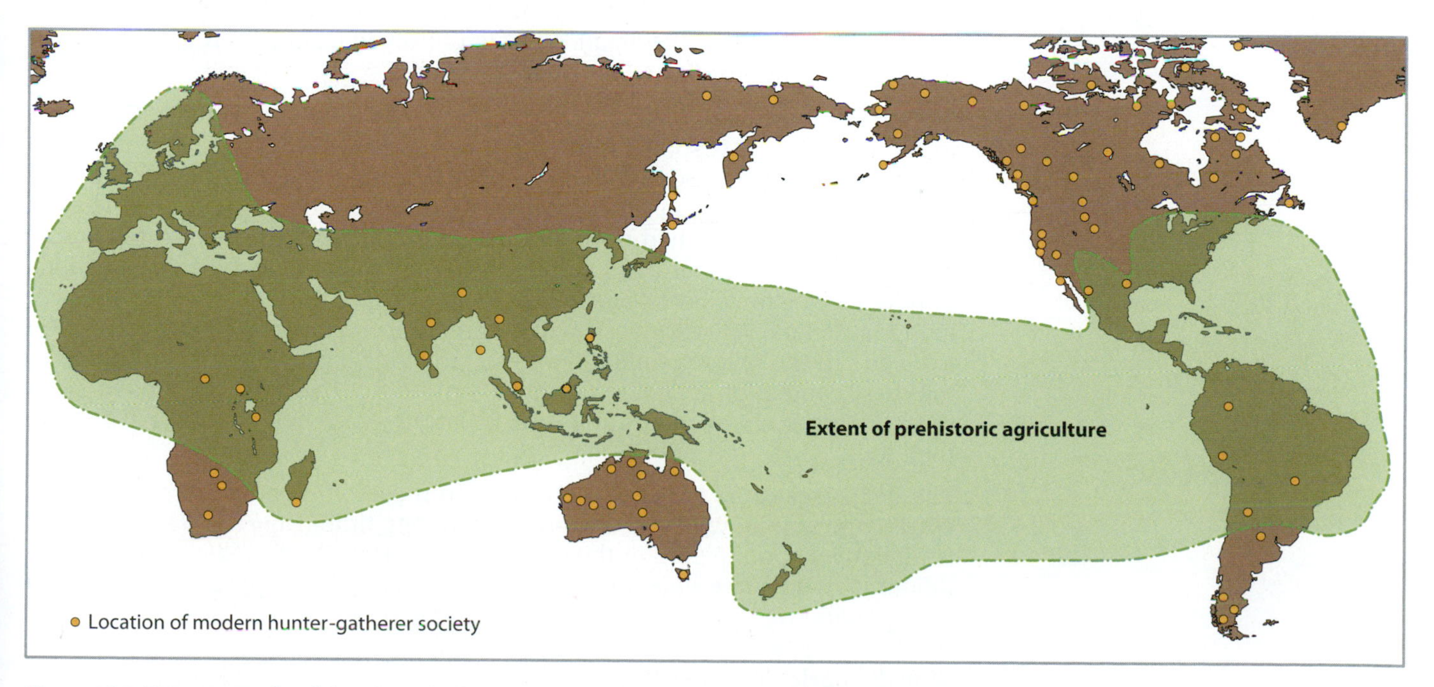

Figure 12.3: The extent of prehistoric agriculture and the marginality of modern hunter-gatherers.
[Data from Bellwood P (2001) *Ann. Rev. Anthropol.* 30, 181; and Kelly RL (1995) The Foraging Spectrum: Diversity in Hunter-Gatherer Lifeways. Smithsonian Institution Press.]

Box 12.1: Sources of evidence in the archaeology of agriculture

Examination of archaeological remains reveals direct and indirect evidence of agricultural activity, and ^{14}C dating (see Box 11.2) can be used to estimate when this activity occurred. At least four kinds of evidence are available:

Human bones: evidence on diet can come from tooth wear, signs of disease, and bone chemistry. Changes in physical activity associated with agriculture can be deduced from skeletal signs of musculature development.

Animal bones (archaeozoology): patterns of animal remains are distinctively different under domestication compared with wild captures. The proportion of remains made up by a single species increases; animals decrease in size as an environmental and genetic consequence of captivity; and the sex ratio and age profile of animal skeletons change, indicating planned culling.

Plant remains (archaeobotany): charred plant remains are chemically and physically stable and can be preserved in archaeological sites where fire was used. They can be separated using the flotation technique, where sediment is mixed with water and the plant material floats on the surface, allowing it to be recovered and microscopically examined. In other contexts, plant remains can be found in bogs, in lake sediments, in frozen soil, or in arid conditions. As well as seeds themselves, useful evidence comes from pollen, starch grains, **phytoliths** (particles of silicon dioxide from leaves or fruits), or **parenchyma** (the storage material of tubers and rhizomes). The recognition of a pattern in which tree pollen levels fall dramatically, followed by a rise in grass and cereal pollen, and finally a return to dominance of tree pollen, has been interpreted as the land clearance effects of slash-and-burn agriculture (*landnam*). However, the activities of foragers, and the effects of tree disease, could also be involved in these shifts. Remains of particular insects have been taken as evidence of storage of particular seed crops, and analysis of human **coprolites** provides a direct way to see what people ate.

Associated cultures, defined by artifacts (including stone tools, ceramics, and rock-art), architecture, burial practices, and the layouts of settlements. For example, a distinctive type of pottery (**linear pottery**, or ***linienbandkeramik***; **LBK**—see Figure 12.2c) is associated with the early European Danubian farming culture, and another type, **Cardial ware**, impressed with the serrated edge of the cardium (cockle) shell, is associated with the spread of farming in Italy, southern France, Spain, Sardinia, and Corsica.

A key point when considering past agricultural activity is that the archaeological evidence has been greatly affected by political contexts: for example, there was work during the 1950s and 1960s in Iraq and Iran but little since then; and for decades huge areas of the Middle East, Africa, Mesoamerica, and East and Southeast Asia have been effectively inaccessible for field research. It is important to distinguish between *evidence of absence* of agriculture in a region, and *absence of evidence*.

new species were domesticated in response to the arrival of agriculture from elsewhere, means that the actual number of truly independent regions of origin remains contentious (**Figure 12.4**).

The best-characterized centers of domestication are the **Fertile Crescent** in the Near East, **Mesoamerica**, and East China, with more recent evidence emerging for lowland South America and highland New Guinea. These areas all had native species suitable for domestication. According to one view, once the "agricultural package" had been assembled it could be exported to regions of similar climate. Without the suitable native species additional domestications of new species may have occurred en route. Dispersal would have been easier along latitudinal (east to west) axes than along longitudinal (north to south) axes due to the greater similarity of the growing seasons.[23] It is worth noting that past centers of agricultural innovation are not the most productive areas today. Climate change and long-term environmental degradation have taken their toll.

The first dates of agricultural innovation in these different centers are spread over six millennia, which to us may seem a long time (**Figure 12.5a**), but in fact is a relatively brief period in the context of the 100 KY of modern human evolution: why do these independent origins of agriculture appear within such a defined window?

Why did agriculture develop?

The Neolithic revolution required no biological changes of humans, but was a cultural phenomenon (albeit with many biological consequences, discussed in

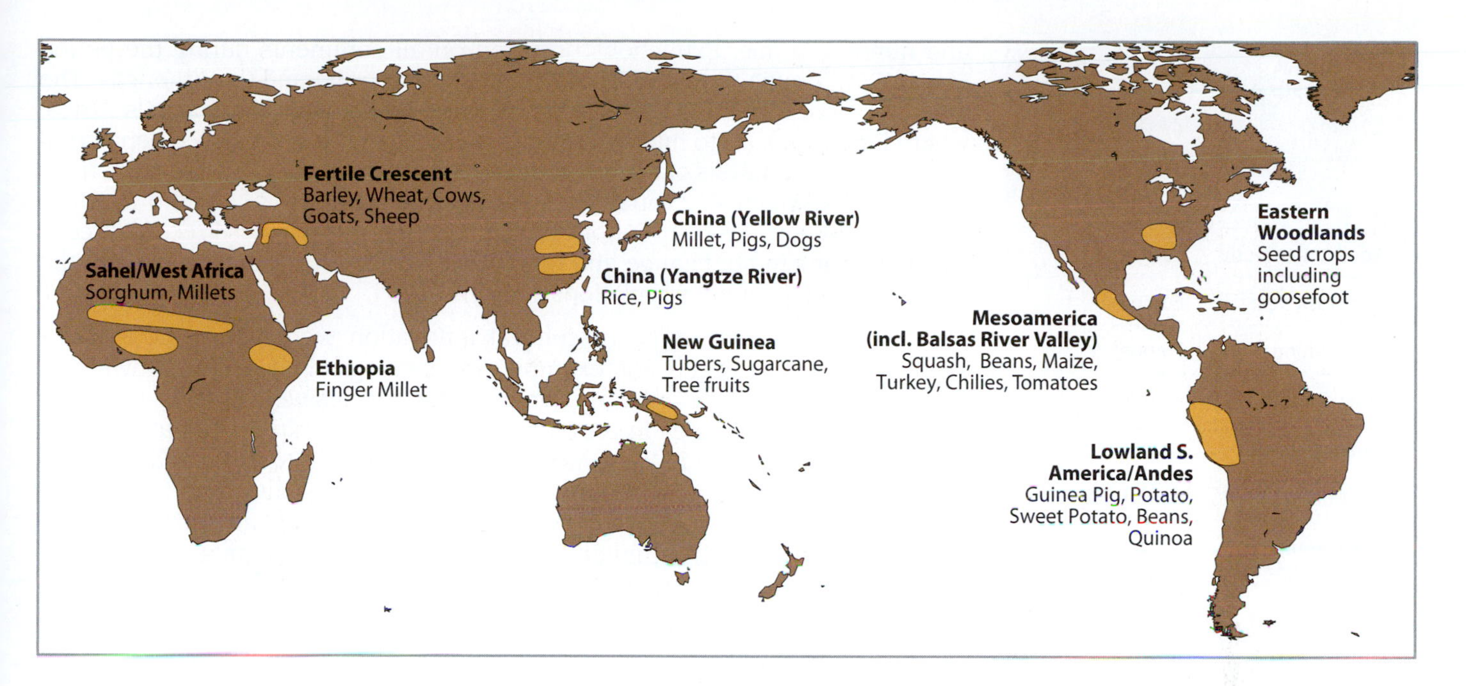

Figure 12.4: Map of centers of agricultural innovation together with some of the plants and animals domesticated within them.

the next section). The traditional view was that it was a step on the natural progression toward "civilized man," and that adopting farming was the only sane option for any self-respecting hunter-gatherer. However, its eventual outcome would have been far from obvious to prehistoric populations. In addition, the view of an inevitable progression does not explain the long delay between the completed development of truly modern behavior by 60 KYA and the advent of farming at 10 KYA.

Climate change was crucial. The end of the last ice age ~14 KYA was followed by relative climatic stability (Figure 12.5b) in which Mesolithic human populations prospered and grew. In particular, the period after ~10 KYA saw particularly stable temperatures, free of major and rapid fluctuations, which would have facilitated the development of farming practices. Another possible factor favoring domestication was the extinction of many large-bodied animals (**Table 12.1**), and there has been much debate over the relative importance of climate change

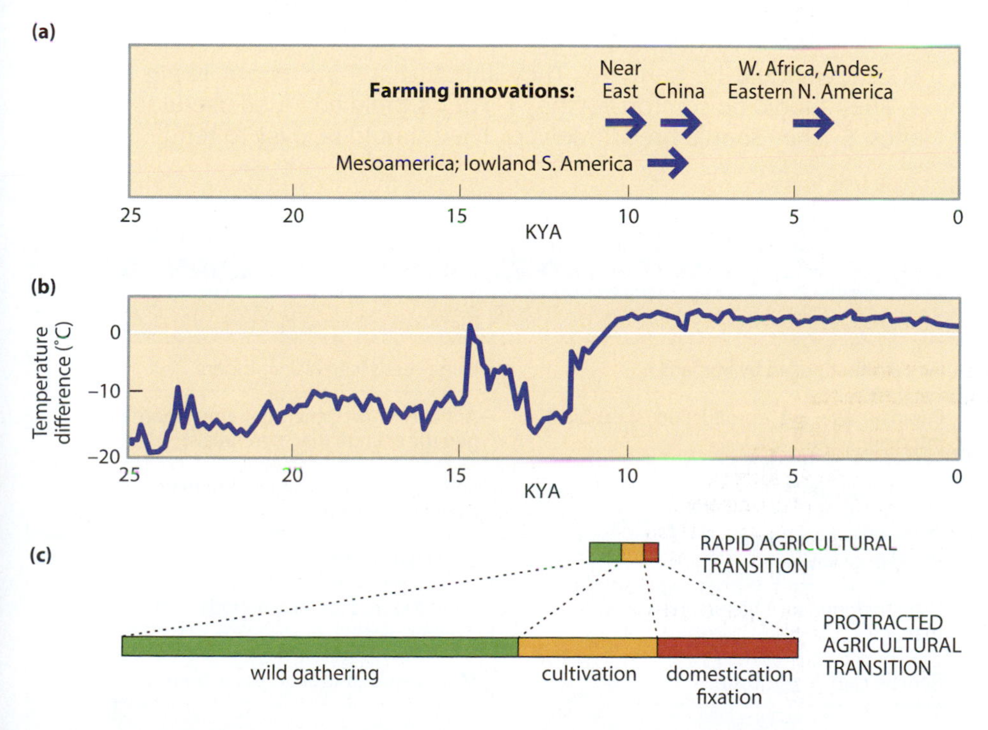

Figure 12.5: Time line of changing temperatures and agricultural development. (a) Periods and places of farming innovation. (b) Climatic variation. The graph shows temperature variation over the last 25 KY with lower temperatures prior to 10 KYA, and very rapid fluctuations of up to 10°C (data from the Vostok ice core in Antarctica: **http://www.amap.no/maps-gra/mg-cc.html**). Temperature difference is shown compared to the current mean surface temperature of −55.5°C. Farming innovations occur during the warmer recent period. (c) The traditional view is of very rapid agricultural transitions, shown here (*top*) for the Near East; below is shown an alternative scheme in which the transition is much more protracted. [Adapted from Allaby RG et al. (2008) *Proc. Natl Acad. Sci. USA* 105, 13982. With permission from National Academy of Sciences.]

TABLE 12.1: EXTINCTIONS OF LARGE-BODIED (>44 KG) MAMMALS BETWEEN 50 AND 10 KYA

Species	Continent
Woolly mammoth	Europe
Giant Irish deer	Europe
Woolly rhino	Europe
Saber-tooth cat	N. and S. America
American camel	N. America
Giant beaver	N. America
Giant ground sloth	N. America
American mastodon	N. America
Big-horned bison	N. America
Glyptodon (giant armadillo)	S. America
Marsupial lion	Australia
Giant short-faced kangaroo	Australia
Giant wombat	Australia

and hunting in this. Extinctions occurred on all continents during the period 50–10 KYA, though they were most marked in Australia and the Americas. The only continent to retain many species of large-bodied wild mammals is Africa, where long exposure to human hunting practices may have allowed behavioral adaptation. An analysis of the distribution and timing of megafaunal extinctions compared with climatic variables and human arrival times,[57] and a study of ancient DNA, species distribution models, and the human fossil record,[42] concur that both climate change and human activity are necessary to explain the changes in megafaunal populations.

The locations of centers of agricultural innovation were largely determined by biogeographic luck.[23] For example, archaeological evidence shows that many of the most important early domesticated species were cereals and other grasses: wheat and barley in the Near East, millet and rice in China, maize in the Americas, sorghum in West Africa, and sugarcane in New Guinea. The distribution of wild grasses with large seeds is highly irregular, with the highest concentration in Southwest Asia. Other "founder" crops and animals that could be domesticated also had limited distributions. Centers of agricultural innovation formed in the areas where these distributions overlapped.

Finally, it is worth noting the irreversibility of farming. It supports a larger population than foraging from the same area of productive land, which could not be sustained after a return to previous subsistence methods. This ratchet-like mechanism ensured that the number of populations practicing agricultural means of food production could only increase over time.

Which domesticates were chosen?

Within each agricultural homeland, the plant and animal species domesticated were much less diverse than those exploited by local hunter-gatherer societies. In addition, after the initial burst of domestication relatively few species were added to most agricultural packages (excepting the Americas). Wheat and rice are amongst the earliest domesticates, yet still occupy the greatest area of agricultural land of all domesticated plants. This suggests that only certain species were amenable to selective breeding for useful traits. **Table 12.2** lists some of their desirable characteristics.

While it is easier to control the reproduction of animals than of plants, it is harder to manage their daily existence. Francis Galton, as with many other subjects, had something to say on the qualities of animals suitable for domestication:

> "1. They should be hardy; 2. They should have an inborn liking for man; 3. They should be comfort-loving; 4. They should be found useful to the savages; 5. They should breed freely; 6. They should be easy to tend.

TABLE 12.2: DESIRABLE CHARACTERISTICS OF SPECIES SUITABLE FOR DOMESTICATION

Animals	Plants
Diet easily supplied by humans	Large seeds (cereals), or tubers
Rapid growth and reproduction (short birth interval)	Annuals (short generation time: more rapid genetic change than perennials)
Placid disposition	Selfing hermaphroditic reproduction, rather than outcrossing
Breed in captivity	Easily harvested
Used to dominance hierarchies—easy to manage	Can be stored for long periods
Remain calm in enclosures	Monogenic control of crucial traits

It would appear that every wild animal has had its chance of being domesticated, that those few which fulfilled the above conditions were domesticated long ago, but that the large remainder, who fail sometimes in only one small particular, are destined to perpetual wildness as long as their race continues. As civilization extends they are doomed to be gradually destroyed off the face of the earth as useless consumers of cultivated produce."

From Inquiries into Human Faculty and Its Development, 1907.

12.3 OUTCOMES OF AGRICULTURE

The outcomes of this cultural transition for the domesticated animal and plant species are considered in **Section 12.7**. Here, we concentrate on the consequences for humans themselves.

Agriculture had major impacts on demography and disease

Rapid demographic growth

The long-term demographic outcomes of the Neolithic transition are obvious. Population sizes increased slowly during the period 100–10 KYA, but today we have a global population hundreds of times greater than that of pre-agricultural hunter-gatherers (**Figure 12.6a**). We also enjoy much greater life expectancy in the modern era compared with pre-agricultural societies. In the short term, demographic growth is signaled in the archaeological record by an abrupt increase in the proportion of 5- to 19-year-old juveniles found in cemeteries[10] at the transition from foraging to farming (Figure 12.6b). In the genetic data, a coalescent-based analysis comparing whole mitochondrial genomes associated with agricultural populations with those associated with hunter-gatherers in Europe (Figure 12.6c), Southeast Asia, and sub-Saharan Africa, shows signs of an approximate fivefold increase in the population growth rate at the local onset of agriculture.[32]

Why did changing the method of food production result in such dramatic population growth? A number of potentially synergistic factors have been proposed. Higher population densities are supported by increased food yields, and improved nutrition leads to earlier **menarche** (first menstrual cycle), resulting in a longer period of female fertility for the same life expectancy; a more stable food supply could also increase fertility through fewer miscarriages, and there would be less need for the "natural" population control methods of infanticide and induced abortion. Greater sedentism allows a shorter interval between births: previously, hunter-gatherers would have carried their young children around with them, limiting the number of offspring that could be raised at any one time. Availability of milk from animals also allows earlier weaning and closer-spaced pregnancies.

Malnutrition and infectious disease

Despite these apparent advantages, population growth rates have been much more dramatic in recent times than during the early years of farming. In fact, **paleopathological** studies indicated that the adoption of agriculture often had a negative impact on human health.[20] Such apparent health disadvantages of farming represent a paradox, given that agriculture clearly underlies the expansion and success of our species in the last 10 KY. Recent studies investigating health and subsistence modes across a range of times, places, and populations have shown a complex relationship between subsistence change and health[53] and, while there is still evidence for a decline in health amongst many populations, the emerging picture is more regionally specific and diverse than previously thought.

Paleopathological evidence for the deleterious consequences of dietary changes comes from the crania and the bones of the eye-sockets of early farmers, which show greater porosity than in hunter-gatherers, indicative of anemia. These and

Figure 12.6: Demographic impact of the Neolithic transition.
(a) Global population size estimates over the past 100 KY. A zone of plausible population sizes between upper and lower bounds is shown on a *logarithmic* scale. Uncertainty in population size estimates is greater in the more distant past.
(b) Age profile of skeletons in cemeteries by relative chronology compared with the local advent of agriculture. The *green* line is fitted using the Loess procedure, which weights the scattered data points (individual cemeteries) by the inverse of their binomial variance to take account of numerical differences between sites. The proportion of young individuals increases at the transition from foraging to farming, signaling a demographic expansion.
(c) Coalescent analysis of whole mtDNA sequences associated with agricultural populations (*blue line*) or hunter-gatherers (*red line*) in Europe shows a rapid increase in female effective population size at the onset of agriculture. [b, from Bocquet-Appel J-P (2008) Explaining the Neolithic demographic transition. In The Neolithic Demographic Transition and its Consequences (J-P Bocquet-Appel & O Bar-Yosef eds). Springer. With permission from Springer Science and Business Media. c, adapted from Gignoux C et al. (2011) *Proc. Natl Acad. Sci. USA* 108, 6044. With permission from Chris Gignoux.]

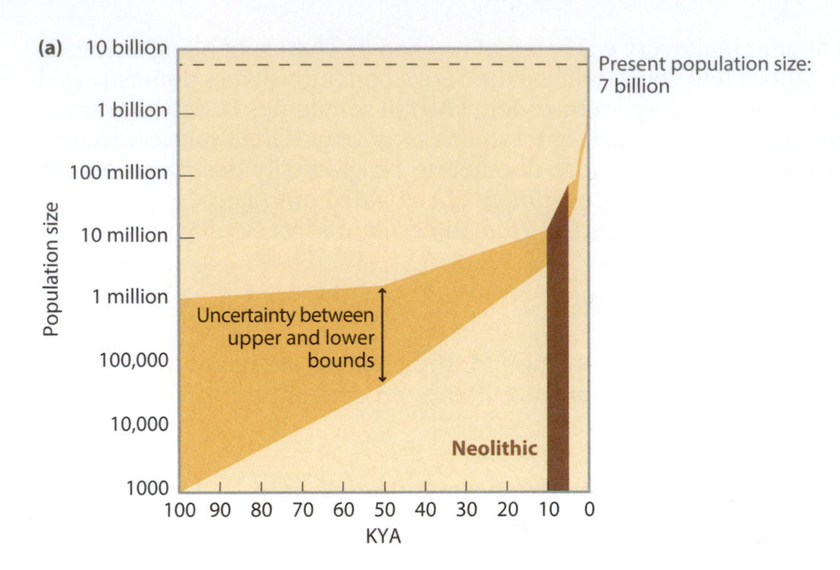

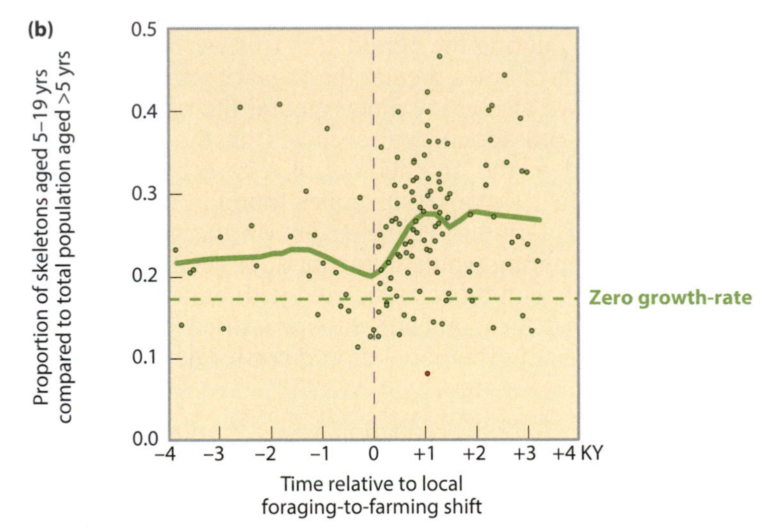

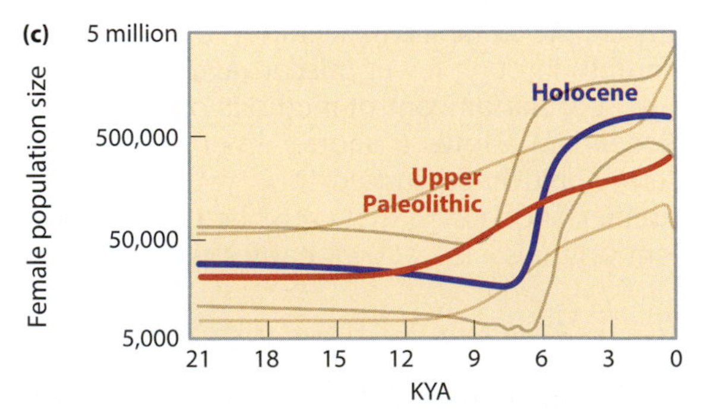

other skeletal indicators suggest that malnutrition was common. Comparative studies of Paleolithic and Mesolithic populations from the same region often show a decline in nutrition over time. While contemporaneous hunters adapted to reduced resources by exploiting a broader spectrum of species in their diet, early farmers, depending on fewer species, were prone to catastrophic crop failure. Even without such failures, reliance on too few crops can result in nutrient and vitamin deficiencies.

Some early farmers also appear to have experienced an increased prevalence of infectious disease, seen as higher frequencies of skeletal lesions, probably connected to their poorer nutrition. Many human infectious diseases result from pathogens transferring from an animal host (zoonosis), and because animal husbandry requires greater daily contact with animals than hunting does, this would have facilitated disease transfer. In addition, higher population densities may well have exacerbated the impact of zoonoses. Many pathogens require high host densities if they are to be sustained in the population (for example, **crowd epidemic diseases** such as smallpox, measles, and mumps[78]). Increased human population densities resulting from post-agricultural population growth and sedentism could have supported pathogens previously confined to herd animals (though some epidemic diseases may have required the even greater densities that followed urbanization). Sedentism also led to accumulations of human and animal wastes, attracting vector-borne diseases such as plague (see Opinion Box 17.1). Increased pressure on supplies of clean water could have resulted in epidemics of cholera. An increased reliance on food storage during leaner times may have led to increased morbidity through spoilage, and attracted rodents from which other zoonoses could be acquired.

The eventual development of resistance to zoonotic pathogens has later relevance for epidemics resulting from contact between previously isolated populations. During the recent era of European colonization, the impact of infectious diseases (for example, smallpox) was often far more severe on indigenous populations than it was on the colonists. Such diseases played a major role in the reduction of the indigenous human population of South and North America from ~70 million in 1492 to ~600,000 by 1800. This differential susceptibility was known at the time, and one of the earliest forms of biological warfare is revealed by the reply, in 1763, of British General Amherst to Colonel Bouquet's suggestion that smallpox-infected blankets be given to Native Americans to precipitate an epidemic: "You will do well to try and inoculate the Indians by means of blanketts (sic), as well as every other method that can serve to extirpate this execrable race." (http://www.umass.edu/legal/derrico/amherst/lord_jeff.html)

Both infectious diseases and dietary changes altered the selective landscape for humans by, for example, selecting for tolerance to high-lactose diets in adults. In Chapters 15–17 we explore some of the genetic consequences of these novel selective pressures.

Agriculture led to major societal changes

Living in an early farming society would have been significantly different from living in a hunting-gathering society. Life was more settled, and spent in larger populations. The dominant social unit became the household rather than the whole group, which may have reinforced the concept of private property. Trading networks were required with outside communities to maintain access to distant resources. Permanent settlement allowed the development of non-portable technology. New tools were required for food harvesting, storage, and preparation. Although the earliest pottery is found among some pre-agricultural societies, clayware is a consistent indicator of the Neolithic transition.

Despite an initial focus on subsistence, it was the ability of agriculture to produce a surplus that transformed human societies. The complex 10,000-year history of these societal developments is beyond the scope of this book, but a simplistic linear account can be summarized as follows. Food surpluses could be used to support full-time craftspeople and a bureaucracy to manage the society. Intensification of agriculture resulted in the new phenomenon of land-owning and wealth, which varied among households, increasing social stratification. Settlement made the fission of groups costly, so new mechanisms for within-group conflict resolution were required. Greater dependence on specific tracts of land also led to the potential for conflict between neighboring settlements, while the central management of multiple settlements represented the first stages of state formation. The development of trade specialization, structured societies, and complex economies eventually precipitated the birth of civilizations—the earliest evidence for writing is largely represented by the administration of economic transactions on clay tablets.

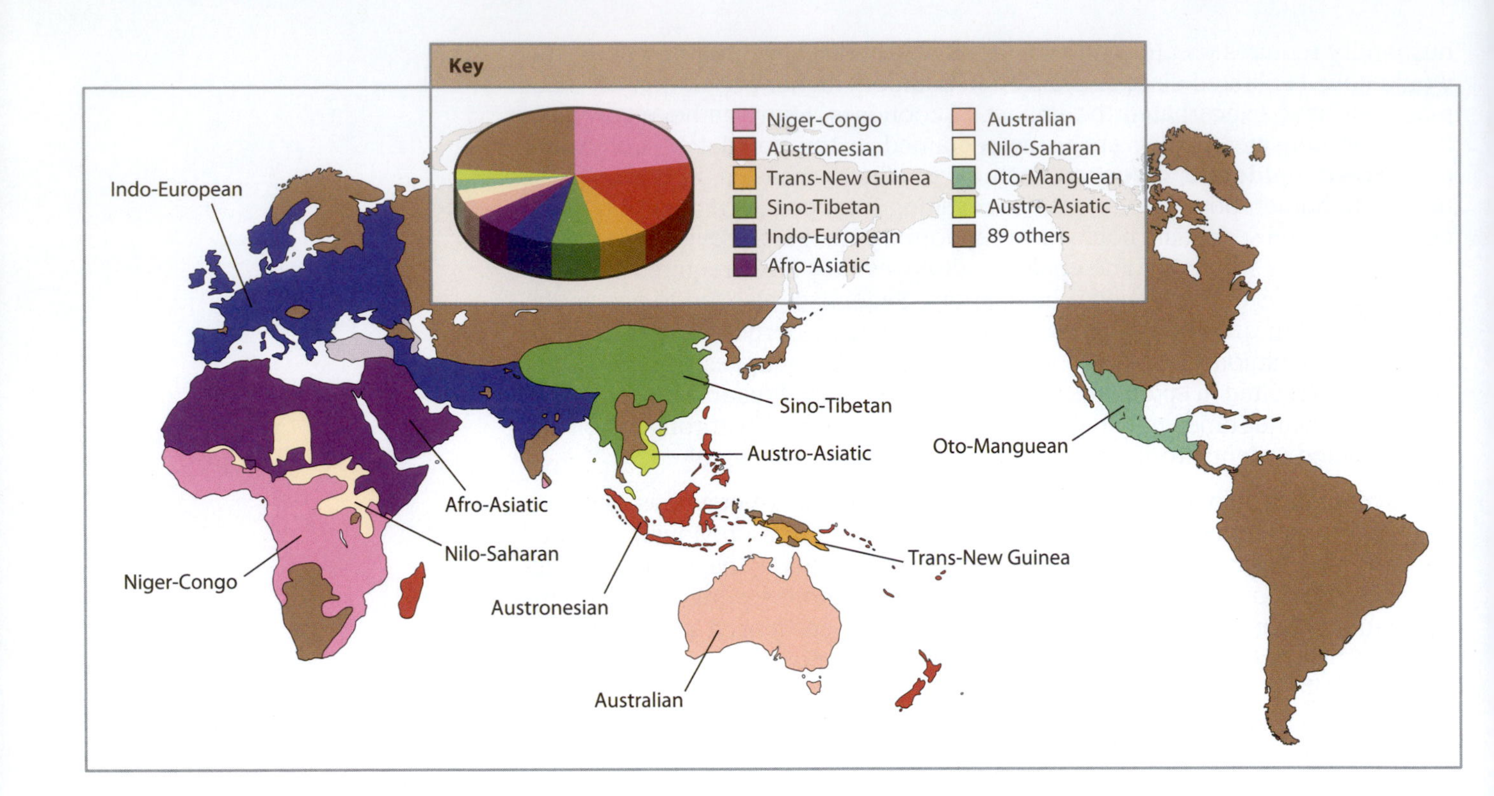

Figure 12.7: Widely spread and rich language families.
A few language families contain disproportionately large numbers of languages, as shown in the pie chart within the upper panel. The map shows the distributions of major language families that each contain more than 100 languages. (Data from **www.ethnologue.com**)

12.4 THE FARMING–LANGUAGE CO-DISPERSAL HYPOTHESIS

Some language families have spread widely and rapidly

The approximately 6900 languages spoken throughout the world today can be grouped into more than 100 families, each on the basis of a hypothetical shared common ancestral language. The ancestral relationships between language families themselves are highly contentious, and probably impossible to reconstruct. A few families contain disproportionately large numbers of languages; for example, two (Austronesian and Niger–Congo) together encompass about 40% of all individual languages. The geographic distribution of such families is highly skewed: while some inhabit areas which contain a high diversity of unrelated languages with limited distributions, others occupy large continental regions in which few, if any, other languages are present (**spread zones**; **Figure 12.7**). Examples of spread-zone language families are Indo-European (including English, Greek, Iranian, Hindi, and most European languages) and Sino-Tibetan, including Chinese.

A single origin of languages within a family that is distributed over a wide area suggests a language range expansion, and (because few linguists believe that any language family can be more than 10 KY old) a recent one. To identify the likely geographic origin two characteristics have been used: the region of greatest linguistic diversity; and the region in which the deepest-rooting languages are spoken, based on reconstruction of their ancestral relationships from a study of shared specific innovations, analogous to the reconstruction of a phylogenetic tree.

The likely homelands of many of the largest language families appear to be situated in and around the centers of agricultural innovation described above. This led to the hypothesis that many of these families moved together with the expansion of agriculture in different areas of the world. Colin Renfrew[60] suggested that the distribution of Indo-European was best explained by the dispersal to Europe of the first farmers from 10 KYA, spreading out of Anatolia and speaking their **Proto-Indo-European** language. This interpretation caused much argument among linguists, because it contradicted the standard view that the first Indo-Europeans were nomadic pastoral horse riders who spread westward from the Steppes north of the Black Sea around 5 KYA. Peter Bellwood independently suggested a farming–language co-dispersal model for the spread of the Austronesian family (including Polynesian languages) in the Pacific.[8] Other spread-zone languages may have been dispersed in the same way, including the Bantu languages, together with the spread of farming out of West Africa (**Section 12.6**).

This view is supported by the finding that areas in which there is little archaeological evidence for agricultural expansions tend to be associated with patchwork-like language distributions. Language families occupy a much more limited space, with less-closely related languages whose origins may go back much deeper in time: these "mosaic zone" families are seen, for instance, in the Caucasus, New Guinea, and North Australia.

Linguistic dating and construction of proto-languages have been used to test the hypothesis

If it were possible to date accurately the spread of languages from linguistic evidence itself, it would be easy to test whether linguistic expansions accompany agricultural expansions. However, despite attempts at linguistic dating (glottochronology), most linguists believe that known differences in rates of change between languages disprove the existence of a "linguistic clock," analogous to the molecular clock (**Section 5.7**) used in genetic dating methods (see Opinion Box 1 in Chapter 1).

An alternative approach comes from attempts to reconstruct the vocabularies of ancestral languages, known as proto-languages. A word present in the same form in all languages within a family indicates that this word was present in the proto-language. By reconstructing a number of different words it should be possible to build up a cultural context for this ancestral language that will allow indirect dating through archaeological evidence. For example, one of the problems with the theory of agricultural origins for Indo-European languages is that it appears to be possible to reconstruct terms associated with horse-drawn chariots in Proto-Indo-European. The first archaeological evidence for these vehicles postdates the spread of agriculture by several millennia, suggesting that the expansion of this language family took place long after the spread of agriculture, and perhaps with the secondary product revolution (5–6 KYA) associated with, amongst other things, the use of horses for transport. However, such a recent date does not agree with lexicostatistical studies described in the next section.

There are alternative explanations for why some families have spread far and wide, while others remain localized. Ecological, geographical, and economic factors can play important roles in determining linguistic diversity. Cultures adapted to coastal environments or arid plains can rapidly spread their languages over large distances. By contrast, highlands provide refuge for remnant languages and often harbor high linguistic diversity.

What are the genetic implications of language spreads?

Cultures can certainly spread without the concomitant spread of genes, but languages are less easy to acquire during adulthood than are other cultural modifications. It has been argued that the improbable nature of this **language shift** in an entire population suggests that the languages must have been spread by the farmers themselves, and that the later extinction of hunter-gatherer languages reflects the demographic superiority of farmers.[8] In such instances we should expect to see clear evidence for gene flow along the same axes as the spread of agriculture and language.

Other considerations suggest that gene flow need not accompany agriculturally associated language expansions. In the previous section we saw the dramatic societal changes that the advent of agriculture entailed. During these changes large elements of the agricultural vocabulary would have to be invented. Under such great social pressures language shift may well have been more likely than assessments of its infrequent modern occurrence might suggest.

The above discussion of language expansion with and without concomitant gene flow relies too heavily upon discrete packages of genes labeled "farmers" and "hunter-gatherers." Admixture could have occurred between these two groups during the long time-scales over which the language expansions took place, resulting in offspring with parents speaking different languages. The adoption by the child of a single language would then necessarily uncouple genes and language to some degree. These complications indicate that we should not expect an absolute correlation between means of food production, language, and genes. Initially, we will examine whether there is any evidence for gene flow along known axes of agricultural expansion.

12.5 OUT OF THE NEAR EAST INTO EUROPE

Nongenetic evidence provides dates for the European Neolithic

Figure 12.8 shows the geographical distribution of the earliest evidence for agriculture throughout Europe, based on 735 archaeological sites.[52] The distributions of the oldest sites (dating to around 11–12 KYA) indicate that agriculture probably originated in the Fertile Crescent in the Near East—the area that today includes northeast Syria, northern Mesopotamia, and part of southeast Turkey. There is a time gradient: sites become consistently younger toward the northwest of Europe, with agricultural practices arriving at the Baltic and the British Isles 5–6 KYA. Although there is debate about interpretation of particular aspects of archaeological evidence, and the small-scale pattern of spread, this overall picture from archaeology has provided a framework for the disputes of genetics.

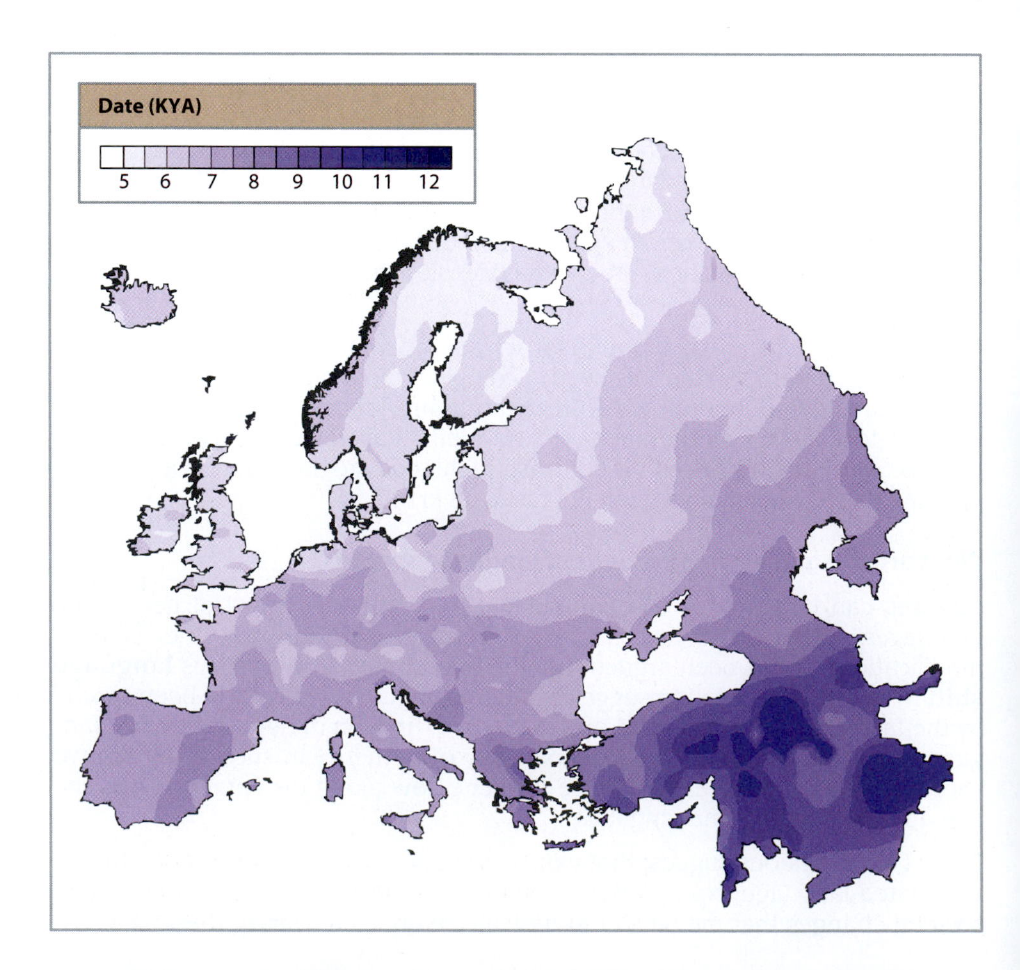

Figure 12.8: Map of the distributions of the earliest archaeological sites in Europe and the Middle East showing evidence of agriculture.
[From Balaresque P et al. (2010) *PLoS Biol.* 8, e1000285. With permission from Public Library of Science. Based on data in Pinhasi R et al. (2005) *PLoS Biol.* 3, e410.]

Figure 12.9: Stonehenge, a megalithic site in England.
(Courtesy of Gareth Wiscombe under the Creative Commons Attribution 2.0 Generic license.)

A simple traditional view of the arrival and spread of agriculture in Europe can be summarized as follows.[6] The crops (wheat, barley, pulses, and flax) and animals (cattle, pigs, sheep, and goats) of the European Neolithic were originally domesticated in the Near East, and brought to Europe by farmers. Farming communities appeared in the Aegean area and Greece around 9 KYA, and there followed two major streams of movement, into southeast Europe, and along the Mediterranean coast. An expansion of farming then occurred, rapid by the prehistoric standards, occupying an area from the Netherlands in the west to the Ukraine in the east, and from the Alps to northern Germany and Poland. From the Mediterranean there were expansions into Italy, France, and the Iberian peninsula. Regional differentiation of Neolithic groups then took place, including the development of the characteristic western **megaliths** (structures built of large stones, including single standing stones; **Figure 12.9**), and dispersal to the extreme west, including Ireland, was completed by 4 KYA. This apparently regular pattern of spread has formed the basis for the genetic studies discussed below.

This traditional view of agricultural spread is by no means universally accepted, however. The importance of rapid maritime transmission of agricultural practices is clear from evidence in Cyprus of imported cereals, domesticated cats and dogs, and practices of house building and stone tool manufacture by 10.6 KYA.[74] The idea of a "Neolithic package," in which all of the elements (permanent village of rectangular houses, domesticated crops and animals, evidence of religion, ground stone tools, and pottery) arrive suddenly, does not always fit the archaeological evidence. The existence of permanent pre-Neolithic settlements, and evidence in some places of the cultivation of crops or the domestication of animals dating many centuries before the arrival of the Neolithic, suggests that the populations already present may have played a considerable role in a slower process of cultural change (**Opinion Box 13**).

As was discussed in **Section 12.4**, the inclusion within the Neolithic package of an Indo-European language has also been challenged. Key to this debate is the time-depth of the language family, but this has been very difficult to agree upon, with some taking the view that 4 KYA or 40 KYA are equally possible. Despite the objections of many historical linguists (see Opinion Box 1 in Chapter 1), attempts have been made to estimate this: one study[13] used a cross-language **Swadesh word list** of ~200 basic items of vocabulary that are relatively resistant to word-borrowing (**Table 12.3** shows the 100-word version of the list) in 103 Indo-European languages, including 20 ancient examples no longer spoken. Application of Bayesian phylogeographic approaches to **cognate** terms shared between languages, together with geographical ranges for each language, strongly favored an origin in Anatolia, rather than the alternative of the

TABLE 12.3: THE SWADESH 100-WORD LIST

I; you; we; this; that; who?; what?; not; all; many; one; two; big; long; small; woman; man; person; fish; bird; dog; louse; tree; seed; leaf; root; (tree) bark; skin; flesh; blood; bone; grease; egg; horn; tail; feather; hair; head; ear; eye; nose; mouth; tooth; tongue; claw; foot; knee; hand; belly; neck; breasts; heart; liver; to drink; to eat; to bite; to see; to hear; to know; to sleep; to die; to kill; to swim; to fly; to walk; to come; to lie (recline); to sit; to stand; to give; to say; sun; moon; star; water; rain; stone; sand; earth; cloud; smoke; fire; ash; to burn; path; mountain; red; green; yellow; white; black; night; hot; cold; full; new; good; round; dry; name

Agriculture was undoubtedly revolutionary in terms of its impact on human demographics and its long-term alteration of environments. However, it is still unclear whether this revolution occurred rapidly, in just a few human generations, or was much slower, taking more than a hundred generations. Debates rage over whether this was a rapid development of an agricultural package based in perhaps just eight or nine focused areas, or whether as many as 24 regions underwent a gradual development process, and eventually coalesced into regional agricultural packages. The alternative time-scales have implications for both why and where agriculture developed. The concept of a package of founder crops originating together from a core area is associated with rapid domestication of those crops. Similarly, a single center of domestication of several species within a short time suggests one main cause for the origin of agriculture, such as rapid climatic change events or periods of social or demographic stress. In contrast, a protracted domestication may be drawn out over space as well as time, with different causal factors at different stages and places in the process. Slow processes of domestication are associated with more dispersed origins, mosaics of habitats, and interacting communities across wider regions. Archaeobotanical evidence for crop domestication increasingly points toward the latter, more complex scenario. Evidence from studies of animal domestication also points toward an extended process that occurred in parallel in several different places, such as in various parts of the Fertile Crescent as well as in other world regions.

Until recently most archaeologists and plant geneticists considered crop domestication a rapid process. A key trait that distinguishes domesticated seed crops, such as cereals, from their wild ancestors is that of seed dispersal, with wild forms naturally shedding mature seeds but domesticates evolving to retain them, and to rely on human harvesting and planting. Experiments in the 1980s in growing and harvesting wild cereals together with simple models of change involving single genetic mutations, suggested that it may have taken only 20–200 years to fully domesticate wild cereals. This meant that domestication could have occurred within a time period similar to the error margin on **radiocarbon** dates, so archaeological evidence could be interpreted as a virtually instantaneous transition from wild to domesticated forms. However, as the collection of archaeobotanical evidence, in particular of **rachises** that are either **shattering** or **non-shattering** (see Figure 12.27), accumulated from the late 1990s onward, a slower process became evident (**Figure 1**). For wheat and barley in the Near East, archaeological plant populations indicate that the non-shattering phenotype took between 2000 and 3000 years (11.5–8.5 KYA) to develop. Recent archaeological data on rice in China suggest a similarly long time-scale (8.5–5.5 KYA).

In another methodological advance, archaeobotanists working in the Near East have begun to recognize an arable weed flora associated with early cereals that are morphologically wild. This suggests that an artificial cultivated environment that supported an arable weed flora arose before morphological change occurred in the cereal, and also that fields were intensively managed and reused. Another gradual change associated with domestication that can be recognized archaeologically is a trend toward larger seed sizes. This trend can extend over a few thousand years in the case of Near Eastern cereals, pea, and lentil, Asian rice, or North American sunflower. Data from rind phytoliths suggest similarly slow improvement in the size of neotropical squashes. In a few crops grain size change may have been more rapid—examples are South Indian mungbean, East Asian melons, and African pearl millet. Such variation in the evolution of a domestication trait suggests somewhat diverse processes affected by genetic differences between crop species or by different cultural practices in different regions. Taken together, advances in the collection and documentation of archaeobotanical data indicate that domestication was a dynamic and diverse evolutionary process over many human generations,[29, 30] rather than a sudden "agricultural revolution" in economy and plant morphology.

Dorian Q Fuller, Institute of Archaeology,
University College London

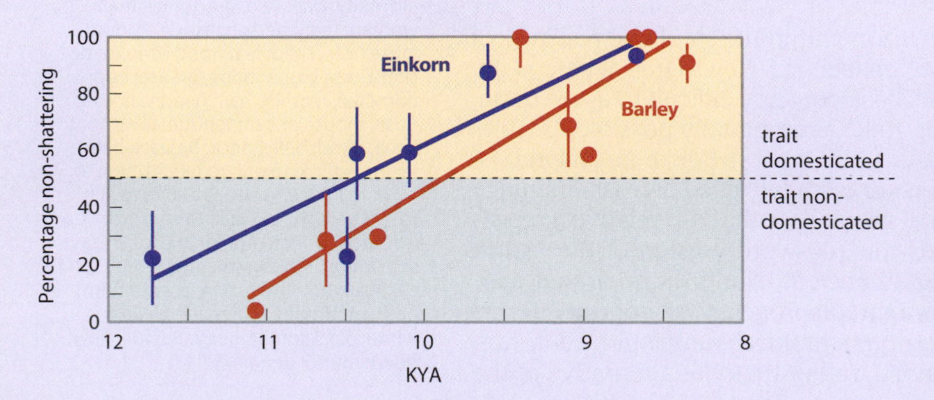

Figure 1: Gradual evolution of a domestication trait based on archaeobotanical evidence. The percentage of the non-shattering trait in Near Eastern archaeological sites of different dates is shown, as estimated from the rachis remains of barley and spikelet fork remains of einkorn wheat. Vertical bars indicate standard deviations. Populations with average percentages falling above the gray area are taken to be domesticated for this trait. [From Fuller DQ et al. (2012) *J. Exp. Bot.* 63, 617. With permission from Oxford University Press.]

steppe region north of the Caspian Sea. The most likely age for Indo-European was estimated to be 7.1–10.4 KY. Thus, both the place and date of origin appear consistent with a farming spread for the language family.

Osteological evidence from **craniometry** (the measurement of skull dimensions) has also contributed to the debate. Analysis of measurements of 116 Mesolithic (hunter-gatherer) and 165 Neolithic (farmer) crania from Southwest Asia and Europe[54] allows a null model of isolation-by-distance (both geographic and temporal) to be rejected. A model with a continuous diffusion of farmers, rather than an indigenous adoption of agricultural technologies by Mesolithic populations, fits the data significantly better. However, extension of the craniometric approach to more outlying regions of Europe[75] suggests that hunter-gatherers in these regions themselves adopted farming practices, and thus that it is impossible to apply a uniform model to this widespread and complicated demographic transition.

Different models of expansion give different expectations for genetic patterns

As stated at the outset of this chapter, two simple models have been used to explain the spread of agriculture: in the first model, acculturation (also known as **cultural diffusion**), the farmers did not move, but the technology and ideas of agriculture did, as they were adopted by indigenous hunter-gatherer populations. The second model involves gene flow, in which the farmers themselves moved, taking agricultural practices with them. While there are many different possible gene-flow models, the most widely discussed is demic diffusion, also known as the wave of advance. Here, spread is stimulated by population growth (itself a result of increasing availability of food) and local migratory activity. What patterns do we expect to find in the genetic diversity of modern European populations under these different models? The answer depends greatly upon the starting conditions (the gene pool carried by the preexisting Paleolithic hunter-gatherer populations compared with that of the Neolithic farmers) and this important issue will be addressed below. Let us first assume for simplicity that the dispersed groups of hunter-gatherers have a uniform composition of genetic variants, completely distinct from that of the farmers.

In a purely demic diffusion model without interbreeding, the migration of farmers would have the effect of replacing the gene pool of the indigenous Europeans with that of the Near Eastern immigrants. We would expect to see strong affinities between the genes of all modern European populations and those of the populations in the Near East, but little genetic differentiation within Europe other than that arising via drift. In a purely acculturation model the spread of farming should not be associated with major genetic changes, and we expect a sharp distinction between the genes of most Europeans and those of Near Eastern populations (Figure 12.1, left panel). An intermediate model, in which farmers spread and interbred with indigenous peoples (Figure 12.1, right panel), should have left its imprint as a **cline**, or gradient, of gene frequencies (**Section 6.8**) from the Near East toward the northwest. The debate that has arisen around the genetic evidence is in fact based not upon either extreme position, which nobody accepts as reasonable, but on versions of a mixed acculturation–demic diffusion model like that espoused by Ammerman and Cavalli-Sforza, and the consequences of such mixed models for modern European genetic diversity.

The early work on these models was notable because of its use of computer simulations as a means to identify the relative importance of different variables, and to estimate the power of different analytical methods to indicate the genetic outcomes of the chosen models. **Figure 12.10** shows an example of a simple simulation of the wave of advance, which used a migration rate of 1 km per year, estimated from the overall archaeological picture of agricultural spread in Europe, and supported by later data from many more sites (0.6–1.3 km per year[52]).

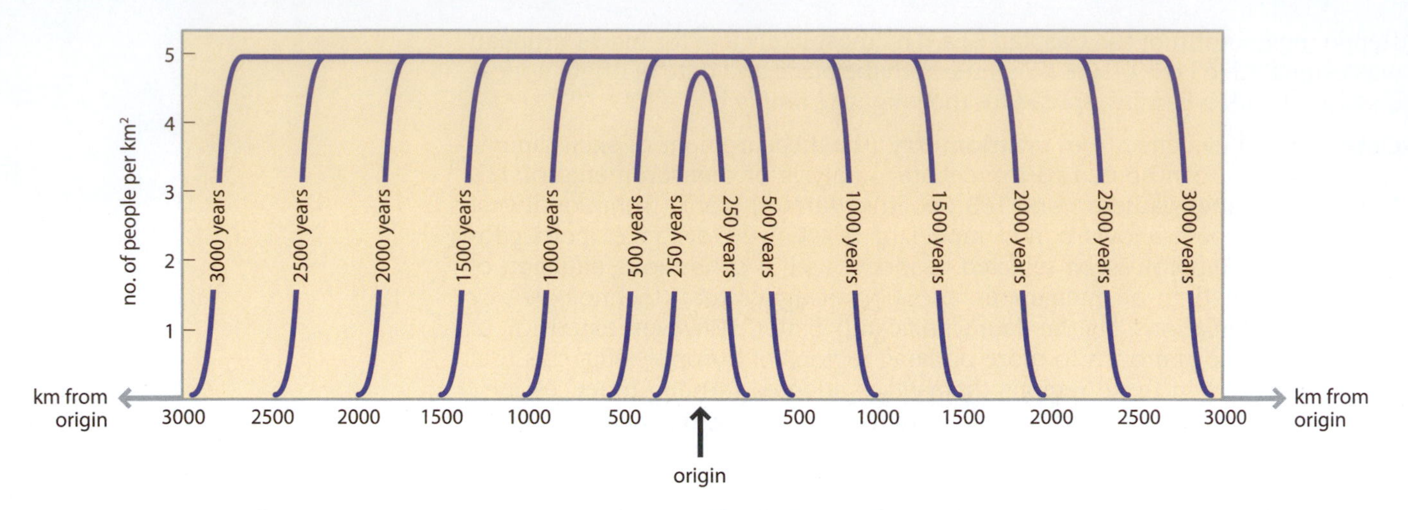

Figure 12.10: Simulating the wave of advance.
Curves showing local population density indicate the position of the wave of advance at 500-year intervals, given a rate of advance of 1 km/year. [From Ammerman AJ & Cavalli-Sforza LL (1984) Neolithic Transition and the Genetics of Populations in Europe. Princeton University Press. With permission from Princeton University Press.]

Models are oversimplifications of reality

So far, we have been considering an unrealistically simplistic situation; in reality, there are many complicating issues. First, most available genetic data refer to modern populations, separated in time by a few hundred generations from those in which agriculture was first established; there has been much opportunity for genetic drift and for subsequent migration events, both small and large scale, to reshape allele frequencies. For example, the so-called barbarian invasions ~1500 years ago involved the immigration of peoples from as far afield as Russia and Mongolia. The genetic impact of such events is unclear.

The genetic composition of indigenous Europeans before the advent of agriculture is unlikely to have been uniform. **Figure 12.11** illustrates the two pre-agricultural demographic events in European prehistory—the original occupation in the Paleolithic, which followed a similar route to agricultural expansion, and the later re-expansion after the last glacial maximum from southerly **glacial refugia** in the peninsulas of Iberia, Italy, and the Balkans. These movements are likely to have produced heterogeneous patterns of allele frequencies among hunter-gatherer populations well before agriculture, and could themselves have resulted in clinal patterns, for example by repeated founder events during the initial Paleolithic expansion that colonized the continent.

Because each ultimately shared a common origin, the indigenous Europeans and the incoming farmers are likely to have shared alleles, some at characteristically different frequencies, and some at similar frequencies. Furthermore, interactions between residents and incomers were probably more complex than simple indifference, intermarriage, or displacement. Factors such as warfare and disease on the one hand, and trade and cooperation on the other, may have been important.

Finally, archaeological evidence suggests that the rate of agricultural advance was far from uniform, and involved episodes of rapid expansion followed by

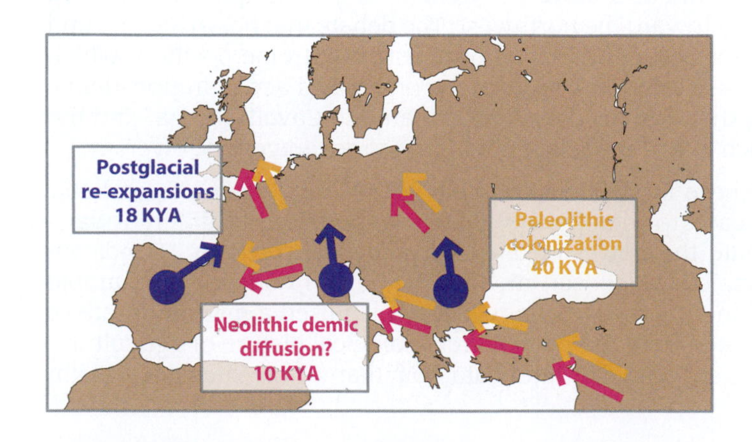

Figure 12.11: Major population movements in European prehistory.
Arrows indicate Paleolithic colonization, postglacial expansion, and Neolithic input from the Middle East.

periods of stasis.[11] Dates for the earliest western Mediterranean Cardial ware sites ranging from central Italy to Portugal are barely distinguishable, indicating that the spread here was very rapid, perhaps taking only six generations,[81] and suggesting colonization by use of boats. These kinds of processes are inadequately described by the simple wave of advance model.

Having stated these caveats, we can now ask what the genetic evidence can tell us. Most data come from analyses of modern populations, but ancient DNA studies are now contributing increasingly to our understanding.

Principal component analysis of classical genetic polymorphisms was influential

Extensive data on European allele frequencies for **classical polymorphisms** (such as blood groups, enzymes detectable in blood, and HLA—see Box 3.1), representing individual loci in the nuclear genome, have been collected and analyzed to test the hypothesis of Neolithic demic diffusion.[18, 46] In all, 94 alleles at 34 loci distributed among 16 different chromosomes were studied, with sample sizes per population varying from a few tens for some loci to almost 3000 for alleles of the ABO blood group system.

Patterns for individual loci are heterogeneous, and must be treated with caution: some might be under natural selection, some might be uninformative (having had no initial difference in frequency between farmers and hunter-gatherers), and some might indicate different population movements. Unlike selection, migration has the same effect on all genes, so in order to circumvent these problems and to discern broad underlying patterns, **principal component analysis** (PCA; **Section 6.3**) was used to analyze data from all alleles simultaneously.

The output of PCA is usually a plot of one principal component (PC) against another, upon which populations are represented by points (see Figure 6.6). Cavalli-Sforza and colleagues developed a more useful method, displaying the output geographically as **synthetic maps** of individual PCs. These should allow common patterns due to single migrations to be abstracted from data on all alleles, and the element of a geographical map is an important **heuristic** device that apparently enables easy recognition of the various patterns, and has led to their discussion in terms of migrations and the archaeological record.

Interpreting synthetic maps

Synthetic maps may appear compatible with particular explanations, but this does not prove that the explanations are correct. Some critics[43] have likened these maps to "Rorschach inkblot" tests, in which the viewer is asked to use their imagination to say what a particular pattern suggests. Further doubt has been cast on their usefulness by simulations demonstrating that the patterns seen in the various PCs, and interpreted to reflect particular population movements, can actually arise as mathematical artifacts under a model of temporally and spatially constant homogeneous short-range migration processes[49].

The first four PCs of European allele frequencies together summarized 68% of the total variance in the data; it is the synthetic map of the first PC, summarizing 28% of the variance, which was taken as providing support for the Neolithic demic diffusion hypothesis (**Figure 12.12**). The map has a strong focus in the Near East, and an approximately concentric gradient extending out to the northwestern fringes of Europe that resembles the map of dates of the arrival of agriculture (Figure 12.8). There need be no simple relationship between the proportion of variance summarized by a PC and the genetic contribution of migrants who contributed to it; indeed, simulation studies[5, 59] suggested that the observed patterns would occur only if the proportion of incoming alleles was greater than 66% at the time of admixture.

Synthetic maps of genetic data may show patterns, but do not indicate when those patterns were established. So the first PC (setting aside possible artifacts)

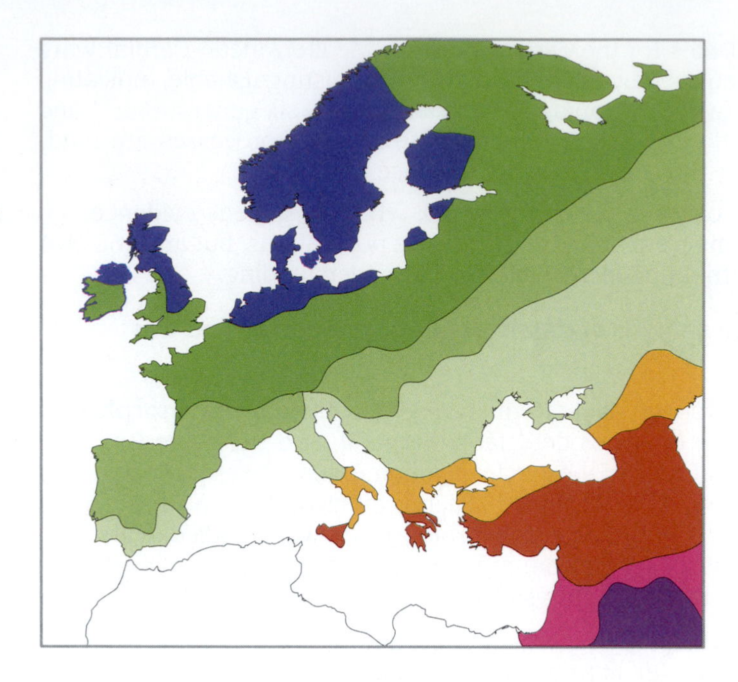

Figure 12.12: Synthetic map of Europe and Western Asia obtained using the first principal component of classical genetic data.
[From Cavalli-Sforza LL et al. (1994) The History and Geography of Human Genes. Princeton University Press. With permission from Princeton University Press.]

may be compatible with the demic diffusion hypothesis, but why should we reject an alternative origin for this pattern in the initial colonization of Europe in the Paleolithic? Support for a Neolithic origin for the gradient of classical gene frequencies has been adduced from Mantel testing (see **Section 6.8**) of partial correlations between three kinds of distance: genetic, geographical, and temporal—the latter based on archaeological evidence for the times of origin of agriculture in different regions. When geographical distances are held constant, genetic distances for several loci show significant correlations with the agriculturally based temporal distances,[72] though this analysis still does not support any particular date range. The other argument that has been used to support a Neolithic origin for the gene frequency patterns is based on the farming–language co-dispersal hypothesis (**Section 12.4**). If Indo-European languages are no more ancient than the Neolithic, as many believe, then the fact that they have spread so widely within (and beyond) Europe suggests a corresponding spread of agriculturalists. The Basques have played an important role here, since their stubbornly non-Indo-European language has suggested to many researchers that they might retain characteristics of a pre-agricultural past (**Box 12.2**).

mtDNA evidence has been controversial, but ancient DNA data are transforming the field

Much of the fiercest debate on the issue of Neolithic demic diffusion in Europe was engendered by the patterns of modern mtDNA diversity and their interpretation. Subsequent arguments became particularly polarized and bitter, and some of the history of the debate is summarized in **Box 12.3**.

Inspection of the frequencies of mtDNA haplogroups in modern European populations (**Figure 12.13**) shows that haplogroup (hg) H dominates the entire continent, with hg U the next most common lineage, and no obvious clinal pattern in these or in any other lineages. It is this lack of clinal patterns, together with the results of **founder analysis** and estimates for the ages of lineages based on HVSI sequences (Box 12.3), that gave rise to the view that mtDNA indicated a major contribution of Paleolithic lineages in Europe, and relatively little influence of incoming farmers. Here, the majority hg H (TMRCA estimated as 19.2–21.4 KYA in the European sample) was seen as a Paleolithic substrate,[62] relatively unaffected by younger incoming lineages from the Near East, which constituted less than 20% of the whole.

Box 12.2: The Basques: a Paleolithic relic in Europe?

Whichever European genetic study you examine, one thing will be clear: the Basques are special. Along with Saami (Lapps), Sardinians, and Icelanders, they are the traditional outliers within the European gene pool. Studies using **paleoecological** information to partition Europe into subregions[62] were happy to free the Basques from the straitjacket of ecology, and consider them separately (Figure 12.13). Admixture approaches to European prehistory often take the Basques as a proxy for the Paleolithic substrate.

While Saami are distinguished by their geographical extremity and lifestyle (reindeer herding), and Sardinians and Icelanders dwell on islands, prone to founder effects and drift, the Basques today are an agricultural community within continental Europe, unencircled by impenetrable geographical barriers (the mountains of the Pyrenees lie *within* the Basque territory). So, why the special status? The key is their language, a remarkable **isolate** within the relatively homogenous Indo-European landscape, and it is this linguistic quirk that has suggested to many that the Basques somehow withstood the introgressions of the Neolithic, and even today preserve some aspects of a Paleolithic population. Mark Kurlansky cites the connection of *aitzur* (hoe), *aizkora* (axe), and *aizto* (knife), with *haitz*, meaning "stone," as a direct link to the Stone Age.[36] In fact, there is evidence from place-names, coins, and inscriptions that other European populations (for example, Etruscans in Italy) also preserved non-Indo-European languages over 2000 years ago, but lost them relatively recently. The supremacy of the Roman Empire was responsible for the **Romanization** of many languages. The Basques dealt well with the arrival of the Romans, serving as soldiers for them as far afield as Hungary, the lower Rhine, and Hadrian's Wall, in Northern England. Linguistic replacement is often more rapid than replacement of genes, and this may suggest that there may be other European populations that could be considered as Paleolithic relics, aside from the Basques.

There is no doubt that Basques have unusual allele frequencies for some classical polymorphisms—they have the highest known frequency of the Rhesus-negative blood group, a finding made way back in the 1940s. However, the combination of linguistic uniqueness, genetic differentiation, and a desire for self-determination may have proved to be a self-fortifying cocktail. Once interest in a particular population has been aroused, it becomes an object for special study, and prophecies may become self-fulfilling. If Galicians or Gascons had preserved non-Indo-European languages, would they have been studied in similar depth, and considered as Paleolithic relics? The assumption that "Basque equals Paleolithic" seems far too simple to be true. A genomewide analysis (from Catalan researchers) of 300,000 SNPs shows that the Basques are not particularly distinct from other populations of the Iberian peninsula, and that French and Spanish Basques are differentiated from each other.[37] However, the controversy continues: an independent genomewide study (from a Basque group)[64] comes to completely opposite conclusions.

Origins are part of today's politics for many populations, but especially so for the Basques. As Roger Collins put it: "a politicization of ... recherché anthropological arguments about the Stone Age is a distinctive feature of the ideological underpinning of modern Basque nationalism, in which the longevity of the people and the continuity of their occupation of their western Pyrenean homelands are arguments of central importance".[21]

The availability of increasing numbers of complete (~16.5 kb) mtDNAs, rather than ~350-bp HVSI sequences, has allowed a more complete assessment of diversity patterns. **Figure 12.14** shows the distribution of pairwise nucleotide differences between coding region sequences.[28] Haplogroup H sequences show an identical mean and mode of 6 nucleotide differences, while the corresponding values for hg U are 18 and 22 differences respectively. These patterns suggest a recent population expansion among hg H sequences, and a more ancient one in hg U. Estimation of past population sizes using a Bayesian coalescent method indicated that the population increase for hg H began around 9 KYA, while hg U expanded 10–20 KYA, followed by a slight decrease 5–6 KYA. Both lineages show similar patterns of growth from 4 KYA to the present. These data suggest that hg U predominated in Paleolithic pre-agricultural populations, and hg H in later-expanding farmers, a picture contrasting with the earlier view of the modern mtDNA evidence, described above, in which most lineages are claimed to show continuity from Paleolithic to Neolithic. The status of haplogroup H, in particular, now shifts from an assumed Paleolithic substrate to a possible product of the Neolithic expansion itself.

Box 12.3: The polarization of debate over European agriculture: Paleolithic United vs. Neolithic Wanderers

The history of the debate over the Neolithic contribution to the gene pool of modern Europeans is interesting, because it exemplifies how arguments become polarized when philosophical and methodological differences combine with salvoes of barbed comments, dividing researchers into bitterly opposed camps. To the outsider (rather like the non-football fan) it may seem ridiculous that people can become so partisan.

The study that triggered the controversy[61] analyzed **HVSI** sequence variation in 821 individuals belonging to 14 populations, one representing the Middle East, the rest European. The analysis took a **phylogeographic** approach: (1) the relationships between haplotypes were displayed phylogenetically using **reduced median networks** (see Section 6.4) and five clusters were identified; (2) geographical distributions of haplotypes were examined, and the most "ancestral" haplotypes within clusters were taken to suggest their sources; (3) TMRCAs were estimated based on the mean pairwise sequence difference between their European members.

The conclusions suggested that farming had been a largely culturally transmitted phenomenon, because: (1) cluster distributions did not show a clinal pattern; (2) all but one of the TMRCAs were >10 KYA, taken to indicate that >80% of mtDNA haplotypes were brought into Europe in the Paleolithic. The star-like shapes of these clusters suggested population growth, ascribed to postglacial expansion; (3) the two youngest subclusters (totaling only 12%) were linked by ancestral haplotypes found only in the Middle East, suggesting that they originated there and were spread as a minor contribution during the Neolithic.

Publication was followed by a stormy debate. Initial attacks criticized the use of mtDNA, on the grounds of the homogenizing effects of **patrilocality** on a maternally inherited locus and possible problems with the mutation rate. Two other criticisms exposed philosophical rifts: (1) the phylogeographic approach was questioned, with Cavalli-Sforza arguing that frequency-based methods (as used for classical polymorphisms) should be employed instead. Later critics have been outspoken, mocking phylogeography as "astrologenetics,"[19] and calling for its reliability to be tested by modeling. (2) The apparent assumption that the age of a cluster equates to the age of a population was also attacked. The analogy of a future European colony on the planet Mars was made[4]—the MRCA of the colonists' mtDNAs might lie in the Paleolithic, but this would not tell us much about the age of the colony (Section 6.6). Counterattacks[63] argued that frequency-based approaches were inappropriate for nonrecombining loci, and that the "lineage = population" assumption had not actually been made, but instead that European founder haplotypes had been used as a baseline from which founder events associated with haplotype clusters could be identified and dated.

Later mtDNA studies increased haplotype resolution and introduced a cladistic classification, but again revealed no obvious geographical patterns (haplogroup H predominating in all populations—Figure 12.13) and mostly Paleolithic TMRCAs.[62] When allowance was made for multiple dispersals of hg H, the modern European Neolithic component was estimated at around 20%, and the major demographic impact again concluded to be postglacial expansion in the late Upper Paleolithic. Opponents of phylogeographic methods used spatial autocorrelation to analyze HVSI data,[70] concluding that there was very little geographical structure to mtDNA diversity in most of Europe, but clinal patterns around the Mediterranean Sea, consistent with their spread during the Neolithic. No evidence was found of northward postglacial expansions. Counter-accusations of methodological errors followed, and the quality of the debate was not improved by an important table being garbled by publisher's formatting errors.

As the main text describes, the field is now being moved on thanks to the increasing availability of whole mtDNA sequences and ancient mtDNA data.

Data from ancient mtDNA

An alternative to indirect deductions from modern population samples is to analyze human remains from the relevant prehistorical periods directly. Because of its relatively high probability of survival in skeletal material (**Section 4.10**), mtDNA is particularly suited to ancient DNA analysis, and suitable datasets are becoming available. **Figure 12.15** summarizes data from several studies on early European Neolithic farmers and hunter-gatherers, based on short (<400 bp) sequences (largely HVSI) from which haplogroups can be deduced.[28, 33, 68] The results need to be interpreted with caution, because sample sizes (for hunter-gatherers in particular) are small. Meaningful conclusions about regional differentiation are not possible, and most of the hunter-gatherers come from central and northern Europe, while the farmers are more widespread. Taken at face value, however, there is a marked discontinuity between the two population

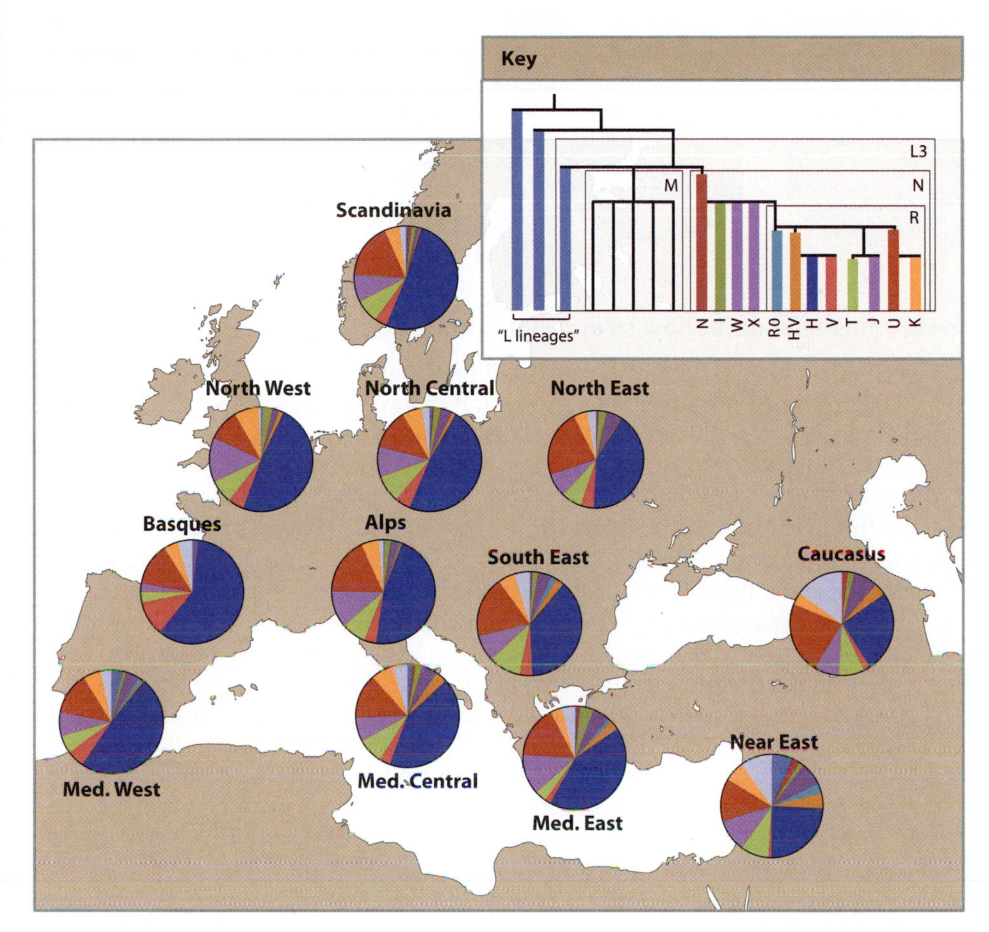

Figure 12.13: Distribution of major mtDNA haplogroups in Europe. Each pie chart shows the frequencies of major haplogroups in each of 12 subregions defined on the basis of paleoecology [Gamble C (1999) The Palaeolithic Societies of Europe. Cambridge University Press]. Note the relatively high frequency of African haplogroups (informally referred to here as "L lineages") in the Near Eastern and south European populations, the marked difference between Near Eastern and Caucasian populations and Europe, and the overall similarity between European populations, with the exception of the Basques, who show markedly higher frequencies of haplogroups H, V, and U than the rest. Sample sizes in Europe range from 156 (Basques) to 456 (North West) and are 1088 for the Near East. The tree above shows the phylogenetic relationships of the different haplogroups. *Pale blue* sectors within pie charts represent a collection of other minor haplogroups. [Data from Richards M et al. (2000) *Am. J. Hum. Genet.* 67, 1251.]

groups. The hunter-gatherers show low diversity, carrying haplogroup U at high frequency (~80%), and hg H at only 7%, while the farmers show greater diversity, carrying hg H at ~28%, and hg U at only 18%. Larger sample sizes, and whole mtDNA genome sequences from ancient material, will be important in strengthening the apparent evidence for a large contribution of Neolithic farmer lineages to the modern mtDNA pool.

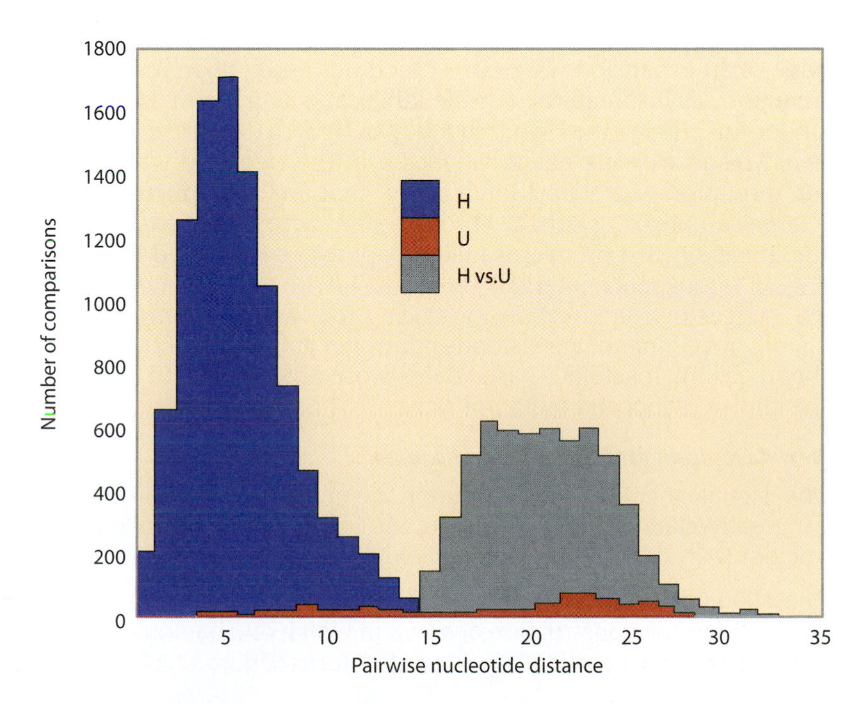

Figure 12.14: Pairwise nucleotide distances for complete mtDNA sequences in haplogroups H and U. Based on 145 hg H and 42 hg U complete sequences from an unbiased set of 353 European mtDNAs. [From Fu Q et al. (2012) *PLoS One* 7, e32473. With permission from Public Library of Science.]

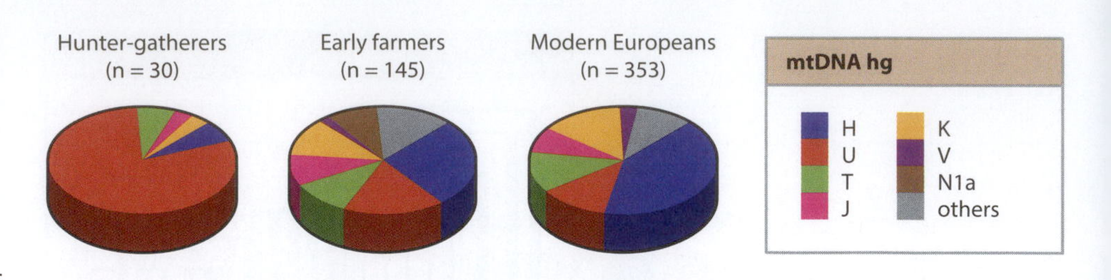

Figure 12.15: mtDNA haplogroups in hunter-gatherers and early European farmers are very different. Data are based on short (<400 bp) segments of mtDNA. [Adapted from Fu Q et al. (2012) *PLoS One* 7, e32473. With permission from Public Library of Science. Additional data from Sanchez-Quinto F et al. (2012) *Curr. Biol.* 22, 1494; and Hervella E et al. (2012) *PLoS One* 7, e34417.]

Y-chromosomal data show strong clines in Europe

The broad picture of Y-chromosome diversity within Europe was established by two large-scale studies[66, 69] analyzing binary polymorphisms, and demonstrating clear southeast–northwest gradients for some major haplogroups (**Figure 12.16**). The two studies differed in their philosophies: one[66] used spatial models to confirm statistically significant clinal patterns, but failed to date any of the lineages it described, while the other[69] rejected a spatially explicit approach, but estimated TMRCAs of lineages based on data from one tetra- and two dinucleotide repeat loci. The latter study followed a phylogeographic path by explicitly linking particular lineages with specific archaeological cultures such as Gravettian (hg I) and Aurignacian (hg R), based on haplogroup distributions and age estimates. Furthermore, it linked postglacial expansions from Ukraine and the Iberian peninsula with two major Y lineages [hg R1a1 and a lineage equivalent to the current R1b1b2 (R1-M269), respectively]. Based on the suggestion that haplogroups G, J2 (J-M172), and E1b1b1 (E-M35) represented probable Neolithic inputs into Europe from the Near East, and a simple measure of their overall proportion, it was suggested that the Neolithic contribution was only ~22%. The only major European-specific lineage, hg I, has widely been proposed to be a signal of Paleolithic contribution to modern European populations.[65] Both studies identified the major Western European haplogroup R1b1b2, carried by some 110 million European men today, as a likely "Paleolithic substrate," present at increasing frequency from the southeast to the northwest. This position thus echoes the original view of the origin of mtDNA haplogroup H. Notably, Y-hg R1b1b2 (like mtDNA hg H) reaches very high frequency (>85%) in the allegedly Paleolithic Basques.

The clines identified by both major studies were taken to indicate the introgression of relatively small numbers of Neolithic Y chromosomes into a Paleolithic substrate. However, a different interpretation comes from considering the nature of a population expansion. Simulations show that a new allele arising at the edge of an expansion wave can sometimes "surf" upon it, increasing rapidly in frequency as it spreads across the landscape away from the mutation source.[26] Under this model, the high frequency of hg R1b1b2 in the northwest of Europe could result from its being caught up in the Neolithic wave of advance. This interpretation was claimed in a study[3] that analyzed microsatellite diversity in a large sample of hg R1b1b2 chromosomes across Europe. Coalescent dating of the lineage based on microsatellite haplotypes estimated its mean age as ~6.5 KY (95% confidence interval: 4.6–9.1 KY), with a per-generation growth rate of 2%, suggesting rapid expansion during the Neolithic period. Subsequent criticism[17] (reflecting the persistently controversial nature of this debate) suggested that these microsatellite-based dates were unreliable, and unduly influenced by the choice of loci and mutation rates.

New developments for the Y chromosome

The significance of the distribution of modern Y diversity in Europe would be greatly clarified by dating methods independent of microsatellites, and by ancient DNA data. Both are now yielding useful insights.

The 1000 Genomes Project's unbiased ascertainment of many Y-chromosomal SNPs allows the construction of a tree in which one haplogroup, R1b1b2, stands out because of its particularly "star-like" form (**Figure 12.17**), indicating a recent population expansion. This tree is difficult to date because the low-coverage sequences used include only a proportion of the SNPs. But when sufficient sequence data become available, it should be possible to estimate the age of the expansion from SNPs themselves, rather than relying on microsatellites.

Retrieving reliable ancient DNA sequences of nuclear loci such as the Y chromosome is more difficult than for mtDNA. However, some data on Neolithic cemeteries and individual specimens have appeared, based on PCR assays targeted at very small (60–80 bp) fragments. These data must be regarded with caution, because males within individual cemeteries may be quite closely related, and the sites yielding data are few and widely separated. However, pooling all the current data from four Neolithic sites in Germany, France, Catalonia, and Italy (**Ötzi**, the Tyrolean Iceman; Box 4.3) provides a picture very different from current western European samples (**Figure 12.18**). The predominant haplogroup is G2a, very rare today, and there are no examples of the now common hg R1b1b2 chromosomes. Clearly, more data are required, but it seems possible that there is a real discontinuity between Neolithic Y lineages and those of

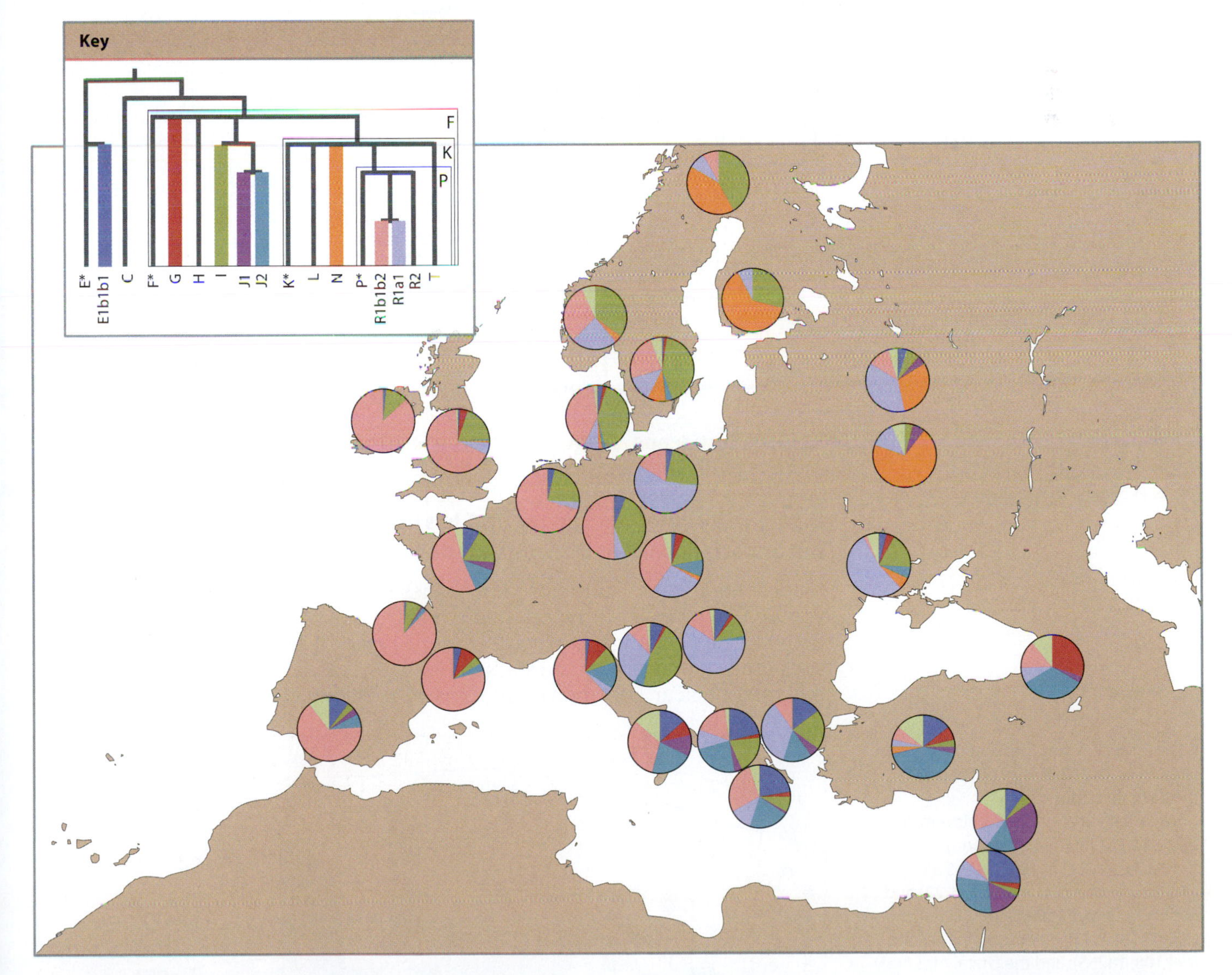

Figure 12.16: Distribution of major Y-chromosomal haplogroups within Europe.
Pie charts show the relative frequencies of different haplogroups, proportional to sector area. The tree above the map shows the phylogenetic relationships and names of the haplogroups. [Most data are from Semino O et al. (2000) *Science* 290, 1155. Additional information as follows: Ireland (Moore L et al. (2006) *Am. J. Hum. Genet.* 78, 334), Sweden (Karlsson A et al. (2006) *Eur. J. Hum. Genet.* 14, 963), Denmark (Børglum A et al. (2007) *Am. J. Phys. Anthropol.* 132, 278), Britain (King T & Jobling MA (2009) *Mol. Biol. Evol.* 26, 1093), and Norway (Jobling MA unpublished data.)]

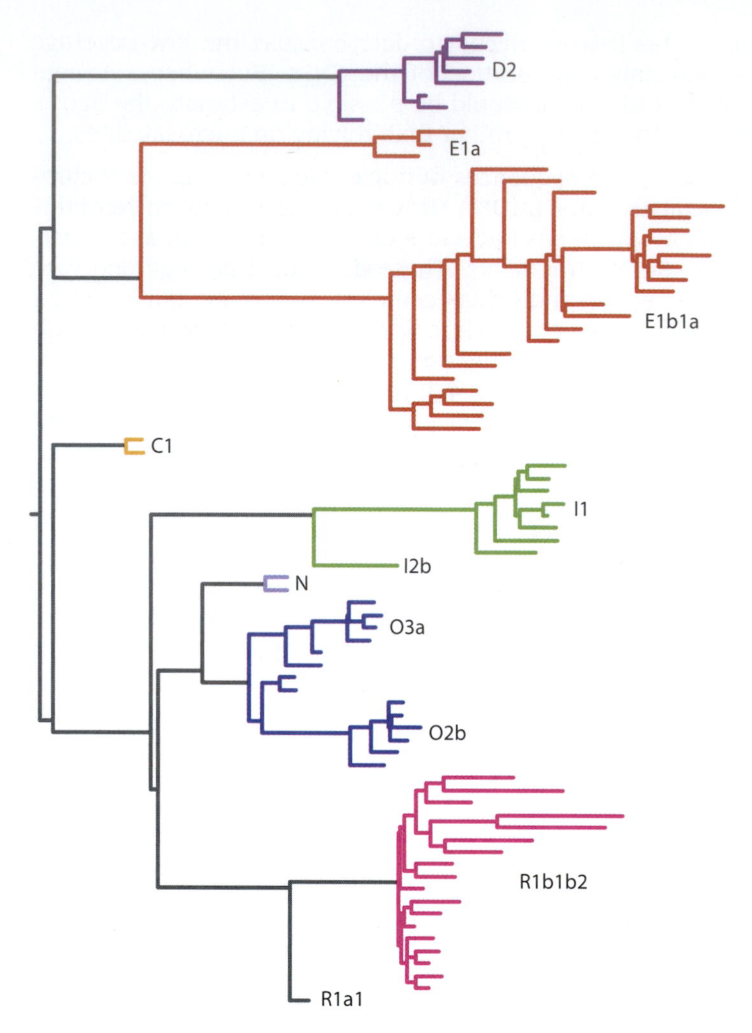

Figure 12.17: Y phylogeny constructed from multiple SNPs without ascertainment bias.
In this maximum-likelihood tree (maximum parsimony, the usual method for the Y chromosome, is precluded because of missing sites in the low-coverage 1000 Genomes Project data) hg R1b1b2 stands out as having short branches indicating a population expansion. [From 1000 Genomes Project Consortium (2010) *Nature* 467, 1061. With permission from Macmillan Publishers Ltd.]

modern Europeans, suggesting that hg R1b1b2 may have expanded in Europe more recently than the arrival of agriculture.

Biparentally inherited nuclear DNA has not yet contributed much, but important ancient DNA data are now emerging

Mitochondrial and Y-chromosomal studies focus on single loci, with the disadvantage that they each reflect single realizations of the evolutionary process; both, too, are strongly subject to the influence of genetic drift, and possibly social selection based on the status of females or males. Turning to biparentally inherited nuclear DNA polymorphisms in the context of European prehistory

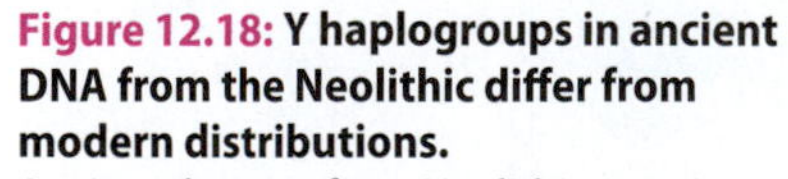

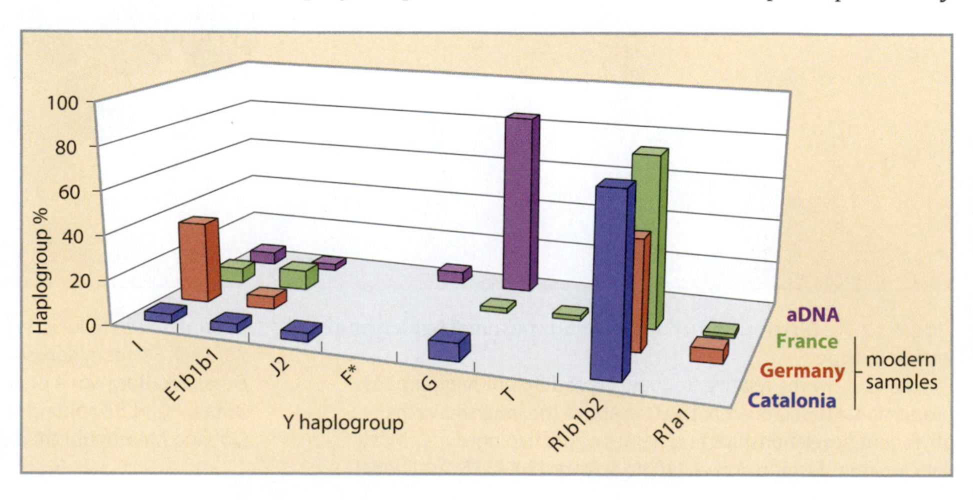

Figure 12.18: Y haplogroups in ancient DNA from the Neolithic differ from modern distributions.
Ancient data are from Neolithic remains dated to ~5–7 KYA in Derenburg, Germany [($n = 3$, Haak W et al. (2010) *PLoS Biol.* 8, e1000536); Aveyron, France ($n = 22$) and Avellaner Cave, Catalonia ($n = 6$, both Lacan M et al. (2011) *Proc. Natl Acad. Sci. USA* 108, 18255); and the Ötztal Alps, Italy ($n = 1$, Keller A et al. (2012) *Nat. Commun.* 3, 698). Modern comparative data on Germany and Catalonia are from Semino O et al. (2000) *Science* 290, 1155; and French data are unpublished observations from MA Jobling].

should in principle provide a more reliable picture. In fact, rather few informative studies have been carried out to date.

Admixture approaches using (as with the Y-chromosome study described above) modern Near-Easterners and Basques as "Neolithic" and "Paleolithic" parentals have been applied to genomewide microsatellite data on 377 loci.[7] European populations were from the CEPH-HGDP panel (Box 10.3), and therefore small in number, and not ideally distributed: French, northern Italians, Sardinians, Tuscans, Orkney Islanders, Russians, and Adygei (from the southern Caucasus). Near Eastern admixture is generally greater than 50%, and a number of different admixture methods show a significant negative correlation between this admixture proportion and distance from the Near East.

Genomewide SNP chips (**Section 4.5**) developed for association studies have been used to analyze European samples, but so far have yielded few insights into prehistory. Two large studies[38, 48] showed rather homogeneous overall patterns within Europe, but nonetheless significant correlations between geographic and genetic distance (see **Section 18.2**). Elevated genetic diversity and shorter linkage disequilibrium in the south[38] is compatible with a population expansion from southern to northern Europe, but this putative expansion cannot be dated—it could reflect postglacial expansion, or an element of Neolithic population spread, or a combination of the two.

Ancient DNA data

Just as the genomic DNA sequences from Neanderthal and Denisovan individuals have revolutionized our understanding of hominid evolution, similar sequences from European hunter-gatherers and early farmers should also be key to advancing the debate over the impact of the Neolithic transition. Because the relevant specimens would be substantially younger than the denizens of the Vindjia and Denisova caves, it ought to be simpler to retrieve sequence data. There are two problems, however: many key southern sites are likely to have maintained substantially higher average temperatures, affecting DNA survival, and proof of authenticity is more difficult, because any retrieved sequences (including mtDNA) are likely to be present in modern individuals, potentially including the excavators or laboratory personnel.

Next-generation sequencing of well-preserved samples from Sweden, where the Neolithic period involved coexistence of hunter-gatherers and agricultural groups for ~1000 years, has yielded ~27–97 Mb of sequence from the genomes of each of three Pitted-Ware Culture hunter-gatherers (4.4–5.3 KYA) and one Funnel Beaker Culture farmer (4.9 KYA) who lived <400 km apart.[71] Support for the idea that these sequences genuinely derive from the specimens themselves comes from a pattern of nucleotide substitution consistent with DNA damage in ancient samples. Comparisons of the different individuals cannot be done directly, because with only a few percent of the genome sequenced, there is very little overlap—a complex statistical procedure therefore has to be used. **Figure 12.19** shows a STRUCTURE analysis of the ancient samples in the context of modern Europeans. The hunter-gatherers most closely resemble modern Finns,

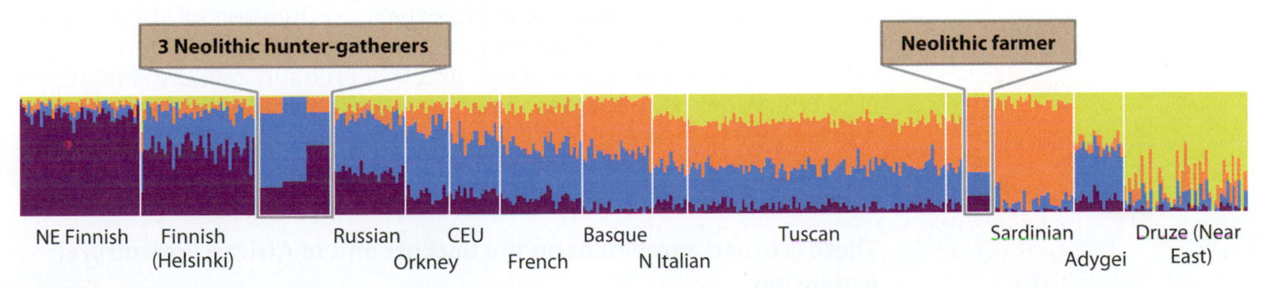

Figure 12.19: Population structure based on partial genome sequences in three Neolithic hunter-gatherers and one farmer, compared with modern populations.
Each individual is shown as a vertical line partitioned into colored segments representing the inferred membership of four genetic clusters. CEU, CEPH Utah residents with ancestry from northern and western Europe. [From Skoglund P et al. (2012) *Science* 336, 466. With permission from AAAS.]

while the Neolithic farmer clusters much more closely with Mediterranean European populations. **Strontium isotope analysis** of the remains indicates that all four individuals had grown up in the local area. These findings suggest that agriculture was brought to northern Europe by people who were genetically distinct from the local hunter-gatherers. Consistent with this, retrieval of ~17–41 Mb of genomic sequence from two hunter-gatherer skeletons from Northern Spain, dated to 7 KYA, shows them to be distinct from the modern inhabitants of southern Europe and the Iberian peninsula.[68]

What developments will shape debate in the future?

Ancient DNA results are having a major impact on this area of study, and new data and insights are likely to emerge. As well as the studies described above on mtDNA, the Y chromosome, and random genomic sequences, targeted studies of nuclear loci will also be useful. Most such studies to date have focused on the DNA-based diagnosis of lactase persistence status in ancient samples, for example, the demonstration that the common European lactase persistence allele (–13910T; **Section 15.6**) exists at only 5% frequency in Pitted-Ware Culture hunter-gatherers, compared with a modern Swedish frequency of 74%,[44] and is absent from a set of eight Neolithic skeletons from central, northeast, and southeast Europe.[16] Modeling of the spread of alleles likely to have been advantageous in adaptation of an agricultural lifestyle, and the analysis of these alleles in ancient samples, should complement each other: for example, the rapid spread of the –13910T lactase allele has likely been influenced by the allele-surfing phenomenon discussed above,[31] as well as positive selection.

Comparisons of modern mtDNA and Y-chromosomal patterns have suggested that sex-biased processes may have been involved in the spread of agriculture—perhaps men from incoming farming populations were at a social or economic advantage that led to the spread of their Y-chromosomal lineages at the expense of hunter-gatherer patrilines, but hunter-gatherer mtDNA lineages were able to persist. Alternatively, the change of lifestyle may have involved a change of mating practices. Modeling approaches to modern genetic data are consistent with a shift to patrilocality on the adoption of agriculture.[58] More work is needed to understand the effects of demographic histories on the distributions of lineages today, and to exclude differential genetic drift as an explanation for any differences.

As whole-genome sequences become available for increasing numbers of modern European samples, comparisons of mtDNA and X- and Y-chromosomal data will give unbiased pictures of diversity and sex-specific processes, linkage-disequilibrium-based approaches will help us to understand the dates at which general clinal patterns were established, and spatial analyses of agriculturally selected alleles will illuminate adaptive processes themselves.

12.6 OUT OF TROPICAL WEST AFRICA INTO SUB-EQUATORIAL AFRICA

The history and complexity of human habitation is much more ancient in Africa than it is in Europe. Hence, we might expect the question of African agricultural expansions to be even more mired in genetic controversy than the European Neolithic. This is not so, however, and this probably reflects: a later spread of agriculture, linked to the expansion of specific populations over the last 3 KY; a reasonable consensus on the linguistic evidence; a paucity of archaeological evidence compared with Europe; and, so far, an absence of vicious exchanges between advocates of opposing philosophies.

There is broad agreement on the background to African agricultural expansion

There is a reasonable consensus that agriculture spread from the Near East into Egypt between 9.5 and 7 KYA. **Emmer wheat** and barley were grown in the fertile Nile Valley, and farming cultures including domesticated pigs and goats

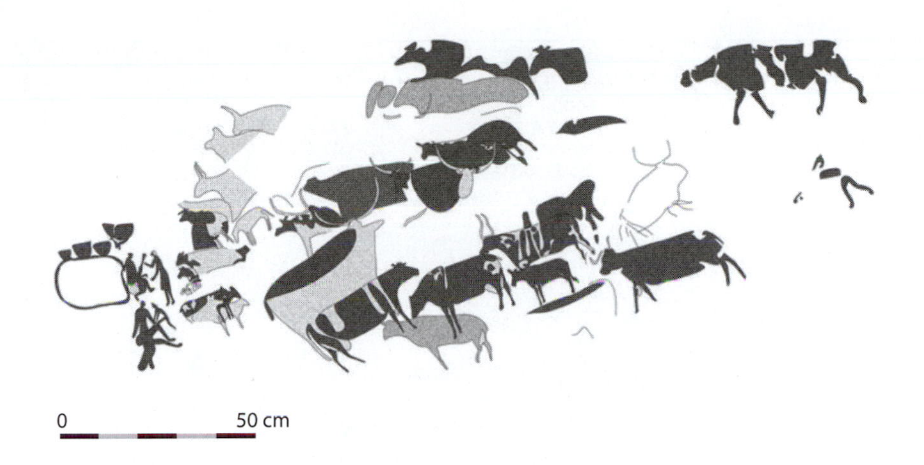

Figure 12.20: Tracing of rock art from Teshuinat II rock shelter, southwest Libya, showing Saharan pastoralists with their pots and cattle.
[From Dunne J et al. (2012) *Nature* 486, 390. With permission from Macmillan Publishers Ltd.]

became firmly established by 5.5 KYA. To the west, ample evidence of early cattle herding in the mountainous regions of the Sahara can be found in the form of striking rock paintings (**Figure 12.20**). Though this kind of evidence is notoriously difficult to date, milk fats have been chemically identified in food residues absorbed into pottery fragments themselves dated to the Middle Pastoral period (8.2–5.8 KYA),[25] thus providing an indirect dating for the cattle-herding practice. At this time the Sahara was not the inhospitable environment it is today, but semi-arid grassland including lakes, and rich in antelope and wild oxen, *Bos primigenius*, the progenitor of modern cattle. Cattle herders may have also been using wild cereal grasses and perhaps cultivating other local crops; they hunted with bows and arrows, and used flaked and polished stone adzes[50]—cutting tools with a flat heavy blade at right angles to the haft.

Around 5.5 KYA the Sahara became drier and lakes disappeared. The northern limit of the area in which the tsetse fly was endemic shifted southward: this fly carries the parasite that causes **trypanosomiasis**, fatal to cattle. Under the influence of these push and pull factors, the pastoralists moved south into the savanna regions, probably cultivating sorghum, millet, and yams. By 3.5 KYA, crops were widespread throughout the belt of savanna south of the Sahara, with farming communities using digging sticks, hoes, and axes in the shifting agriculture of woodland soils ("slash and burn"). Soil had to be managed carefully, and required different practices for different places, so this cannot be regarded as an unsophisticated farming culture.

Rapid spread of farming economies

Africa is different from other parts of the Old World in that there was (except in Egypt and some other northern areas) no distinct Bronze or Copper Age during which these softer metals were utilized. Particularly in sub-Saharan Africa, the first metal to be introduced was iron, possibly from northern Africa west of Egypt, or from Egypt itself. Its arrival south of the Sahara around 2.7 KYA coincided with the rapid expansion of farming economies throughout sub-Saharan Africa. This spread, a complex and poorly understood series of population movements, is known as the **Bantu expansion**, since Bantu languages are suggested to have been co-dispersed with agriculture over much of east, central, and southern Africa from a homeland in eastern West Africa. The linguistic evidence is the subject of the next section. Some archaeologists regard the term Bantu expansion as misleading, because it implies that the traits of iron-working, farming, and Bantu-speaking were spread together by the same people. In fact, there was a great variety of regional and local events of agricultural expansion linked to social, political, and cultural change. Though there is evidence that farming practices were established over a considerable area before the arrival of iron, its availability certainly made farming much more efficient; by 1.3 KYA domesticated species originating in equatorial Africa, stock-breeding, iron-working, and new forms of social organization were spread widely through the center and south of the continent.

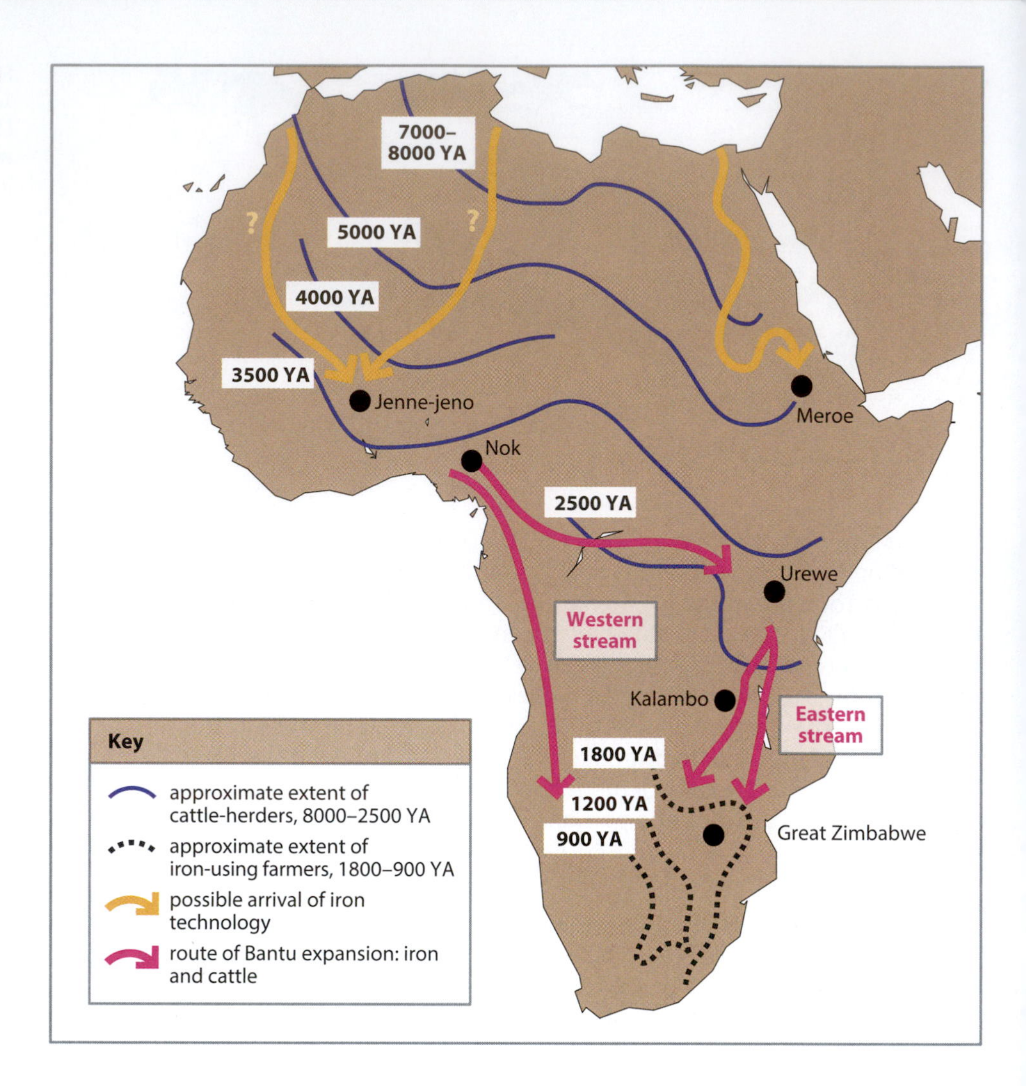

Figure 12.21: The spread of agriculture and iron-working in Africa.

There is much archaeological evidence for early iron-working, for example, in the remains of furnaces at Nok, in Nigeria. However, the details of the origins and pattern of spread of the expansion from this center southward, and possibly to another center in the east, Urewe, are far from clear. A simple picture of the probable routes and relevant dates is shown in **Figure 12.21**. Because of their high degree of homogeneity in the archaeological record, the early iron-working communities of eastern and southern Africa can be considered as a single archaeological culture, known as the **Chifumbaze complex**.[50]

There are serious practical difficulties in studying the archaeological evidence for agriculture in tropical Africa; the area is covered with trees, which makes fieldwork difficult; the important early crops were tubers, and the methods to study them (see Box 12.1) have only recently been introduced; and, last but not least, the political situation in many tropical African countries is a severe hindrance. The apparent simplicity of the archaeology of sub-Saharan Africa may simply reflect the relative paucity of evidence obtained so far.

Bantu languages spread far and rapidly

The distribution of language families spoken within Africa is shown in **Figure 12.22a**. The dominance of the Bantu subgroup of the Niger–Congo family in central and southern Africa is striking in comparison with the diversity of subgroups of Niger–Congo and of other families in the equatorial region and parts of the north. The ~600 languages within the Bantu group are closely related, with high mutual intelligibility, suggesting a recent common origin, and their distribution coincides well with that of the Chifumbaze complex. This has led to the idea that the modern speakers of Bantu, now numbering over ~220 million people (about

28% of Africans) spread over almost 9 million km^2, owe their wide distribution to the expansion of iron-working agriculturalists. Linguistically speaking, at least, this expansion has its origin in the Benue River Valley, between southeast Nigeria and western Cameroon, where linguists believe that the ancestral Bantu language was spoken.

Modern Bantu languages can be split into two major groups, spoken in the eastern and western parts of Bantu territory respectively. The boundary between the two follows the eastern edge of the equatorial forest, and the western branch of the Rift Valley, but is less clearly defined further south. The earliest dispersal of Bantu speakers seems to have been that of the western group in the equatorial forest (Figure 12.22b), with several distinct stages suggested by lexicostatistical studies. The final expansion took place into much of the southern savanna, where subsequent interaction took place between speakers of Eastern and Western Bantu.

Finding specific evidence to link the Bantu languages with the spread of agriculture is not simple, partly because the agricultural practices in the equatorial west were very different from those in the eastern savanna, and so we expect linguistic differences in agricultural terms from these different regions. However, some

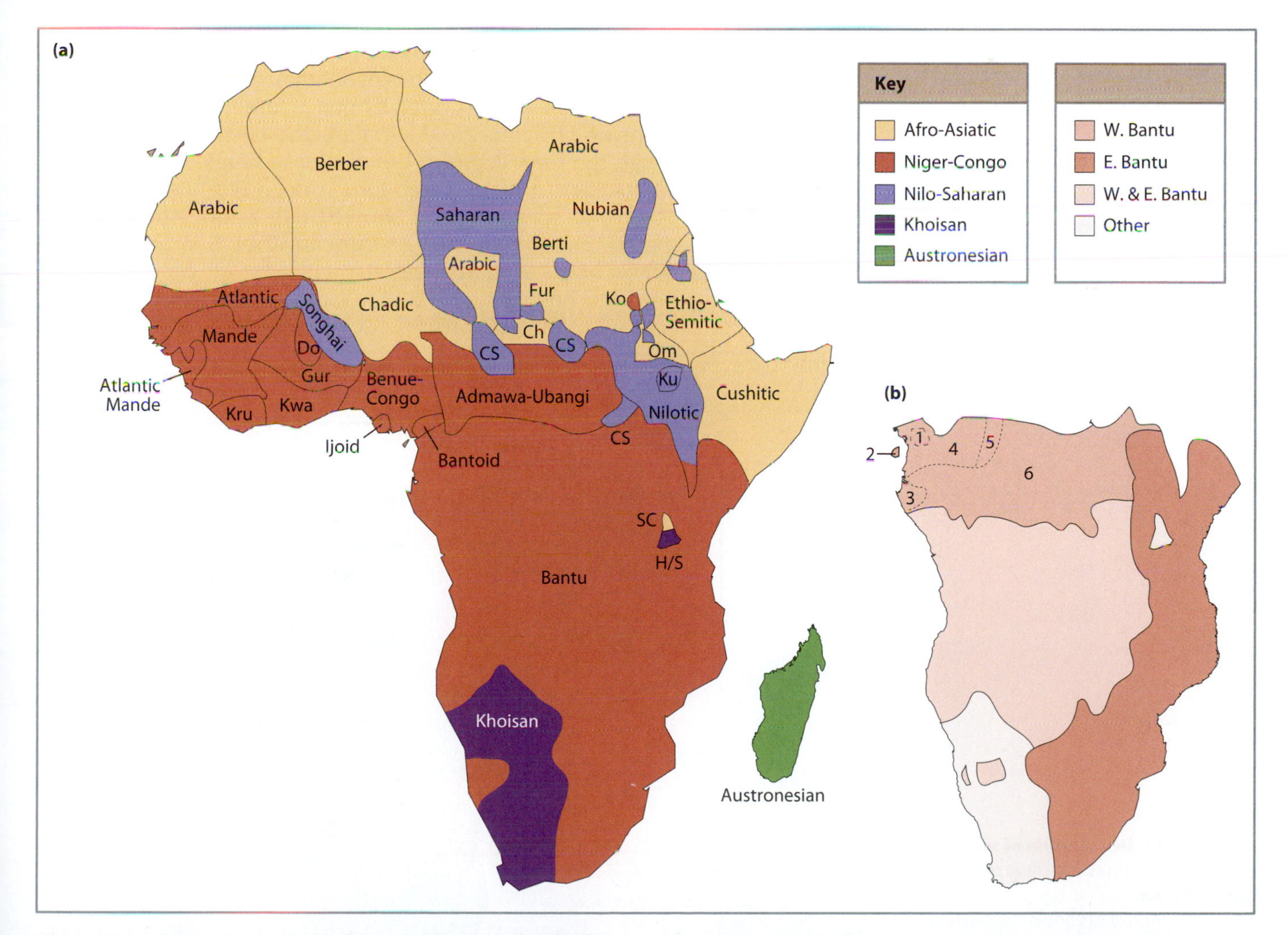

Figure 12.22: African language families, showing the distribution of Bantu languages.
(a) Bantu languages are widespread in central and southern Africa. Not all language families are named here. Do: Dogon; Ch: Chadic; CS: Central Sudanic; H/S: Hadza/Sandawe; Ko: Kordofanian; Ku: Kuliak; Om: Omotic; SC: South Cushitic. (b) The distribution of Bantu subfamilies. The numbers 1 to 6 refer to the successive stages of dispersal based on lexicostatistical evidence. [a, from Blench R (1993) In The Archaeology of Africa: Food, Metals and Towns (T Shaw et al. eds). Routledge. b, from Phillipson DW (2005) African Archaeology, 3rd ed. Cambridge University Press.]

words (for example, that for *goat*) can be traced to a common ancestral form and the geographical distribution in Bantu of many loan words for cattle, sheep, and cereal cultivation suggests that the people who spread the farming practices were largely Bantu speakers. A maximum parsimony approach to Bantu language classification led to well-supported trees that were claimed to be highly consistent with archaeological evidence for the spread of agriculture,[34] supporting the idea that the primary Western and Eastern Bantu languages split ~4 KYA. However, there is disagreement here, with other linguists disputing the early split, and considering Eastern Bantu as a later development, only ~2 KYA.[27]

Genetic evidence is broadly consistent, though ancient DNA data are lacking

Understandably, many studies of human genetic diversity within Africa have focused upon the issue of the origin of modern humans and early events in prehistory. However, genetic evidence relating to the spread of farming practices has also been presented.

Genomewide evidence

A genomewide study has been carried out in a set of 113 African populations, using a panel of 1327 polymorphisms, including 848 microsatellites and 476 indels.[73] In a Bayesian clustering analysis (**Figure 12.23**) the most geographically widespread cluster extends from the Mandinka of West Africa through central Africa to the Bantu-speaking Venda and Xhosa of South Africa. This corresponds closely to the distribution of the Niger–Congo language family, and seems compatible with the expansion of Bantu-speaking populations within the past few thousand years, though the time-depth of the cluster is unknown. STRUCTURE analysis also revealed substructure between East African Bantu speakers and West Central African Bantu speakers, and showed that South African Bantu have substantial shared ancestry with South African Khoe-San and western African Bantu, and lower levels of shared ancestry with East African Bantu. This picture is compatible with linguistic and archaeological evidence for distinct East and West African Bantu migrations into southern Africa, and admixture with local hunter-gatherer populations.

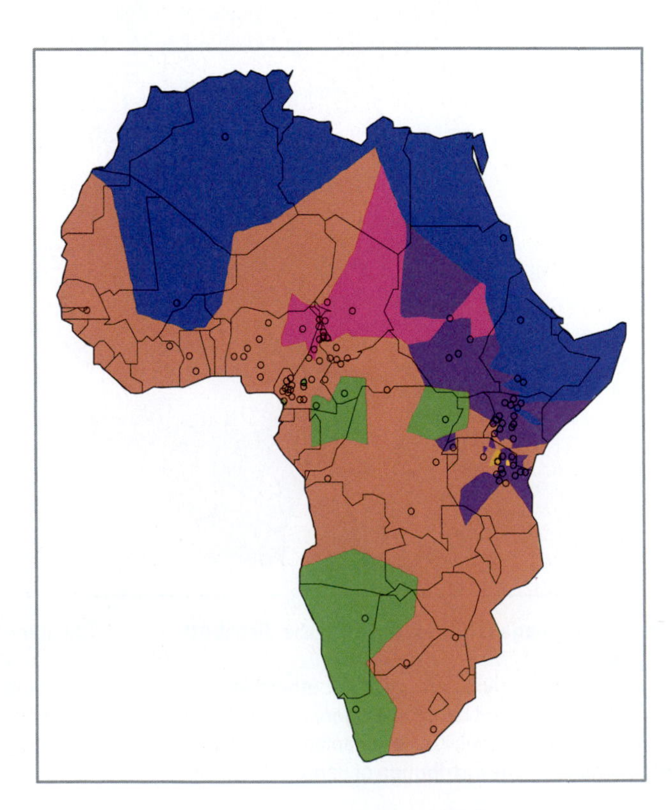

Figure 12.23: Distribution of six clusters based on genomewide diversity data within Africa.
Each color represents one of six clusters produced using a Bayesian clustering analysis assuming no admixture. The *brown* cluster corresponds well with the distribution of Niger–Congo languages. [From Tishkoff SA et al. (2009) *Science* 324, 1035. With permission from AAAS.]

Evidence from mtDNA and the Y chromosome

A comparison of linguistic and geographical variation with the patterns of diversity in mtDNA and the Y chromosome has suggested a sex-biased process in the expansion of Bantu speakers.[79] mtDNA HVSI sequences, and Y haplogroups defined by 50 SNPs, were analyzed in ~40 African populations. Mantel tests showed a strong partial correlation between Y-chromosomal genetic distances and linguistic distances, but no correlation between genetic and geographic distances. By contrast, mtDNA variation was correlated weakly with both language and geography. When Bantu speakers (but not speakers of other language groups) were removed, the correlation with linguistic variation disappeared for the Y chromosome but strengthened for mtDNA. This suggests that patterns of gene flow in Africa have differed for men and women, and specifically that Y-chromosome lineages spread by Bantu speakers act to homogenize Y diversity across large geographical distances. This process could have been driven by the introduction of **polygyny**, a practice that is more common among farming than foraging populations, and the consequent replacement of indigenous Y chromosomes (but not mtDNAs) by those of the incoming farmers. In a separate study,[51] resequencing data based on 780 bp of mtDNA coding region and 6.6 kb of the Y chromosome fitted a model of population growth in farming populations for mtDNA, but not for the Y chromosome. This could be explained by reduced effective population size for the Y or reduced effective migration rate, perhaps due to the respective practices of polygyny and patrilocality, associated with agricultural lifestyles.

Surveys of mitochondrial and Y-chromosomal diversity have also identified particular lineages that may have been spread with the Bantu expansion. In agreement with the studies described above, the candidate paternal lineages are much more dominant than the maternal lineages. Africa is notable for being home to the two deepest-rooting Y haplogroups, A and B, but these are present at only 13% overall in a sample of 3255 chromosomes (**Figure 12.24**); there is one predominant and derived lineage, haplogroup E1b1a, that accounts for over 59% of African chromosomes overall, and about 80% of chromosomes in Bantu speakers from Cameroon and West Africa. It is found at lower frequencies among Pygmy populations who have retained their hunter-gatherer lifestyles, but experienced some admixture from agriculturalists. From its distribution this seems a strong candidate for a "Bantu expansion lineage" (and is grouped with haplogroup B2a as another). If this were so, its age should be consistent with the archaeological dates for the expansion. A microsatellite network shows a star-like pattern, consistent with a population expansion, and the mean TMRCA estimate from these data is 5.8 KYA.[9]

Continent-wide analyses of mtDNA diversity suggest haplogroups L0a, L2a, L3b, and L3e as signature lineages of the Bantu expansion, because they are found in the highest frequencies in modern Bantu-speaking populations (**Figure 12.25**). Coalescent analysis of whole mtDNA sequences associated with agricultural populations or hunter-gatherers in Africa shows a rapid increase in female effective population size in the former, approximately at the onset of agriculture, 4.6 KYA (95% CI: 3–10 KYA).[32]

A recent analysis of genetic distances based on autosomal, mtDNA, and Y-chromosomal data on modern populations, together with geographical and linguistic distances,[22] finds evidence for demic diffusion in reasonable agreement with archaeological and linguistic evidence for the Bantu expansion. However, it finds little support for an early split between Western and Eastern Bantu speakers. As discussed in **Section 12.5**, in Europe ancient DNA data are contributing greatly to our understanding of the spread of agriculture. In Africa, though the key period is more recent than the European Neolithic, no ancient DNA data have been published. Even if relevant skeletal remains can be found, DNA survival may be poor under tropical conditions, so it seems likely that inferences from modern populations will remain the sole source of evidence in Africa.

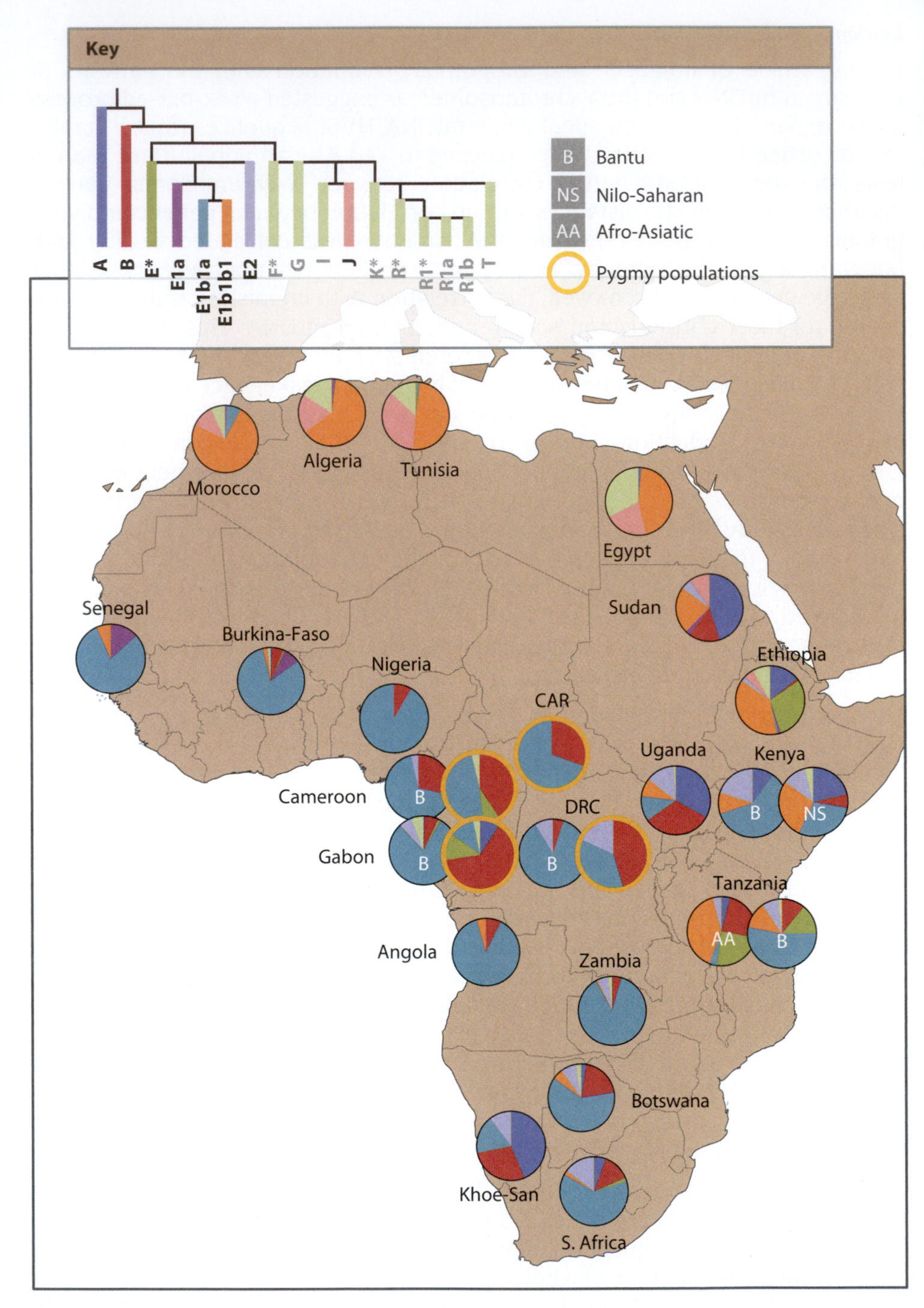

Figure 12.24: Distribution of major Y-chromosomal haplogroups in Africa. Pie charts show the frequencies of different haplogroups represented as sector areas. The tree shows the phylogenetic relationships of the haplogroups, with color coding corresponding to the pie charts. For some regions, both Bantu-speaking (B) and non-Bantu-speaking (NS, AA) populations are shown. Pygmy populations, speaking non-Bantu languages in the Niger–Congo family, are indicated by *yellow* edging. [Data from Arredi B et al. (2004) *Am. J. Hum. Genet.* 75, 338; and de Filippo C et al. (2011) *Mol. Biol. Evol.* 28, 1255.]

12.7 GENETIC ANALYSIS OF DOMESTICATED ANIMALS AND PLANTS

Advances in animal and plant genomics, and in sequencing and genotyping technologies, make it possible to address questions about the genetic basis of artificial selection in domesticated species. Reference genome sequences (see www.ensembl.org) have been generated for many key species of animals (cattle, pig, sheep, horse, and dog) and plants (rice, maize, foxtail millet, sorghum, tomato, grape, and soybean). These facilitate diversity studies of modern animal and plant populations, which, together with advances in ancient DNA methods, provide information about the number and geographical origin of domestication events. In turn, this informs our understanding of the whole process of the development of agriculture.

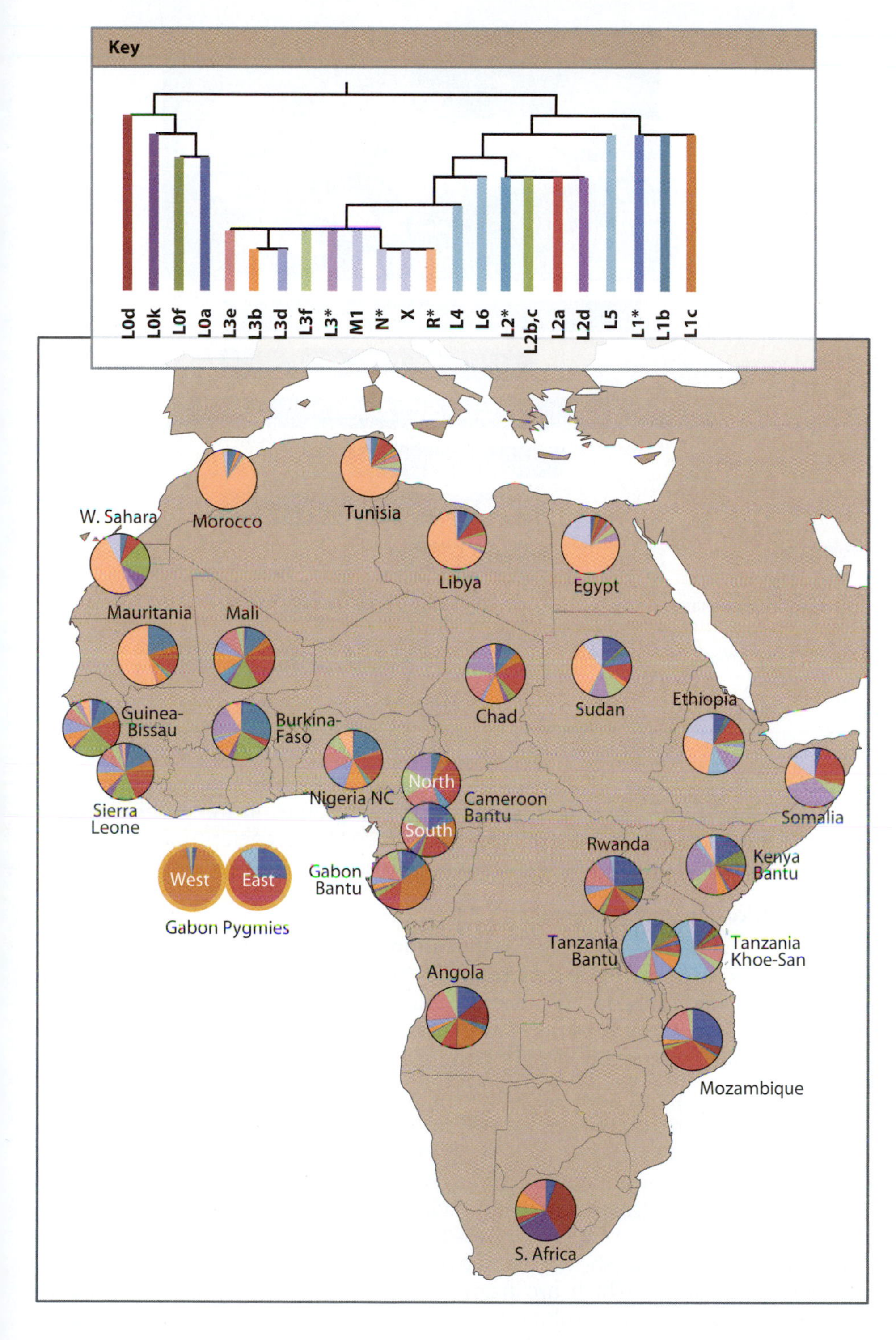

Figure 12.25: Distribution of major mtDNA haplogroups in Africa.
Pie charts show the frequencies of different haplogroups represented as sector areas. The tree shows the phylogenetic relationships of the haplogroups, with color coding corresponding to the pie charts. NC, Niger-Congo speakers. Haplogroups M1, N*, and X are pooled, as are haplogroups L4, L5, and L6. [Data from Rosa A & Brehm A (2011) *J. Anthropol. Sci.* 89, 25.]

Selective regimes had a massive impact on phenotypes and genetic diversity

Since the beginning of agriculture, people have selected desirable traits to improve product quality and manageability of animal and crop species. Modern animal and plant domesticates differ significantly in their morphology and biochemistry from the wild species from which they were selectively bred. The propagation of individuals with desirable characteristics can lead to dramatic phenotypic changes, as demonstrated by the diversity of pedigree dogs bred for different purposes compared with the gray wolf, or between wild and cultivated sunflowers first domesticated ~4 KYA in the Americas. A chihuahua and a wolf, or a cultivated and wild sunflower, are interfertile and in the latter case considered as a single species, despite their evident differences (**Figure 12.26**). These

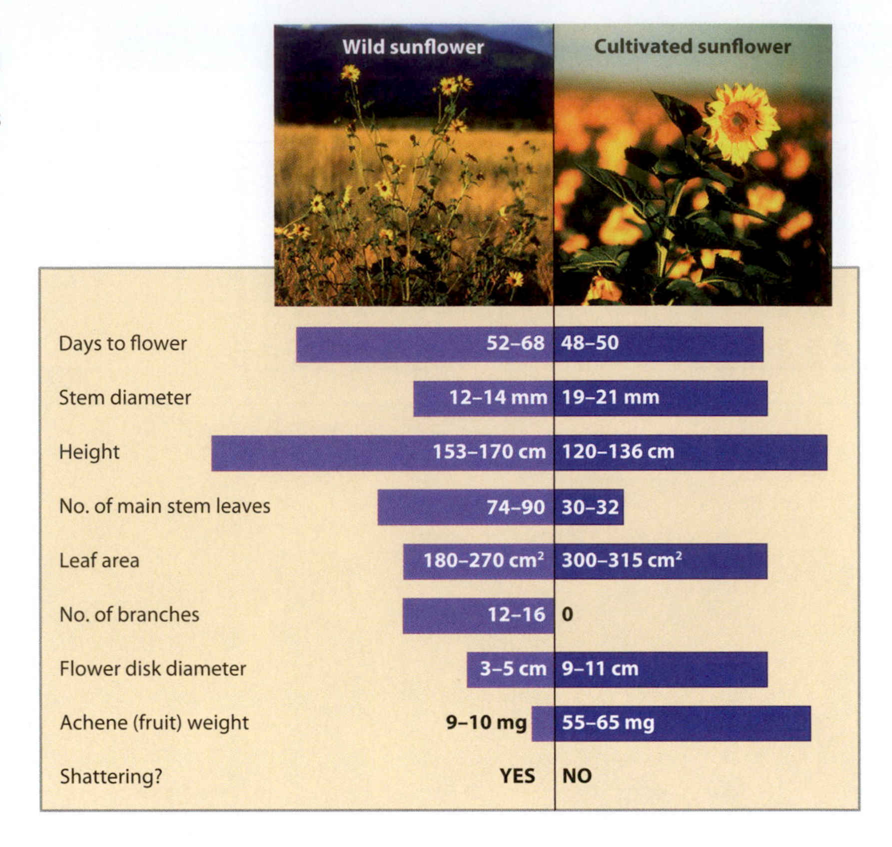

Figure 12.26: Some phenotypic differences between wild and cultivated sunflower.
Schematic illustration of 9 of 18 major traits that differ. Lengths of bars represent mean values. Note that time to flowering is not only shorter in cultivated sunflower, but also more predictable. These two plants are interfertile, and regarded as the same species. [Data from Burke J et al. (2002) *Genetics* 161, 1257. Image from Doebley J et al. (2006) *Cell* 127, 1309. With permission from Elsevier.]

artificial adaptations become apparent at different times in the prehistory of a domesticated species. In the archaeological record it is easier to identify morphological changes than biochemical alterations.

Key domestication changes in crops

Crops share a number of changes resulting from domestication, sometimes known as the **domestication syndrome**. These typically include loss of seed dormancy, synchronized flowering, a reduction in unpalatable substances in edible parts, larger fruits or grains, and a loss of natural seed dispersal (shattering) so that seeds remain attached to the plant for easier harvest (**Figure 12.27**).

Because the key plant domestication traits are mostly quantitative in nature, the underlying genetic changes are assumed to involve a number of **quantitative trait loci** (**QTLs**; see Box 15.3). These can be identified through analysis of the segregation of phenotypic traits in the offspring of crosses (for example, between domesticated strains and wild relatives); examples of some genes identified in this way are shown in **Table 12.4**. QTL mapping in different domesticated cereal species has shown that the genes controlling the critical phenotypic differences, many of which are transcriptional regulators, are rather few in number. For example, the striking morphological differences between domesticated maize and its wild progenitor, **teosinte**, can be ascribed to as few as five QTLs, each having a major effect. Some QTLs are orthologous in different plant domesticates; for example, shattering behavior in sorghum is controlled by a single gene, *Shattering1* (*sh1*), and comparative genomic analysis shows that its orthologs in rice and maize are also involved in the same phenotype.[40] Note that such comparative genomic approaches are difficult in the *Triticeae* cereals, including wheat and barley, because of their enormous polyploid genome sizes. Modern bread wheat, for example, is a hexaploid species formed from the hybridization of the tetraploid wheat *Triticum turgidum* with the diploid *Aegilops tauschii* (goat grass), and has a genome rich in repeated sequences and, at ~17,000 Mb, nearly three times the size of the human genome. In non-cereal species, too, it can be found that only a few QTLs are responsible for major

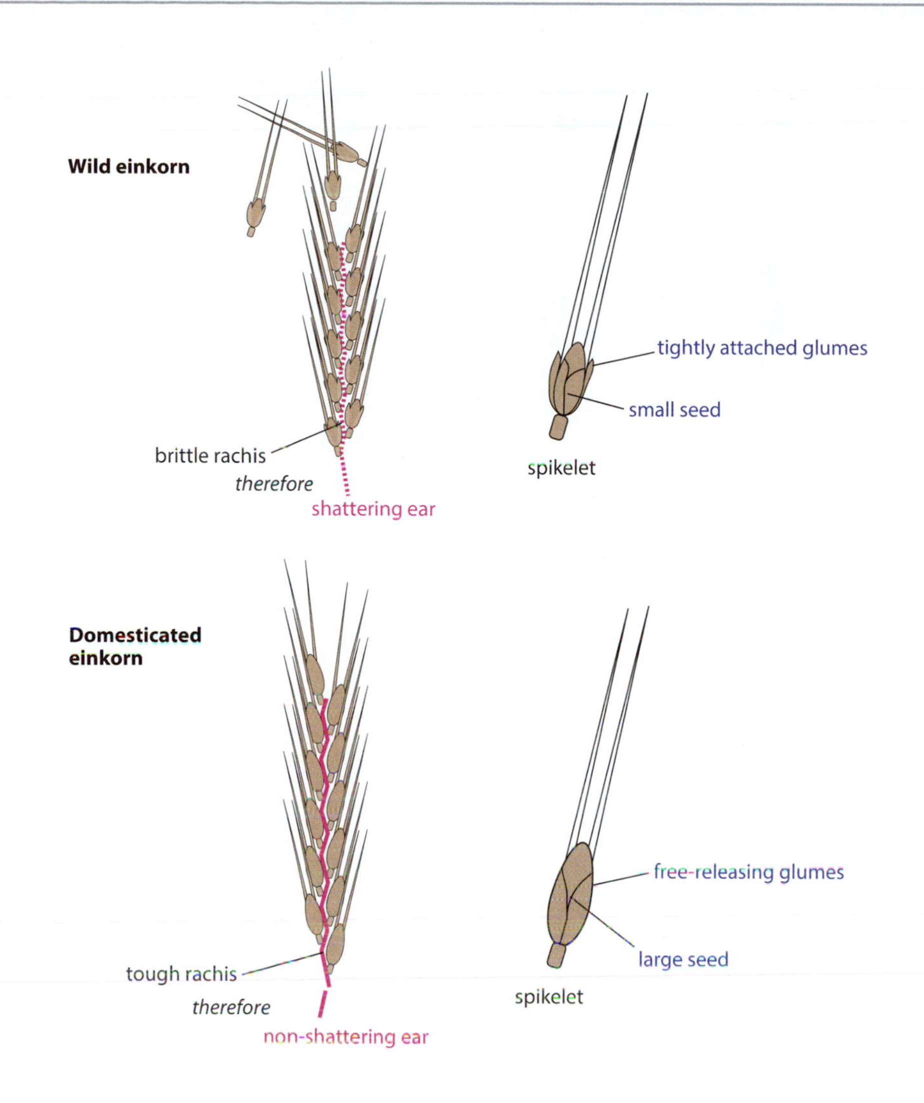

Figure 12.27: The three key domestication traits of cereals. The tough rachis of domesticated wheats allows the ear to be harvested as a whole, rather than the natural process of shattering, where the spikelets are released individually on maturity. Non-shattering mutants in the wild will be strongly selected against. The seeds are considerably larger than in the wild progenitor (often accompanied by polyploidy—genome copy number >2), and they are freely released during threshing from their glumes (leaf-like structures that protect them).

selected changes. In the eggplant (aubergine), just six loci control many aspects of fruit size, color, and shape and also plant prickliness, and some of these are orthologous to loci having the same effects in other members of the *Solanaceae*: tomato, pepper, and potato.[24] Unsurprisingly, consideration of a wider range of

TABLE 12.4: SOME IMPORTANT GENES IN CROP DOMESTICATION

Gene	Species	Function	Selection evidence?	Causative change
tb1	maize	transcriptional regulator; plant and inflorescence structure	Y	regulatory
tga1	maize	transcriptional regulator; seed casing (**glume**) form	Y	amino acid
qSH1	rice	transcriptional regulator; shattering	Y	regulatory
Rc	rice	transcriptional regulator; seed color	Y	coding sequence disruption
sh4	rice	transcriptional regulator; shattering	Y	regulatory/amino acid
fw2.2	tomato	cell signaling; fruit weight	NT	regulatory
Q	wheat	transcriptional regulator; inflorescence structure	NT	regulatory/amino acid
Sh1	sorghum, rice, maize	transcriptional regulator; shattering	Y	regulatory/truncation/splice-site variant

NT, not tested. [Adapted from Doebley J et al. (2006) *Cell* 127, 1309. With permission from Elsevier. Additional information from Gross B et al. (2010) *Mol. Ecol.* 19, 3380; Lin Z et al. (2012) *Nat. Genet.* 44, 720; Zhang L et al. (2009) *New Phytol.* 184, 708.]

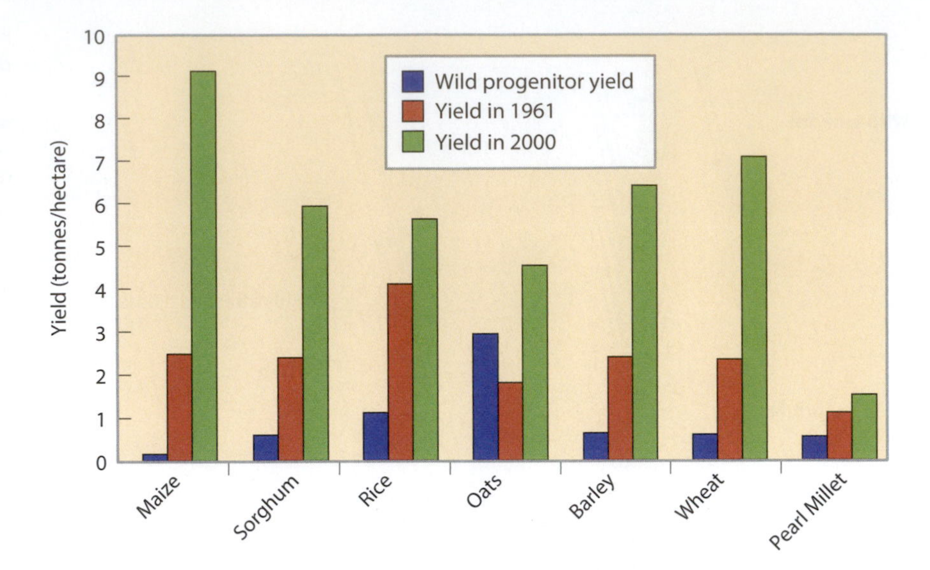

Figure 12.28: Increases in yield in domesticated cereals compared with wild progenitors.
[Data from Buckler ES et al. (2001) *Genet. Res.* 77, 213.]

domesticated plants[47] presents a more complex picture with many departures from the domestication syndrome picture presented above.

Since the establishment of the basic three cereal traits shown in Figure 12.27, selective breeding has concentrated on increasing yield. **Figure 12.28** illustrates the dramatic increases in cereal yield since domestication, and also over the last 50 years; recent increases have a substantial genetic component due to advances in plant breeding. In some species, selective breeding has also concentrated on making the edible parts more palatable by removing the chemical and physical defenses that protect them against overgrazing in the wild. Plants belonging to the *Cucurbitaceae* (for example, pumpkins, melons, and cucumbers) have lost their natural bitterness due to reduced production of the bitter compound cucurbitacin, and also have fewer phytoliths in their rinds.[55] Similarly, cultivated tomatoes (*Solanum lycopersicum*) have down-regulated production of the unpleasant-smelling 2-phenylethanol and phenylacetaldehyde compared with their wild relatives (*S. pennellii*).

Effects on crop genetic diversity

It might be imagined that continuous selection for advantageous traits among cereal species would have led to low genetic diversity. Surprisingly, this is not always the case. General estimates for maize, sorghum, rice, oats, wheat, and pearl millet indicate that the domesticates have retained on average two-thirds of the nucleotide diversity of their wild progenitors, and cultivated barley has even more diversity than wild. This high diversity is probably a result of outcrossing with other strains during and after domestication, and also of the very large numbers of plants that would have to be maintained in a population while a crop was being used for subsistence.[15]

However, when genetic studies focus upon specific loci controlling important selected traits, reduced diversity is apparent. One major difference between domesticated maize and its progenitor is that teosinte has many secondary branches, each bearing 10–12 kernels, while maize has only one or two short branches, each bearing ~300 kernels. This difference is strongly influenced by the gene *teosinte branched1* (*tb1*) encoding a transcription factor that regulates the cell cycle, acting as a repressor of organ growth and thereby repressing branch outgrowth. A study of genetic diversity at this locus[76] shows that, while the coding region of the gene displays no evidence of the action of strong recent selection, with maize carrying 39% of the nucleotide diversity of teosinte, a 1.1-kb region containing the promoter carries only 3% of the teosinte diversity (**Figure 12.29**). In a phylogenetic analysis of the coding region, maize sequences fall into many different clades, consistent with neutrality, but by contrast the promoter region sequences are monophyletic. An **HKA test** for selection (see

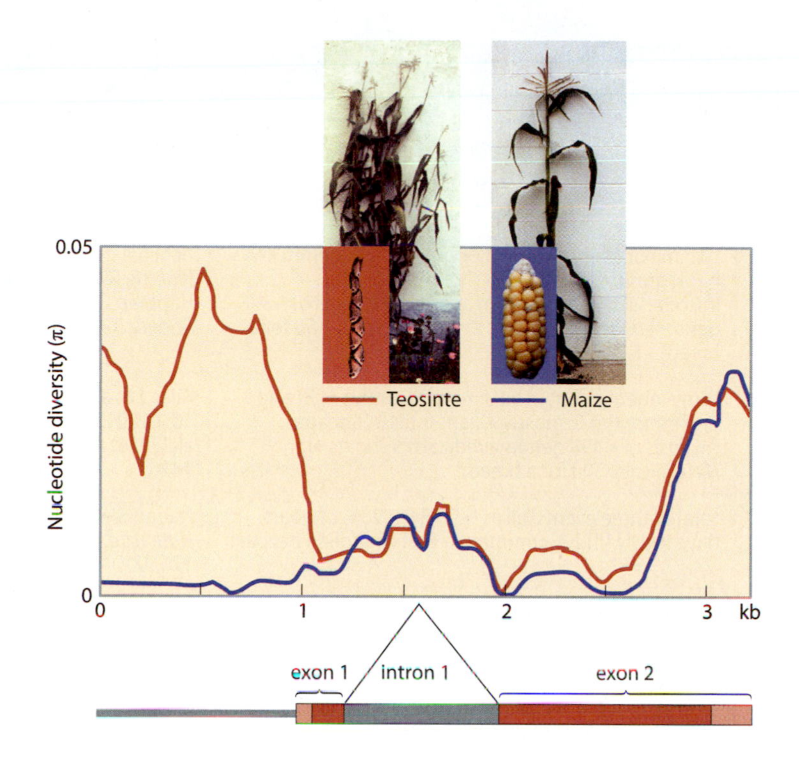

Figure 12.29: The effects of artificial selection on the *teosinte branched1* (*tb1*) gene.
Nucleotide diversity is greatly reduced in the domesticated maize promoter region compared with the same region in its wild progenitor, teosinte. [From Wang RL et al. (1999) *Nature* 398, 236. With permission from Macmillan Publishers Ltd. Images from Doebley JF et al. (2006) *Cell* 127, 1309. With permission from Elsevier.]

Section 6.7) is highly significant, and thus the promoter region has been under strong selection during domestication. Similar evidence for selection exists for many other genes in other species (**Table 12.5**).

Phenotypic and genetic change in animals

In comparison with plants, morphological changes in domesticated animals are more species-specific. Size changes are more difficult to discern due to the changes in population structure that may result from slaughter at a younger age. However, in cattle, for example, there is clear evidence of reduction of size from the Neolithic to the Middle Ages, as well as the reduction of large horns. Behavioral changes were bred into animal stocks, making them less aggressive and more easily managed. There were often changes in skull shape due to artificial selection for retention of juvenile features in adults, a characteristic known as **neoteny**, and reductions in brain size and acuity of the senses compared with wild progenitors—both essential in the wild, but a waste of energy under domestication. Changes in features exploited during the secondary product revolution appear later in prehistory, for example, sheep hair becomes finer and less pigmented. Most animal QTL studies have concentrated not on identifying genes important in early domestication, but on traits that are thought useful in improving yield and quality of animal products, in particular meat and milk.

Many species show evidence of strong selection at the *MC1R* gene (Section 15.3), indicating that coat color and pattern were important for early farmers and breeders, probably in order to distinguish their own animals from others, or for cultural reasons. There is evidence in cattle for selection on the myostatin gene, associated with muscle composition, and also the growth hormone receptor gene, associated with growth rate and various other traits. In dogs, there is evidence of strong selection on a number of genes associated with growth (for example, *insulin-like growth factor 1*) and skeletal traits, many of which are related to breed-specific characteristics.[77] Whole-genome resequencing approaches in chickens[67] have identified many candidate loci under selection. One, the thyroid stimulating hormone receptor (*TSHR*) gene, has undergone a massive selective sweep such that almost all domestic chickens are homozygous, despite an overall high genetic diversity in this species. The sweep haplotype includes a nonconservative amino acid substitution, and the

TABLE 12.5:
EVIDENCE ON ANIMAL DOMESTICATION EVENTS FROM GENETIC DATA

Domesticate	Progenitor	Earliest evidence of domestication	Conclusions from genetic evidence	Reference
Cattle, *Bos taurus, Bos indicus*	aurochs, *Bos primigenius* (**extinct**)	10.3–10.8 KYA, Fertile Crescent	two major events: distinct origins for *taurus* (mtDNA haplogroup T) and *indicus* (hg I). Some later introgressed lineages from aurochs. Analysis of ~37K genomewide SNPs shows low N_e (typically 150 for a breed)	Achilli A et al. (2008) *Curr. Biol.* 18, R157; Bovine HapMap Consortium (2009) *Science* 324, 528.
Sheep, *Ovis aries*	Asiatic mouflon, *Ovis orientalis*	10–11 KYA, Fertile Crescent	three independent events: two major (hg A, B) and one minor (hg C, mostly Asian) mtDNA lineages. Analysis of ~49K genomewide SNPs shows high N_e (typically 300 for a breed)	Kijas J et al. (2012) *PLoS Biol*, 10, e1001258; Pereira F et al. (2006) *Mol. Biol. Evol.* 23, 1420.
Goat, *Capra hircus*	bezoar goat, *Capra aegagrus*	~10.5 KYA, S. E. Anatolia	total of three events, all in Near East: 90% of goats have mtDNA hg A, common in East Anatolian bezoars	Naderi S et al. (2008) *Proc. Natl Acad. Sci. USA* 105, 17659.
Pig, *Sus scrofa*	wild boar, *Sus scrofa*	~9 KYA, Near East	multiple domestications including independent events in Near East, Europe, and Island S. E. Asia: mtDNA control region sequences show 14 major clusters	Larson G et al. (2005) *Science* 307, 1618.
Horse, *Equus caballus*	wild horse, *Equus ferus* (**extinct**)	6–7 KYA	many events across Eurasia: high mtDNA diversity (17 hgs). Low Y-chromosome diversity in modern horses (greater in ancient DNA) indicating highly controlled breeding of males	Achilli A et al. (2012) *Proc. Natl Acad. Sci. USA* 109, 2449; Lippold S et al. (2011) *Nat. Commun.* 2, 450.
Dog, *Canis familiaris*	gray wolf, *Canis lupus*	~11.5 KYA, Levant (modern Israel)	unclear. Haplotype sharing (based on ~48K genomewide SNPs) greatest with Middle Eastern wolves, suggesting major origin there, plus secondary events in Europe and E. Asia	Vonholdt B et al. (2010) *Nature* 464, 898.
Chicken, *Gallus gallus domesticus*	red jungle-fowl, *Gallus gallus*	~8 KYA, N. E. China; >4 KYA Indus Valley	many events in Asia: mtDNA HVSI sequences show multiple clusters. Sequence data from 30 nuclear loci indicate recent introgressions from junglefowl	Liu Y et al. (2006) *Mol. Phylogenet. Evol.* 38, 12; Sawai H et al. (2010) *PLoS One* 5, e10639.

Note that identically named mtDNA haplogroups (for example, hg A) referred to in different species are unrelated to each other.

basis of selection is suggested to be related to the absence of the strict regulation of seasonal reproduction that is found in natural populations.

How have the origins of domesticated plants been identified?

Attempts to locate the origin of plant domestication events have involved two major lines of enquiry:

- The identification of seeds of the wild species in early archaeological sites (**Figure 12.30**), with seeds of domesticated species succeeding in later strata.
- The identification and location of likely wild progenitor species still existing, and the analysis of their progenitor status, for example by typing of genetic polymorphisms, and phylogenetic comparisons with domesticated varieties. This approach may be unreliable if, as stated above, crops retain high diversity because of gene flow.

In the case of the European Neolithic, because the progenitor range of **einkorn**, rye, and some legumes—including lentil, pea, and chickpea—overlap in the foothills of the Karaçadag mountains in southeastern Turkey, it was proposed that this region within the Fertile Crescent was a "cradle of agriculture" in which many innovations occurred.[39] However, this picture reflects the once widespread view that domestication and the associated morphological changes took place in single locations over only a few human generations, triggered by intense intervention and harvesting practices. More recently there has been

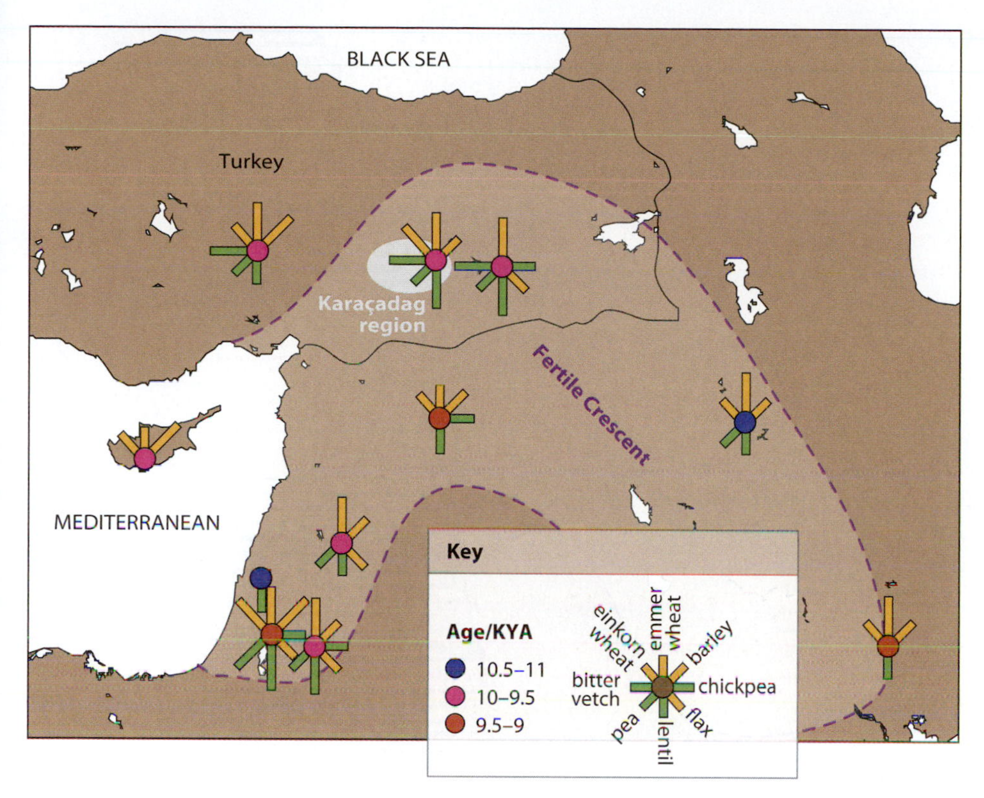

Figure 12.30: Distribution of Southwest Asian archaeological sites showing the first evidence for domesticated crops. [From Zohary D et al. (2012) Domestication of Plants in the Old World, Oxford University Press. With permission from Oxford University Press.]

a shift, based on archaeological findings, toward viewing domestication as a more gradual process starting as early as 12.5 KYA, and extending over a longer period, probably several millennia (Opinion Box 13).

If there were a single domestication event in the history of a species, phylogenetic trees would be expected to show the domesticated species lying within a single clade. Earlier claims that this pattern was demonstrable for barley, for example, were based firmly in the cradle-of-agriculture view, and are now discounted.[14] However, the pattern does appear to apply to maize. In a phylogeny constructed from genetic distances based on the proportion of shared alleles at 99 microsatellites (**Figure 12.31**), maize falls into a **monophyletic** group with strong **bootstrap** support, and teosinte branches basal to the maize clade all come from the Balsas river region of southwest Mexico.[45] The microsatellite data can also be used to date the divergence of the teosinte and maize lineages, and at 9.2 KYA (with 95% confidence interval 5.7–13.1 KYA), this is consistent with archaeological evidence showing maize starch grains on grinding stones from the Xihuatoxtla shelter, in the Central Balsas Valley, dated to 8.7 KYA.[56] There has been controversy on whether the divergent *indica* and *japonica* strains of rice (*Oryza sativa*) underwent a single domestication from the wild progenitor *O. rufipogon*, or two independent domestications. Phylogenetic analysis of high-throughput RNA sequence data[80] supports two domestications, but also provides evidence for transfer of genes from *japonica* to *indica* rice.

Ancient DNA evidence has not been greatly used to study plant origins, largely because the widely available remains, such as phytoliths or charred seeds, preserve little DNA. Analysis of three domestication genes, including *tb1*, in archaeological corn cobs showed that alleles typical of modern maize rather than wild teosinte were present in Mexican maize by 4.4 KYA.[35]

How have the origins of domesticated animals been identified?

Locating domestication events for animals differs from the situation for plants in a number of respects. Unlike plants, which if left to their own devices and not subject to extreme climate changes will stay where they are, animals are mobile, and can be hunted to extinction if not already domesticated. The last

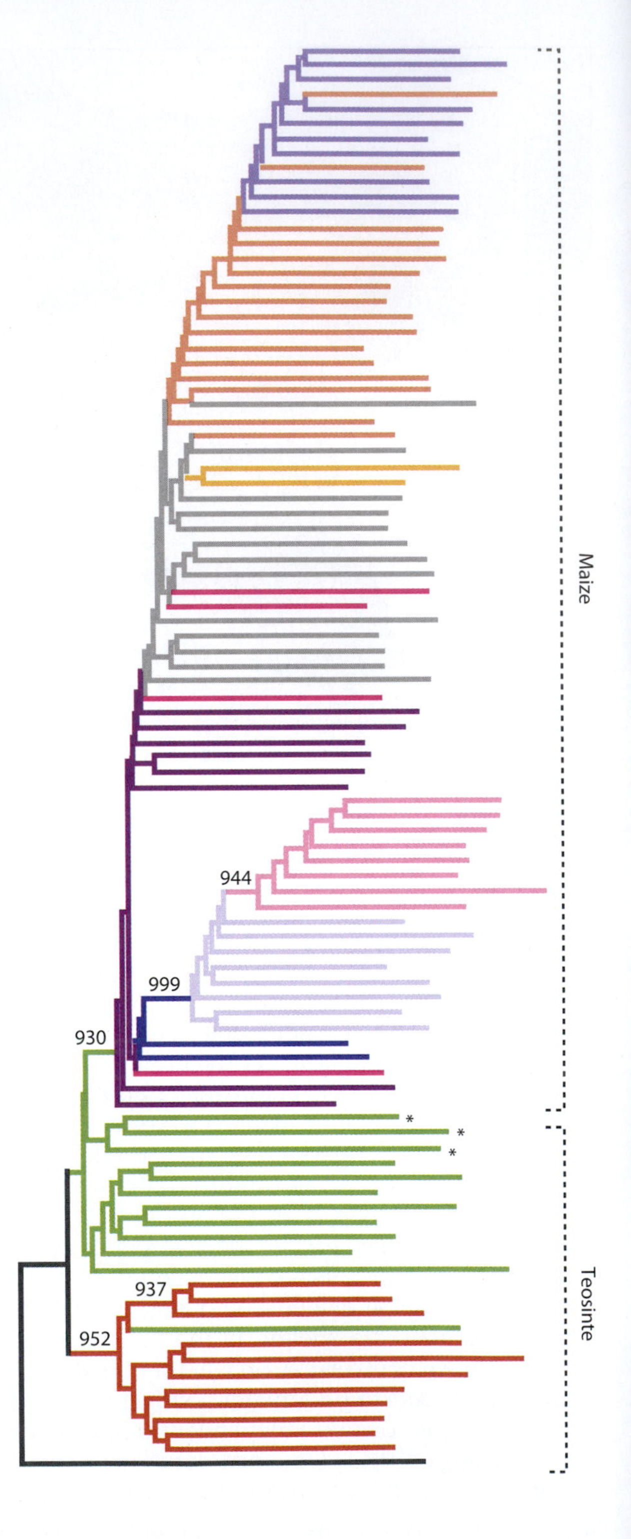

Figure 12.31: Phylogenetic tree showing a single origin for domesticated maize. Tree based on microsatellite data from 99 loci in 95 ecogeographically defined groups of maize and teosinte. Numbers on branches indicate the number of times a clade appeared among 1000 bootstrap samples, with only bootstrap values >900 shown. Asterisks identify those populations of ssp. *parviglumis* basal to maize, all of which are from the central Balsas river region. [From Matsuoka Y et al. (2002) *Proc. Natl Acad. Sci. USA* 99, 6080. With permission from National Academy of Sciences.]

aurochs (this is the singular noun, the plural is *aurochsen*), progenitor of modern cattle, is said to have been killed in Poland in 1627, and the wild ancestors of the horse, the Old World camel, and the goat have also disappeared. Ancient DNA studies clearly can make a contribution to resolving this problem.

Most genetic studies of the origins of animal domestication have employed mtDNA, though this is now changing with more genomewide approaches, often driven by ongoing interest in breed improvement. Across species, mtDNA has sufficient sequence conservation to allow the design of universal PCR primers, but enough diversity to yield a large number of informative haplotypes.

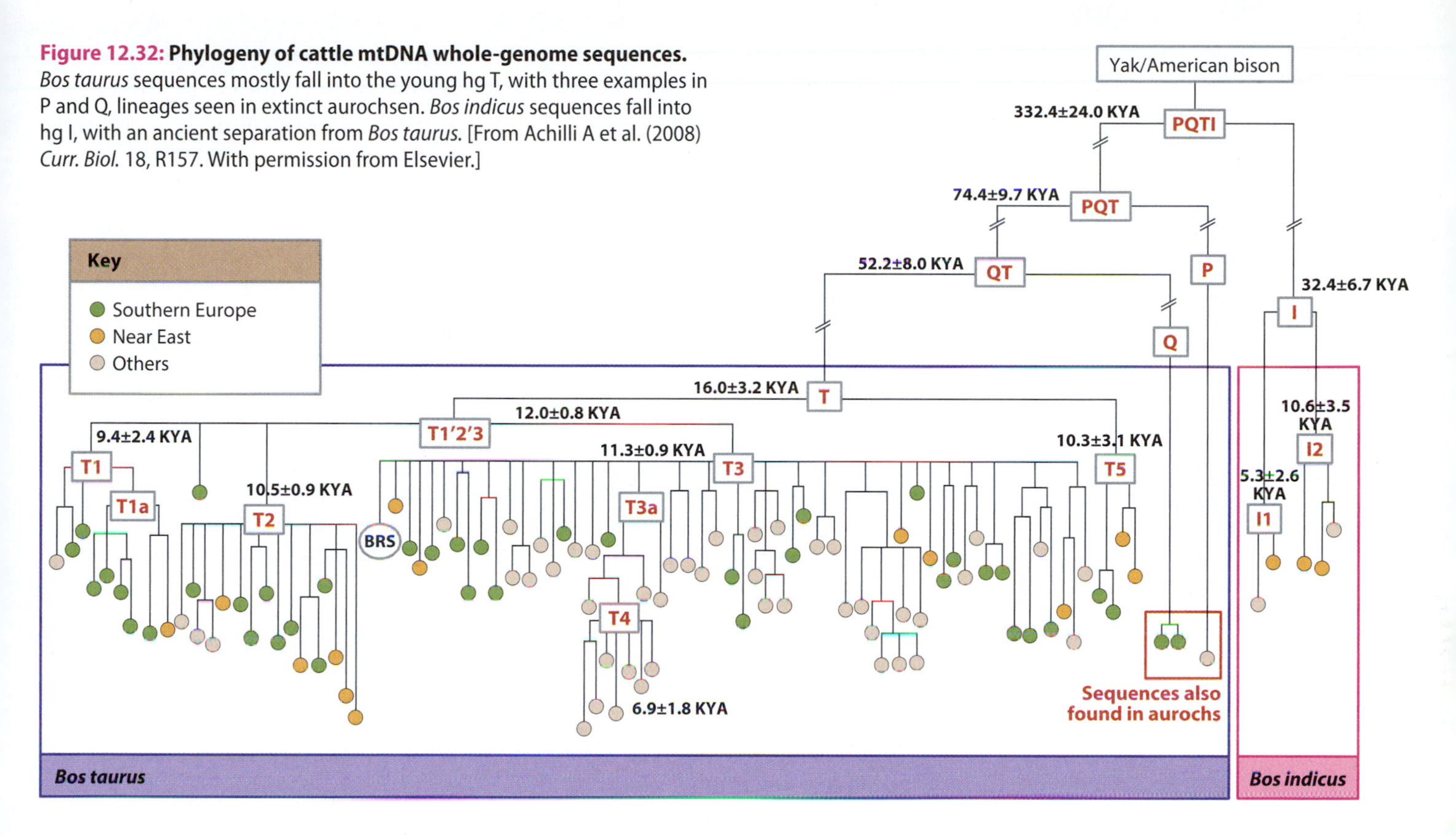

Figure 12.32: Phylogeny of cattle mtDNA whole-genome sequences. *Bos taurus* sequences mostly fall into the young hg T, with three examples in P and Q, lineages seen in extinct aurochsen. *Bos indicus* sequences fall into hg I, with an ancient separation from *Bos taurus*. [From Achilli A et al. (2008) *Curr. Biol.* 18, R157. With permission from Elsevier.]

Also, ancient DNA work is most straightforward when mtDNA is being studied (see **Section 4.10**). Another advantage comes from the mating structures of domesticated animal populations: under managed breeding, males have many more offspring than females do, and this may mean that past spatial patterns of paternally and biparentally inherited loci are obscured; indeed, Y-chromosome diversity is often extremely low and uninformative, as in the horse.[41] The maternally inherited mtDNA is therefore expected to show relative continuity with populations of the past. Since all domestic livestock species can interbreed with their wild relatives, introgression is a potential problem in interpreting domestic animal phylogenies; however, introgression events might be biased toward wild males, rather than females, so mtDNA may be relatively insensitive to this problem.

Cattle domestication

Cattle domestication history provides a good example where evidence from both ancient and modern sequences has been combined. There are two major types of cattle, the European humpless *Bos taurus* and the more arid-adapted Indian **zebu** *Bos indicus*, considered different species even though they are completely interfertile. Archaeological evidence suggests that taurine cattle were domesticated from the aurochs in the Near East between 10.8 and 10.3 KYA. mtDNA analysis suggests that they derive from two independent domestication events. A tree of complete mtDNA sequences (**Figure 12.32**) shows a deep divergence between *Bos taurus* [mostly in haplogroup (hg) T] and *Bos indicus* (all hg I), estimated as ~330 KYA.[1] The hg T taurine sequences are very closely related, with a TMRCA of 16.0 ± 3.2 KYA, but surprisingly there are also three mtDNAs among taurine samples that lie within hgs P and Q, lineages found in ancient DNA analysis of northern European aurochsen. This suggests some introgressions from the aurochs after domestication. Coalescent simulation and approximate Bayesian computation applied to a sample of 15 Neolithic to Iron Age Iranian domestic cattle mtDNA HVSI sequences[12] estimates that only ~80 female aurochsen were initially domesticated. This suggests the domestication process may have been limited by the difficulty and danger of managing and breeding the large, aggressive, and territorial wild progenitors of *Bos taurus*.

This picture of a small number of independent domestications is a common theme for sheep and goats (Table 12.5), while others, including pigs, horses, dogs, and chickens, present more complex pictures including multiple centers of domestication, plus some introgression from wild relatives.

SUMMARY

- The development of agriculture over the last 10 KY was an important cultural innovation that allowed human populations to expand dramatically. There is a major debate on the extent to which the spread of agricultural practices was accompanied by gene flow from farming populations.
- Agriculture involved the domestication of animal and plant species, and is recognizable by direct evidence of this and by changes in associated material culture including stone tools and pottery. These changes have been referred to as the "Neolithic revolution," but there has been a shift toward a more gradualist view.
- Agriculture began independently in different times and places. Its development was probably triggered by the needs of a growing human population in a period of relative climatic stability.
- As well as the eventual population growth enabled by agriculture, there were problems, including initial poor nutrition, and spread of disease from animals to humans.
- The geographical correlations of agricultural origins with centers of large language families have suggested that agriculture and language may have dispersed together.
- Principal component analysis of classical gene frequencies showed a gradient from the southeast to the northwest of Europe, interpreted as evidence for Neolithic gene flow. However, the age of the gradient is unknown, and there are doubts about mathematical artifacts arising from the methods.
- Patterns of mtDNA diversity in Europe show no strong clines, and this has been taken to reflect a largely Paleolithic contribution. By contrast, ancient DNA evidence, though scanty, suggests a discontinuity between hunter-gatherers and farmers.
- Major Y-chromosome lineages show strong clines, and their interpretation has been controversial. Again, emerging ancient DNA data suggest a lack of continuity with modern patterns, suggesting possible large Neolithic contributions. Large-scale Y-chromosome sequence data indicate a recent expansion for the major European lineage.
- Genomewide data on modern populations have not been very informative, but genome sequence from ancient samples suggests that agriculture was brought to northern Europe by people who were genetically distinct from the local hunter-gatherers.
- The spread of iron-working, Bantu languages, and agriculture from West Africa into central and southern Africa (the Bantu expansion) occurred over the last 3 KY. Patterns of modern genetic diversity seem broadly consistent with the archaeological and linguistic framework, but ancient DNA data are not yet available.
- Genetic analyses of domesticated animals and plants can locate the origins of agriculture. Selected changes in animals were mostly behavioral, and have received little attention from geneticists. Changes in plants focused on improved ease of harvesting and yield. The same genes have been the target of selection in many plant species, but at the same time a high level of overall genetic diversity has been maintained.

- Phylogeographic studies of plant diversity can suggest the locations of domestication events. Several species seem to have been domesticated in southeastern Turkey. Cattle, sheep, and goats show patterns consistent with small numbers of domestication events, while other animals were domesticated many times.

QUESTIONS

Question 12–1: Put the following domesticated species in order of the time of first domestication: chilli pepper, goat, dog, maize, chicken, rice, emmer wheat, guinea pig, pearl millet, turkey.

Question 12–2: Can the transition from foraging to farming reasonably be characterized as a revolution?

Question 12–3: List some of the advantages and disadvantages of making this transition. Why did it occur when it did?

Question 12–4: What evidence has been used to support the co-dispersal of agriculture and language?

Question 12–5: Compare and contrast the processes of plant and animal domestication.

Question 12–6: Giving examples based on genetic data, describe how the adoption of agriculture has changed the selective environment for humans. Refer also to Chapters 15–17.

Question 12–7: You obtain high-quality, high-coverage genome sequence data from 10 ancient hunter-gatherers and 10 early farmers distributed across Europe. What analyses would you carry out to test ideas about the transition from foraging to farming?

Question 12–8: Over lunch, Professor Pangloss (who is fond of exotica) asserts that the steak pies from his local butcher are pure, finest, imported zebu. You are not so sure, but how would you test for adulteration with boring British beef?

REFERENCES

The references highlighted in purple are considered to be important (for this chapter) by the authors.

1. **Achilli A, Olivieri A, Pellecchia M et al.** (2008) Mitochondrial genomes of extinct aurochs survive in domestic cattle. *Curr. Biol.* **18**, R157–158.
2. **Ammerman AJ & Cavalli-Sforza LL** (1984) Neolithic Transition and the Genetics of Populations in Europe. Princeton University Press.
3. **Balaresque P, Bowden GR, Adams SM et al.** (2010) A predominantly Neolithic origin for European paternal lineages. *PLoS Biol.* **8**, e1000285.
4. **Barbujani G, Bertorelle G & Chikhi L** (1998) Evidence for Paleolithic and Neolithic gene flow in Europe. *Am. J. Hum. Genet.* **62**, 488–491.
5. **Barbujani G, Sokal RR & Oden NL** (1995) Indo-European origins: a computer-simulation test of five hypotheses. *Am. J. Phys. Anthropol.* **96**, 109–132.
6. **Barker G** (2006) The Agricultural Revolution in Prehistory: Why Did Foragers Become Farmers? Oxford University Press.
7. **Belle EM, Landry PA & Barbujani G** (2006) Origins and evolution of the Europeans' genome: evidence from multiple microsatellite loci. *Proc. Biol. Sci.* **273**, 1595–1602.
8. **Bellwood P** (2001) Early agriculturalist population diasporas? Farming, language and genes. *Ann. Rev. Anthropol.* **30**, 181–207.
9. **Berniell-Lee G, Calafell F, Bosch E et al.** (2009) Genetic and demographic implications of the Bantu expansion: insights from human paternal lineages. *Mol. Biol. Evol.* **26**, 1581–1589.
10. **Bocquet-Appel J-P** (2008) Explaining the Neolithic demographic transition. In The Neolithic Demographic Transition and its Consequences (J-P Bocquet-Appel & O Bar-Yosef eds). Springer.
11. **Bocquet-Appel J-P, Naji S, Vander Linden M & Kozlowski JK** (2009) Detection of diffusion and contact zones of early farming in Europe from the space-time distribution of 14C dates. *J. Archaeol. Sci.* **36**, 807–820.
12. **Bollongino R, Burger J, Powell A et al.** (2012) Modern taurine cattle descended from small number of Near-Eastern founders. *Mol. Biol. Evol.* 29, 2101–2204.
13. **Bouckaert R, Lemey P, Dunn M et al.** (2012) Mapping the origins and expansion of the Indo-European language family. *Science* **337**, 957–960.
14. **Brown TA, Jones MK, Powell W & Allaby RG** (2009) The complex origins of domesticated crops in the Fertile Crescent. *Trends Ecol. Evol.* **24**, 103–109.
15. **Buckler ES, Thornsberry JM & Kresovich S** (2001) Molecular diversity, structure and domestication of grasses. *Genet. Res.* **77**, 213–218.
16. **Burger J, Kirchner M, Bramanti B et al.** (2007) Absence of the lactase-persistence-associated allele in early Neolithic Europeans. *Proc. Natl Acad. Sci. USA* **104**, 3736–3741.
17. **Busby GB, Brisighelli F, Sanchez-Diz P et al.** (2012) The peopling of Europe and the cautionary tale of Y chromosome lineage R-M269. *Proc. Biol. Sci.* **279**, 884–892.
18. **Cavalli-Sforza LL, Menozzi P & Piazza A** (1994) The History and Geography of Human Genes. Princeton University Press.
19. **Chikhi L** (2009) Update to Chikhi et al.'s "Clinal variation in the nuclear DNA of Europeans" (1998): genetic data and storytelling–from archaeogenetics to astrologenetics? *Hum. Biol.* **81**, 639–643.
20. **Cohen MN & Armelagos GJ** (1984) Paleopathology at the Origins of Agriculture. Academic Press.
21. **Collins R** (1986) The Basques. Blackwell.
22. **de Filippo C, Bostoen K, Stoneking M & Pakendorf B** (2012) Bringing together linguistic and genetic evidence to test the Bantu expansion. *Proc. Biol. Sci.* **279**, 3256–3263.

23. **Diamond J** (1998) Guns, Germs and Steel: A Short History of Everybody for the Last 13,000 years. Vintage.
24. **Doganlar S, Frary A, Daunay MC et al.** (2002) Conservation of gene function in the Solanaceae as revealed by comparative mapping of domestication traits in eggplant. *Genetics* **161**, 1713–1726.
25. **Dunne J, Evershed RP, Salque M et al.** (2012) First dairying in green Saharan Africa in the fifth millennium BC. *Nature* **486**, 390–394.
26. **Edmonds CA, Lillie AS & Cavalli-Sforza LL** (2004) Mutations arising in the wave front of an expanding population. *Proc. Natl Acad. Sci. USA* **101**, 975–979.
27. **Ehret C** (2001) Bantu expansions: re-envisioning a central problem of early African history. *Int. J. Afr. Hist. Stud.* **34**, 5–27.
28. **Fu Q, Rudan P, Pääbo S & Krause J** (2012) Complete mitochondrial genomes reveal Neolithic expansion into Europe. *PLoS One* **7**, e32473.
29. **Fuller DQ** (2010) An emerging paradigm shift in the origins of agriculture. *General Anthropol.* **17**, 8–12.
30. **Fuller DQ, WIllcox G & Allaby RG** (2011) Cultivation and domestication had multiple origins: arguments against the core area hypothesis for the origins of agriculture in the Near East. *World Archaeol.* **43**, 628–652.
31. **Gerbault P, Liebert A, Itan Y et al.** (2011) Evolution of lactase persistence: an example of human niche construction. *Philos. Trans. R. Soc. Lond. B Biol. Sci.* **366**, 863–877.
32. **Gignoux CR, Henn BM & Mountain JL** (2011) Rapid, global demographic expansions after the origins of agriculture. *Proc. Natl Acad. Sci. USA* **108**, 6044–6049.
33. **Hervella M, Izagirre N, Alonso S et al.** (2012) Ancient DNA from hunter-gatherer and farmer groups from Northern Spain supports a random dispersion model for the Neolithic expansion into Europe. *PLoS One* **7**, e34417.
34. **Holden CJ** (2002) Bantu language trees reflect the spread of farming across sub-Saharan Africa: a maximum-parsimony analysis. *Proc. R. Soc. Lond. B* **269**, 793–799.
35. **Jaenicke-Despres V, Buckler ES, Smith BD et al.** (2003) Early allelic selection in maize as revealed by ancient DNA. *Science* **302**, 1206–1208.
36. **Kurlansky M** (1999) The Basque History of the World. Jonathan Cape.
37. **Laayouni H, Calafell F & Bertranpetit J** (2010) A genome-wide survey does not show the genetic distinctiveness of Basques. *Hum. Genet.* **127**, 455–458.
38. **Lao O, Lu TT, Nothnagel M et al.** (2008) Correlation between genetic and geographic structure in Europe. *Curr. Biol.* **18**, 1241–1248.
39. **Lev-Yadun S, Gopher A & Abbo S** (2000) The cradle of agriculture. *Science* **288**, 1602–1603.
40. **Lin Z, Li X, Shannon LM et al.** (2012) Parallel domestication of the *Shattering1* genes in cereals. *Nat. Genet.* **44**, 720–724.
41. **Lippold S, Knapp M, Kuznetsova T et al.** (2011) Discovery of lost diversity of paternal horse lineages using ancient DNA. *Nat. Commun.* **2**, 450.
42. **Lorenzen ED, Nogues-Bravo D, Orlando L et al.** (2011) Species-specific responses of Late Quaternary megafauna to climate and humans. *Nature* **479**, 359–364.
43. **MacEachern S** (2000) Genes, tribes and African history. *Curr. Anthropol.* **41**, 357–384.
44. **Malmstrom H, Linderholm A, Liden K et al.** (2010) High frequency of lactose intolerance in a prehistoric hunter-gatherer population in northern Europe. *BMC Evol. Biol.* **10**, 89.
45. **Matsuoka Y, Vigouroux Y, Goodman MM et al.** (2002) A single domestication for maize shown by multilocus microsatellite genotyping. *Proc. Natl Acad. Sci. USA* **99**, 6080–6084.
46. **Menozzi P, Piazza A & Cavalli-Sforza LL** (1978) Synthetic maps of human gene frequencies in Europeans. *Science* **201**, 786–792.
47. **Meyer RS, DuVal AE & Jensen HR** (2012) Patterns and processes in crop domestication: an historical review and quantitative analysis of 203 global food crops. *New Phytol.* **196**, 29–48.
48. **Novembre J, Johnson T, Bryc K et al.** (2008) Genes mirror geography within Europe. *Nature* **456**, 98–101.
49. **Novembre J & Stephens M** (2008) Interpreting principal component analyses of spatial population genetic variation. *Nat. Genet.* **40**, 646–649.
50. **Phillipson DW** (2005) African Archaeology, 3rd ed. Cambridge University Press.
51. **Pilkington MM, Wilder JA, Mendez FL et al.** (2008) Contrasting signatures of population growth for mitochondrial DNA and Y chromosomes among human populations in Africa. *Mol. Biol. Evol.* **25**, 517–525.
52. **Pinhasi R, Fort J & Ammerman AJ** (2005) Tracing the origin and spread of agriculture in Europe. *PLoS Biol.* **3**, e410.
53. **Pinhasi R & Stock JT** (eds) (2011) Human Bioarchaeology of the Transition to Agriculture. Wiley-Blackwell.
54. **Pinhasi R & von Cramon-Taubadel N** (2009) Craniometric data supports demic diffusion model for the spread of agriculture into Europe. *PLoS One* **4**, e6747.
55. **Piperno DR, Holst I, Wessel-Beaver L & Andres TC** (2002) Evidence for the control of phytolith formation in Cucurbita fruits by the hard rind (Hr) genetic locus: Archaeological and ecological implications. *Proc. Natl Acad. Sci. USA* **99**, 10923–10928.
56. **Piperno DR, Ranere AJ, Holst I et al.** (2009) Starch grain and phytolith evidence for early ninth millennium B.P. maize from the Central Balsas River Valley, Mexico. *Proc. Natl Acad. Sci. USA* **106**, 5019–5024.
57. **Prescott GW, Williams DR, Balmford A et al.** (2012) Quantitative global analysis of the role of climate and people in explaining late Quaternary megafaunal extinctions. *Proc. Natl Acad. Sci. USA* **109**, 4527–4531.
58. **Rasteiro R, Bouttier PA, Sousa VC & Chikhi L** (2012) Investigating sex-biased migration during the Neolithic transition in Europe, using an explicit spatial simulation framework. *Proc. Biol. Sci.* **279**, 2409–2416.
59. **Rendine S, Piazza A & Cavalli-Sforza LL** (1986) Simulation and separation by principal components of multiple demic expansions in Europe. *Am. Nat.* **128**, 681–706.
60. **Renfrew C** (1988) Archaeology and Language: the puzzle of Indo-European origins. Cambridge University Press. (Reissued 2000, Picador.)
61. **Richards M, CorteReal H, Forster P et al.** (1996) Paleolithic and neolithic lineages in the European mitochondrial gene pool. *Am. J. Hum. Genet.* **59**, 185–203.
62. **Richards M, Macaulay V, Hickey E et al.** (2000) Tracing European founder lineages in the near eastern mtDNA pool. *Am. J. Hum. Genet.* **67**, 1251–1276.
63. **Richards M & Sykes B** (1998) Evidence for Paleolithic and Neolithic gene flow in Europe – Reply. *Am. J. Hum. Genet.* **62**, 491–492.
64. **Rodriguez-Ezpeleta N, Alvarez-Busto J, Imaz L et al.** (2010) High-density SNP genotyping detects homogeneity of Spanish and French

Basques, and confirms their genomic distinctiveness from other European populations. *Hum. Genet.* **128**, 113–117.

65. **Rootsi S, Magri C, Kivisild T et al.** (2004) Phylogeography of Y-chromosome haplogroup I reveals distinct domains of prehistoric gene flow in Europe. *Am. J. Hum. Genet.* **75**, 128–137.

66. **Rosser ZH, Zerjal T, Hurles ME et al.** (2000) Y-chromosomal diversity within Europe is clinal and influenced primarily by geography, rather than by language. *Am. J. Hum. Genet.* **67**, 1526–1543.

67. **Rubin CJ, Zody MC, Eriksson J et al.** (2010) Whole-genome resequencing reveals loci under selection during chicken domestication. *Nature* **464**, 587–591.

68. **Sanchez-Quinto F, Schroeder H, Ramirez O et al.** (2012) Genomic affinities of two 7000-year-old Iberian hunter-gatherers. *Curr. Biol.* **22**, 1494–1499.

69. **Semino O, Passarino G, Oefner PJ et al.** (2000) The genetic legacy of Paleolithic *Homo sapiens sapiens* in extant Europeans: a Y chromosome perspective. *Science* **290**, 1155–1159.

70. **Simoni L, Calafell F, Pettener D et al.** (2000) Geographic patterns of mtDNA diversity in Europe. *Am. J. Hum. Genet.* **66**, 262–278.

71. **Skoglund P, Malmstrom H, Raghavan M et al.** (2012) Origins and genetic legacy of Neolithic farmers and hunter-gatherers in Europe. *Science* **336**, 466–469.

72. **Sokal RR, Oden NL & Wilson C** (1991) Genetic evidence for the spread of agriculture in Europe by demic diffusion. *Nature* **351**, 143–145.

73. **Tishkoff SA, Reed FA, Friedlaender FR et al.** (2009) The genetic structure and history of Africans and African Americans. *Science* **324**, 1035–1044.

74. **Vigne JD, Briois F, Zazzo A et al.** (2012) First wave of cultivators spread to Cyprus at least 10,600 y ago. *Proc. Natl Acad. Sci. USA* **109**, 8445–8449.

75. **von Cramon-Taubadel N & Pinhasi R** (2011) Craniometric data support a mosaic model of demic and cultural Neolithic diffusion to outlying regions of Europe. *Proc. Biol. Sci.* **278**, 2874–2880.

76. **Wang RL, Stec A, Hey J et al.** (1999) The limits of selection during maize domestication. *Nature* **398**, 236–239.

77. **Wiener P & Wilkinson S** (2011) Deciphering the genetic basis of animal domestication. *Proc. Biol. Sci.* **278**, 3161–3170.

78. **Wolfe ND, Dunavan CP & Diamond J** (2007) Origins of major human infectious diseases. *Nature* **447**, 279–283.

79. **Wood ET, Stover DA, Ehret C et al.** (2005) Contrasting patterns of Y chromosome and mtDNA variation in Africa: evidence for sex-biased demographic processes. *Eur. J. Hum. Genet.* **13**, 867–876.

80. **Yang CC, Kawahara Y, Mizuno H et al.** (2012) Independent domestication of Asian rice followed by gene flow from *japonica* to *indica*. *Mol. Biol. Evol.* **29**, 1471–1479.

81. **Zilhão J** (2001) Radiocarbon evidence for maritime pioneer colonization at the origins of farming in west Mediterranean Europe. *Proc. Natl Acad. Sci. USA* **98**, 14180–14185.

INTO NEW-FOUND LANDS

CHAPTER THIRTEEN

In Chapter 11 we discussed the evidence for a recent origin of **anatomically modern humans** in Africa. We also saw how, 60–45 KYA, modern humans settled Eurasia and Australia. Several areas currently occupied by humans were not settled during this primary exodus from Africa, but were discovered and peopled more recently. These "new-found lands" include the Americas, islands in the Pacific and elsewhere, and the extreme northerly latitudes of Eurasia. In contrast, some areas previously settled by humans are now no longer inhabited. As we shall see, this is primarily due to rising sea levels submerging low-lying **continental shelves**.

In this chapter we will explore the evidence, both genetic and nongenetic, for the settlement of these new lands. We will be asking when it was accomplished, by which route, and by whom. We will investigate the role of changing sea levels in influencing human dispersals, and then consider two case studies that have captured the imagination of generations of prehistorians: the peopling of the Americas and the settlement of the Pacific. These new-found lands, and the act of reaching them, would have exerted novel selective pressures on migrating populations. In addition, the **founder effects** that inevitably accompanied such migrations have also shaped modern genetic diversity. We will encounter the evidence for these two processes, as well as their implications, in this chapter.

13.1 SETTLEMENT OF THE NEW TERRITORIES

Sea levels have changed since the out-of-Africa migration

In **Section 11.1** we saw that over the past 100 KY the global climate fluctuated considerably and was generally appreciably cooler than the past 10 KY of stable warmth. We also saw that during this colder period, far more of the world's water was tied up in huge **ice sheets** at extreme latitudes, and that sea levels were therefore much lower than they are at present. Since sea levels determine both the amount and accessibility of available land, we will now explore the nature of sea level change in more detail.

The proportion of the world's water tied up in ice sheets is only one (**eustatic**) factor that determines sea levels (**Figure 13.1**). Two other important factors are regional (**isostatic**) changes to the Earth's crust and thermal expansion of water. Isostatic changes can either alter the volume of an oceanic basin, or the height of a landmass. The Earth's crust lies on top of the liquid interior of our planet in a set of tiled tectonic plates. Consequently the weight of water in oceanic basins and ice caps on landmasses causes the crust to be depressed. This effect is illustrated by the "rebound" of land previously covered by the ice caps. For example, Chesapeake Bay in northeast USA is still rising by as much as

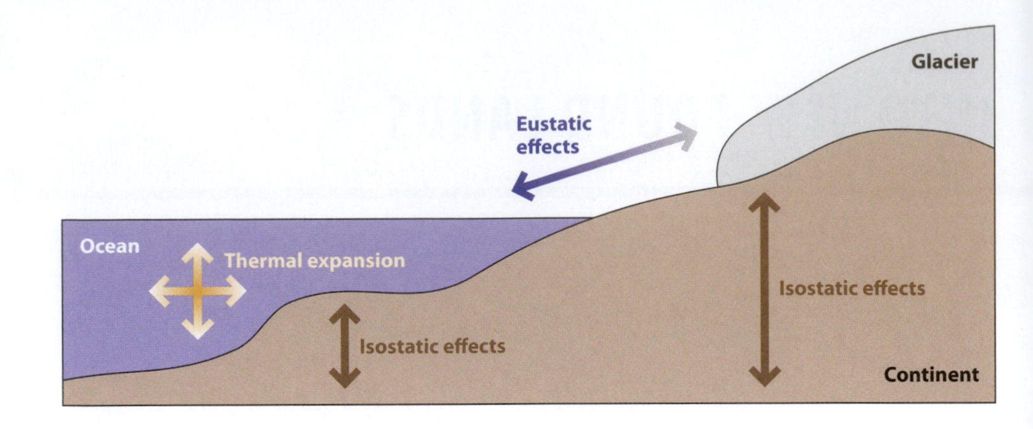

Figure 13.1: Three factors influencing sea levels.

3–4 mm per year as a result of the removal of a former ice cap. The thermal expansion of water affects global sea levels because at higher temperatures water becomes less dense and the same mass of water occupies a larger volume.

Figure 13.2 shows the approximate sea levels over the past 40 KY. The lowest sea levels existed during the **last glacial maximum** (**LGM**), ~20 KYA. Sea level rises since the LGM have not been smooth, because sudden thaws of the ice sheets and breakage of dams retaining inland lakes released huge volumes of water that resulted in rapid sea level rises known as **meltwater pulses**. These can be observed to occur at the same time in sea-level records from around the world (for example, **radiocarbon**-dated remains of submerged corals or mangrove swamps, which were previously close to sea level). Two meltwater pulses are thought to have bracketed a cold period, known as the Younger Dryas, that occurred between 13 and 11.5 KYA. Rising sea levels finally tailed off about 6 KYA, after having risen approximately 120 m over 14 KY.

To put these changes into perspective, it has been predicted that if all remaining glaciers (>160,000), ice caps (70), and ice sheets (two) were to melt, present sea levels would only rise by about 70 m.[27] The global average sea levels shown in Figure 13.2 can be used to predict the approximate positions of coastlines at any time in the past 40 KY from an accurate map of modern **seafloor topography**. More precise reconstructions would require that regional isostatic effects are taken into account, but these are only known accurately for a few locations worldwide. **Figure 13.3** shows the approximate extent of additional exposed land during the LGM, when sea levels were 120 m lower than today, and at the end of the Younger Dryas ~11.5 KYA, when sea levels were about 50 m lower. Rising sea levels would have had the greatest effect on continental shelves with shallow gradients, where a small change in sea level could cause a major change in landmass.

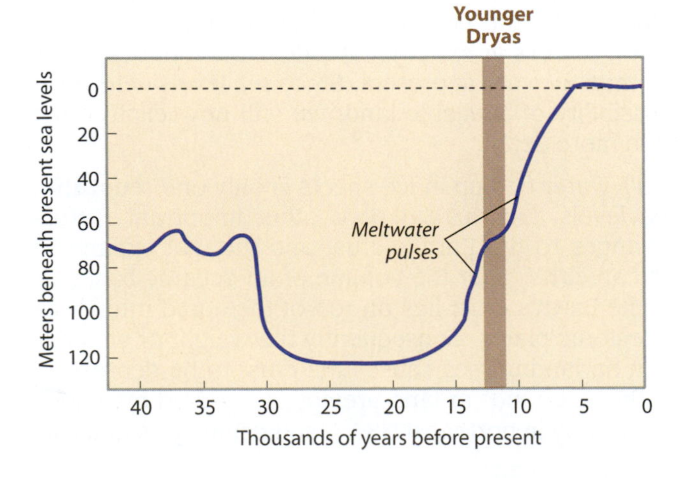

Figure 13.2: Sea levels since 40 KYA. The Younger Dryas cold period interrupts the two meltwater pulses at the end of the last ice age. [From IPCC (2001). Climate Change 2001: The Scientific Basis. Geneva: Intergovernmental Panel on Climate Change. And from Lambeck K et al. (2002) *Nature* 419, 199. With permission from Macmillan Publishers Ltd.]

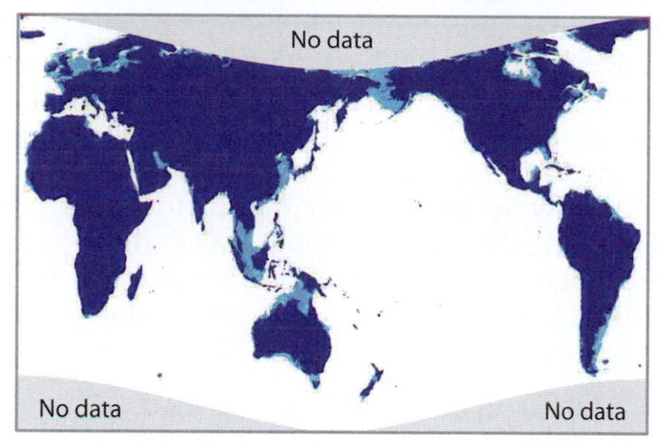

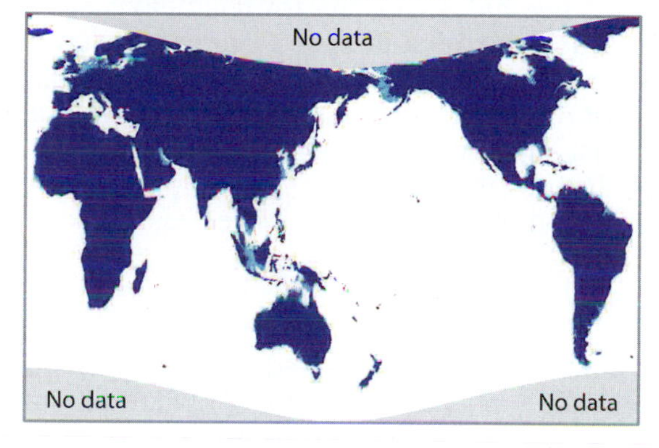

Figure 13.3: Approximate coastlines during the LGM and at the end of the Younger Dryas.
Present-day continents are shown in *dark blue*; *light blue* denotes the extra landmass at lower sea levels. [Bathymetric (sea depth) data from Smith WHF & Sandwell DT (1997) *Science* 277, 1956.]

Four areas of the world were especially prone to large changes in landmass as a result of rising sea levels:

- The **Sahul** landmass, which consisted of Australia, Tasmania, and New Guinea.
- Island Southeast Asia, which during the LGM formed a large landmass known as **Sunda**.
- The Bering Straits between Alaska and Eastern Siberia, which formed a land bridge between the Old and New Worlds, known as **Beringia**.
- The English Channel and the North Sea which, at the time of the LGM and until about 7–8 KYA, formed a land bridge between Britain and the European continent.

These dramatic changes in sea levels would have precipitated rapid changes in settlement patterns and perhaps population crashes; many **hunter-gatherers** occupied coastal niches. Despite the apparently recent malaria-driven retreat of some tropical populations to higher altitudes, modern populations are still disproportionately skewed toward coastal regions.[5]

About 14 KYA, sea levels may have risen by as much as 16 m per century.[22]. By comparison, in our admittedly more densely populated modern world, a sea level rise of only 1 m over the next century (or even greater by some predictions) would result in the likely relocation of 140 million people (10% of the population) in China and Bangladesh alone.

What drives new settlement of uninhabited lands?

The agricultural dispersals described in Chapter 12 were facilitated by a period of unprecedented climatic stability, and expanded into lands previously settled by hunter-gatherers. By contrast, since the out-of-Africa migrations, changing environments facilitated the settlement of previously uninhabited lands by a number of means. First, retreating ice sheets revealed land bridges leading to uninhabited landmasses; however, in some cases sea level rises meant that these bridges were only passable during a limited time window (for example, the **Beringian** land bridge between Northeast Siberia and North America). Second, warming climates allowed flora and fauna to follow retreating ice sheets into more extreme latitudes and altitudes. Humans could then have extended their range without needing to change their diet. In theory, this could allow the mixing of populations that were previously confined to geographically restricted **glacial refugia**. However, there are differences of opinion as to whether these refugia were indeed isolated from one another. Some environments could have been buffered against climate change and certain topographical features in more northerly latitudes might have formed hospitable microclimates.

Thus far we have primarily considered external factors that determine when and where humans might migrate. Clearly, however, humans occupy a far broader range of environments than most other mammalian species by virtue of behavioral adaptability and a capacity for technological innovation. These cultural changes can facilitate the settlement of previously uninhabited lands by allowing previously inaccessible lands to be reached, perhaps by use of a new means of transport or navigation (for example, the double-hulled ocean-going canoes of Polynesians); and, enabling previously inhospitable conditions to be tolerated, through changes in diet, lifestyle, and construction of shelters (for example, the construction of stone houses on treeless islands).

13.2 PEOPLING OF THE AMERICAS

The origins of the Americans have attracted enormous attention and controversy, focusing particularly on two questions: (1) when did humans first enter the Americas and where did they come from? And (2), how many major migrations were there? Despite the continuing controversy about the time of the first entry, which is discussed below, it is agreed that the Americas were the last of the inhabited continents to be settled by humans. Current genetic and archaeological evidence implies a dispersal from a Siberian source toward the Bering land bridge no earlier than 32 KYA (Yana Rhinoceros Horn Site, **Figure 13.4**) followed by a migration from Beringia into the Americas after 18 KYA.[16] Two routes have been proposed (see Figure 13.4): one by land from Siberia across the Beringian land bridge and down an ice-free corridor in North America; and the other by sea along the coast of Siberia, Beringia, and North America. The question of which route was used is important, because some sections of the land route were only passable during limited periods, and so this constrains the possible dates of entry. In contrast, the coastal route does not impose such obvious constraints. The Beringian land bridge was most extensive when most ice was present in ice caps, but this same ice abundance impeded movement further into North America; the simultaneous presence of both the land bridge and the ice-free corridor was thus limited to the time before 36 KYA and a short period near the end of the ice age (**Figure 13.5**). People could have moved into Alaska in between these dates, and could thus have been present in North America, but, according to this model, would not have been able to penetrate far inland at this time.

Analysis of the northwest coastal environment is complicated by the sea level changes that have taken place: the late Pleistocene coastline is now under water and difficult to investigate for archaeological sites. Nevertheless, human remains from near the coast dated to 11 KYA (see below), and stone tools recovered from under water off the coast of British Columbia from deposits of 10.6 KYA, demonstrate human presence near the coast in the early postglacial period. Earlier use of this route during the glacial period seems possible and is an important focus for future research.

Other alternatives, such as travel across the Northern Atlantic, have been suggested, but are much less plausible. Sources of evidence for the peopling of the Americas, as for other parts of the world (Chapter 9), are information about

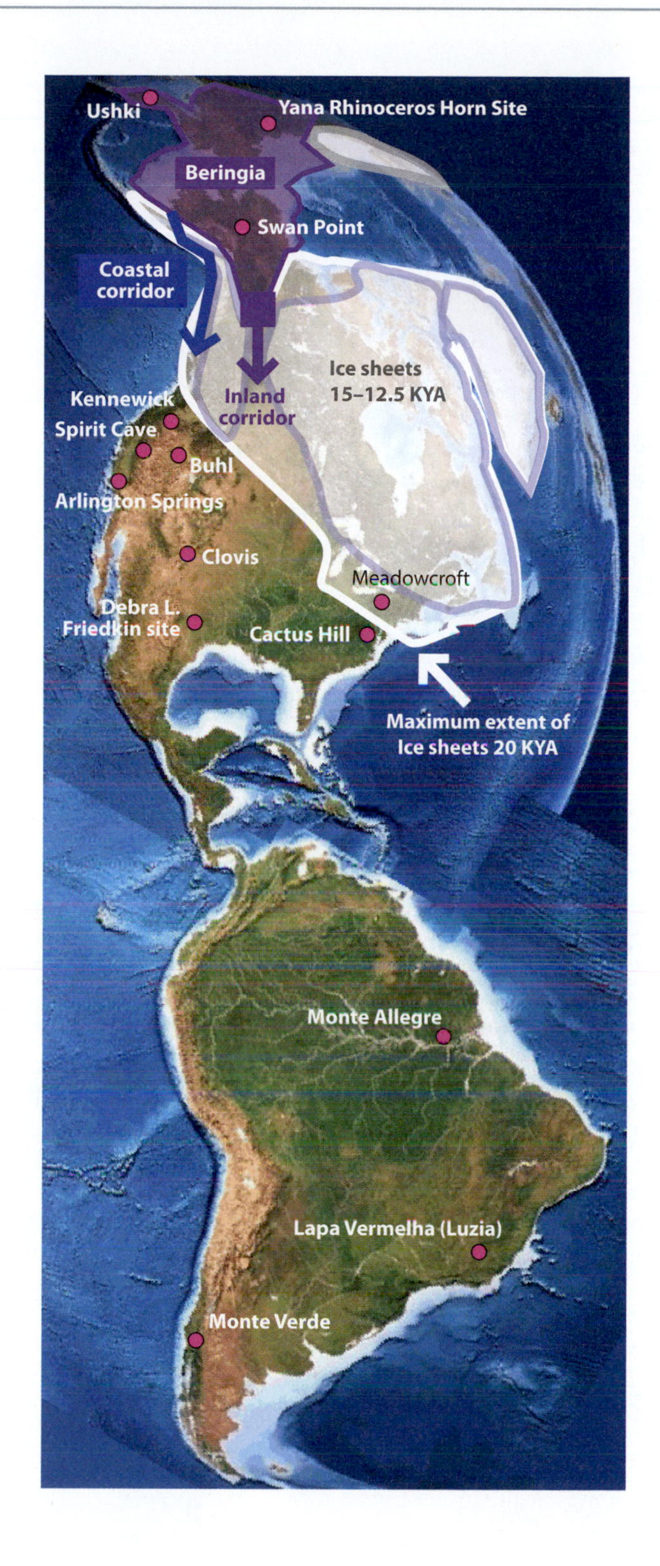

Figure 13.4: Suggested routes and locations of selected fossil and archaeological sites in the Americas. The first settlers could have traveled by land (*purple arrows*) across the Beringian land bridge into Alaska and then down the ice-free corridor, although these two routes were not usually available at the same time. Alternatively, they could have traveled along the coast (*blue arrows*), using land or sea. Scenarios in which travel is inland for some of the distance and coastal for the rest are also possible. Surprisingly, the earliest widely accepted site (Monte Verde) and human fossil (Luzia) are both in South America.

the environmental conditions, linguistics, paleontology and archaeology, and genetics; indeed, the development and first applications of several methods for analyzing genetic data have been driven by the desire to answer questions about human entry into the Americas.

The changing environment has provided several opportunities for the peopling of the New World

Climate during the period of interest was dominated by the ice age, with the full glacial period or LGM around 18–21 KYA; for a discussion of the difference between calibrated and radiocarbon years, see **Figure 13.6** and Box 11.2.

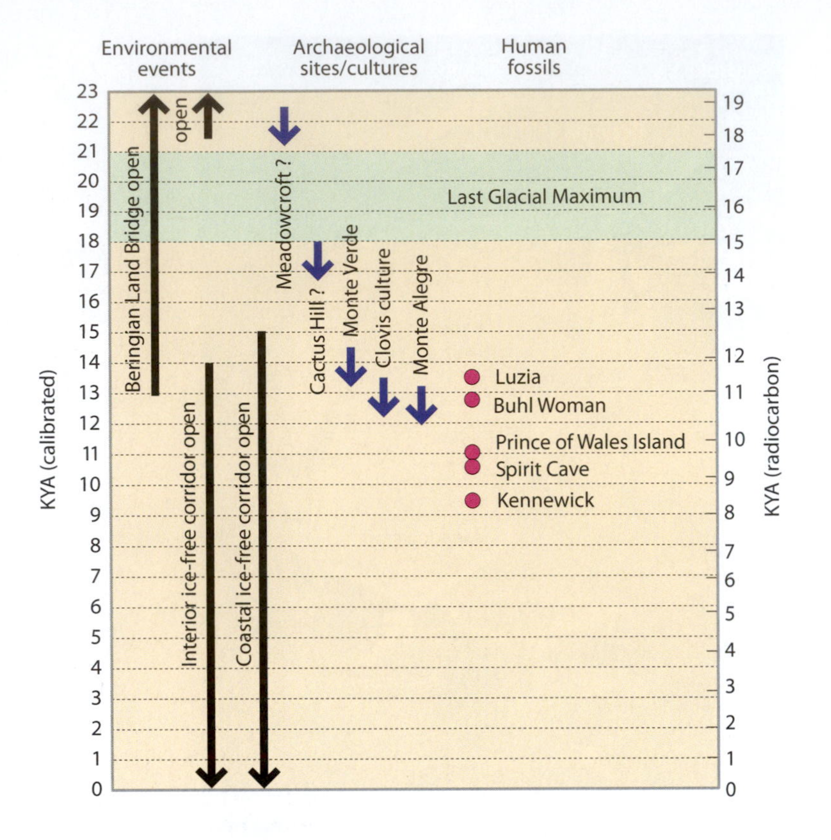

Figure 13.5: Chronology of events and sites relevant to the peopling of the Americas. Geological and climatic events determined the environmental background. Selected archaeological sites and human fossils are included. Note the different radiocarbon (*right*) and calibrated (*left*) time-scales.

Dates in this chapter will all be calibrated unless otherwise stated. In addition to the low temperature, and its consequences for plant and animal life, there were significant effects on sea level (Figure 13.3). The Beringian land bridge was present between about 65 and 36 KYA, and again between 30 KYA and perhaps as recently as 13 KYA (Figure 13.5), and reached a maximum width of about 2000 km (~1200 miles). The hearth and dwelling features associated with human occupation at the Kamchatkan site Ushki and distinctive **micro-blades** found at Swan Point are the earliest (13–14 KYA) archaeological sites recovered so far from Beringia.[16] The land in Beringia was not covered by an ice cap, but was probably tundra, dominated by grass, sedge, and *Artemisia* (a genus that includes sagebrush) and inhabited by a rich fauna including mammoths, bison, horses, and camels. Further south, for much of the **Pleistocene**,

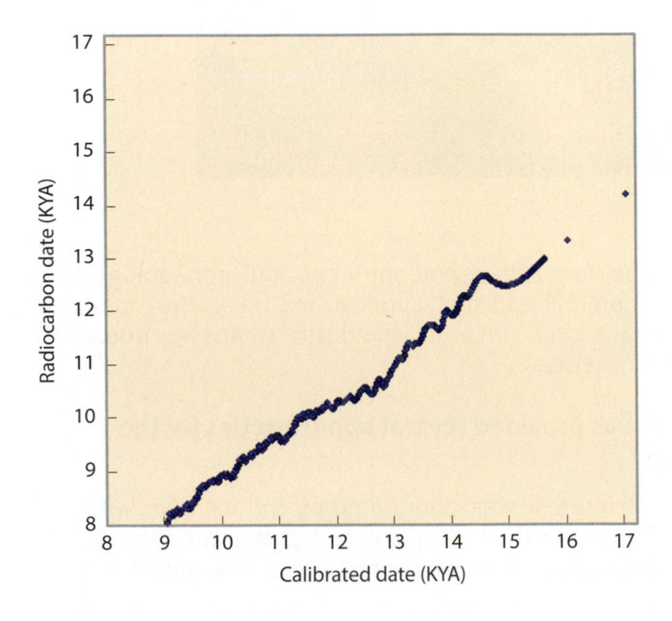

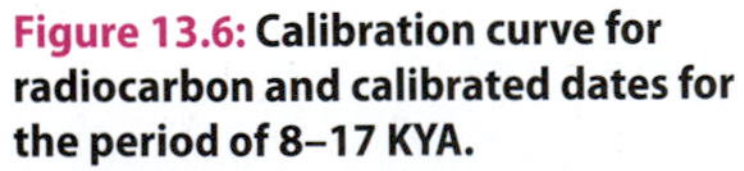

Figure 13.6: Calibration curve for radiocarbon and calibrated dates for the period of 8–17 KYA. Dates for fossils or archaeological remains are often given in "radiocarbon years" rather than "calibrated" or "calendar years" (Box 11.2). The difference for the time period discussed here can be as much as 2 KY, so that a radiocarbon date of 11 KYA corresponds to a calibrated date of 13 KYA. Note that it may often be unclear whether dates encountered in the literature are calibrated or not.

any travelers from Beringia would then have encountered a major obstacle: an ice sheet covering a large part of North America, which would have formed an impenetrable barrier. At times, it was divided into eastern (Laurentide) and western (Cordilleran) sections by an ice-free corridor stretching from present-day Yukon, through Canada, to Montana. While the environment in this corridor would still have been inhospitable, it constitutes one possible route into the rest of the Americas. It is thought that this corridor was open before the LGM until 24 KYA and again after 14 KYA.[16] The coastal ice-free corridor emerged probably before the interior corridor, approximately 15 KYA (see Figure 13.5). Since archaeological evidence is still lacking, it is difficult to judge which route(s) the first Americans took. Approximately 13-KY-old human remains from Arlington Springs, Santa Rosa Island, California, however, suggest that the early settlers of North America used watercraft.[16]

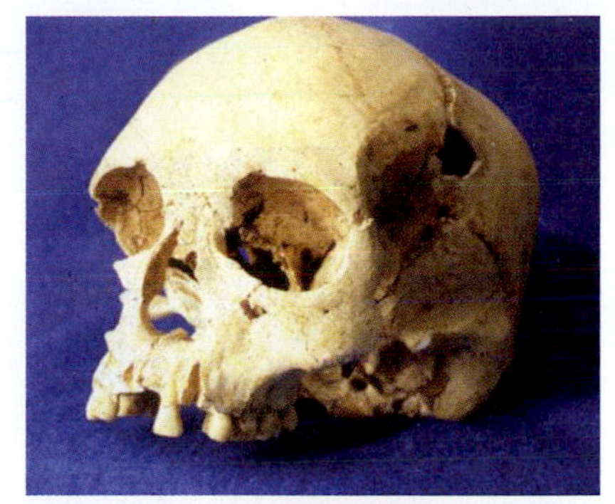

Figure 13.7: Luzia skull. The female skeleton, considered to be one of the oldest Native Americans, was found in Lapa Vermelha, Brazil, in 1975 by archaeologist Annette Laming-Emperaire. (Facial reconstruction by Richard Neave of Manchester University.)

Fossil and archaeological evidence provide a range of dates for the settlement of the New World

Fossils

Human fossils provide unequivocal evidence for the presence of humans in the Americas, and those associated with reliable dates give valuable information about the timing of the early peopling of these regions and the morphology of the settlers. Their study, however, also raises ethical and, in some countries, legal issues, which can be complicated by the difficulty of establishing whose ancestors any fossil is most likely to represent. The number of early fossils is small, but they often attract much attention, so that some are now well known.

"Luzia," possibly the earliest fossil in the Americas, was named after the famous *Australopithecus afarensis* fossil "Lucy" (**Section 9.1**). Luzia's skull (**Figure 13.7**) and some additional bones were found in 1975 in the Lapa Vermelha rock shelter in Minas Gerais, Brazil. Single-grain **optically stimulated luminescence** dates of sediments closest to where Luzia was recovered range from 12.7 to 16.0 KYA,[13] sustaining the claims that Luzia could represent the oldest modern human remains from the New World.

"Buhl Woman" was found in a quarry near Buhl, Idaho, in 1989 with artifacts including an obsidian **biface** and a bone needle, suggesting a burial. She was judged to be a healthy 17–21-year-old and the cause of death was unclear. Dating of bone collagen gave ~12.9 KYA. She was reburied in 1991, after only limited study, in accordance with the Native American Graves Protection and Repatriation Act.

"Prince of Wales Island Man," discovered in 1996, is represented by a lower jaw, vertebrae, and pelvis found with a stone point in the On Your Knees Cave near the coast of Prince of Wales Island, Alaska. The remains were studied by scientists in association with the Tlingit and Haida people, and a fragment of the jaw was dated to ~11 KYA. Ancient DNA analyses performed on these remains[32] revealed that the Y chromosome haplogroup Q1a3-M3 of this man is that carried by a majority of Native Americans living today while his mtDNA belongs to a rare lineage now called D4h3 (see Figures 13.12 and 13.13) that is found at low frequencies along the Pacific coast from Mexico down to Chile and which may have spread together with an early coastal dispersal of the Paleoindians.[41]

"Spirit Cave Man" is represented by a burial of a short man aged 40–45 years with signs of dental abscesses and several injuries, partially mummified by the dry conditions and thus providing unusual insights into ancient perishable materials. He was found in 1940 in a dry rock shelter in Nevada and was wearing moccasins and wrapped in a skin robe and two tule (bulrush) mats. His bones were dated to 10.6 KYA. Examination of his stomach contents showed that his last meal included fish.

"Kennewick Man" (**Figure 13.8**) was discovered in 1996 eroding from the banks of the Columbia River in Washington State and has been the subject of considerable controversy and legal action, which have led to his wide notoriety. Initially classified as a modern skeleton, it was only after a stone projectile point was discovered in the right pelvis bone that radiocarbon dating and DNA analysis were initiated. The skeleton, dated to 9.3–9.5 KYA, was almost complete, and was that of a 40–55-year-old man who had suffered numerous injuries in life, including fractures of six ribs and atrophy of the left arm. Controversy focused on his "European" skeletal morphology: was he truly ancestral to modern Native Americans? Attempts to amplify DNA by several labs have so far been unsuccessful.

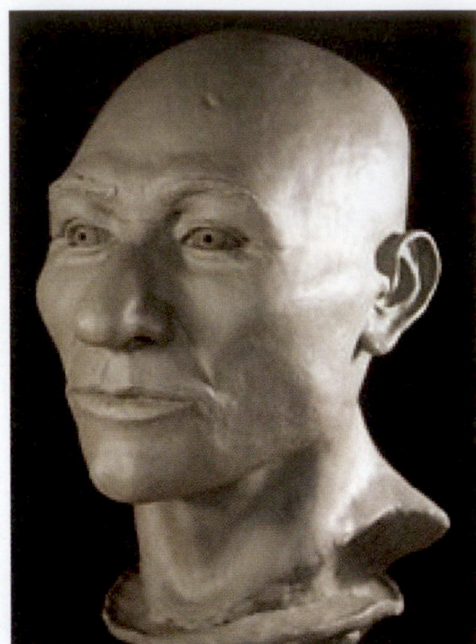

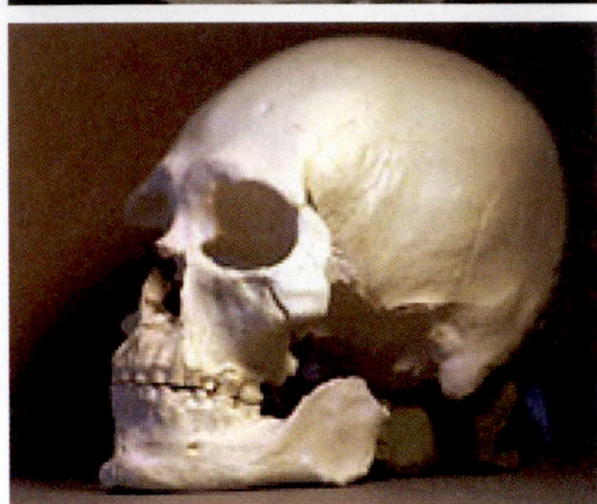

Figure 13.8: Kennewick Man. The skull belongs to a near-complete skeleton of a 40–55-year-old man dating to 9.5–9.3 KYA. Considerable controversy has accompanied this find, including disagreements about the morphological similarities to modern populations. Attempts to analyze DNA have so far been unsuccessful. (With permission from Associated Press.)

Archaeological remains

The **Clovis** culture is universally accepted as clear evidence for humans in the Americas by about 13.5 KYA. There have been many claims for "pre-Clovis" remains, and these have often been highly controversial, although some pre-Clovis sites are now widely accepted. We will therefore start by considering Clovis.

Clovis and the Paleoindians

By 14 KYA, the North American ice sheet was receding and parts of central North America supported an extensive **megafauna** including mammoths and bison. Soon after, the Clovis cultural complex appeared in the archaeological record. It is named after the town of Clovis in New Mexico, near which one of the first sites to be investigated, Blackwater Draw, is located (Figure 13.4). It is characterized by **Clovis points** (**Figure 13.9**), fluted (grooved) projectile points, constructed from a variety of stone types. The earliest Clovis remains date from around 13.5 KYA, and the culture appears to have spread over much of the nonglaciated part of North America within a few hundred years, but was soon replaced by other styles of points. These include Folsom (~12.9–12 KYA), distinguished by their smaller size and longer flutes.

The Clovis people were undoubtedly big game hunters and their points are found associated with, or occasionally embedded in, the bones of large animals. Mammoths were a common prey species, but **mastodons** and smaller animals were also hunted. The Folsom people are particularly associated with bison. However, it is likely that all these **Paleoindians** also ate a wide variety of foods that could have included small game, fish, shellfish, and plants. These may be less well preserved in archaeological sites, so the relative importance of the different resources can be difficult to assess. It has been suggested that the wide and rapid spread of the Clovis people was due to their pursuit of prey, and that the Paleoindians may have been responsible for the extinction of some prey species (**Box 13.1**), necessitating a subsequent change in lifestyle.

Clovis points are not found outside North America, but Paleoindian remains are also known from many sites in South America and these sites are characterized by a diversity of artifacts including fishtail, willow-leaf, and triangular stemmed points. Some are contemporaries of Clovis or Folsom, including those from the cave "Caverna da Pedra Pintada" ("Cave of the Painted Rock") near Monte Alegre in Amazonian Brazil. This site contains rock paintings and yielded biological remains indicating the use of Brazil nuts, fish, turtles, and mussels, as well as triangular points made of quartz and **chalcedony** that could have been used for hunting larger animals. Radiocarbon dates from single plant specimens suggested that the initial occupation occurred from ~13.2 to 12.5 KYA, only ~300 years after the earliest Clovis dates. Thus by ~13 KYA, there were distinct Paleoindian populations with varied lifestyles in both North and South America.

Pre-Clovis sites

The abundance of Clovis and other Paleoindian remains, contrasted with the difficulty of reliably identifying pre-Clovis sites, led to the "Clovis-first" hypothesis: the idea that the Clovis people were the first humans in the Americas. Work over the last few decades, however, has persuaded the majority of investigators that earlier sites do exist. The most widely accepted of these is Monte Verde in southern Chile, a wet site in an upland bog containing stone implements and also organic remains such as charcoal and wooden tools, and even chewed leaves, and also bones with soft tissue adhering. The earliest dates obtained were around 14.5 KYA, a thousand years earlier than Clovis. In North America, the Meadowcroft rock shelter in Pennsylvania has yielded some 20,000 stone flakes, 1 million animal remains, and 1.4 million plant remains. The **stratigraphy** dates to >30 KYA, with the oldest signs of human occupation apparently at 22–23 KYA, but critics suggest that contamination of the dated material with natural carbon may produce artifactually old dates. Other claims for early sites include Cactus Hill, Virginia (~18 KYA), but again critics are not convinced that the site is free from disturbance that might associate the remains signaling human occupation with the earlier material used for dating.[16] However, another, more recently discovered assemblage of small and lightweight artifacts beneath a typical Clovis complex at Debra L. Friedkin site from Texas provides strong support for the view that North America was settled before Clovis. The more than 15,000 pre-Clovis artifacts, known also as the Buttermilk Creek Complex, date to 15.5–13.2 KYA.[50]

Figure 13.9: Clovis points. After ~13.5 KYA, finely made points were manufactured in North America. These Clovis points are housed in the Southern Oregon University Laboratory of Anthropology.

Unresolved issues

More work is required to validate these and other pre-Clovis sites, and major uncertainty still surrounds the question of when humans first entered the Americas. Was it a short time before Clovis, or much earlier? The main reason why a very early date (before 20 KYA) appears unlikely is that humans entering such a favorable environment would be expected to increase rapidly in numbers and leave obvious traces. Few, if any, such traces have been found, although it is of course possible that future discoveries will reveal more. However, the

Box 13.1: Why did the megafauna go extinct?

At ~14 KYA, some 35 genera of large American mammals became extinct, including mammoths, mastodons, camels, horses, giant ground sloths, bears, and saber-toothed cats. Suggested explanations include

- Climate change
- Disease, perhaps introduced by humans
- Overkill by human hunters

It has not been easy to distinguish between these possibilities because climatic warming and human entry (or expansion in numbers to become visible in the archaeological record) occurred at about the same time. It has been suggested that extraterrestrial objects may have struck northern North America 13 KYA, triggering the **Younger Dryas** major cooling event, which could be responsible for the extinction of the North American megafauna.[14] According to this model, the climate change would also reduce the size of human populations. Analyses of 1500 radiocarbon-dated archaeological sites in Canada and the USA associated with human occupation have, however, not supported the evidence of a major bottleneck in Paleoindian population size at this period.[3]

However, there have been many fluctuations in the climate over the last few million years which did not lead to mass extinctions and there is no good reason why the Pleistocene–Holocene change should have been different, except for the presence of humans. Disease rarely kills all members of a species, as illustrated by the ability of Australian rabbits to survive myxomatosis, and would be unlikely to affect so many different species. In contrast, highly skilled hunters, encountering naive prey, could rapidly exterminate a large proportion of the species, according to the "blitzkrieg" overkill hypothesis. Modeling by Alroy[1] examined the effects of varying parameters such as dispersal rate and competition among prey species, human population growth rate, and hunting ability. Most simulations led to a major mass extinction. Thus, from both qualitative and quantitative considerations, it seems likely that humans caused the extinction of the megafauna. Indeed, wherever modern humans have encountered naive animals, including in Australia, Madagascar, and New Zealand, mass extinction has followed. Only the African megafauna, which evolved in contact with hominids for millions of years, avoided mass extinction: they probably adapted to the gradually increasing hunting skills.

In Alroy's scenarios, extinction follows rapidly after the appearance of humans: the median time in these models from the introduction of 100 humans to the extinction of the prey was 1229 years. If this work is a reliable guide to the consequences of hunters entering a new territory, mass extinction would itself provide a recognizable marker in the fossil record for the appearance of modern humans.

acceptance of even a single pre-Clovis site, Monte Verde, has major implications. It would have taken a significant amount of time for humans to travel the 10,000 miles from Alaska to Chile (Figure 13.4), since people adapted to a cold Arctic environment would have had to move through temperate and tropical regions. How much time would this have required? If it was a few thousand years, the ice-free corridor might not have been open. Did these people travel by a fast coastal route, the so-called "Kelp Highway"?[11] These questions remain unanswered.

Did the first settlers go extinct?

The early (>8 KYA) Paleoindian skeletal material is often morphologically distinct from modern Native Americans: an observation reflected in the media by descriptions of Kennewick Man as "European" in appearance, or Luzia as "Australian." Attempts have been made to summarize the differences by describing the early individuals as having long (measured front to back) skulls with narrow faces, prominent noses, and inconspicuous cheekbones, in contrast to modern Native Americans who have more rounded skulls, flatter broader faces, smaller noses, and more widely flaring cheekbones. However, there are so few reliably dated ancient crania, and they are so diverse, that it may be unreasonable to regard them as a single group. Morphometric comparisons (see **Figure 13.10**) of modern population samples with ancient fossils illustrate the diversity of the four oldest crania used in that study (which included Spirit Cave), and, for three of them, their difference from most modern Native American populations: they show more similarity to Inuits, Polynesians, or Ainu. There is evidence that some of these morphologically diverse populations survived into historical times, for example, in the isolated Baja California Peninsula in Mexico, where skulls dating between 2.8 and 0.3 KYA resembled Paleoindians.[17]

It thus appears that there are real differences in morphology, but what do these tell us? Cranial morphology is influenced by environmental as well as genetic factors (see **Section 15.2**). Lifestyle and diet differed significantly between the late Pleistocene/early Holocene and more recent times, even before European contact. It would thus be premature to conclude that the initial settlers were genetically distinct: ancient DNA analysis is needed for this, and would be of considerable scientific interest. It would be even more rash to deduce that there were different origins or migrations for these first settlers. A comprehensive

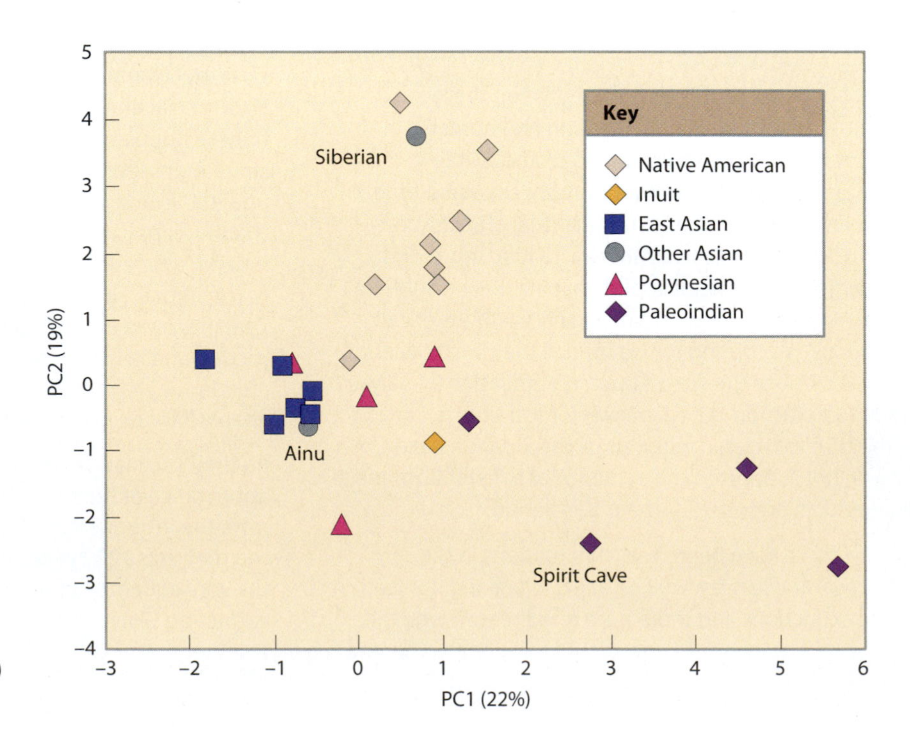

Figure 13.10: Morphometric analysis comparing individual early North American crania with modern population samples.
[Data from Jantz RL and Owsley DW (2001) *Am. J. Phys. Anthropol.* 114, 146.]

analysis of early American and European skulls together with contemporary data concluded that the continental differentiation of modern human morphology is probably a late phenomenon that occurred after the initial settlement of the Americas.[25]

A three-migration hypothesis has been suggested on linguistic grounds

In 1986, Greenberg, Turner, and Zegura published a controversial proposal that has been at the center of much of the subsequent debate: in a synthesis of linguistic, dental, and genetic evidence, they suggested that three separate migrations had contributed to the settlement of the Americas.[20] Languages were classified into three families:

- **Eskimo-Aleut** (the name "Inuit" is used for the people, but "Eskimo" for the language), spoken in the far north, as well as in Greenland and parts of Siberia. This family is widely recognized and accepted.
- **Na-Dene**, spoken in parts of North America. This family has deeper divisions than Eskimo-Aleut, and is considered by many linguists to be related to the Yeniseian family of languages of Central Siberia.
- **Amerind**, a vast family containing the remaining indigenous languages spoken in North America, and all endemic languages from Central and South America. In contrast to the two previous families, the grouping of these diverse languages into a single family, with the implication of a common descent from a hypothetical proto-Amerind, is highly controversial and not supported by most linguists. Some conservative linguists find the evidence for universal features that would support the existence of a single or even a few major language families in the Americas lacking and would place them in hundreds of different families, with as many languages remaining as unclassified isolates.[4, 44] However, because of historical reasons and its simplicity, the Greenberg classification scheme has remained widely used by geneticists as a test hypothesis for assessing the number of distinct dispersal events to the Americas.

Greenberg et al. hypothesized that the first Amerind dispersal dated to >11 KYA, the second Na-Dene dispersal occurred around 9 KYA, and the third Eskimo-Aleutian dispersal was dated to around 4 KYA. In the following sections, therefore, the reader is advised to remember that this classification scheme is not widely supported by linguists.

Genetic evidence has been used to test the single- and the three-wave migration scenarios

Greenberg et al. reviewed the genetic evidence available to them, mainly from **classical polymorphisms**, and concluded that it was compatible with their three-migration hypothesis, but provided only "*supplementary rather than primary*" support. Considerably more genetic data are now available and can be used to ask: Which Old World populations are most similar to Native Americans? How diverse are Native Americans compared with other continental populations? Are there distinct genetic subgroups within Native Americans? What insights into their demographic history can genetic analyses provide?

A problem in interpreting the results of genetic analysis in Native American populations, particularly those from North America, is that of pervasive **admixture** with other populations, including Europeans and Africans, over the last few hundred years (see Chapter 14). This admixture has important consequences: it increases genetic diversity, and it introduces foreign lineages that could lead to incorrect conclusions if their origins are not identified. In addition, the definition of population membership is perhaps less straightforward than in other parts of the world (see Box 1.1 and **Section 10.1**).

Overall, classical polymorphism analyses suggested that Native American populations were most similar to North Asian populations, consistent with entry

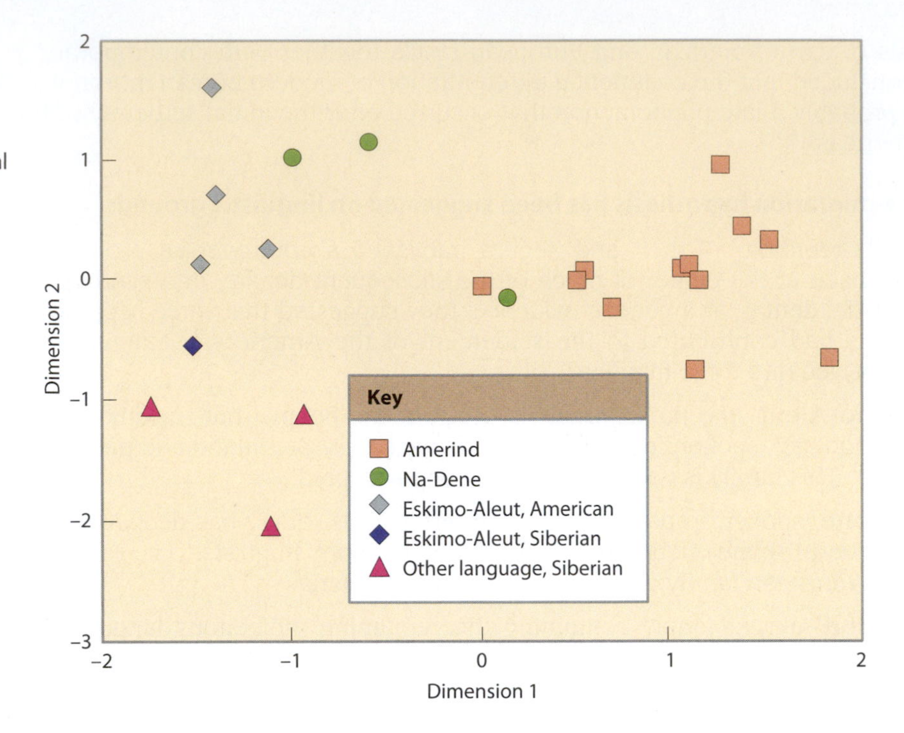

Figure 13.11: Multi-dimensional scaling analysis of population pairwise F_{ST} values calculated from classical polymorphism frequencies. For more information on multi-dimensional scaling see Section 6.3. [Data from Cavalli-Sforza LL (1994) The History and Geography of Human Genes. Princeton University Press.]

into the Americas through Beringia. A more extensive analysis of classical polymorphism data was presented by Cavalli-Sforza and colleagues, who calculated **pairwise** F_{ST} values (Box 5.2) between 20 American populations or groups of populations, and three Siberian populations. Their data (**Figure 13.11**) showed the distinction between the American and Siberian populations, and revealed two clusters within America: one containing the Arctic populations, and the other the remaining North American and all the South American populations. There was thus some correlation with language, since all Eskimo-Aleut speakers fell into the Arctic cluster and all Amerind speakers into the second cluster. However, the Na-Dene speakers did not form a separate third cluster, but resembled their geographical neighbors: the northern Na-Dene fell into the Arctic cluster and the southern Na-Dene into the Amerind cluster. The authors concluded that "*genetic analysis fully confirms the division of American natives into three major clusters*," but this required the assumption of considerable admixture between the Na-Dene speakers and their neighbors, and/or similar source populations for Eskimo-Aleut and Na-Dene. An alternative conclusion would be that there are two genetic groups and the Na-Dene speakers do not form a single genetic unit.

Mitochondrial DNA evidence

One of the first mtDNA studies in Native Americans identified four **haplogroups** that captured most of the variation observed in the New World.[48] The four haplogroups, which also appeared to be common among Asian populations, were designated A, B, C, and D. Since this study, improved resolution based on the analysis of complete sequences of mtDNA (http://www.phylotree.org/) has led to the recognition of subdivisions within the A, B, C, and D groups, and an additional rare haplogroup X, so that at least 15 subclades can claim founder status in the peopling of the Americas (**Figure 13.12**).

The distributions and times to the most recent common ancestor (**TMRCAs**) of the American founder-haplogroups have been of considerable interest. It must be remembered that a TMRCA does not date a migration, but the TMRCA of a lineage shared between two locales must pre-date a migration between them. Eskimo-Aleut populations have characteristically high frequencies of A2a and A2b, and lower incidences of D2a and D3 founder lineages. Because these four founder haplogroups have younger coalescent times than the other founders,

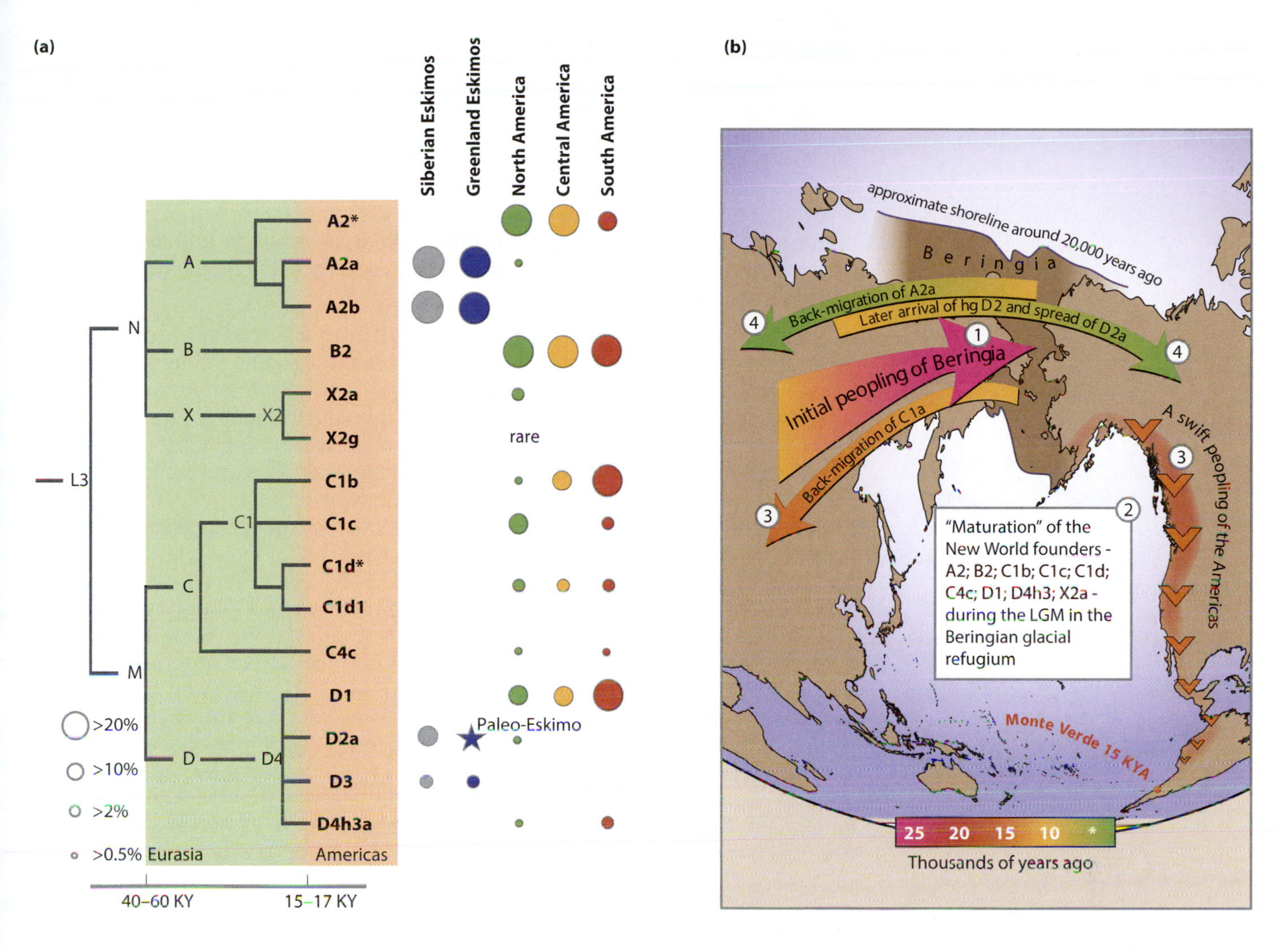

Figure 13.12: Phylogenetic tree and dispersal map of Native American mtDNA founder haplogroups. (a) Phylogenetic tree of mtDNA founder haplogroups and their frequencies. (b) A schematic map of the founder dispersals into and out of Beringia. Colors of the arrows correspond to approximate timing of the events and are decoded in the colored time-bar. The initial peopling of Beringia (1) was followed by a standstill (2) after which the ancestors of the Native Americans spread swiftly all over the New World while some of the Beringian maternal lineages—for example C1a—spread westward (3). More recent (4) genetic exchange is manifested by back-migration of A2a into Siberia and the spread of D2a into northeastern America that postdated the initial peopling of the New World. [a, adapted from Perego UA et al. (2010) *Genome Res.* 20, 1174. With permission from Cold Spring Harbor Laboratory Press. Additional data from Gilbert MT et al. (2008) *Science* 320, 1787; Helgason A et al. (2006) *Am. J. Phys. Anthropol.* 130, 123; Malhi RS et al. (2009) *Am. J. Phys. Anthropol.* 140, 203. b, from Tamm E et al. (2007) *PLoS One* 2, e829. With permission from Public Library of Science.]

and because they are closely related to mtDNA lineages of the Northeast Asian Chukchi, Koryak, and Eskimo-Aleutian populations (for example, the Siberian Eskimos in Figure 13.12) while not being found elsewhere in the Americas, it is most plausible that Inuits made their way to North America and to Greenland long after most of the Americas were already settled by other Native American groups. Ancient DNA evidence from Greenland even suggests that the Paleo- (~4.5 KYA) and Neo-Eskimo (~1 KYA) dispersals represent at least two separate events. On the other hand, no uniquely Na-Dene-specific founder haplogroups have been identified so far: subarctic speakers of Athapascan, which forms a major group of Na-Dene languages, are similar in their haplogroup composition (A2a and D2a) to Eskimo-Aleutians, whereas southern Athapascans such as Navajos and Apaches have become more similar to their neighboring Zuni- and Yuman-speaking populations, probably because of recent admixture.[36]

Interpretation of the mtDNA data

While the distribution of mtDNAs in the Americas and potential source regions is reasonably well established, the insights they provide into the number and timing of the migrations remain contentious. The starting assumption is that the modern population data provide useful insights into ancient populations from the same area, but this assumption may not be reliable if there has been extensive drift, admixture, or movement of people. A second factor is that information about molecular lineages provides only indirect information about population origins: the TMRCA of a lineage, for example, does not itself tell us whether the lineage arose in Asia or America.

A glance at Figure 13.12a shows that, apart from the Eskimo-Aleutian lineages, most other founder haplogroups are spread at variable frequencies across North, Central, and South America. Because of their wide distribution and similar coalescent times (15–17 KYA) these Pan-American founder groups have been associated in computer simulations with the model assuming a fast coastal dispersal route in the initial settlement of the Americas. Three exceptions to this pattern are the rare founder haplogroups X2a, X2g, and C4c, which are locally frequent in some Algonquin-speaking groups and have been suggested to derive from the same Beringian gene pool but via a dispersal along the inland ice-free corridor. Taken together, the mtDNA evidence is by and large consistent with a formation of a Beringian mtDNA gene pool by around 18 KYA, followed by two almost simultaneous dispersals into the Americas: (1) one along the coastal ice-free corridor from which the majority of current Native American mtDNA diversity derives, and (2) possibly an additional minor migration through the inland corridor. After the initial settlement of the Americas the Beringian gene pool may have been enriched or even replaced by additional gene flow from Siberia becoming, in the Holocene, the source of additional secondary migrations to North America and Greenland in association with the Eskimo-Aleutian languages. Also, some mtDNA lineages, such as C1a, may have dispersed from Beringia back to Siberia (Figure 13.12b).

Evidence from the Y chromosome

Y chromosomes belonging to one particular haplogroup, Q1a3a (Q1a-M3), are present in all three linguistic groups in the Americas, but are very rare elsewhere, being found only in a few Siberian populations who live near the Bering Strait. This observation does not fit easily with the three-migration hypothesis and, like some interpretations of the mtDNA data, has led to suggestions of a single Beringian gene pool from which all Native American populations have derived their genetic variation. The time and place of origin of this lineage are thus of some interest, and are discussed below.

While this lineage is frequent in the Americas, making up 58% of the Y chromosomes according to the combined results of two substantial surveys including both North and South American populations,[28, 35] additional lineages are also present. Do these represent recent admixture, which is likely to be predominantly male-mediated and thus contribute more to the pool of Y chromosomes than to the mtDNAs, or other indigenous lineages? It is assumed that admixture will be mainly of European (colonist) or African (slave) origin, so lineages found in Native Americans that are frequent in these external populations are ascribed to admixture, while those that are rare or absent from these known sources are candidates for further indigenous lineages. According to these criteria, additional founding lineages are likely to include one or more within each of haplogroups Q1a-P36 and C3-M217 (**Figure 13.13**). The status of other lineages remains uncertain, but some of them, for example, from haplogroup N, may represent further rare founders.

As with the mtDNA lineages, we can now ask what the most likely geographical sources for the founding Y lineages are. Since Q1a3a-M3 seems to have an origin within Beringia or the Americas, this question comes down to finding sources for the precursor to Q1a3a, any additional Native American

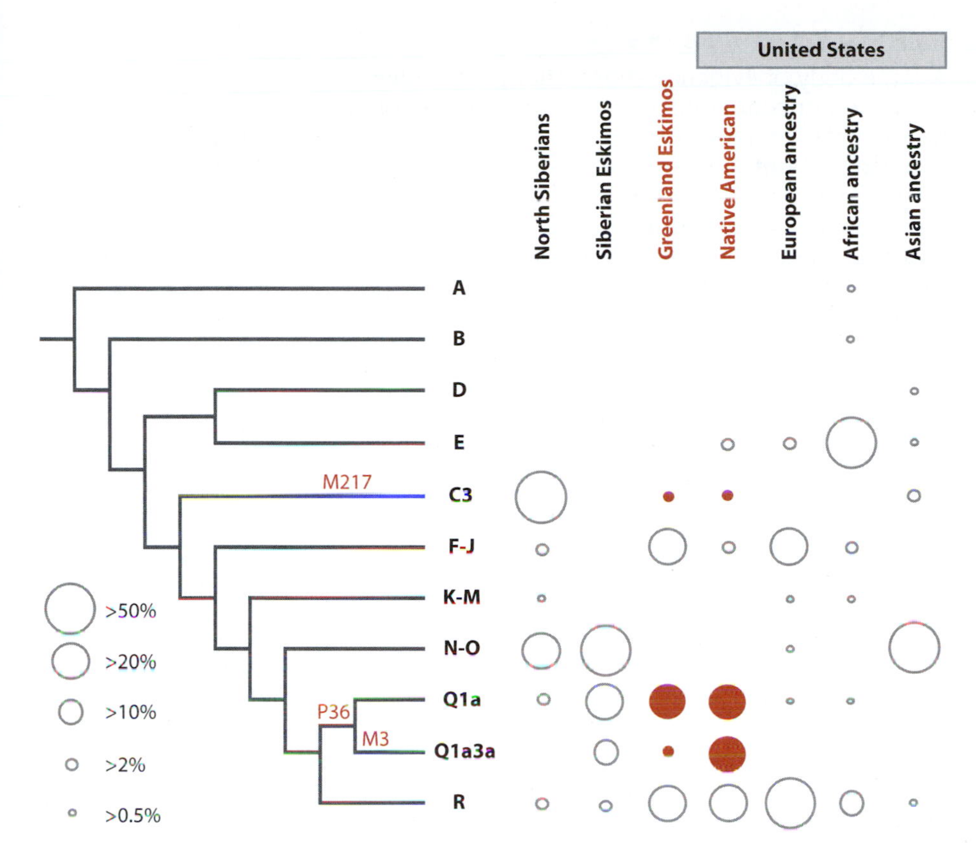

Figure 13.13: Y-chromosomal phylogeny relevant to the peopling of the Americas. Indigenous Native American haplogroups are highlighted in *red*. Haplogroup N may represent a further rare founder lineage. Note that additional polymorphisms can subdivide some of these haplogroups further.

subdivisions of Q, and C3. For haplogroup C3, this is relatively simple. C3 is common throughout much of East Asia, but when additional information about microsatellite subtypes is taken into account, the most likely source is from the region of Lake Baikal.[28, 35] Haplogroup Q has characteristically high frequencies in certain populations of Siberia (Altaians, Selkups, and Kets). Its short tandem repeat (STR)-based age estimates in Asian populations are in the 15–18 KYA range, and while the Native American STR variation is comparable to that of Asian Q lineages, these data are in line with the model implying postglacial dispersal into the Americas. While admixture in the Americas cannot be overlooked, all analyses aiming to infer population ancestry from distinct sources in Asia are based on present-day allele frequencies and mostly assume that lineage distributions in modern Siberian populations are representative of those in the same area prior to the settlement of the Americas. Yet there is likely to have been much subsequent change due to migration, admixture, drift, and other causes. Indeed, as climate has improved since the first migration from Asia to the Americas, be it ~14 KYA or >20 KYA, it is possible that subsequent repopulation of northern Asian latitudes has obscured the ancestral gene pool of Native Americans. If this has happened, it will be very difficult to identify with any reasonable precision a distinct source region in Asia as a founder of the Native American gene pool from modern population data.

The next question is, what can be deduced about the timing of the migrations from the Y data? The wide geographical distribution and high frequency of the Q1a*/Q1a3a lineages in the Americas suggests that they or their precursors were carried by the earlier (or only) migration, with the C3 lineages entering a little later or much later. If Q1a* lineages are found in Siberia and the Americas, but Q1a3a-M3 only in the Americas, the dates of the two mutations defining these lineages bracket the migration. Estimates of these dates have wide error margins, but taken at face value would place the migration after ~20 KYA but before ~10 KYA[9]: consistent with the archaeological data, but doing little to refine it.

Evidence from the autosomes

A large body of evidence shows that genetic diversity is lower in the Americas than in other continents. For example, the use of 377 autosomal microsatellites to characterize 1056 individuals from the HGDP diversity panel (**Section 10.1**) showed that **heterozygosity** was, on average, lowest in the Americas. One of the models that have been put forward to explain decreasing heterozygosity pattern globally is the **serial founder model** by which the peopling of the world was a process involving a series of consecutive founder effects, each decreasing the genetic variation within the daughter populations (Chapter 9). In the case of the peopling of the Americas, the low genomic heterozygosity estimates (Figure 9.15) which show north-to-south declining patterns[43, 54] are in line with the model of one primary wave of dispersal associated with a severe bottleneck at the founding stage. Based on the analyses of multiple nuclear loci, the effective population size at the founding stage of the New World has been estimated within the range of tens,[24] hundreds,[12] or a few thousands[33] of individuals.

Analyses of sequence data from multiple autosomal loci have produced estimates of the divergence times of Native American and Asian populations that range between 10.5 KYA[12] and 22 KYA[21] (95% CI: 7.5–27 KYA). These results, as well as the ~13 KYA age of the variation of an autosomal microsatellite D9S1120, have been interpreted in support of the model according to which all Native American populations derive from a single Asian source.[45] Notably, short alleles at D9S1120 are specifically shared by all Native American populations and Northeast Siberians, consistent with a single Beringian ancestral population. However, support for two separate Asian sources and a two-phase migration scenario has been inferred by analyses of the SNP frequency and allelic linkage patterns in the 51 **HGDP** populations.[23] Multiple dispersals from a Beringian source were also concluded in an ancient DNA study where both Paleo- and Neo-Eskimos from Greenland were inferred to carry distinct ancestry components that they shared with Northeast Siberians but not with other Native American populations. The Na-Dene group appeared as intermediate between the Inuit and "Amerind" groups, likely reflecting local admixture (**Figure 13.14**).

On the basis of analyses of f_4 statistics (**Section 14.3**) on genomewide SNP data from 52 Native American and 17 Siberian populations it has been argued that the majority of the New World populations stem solely from the first dispersal whereas a minority of populations including those speaking Na-Dene and Eskimo-Aleutian languages carry proportions of their ancestry from two

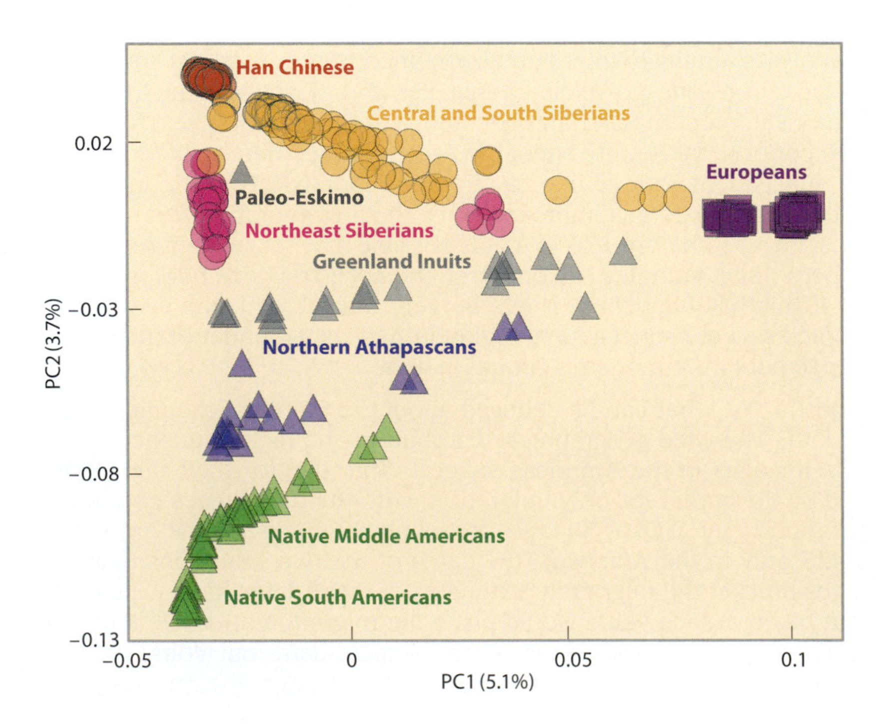

Figure 13.14: Principal component analysis plot of Eurasian and American populations.
[Adapted from Rasmussen M et al. (2010) *Nature* 463, 757. With permission from Macmillan Publishers Ltd.]

additional migration waves.[43] The Na-Dene-speaking Chippewas from Canada were estimated to have approximately one-tenth of their ancestry inherited from the second wave of the Paleo-Eskimo dispersal and the present-day Eskimo-Aleutians were concluded to derive from yet another ancestry component that was formed by a later admixture of the first- and second-wave components (**Figure 13.15**).

Patterns of genetic admixture with European populations, as noted above in the Y-chromosomal context, are also clearly detectable in ancestry inferences based on genomewide data—in the PCA plots (see Chapter 6) an almost continuum of admixed ancestry falls between the European, Native American, and Northeast Asian clusters (Figure 13.14).

Conclusions from the genetic data

Overall, the genetic data tend to support the hypothesis of (at least) three independent waves of dispersal into the Americas. Analyses of genomewide data reveal the similarity of most American populations to one another and their distinction from the rest of the world, supporting their descent from a unique Beringian source. The majority of Native Americans trace their ancestry solely to this first dispersal whereas the Na-Dene and Eskimo-Aleutian populations draw their genetic diversity from a mixture of dispersal waves: the early dispersal and two more recent expansions from the (circum-) Beringian gene pool currently provide the best explanation for the available data (Figure 13.15).

The initial migration at the time suggested by the archaeological data, ~20–15 KYA, would provide a coherent explanation for our current observations. This migration could have established the basic genetic patterning seen throughout most of the New World today. Two additional migrations, at later times, may then have modified this pattern specifically in North America. It seems likely that all dispersal events had their source in or around Beringia and involved elements of the same gene pool sampled at different times. It is also plausible, however, that before the later dispersal events occurred, the (circum-) Beringian gene pool had been enriched by additional gene flow from other Siberian populations that had meanwhile set foot in the neighborhood. Due to advances in technology, it will become feasible in the near future to address the questions of Native American origins at the scale of sequence data from the entire genome. New technological advances also make it likely that ancient DNA evidence will prove crucial in solving many of the current debates.

13.3 PEOPLING OF THE PACIFIC

The origins of Pacific Islanders have intrigued researchers from many disciplines since the voyages of Captain Cook in the late eighteenth century. Cook himself was struck by the mutual intelligibility of languages spoken on islands separated by thousands of kilometers, and by the navigational sophistication necessary to voyage between them.

As we shall see, the likely origins of Pacific Islanders from the west, in Island Southeast Asia, require that we examine in more detail the changing environment over the past 40 KY. Alfred Wallace (the co-discoverer of natural selection) was one of the first to notice the sharp faunal differences that exist within this region. Consequently, the division between the typically Asian ecology in the west and the distinctly different ecology in the east is known as the **Wallace line**. This line closely corresponds with the Sunda landmass in existence during the LGM, described earlier in this chapter (see **Figure 13.16**). It is even possible to reconstruct a major drainage system on the Sunda continent, known as the Molengraaff river, of which many of the major Indonesian rivers on different islands would have been tributaries. The islands that reside between the Sunda and Sahul landmasses are known collectively as **Wallacea**.

The land bridges that joined these islands into their respective landmasses would have been present throughout most of the period of the out-of-Africa dispersal

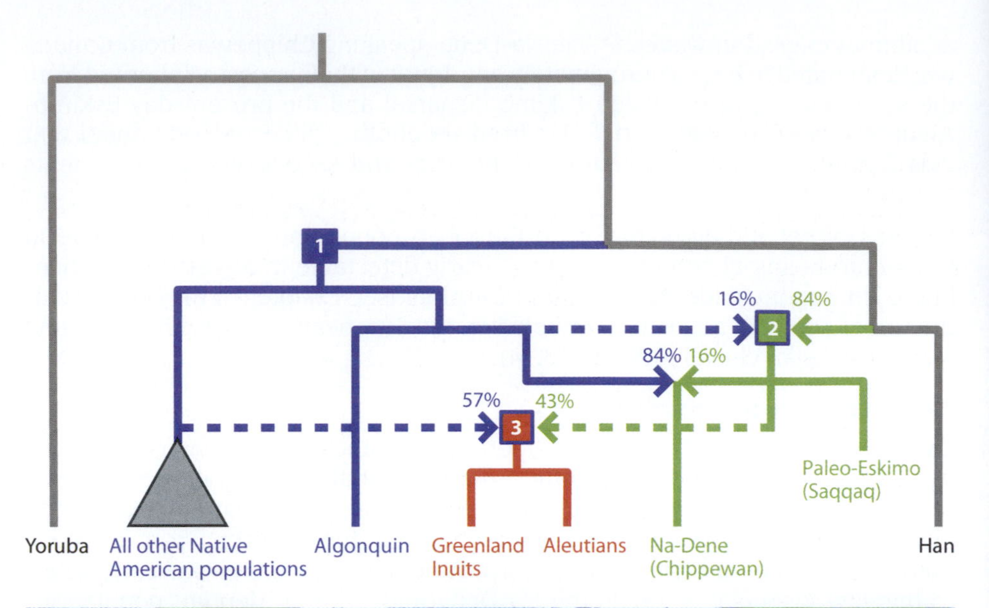

Figure 13.15: Peopling of the Americas from Beringia in three waves. The three Native American dispersal waves with their estimated admixture contributions to extinct and presently living Native American populations were inferred from analyses of genomewide SNP data [Reich D et al. (2012) *Nature* 488, 370]. The dates of the respective migration waves inferred from mtDNA data are shown according to M.T. Gilbert et al. [Gilbert MT et al. (2008) *Science* 320, 1787].

events and would only have been submerged about 8 KYA. Now we turn to the fossil and archaeological evidence for the first settlement of the Pacific.

Fossil and archaeological evidence suggest that Remote Oceania was settled more recently than Near Oceania

The islands of the Pacific are classified into **Near Oceania**, first settled approximately 45 KYA, and **Remote Oceania**, first settled within the last 3.5 KYA. Near Oceania includes New Guinea and islands lying off its northeast coast. New Guinea formed part of the Sahul landmass when it was first settled, as discussed in **Section 11.2**. However, the further reaches of Near Oceania, which includes some islands in what are now the Solomon Islands, would have required substantial voyages of 50–100 km. The earliest evidence for human occupation of the Solomon Islands, the majority of which may have been combined into a single landmass at the time, dates to 29 KYA. There are earlier finds on New Ireland that date to about 35 KYA, which support this early migration eastward from the Sahul continent. These earliest Pacific Islanders would have been hunter-gatherers, dependent on local wild resources. The islands of Remote Oceania have substantially fewer floral and faunal resources (in terms of genera and species) than those in Near Oceania.

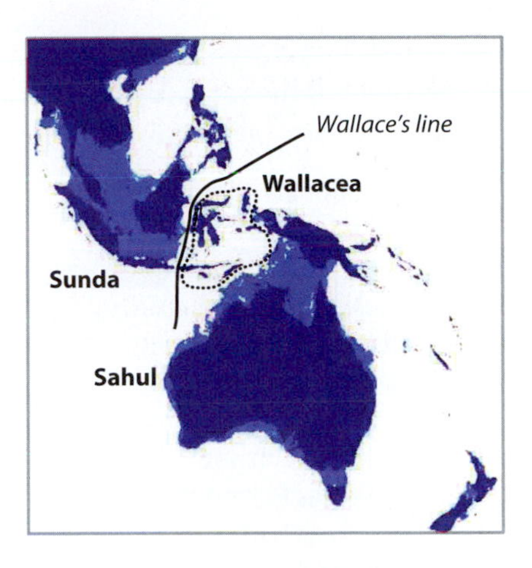

Figure 13.16: Sunda, Sahul, and the Wallace line.
The islands of Wallacea lie between the two great landmasses of the last ice age.

There is no evidence for human occupation further to the east, where distances between islands increase markedly (>350 km), until some 25 KY later. One factor that has been emphasized in determining the distinction between Near and Remote Oceania is "intervisibility"; in other words, the ability to see the destination from the island of origin, albeit only from an elevated viewpoint in some cases. Alternatively, the destination may be seen from the water, before the island of origin has disappeared from view. It has been estimated that all of the sea voyages accomplished during the settlement of Near Oceania would have been toward visible destinations, whereas this is certainly not true of voyages in Remote Oceania. Voyaging continued among the populations of Near Oceania as shown by trade in **obsidian** from New Britain to other islands from 12 KYA onward.

The earliest settlements of islands further to the east, in Remote Oceania, are all associated with the same cultural package, known as **Lapita**. This package included the means for agricultural food production, characteristic pottery and tools, and superior voyaging and navigational technologies. The Lapita complex appears to have its origins 3.5 KYA in the islands just off the north coast of New Guinea, in the Bismarck Archipelago. From here, there is a rapid eastward spread of this culture, taking only 500–600 years to reach Hawaii, Easter Island, and New Zealand (**Figure 13.17**).

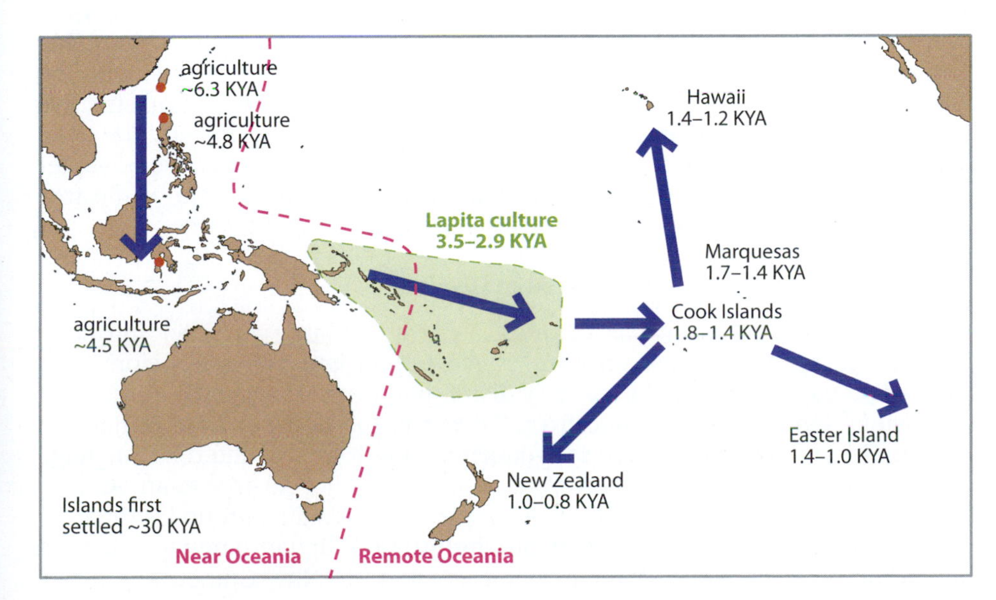

Figure 13.17: Near Oceania, Remote Oceania, and the Lapita culture.
The *pink* dashed line divides the islands of the Pacific Ocean into Near and Remote Oceania. Labels indicate the earliest dates for agriculture in different locations, and arrows indicate the direction of movement inferred from archaeological dates.
The shaded area represents the region occupied by the Lapita culture which spans the boundary between Near and Remote Oceania.

Box 13.2: Is the category "Melanesian" meaningless?

The traditional tripartite classification of the islands of the Pacific into Polynesia, Melanesia, and Micronesia was based partially on dramatic differences in human phenotypes. Melanesians share darker skin and "frizzy" hair, whereas Polynesians and Micronesians have a more "**mongoloid**" appearance. As discussed in Section 15.3, skin color can be an unreliable marker of population prehistory due to its interaction with environmental factors. In addition, only Polynesia appears likely to be a coherent grouping with regard to the archaeological and linguistic evidence. All languages within this region belong to a single subfamily, Polynesian, to the exclusion of other Oceanic languages. By contrast, Melanesia contains islands settled 30 KYA and others 3 KYA, and peoples speaking many different language families. Micronesia contains islands settled from the west and islands settled from the south, and although all are Austronesian speakers, they speak languages belonging to different subfamilies; Oceanic languages are only spoken in central and eastern Micronesia. Peoples residing within the geographical area "Melanesia" are so heterogeneous as to make this description highly ambiguous and capable of almost any interpretation. This classification has also focused attention on Polynesians to the detriment of other inhabitants of Remote Oceania.

Pacific islands have traditionally been divided into three groups: **Polynesia** ("many islands") is the area within a triangle with corners at Hawaii, New Zealand (**Aotearoa**), and Easter Island (**Rapanui**); **Melanesia** ("black islands") is the area to the west of Polynesia, up to and including New Guinea (see **Box 13.2**); and **Micronesia** ("small islands") are the low-lying coral atolls to the north of Melanesia. After the initial spread of the Lapita culture to the western fringe of Polynesia there was a time lag of about 1000 years before a distinctive Polynesian culture developed and spread to previously uninhabited islands, first into central Polynesia, and then finally dispersed to the periphery (Hawaii, Aotearoa, and Rapanui) by about 800 YA (see Figure 13.17). There is also evidence of back-migration of Polynesian cultures to some of the islands to the west, from whence they came. The timing and origins of first migrations to Micronesia are less well characterized. The western fringe of Micronesia appears to have been settled directly from Island Southeast Asia around 3 KYA, whereas most of Micronesia was settled slightly later, probably from islands to the south.

After their initial settlement, and despite the great distances between them, the island groups of Polynesia did not become isolated from one another. Archaeological evidence for long-distance voyaging comes from the frequent identification of stone tools in one **archipelago** with isotope signatures diagnostic of rocks from other archipelagos.[6]

As with the settlement of the Americas, the archaeological record details the destructive ecological impact of these new arrivals on their various islands. Indeed, dramatic ecological changes are commonly understood on Pacific islands as one of the first markers of human settlement. These impacts result from a number of human practices, including habitat modification (for example, land clearance); introduction of **commensal** and domesticated animals and plants that compete for resources with indigenous species (for example, rats); and hunting to extinction (for example, of flightless birds in New Zealand).These impacts could be very rapid. The absence of ground-dwelling mammals in New Zealand resulted in several bird species evolving to efficiently exploit this ecological niche. Loss of the ability to fly was one common consequence for many of these species. This is best exemplified by the 11 species of Moa that existed on New Zealand at the time of the first Polynesian settlements. Some of these birds were over 2 m tall. Within a few hundred years, all of these species had become extinct, over an area of 270 km^2.

A fully developed agricultural package utilizing diverse domesticated species, pottery, and complex navigational skills does not tend to develop in isolation on small islands. This raises the question: what are the origins of the Lapita culture? Agriculture in Island Southeast Asia pre-dates Lapita, and appears to arrive from the north, from continental Southeast Asia. The earliest **Neolithic** remains in this region have produced a southward gradient of dates, with the oldest in Taiwan. Elements of the Lapita culture (for example, the distinctive pottery) clearly owe ancestry to these older Neolithic assemblages. However, **horticulture** has an ancient independent ancestry in New Guinea (see Chapter 12) and other elements of the Lapita culture (for example, some of the tree crops) appear to have been derived from these more local practices.

Two groups of languages are spoken in Oceania

Languages spoken on Pacific islands are traditionally classified into two groups: **Austronesian** and **Papuan**. Austronesian languages all belong to a single language family and are widely distributed from Madagascar in the west to Easter Island in the east, and from Taiwan in the north to New Zealand in the south. By contrast, Papuan languages belong to many different language families, lumped together primarily because they are not Austronesian. As with Amerind languages, Papuan languages represent a residual group that may or may not derive from a common **proto-language**. Papuan languages have a much more restricted distribution, being confined to New Guinea and a few

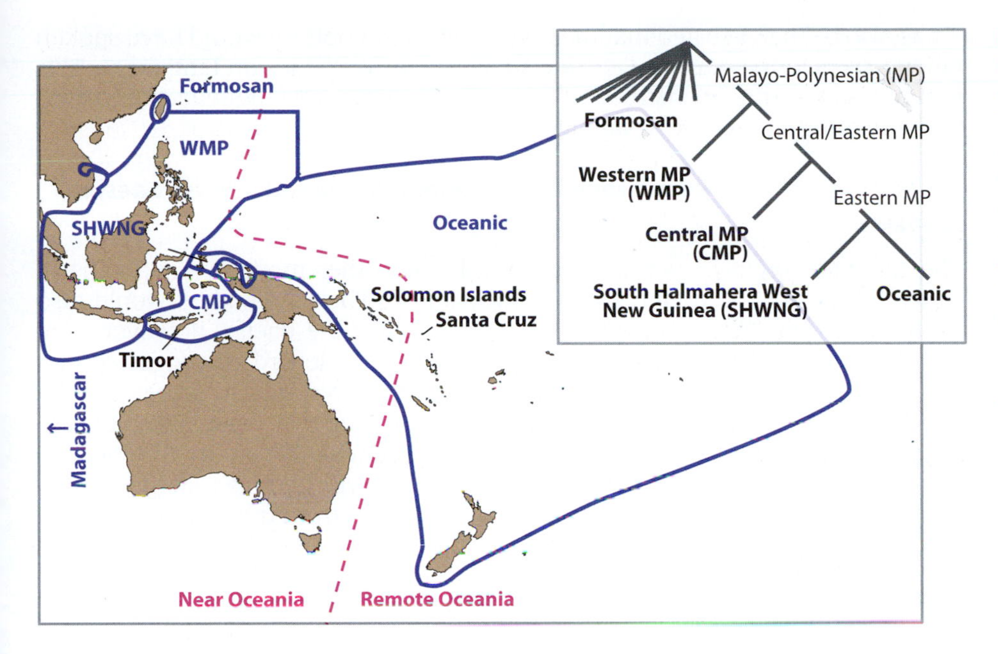

Figure 13.18: Tree of Austronesian languages and their geographical distribution. The map shows the distribution of different Austronesian language groupings in *blue*, relative to the boundary between Near and Remote Oceania. The inset shows the ancestral relationships between these groupings according to Blust R (1999) *Symp. Ser. Inst. Ling. Acad. Sin.* 1, 31; and Pawley A and Ross M (1993) *Annu. Rev. Anthropol.* 22, 425. Timor and Santa Cruz (*black* labels) indicate the western and eastern limits of Papuan languages.

nearby islands, from Timor in the west to the Solomon Islands in the east. The eastern limit of Papuan languages fits exactly with the boundary of Near Oceania, which lies between the Santa Cruz group and the rest of the Solomon Islands to the west (**Figure 13.18**). Papuan languages predominate in New Guinea, where Austronesian languages are restricted to coastal areas. The vast diversity of Papuan languages shares common elements of language structure, including sound system and grammar, and may derive from a late Pleistocene dispersal, significantly pre-dating the inflow of Austronesian-speaking populations in the area.[10]

The fact that all Austronesian languages belong to a single family, and therefore have a recent common origin, is of itself evidence for a recent spread of these languages over their wide geographical range. By examining the pattern of linguistic diversity within Austronesian languages, we can gain further information as to the nature of that spread. Applying **Bayesian** phylogenetic analyses to lexical data, the age of the Austronesian languages has been estimated at approximately 5 KYA.[18] Investigating which Austronesian languages share unique innovations allows subfamilies of related languages to be defined, and a branching pattern similar to a phylogenetic tree to be reconstructed. From this evidence it has become apparent that Taiwan contains by far the most diverse set of Austronesian languages found in a single locale, with nine out of the ten major Austronesian subfamilies confined to this island. All remaining Austronesian languages belong to a single subfamily, Malayo-Polynesian. This subfamily can itself be divided into smaller groups of related subfamilies. The branching structure and geographic distribution of these subfamilies is shown in Figure 13.18. It can be seen that almost all of the languages of Remote Oceania belong to the Malayo-Polynesian subgroup of Austronesian languages.

The close fit between the archaeological and linguistic evidence has led to attempts to meld the evidence into a coherent "archaeo-linguistic" perspective. It has been suggested that the original inhabitants of Near Oceania spoke non-Austronesian languages, and that the Austronesian languages arrived with farming, which was ultimately from Southeast China, but came via Taiwan. The congruence, in Taiwan, of the greatest diversity of Austronesian languages, together with the oldest farming remains in Island Southeast Asia, supports this scenario, as does the similarity between the limits of Papuan languages and the settlement of Near Oceania. Furthermore, reconstructing elements of the vocabulary that must have been present in the ancestral "proto-Austronesian" language reveals that its speakers were indeed farmers and not hunter-gatherers.

It has also been suggested that the makers of Lapita pottery would have spoken languages belonging to the Oceanic subgroup of Austronesian languages. This is supported by the reconstruction in the "proto-Oceanic" language of linguistic terms relating to oceanic voyaging.

Several models have been proposed to explain the spread of Austronesian speakers

The ultimate origin of Pacific Islanders, like all other modern humans, is in Africa. There is little doubt that the ancestors of Pacific Islanders would also have had to pass through continental Asia to reach the Pacific; however, here the different models for Pacific origins diverge. Thor Heyerdahl suggested that the first settlers in the Pacific would have come from the east, somewhere in South America. He backed up his hypothesis by taking the unusual step of sailing a vessel, similar to those made by Native Americans, from the Americas to Polynesia, to prove that it could be done. Drifting with the prevailing westerly winds, Heyerdahl argued that peopling of the Pacific from the east would have to have been accomplished in the teeth of these winds. He also put forward as evidence the likely American origins of some important Pacific crops, and similarities between stone facing designs on monuments in Polynesia and South America. The American origin of sweet potato (*Ipomoea batatas*) has since been confirmed; however, in the face of overwhelming evidence of Asian origins for Pacific languages and cultures, more recent interpretations of this connection have suggested at most limited prehistoric trading contacts between Polynesians and Native Americans.[19]

All other models envisage a movement into the Pacific from the west; however, there is considerable diversity among them. As we have seen above, there is good evidence for two major cultural movements into the islands southeast of the Asian continent, the first associated with first settlement of the region by modern humans, and the second with the spread of agriculture. If we are to distinguish between the relative contributions to Pacific Islanders of ancestral peoples whose migrations may have accompanied these different cultural expansions, it is necessary to distinguish between *proximate* and *ultimate* origins. The *ultimate* origins, say 45 KYA, of genetic lineages from either source population, would be in continental Asia. However, from 29 to 6 KYA, the *proximate* origins of genetic lineages derived from either original settlers or farmers, would be in Near Oceania or continental Asia respectively. Thus we must consider not only the geographical origin of a lineage, but also its time of origin.

Essentially there are two models that are polar opposites, between which lie further models in what archaeologist Peter Bellwood calls "the continuum of reality." Metaphors have unfortunately run riot in this arena of prehistory. At one end of the continuum, the "Express Train" model has Austronesian speakers arrive from the north and, with negligible admixture with indigenous Near Oceanians, disperse into the Pacific.[8] Under this model we should expect proximate origins for Pacific lineages in Taiwan and continental Asia. At the other end of the continuum, the "Entangled Bank" model sees the movement into Remote Oceania as being the culmination of 40 KY of interactions, in a "voyaging corridor," among indigenous Near Oceanians.[47] By contrast, this would suggest genetic lineages in Remote Oceania have a proximate origin in Near Oceania. Between these opposing views are a number of intermediate models that are summarized in **Figure 13.19**. While the spread of agriculture has commonly been seen as the engine driving population dispersals, some of the models emphasize the role of the rising sea levels and the sinking of the Sunda continent in driving population dispersals.[40]

A few comments are needed to put these different models for the origins of human settlement of Remote Oceania into context. One model may not fit all the evidence. Whereas a certain model may be more appropriate for the spread of languages, another may fit better to the spread of cultures and genetic lineages. Indeed, many of these models originate in different disciplines, and their

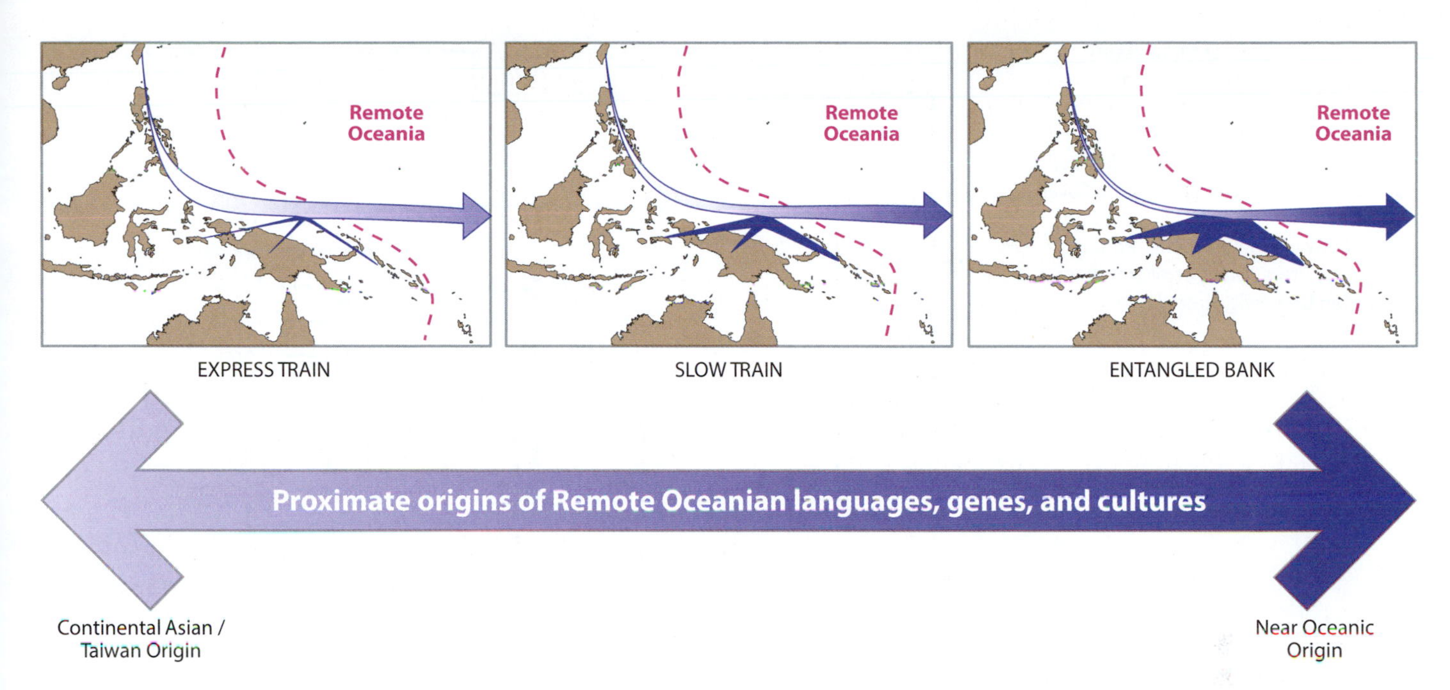

Figure 13.19: Spectrum of alternative models for the origins of the inhabitants of Remote Oceania. The width of the arrows conveys the relative ancestral contributions of Island Southeast Asian and Near Oceanian populations to the inhabitants of Remote Oceania. The dashed line is the boundary between Near and Remote Oceania. The spectrum beneath represents the continuum of admixture proportions encapsulated within each model.

original proponents may not have intended them to have any validity beyond that discipline. It may be the case, for example, that the spread of languages occurred in one direction at one given time point and the spread of certain genetic lineages preceded or followed at different times and from different sources. These models are not complete: they only describe the contributions of the first settlers of Remote Oceania, and take no account of post-settlement gene flow, which also plays a role in shaping modern genetic diversity. At least some secondary movements of non-Austronesian speakers into Remote Oceania must have taken place, if only to account for the Papuan languages on the Santa Cruz islands. These two important characteristics of models—that one size does not necessarily fit all, and that they lack sufficient complexity—are common to many different debates for the origins of individual peoples.

Austronesian dispersal models have been tested with genetic evidence

While in this chapter our focus is on evidence for initial settlement, because we draw our inferences from modern genetic diversity we must tease apart the conflated influence of post-settlement gene flow from the signal of the first migrations. We must also take account of time-depth in order to distinguish between proximate and ultimate origins. For these reasons we might not expect classical polymorphisms to be highly informative.

Classical polymorphisms

When analyzed using classical polymorphisms, Australian and New Guinean populations exhibit few similarities, suggesting that later region-specific migrations (that is, to New Guinea but not Australia, or vice versa) have played an important role in shaping modern diversity. An east–west gradient of gene frequencies across New Guinea has been taken as evidence of the predominance of longitudinal migrations across this island; however, these patterns may derive from multiple migrations, and remain undated.

A tree of Near and Remote Oceanic populations constructed from genetic distances between them shows little correspondence with the traditional (Polynesian, Melanesian, and Micronesian) classification of these islands (**Figure 13.20**). The authors attributed this to the influence of high levels of recent gene flow from Near to Remote Oceania that has obscured the patterns of genetic diversity associated with the initial settlement. However, there is some clustering of Eastern Polynesian populations, perhaps indicating that these populations were less prone to post-settlement gene flow by virtue of their isolation.

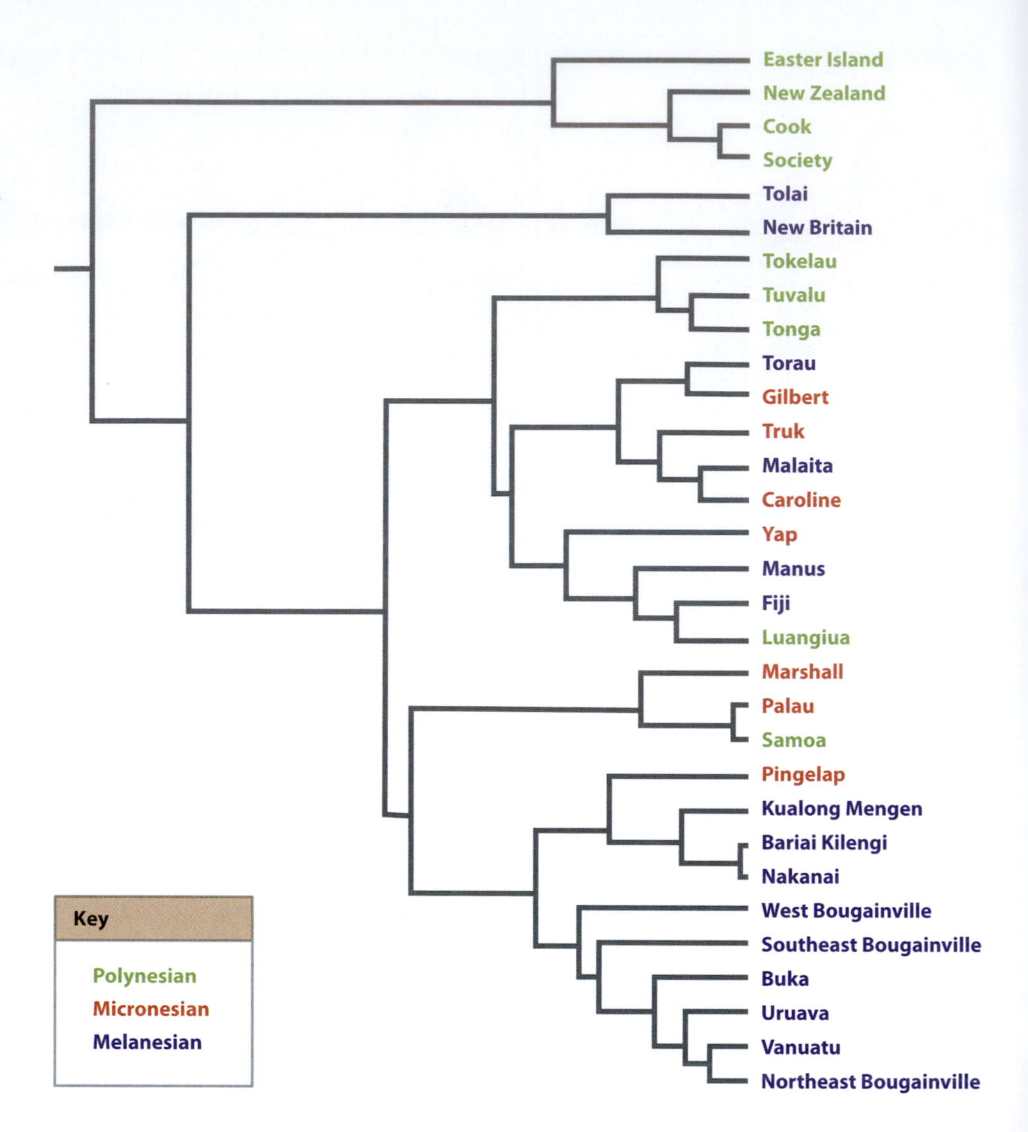

Figure 13.20: Tree of Melanesian, Micronesian, and Polynesian populations. A neighbor-joining tree of Pacific populations constructed from classical polymorphism frequencies. [From Cavalli-Sforza LL et al. (1994) The History and Geography of Human Genes. Princeton University Press. With permission from Princeton University Press.]

It is worth noting that population differentiation apparent within trees of populations is generally attributed to the operation of genetic drift. The successive founder effects implicit within the peopling of the Pacific would lead to variable **effective population sizes** and as a consequence different rates of divergence between different islands. These imbalances might be expected to cause such trees to assume unusual topologies. This is indeed the case in Figure 13.20, where the Eastern Polynesian populations appear as an outgroup to all the other populations in Near and Remote Oceania, rather than being a subclade within other populations of Remote Oceania as might be expected. Selection might be expected to affect the diversity of individual loci, but is unlikely to significantly perturb a multi-locus tree such as that shown in Figure 13.20.

Globin gene mutations

One set of autosomal polymorphisms has been particularly informative about prehistoric migrations in the Pacific, and the influence of natural selection. These polymorphisms are deletions of the α-globin genes that result in **α-thalassemia** (**Section 16.4**). Region-specific polymorphisms are often only present at low frequencies due to a recent mutational origin; however, globin deletions are present at appreciable frequencies because they protect against malaria, a disease prevalent in coastal regions from New Guinea to Vanuatu, but absent from more easterly locations in Remote Oceania.

The α-globin gene lies on chromosome 16, where there are two copies that have arisen through an ancient **gene duplication**. The high frequencies in the

Pacific of two different deletions removing a single α-globin gene means that in some locations individuals with the normal copy number of four are actually quite rare, and many more have two or three copies.

Recurrent 4.2- and 3.7-kb deletions ($\alpha^{3.7}$ and $\alpha^{4.2}$) that lead to the removal of a single α-globin gene result from **non-allelic homologous recombination** between **paralogous** sequences (see **Section 3.6**) and are found worldwide among tropical and subtropical populations. In Remote Oceania, the two major α-globin deletions are an $\alpha^{3.7}$ deletion of the III subtype ($\alpha^{3.7III}$) and an $\alpha^{4.2}$ deletion on an RFLP haplotype designated IIIa. Whereas the $\alpha^{3.7III}$ deletion predominates all over Remote Oceania, the $\alpha^{4.2}$ deletion is only present at appreciable frequencies on the western fringes and at low frequency in Tonga (see **Figure 13.21**). Neither of these particular deletions has been found further west than the Wallace line. Although deletions of the same size are found at high frequencies throughout Island Southeast Asia, these occur on different RFLP haplotypes and appear to be independent events. In addition, at least three independent double α-globin deletions are found in Island Southeast Asia at high frequencies, but these are not observed in either Near or Remote Oceania.

The inference most commonly drawn from the geographical distributions of α-globin deletions given above is that the inhabitants of Remote Oceania have dual genetic ancestry from Island Southeast Asia and Near Oceania. The analysis of mutations in other globin genes also supports this interpretation. The $\alpha^{3.7III}$ and $\alpha^{4.2}$ deletions both arose on a IIIa RFLP haplotype in Near Oceania, so why are their distributions in Oceania so different? There are two possible explanations: only the $\alpha^{3.7III}$ deletion was picked up during the initial Austronesian dispersal, and the movement of the $\alpha^{4.2}$ deletion into Remote Oceania resulted from later migrations; or, the two deletions were dispersed by the same migration, but the cumulative impact of sequential founder effects skewed their relative frequencies. The $\alpha^{4.2}$ deletion predominates among populations of New Guinea, whereas the $\alpha^{3.7III}$ deletion is more frequent in the Bismarck Archipelago off the New Guinean coast, thought to be the Lapita homeland. This differentiation in present-day Near Oceania has led most people to favor the first of the two options detailed above.

Mitochondrial DNA

Almost all mitochondrial DNA sequences within Remote Oceania can be apportioned into one of a few haplogroups that are nested within the major "Out of Africa" haplogroups (M and N), and can be related together by a phylogeny based on complete mtDNA sequences, shown in **Figure 13.22**. All of these lineages can

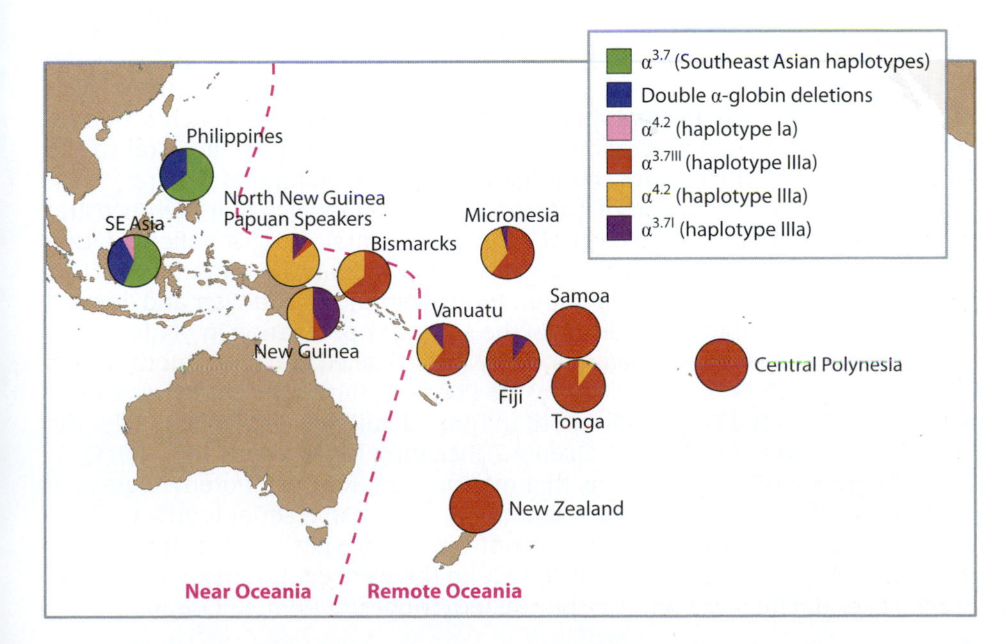

Figure 13.21: Map of α-globin deletion frequencies.
[Adapted from Oppenheimer S & Richards M (2001) *Sci. Prog.* 84, 157. With permission from Science Reviews 2000 Ltd. Additional data from O'Shaughnessy DF et al. (1990) *Am. J. Hum. Genet.* 46, 144.]

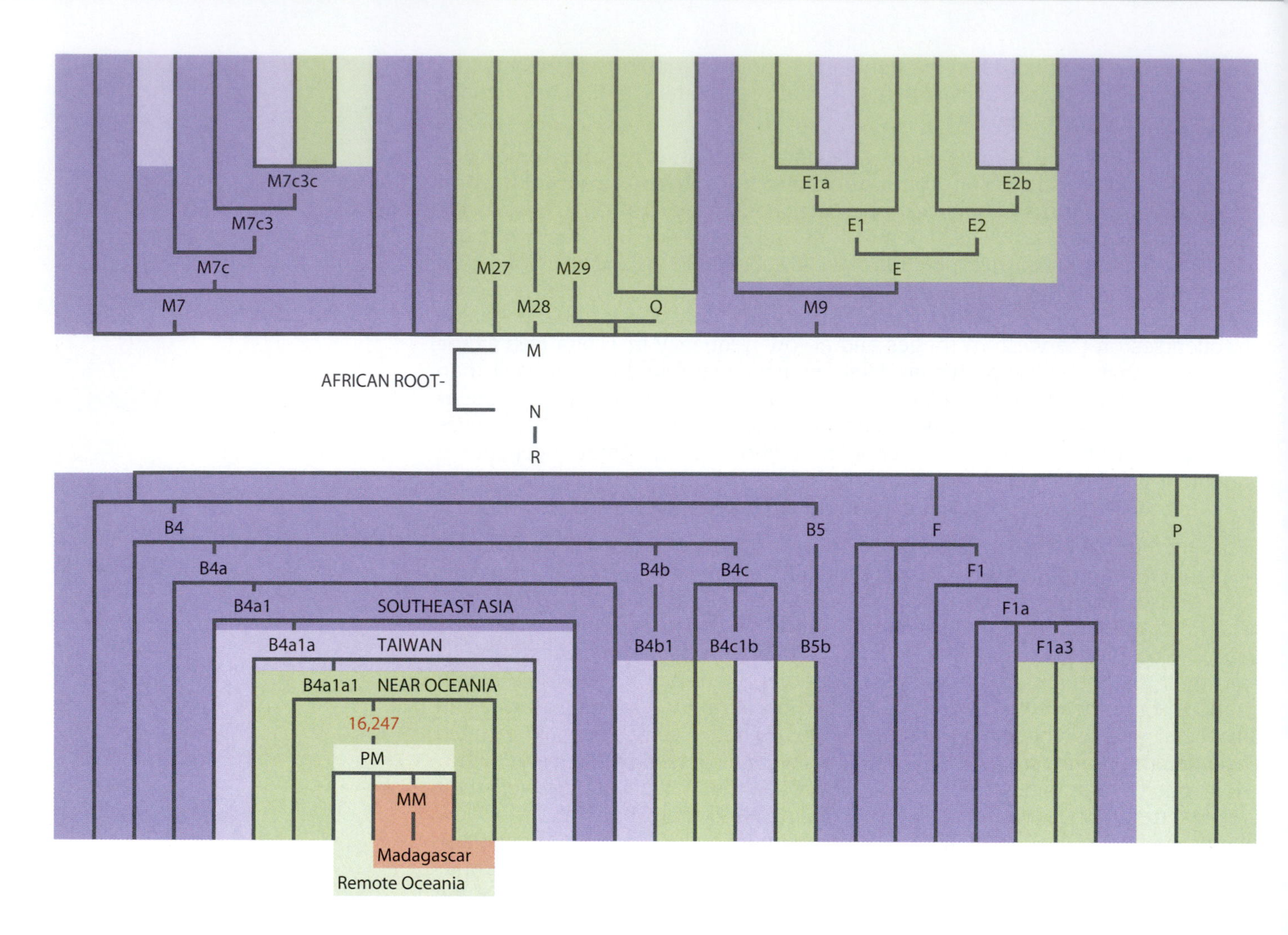

Figure 13.22: Phylogeny of mtDNA lineages in Near and Remote Oceania. Haplogroup names reflect the Phylotree nomenclature which is based on complete mtDNA sequence studies. Geographic spread of haplogroups is indicated by color shades. PM refers to "Polynesian motif" defined by 16,247 mutation (discussed in text and shown in *red* font) and MM to "Malagasy motif."

be found in populations to the west of Remote Oceania. To examine the origins of individual lineages in more detail, it is necessary to consider the wider geographical distributions of these lineages among modern populations in Island Southeast Asia, Near Oceania, and Remote Oceania (see **Figure 13.23**).

Lineage diversity in Remote Oceania decreases from west to east, until a single lineage becomes almost fixed in some Eastern Polynesian islands. This single lineage was first defined by four characteristic mutations at positions 16,189, 16,217, 16,247, and 16,261 within the **hypervariable segment I (HVSI)** of the mtDNA control region. The wide spread of this lineage across Oceania has led to it being dubbed the **Polynesian motif**. A phylogenetic approach based on complete mtDNA sequences allows us to identify haplotypes ancestral to the Polynesian motif, which share some but not all of its mutations (Figure 13.22). The geographical distribution of haplogroup B4a haplotypes that are ancestral to the Polynesian motif stretches back into continental Asia. Genetic variation of B4a1a1 lineages in Remote Oceania is nested within the variation of Near Oceania, which in turn is nested within B4a1a variation in Taiwan and Eastern Indonesia.[46, 49] Notably, the majority of the Malagasy population in Madagascar, who are also Austronesian speakers, carry the so-called Malagasy motif, which is nested within the branch defined by the Polynesian motif.[42] Before we examine the origins of the Polynesian motif in more detail, we will first consider the other lineages found in Remote Oceania, including haplogroups M27–29, Q, E, and P. There are two explanations that may account for the different degrees of penetration of these lineages into Remote Oceania. First, serial founder effects may have resulted in the near-fixation of the Polynesian motif in the east by drift. Second, post-settlement gene flow from the west could explain the occurrence of certain lineages only on the western fringes of Remote Oceania.

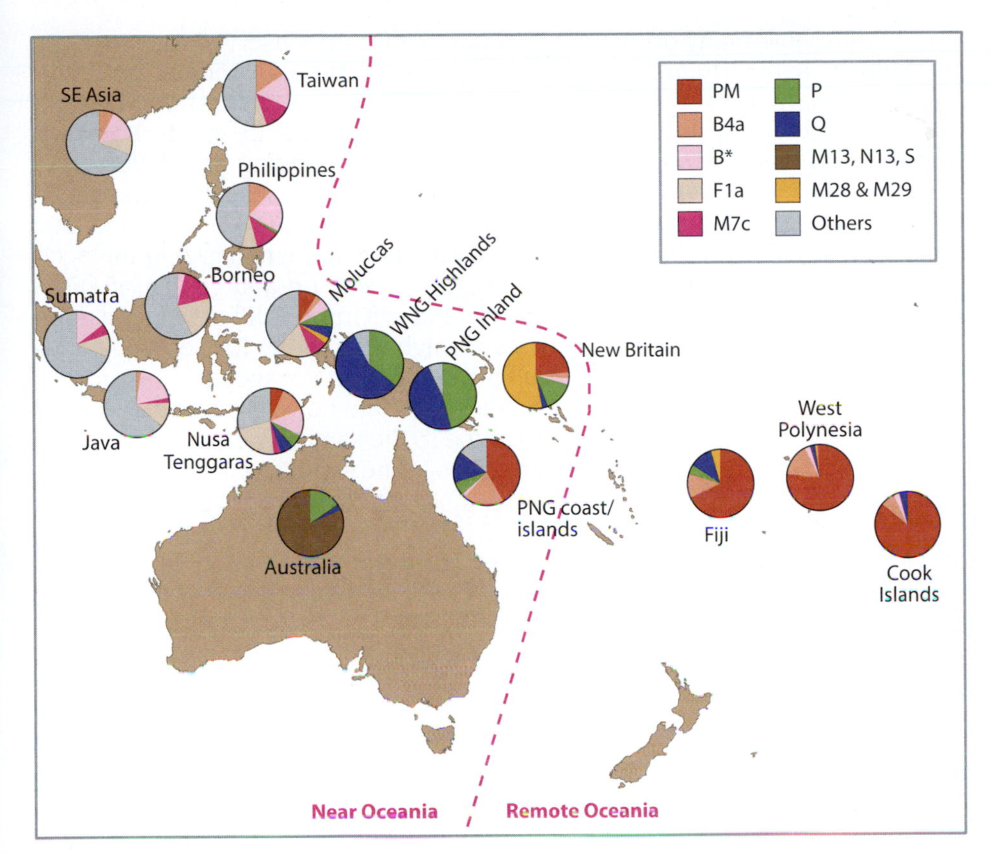

Figure 13.23: Geographical distribution of the major mtDNA lineages in Southeast Asia and the Pacific. WNG, West New Guinea; PNG, Papua New Guinea. [Based on data from Kayser M et al. (2006) *Mol. Biol. Evol.* 23, 2234; Hudjashov G et al. (2007) *Proc. Natl Acad. Sci. USA* 104, 8726; Trejaut JA et al. (2005) *PLoS Biol.* 3, e247; and Tabbada KA et al. (2010) *Mol. Biol. Evol.* 27, 21.]

Several haplogroups, such as M7c and B4a1a that are found in Remote Oceania are also found in Island Southeast Asian, but not highland New Guinean, populations. By contrast, two of the lineages found in Remote Oceania (P and Q) are also found at high frequency among highland New Guinean populations, but not among Island Southeast Asian populations. While haplogroup Q can be found at low frequencies throughout Remote Oceania, haplogroup P is confined to the western fringes.

The clear distinctions between the lineage distributions found in Island Southeast Asia and Near Oceania greatly facilitate discussions of lineage origins. However, no modern population should be considered a "fossilized" remnant of ancient diversity. Despite this fact, isolated populations are less prone to recent admixture and may have retained lineages derived from more ancient migrations. For these reasons, a contrast is often drawn between highland and coastal New Guinean populations. The assumption is that the more isolated highland populations are more likely to have retained lineages derived from the earliest settlement of Sahul. Support for this assumption comes from the observation that the lineages found in coastal New Guinean populations at higher frequencies than in the highlands are also found in other Southeast Asian populations.

From analysis of the geographical distributions of mtDNA lineages in Remote Oceania, it appears that there are at least two proximate origins for lineages found in this region, in Near Oceania and Island Southeast Asia. Could it be that there was one origin for the lineages carried by the initial settlers, and that the other contribution to modern populations results from post-settlement gene flow? If this were true we might expect to see that all lineages from one proximate origin were confined to the western fringes of Remote Oceania. This appears not to be the case: haplogroup Q and haplogroup B4a1a1 are both found throughout Remote Oceania, and have different origins. Thus it appears likely that both proximate origins of modern mtDNA lineages were represented among the initial settlers of Remote Oceania.

Considering how many distinct mtDNA haplotypes are observed today among the **Māoris**, some researchers have tried to assess through computer simulations how many women might have been in the canoes (wakas) that brought the initial settlers to Aotearoa (the Māori name for New Zealand). **Forward simulations** have been carried out using different founding population sizes and growth parameters, to find the size that best re-creates the modern genetic diversity. One study estimated the wakas to contain between 50 and 100 women, a small figure that agrees with **oral histories** emphasizing the accidental discovery of Aotearoa.[39] Another study, based on a larger set of Māori samples, identified further founder types, and relying on different population growth models, yielded estimates twice as high than in the former study, thus supporting a concept of planned multiple settlement voyages to Aotearoa by Polynesian navigators.[51] These two studies illustrate how sensitive the estimation of past demographic parameters can be to the sampling of modern genetic variation and the choice of parameter values in the simulations.

The Y chromosome

As in America, the male bias of Europeans visiting Remote Oceania (sailors, whalers, traders, missionaries) over the past 250 years means that, in addition to prehistoric post-settlement gene flow, historic admixture with Europeans may complicate the signal of initial settlement present in the paternal lineages of modern Oceanic populations. However, in practice the high geographic differentiation of the Y chromosome means that European admixture can be easily identified and discounted before further analysis.[26] A number of studies show that European admixture varies greatly between different islands of Remote Oceania, with perhaps the greatest influence in the Cook Islands where approximately one-third of all Y chromosomes belong to three diagnostically European lineages. The relative abundance of these admixed lineages can be used to pinpoint their likely origin within Europe, assuming that they all come from the same region. These considerations support an origin for these lineages in northwest Europe, in agreement with the historical evidence. Having discounted any European admixture, the remaining indigenous paternal lineages of Remote Oceania can be related by the phylogeny shown in Figure 13.24.

The distribution of these paternal lineages is shown in Figure 13.25. The pattern of lineage distribution is strikingly reminiscent of the mtDNA lineage pattern. The lineage distributions of Near Oceania and Island Southeast Asia are quite distinct and lineages found in Remote Oceania can be traced to both source areas. Again, the predominant lineage in Remote Oceania, haplogroup C2, can

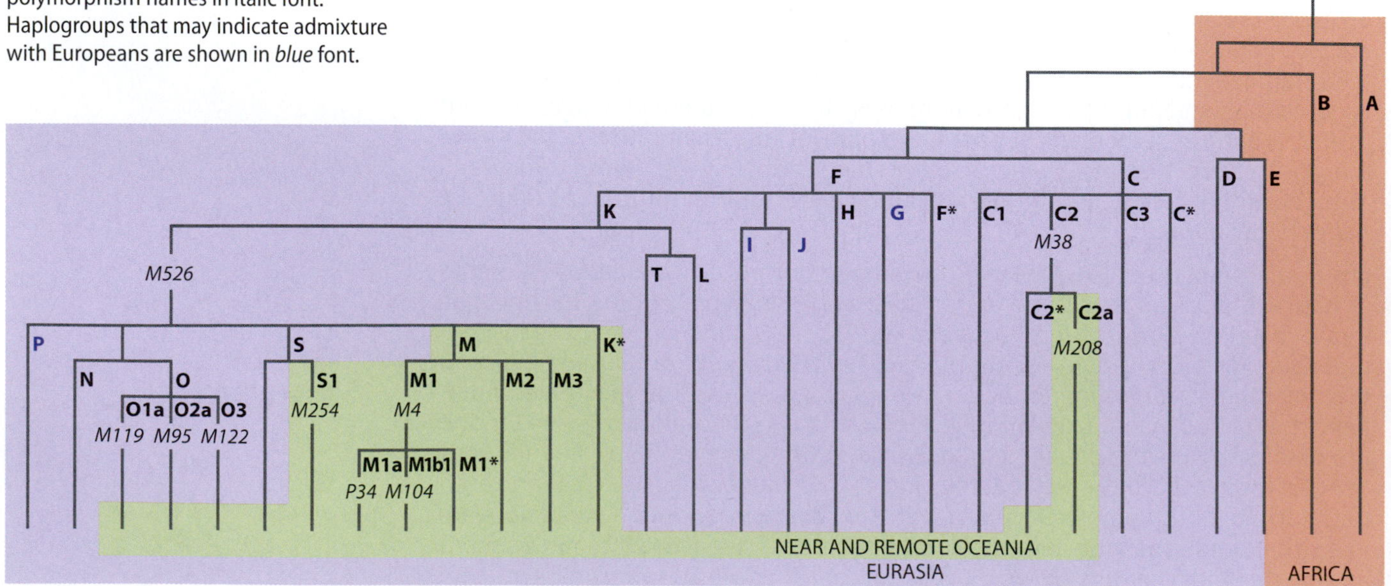

Figure 13.24: Phylogeny of Y-chromosomal lineages in Near and Remote Oceania.
Haplogroup names are shown in bold, polymorphism names in italic font. Haplogroups that may indicate admixture with Europeans are shown in *blue* font.

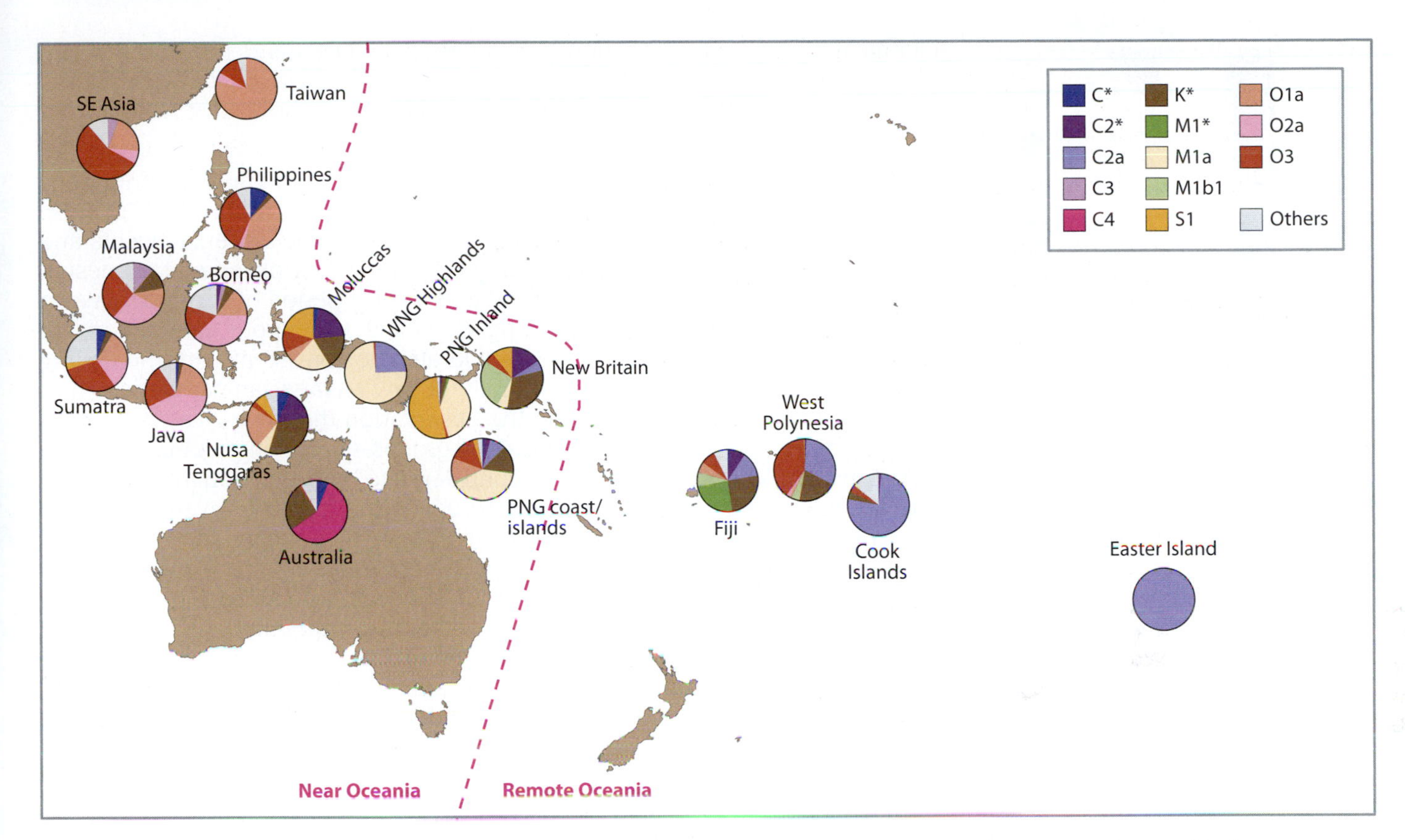

Figure 13.25: Geographical distributions of major Y-chromosomal lineages in Oceania.
WNG, West New Guinea; PNG, Papua New Guinea. [Based on data from Kayser M et al. (2006) *Mol. Biol. Evol.* 23, 2234; and Cox MP et al. (2007) *Hum. Biol.* 79, 525.]

be traced as far west as Wallacea, the Moluccas, and Nusa Tenggaras (Figure 13.25). Its major subclade C2a is found only in Remote and Near Oceania. Most Polynesian C2a chromosomes carry yet another variant allele, P33, which is absent in Near Oceania.[7] Only in Wallacea and West New Guinea, though, is C2 found together with other haplogroup C chromosomes that do not belong to this lineage. Estimates for the age of the extant Y chromosomes belonging to haplogroup C2 pre-date the spread of farming. For example, its TMRCA estimate in Northwest New Guinea is ~12 KYA.[38]

Similarly to the mtDNA haplogroup B4a sharing between Island Southeast Asia and Polynesia, Y chromosome haplogroup O variation also reflects such links. The subclade O1a2-M110 that is specifically frequent and diverse in Taiwan is found among some Near and Remote Oceanian populations, albeit at low frequency. Among most Polynesian populations, however, O1a lineages are absent and the O3a-M324 lineages that are common there do not seem to have ancestry in Taiwan but more generally in mainland Southeast Asia.[29]

In conclusion, the predominant maternal and paternal lineages in Remote Oceania appear to have an origin in Wallacea that pre-dates the spread of farming into this region. However, in addition to these lineages, there are contributions from Island Southeast Asian and New Guinean sources, indicating that the initial settlers of the remote islands of the Pacific were an admixed population with at least three sources of genetic input. As a summary of mtDNA and Y-chromosome studies, it has been estimated that the proximate origins of 94% of Polynesian maternal lineages are in East Asia whereas 66% of their paternal DNA is inherited from admixture with Melanesians.[30] These contrasting estimates (**Figure 13.26**) suggest that the admixture would have been sex-specific and several authors have interpreted these patterns as a reflection of **matrilocal** marital residence customs of the proto-Austronesians.

Autosomal evidence

Analyses of genomewide microsatellite and SNP data provide further insights into the debate on dispersal routes and times in Near and Remote Oceania. Scans of hundreds of microsatellite repeat loci and **indel** polymorphisms have

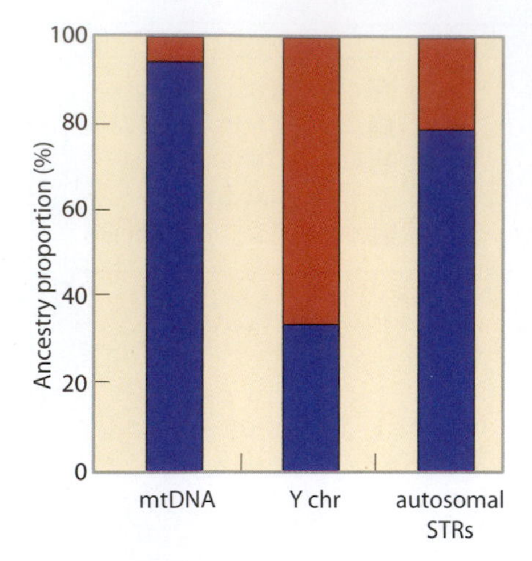

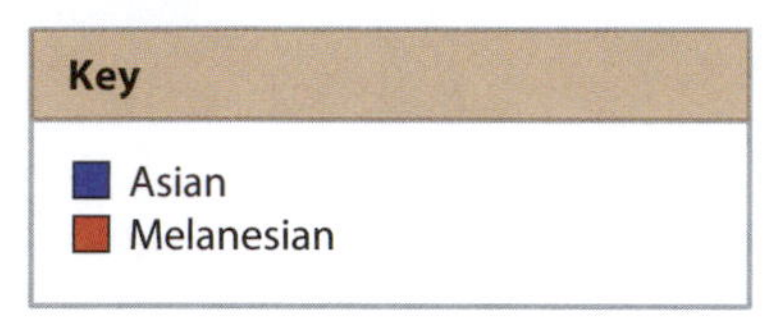

Figure 13.26: Estimated Asian and Melanesian ancestry proportions in Polynesians.
[Adapted from Kayser M et al. (2008) *Am. J. Hum. Genet.* 82, 194. With permission from Elsevier.]

identified low genetic variation among Remote Oceanian populations and relatively high diversity in Melanesia[15]. In line with mtDNA data, the majority (~80%) of the Polynesian autosomal genome appears to have East Asian rather than Melanesian (~20%) affiliation[31] (Figure 13.26).

While the Remote Oceanian populations appear to have larger contributions from East Asia than from Melanesia in their genomes (**Figure 13.27**), these analyses do not inform us about the pace and time of the dispersal events and admixture. On one hand, it has been argued that the low observed Melanesian genetic contribution implies fast spread and a short period of contact between dispersing Austronesian speakers and the local indigenous populations of Melanesia.[15] The fast spread model would also be consistent with the finding of only a minor East Asian contribution in the few Melanesian populations that speak Austronesian languages, and the observation that this ancestry component is entirely lacking in Papuan-speaking groups. On the other hand, the ~20% of Melanesian contribution has been considered to be large enough to corroborate the "slow boat" model according to which the dispersal event was slow enough to capture local admixture at this scale.[31] The admixture apparently occurred predominantly via the male-specific contribution from Melanesians to the ancestors of Remote Oceanians. Analyses of genomewide SNP data have supported an even greater contribution of East Asian populations (87%) to the gene pool of Remote Oceanians and an admixture event dated to ~3 KYA.[52] Considering the continuum of models between the "fast" and the "slow," as represented in Figure 13.19, the autosomal evidence would therefore not support either of the extremes but would be consistent with a smaller range of models in between them.

Evidence from other species has been used to test the Austronesian dispersal models

Studies of the biological impact of the last ice age have been greatly informed by noting common patterns of genetic diversity among several extant floral and faunal species. Each species provides independent evidence on a common evolutionary process, in this case, the settlement of previously inhospitable lands. The term **comparative phylogeography** has been coined to describe this comparison of geographical distributions of lineages from different species that have undergone the same evolutionary process.

When humans settle previously uninhabited lands they bring with them (intentionally and unintentionally) a panoply of other species, including infectious microbes, domesticated plants and animals, and non-domesticated species that coexist with humans (**commensals**). Thus the geographical patterns of genetic diversity in these species have been shaped, at least in part, by a common process—human migration. Can a similar comparative approach to the one described above allow the analysis of genetic diversity in nonhuman species to shed light on human migrations? In principle, such an approach to studying human migration has several advantages over using human samples. In particular, ancient DNA studies are often facilitated by a relative abundance

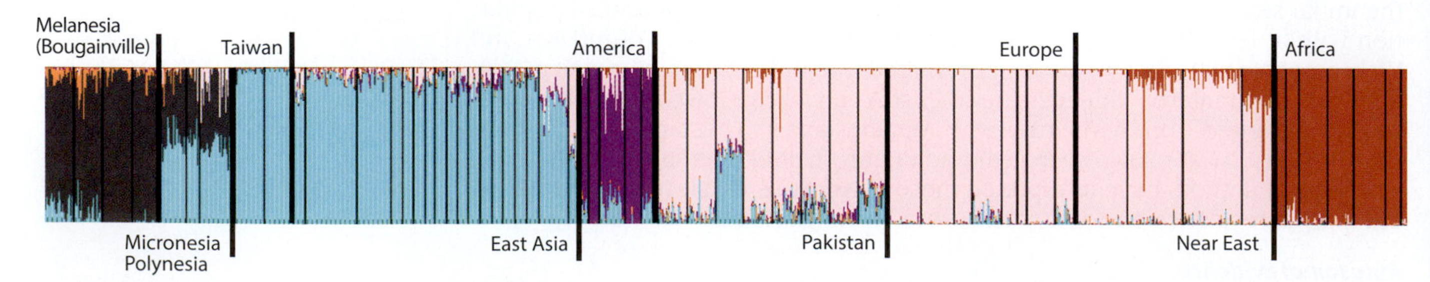

Figure 13.27: Analysis of microsatellite- and indel-based population structure in Pacific and worldwide populations. Analyses have been performed using STRUCTURE program at $K = 6$ and combine 687 microsatellites and 203 indels across the genome. [Adapted from Friedlaender JS et al. (2008) *PLoS Genet.* 4, e19. With permission from Public Library of Science.]

of samples (especially from domesticated species) and the absence of problems caused by human DNA contamination (see **Section 4.10**).

The potential for gaining insight into human migration from the genetics of nonhuman species has been explored in the settlement of the Pacific. Several domesticates have been studied (for example, pigs, and the Polynesian food rat, *Rattus exulans*, shown in **Figure 13.28**), as has a commensal lizard (*Lipinia noctua*) that was presumably a stowaway on ocean-going canoes, and an infrequently **pathogenic** virus (human polyomavirus JC).

Despite the relative paucity of studies on individual species at present, there is support from analysis of all the species mentioned above for settlement of the Pacific from the West. Analysis of whole genome sequences (5.1 kb) of the JC virus in Remote Oceania suggests that the settlers carried at least two distinct viral subtypes, one with East Asian origins and the other from New Guinea.[53] Similarly, two distinct types of the bacterium *Helicobacter pylori* provide further clues about the peopling of Oceania: one haplotype called hpSahul is found only in Melanesia and Australia while the second, hspMaori, is associated with Austronesian-speaking populations and supports the out-of-Taiwan dispersal model (**Figure 13.29**). In addition, mtDNA phylogenies of *Lipinia noctua* exhibit

Figure 13.28: Polynesian or Pacific food rat (*Rattus exulans*).
(Courtesy of Cliff under the Creative Commons Attribution 2.0 Generic license.)

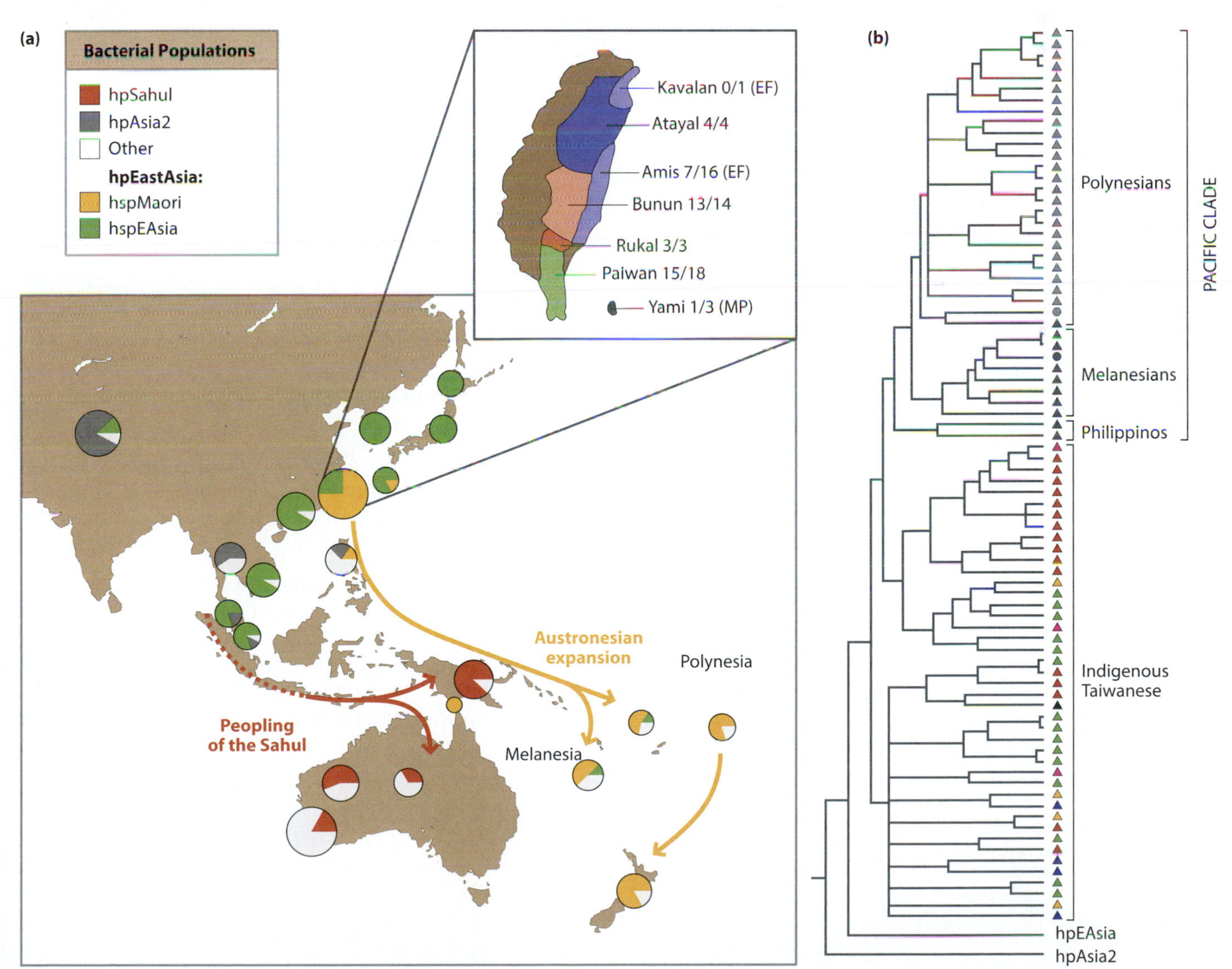

Figure 13.29: Geographic distribution and phylogeny of *H. pylori* haplotypes in Asia, Near Oceania, and Remote Oceania.
(a) Geographic spread of *H. pylori* haplotypes and inferred dispersal events. (b) Phylogenetic relationships between the hspMaori haplotypes are presented as a consensus tree, using hspAsia2 and hspEAsia as outgroups. The population affiliations of Taiwanese individuals (triangles) are indicated in the inset of Figure 13.29a. [From Moodley Y et al. (2009) *Science* 323, 527. With permission from AAAS.]

a clear signal of the rapid transport into Remote Oceania of this native New Guinean lizard.[2] More detailed studies of mtDNA diversity of the Polynesian food rat in Remote Oceania suggest that substantial levels of inter-island voyaging opposed the isolation of previously settled islands.[37] Also, these data show that islands in Remote Oceania now uninhabited (Kermadecs) were formerly settled by humans either full time or on a seasonal basis.

Analyses of pig mtDNA have provided evidence for an early human-mediated introduction of the Sulawesi warty pig (*Sus celebensis*) to Flores and Timor while the spread of domestic pig (*Sus scrofa*) through Island Southeast Asia into Oceania can be ascribed to two later dispersals. One of the inferred dispersal routes of *S. scrofa* correlates with the spread of Neolithic (Lapita) and later Polynesian migrations while linking Oceanic pigs with those from mainland Southeast Asia. Interestingly, it appears that the so-called "wild" pigs of New Guinea have descended from the domestic pigs introduced by early agriculturalists.[34]

In conclusion, genetic studies of nonhuman species support the rapid peopling of Remote Oceania from the west by an admixed population. In addition, initial settlement was followed by high levels of inter-island voyaging that indicate that rather than being a barrier to gene flow, the large oceanic distances between islands could more properly be considered as well-trodden highways.

SUMMARY

- Rising sea level has both driven human migration by submerging previously settled lands, and restricted it by inundating land bridges.
- Technological ability and environmental opportunity both govern when and from where new-found lands could be settled in prehistory.
- Evidence from human fossils and archaeology suggests that humans were present in the Americas by 14.5 KYA and soon spread over both continents, probably exterminating most of the megafauna.
- The genetic evidence suggests that most Native American diversity can be accounted for by one major migration into Beringia, with dispersal into the rest of the continents 20–15 KYA. Later, two or more minor dispersals may have contributed to the North American gene pool.
- In the Pacific, genetic evidence supports the archaeological and linguistic evidence for the origins of Pacific Islanders from the west and not from the Americas.
- The genetic lineages that predominate on recently settled Pacific islands suggest that the peoples who first settled these remote islands were admixed, with contributions from Southeast Asia and the indigenous people of Wallacea and Melanesia. Comparisons of autosomal and uniparentally inherited loci suggest that more than 80% of the autosomal loci of Polynesians have a Southeast Asian origin and that the genetic contributions from Melanesia were mainly male-specific.
- Analyzing the genetic diversity of pathogens, commensals, and domesticated species that owe their geographic distribution to human migration can itself be informative about patterns of human migration.

QUESTIONS

Question 13–1: What types of genetic data and analysis methods would be required to solve the question of whether peopling of the Americas involved one or three dispersal events?

Question 13–2: What would be the key advantages and disadvantages of using genomewide genotype data as opposed to high-resolution sequence information from a few selected loci and sequence data from the entire genome in addressing questions of genetic origins of the New World and Remote Oceanian populations?

Question 13–3: Suppose an ancient DNA study on the remains of Kennewick Man would recover mtDNA sequence from haplogroup E and nuclear DNA signatures similar to modern-day Central Asian populations. How would this evidence change our current interpretation of Native American origins?

Question 13–4: Propose at least three reasons why the date estimate of the Austronesian language family would be more recent than coalescent date estimates from genetic studies based on a single gene or a locus.

Question 13–5: For the eight Pacific islands: Aotearoa, Rapa Nui, Guadalcanal, Fiji, Honshu, Tonga, Oahu, Taiwan, a) Place them in order of increasing longitude; b) Place them in order of settlement time, from most ancient to most recent; c) Use the online language database www.ethnologue.com to see what their majority indigenous languages are, and consider the relationships of these languages.

Question 13–6: Suppose that human remains dating back to 12 KYA are found in Antarctica, their DNA is successfully analyzed and contamination with modern DNA can be excluded. The recovered DNA shows genomewide haplotype heterozygosity of 0.68. Given the data plotted in Figure 9.15, consider critically the possibility of South American origin of the Antarcticans.

REFERENCES

The references highlighted in purple are considered to be important (for this chapter) by the authors.

1. **Alroy J** (2001) A multispecies overkill simulation of the end-Pleistocene megafaunal mass extinction. *Science* **292**, 1893–1896.
2. **Austin CC** (1999) Lizards took express train to Polynesia. *Nature* **397**, 113–114.
3. **Buchanan B, Collard M & Edinborough K** (2008) Paleoindian demography and the extraterrestrial impact hypothesis. *Proc. Natl Acad. Sci. USA* **105**, 11651–11654.
4. **Campbell L** (1997) American Indian Languages: The Historical Linguistics of Native America. Oxford University Press.
5. **Cohen JE & Small C** (1998) Hypsographic demography: the distribution of human population by altitude. *Proc. Natl Acad. Sci. USA* **95**, 14009–14014.
6. **Collerson KD & Weisler MI** (2007) Stone adze compositions and the extent of ancient Polynesian voyaging and trade. *Science* **317**, 1907–1911.
7. **Cox MP, Redd AJ, Karafet TM et al.** (2007) A Polynesian motif on the Y chromosome: population structure in remote Oceania. *Hum. Biol.* **79**, 525–535.
8. **Diamond J** (1988) Express train to Polynesia. *Nature* **336**, 307–308.
9. **Dulik MC, Zhadanov SI, Osipova LP et al.** (2012) Mitochondrial DNA and Y chromosome variation provides evidence for a recent common ancestry between Native Americans and Indigenous Altaians. *Am. J. Hum. Genet.* **90**, 229–246.
10. **Dunn M, Terrill A, Reesink G et al.** (2005) Structural phylogenetics and the reconstruction of ancient language history. *Science* **309**, 2072–2075.
11. **Erlandson JM, Graham MH, Bourque BJ et al.** (2007) The kelp highway hypothesis: marine ecology, the coastal migration theory, and the peopling of the Americas. *J. Isl. Coast. Archaeol.* **2**, 161–174.
12. **Fagundes NJ, Ray N, Beaumont M et al.** (2007) Statistical evaluation of alternative models of human evolution. *Proc. Natl Acad. Sci. USA* **104**, 17614–17619.
13. **Feathers J, Kipnis R, Piló L et al.** (2010) How old is Luzia? Luminescence dating and stratigraphic integrity at Lapa Vermelha, Lagoa Santa, Brazil. *Geoarchaeology* **25**, 395–436.
14. **Firestone RB, West A, Kennett JP et al.** (2007) Evidence for an extraterrestrial impact 12,900 years ago that contributed to the megafaunal extinctions and the Younger Dryas cooling. *Proc. Natl Acad. Sci. USA* **104**, 16016–16021.
15. **Friedlaender JS, Friedlaender FR, Reed FA et al.** (2008) The genetic structure of Pacific Islanders. *PLoS Genet.* **4**, e19.
16. **Goebel T, Waters MR & O'Rourke DH** (2008) The late Pleistocene dispersal of modern humans in the Americas. *Science* **319**, 1497–1502.
17. **González-José R, Gonzalez-Martin A, Hernandez M et al.** (2003) Craniometric evidence for Palaeoamerican survival in Baja California. *Nature* **425**, 62–65.
18. **Gray RD, Drummond AJ & Greenhill SJ** (2009) Language phylogenies reveal expansion pulses and pauses in Pacific settlement. *Science* **323**, 479–483.
19. **Green RC** (2000) A range of disciplines support a dual origin of the bottle gourd in the Pacific. *J. Polynesian Soc.* **109**, 191–197.
20. **Greenberg JH, Turner II CG & Zegura SL** (1986) The settlement of the Americas: a comparison of the linguistic, dental and genetic evidence. *Curr. Anthropol.* **27**, 477–497.
21. **Gutenkunst RN, Hernandez RD, Williamson SH & Bustamante CD** (2009) Inferring the joint demographic history of multiple populations from multidimensional SNP frequency data. *PLoS Genet.* **5**, e1000695.
22. **Hanebuth T, Stattegger K & Grootes PM** (2000) Rapid flooding of the Sunda shelf: a late-glacial sea-level record. *Science* **288**, 1033–1035.
23. **Hellenthal G, Auton A & Falush D** (2008) Inferring human colonization history using a copying model. *PLoS Genet.* **4**, e1000078.
24. **Hey J** (2005) On the number of New World founders: a population genetic portrait of the peopling of the Americas. *PLoS Biol.* **3**, e193.

25. **Hubbe M, Harvati K & Neves W** (2011) Paleoamerican morphology in the context of European and East Asian late Pleistocene variation: implications for human dispersion into the New World. *Am. J. Phys. Anthropol.* **144**, 442–453.

26. **Hurles ME, Irven C, Nicholson J et al.** (1998) European Y-chromosomal lineages in Polynesians: a contrast to the population structure revealed by mtDNA. *Am. J. Hum. Genet.* **63**, 1793–1806.

27. **IPCC.** Climate Change 2001: The Scientific Basis. Geneva: Intergovernmental Panel on Climate Change, 2001.

28. **Karafet TM, Zegura SL, Posukh O et al.** (1999) Ancestral Asian source(s) of new world Y-chromosome founder haplotypes. *Am. J. Hum. Genet.* **64**, 817–831.

29. **Kayser M** (2010) The human genetic history of Oceania: near and remote views of dispersal. *Curr. Biol.* **20**, R194–201.

30. **Kayser M, Brauer S, Cordaux R et al.** (2006) Melanesian and Asian origins of Polynesians: mtDNA and Y chromosome gradients across the Pacific. *Mol. Biol. Evol.* **23**, 2234–2244.

31. **Kayser M, Lao O, Saar K et al.** (2008) Genome-wide analysis indicates more Asian than Melanesian ancestry of Polynesians. *Am. J. Hum. Genet.* **82**, 194–198.

32. **Kemp BM, Malhi RS, McDonough J et al.** (2007) Genetic analysis of early Holocene skeletal remains from Alaska and its implications for the settlement of the Americas. *Am. J. Phys. Anthropol.* **132**, 605–621.

33. **Kitchen A, Miyamoto MM & Mulligan CJ** (2008) A three-stage colonization model for the peopling of the Americas. *PLoS One* **3**, e1596.

34. **Larson G, Cucchi T, Fujita M et al.** (2007) Phylogeny and ancient DNA of *Sus* provides insights into neolithic expansion in Island Southeast Asia and Oceania. *Proc. Natl Acad. Sci. USA* **104**, 4834–4839.

35. **Lell JT, Sukernik RI, Starikovskaya YB et al.** (2002) The dual origin and Siberian affinities of Native American Y chromosomes. *Am. J. Hum. Genet.* **70**, 192–206.

36. **Malhi RS, Mortensen HM, Eshleman JA et al.** (2003) Native American mtDNA prehistory in the American Southwest. *Am. J. Phys. Anthropol.* **120**, 108–124.

37. **Matisoo-Smith E, Roberts RM, Irwin GJ et al.** (1998) Patterns of prehistoric human mobility in Polynesia indicated by mtDNA from the Pacific rat. *Proc. Natl Acad. Sci. USA* **95**, 15145–15150.

38. **Mona S, Grunz KE, Brauer S et al.** (2009) Genetic admixture history of Eastern Indonesia as revealed by Y-chromosome and mitochondrial DNA analysis. *Mol. Biol. Evol.* **26**, 1865–1877.

39. **Murray-McIntosh RP, Scrimshaw BJ, Hatfield PJ & Penny D** (1998) Testing migration patterns and estimating founding population size in Polynesia by using human mtDNA sequences. *Proc. Natl Acad. Sci. USA* **95**, 9047–9052.

40. **Oppenheimer SJ & Richards M** (2001) Polynesian origins. Slow boat to Melanesia? *Nature* **410**, 166–167.

41. **Perego UA, Achilli A, Angerhofer N et al.** (2009) Distinctive Paleo-Indian migration routes from Beringia marked by two rare mtDNA haplogroups. *Curr. Biol.* **19**, 1–8.

42. **Razafindrazaka H, Ricaut FX, Cox MP et al.** (2010) Complete mitochondrial DNA sequences provide new insights into the Polynesian motif and the peopling of Madagascar. *Eur. J. Hum. Genet.* **18**, 575–581.

43. **Reich D, Patterson N, Campbell D et al.** (2012) Reconstructing Native American population history. *Nature* **488**, 370–374.

44. **Renfrew C** (ed.) (2000) America Past, America Present: Genes and Languages in the Americas and Beyond. Cambridge University Press.

45. **Schroeder KB, Jakobsson M, Crawford MH et al.** (2009) Haplotypic background of a private allele at high frequency in the Americas. *Mol. Biol. Evol.* **26**, 995–1016.

46. **Soares P, Rito T, Trejaut J et al.** (2011) Ancient voyaging and Polynesian origins. *Am. J. Hum. Genet.* **88**, 239–247.

47. **Terrell J, Hunt TL & Gosden C** (1997) The dimension of social life in the Pacific: human diversity and the myth of the primitive isolate. *Curr. Anthropol.* **38**, 155–195.

48. **Torroni A, Schurr TG, Cabell MF et al.** (1993) Asian affinities and continental radiation of the four founding Native American mtDNAs. *Am. J. Hum. Genet.* **53**, 563–590.

49. **Trejaut JA, Kivisild T, Loo JH et al.** (2005) Traces of archaic mitochondrial lineages persist in Austronesian-speaking Formosan populations. *PLoS Biol.* **3**, e247.

50. **Waters MR, Forman SL, Jennings TA et al.** (2011) The Buttermilk Creek complex and the origins of Clovis at the Debra L. Friedkin site, Texas. *Science* **331**, 1599–1603.

51. **Whyte AL, Marshall SJ & Chambers GK** (2005) Human evolution in Polynesia. *Hum. Biol.* **77**, 157–177.

52. **Wollstein A, Lao O, Becker C et al.** (2010) Demographic history of Oceania inferred from genome-wide data. *Curr. Biol.* **20**, 1983–1992.

53. **Yanagihara R, Nerurkar VR, Scheirich I et al.** (2002) JC virus genotypes in the western Pacific suggest Asian mainland relationships and virus association with early population movements. *Hum. Biol.* **74**, 473–488.

54. **Yang NN, Mazieres S, Bravi C et al.** (2010) Contrasting patterns of nuclear and mtDNA diversity in Native American populations. *Ann. Hum. Genet.* **74**, 525–538.

WHAT HAPPENS WHEN POPULATIONS MEET

CHAPTER FOURTEEN

The previous four chapters presented the evidence for the first settlement of the Old and the New World by anatomically modern humans. In Chapter 9 we learned that the majority of the genetic variation observed outside Africa today derives from an out-of-Africa dispersal event 50–70 KYA with a minor addition of gene flow from archaic hominins, including the Neanderthals. In the succeeding Chapters 10–13 we saw how, during and after the initial colonization of the continents, populations diversified due to genetic drift and accumulation of new mutations. As a result, we can infer the continental origins of individuals easily, even from a small number of carefully chosen genetic polymorphisms. However, populations are not discrete entities: we know from archaeological and historical evidence that they are often in flux, generating hybrid populations and individuals of mixed ancestry. This chapter discusses this process of mixing, known as **admixture**, and how we can detect and measure the extent of admixture from genetic evidence.

14.1 WHAT IS GENETIC ADMIXTURE?

Neighboring populations frequently exchange individuals that contribute to an ongoing process of bidirectional **gene flow** between them. However, a third, hybrid population does not usually result from this kind of exchange. The term admixture is reserved for the formation of a hybrid population from the mixing of ancestral populations that have previously been relatively isolated from one another. The range expansion or migration of one population into a region inhabited by a previously isolated population is one such scenario. Thus, admixture can be thought of as being initiated at a specific point in time, when the populations first came into contact.

As with most studies described in this book, we are almost exclusively limited to examining modern genetic diversity. When we examine modern populations, we detect not simply the proportions of admixture established when the populations first met, but the summation of cumulative gene flow from the time when they first met to the present day. Thus the consequences of admixture and gene flow may be difficult to distinguish. Of course, the imprint of past admixture in modern populations has also been modified by the **drift**, selection, and mutation processes that shape all genetic diversity.

Many different issues of population prehistory can be viewed as questions about admixture. All that is required is that alternative ancestral populations can be differentiated from one another in either time or space. For example, the relative contributions to the modern gene pool of several migrations to the same location can be thought of as admixture between the source populations for each migration. It is in this framework that the relative contributions of **Paleolithic**

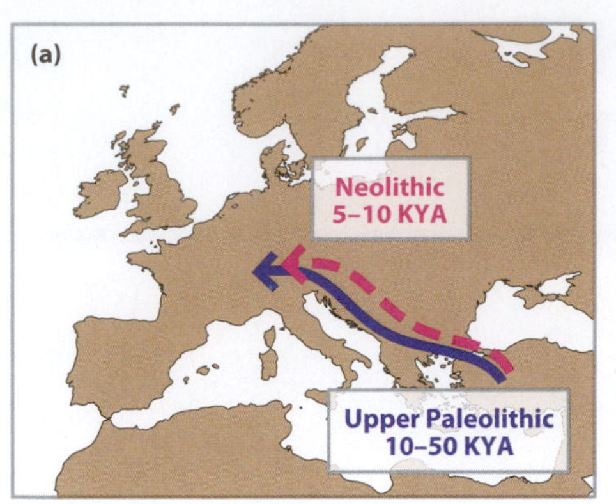

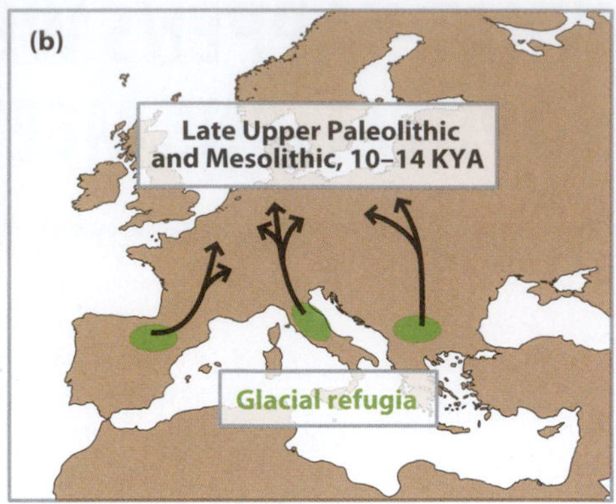

Figure 14.1: Maps showing potential sources of past admixture in Europe. (a) Neolithic and Paleolithic peoples migrated into Europe at different times, although by similar routes, and both are thought to have contributed to the modern European gene pool. (b) Peoples from different glacial refugia (*green*) may have been isolated from one another during the ice age, and as conditions improved northward migration would have presented opportunities for admixture.

hunter-gatherers and **Neolithic** farmers to modern European diversity have been considered (see **Figure 14.1**). Whilst the two ancestral populations spread from largely similar geographical origins in the Near East, they are separated in time by thousands of years. This particular example is explored in greater detail in **Section 12.5**.

The processes of isolation and range expansion that result in subsequent admixture can be driven by environmental changes. During the recent **ice ages**, the environment in more northerly latitudes became uninhabitable. Humans and other plant and animal species found refuge in pockets of more hospitable climate, known as **glacial refugia**. These refugia were often isolated from one another. For example, three major European glacial refugia were the Iberian Peninsula, Italy, and the Balkans/Greece.

After the end of the last ice age, about 14 KYA, many species started the long process of recolonizing the more northerly latitudes from these refugia. During this process, previously isolated populations were often brought back into contact with one another, the genetic consequences of which can be analyzed through a consideration of admixture.

More recent historical events that can be studied through an appreciation of admixture processes include episodes of enforced migration. These episodes have often been motivated by colonialism and/or the creation of a subjugated labor force—in other words, a slave trade. Although the eighteenth century Atlantic slave trade has received most attention, slavery was widespread throughout the ancient world including the Egyptian, Greek, and Roman empires, among Arabs, in Iceland, in the Pacific, and in Africa itself.

Historically, some of the first studies of genetic admixture at the molecular level were those that analyzed the frequencies of different blood group protein alleles in African-Americans, comparing them to European-Americans and Africans.[12] The aim was primarily to quantify European admixture among African-Americans. Both marital records and a supposed lightening of African-American skin color provided external evidence of admixture. A number of studies of different blood groups in different US populations were published throughout the 1950s and 1960s (for example by T. E. Reed[30]). They demonstrated that the extent of European admixture varied considerably among the different regional populations of African-Americans over the 10 generations or so since the peak of the major period of African slavery. The proportion of European genes within different African-American populations was shown to vary from ~4% to ~30%, with southern populations having consistently lower levels of European admixture. More recent studies use DNA polymorphisms rather than protein polymorphisms, but the conclusions remain the same.

Often genetic studies of prehistoric admixture events are initiated when evidence from nongenetic sources indicates that admixture might have occurred. This is because, unsurprisingly, the meeting of previously isolated populations

Box 14.1: The ever-changing terminology of people with mixed ancestry

Societies undergoing admixture have often sought to classify individuals on the basis of their proportions of admixture. Many of these largely historical terms introduced fine gradations of admixture, but the number of terms required to cover all possible proportions doubles each generation, and it quickly becomes impractical to maintain a word for each fraction. As a result many of these words have all but died out, whilst others have been retained in common usage as general terms for people of mixed ancestry.

As with all terminologies associated with contentious societal issues, many of these words have been considered offensive at some point in time. It could be argued that the rapid turnover of names for individuals of mixed ancestry is driven by society's need to find neutral words free from negative connotations. However, as the societal inequalities remain, these new terms attract derogatory associations and fresh terms need to be invented at regular intervals.

Term	Parents	Proportion
Mulatto	Black and White	1/2 Black
Quadroon	Mulatto and White	1/4 Black
Octoroon	Quadroon and White	1/8 Black
Mustifee	Octoroon and White	1/16 Black
Mustifino	Mustifee and White	1/32 Black
Cascos	Mulatto and Mulatto	1/2 Black
Sambo	Mulatto and Black	3/4 Black
Mango	Sambo and Black	7/8 Black
Metisse	White and Native American	General
Mestizo	White and Native American	General
Griffe	Black and Native American	General
Hapa	Asian/Polynesian and White	General

has effects beyond the realm of genetics (see **Box 14.1**). But can genetic admixture be recognized in the absence of corroborative historical or prehistorical information? This question will be addressed in Section 14.3.

Admixture has distinct effects on genetic diversity

The process of admixture shapes genetic diversity in a number of different ways. In this chapter we will explore how seeking these different imprints in modern genetic diversity can lead to the inference that admixture occurred some time in the past. Our ability to detect admixture depends in part on the number of polymorphisms used, and how differentiated the source populations were from one another. As we shall see, the more different the ancestral populations were, the easier it is to detect and quantify admixture.

While admixture at the population level can be detected in a single genetic locus, multiple loci will be required to infer admixed ancestry in a single individual. This raises the additional complication that some alleles may have their ancestry in one parental population while other alleles have their ancestry in another. This is an inevitable consequence of sexual reproduction and **diploidy**. In fact, it becomes increasingly unlikely over time that any individual from an admixed population will be able to trace all their genes to a single source population (see **Figure 14.2** and **Table 14.1**). Different genomes within an admixed population, though, are likely to exhibit differing amounts of admixture. In **panmictic** populations this difference decreases rapidly in time, but admixed populations are not always panmictic. Nonrandom mating could be due to geographic structure or socioeconomic factors, both of which can contribute to the variation in admixture proportions at the individual level. Although an estimate of population admixture is typically presented as an average of the admixture among the individual genomes within it, the range of variation among individuals can inform us both about the time since the admixture event, and the nature of the admixture event. For example, analysis of 181 Mexicans, representing a relatively young admixed population, showed the full range of 0–100% European ancestry.[9] In contrast, analysis of Polynesians, who derive from an admixture

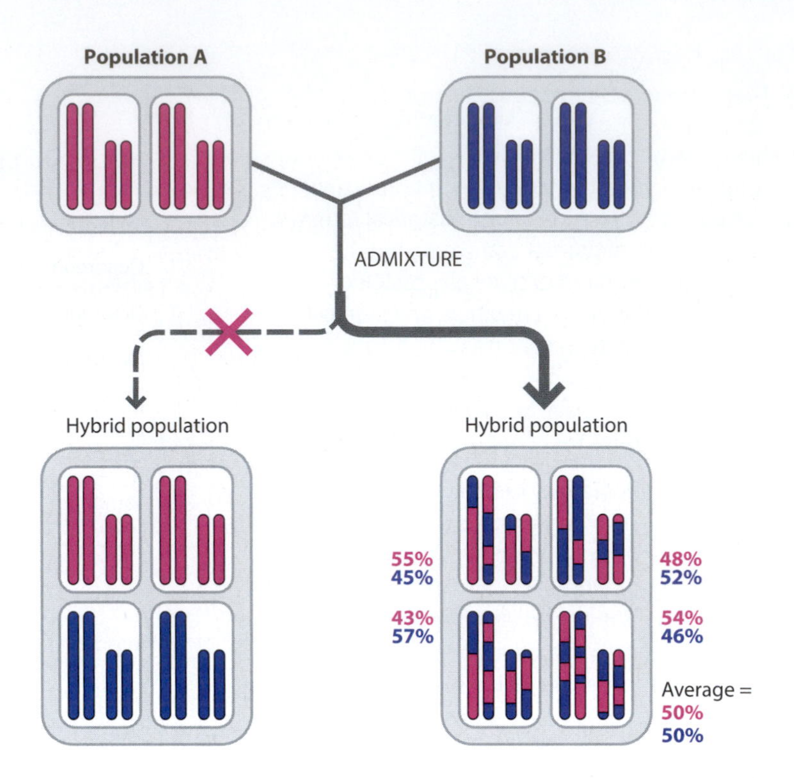

Figure 14.2: Admixture within individual genomes. Diploid genomes comprising two autosomes are shown schematically in ancestral and hybrid populations. There are two individuals (*white rounded rectangles*) in each ancestral population (*gray rounded rectangles*), and four individuals in each hybrid population. Admixture does not result in a population in which individuals can trace the ancestry of their entire genome to one of two ancestral populations, but rather a population in which all individuals have genomes of mixed ancestry.

event more than 100 generations ago, showed a narrow range of 76–84% of Asian contribution[19] (see **Section 13.3** for details).

The complex genetic ancestry of a genome often contrasts with the simplicity of an individual's perceived identity. Because of this, public dissemination of the results from admixture studies needs to be undertaken responsibly and with due care for their potential impact. It is worth remembering that ancestral populations themselves are likely to carry mixed ancestry of genes some of which, as we saw previously (**Section 9.5**), can be traced back to admixture between modern humans and Neanderthals.

In this chapter we will consider a variety of statistical methods that have been used to study this important issue. Then we will examine some case studies that illustrate the power of these methods and demonstrate the diverse outcomes of population encounters in the past. But first, we will consider the impact of admixture on other features of the populations involved.

TABLE 14.1: PROBABILITY OF ALL GENES OF AN INDIVIDUAL DERIVING FROM A SINGLE ANCESTRAL POPULATION IN A PANMICTIC HYBRID POPULATION

ADMIXTURE 50:50					ADMIXTURE 1:20				
t	A	B	A+B	AB	t	A	B	A+B	AB
0	0.5	0.5	1	0	0	0.05	0.95	1	0
1	0.25	0.25	0.5	0.5	1	0.003	0.903	0.905	0.095
2	0.063	0.063	0.125	0.875	2	6×10^{-6}	0.815	0.815	0.185
3	0.004	0.004	0.008	0.992	3	4×10^{-11}	0.663	0.663	0.337
4	2×10^{-5}	2×10^{-5}	3×10^{-5}	1	4	2×10^{-21}	0.44	0.44	0.56
5	2×10^{-10}	2×10^{-10}	5×10^{-10}	1	5	2×10^{-42}	0.194	0.194	0.806

t, number of generations since admixture. A, B, probability of observing individuals with full ancestry in either of the two source populations A and B; $A_t = (A_{t-1})^2$; $B_t = (B_{t-1})^2$. A+B, probability of observing non-admixed individuals. AB, probability of observing admixed individuals; $1 - (A + B)$.

14.2 THE IMPACT OF ADMIXTURE

Different sources of evidence can inform us about admixture

Genetic admixture is not the only consequence of the meeting of populations. Such events often impact significantly upon the cultural features of the populations involved. Thus many episodes of prehistoric admixture may well be detected using other records of prehistory, although it should be remembered that there need not necessarily be an archaeological or linguistic correlate for every genetic episode, and vice versa.

In the first few generations of admixture, individuals that descend primarily from one of the ancestral populations are often easily identifiable through their language or appearance. Individuals apportioned to the different ancestries are rarely on an equal footing in the nascent society. For example, the status of African slaves, European settlers, and Native Americans within the Americas was far from equal. Genetics cannot be divorced from these sociological considerations, since they directly influence the nature of the admixture. Rather, integrating genetic evidence with other prehistorical and historical records allows a richer appreciation of population encounters in prehistory.

Consequences of admixture for language

Populations that have been isolated from one another will accumulate linguistic differences relatively rapidly, and perhaps even speak different, mutually unintelligible, languages. What kinds of linguistic changes in a hybrid population might we expect to see as a result of their admixture? Bilingualism can be one short-term outcome, involving little change to either language. Alternative outcomes can be language mixing leading to highly dynamic **pidgin** languages that can be specific to a given location where admixture occurred (for example, Spanglish, referring to different blends of Spanish and English in Central America), or the establishment of a **lingua franca**—a language that would be widely understood in a broader geographic region where many languages meet. For example, during the Renaissance era commerce and diplomacy in the eastern Mediterranean was mediated largely by a mixed language that was based on Romance languages (Italian, Spanish, and French) enriched with specific loan-words from the Arabic, Turkish, and Greek languages.

The first point to appreciate is that a language is unlike a genome in several important ways. Whereas it is perfectly possible to assemble a fully functioning hybrid genome from several ancestral genomes with no associated costs, the same is not generally true of languages, because they must maintain a certain level of coherence to function adequately. Pidgin languages can arise and be erased over the span of a single generation. When the next generation of children of the pidgin speakers starts speaking it, the hybrid language becomes fixed. Linguists call native languages that represent fixed hybrids of parental languages **creoles**. A number of well-known modern creoles derive from European languages (French, Spanish, Portuguese, and English) and indigenous languages brought together by the actions of European colonial powers over the past few hundred years. For example, the Cajun language of Louisiana is a creole derived from French and languages spoken by African slaves. Creolization is considered to be a relatively rare process in language evolution.

Much more common than the development of creoles is the limited incorporation of certain features of one language into a dominant substrate from another language. These linguistic borrowings can affect different aspects of the language. The simplest example is the incorporation of outside words into a language. For example, among Polynesian languages, the term for the sweet potato (*kumara* in New Zealand Maori) derives from the word *kumar* from the Quechuan languages spoken in South America. As well as words, elements of structure can also be borrowed. For example, the order of subject, verb, and object within a sentence is often different between languages. Some **Austronesian** languages spoken in areas with neighboring **Papuan** speakers have adopted the "verb last" order of these Papuan languages (subject–object–verb: "he it hit"), as opposed to the more common "verb medial" organization (subject–verb–object: "he hit it") found among closely related Austronesian languages. These types of structural change result from what linguist Malcolm Ross calls:

> ... the natural pressure to relieve the bilingual speakers' mental burden by expressing meanings in parallel ways in both languages.

There are thus a wide variety of possible linguistic consequences of contact between populations speaking different languages, and which particular consequence follows in which situation is—as with the genetic outcomes—largely determined by the social context of this contact.

Spoken languages are not the only linguistic source of evidence for admixture: the names of people (surnames in particular) and of places (together referred to as **onomastic** evidence) are both capable of revealing the hybrid nature of a population. For example, English towns have names derived from Celtic, French (Norman), and Scandinavian languages as well as from Anglo-Saxon. It is worth noting, however, that evidence of past contact from sources such as place names does not imply that significant genetic admixture will be found in the current inhabitants. The Caribbean island populations of today, for example, may contain little genetic input from the original inhabitants.

In societies where surnames follow clear lines of inheritance, they have often been used in population genetic analyses, and admixture studies are no exception. Patterns of surname introgression have been shown to be correlated with levels of admixture in a number of different populations.[7] These conclusions have been reinforced by genetic analysis. Nevertheless, such surname studies have been dubbed the "poor man's population genetics,"[11] and are of real use only where genetic data are unavailable, and when admixture has occurred within the time frame of surname usage: this varies greatly from population to population, and may be very recent. However, if records are sufficiently detailed, surname analysis can reveal how admixture processes may have changed over time.

Archaeological evidence for admixture

The answers that archaeology provides to the question, What happens when populations meet? are primarily cultural in nature. The temporal and spatial distribution of archaeological sites can be used to demonstrate contact between cultures, and subsequent cultural change. However, such evidence only establishes the *potential* for genetic admixture. Before genetic admixture can be inferred, it must be assumed that populations with different **material cultures** are also different genetically and that the movement of artifacts is mirrored by the movement of people. In other words, artifacts are not being distributed by a set of sequential trading exchanges. New cultures can be adopted wholesale, or elements of individual cultures can be combined together in a process of integration. The integration of two cultural traditions may or may not be accompanied by genetic admixture. Similarly, a wholesale replacement of cultural practices may or may not be associated with a similar replacement of genes. With these caveats in mind it is worth noting that the spatiotemporal spread of an archaeological culture does indicate the geographical location of likely ancestral populations. In addition, the precision of archaeological dating provides good estimates for the time-scale of potential admixture processes.

Approaches based on **physical anthropology** have been adopted to seek phenotypic changes associated with genetic admixture in skeletal remains. Such work is often contentious, as human populations can rarely be well differentiated on skeletal evidence alone. In addition, alterations in cultural practices, for example, specific foot-binding or skull-compression traditions, or differences in diet, may cause significant morphological changes through **developmental plasticity**, rather than any change of genes[18] (see **Section 15.2**).

The biological impact of admixture

Our focus in this chapter is on identifying and quantifying past admixture. While in cases of highly differentiated ancestral populations, a small number of variants are sufficient to detect admixture, quantification of admixture proportions requires a larger number of polymorphisms across the whole genome so that specific selection processes can be excluded as explanations for modern patterns of genetic diversity. Nonetheless, all genetic admixtures will lead to a variety of phenotypic effects. Any quantitative trait that is genetically encoded and well differentiated between populations will be altered in admixed populations. Obvious physical examples include pigmentation, body proportions, and stature. In the past, these phenotypic data have been used to calculate admixture proportions, and, indeed, protein-coding genes associated with skin color show F_{ST} (see Box 5.2) values that are higher than the genome average, and can therefore give information about ancestry in admixed populations (Chapter 15).

Disease **prevalences** are often different between ancestral populations (see Chapter 16 for a discussion). An obvious medical consequence of admixture is that the hybrid population is expected to have disease prevalences for **Mendelian** disorders that are intermediate between those of the ancestral populations. When the most frequent diseases differ between the populations, this can lead to an overall lowering of burden of these diseases through a reduction in the probability of having two parents carrying the same deleterious **recessive** allele (**Table 14.2**).

Given the variation in degree of admixture among individuals in an admixed population, the proportion of admixture can be correlated with susceptibility to certain diseases more prevalent in one or other of the ancestral populations. It has been proposed that the prevalence of type 2 diabetes (OMIM 125853) in different Native American populations is positively correlated with the proportion of Native American genes, irrespective of whether this proportion is assessed phenotypically, genealogically, or genetically.[7] In practice, many studies have confirmed that non-diabetic Native Americans have on average significantly higher European admixture in their genomes than those diagnosed with diabetes. However, greater European ancestry also correlates with higher socioeconomic status which can, at least partly, explain the relationship between the disease and ancestry. While it is unlikely that assessing overall levels of individual admixture will have major predictive value of disease for individuals or for drug design (see Box 17.5), inferring the ancestry of particular genomic regions in admixed populations has proved to be a successful method for identifying disease loci (see **Section 14.4**).

For complex diseases, the possibility remains that each ancestral population contains individuals with co-adapted combinations of alleles that will be disrupted by admixture, resulting in a higher burden of disease in the hybrid

TABLE 14.2:
ADMIXTURE CAN REDUCE THE DISEASE BURDEN OF RECESSIVE SINGLE-GENE DISORDERS

Population	Carrier frequency of allele A	Carrier frequency of allele B	Incidence of disease A	Incidence of disease B	Total disease incidence
A	1/10	0	1/400	0	1/400
B	0	1/15	0	1/900	1/900
1:1 admixture	1/20	1/30	1/1600	1/3600	~1/1100

Only a quarter of children with carrier parents will be affected by a recessive disease.

population than in either ancestral population. No example of this **outbreeding depression** has yet been demonstrated in mixed human groups. However, analyses of numbers of children born to Icelandic couples over the past two centuries revealed a positive correlation between kinship and fertility: the highest reproductive success was observed in couples who were related at the level of third and fourth cousins.[15] Given the relatively minor socioeconomic differences among the mostly non-admixed families of Iceland, this correlation may have a biological basis.

There remains another mechanism by which admixture can result in an increased disease burden. Human populations often harbor their own populations of pathogens to which they have previously developed resistance. The release of these pathogens into previously unexposed populations could result in a substantial increase in the incidence and severity of infectious disease. This type of episode is exemplified by the population crashes witnessed in Polynesia and the Americas on first contact with Europeans bearing novel pathogens, such as those causing smallpox and measles. In such cases the resulting selective pressures are expected to result in a substantial bias toward contributions from the resistant ancestral population in the admixed population. In the first generations after admixture this bias would extend toward all genomic loci irrespective of their linkage to the locus conferring disease resistance, although in later generations, as the admixture patterns become more and more fragmented across the genome, this bias would be confined to linked loci.

14.3 DETECTING ADMIXTURE

Methods based on allele frequency can be used to detect admixture

Allele frequency-based methods were among the first tools developed to detect admixture from protein data. The simplest scenario occurs when no alleles are shared between the ancestral populations. Each allele in the hybrid population can then be unambiguously assigned to an ancestral population, and the proportion of admixture calculated by simply counting up the number of alleles assigned to each population. However, an absolute distinction between ancestral populations is rare; more often, alleles are found within many populations at differing frequencies. In principle, it is easy to estimate the proportion of admixture in a hybrid population formed from two ancestral populations (see **Figure 14.3**).

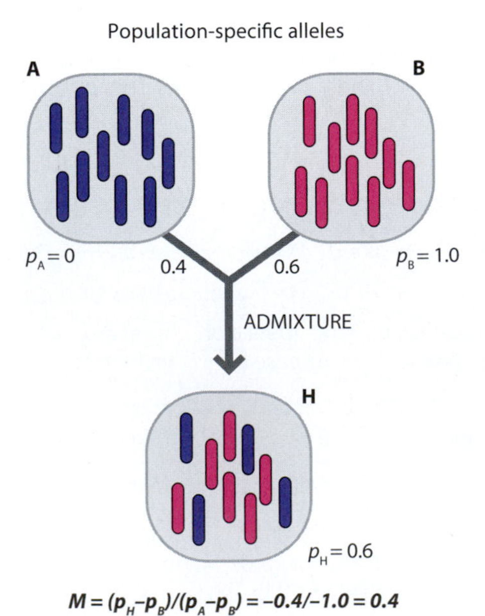

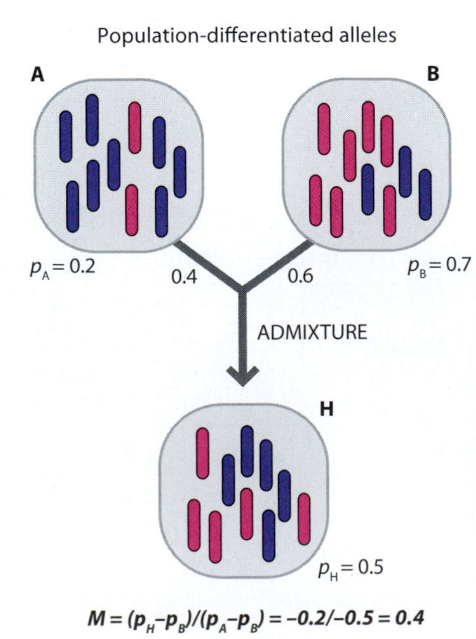

Figure 14.3: Calculating admixture proportions (*M*) when alleles are population-specific, and when they are present in both populations but at different frequencies. p_A, p_B, and p_H are the frequencies of an allele in the two parental populations A and B, and the hybrid population, H, respectively. The admixture proportions (*M*) are calculated using the equation described in the text, for two different scenarios.

For any given allele, if we know its frequency in the ancestral populations A and B (p_A and p_B) and in the hybrid population (p_H), we can estimate the proportion (M) that ancestral population A contributed to the admixed population by re-arranging the equation[5]

$$p_H = Mp_A + (1 - M)p_B$$

to give

$$M = (p_H - p_B)/(p_A - p_B)$$

Obviously, this approach requires the unambiguous identification of the ancestral populations as well as a number of assumptions, such as a lack of subsequent gene flow, which are either unrealistic or difficult to test in practice. Further complicating questions include:

- If more than one locus is being studied, how should they be averaged?
- What have the effects of genetic drift and selection upon allele frequencies been in all three populations since admixture?
- What if we misidentify the ancestral populations?
- What if there were more than two ancestral populations?
- What if an allele in the admixed population is not found in either ancestral population?

These complications have led to the development of a series of different admixture estimation procedures, which can be classified on the basis of the type of data used (genomewide or locus-specific), the assumed model for admixture, and whether they seek to estimate admixture at the level of the population or the individual. **Figure 14.4** illustrates a number of different admixture models.

Given a set of multi-locus allele frequencies in ancestral and hybrid populations, what is the best way of getting a single estimate of admixture proportions from

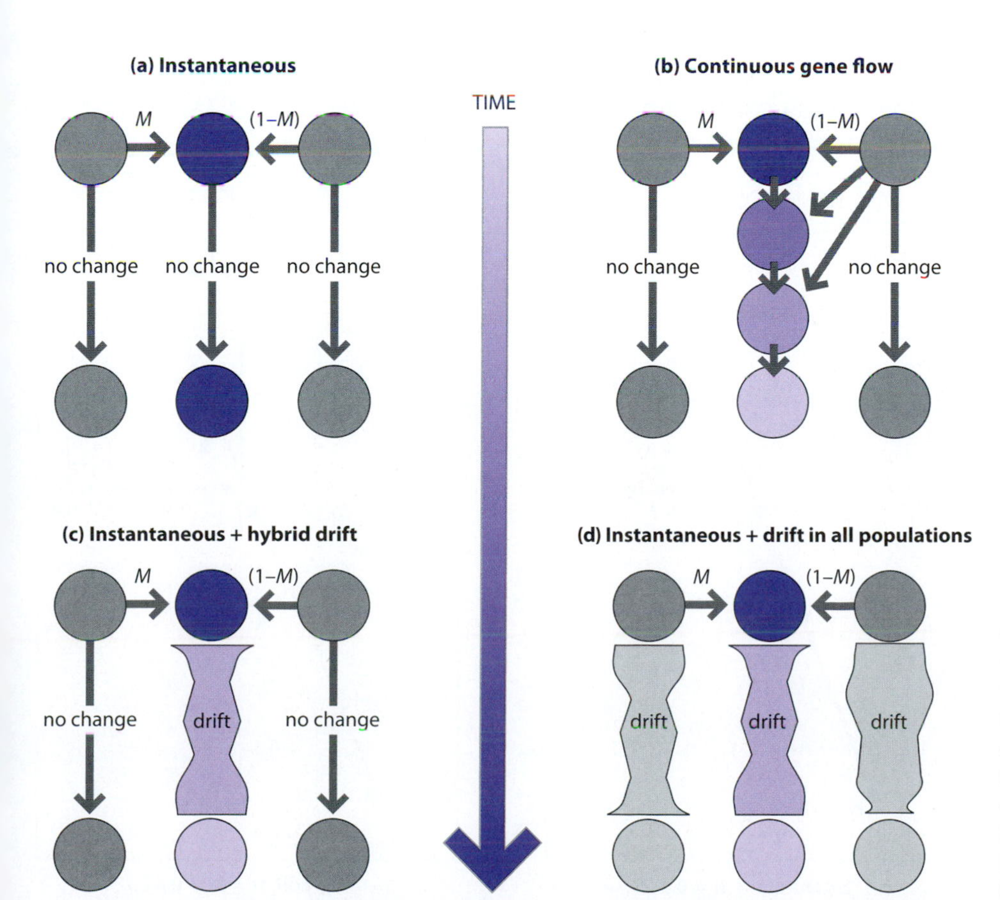

Figure 14.4: Different admixture models of varying complexity. (a) Instantaneous admixture, (b) cumulative effect of gene flow across many generations, (c) instantaneous admixture allowing for drift in the hybrid population, and (d) instantaneous admixture allowing for drift in all three populations. *Gray* circles, parental populations; *blue* circles, hybrid populations.

these data? Admixture estimates can be calculated for each allele, or locus, individually and then averaged. There are a number of different ways of averaging this information across loci and assessing ancestry from the data.

The equation for M given above suggests that estimates of admixture proportions from different alleles should be related linearly. In other words plotting ($p_H - p_B$) against ($p_A - p_B$) for different alleles should give a straight line of gradient M. However, drift, selection, and imprecision of allele frequency estimation can lead to deviations from the linear relationship in real data. **Figure 14.5** shows this property for an idealized admixture situation where all alleles give the same estimate of M.

The plot in Figure 14.5 immediately suggests one method of averaging information from different estimates, namely to plot the least-squares regression line between the points and take its gradient as the multi-locus estimate of admixture.[32] This **estimator** of admixture is often known as $\mathbf{m_R}$.

The above method assumes that the allele frequencies are known without error; it does not take account of the different levels of precision associated with each individual estimate. This can be considered by averaging the different estimates weighted according to their precision, as assessed by their **variances**. These variances depend on the size of the samples. The weighting factor commonly used is the inverse of the variance of the estimate. In other words, the higher the variance of the estimate, the less we are sure that it is accurate and the lower the weight we give it.[6] This weighted average approach takes into account sampling effects on all allele frequency estimates.

(a) Artificial population

Allele	p_A	p_B	p_H	$p_H - p_B$	$p_A - p_B$
A	0.6	0.5	0.54	0.04	0.1
B	0.1	0.8	0.52	−0.28	−0.7
C	0.3	0.6	0.48	−0.12	−0.3
D	0.45	0.7	0.6	−0.1	−0.25
E	0.35	0.15	0.23	0.08	0.2
F	0.8	0.3	0.5	0.2	0.5
G	0.2	0.25	0.23	−0.02	−0.05
H	0.4	0.6	0.52	−0.08	−0.2

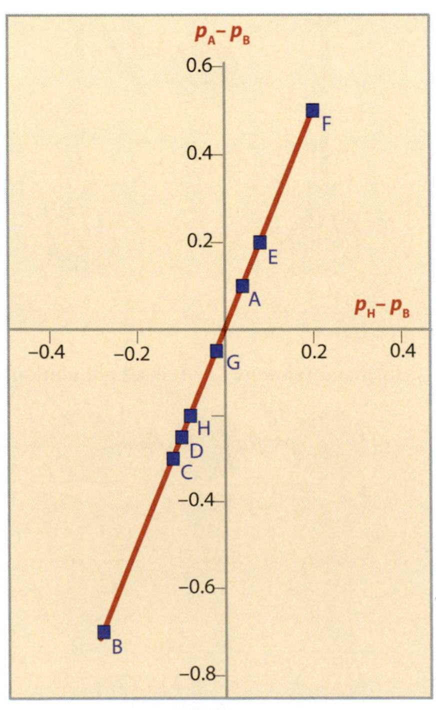

(b) Real population

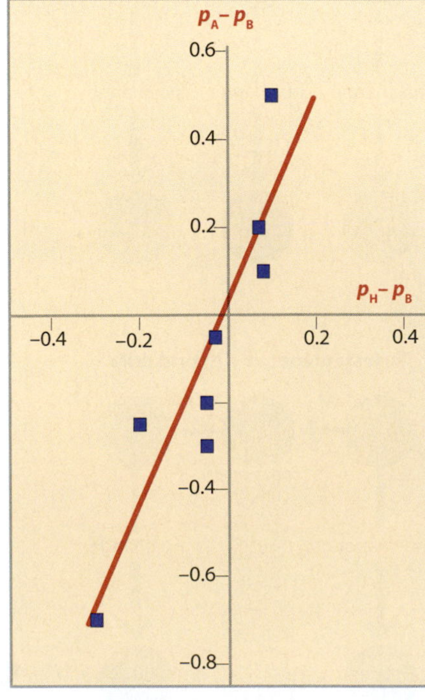

Figure 14.5: Linearity of ($p_H - p_B$) against ($p_A - p_B$). (a) An idealized case of admixture in which eight alleles (A–H) give exactly the same value for M, and (b) a more realistic set of variant estimates of M from different alleles; M is estimated by fitting a best fit line through the points.

More recently, simulation modeling methods have been developed to infer multiple parameters of population histories in the presence of admixture.[10] However, all methods designed for admixture estimation will produce estimates of M even if the ancestral populations have been grossly misidentified, and so care will always be necessary to identify them correctly. This often requires the support of historical, archaeological, and linguistic evidence.

Admixture proportions vary among individuals and populations

A positive estimate of admixture proportions within a hybrid population does not mean that all individuals have the same ancestry ratios. Even if several generations have passed since the admixture event, populations often remain heterogeneous in terms of individual admixture proportions. Similarly, different sampling locations may yield variable estimates if admixture is geographically structured. Besides these inter-individual and interpopulation variations in the amount of *global* genomic estimates of admixture, representing averages over individual genomes, there is also *local* variation at various genes and noncoding loci (**Section 14.4**). Substructure within the admixed population, which can be caused by geographic partitioning, socioeconomic factors, natural selection, and potentially also by **assortative mating**, may reveal itself in unusually large variation in individual admixture estimates as seen, for example, in a number of studies examining admixture in African-Americans (**Table 14.3**). Levels of population admixture can be determined from individual admixture estimates, but not vice versa. How can we investigate this finer-grained admixture?

Calculating individual admixture levels using multiple loci

The actual calculation of individual admixture levels is statistically complex, because information has to be incorporated from many alleles at many loci, and most approaches necessarily have to make simplifying assumptions about past population structure. These include reduction of the number of ancestry groups, ideally down to two parental populations, and reduction of the observed genetic diversity to a minimum number of components. An appreciation of genomewide admixture can only be gained by inferring the ancestry of multiple unlinked loci within that genome, and thus requires substantial genotyping effort. Some alleles may not be particularly well differentiated between different ancestral populations, and thus may not be particularly informative. To reduce the genotyping load many studies have focused on what are called **ancestry informative markers** (AIMs). These are polymorphisms where the allele frequency in the ancestral populations differs considerably (>45%;[37] this work is discussed in a forensic context in **Section 18.2**). Analysis of AIMs has the additional advantage of giving a more accurate estimate of the amount of admixture compared to the same number of less well-differentiated alleles. The high degree of population differentiation exhibited by mtDNA and the nonrecombining portion of the Y chromosome makes them valuable sources of such AIMs. However, both mtDNA and the entire male-specific region of the Y chromosome each represent a single locus, with a single evolutionary history that is not necessarily representative of the rest of the genome.

There are potential problems in focusing solely on AIMs. Because of the low degree of **genetic differentiation** among modern humans, most alleles that exist at moderate to high frequency are not highly population-specific (see **Section 10.2** for more details). Besides drift there are at least two reasons why some alleles do show appreciable population specificity: (1) they result from relatively recent mutations that have not had sufficient time to disperse to any great degree; (2) they result from the action of selection, which influences allele frequency differently in different selective environments. It should also be noted that because of the development of high-throughput genotyping methods there is less need to focus on AIMs and it may be easier and more cost-effective to genotype samples for many polymorphisms genomewide (for example, using a SNP chip) rather than using a small number of custom AIMs.

TABLE 14.3:
ADMIXTURE ESTIMATES IN SOUTH AMERICAN POPULATIONS FOR LOCI WITH DIFFERENT INHERITANCE PATTERNS

Population/locus	% African	% European	% Native American	% Other
Afro-Uruguayan[a]				
Y-chromosomal	30	64	6	
Autosomal	47	38	15	
Mitochondrial	52	19	29	
Brazilian Whites[b]				
Y-chromosomal	3	97	0	
Mitochondrial	28	39	33	
Colombians[c]				
Y-chromosomal	9	79	12	
Autosomal	11	42	47	
Mitochondrial	6	4	90	
US Hispanic[d]				
Y-chromosomal	21	69	8	2
Autosomal	12	61	15	12
Mitochondrial	15	24	49	12
Argentinians[e]				
Y-chromosomal	1	94	5	
Autosomal	4	80	15	
Mitochondrial	2	44	54	

Data from:
[a] Sans M et al. (2002) *Am. J. Phys. Anthropol.* 118, 33.
[b] Alves-Silva J et al. (2000) *Am. J. Hum. Genet.* 67, 444; and Carvalho-Silva DR et al. (2001) *Am. J. Hum. Genet.* 68, 281.
[c] Rojas W et al. (2010) *Am. J. Phys. Anthropol.* 143, 13.
[d] Lao O et al. (2010) *Hum. Mutat.* 31, E1875.
[e] Corach D et al. (2010) *Ann. Hum. Genet.* 74, 65.

Calculating individual admixture levels using genomewide data

High-resolution genomewide genotype data are now available for many populations throughout the world and a number of approaches have been developed to simultaneously assess population structure (**Sections 10.2 and 10.3**) and admixture from such data. One method of reconstructing genetic ancestry from given genotype data, employed by programs such as EIGENSTRAT,[26] is based on **principal component analysis** (PCA; **Section 6.3**). PCA can be used to assess the clustering of individuals or populations at low-dimensional projections of the data. As populations diverge from each other over time by genetic drift, the clusters detected on the low-dimensional scatter plots of PCA are expected to become more and more distinct with decreasing levels of overlap between them. Individuals representing admixture of two distinct ancestral populations would be expected to lie on a cline between the clusters: the exact position of admixed individuals on this cline would be determined by the ancestral contributions of the two parental populations.

Computer simulations show that PCA can also be used to predict contributions to an admixed population from ancestral populations that are either extinct or not available for genotyping for other reasons. **Figure 14.6** illustrates a two-dimensional PCA plot applied to the results of a computer simulation study where hybrid population C was derived from ancestral populations A (extinct) and B (extant) while population D was left to drift without admixture. As a result, individuals from the two non-admixed populations B and D cluster tightly together whereas individuals from the hybrid population C are dispersed between the two ancestral populations. Although population A was not sampled it can be reconstructed as a source of admixture because some individuals in the hybrid group C are still characterized by a high genetic contribution from A. This outcome is important because when a hybrid population is formed from two distinct parental groups, then for a number of generations individuals in the hybrid group will carry different proportions of admixture. This proportion scales linearly with the genetic distance of each individual from their parental populations.

However, accurate and reliable assessment of individual admixture proportions would certainly benefit if data from parental populations were available, particularly in cases where one of them has made a low contribution. Application of PCA methods for inferring admixture to real data should also be carried out with care because alternative demographic models can produce similar patterns: for example, population substructure and higher effective population size of population C in Figure 14.6 could have generated the observed pattern even if ancestral population A never existed. It should be also noted that, when data from only a single hybrid population are available, this method is capable of identifying admixture only where individuals differ from each other in admixture proportions. If admixture is ancient, all individuals are likely to have highly similar admixture proportions and subsequently do not allow admixture detection by PCA.

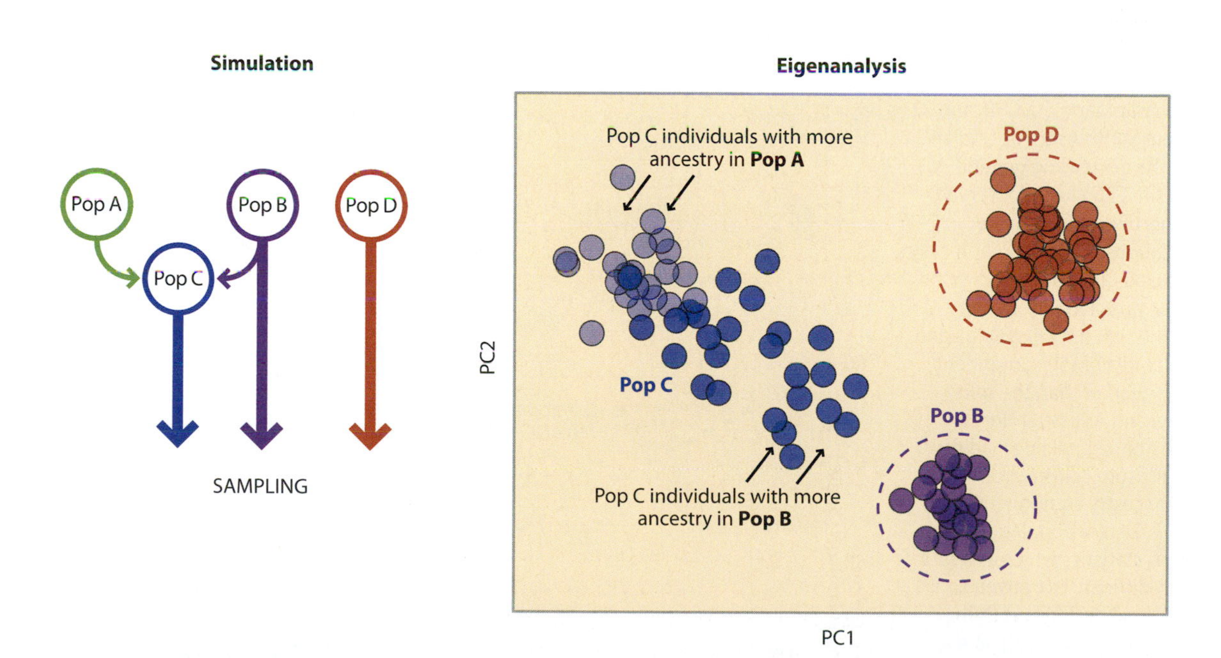

Figure 14.6: Assessment of admixture by PCA.
The plot is based on a computer simulation of multi-locus data (10,000 unlinked variants) where population C was derived by recent admixture from populations A (70%) and B (30%). Individuals from population A were not sampled and included in the PCA. In the PCA each individual is plotted on a scatter plot, with the *x* and *y* coordinates of each individual point representing the values of the first two PCs of that individual's genotype data. Individuals from population C have variable levels of admixture and are dispersed along a line joining the two parental populations. Their position on this line is determined by the admixture proportions and does not change when incorporating population A back in the analyses. [Adapted from Patterson N et al. (2006) *PLoS Genet.* 2, e190. With permission from Public Library of Science.]

Calculating admixture levels from estimated ancestry components

To address admixture events that have occurred many hundreds of generations ago, multi-population ancestry assessment algorithms have been proposed.[31] These methods should be regarded with caution because they rely on specific models of population history, whose assumptions are not always testable. Like the individual ancestry assessment from PCA, these algorithms relate the ancestry proportions in admixed populations to coordinates of a regression line at a low-dimensional projection of the data. As an example, consider PCA of South Asian populations in the context of data from European and East Asian populations (**Figure 14.7a**). While Europeans and East Asians form tight clusters, the individuals from South Asia are dispersed more widely in the plot. Within each population of South Asia the variation is minimal, but when plotted together they form an inverted v-shaped cluster. One arm of this cluster, dubbed the "Indian Cline," stretches out toward European populations. Because North Indian and Pakistani populations appear to be closest to Europe on this cline, one possible interpretation is that these populations have received a higher admixture proportion from a hypothetical Ancestral North Indian (ANI) population that shares its ancestry with populations of Europe and northern Caucasus. In contrast, South Indian populations would be derived largely from a separate, Ancestral South Indian (ASI) source (Figure 14.7b). Assuming such ancestral populations existed, how can we quantify these admixture proportions given that neither of the ancestral groups can be sampled today? One such method,[31] based on the regression line estimated from a four-population statistic, f_4, is illustrated in Figure 14.7c. This method estimated the ancestral ANI contribution in Indian populations as 40–80%.

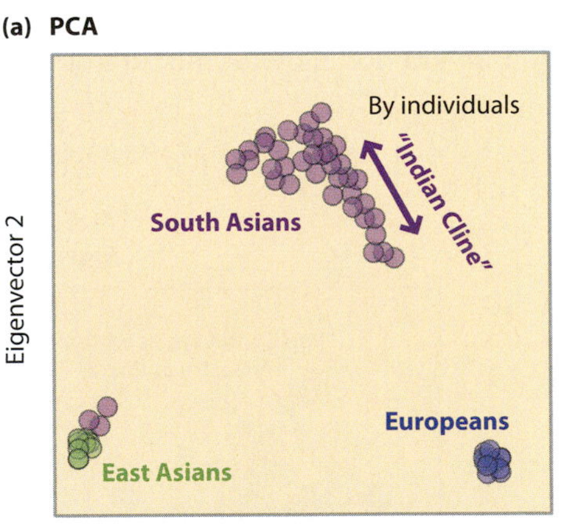

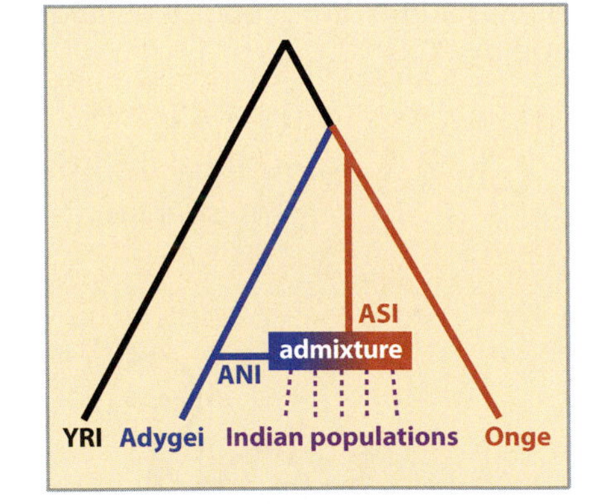

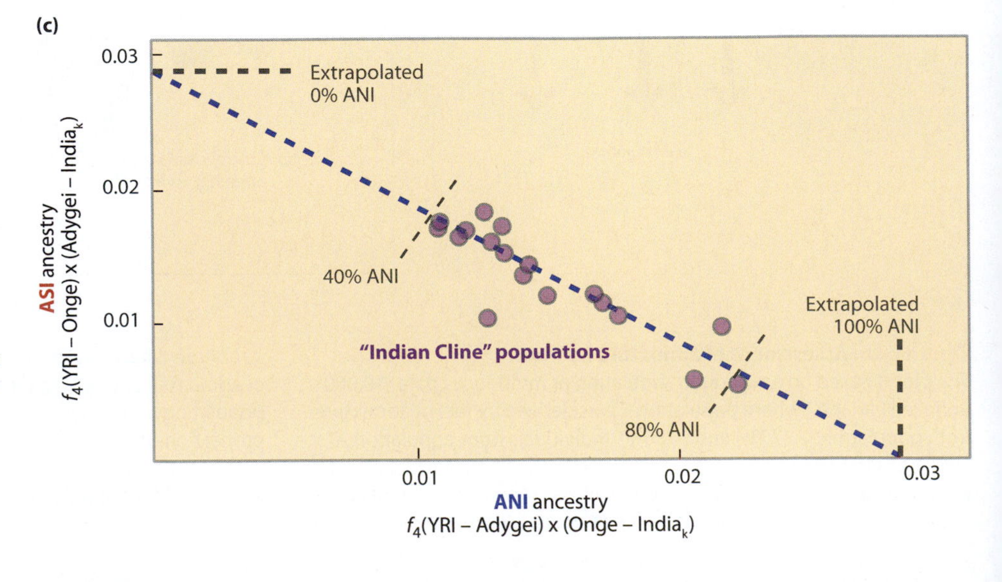

Figure 14.7: Ancestry estimation along the Indian Cline. (a) PCA plot based on genomewide SNP data assessed in HapMap CEU, CHB, and 22 populations from South Asia. The "Indian Cline" refers to the decreasing genetic distances from the CEU cluster observed from south to northwest Indian and Pakistani populations. (b) Demographically explicit model explaining the extent of admixture of the populations on the "Indian Cline." The split between Onge and Ancestral South Indian (ASI) component is expected to be earlier than the split between Adygei and Ancestral North Indian (ANI). The horizontal line leading to ANI reflects admixture. (c) The proportions of ANI and ASI ancestry components in the Indian and Pakistani populations assessed using f_4 statistics which assesses allele frequency differences between pairs of populations: Yoruban Africans (YRI), Adygei, Onge (Andaman Islands), and a number of South Asian populations tested for admixture. The projections of the Indian and Pakistani population on the regression line are informative about their relative admixture proportions in ANI and ASI. [Adapted from Reich D et al. (2009) *Nature* 461, 489. With permission from Macmillan Publishers Ltd.]

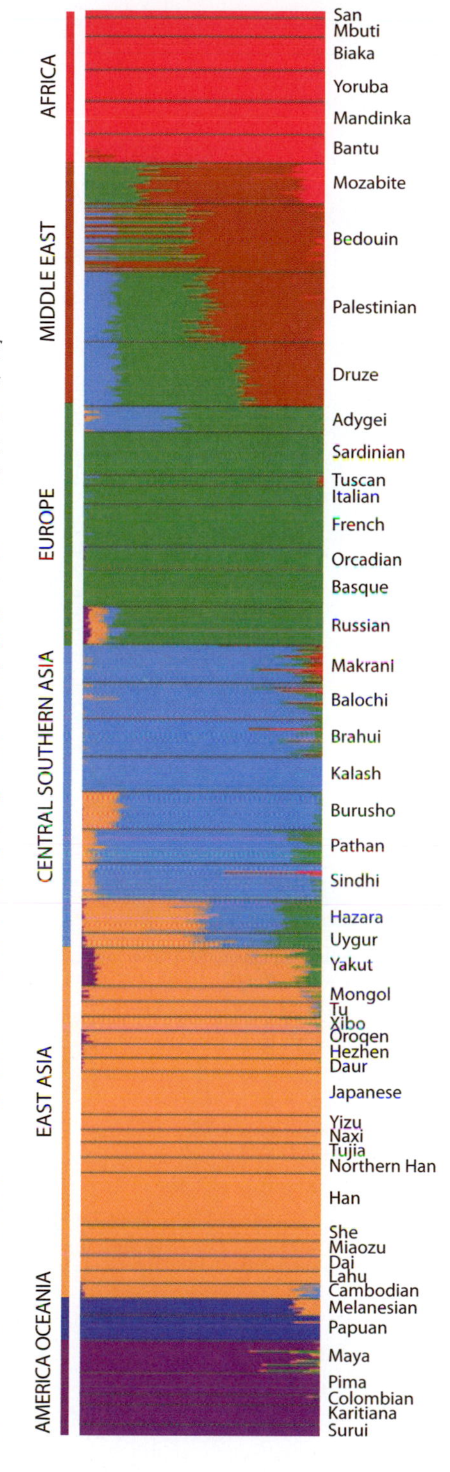

Figure 14.8: Global ancestry profiles of individuals in the HGDP-CEPH panel.
Ancestry as inferred using FRAPPE program at $K = 7$, that is, assuming there are seven clusters. The plot is based on 938 individuals from the HGDP-CEPH panel genotyped for 650,000 SNPs over the genome. Each individual is represented by a horizontal line partitioned into colored segments whose lengths correspond to the ancestry coefficients in the seven ancestry components. Population labels have been added after individual ancestry assessment. [From Li JZ et al. (2008) *Science* 319, 1100. With permission from AAAS.]

In principle, it should be possible in the future to go beyond identification of ancestral populations on PCA plots, and to reconstruct with a certain probability the ancestral genomes themselves from the fragments that survive in modern populations. However, this approach will be feasible only for populations that have gone extinct through processes involving admixture, as is the case, for example, for Tasmanians and many Native American populations.

Clustering methods can take multi-locus genotypes of individuals from several populations and apportion them into well-resolved clusters that are clearly differentiated from one another. One fundamental problem is deciding upon the most likely number of clusters. A number of model-based clustering methods (STRUCTURE, FRAPPE, ADMIXTURE) have been devised that determine the cluster number, the frequency of any given allele in each cluster, and the proportion of each individual's genome that owes ancestry to each cluster.[1, 28, 39] Thus, under the assumption that all loci are in Hardy–Weinberg equilibrium in all populations, STRUCTURE-like approaches (**Section 6.3**) are capable of calculating individual admixture proportions from genetic clusters estimated from the data themselves, rather than from allele frequencies in sampled populations, whose definitions may often rely on external evidence (for example, a shared language or nationality). An example of such a plot where 938 individuals from the 51 populations of the **HGDP-CEPH panel** (Table 10.1, Box 10.3) are allocated to seven ancestry components is shown in **Figure 14.8**. Simulated evolution of two and three populations in the absence of admixture has shown that the clusters formed correspond to the populations themselves and that individuals owe all their ancestry to a single cluster. By contrast, when admixture is included within the simulation, again these clusters are formed but many individuals show some ancestry in between them (**Figure 14.9**).

Although diversity and fine-scale population structure exist both within Africa and Europe, and within Native American populations, it is fairly straightforward to estimate the admixture proportions of each of the three ancestral continental sources in present-day American populations. Earlier, in Chapter 9, we saw how even admixture with Neanderthals is embedded in the genomes of all non-African populations. Thus, a consideration of individual genetic diversity can also reveal cryptic **population structures** that were not previously known to exist. Each individual genome can, in a sense, be considered to be like a palimpsest of multilayered history of successive periods of drift and admixture events in multiple ancestral populations, each of which, in turn, may have had its own multifaceted history. The identification of cryptic population structure is important for other applications. For example, if undetected it can cause spurious associations when hunting disease genes (see Chapter 17) and unreliable match probabilities in forensic situations (see Chapter 18). **Table 14.4** lists software for estimating admixture proportions.

Problems of measuring admixture

Any analysis of admixture is, however, an oversimplification of the population history because of the large number of parameters that are used, some of which need to be fixed at assumed values. Even in cases where it seems that we have succeeded in reducing the ancestry of present populations down to two or a few ancestral populations, it would be naive and wrong to consider the reconstructed ancestral groups as pure types akin to the nineteenth-century concept of "race." Most clustering algorithms for ancestry detection have been developed to correct for population stratification as a confounding factor in association studies. For such purposes it is useful to detect hidden structuring of the data among **cases and controls** regardless of the meaning of the revealed ancestry components. Their interpretation in terms of demographic histories of populations should be considered with caution.[43] Different opinions have been expressed about whether the ancestry components revealed by the clustering algorithms really reflect past population structure, or are due to sampling at discrete points within an underlying clinal space of variation of human genetic diversity.[14, 34, 36] Some of the admixture components shown in Figure 14.8, for example, would not be compatible with any demographic scenario that would be supported by historical or archaeological evidence: for example, the Native American component detected in Russians.

Natural selection can affect the admixture proportions of individual genes

If all tested loci are evolving neutrally, admixture should affect allele frequencies to an equal degree, since all depend upon the same parameter (M, introduced at the beginning of this section). However, selection can bias the frequency of alleles in an admixed population, which often inhabits an environment exerting different selective pressures from that inhabited by either ancestral population. A change in selective pressures will cause a change in allele frequencies at those loci irrespective of any admixture event. Consider an allele that was previously maintained in one of the ancestral populations by **balancing selection** that might now be present in an admixed population in which it has no **heterozygote advantage**. For example, the sickle-cell disease (OMIM 603903) allele (Hb^S) in heterozygotes protects against malaria in Africa (**Section 16.4**), but in the admixed population of African-Americans in the USA, where malaria is absent, is simply deleterious in homozygotes. Consequently, the frequency of

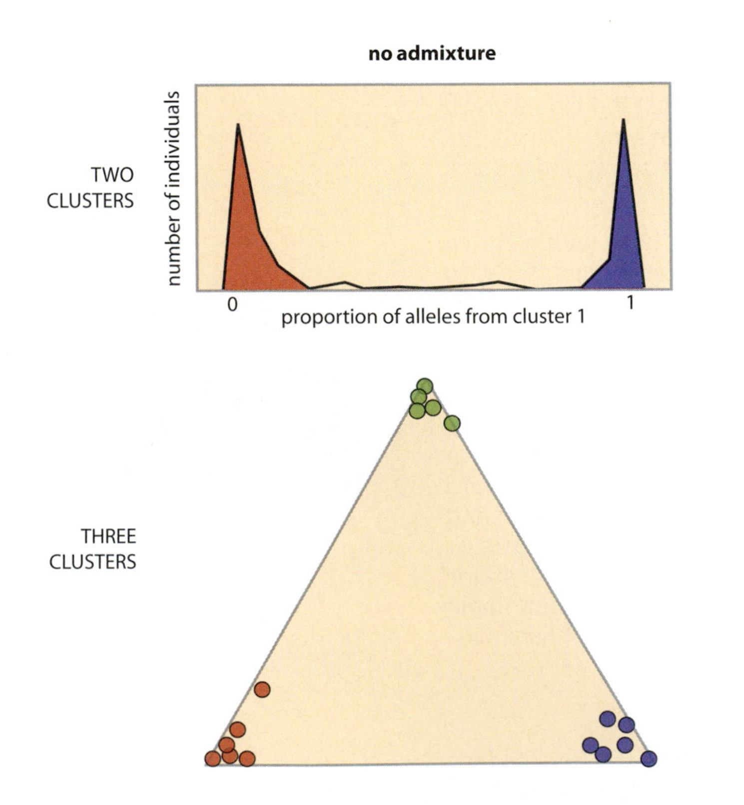

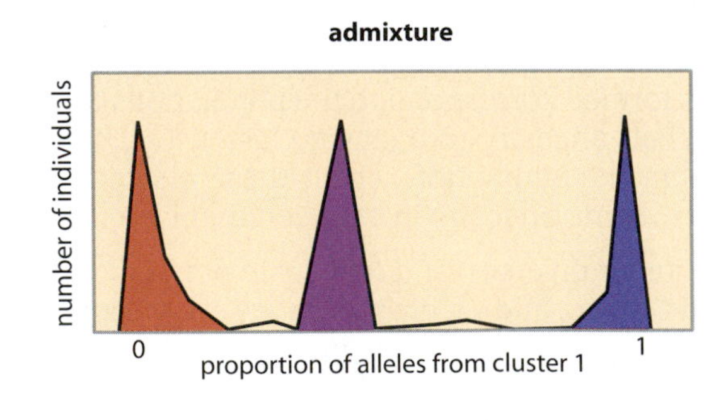

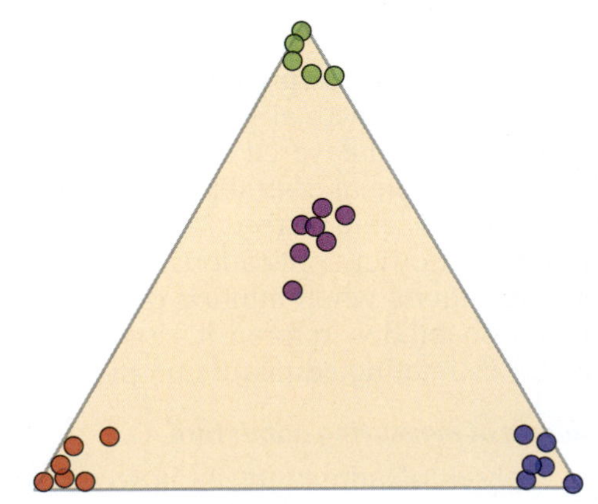

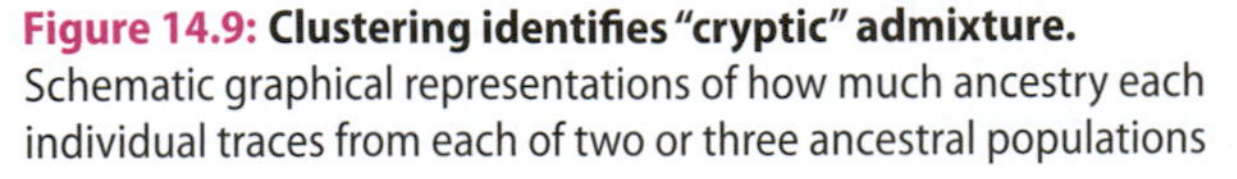

Figure 14.9: Clustering identifies "cryptic" admixture. Schematic graphical representations of how much ancestry each individual traces from each of two or three ancestral populations (clusters). In the absence of admixture the majority of alleles in each individual can be traced to a single cluster. Admixture can be identified when groups of individuals are found to fall between clusters.

TABLE 14.4: SOFTWARE FOR ADMIXTURE ANALYSIS

Method	Software	URL
Gene identity	ADMIX95	http://www.genetica.fmed.edu.uy/software.htm
Bayesian individual admixture	ADMIXMAP	http://homepages.ed.ac.uk/pmckeigu/admixmap/
Identification of population structure	STRUCTURE	http://pritch.bsd.uchicago.edu/
	FRAPPE	http://med.stanford.edu/tanglab/software/frappe.html
	ADMIXTURE	http://www.genetics.ucla.edu/software/admixture/publications.html
	FINESTRUCTURE	http://www.maths.bris.ac.uk/~madjl/finestructure/chromopainter_info.html
Principal component analysis	EIGENSTRAT	http://genepath.med.harvard.edu/~reich/EIGENSTRAT.htm
Population trees with admixture	TREEMIX	http://pritch.bsd.uchicago.edu/
Coalescent-based simulation of population histories with admixture	SPLATCHE	http://www.splatche.com/
Admixture mapping of complex traits	MALDSOFT	http://pritch.bsd.uchicago.edu/
Identification of ancestry segments in admixed individuals	ANCESTRYMAP	http://genepath.med.harvard.edu/~reich/Software.htm
	SABER	http://med.stanford.edu/tanglab/software/
	HAPMIX	http://www.stats.ox.ac.uk/~myers/software.html
	CHROMOPAINTER	http://www.maths.bris.ac.uk/~madjl/finestructure/chromopainter_info.html

Hb^S in African-Americans is much closer to that in European-Americans than we might otherwise expect. Admixture proportions calculated solely on the basis of this allele give much higher estimates for the contribution of European genes to African-Americans than do other, neutral, loci. Similarly, HLA (Box 5.3) alleles in Puerto Ricans reflect an excess of African and deficiency of European ancestry when compared with global genomewide averages, suggesting adaptive advantage of the African alleles (**Figure 14.10**).

This finding can be turned on its head to provide a means for identifying selection acting upon specific alleles. Heterogeneity among admixture estimates derived from the frequencies of different alleles can be used to pinpoint those alleles whose admixture estimates deviate significantly from those of most others (see **Figure 14.11**).

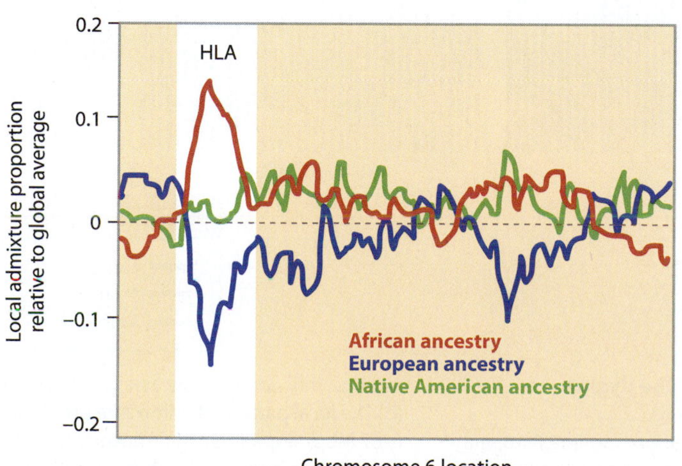

Figure 14.10: Deviations of admixture proportions due to natural selection on HLA alleles.
An excess of African and deficiency of European ancestry at the HLA locus (*white* bar) on chromosome 6. [Adapted from Tang H et al. (2007) *Am. J. Hum. Genet.* 81, 626. With permission from Elsevier; and Oleksyk TK et al. (2010) *Philos. Trans. R. Soc. Lond. B Biol. Sci.* 365, 185. With permission from Royal Society Publishing.]

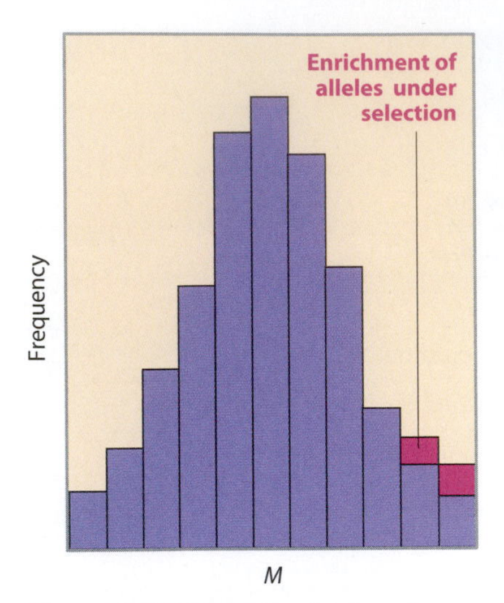

Figure 14.11: Deviant estimates of admixture (*M*) reveal outlier alleles under selection.
A change of selective environments between ancestral and hybrid populations can lead to selection pressures on some alleles resulting in estimates of *M* that are outliers when compared with neutral alleles.

However, selection is not the only evolutionary process that can distort allele frequencies and the admixture estimates derived from them. **Genetic drift** also influences these estimates. Distinguishing between systematic biases in admixture estimates resulting from selection, on the one hand, and random biases introduced by sampling effects and genetic drift, on the other, is far from easy. When searching for outliers of *M* from empirical distributions, as shown in Figure 14.11, caution is required because we do not know what proportion of the genome has been affected by selection and how much variation is due to drift.

14.4 LOCAL ADMIXTURE AND LINKAGE DISEQUILIBRIUM

As we saw in the previous section, admixture proportions can vary in populations due to inter-individual differences, and in addition the genome of each individual can be a composite of clusters of segments that differ in admixture contributions because both selection and drift affect the frequencies of alleles at unlinked loci independently. Over time, the signal of admixture will be divided by recombination into smaller and smaller segments of the genome. **Figure 14.12** illustrates the distinction of the global and local patterns of admixture in the example of the Uyghur population from Central Asia, which has been estimated to derive from an admixture event involving East and West Eurasian sources more than 100 generations ago.

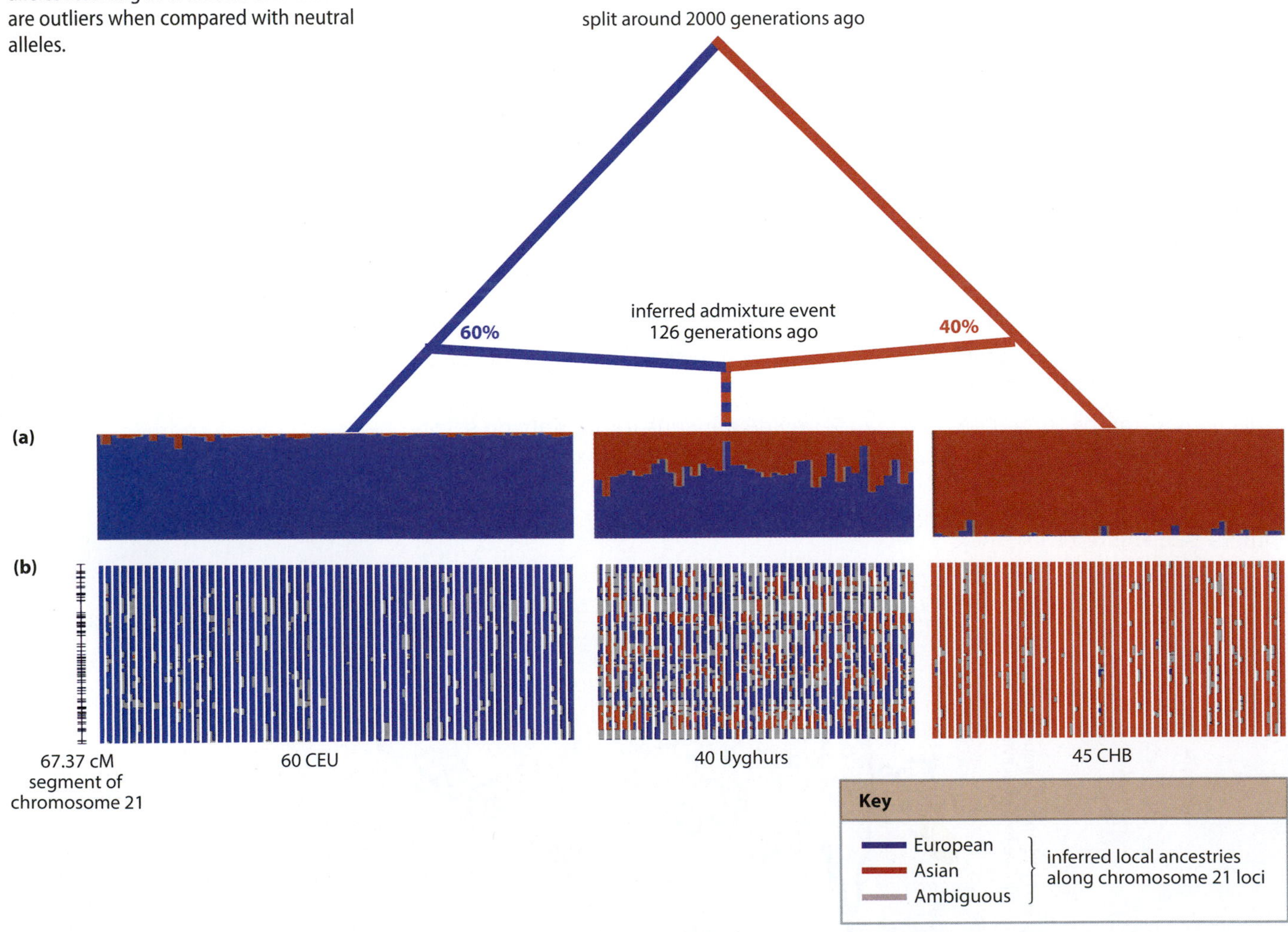

Figure 14.12: Local and global admixture in the Uyghur population.
(a) The global, genomewide admixture contribution from the European source was estimated in the Uyghur population to be around 60% on average. Individuals show a range of variation between 40 and 85%. (b) The local pattern of admixture profile is illustrated on a 67.37 cM (~67 Mbp) segment of chromosome 21. [Adapted from Xu S et al. (2008) *Am. J. Hum. Genet.* 82, 883. With permission from Elsevier.]

How does admixture generate linkage disequilibrium?

In Chapters 3 and 5 we encountered the phenomenon of **linkage disequilibrium** (LD), whereby alleles at different loci tend to be co-inherited more often than we might otherwise expect. Admixture events generate LD between all loci for which differences in allele frequency exist between the two ancestral populations (see **Figure 14.13**). Over the first few generations, LD can be detected even between physically unlinked loci, for example on different chromosomes. However, the detectable association between unlinked loci (self-contradictory though this may sound) dissipates rapidly over a few generations as a result of **chromosomal assortment**. LD at physically linked loci decays more slowly due to recombination events. As a result, recently admixed populations should exhibit LD over greater genetic distances than non-admixed populations. This makes them of potential use for mapping traits and disease genes.

A number of factors affect the extent of LD exhibited by an admixed population. These include:

- The time since admixture
- The admixture dynamics, for example, instantaneous or continuous gene flow
- The relative contributions of different ancestral populations
- The allele frequency differences between ancestral populations
- The pattern of recombination in the human genome

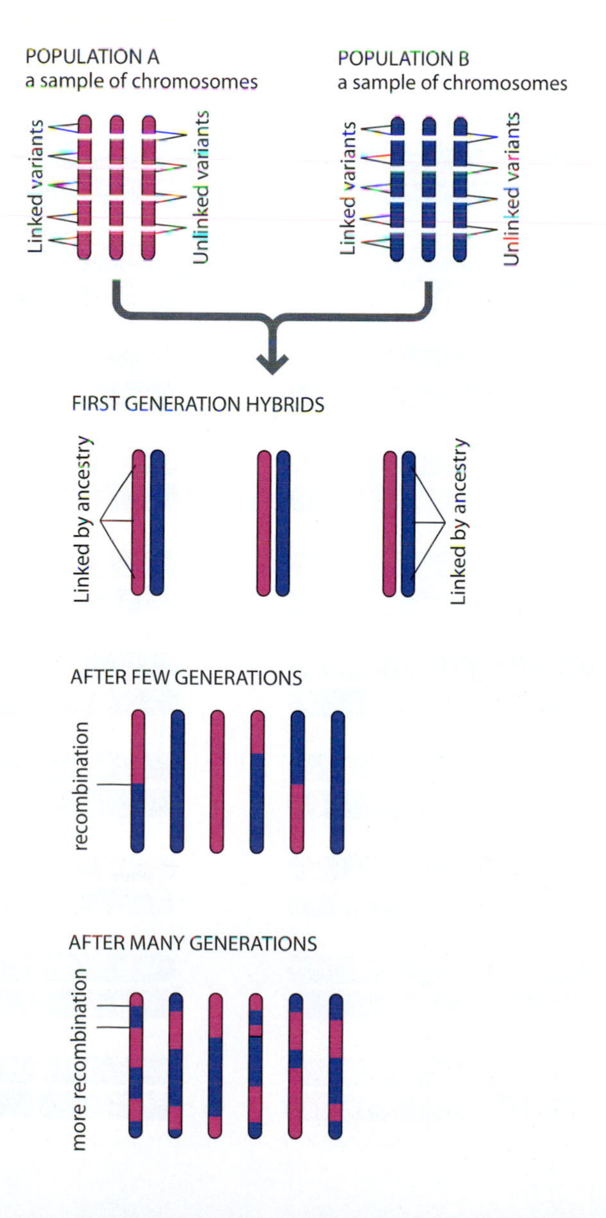

Figure 14.13: Admixture generates linkage disequilibrium (LD). Decline in extent of LD over the generations after admixture. Two ancestral populations (A and B) contribute a sample of three chromosomes to a hybrid population. Three first-generation hybrids show complete linkage of all polymorphic variants on the same chromosome and because of different allele frequencies in A and B the ancestry of each chromosome can be wholly traced back to only one ancestral population. Over time the LD breaks down due to recombination events and the ancestry becomes fragmented, until only closely positioned alleles on the same chromosome are in LD.

Admixture mapping

The approach of mapping genes underlying phenotypic traits by assessing their association with LD caused by admixed ancestry is generally referred to as Mapping Admixture by LD (MALD).[8, 38] A number of MALD algorithms have been developed as statistically powerful alternatives to genome-wide association studies (GWAS) (Table 14.4). The idea behind the approach is that genes associated with a phenotypic trait (or disease) that displays a substantial frequency difference between two parental populations are expected to show pronounced admixture LD among the cases sampled from the hybrid population (**Figure 14.14**). This approach can be successfully applied even as a cases-only study design with a few thousand AIMs and has the power to detect associations with modest odds ratios with sample sizes of only few thousands of individuals.[22, 25] Because the length of the haplotypes resulting from admixture decays

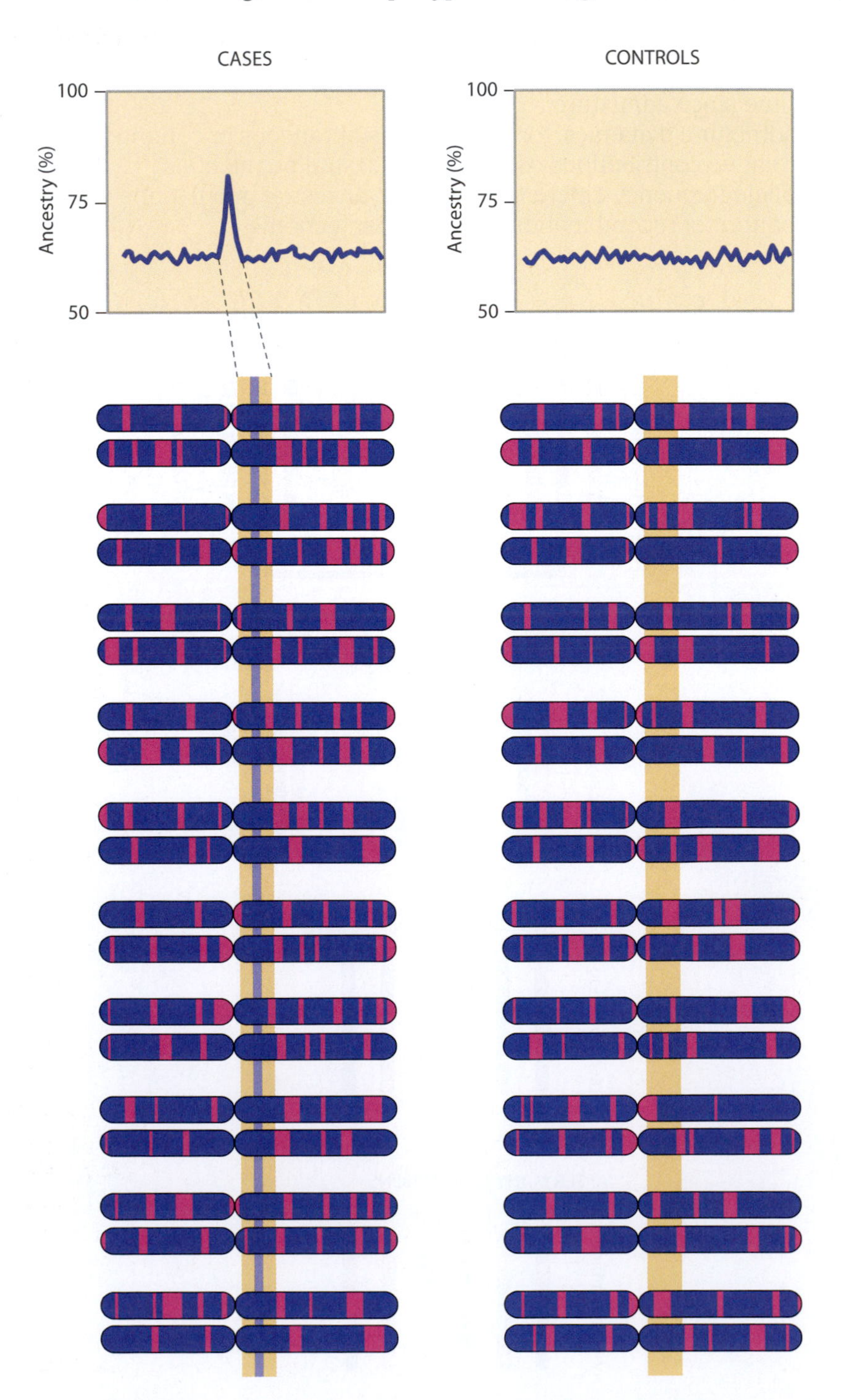

Figure 14.14: Mapping admixture by LD (MALD).
Admixture mapping will highlight genetic loci that show allele frequency differences between two parental populations in association with a disease. Ancestry in the population with the higher disease prevalence is shown in *dark blue*. Individuals with the disease (cases) show higher ancestry at associated loci whereas randomly taken controls from the same hybrid population display an ancestry profile consistent with the genome-average admixture proportions. [From Winkler CA et al. (2010) *Annu. Rev. Genomics Hum. Genet.* 11, 65. With permission from Annual Reviews.]

over time, different polymorphism densities are required for different admixed populations. In the case of African-Americans, for example, where admixture dates mostly to <10 generations ago, admixture LD extends on average for 17 cM,[25] while in Uyghurs (Figure 14.12), where admixture dates to >100 generations ago, the blocks extend only 2–4 cM on average.[45]

Admixture mapping can also be applied to cases involving more than two ancestral populations. Admixture in Latin America is often three-way (as already shown in Figure 14.10) as it involves mixtures of Native American, European, and African ancestries. Individuals from the "colored" population of South Africa can represent five-way admixture combining ancestries from Bantu, Khoe-San, European, South Asian, and Indonesian sources.[44] The greater the number of parental populations and the more ancient the admixture event, the more polymorphisms will be needed to distinguish the ancestry profiles in MALD. Extension of the existing AIM sets is likely to result from the application of new sequencing technologies, for example in the **1000 Genomes Project** (Box 3.2). With current high-throughput genotyping methods it is feasible to genotype cost-effectively sufficient numbers of SNPs across the genome to allow the recovery of even short-range haplotypes for admixture mapping.

A number of genes have been successfully mapped using the MALD approach. These include variation in the *DARC* gene promoter associated with white blood cell counts. The FY^*B^{ES} allele is almost fixed in sub-Saharan Africans while its frequency is close to zero in Europe. Without the European-African admixture in the Americas, the mapping of the locus to a 30 cM region on chromosome 1 would have been challenging when it was achieved[20] in 2000. While MALD is most effective in mapping diseases involving a single gene or a few genes, it has also proven successful in mapping candidate regions for some complex disorders, for example prostate cancer. The lack of MALD success so far in mapping genes associated with hypertension and aggressive breast cancer in African-Americans, despite the existence of significant disease prevalence differences among Africans and Europeans, has been attributed to the involvement of many genes with minor impact and/or the effect of nongenetic factors.[44]

Admixture dating

As we saw in the preceding sections, admixture events leave behind a specific imprint on the genomes of the hybrid population in the form of long-range LD patterns. Because the break-up of LD is time-dependent, it is possible to infer with some accuracy the date of admixture from the lengths of the "migrant tracts," that is, chromosome segments with recent migrant ancestry.[27] Several methods have been developed to further examine the relation between time since admixture and the extent of LD decay, while also considering the confounding factor of the background LD that existed in the two parental groups before the admixture. One such method, ROLLOFF, assesses the decline of LD between SNPs at distances greater than 0.5 cM.[23] It examines the decline of correlation between allele frequency differences and LD in specified parental populations for pairs of polymorphisms at increasing genetic distances. The date estimates are obtained by fitting the exponential distribution to these correlations and solving it as a function of the number of generations and genetic distance.[23] Using this method, the 1–3% of African admixture observed throughout southern Europe was dated to around 55 generations, consistent with the historically attested African slave trade practiced by the Roman Empire. Slightly older, 72-generations-old African admixture was detected by the same method among eight Jewish groups.[23] Computer simulations suggest that the method is accurate up to a time-depth of 300 generations (~9 KY) in cases of single and discrete events of admixture. Other methods have been developed to estimate the time since admixture, such as the sliding window wavelet decomposition method for assessing recombination breakpoints that have occurred on chromosomes after admixture.[29] These methods appear to give accurate dates of admixture up to a few hundred generations, after which the LD signal becomes inaccurate for dating. When gene flow between populations occurs continuously over time or in multiple pulses, these methods are likely to yield date estimates that are more recent than the actual initiation of admixture.

14.5 SEX-BIASED ADMIXTURE

What is sex-biased admixture?

A phenomenon known as **sex-biased admixture** refers to cases where sex-specific loci give different estimates of admixture proportions. Males or females may contribute disproportionate amounts of admixture. In the extreme case, admixture may be restricted to one sex only, so-called **sex-specific admixture**. Sex-biased admixture can result from a sex bias in the makeup of one of the ancestral populations. There are clear examples of this from recent colonial admixture. The European explorers, traders, and missionaries who traveled the world over the past 500 years were predominantly male. As a consequence, admixed populations resulting from these contacts are likely to exhibit male-biased admixture. We saw an unambiguous example of this in Chapter 13, where the paternally inherited Y chromosomes of Cook Islanders in the Pacific are one-third European, but no European admixture appears among their maternally inherited mitochondrial DNA.[16] Sex-biased admixture is not restricted to the scenario of significant sex biases in one ancestral group as outlined above; it may also result from admixture between ancestral populations, neither of which is sex-biased (see **Figure 14.15**). As emphasized in previous sections, we cannot divorce genetic admixture from the wider social context in which it occurs. Ancestral populations rarely have equal status when they encounter one another for the first time. The colonial situation illustrates this clearly. In addition, human populations rarely exhibit random mating, especially across perceived "racial" or socioeconomic boundaries. Such boundaries are often more permeable to one sex than the other, an imbalance that is sometimes dependent on whether the individual is mating "above" or "below" themselves in social status terms. For example, in the Indian caste system, it is easier for a woman to marry a man from a higher caste than vice versa. Directional mating is also apparent in Western societies. In England, the frequency of marriages between white females and African-Caribbean males is greater than that between African-Caribbean females and white males.[21] The social treatment of mixed marriages is an important factor: how are offspring from unions across

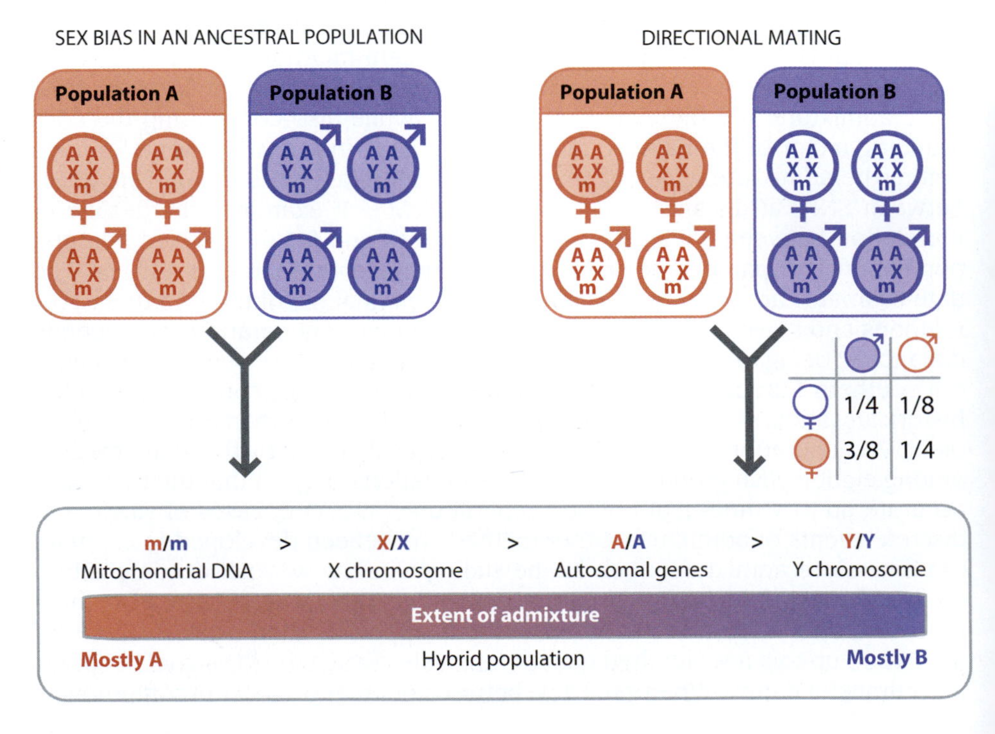

Figure 14.15: Scenarios under which sex-biased admixture may occur. Two different scenarios that lead to sex-biased admixture are shown. Each individual is represented by a pair of autosomes (AA), a pair of sex chromosomes (XX or YX), and a mitochondrial genome (*circle*) colored according to their population affiliation. In each scenario two ancestral populations (A and B) contribute to the admixed population. In the first scenario, population B consists only of males and so contributes no mitochondrial genomes, many Y chromosomes, and intermediate levels of autosomes and X chromosomes to the admixed population. In the second scenario, mating of males from population B with females from population A represents three-quarters of the mixed matings that contribute to the admixed population, and outweighs matings of females from population B with males from population A, leading to a sex bias in the contributions to the admixed population.

status boundaries incorporated into the population? Mixed unions can be stigmatized and therefore incorporated into the lower status group, or ostracized from all groups. All these factors potentially skew the contributions of males and females to all admixed populations. Thus sex-biased admixture may be a relatively common feature of admixture, but how can it be detected?

Detecting sex-biased admixture

Sex-biased admixture will cause admixture estimates from loci with different patterns of inheritance to differ markedly. Thus, admixture estimates are not pooled from all loci, but are compared between loci with different patterns of inheritance. There are four different modes of inheritance in the human genome:

- Exclusively maternal inheritance—mitochondrial DNA
- Exclusively paternal inheritance—Y chromosome
- 2:1 female-biased inheritance—X chromosome
- Equal, **biparental inheritance**—autosomes

If females contributed more than males to an admixed population, estimates of admixture would lie on a gradient (Figure 14.15):

mitochondrial DNA > X chromosome > autosomes > Y chromosome

By contrast, if males contributed a greater proportion, the gradient would be reversed:

Y chromosome > autosomes > X chromosome > mitochondrial DNA

In principle, comparisons between any two of these types of loci can reveal sex-biased admixture. However, for historical and technical reasons, the most common comparisons made in practice have been between either mtDNA and autosomal estimates, or mtDNA and Y-chromosomal estimates. Given the often substantial variance around admixture estimates described above, it makes sense to compare loci at which the greatest differences in admixture estimates should be expected. It is important to take drift into account when comparing mtDNA and Y-chromosomal admixture estimates, which, due to the small effective population sizes of these loci, are prone to **stochastic** fluctuations.

As with all admixture studies, an additional factor to be considered when contrasting admixture estimates from different loci is the allele frequency difference between ancestral populations; the larger this difference, the more power admixture estimation has to detect sex bias. We saw in Chapter 10 that the level of population differentiation differs between loci with different inheritance patterns, with Y-chromosomal polymorphisms exhibiting by far the highest levels of differentiation. This means that, on average, Y-chromosomal polymorphisms will exhibit the greatest difference in frequency between ancestral populations. This makes the inclusion of these polymorphisms particularly attractive for studying sex-biased admixture.

Care must always be taken when comparing loci with different patterns of inheritance because the polymorphisms being analyzed often have different mutational dynamics. Any differences between them may result from the mutation-rate differences rather than the inheritance differences. For example, comparisons are commonly made between mtDNA sequences and autosomal microsatellites (an example is Seielstad et al.[35]), which by virtue of dissimilar mutation dynamics may be more unreliable than comparisons of more similar loci.

Sex-biased admixture resulting from directional mating

A more complex pattern of admixture than the simple model described above is revealed when examining the origins of South American populations. Three major ancestral groups have contributed to modern genetic diversity on this continent: Native Americans, European colonists, and African slaves. Contributions occurred at different times and were from populations of different

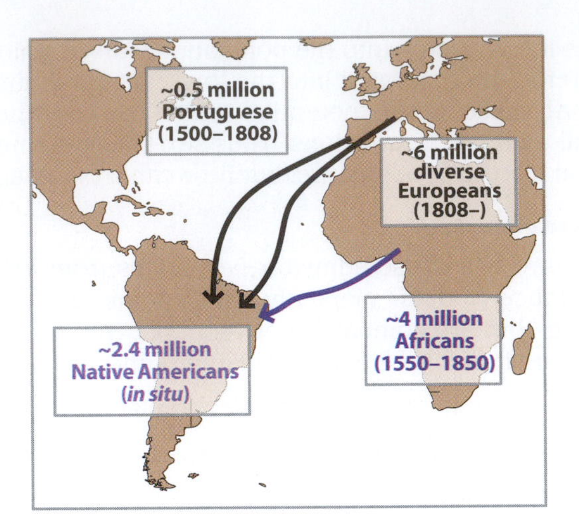

Figure 14.16: Map of genetic sources of Brazilian populations.

sizes. For example, the Native American population of Brazil at the time of its "discovery" by the Portuguese in ~1500 AD was thought to number ~2.4 million. Some half a million European colonists, predominantly male Portuguese, had arrived by 1808. The next two centuries saw the arrival of ~6 million diverse settlers, of which 70% came from Portugal and Italy (other sources included Spain, Germany, Syria, Lebanon, and Japan). Meanwhile, in the 300 years between the mid-sixteenth and nineteenth centuries, some 4 million African slaves were imported into the country. These movements are summarized in **Figure 14.16**.

The present-day populations of Brazil, and the rest of South America, are far from homogeneous. There is substantial population structure, with groups tracing predominant ancestry to different source populations, each with a very different socioeconomic status. A predominantly white middle class tends to occupy a position within society "above" groups with more apparent ancestry from Native Americans or Africans, or both. However, levels of segregation differ greatly between different South American countries. Genetic studies reveal a striking picture of sex-biased admixture in all groups, with directional mating between European males and Native American and African females. Having said that, the male-biased demography of the earliest settlers is also likely to have played a role in establishing the sex-biased admixture among modern populations.

Table 14.3 gives admixture estimates from Y-chromosomal, mitochondrial, and autosomal loci in South and North American populations: an Afro-Uruguayan population claiming predominantly African ancestry, a countrywide Brazilian "white" population, and average countrywide estimates for Colombian and Argentinian populations and for individuals describing themselves as Latinos from the USA. Admixture estimates in these populations were obtained by various methods, including the gene identity method, allele counting, the weighted least-squares approach, STRUCTURE-like analyses, and summary assignments of mtDNA and Y chromosome by their continental affiliations.

Regardless of the different methods being used, a number of general conclusions can be drawn from these results:

- All populations show some level of ancestry from at least three continental sources.
- As expected, autosomal values for admixture lie between mitochondrial and Y-chromosomal estimates.
- European ancestry is found to be consistently greater among Y chromosomes than among mtDNAs.
- African and Native American ancestry is greater among mtDNAs than among Y chromosomes.

Thus all the Latin American groups exhibit the same pattern of directional mating, despite their different socioeconomic status within society. Studies of African-American populations from the USA also demonstrate that European admixture is biased toward males. More surprisingly, low levels of female-biased Native American admixture have been identified in these populations, revealing them to have three ancestral populations, rather than simply being a European-African mix. The events of the past 500 years have clearly had a much more detrimental effect on the frequency of Native American paternal lineages than upon the frequency of their maternal lineages.

The effect of admixture on our genealogical ancestry

As we learned in previous sections, when populations meet, the common outcome is that they admix. Because even low levels of gene flow over generations can lead to substantial gene exchange, it is likely that two individuals even from distant corners of the world can share some genetic variants by descent because of shared relatives within the past few generations. But exactly how closely related are we? Theoretical predictions are that in a panmictic population the individual genealogies coalesce to at least one common ancestor $T \approx \log_2 N$ generations ago. Assuming that the historical population size of Yorkshire and County Durham in the UK was between 1 million and 10 million, everyone from this region is expected to share a common ancestor with Kate Middleton (now Duchess of Cambridge) within the last 20–23 generations (~600–700 years) because her paternal and maternal ancestors were from that area. By the same formula, anyone having relatives in Britain is expected to be related to everyone in Britain, including the British Royal Family, via at least a single connection within the last 26 generations (~800 years). But the human species is certainly not a panmictic population: before the era of air and rail travel people were more sedentary and the proportion of international and intercontinental marriages was low. A computer simulation study attempting to model past gene flow in the world through a complex network of intra- and intercontinental migration rates inferred from historical and archaeological evidence estimated that the most recent common ancestor of all humans, considering all individual genealogies, may have lived as recently as 1415 BC—only 114 generations ago.[33] This estimate suggests that for any individual living today anywhere in the world, including the Tasmanians who were physically isolated from the rest of the world for ~12 KY but experienced European admixture over the past six generations, there would be a common ancestor only a couple of thousand years ago, shared with everyone else in the world. On the other hand, how much genetic information have we inherited from such a distant common ancestor? The answer is not much: any one particular ancestor who lived 20 generations ago was one out of 2^{20} ancestors, which means that he or she has passed down to us any particular gene with a probability less than one in a million. Given the limited number of recombination events over this short time period there is also a substantial variance in the amount of genome contributed by the set of ancestors. It is quite likely that even relatively recent ancestors can have left no genomic trace in an individual genome.

14.6 TRANSNATIONAL ISOLATES

When populations meet, the extent of admixture measurable from allele frequency patterns can vary quite substantially. In contrast to the populations that were the focus of the above sections, **isolated populations** are those that by virtue of their geography, history, and/or culture have experienced little gene flow with surrounding populations. Isolation due to geographic barriers has led some island populations from the Indian Ocean, for example the Andaman Islanders, to develop and maintain their unique allele frequency and phenotypic characteristics.[31, 40] In the case of mainland isolates, the term "isolation" is relative. No threshold of per-generation gene flow has been set that defines a population isolate; rather, a population is isolated when its surrounding populations more readily exchange genes with one another than with the isolate. This isolation can be revealed by unusual allele frequencies within the population, compared with surrounding populations, and is often associated with linguistic and geographical boundaries. For example, Basques and Finns are often regarded as population isolates within Europe. As a result of their isolation, population isolates often have unusually high frequencies of some typically rare genetic diseases. The Finnish genetic **disease heritage** is discussed in greater detail in Chapter 16.

Population isolates are commonly restricted to defined geographical regions. Luba Kalaydjieva has coined the term **transnational isolates** to refer to groups which, despite a widespread geographical distribution, remain isolated, largely through the social practice of **endogamy**. While the word was first used to describe the particular situation of European Roma, here it is extended to other groups. As with traditionally defined population isolates, the genetic coherence of transnational isolates is readily apparent in their common disease heritage.

The paradox between the genetic coherence and geographical dispersal of these transnational isolates could result from one of two processes: either coherence is actively maintained through mating over large distances, but within the group, or coherence results from a recent migration from a common point of origin, but is decaying over time. As we shall see, the latter process is the more frequent explanation.

Roma and Jews are examples of widely spread transnational isolates

European Roma

European Roma, often called Gypsies, represent a population of about 8 million spread over the European continent. They are found in highest concentrations in southeastern Europe and the Iberian Peninsula. Historical records indicate they entered Europe about 1000 YA, gradually spreading across the continent from the southeast (see Figure 14.17). The Roma speak a variety of Romani dialects although some have adopted the languages of surrounding populations. The linguistic affinities of Romani languages indicate an origin somewhere on the Indian subcontinent.

The social structure of the Roma is orientated around small, endogamous groups, often associated with a specific trade and religion. However, these religions differ greatly between groups. Islam, Roman Catholicism, Protestantism, and the Eastern Orthodox Church are all represented among European Roma.

A wide variety of Mendelian disorders are found among the Roma, such as Lom type of motor and sensory neuropathy (OMIM 601455) and spinal muscular atrophy (OMIM 253300). These disorders are characterized by homogeneity, with single mutations underlying most cases. These mutations each tend to exist on a single **haplotypic background**, thus indicating a single common and recent origin. Some of these disorders are common in other European populations, while others are specific to the Roma. Some of the Roma-specific mutations are found only in certain groups, whereas others are spread across all European Roma.[17] Thus, while there is an obvious **founder effect** resulting from a common origin, there is also substantial heterogeneity between groups resulting in considerable internal diversity. Genetic distances between Roma groups both within and between countries are typically larger than those between the surrounding European populations. Thus individual Roma groups can be thought of as isolates within a larger isolate. Three processes could cause this population differentiation:

- High levels of drift due to endogamous practices in small populations
- Different levels of admixture between Roma groups and the surrounding populations
- Original substructure in the ancestral Roma population, maintained over time

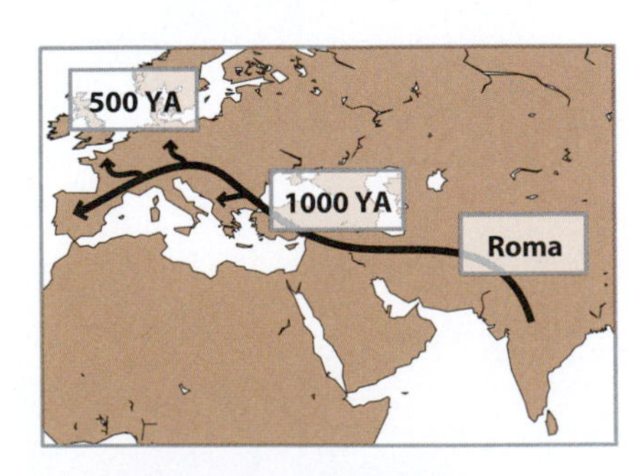

Figure 14.17: The Roma diaspora. The recent Indian origins of the Roma are supported by historical, linguistic, and genetic evidence. Approximate dates derive from historical records.

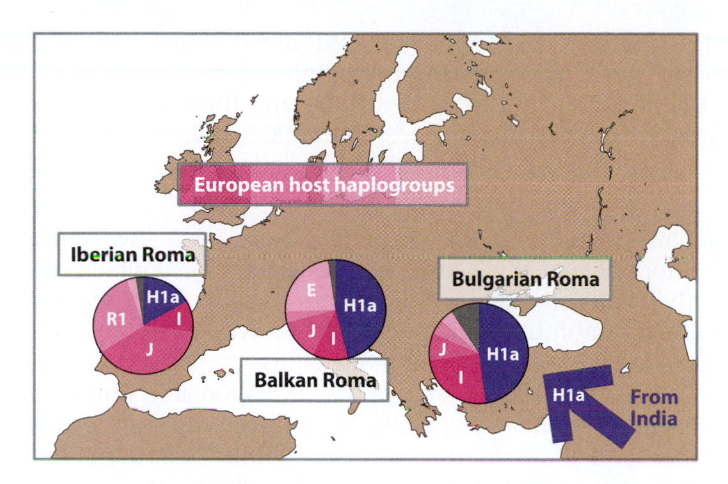

Figure 14.18: Frequencies of Indian and European-specific Y-chromosomal lineages among different Roma groups. [Data from Gresham D et al. (2001) *Am. J. Hum. Genet.* 69, 1314; Gusmao A et al. (2008) *Ann. Hum. Genet.* 72, 215; Regueiro M et al. (2011) *Am. J. Phys. Anthropol.* 144, 80.]

As a consequence of these issues, population genetic studies have tended to focus on similarities between the different Roma groups and their admixture with surrounding European populations after dispersal from their common source. Genetic evidence from classical, Y-chromosomal, and mtDNA studies has shown that the Roma share alleles and lineages with populations on the Indian subcontinent that are not found in other European populations. Nevertheless, it has not been possible to identify a single likely ancestral population. This inability severely hampers our ability to perform the kinds of quantitative admixture analyses examined previously in this chapter.

However, by observing the frequencies of European-specific Y-chromosomal lineages among different Roma populations, some conclusions can be drawn about the nature of the admixture (**Figure 14.18**). First, the degree of admixture is highly variable between different populations, while showing a general decline of the otherwise Indian-specific haplogroup H1a frequency from southeast to western Europe. Second, the admixed lineages reflect the lineage distributions within surrounding populations. It can therefore be inferred that multiple independent admixture events have occurred in the different populations, and that admixture has played a significant role in population differentiation among the different Roma groups.

The Jews

A common religion, language, and traditions unite the Jewish people. Historical and linguistic evidence attests to their Bronze Age origins in the Middle East. The ~14 million modern Jews reside mostly in the USA (~6 million) and Israel (~5 million). Despite the maternal inheritance of "Jewishness" prescribed by religious law, the practice of endogamy ensures a degree of both paternal and maternal genetic continuity; the Jewish religion does not seek converts with the same enthusiasm as some other faiths. Jews are generally classified into three groups on the basis of their ancestral migrations (see **Figure 14.19**).

Ashkenazi Jews ("Ashkenaz" is the medieval Hebrew name for the land around the Rhine Valley) had migrated from the Middle East into Central Europe during the early Middle Ages and subsequently moved within northern Europe, often attempting to avoid persecution. During the nineteenth and twentieth centuries many Ashkenazi Jews left Europe for the Americas, Australia, and South Africa, and as a consequence they now make up ~90% of the US Jewish population.

Sephardic Jews ("Sepharad" in Hebrew meaning "Spanish") had resided in the Iberian Peninsula for centuries prior to being persecuted by the Spanish Inquisition during the fifteenth century. This led to their dispersal to mainly Mediterranean countries (Italy, the Balkans, North Africa, Turkey, and Lebanon), Syria, and the Americas.

Middle-Eastern (Oriental) Jews remained in the Levant (lands on the Eastern edge of the Mediterranean Sea) and surrounding countries (Iran, Iraq, and the Arabian Peninsula).

Based on the analyses of genomewide SNP allele frequencies and autosomal population structure, most Jewish populations carry a composite of ancestry components characteristic of the Near East and their historical host populations (Figure 14.19) while in a fine-resolution PCA all Jewish groups, except for Ethiopian and Indian Jews (additional small groups not listed above), formed a tight cluster overlapping with the Druze and Cypriot populations.[4]

The Mendelian disorders of Jewish people have been investigated in great depth. A plethora of genetic diseases have been identified which typically result from one or two common founder mutations. As with the Roma, some of these mutations are common to many Jewish groups whereas others are specific to certain populations. For example, Tay-Sachs disease (OMIM 272800) is prominent only among Ashkenazi Jews where the responsible mutation reaches a heterozygote carrier frequency of 1/25, whereas the carrier frequency of alleles responsible for familial Mediterranean fever (OMIM 249100) is between 1/10 and 1/5 in populations from all three major Jewish groups. The likelihood that an Ashkenazi Jew is a carrier for one of the eight most common disease alleles is 1/4. It has been suggested that either the common mutations derive from mutational events at different times during the Jewish diaspora, or that genetic drift has resulted in the loss of the disease alleles from some populations, perhaps as a result of founder effects. The dating of different disease alleles to different times supports the former explanation.[24] However, as an additional complication, some disease alleles appear to have been recent introductions via admixture, as they are common in surrounding populations, but not in other Jewish groups. The utility of population isolates in identifying disease-related genes is explored in greater detail in **Section 16.2**.

Many haplotypes have been identified that are shared by all three Jewish populations but not with their surrounding populations. This provides considerable

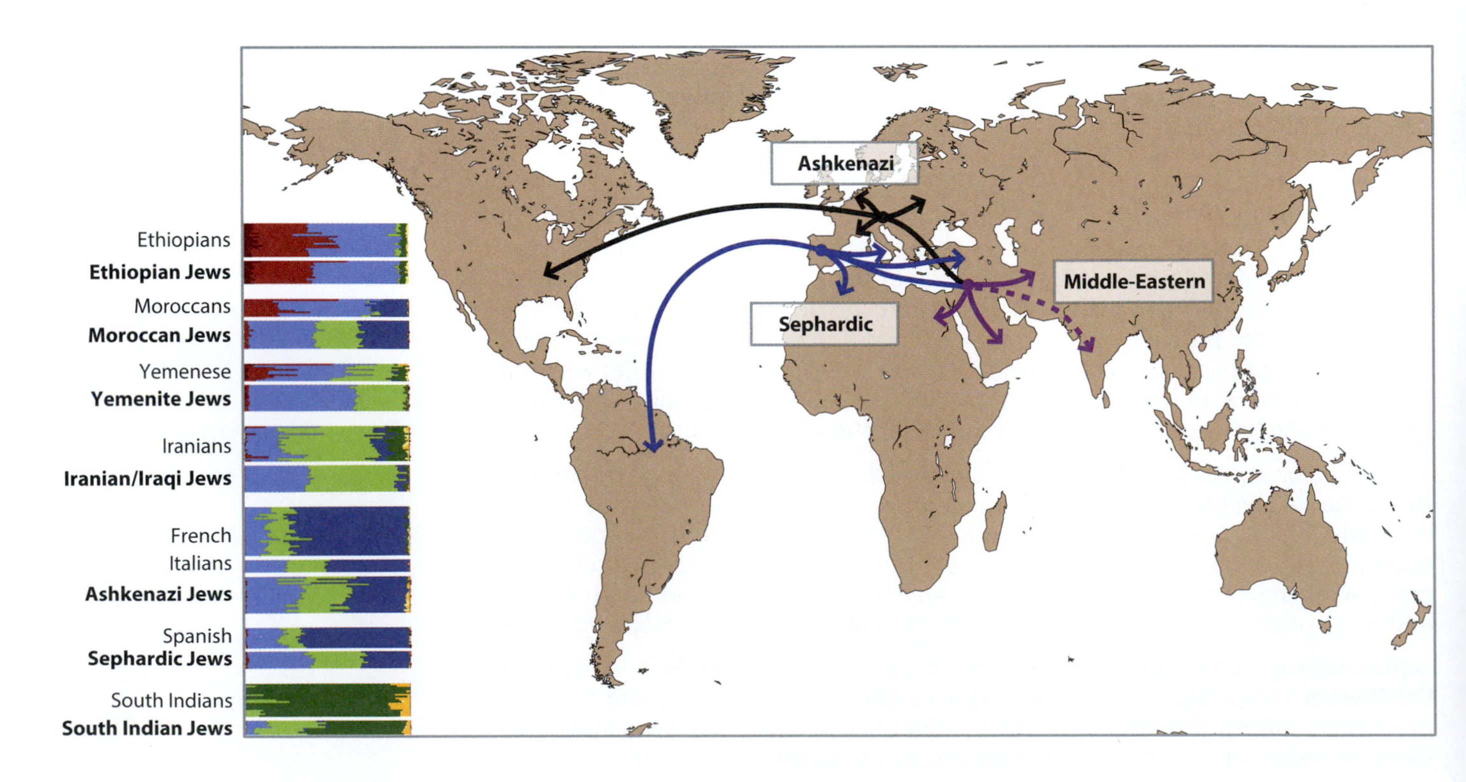

Figure 14.19: The Jewish diaspora.
The tripartite division of Jewish peoples based on their history of migrations is indicated by arrows. Population structure inferred with the ADMIXTURE program at $K = 8$ is shown at the bottom-left corner of the figure. *Light blue* and *light green* colored components, characteristic to Near Eastern populations, can be found, to various degrees, being detected in most Jewish groups. [Adapted from Behar DM et al. (2010) *Nature* 466, 238. With permission from Macmillan Publishers Ltd.]

TABLE 14.5:
GENETIC DIVERSITY AMONG DIVERSE JEWISH POPULATIONS AND THEIR HOST, NON-JEWISH, POPULATIONS

Jewish population	Host population	Less genetic diversity in Jewish compared with host population?	
		Y-chromosomal	Mitochondrial
Ashkenazi	German	no	yes
Moroccan	Berber	yes	yes
Iraqi	Syrian	no	yes
Georgian	Georgian	yes	yes
Bukharan	Uzbekistani	yes	yes
Yemeni	Yemeni	no	yes
Ethiopian	Ethiopian	no	yes
Indian	Hindu	no	yes

Data from Thomas MG et al. (2002) *Am. J. Hum. Genet.* 70, 1411.

support for the common origin of these groups and the partial maintenance of their genetic integrity through endogamous practices over the past 2 KY. One specific Y-chromosomal haplotype that appears to have been co-inherited with the paternally inherited Cohanim priesthood[13, 41] is discussed in Chapter 18.

For calculating admixture estimates for Jewish populations, one ancestral population, that of the Middle East, can be assigned with relative confidence. Given the different migratory histories of Ashkenazi and Sephardic Jews, other genetic contributions could come from a number of other ancestral host populations. The maternal inheritance of "Jewishness" suggests that admixture might be lower in maternal lineages than in paternal lineages, and thus it is of interest to compare admixture estimates for Y-chromosomal and mtDNA polymorphisms. A comparison of genetic diversity among diverse Jewish and non-Jewish host populations showed that whereas Y-chromosomal diversity was not generally lower than that in the host population, mtDNA diversity was frequently significantly lower in Jewish populations[42] (see **Table 14.5**). This suggests either (1) that founder effects and other causes of genetic drift were stronger in maternal lineages; or (2) that admixture has been consistently male-biased, resulting in the introduction of paternal lineage diversity. As we saw in **Section 10.2**, higher drift in maternal than in paternal lineages is an unusual finding; globally, the Y chromosome shows the greatest genetic differentiation of all loci. Ashkenazi Jews, unlike Sephardic or Oriental Jews, are indeed characterized by severe founder effect(s) as evidenced by their characteristically high frequency of mtDNA haplogroup K.[2, 3]

SUMMARY

- Genetic admixture is the process by which a hybrid population is formed from contributions by two or more parental, or ancestral, populations. It is likely that every human population has been influenced by admixture, and many important issues in recent human evolution can be considered to be questions of admixture.
- Past admixture events can be identified in the historical, linguistic, and archaeological records as well as through their impact on patterns of genetic diversity.

- The genetic contribution of an ancestral population to the hybrid population can be estimated by considering the frequencies of a given allele in the hybrid population and both ancestral populations. Admixture estimates require the correct identification of the ancestral populations and are most accurate when an allele is present at very different frequencies in the two ancestral populations.
- A number of different methods for estimating admixture proportions have been devised to best combine information from multiple alleles into a single estimate, and that take into account some of the confounding factors, such as genetic drift in both ancestral and hybrid populations since the admixture event.
- In an admixed population, individual genomes are themselves admixed to a greater or lesser degree. By typing many polymorphisms, levels of individual admixture can be estimated and compared between individuals and regions of the genome.
- Through admixture mapping, different genes and genomic regions can be studied to reveal associations with phenotypes and disease.
- Admixture results in elevated levels of linkage disequilibrium, which decays over time. This relationship between the extent of LD and time can be used for dating admixture events.
- Under a number of different admixture scenarios, the contributions of males and females from ancestral populations may not be equal. This sex-biased admixture reveals itself in discrepant admixture estimates from loci with different patterns of inheritance (that is, mtDNA, Y chromosomes, X chromosomes, and autosomes). Sex-biased admixture is commonly observed in the formation of admixed populations, for example, as a result of the Atlantic slave trade.
- Admixture cannot be divorced from its social context. Social practices such as endogamy restrict admixture with other populations, and the relative socioeconomic status of the different ancestral populations can lead to directional mating between males and females of the two groups.
- Populations that live and breed in isolation due to geographic, cultural, or socioeconomic barriers experience less admixture than others. Transnational isolates (for example, the Roma and the Jews) are populations that maintain genetic coherence over vast geographical distances as a result of recent dispersal from a common origin and endogamous mating practices that restrict admixture.

QUESTIONS

Question 14–1: Consider a situation where a group of individuals from population A moves into the territory of population B with much larger population size and remains relatively isolated there. Each generation 95% of the individuals in the isolated group will mate with each other and only 5% will mate with individuals from population B and remain in the isolated community. What will be the expected ancestry proportions of A and B in that deme after (a) 10 generations, and (b) 100 generations?

Question 14–2: Smokers with A/A genotype (rs762551) in the *CYP1A2* gene metabolize caffeine 1.6 times faster than other genotypes. The A allele frequency is 70% in Europe and 50% in India. You collect samples from second-generation Indians from a number of European cities and determine their A-allele frequency to be 52% on average. What would be your estimate of admixture? If, instead, the estimated allele frequency in your sample of European Indians was 48%, how would you explain the result?

Question 14–3: What are the key intrinsic and technology-related limitations to admixture inference and dating?

Question 14–4: Discuss the benefits and limitations of admixture mapping.

Question 14–5: Using the HapMap browser (http://hapmap.ncbi.nlm.nih.gov/ choose Phase 1, 2, and 3 data merged) and the formula for determining admixture rate M (Section 14.3) estimate the proportion of European admixture in African Americans (ASW) and Mexicans (MEX) on the basis of allele frequency data of the following two SNPs, rs1426654 and rs2227282. For parental sources consider data from the CEU, YRI, and CHB samples.

Question 14–6: Below are mtDNA and Y-chromosome haplogroup (hg) data for populations from three places: Greenland, Madagascar and the Cook Islands. Use information in the Appendix to deduce: a) which population is which, and b) whether the data suggest a history of admixture, and if so, of what nature.

	mtDNA haplogroup %										Y haplogroup %									
	A	B	E	F	L0/ L1	L2	L3	M	Q	R	B	C	E	I	J	K	L	O	Q	R
Pop. A		80							20			50		10		12				28
Pop. B	100											3	1	29	1				39	27
Pop. C		27	4	4	8	14	15	26		1	8		65		2		1	21		3

REFERENCES

The references highlighted in purple are considered to be important (for this chapter) by the authors.

1. **Alexander DH, Novembre J & Lange K** (2009) Fast model-based estimation of ancestry in unrelated individuals. *Genome Res.* **19**, 1655–1664.
2. **Behar DM, Metspalu E, Kivisild T et al.** (2006) The matrilineal ancestry of Ashkenazi Jewry: portrait of a recent founder event. *Am. J. Hum. Genet.* **78**, 487–497.
3. **Behar DM, Metspalu E, Kivisild T et al.** (2008) Counting the founders: the matrilineal genetic ancestry of the Jewish Diaspora. *PLoS One* **3**, e2062.
4. **Behar DM, Yunusbayev B, Metspalu M et al.** (2010) The genome-wide structure of the Jewish people. *Nature* **466**, 238–242.
5. **Bernstein F** (1931) Comitato Italiano per o Studio dei Problemi Della Populazione, pp 227–243. Roma: Instituto Poligrafico dello Stato.
6. **Cavalli-Sforza LL & Bodmer WF** (1971) The Genetics of Human Populations, pp 227–243. W.H. Freeman.
7. **Chakraborty R** (1986) Gene admixture in human populations: models and predictions. *Yearbk Phys. Anthropol.* **29**, 1–43.
8. **Chakraborty R & Weiss KM** (1988) Admixture as a tool for finding linked genes and detecting that difference from allelic association between loci. *Proc. Natl Acad. Sci. USA* **85**, 9119–9123.
9. **Choudhry S, Coyle NE, Tang H et al.** (2006) Population stratification confounds genetic association studies among Latinos. *Hum. Genet.* **118**, 652–664.
10. **Cornuet JM, Santos F, Beaumont MA et al.** (2008) Inferring population history with DIY ABC: a user-friendly approach to approximate Bayesian computation. *Bioinformatics* **24**, 2713–2719.
11. **Crow JF** (1983) Surnames as markers of inbreeding and migration. Discussion. *Hum. Biol.* **55**, 383–397.
12. **Glass B & Li CC** (1953) The dynamics of racial intermixture; an analysis based on the American Negro. *Am. J. Hum. Genet.* **5**, 1–20.
13. **Hammer MF, Behar DM, Karafet TM et al.** (2009) Extended Y chromosome haplotypes resolve multiple and unique lineages of the Jewish priesthood. *Hum. Genet.* **126**, 707–717.
14. **Handley LJ, Manica A, Goudet J & Balloux F** (2007) Going the distance: human population genetics in a clinal world. *Trends Genet.* **23**, 432–439.
15. **Helgason A, Palsson S, Gudbjartsson DF et al.** (2008) An association between the kinship and fertility of human couples. *Science* **319**, 813–816.
16. **Hurles ME, Irven C, Nicholson J et al.** (1998) European Y-chromosomal lineages in Polynesians: a contrast to the population structure revealed by mtDNA. *Am. J. Hum. Genet.* **63**, 1793–1806.
17. **Kalaydjieva L, Gresham D & Calafell F** (2001) Genetic studies of the Roma (Gypsies): a review. *BMC Med. Genet.* **2**, 5.
18. **Kaplan B** (1954) Environment and human plasticity. *Am. Anthropologist* **56**, 781–799.
19. **Kayser M, Lao O, Saar K, et al.** (2008) Genome-wide analysis indicates more Asian than Melanesian ancestry of Polynesians. *Am. J. Hum. Genet.* **82**, 194–198.
20. **Lautenberger JA, Stephens JC, O'Brien SJ & Smith MW** (2000) Significant admixture linkage disequilibrium across 30 cM around the FY locus in African Americans. *Am. J. Hum. Genet.* **66**, 969–978.
21. **Model S & Fisher G** (2002) Unions between blacks and whites: England and the US compared. *Ethnic Racial Stud.* **25**, 728–754.
22. **Montana G & Pritchard JK** (2004) Statistical tests for admixture mapping with case-control and cases-only data. *Am. J. Hum. Genet.* **75**, 771–789.
23. **Moorjani P, Patterson N, Hirschhorn JN et al.** (2011) The history of African gene flow into Southern Europeans, Levantines, and Jews. *PLoS Genet.* **7**, e1001373.
24. **Ostrer H** (2001) A genetic profile of contemporary Jewish populations. *Nat. Rev. Genet.* **2**, 891–898.
25. **Patterson N, Hattangadi N, Lane B et al.** (2004) Methods for high-density admixture mapping of disease genes. *Am. J. Hum. Genet.* **74**, 979–1000.
26. **Patterson N, Price AL & Reich D** (2006) Population structure and eigenanalysis. *PLoS Genet.* **2**, e190.
27. **Pool JE & Nielsen R** (2009) Inference of historical changes in migration rate from the lengths of migrant tracts. *Genetics* **181**, 711–719.
28. **Pritchard JK, Stephens M & Donnelly P** (2000) Inference of population structure using multilocus genotype data. *Genetics* **155**, 945–959.
29. **Pugach I, Matveyev R, Wollstein A et al.** (2011) Dating the age of admixture via wavelet transform analysis of genome-wide data. *Genome Biol.* **12**, R19.
30. **Reed TE** (1969) Caucasian genes in American Negroes. *Science* **165**, 762–768.

31. **Reich D, Thangaraj K, Patterson N et al.** (2009) Reconstructing Indian population history. *Nature* **461**, 489–494.

32. **Roberts DF & Hiorns RW** (1962) The dynamics of racial intermixture. *Am. J. Hum. Genet.* **14**, 261–277.

33. **Rohde DL, Olson S & Chang JT** (2004) Modelling the recent common ancestry of all living humans. *Nature* **431**, 562–566.

34. **Rosenberg NA, Mahajan S, Ramachandran S et al.** (2005) Clines, clusters, and the effect of study design on the inference of human population structure. *PLoS Genet.* **1**, e70.

35. **Seielstad MT, Minch E & Cavalli-Sforza LL** (1998) Genetic evidence for a higher female migration rate in humans. *Nat. Genet.* **20**, 278–280.

36. **Serre D & Pääbo S** (2004) Evidence for gradients of human genetic diversity within and among continents. *Genome Res.* **14**, 1679–1685.

37. **Shriver MD, Smith MW, Jin L et al.** (1997) Ethnic-affiliation estimation by use of population-specific DNA markers. *Am. J. Hum. Genet.* **60**, 957–964.

38. **Stephens JC, Briscoe D & O'Brien SJ** (1994) Mapping by admixture linkage disequilibrium in human populations: limits and guidelines. *Am. J. Hum. Genet.* **55**, 809–824.

39. **Tang H, Peng J, Wang P & Risch NJ** (2005) Estimation of individual admixture: analytical and study design considerations. *Genet. Epidemiol.* **28**, 289–301.

40. **Thangaraj K, Chaubey G, Kivisild T et al.** (2005) Reconstructing the origin of Andaman Islanders. *Science* **308**, 996.

41. **Thomas MG, Skorecki K, Ben-Ami H et al.** (1998) Origins of Old Testament priests. *Nature* **394**, 138–140.

42. **Thomas MG, Weale ME, Jones AL et al.** (2002) Founding mothers of Jewish communities: geographically separated Jewish groups were independently founded by very few female ancestors. *Am. J. Hum. Genet.* **70**, 1411–1420.

43. **Weiss KM & Long JC** (2009) Non-Darwinian estimation: my ancestors, my genes' ancestors. *Genome Res.* **19**, 703–710.

44. **Winkler CA, Nelson GW & Smith MW** (2010) Admixture mapping comes of age. *Annu. Rev. Genomics Hum. Genet.* **11**, 65–89.

45. **Xu S, Huang W, Qian J & Jin L** (2008) Analysis of genomic admixture in Uyghur and its implication in mapping strategy. *Am. J. Hum. Genet.* **82**, 883–894.

SECTION 5

HOW IS AN EVOLUTIONARY PERSPECTIVE USEFUL?

Having an evolutionary perspective on human genetic diversity not only illuminates the origins of our species and the way that early humans spread across the world, but has practical applications too. It gives us an essential framework for understanding "normal" and pathogenic phenotypic variation among people, and is important in making reasonable deductions about individuals from their DNA.

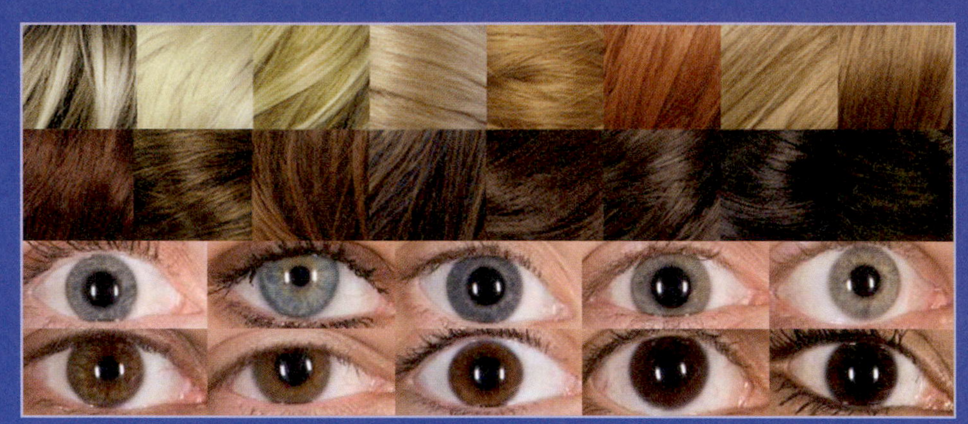

Eyes from Walsh S et al. (2011) *Forensic Sci. Int. Genet.* 5,170. With permission from Elsevier.

CHAPTER 15
UNDERSTANDING THE PAST, PRESENT, AND FUTURE OF PHENOTYPIC VARIATION

In this chapter we discuss "normal" phenotypic variation, including the complex phenotypes of morphology, pigmentation, and dietary differences. We ask to what extent these can be understood in terms of adaptation, drift, and sexual selection.

CHAPTER 16
EVOLUTIONARY INSIGHTS INTO SIMPLE GENETIC DISEASES

Many aspects of our health are influenced by our genomes, and since these are products of our history, evolutionary genetics has important implications for diagnosis and treatment. Here we examine genetic influences on the distributions and frequencies of simple genetic diseases, including the roles of demographic history and selection.

CHAPTER 17
EVOLUTION AND COMPLEX DISEASES

In this chapter we use an evolutionary perspective to illuminate the genetic influences on complex disease, including infection. We also describe the use of evolutionary information in identifying disease genes, and its importance for medical treatments, including the genetic basis for different responses to drugs.

CHAPTER 18
IDENTITY AND IDENTIFICATION

Each copy of the human genome is different, and exploiting these differences allows the identification of individuals from their DNA, in forensic analysis. This chapter explains how this is done, and also examines the use of DNA in determining family relationships and phenotypic characteristics. We also introduce the opportunities and concerns of "personal genomics," where individuals can investigate their own genomic DNA sequences.

UNDERSTANDING THE PAST, PRESENT, AND FUTURE OF PHENOTYPIC VARIATION

CHAPTER FIFTEEN

In previous chapters, we saw that humans expanded out of Africa and moved across continents, encountering new environments. In addition, humans modified their own environment, most dramatically by the development of agriculture (Chapter 12). Adaptations to such new environments can be genetic, physiological, or cultural, or any combination of these three. In this chapter we will provide a broad overview of the evidence for genetic adaptation of particular traits to different environments, and will select five examples to discuss in depth, reflecting adaptation to changes in climate, altitude, and diet. Finally, in **Section 15.7**, we will move from a consideration of the influence of our evolutionary past to a consideration of our evolutionary future. As our species continues to grow in number and appears increasingly dominated by cultural rather than biological influences, have we become insulated from the forces of natural selection? Can we use genetics to predict anything about the future or survival of our species?

So far, the emphasis of this book has been upon genetic variation that is thought to be selectively neutral, and to have little or no influence on our phenotype. Here, and in the following chapters, we change our focus, and begin an exploration of the phenotypic diversity of modern human populations from an evolutionary genetic perspective. There are three key questions that we address: To what extent have humans adapted to different environments? How much of this adaptation has been driven by natural selection? And what is the genetic basis of this adaptation?

15.1 NORMAL AND PATHOGENIC VARIATION IN AN EVOLUTIONARY CONTEXT

This chapter examines the genetic basis of normal nonpathogenic phenotypic variation, while Chapters 16 and 17 will focus on variation influencing disease, but this is not always a simple distinction to make. For example, everyone would agree that type 2 diabetes is a disease, while having red hair and fair skin is considered to be normal variation, and regarded as a trait. However, having red hair and fair skin increases the risk of developing the serious cancer melanoma,[9] and, on the other hand, having a higher than average body mass index increases the susceptibility to type 2 diabetes. In this chapter we will discuss the genetic bases of phenotypic variation among normal healthy people, but will point out how these may influence the risk of pathogenic consequences. Similarly, in Chapters 16 and 17 we will examine phenotypes that are usually detrimental, but can be neutral or advantageous under some circumstances. Some of this normal variation represents adaptations to the environment, such as climate and diet, and at the same time its expression can be environmentally influenced. This environmental influence does not remain constant, and there may be examples of adaptations that were beneficial in our ancestors that are detrimental today in modern populations (a possible example, the thrifty genotype, is discussed in Chapter 17). It is important to bear in mind that both normal and pathogenic genetic variation reflects the interaction of our genes with the wider environment around us.

Human phenotypic variation is influenced by many aspects of our evolutionary past:

- How our species originated, and the effect of this upon our genetic diversity.
- The early differentiation and spread of human populations.
- The development of agriculture, recent demographic expansions, change in diet, and close contact with animal species.
- The recent admixture of populations with different histories, and recent migrations.

Humans are a young species that has undergone recent dispersal out of Africa followed by explosive population growth beginning in the Neolithic, and, only very recently, extensive remixing by intercontinental migrations. The consequence of this history is relatively low genetic diversity, a preponderance of alleles that have arisen recently and are rare in the population, and little differentiation between populations in different places (Chapter 10). Unlike the other great apes, humans have moved into a wide range of very different environments, with different climates, altitudes, food sources, and pathogens. Selective regimes are therefore very diverse for humans.

In attempting to understand phenotypic variation in terms of our evolutionary history, uncertainty about particular questions has an important influence on our choice of explanations. For example, the debate over the extent of gene flow between pre-Neolithic continental groups (**Section 9.3**) leads to uncertainty over whether those of our phenotypic characteristics that do show strong geographical differentiation are of ancient or more recent origin. If there had been little gene flow, an ancient origin is more likely, but if gene flow had been greater, this would favor a more recent origin of regional phenotypic differentiation.

15.2 KNOWN VARIATION IN HUMAN PHENOTYPES

What is known about human phenotypic variation?

Before we begin a discussion of phenotypic variation we need to distinguish between a number of different underlying mechanisms. First, there are physiological mechanisms representing short-term and sometimes rapidly reversible responses to environmental change, known as **acclimatization**. Examples are tanning, where skin pigmentation increases as a response to increased sun exposure, and the increased production of hemoglobin in response to reduced oxygen concentrations at high altitude. Although there are genetic influences acting on the efficiency of these responses, we will not consider acclimatization further here. Second, there is developmental plasticity, in which the environment influences the long-term development of an individual. Examples are the possible influence of changed environment on the proportions of the skull (see later in this section); the possible influence of poor nutrition on aspects of metabolism (**Section 17.2**); and some adaptations to high altitude (**Section 15.4**). Again, these mechanisms are expected to have genetic influences, though any such links are poorly understood. Finally, there are genetic mechanisms directly responsible for phenotypic variation, which are the primary subject of this chapter. The distribution of alleles underlying these phenotypes is governed by the principles outlined in Chapter 5: mutation, genetic drift, founder effects, gene flow between populations, and adaptation through natural selection. Until recently, studies of the basis of human phenotypic variation were almost exclusively the preserve of biological anthropologists. Major efforts went into describing the patterns of variation among human populations in pigmentation, morphology, physiology, behavior, and life history. Hypotheses have been suggested for the adaptive (or non-adaptive) significance of many different phenotypes. Some of these will be described below, and later we will focus on what evidence is available from genetic studies.

As has been pointed out already, humans inhabit an unusually broad range of environments. Adaptive responses to similar environments are likely to result in similar phenotypes: an often-used example is dark skin in Africans, South Indians, Australian Aborigines, and Melanesians. Because these physical characteristics are likely to be responsive to the environment, the apparent similarities they present are unlikely to be representative of the rest of the genome, and we might expect them to be poor indicators of population affinity. However, arguments persist that particular shared phenotypes indicate some relatedness; this is connected to the issue, discussed above, of uncertainty about the extent of gene flow between populations before the Neolithic.

What kinds of normal phenotypic variation have been studied? One of the most obvious differences between populations is pigmentation of skin, hair, and eyes; another striking difference is in digestion, in particular the degree to which lactose (a sugar from milk) is tolerated in adult life. Because genetics has contributed heavily to these two areas, they will form the basis of two subsequent sections of this chapter (**Sections 15.3 and 15.6**). Human adult height varies between populations and between individuals, and the genetic basis of this variation has been well studied (**Box 15.1**). However, here we will consider briefly some of the other differences studied by physical anthropologists, most of which have been little investigated from a genetic point of view.

Morphology and temperature adaptation

Body size and proportions vary greatly among human populations, and this is generally interpreted as an adaptive response to different thermal environments. Studies of other mammals, and also of birds, have led to the recognition of a general relationship between climate and morphology—body size increases with distance from the equator. Physical anthropologists refer to two rules that describe the relationships between body morphology and climate; both are essentially renderings of the same simple physical principle. **Bergmann's rule** states that body size increases as climate becomes colder, because as mass increases, surface area does not increase proportionately. Heat is lost at the surface, so increased body mass means better heat retention. **Allen's rule** tells us that shorter appendages (limbs) are favored in colder climates because they have higher ratios of mass to surface area, and therefore retain heat. Conversely, longer appendages are favored in hotter climates. According to these principles, people are best adapted to hot climates when they have a thin body shape with long limbs, and best adapted to cold climates when they are stocky and short-limbed. Studies on many populations generally confirm this, with East African pastoralists and Inuit from the Arctic often being used as typical examples[68] (**Figure 15.1**). However, not all populations conform to the rules. The small body size of African Pygmies is sometimes explained as an adaptation to the hot climate of central Africa, but there is no direct evidence to support this. Alternatively, it could be an adaptation to moving through dense vegetation, or a result of sexual selection, or selection for earlier reproduction.

Facial features

Differences between populations have long been used as part of the traditional descriptions of races.[33] Degree of flatness of the face, shapes of ears, nose, and lips, and the distribution and texture of hair (color is considered below) vary among populations, and the genetic component of this variation is just beginning to be investigated.[49] Adaptive explanations have been put forward for some of these differences. Nose shape shows some association with climate, and it has been suggested that a narrow nose may warm and moisten air before it reaches the lungs more efficiently than a broad one. The internal skin-fold of the eyelid (**epicanthal fold**) seen in many Asian populations has been explained

Box 15.1: Genomewide association studies of adult height

In 1885 Francis Galton published one of the first studies on hereditary traits in humans, showing that the average height of the two parents is correlated with the height of the child in adulthood. As Galton acknowledged, adult height is an ideal quantitative trait (see Box 15.3) because it is easy to measure accurately, and subsequent data have shown that it has a broad-sense heritability of around 80%, meaning that four-fifths of the variation observed in height is due to genetic variation.

Initial studies using genomewide linkage to identify quantitative trait loci (QTLs) produced inconsistent results, and it was only by pooling all available data from over 6000 twins that three significant QTLs were found, each contributing only a small amount to the total genetic variation.[61] Genomewide association studies (GWASs) have been more successful, because of the smaller effects that association studies can detect, and the huge sample sizes analyzed—these studies aimed to understand the genetic basis of other, more medically important, phenotypes, and height was recorded incidentally. By 2008, there were five published studies each with around 30,000 samples, which in total identified 54 loci in which variation is associated with adult height. These loci included regions where there is a strong candidate gene that, for example, affects growth when deleted in mice; and also regions where there are no good candidate genes or no genes at all. Only about 5% of the total variation in height was explained by the common variants identified in these studies, with the most strongly associated single variant explaining just 0.3% of the total variation in height.[81] Indeed, Galton's original methods for predicting the height of a child in adulthood are more effective than methods based on the presence of certain alleles at these 54 genetic loci: [2] knowledge of both parents' heights allows 40% of the total variation in height to be explained. However, analysis of a larger sample set (184,000 people) has found a total of 180 variants explaining 10% of the total variation in height.[1] These variants have, individually, extremely small effect sizes, and so even larger studies are likely to describe even more of the observed variation in height. Recessive genetic variants are also likely to be important, since there is an inverse correlation between genomewide homozygosity and adult height.[52] At an extreme, the offspring of first cousins are predicted to be 3 cm shorter on average than the offspring of unrelated parents. This is an example of inbreeding depression (Section 5.3).

How does this new knowledge contribute to our understanding of the evolution of adult height and the variation in height that we see in different populations today? It must be remembered that broad-sense heritability defines the amount of genetic variation *in a trait in a given population at a given time*, and is critically dependent on the range of environments in that population and time. Therefore heritability is not fixed, but will vary across time and between populations. For example, we know that height is dependent on nutrition, yet the estimates for heritability are based on Western populations which are not generally subject to undernutrition and have had limited exposure to infectious disease. In historical and contemporary populations where access to and utilization of food, and exposure to infection, is less uniform, we would expect an increase in the environmental influence on adult height variation and consequent reduction in broad-sense heritability. In addition, different genes may contribute to height variation in these populations (such as genes involved in nutrition), and in differing amounts. So, for evolutionary analysis, the current crop of GWASs is useful for identifying pathways involved in the development of adult stature, but may not necessarily identify all the genes involved in the evolution of adult height in modern humans.

as a protection from cold air and snow glare.[18] However, these are **"Just So" stories**, with little or no direct evidence, and again the features do not show geographical distributions compatible with simple adaptive explanations.

Tooth morphology and cranial proportions

Complex tooth crown shape with a particular upper incisor morphology (**shoveled incisors**) is prevalent in some Asian and Native American populations (**Figure 15.2**). The geographic distribution of shoveling is highly correlated with the distribution of a SNP in the *EDAR* gene (**Section 10.3**). The shape of the cranium has been the focus of special interest from anthropologists for many years. One type of craniometric trait, the **cephalic index** (**CI**, also called cranial index), is the ratio of the breadth to the length of the skull multiplied by 100, and provides a single, if crude, figure to describe the proportions of a skull. Individuals with CIs below 75 have long, narrow heads and are termed **dolichocephalic**, while those with CIs over 80 have broad heads and are **brachycephalic**. Those with CIs between 75 and 80 are **mesocephalic**. Many data were gathered on

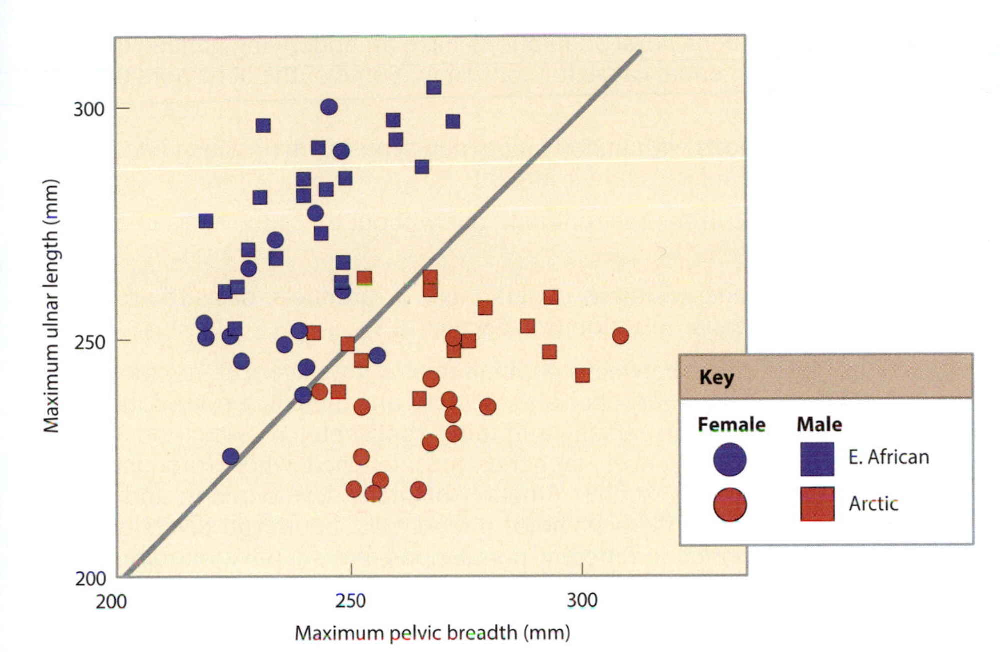

Figure 15.1: Differing body proportions in East African and Arctic populations. Maximum length of ulna (the larger of the two bones of the forearm) plotted against bi-iliac (maximum pelvic) breadth for modern East Africans and modern samples from Alaskan Inupiat and Aleut. East Africans generally have long arms and narrow pelvises, while most of the Arctic peoples have short arms and broad pelvises. [Adapted from Ruff C (2002) *Annu. Rev. Anthropol.* 31, 211. With permission from Annual Reviews.]

the CIs within different populations. However, highly influential work by Franz Boas, an American anthropologist, indicated that the CIs of children born in the USA of European immigrant parents differed from those of their European-born siblings.[8] In other words, inherited aspects of the CI were not as important as early environmental influence, and this suggested that cranial proportion was a developmentally plastic trait, and therefore not a reliable phenotype to use as a classificatory tool, either for modern or for ancient samples. Boas' conclusions were an important part of the refutation of racism. Despite 90 years of acceptance, reanalyses of Boas' original data using modern statistical methods and models from quantitative genetics, find that there is high heritability of CI.[72] Further analysis and new data suggest that, although there is some developmental plasticity in craniometric traits, they do represent underlying genetic differences,[66] and in fact several craniometric traits show a close correlation between diversity and distance from Africa, with the most diverse craniometric measurements in populations in Africa[51] (**Section 9.1**). This closely mirrors the loss of diversity seen in DNA polymorphisms with increasing distance from Africa (**Section 9.4**).

Behavioral differences

Sociobiology, and its outgrowth, evolutionary psychology, are concerned with finding explanations in our primate past and our adaptive human past for complex modern behaviors and traits. These are contentious fields, in which the unit of investigation tends to be the human species as a whole. Discovering meaningful behavioral differences between populations that may have adaptive (and therefore underlying genetic) causes is not trivial. Human behavioral ecologists have studied important aspects of behavior such as inter-birth intervals, marriage practices, and foraging strategies.[44] However, while a phenotype like pigmentation may be easy to study and quantify (though not so easy to understand—see **Section 15.3**) in any population, these behavioral studies are difficult and are focused on a small number of "traditional" populations. Modern industrialized populations are considered less suitable material, as if they are somehow insulated from the evolutionary forces that might have shaped these behaviors originally (**Section 15.7**). It does not take an anthropologist to realize that there are, for example, differences in mating strategies between populations. Humans are supremely flexible in their responses to different physical and social environments, and our chief adaptations to the environment are socially transmitted cultural ones (Box 15.5)—for example, without the wearing of warm clothes, Inuit people could certainly not survive in the Arctic. Even if

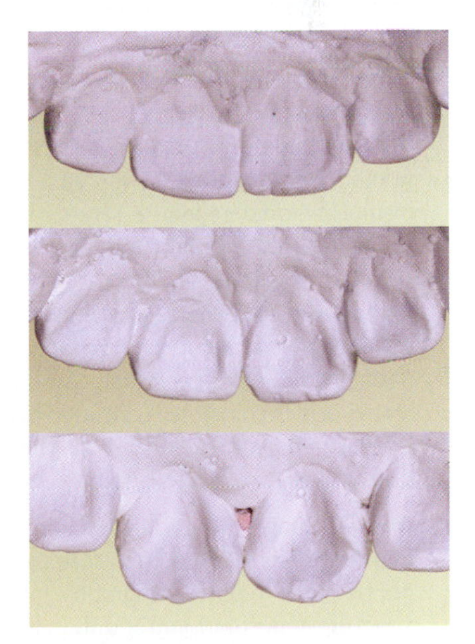

Figure 15.2: Different degrees of "shoveling" on the lingual side of human incisors. Top to bottom: increasing amount of shoveling; grade 1, grade 3, grade 5. [Adapted from Kimura R et al. (2009) *Am. J. Hum. Genet.* 85, 528. With permission from Elsevier.]

we do assume that behavioral adaptations have an underlying genetic component, there is still potential cause for confusion. Some of the behaviors that are observed could be:

- Ancient adaptations with underlying genetic causes, maintained because of continuity of the selective environment
- Adaptive to the current environment, but without the same basis in ancient adaptation
- Non-adaptive traits arising as evolutionary by-products, or as the result of forces such as sexual selection (see **Section 15.3**)

Studies of differences between populations in intelligence, as measured through **IQ tests**, have a very checkered history, and there is serious doubt over whether IQ testing is a universally and fairly applicable cross-cultural assessment of intelligence. However, as genes are identified whose products play specific roles in neurotransmitter function or brain development, and which are associated with specific behavioral phenotypes, haplotype diversity studies have been performed in different populations and adaptive interpretations have been put forward. Two controversial examples are considered in **Box 15.2** which discusses the diversity of haplotypes at the *ASPM* and *MCPH1* genes,[20, 53] genes involved in brain development. There are several other examples of candidate genes whose products may influence behavioral disorders, including monoamine oxidase A (*MAOA*), which metabolizes biogenic amines such as

Box 15.2: *ASPM* and *MCPH1*: selection at loci affecting brain size, or "Just So" stories?

Over-interpreting patterns of genetic variation to fit with poorly substantiated hypotheses, whether in studies of brain size or other phenotypes, has increased in parallel with the availability of genetic variation data. An example is the case for recent natural selection at two genes involved in brain development: Microcephalin (*MCPH1*) and Abnormal Spindle-like Microcephaly associated (*ASPM*).[20, 53] Homozygous null mutations at these two genes cause primary **microcephaly** in humans, a clinical condition which results in moderate mental retardation and a small, but normally formed, brain. Analyzing haplotype frequencies of both these genes in different populations showed that, for both genes, there was an unusually high frequency of a young, derived haplotype in European populations. These derived haplotypes carried at least one allele that altered an amino acid in the protein encoded by the gene, suggesting a possible functional effect of that haplotype, although function was not investigated.

The key phrase, to a population geneticist, is "unusually high frequency." Spelled out, this means a frequency that would not be produced by demographic forces such as population structure and expansion, and therefore likely to have been produced by natural selection. So to prove that a haplotype is at "unusually high frequency," various models of what is "usual" were made, making assumptions about human demography. The problem is that we do not know much about human demographic history so the assumptions that are made for the model define what is a "usual" frequency, and hence what is "unusual." The assumptions made in the original papers describing the findings, and in a subsequent paper refuting the interpretation of natural selection,[16] are all justified given the current knowledge of ancient human demography; but while one set of assumptions shows the haplotype frequency to be "unusual," the other set of assumptions shows it to be "usual" thus suggesting that the pattern of haplotype frequency can be explained by human demography.

The agent for any natural selection for brain size that was acting on these genes was not explicitly stated in the original papers, but the implication was that differences in intelligence, of some sort, were being selected for. Indeed, according to the Wall Street Journal, the university where the research was conducted promptly filed a patent for the use of the work as a DNA-based intelligence test. This implication, combined with the data showing that sub-Saharan Africans have the lowest frequency of the putatively selected haplotype for both genes, led to appropriation of the research by racists and considerable criticism from the scientific community. More recent results show that there is no link between variation at these genes and intelligence or general cognitive ability.[54, 76] It seems this "Just So" story is a cautionary tale against over-interpretation of genetic diversity data, a tale that should be remembered when we begin to have access to vast amounts of variation data from many different human genomes.

the neurotransmitter serotonin; tryptophan hydroxylase (*TPH*), the rate-limiting enzyme for serotonin biosynthesis; and the serotonin transporter (*SLC6A4*), important in serotonin uptake at synapses.

How do we uncover genotypes underlying phenotypes?

As should be clear from the brief discussion of the debate surrounding Boas' research, anthropologists want to be able to distinguish between an innate (genetic) basis for a characteristic, and an environmental one. This dissection of genes and environment is a central part of the study of any complex trait, and is addressed by **quantitative genetics** (**Box 15.3**). Boas asked about the environmental effect on cephalic index by comparing immigrants' offspring born in the USA with those born in the country of origin; such studies are still of use. However, a more widely used approach to separating the effects of genes and environment is twin studies, the value of which was initially pointed out by Francis Galton. Identical (**monozygotic**, **MZ**) twins have exactly the same genotype, so their many phenotypic similarities might be assumed to have a genetic basis. However, they also shared a uterus, and normally their postnatal environment too. Nonidentical (**dizygotic**, **DZ**) twins share their environments, but genetically are no more similar to each other than are ordinary siblings. A comparison of the two kinds of twins should therefore identify genetic components responsible for any greater degree of resemblance between MZ than between DZ twins.

There are some problems with a simple interpretation of twin studies. MZ twins have more similar DNA methylation patterns (**Section 2.2**) than DZ twins do, confounding **epigenetic** effects with genetic effects.[7] The assumption that the extent of environmental sharing is identical for the two twin types is not entirely valid: MZ twins are always of the same sex, while on average half of DZ twins

Box 15.3: Heritability, quantitative genetics, and complex traits

The discipline of quantitative genetics is fundamental to the identification of quantitative trait loci (QTLs) contributing to complex traits. It aims to decompose the total variance in a phenotype (V_P) into its genetic (V_G) and environmental (V_E) variance components, such that:

$$V_P = V_G + V_E$$

This makes two assumptions:

- There is no genotype by environment covariance ($V_{G,E}$), such as parents with high intelligence providing an intelligence-stimulating environment for their children.
- There is no genotype–environment interaction (V_{G*E}), where the effect of the genotype depends on the environment.

The genetic component (V_G) can be further decomposed. Genetic variance components can be:

- Additive (V_A) (due to homozygous alleles)
- Dominant (V_D) (due to heterozygous alleles)
- Epistatic (V_I) (due to interactions between genes)

Environmental variance components can be due to specific, identified elements in the environment, or to unidentified factors.

The proportion of variance in a trait that is explained by genetic factors is known as its **heritability**. Heritability is subdivided into **broad-sense heritability** (the proportion of genetic variance that can be attributed to all the genetic effects listed above, that is, V_G/V_P), and **narrow-sense heritability** (only that proportion of genetic variance attributable to additive effects, that is, V_A/V_P).[79]

The ability to identify QTLs depends on sample size, the nature of the units (for example, individuals in a population, pedigrees, or sib-pairs) being studied, and the strength of the effect of the QTL itself. QTLs can be identified using GWASs, or by linkage approaches. Traditional linkage methods are based on a specified genetic model that incorporates a defined mode of inheritance, **penetrance** (probability of manifesting the trait given possession of the allele), and frequency of the underlying allele. In some cases this model can be successfully applied to a trait—an example is the tasting/non-tasting of phenylthiocarbamide (Section 15.5). However, most complex traits are governed by many loci for which such parameters are not known, as well as by environmental influences, so the application of these traditional models is inappropriate. More sophisticated models have been developed.[67]

TABLE 15.1:
SOME FREQUENT HUMAN PHENOTYPIC TRAITS WITH A SIMPLE KNOWN GENETIC BASIS

Trait	Phenotypes	Gene	Polymorphism	Molecular evidence for selection	Variance explained %	Reference
Cerumen (earwax)	dry/wet	*ABCC11*	rs17822931	high population differentiation	100	Yoshiura K et al. (2006) *Nat. Genet.* 38, 324.
Hair thickness	small/large diameter	*EDAR*	rs3827760	high population differentiation, extended haplotype homozygosity	–	Fujimoto A et al. (2008) *Hum. Mol. Genet.* 17, 835.
Tooth morphology	degree of "shoveling" (Figure 15.2)	*EDAR*	rs3827760	high population differentiation, extended haplotype homozygosity	19	Kimura R et al. (2009) *Am. J. Hum. Genet.* 85, 528.
Androstenone smell	pleasant/offensive/no smell	*OR7D4*	rs61729907 and rs5020278	none	19–39	Keller A et al. (2007) *Nature* 449, 468.
Color vision	protanomaly	*OPN1LW* *OPN1MW*	5′ *OPN1LW*- 3′ *OPN1MW* fusion gene	none	–	Deeb S (2006) *Curr. Opin. Genet. Dev.* 16, 301.
Color vision	deuteranomaly	*OPN1LW* *OPN1MW*	5′ *OPN1MW*-3′ *OPN1LW* fusion gene	none	–	
Color vision	other fine red–green spectral tuning	*OPN1LW*	rs949431	none	–	

are brother–sister pairs, and MZ twins may be treated more similarly because they look more alike. Studies of separated MZ twins are also problematic because sample sizes tend to be very small, and because of early environmental sharing. Notwithstanding these problems, once a genetic basis for a phenotype has been confirmed, there are different approaches that can be taken to identify the genes responsible.

Twin studies, and studies of inheritance of phenotypes in families, often show a genetic basis of the phenotype in question, but only a few examples exist where genetic variation in a *single gene* causes all or a substantial part of a common phenotypic variation (see **Table 15.1** for some examples that are not discussed below). One thing seems clear—almost all quantitative phenotypes have a multigenic basis (see discussion of adult height in Box 15.1). Indeed, none of the most frequently quantified traits exhibit Mendelian inheritance, including those often used to illustrate supposedly simple Mendelian inheritance in humans, such as tongue-rolling, attached earlobes, and presence of a widow's peak (see http://udel.edu/~mcdonald/mythintro.html). This means that mapping approaches using **genetic linkage** will not usually work.

As more becomes known about the physiological or biochemical roles of particular gene products, this knowledge allows a **candidate gene** approach to be taken to understanding the genetic basis of a particular complex phenotype. The candidate gene approach has helped to delineate the biochemical pathways involved in some phenotypes, but has failed to yield many alleles with large phenotypic effects. One reason may be that many candidates have been identified as a result of the phenotypes of naturally occurring mutants or **gene knockouts** in mice, and the evolutionary distance between humans and mice may be sufficiently large for functional divergence to have occurred. A further reason is that some human genes have been chosen as candidates for affecting normal variation in a trait because mutations in these genes can give rise to specific abnormal human phenotypes. The connection between normal and abnormal variation is often not straightforward. An example of this is the limited success of using variation in genes underlying pigmentation phenotypes such as **albinism** to explain normal pigmentation variation (**Section 15.3**).

An alternative, and less biased, approach to the problem is a genomewide association study (GWAS, Box 17.2) for alleles involved in a phenotype, in which genome variation is surveyed using a large number of SNPs spread throughout the genome (see Box 15.1 for a discussion of this approach to studying human height).

What have we discovered about genotypes underlying phenotypes?

Despite the fact that biological anthropologists have gathered huge amounts of descriptive data on morphological, physiological, and behavioral variation, and formulated theories to explain its adaptive significance, we are still only beginning to uncover the molecular genetic basis of differences in these phenotypes between populations.

The phenotypes we discuss below are, at the time of writing, perhaps the most-studied examples in the search for the genetic basis for adaptation. Much remains to be discovered; in some cases the environmental change causing the adaptation is not clear, and in most cases how the variation at the genes identified actually alters the phenotype in question remains to be demonstrated, although some plausible models exist. New genotyping technologies have allowed rapid progress in gathering genetic data over the past few years, but to conclusively demonstrate the link between genotype and phenotype, time-consuming and expensive functional experiments are needed. An emerging theme running through this chapter is the concept of convergent evolution, where different populations have evolved similar phenotypes to adapt to an environmental challenge, but the alleles underlying those adaptations are different (see **Sections 15.4 and 15.6** in particular).

For other traits where the pattern of selection is less obvious, such as adult human height or obesity, GWASs are identifying hundreds of alleles that individually have tiny effects in determining the phenotype within a population (Box 15.1). Whether variation at these same loci is responsible for interpopulation variation in height remains to be seen; it has been suggested that the alleles that influence adult human height in Europeans have been under recent selection,[78] because of a very subtle cline in allele frequency between southern and northern Europeans. Further studies will tease apart the subtle effects of selection and population history detectable when many thousands of individuals are analyzed.

15.3 SKIN PIGMENTATION AS AN ADAPTATION TO ULTRAVIOLET LIGHT

Skin color is one of the most obvious ways in which humans differ, and has been widely used in attempts to define "races" (Box 10.1). It shows a highly nonrandom geographical distribution (**Figure 15.3**), particularly in the indigenous populations of the Old World. We find populations with the darkest skin colors in the tropics, and those with lighter pigmentation in more northerly latitudes. A number of adaptive explanations have been put forward for this variation in pigmentation. However, as with many apparently adaptive phenotypes in humans, there are populations who do not fit into this neat adaptive pattern, and additional explanations must be sought.

If we consider pigmentation as a quantitative trait, its proportion of variation between populations is high. One estimate[65] gives it an F_{ST} of 0.6, as opposed to 0.15 for neutral autosomal variants. A key early study on the genetics of human "races" stated that results on neutral polymorphisms did not apply to "*those genes which control morphological characters such as pigmentation and facial structure*."[56] This reinforces the point that the easily observed differences between populations are probably the result of adaptation or other selective forces, and are therefore a poor reflection of the relationships between populations. It undermines any attempt to construct racial groups on traditional lines of observable phenotypic differences (**Section 10.1**).

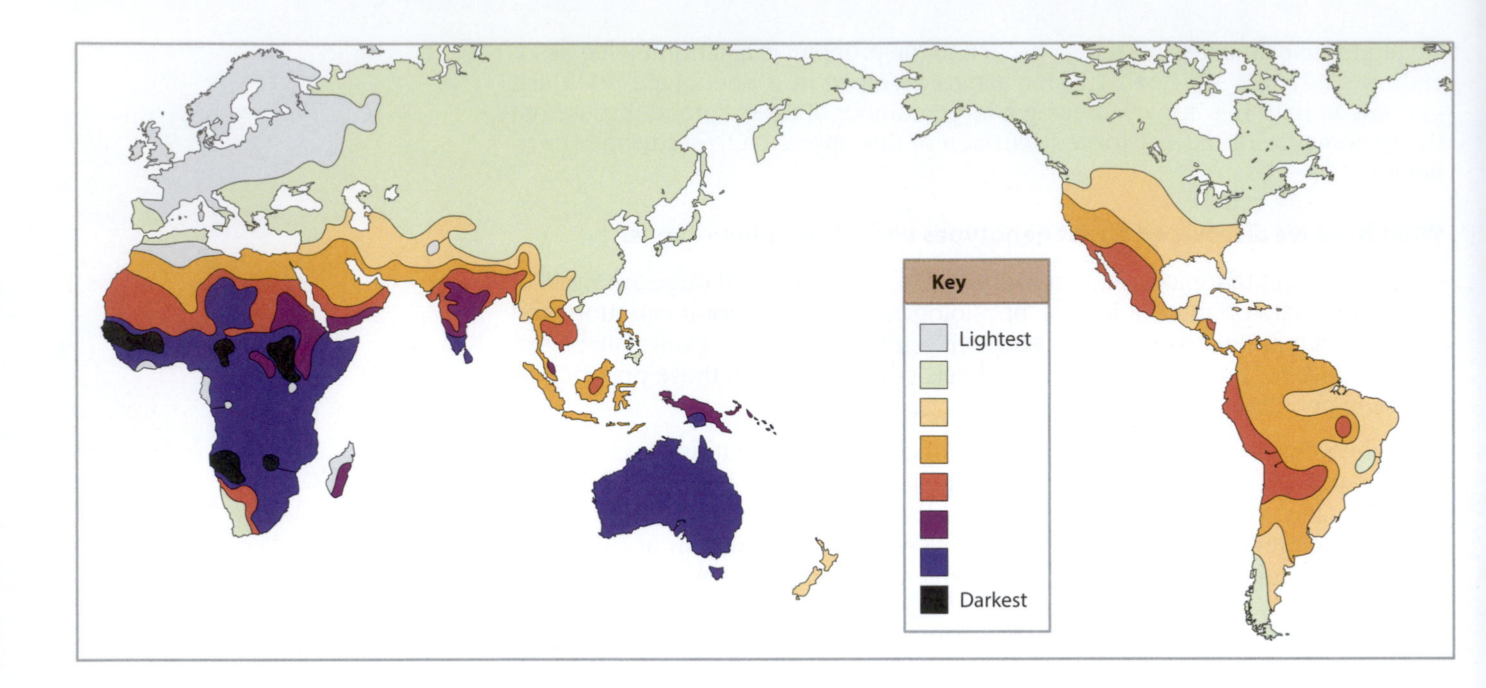

Figure 15.3: Distribution of different skin colors in indigenous populations of the world.
[From Jurmain R et al. (2000) Introduction to Physical Anthropology, 8th ed. Wadsworth/Thomson Learning. With permission from Wadsworth/Thomson Learning.]

Melanin is the most important pigment influencing skin color

Melanin is a granular substance produced by specialized cells called **melanocytes** that are at the boundary between the dermis and epidermis. It is concentrated in vesicles known as melanosomes, and these are transported into keratinocytes. Melanin is responsible not only for the color of skin, but also that of hair and (to a large extent) eyes. Hair follicles and surrounding keratinocytes receive melanosomes from melanocytes in a similar way to skin pigmentation, while in the iris the melanosomes are retained within the melanocytes themselves.

The number of melanocytes is about the same in different individuals, and variation in skin pigmentation results from differences in the number, size, and distribution of melanosomes within the keratinocytes. Dark skin contains many large, very dark, melanosomes, while lighter skin contains smaller and less dense melanosomes (**Figure 15.4**); these differences exist at birth, though subsequent exposure to the sun can alter **melanosome** size and clustering, and result in changed pigmentation (tanning).

As well as the size and distribution of melanosomes, the type of melanin within them also influences pigmentation. The chemical structures of different classes of melanin polymer are complex and difficult to define, but it is clear that black/brown pigments are formed by synthesis of eumelanin, while red/yellow pigments result from synthesis of the sulfur-containing pheomelanin. A melanocyte can synthesize both types of melanin.

Variable ultraviolet light exposure is an adaptive explanation for skin color variation

Anthropologists have addressed the question of the probable ancestral state of human skin color. It is likely that the skin of the earliest hominins was similar to that of the chimpanzee—white, but covered with dark hair. Indeed, the hair-covered skin of most primates is white, suggesting that this is the primate ancestral state, but exposed skin in all primates is pigmented to some extent, indicating that the potential for melanogenesis is also probably ancestral. Bipedalism and later brain expansion in early hominin evolution (Chapter 8) is thought to have required the evolution of a sensitive whole-body cooling mechanism capable of regulating brain temperature precisely. This involved the loss of hair, and a concomitant increase in number of sweat glands. Bare skin is at risk from

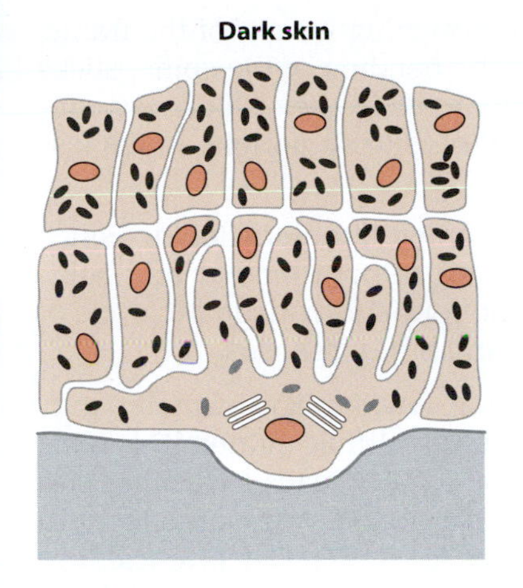

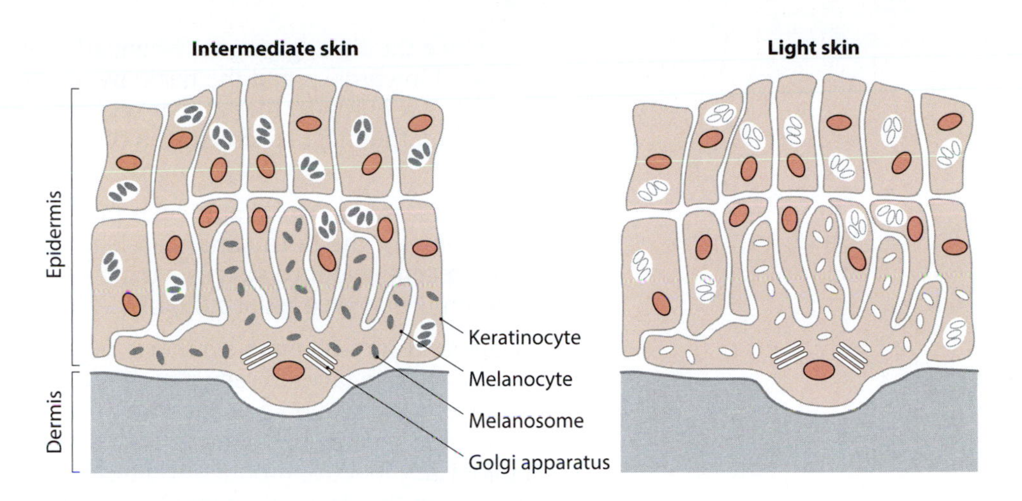

Figure 15.4: Different skin colors are determined by melanosome type and distribution.
In dark skin, melanosomes accumulate eumelanin and remain as single particles, while in lighter skins they contain increasing amounts of pheomelanin and cluster in membrane-bound organelles.

damage from ultraviolet (UV) radiation (**UVR**), in particular UV-B (wavelength 290–315 nm), the most energetic form of UVR that normally reaches the Earth's surface. Melanin has a twofold role in protecting us from UVR. First, it reduces the amount of radiation entering the deeper layers of the epidermis by absorbing or scattering it. Second, it acts as a filter, absorbing chemical by-products of UVR damage that would otherwise be toxic or carcinogenic. The immediate ancestors of all modern humans therefore must have developed dark skins, and the issue of skin color variation becomes a question of explaining the different degrees of subsequent loss of pigmentation.

What are the potential effects of UVR exposure that might have acted as selective pressures upon melanization?

Short-term UVR exposure causes sunburn: almost everyone with light skin knows the unpleasant after-effects of too much sun. Apart from discomfort and lowering of the pain threshold, more importantly there is damage to sweat glands and suppression of sweating which disrupts thermoregulation. Severe sunburn can be fatal. Light skin is therefore likely to have been strongly disadvantageous in UVR-exposed foraging societies.

Long-term UVR exposure causes cancers: degenerative changes in the dermis and epidermis due to UVR can eventually lead to basal cell and squamous cell carcinomas, and also the often-fatal melanomas. However, these cancers are usually manifested after reproductive age, and are therefore not thought to be a strong selective force.

UVR causes nutrient photodegradation in the skin: important micronutrients such as **flavins**, **carotenoids**, **tocopherol**, and **folate** are sensitive to photodegradation, and it has been suggested that this negative effect of UVR has formed part of the selective basis for skin pigmentation.

The effects of UVR are almost universally harmful, but it does have one benefit: the synthesis of vitamin D. This vitamin plays an essential role in the mineralization and normal growth of bone during infancy and childhood. It is present in a number of foodstuffs, including liver, fish oils, egg yolk, and milk products, and these foods comprise a different proportion of the diet in different populations. For most populations, the majority of the vitamin D requirement comes from the action of UVR on a steroid precursor in the skin, 7-dehydrocholesterol. In its active form, vitamin D acts as a hormone to regulate calcium absorption from the intestine and to regulate levels of calcium and phosphate in the bones. Insufficient vitamin D results in **rickets** in childhood, and **osteomalacia** in adulthood, conditions in which bones are soft, and become distorted, particularly in the weight-bearing parts of the skeleton, the legs and pelvis. Rickets was described in 1645 by Daniel Webster:

> ... the whole bony structure is as flexible as softened wax, so that the flaccid and enervated legs can hardly support the superposed weight of the body;

> hence the tibia, giving way beneath the overpowering weight of the frame, bend inwards ... and the back, by reason of the bending of the spine, sticks out in a hump in the lumbar region ... the patients in their weakness cannot (in the most severe stages of the disease) bear to sit upright, much less stand ....

Pelvic deformities are a particular problem for women during childbirth, since narrowing of the birth canal can lead to the death of both mother and baby. More recent evidence suggests that besides bone formation adequate levels of vitamin D may also reduce the risk of infection and other diseases.

A demand for adequate synthesis of vitamin D in low-sunlight climates may therefore have favored reduction in skin pigmentation. While vitamin D is toxic in excess, there is no evidence that overexposure to UVR leads to synthesis to toxic levels, since pre-vitamin D3 (a vitamin D precursor) can isomerize into biologically inactive products. Cultural adaptations to vitamin D deficiencies are dietary: in the Arctic they involve the traditional consumption of plentiful oily fish, and in many countries a daily spoonful of cod-liver oil or vitamin supplementation of milk and other foods.

Figure 15.5 illustrates how these various factors could interact to influence skin pigmentation. In order to evaluate the different hypotheses, it is necessary to have reliable data on pigmentation in different populations as well as accurate information on the amount of UVR exposure that these populations experience. Skin color is measured using a reflectometer, which assesses how much light

Skin color	Sunburn	Micro-nutrients	Vitamin D	Sunburn	Micro-nutrients	Vitamin D	Sunburn	Micro-nutrients	Vitamin D
DARK	OK	OK	OK	OK	OK	↓ increased risk of disease	OK	OK	↓ increased risk of disease
	↑	↓ NTDs, male infertility	OK	OK	OK	OK	OK	OK	↓ increased risk of disease
LIGHT	↑	↓ NTDs, male infertility	OK	↑	↓ NTDs, male infertility	OK	OK	OK	OK

Figure 15.5: Balancing the harmful and beneficial effects of UVR through skin color adaptation. The harmful effects of sunburn and micronutrient (including folate) photodegradation are balanced against the benefit of vitamin D synthesis through UVR, and together these form an adaptive explanation for the distribution of skin color variation. Note that cultural adaptations (protection of skin by clothing and shade, and dietary folate and vitamin D) are also important—see text. NTDs, neural tube defects.

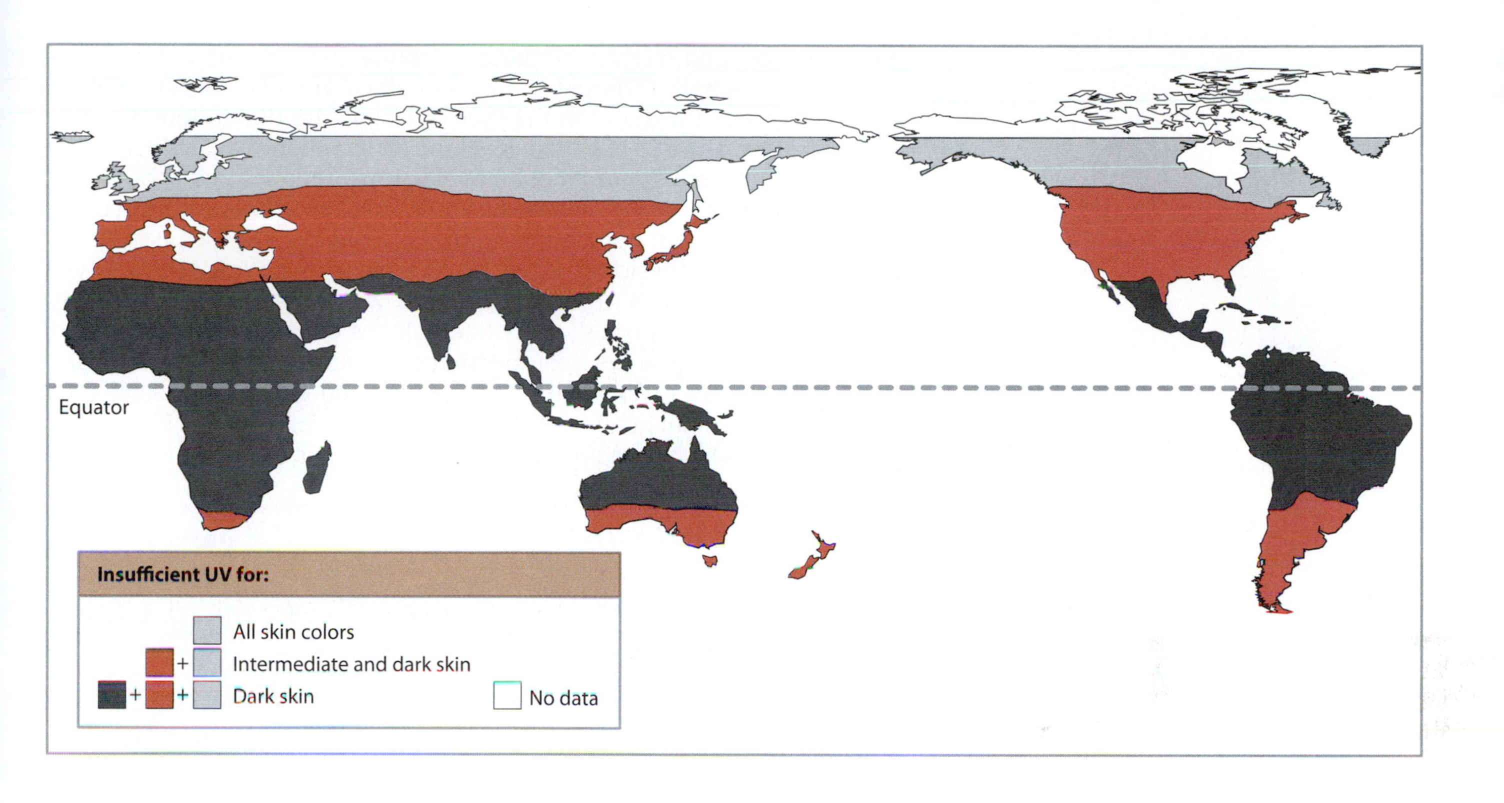

Figure 15.6: Estimated areas in which UVR is insufficient for vitamin D synthesis in different skin colors. In northern Europe, for example, there is insufficient UVR for the adequate synthesis of vitamin D. The dose of UV considered is the average annual UV minimal erythemal dose, that is, the amount of UVR causing a barely perceptible reddening of the skin. [From Jablonski NG & Chaplin G (2000) *J. Hum. Evol.* 39, 57. With permission from Elsevier.]

of a particular wavelength is reflected from the skin, usually in a region that is little exposed to the sun, such as the underarm area. A comparison of UVR data with extensive skin reflectance data from different populations illustrated the different potentials for UV-induced vitamin D synthesis in different regions and in different skin colors.[36] The UV dose considered in this study is the minimal erythemal dose (UVMED), which is the amount of UVR required to produce a barely perceptible reddening of lightly pigmented skin. **Figure 15.6** shows the three latitudinal regions defined in this way; the correlation observed between skin color distribution and predicted vitamin D synthetic potential is taken to support the vitamin D hypothesis. Although the amount of data analyzed in this study is impressive, it does not seem to represent a truly objective test of opposing hypotheses. Cultural adaptations are complicating factors that are difficult to take into account: dietary practices can cope with a lack of vitamin D (or, indeed, folate), and shade-seeking or the wearing of clothes can provide protection from the damaging effects of UVR. Also, there is little paleoanthropological evidence for rickets or osteomalacia in ancient populations. It seems likely that a need for vitamin D synthesis and photo-protection have both played a role in the distribution of skin color, together with cultural adaptations, and, possibly, sexual selection.

Several genes that affect human pigmentation are known

The usefulness of other organisms for identifying pigmentation genes and pathways in humans is an example of **comparative genomics**, which relies on the roles of particular genes being conserved across vertebrate evolution. The biochemical pathways underlying melanin synthesis have been elucidated mostly by genetic analysis of mouse coat color mutations. Over 370 genes have been identified that affect mouse coat color—many of their phenotypes were once highly prized by collectors of "fancy mice." Now over 170 of the genes underlying these traits have been identified, and for many of them mutations in their human orthologs show corresponding phenotypes; **Table 15.2** lists a selection. Mutations in some of these genes result in Mendelian forms of traits that usually show more complex inheritance; for example disruption of melanin synthesis resulting in albinism. Other mutations in the same genes may have a milder effect on phenotype, and might contribute to normal skin pigmentation variation as a quantitative trait. Different forms of oculocutaneous albinism (**OCA**), in which there is lack of pigment in the eyes and skin, are due to mutations in the *TYR*, *OCA2*, *TYRP*, and *SLC45A2* genes. As the albino phenotype reduces evolutionary fitness, because of photophobia, **nystagmus**, reduced acuity of vision, and increased susceptibility to melanoma, none of these mutations occur as common variants in human populations. Other pigmentation genes have even wider developmental or physiological effects: an example is Waardenburg syndrome type 1 (OMIM 193500), in which a frontal white blaze of hair, variably pigmented irises, white eyelashes, and white patches of skin are accompanied by deafness. The association between pigmentation and deafness here results from another function of melanocytes: they are needed within the cochlea for the development of normal hearing. The gene defect underlying this phenotype is in the *PAX3* gene, which is required for the correct migration of melanocytes

TABLE 15.2:
EXAMPLES OF HUMAN GENES ASSOCIATED WITH PIGMENTATION PHENOTYPES

Gene	Protein, function	Mutant human phenotype	OMIM no.	Mouse coat color or other phenotype
Melanocyte function				
TYR	tyrosinase, oxidation of tyrosine, dopa	oculocutaneous albinism (OCA) type 1	203100	albino
OCA2	P-protein, regulating pH of melanosome	OCA type 2	203200	pink-eyed dilution
TYRP1	oxidation of DHICA, stabilization of tyrosinase	OCA type 3	203290	brown
SLC45A2	MATP calcium transporter	OCA type 4	606574	underwhite
SLC24A5	NCKX5 calcium transporter	association with fair skin	609802	ocular albinism
ASIP	agouti signal protein, pheomelanogenic stimulation	association with dark hair and brown eyes	600201	agouti
MC1R	melanocyte-stimulating hormone (MSH) receptor	red hair or blond/brown hair	155555	extension
POMC	pro-opiomelanocortin (from which MSH and adrenocorticotropic hormone are produced), pheomelanogenesis	red hair with severe early-onset obesity and adrenal insufficiency	609734	pomc1
OA1	G-protein-coupled receptor	ocular albinism type 1	300500	oa1
Melanosome transport/uptake by keratinocyte				
MYO5A	myosin Va motor protein	Griscelli syndrome (includes partial albinism)	214450	dilute
RAB27A	Ras family protein	Griscelli syndrome	214450	ashen
Developmental				
KIT	c-kit receptor, melanoblast migration	piebaldism	172800	dominant *PAX3*
	transcription factor in neural tube development	Waardenburg syndrome type 1	193500	splotch
MITF	transcription factor	Waardenburg syndrome type 2	193510 microphthalmia	–

DHICA, dihydroxyindole carboxylate. [Adapted from Sturm R et al. (2001) *Gene* 277, 49. With permission from Elsevier. Additional data from http://www.espcr.org/micemut/.]

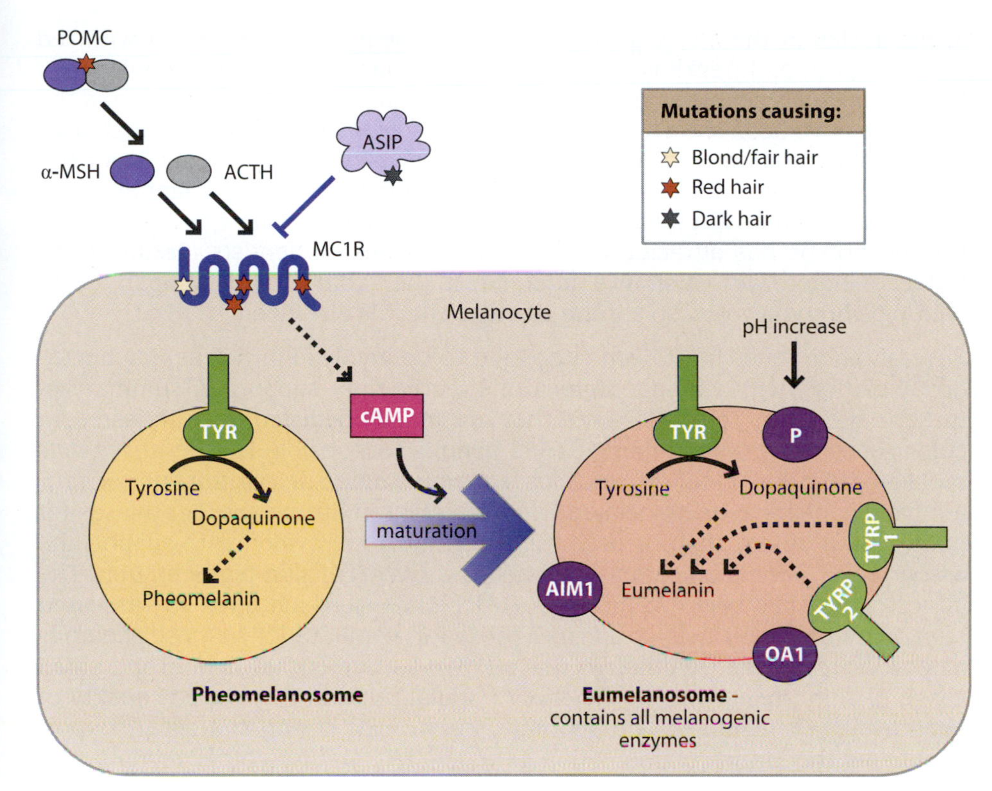

Figure 15.7: The control of pigment type switching in the melanocyte. The POMC precursor is cleaved to give α-MSH and ACTH (adrenocorticotropic hormone), which stimulate the MC1R protein. The ASIP protein is an antagonist of stimulation. Stimulated MC1R elevates intracellular cAMP, which acts to stimulate melanosome maturation. The mature eumelanosome contains all the enzymes necessary for synthesis of the dark pigment eumelanin. In the absence of MC1R stimulation, tyrosine is converted to the red/yellow pheomelanin in the pheomelanosome. Some key mutations affecting normal pigmentation are shown by stars. A subset of proteins are shown, and the pathway of conversion of tyrosine is not shown in detail. Also see Table 15.2 for further information. [Adapted from Sturm R et al. (2001) *Gene* 277, 49. With permission from Elsevier.]

from the neural crest, within which they originate. **Figure 15.7** shows the subcellular location and function of some of the gene products affecting pigmentation.

Analysis of pigmentation mutants in other model organisms, apart from the mouse, has also provided insights into human pigmentation. The skin of the zebrafish (*Danio rerio*) mutant *golden* contains fewer and smaller melanosomes than the wild type. The mutated gene (*slc24a5*) encodes a cation transporter which is involved in the biosynthesis of melanosomes in the skin, although its exact role is not clear. In humans, the orthologous gene is *SLC24A5*, and human mRNA encoding *SLC24A5* can restore wild-type zebrafish pigmentation when injected into *golden* zebrafish embryos. A nonsynonymous SNP (Ala111Thr, rs1426654) correlates with skin color, as measured by reflectometry, in African-Americans, African-Caribbeans,[46] and South Asians;[74] it accounts for about 30% of the phenotypic variation in skin color, at least in the first two of these populations. The **derived** threonine-111 allele is associated with paler skin, and is almost **fixed** in some European populations (**Figure 15.8**); it has also been shown to disrupt the biological activity of the SLC24A5 protein suggesting that this is the functionally relevant allele.

Mutations in the melanocortin 1 receptor gene (***MC1R***) affect not only human pigmentation, but also that of wild animals, lab mice, and many domestic animals, including cattle, horses, sheep, pigs, dogs, and chickens. The *MC1R* gene product lies in the cell membrane of the melanocyte (Figure 15.7), and is the receptor for α-melanocyte-stimulating hormone (α-MSH). Receptor stimulation leads to elevation of intracellular concentration of a signaling molecule cyclic AMP (**cAMP**), thus inducing changes in protein activity through phosphorylation, altered gene expression, and ultimately the generation of a mature eumelanogenic melanosome. In the absence of a signal via MC1R, the eumelanosome cannot form, and instead the immature pheomelanosome persists, with a consequent phenotype of reduced pigmentation. Indeed, red hair has been shown to be the phenotype of a homozygous null *MC1R* human, lacking any melanocortin 1 receptor activity.[6]

Three alleles of the *MC1R* gene have been shown to be associated with red hair, fair skin, and freckling and a fourth allele has been associated with fair/blond hair.[27] These alleles map to the intracellular part of the protein, and the first three have been shown to reduce the ability of MC1R to stimulate increases in the concentration of cAMP. The mode of inheritance of red hair has been thought to be autosomal recessive, and the inheritance of the variant alleles is broadly consistent with this.[23] This apparent simple inheritance of a pigmentation phenotype has attracted the interest of forensic scientists (**Section 18.2**). However, heterozygotes show a difference in their ability to tan compared with wild-type homozygotes, so a **gene dosage** effect is also likely.

Several other genes have been suggested to be involved in normal pigmentation variation, with varying degrees of experimental support. A summary of these is shown in Table 15.2, and they are mostly deduced from mouse coat color genes: for example the *SLC45A2* gene is responsible for the *underwhite* mouse coat color mutation, encodes a melanosome-specific transporter protein, and appears to be associated with skin color in Europeans. For this, as for many of the others, there is as yet limited functional evidence to support the association. A complementary approach is a GWAS of skin pigmentation. The largest so far has been on nearly 3000 Icelanders, which used self-assessed high or low skin sensitivity to sun as a proxy for pigmentation in these generally pale-skinned people.[75] Although this study was limited because of the small range of skin phenotypes in the study population, and the unclear quality of phenotyping, it identified one genomic region with strong statistical support containing the known pigmentation gene (*MC1R*), and two other regions, one containing another known pigmentation gene tyrosinase (*TYR*). Future genomewide studies on more phenotypically diverse populations treating skin color as a quantitative trait may both confirm existing results and identify new pigmentation genes.

Genetic variation in human pigmentation genes is consistent with natural selection

If selection has been acting on human pigmentation genes, then the signal of that selection may be detected in patterns of allele and haplotype diversity

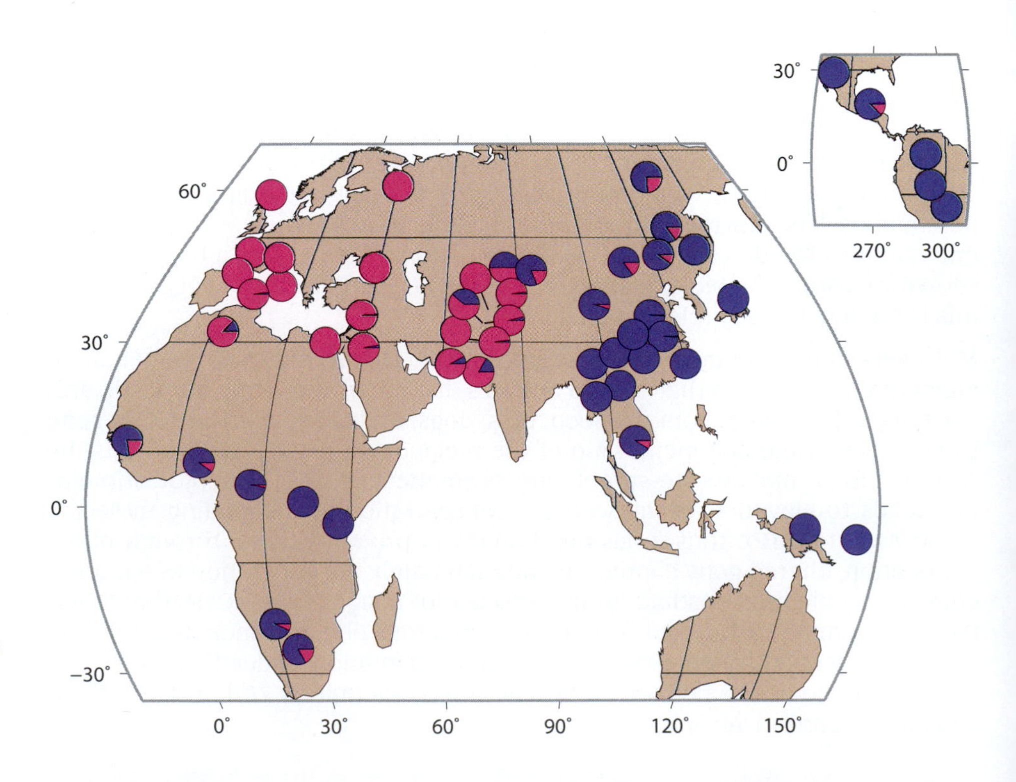

Figure 15.8: Global distribution of *SLC24A5* alleles.
Each pie represents an individual population from the CEPH-HGDP collection (Box 10.2). The threonine-111 allele frequency is shown in *red* and the alanine-111 frequency in *blue*. [From Coop G et al. (2009) *PLoS Genet.* 5, e1000500. With permission from Public Library of Science.]

within and between populations. For *SLC24A5*, there are several lines of evidence suggesting recent positive selection of an allele contributing to a lighter skin phenotype in Europeans.[35, 58] First, the SNP associated with skin color at *SLC24A5* shows the largest allele frequency difference of any functional SNP between Europeans and other populations, from whole-genome sequencing data (Table 10.4). This high level of population differentiation is also seen in other pigmentation genes[63] (for example, *SLC45A2*, Table 10.4). Second, the derived allele (Thr111) associated with lighter skin lies on a large extended haplotype in Europeans (see Chapter 6). However, in East Asian populations the ancestral Ala111 allele is almost universal in the population, yet East Asian individuals generally have paler skin than other populations with high frequency of the Ala111 allele. This demonstrates that the genetic basis for skin color variation differs between populations, and clearly further work is needed.

A study of *MC1R* variation based on resequencing of the gene revealed a striking difference in the distribution of haplotypes between African and Eurasian populations.[28] There are only five African haplotypes in the sample, and a complete absence of nonsynonymous nucleotide substitutions. In Eurasia, by contrast, there are 14 haplotypes, with 10 nonsynonymous mutations and three synonymous ones. Comparing these findings with the known number of synonymous and nonsynonymous changes between the human consensus sequence and the chimpanzee sequence shows that the high degree of amino acid sequence conservation in Africa is highly unlikely to have arisen by chance. It probably reflects strong functional constraint in Africa, where any diversion from eumelanin production is strongly deleterious. The pattern in Eurasia seems compatible with low selective constraint, rather than selective enhancement of diversity, and this is supported by HKA tests (**Section 6.7**) comparing *MC1R* and the β-globin gene between human populations and between humans and chimpanzees. On the face of it, this represents a challenge to the vitamin D hypothesis, but it must be remembered that the power of these tests to detect selection is limited.

MC1R and *SLC24A5* have both been heavily studied, yet appear to present a paradox: strong positive selection for pale skin in Europeans revealed by *SLC24A5* diversity, yet strong purifying selection in Africans combined with relaxation of functional constraint in Europeans revealed by *MC1R* diversity. However, we must understand the difference between examining patterns of selection at individual genes involved in skin pigmentation and patterns of selection for the skin pigmentation phenotype itself. Because many genes contribute to a molecular pathway that generates the phenotype, examination of each gene in isolation will only reveal part of the selective history of the phenotype.

Does sexual selection have a role in human phenotypic variation?

An adaptive model for human skin color variation is described above, and, as has been mentioned at various points in this chapter, there have been adaptive explanations proposed for many other differences between human populations. However, two major problems remain. The first is that, even when an adaptive explanation seems reasonable and persuasive, there are often populations that are in the "wrong" environment and therefore do not fit with the adaptive explanation. The second is that some adaptive explanations are far from persuasive in the first place. An alternative to adaptation through natural selection is to turn instead to the peacock's tail as a model. This spectacular ornament is nothing to do with natural selection, and everything to do with sexual selection. Females choose mates on the basis of the quality of their tails, which leads to an increasing degree of tail ornamentation, to the extent that this can actually become maladaptive, endangering the survival of its bearer. For skin color a model of constraint in Africa and neutrality in northern regions, coupled with sexual selection, would be sufficient to explain its geographical variation. It has been suggested that a model of sexual selection might be applicable to other phenotypic variation among human populations.[17, 18, 29] Some of Charles Darwin's views on this are summarized in **Box 15.4**.

Under a sexual selection hypothesis, human skin pigmentation dimorphism, where females generally are paler skinned than males, is proposed to be driven by mate choice where paler females are more attractive to males, or darker males are more attractive to females. Similarly, sexual selection would

Box 15.4: Darwin's "The Descent of Man and Selection in Relation to Sex"

In *The Descent of Man*, published in 1871, Darwin turns his attention from a general consideration of the relationships and origins of species to the origin and diversity of one species, *Homo sapiens*. In a series of chapters in the middle of the book he establishes the importance of sexual selection in evolution by describing examples among animals, and in the last few chapters turns his attention to humans:

> During many years it has seemed to me highly probable that sexual selection has played an important part in differentiating the races of man; but in my Origin of Species I contented myself by merely alluding to this belief. ...let us see how far the men are attracted by the appearance of their women, and what are their ideas of beauty.

Darwin gives many examples, drawn from the experiences of travelers and early anthropologists. These examples demonstrate his simple yet powerful observation that physical traits deemed beautiful (for example, hairiness, facial features, and stature) vary hugely among different populations, yet are always found at high frequency within the population that favors them. From our perspective at the beginning of the twenty-first century, many of these anecdotal comments might seem offensive; in particular, white male anthropologists in the past have come in for justified criticism because of their attention to sexual characteristics of the women of "native" populations. Here are a few of the examples used by Darwin:

> Pallas, who visited the northern parts of the Chinese empire, says, 'those women are preferred who have the Mandschu form; that is to say, a broad face, high cheek-bones, very broad noses, and enormous ears'; and Vogt remarks that the obliquity of the eye, which is proper to the Chinese and Japanese, is exaggerated in their pictures for the purpose, as it 'seems, of exhibiting its beauty, as contrasted with the eye of the red-haired barbarians.' It is well known, as Huc repeatedly remarks, that the Chinese of the interior think Europeans hideous, with their white skins and prominent noses. The nose is far from being too prominent, according to our ideas, in the natives of Ceylon; yet 'the Chinese in the seventh century, accustomed to the flat features of the Mongol races, were surprised at the prominent noses of the Cingalese; and Thsang described them as having "the beak of a bird, with the body of a man."'

> It is well known that with many Hottentot women the posterior part of the body projects in a wonderful manner; they are steatopygous; and Sir Andrew Smith is certain that this peculiarity is greatly admired by the men. ... some of the women in various negro tribes have the same peculiarity; and, according to Burton, the Somal men are said to choose their wives by ranging them in a line, and by picking her out who projects farthest *a tergo*. Nothing can be more hateful to a negro than the opposite form.

> With mammals the general rule appears to be that characters of all kinds are inherited equally by the males and females; we might therefore expect that with mankind any characters gained by the females or by the males through sexual selection would commonly be transferred to the offspring of both sexes. If any change has thus been effected, it is almost certain that the different races would be differently modified, as each has its own standard of beauty.

Darwin surmises that the strength of sexual selection among developed nations may be diminished:

> Civilised men are largely attracted by the mental charms of women, by their wealth, and especially by their social position; for men rarely marry into a much lower rank. The men who succeed in obtaining the more beautiful women will not have a better chance of leaving a long line of descendants than other men with plainer wives.

However, social stratification may mean that there are subpopulations in which sexual selection is still an active force:

> Many persons are convinced, as it appears to me with justice, that our aristocracy, including under this term all wealthy families in which primogeniture has long prevailed, from having chosen during many generations from all classes the more beautiful women as their wives, have become handsomer, according to the European standard, than the middle classes.

Thus, Darwin had a strong belief that sexual selection was a major force in shaping human phenotypes, and in particular the phenotypic differences between different human populations.

mean that the pale skin color seen at high latitudes is due to absence of natural selection for dark skin mediated by protection against UVR. If true, then at low latitudes natural selection for dark skin pigmentation will have overridden sexual selection and therefore males and females will have more similar skin pigmentation. Conversely, at higher latitudes, natural selection will be weaker and sexual selection more powerful, resulting in a greater difference in skin pigmentation between males and females. Using results collated from the published literature, one study found no association of sex difference in pigmentation with latitude, failing to support this particular sexual selection hypothesis.[50]

15.4 LIFE AT HIGH ALTITUDE AND ADAPTATION TO HYPOXIA

At an altitude of 4000 m, every intake of breath contains 60% fewer oxygen molecules than the same breath at sea level. People from low altitudes can become acclimatized to some extent at high altitudes through physiological mechanisms. However, there are about 25 million people who live permanently in regions over 3000 m in altitude (in the Andes, Ethiopia, and the Himalayas). In these people there is evidence of adaptation to high altitude, and this adaptation is biological because traditional technology or changes in behavior do not offer protection against **hypoxia**. For example, some populations of Tibet do not show a reduced birth weight, characteristic of other high-altitude populations or recent immigrants, and this may reflect an adaptive difference in maternal blood flow to the placenta during pregnancy.[38]

Natural selection has influenced the overproduction of red blood cells

Chronic mountain sickness is seen in some individuals who move from low-altitude regions to areas of high altitude, and is caused by an overproduction of hemoglobin in response to hypoxia. Tibetans are an exception, and show very little increase in hemoglobin, as a result of a biological adaptation. A candidate gene analysis identified the Endothelial Pas domain protein 1 gene (*EPAS1*) which encodes the HIF-2α transcription factor, a member of the hypoxia-inducible factor family of proteins that regulate the cellular response to hypoxia. Tibetans show a very high level of differentiation from lowland Han Chinese in this gene (**Figure 15.9**), and extended haplotype tests (**Section 6.7**) show evidence of recent strong positive selection.[5, 70] In addition, a direct association of *EPAS1* variants with hemoglobin concentration has been shown: individuals carrying the Tibetan-specific haplotype have a lower hemoglobin concentration at high altitude than those who do not carry the haplotype. The *EPAS1* haplotype accounts for a large proportion of the variation in hemoglobin levels in Tibetans, and in terms of absolute amounts of hemoglobin, has a much larger effect than any single allele within low-altitude populations.

Although an *EPAS1* haplotype that affects the response to hypoxia has been identified, the variant causing this effect has not yet been identified. Indeed, alleles of intronic SNPs showed the strongest association with hemoglobin concentration and showed the highest level of differentiation between populations. As in the case of lactase persistence and the lactase gene (**Section 15.6**), it is likely that the causative allele affects expression levels of the gene. *EGLN1* is another gene, identified by several studies, where variation is associated with hemoglobin levels and which has been subject to natural selection at high altitude. It encodes a protein that negatively regulates HIF-2α by targeting it for degradation under normal oxygen conditions; under hypoxic conditions it reduces this targeting allowing HIF-2α to initiate expression of downstream genes as part of the hypoxic response.[60]

Several other genes involved in the body's response to hypoxia have shown evidence of natural selection among Tibetans in some, but not all, studies. Some of these are very promising (for example, β-globin) but await further studies to confirm the nature of the genetic variation and its functional role.

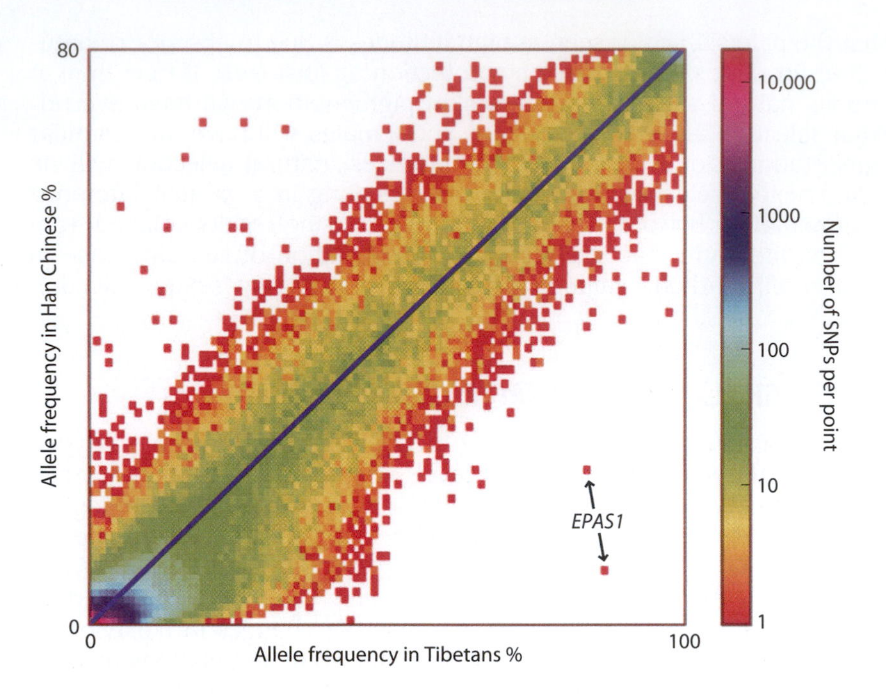

Figure 15.9: Allele frequency differences between genomes of Han Chinese and Tibetan populations. Each point on the graph corresponds to a derived allele frequency in Tibetans (*x* axis) and Han Chinese (*y* axis), with the number of SNPs showing those particular two allele frequencies indicated by the color of the point (legend on the right of the graph). Derived alleles of two SNPs in *EPAS1* highlighted on the graph are at unusually high frequency in Tibetans and low frequency in Han Chinese. [From Yi X et al. (2010) *Science* 329, 75. With permission from AAAS.]

High-altitude populations differ in their adaptation to altitude

Most research to date has focused on the Tibetan population, and the adaptations to hypoxia shown by Ethiopian and Andean populations are less clear. A genomewide scan for extended haplotype homozygosity in Andean populations identified several genes involved in the response to hypoxia that were different from those identified in Tibetans. This may seem at first surprising, but it is known that the adaptation to hypoxia is physiologically different between Andeans and Tibetans. For example, Tibetans show characteristically low hemoglobin concentration (which is associated with the genetic variation in *EPAS1*), and high blood capillary density, not shown by Andeans.[4]

Given this difference in physiological adaptation to the same environmental trait, it is perhaps less surprising that the selected genes can be different between the Tibetans and Andeans. This is an example of convergent evolution. Very little research has been done on Ethiopians who live at high altitude.

15.5 VARIATION IN THE SENSE OF TASTE

Adaptation to different foods allowed ancestral hominins to expand into new environments and exploit the new food sources found there. Consumption of different plants in these new environments would have exposed humans to a variety of plant chemicals with potentially toxic effects, including **alkaloids**. For example, the legume plant family (Family *Fabaceae*) contains not only familiar edible legumes like peas and beans, but also many toxic species such as brooms, which contain high levels of alkaloids. The ability to detect such chemicals as a bitter taste would have allowed avoidance of plant species, or parts of plants, that contained high levels of these chemicals.

Is it likely that the taste variation we observe is due to adaptation to exploiting new food sources, and avoiding poisons? Alleles causing functional differences in response to chemicals present in food are interesting, and are important contributions to the genetic variation in human phenotypes. Although attractive adaptive hypotheses have been proposed for this variation, there is little evidence of positive selection in most cases. Like the variation in the sense of smell (**Opinion Box 14**), it may be more likely that the phenotypic variation we observe is simply due to removal of selective constraint during human evolution, allowing these phenotypic variants to become a component of **neutral variation**.

While color blindness is restricted to a few percent of humans, odor blindness (specific anosmia or hyposmia) affects us all: practically every human is incapable of sensing a few of the many thousands of odorants perceived by our species. Furthermore, pronounced inter-individual variation also occurs in perception of odor quality. A prominent example is the steroid odorant androstenone, to which 30% of humans are completely insensitive, while the rest show a strong dichotomy between pleasant and utterly unpleasant perception. Underlying missense variations in an olfactory receptor (OR) gene, *OR7D4*, have been identified (Table 15.1) and linked to odorant sensitivity by a combination of sensory evaluation and *in vitro* screens of olfactory receptors expressed in experimental cell lines. This gene is one of ~400 active OR genes, a repertoire that allows a huge inventory of volatile odorous compounds to be recognized. Most odorants bind to several OR proteins, and exact identification stems from a combinatorial code, much like color vision. Consequently, olfaction is often a polygenic trait, but it is conjectured that losing the highest affinity OR for a given odorant would manifest nearly Mendelian inheritance. Recent research has shown that the olfactory repertoire shows the most extreme variation in gene content of any human gene family. An average human individual shows heterozygosity at one-third of all ~400 OR loci; hence the effective personal protein repertoire may be as large as 600. Curiously, the wiring diagram of the chemosensory neurons suggests that the brain receives distinct odor information from each of these individual alleles. Surveys to date have found a whopping ~4000 missense alleles. There are indications that some of this variability is neutral but it may also be shaped in some ORs by the selective advantage of enhanced allele repertoire, that is, by balancing selection, as seen in some major histocompatibility complex proteins. In parallel, those ORs addressing essential, stereotyped behavior (for example, sex attractants or poisons) might be subject to purifying selection. Selection will thus depend on the nature of the relevant odorants, and it is quite likely that each different OR repertoire member would show a different evolutionary signature.

Because of the redundant receptor–odorant relationships, many individual OR genes appear to be non-essential for survival. This gave rise to a very prevalent phenomenon of loss-of-function mutations that readily spread in the population. Functional inactivation may be by nonsense mutations, frame-disrupting indels, or deletion copy-number variants (CNVs), sometimes affecting contiguous OR subclusters.[30, 55] Our recent data suggest that such repertoire diminution events are highly prevalent, affecting as many as 60% of the active OR collection, generating a complex pattern of inter-individual deletion variability (**Figure 1**). This further accentuates the great degree of inter-individual genomic diversity in this functional pathway, including cases in which an interlocus in-frame deletion CNV results in a novel OR sequence. So far, one example of a genetic association between an olfactory phenotype and such a loss-of-function event has been discovered, whereby individuals homozygous for an intact *OR11H7P*, a reference genome stop-SNP pseudogene, are more sensitive to the sweaty odorant isovaleric acid. Olfaction thus stands out as a prime example of widespread, functionally relevant genetic variability, but much further work will be required for complete genotype–phenotype mapping in this pathway.

Doron Lancet and Tsviya Olender,
Department of Molecular Genetics, Weizmann Institute of Science, Rehovot, Israel

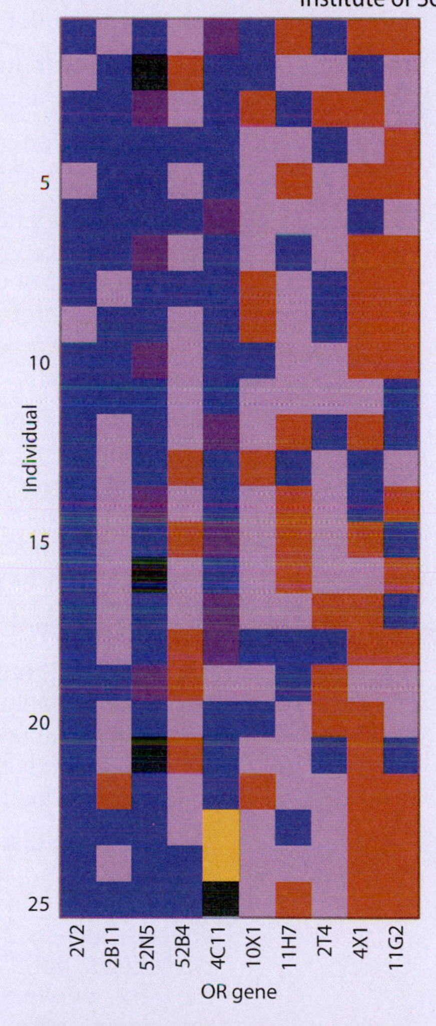

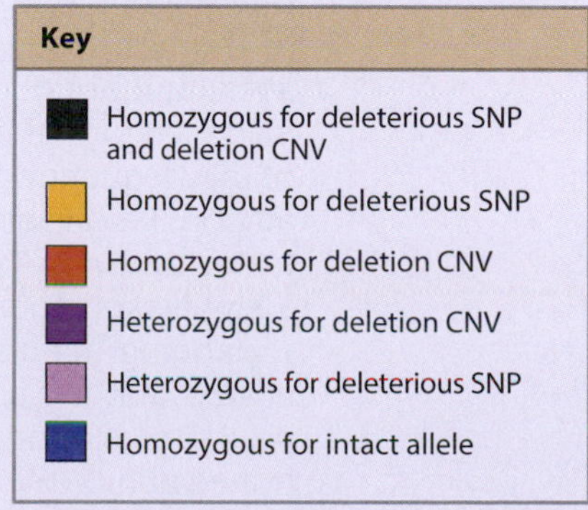

Figure 1 Genetic variation in some OR genes. Genotype calls of several OR genes, for which both an intact and inactive allele are present in the population. Each row represents an individual, and every column represents a gene.

Variation in tasting phenylthiocarbamide is mostly due to alleles of the *TAS2R38* gene

There are several reports of phenotypic variation in sensitivity to bitter **tastants**, including quinine, tetracycline, and chloramphenicol. The best-characterized taste polymorphism is related to the synthetic tastant phenylthiocarbamide (PTC). It is important to remember that PTC, unlike the other tastants mentioned, does not occur in nature and would not have been encountered in the diet of ancestral humans. However, variation in PTC-tasting is very closely correlated with variation in tasting other thiocyanate-containing compounds, and such compounds are produced during breakdown of **glucosinolates** present in the *Brassica* order of plants, for example.[12]

The human variation in PTC-tasting was discovered by accident in 1931, when, in an incident illustrating the lack of modern safety procedures, Dr Arthur Fox

> had occasion to prepare a quantity of phenylthiocarbamide, and while placing it in a bottle the dust flew around in the air.[24]

Another occupant of the lab tasted the dust as bitter, whilst Dr Fox did not, and this dichotomous taster/non-taster phenotype was confirmed, first on a larger number of unrelated individuals, then on a number of families, suggesting that the non-taster phenotype was inherited as an autosomal recessive trait (OMIM 171200). It took another 70 years before the genetic basis of this trait was identified.

Although it is normally treated as a dichotomous trait, the threshold for PTC testing is in fact quantitative (Box 15.3). Genomewide QTL linkage analysis identified a gene that was subsequently named taste receptor 2 member 38 (*TAS2R38*), which encodes a G-protein signaling receptor and accounts for 50–80% of the phenotypic variation in PTC-tasting.[39] Surprisingly, the non-taster allele appears to produce a functional receptor which differs from the taster allele by three amino acids at positions 49, 262, and 296 (rs714598, rs1726866, rs10246939). Taster haplotypes have a proline, alanine, and valine at those positions (PAV haplotype) while non-tasters have an alanine, valine, and isoleucine (AVI haplotype). This suggests that the PTC non-taster allele may in fact be a functional taster allele for another compound; indeed, fruits of the bignay tree *Antidesma bunius* (Family Euphorbiaceae) taste bitter to PTC *non-tasters* but sweet to PTC-tasters, although the active compound has not yet been identified.

Following the discovery that chimpanzees also showed this dichotomous tasting/non-tasting trait, it was thought initially that long-standing balancing selection had maintained this variation in both *Homo* and *Pan* since their divergence. Analysis of the *TAS2R38* gene in chimpanzees confirms that there are two alleles, but the non-taster allele is due to a mutation of the start codon of the gene, preventing translation of the gene into a protein product. This separate origin of the non-tasting allele in chimpanzees shows that the generation of the non-tasting trait is an independent event in both *Homo* and *Pan* lineages, and not due to long-term balancing selection.[82] However, there is some evidence for balancing selection (**Section 5.4**, Box 6.6) in humans from analysis of the frequency distribution of *TAS2R38* haplotypes in different populations. In all populations analyzed, with the notable exception of Native Americans, both taster and non-taster haplotypes had intermediate frequencies, and showed a significantly positive value for Tajima's *D* statistic (**Section 6.7**), assuming an expanding population size.[83] Sequence analysis of a single Neanderthal suggested that it was a heterozygote, supporting the idea that the polymorphism pre-dates splitting of Neanderthal and modern humans 500 KYA.[45] As mentioned above, the selective agents driving balancing selection are not clear. It is possible that heterozygotes are at an advantage because of their broader bitter-taste sensitivity, as compared to PTC-taster and PTC-non-taster homozygotes. An alternative hypothesis is that pathogens are the selective agents, following the observation that *TAS2R38* is expressed in the respiratory tract and binds

quorum-sensing molecules from the pathogenic bacterium *Pseudomonas aeruginosa*, and that the AVI haplotype encodes for a TAS2R38 protein that shows a weaker response to these molecules.[48] Indeed, the observation of many nonsynonymous alleles of *TAS2R38* in Africa, together with a lack of correlation between genotype frequency and diet across African populations, supports a role of pathogens as a selective agent.[13] *TAS2R38* is clearly a **pleiotropic** gene, and disentangling its evolutionary history remains a fascinating challenge.

There is extensive diversity of bitter taste receptors in humans

TAS2R38 is one of a family of 25 bitter taste receptor genes in humans. They encode G-protein-coupled receptors embedded in the cell membrane that respond to a chemical **agonist** by stimulating a downstream signaling cascade within the cell. Some have been linked to their cognate tastants, for example the *TAS2R16* gene encodes a receptor for β-glucopyranosides[10] and there is some evidence favoring positive selection for a high-sensitivity tasting allele, suggesting that increased sensitivity toward these compounds has been advantageous.[71] Indeed, a subclass of β-glucopyranosides are the cyanogenic glycosides, which generate highly toxic cyanide ions following digestion, are present in many plant species as a defense against herbivores, and are also used as defense against insectivores by several types of arthropod, including beetles, centipedes, and butterflies.[84] The high-sensitivity tasting allele carries an asparagine at amino acid 172 (rs846664), and **transfecting** human cells with this high-sensitivity receptor results in an increased influx of calcium ions following stimulation with several naturally occurring β-glucopyranosides. This influx indicates activation of the G-protein signaling pathway within the cell, which, in taste receptor cells, would activate the cell and stimulate the appropriate neuron. However, taste tests of different bitter substances in humans with different *TAS2R16* genotypes have not yet been performed.

Two other bitter taste receptors, TAS2R43 and TAS2R44, respond to the synthetic sweetener saccharin and the natural compounds aloin (from the plant genus *Aloe*) and the **nephrotoxin** aristolochic acid (found in wild ginger, genus *Asarum*, not to be confused with ginger, genus *Zingiber*). *TAS2R43* shows a deletion polymorphism and an amino acid polymorphism changing a serine to a tryptophan residue at position 35 (rs68157013). The tryptophan allele shows higher sensitivity to aloin both in taste tests on human volunteers and analysis of calcium influx in cells transfected with the *TAS2R43* receptor.[64]

The tastants for most other TAS2Rs have not yet been identified, and this remains a limiting factor in further genotype–phenotype studies of bitter taste. The *TAS2R* family of genes shows a large number of coding polymorphisms and possible copy-number polymorphisms, suggesting that much more phenotypic variation remains to be identified.

Sweet, umami, and sour tastes may show genetic polymorphism

Bitter taste is clearly linked to the rejection of food, but the other three tastes, particularly umami (Japanese for "delicious flavor"—often translated as "savory taste") and sweet, are appetitive sensations in food recognition. These non-bitter tastes are also mediated by transmembrane G-protein-coupled receptors. Both sweet and umami tastes are detected by the taste receptor family TAS1R, which, in humans, has three members TAS1R1, TAS1R2, and TAS1R3. Sucrose sensitivity differs between individuals as a quantitative trait, and two SNPs (rs307355 and rs3574481) account for 16% of its variability. Both SNPs affect expression of the *TAS1R3* gene, suggesting that the variation in the amount of receptor is responsible, with high-sensitivity taste alleles of both SNPs most frequent in Western Europe.[26] It is hypothesized that the high-sensitivity allele has been driven to high frequencies by selection favoring high-sensitivity taste to sucrose-containing plants in cold climates. This suggestion is attractive, but as yet there is no genetic evidence for selection acting at these alleles. Umami tasting (as measured by sensitivity to monosodium glutamate, MSG) shows a

bimodal distribution in humans, with 10% of Europeans showing low sensitivity to MSG. The underlying genetic variants are not known, although the receptor that is responsible for at least part of umami sensitivity is a heterodimer of the **peptides** encoded by *TAS1R1* and *TAS1R3* genes, and different alleles in these genes may cause at least part of the observed variation.[69]

The nature of sour taste is less well characterized, and there appears to be very little variation in taste sensitivity between individuals, at least in Europeans. A variety of acids are used as tastants to measure sour taste, and it is usually accepted that detectors of H^+ ions are responsible for sour taste; however, several other mechanisms have been proposed. Given that organic acids are more likely to have been responsible for naturally occurring sour taste rather than inorganic acids such as HCl, frequently used as a tastant in experiments, sensitivity studies using a variety of different organic acids may provide further insight.

15.6 ADAPTING TO A CHANGING DIET BY DIGESTING MILK AND STARCH

In most of the world's population the ability to digest lactose, the major sugar of milk (**Figure 15.10**), declines rapidly after weaning, because of falling levels of the enzyme lactase in the small intestine. Consequently, in the majority of people, ingestion of more than a small quantity of milk (containing 4–8% lactose) in adult life causes abdominal pain, flatus, and diarrhea, since lactose causes osmotic transport of water into the small intestine, and is fermented in the colon by gut bacteria. This permanent reduction in lactase levels is common to all mammals, and may be important in encouraging young animals to be weaned toward an adult diet. In lactose-intolerant people, milk products that are soured or otherwise treated, such as cheeses and yogurts, cause few problems because they contain relatively low levels of lactose, or even provide bacteria that themselves secrete lactases (for example, *Lactobacillus acidophilus*). This can be seen as a cultural adaptation to the trait of **lactose intolerance**.

However, substantial numbers of people worldwide can continue to drink fresh milk as adults without problems, because lactase activity persists into adulthood (**lactase persistence**). In such adults, prolonged avoidance of lactose does not result in lactose intolerance: the trait is an innate one, representing a genetic adaptation.

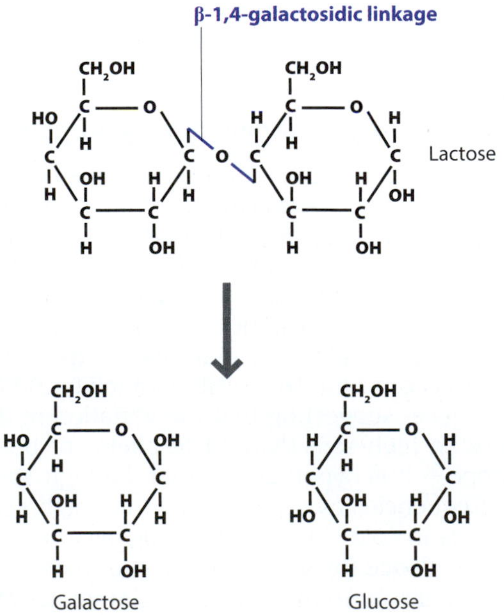

Figure 15.10: Cleavage of the disaccharide lactose by the enzyme lactase.

There are several adaptive hypotheses to explain lactase persistence

The geographical distribution of lactase persistence is very non-uniform (**Figure 15.11**): it is highly prevalent (>70% population frequency) in Europeans and in certain African pastoralists, such as the Beja of Sudan. It is found at intermediate (30–70%) frequency in the Middle East, around the Mediterranean, and in south and central Asia, and at low frequency in Native Americans and Pacific Islanders, as well as in much of sub-Saharan Africa and Southeast Asia. The phenotype thus seems to be most frequent among people who have a history of drinking fresh milk, and least frequent among people who have no such history. Lactase persistence has therefore long been regarded as an adaptation to dietary change brought about by the development of agriculture and animal domestication in the Near East and Africa (**Section 12.2**). Before discussing this and other hypotheses in more detail, it is important to note that data are very patchy for some regions of the world, and that several different tests are used to determine the lactase-persistence phenotype, each with different advantages and disadvantages.[32] It is also worth noting that these tests measure the ability to digest 50 g of lactose, which is equivalent to about a liter of fresh milk, more than most people drink at any one time.

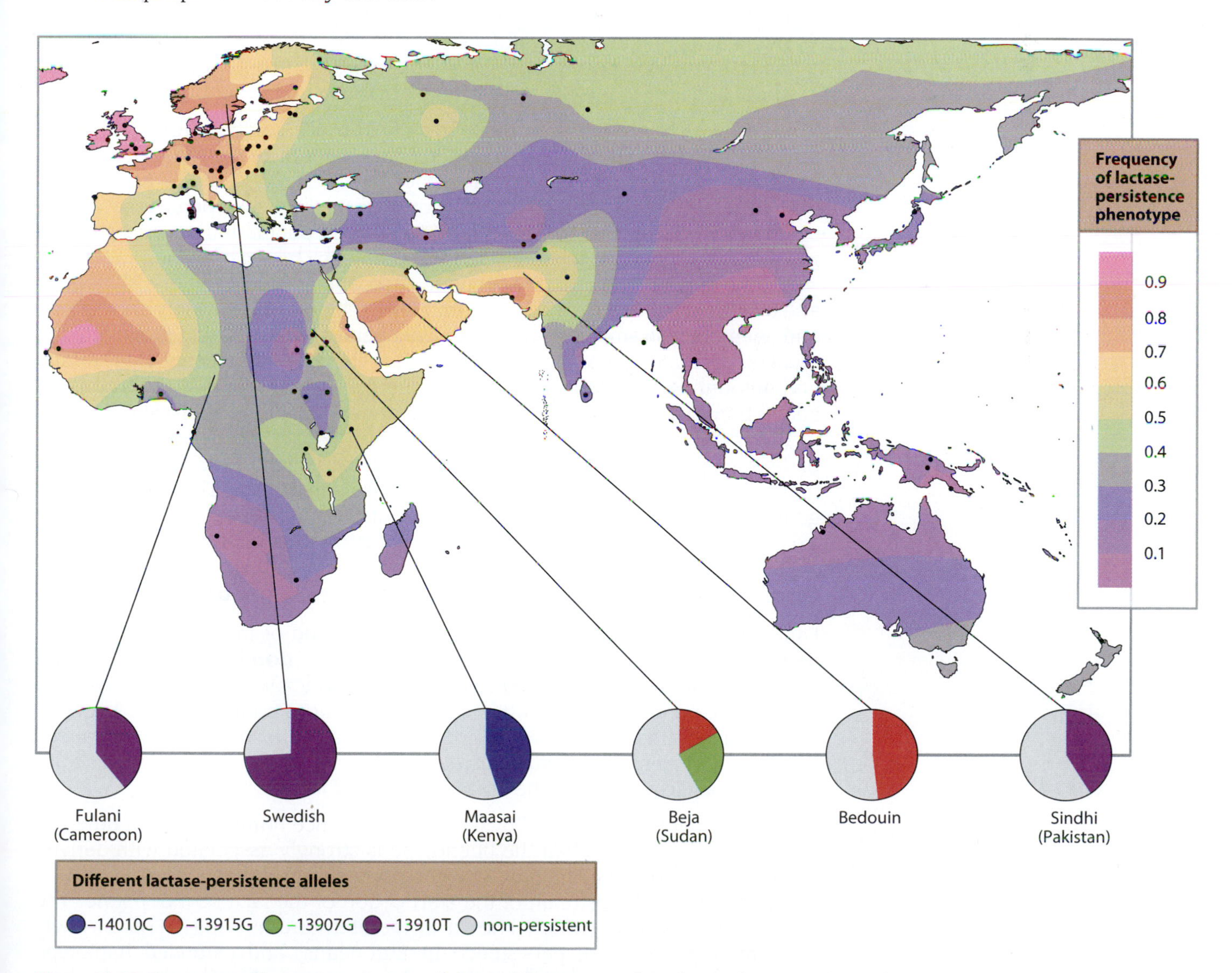

Figure 15.11: Frequency of lactase-persistence alleles in different populations.
The map, together with the key on the right, shows the frequency of the lactase-persistence phenotype, with sampling locations shown as *black* dots. The pie charts, together with the key below, show the frequencies of different lactase-persistence alleles in different populations. [Adapted from Itan Y et al. (2010) *BMC Evol. Biol.* 10, 36. With permission from BioMed Central.]

The adaptive hypotheses make the reasonable assumption that lactase non-persistence is the ancestral state because other mammals are non-persistent. Several selective explanations have been proposed for the variation in lactase persistence, the three most persuasive being based on the increase in frequency of persistence alleles in populations practicing dairying:

- *Food value*: the lactose component of fresh milk provides a valuable extra source of nutrition.
- *Water content*: milk provides important additional fluid in arid regions. For some desert nomads, milk may be the only source of water at some times of year, and diarrhea as a consequence of lactose intolerance would be strongly disadvantageous. In addition, it can provide a pathogen-free drink anywhere.
- *Improved calcium absorption*: this hypothesis was put forward to explain the rapid establishment of a high frequency of the phenotype in northern Europe, even though mixed farming meant lower reliance upon milk than elsewhere. As was discussed in **Section 15.3**, populations in northern latitudes may be susceptible to rickets and osteomalacia because of low sunlight and consequent low vitamin D levels. Calcium can help to prevent rickets, probably by reducing the breakdown of vitamin D in the liver. Milk contains calcium, and lactose promotes its absorption; therefore lactase-persistent individuals should be able to absorb more calcium than non-persistent individuals.

How can we choose between these hypotheses? We could try to examine the correlations between pastoralism, aridity, and amount of sunshine on the one hand, and lactase persistence on the other. The difficulty with this approach is that, in effect, it regards each population with high lactase persistence as an independent occurrence of the trait, and ignores population co-ancestry. An attempt has been made to surmount this problem by taking population relationships into account using a phylogenetic approach.[31] Both a genetic phylogeny and a cultural phylogeny were used (but note that they are strongly correlated), and within the framework of these phylogenies, pastoralism explained the greatest amount of variance in lactase persistence levels, rather than aridity and amount of sunshine. This supports the first hypothesis that the trait is an adaptation to dairying. Furthermore, a maximum likelihood approach (Box 6.3) to the possible pathways of change between the four possible combinations of lactase persistence/non-persistence and dairying/non-dairying indicated that dairying probably originated before lactase persistence. This linkage of the genetic trait of lactase persistence to the cultural trait of milk production is an excellent example of **gene–culture co-evolution** (**Box 15.5**).

Lactase persistence is caused by SNPs within an enhancer of the lactase gene

Family and twin studies have shown that the mode of inheritance of lactase persistence is consistent with an underlying single **dominant** mutation with high penetrance (OMIM 223100). Therefore, in principle it should be easier to identify the genetic factors in this case than in more complex traits. Lactase-non-persistent individuals are homozygous for an autosomal recessive allele, and lactase-persistent individuals are either heterozygous or homozygous for a dominant allele preventing the normal decline of lactase activity.

Analysis of haplotypes shared by lactase-persistence homozygotes in different populations has shown that the phenotype is strongly associated with derived alleles of four SNPs which differ between populations, but all cluster in a small region about 14 kb upstream of the start codon of the lactase (*LCT*) gene. The T allele at rs4988235 (widely known in the literature as –13910T) is strongly associated with lactase persistence in Europeans, and extended haplotype tests (Figure 6.25) provide evidence of a **hard sweep**, showing that this allele has been strongly selected in recent history. Indeed, in ancient DNA analysis this variant was not found in eight individuals living in the early Neolithic (7–8 KYA) suggesting it was absent, or at low frequencies, when farming was first introduced into Europe.[11] Another variant (–14010C) is associated with lactase persistence in pastoralist populations from Kenya and Tanzania, including the Maasai. Again, analysis of extended haplotypes suggests recent positive selection for this allele. This provides strong support for the hypothesis that there has been strong positive selection for lactase persistence in response to dairying. Alleles at two other SNPs in this small genomic region are also associated with lactase persistence—the G allele of rs41480347 (–13915G) in Bedouin and Somali camel herders and the G allele of rs41525747 (–13907G) in Somali camel herders, together with several other rare variants within the same region. This is another example of convergent evolution of a favorable trait in different human populations: there are different molecular causes of lactase persistence in different parts of the world, arising from alleles of different closely linked SNPs (Figure 15.11).[34, 77] We can infer that selection began operating when the culture of milking domestic livestock arrived, or was developed, in each population. The genetic signature of selection differs between different populations: in Europeans and Maasai there is a signature of a hard selective sweep detectable by extended haplotype tests, while in the camel herders of Arabia and the Horn of Africa the existence of several rare variants that cause lactase persistence is evidence of a **soft sweep** in these populations (Figure 6.21).

All the alleles described that are associated with lactase persistence are neither in the coding region nor in a promoter of the lactase gene, but are in an enhancer which is part of a LINE within an intron of another gene, 14 kb away.

Box 15.5: Gene–culture co-evolutionary theory

We inherit two kinds of information from our ancestors, genetic and cultural, and both genetic and cultural adaptations are important in human evolution. The study of gene–culture co-evolution investigates the interaction of these two spheres and their effects on the evolutionary process.[44] Culture is an evolving set of beliefs, ideas, values, and knowledge that is learned and can be socially transmitted between individuals. Sometimes, selective pressures acting upon genes can be modified by culture (such as sexual selection with culturally transmitted mating preferences, see below). On the other hand, whether or not culture is adopted by an individual will depend on their genetic constitution (the example of lactase persistence is described below).

The quantitative study of gene–culture co-evolution was originated by Feldman and Cavalli-Sforza.[21] Their models incorporated both the differential transmission of genes from one generation to the next, and the diffusion of cultural information, allowing the evolution of the two to be mutually dependent. While gene transmission is an exclusively "vertical" process (from parents to children), cultural transmission can be "vertical," "oblique" (for example, from teachers to children), or "horizontal" (among siblings or friends within the same generation). Oblique or horizontal transmission is faster than vertical transmission, which is necessarily measured in generations.

According to gene–culture co-evolutionary theory, sexual selection can alter allele frequencies even if mating preferences are entirely learned and culturally transmitted. Mathematical modeling suggests that if a culturally transmitted mating preference increases in the population by learning or just by chance, it can drive the frequency of a genetic trait to fixation, even if that trait is maladaptive in other ways. Given the increasing evidence for social learning of mate-choice, the difference in perception of female attractiveness between groups (see Box 15.4), and how such preferences change rapidly over time, sexual selection may have even more drastic consequences on allele frequencies than previously imagined.[43]

Gene–culture co-evolutionary theory can also be applied to the question of whether or not selection pressures following the adoption of dairying led to the spread of lactase persistence.[22] This analysis has shown that the persistence allele (see Section 15.6) could increase to high frequencies within ~300 generations (the interval between the origin of dairying and the present) only if there is strong vertical cultural transmission of milk consumption. If a significant proportion of the offspring of milk consumers did not themselves drink milk, it would require an unrealistically high selective advantage of milk drinking for the allele frequency to rise. The implication of this is that differences between cultures in the strength of cultural transmission might have had a large effect on the current distribution of the persistence trait. In this example the complicating effect of culture means that the predictions of traditional genetic models could arrive at the wrong answer.

They would not be considered good candidates for having the effect of lactase expression without functional analysis. Evidence comes from using a **reporter gene** construct, where a reporter gene (encoding luciferase or β-galactosidase) was **ligated** to the lactase promoter and different alleles of the region associated with lactase persistence. When introduced into a human intestinal cell line that expresses lactase, the alleles associated with persistence increased expression of the reporter gene, supporting the idea that these alleles increase expression of lactase in the adult human intestine.

Increased copy number of the amylase gene reflects an adaptation to a high-starch diet

The agricultural revolution was accompanied by major changes in diet for many human populations, and it would be strange if the ability to digest lactose were the only example of a dietary adaptation. Adaptations to increased levels of carbohydrate in the diet are also strong candidates. The domestication of plants such as wheat and rice is likely to have led to a large increase in dietary starch. For example, rice domestication occurred between 6 and 7.5 KYA in the lower Yangtze region of China, and archaeological evidence suggests that a large increase in rice consumption followed.[25] Even as recently as 40 years ago, rice accounted for 40% of the calorie intake for the Chinese and Japanese, so if this reflects the proportion of calorie intake due to rice after domestication, the selective advantage of efficient digestion of rice must have been significant. Rice is 80% starch, and starch is mostly comprised of the polysaccharide amylose, which is digested by amylases in the saliva and pancreatic bile to its constituent disaccharides, maltose and isomaltose. These are then cleaved into the monosaccharide glucose by the enzymes maltase-glucoamylase and sucrase-isomaltase (**Figure 15.12**).

Although a rich source of calories, a diet rich in starches would have placed stress on the enzymatic capacities of the digestive system in agricultural populations. The salivary amylase gene (*AMY1*) is polymorphic in copy number, with copy numbers per diploid genome ranging from 2 to 15 (**Figure 15.13**). Individuals with a higher gene copy number of *AMY1* produce more amylase in their saliva, and are therefore able to digest starch more efficiently. This increase in *AMY1* copy number is common to all human populations, in contrast to chimpanzees, where the copy number of 2 per diploid genome is not variable. This increase

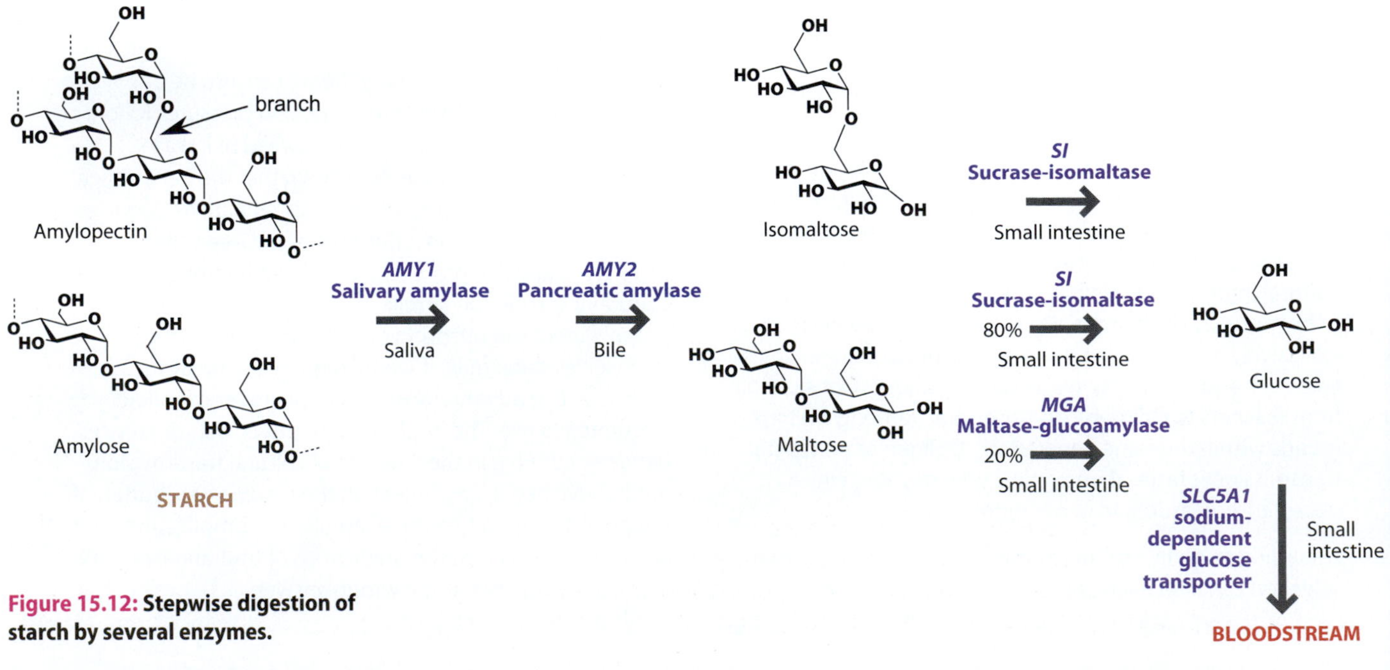

Figure 15.12: Stepwise digestion of starch by several enzymes.

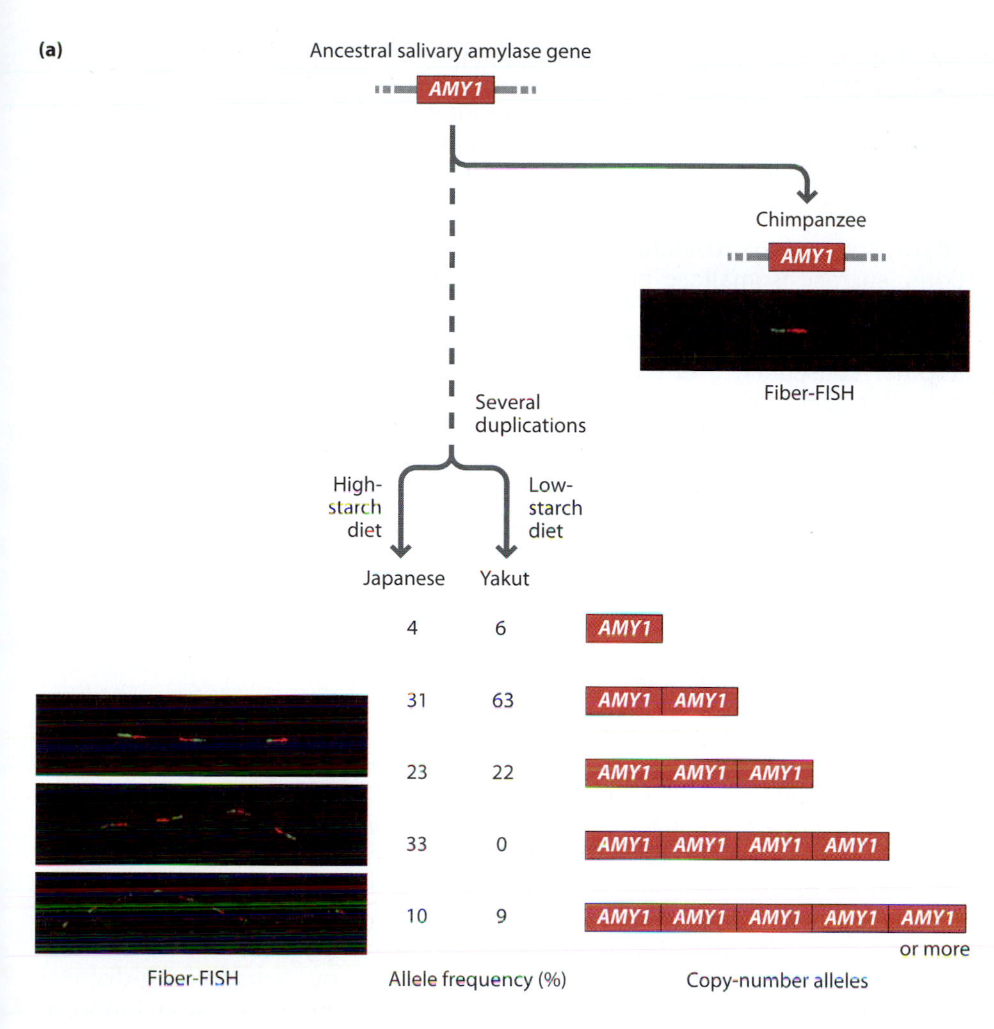

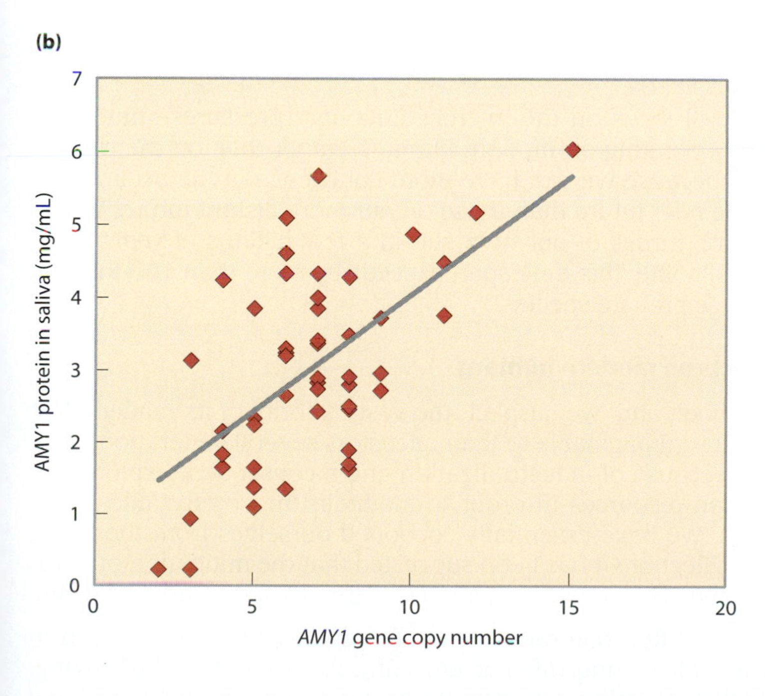

Figure 15.13: Copy-number variation of the salivary amylase gene and starch consumption. Part (a) shows that amplification of the salivary amylase gene *AMY1* occurred only in the human lineage, and higher copy-number alleles predominate in the Japanese, who eat a starch-rich diet, compared with the Yakut population. Fiber-FISH shows examples of the multiple copies of the gene on stretched DNA fibers. Part (b) shows the relationship between *AMY1* gene copy number and amount of the salivary amylase protein in saliva. [Adapted from Perry GH et al. (2007) *Nat. Genet.* 39, 1256. With permission from Macmillan Publishers Ltd.]

in *AMY1* copy number in humans has been suggested to be an adaptation to increased starch in the diet due to increased consumption of plant tubers by early hominins. Furthermore, individuals from a population that traditionally has a low-starch diet, such as the Yakut of northeastern Siberia, tend to have lower *AMY1* gene copy number than people with a high-starch diet, such as the Japanese.[62] This difference within modern humans may be due to selection in high-starch-diet populations favoring an increased digestive capacity of amylase, to make best use of dietary calories when healthy, and when ill, such as during episodes of diarrhea, where oral digestion of starch can be very important.[47]

There is suggestive evidence that, in the Han Chinese and Japanese populations, sucrase-isomaltase has undergone strong recent natural selection.[80] It is not known, at the moment, whether this selection is for an allele coding for a more active variant of sucrase-isomaltase. If so, it would suggest that this is another consequence of the selective pressure for coping with a high-starch diet. Deficiency of sucrase-isomaltase, caused by alleles in the sucrase-isomaltase gene that inactivate the encoded enzyme, reaches 10% frequency in Inuit of Greenland, a population whose calories traditionally come not from starch or sugar, but from animal products. This suggests inactivating alleles of sucrase-isomaltase in Inuits have reached a high frequency by genetic drift because of the absence of purifying selection.

15.7 THE FUTURE OF HUMAN EVOLUTION

Have we stopped evolving?

The human population is clearly not in a state of equilibrium. Recent patterns of migration and population growth are far removed from those that have predominated over much of human prehistory. But while the human species is undoubtedly in a state of flux, *have we stopped evolving?* The motivation for this question derives from the observation that humans occupy a broad range of environments and respond to environmental changes by evolving predominantly in a cultural rather than biological fashion. While we may yet uncover genetic changes that have aided their colonization of isolated islands, Polynesians have prospered in the South Pacific by virtue of their innovations in long-distance navigation and vessel construction, and ability to support large populations in limited environments through fishing and agriculture. Likewise, despite conforming to Bergmann and Allen's rules (**Section 15.2**), Inuit would not survive long in the Arctic without the cultural acquisitions of clothing, shelter, and fire.

In this section, we will focus on the microevolutionary pressures—mutation, drift, and selection—operating on modern humans, rather than on the macroevolutionary future, because we can have more confidence in the accuracy of predictions about the near future than those on our more distant future. Whilst the catastrophic death throes of our own sun in a few billions of years' time exercise the minds of some, this time span exceeds by more than 100-fold the life span of an average primate species.

Natural selection acts on modern humans

The authors of this book and, we suspect, the vast majority of its readers, live in a degree of comfort unimaginable to their ancestors several generations ago. This has happened because of industrialization and a consequent exploitation of natural and human resources in order to maintain the comfortable life in developed countries. We have essentially cocooned ourselves from the worst excesses of natural selection; it has been suggested that the mitigation of natural selection is a defining feature of civilization (see Box 15.4, for Darwin's view).

Nevertheless, it is clear that natural selection is ongoing in humans, even in developed countries. Measuring this current natural selection is challenging, but can be approached by using data from large, long-term, multigenerational **longitudinal studies**, established initially to measure social and genetic risk factors for disease. In developed countries, where almost all such studies are carried out, child mortality is low and most individuals survive from birth through child-bearing age. In these populations, evolutionary fitness can be estimated by measuring completed family size, or lifetime reproductive success. By examining how different traits, such as adult height, are correlated with lifetime reproductive success, the amount of natural selection acting on those traits can be estimated. If a trait's heritability is high, indicating that a large amount of the variation within it is genetic, then natural selection on that trait is likely to have a substantial impact on the frequencies of associated alleles.[73]

This approach has already yielded interesting results, such as selection for tall men and short women, and is likely to be a powerful approach in analyzing even larger longitudinal studies in the future. However, it will be difficult to predict the effect of current natural selection on allele frequencies, given the long time-scale required to significantly change allele frequencies, even in a trait caused by a single gene. In addition, many of the traits have a substantial cultural component, which can vary drastically even over two or three generations.

Can we predict the role of natural selection in the future?

The need for adaptation to a sustainable human economy (see **Box 15.6**) will impose unknown demands on human adaptability, and whether these demands can be satisfied purely by cultural evolution alone remains to be seen. Our knowledge of human evolution and what has driven adaptation in the past (see also Chapters 16 and 17) suggests that any future biological adaptation is most likely to occur from the following three aspects of environmental change.

Climate change

As we have seen in this chapter, and preceding chapters, environmental changes in the past have been instrumental in facilitating the spread of humans and in determining the pattern of land use. The past 10 KY have seen a far more stable global climate than in any period of similar length in the previous 100 KY. This has allowed humans to develop complex societies, but the current century will see major changes in how humans interact with the global environment. The scientific consensus is that climate change is inevitable, with an increase of 2.5–4°C predicted within the next 100 years, resulting in a sea level rise of up to 50 cm. The outcomes for different regions are likely to be very different, including increases and decreases in temperatures, and greater frequency and magnitude of extreme events such as droughts and floods. These changes will have different impacts on individual countries which themselves vary in their abilities to adapt culturally, but in general will lead to greater adverse impacts on people in developing countries. Sea level rises resulting from melting glaciers and ice sheets will potentially change settlement patterns in flood-prone areas, resulting in the migration of large numbers of people.

Dietary change

Response of food crops to climate change in the short term is unclear. Small increases in temperature may increase crop yields, but this is likely to be offset by extreme weather events and restrictions in water availability for irrigation. Instances of malnutrition because of crop failure are likely to increase, and, combined with infectious disease, are likely to increase mortality, especially in children.

Infectious disease

Infectious disease plays a dominant role in mortality in the developing world (**Figure 15.14**), and genetic variation is important in mediating variable resistance to infectious diseases (**Section 16.4** and Chapter 17). The spread of infectious disease has been greatly facilitated by increases in mobility, and is likely to be exacerbated by climate change. The aftermath of an extreme weather event such as a flood often includes disease epidemics, for example outbreaks of diarrheal diseases. Even in the absence of such events, the geographical distributions of certain infectious diseases are already changing because of climatic oscillations.[59] The movement of infectious disease into new areas, both directly by the pathogen and indirectly by expansion of the **vector** range, will expose

Box 15.6: The oversized ecological footprint of *Homo sapiens*

The consumption of environmental resources by the 7 billion human inhabitants of planet Earth is unsustainable. A consideration of our species' "ecological footprint" is perhaps the most effective method for demonstrating this. An ecological footprint is "*the area of biologically productive land and water that is required to produce the resources consumed and wastes generated by humanity,*" and can be compared with the amount of biologically productive land available on our planet. Seven major activities determine our land requirements:

- Growing crops
- Grazing animals
- Harvesting timber
- Marine and freshwater fishing
- Accommodation of residential and industrial facilities
- Burning fossil fuels
- Waste disposal

From this list it can be appreciated that an individual's ecological footprint is dependent on their lifestyle. Thus footprints can be calculated for the average inhabitant of any given country to compare the equality of land usage and predict future ecological footprints on the basis of demography and development. Calculations based primarily on the first six of the activities listed above reveal that humanity's ecological footprint already exceeds the total biologically productive land area available on our planet, and has done so for over two decades (**Figure 1**). Driven primarily by increases in carbon footprint allied to the growing human population, our ecological footprint will expand even faster.

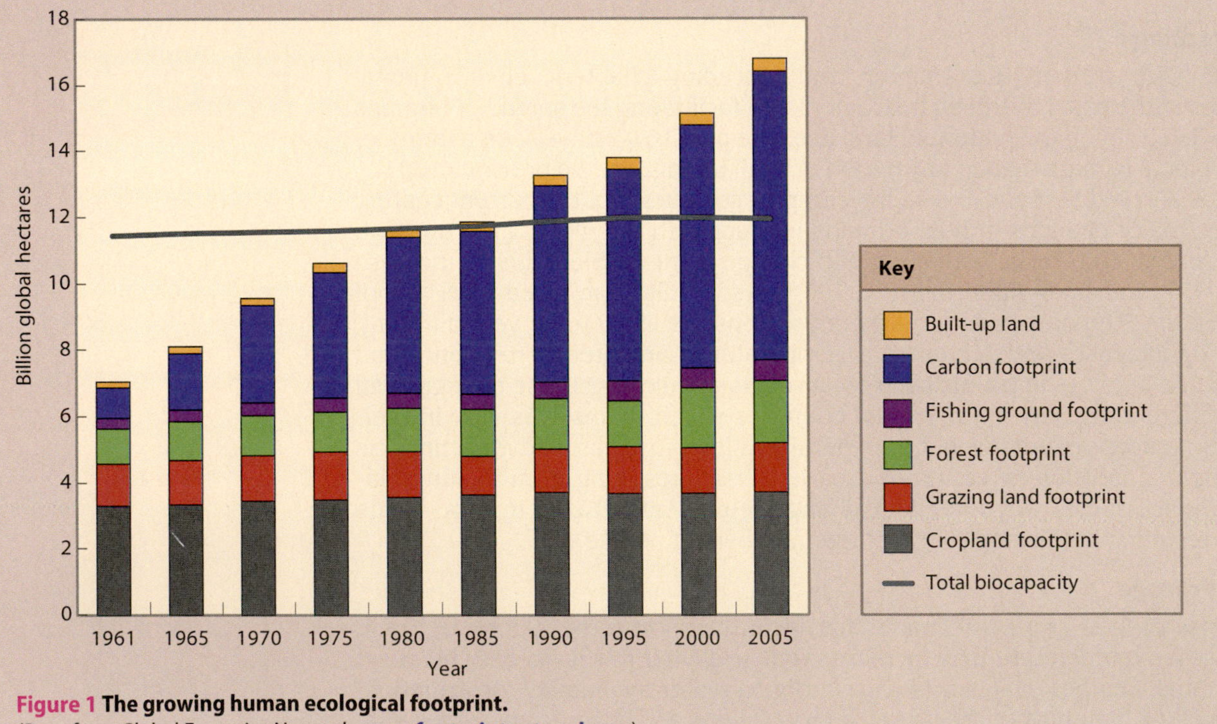

Figure 1 The growing human ecological footprint.
(Data from Global Footprint Network **www.footprintnetwork.org**)

populations who do not have even the partial genetic resistance given by long-term exposure. New pathogens will continue to evolve and emerge, adapting to new hosts and circumventing both immune and chemical defenses; seven billion large-bodied mammals with limited genetic diversity and high mobility represents a host environment of almost unparalleled richness for a new pathogen. Taken together, mortality rates from infectious disease will increase, and perhaps provide the most immediate danger of catastrophic mortality.

What will be the effects of future demographic changes?

It is clear that, in general, improved nutrition and sanitation have led to dramatic reductions in infant mortality and increases in life expectancy over the

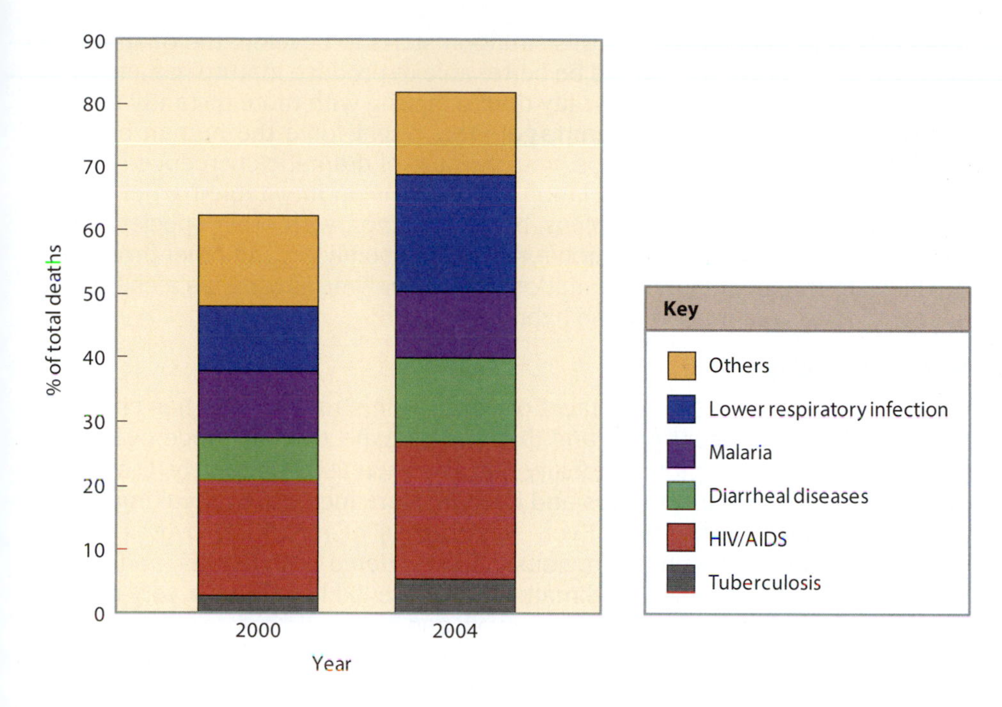

Figure 15.14: An increasing burden of infectious disease in Africa. Figures show the percentage of all deaths attributable to different infectious diseases, including respiratory disease. Note that the total number of reported deaths from all causes also rose from 10,200,000 in the year 2000 to 11,200,000 in 2004. (Data from the World Health Organization.)

past few centuries. This has resulted in an exponential increase in population size: our present census population size of over 7 billion contrasts sharply with our long-term effective population size of only about 10,000 individuals (**Section 6.6**), and the human population is predicted to keep growing in size for some time yet. Although alternative models of future population growth differ substantially in their projections, most agree that the rate of population increase is finally slowing, perhaps for the first time since the advent of agriculture some 10 KYA.

Increasing population size

In larger populations, as genetic drift lessens, smaller selection pressures become able to influence allele frequencies (**Section 5.6**). An increasing effective population size should allow selection pressures of lesser magnitude that were previously overridden by stochastic allele frequency changes to make their presence felt. This feature of future human populations may interact synergistically with the different selective environments discussed in the previous sections of this chapter.

Effective population size determines the amount of neutral genetic diversity within a species. As genetic drift lessens, the balance between mutation introducing new alleles and drift removing them will be adjusted (**Section 5.6**). Thus, if the population remains at a large and constant size for a long period of time, we can expect much higher levels of specieswide genetic diversity in the future. When comparing two European DNA sequences, we will see a difference about every 1250 bp on average, which reflects the rarity (in a given DNA sequence) of variants with common alleles in the population. However, sequencing studies detect a large number of very rare variants, reflecting the recent expansion of human populations. Indeed, one study sequencing 864 kb of DNA in 14,000 individuals detected a variable site every 17 bp.[57] If the human population establishes mutation–drift equilibrium at even a small fraction of the current population size, some of these rare variants will become more common due to genetic drift, and this level of diversity will be likely to disrupt homologous recombination. The result would be repression of recombination during meiosis, although it is unclear what effect islands of sequence identity resulting from regions of selective constraint would have on this process. This repression would in turn result in frequent chromosomal aneuploidy, leading to a dramatic reduction in human fertility. As this situation starts to develop, the offspring of closely related individuals would be better able to produce mature gametes and would therefore have greater fertility than offspring with more distantly related parents (**heterozygote disadvantage**). This might force the human population into isolated breeding groups, with barriers of dramatically reduced hybrid fertility leading to eventual speciation. This seems an inevitable barrier to the long-term integrity of *Homo sapiens*, but in common with other species, environmental change, sexual selection for courtship behavior, and other processes are likely to fragment larger populations before polymorphism has a chance to accrue to levels required to cause hybrid sterility.[15]

Increased mobility

Population size is not the only facet of population structure that has changed significantly in the recent past, and that can be expected to change over subsequent centuries. Individual mobility has also increased massively. Distances between the birthplaces of wives and husbands are increasing, breaking down prior population substructure. Even small levels of migration over multiple generations can homogenize previously differentiated populations, and large-scale migrations as a result of climate change are likely to have a very similar homogenizing effect. Over time, phenotypic and genotypic differences between regional populations can be expected to lessen as a result of this increased mobility. A greater proportion of global genetic diversity will be found within any single population. Measures of apportionment of diversity, such as F_{ST}, will tend to zero if our species ever approaches **panmixis**.

Differential fertility

The phenomenon of differential fertility is easy to demonstrate, but showing that it has any genetic basis is far more difficult. A proof of its heritability is not sufficient, because there is much potential for social as well as biological inheritance (see Box 15.1). Given these challenges, it is difficult to determine how much natural selection is acting on differential fertility.

Differential fertility is a common feature of social inequalities, although it appears that a shift in direction has occurred in recent times. Ethnographic studies of indigenous societies often show that high status affords better access to resources and more offspring. By contrast, in the developed world it is the lower socioeconomic sections of society that consistently out-reproduce the higher groups.

A long-term study of twins suggests that there is an appreciable genetic basis underlying differential fertility.[40] The researchers compared the family sizes of monozygotic and dizygotic twins and found that about 40% of the variance in reproductive success could be attributed to genetic factors, even after taking into account differences in education background and religious affiliation amongst the women. These environmental factors do play a significant role: Roman Catholic women had 1.4 times more reproductive success than non-religious women did, and those educated the least also had 1.4 times more reproductive success than those with a university education. The latter statistic is the kind that disturbs **eugenicists**.

Such a high degree of genetic heritability could not be sustained over the long term, as beneficial alleles would quickly become fixed. The suggestion was therefore made[40] that selection might be acting on different traits than in previous generations. This provides a mechanism linking cultural and biological evolution. A constantly changing cultural world allows previously unimportant genetically encoded traits to determine reproductive fitness, thus maintaining a high level of genetic heritability. For example, it might be that in pre-industrial societies, reproductive biology (for example, wide birth canal) was the most important determinant of reproductive fitness, whereas in modern developed societies, behavioral traits (for example, desire for a large family) may have become more important. Over the long term, if the reproductive culture remains relatively stable, the heritability of differential fitness is likely to decline as a result of the spread of the beneficial alleles.

In the meantime, natural selection on differential fertility might be driving dramatic changes in specific alleles responsible for greater reproductive fitness. This appears to be happening at a large polymorphic inversion on chromosome 17, where females carrying the inversion allele have more children than non-carriers, providing a strong force that will increase inversion allele frequency (**Box 15.7**). The reason for this is unclear, but inversion carriers have a higher genomewide recombination rate than non-carriers, and differential recombination rate has been directly linked with differential fertility.

Differential generation time

An increase in generation time will result in an increase in the number of male germ-line replications per generation, while the number in females will stay the same. This is because in males, a fixed number of replications occur prior to puberty (~35) and then replication accelerates to a much faster rate (~23 per year). If generation time increases, then the length of time spent at this faster rate increases, and the result is an increase in mutation rate driven by increased paternal age (**Section 3.7**). This has a particular effect on the male-specific Y chromosome compared to autosome; conversely this phenomenon will affect the X chromosome less than autosomes, because an X chromosome spends two-thirds of its time in females.

It is generally assumed that the generation time has increased recently, and will continue to increase, as the age of the female at first birth is higher now than was likely in the past. For example, the average age of the mother at first birth in the Organisation of Economic Co-operation and Development (OECD) countries in 2005 ranged from 21.3 (Mexico) to 29.8 (UK), with an average of 27.7; this is high compared to hunter-gatherer humans (19.5 years). However, the average age of the mother at last birth is also important for determining generation time, and for this statistic the data are patchy. Indeed, generation times are remarkably consistent in industrial and hunter-gatherer populations (Table 6.6), and analysis of Icelandic genealogical records[42] has shown remarkable constancy since 1650 of paternal age at conception of child of about 35 years of age, with a temporary reduction between 1950 and 2000.

Box 15.7: Recombination rate, maternal age, and fertility

Analysis of a large dataset from the Icelandic population has shown that maternal meiotic recombination rate increases with maternal age.[41] This is surprising because if, as is generally accepted, all the mother's eggs are produced when she herself was a fetus, then all the meiotic recombination would have occurred at that time.

In this study meiotic recombination events were measured by analysis of living offspring, and meiotic recombination events from failed pregnancies (due to chromosomal rearrangements and aneuploidies) could not be measured. So perhaps this increase in recombination with maternal age is due to selection bias: mothers with higher maternal recombination rates are more likely to produce viable offspring when older. Perhaps higher recombination rates protect the fetus against maternal age-related **nondisjunction**, which causes aneuploidy and is therefore responsible for the larger number of fetuses not carried to term in older women. Further analysis of the Icelandic data supports this: after accounting for the effect of maternal age, mothers with higher recombination rates had more children. Analysis of a smaller dataset also confirmed the association of maternal age and recombination rate, and recombination rate with family size.[14]

There is substantial genetic variation in recombination rate between different individuals, with a heritability estimate of 30%. If higher recombination rate has a positive effect on reproductive success, then alleles which increase the genomewide recombination rate will be strongly selected in the population. Alleles that affect the rate of recombination at particular **recombination hotspots** are known,[37] and locations of these hotspots are affected by the alleles of the *PRDM9* gene (Section 3.8).

Will the mutation rate change?

Both endogenous and exogenous mutagenic factors (**Section 3.2**) are likely to determine whether mutation rates remain constant in the near future. Endogenous mutagenic factors include DNA replication errors, reactive by-products of metabolism, and DNA repair efficiency, which are extremely unlikely to change in the short to medium term. In contrast, levels of exogenous mutagenic factors, including chemical mutagens, ultraviolet (UV) radiation, and ionizing radiation, have the potential to change rapidly through anthropogenic causes, and we should be aware of their potential medium- and long-term effects.

The number of synthetic chemicals in the environment is increasing but their mutagenic potential is assessed prior to their release and levels are regulated, at least in developed countries. The monitoring of predetermined safety levels by environmental agencies has thus far ensured that global chemical mutagenicity has not altered significantly, although it is unclear how much local releases of large amounts of potential mutagens may affect local populations in the medium term (for example, industrial release of methyl isocyanate in Bhopal, India, in 1986).

Nuclear weapons testing and radiation leaks have increased levels of background radiation, albeit minimally. The global increase in radiation exposure as a result of our nuclear activities is far outweighed by regional variation in levels of natural radioactivity, principally from rocks. However, species have never (prior to nuclear weapons, nuclear accidents, and radiotherapy to treat cancer) been exposed to such high-energy ionizing radiation (IR) before, and there may well be unpredictable consequences in the way that cells behave. The mutation rate at human minisatellites has been shown to be elevated by radiation exposure.[19] Work in mice has suggested that increased mutation rates as a result of exposure to high-energy IR are not restricted to the exposed generation, but can be inherited by future unexposed generations.[3]

Cosmic radiation with mutagenic capabilities is relatively constant in magnitude. Increased UV radiation has reached extreme latitudes through the ozone holes formed by **anthropogenic** degradation, resulting in an elevated prevalence of skin cancer, especially in Australasia. However, the low penetration of UV radiation means that it does not affect germ-line mutation rates.

SUMMARY

- There is much variation in morphology among humans. The contribution of genetic variation to this morphological variation is incompletely understood, and the adaptive significance of much of this is hard to prove.
- Variation in pigmentation has been widely studied, and several hypotheses have been put forward to explain it. The most likely explanation is that it results from a balance between the opposing needs for UV exposure for vitamin D synthesis, and protection from UV damage.
- Variation at several genes is known to contribute to pigmentation. Variation at *MC1R* can cause red hair and fair skin, and one SNP in the *SLC24A5* gene accounts for about 30% of the phenotypic variation in skin color in certain populations, with the derived allele fixed in European populations.
- *EPAS1* and *EGLN1* genes have been identified in high-altitude populations as targets of recent positive selection. Tibetan, Andean, and Ethiopian populations are characterized by distinct physiological adaptations to high altitude and this is reflected in the convergent evolution of the adaptive mechanisms.
- Major dietary changes in the course of human evolution have occurred due to encounters with new plant species in human dispersals, and the transition from hunter-gathering to settled agricultural lifestyle. These changes are likely to have caused genetic adaptation in several genes encoding digestive enzymes and taste receptors.

- Lactase persistence is an adaptation to milk-drinking and represents an example of "gene–culture co-evolution" that has developed since the beginning of agriculture 10 KYA. One allele is responsible for lactase persistence in Europeans, but several other alleles cause the trait in other populations. The alleles are all within a ~100 bp region that contains an enhancer that controls expression of the lactase gene.
- Natural selection continues to act on modern human populations. As in the past, climatic factors, dietary change, and infectious disease are very likely to be the selective pressures that cause continued human evolution.
- An increasing population size will enable weaker selection to alter allele frequencies, and will mean genetic differences between individual humans will increase due to rare or even unique alleles. In contrast, increased mobility will reduce differences between human populations.

QUESTIONS

Question 15–1: Both adaptation to milk-drinking and adaptation to high altitude show evidence of convergent evolution at the genetic level. What are the similarities and the differences in the genetic changes underlying these adaptations?

Question 15–2: Lactase persistence is regarded as the classic example of gene–culture co-evolution. Could any of the other adaptations described in this chapter be subject to this form of evolution, and how might your hypotheses be tested?

Question 15–3: Compare and contrast the relative importance of genetic and cultural adaptation to:
(a) The availability of milk from domestic animals.
(b) Low levels of ultraviolet light at higher latitudes.

Question 15–4: One of the assumptions in determining broad-sense heritability is that there is no genotype–environment interaction, where the effect of the genotype depends on the environment. Ignoring genotype–environment interactions will inflate the value of environmental variance (V_E). Giving examples, discuss how realistic this assumption is for human populations, and how it could affect estimates of broad-sense heritability.

Question 15–5: A serine–tryptophan polymorphism in the *TAS2R43* gene affects aloin taste sensitivity, with a tryptophan at position 35 being the sensitive allele.
(a) Using the UCSC Genome Browser, identify the ancestral amino acid at this polymorphism.
(b) The plants of the genus *Aloe* contain aloin and grow in semi-arid conditions. Given your answer to part (a), suggest a hypothesis for the evolution of this polymorphism.

Question 15–6: The United Nations predicts that the human global population is most likely to stabilize at 9 billion within the next 100 years. If the population level remains at this level for the next 10 KY, what would you predict about the following parameters, compared to today's values?
(a) Effective population size.
(b) Average heterozygosity.
(c) The strength of selection required to overcome effects of genetic drift.
(d) F_{ST} between populations.
Over the same time-scale, the chimpanzee population declines to 400 individuals, and remains stable thereafter. 200 chimpanzees are kept as an interbreeding population across the world's zoos, while 200 remain as wild in a nature reserve. In these instances, how will the four parameters (a–d) change?

REFERENCES

The references highlighted in purple are considered to be important (for this chapter) by the authors.

1. **Allen HL, Estrada K, Lettre G et al.** (2010) Hundreds of variants clustered in genomic loci and biological pathways affect human height. *Nature* **467**, 832–838.
2. **Aulchenko YS, Struchalin MV, Belonogova NM et al.** (2009) Predicting human height by Victorian and genomic methods. *Eur. J. Hum. Genet.* **17**, 1070–1075.
3. **Barber R, Plumb MA, Boulton E et al.** (2002) Elevated mutation rates in the germ line of first- and second-generation offspring of irradiated male mice. *Proc. Natl Acad. Sci. USA* **99**, 6877–6882.
4. **Beall CM** (2007) Two routes to functional adaptation: Tibetan and Andean high-altitude natives. *Proc. Natl Acad. Sci. USA* **104**, 8655–8660.
5. **Beall CM, Cavalleri GL, Deng L et al.** (2010) Natural selection on *EPAS1* (HIF2α) associated with low hemoglobin concentration in Tibetan highlanders. *Proc. Natl Acad. Sci. USA* **107**, 11459–11464.
6. **Beaumont KA, Shekar SN, Cook AL et al.** (2008) Red hair is the null phenotype of *MC1R*. *Hum. Mutat.* **29**, E88–E94.
7. **Bell JT & Spector TD** (2011) A twin approach to unraveling epigenetics. *Trends Genet.* **27**, 116–125.
8. **Boas F** (1910) Changes in bodily form of descendants of immigrants. United States Immigration Commission, Senate document no.208, 61st Congress. Washington, D.C.: Government Printing Office.

9. **Box NF, Duffy DL, Chen W et al.** (2001) *MC1R* genotype modifies risk of melanoma in families segregating *CDKN2A* mutations. *Am. J. Hum. Genet.* **69**, 765–773.

10. **Bufe B, Hofmann T, Krautwurst D et al.** (2002) The human *TAS2R16* receptor mediates bitter taste in response to β-glucopyranosides. *Nat. Genet.* **32**, 397–401.

11. **Burger J, Kirchner M, Bramanti B et al.** (2007) Absence of the lactase-persistence-associated allele in early Neolithic Europeans. *Proc. Natl Acad. Sci. USA* **104**, 3736–3741.

12. **Burow M, Bergner A, Gershenzon J & Wittstock U** (2007) Glucosinolate hydrolysis in *Lepidium sativum*—identification of the thiocyanate-forming protein. *Plant Mol. Biol.* **63**, 49–61.

13. **Campbell MC, Ranciaro A, Froment A et al.** (2012) Evolution of functionally diverse alleles associated with PTC bitter taste sensitivity in Africa. *Mol. Biol. Evol.* **29**, 1141–1153.

14. **Coop G, Wen X, Ober C et al.** (2008) High-resolution mapping of crossovers reveals extensive variation in fine-scale recombination patterns among humans. *Science* **319**, 1395–1398.

15. **Coyne JA & Allen Orr H** (1998) The evolutionary genetics of speciation. *Philos. Trans. R. Soc. Lond. B Biol. Sci.* **353**, 287–305.

16. **Currat M, Excoffier L, Maddison W et al.** (2006) Comment on "Ongoing adaptive evolution of *ASPM*, a brain size determinant in *Homo sapiens*" and "Microcephalin, a gene regulating brain size, continues to evolve adaptively in humans." *Science* **313**, 172.

17. **Darwin C** (1871) The Descent of Man and Selection in Relation to Sex. John Murray.

18. **Diamond J** (1991) The Rise and Fall of the Third Chimpanzee. Vintage.

19. **Dubrova YE, Nesterov VN, Krouchinsky NG et al.** (1996) Human minisatellite mutation rate after the Chernobyl accident. *Nature* **380**, 683–686.

20. **Evans PD, Gilbert SL, Mekel-Bobrov N et al.** (2005) Microcephalin, a gene regulating brain size, continues to evolve adaptively in humans. *Science* **309**, 1717–1720.

21. **Feldman MW & Cavalli-Sforza LL** (1976) Cultural and biological evolutionary processes: selection for a trait under complex transmission. *Theor. Popul. Biol.* **9**, 238–259.

22. **Feldman MW & Cavalli-Sforza LL** (1989) On the theory of evolution under genetic and cultural transmission with application to the lactose absorption problem. In Mathematical Evolutionary Theory (MW Feldman ed), pp 145–173. Princeton University Press.

23. **Flanagan N, Healy E, Ray A et al.** (2000) Pleiotropic effects of the melanocortin 1 receptor (*MC1R*) gene on human pigmentation. *Hum. Mol. Genet.* **9**, 2531–2537.

24. **Fox AL** (1932) The relationship between chemical constitution and taste. *Proc. Natl Acad. Sci. USA* **18**, 115–120.

25. **Fuller DQ, Qin L, Zheng Y et al.** (2009) The domestication process and domestication rate in rice: spikelet bases from the Lower Yangtze. *Science* **323**, 1607–1610.

26. **Fushan AA, Simons CT, Slack JP & Drayna D** (2010) Association between common variation in genes encoding sweet taste signaling components and human sucrose perception. *Chem. Senses* **35**, 579–592.

27. **García Borrón JC, Sánchez Laorden BL & Jiménez Cervantes C** (2005) Melanocortin 1 receptor structure and functional regulation. *Pigment Cell. Res.* **18**, 393–410.

28. **Harding RM, Healy E, Ray AJ et al.** (2000) Evidence for variable selective pressures at *MC1R*. *Am. J. Hum. Genet.* **66**, 1351–1361.

29. **Harpending H & Rogers A** (2000) Genetic perspectives on human origins and differentiation. *Annu. Rev. Genomics Hum. Genet.* **1**, 361–385.

30. **Hasin-Brumshtein Y, Lancet D & Olender T** (2009) Human olfaction: from genomic variation to phenotypic diversity. *Trends Genet.* **25**, 178–184.

31. **Holden C & Mace R** (1997) Phylogenetic analysis of the evolution of lactose digestion in adults. *Hum. Biol.* **69**, 605–628.

32. **Hollox EJ & Swallow DM** (2002) Lactase deficiency—biological and medical aspects of the adult human lactase polymorphism. In Genetic Basis of Common Diseases (RA King, JI Rotter & AG Motulsky eds), 2nd ed. pp 250–265. Oxford University Press.

33. **Howells WW** (1992) The dispersion of modern humans. In The Cambridge Encyclopedia of Human Evolution (S Jones, R Martin & D Pilbeam eds), pp 389–401. Cambridge University Press.

34. **Itan Y, Jones BL, Ingram CJE et al.** (2010) A worldwide correlation of lactase persistence phenotype and genotypes. *BMC Evol. Biol.* **10**, 36.

35. **Izagirre N, García I, Junquera C et al.** (2006) A scan for signatures of positive selection in candidate loci for skin pigmentation in humans. *Mol. Biol. Evol.* **23**, 1697–1706.

36. **Jablonski NG & Chaplin G** (2000) The evolution of skin coloration. *J. Hum. Evol.* **39**, 57–106.

37. **Jeffreys AJ & Neumann R** (2009) The rise and fall of a human recombination hot spot. *Nat. Genet.* **41**, 625–629.

38. **Julian CG, Wilson MJ & Moore LG** (2009) Evolutionary adaptation to high altitude: a view from *in utero*. *Am. J. Hum. Biol.* **21**, 614–622.

39. **Kim U, Jorgenson E, Coon H et al.** (2003) Positional cloning of the human quantitative trait locus underlying taste sensitivity to phenylthiocarbamide. *Science* **299**, 1221–1225.

40. **Kirk KM, Blomberg SP, Duffy DL et al.** (2001) Natural selection and quantitative genetics of life-history traits in Western women: a twin study. *Evolution* **55**, 423–435.

41. **Kong A, Barnard J, Gudbjartsson DF et al.** (2004) Recombination rate and reproductive success in humans. *Nat. Genet.* **36**, 1203–1206.

42. **Kong A, Frigge ML, Masson G et al.** (2012) Rate of *de novo* mutations and the importance of father's age to disease risk. *Nature* **488**, 471–475.

43. **Laland KN** (2008) Exploring gene–culture interactions: insights from handedness, sexual selection and niche-construction case studies. *Philos. Trans. R. Soc. Lond. B Biol. Sci.* **363**, 3577–3589.

44. **Laland KN & Brown GR** (2002) Sense and Nonsense: Evolutionary Perspectives on Human Behaviour. Oxford University Press.

45. **Lalueza-Fox C, Gigli E, De la Rasilla M et al.** (2009) Bitter taste perception in Neanderthals through the analysis of the *TAS2R38* gene. *Biol. Lett.* **5**, 809–811.

46. **Lamason RL, Mohideen MAPK, Mest JR, et al.** (2005) SLC24A5, a putative cation exchanger, affects pigmentation in zebrafish and humans. *Science* **310**, 1782–1786.

47. **Lebenthal E** (1987) Role of salivary amylase in gastric and intestinal digestion of starch. *Dig. Dis. Sci.* **32**, 1155–1157.

48. **Lee RJ, Xiong G, Kofonow JM et al.** (2012) T2R38 taste receptor polymorphisms underlie susceptibility to upper respiratory infection. *J. Clin. Invest.* **122**, 4145–4159.

49. **Liu F, van der Lijn F, Schurmann C et al.** (2012) A genome-wide association study identifies five loci influencing facial morphology in Europeans. *PLoS Genet.* **8**, e1002932.

50. **Madrigal L & Kelly W** (2007) Human skin color sexual dimorphism: a test of the sexual selection hypothesis. *Am. J. Phys. Anthropol.* **132**, 470–482.

51. **Manica A, Amos W, Balloux F & Hanihara T** (2007) The effect of ancient population bottlenecks on human phenotypic variation. *Nature* **448**, 346–348.

52. **McQuillan R, Eklund N, Pirastu N et al.** (2012) Evidence of inbreeding depression on human height. *PLoS Genet.* **8**, e1002655.

53. **Mekel-Bobrov N, Gilbert SL, Evans PD et al.** (2005) Ongoing adaptive evolution of *ASPM*, a brain size determinant in *Homo sapiens*. *Science* **309**, 1720–1722.

54. **Mekel-Bobrov N, Posthuma D, Gilbert SL et al.** (2007) The ongoing adaptive evolution of ASPM and Microcephalin is not explained by increased intelligence. *Hum. Mol. Genet.* **16**, 600–608.

55. **Menashe I, Abaffy T, Hasin Y et al.** (2007) Genetic elucidation of human hyperosmia to isovaleric acid. *PLoS Biol.* **5**, e284.

56. **Nei M & Roychoudhury A** (1982) Genetic relationship and evolution of human races. *Evol. Biol.* **14**, 1–59.

57. **Nelson MR, Wegmann D, Ehm MG et al.** (2012) An abundance of rare functional variants in 202 drug target genes sequenced in 14,002 people. *Science* **337**, 100–104.

58. **Norton HL, Kittles RA, Parra E et al.** (2007) Genetic evidence for the convergent evolution of light skin in Europeans and East Asians. *Mol. Biol. Evol.* **24**, 710–722

59. **Patz JA & Kovats RS** (2002) Hotspots in climate change and human health. *BMJ* **325**, 1094–1098.

60. **Peng J, Zhang L, Drysdale L & Fong GH** (2000) The transcription factor EPAS-1/hypoxia-inducible factor 2 plays an important role in vascular remodeling. *Proc. Natl Acad. Sci. USA* **97**, 8386–8391.

61. **Perola M, Sammalisto S, Hiekkalinna T et al.** (2007) Combined genome scans for body stature in 6602 European twins: evidence for common Caucasian loci. *PLoS Genet.* **3**, e97.

62. **Perry GH, Dominy NJ, Claw KG et al.** (2007) Diet and the evolution of human amylase gene copy number variation. *Nat. Genet.* **39**, 1256–1260.

63. **Pickrell JK, Coop G, Novembre J et al.** (2009) Signals of recent positive selection in a worldwide sample of human populations. *Genome Res.* **19**, 826–837.

64. **Pronin AN, Xu H, Tang H et al.** (2007) Specific alleles of bitter receptor genes influence human sensitivity to the bitterness of aloin and saccharin. *Curr. Biol.* **17**, 1403–1408.

65. **Relethford JH** (1992) Cross-cultural analysis of migration rates: effects of geographic distance and population size. *Am. J. Phys. Anthropol.* **89**, 459–466.

66. **Relethford JH** (2004) Boas and beyond: migration and craniometric variation. *Am. J. Hum. Biol.* **16**, 379–386.

67. **Rogers J, Mahaney MC, Almasy L et al.** (1999) Quantitative trait linkage mapping in anthropology. *Yearbk Phys. Anthropol.* **42**, 127–151.

68. **Ruff C** (2002) Variation in human body size and shape. *Annu. Rev. Anthropol.* **31**, 211–232.

69. **Shigemura N, Shirosaki S, Ohkuri T et al.** (2009) Variation in umami perception and in candidate genes for the umami receptor in mice and humans. *Am. J. Clin. Nutr.* **90**, 764S–769S.

70. **Simonson TS, Yang Y, Huff CD et al.** (2010) Genetic evidence for high-altitude adaptation in Tibet. *Science* **329**, 72–75.

71. **Soranzo N, Bufe B, Sabeti PC et al.** (2005) Positive selection on a high-sensitivity allele of the human bitter-taste receptor *TAS2R16*. *Curr. Biol.* **15**, 1257–1265.

72. **Sparks CS & Jantz RL** (2002) A reassessment of human cranial plasticity: Boas revisited. *Proc. Natl Acad. Sci. USA* **99**, 14636–14639.

73. **Stearns SC, Byars SG, Govindaraju DR & Ewbank D** (2010) Measuring selection in contemporary human populations. *Nat. Rev. Genet.* **11**, 611–622.

74. **Stokowski RP, Pant P, Dadd T et al.** (2007) A genomewide association study of skin pigmentation in a South Asian population. *Am. J. Hum. Genet.* **81**, 1119–1132.

75. **Sulem P, Gudbjartsson DF, Stacey SN et al.** (2007) Genetic determinants of hair, eye and skin pigmentation in Europeans. *Nat. Genet.* **39**, 1443–1452.

76. **Timpson N, Heron J, Smith GD & Enard W** (2007) Comment on papers by Evans et al. and Mekel-Bobrov et al. on evidence for positive selection of *MCPH1* and *ASPM*. *Science* **317**, 1036.

77. **Tishkoff SA, Reed FA, Ranciaro A et al.** (2006) Convergent adaptation of human lactase persistence in Africa and Europe. *Nat. Genet.* **39**, 31–40.

78. **Turchin MC, Chiang CWK, Palmer CD et al.** (2012) Evidence of widespread selection on standing variation in Europe at height-associated SNPs. *Nat. Genet.* **44**, 1015–1019.

79. **Visscher PM, Hill WG & Wray NR** (2008) Heritability in the genomics era—concepts and misconceptions. *Nat. Rev. Genet.* **9**, 255–266.

80. **Voight BF, Kudaravalli S, Wen X & Pritchard JK** (2006) A map of recent positive selection in the human genome. *PLoS Biol.* **4**, e72.

81. **Weedon MN & Frayling TM** (2008) Reaching new heights: insights into the genetics of human stature. *Trends Genet.* **24**, 595–603.

82. **Wooding S, Bufe B, Grassi C et al.** (2006) Independent evolution of bitter-taste sensitivity in humans and chimpanzees. *Nature* **440**, 930–934.

83. **Wooding S, Kim U, Bamshad MJ et al.** (2004) Natural selection and molecular evolution in ptc, a bitter-taste receptor gene. *Am. J. Hum. Genet.* **74**, 637–646.

84. **Zagrobelny M, Bak S, Rasmussen AV et al.** (2004) Cyanogenic glucosides and plant-insect interactions. *Phytochemistry* **65**, 293–306.

EVOLUTIONARY INSIGHTS INTO SIMPLE GENETIC DISEASES

CHAPTER SIXTEEN

In the previous chapter we discussed the evolution of particular human phenotypes. In this and the next chapter, we discuss the evolutionary genetics of human disease. The line between normal variation and disease can be blurred, and the precise boundary is often defined differently depending on the environment, society, and the ease of access to medical care.

In this chapter we discuss simple human genetic diseases that are inherited in a Mendelian manner, and in the next chapter we discuss complex disease. Just as there is no complete division between normal and pathogenic variation, so there is no clear distinction between simple and complex genetic diseases. For all of the simple genetic diseases discussed, other modifier genes, as well as the environment, affect their severity and nature. Nevertheless, the division remains helpful, and one that will be used here: simple genetic diseases are those where individuals carrying one disease allele (for dominant diseases) or two alleles (for recessive diseases) generally manifest the disease phenotype, despite variation due to **stochasticity** in the development of the phenotype, environmental effects, and the effects of other genes. Simple genetic diseases are also usually rare, although there are exceptions, as we shall see. Complex genetic diseases are those where a single disease allele (heterozygous or homozygous) does not cause the disease, and can be common in the population. In practice, the most useful distinguishing criterion is the inheritance pattern in families: if Mendelian inheritance can be discerned, the disease is a simple genetic disease.

We now know the molecular basis of over 3000 simple genetic diseases, and the rate of discovery is unlikely to slow given the variety of powerful methods that can now be used to discover the genes responsible (**Box 16.1**). The genetic diseases we will mention in this chapter are summarized in **Table 16.1**, together with details on their **incidence**, mode of inheritance, and reference to the database Online Mendelian Inheritance in Man (OMIM, omim.org). Table 16.1 includes some of the most frequent simple genetic diseases, and we refer to their incidence as high, typically meaning a fraction of one percent; however, this is relative to other simple diseases, and not relative to many complex diseases, which can manifest in a substantial proportion of the population. In contrast, simple genetic diseases, and the alleles that cause them, are very rare.

We hope that this chapter, and the next, will be valuable to those whose main interest is medicine, as well as to anthropologists and geneticists. We can respond to a question about why a patient has glucose-6-phosphate dehydrogenase (**G6PD**) deficiency, for example, by answering that G6PD deficiency is caused by mutant alleles of the *G6PD* gene. This is sometimes known as the "proximate" explanation. But a doctor might ask, why is G6PD deficiency not seen in my European-American patients? And why is Huntington's disease not

Box 16.1: Identifying genes underlying Mendelian disorders

One of the great successes of human genetics has been the identification of the genes responsible for many simple genetic disorders. Linkage analysis is most readily applicable to Mendelian disorders (those that show the classic patterns of autosomal-dominant, autosomal-recessive, or sex-linked inheritance), and the principle behind it is to observe the co-segregation of a disease phenotype with DNA polymorphisms in pedigrees. Recombination events are likely to separate an allele from the mutant allele in the causative gene if the two are far apart, and are less likely to separate them if they are close together. The results are commonly expressed as a **LOD score**, which is the logarithm (in base 10) of the odds of linkage—the ratio of the likelihood that loci are linked to the likelihood that they are not linked. For example, a LOD score of 3 represents odds of a thousand to one in favor of linkage, and is normally taken as the minimum value indicating linkage with a 5% chance of error. Even when the underlying molecular cause of the disease was unknown, genetic linkage analysis in pedigrees segregating the disease could be used to identify a critical region of the genome, defined by flanking recombination events, within which the gene must lie. Then chromosomal rearrangements, deletions, testing of candidate genes, or simply a systematic evaluation of all genes within the critical region for candidate causative mutations would reliably lead to the identification of the relevant gene. This procedure could be complicated by factors such as the degree of **penetrance** (does an individual carrying the mutated gene always exhibit the phenotype?) or **heterogeneity** (do mutations in more than one gene result in the same phenotype?), but has nevertheless led to the successful identification of over 3000 important genes, with the immediate possibility of offering diagnosis and the more distant prospect that improved treatment may result from an understanding of the functional effect of the mutant allele.

Nevertheless, an understanding of population history can be important in the selection of the individuals to be studied and the strategy used. An example of this is the use of isolated populations, particularly those established from a small number of founders, such as the Finns. Because of genetic drift in a small population, the genetic disease spectrum may be very different from that in surrounding populations, as described in Section 16.2. These "Finnish" diseases are often derived mainly from a single founder, which reduces the genetic heterogeneity, and this assumption allows the use of linkage disequilibrium (LD) analyses in the search for the causative gene.

Another example of the importance to gene mapping of population history, albeit on a very recent time-scale, is the use of **autozygosity mapping**.[51] This relies on identification of large tracts of the genome that are identical-by-descent; that is, the same section of both homologous chromosomes in one individual has been inherited from the same individual chromosome in recent history. The large tracts are identified as long runs of homozygosity for SNPs, but the term autozygous is used to imply a very recent shared ancestor for these regions. Such autozygosity is most frequently due to consanguinity in a family, and consanguineous families can be used to map regions of autozygosity that may contain a gene causing a recessive disease. High rates of consanguinity in some populations can allow the identification of disease genes using autozygosity mapping, as exemplified by the identification of mutations in the *SCN9A* gene as the cause of the congenital inability to feel pain in families from northern Pakistan.[10]

The development of whole exome, and whole genome, sequencing (see Section 4.4) has allowed the analysis of Mendelian diseases where only one or a handful of families show the disease, or where disease occurrence in a particular family is not explained by a mutation in the known disease gene. The rationale is that, by sequencing all exons and examining the inheritance pattern of potentially pathogenic mutations in the family, the causative mutation can be identified. However, this is a challenging task given the recent population expansion and high mutational load of humans (see Box 16.2). Indeed, given that every person is likely to be heterozygous for about 100 variants predicted to abrogate function for that gene, and homozygous for several such variants,[29] distinguishing the mutations that actually cause the disease will require direct biological evidence.

seen in my African-American patients? These questions can be explored and sometimes given an answer rooted in evolutionary genetics and biology—an "ultimate" answer. For most genetic diseases, there is no satisfying ultimate explanation for their presence or variation, they "just are"—that is, new mutations are generated spontaneously and the diseases they cause are selected against, removing those alleles from the population. We discuss this evolutionary process in **Section 16.1**. Exceptions to this rule are interesting and informative both to those involved in medicine, and to those more interested in human evolution in general. We devote **Sections 16.2–16.4** to these exceptions, and what can be learnt from them.

TABLE 16.1:
GENETIC DISEASES DISCUSSED IN THE TEXT (IN ORDER OF APPEARANCE)

Disease	Gene symbol	Gene name	Inheritance	Incidence	Clinical phenotype	OMIM	Section
Hutchinson-Gilford progeria syndrome	*LMNA*	lamin A	autosomal dominant	1 in 4 million	"early aging," short stature, loss of hair, low body weight	176670	16.1
Hereditary hemorrhagic telangiectasia	*ENG* and *ACVRL1* (mostly; two other loci known)	endoglin activin receptor-like kinase 1	autosomal dominant	1 in 5000	blood spots (telangiectases) on mucosa, hemorrhaging	187300	16.1
Duchenne muscular dystrophy	*DMD*	dystrophin	X-linked recessive	1 in 3000 males	muscle weakness, respiratory failure	310200	16.1
Huntington's disease	*HTT*	huntingtin	autosomal dominant	1 in 20,000 (western Europeans) 1 in 500,000 (Japanese)	movement abnormalities, personality changes, cognitive loss	143100	16.1
Tay-Sachs disease	*HEXA*	hexosaminidase A	autosomal recessive	1 in 3600 (Ashkenazi Jews) 1 in 360,000 (other North Americans)	loss of movement, visual loss, death between 2 and 4 years of age	272800	16.2
Charcot-Marie-Tooth disease type 1A	*PMP22* (mostly)	peripheral myelin protein 22	autosomal dominant	no estimates for incidence	atrophy of leg muscles, sensory impairment	118220	16.3
17q21.31 microdeletion syndrome	several	several	autosomal dominant	1 in 17,000	psychomotor retardation, communication difficulties	610443	16.3
Sickle-cell anemia	*HBB*	β-globin	autosomal recessive	1 in 700 (African-Americans) 1 in 160,000 (European-Americans)	anemia, repeated infections, loss of vision, stroke	603903	16.4
β-thalassemia	*HBB*	β-globin	autosomal recessive	1 in 250 (Cyprus) 1 in 3 million (native English)	anemia, severity dependent on genotype	613985	16.4
α-thalassemia	*HBA*	α-globin	autosomal recessive (in most cases)	1 in 12 (Greece)	anemia, severity dependent on genotype (Figure 16.7)	604131	16.4
Glucose-6-phosphate dehydrogenase deficiency	*G6PD*	glucose-6-phosphate dehydrogenase	X-linked recessive	1 in 30 (Iran)	neonatal jaundice, acute hemolytic anemia	305900	16.4
Cystic fibrosis	*CFTR*	cystic fibrosis transmembrane conductance regulator	autosomal recessive	1 in 2400 (United Kingdom)	lung infections, infertility, poor growth	219700	Box 16.3
Hemolytic disease of the newborn (Rh-mediated)	*RHD*	Rhesus blood group D antigen	autosomal recessive	1 in 250 (USA, prior to medical treatment)	hemolytic disease in neonates	111680	Box 16.4

16.1 GENETIC DISEASE AND MUTATION–SELECTION BALANCE

What distribution of frequencies would we expect for genetic diseases? Diseases are, by definition, disadvantageous, and disease-causing alleles will be selected against in the population if they affect reproductive fitness, and thus will never become common. Alleles causing simple diseases are therefore predicted to be in mutation–selection balance (see **Section 5.6** for discussion on mutation–selection balance, Table 3.1 for current estimates of mutation rates, and **Box 16.2** for the consequences of mutation–selection balance for the population as a whole). This prediction allows us to make some inferences about the expected frequency of genetic diseases. If a disease is dominant—that is, individuals carrying only one mutant allele have the disease—then the frequency of the disease allele is simply a balance between the mutation rate and the strength of selection acting against it:

$$P = \frac{u}{S}$$

where P is the frequency of the disease allele, u is the **deleterious** mutation rate of the gene, and S the strength of selection against the disease allele (Chapter 5). Note that here we refer to the deleterious mutation rate per gene per generation, rather than mutation rate per nucleotide per generation, which is used through the rest of the book. This is because for genetic diseases we measure mutations in the gene that cause the disease (deleterious mutations) rather than any mutation at a single nucleotide.

If the disease is recessive—that is, only homozygotes or **compound heterozygotes** for mutant alleles suffer from the disease—then the disease alleles will be at a higher frequency than the disease itself in the population because most are "hidden" in heterozygotes. The frequency of the disease allele (P) can then be calculated as

$$P = \sqrt{\frac{u}{S}}$$

So, we can see that in both dominant and recessive diseases, the deleterious mutation rate and the strength of selection can affect the frequency of the disease allele.

How does this affect the incidence of a particular disease? We can use the Hardy–Weinberg equation (**Section 5.1**) to calculate that for a dominant disease, almost all patients will be heterozygotes for the disease allele because $2P(1 - P) > P^2$, particularly if the disease is rare and therefore P is small. Given that P is small, $1 - P$ will be almost 1 so, for a dominant genetic disease, the incidence at birth (including new mutations and mutant alleles transmitted from the parents) is:

$$\text{incidence at birth} \approx \frac{2u}{S}$$

For a recessive disease, all patients will be homozygous for the disease allele, so, following the Hardy–Weinberg law:

$$\text{incidence at birth} \approx \frac{u}{S}$$

Variation in the strength of purifying selection can affect incidence of genetic disease

In the most extreme cases a fully penetrant dominant disease prevents reproduction of affected individuals because they die in childhood or are infertile. So, all mutant alleles in the population will produce affected individuals, who will then fail to transmit the mutant allele. In this case, $S = 1$, all cases of the disease will be due to independent de novo mutations, and the frequency of the disease allele will equal the mutation rate. An example is **Hutchinson-Gilford Progeria Syndrome** (**HGPS**, OMIM 176670), an autosomal dominant disease with very early-onset symptoms that resemble aging. The disease is very rare (incidence about 1 in 4 million) and almost always caused by particular mutations in the lamin A gene (*LMNA*) that cause truncation of the protein. Most individuals with this disease die before 13 years of age.[22]

Box 16.2: The human mutational load

What effect does the occurrence of simple genetic diseases have upon population fitness? This question was raised independently by the geneticists J. B. S. Haldane (1892–1964) and H. J. Muller (1890–1967) and answered by the Haldane–Muller principle, which states that the effect of mutation–selection balance on the average fitness of a population depends only on the overall deleterious mutation rate and not on the severity of the mutations. This is because the effects of severe and mild mutations balance out: the more severe the mutation, the lower equilibrium frequency in the population. The effect on fitness of deleterious alleles in a population is called its mutation load.

Muller was particularly concerned about the number of alleles for recessive conditions that may be mildly deleterious when heterozygous in an individual, and their effect on the average fitness of the population. He published his concerns hidden deep in population genetic theory, in 1950, in an influential paper.[31] In a particularly vivid passage he described the human mutational load:

> We tend to carry our burden more or less unconsciously at first, and then for a time rather zealously ..., but we usually become more weighed down by it in our later years, as all our powers gradually dwindle.

Muller then discusses the relaxation of natural selection since industrialization, and suggests that without some kind of artificial selection the accumulation of deleterious alleles as heterozygotes would endanger the survival of the human species. As a practical solution Muller suggested some kind of reproductive self-control for carriers of these mutations "motivated by their own desire to contribute to human benefit." Ignoring, for a moment, the ethical implications, was Muller right to be worried?

We have seen in the previous chapter, and will see again in the next, that although the process of natural selection has changed for some populations, it remains a force in human evolution, and is likely to remove deleterious alleles from the population. Nonetheless, the accumulation of recessive disease alleles as a result of population processes could be important. It has been suggested that endogamy and consanguinity, although increasing the frequency of autosomal recessive disease (Section 16.2), may be expected to eventually reduce the frequency of autosomal recessive disease *alleles* as they are more likely to be present as homozygotes and therefore allow selection to act. Therefore, while consanguinity may not be good for the individual with a genetic disease, it might be good for the population. The corollary of this prediction is that increasing **panmixia** will cause a reduction in genetic disease but allow deleterious recessive alleles to be maintained for longer in the human population, thereby increasing the mutation load. There are data to support this: in the USA, a study measured the amount of **autozygosity** in genomes of individuals born at different times since the beginning of the twentieth century. There was a negative correlation with age, so that younger people showed less genomewide autozygosity, presumably as a result of the known increase in outbreeding in the past 100 years (**Figure 1**).

Muller estimated that "the average individual is probably heterozygous for at least 8 genes, and possibly for scores, each of which produces a significant but usually slight detrimental effect," while Morton and colleagues calculated from a consideration of consanguineous marriages that "the average person carries heterozygously the equivalent of 3–5 recessive lethals." Resequencing of thousands of genomes is already discovering a large number of individually rare variants at particular loci,[9, 48] and providing direct measurements of the numbers of disease alleles per individual, which are in good agreement with the early estimates.

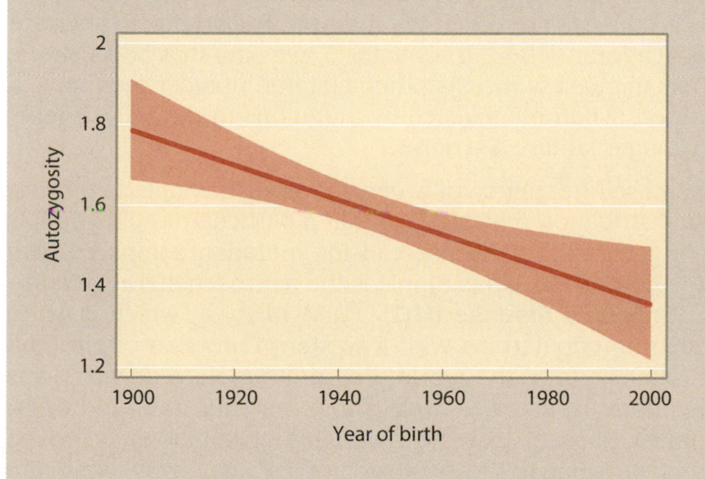

Figure 1: Decreasing genome autozygosity in more recent births.
Individuals born more recently show reduced levels of autozygosity compared with those born at the start of the twentieth century. Autozygosity was measured as contiguous regions of homozygosity from genomewide SNP genotypes, from 809 North Americans of European ancestry. The *red* line shows the predicted trend derived from these data, and the *orange* shaded region shows the 95% confidence interval for this estimate. [From Nalls MA et al. (2009) *PLoS Genet.* 5, e1000415. With permission from National Academy of Sciences.]

If the disease is less severe or occurs later in life, after reproductive age, then the prevalence can be higher due to a reduction in the strength of purifying selection. An example is **hereditary hemorrhagic telangiectasia** (HHT, OMIM 187300), which is a fully penetrant dominant disease that leads to recurrent and often severe nosebleeds and gastrointestinal bleeding. It is caused by over 600 different mutant alleles in the endoglin (*ENG*) and activin receptor-like kinase 1 (*ACVRL1*) genes, with most mutant alleles being unique to one family. It is one of the most frequent simple dominant Mendelian diseases, with an incidence at birth of about 1 in 5000. The true incidence may be considerably higher, as HHT is thought to be under-diagnosed, because diagnosis often requires thorough patient investigation and often only follows rare, severe life-threatening hemorrhages.[44] The high incidence is partly due to the large **mutational target**: many different amino acid sites in the two genes which, when mutated, can cause the disease (see next section). HHT can be caused by 600 different mutant alleles while only a few truncating mutations in *LMNA* cause HGPS. However, the severity of HHT symptoms are mild and develop with age, so that, although there is a small increased risk of stroke in children and a small increased risk of complications during childbirth, the effect on reproductive fitness is likely to be minimal. This leads to two differences when compared with HGPS. First, HHT disease alleles can be inherited, so most new cases which contribute to the incidence of the disease are in fact not due to de novo mutations but due to a mutant allele inherited from a parent. Second, the life expectancy of HHT patients is not much less than the population average, so the long life span of HHT patients contributes to the high prevalence of the disease.

Variation in the deleterious mutation rate can affect incidence of genetic disease

In the section above, we introduced the concept of a mutational target, where the difference in incidence of HHT and HGPS is due in part to the number of amino acid sites which, when mutated, cause the disease. Another well-known example of this is **Duchenne Muscular Dystrophy** (**DMD**, OMIM 310200), which is a progressive muscle-weakening disease caused by mutations of the dystrophin gene (*DMD*). It is X-linked recessive in inheritance, so that it is most common in males (who have one X chromosome) and extremely rare in females. Death occurs around 18 years of age, and patients very rarely reproduce. Given the strong selective disadvantage of the disease, it is at first surprising that it is so common, with an incidence at birth of around 1 in 3000 males. The *DMD* gene is very large, with its 79 exons spanning over 2 Mb of the X chromosome, so deletions, if we assume that they occur with an equal probability across the genome, are 20 times more likely to affect a gene of this size compared with a gene whose 79 exons are spread over 100 kb. Indeed, deletions account for two-thirds of all mutations causing DMD. Furthermore, the genomic region containing the *DMD* gene seems more prone to deletion than an average part of the genome. The reasons for this are not yet clear—it may be in part due to aberrant activation of DNA replication origins.[3] In Chapter 3, we saw that because the mutation rate in fathers increases with age, then children of older men have a higher frequency of Apert syndrome, caused by mutations in the *FGFR2* gene, than the children of younger fathers (**Section 3.7**).

In some cases, an increased frequency of a genetic disease coupled with differential geographical distribution may suggest that a particular haplotype has a higher probability of generating a disease-causing mutation compared with other haplotypes in the population. An example is the neurodegenerative autosomal dominant **Huntington's disease** (**HD**, OMIM 143100), which is 10 to 100 times more frequent in populations with a western European origin than in populations from other parts of the world. Before its molecular basis was elucidated, it was therefore expected to have a low mutation rate and originate from a single founder. Subsequently, however, this explanation has proved inadequate. The phenotype is due to the expansion of a $(CAG)_n$ array within the first exon of the huntingtin gene (*HTT*) that codes for a polyglutamine tract in the protein. The length of the array is highly polymorphic, ranging from ~10 to 35 CAGs in unaffected individuals, with 17 CAGs being the most common. If the array exceeds 35 CAGs then there are pathogenic consequences in striatal cells (a subset of neurons) in adults.

Pedigree studies have shown that a significant proportion of families (~3%) manifest the disease as a result of new mutations, so the mutation rate is not low and, for most populations, there is no single founder. But if there are many recurrent HD mutations, why should they occur in Europeans so much more often than in other populations? Examining the SNP haplotype background in the *HTT* chromosomal region from European HD patients showed that two closely related haplotypes (A1 and A2) are present on 80% of chromosomes with a HD disease-length $(CAG)_n$ array, yet they are at a frequency of 25% in the European population as a whole (**Figure 16.1**).[52] This strongly suggests that these haplotypes affect the mutation rate of the $(CAG)_n$ by predisposing it to array expansion. In addition, haplotypes A1 and A2 are only present in European populations, suggesting that they are responsible for the geographical differences in HD incidence (Figure 16.1).

How haplotypes A1 and A2 actually affect the mutation rate of the $(CAG)_n$ array is not known. There are two competing models:

1. Alteration of a regulatory element. Haplotypes A1 and A2 carry SNP alleles that alter the sequence of a regulatory element, which in turn affects binding of a protein that regulates mutation at the $(CAG)_n$ array. There is no direct evidence for such a regulatory element, although such elements are known for other trinucleotide repeat arrays, such as in the spinocerebellar ataxia 7 gene (*SCA7*).[26]
2. A longer, nonpathogenic, $(CAG)_n$ array. A significant number of haplotypes A1 and A2 chromosomes carry a longer-than-average $(CAG)_{20}$ array. While not pathogenic in itself, it has been shown that repeat arrays within the normal range have a mutation rate that is correlated with array length, so that, for example, a $(CAG)_{20}$ array has a mutation rate that is almost four times higher than a $(CAG)_{17}$ array. Examining the haplotypes A1 and A2 that do not carry a HD disease-length $(CAG)_n$ array shows that they do carry a higher number of CAG trinucleotides in the array compared with other haplotypes, so we would expect those arrays to show a higher mutation rate.[12]

Whatever the mechanism, it is likely that haplotypes A1 and A2 rose to high frequencies in Europeans because of genetic drift, and so drift combined with a high mutation rate is responsible for the relatively high incidence of HD in European populations. The importance of drift in increasing frequency of disease-causing alleles—and therefore the incidence of genetic disease—even in the face of strong purifying selection, is discussed in the next section.

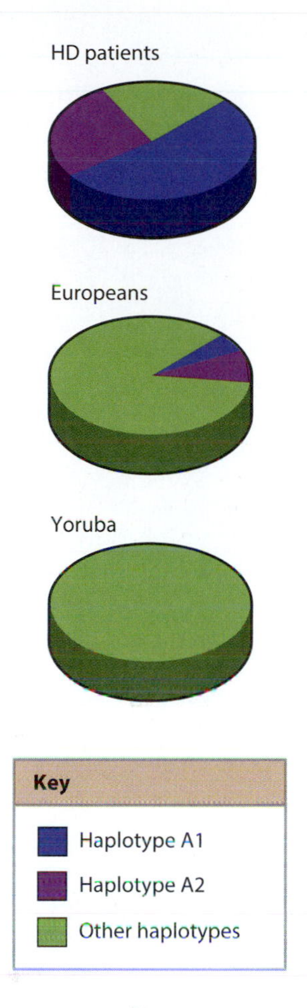

Figure 16.1: Distribution of huntingtin gene haplotypes in Huntington's disease (HD) patients and healthy individuals.
Pie charts showing huntingtin haplotypes A1 and A2 as a proportion of all haplotypes in different populations. HD patients are individuals with a $(CAG)_n$ array longer than 35 CAGs, compared with healthy Europeans and healthy Yoruba, from Nigeria. [From Warby SC et al. (2009) *Am. J. Hum. Genet.* 84, 351. With permission from Elsevier.]

16.2 GENETIC DRIFT, FOUNDER EFFECTS, AND CONSANGUINITY

In this section we discuss the role of genetic drift in affecting the distribution and incidence of Mendelian diseases.

We have seen in the previous section that, in general, if selection is strong and the mutation rate uniform, then for both dominant and recessive diseases, mutations will probably occur with equal frequency in different populations, and each disease will be rare and have a relatively uniform geographical distribution. It is departures from this paradigm that interest evolutionary geneticists. When a simple genetic disease is either more frequent than expected, or has a particular geographical distribution, then this may be due to the effects of natural selection (**Section 16.4**) or genetic drift. A concentration of a set of normally rare disorders within a population is known as a **genetic disease heritage**, and, because of the unlinked nature of the causal loci underlying these disorders, must result from some evolutionary process that affects many genomic regions simultaneously.

Unlike selection or locus-specific mutation rates, genetic drift affects all alleles within a population, and past demography that resulted in a low effective population size is a frequent cause of a recognizable genetic disease heritage. These demographic processes include founder events, population bottlenecks, and long-term endogamy. It is often found that populations with distinct genetic disease heritages also harbor unusual frequencies of neutral alleles as well, further supporting the causal role of genetic drift. Furthermore, there appears to be an inverse relationship between disease frequency and allelic heterogeneity. In a population where a particular Mendelian disorder is common as a result of genetic drift, the disorder is often caused by fewer mutant alleles than in populations in which the disease is less frequent.

Table 16.2 gives examples of population isolates that have been used in genetic studies, and below we discuss particular examples where founder effects, endogamy, and consanguinity have affected the genetic disease heritage of particular populations.

Jewish populations have a particular disease heritage

Over 40 Mendelian disorders have been identified that are present at markedly higher frequencies in Jewish populations than in surrounding populations, and some also occur at higher frequencies in particular Jewish groups.[35] This disease heritage relates in part to their complex history of **diaspora** (see **Section 14.6**).

Lysosomal storage disorders are a group of genetic diseases that all involve defects of hydrolase enzymes in the lysosome, a cellular organelle involved in the degradation of different molecules. **Tay-Sachs disease** (OMIM 272800) is the most frequent of four such disorders in the Ashkenazi Jewish population, the others being Gaucher disease (OMIM 230800), Niemann-Pick disease (OMIM 257220), and mucolipidosis type IV (OMIM 252650). Tay-Sachs disease is characterized by paralysis, dementia, and blindness, resulting in death usually by the time the patient is 3 years old. It is a recessive disease, caused by mutations in the hexosaminidase gene *HEXA*, and is 100 times more common in

TABLE 16.2:
SOME POPULATION ISOLATES USED IN GENETIC STUDIES

Genetic isolate	Country	Example of a mapped disease gene			
		Disease	OMIM reference number	Gene symbol	Reference
Finnish (see Section 16.2)	Finland	congenital chloride diarrhea	214700	*SLC26A3*	Höglund P et al. (1996) *Nat. Genet.* 14, 316.
Jewish communities (see Section 16.2)	Various	familial dysautonomia	223900	*IKBKAP*	Slaugenhaupt SA et al. (2001) *Am. J. Hum. Genet.* 68, 598.
Old Order Amish	USA	Ellis-van Creveld syndrome	225500	*EVC*	McKusick VA (2000) *Nat. Genet.* 24, 203.
Hutterite	USA	Bowen-Conradi syndrome	211180	*EMG1*	Armistead J et al. (2009) *Am. J. Hum. Genet.* 84, 728.
Sardinian	Italy	early onset Parkinson's disease	605909	*PINK1*	Valente EM et al. (2004) *Science* 304, 1158.
Margarita Island	Venezuela	cleft lip/palate-ectodermal dysplasia	119530	*CLPED1*	Suzuki K et al. (2000) *Nat. Genet.* 25, 427.
Costa Rica central valley	Costa Rica	inherited deafness	124900	*DIAPH1*	Lynch ED et al. (1997) *Science* 278, 1315.

Adapted from Arcos-Burgos M & Muenke M (2002) *Clin. Genet.* 61, 233. With permission from John Wiley & Sons, Inc.

Ashkenazi Jews than in other populations. Given that all four lysosomal storage disorders involve a particular physiological process it has been argued that the high frequency of such diseases might be due to heterozygous carriers having some, unknown, selective advantage.

An alternative explanation, however, is that lysosomal storage disorder alleles have risen to high frequency by genetic drift due to founder effects and endogamy within the Ashkenazi Jewish community.[45] Coalescent analysis (**Section 6.5**) of the most common disease-causing mutations of lysosomal storage disorders and other diseases prevalent in Jews suggests that most originated around 50 generations (~1300 years) ago, which coincides with historical evidence suggesting that the Ashkenazi Jews moved and expanded into central Europe at this time.[40]

Finns have a disease heritage very distinct from other Europeans

Finns have a dramatically different spectrum of inherited, mostly recessive, diseases, when compared with their European neighbors. Each of these diseases shows reduced **allele heterogeneity**, so that in recessive diseases over 50% of cases are due to homozygosity for one allele, which is a direct consequence of a founder effect or bottleneck in the population history of the Finns. This lack of allelic heterogeneity, together with the long stretches of linkage disequilibrium, has the advantage of allowing efficient mapping and identification of the genes causing the genetic diseases (see Box 16.1). The lack of heterogeneity also allows accurate identification of heterozygotes for the recessive disease allele (carriers), since only one or a small number of alleles need to be tested. Twenty percent of Finns are known to be carriers for one of the recessive diseases regarded as part of the Finnish disease heritage.[33, 36] Another consequence of the founder effect is that recessive diseases that are generally common in European populations, such as cystic fibrosis, are rare or absent in Finland.

The Finnish founder effect is a result of the small initial population size of the Finns followed by subsequent population growth, and isolation of the Finnish population from other neighboring populations. Finland was covered by a thick ice cap during the last glacial period (up to ~10 KYA), and human settlement is likely to have occurred shortly after the melting of the ice. Archaeological data provide signs of human activity before 9 KYA, followed by the arrival of agriculture at about 4–3.5 KYA. Given the limitations of agriculture in the Finnish climate, it is likely that any settlements remained small and isolated, relying on hunting, and evidence of permanent settlement of the southern and western coasts of Finland dates back only about 2 KY. Long-term genetic drift, or recent population movements, or a combination of both, has led to distinct geographical patterns in disease mutations (**Figure 16.2**).[32] Small founder populations

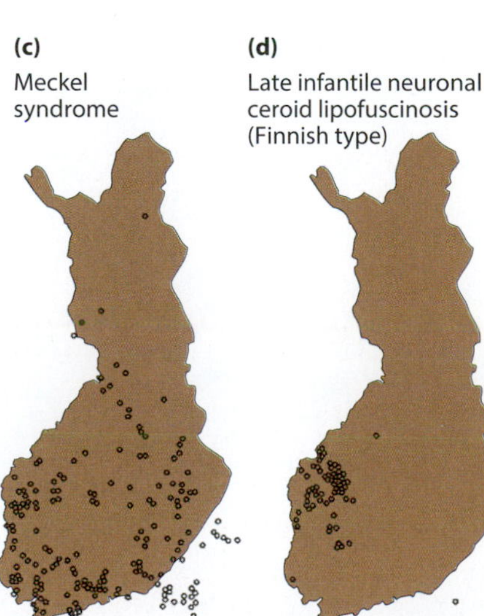

Figure 16.2: Distributions of four "Finnish Disease Heritage" diseases. The inset map on the left shows Finland in the context of Scandinavia in northwest Europe. The four maps (a–d) show different geographic patterns of the distribution of the birthplaces of the four grandparents of patients with the particular recessive disease. (a) Concentration in an area of late settlement of Finland, populated since the 1500s. (b) Dispersed pattern throughout Finland. (c) Concentration in the west of Finland where population density is the highest. (d) Restricted geographic pattern where most of the grandparents of all patients were born within one county. [From Norio R (2003) *Hum. Genet.* 112, 470. With permission from Springer Science and Business Media.]

from the southwestern coastline expanded toward the northeastern wilderness after the sixteenth century.

Language is likely to have been a barrier to immigration and therefore to admixture following the population expansion of the Finns. The Finnish language belongs to a small group of Uralic languages, which differ drastically from most languages spoken in Europe, which belong to the Indo-European family. Other Uralic-speaking populations are the Saami (~50,000–80,000), the Estonians (~1.5 million), and the Hungarians (~12 million), but there are several smaller regions in Russia and Siberia where Uralic languages are also spoken. Genetic evidence from the Y chromosome supports the concept of Finnish isolation, based on low haplotype diversity,[55] but mitochondrial DNA diversity is similar to that seen in other European populations.[19]

The Finns are now one of the best genetically characterized populations in the world, an effort justified by the promise of revealing the genes behind their genetic disease heritage. A modern healthcare system allied to the presence of genetic researchers of international quality has enabled these genes to be identified, with benefits not only for the Finnish population, but also for our understanding of the role of genes in disease. It provides a good example for other population isolates intending to investigate their own genetic disease heritage.

Consanguinity can lead to increased rates of genetic disease

Mendelian genetic diseases and birth defects (many of which have a genetic cause) are particularly frequent in communities with a high rate of marriage between relatives (consanguinity) (**Figure 16.3a** and **Section 5.3**). Some autosomal recessive diseases cluster in particular Arab communities, such as Canavan's disease (OMIM 271900) in Bedouin of Saudi Arabia.[2]

The description "Arabic" covers people from a large part of the world, ranging from Morocco in the west to Qatar in the East. These populations have different genetic histories but share socio-cultural factors such as language. The Islamic religion is also a shared cultural factor among most Arabs, although many of the socio-cultural factors apply to non-Islamic Arabs (such as those who are Christian) and Muslims outside the Arab world, such as those in Iran and Pakistan. Consanguineous marriages (second cousins or closer, coefficient of kinship $f \geq 0.0156$, **Section 5.3**) are common in many Arab countries, with many showing rates between 20 and 40% (Figure 16.3b) in contrast to European countries which typically show rates of 0.5–1%.[4, 47] This results in inbreeding depression (**Section 5.3**) as a result of frequent manifestations of recessive diseases, and poses a particular problem for the healthcare services in these countries. A study of the Pakistani community living in the UK has shown that for every increase of 0.01 in the coefficient of kinship between parents f, the incidence of autosomal recessive disorders would increase by ~0.7%.[5] For consanguineous families, autozygosity mapping approaches can be used to identify the gene causing the disease (see Box 16.1).

16.3 EVOLUTIONARY CAUSES OF GENOMIC DISORDERS

A genomic disorder is a disease caused by a genomic rearrangement, such as a deletion, duplication, or inversion, which results directly from the architecture of the genome.[6] This architecture can encompass segmental duplications, inversions, or more complex structural variation (**Section 3.6**). Genomic disorders include simple Mendelian and **sporadic** genetic diseases like Charcot-Marie-Tooth disease and **Williams-Beuren syndrome** (OMIM 194050), but also some cases of more complex phenotypes such as developmental delay, and neuropsychiatric syndromes such as autism, schizophrenia, and bipolar disorder. Genomic rearrangements are often individually rare—a particular deletion or duplication will be the cause of disease in a very small number of schizophrenia patients, for example—but together may explain a large proportion of the genetic component of these diseases. For these neuropsychiatric syndromes, at least, there is an overlap between simple genetic diseases and more complex multifactorial disease: the common complex disease may in fact be a multitude of rare simple diseases with indistinguishable phenotypes. The simple genetic diseases caused by large genomic rearrangements are usually highly deleterious, so our expectation is that they are in mutation–selection balance in the population.

Segmental duplications allow genomic rearrangements with disease consequences

The term genomic disorder was first used to describe **Charcot-Marie-Tooth disease type 1A** (CMT1A; OMIM 118220), a dominant genetic disease characterized by progressive motor and sensory neuropathy. The disease is caused by

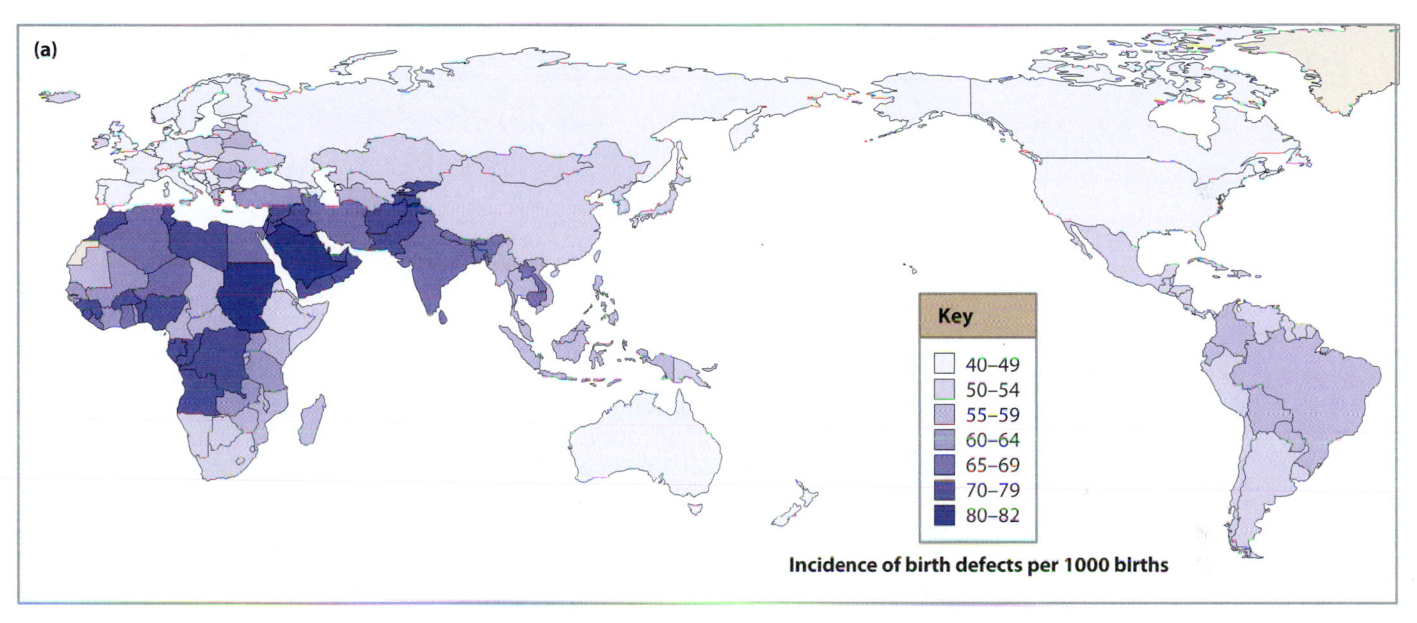

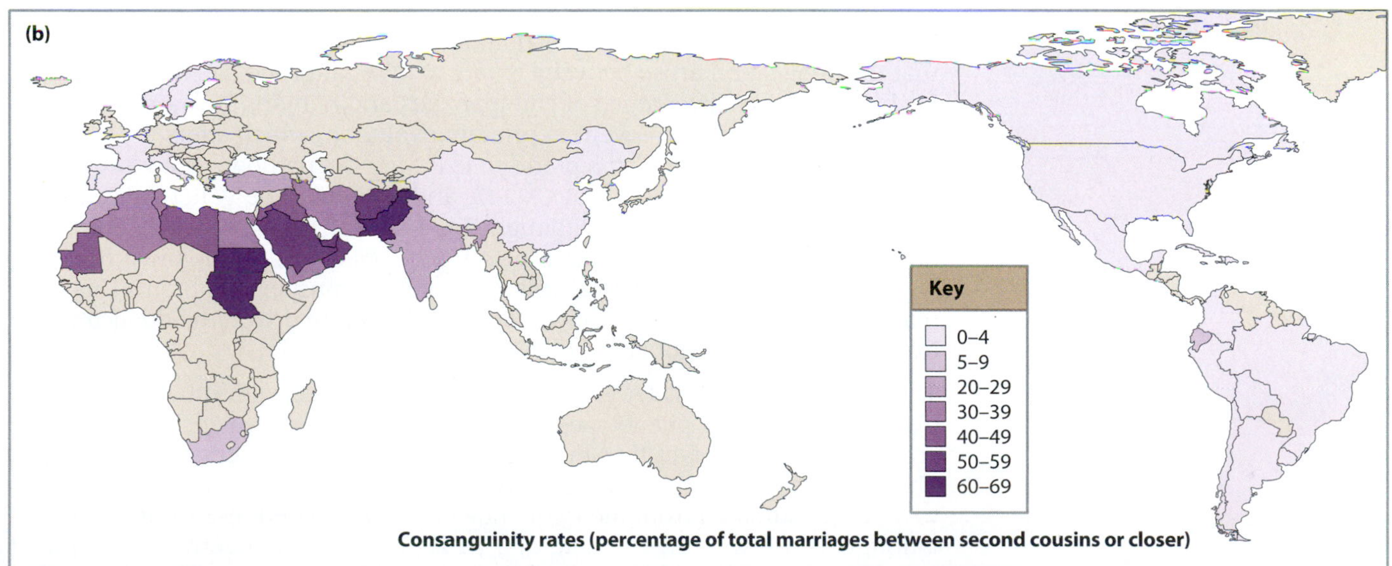

Figure 16.3: Levels of consanguinity and incidence of birth defects.
(a) The incidence of birth defects per 1000 births, by country. A birth defect is defined as any abnormality affecting body structure or function that is present from birth, and includes simple genetic diseases, chromosomal disorders, and nongenetic disorders, such as those resulting from exposure to environmental teratogens. (b) Consanguinity rates per country, where data are available. The proportion of total marriages between second cousins or closer is shown (corresponding to a coefficient of kinship $f \geq 0.0156$). [a, data from Christianson AL et al. (2006) March of Dimes Birth Defects Foundation, White Plains, NY; b, data from Tadmouri GO (2006) Genetic Disorders in the Arab World, Dubai.]

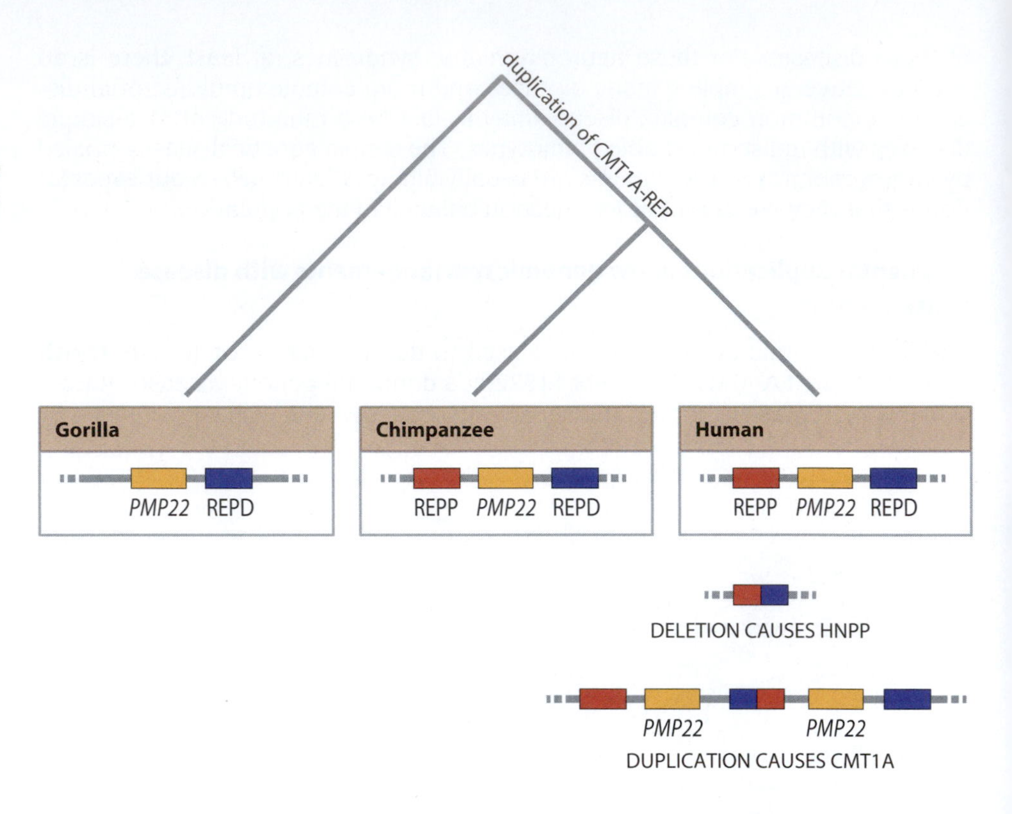

Figure 16.4: Evolutionary history of CMT1A repeats, and Charcot-Marie-Tooth disease.
A duplication of a 24-kb region (REPP in *red*, REPD in *blue*) occurred after divergence of the gorilla lineage but before divergence of the chimpanzee and human lineages. Recurrent non-allelic homologous recombination events between these duplicated regions cause copy-number variation of the *PMP22* gene and the genetic diseases HNPP (deletion) and CMT1A (duplication) in humans.

duplication of the peripheral myelin protein 22 gene (*PMP22*) on chromosome 17. The pathogenic duplication is caused by non-allelic homologous recombination (NAHR, **Section 3.6**) between two 24-kb segmental duplications in the human genome flanking the gene, termed CMT1A-REPs. The reciprocal deletion generated by NAHR between these two elements causes a different genetic disease, which is called hereditary neuropathy with liability to pressure palsies (HNPP).

Two copies of the CMT1A-REP sequence, which sponsor NAHR between the two elements, are found only in humans and chimpanzees, which suggests that the duplication arose between 6 and 10 MYA; after the divergence of gorillas but before the divergence of chimpanzees and humans[25] (**Figure 16.4**). This shows that a duplication event in primate evolution, which alters the genomic architecture of the region, can predispose to rare but pathogenic consequences in humans. It also predicts that CMT1A and HNPP will be genetic diseases in chimpanzees and bonobos, although this has not been reported. The original duplication event generated a novel gene, *FAM18B*, of unknown function, which is expressed in bronchial epithelia, skeletal muscle, and brain. This gene may have offered a selective advantage and therefore driven the duplication to fixation, but this has not been proven.[24]

The prevalence of CMT1A is relatively high compared with other single-gene diseases with severe phenotypes. Estimates of CMT1A prevalence vary, and are complicated by the fact that many patients reproduce successfully, so not all observed mutatant alleles are de novo. However, a prevalence of 1 in 23,000 to 41,000, estimated from the frequency of the diagnosed disease in the population, concurs with an estimate of 1 in 23,000 to 79,000 derived from direct measurement of the *PMP22* gene duplication rate in sperm. This shows that the high prevalence is caused by the mutation rate, which is three orders of magnitude higher than the mutation rate of a single nucleotide base[49] (see **Section 3.2**).

Another example of segmental duplications sponsoring recurrent pathogenic deletions and duplications is the case of the rare **17q21.31 microdeletion syndrome** (OMIM 610443), where patients have learning disability. The locus is polymorphically inverted in humans, with haplotype H1 carrying the non-inverted orientation and haplotype H2 carrying the inversion. The orientation of the segmental duplications in the inverted H2 haplotype allows deletion of the region by NAHR, causing the disease, while deletions on the H1 haplotype are not seen. The frequency of haplotypes H1 and H2 differs between populations because of either genetic drift or selection (**Section 6.7**), with haplotype H2 reaching a frequency of 30% in Europeans yet being very rare in non-Europeans, making 17q21.31 microdeletion syndrome a European-specific disease.[43, 56]

Duplications accumulated in ancestral primates

Are the examples discussed above unusual cases, or do they indicate a general principle explaining the existence of pathogenic deletions and duplications in the human genome? Comparing segmental duplication maps of ape and macaque genomes suggests that most human segmental duplications arose after the divergence of the Old World Monkey and hominoid lineages, with a significant burst between the human–orangutan ancestor and the human–chimpanzee ancestor.[30]

The reason for the accumulation of segmental duplications in ancestral primates (particularly the ancestor to gorilla, chimpanzee, and humans) is not yet clear. Some segmental duplications may have increased in number due to positive selection throughout recent primate evolution. However, it seems likely that many of the complex segmental duplication-rich regions were generated by a combination of a higher duplication rate, duplication shadowing (where existing duplications sponsor further duplications in the same region), and fixation by genetic drift.

16.4 GENETIC DISEASES AND SELECTION BY MALARIA

If a simple genetic disease is more frequent than expected given mutation–selection balance, or has a particular geographical distribution which is not readily explained by demographic history, then the effects of natural selection may be invoked to explain it: an example is the lysosomal storage diseases in Ashkenazi Jews discussed above (**Section 16.2**). However, just as a selective explanation for allele distributions in non-disease phenotypes can be unfounded (Box 15.2) because the pattern of distribution can readily be explained by genetic drift, so the same explanation can apply to diseases. Many selective explanations for differences in distribution of genetic diseases are just hypotheses with very little genetic or epidemiological data to support them. In these cases, genetic drift is not just the null hypothesis, but the hypothesis that is best supported by the available data. In fact, unusual distributions of Mendelian simple diseases can almost always be explained by genetic drift.

With this caveat in mind, we illustrate in this section some exceptional but well-founded cases of malaria resistance where natural selection for heterozygous carriers of genetic disease maintains a high frequency of the disease allele, and thus a high incidence of disease, in certain parts of the world. We critically discuss two more unusual cases in this chapter (cystic fibrosis, **Box 16.3**; and Rh hemolytic disease, **Box 16.4**), because although they have been extensively debated in the literature and in other textbooks as cases where natural selection may have maintained a high frequency of a disease allele, there is no compelling evidence for this.

For all three examples of malaria resistance discussed below, there is strong evidence that a high frequency of the disease allele, and hence high incidence of an autosomal recessive disease, is maintained by natural selection favoring heterozygotes who show either increased resistance to infection by malarial protozoan parasites (usually *Plasmodium falciparum* in Africa and *Plasmodium vivax* outside Africa) or milder symptoms following malaria infection. These are convincing cases of heterozygote advantage, a form of balancing selection (**Section 5.4**).

Box 16.3: Why is cystic fibrosis (CF) so common in Europe?

Cystic fibrosis (CF) has a high incidence in Europeans (for example, 1 in 2400 in the United Kingdom), where it is the most common severe autosomal recessive disease. Affected individuals carry two inactive copies of the *CFTR* (cystic fibrosis transmembrane conductance regulator) gene. One allele, ΔF508, a deletion of 3 bp which removes the phenylalanine at amino acid 508 and leads to a misfolded, nonfunctional protein, accounts for about two-thirds of CF disease alleles in the population, although a total of over 1000 disease alleles are known, most of which are very rare. It has been proposed that the high frequency of the disease is due to a selective advantage of the heterozygote.[41] This is supported by the high F_{ST} value found for SNPs around the *CFTR* gene in a genomewide scan, indicating high population differentiation between Europeans and other populations.[1] Evidence for selection is less clear than for sickle-cell anemia, but heterozygotes have variously been proposed to be more resistant to diseases such as cholera, typhoid fever, tuberculosis, or bronchial asthma.

The CFTR protein is located in the membrane of several cell types, including the epithelial cells lining the small intestine, where it regulates chloride ion transport, in turn linked to fluid secretion. Untreated cholera leads to excessive chloride ion and fluid secretion that can be fatal. In a mouse model, animals carrying one copy of the *CFTR* gene secreted only half the amount of fluid and chloride ions in response to cholera toxin compared with animals carrying two copies,[17] so it seems reasonable to think that the same would happen in humans. Measurement of intestinal chloride secretion in humans in response to prostaglandin stimulation (considered to be a model for cholera toxin exposure), however, revealed no difference between unaffected individuals and CF heterozygotes,[23] so some doubt remains about whether this scenario provides an adequate explanation. In addition, cholera was first reported in Europe in 1832, but the ΔF508 mutation must be much older than this because of its Europe-wide distribution and high frequency. Indeed, the ΔF508 mutation has been found from 2000-year-old skeletons in Brittany,[13] and dating the origin of the ΔF508 mutation, using population genetic methods described in Section 6.7, suggests a best guess of around 15 KYA (**Table 1**). This shows that cholera was not the selective agent.

An alternative possibility comes from work demonstrating that CFTR is used by *Salmonella typhi*, the bacterium that causes typhoid fever, as a receptor to enter epithelial cells.[38] Cultured mouse cells take up little *S. typhi*, but will do so if they express wild-type human *CFTR*. If, however, they express *ΔF508 CFTR*, they take up only the basal level. Thus human CF heterozygotes might be partially protected against typhoid fever by reduced uptake of the bacterium. Another suggestion has been that CF heterozygotes are resistant to tuberculosis, but as yet there is no strong evidence to support this.

Statistical analysis of the pattern of variation around the *CFTR* locus can also be used to assess whether ΔF508 heterozygotes are likely to have had a selective advantage, and some studies have indicated such an effect, although this conclusion depends greatly on the population genetic assumptions made.

Large-scale datasets rigorously testing the selection hypothesis have not yet appeared, and neither has convincing epidemiological evidence for an advantage of CF carriers. For the moment the case for selection is not proven, and we should remember how demographic factors can apparently generate counterintuitive patterns of diversity in the face of natural selection, as may have happened at the Rh blood group locus (see Box 16.4).

TABLE 1:
DATE ESTIMATES OF THE ORIGIN OF THE ΔF508 *CFTR* MUTATION

Time (KYA), 27 years/ generation	Time (generations)	Basis	Reference
5.4	200	linkage disequilibrium pattern	Serre J et al. (1990) *Hum. Genet.* 84, 449.
>70	>2627	variation at 3 microsatellites	Morral N et al. (1994) *Nat. Genet.* 7, 169.
>24	>900	reanalysis of Morral et al. data	Kaplan N et al. (1994) *Nat. Genet.* 8, 216.
>13.5	>500	variation at 3 microsatellites	Slatkin M & Bertorelle G (2001) *Genetics* 158, 865.
>15	>580	coalescent analysis of Morral et al. data	Wiuf C (2001) *Genet. Res.* 78, 41.
>4.9 (no population expansion)	>183 (no population expansion)	coalescent analysis of SNP data	Morris A et al. (2002) *Am. J. Hum. Genet.* 70, 686.
>2.5 (rapidly expanding population)	>92 (rapidly expanding population)		

Box 16.4: The evolutionary puzzle of the Rhesus negative blood group

In 1939, Drs Philip Levine and Rufus Stetson reported a case of an unusual complication during childbirth. A woman's second pregnancy had ended with a stillbirth; and she had reacted severely to the resulting transfusion of her husband's blood. Her blood serum was shown to contain an antibody that reacted to her husband's red blood cells, and to the red blood cells of most other European-Americans. The following year, Landsteiner and Wiener discovered an antibody that also reacted to the red blood cells from most people, and was unrelated to other known blood groups, such as ABO. They called the antibody anti-Rh and the new blood group Rh (or Rhesus, because the antibody was produced by immunizing rabbits with rhesus macaque blood). People who react to this antibody are Rh-positive, and those who do not are Rh-negative.

During an Rh-negative mother's first pregnancy, an Rh-positive child immunizes her against the Rh antigen, so that if subsequent pregnancies carry an Rh-positive child, the fetus will receive anti-Rh antibodies causing hemolysis of the fetal red blood cells. This causes either death of the fetus and subsequent stillbirth, or a severe hemolytic anemia of the newborn infant. This genetic disease has fortunately become rare since the development of Rho(D) antibody, which can be given to Rh-negative mothers around 28 weeks into pregnancy, and prevents them becoming immunized against Rh antigen from fetal cells that have entered the maternal bloodstream.

Before the development of this routine treatment, hemolytic disease of the newborn was quite common. For example, the incidence at the Chicago Lying-In Hospital between 1940 and 1946 was estimated to be 1 in 250 women, with two-thirds of these cases resulting in newborn death or stillbirth. Given that the Rh blood group has been estimated to be responsible for about 90% of cases of stillbirth, it must surely have been subject to natural selection. In this example, where the selection is against heterozygotes (children who are Rh-positive but have an Rh-negative mother), we would expect selection against the rarer allele, in this case the Rh-negative allele. However, in Europe, the Rh-negative allele is typically at a frequency of 40%. Why is it so common?

Explanations have been a source of pub discussions (for evolutionary geneticists, at least) for over 70 years.[15, 18] One suggestion, from the population geneticist R. A. Fisher among others, is that of reproductive compensation, where Rh-negative mothers compensate by having more children than Rh-positive mothers. Despite much analysis, there is no evidence to support this.[39]

At the molecular level, the Rh blood group is determined by the Rhesus blood group D antigen (*RHD*) gene. In Europeans the homozygous deletion of the *RHD* gene causes the individual to be Rh-negative, although other rarer mutant alleles in the gene can cause the Rh-negative blood group in other populations. Heterozygous deletion, or homozygous presence, of the *RHD* gene causes the individual to be Rh-positive. Could there be a benefit of loss of the *RHD* gene? Unfortunately, the function of this gene is still not known for certain, although it is expressed uniquely on red blood cells, and by comparison with similar genes of known function in other species it is likely to be involved in transporting carbon dioxide through the membrane. It may be that the RhD protein is a receptor for an unknown or extinct parasite, and deletion provided resistance to that parasite.

Molecular analysis of the *RHD* gene has not provided support for recent positive selection for the deletion allele, suggesting either that current methods cannot detect the signature of older episodic selection at this locus, or that the deletion rose to high frequencies in Europe by genetic drift.[37] If such an apparently deleterious allele did rise by genetic drift to such significant levels, perhaps during migration and settlement of Europe, it provides an example where natural selection is not needed to explain a high frequency of a deleterious allele.

We focus on the evidence from three types of study:

1. Geographical correlation of disease frequency and agent of selection
2. Clinical observation of different **morbidity**/mortality of heterozygous carriers
3. Population genetic evidence of natural selection.

Sickle-cell anemia is frequent in certain populations due to balancing selection

Sickle-cell anemia (OMIM 603903) is a severe anemia with characteristic "sickling" of red blood cells, particularly under lower oxygen conditions (**Figure 16.5**). This is caused by homozygosity for the Hb^S allele of the β-globin gene (*HBB*), in which a single nucleotide substitution changes glutamic acid to valine at the sixth amino acid (Glu6Val, rs334). The resulting hemoglobin molecule, which is

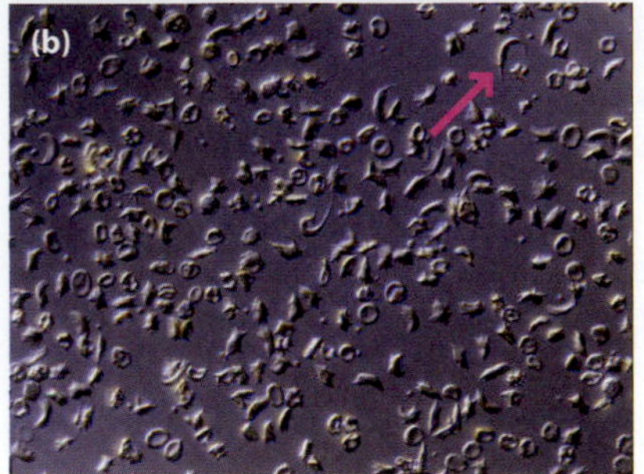

Figure 16.5: Sickling of red blood cells in sickle-cell anemia. Top, healthy red blood cells are biconcave discs with a diameter of about 8 μm. Under normal oxygen levels in the blood, an individual has normal red blood cells, as shown. Bottom, in individuals with sickle-cell disease, under low oxygen levels in the blood, blood cells show characteristic sickling morphology. An example is highlighted. (With permission from Wellcome Trust Image Library.)

a tetramer of two α-globin and two β-globin molecules and transports oxygen in the blood, aggregates in deoxygenated blood resulting in insoluble protein fibers which cause the sickling phenomenon. Sickle-cell anemia is a recessive disease because Hb^S heterozygotes are clinically normal, although red blood cells from heterozygotes can show sickling under very low oxygen conditions *in vitro*.

The Hb^S allele is particularly frequent in regions where malaria is endemic (**Figure 16.6a,b**). This suggests that balancing selection is operating, in that HbS heterozygotes may show protection against malaria and therefore maintain the Hb^S allele at high frequencies. This does seem to be the case—heterozygotes have been shown clinically to show less severe malaria symptoms than homozygotes for the non-sickle allele Hb^B. (**Table 16.3**).

Determining the ages of alleles protective against malaria allows comparison with information from other sources (for example, archaeology), either to

TABLE 16.3:
PROTECTION AGAINST SEVERE MALARIA BY GENETIC DISEASE ALLELES

Study population	Gene	Disease allele	Non-malaria controls		Severe malaria cases[a]		*p* value	Reference
			Number	% individuals carrying allele	Number	% individuals carrying allele		
Gambia and Kenya, female	*G6PD*	A–	325	19.7	388	10.8	0.006	Ruwende C et al. (1995) *Nature* 376, 246.
Gambia and Kenya, male			388	10.9	396	4.3	0.004	
Afghan refugees in Pakistan, female	*G6PD*	Med	154	18.2	78*	7.7	0.037	Leslie T et al. (2010) *PLoS Med.* 7, e1000283.
Afghan refugees in Pakistan, male			124	12.1	62*	1.6	0.041	
Gambia	*HBB*	Hb^S	3875	7.5	2488	1.3	<0.001	Clark TG et al. (2009) *Eur. J. Hum. Genet.* 17, 1080.
Ghana	*HBB*	Hb^S	2048	14.8	2591	1.4	<0.001	May J et al. (2007) *JAMA* 297, 2220.
Ghana	*HBA*	α–	2048	27.3	2591	25.2	0.03	
Ghana, children <5 years old	*HBA*	α–	1093	32.7	261	23.7	0.04	Mockenhaupt FP et al. (2004) *Blood* 104, 2003.
Kenya	*HBA*	α–	648	50.6	655	47.0	0.013	Williams TN et al. (2005) *Blood* 106, 368.

[a]Studies examined infection by *Plasmodium falciparum* except * where *P. vivax* infection was measured.
For α-thalassemia studies, figures show percentage of heterozygotes for α–, not homozygotes.

Figure 16.6 *(right)*: Geographical distribution of malaria and frequencies of alleles maintained by malarial selection. (a) Current geographical distribution of *Plasmodium falciparum* malaria endemicity. API, annual parasite incidence. (b) Frequency distribution of the sickle-cell hemoglobin allele HbS. (c) Frequency distribution of G6PD deficiency (G6PDd) alleles. Note that only countries with endemic malaria are shown with G6PD deficiency data. (Courtesy of Fred Piel and Simon Hay of the Malaria Atlas Project.)

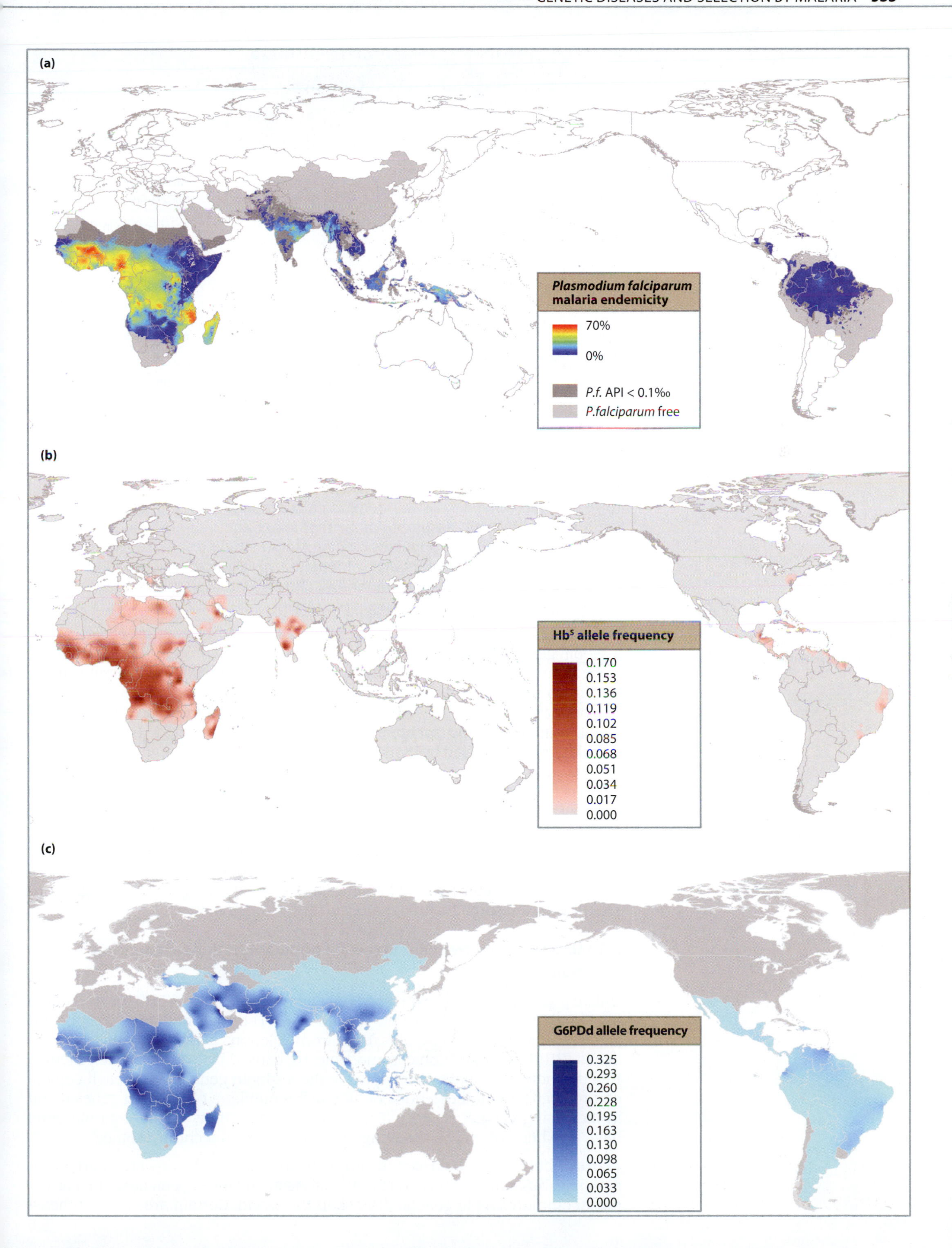
(a)
Plasmodium falciparum malaria endemicity
70%
0%
P.f. API < 0.1‰
P.falciparum free
(b)
Hb^S allele frequency
0.170
0.153
0.136
0.119
0.102
0.085
0.068
0.051
0.034
0.017
0.000
(c)
G6PDd allele frequency
0.325
0.293
0.260
0.228
0.195
0.163
0.130
0.098
0.065
0.033
0.000

TABLE 16.4:
α- AND β-GLOBIN VARIANTS THAT CONFER RESISTANCE TO MALARIA

Gene	Allele	Molecular basis	Anemia in homozygotes	History
HBB	Hb^S	Glu6Val	severe	single origin in West Africa; multiple origins elsewhere
	Hb^E	Glu26Lys	mild	single origin in Thailand; multiple origins elsewhere
	other β-thalassemia	many alleles	variable	usually very population-specific
HBA	α-thalassemia	α– (one gene deleted)	very mild	recurrent mutation
		–– (both genes deleted)	severe or fatal	recurrent mutation

Adapted from Hedrick P (2011) *Heredity* 107, 283. With permission from Macmillan Publishers Ltd.

evaluate the hypothesis of malarial selection, or (if this is assumed to be true) to assess the reliability of the genetic calculations. Modeling selection under various demographic assumptions at this locus and using SNP data from the West African Mandinka population suggested a single origin for the Hb^S allele between 700 and 2700 years ago, at least in this particular population. This is consistent with the spread of malaria occurring with agriculture, although the interpretation is dependent on a simple model of population growth.[11]

β-Thalassemias (OMIM 613985) are anemias caused by mutant alleles in the β-globin gene that lead to decreased levels of β-globin protein, and are thought to be maintained in malaria endemic regions by selection, in a similar manner to Hb^S. The geographical overlap between β-thalassemia and malaria is not perfect, mainly because β-thalassemia is still common in areas, such as Mediterranean Europe, where malaria has been eradicated. In most cases, β-thalassemia is caused by individually rare, population- or region-specific alleles, but an exception is the Hb^E allele (Glu26Lys, rs33950507, **Table 16.4**). This allele is frequent in Southeast Asia, occurring at up to 25% in some groups; in homozygous form it causes anemia, but in heterozygous form offers protection against malaria caused by *P. falciparum*.[7] Analysis of SNP diversity surrounding the β-globin gene in Thai individuals confirmed that the Hb^E allele exists on a long extended haplotype, a sign of positive selection (**Section 6.7**). The age of this allele is between 1000 and 4500 years, again showing a recent origin consistent with the spread of malaria occurring with agriculture.[34]

α-Thalassemias are frequent in certain populations due to balancing selection

Unlike β-globin, which is expressed only after birth, α-globin is expressed in the adult and fetus, so α-thalassemia (OMIM 604131) causes both pre- and postnatal disease. Most α-thalassemias result from deletions of one of the two α-globin genes (*HBA1* and *HBA2*) on chromosome 16, caused by NAHR between them.[16] Healthy individuals have four copies of the α-globin gene (two on each copy of chromosome 16) and deletions of increasing numbers of α-globin genes result in increasing clinical severity (**Figure 16.7**). Duplications of the α-globin gene have also been observed, resulting in three copies on a chromosome.[27]

The distribution of α-thalassemia mirrors that of malaria, being particularly common in sub-Saharan Africa and Southeast Asia, and it has been described as the most common simple genetic disease in the world. Certain areas of Southeast

Clinical phenotype	α-globin gene arrangement	α-globin genotype	Number of functional α-globin genes
Normal	α2 α1	αα/αα	4
Normal	α1/α2	ααα/αα	5
Normal (silent carrier)	α2/α1	αα/α–	3
α-thalassemia trait (mild anemia)	or	α–/α– αα/––	2
Moderately severe hemolytic anemia		α–/––	1
Hydrops fetalis or homozygous α-thalassemia		––/––	0

Figure 16.7: Clinical phenotypes caused by different copy numbers of α-globin genes.

Asia have very high levels of the disease (for example, 33.6% prevalence in the Kadazan-Dusun population of Malaysian Borneo) and gene deletions (α– alleles) have been used as genetic markers to answer questions about the peopling of the Pacific Islands (**Section 13.3**).

Comparison of the frequency of the α-thalassemia deletion genotypes in malaria patients and controls confirms that individuals heterozygous for the deletion (α–/αα) are significantly protected against severe malaria, but also allows population genetic estimation of the selection coefficient for the α– allele in a malarial environment.[20, 54] The value is between 0.05 and 0.08, which would normally lead to rapid fixation of the α– allele if the selection against the α-thalassemia in α–/α– individuals were not taken into account.

In sub-Saharan populations where malaria is endemic, the frequency of the sickle-cell Hb^S allele (Figure 16.6b) and α– thalassemia alleles are high (0.33–0.4 in Nigeria compared with <0.01 in the UK). When both alleles are present in the same individual, that is, the individual is heterozygous for the Hb^S allele and homozygous for the α– allele, then the protective effect against malaria is abolished. This is an example of **negative epistasis**, and may be another reason why, in addition to sickle-cell anemia and α-thalassemia, the malaria-protective alleles Hb^S and α– do not rise to even higher frequencies in malaria-endemic regions.[53]

In northern European populations, where malaria is not endemic, we would expect the α– allele to be in mutation–selection balance. The frequency of the α– allele is less than 1%, and direct measurement of α-globin mutation rate in sperm gives a deletion mutation rate of 4.2×10^{-5} per generation. If we assume maternal and paternal mutation rates to be the same, this suggests that either the selection against α-thalassemic individuals homozygous for α– deletion in northern Europe has been very strong or that there is selection against α-thalassemic individuals combined with weak selection against heterozygotes.[21]

A summary of α- and β-globin variants that confer resistance to malaria is given in Table 16.4.

Glucose-6-phosphate dehydrogenase deficiency alleles are maintained at high frequency in malaria-endemic populations

Glucose-6-phosphate dehydrogenase (G6PD) is a **housekeeping** enzyme, which is found in all cells and catalyzes the reaction of glucose-6-phosphate with **NADP** to form 6-phospho-gluconate and **NADPH**, in a pathway leading eventually to the production of ribose. The production of NADPH is of particular significance because this chemically reduced (that is, NADPH instead of NADP) form is necessary to avoid oxidative damage in cells, and G6PD is the only enzyme in red blood cells that can produce it.

Complete absence of G6PD enzyme activity is unknown, and is probably lethal, but a condition with reduced levels of G6PD is quite common in Africa and South Asia, and known as glucose-6-phosphate dehydrogenase deficiency (Figure 16.6c, OMIM 305900). It is an X-linked recessive disease, caused by several different alleles of the *G6PD* gene that either reduce the catalytic activity or the stability of the resulting enzyme. Because the capacity of the red blood cells to resist oxidative damage is reduced in G6PD-deficient individuals, the individual is vulnerable to drugs (such as the anti-malarial compound primaquine) or naturally occurring substances that increase oxidative stress, with neonatal jaundice or hemolytic anemia the result. One such naturally occurring compound is vicine, an alkaloid glycoside which is present in broad beans (*Vicia faba*), also known as fava beans, hence the alternative disease name **favism**.

A high frequency of G6PD deficiency alleles is coincident with a high prevalence of *Plasmodium falciparum* malaria across the world (Figure 16.6c). This suggests that G6PD deficiency may protect against malaria, as confirmed by epidemiological studies which have shown that in heterozygous females and hemizygous males, the risk of severe malaria is reduced by about 50% (Table 16.3; female homozygotes were not evaluated in the Gambian and Kenyan study because they were too rare). Also, *in vitro* studies have shown that *P. falciparum* grows poorly in G6PD-deficient red blood cells, at least until it adapts by producing its own G6PD. It is thought that the parasite depletes the red blood cell of NADPH, leading to peroxide-induced hemolysis in G6PD-deficient individuals and thus loss of a few cells, but curtailment of the parasite's development.

Population genetic analysis also supports the view that selection maintains G6PD deficiency alleles in malaria-endemic populations. The ancestral *G6PD* allele, determined by comparison with ape sequences, is B. Several hundred G6PD deficiency alleles have been identified, but few of them are common. In the Mediterranean region, the Middle East, and India, the Med allele (Ser188Phe, rs5030868) with ~3% of B enzyme activity is present at a frequency of ~2–20% in different populations, and is associated with protection against infection by the malaria parasite *P. vivax*. In Southeast Asia the Mahidol (Ser163Gly, rs137852314) variant has between 5 and 30% of B enzyme activity and is present at frequencies of around 10%. This variant has been shown to be associated with a reduced number of *P. vivax* parasites in the blood, and shows extended haplotypes characteristic of recent strong positive selection (**Figure 16.8**, **Section 6.7**).[28]

In sub-Saharan Africa, most G6PD-deficient individuals carry the A or A– alleles. The A allele (Asn126Asp, rs1050829) can represent up to 40% of a population and has ~85% of B activity, but does not protect against severe malaria and should not be considered in the same way as other deficiency alleles; it remains unclear whether it conferred a selective advantage in the past. There are several deficiency alleles classified as A– in sub-Saharan Africa[8] but the most studied is A– (Val68Met, rs1050828), which has ~12% of B activity, and can be present at a frequency of up to 25%. This allele has been shown by extended haplotype tests to have undergone recent positive selection (Figure 16.8, see also **Section 6.7**).[42]

When did these alleles originate? Estimates of the TMRCAs of subsets of the chromosomes have been obtained in several ways. Verrelli et al. used coalescent-based dating to calculate that the TMRCA for the observed G6PD variation in Africa was 620 (480–760) KYA, a time that is well within the range expected for a neutral X-linked segment of DNA.[50] The coalescence time for A alleles is also ancient (316 ± 244 KYA), but the coalescence time for the A– allele (45 ± 20 KYA) is more recent. A joint estimation of age and selection coefficient for the A– allele, using linkage disequilibrium,[46] gave values of between 0.5 and 2

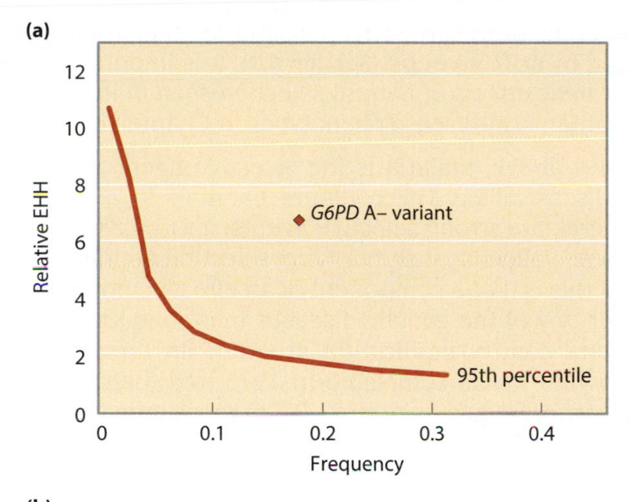

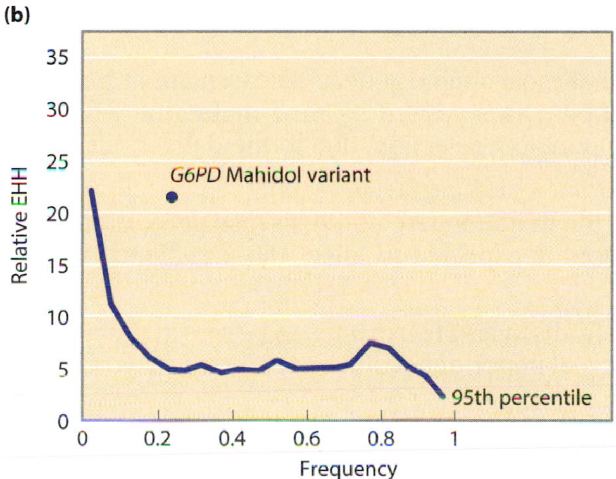

Figure 16.8: Extended haplotypes generated by selection at the *G6PD* gene. SNPs are plotted on each graph showing relative extended haplotype homozygosity (REHH, see Section 6.7) on the *y* axis and allele frequency on the *x* axis. (a) The SNP defining the G6PD A– allele is highlighted, showing an unusually high REHH indicative of selection. (b) The SNP defining the G6PD Mahidol allele is highlighted, showing an unusually high REHH indicative of selection in Thais. [a, empirical SNP data from YRI HapMap population, from Sabeti PC et al. (2002) *Nature* 419, 832; b, empirical data from CHB HapMap population, from Louicharoen C et al. (2009) *Science* 326, 1546.]

KYA and $s \approx 0.26$. For the Mahidol allele, analysis of linkage disequilibrium suggests a date of origin of 1–2.5 KYA (assuming a 27-year generation time) and a selection coefficient of $s \approx 0.2$, and for the Med allele a date of 10 ± 25 KYA was estimated using a coalescent approach.[28, 50] The difference in age estimates between the linkage disequilibrium approaches and the coalescent approach is likely to reflect the fact that, although the alleles themselves are old (as measured by coalescent analysis), the selective sweeps increasing their frequency (as measured by the linkage disequilibrium analysis) are more recent.

As usual, there are very large uncertainties associated with these age estimates, and the selection coefficients seem remarkably high. However, the deficiency alleles are clearly recent, and thus consistent with malarial selection and with the idea that meaningful (although imprecise) estimates are obtained from genetic data. The ancient date for the A allele suggests that it is neutral, or at least not involved in resistance to malaria.

What can these examples tell us about natural selection?

The cases discussed above show how studying the ultimate cause of disease can yield important and interesting insights into human evolution and medicine.

A message from the examples above comes from the nature and frequencies of the different alleles that are responsible for the disease. This is also known as the **allelic spectrum**, **allelic architecture**, or **genetic architecture** of disease, and has consequences for mapping genes involved in complex disease, discussed in the next chapter. The diseases have either different causative alleles, or multiple occurrences of the same allele, or both. Strong positive selection has operated on one or a handful of these disease alleles in certain geographical areas, with different rarer alleles responsible for the disease in other areas, possibly affected by **soft sweeps** (Section 6.7). It is important to note the similarities in the genetic effects of natural selection seen in this section with those described on human non-disease phenotypes in Chapter 15.

For the examples above, malaria is the selective agent maintaining the high frequency of a disease allele. This confirms that malaria, and perhaps infectious diseases in general, are strong selective agents, if they can maintain a high frequency of a disease allele against purifying selection acting on the same allele. The corollary of this is that a component of an effective population-based treatment regime for any of the genetic diseases mentioned in this section should include a malaria control/eradication program. In the Maldives, for example, such a program[14] was completed in 1975 and there has been a drop in β-thalassemia carrier frequency from 21.3% in those individuals born in 1970 to 16% for those born in 1989.

SUMMARY

- We expect that most simple genetic diseases are in mutation–selection balance, with new cases caused by new mutations which are subsequently removed by purifying selection, due to the deleterious consequences of the disease.
- Variation in the mutation rate of the disease gene, or in the strength of purifying selection, is expected to affect the prevalence of the simple genetic disease.
- Simple genetic diseases are expected to be rare in the population, and exceptions may reveal important evolutionary forces at work, such as genetic drift or selection.
- Demographic processes such as founder effects and consanguinity can affect the prevalence of simple genetic diseases in a population, increasing some and decreasing others.
- Segmental duplications allow large-scale genomic rearrangements. If these genomic rearrangements cause a disease, they are called genomic disorders.
- The maintenance of certain disease-causing alleles by their selective advantage in malaria-endemic regions provides the best examples of positive selection increasing the prevalence of genetic diseases.

QUESTIONS

Question 16–1: For the following four diseases, explain the molecular basis of their dominant modes of inheritance: Huntington's disease, Hutchinson-Gilford progeria syndrome, achondroplasia, hereditary hemorrhagic telangiectasia.

Question 16–2: Distinguish prevalence of a disease from incidence of a disease. What factor influences the prevalence but not the incidence of a disease?

Question 16–3: In the Maldives there has been a drop in β-thalassemia carrier frequency from 21.3% in those individuals born in 1970 to 16% for those born in 1989. Calculate the corresponding fall in the numbers of new β-thalassemia patients between 1970 and 1989, given these rates.

Question 16–4: The incidence at birth of an autosomal dominant disease is 1 in 50,000. The clinical genetics lab has confirmed that all patients have a unique mutation in the causative gene. Assuming mutation-selection balance, calculate the rate of new mutations at this gene if

(a) All patients die before age 12

(b) All patients live a normal life span, but males with the disease are sterile

Question 16–5: Discuss how the development of agriculture has shaped our genome through malaria.

Question 16–6: Using dbSNP, determine the allele frequency of the A– G6PD allele (rs1050828) in the HapMap populations. Plot the frequencies on a sketch map of the world. Does your allele frequency map agree with Figure 16.6c?

REFERENCES

The references highlighted in purple are considered to be important (for this chapter) by the authors.

1. **Akey JM, Zhang G, Zhang K et al.** (2002) Interrogating a high-density SNP map for signatures of natural selection. *Genome Res.* **12**, 1805–1814.
2. **Al-Gazali L, Hamamy H & Al-Arrayad S** (2006) Genetic disorders in the Arab world. *BMJ* **333**, 831–834.
3. **Ankala A, Kohn JN, Hegde A et al.** (2012) Aberrant firing of replication origins potentially explains intragenic nonrecurrent rearrangements within genes, including the human *DMD* gene. *Genome Res.* **22**, 25–34.
4. **Bittles AH & Hamamy H** (2010) Endogamy and consanguineous marriage in Arab populations. In Genetic Disorders Among Arab Populations, 2nd ed (Teebi AS ed), pp 85–108. Springer.
5. **Bundey S & Alam H** (1993) A five-year prospective study of the health of children in different ethnic groups, with particular reference to the effect of inbreeding. *Eur. J. Hum. Genet.* **1**, 206–219.
6. **Carvalho C, Zhang F & Lupski JR** (2010) Genomic disorders: a window into human gene and genome evolution. *Proc. Natl Acad. Sci. USA* **107** Suppl 1, 1765–1771.
7. **Chotivanich K, Udomsangpetch R, Pattanapanyasat K et al.** (2002) Hemoglobin E: a balanced polymorphism protective against high parasitemias and thus severe *P. falciparum* malaria. *Blood* **100**, 1172–1176.
8. **Clark TG, Fry AE, Auburn S et al.** (2009) Allelic heterogeneity of G6PD deficiency in West Africa and severe malaria susceptibility. *Eur. J. Hum. Genet.* **17**, 1080–1085.
9. **Coventry A, Bull-Otterson LM, Liu X et al.** (2010) Deep resequencing reveals excess rare recent variants consistent with explosive population growth. *Nat. Commun.* **1**, 131.
10. **Cox JJ, Reimann F, Nicholas AK et al.** (2006) An *SCN9A* channelopathy causes congenital inability to experience pain. *Nature* **444**, 894–898.
11. **Currat M, Trabuchet G, Rees D et al.** (2002) Molecular analysis of the β-globin gene cluster in the Niokholo Mandenka population reveals a recent origin of the β-S Senegal mutation. *Am. J. Hum. Genet.* **70**, 207–223.
12. **Falush D** (2009) Haplotype background, repeat length evolution, and Huntington's disease. *Am. J. Hum. Genet.* **85**, 939–942.
13. **Farrell P, Le Marechal C, Ferec C et al.** (2007) Discovery of the principal cystic fibrosis mutation (F508del) in ancient DNA from Iron Age Europeans. *Nature Precedings,* http://hdl.handle.net/10101/npre.2007.1276.1
14. **Firdous N, Gibbons S & Modell B** (2011) Falling prevalence of β-thalassaemia and eradication of malaria in the Maldives. *J. Community Genet.* **1**, 173–189.
15. **Fisher R, Race R & Taylor G** (1944) Mutation and the Rhesus reaction. *Nature* **153**, 106.
16. **Flint J, Harding RM, Boyce AJ & Clegg JB** (1998) The population genetics of the haemoglobinopathies. *Baillières Clin. Haematol .* **11**, 1–51.
17. **Gabriel SE, Brigman KN, Koller BH et al.** (1994) Cystic fibrosis heterozygote resistance to cholera toxin in the cystic fibrosis mouse model. *Science* **266**, 107–109.
18. **Haldane J** (1941) Selection against heterozygosis in man. *Ann. Hum. Genet.* **11**, 333–340.
19. **Hedman M, Brandstätter A, Pimenoff V et al.** (2007) Finnish mitochondrial DNA HVS-I and HVS-II population data. *Forensic Sci. Int.* **172**, 171–178.
20. **Hedrick P** (2011) Population genetics of malaria resistance in humans. *Heredity* **107**, 283–304.
21. **Hedrick PW** (2011) Selection and mutation for α-thalassemia in nonmalarial and malarial environments. *Ann. Hum. Genet.* **75**, 468–474.
22. **Hennekam R** (2006) Hutchinson–Gilford progeria syndrome: Review of the phenotype. *Am. J. Med. Genet. A* **140**, 2603–2624.
23. **Högenauer C, Santa Ana CA, Porter JL et al.** (2000) Active intestinal chloride secretion in human carriers of cystic fibrosis mutations: an evaluation of the hypothesis that heterozygotes have subnormal active intestinal chloride secretion. *Am. J. Hum. Genet.* **67**, 1422–1427.
24. **Inoue K, Dewar K, Katsanis N et al.** (2001) The 1.4-Mb CMT1A duplication/HNPP deletion genomic region reveals unique genome architectural features and provides insights into the recent evolution of new genes. *Genome Res.* **11**, 1018–1033.
25. **Keller MP, Seifried BA & Chance PF** (1999) Molecular evolution of the CMT1A-REP region: a human-and chimpanzee-specific repeat. *Mol. Biol. Evol.* **16**, 1019–1026.
26. **Libby RT, Hagerman KA, Pineda VV et al.** (2008) *CTCF* cis-regulates trinucleotide repeat instability in an epigenetic manner: a novel basis for mutational hot spot determination. *PLoS Genet.* **4**, e1000257.
27. **Liu Y, Old J, Miles K et al.** (2000) Rapid detection of α-thalassaemia deletions and α-globin gene triplication by multiplex polymerase chain reactions. *Br. J. Haematol.* **108**, 295–299.
28. **Louicharoen C, Patin E, Paul R et al.** (2009) Positively selected G6PD-Mahidol mutation reduces *Plasmodium vivax* density in Southeast Asians. *Science* **326**, 1546–1549.
29. **MacArthur DG, Balasubramanian S, Frankish A et al.** (2012) A systematic survey of loss-of-function variants in human protein-coding genes. *Science* **335**, 823–828.
30. **Marques-Bonet T, Kidd JM, Ventura M et al.** (2009) A burst of segmental duplications in the genome of the African great ape ancestor. *Nature* **457**, 877–881.
31. **Muller HJ** (1950) Our load of mutations. *Am. J. Hum. Genet.* **2**, 111–176.
32. **Norio R** (2003) Finnish disease heritage I. *Hum. Genet.* **112**, 441–456.
33. **Norio R** (2003) The Finnish disease heritage III: The individual diseases. *Hum. Genet.* **112**, 470–526.
34. **Ohashi J, Naka I, Patarapotikul J et al.** (2004) Extended linkage disequilibrium surrounding the hemoglobin E variant due to malarial selection. *Am. J. Hum. Genet.* **74**, 1198–1208.
35. **Ostrer H** (2001) A genetic profile of contemporary Jewish populations. *Nat. Rev. Genet.* **2**, 891–898.
36. **Peltonen L, Jalanko A & Varilo T** (1999) Molecular genetics of the Finnish disease heritage. *Hum. Mol. Genet.* **8**, 1913–1923.
37. **Perry GH, Xue Y, Smith RS et al.** (2012) Evolutionary genetics of the human Rh blood group system. *Hum. Genet.* **131**, 1205–1216.
38. **Pier GB, Grout M, Zaidi T et al.** (1998) *Salmonella typhi* uses CFTR to enter intestinal epithelial cells. *Nature* **393**, 79–82.
39. **Reed T** (1971) Dogma disputed. Does reproductive compensation exist? An analysis of Rh data. *Am. J. Hum. Genet.* **23**, 215–224.

40. **Risch N, Tang H, Katzenstein H & Ekstein J** (2003) Geographic distribution of disease mutations in the Ashkenazi Jewish population supports genetic drift over selection. *Am. J. Hum. Genet* .**72**, 812–822.

41. **Romeo G, Devoto M & Galietta LJV** (1989) Why is the cystic fibrosis gene so frequent? *Hum. Genet.* **84**, 1–5.

42. **Sabeti PC, Reich DE, Higgins JM et al.** (2002) Detecting recent positive selection in the human genome from haplotype structure. *Nature* **419**, 832–837.

43. **Shaw-Smith C, Pittman AM, Willatt L et al.** (2006) Microdeletion encompassing *MAPT* at chromosome 17q21.3 is associated with developmental delay and learning disability. *Nat. Genet.* **38**, 1032–1037.

44. **Shovlin CL** (2010) Hereditary haemorrhagic telangiectasia: pathophysiology, diagnosis and treatment. *Blood Rev.* **24**, 203–219.

45. **Slatkin M** (2004) A population-genetic test of founder effects and implications for Ashkenazi Jewish diseases. *Am. J. Hum. Genet.* **75**, 282–293.

46. **Slatkin M** (2008) A Bayesian method for jointly estimating allele age and selection intensity. *Genet. Res. (Camb.)* **90**, 129–137.

47. **Tadmouri GO, Nair P, Obeid T et al.** (2009) Consanguinity and reproductive health among Arabs. *Reprod. Health* **6**, 17.

48. **The 1000 Genomes Project Consortium** (2012) An integrated map of genetic variation from 1092 human genomes. *Nature* **491**, 56–65.

49. **Turner DJ, Miretti M, Rajan D et al.** (2007) Germline rates of de novo meiotic deletions and duplications causing several genomic disorders. *Nat. Genet.* **40**, 90–95.

50. **Verrelli BC, McDonald JH, Argyropoulos G et al.** (2002) Evidence for balancing selection from nucleotide sequence analyses of human *G6PD*. *Am. J. Hum. Genet.* **71**, 1112–1128.

51. **Wang S, Haynes C, Barany F & Ott J** (2009) Genome-wide autozygosity mapping in human populations. *Genet. Epidemiol.* **33**, 172–180.

52. **Warby SC, Montpetit A, Hayden AR et al.** (2009) CAG expansion in the Huntington disease gene is associated with a specific and targetable predisposing haplogroup. *Am. J. Hum. Genet.* **84**, 351–366.

53. **Williams TN, Mwangi TW, Wambua S et al.** (2005) Negative epistasis between the malaria-protective effects of α+-thalassemia and the sickle cell trait. *Nat. Genet.* **37**, 1253–1257.

54. **Williams TN, Wambua S, Uyoga S et al.** (2005) Both heterozygous and homozygous α+ thalassemias protect against severe and fatal *Plasmodium falciparum* malaria on the coast of Kenya. *Blood* **106**, 368–371.

55. **Zerjal T, Dashnyam B, Pandya A et al.** (1997) Genetic relationships of Asians and Northern Europeans, revealed by Y-chromosomal DNA analysis. *Am. J. Hum. Genet.* **60**, 1174–1183.

56. **Zody MC, Jiang Z, Fung HC et al.** (2008) Evolutionary toggling of the *MAPT* 17q21. 31 inversion region. *Nat. Genet.* **40**, 1076–1083.

EVOLUTION AND COMPLEX DISEASES

CHAPTER SEVENTEEN

In this chapter we will ask how an evolutionary perspective can help illuminate the genetic influences on complex disease incidence or prevalence. Understanding these influences is difficult, for reasons that are summarized in **Figure 17.1**.

In a complex disease, the presence of a particular allele in an individual contributes to disease risk, but is not enough to cause the disease. If we want to investigate the influence of such an allele upon a disease phenotype in affected individuals, we determine genotypes. The causative allele itself is rarely among the genotyped variants, so these usually act as proxies. Genotypes are shown to be associated with the causative allele, or not, by genotyping technologies (which have a low error rate) and statistical methods that consider, for example, the frequency of each genotype in cases and controls. The strength of the allele's effect and the sample size determine the power of the study to discover the association.

However, this picture is complicated by many other factors. First, describing the phenotype accurately can be difficult—there may be many other similar phenotypes that have distinct **etiologies**. This problem can be addressed to some extent by dividing the disease phenotypes into more readily definable quantitative parameters called **endophenotypes**. An example is the use of cholesterol levels and blood pressure in cardiovascular disease. Second, the environment may have a strong influence on the phenotype or endophenotype, and this environment can be individual-specific, shared within families, or shared within populations. The environment itself can be measured, but not without error. Third, many alleles at many genes contribute to the heritability of the endophenotypes, and there are interactions between alleles, and also between genes and environment. Allele frequencies are influenced by the forces described in Chapter 5—natural selection, mutation, and drift—contributing to variation between individuals of the same population and variation among populations. Thinking in terms of evolution can help illuminate the genetic landscape of complex diseases, an approach that has been called **evolutionary medicine**.[9, 33]

17.1 DEFINING COMPLEX DISEASE

Many aspects of our health are influenced by our genotype, and since our genotype is the product of our evolutionary history, evolutionary genetics can have important implications for diagnosing, understanding, and treating medical conditions. We have seen in Chapter 15 that, at the level of the genome, there is no clear distinction between "normal" genetic variation and variation associated with disease. We have also seen in Chapter 16 the imperfect but useful nature of the distinction between the simple genetic diseases that were discussed there and the complex genetic diseases that are the subject of this chapter. To reiterate, simple genetic diseases are almost always rare and are generally caused by a single high-penetrance disease allele within any particular family. Complex genetic diseases are those where a single disease allele (heterozygous or homozygous) is not sufficient to cause the disease; these diseases may be common in the population. **Table 17.1** lists the incidence of a selection of complex diseases in different populations, and it is instructive to compare the incidences of these diseases with those of simple diseases shown in Table 16.1. In practice, the most useful criterion distinguishing simple from complex disease is the inheritance pattern in families: if Mendelian inheritance can be discerned, the disease is a simple genetic disease.

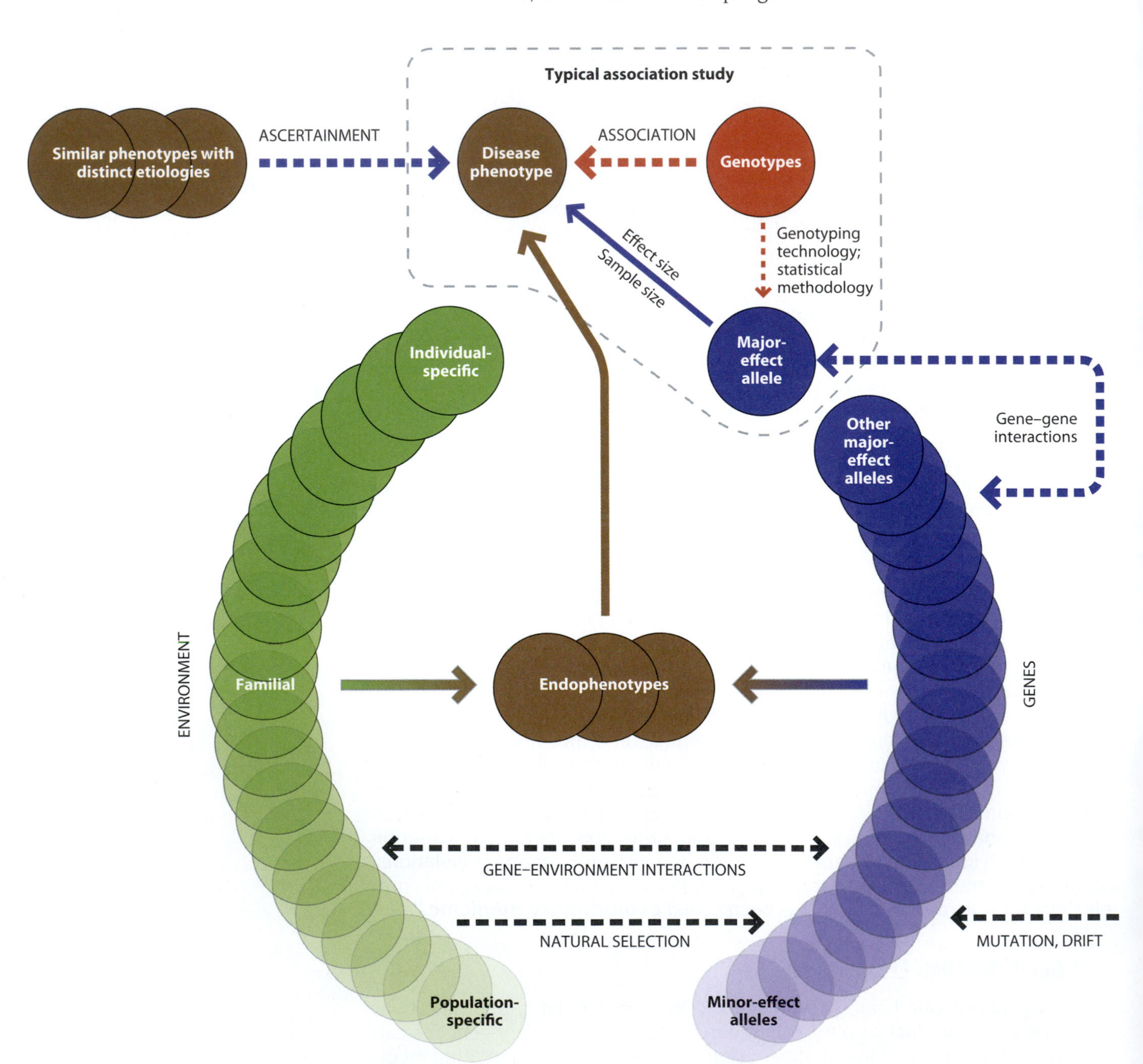

Figure 17.1: Schematic representation of the complexity of complex disease.
The dotted box contains a simple scheme for a typical study of the association between an allele and a disease phenotype. Outside this box are illustrated various other factors that complicate the task; for further discussion, see the text.

TABLE 17.1:
INCIDENCE OF A SELECTION OF COMPLEX DISEASES IN DIFFERENT POPULATIONS

Disease	OMIM	Phenotype	Population	Incidence per 100,000 per year	Section
Chronic obstructive pulmonary disease	606963	irreversible obstruction of airways caused by bronchitis, emphysema, or other disease	England	200	–
Type 2 diabetes	125853	high blood sugar; major cause of adult blindness; two- to fourfold increase in cardiovascular mortality	Italy	760	17.2
			United States	1000–1500	
			Pima Indians	~2400	
Age-related macular degeneration	603075	progressive degeneration of cells in the macula region of the retina	Latino-Americans in Los Angeles	50	17.3
			European-Americans in Wisconsin	180	
Crohn's disease	266600	chronic intestinal inflammation, discontinuous; inflammation of skin, eyes, or joints	England	6–11	17.3, Figure 17.9
			China	0.3	
Cardiomyopathy	192600 (rare familial hypertrophic cardiomyopathy)	deterioration of the heart muscle, for any reason, usually leading to heart failure	UK (male)	440	17.4
			UK (female)	390	
Kidney failure	n/a	loss of kidney function, known as end-stage renal disease	African-Americans	100	17.4
			European-Americans	25	

In this chapter we will also provide an overview of how genetic factors in complex disease are discovered, and how evolutionary information can help us to investigate the genetics of complex disease. We provide several key examples of how the allelic architecture or allelic spectrum (**Figure 17.2**) of a disease is influenced by evolution, with implications for understanding how the disease occurs and potential for treatment. We also discuss the importance of evolutionary information for medical treatments, particularly in the use of drugs to treat disease.

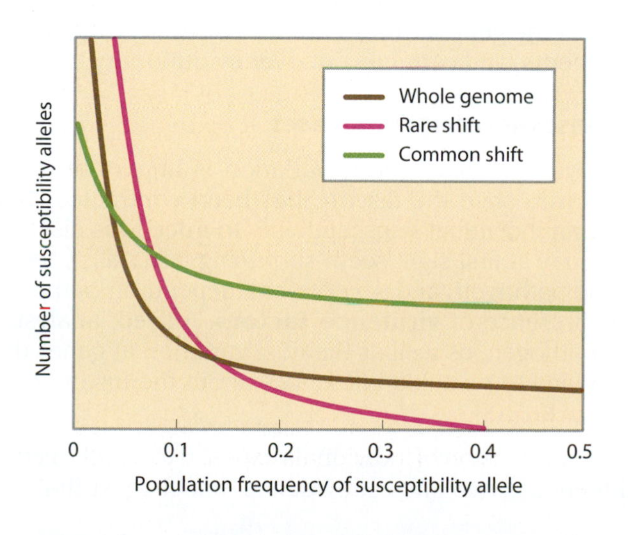

Figure 17.2: Three different models predicting the allelic spectra of complex disease.
This graph relates the number of susceptibility alleles for a complex disease with their frequency in the population. The *brown* line represents a model where the spectrum of disease susceptibility alleles is the same as the allelic spectra for all variation in the genome, where most alleles are rare. The *green* line represents the common-disease common-variant model (Box 17.3), where disease-susceptibility alleles are more likely to be common than the genome average. The *pink* line represents the mutation–selection model, where disease susceptibility alleles are more likely to be rare than the genome average.

TABLE 17.2:
ESTIMATES OF GENETIC CONTRIBUTIONS TO VARIATION IN COMPLEX DISEASE RISK FROM A SELECTION OF STUDIES

Disease	Approximate heritability (population)	Sibling recurrence risk (population)
Type 2 diabetes	0.26 (Denmark) 0.31 (Sweden)	1.2–1.8 (United States)
Crohn's disease	0.34–1 (Sweden)	35 (England)
Schizophrenia	0.82 (UK) 0.84 (Finland)	8.6 (Sweden)
Malaria	~0.25 (Kenya)	–
Helicobacter pylori infection	0.5–0.6 (Sweden)	–
Leprosy	0.57 (Kembanglemari, Indonesia)	1.8–2.5 (Malawi)
Tuberculosis (measured by tuberculin skin test)	~0.5 (Gambia)	–

The genetic contribution to variation in disease risk varies between diseases

How do we determine whether genetic variation is important in the variation in risk of developing a complex disease? In Chapter 15 we introduced heritability, a statistic used to determine the amount of total variation in a trait that could be accounted for by genetic variation (Box 15.3). Disease is simply a medically important trait, so measuring the heritability of a disease by twin or population studies can give a measure of the importance of genetic variation in the susceptibility of an individual to a complex disease.

Another, perhaps more intuitive, approach is the observation of familial clustering. If a disease runs in a family, even if there is no clear Mendelian inheritance, then we might suspect that the disease has a large genetic component. Familial clustering can be quantified using the **risk ratio** (λ). A common form of this is the **sibling risk ratio** (λ_S) which is the probability (relative to the population level of risk) that, if a certain individual has a particular disease, their sibling will also have that disease.

There are different levels of genetic contribution to the risk of developing different complex diseases (Figure 17.1, **Table 17.2**). This is because diseases differ in their etiology, and have different environmental components. Whereas many simple diseases are congenital or early-onset, most complex diseases are late-onset and their incidence is dependent upon nongenetic factors. Indeed we would not expect the genetic contribution to disease risk to be the same in different environments, and certainly not over evolutionary time periods.

Infectious diseases are complex diseases

In Chapter 16 we saw that genetic variation is important in determining the severity of malaria, once the disease has been contracted. Genetic variation can also affect an individual's susceptibility to infectious disease, as well as its severity.[4] This may at first sight seem surprising; after all, infection depends on exposure to the pathogen, and severity can depend on pathogen genetic variation such as presence of **virulence factors**. Indeed, analysis of the genetic variation of a pathogen, as well as the host variation at genes that respond to a pathogen, can lead to valuable inferences about the history of human disease (Box 7.3, **Opinion Box 15**).

However, only a proportion of individuals exposed to a pathogen develop clinical disease and there are two lines of evidence that suggest that genetic variation

Most bacterial taxa are genetically diverse, representing a long history of evolution, in some cases even co-evolution with humans.[30] But some are simpler, and this group of so-called genetically monomorphic pathogens is a common cause of dramatic diseases, including leprosy, anthrax, plague, typhoid fever, cholera, whooping cough, and glanders.[1] For these bacteria, multiple genome sequences have revealed only a few thousand SNPs per taxon, or even fewer, and their molecular clock rates are so fast that coalescence times can be very recent. Thus the argument in Box 16.3 that cystic fibrosis may have been selected by cholera is probably invalid because the causative organism, pandemic *Vibrio cholerae*, probably did not exist 200 years ago. The causative agents of tuberculosis and typhoid fever are also young, but current estimates of their ages are unreliable because they pre-date our new understanding of just how fast bacteria can accumulate SNPs.

Our understanding of the potential selective pressures on humans might profit from an improved understanding of pathogen history. A good example is plague, caused by *Yersinia pestis*, where prior pandemics and population expansions have been illuminated by analyses of genomes from both extant bacterial isolates (Figure 1) and ancient DNA.[1] Historians agree that (at least) three major plague pandemics have swept through much of the world: the Justinianic Pandemic (541 to late eighth century), the Black Death (1347 to mid-eighteenth century), and the ongoing Third Pandemic, which began in China in the mid-nineteenth century and spread from Hong Kong in 1894. Modern human plague is a zoonosis, in which humans are infected from rodents or other animals. In Central Asia and elsewhere, infected rodent populations exist, within which transmission of *Y. pestis* occurs via rodent-species-specific fleas. Human outbreaks result when *Y. pestis* is transmitted to local potential hosts, such as rats. The Third Pandemic is thought to have resulted from the development of rapid global communications via ships, whose voyages were quick enough to allow dissemination of infected rats to new foci. Subsequent host jumps to indigenous rodent populations have resulted in foci in multiple areas within Africa, and North and South America.

The Black Death is thought to have killed one-third of the European population. Until recently, there was a lack of molecular genetic data to prove that this epidemic was caused by *Y. pestis*. Historians and epidemiologists have repeatedly commented on discrepancies between modern plague and the Black Death regarding the local speed of expansion of the disease, its symptoms, and the number of deaths. This has now been addressed through the repeated demonstration of ancient *Y. pestis* DNA in the tooth pulp of affected skeletons[13] and the successful sequencing of genomes from such sources.[3] The discrepancies in epidemiology and symptoms remain unexplained.

Overlapping molecular clock rates were estimated from population genetic studies of *Y. pestis* isolated over almost 100 years in Madagascar, and from skeletons of known burial dates during the Black Death. These rates are orders of magnitude slower than for many other pathogenic bacteria,[1] and indicate that the MRCA of *Y. pestis* existed about 2000 YA. That ancestor in turn represents an asexual clone of a more diverse species, *Y. pseudotuberculosis*, which is transmitted by the oral–fecal route and causes gastroenteritis rather than plague. Genetic data from modern isolates indicate that *Y. pestis* originated in the vicinity of China, and has spread globally many times. During the Third Pandemic, variants that differed by only one or a few SNPs seeded different countries, including Madagascar and the continental USA, and founded all extant isolates in those areas.[2] Similar patterns of spread were responsible for populations currently found in Western Asia and Africa. This historical reconstruction acts as a paradigm for similar reconstructions of the spread of epidemic disease that need to be carefully matched against genetic changes in their natural hosts.

Mark Achtman, Environmental Research Institute, University College Cork, Ireland

(a)

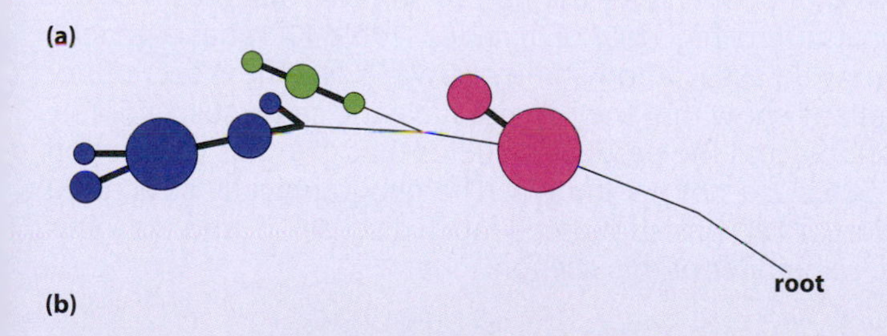

(b)

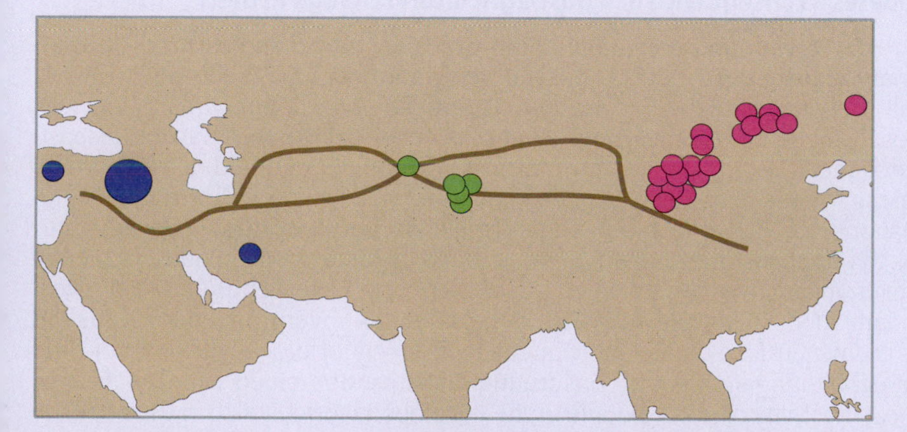

Figure 1: Postulated spread of *Y. pestis* along the Silk Road.
(a) Phylogeny of *Y. pestis* (biovar Medievalis) from SNP data. Haplotypes are colored to correspond with the sampling sites in (b). *Thick black* lines represent an inter-haplotype distance of one SNP allele, *thin black* lines represent an inter-haplotype distance of more than one SNP allele. Circle sizes are proportional to haplotype frequency. (b) Sampling sites of *Y. pestis*, with colors corresponding to the haplotypes shown in (a). The Silk Road is shown as a *brown* line. (a, adapted from Morelli G et al (2010) *Nat. Genet.* 42, 1140. With permission from Macmillan Publishers Ltd.)

makes a substantial contribution to an individual's susceptibility. First, familial clustering and a high concordance rate in monozygotic twins are observed for leprosy and tuberculosis, for example, and in Table 17.2 we have highlighted four infectious diseases where the heritability of susceptibility to that disease has been measured. Second, alleles at single genes can confer protection against certain infections, or susceptibility to other infections. Examples include the *CCR5Δ32* allele discussed in **Section 17.4**, where homozygotes are resistant to HIV infection; and rare alleles at certain genes involved in the immune response (for example, interferon gamma receptor 1; *IFNGR1*, OMIM 107470), which lead to a very high susceptibility to mycobacterial infection.

17.2 THE GLOBAL DISTRIBUTION OF COMPLEX DISEASES

Complex diseases are spread widely across the world and occur at frequencies that can be orders of magnitude higher than those of simple Mendelian disorders. As discussed above, both genetic and environmental factors contribute to the risk of their development during an individual life span. Some complex diseases, such as schizophrenia, have similar prevalence in most populations: about 0.4% in this case. In contrast, diseases such as obesity or infectious and parasitic disease clearly vary in their distribution across the world (**Figure 17.3**). This is primarily due to environmental variation; in particular, for obesity the availability of cheap calorific food in certain areas of the world combined with the lack of physical activity, and for infectious diseases the geographical range of the pathogen or its vector.

But how much is the incidence of complex disease determined by variation in frequency of different susceptibility alleles? In the previous chapter, we saw that malaria resistance alleles occur in malarial regions for good evolutionary reasons, namely natural selection by the disease (**Section 16.4**). It is not the case, of course, that absence of malaria in Northern Europe is due to genetic variation, but rather a consequence of the absence of the pathogen. Indeed, northern Europeans, lacking protective alleles, may be more likely than local people to suffer from severe malaria if they move to a malaria-endemic region.

Genetic variation has played a role in the prevalence of other complex diseases, such as type 2 diabetes (T2D; OMIM 125853) in westernized Pacific Islanders, Native Americans, and Australian Aborigines. For example, on the island of Nauru in Micronesia, 40% of people over the age of 15 have diabetes. For the Pima Indians in Arizona, the very high rate of diabetes (Table 17.1) has a genetic component, but genomewide association studies (GWASs; see next section) of diabetes in the Pima Indians show that they do not have the same susceptibility alleles as other populations, and the particular alleles that give this population such a high risk of diabetes are not yet known. The geographically structured and changing distribution of T2D has stimulated much discussion, and we focus on this example for the remainder of this section.

Is diabetes a consequence of a post-agricultural change in diet?

Could the varying and sometimes high prevalence of diabetes have an evolutionary explanation? Several possible explanations have been put forward, but all remain speculative at present. The most well-known explanation was proposed in 1962 by James Neel (**Figure 17.4**). He suggested that the disease diabetes mellitus was the result of "a thrifty genotype rendered detrimental by 'progress'".[32] His hypothesis was that the rapid release of the hormone insulin in response to elevated blood-sugar levels (**hyperglycemia**) was advantageous to our ancestors, allowing them to build up fat deposits in times of plenty. However, in an environment where there is continuously plentiful or even overabundant food, this rapid response is detrimental—overproduction of insulin leads to insulin resistance, subsequent high levels of blood glucose, and the set of debilitating symptoms constituting diabetes. Combined with the relative physical inactivity of many people in the developed world, it also causes obesity

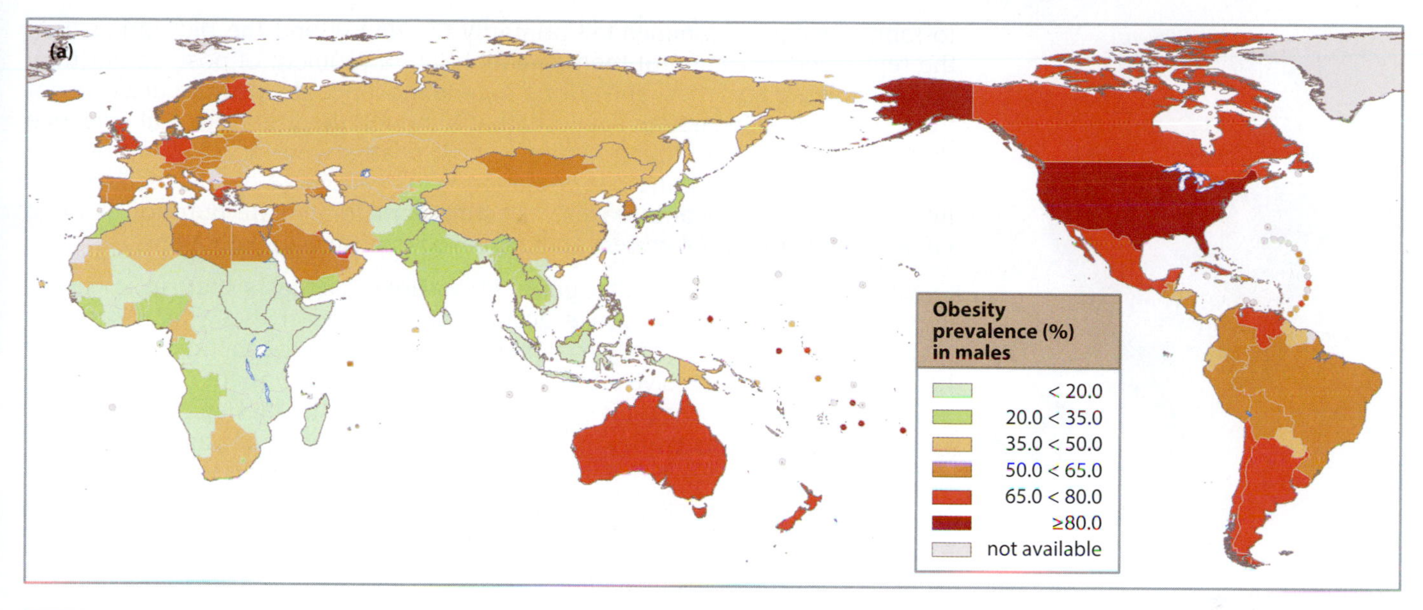

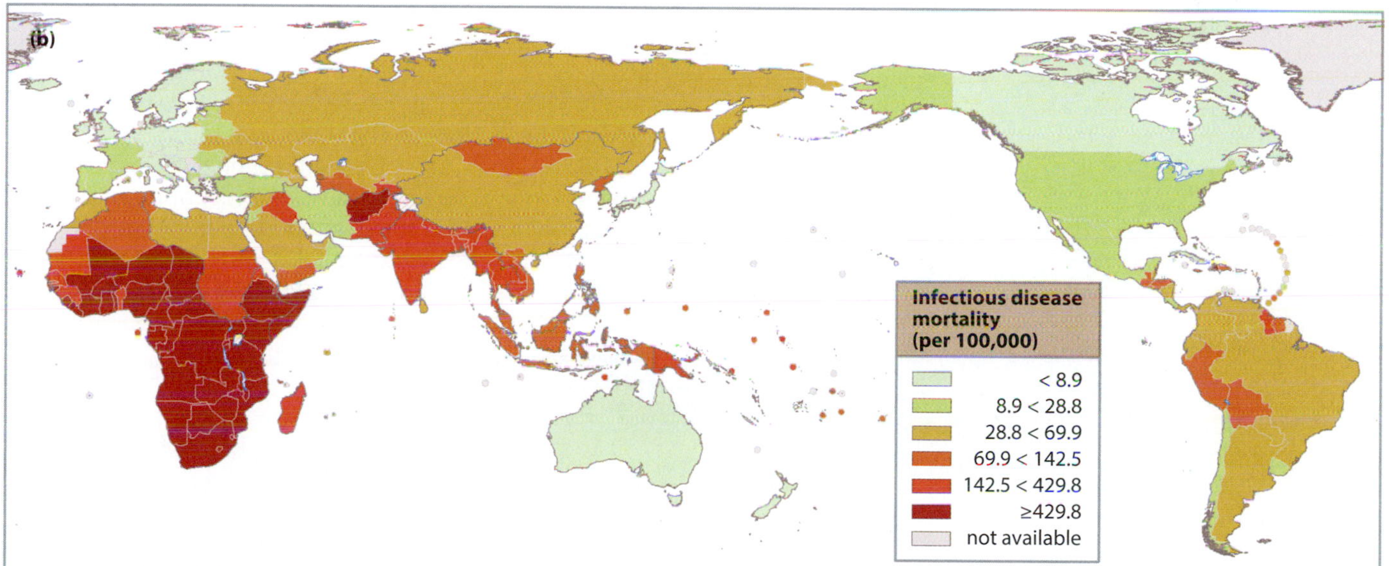

Figure 17.3: Geographical distributions of complex diseases. (a) Obesity. Estimated overweight and obesity (BMI > 25) prevalence (%) in adult males. (b) Infectious disease. Estimated age-standardized death rate (per 100,000 individuals) due to infectious and parasitic diseases in adults. (Data from WHO Global Infobase, **https://apps.who.int/infobase/**.)

(**Figure 17.5**, **Box 17.1**). Neel later refined his hypothesis so that it applied specifically to T2D. This **thrifty genotype** hypothesis has been widely used to explain the high incidences of the disease among westernized Native Americans, Australian Aborigines, and Pacific Islanders. It is argued that, because these regions of the world were settled under difficult circumstances, the thrifty genotype was strongly favored. However, it seems likely that food shortages were an ever-present threat for our ancestors, and unlikely that this threat suddenly disappeared with the advent of agriculture; indeed, paleopathological evidence suggests that nutrition among early farmers was often poorer than among hunter-gatherers (**Section 12.3**), and so the thrifty genotype could have been favorable in a wider group of populations until very recently.

The drifty gene hypothesis

Neel's thrifty genotype hypothesis has not gone unchallenged. A major criticism has been the lack of evidence for differential survival of obese and lean individuals in famines, necessary for the spread of thrifty alleles. According

Figure 17.4: James V. Neel. (Courtesy of University of Michigan.)

to John Speakman, famines kill primarily the young and the old, rather than the reproductive section of the population, are a problem of post- rather than pre-agricultural societies, and during famine most people die from disease, not from starvation. Speakman suggests that although there is historical evidence for frequent and devastating famines, most deaths were not due to absolute starvation, that is, running out of energy.[45] Instead, most deaths were due to infectious and diarrheal diseases, reflecting poor immediate nutrition (lack of vitamins rather than fat stores) and increased exposure to pathogens.

Speakman's alternative **drifty gene hypothesis** suggests that, in ancestral populations, lipid storage genes accumulated variation that was essentially neutral and influenced mainly by genetic drift, because the metabolism did not have to cope with excessive fat intake.[46] However, these variants revealed their functional consequences under the pressure of a high-fat diet. According to this view of **ancestral neutrality**, the variation of T2D prevalence among present-day populations is unsurprising, whereas long-term selection for thrifty alleles would have led to the fixation of the obesity and T2D risk alleles in all ancestral groups assuming the universality of the selection pressures.

Evidence from genomewide studies

If the thrifty genotype hypothesis is correct, we might expect haplotypes associated with an increase in T2D risk to have undergone positive selection. If this selective event was in early *Homo* (~1 MYA), then T2D susceptibility haplotypes would be ancestral and fixed in all humans. If, however, a recent selective event (between 5 and 50 KYA) favored the thrifty genotype, we would expect T2D susceptibility haplotypes to be derived and have reached a higher frequency than would be expected for their age. Do we see any evidence of recent natural selection on T2D susceptibility haplotypes? The most-studied example is the gene encoding a transcription factor (*TCF7L2*), where alleles at a SNP (rs7903146) contribute to T2D susceptibility.[17] The low-risk haplotype is derived and shows

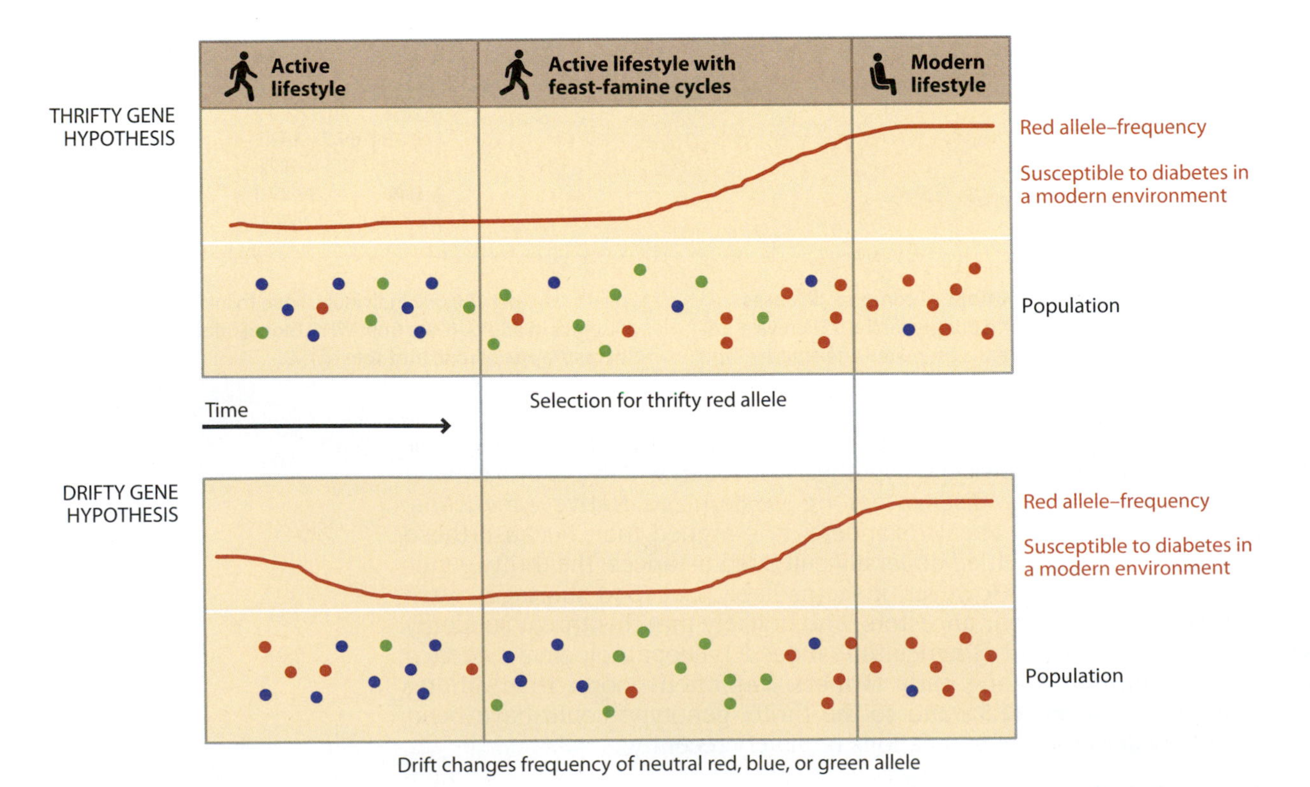

Figure 17.5: The thrifty gene and drifty gene hypotheses. A comparison of two models that might explain the existence of diabetes susceptibility alleles in modern populations. The frequency of diabetes susceptibility alleles in the population is represented by a *red* line; these alleles are also shown schematically as *red* dots in the box below, with nonsusceptibility alleles as *green* or *blue* dots. See text for more details.

Box 17.1: Should we all eat a hunter-gatherer's diet?

The fact that our diet has changed since the advent of agriculture, with consequent effects on our genotype, leads to the wide perception that westernized modern humans are eating the wrong diet. Some suggest that we should be returning to the natural diet of our Paleolithic hunter-gatherer ancestors, rejecting the products of the agricultural Neolithic, and thus avoiding the diseases of civilization, such as coronary heart disease, diabetes, and cancers.[10] Analysis of the diets of 229 modern hunter-gatherer populations indicates that they obtain 19–35% of their energy from protein, 22–40% from carbohydrate, and 28–58% from fat.[48] The equivalent figures for US adults are 15.5% from protein, 49% from carbohydrate, and 34% from fat (and 3% from alcohol). Thus, the US diet's energy from protein and carbohydrate is outside the range of the hunter-gatherer populations. A paleo-diet called NeanderThin ("Eat like a caveman to achieve a lean, strong, healthy body") is followed by some neo-Paleolithics, and includes a lot of meat, and no Neolithic products. An alternative view[29] takes into account the fact that modern-day hunter-gatherers are largely free of the diseases of civilization regardless of whether they eat largely animal or plant foods, or even if, like the Yanomamo of South America, they rely on plant foods taken from a single species. Also, a consideration of our great ape relatives and the likely progenitor species of chimpanzees and humans suggests that our distant ancestors had a strongly plant-based diet. While it is certainly true that many of us nowadays do have very unhealthy diets, humans are successful omnivores, and it will take careful study rather than supposition to disentangle the evolutionary web of diet, adaptation, and disease.

a high degree of population differentiation, being rare in the YRI HapMap population, at intermediate frequency in the CEU HapMap population, and frequent (95%) in the CHB and JPT HapMap populations. Haplotype selection tests using **EHH** (**extended haplotype homozygosity**; **Section 6.7**) show that in the Chinese and Japanese populations this low-risk haplotype has risen to a high frequency by positive selection. This result is the opposite of that predicted by the thrifty gene hypothesis, and has not yet received a biological explanation.

The thrifty phenotype hypothesis

While the thrifty and drifty gene hypotheses try to explain our current state of obesity (Figure 17.5) and T2D epidemics through long-term evolutionary processes, there is growing evidence supporting the view that short-term response to environmental variation, through developmental plasticity, can significantly modulate the individual risk of T2D. It has been proposed that poor nutrition ***in utero*** and in early infancy may be a major cause of increased risk of T2D. In this **thrifty phenotype** hypothesis,[14] the fetus adjusts its development to maternal malnutrition by itself becoming nutritionally thrifty, resulting in decreased growth, hormonal and metabolic adaptations, and changes in the cells of the pancreas responsible for insulin secretion. In effect, the fetus programs itself to survive in a poor environment. A transition to over-nutrition later in life makes this developmental adaptation disadvantageous, as it predisposes to T2D through reduced secretion of insulin, or insulin resistance (**Figure 17.6**). The thrifty phenotype hypothesis has been supported by numerous epidemiological and animal experimental studies.[50]

17.3 IDENTIFYING ALLELES INVOLVED IN COMPLEX DISEASE

Genetic association studies are more powerful than linkage studies for detecting small genetic effects

Genetic linkage analysis, as we have seen, is very successful in identifying genes that cause simple Mendelian diseases (Chapter 16), where the mode of inheritance (dominant, recessive, and so forth) can be determined. Linkage analysis can also be used in non-Mendelian complex diseases, where there is no model specifying how the disease is inherited in the family, and this is called **nonparametric linkage analysis**. The aim is to identify chromosomal regions more likely to be shared by family members with the disease than family members without the disease.

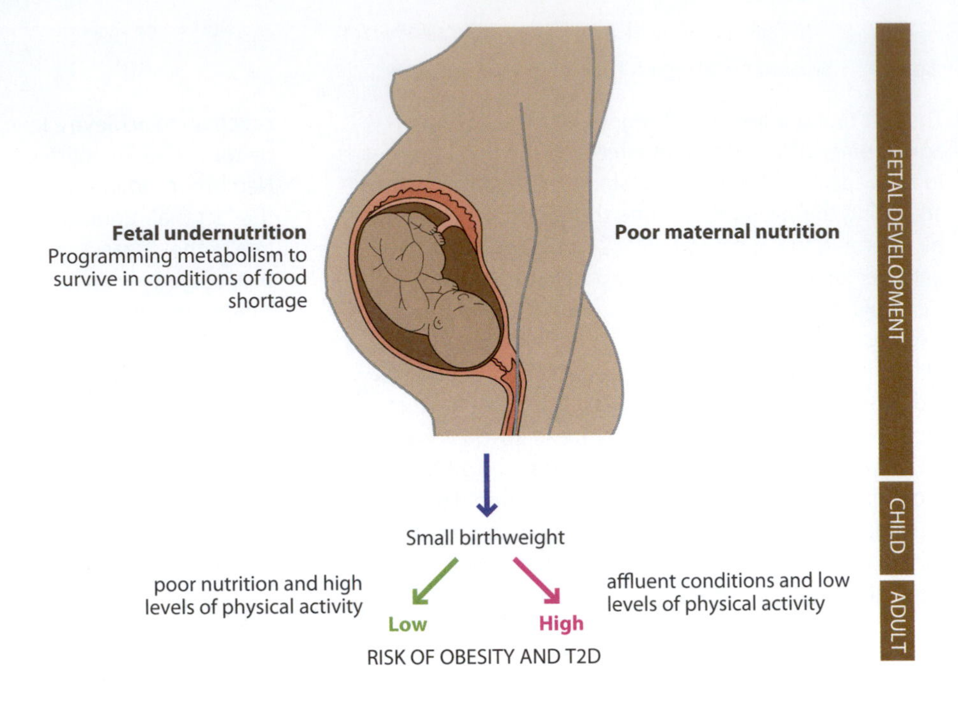

Figure 17.6: The thrifty phenotype hypothesis.
An alternative model explaining different individual risks of obesity and type 2 diabetes (T2D).

A widely used example of nonparametric analysis is the **affected-sib-pair** study design. On average, two siblings are expected to share 50% of their genome. If both siblings have a common complex disease, we might expect genomic regions that contain a susceptibility allele to that disease to be more likely to be shared between those siblings. By examining the alleles at SNPs or microsatellites across the genome in a large collection of affected siblings and their parents, we might be able to identify those regions that are significantly more likely to be shared than expected by chance (that is, >50%). These studies have been limited in their success, partly because they define very large candidate regions containing many genes, and partly because a large number of families are needed to confidently detect a linkage of a region with susceptibility to disease (**Table 17.3**).

Genetic association is an alternative to genetic linkage which tests whether individuals carrying a particular allele at a polymorphic site (usually a SNP) are more or less likely to have a certain disease than those individuals who do not carry that allele. The affected-sib-pair design, with two parents and two affected

TABLE 17.3:
A COMPARISON OF THE POWER OF LINKAGE AND ASSOCIATION STUDIES TO IDENTIFY A COMPLEX DISEASE SUSCEPTIBILITY GENE

Genotypic relative risk	Frequency of disease allele	Number of families needed	
		Linkage[a]	Association[b]
2.0	0.01	296,710	5823
	0.1	5382	695
	0.5	2498	340
1.5	0.01	4,620,807	19,320
	0.1	67,816	2218
	0.5	17,997	949

[a] Each family consisted of two parents and two affected children.
[b] Each family consisted of two parents and one affected child.
(Data from Risch N & Merikangas K (1996) *Science* 273,1516.)

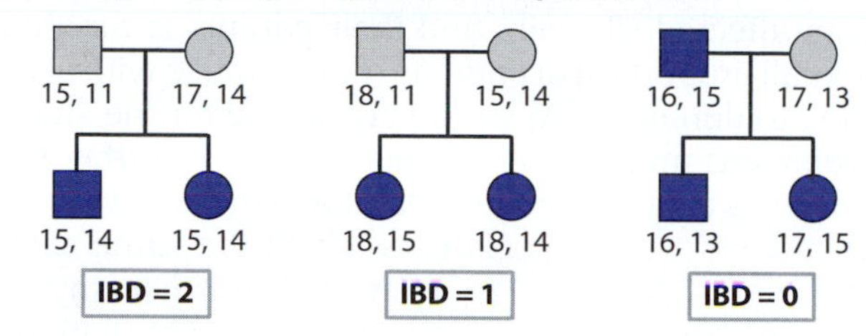

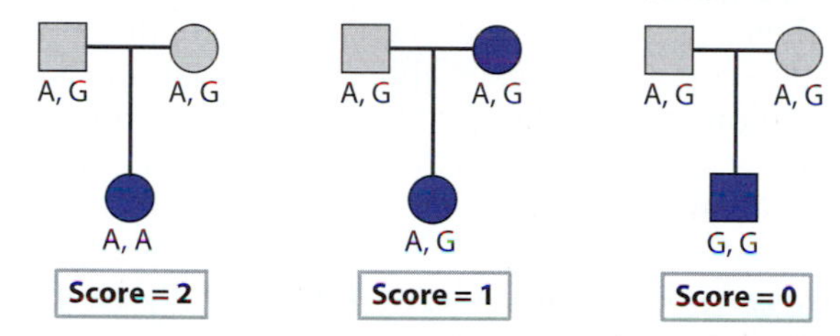

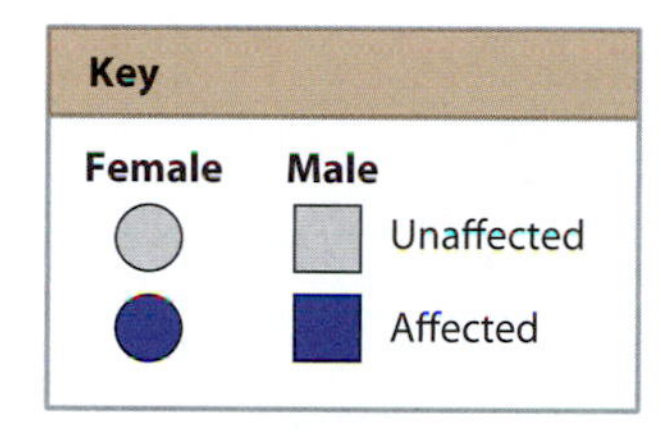

Figure 17.7: Comparison of linkage and association mapping of a complex disease.
(a) The linkage study uses families consisting of two affected children (*blue*; an affected sib-pair) and their parents (who may or may not be affected). All individuals are typed with a genomewide set of microsatellites. Here, the results for one locus in three families are shown. The number of alleles at each locus that are shared identical by descent (IBD) by the affected sibs is counted: this number can be 2, 1, or 0. Note that in the family on the left, affected sibs both typed as 17, 15; 17, 11; or 14, 11 would also score 2; also, not all families will be informative for every locus. The expected null distribution of 2, 1, 0 scores is 0.25, 0.5, 0.25 and each locus is tested for a departure from this distribution. (b) The association study uses an affected child and his or her parents. All individuals are typed with a genomewide set of SNPs and the transmission of each allele (here, the A allele) to the affected child is scored. The null expectation is that the distribution of transmissions: nontransmissions will be 0.5:0.5 and each SNP is tested for departure from this distribution.

sib-pairs, can also be used to test for genetic association using the **transmission disequilibrium test** (**TDT**).

The increased power of association, when compared with linkage, to detect small genetic susceptibilities to disease was illustrated by a simulation study that contrasted linkage using affected sib-pairs with TDT.[39] First, a linkage study was simulated with a genome screen using 500 microsatellites, with a >95% chance of no false positives and an 80% power of detecting linkage. Second, an association study using a TDT design (**Figure 17.7**) was simulated, again with a >95% chance of no false positives and an 80% power of detecting linkage. In this simulation, all individuals had genotypes from a genomewide set of SNPs and the transmission of each allele (here, the A allele) to the affected child was scored. In a TDT, the null expectation is that the distribution of transmissions to nontransmissions will be 0.5:0.5 and each SNP is tested for departure from this distribution.

The simulations also needed to define the quantitative effect of a susceptibility allele, the **genotypic relative risk** (**GRR**), as the increased chance that an individual with a particular genotype would develop the disease. GRR values of 1.5 and 2.0 were considered, among others. An individual can carry zero, one, or two copies of the susceptibility allele, and the risk associated with two copies was multiplicative, that is, individuals homozygous for the susceptibility allele had GRRs of 1.5^2 (= 2.25) or 2.0^2 (= 4.0), respectively. The necessary number of families was then calculated, based on either linkage or association approaches. This number depended strongly on the **relative risk** and frequency of the susceptibility allele, but was always smaller for the association study than for linkage (Table 17.3). A major reason for this is that TDT uses more information than does linkage: it takes into account *which* allele is transmitted, while linkage only takes into account that *an* allele is shared identical by descent, without considering which allele it is. While these conclusions were derived using only a single design of linkage and association study, they have proved generally applicable.

TDT is not often used for association studies, for the practical reason that collecting DNA from affected sib-pairs and their parents is time-consuming, and because only a small proportion of patients (that is, those with an affected sib and parents all willing to donate DNA) would be eligible for the study. Case-control association studies are much more popular, because in theory any unrelated patient can be used as a case, and are potentially more powerful than affected-sib-pair studies in identifying associations, given the same sample size.[31] For these studies, care must be taken to determine whether any association discovered is due to true association with the disease or population structure, also referred to as **population stratification**. Population structure can be present in any population, but we can illustrate the problem by considering an admixed population, which is a population with high levels of population structure. In an admixed population a polymorphism and disease that were both present at a higher frequency in one parental population than the other may show association. An advantage of family-based association studies, like TDT, is that they are not affected by population structure.

Candidate gene association studies have not generally been successful in identifying susceptibility alleles for complex disease

Until the early 2000s, typing a large number of SNPs on a large number of samples was a time-consuming and expensive task. By selecting SNPs within and around candidate genes, an association study would become feasible, but could only successfully detect association if relevant genes had been identified. This required knowledge of the function of the genes, and the etiology of the disease, so that genes whose protein products played a role in the disease could be tested. It seemed an attractive approach, although we know from simple Mendelian disease genetics that the gene causing a disease may not be obvious—consider, for example, the *HTT* gene causing Huntington's disease (**Section 16.1**), which is expressed in most tissues and does not have a known function. Many early candidate gene studies suffered from inadequate matching of cases and controls, low statistical power to detect an effect, and **publication bias** (where negative results were not reported), and their findings have often not been **replicated**. There have, however, been a few robust, replicated associations of alleles with disease identified in this way. An example is the association of alleles at the complement factor H (*CFH*) gene with age-related macular degeneration (Table 17.1), where a study analyzing *CFH* as a candidate gene within a very large genomic region previously identified by genomewide linkage analysis has been confirmed by genomewide association studies.[8, 25]

Genomewide association studies can reliably identify susceptibility alleles to complex disease

A genomewide association study (GWAS, **Box 17.2**) types a large number of SNPs (hundreds of thousands to millions) across the whole genome to test for association with a disease, and such studies were made possible by the development of SNP chips (**Section 4.5**).

The strength of GWASs is that they are not biased by prior assumptions about the identity of the disease locus, unlike candidate gene association studies, and can detect relatively weak effects of an allele on disease susceptibility, unlike genomewide linkage studies. GWASs do, however, make the assumption that the SNPs being tested will either include the causative allele itself or a SNP in strong linkage disequilibrium (LD) with (**tagging**) the causative allele. The existence of a haplotype block structure in the human genome (**Section 3.8**) supports this assumption, because much of the common genetic variation within a haplotype block will be tagged by alleles at a few SNPs. Exactly how many SNPs can effectively tag a block depends on the LD structure of the block, which is a consequence of its evolutionary history in the population, and how a haplotype block is defined. A rule-of-thumb is that a third of all common SNPs

Box 17.2: What is a GWAS?

A genomewide association study (GWAS, pronounced gee-wazz, plural GWASs) is a mainstay of human genetics research (**Figure 1**). Most are **case-control studies**, such as those analyzing Crohn's disease described in this chapter, where many different SNPs are genotyped and each allele tested for whether it is more frequent in cases or controls, and hence shows association with a disease. Controls that are well-matched for age, sex, and population are critical for case-control studies. Others measure association of alleles with a quantitative trait, such as adult height (Box 15.1).

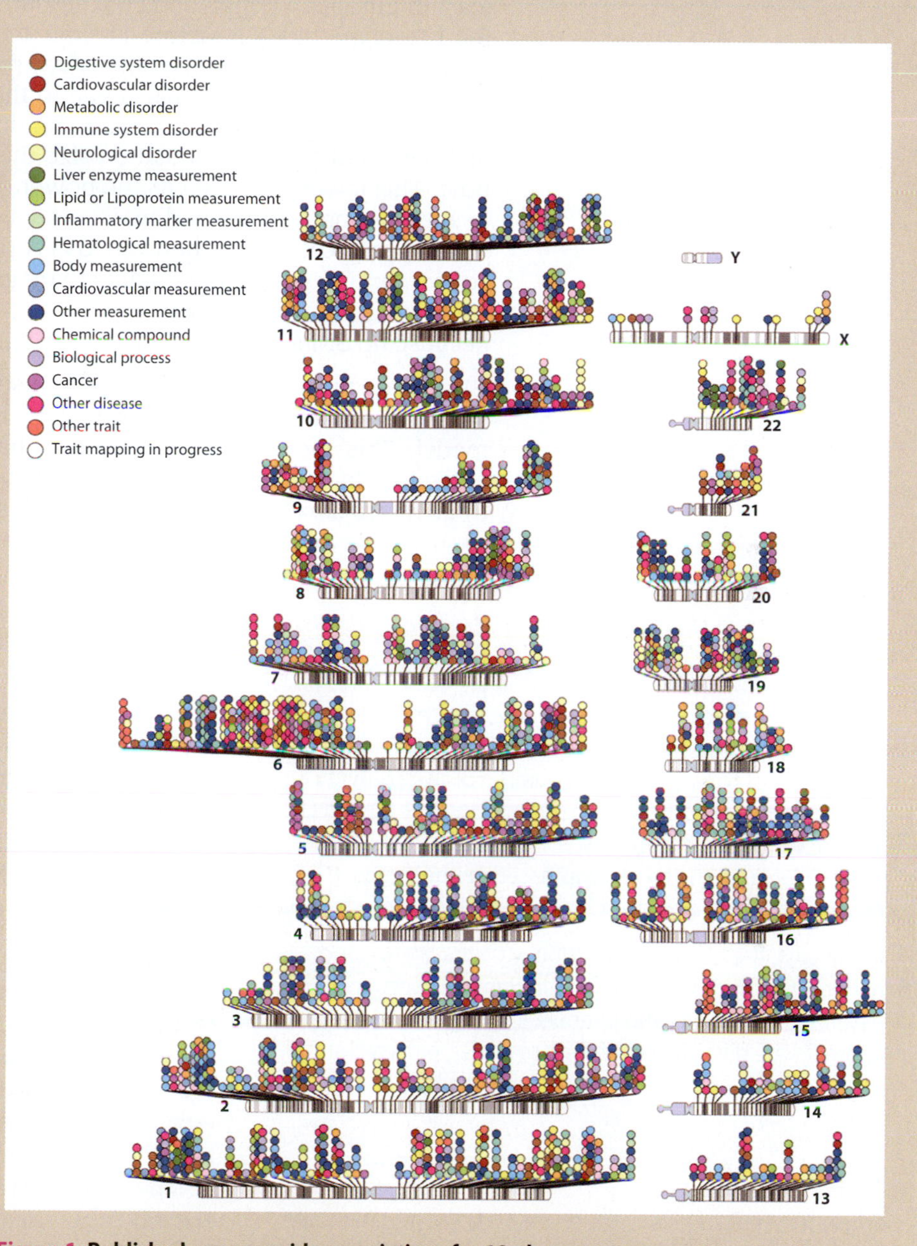

Figure 1: Published genomewide associations for 18 phenotype categories. Positions of loci associated (at $p \leq 5 \times 10^{-8}$) with diseases/non-disease phenotypes identified by GWASs up to July 2012, placed on a human karyotype idiogram. Different colored spots represent different phenotypes (From www.genome.gov/gwastudies/)

Early studies confirmed the idea that most genetic effects on disease or non-disease phenotypes are weak. This has the consequence that large numbers of samples (typically thousands of cases and controls) are required to convincingly demonstrate a true association of an allele with a disease, rather than an association that could arise by chance (false positive). In addition, because so many individual loci are tested, GWASs are vulnerable to the **multiple testing** problem, where an apparently statistically significant association will be found by chance if enough SNPs are tested. This danger needs to be counteracted by very large sample sizes and the use of an appropriately stringent level of significance, typically $p < 5 \times 10^{-8}$, to identify a real association of a SNP allele with a disease. It can be intuitively understood that the value is based on the idea that a million independent genomic regions are being tested and therefore a **Bonferroni correction** of 0.05/1,000,000 is applied. However, a more flexible Bayesian (Box 6.4) justification for the level of significance is often used (see Box 1 of reference 51 for further details).

The number of samples required for a GWAS is usually well beyond the number obtainable from a single research center or hospital, and large collaborative consortia are the norm. Indeed, constructing, organizing, and maintaining such consortia is perhaps the most difficult aspect of a GWAS.

For many diseases, several competing consortia combine their results after their initial publications to allow a more powerful analysis to be performed, called a meta-analysis. This uses the genotypes already gathered to identify alleles with even weaker associations with the disease or quantitative trait.

For further reading on GWASs and their role in disease research, we recommend Strachan and Read's *Human Molecular Genetics*.[47]

within a block can tag almost all of the common variation in that block, at least in European populations.[23] The HapMap Project (Box 3.6) has shown that the haplotype block structure is similar between populations, but less block-like in the YRI population from Africa, reflecting a lower level of LD between SNPs in that population. This is due to the demographic history of the Yoruba, who did not experience the genetic bottleneck of the movement out of Africa, and have had a larger population size than non-African populations since that time. This lower LD weakens the power of GWASs for a given number of SNPs in African populations, and may go some way toward explaining the unusual results when GWASs are performed in African populations. For example, a large study examining the susceptibility to malaria did not detect an association with *G6PD*, which carries a protective variant against severe malaria (**Section 16.4**), because no SNPs typed by the genotyping chip tagged the A– variant of *G6PD*.[19]

GWASs are most powerful when the causal allele is common and thus the allelic spectrum of the disease follows the **common-disease common-variant** model. This model, which proposes that common diseases will usually be influenced by one or a small number of susceptibility alleles with relatively high frequencies (**Box 17.3**), provided an early stimulus for the GWAS field.

Box 17.3: The common-disease common-variant model

The common-disease common-variant (CDCV) model is a hypothesis that proposes that common diseases will usually be influenced by one or a small number of susceptibility alleles at each locus. In the late 1990s, very little empirical evidence on the susceptibility alleles for common diseases was available, so the model was based on theoretical expectations. The argument involved the following steps.

In the past, the human population was small in size and had reached mutation–drift equilibrium (see Section 5.6). Assuming that it was also panmictic and of constant size, the approximate diversity (*D*) of disease alleles, or its reciprocal (*n*, the effective number of alleles), is given by:

$$D_{\text{disease}} \approx \frac{1}{1 + 4N_e\mu(1 - f_0)}$$

where N_e is effective population size, μ is the mutation rate for a non-disease allele mutating into a disease allele, and f_0 is the combined equilibrium frequency of all the disease alleles at a locus.

Assuming $N_e = 10^4$ and the mutation rate per gene per generation as 3.2×10^{-6}, Reich and Lander compared the number of alleles expected for a rare disease ($f_0 = 0.001$) and a common disease ($f_0 = 0.2$). For both, the expression $4N_e\mu(1 - f_0)$ is small and the effective number of alleles, *n*, is close to 1: about 1.1, with a single disease allele accounting for ~90% of the total.

We know that the human population is now large; they consider the value $N_e = 6 \times 10^9$. This is the current census size, not the current N_e, but it may represent a future N_e, and the calculations do not depend critically on this value. The expected number of alleles for a rare disease and a common disease at equilibrium are again similar to one another, but because of the larger N_e, the expression $4N_e\mu(1 - f_0)$ is large, and *n* is about 77,000 for the rare class and 61,000 for the common class. This should not be a surprise: large populations contain more diversity than small ones; it also predicts that the allelic spectrum of a common disease will be very diverse at mutation–drift equilibrium.

The human population, however, is not at equilibrium because it has expanded in the recent past, and time is required to reach equilibrium. There is a crucial difference between the two types of disease in the rate at which equilibrium is attained after an expansion. Initially, if the expansion was instantaneous, *n* would still be ~1.1 for both. The subsequent increase in diversity will, however, be rapid for the rare class, but slow for the common class. This is because, although the mutation rate is assumed to be the same for both, a new mutation in a rare disease adds significant diversity to the small pool of disease alleles, while a new mutation in a common disease adds little to the large pool of disease alleles that already exists. In addition, there is stronger selection against the rare disease, removing some of the alleles that are already present, while selection against the common disease is weaker and the original disease allele decreases less rapidly.

Thus, for a population that expanded some 50 KYA, *n* is barely above 1 for the common disease, but it is approaching its equilibrium value for the rare disease. Therefore rare diseases will have many alleles (as is often observed, Section 16.1), but common diseases are predicted to have far fewer alleles at each locus. An effective number of alleles of 10 or fewer was adopted as a criterion for a simple allelic spectrum.

GWASs have varied in their success in identifying susceptibility alleles to common disease. For example, in similarly powered studies, nine loci were identified as being strongly associated with Crohn's disease (Table 17.1) but none was identified as being strongly associated with hypertension (raised blood pressure). In addition, common genetic variation has been shown not to account for much genetic risk in neurological diseases such as bipolar disorder and schizophrenia (**Box 17.4**). A summary of results from the first tranche of GWASs shows that most common susceptibility alleles have a small (<1.5) individual effect on the risk of disease, as measured by the **odds ratio** (**OR**).[18] In this summary, two of the three alleles that are common and have a very large odds ratio (>10) are for non-disease phenotypes (hair color and eye color), and are likely to have been driven to high frequencies by positive selection (**Figure 17.8**, see also Chapter 15). Larger studies, combining several GWASs for the same disease in a meta-analysis, can identify even weaker effect alleles at a stringent significance

Box 17.4: Selection for shamans?

Neuropsychiatric disorders, such as schizophrenia, depression, and obsessive-compulsive disorder, are debilitating conditions. Genetic variation is known to play an important role in the development of these common disorders; for example, schizophrenia has an estimated heritability of 0.8 (Table 17.2) and a prevalence[42] of around 1 in 250. Given its clear selective disadvantage, where individuals often are childless or commit suicide,[2, 28] why have schizophrenia susceptibility alleles not been removed by purifying selection?

Three explanations have been put forward:[24]

1. **Ancestral neutrality.** Susceptibility alleles were not disadvantageous in the past, and have become disadvantageous too recently for purifying selection to have had a significant effect on the frequency of susceptibility alleles.
2. **Balancing selection.** Susceptibility alleles are disadvantageous, but only at certain times, and may also have beneficial effects that allow them to be maintained in the population at appreciable frequencies through balancing selection.
3. **Mutation–selection balance.** Susceptibility alleles are disadvantageous, and are removed by purifying selection, but new susceptibility alleles are repeatedly generated by recurrent mutation.

Explanations 1 and 2 are related, in that they suggest that susceptibility alleles are not now, or have not always been, disadvantageous. Does the modern environment lead to expression of these alleles as schizophrenia, while in the past they had a mild phenotype? There is little evidence for this; although the clinical description of schizophrenia dates only from the nineteenth century, it is very likely that schizophrenic traits were recorded in earlier writings, where they were described as mania or generic madness. The earliest description of schizophrenia may be that of Opicinus de Canistris (1296–1350), a cleric and writer, who described his mirrored visions and reflections of himself as the embodiment of a map of Europe, such that, for example, his constipation represented political problems in northern Italy—the "abdomen" of Europe.[16]

Could susceptibility alleles for symptoms displayed by schizophrenics even have been favored in the past—the idea of the "selection for shamans" introduced in the box title? Any form of positive selection for these traits must have been particularly strong to balance the negative effects; however, given that there are cognitive affinities between those successful in creative endeavors and those diagnosed with a mental illness, it is at least plausible that "creativity alleles" have been favored by natural or sexual selection, and that perhaps too many such alleles can increase susceptibility to a mental illness.

What can genetics tell us about schizophrenia? GWASs have had limited success, suggesting a limited role for common variation in schizophrenia, although an interesting but puzzling link has been established with variation at the major histocompatibility complex (Box 5.3), suggesting a possible link with infectious and/or inflammatory disease.[43] However, large-scale deletions and duplications have been associated with schizophrenia. These are individually very rare but highly penetrant and together account for a measurable portion of the genetic variation in susceptibility. These deletions and duplications are recurrent and mediated by complex duplication-rich regions of the genome. This supports a model of recurrent de novo mutation generating large deletions and duplications of major effect, which are subsequently removed by purifying selection. The observation of a **paternal age effect** (Section 3.7) for schizophrenia further supports the role of de novo mutation in developing the disease. Taken together, the genetic data favor explanation 3.

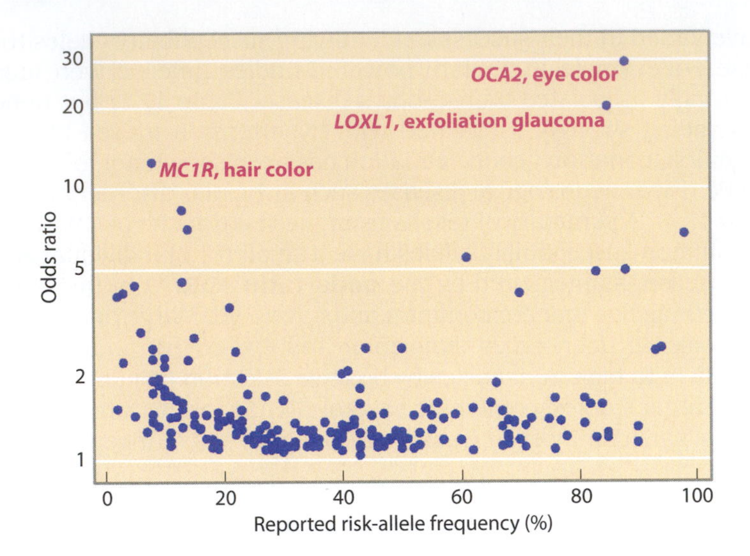

Figure 17.8: Relationship between frequency of susceptibility alleles for complex disease and odds ratio of developing that disease given presence of the susceptibility allele. Most alleles have low odds ratios and therefore have a very small individual effect on disease risk. Alleles with large odds ratios tend to occur at lower frequency, but there are exceptions, for example in two of the three cases highlighted where the SNP is in a gene which may have undergone selection. [From Hindorff LA et al (2009) *Proc. Natl Acad. Sci. USA* 106, 9362. With permission from Teri Manolio, National Human Genome Research Institute, NIH, USA.]

level, and imputation of SNPs that were not actually typed experimentally can also increase power. However, it is clear that even GWASs using many thousands of samples and complete ascertainment of variants will not account for the entire genetic component of complex disease susceptibility and this has led to discussions about what strategy to take to identify the remaining genetic component (**Figure 17.9** and **Opinion Box 16**). Given that GWASs have not identified most of the genetic variation underlying susceptibility to disease, it seems likely that the common-disease common-variant model is wrong—or at least not applicable for most variants and most diseases—and that rare variants may contribute, along with interactions between genes and also of genes with the environment. We discuss the alternative model that disease alleles are mostly in mutation–drift equilibrium in **Section 17.4**.

GWAS data have been used for evolutionary genetic analysis

The huge amount of data generated from a GWAS can be used for population genetic analysis. Indeed, analysis of allele frequencies across the genome between cases and controls can give an indication of population stratification. This is based on the assumption that allelic differences between cases and controls due to susceptibility differences will only affect a relatively small number of SNPs, whilst population stratification will affect a much larger number of SNPs. There are several different approaches for analyzing the pattern of variation across the thousands of SNPs typed by a SNP chip in a case-control study and correcting any genetic associations of alleles with disease for the confounding effect of population structure. These methods include approaches based on principal component analysis (PCA; **Section 6.3**).[36]

An example is the analysis of control individuals from Britain used for the Wellcome Trust Case-Control Consortium analysis.[51] The main role of this analysis was to test for any significant population stratification resulting from sampling from different regions of the UK. This was not found, justifying the grouping of all the controls together as one control set for use in the association studies. However, PCA of the full genomewide dataset (500,000 SNPs) showed that different regions could be distinguished (**Figure 17.10a**). In addition, alleles of a number of different SNPs showed particular geographical differentiation, such as one found in a likely regulatory element downstream of the Toll-like receptor 6 (*TLR6*) gene (Figure 17.10b). The TLR6 protein is involved in innate immune recognition of microorganisms, and may have been subject to positive selection in the past.

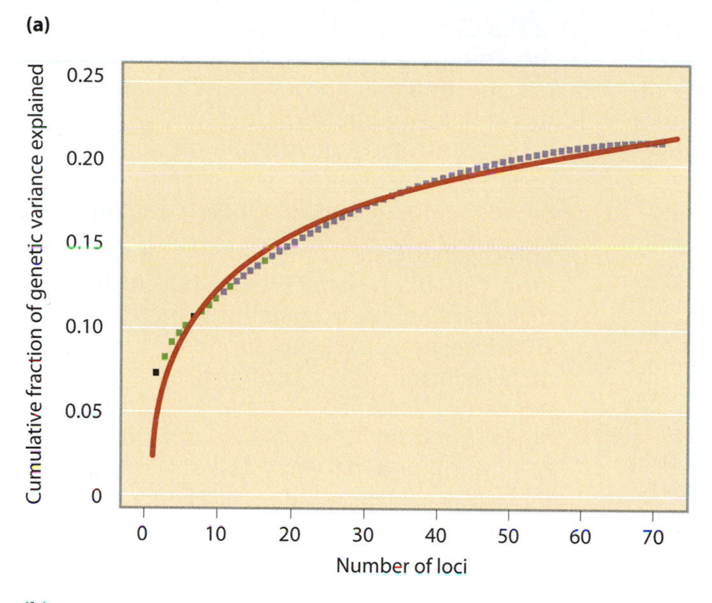

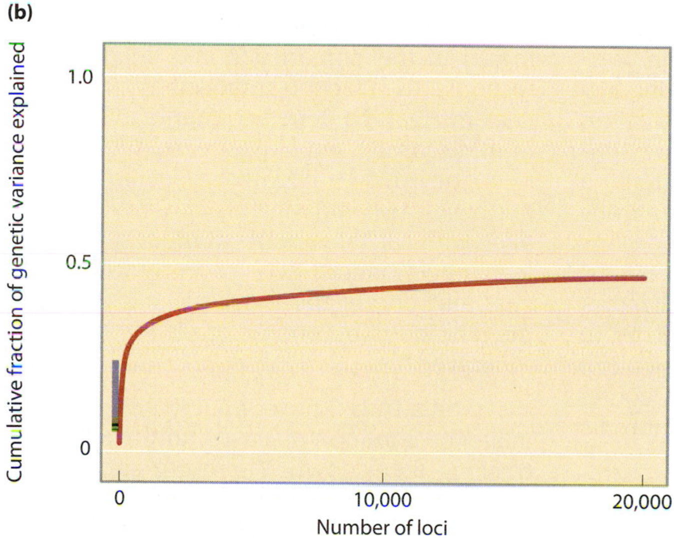

Figure 17.9: Proportion of genetic variation of Crohn's disease currently explained.
(a) Just over 20% of the genetic variation in susceptibility to Crohn's disease, a complex inflammatory disease of the gut, is explained by 71 loci. *Black* points indicate loci identified before GWASs, *green* points indicate loci identified in individual GWASs, and *blue* those loci identified in meta-analyses of several GWAS experiments together. The *solid red* line shows a logarithmic model fitting the data.
(b) Extending the logarithmic model fitting the data in (a) predicts that less than 50% of the total genetic variation in susceptibility to Crohn's disease will be found by ever-larger GWASs, even in a situation where 20,000 independent common alleles contribute to an individual's susceptibility to the disease. [From Franke A et al (2010) *Nat. Genet.* 42, 1118. With permission from Macmillan Publishers Ltd.]

17.4 WHAT COMPLEX DISEASE ALLELES DO WE EXPECT TO FIND IN THE POPULATION?

Since disease is, if it affects reproductive fitness, evolutionarily disadvantageous, we expect that alleles that increase disease susceptibility will be removed from the population by purifying selection, while alleles that increase disease resistance will increase in frequency. In this section, we consider these two scenarios, and then some examples that illustrate them.

Negative selection acts on disease susceptibility alleles

New susceptibility alleles are continually generated by mutation, and the rate at which they are removed—the strength of the purifying selection—will depend on how disadvantageous they are. In evolutionary terms, this means how much they decrease the number of offspring of the carrier, rather than their perhaps more obvious effects on the health of the individual. As a consequence of this **mutation–selection balance**, alleles which lead to large increases in complex disease susceptibility are kept at low frequencies in the population, and blend into Mendelian disease alleles with variable penetrance. Indeed, for many complex diseases, there are Mendelian forms which are distinguished by their inheritance

OPINION BOX 16: The "missing heritability" of complex disease

GWASs have been remarkably successful at identifying associations with common diseases that are well replicated and that nearly everyone agrees represent real associations with risk of disease or with variation in other traits. For most of the traits studied, however, the associated variants have small apparent effects on risk, and collectively the associated variants usually explain only a small proportion of the total heritability of the disease or trait. This observation has led to the so-called "**missing heritability**" problem: the diseases themselves are known to be heritable, but we are unable to explain most of this heritability. For example, even the combined effects of all associated variants from a GWAS of 184,000 people explain only 10% of the variation in human height[26] (Box 15.1). An additional complication in the interpretation of GWAS results is the inability to identify the causal variants responsible for the associations.

The characterization of these results as constituting a missing heritability problem is perhaps overblown. GWASs are designed to represent most common variants in the human genome. Therefore, virtually any common variant with a large or even moderate effect on risk of disease would be identified by GWASs. The simplest interpretation of these results, therefore, is that the variants that influence common diseases are generally too rare to be detected by GWASs. Most likely, for many traits, the heritability is "hiding" in the most obvious place: variants that are too infrequent to be well represented in a GWAS (**Figure 1**). Indeed, in sharp contrast to the common variants analyzed by GWAS, rare, causal copy-number variants have been identified that confer high risk for common diseases such as schizophrenia and autism.

A clear motivation for GWASs was supplied by theoretical arguments that supported the common-disease common-variant (CDCV) hypothesis (Box 17.3). An influential paper by Reich and Lander[38] argued that due to the recent increase in the size of the human population, variants that cause common diseases could have risen to relatively high frequencies. Other modeling studies, however, showed that rare variants were more likely to be influencing common diseases.[37] The early popularity of the CDCV hypothesis was likely a simple reflection of the emerging accessibility of common variants to systematic analysis through GWAS as opposed to clear theoretical arguments in its favor. Now that GWASs have been performed for virtually all common diseases, the results are most easily reconciled with the rare variant model, although this is not (yet) a unanimous view.

Some geneticists argued against a major role for common variants before GWASs, but in the post-GWAS era this point of view is gaining increasing traction. The GWAS results also bring us closer to the model that was once referred to as the classical school, where most genes are assumed to have a "wild-type" version, and deviation from this type occurs primarily in the form of harmful genetic variants kept in the population by a mutation–selection balance. This is in contrast with the balance school, positing that much of the functional variation in genomes is maintained by some form of balancing selection favoring functional alternative alleles. There are certainly clear exceptions to the classical model in the human genome, in particular related to response to selection imposed by pathogens. However, it seems that the genetic variants influencing most common diseases have almost certainly been kept at a low frequency due to negative selection and that the total burden of such variants reflects a mutation–selection balance with causal variants at modest frequency at best. A leading genomicist even recently went so far as to suggest that one of the most striking things to emerge in the post-human-genome era is the concept of a human "wild type."[35]

We won't truly know what types of genetic variants influence common diseases until we have succeeded in tracking them down. There are currently thousands of secure associations that have been identified by GWASs, but very few of them have been tracked to causal variants. The next phase of study, the sequencing of whole genomes, should finally allow researchers to identify causal variants. This next stage will explain some portion of the original GWAS signals and will also identify new associations. It is to be hoped that in the coming years we will be able to explain the problem of the missing heritability.

Elizabeth T. Cirulli and David B. Goldstein
Center for Human Genome Variation,
Duke University Medical School

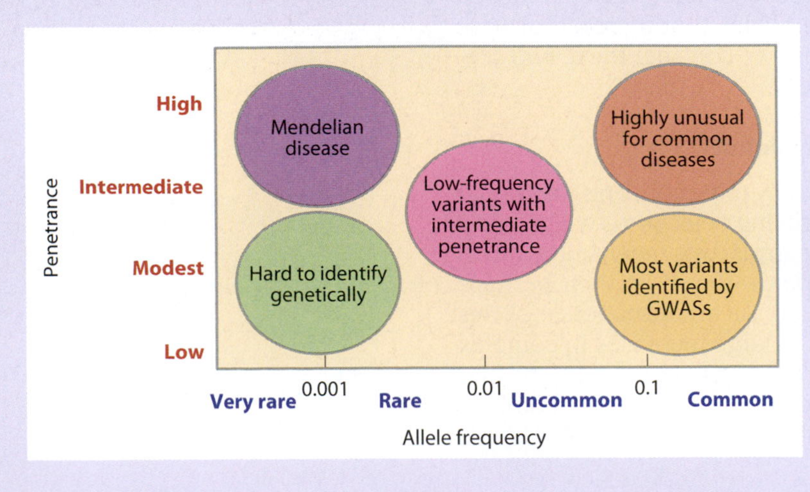

Figure 1: Frequency and penetrance of genetic variants contributing to disease. A significant proportion of the 'missing heritability' of complex disease is likely to be attributable to low-frequency variants with intermediate penetrance (center). [From McCarthy MI et al. (2008) *Nat. Genet.* 9, 356.]

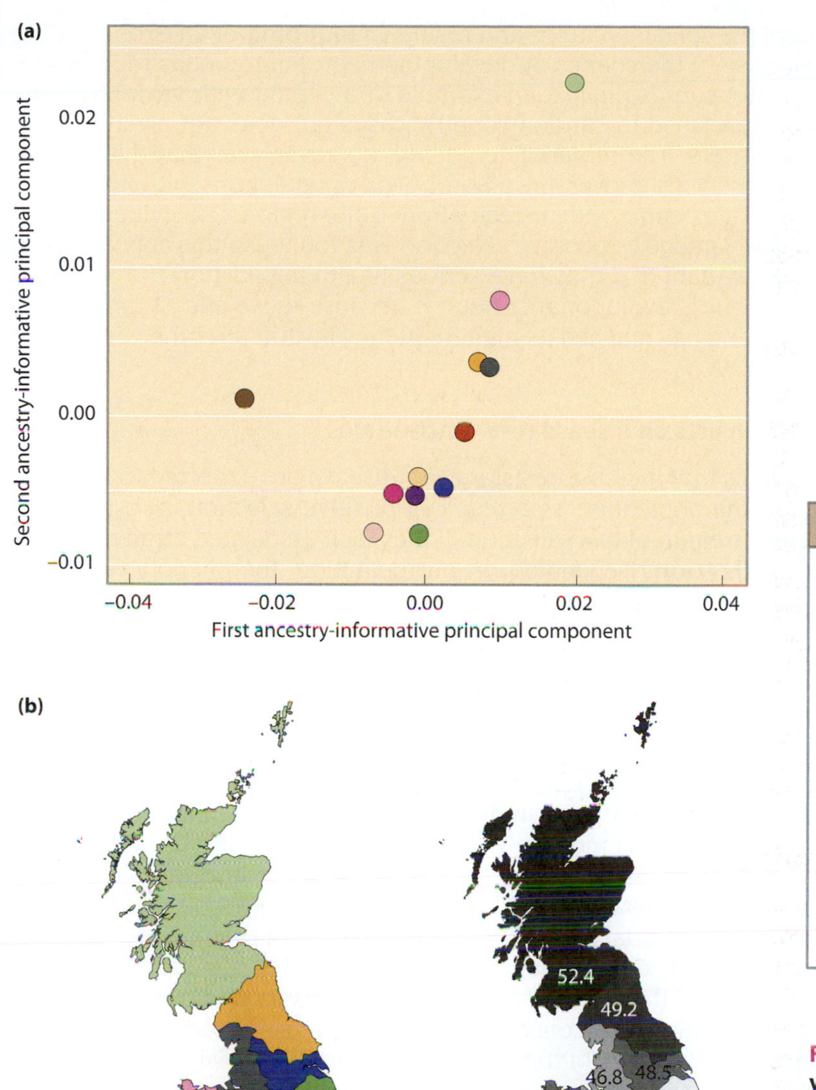

Figure 17.10: Analysis of British regional genetic variation using data from a large case-control study. (a) Principal component analysis of British regions, defined according to the map in (b), using genomewide SNP data. (b) Frequency of a SNP allele downstream of the Toll-like receptor 6 gene (*TLR6*) that shows a major frequency cline across Britain. [Data from Wellcome Trust Case-Control Consortium (2007), *Nature* 447, 661.]

pattern and age of onset rather than their disease phenotype. For example, we introduced above (**Section 17.1**) Mendelian susceptibility to mycobacterial diseases, but other common complex diseases such as cardiomyopathy can also have Mendelian forms (Table 17.1, and see below). Correspondingly, alleles that lead to only small increases in disease susceptibility can become frequent in the population, and blend into the middle of a susceptibility spectrum that ranges from susceptibility alleles through neutral alleles to protective alleles.

An important factor in determining the strength of purifying selection is the age of onset of the disease. Alleles that lead to disease late in life may be responsible for the death of the carrier, but be nearly neutral in the population: grandparents may contribute to the survival of their grandchildren, but the evolutionary consequences of this are not strong. Consequently, many complex diseases, including heart disease, cancer, and diabetes, are largely diseases of old age. **Cardiomyopathy** is a leading cause of heart failure. While Mendelian forms are known, these do not account for the majority of cardiomyopathy cases, but studies have identified a set of relevant genes. One of these is *MYBPC3*, which codes for a constituent of the heart muscle. A 25-bp deletion within an intron of this gene alters the splicing pattern and results in **skipping** of an exon, leading to a large increase in the chance of developing heart failure (odds ratio of ~8). This in itself is not surprising: just an example of a variant with **variable penetrance** that is enriched in cardiomyopathy cases but does not always cause disease in its carriers. The remarkable finding, however, was that this variant has reached frequencies of over 4% in South Asian populations, although it has not been reported in people with ancestry from other regions. No evidence for a compensating advantage or positive selection was found, so this appears to be an example of a variant that may have serious health implications for its 40–50 million carriers, but is evolutionarily neutral because these effects are usually seen in old age, and has reached its high frequency by drift after the peopling of South Asia ~50 KYA.[7]

Positive selection acts on disease resistance alleles

In contrast, alleles that increase resistance to disease are expected to increase in frequency in the population as a result of positive selection, at a rate that depends on the strength of this selection. If they lack associated disadvantages, they will reach **fixation**. Depending on our definition, the phenotypic consequences of moving to high altitude, or lactose intolerance in adults, might be considered diseases, but these are discussed elsewhere (Chapter 15), and here we consider three examples of alleles that increase resistance to more clear-cut diseases.

Severe sepsis and CASP12

Severe **sepsis** is an over-reaction of the immune system, usually triggered by infection, and associated with organ dysfunction and a high mortality rate (20–35% within 30 days) even in modern Western hospitals. The lethal effects of the major killers of children—pneumonia, diarrhea, malaria, and measles—often occur via a common pathway leading to sepsis. CASP12 influences this pathway, and humans or mice that lack CASP12 are less likely to develop severe sepsis. A stop codon within the *CASP12* gene (rs497116) leads to its inactivation and a single inactive form of the gene has increased to an estimated ~95% frequency in the human population as a consequence of positive selection within the last 60–100 KY.[52] The active form of the gene is now largely restricted to sub-Saharan African populations, particularly hunter-gatherers. It was suggested that the driving force for this selection was sepsis resistance in populations as their numbers increased and individuals encountered more infectious diseases.

Malaria and the Duffy antigen

Malaria and its consequences for human evolutionary genetics were introduced in **Section 16.4**. Here, we consider the **Duffy (FY) antigen**, a classical blood group polymorphism with three main alleles *FY*A*, *FY*B*, and $FY{*}B^{ES}$ (also called *FY*O*). This antigen lies on a cell-surface protein called the Duffy antigen receptor for chemokines (DARC), which is present on the surface of many tissues, including spleen, brain, and red blood cells. The protein serves as the sole receptor for the human malarial parasite *Plasmodium vivax*. $FY{*}B^{ES}/FY{*}B^{ES}$ individuals lack DARC on the surface of their red blood cells (erythrocytes, hence the ES superscript standing for erythrocyte-silent), but express DARC in other tissues. This is because a T to C change (rs2814778) 46 bp upstream of the *DARC* transcription start site abolishes transcription of the gene in red blood cells, probably by preventing binding of a transcription factor (GATA-1) required specifically for expression in erythrocytes. These individuals are consequently almost completely resistant to *P. vivax* infection, and, as might be expected, the $FY{*}B^{ES}$ allele is either fixed or almost fixed in sub-Saharan African populations where *P. vivax* malaria is present, yet the allele is absent outside Africa (**Figure 17.11**).

While the beneficial effects of the $FY{*}B^{ES}$ allele in protecting against *P. vivax* malaria are unequivocal, this allele has been found to have deleterious consequences for other diseases. $FY{*}B^{ES}/FY{*}B^{ES}$ homozygotes are more likely to become infected by HIV-1, the virus that causes AIDS, although they then

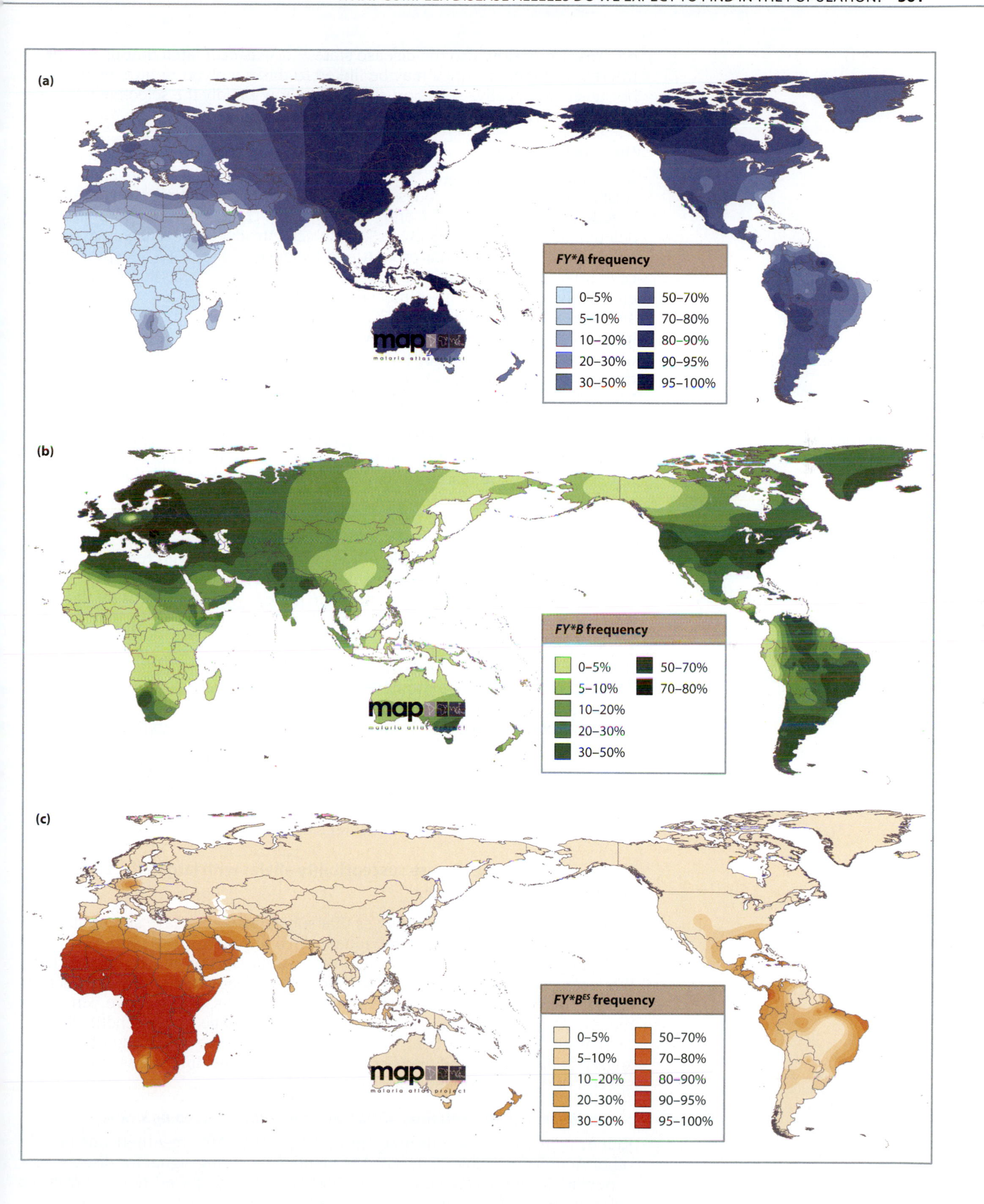

Figure 17.11: Global distribution of Duffy alleles. Interpolated maps of (a) *FY*A*, (b) *FY*B*, and (c) *FY*B*^{ES} allele frequency.

[From Howes RE, et al. (2011) *Nat. Commun.* 2, 266. With permission from Macmillan Publishers Ltd.]

progress more slowly to the disease state.[15] It was estimated that up to 11% of the HIV-1 burden in Africa may be linked to this genotype, so it may now also be subject to negative selection, illustrating how rapidly the selective forces on a variant can change.

HIV-1 and CCR5Δ32

Some individuals are highly resistant to HIV infection and the development of AIDS. The *CCR5* gene encodes a cell surface receptor that acts as a co-factor, together with CD4, for HIV entry into T cells. There is a polymorphic 32-bp deletion in this gene (*CCR5Δ32*, rs333), which introduces a frameshift mutation, leading to a nonfunctional protein and absence of the CCR5 receptor on the cell surface in individuals homozygous for the allele. These homozygotes are effectively resistant to HIV-1 infection, and heterozygotes (*CCR5Δ32/+*) take on average 2–3 years longer than *CCR5+/+* homozygotes to progress from HIV-1 infection to the development of symptoms of AIDS. Lack of a CCR5 receptor does not cause any obvious pathology, although there is evidence suggesting that *CCR5Δ32* homozygotes are more vulnerable to severe symptoms following West Nile virus infection.[12] The *CCR5Δ32* allele is only at appreciable frequencies in Europeans, with the highest frequency of 14% in Scandinavia. HIV is a recent human pathogen, and it is most prevalent in Africa, so the high frequency and global prevalence of the *CCR5Δ32* allele could not have been caused by natural selection for resistance to AIDS. It has been suggested that selection by another infectious disease in the past, such as smallpox or plague, is responsible for the high frequency of *CCR5Δ32* in Northern Europe, but there is no evidence of recent directional selection from extended haplotype tests.[41] An older selective event, or drift via **allele-surfing** (**Section 12.5**) remain plausible explanations.[34] Despite this ambiguity about the population-genetic evidence for positive selection in the past, the allele is clearly advantageous for individuals in populations exposed to the HIV epidemic and might even be experiencing positive selection now.

It is striking that these three examples of variants that increase resistance to disease are therefore likely to have been positively selected now or in the past, and all lead to loss of function. This may reflect the large effects of such variants on protein product levels and the ease of linking these to their phenotypic consequences rather than a lack of other classes of variant that increase disease resistance. Indeed, a rare allele of the *APP* gene (rs63750847-A, minor allele frequency <0.5%) results in an alanine to threonine substitution at position 673 which reduces its protease activity by 40%, and carriers of this allele have increased resistance to Alzheimer's disease.[20]

Unexpectedly, some disease susceptibility alleles with large effects are observed at high frequency

It may seem paradoxical that alleles that substantially increase susceptibility to disease should be found at high frequency in a population, but we have already encountered examples in the section on malaria (**Section 16.4**) where variants that increase resistance to this pathogen also lead to the genetic diseases thalassemia or sickle-cell disease. In malarial regions, the selective benefits of malaria resistance outweigh the selective disadvantage of the genetic disease. Similar scenarios have been found to underlie some other high-frequency susceptibility alleles, and may account for some of the differences in the allelic spectra of different common diseases.

Susceptibility to kidney disease, APOL1, and resistance to sleeping sickness

Kidney failure is about four times more frequent in African-Americans than in other Americans (Table 17.1). GWASs identified a strong signal for disease susceptibility in the *APOL1* gene, peaking at two closely spaced nonsynonymous variants in perfect LD [rs73885319 (Ser342Gly) and rs60910145 (Ile384Met)], present in 52% of the cases. After allowing for this signal, a second nearby signal was detected (rs71785313, an in-frame 6-bp deletion removing the amino acids Asn388 and Tyr389), present on a different haplotype and found in 23% of cases. The two risk haplotypes were present at a combined frequency of ~33% in controls (geographically matched African-Americans), and the OR of developing kidney disease for an individual carrying two (compared with zero or one) risk alleles was 10.5. How could such harmful alleles be present in one-third of the general population? The haplotype carrying the two nonsynonymous variants showed evidence of positive selection in YRI HapMap samples from its integrated haplotype score (**iHS**; **Section 6.7**), and the deletion haplotype was at too low a frequency to test. APOL1 was already known to lyse and thus protect against *Trypanosoma brucei*, which causes sleeping sickness. Both disease-associated variants of APOL1 were able to lyse a strain of this pathogen, *T. b. rhodesiense*, while the ancestral (non-disease) form was not. Thus there is strong evidence that selection for resistance to sleeping sickness has, as an unfortunate consequence, resulted in the high frequency of variants that strongly increase risk of developing kidney disease. These variants together account for the excess risk of kidney failure in African-Americans.[11]

Implications for other GWAS results

Malaria and sleeping sickness are severe diseases responsible for high mortality and thus act as strong selective forces on human populations. Are they exceptional in selecting for resistance alleles that increase susceptibility to other diseases, or just the most obvious examples of a more common phenomenon? A number of lines of evidence support the latter conclusion. For example, a GWAS of the genetic influences on blood cell number and volume in Europeans identified a 1.6-Mb region with extended LD on chromosome 12q24 associated with platelet number.[44] This region contained a haplotype that had risen to high frequency, most likely because of positive selection beginning ~3.4 KYA, but where the derived (selected) state was associated with increased risks of coronary artery disease (OR 1.1–1.2), type 1 diabetes, and celiac disease. The region contained 15 genes, and the target of selection and selective agent remain unknown. Nevertheless, one of the genes, *SH2B3*, showed an amino acid difference between the selected and non-selected haplotypes (corresponding to the SNP rs3184504), and has been implicated in the T-cell-mediated immune response. It is therefore possible that this disadvantageous haplotype was driven to high frequency in Europeans, but not in other populations, because it increased resistance to an unidentified pathogen present in this geographical region within historical times.

More generally, there is a striking difference in the success of similarly powered GWASs in identifying hits in different diseases (**Section 17.3**). Diseases with an autoimmune component usually yield plentiful hits and, furthermore, a study of 107 risk SNPs in seven diseases (celiac disease, Crohn's disease, multiple sclerosis, psoriasis, rheumatoid arthritis, systemic lupus erythematosus, and type 1 diabetes) found that 44% of the SNPs were shared between diseases.[5] They are likely to share the same functional basis for the susceptibility to disease; for example, a region on chromosome 5p13.1 contains eight SNPs where alleles are associated with different autoimmune diseases and several are within or near functional elements that are active in **T cells**.[49] It seems possible that selection for a highly active immune system in past pathogen-rich environments was advantageous, but in current Western societies manifests itself in a set of autoimmune disease susceptibilities.[21] The repeated confirmation of the importance of HLA alleles (Box 5.3) in susceptibility to different infectious and autoimmune diseases further supports this idea.

17.5 GENETIC INFLUENCE ON VARIABLE RESPONSE TO DRUGS

In this final section we discuss the subject of **pharmacogenetics**, which focuses on genetic variation affecting therapeutic response to drugs. While not exactly a complex disease, differential response to a drug is certainly a complex trait, and can have serious medical consequences.

Throughout evolution, organisms have developed biochemical systems for dealing with the endogenous toxic substances generated as by-products of metabolism. In addition, the diet introduces exogenous toxins that must be neutralized. For humans, shifts in diet through time and in novel environments have placed different challenges on these systems, so it might be expected that the genes underlying them vary not only among individuals, but also among populations with different histories.

In today's medicalized world, the many drugs we are prescribed are processed by these same biochemical pathways, and therefore variability in the underlying genes, collectively referred to as **ADME** (absorption, distribution, metabolism, and excretion) genes, also affects drug responses. "If it were not for the great variability among individuals medicine might as well be a science and not an art," wrote Sir William Osler in 1892, a view that is still commonly held today. However, optimists believe that medicine may be changed from an art to a science by developments in the field of pharmacogenetics. This is related to **pharmacogenomics**, which takes a more global view of the effects of drugs on patterns of gene expression, and of the genetic influences on drug response. The ultimate goal of pharmacogenetics is **personalized medicine**—tailored treatment of individuals based on their genotypes.

Apart from the many potentially variable genes that can contribute to the fate of a drug in the body, there are environmental influences, for example a patient's age, weight, diet, disease state, and use of other medications. However, studies in the 1960s demonstrated clear heritability of the half-lives of some drugs (**Figure 17.12**). The most medically important effects are those that involve drugs with a narrow **therapeutic index**—the difference between the dose needed to achieve the desired effect and the toxic dose.

The most productive approach for elucidating the genetic basis of variable drug response has been investigation of candidate genes, based on biochemical knowledge of the pathways involved; **Table 17.4** includes some examples. GWASs, which assume nothing about the underlying biology, have also been carried out.[6] Such studies of severe adverse drug reactions suffer from small sample sizes, since such reactions are (fortunately) rare, and the studies are therefore difficult to replicate. Significantly associated genes are often ones already known from candidate approaches, but some novel genes have also been identified. An example[6] is the identification from GWASs of *VKORC1* and *CYP2C9*, already known to harbor the major genetic influences on warfarin sensitivity (Table 17.4), as well as a SNP in a previously unknown third gene, *CYP4F2*, that explains only ~1.5% of warfarin dose variability.

Population differences in drug-response genes exist, but are not well understood

Researchers studying interpopulation differentiation of ADME genes have proposed the unwieldy terms "pharmacoanthropology" or "ethnopharmacology" for their subject. Differences have been observed for specific alleles in particular ADME genes (Table 17.4), though it is usually difficult to say whether they reflect genetic drift, or differential past selection. In cases where a pharmacological difference exists between population groups but the genetic basis is unknown, ethnic origin can be used as a dangerous proxy (**Box 17.5**).

One example that has received particular attention is the arylamine N-acetyltransferase 2 (*NAT2*) gene, involved in responses to a wide range of **xenobiotics**. Polymorphisms give rise to either a slow- or fast-acetylator phenotype, influencing individual differences in cancer susceptibility through variation in the metabolism of carcinogens, and the effectiveness of drugs including the anti-tuberculosis agent isoniazid. There is wide variation in the frequencies of alleles giving rise to the two phenotypes (**Figure 17.13a**), with populations from the Americas and East Asia showing relatively high frequencies of the fast-acetylator phenotype compared with Africa and West Eurasia.

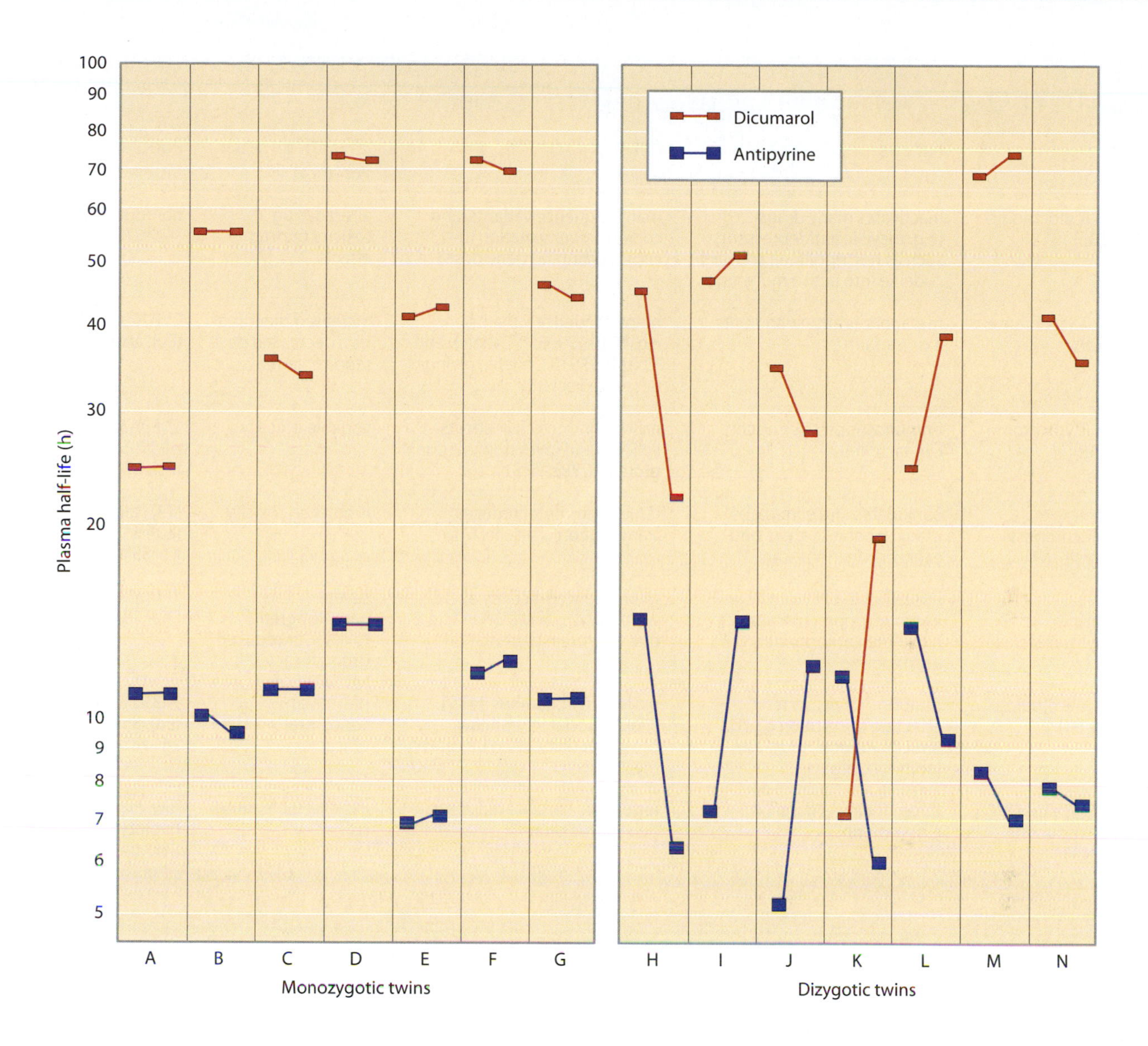

Figure 17.12: Plasma half-lives of two drugs in monozygotic and dizygotic twin-pairs.
The difference between monozygotic and dizygotic twin-pairs shows strong heritability for the half-lives of two drugs. [From Weinshilboum RM & Wang L (2006) *Annu. Rev. Genomics Hum. Genet.* 7, 223. With permission from Annual Reviews.]

When the subsistence strategies practiced by these populations are considered on a global scale, hunter-gatherers show significantly higher levels of the fast-acetylator phenotype (77.6%) compared with pastoralists (51.8%) and agriculturalists (54.9%).[40] Figure 17.13b shows the frequencies of fast-acetylator phenotypes in populations from the three groups within Africa. These findings suggest that *NAT2* may be an example of a gene that has been influenced by culturally driven selective pressures arising from xenobiotic exposure through dietary change. However, the mechanism whereby a slower rate of acetylation may have gained a selective advantage in populations shifting from foraging to pastoralism or agriculture remains to be discovered.

In a systematic study[27] that explored interpopulation differentiation of 283 ADME genes in 62 populations of the CEPH-HGDP and **HapMap Phase III** panels the weighted average F_{ST} of multiple sites from haplotypes of each gene was calculated. The distribution of these genes was compared against a random set of genes and nongenic regions. The range of values for the ADME genes is significantly wider than that for the controls, and particularly so for a subset of 31 genes for which the evidence of involvement in drug response is strongest. Many of these genes showed evidence of recent positive selection using the CLR test (**Section 6.7**) and the Ln(Rsb) extended haplotype test (Table 6.7), with signals of selection in the non-African populations depending highly on geographic region.

TABLE 17.4:
EXAMPLES OF GENES IN WHICH POPULATION VARIATION AFFECTS DRUG RESPONSE

Gene	Function, and drugs affected	Polymorphisms	Effect of genetic variation	Population differences
CYP2D6 (Cytochrome P450 2D6)	inactivates many drugs (e.g. tricyclic antidepressants such as nortriptyline); converts codeine into active morphine	many sequence variants, and copy-number variation (0–13 copies changing gene dose)	altered drug toxicity or drug effects	nonfunctional alleles: ~25% (Eu), ~5% (As, Af)
CYP2C9 (Cytochrome P450 2C9)	metabolizes anti-coagulants, e.g. warfarin	many sequence variants, e.g. *2 (rs1799853 T allele), and *3 (rs1057910 C allele), giving decreased activity	increased bleeding risk; decreased dose requirements	CYP2C9*2 allele: 10% (Eu), absent in As and Af
CYP3A5 (Cytochrome P450 3A5)	immunosuppressive agents, e.g. tacrolimus	nsSNPs; intronic SNP affects splicing and gives nonfunctional protein (CYP3A5*3)	variable efficacy	CYP3A5*3 allele: ~85–95% (Eu), 60–73% (As), 27–50% (Af)
UGT1A1 (UDP-glucuronosyl-transferase 1A1)	conjugates and eliminates many xenobiotics, e.g. anti-cancer drug irinotecan	$(TA)_7$ in promoter reduces transcription cf. usual $(TA)_6$	irinotecan toxicity	$(TA)_7$ allele: 28–38% (Eu), 8–35% (As), 43–55% (Af)
TPMT (Thiopurine S-methyl-transferase)	methylation of cytotoxic and immunosuppressive agents, e.g. 6-mercaptopurine, azathioprine	many reduced-activity alleles with nsSNPs; TPMT*3A inactive due to protein degradation	severe hematopoietic toxicity, or reduced drug efficacy	TPMT*3A allele: ~5% in Eu, <1% in As and Af
ABCB1 (ATP binding cassette B1)	multidrug transporter affecting absorption of e.g. anticonvulsants, protease inhibitors, digoxin	nsSNPs; synonymous 3435T allele associated with low protein activity by an unknown mechanism	plasma drug concentration and efficacy	3435T allele: 37–57% (Eu), 35–62% (As), 14–28% (Af)
VKORC1 (Vitamin K epoxide reductase complex 1)	target for anti-coagulants, e.g. warfarin	regulatory variants reducing gene expression	increased bleeding risk	low-expressing haplotypes: 37% (Eu), 89% (As), 14% (Af)

ns, nonsynonymous; Eu, Europeans; Af, Africans; As, Asians
[Data from Johnson JA (2003) *Trends Genet.* 19, 660 and several other sources.]

Box 17.5: "Race-specific" drugs—the dangerous example of BiDil

Examples have been given in Table 17.4 and in the main text in which differences between populations in drug response are well understood at the biochemical level. In principle, genetic testing can be used in these cases to stratify patients for drug prescription. However, when a pharmacological difference exists but the molecular mechanism is not known, a genetic test cannot be used to stratify by genotype. A dangerous proxy might then be employed—the patient's ethnic origin itself.

Particular controversy arose from a decision by the US Food and Drug Administration in 2005 to approve a fixed-dose combination tablet formulation of isosorbide dinitrate and hydralazine hydrochloride (under the trade name BiDil), as an addition to the standard treatment for congestive heart failure in patients who self-identified as "Black."[22] Differences in the incidence of hypertension among Americans of African and non-African descent are well established, but their causes remain controversial. A large study demonstrated the efficacy of BiDil in African-Americans with advanced heart failure, though it did not compare these patients with others of different ancestry. Nonetheless, "race-specific" approval was given, and led to a media reaction including some favorable headlines, but others such as "Is This the Future We Really Want? Different Drugs for Different Races?" that were much less positive.

Sales of the drug were not as strong as predicted, and the original manufacturer ceased marketing in 2008. This may reflect a lack of general enthusiasm for drugs targeted at only one population group. In any case, the licensing of such drugs is regrettable, since it tends to reinforce the view that "race" exists as a meaningful biological entity, as well as discouraging research into the actual genetic and biochemical basis of drug response variation.

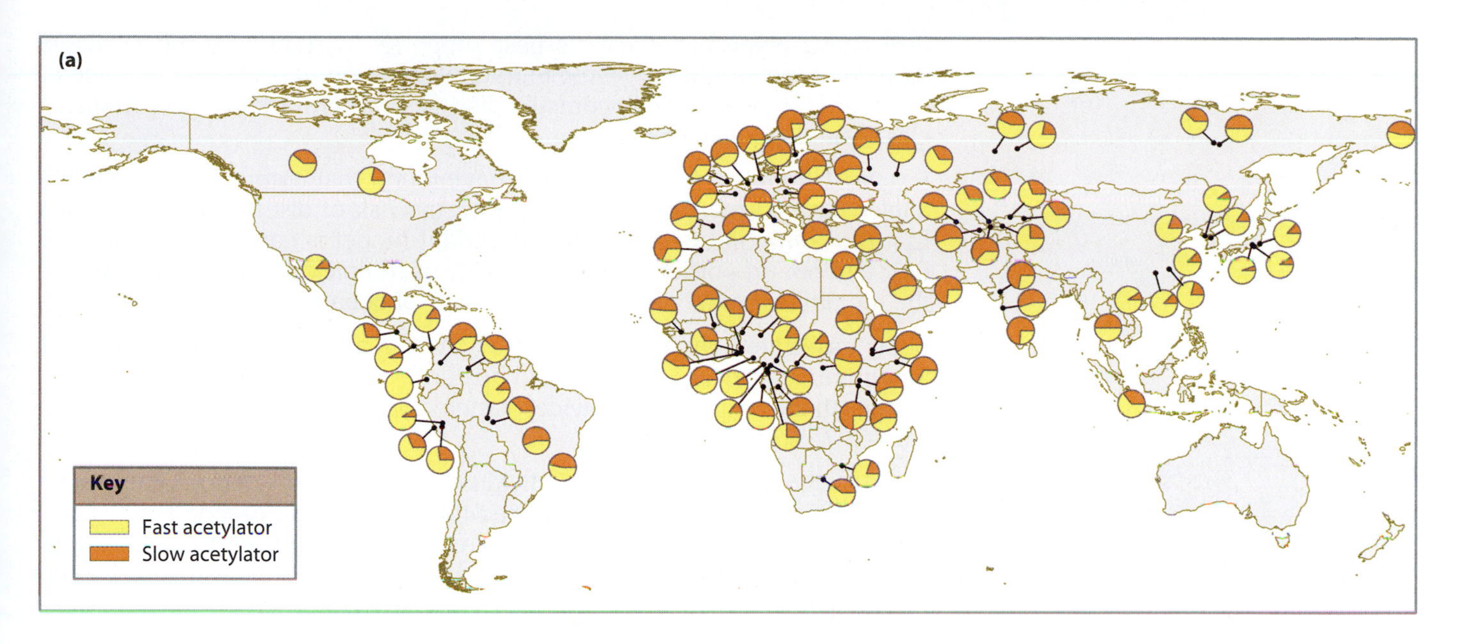

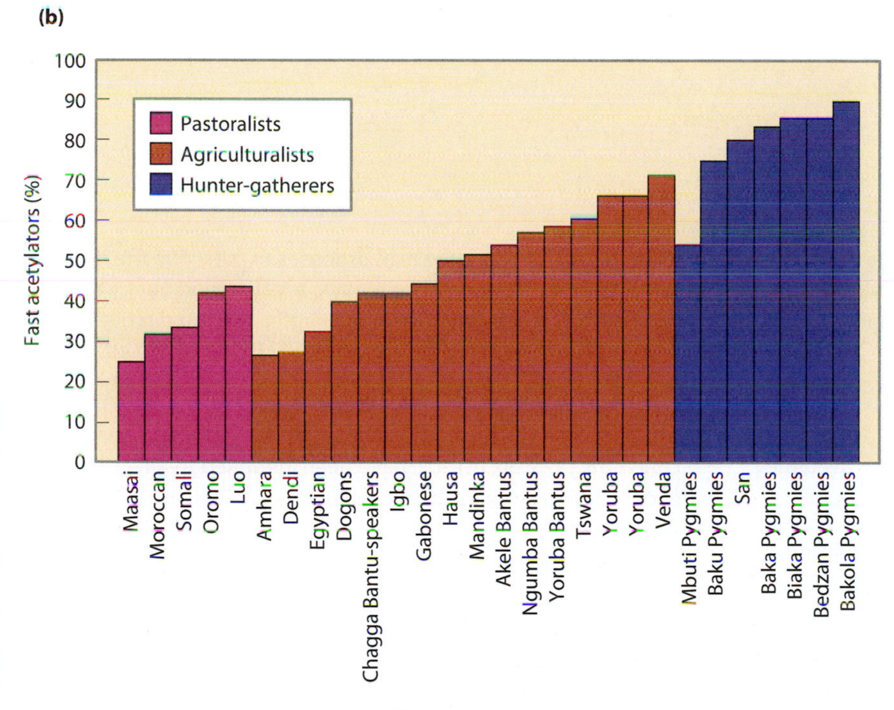

Figure 17.13: Population variation of the slow- and fast-acetylator phenotypes of *NAT2*. (a) Global distribution of slow and fast acetylators, corresponding respectively to three and five SNP-defined haplotypes of the *NAT2* gene. (b) Frequency of the fast-acetylator phenotype in African populations practicing different subsistence strategies. The mean frequency difference between hunter-gatherers and the other two categories is significant. [From Sabbagh A et al (2011) *PloS One* 6, e18507. With permission from Public Library of Science.]

SUMMARY

- Simple Mendelian diseases are often rare, often have an early onset, and are determined almost completely by genetic variation. Complex diseases are more frequent, have a late onset, and, to a significant extent, are subject to environmental variation.
- The genetic contribution to variation in disease risk varies between diseases and familial clustering can be used to quantify the genetic component of the disease risk.
- Although infectious diseases are caused by pathogens circulating in the environment, individual susceptibility to infectious diseases in humans is complex in its genetic basis and is subject to evolutionary processes.
- A number of evolutionary hypotheses have been put forward to explain the high incidence of complex diseases such as type 2 diabetes. Of these, the

thrifty phenotype hypothesis is best supported currently by the available data. More generally, polygenic mutation–selection balance and late age of onset are the reasons why complex diseases can prevail at higher frequencies than Mendelian disorders.

- For most complex diseases GWASs have found that common susceptibility alleles have a small individual effect on the risk of disease and concluded that the common-disease common-variant model is capable of explaining only a minor proportion of the heritability of complex diseases. The cause of the remaining heritability remains a matter of debate.
- Many associated variants detected by GWASs are shared among different autoimmune diseases, which could be due to past selection for an immune system that is exposed to a wider variety of pathogens than it is today, at least in Western countries.
- Genetic variation affects interindividual drug response, and some of the underlying genes have been identified through candidate-gene studies and GWASs. Some interpopulation allele-frequency differences may be due to long-term differences in subsistence strategies, mediated through unknown mechanisms.

QUESTIONS

Question 17-1: The frequency of the *CCR5Δ32* allele varies between populations. Calculate the probability that someone is homozygous for this allele, and therefore HIV-resistant, in

(a) Umea, Sweden (*CCR5Δ32* frequency 14.2%)

(b) Moscow, Russia (*CCR5Δ32* frequency 12.2%)

(c) Seville, Spain (*CCR5Δ32* frequency 3.8%)

Question 17-2: Has modern molecular genetics disproved the thrifty gene hypothesis?

Question 17-3: A clinician from a city with a large ethnic Indian population is interested in developing a molecular genetic test for the 25-bp deletion in the *MYBPC3* gene.

(a) How would you develop an assay for the clinician? She will have saliva from each individual as a source of DNA, and is keen for the test to be cheap (Chapter 4 will help).

(b) Professor Pangloss has used your assay on DNA extracted from a Neanderthal skeleton found in Europe, and has found that the Neanderthal was homozygous for the deletion (Chapter 4 will help, again). Pangloss suggests that the Neanderthal therefore died of heart failure and that there is extensive Neanderthal admixture in the modern Indian population. Do you agree? Explain your reasoning.

Question 17-4: Your physician proposes to prescribe the drug warfarin, and you first decide to have a test carried out by a personal genomics company (see Chapter 18). You discover that you carry the allele combination CYP2C9*3/*3, VKORC1 –1639/3673 AA. What would you say to your physician when he offers you the drug?

Question 17-5: Six risk loci have been identified to affect the risk of developing chronic obstructive pulmonary disease (COPD, see Table 17.1).

(a) Calculate the proportion of people who have all 12 risk alleles (that is, homozygous for the risk allele at all six loci) in the HapMap populations using dbSNP. The risk alleles are: T at rs2571445, A at rs10516526, T at rs12504628, A at rs3995090, C at rs2070600, and A at rs12899618.

(b) With all 12 risk alleles, Europeans have about a 1.6 times greater risk of developing COPD. Assuming that the risk per allele is directly comparable between populations, calculate the number of individuals of the highest risk class in Beijing, Tuscany (Italy), and of the Maasai people in Kenya. Use Wikipedia as a source of population sizes.

(c) Which environmental factors are likely to influence COPD rates in different countries?

REFERENCES

The references highlighted in purple are considered to be important (for this chapter) by the authors.

1. **Achtman M** (2012) Insights from genomic comparisons of genetically monomorphic bacterial pathogens. *Philos. Trans. R. Soc. Lond. B Biol. Sci.* **367**, 860–867.
2. **Allebeck P** (1989) Schizophrenia: a life-shortening disease. *Schizophr. Bull.* **15**, 81–89.
3. **Bos KI, Schuenemann VJ, Golding GB et al.** (2011) A draft genome of *Yersinia pestis* from victims of the Black Death. *Nature* **478**, 506–510.
4. **Chapman SJ & Hill AVS** (2012) Human genetic susceptibility to infectious disease. *Nat. Rev. Genet.* **13**, 175–188.
5. **Cotsapas C, Voight BF, Rossin E et al.** (2011) Pervasive sharing of genetic effects in autoimmune disease. *PLoS Genet.* **7**, e1002254.
6. **Daly AK** (2010) Genome-wide association studies in pharmacogenomics. *Nat. Rev. Genet.* **11**, 241–246.
7. **Dhandapany PS, Sadayappan S, Xue Y et al.** (2009) A common *MYBPC3* (cardiac myosin binding protein C) variant associated with cardiomyopathies in South Asia. *Nat. Genet.* **41**, 187–191.
8. **Edwards AO, Ritter III R, Abel KJ et al.** (2005) Complement factor H polymorphism and age-related macular degeneration. *Science* **308**, 421–424.
9. **Elton S & O'Higgins P** (eds) (2008) Medicine and Evolution: Current Applications, Future Prospects. CRC Press.
10. **Frassetto LA, Schloetter M, Mietus-Synder M et al.** (2009) Metabolic and physiologic improvements from consuming a Paleolithic, hunter-gatherer type diet. *Eur. J. Clin. Nutr.* **63**, 947–955.
11. **Genovese G, Friedman DJ, Ross MD et al.** (2010) Association of trypanolytic *APOL1* variants with kidney disease in African Americans. *Science* **329**, 841–845.
12. **Glass WG, McDermott DH, Lim JK et al.** (2006) CCR5 deficiency increases risk of symptomatic West Nile virus infection. *J. Exp. Med.* **203**, 35–40.
13. **Haensch S, Bianucci R, Signoli M et al.** (2010) Distinct clones of *Yersinia pestis* caused the Black Death. *PLoS Pathog.* **6**, e1001134.
14. **Hales CN & Barker DJ** (1992) Type 2 (non-insulin-dependent) diabetes mellitus: the thrifty phenotype hypothesis. *Diabetologia* **35**, 595–601.
15. **He W, Neil S, Kulkarni H et al.** (2008) Duffy antigen receptor for chemokines mediates *trans*-infection of HIV-1 from red blood cells to target cells and affects HIV-AIDS susceptibility. *Cell Host Microbe* **4**, 52–62.
16. **Heinrichs RW** (2003) Historical origins of schizophrenia: two early madmen and their illness. *J. Hist. Behav. Sci.* **39**, 349–363.
17. **Helgason A, Palsson S, Thorleifsson G et al.** (2007) Refining the impact of *TCF7L2* gene variants on type 2 diabetes and adaptive evolution. *Nat. Genet.* **39**, 218–225.
18. **Hindorff LA, Sethupathy P, Junkins HA et al.** (2009) Potential etiologic and functional implications of genome-wide association loci for human diseases and traits. *Proc. Natl Acad. Sci. USA* **106**, 9362–9367.
19. **Jallow M, Teo YY, Small KS et al.** (2009) Genome-wide and fine-resolution association analysis of malaria in West Africa. *Nat. Genet.* **41**, 657–665.
20. **Jonsson T, Atwal JK, Steinberg S et al.** (2012) A mutation in *APP* protects against Alzheimer's disease and age-related cognitive decline. *Nature* **486**, 96–99.
21. **Jostins L, Ripke S, Weersma RK et al.** (2012) Host–microbe interactions have shaped the genetic architecture of inflammatory bowel disease. *Nature* **491**, 119–124.
22. **Kahn J** (2005) Misreading race and genomics after BiDil. *Nat. Genet.* **37**, 655–656.
23. **Ke X, Hunt S, Tapper W et al.** (2004) The impact of SNP density on fine-scale patterns of linkage disequilibrium. *Hum. Mol. Genet.* **13**, 577–588.
24. **Keller MC & Miller G** (2006) Resolving the paradox of common, harmful, heritable mental disorders: which evolutionary genetic models work best? *Behav. Brain Sci.* **29**, 385–404.
25. **Klein RJ, Zeiss C, Chew EY et al.** (2005) Complement factor H polymorphism in age-related macular degeneration. *Science* **308**, 385–389.
26. **Lango Allen H, Estrada K, Lettre G et al.** (2010) Hundreds of variants clustered in genomic loci and biological pathways affect human height. *Nature* **467**, 832–838.
27. **Li J, Zhang L, Zhou H et al.** (2011) Global patterns of genetic diversity and signals of natural selection for human ADME genes. *Hum. Mol. Genet.* **20**, 528–540.
28. **McGrath JJ, Hearle J, Jenner L et al.** (1999) The fertility and fecundity of patients with psychoses. *Acta Psychiatr. Scand.* **99**, 441–446.
29. **Milton K** (2000) Hunter-gatherer diets—a different perspective. *Am. J. Clin. Nutr.* **71**, 665–667.
30. **Moodley Y, Linz B, Yamaoka Y et al.** (2009) The peopling of the Pacific from a bacterial perspective. *Science* **323**, 527–530.
31. **Morton N & Collins A** (1998) Tests and estimates of allelic association in complex inheritance. *Proc. Natl Acad. Sci. USA* **95**, 11389–11393.
32. **Neel JV** (1962) Diabetes mellitus: a "thrifty" genotype rendered detrimental by "progress"? *Am. J. Hum. Genet.* **14**, 353–362.
33. **Nesse RM & Stearns SC** (2008) The great opportunity: Evolutionary applications to medicine and public health. *Evol. Appl.* **1**, 28–48.
34. **Novembre J, Galvani AP & Slatkin M** (2005) The geographic spread of the CCR5 Δ32 HIV-resistance allele. *PLoS Biol.* **3**, e339.
35. **Olson MV** (2011) Genome-sequencing anniversary. What does a "normal" human genome look like? *Science* **331**, 872.
36. **Price AL, Zaitlen NA, Reich D & Patterson N** (2010) New approaches to population stratification in genome-wide association studies. *Nat. Rev. Genet.* **11**, 459–463.
37. **Pritchard JK** (2001) Are rare variants responsible for susceptibility to complex diseases? *Am. J. Hum. Genet.* **69**, 124–137.
38. **Reich DE & Lander ES** (2001) On the allelic spectrum of human disease. *Trends Genet.* **17**, 502–510.
39. **Risch N & Merikangas K** (1996) The future of genetic studies of complex human diseases. *Science* **273**, 1516–1517.

40. **Sabbagh A, Darlu P, Crouau-Roy B & Poloni ES** (2011) Arylamine *N*-acetyltransferase 2 (*NAT2*) genetic diversity and traditional subsistence: a worldwide population survey. *PLoS One* **6**, e18507.

41. **Sabeti PC, Walsh E, Schaffner SF et al.** (2005) The case for selection at *CCR5-Δ32*. *PLoS Biol.* **3**, e378.

42. **Saha S, Chant D, Welham J & McGrath J** (2005) A systematic review of the prevalence of schizophrenia. *PLoS Med.* **2**, e141.

43. **Shi J, Levinson DF, Duan J et al.** (2009) Common variants on chromosome 6p22.1 are associated with schizophrenia. *Nature* **460**, 753–757.

44. **Soranzo N, Spector TD, Mangino M et al.** (2009) A genome-wide meta-analysis identifies 22 loci associated with eight hematological parameters in the HAEMGEN consortium. *Nat. Genet.* **41**, 1182–1190.

45. **Speakman JR** (2006) Thrifty genes for obesity and the metabolic syndrome—time to call off the search? *Diab. Vasc. Dis. Res.* **3**, 7–11.

46. **Speakman JR** (2008) Thrifty genes for obesity, an attractive but flawed idea, and an alternative perspective: the 'drifty gene' hypothesis. *Int. J. Obes. (Lond.)* **32**, 1611–1617.

47. **Strachan T & Read AP** (2010) Human Molecular Genetics 4. 4th ed. Garland Science.

48. **Ströhle A, Hahn A & Sebastian A** (2010) Estimation of the diet-dependent net acid load in 229 worldwide historically studied hunter-gatherer societies. *Am. J. Clin. Nutr.* **91**, 406–412.

49. **The ENCODE Project Consortium** (2012) An integrated encyclopedia of DNA elements in the human genome. *Nature* **489**, 57–74.

50. **Vaag A, Grunnet L, Arora G & Brøns C** (2012) The thrifty phenotype hypothesis revisited. *Diabetologia* **55**, 2085–2088.

51. **Wellcome Trust Case-Control Consortium** (2007) Genome-wide association study of 14,000 cases of seven common diseases and 3,000 shared controls. *Nature* **447**, 661–678.

52. **Xue Y, Daly A, Yngvadottir B et al.** (2006) Spread of an inactive form of caspase-12 in humans is due to recent positive selection. *Am. J. Hum. Genet.* **78**, 659–670.

IDENTITY AND IDENTIFICATION

CHAPTER EIGHTEEN

Genetic diversity data are collected from individuals, but these data are traditionally interpreted in terms of groups of interbreeding individuals (populations), and their population structure. Such interpretations have been the basis of this book so far. But everyone's genome is unique (apart from those of monozygotic twins, although even these may differ because of **somatic mutations**)—each contains millions of SNPs and hundreds of thousands of other variants, including microsatellites, minisatellites, and CNVs that differentiate it from the rest (Chapter 3). The variation is generated by mutation and, while most is neutral, some affects visible traits. As the resolution of data increases with advances in technology, we can gather more and more information from individual diploid genomes that can tell us about their carriers.

In this, the final chapter, we focus on the individual instead of the population. We ask four major questions: first, how can we best use genetic variation for **individual identification** of any one of the >7 billion human beings with very high probability? Second, what can we determine about an individual from their DNA? This includes physical characteristics such as sex and pigmentation, and other classifications such as population of origin. Third, how can we decide from genetic variation if two individuals are close relatives, such as parent and child, or if they share membership of a larger grouping such as a genealogy, a clan, or some other genetically defined group? And finally, as affordable analysis increases to the ultimate level of the whole genome sequence, what will the **personal genome** offer to individuals who choose to access it?

Many of these questions are of great social interest, and, in the form of forensic DNA analysis and paternity testing, are key issues in law enforcement and social policy.[31] Questions about the genetics of an individual cannot be asked in isolation, however, because the answers—such as how certain we can be that two DNA samples derive from the same person, or whether an individual can be confidently assigned to a particular population from their DNA sequence—depend on the degree and structure of genetic variation in the population to which that person belongs. Thus, studies of individual variation must be set in the context of population variation, and appropriate statistical methods applied to evaluate different hypotheses or make predictions about population membership. These statistical issues may be complex, but at the same time they must often be presented to laypeople in criminal cases, or to the customers of personal genomics companies, in a way that is comprehensible and not misleading. Aside from statistical issues, there are practical issues of cost, data quality, and error rates in DNA analysis, and serious ethical questions related to the custodianship and appropriate use of databases containing DNA data on individuals.

18.1 INDIVIDUAL IDENTIFICATION

The diversity of the human genome, and the mutation processes that underlie this diversity, have been discussed in Chapter 3. On average, we expect to find a base-substitutional difference (a SNP) between two copies of the genome about every 1250 bp (see **Section 3.2**). This means that there will be a few million of these differences between the diploid genome of one individual and that of another. There are fewer polymorphic microsatellites and minisatellites than there are SNPs, but most of these have many more alleles and much higher heterozygosities, so among the few tens of thousands of potentially useful loci, we expect the majority to reveal differences between two individuals chosen at random. There are thus an enormous number of differences between the DNA carried by one person and that carried by another, and easy access to this variation would make it a simple matter to individually identify someone from a DNA sample, or to show with very high probability that two samples came from the same person. Developments in high-throughput genotyping and DNA sequencing (**Sections 4.4** and **4.5**) now allow this in principle. In practice, however, there are methodological and financial constraints that influence the kinds and numbers of polymorphisms that are used for analysis. The situation in a forensic laboratory is very different from that in a research context.

First, methods for individual identification evolve slowly, because their acceptance depends on the steady accumulation of confidence based upon their successful use in the courtroom (through legal precedent) and robust standards in testing (validation). Large **forensic identification databases** are expensive and time-consuming to set up, so frequent changes in established DNA typing systems are undesirable. Second, DNA samples are often degraded or present in trace amounts, so methods must be very sensitive—there is an analogy with ancient DNA studies here, including the concern about possible contamination (**Box 18.1**). Third, genotyping errors can have serious consequences, so all methods and procedures must be reliable and accurate. A false inclusion (**type 1 error**)—an incorrect statement that a person's DNA matches that left at a crime scene by a perpetrator, for example—could lead to the conviction of an innocent person. A false exclusion (**type 2 error**) could lead to a guilty

Box 18.1: The analogy between forensic and ancient DNA analysis

With the introduction of highly sensitive PCR-based methods, forensic scientists gained access to DNA profiles from a wide range of samples that were previously inaccessible: single hairs, cigarette butts, saliva on postage stamps, bite marks, fingerprints (the conventional kind), and so forth. Section 18.3 describes successful DNA analysis of the skeletal remains of the Romanovs. Analysis of such DNA resembles ancient DNA studies (**aDNA**; see Section 4.10): template DNA is present in very small quantities and is often extensively degraded, so that it is difficult to amplify. Another analogy is the ever-present danger of contamination by modern DNA, for which reason personnel involved in collecting and processing samples have their own profiles analyzed so that these can be recognized if they occur among the obtained profiles. Purpose-built facilities for analysis are similar to aDNA laboratories, with dedicated equipment, filtered air, and protective clothing.

The sensitivity of forensic trace DNA methods is impressive.[56] Touched surfaces (such as a phone, door-handle, or keys) can yield full microsatellite profiles, and profiles can also be obtained from single cells by increasing PCR cycle number. Problems associated with profiles from such samples (<100 pg) are **allele drop-out**, where expected peaks on an **electopherogram** fail to appear, artifact formation giving spurious bands (**drop-in**), difficulty of mixture interpretation, and unavoidable low-level laboratory contamination. Guidelines for this work include replicated detection of every allele before reporting the profile. The use of such profiles in criminal casework needs effective monitoring, because DNA can be transferred easily from one individual to another by contact, such as a handshake. The original intention was for the method to be used only as an investigative tool; when it entered the courtroom in a high profile 2007 case in Northern Ireland, controversy over the interpretation of trace DNA evidence led to criticism by the judge, and a review of the use of the method (Caddy Report, 2008: www.bioforensics.com/articles/Caddy_Report.pdf).

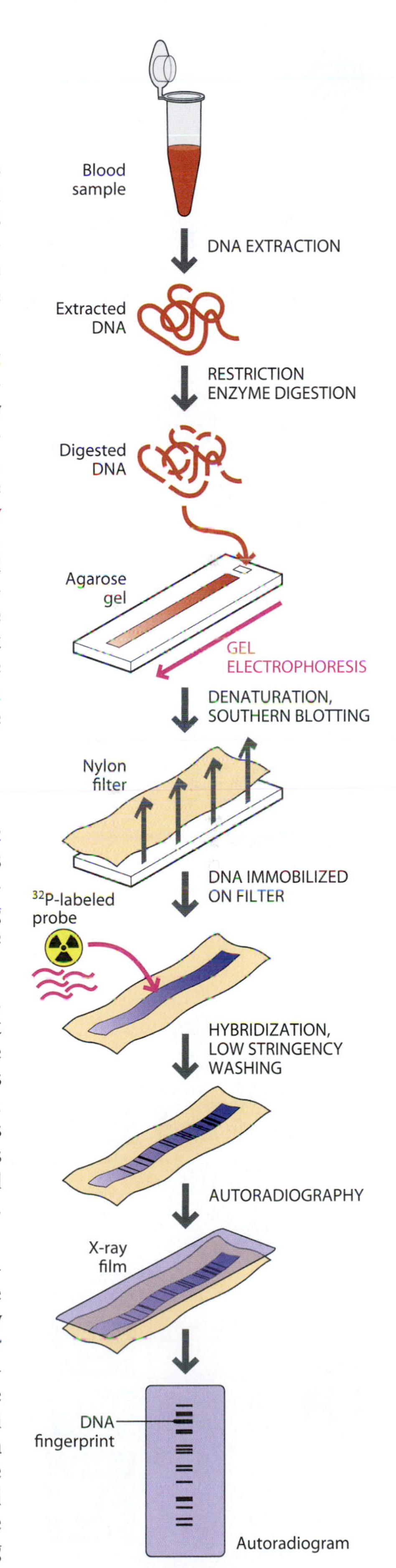

Figure 18.1: Procedure for generating a minisatellite DNA fingerprint. Illustrated here is the original DNA fingerprinting procedure, used from the mid-1980s to the 1990s, and now superseded by microsatellite profiling.

person being judged innocent, and free to commit further offenses. The latter is much the more likely outcome of errors in DNA profiling, or data handling. Such errors therefore tend to favor the suspect in a criminal case. Finally, questions about methods and statistics are often raised in the adversarial and conservative environment of the courtroom, rather than the more relaxed realm of the scientific seminar. This does not promote a productive and objective debate about the issues.

At this stage we must emphasize an important distinction between different uses of DNA-based evidence. The discussion above deals primarily with evidence that will come to court, be evaluated by a judge and jury and, in many jurisdictions, be debated by defense and prosecuting counsel. However, in principle, evidence used only for investigative purposes need never come to court. For example, a match between the DNA of a suspect and a sample from a crime scene will lead to the investigation of that suspect; their guilt or innocence may then be confirmed by other, non-DNA-based means (for example, a confession, a blood-stained hammer, or an irrefutable alibi), and so the DNA evidence need not be discussed in the courtroom. The kind of DNA evidence used in investigations can therefore be more novel and specialized than the points made above might suggest. In addition, even if a crime-scene sample produces a perfect DNA profile, this does not necessarily show that the person who left that profile was the perpetrator of the crime. It only demonstrates that, at some point in time, his or her sample was deposited at that particular place. The rest is for the court to decide.

The first DNA fingerprinting and profiling methods relied on minisatellites

The technique of DNA fingerprinting was developed[30] in Leicester, UK, by Alec Jeffreys in 1984, and rapidly became established as a highly effective means of distinguishing between DNA samples from different individuals, and establishing family relationships. Prior to the use of DNA, poorly discriminating protein-based classical polymorphisms such as blood groups (see Box 3.1) were the only available tools.

The original DNA-based method used a radioactively labeled probe (a short, specific DNA sequence) containing a minisatellite sequence to detect (in at least 250 ng genomic DNA) a collection of 15–20 highly polymorphic minisatellite loci containing shared sequence motifs (**Figure 18.1**). Variable-length fragments corresponding to these minisatellites were visualized as **DNA fingerprints**. The probability of two unrelated individuals sharing a DNA fingerprint was $\sim 10^{-11}$, and in practice the only individuals shown to share DNA fingerprints were monozygotic twins. The later refinement of **single-locus profiling** used sequential analysis of at least four individual hypervariable minisatellites, requiring only 10 ng of genomic DNA, and simplifying interpretation.

These methods were first applied in a criminal case in 1986, to investigate a double rape and murder in Enderby (Leicestershire, UK). First, it was shown that the two crimes, occurring 3 years apart, had been committed by the same man by finding identical profiles in semen samples taken from both victims (a **match**, or an **inclusion**). This connection of one crime scene to another is of great importance to the police. Second, a prime suspect who had confessed to one of the killings was shown to be innocent because his DNA profile was different from that found in the semen samples (an **exclusion**). The first ever **mass screen** was then organized: ~5000 blood samples were taken from men living in the locality, and DNA was extracted from the 500 samples not excluded by an initial blood-group-based test. The true killer, Colin Pitchfork, was apparently aware of the potential of DNA methods, because he evaded the screen by persuading

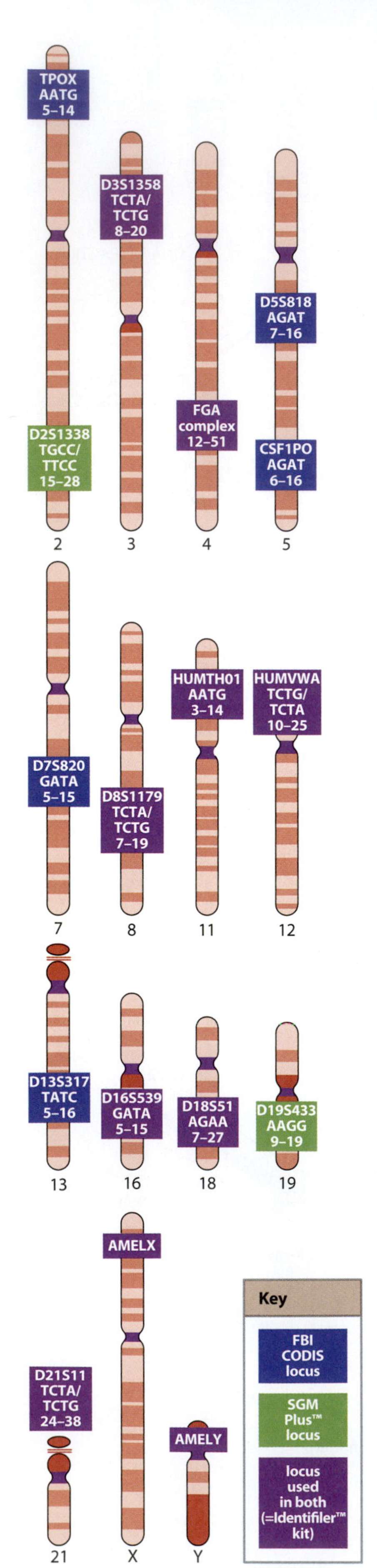

Figure 18.2: Microsatellites commonly used in multiplex DNA profiling, and their chromosomal locations.
Loci forming the US FBI's Combined DNA Index System (CODIS) and the SGM (Second Generation Multiplex) plus system used in the UK are shown. Note that two of the CODIS loci are on the same arm of chromosome 5, though separated by a large physical and genetic distance. The repeat sequence given for each locus is the predominant repeat; some contain additional or variant repeat types. (Data from STRBase: http://www.cstl.nist.gov/biotech/strbase/index.htm.)

a colleague to provide a blood sample on his behalf. Shortly afterward, the colleague talked about this exploit to friends who tipped off the police; Pitchfork was arrested, his single-locus profile and DNA fingerprint were shown to match the samples from the crime scene, and he pleaded guilty and was imprisoned.

PCR-based microsatellite profiling superseded minisatellite analysis

DNA profiling entered the PCR era in the early 1990s, with the now-universal adoption of microsatellites[9]—referred to in the forensic community by their alternative name, short tandem repeats (STRs). STR profiling has several advantages over earlier minisatellite-based methods. PCR allows the determination of a profile from very small (<1 ng) amounts of DNA[56] (though at the same time this sensitivity introduces potential problems with DNA contamination; see Box 18.1). Furthermore, small PCR amplicon sizes (~90–420 bp) allow profiling from badly degraded DNA samples. So-called **mini-STR** multiplexes use redesigned PCR primers to reduce amplicon sizes for a subset of the same microsatellites to ~70–280 bp, thus allowing recovery of profiles in more extreme cases of DNA degradation. Semi-automated analysis using fluorescently labeled primers on a capillary electrophoresis apparatus (see Figure 4.14) allows high-throughput profiling and the generation of large databases. Using this technology, discrete allele lengths (and hence repeat numbers) for the microsatellite loci can be measured accurately, which facilitates database generation and comparisons between profiles in digital form. Fragments are sized with reference to **allelic ladders**—sets of PCR products corresponding to common alleles across the known size range.

Microsatellites are generally chosen from different chromosomes so that they are uncorrelated and each contributes independently to the variability of a profile (**Figure 18.2**), and are assumed to be selectively neutral. They have high heterozygosities (each >0.7), and give specific and robust PCR amplification. Although allele length variability (and consequently the total number of alleles at a locus) is much lower than that for minisatellites, when several microsatellites are simultaneously amplified (multiplexed) they give highly informative profiles. The multiplex currently in use in the UK (**SGM+**) amplifies 10 loci, the system used by the FBI in the USA (**CODIS**) uses 13 loci, and most commercially available multiplexes, such as the Identifiler® system from Applied Biosystems (covering both SGM+ and CODIS; **Figure 18.3**), contain at least 15 loci. One measure of the discriminatory power of profiles is the **probability of identity** (P_I), the chance of two individuals selected at random having an identical profile, calculated by summing the square of the genotype frequencies. For systems widely in use, P_I values are extremely small, typically ranging from ~10^{-12} to 10^{-13} for the ten-locus SGM+ system, and ~10^{-17} to 10^{-18} for the fifteen-locus Identifiler® system.

How do we interpret matching DNA profiles?

When a crime-scene **DNA profile** matches that of a suspect, we need to ask whether this means that the sample was really deposited by the suspect, or by some other person whose profile happens to be the same. These possibilities can be considered using a **likelihood ratio**, which compares the probability of the evidence under the two hypotheses. This can also be presented as a **match**

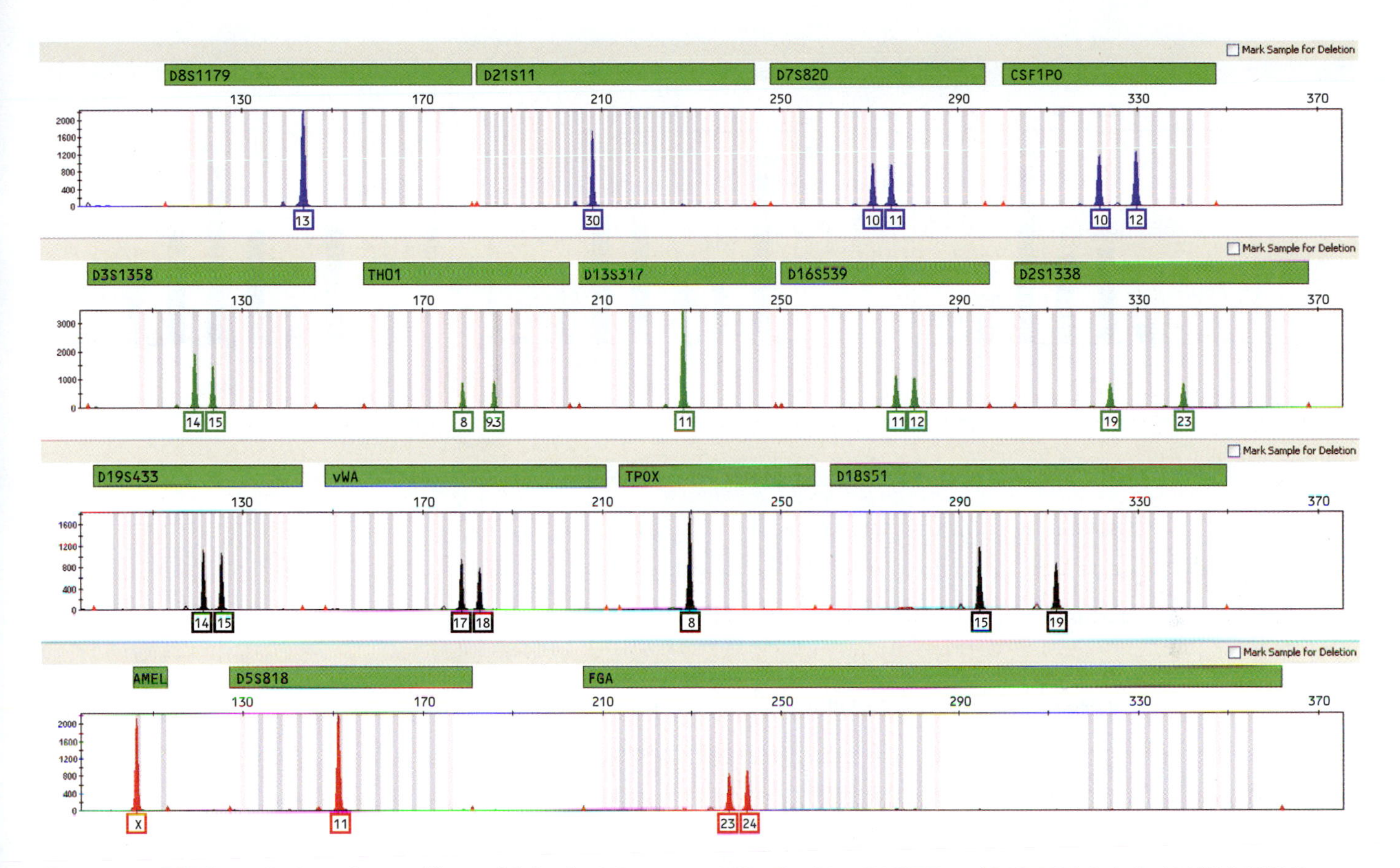

Figure 18.3: A high-resolution microsatellite multiplex for DNA profiling.
Result of amplifying a single 1 ng female genomic DNA sample using the Identifiler® system from Applied Biosystems, and detection using a capillary electrophoresis system, the ABI PRISM® 3130xl Genetic Analyzer. From top to bottom, the panels display the loci: D8S1179, D21S11, D7S820, and CSF1PO labeled with the dye 6-FAM; D3S1358, TH01, D13S317, D16S539, and D2S1338 labeled with VIC; D19S443, vWA, TPOX, and D18S51 labeled with NED; Amelogenin (sex test), D5S818, and FGA labeled with PET. A size standard (labeled with LIZ; not shown) is incorporated in each capillary run. Numbers on the y-axis refer to fluorescent units, and numbers on the x-axis (at the top of each panel) to bp. (From AmpFlSTR® Identifiler® PCR Amplification Kit User Guide. Courtesy of Applied Biosystems.)

probability—the chance that a profile chosen at random from the population would match the profile found at the crime scene. Because no population has been comprehensively profiled, a relatively small **forensic population database** (typically 100–200 unrelated individuals) is used. The match probability is then calculated using the **product rule**, by multiplying together the genotype frequencies at each locus calculated using the population frequency of each of the profile's component microsatellite alleles (equivalent to the estimated population frequency of the profile—Figure 18.4). Rare alleles are treated conservatively: for example, if an allele is observed fewer than five times in a sample (N) of 100 individuals, then instead of the observed frequency, its minimum frequency is adjusted up to $5/2N$, or 2.5%. Resulting match probabilities are low—for example, typically ~10^{-10}–10^{-11} for SGM+.

Which population should be chosen to form the forensic population database? One option would be to use the available databases that represent the large groups (for example, "US **Caucasians**" or "Hispanics") familiar to police forces, though not well-recognized by anthropologists and population geneticists (see Box 1.1). However, if a suspect comes from a subpopulation that is distinct from others (that is, there is population substructure), then some alleles may be present at higher frequency than expected. Choice of an inappropriate database population might then lead to a misleadingly low match probability. Much debate took place on this issue in the minisatellite era, and a conservative parameter (equivalent to F_{ST}—see Box 5.2) was incorporated to account for population substructure. This had the effect of reducing the match probability

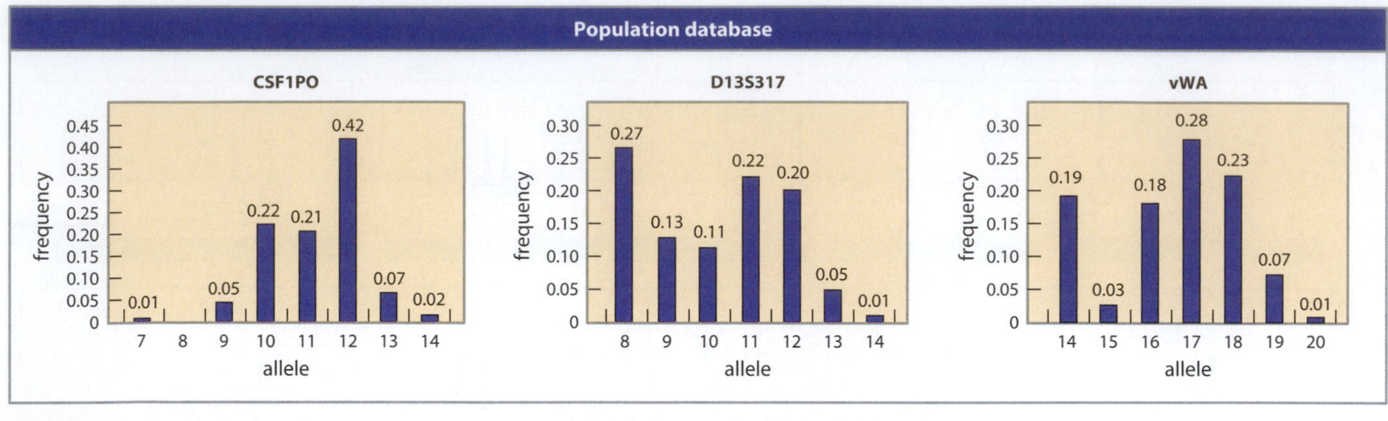

	p *q*	*p* *q*	*p* *q*
Suspect's profile	**9, 12**	**10, 13**	**15, 17**
Frequency of each allele	**0.05 0.42**	**0.11 0.05**	**0.03 0.28**
Per-locus genotype frequency = 2*pq*	**0.0420**	**0.0110**	**0.0168**
Multiply to give profile frequency	**0.0420 x 0.0110 x 0.0168 = 7.76×10^{-6} or 1 in ~129,000**		

Figure 18.4: Estimating the population frequency of a DNA profile. The bar charts show the frequencies of alleles at three forensic microsatellites (CSF1PO, D13S317, and vWA) in a hypothetical population database. Estimation of per-locus genotype frequencies by multiplication assumes that the allele frequencies are independent; likewise, genotype frequencies are assumed to be independent when they are multiplied to give profile frequencies (the product rule).

by up to two orders of magnitude. The same approach has been taken to microsatellite profiles, but in practice typical match probabilities are so low that adjusting for population substructure has little influence on their evidential weight. It is now standard practice for forensic geneticists in particular jurisdictions to profile their local populations to provide relevant databases.

Despite the high discriminatory power of DNA evidence, the way that it is dealt with in court can be distorting. The **prosecutor's fallacy** is a commonly cited example. Suppose a DNA match at a crime scene has a probability of occurrence in a population of 1 in a million. The prosecutor has found that the accused has this match and says "the odds are a million to one in favor of the suspect being guilty." It ought to be impossible for the jury to acquit against such apparently overwhelming odds. Unfortunately, however, there were 10 million people in the city on the night of the crime, so the defense counsel argues that 10 people among the population of the city on the night of the crime could have matched the DNA sample (the "defendant's fallacy"). If the DNA match is the only evidence available, then the odds are 10 to 1 in favor of the suspect being innocent! What the prosecutor should have said was: "The probability of a match if the crime-scene sample came from someone other than the suspect, is one in a million." The discussion should be about comparing the likelihood of observing the evidence given that the defendant is guilty, to the likelihood of observing the evidence given that someone else is guilty.

Complications from related individuals, and DNA mixtures

Match probabilities can also be affected by other factors. The discussion above considers chance matches between individuals who are unrelated. However, if a crime-scene sample were actually deposited by a relative of the suspect, the situation would be very different. Relatives are expected to share alleles with the suspect, and are also much more likely than people taken at random to share an environment, and therefore an opportunity to commit the crime. These possibilities can be incorporated within the statistical interpretation of the evidence using likelihood ratios. Considering, for example, a suspect's brother as an alternative culprit has a marked effect on match probability: for SGM+ the value increases by six orders of magnitude[16] to $\sim 10^{-4}$–10^{-5}. A technical factor that can also increase the match probability is the fact that crime-scene samples can often comprise a mixture from multiple individuals and therefore be difficult to interpret, or be incomplete because of DNA scarcity and degradation, with only a subset of the microsatellite loci or alleles providing reliable data. This is a particular problem in so-called **low-copy-number** profiling, where PCR cycle numbers are increased to produce profiles from extremely small amounts of DNA (Box 18.1).

Large forensic identification databases are powerful tools in crime-fighting

If no suspect is available, it may be possible to match a crime-scene sample to a database of profiles from individuals and thus to identify a suspect in a "cold hit"; if the database also contains crime-scene samples, then the profile might match one of these, thus providing a link. The rules and procedures for constructing forensic identification databases and the legal restrictions on how they may be used vary greatly from country to country, and, in the USA, from state to state.

Here we discuss the National DNA Database (**NDNAD**) of England and Wales as an exemplar, because it is the longest established (since 1995), and until recently the largest in the world (www.npia.police.uk/en/8934.htm). Acts of Parliament passed at its inception allowed the police to take non-intimate samples (that is, **buccal** swabs and hair samples with roots) for profiling without consent (sometimes referred to as informed non-consent) from people in police custody who were charged with a recordable offense (one subject to a term of imprisonment). Subsequent changes in the law permitted similar sampling from people who had been previously convicted of burglary, or violent or sexual offenses and who were still in custody. The rapid growth of the database was aided by changes allowing profiles to be retained, even if arrestees are acquitted or charges dropped. Profiles taken from volunteers in mass screens could also be retained, provided that the volunteers consented in writing.

By December 2012 the NDNAD (England & Wales) contained 401,813 crime-scene profiles, plus profiles from 5,733,486 individuals—about 10% of the population. This remarkably high proportion compares with a figure of just 1.7% for the USA. The performance of the NDNAD is certainly impressive. By the same date it had provided 428,097 matches of crime scene to suspect, of which 2716 were in murder/manslaughter cases, and 6085 in rape cases. Because of the sensitivity of the PCR-based profiling and the size of the database, "cold" cases can now be readdressed provided that crime-scene samples have not been discarded. A recent example was the conviction of a man in 2011 for a rape committed in 1988; a microsatellite profile was obtained from a stored crime-scene swab, and matched a profile in the database, deposited after the attacker had been investigated for a later offense. The NDNAD has also been used as an investigative tool, particularly in **familial searching**.[20] Here, a crime-scene profile finds no perfect match in the database, but identifies some profiles similar enough to belong to close relatives due to familial allele sharing. These people are then interviewed, which can lead to a suspect whose profile matches the crime scene perfectly. David Lloyd, a serial rapist in the 1980s, was caught in this way in 2006 using the profile of his sister, investigated for a drink-driving offense.

Controversial aspects of identification databases

Forensic identification databases are seen by many as an absolute good. For any offender whose profile is in such a database, the risk of being detected in further criminal activity is high. If this acts as a deterrent to further offending, this is a benefit, though since many serious crimes are committed on impulse the deterrence may not be strong, and the database's primary advantage will be the speedy apprehension of culprits. However, there are also potential drawbacks. No database is perfect—all will contain some errors, despite the requirement for laboratories contributing data to be properly accredited; the issue is what the error rate is, and what the consequences of the errors may be. False exclusions (type 2 errors) are the most obvious outcome. The error rates of large databases are not discussed, but should be; any assurance that there are no errors would be a political, rather than a scientific, claim.

The enormous NDNAD of England and Wales continues to increase in size; the ~5.6 million individuals whose profiles it contains (December 2012) greatly exceed the police estimate of the number of active criminals, which is ~3 million. The fact that it contains the profiles of many, but not all, innocent people, and the over-representation of some ethnic groups [for example, 7.3% "African-Caribbean," compared with a population frequency of 2.9% "Black British" (2001 census data)], discriminates against such groups. Databases in many other countries require a certain degree of demonstrable connection to a crime for a profile to be retained, and the NDNAD's use for familial searching would in most jurisdictions be regarded as an unacceptable infringement of privacy. Databases also inadvertently contain potential information about **sex-reversal** syndromes, due to the inclusion of a **sex test** in the profiles (see **Section 18.2**) private genetic information about the phenotypes of individuals that might be regarded to be an infringement of medical privacy. A 2008 case was brought by two men whose DNA profiles were retained in the NDNAD, but who had been respectively acquitted, and not prosecuted (http://www.bailii.org/eu/cases/ECHR/2008/1581.html). The European Court of Human Rights ruled that the retention of profiles of innocent people in the NDNAD was an unlawful breach of their human rights, and that such data should be destroyed. The UK government has not complied, and instead has proposed retention for 6–12 years; meanwhile, profiles continue to be collected, though it is possible for innocent individuals to request removal of their profiles. Some people have proposed a UK populationwide database held by an independent body that would consider police requests for access, or even a global DNA database to help fight global crime and terrorism. Civil liberties debates on these issues are set to continue.

Before moving on from forensic identification databases, it is worth considering the future of medical and research genetic databases, and the possible confluence between them. An individual's genotype for the hundreds of thousands of SNPs generated using a SNP chip is expected to be unique, unless they have an identical twin. In principle, medical SNP databases could therefore be used for individual identification. Current SNP chip technologies require DNA in an amount and quality that is greater than that commonly encountered in the forensic setting, but this may change. It has been claimed that individual genomewide SNP genotypes can be de-convoluted from DNA mixtures,[26] although the approach has been criticized on statistical and practical grounds.[14] Perhaps in future, when all members of a population have genetic data recorded at birth for medical purposes, the temptation to exploit this information for identification purposes will be difficult to resist.

The Y chromosome and mtDNA are useful in specialized cases

DNA profiles based on a set of autosomal microsatellites owe their enormous variability and discriminatory power to three processes: mutation, which generates new alleles, and independent chromosomal assortment and recombination, which together re-assort the alleles each generation. Because of this, the only person expected to share someone's profile is their identical (monozygotic) twin. Since the frequency of such twins is rare (typically ~1 in 300 births) and their existence usually known about, this is not a practical problem. Now consider DNA profiles based on nonrecombining segments of DNA—the Y chromosome and mtDNA. First, genetic variability in these systems is due only to mutation, and this greatly reduces haplotype variability. Second, all members of a man's close **patriline** (his father, brother, son, paternal uncle, and so on) are expected to share his Y-chromosomal haplotype, and all members of a person's **matriline** are expected to share their mtDNA haplotype. Obviously, some patrilineal or matrilineal relatives will differ from a sample donor due to mutation, and the more distantly related they are, and the more polymorphisms we type, the more likely this becomes. Nonetheless, we expect that a sizeable number of relatives will be indistinguishable from the donor of a sample using this kind of DNA evidence.

An additional problem for both the Y chromosome and mtDNA is the possible degree of geographic substructuring, which can affect our interpretation of an inclusion, where there is a need for statistical calculations. We have already considered the problem of population structure in autosomal profiling, but it is potentially much more severe for these nonrecombining pieces of DNA, each of which is a single genetic locus. Past mating practices and genetic drift (see **Section 5.3** and **Appendix**) are strong potential forces that could lead to local concentrations of particular Y-chromosomal or mtDNA haplotypes. An example of a high-frequency Y haplotype defined using SNPs and 16 microsatellites is described in **Section 18.3**

These factors are not a problem for exclusions. For example, if a suspect fails to share a Y-chromosomal profile with a crime-scene sample, this fact excludes him, regardless of how common his patrilineal haplotype may be. However, the sharing of haplotypes within patrilines or matrilines might seem to preclude the use of the Y chromosome or mtDNA in matching samples to suspects. But this is not entirely so: they have both come to play useful, if specialized roles, which are explored below, and recent technical advances in genotyping and sequencing promise to further increase their usefulness.

Y chromosomes in individual identification

The first consideration for Y-chromosomal profiling is that a high proportion of criminals, and >90% of violent offenders, are male. If a perpetrator leaves some of his cells behind at a crime scene, these will usually contain a Y chromosome. In rape cases the normal practice is to attempt to obtain an assailant-specific autosomal DNA profile from a vaginal swab by the method of **differential lysis**, which allows sperm cells to be selectively enriched.[22] However, in cases where this fails (for example when the rapist is **azoospermic** and has no sperm in his semen[3] or where there are other mixtures, such as blood mixed with blood or finger-nail scrapings from a victim), a Y-chromosomal microsatellite profile provides a sensitive means to gain specific information about the assailant. In multiple rapes it may be possible to estimate the likely number of assailants.

A set of between 9 and 17 mostly tetranucleotide Y-chromosomal microsatellites is widely used: as with autosomal profiling, commercial kits dominate forensic practice. When a Y-chromosomal profile from a crime-scene sample matches that of a suspect, the significance of the match cannot be assessed in the same way as for an autosomal profile—the product rule referred to above (see Figure 18.4) is clearly not appropriate for Y-chromosomal microsatellites, because the alleles are not independently inherited. Instead, the profile is compared with a forensic population database, for example the large online Y-Haplotype Reference Database (www.yhrd.org), which currently (Release 43; 2013) contains 112,005 haplotypes from 834 different populations. There are many contributing laboratories, and all are required to undertake a blind proficiency test. Profile frequencies can be determined simply by counting; if a haplotype is absent from the database, the conservative (albeit mathematically unrealistic) assumption that it would be the next one added can be made. Haplotype frequencies observed in or extrapolated from such databases are very high compared with those of autosomal genotypes—for example, for a nine-microsatellite profile[47] they typically range from 10^{-5} to 10^{-3}.

The microsatellites employed in Y-chromosomal analysis are a heterogeneous set discovered before the Y reference sequence was available. With the sequence in hand, it was possible to carry out a systematic survey, revealing 167 novel potentially useful loci.[33] A survey of mutation at 186 microsatellites in 2000 father–son pairs[2] revealed that some have extremely high mutation rates (up to 7.44×10^{-2} per transmission of the Y chromosome), and showed that a set of the most mutable 13 microsatellites could distinguish between closely related males; empirically, 70% of father–son pairs, 56% of brothers, and 67% of cousins could be differentiated by mutations at these so-called **rapidly mutating STRs**. With such discriminatory power, the problems of shared haplotypes among patrilineal relatives are reduced, and this should increase the utility of the Y chromosome in forensic genetics.

mtDNA in individual identification

The major virtue of mtDNA for forensic analysis is the same as that for the analysis of ancient samples (see **Section 4.10**). Because mtDNA has a far higher copy number per cell than nuclear DNA, it has a correspondingly greater chance of survival, and is therefore useful in the analysis of forensic samples that contain little DNA (for example, hair shafts), are old, or have received severe environmental damage. Also, because of its matrilineal inheritance, a matrilineal relative can be used as reference material for matching. As with Y-chromosomal haplotypes, the population frequency of an mtDNA sequence cannot be estimated by multiplication of the frequencies of individual variant sites, and the normal practice is simply to count the number of sequence matches in a database. EMPOP (empop.org), a database of mtDNA control region sequences, currently (Release 9; January 2013) contains 29,444 sequences. Quality control is very stringent, and primary sequence data are linked to sequences; this follows a number of controversies over the quality of sequence data in forensic databases. The random match probability is again high compared with autosomal profiles; for example, for the whole control region in a population of 273 Austrians[6] it was 0.8%, that is, 1 in ~120.

In samples that have undergone extreme treatments such as burning, mtDNA may be the only recoverable DNA suitable for analysis. For the victims of accidents, warfare or disasters, or of the 9/11 terrorist attacks, therefore, this kind of analysis may be the only hope for identification. In such a case, matrilineal relatives provide a reference source that allows lineage matches to be made. The identification of Michael J. Blassie, interred for 14 years in Arlington Cemetery as the Vietnam Unknown Soldier, was an example of the success of this approach. In events such as air crashes, the population of potential sources of the mtDNA sequences at the crash scene may be large, but it is "closed"—that is, confined to a known set of individuals, and hopes are high that full identification, at least to the level of the matriline, can be made.

18.2 WHAT DNA CAN TELL US ABOUT JOHN OR JANE DOE

In this section we describe attempts to deduce information about aspects of phenotype from DNA. In practice, the police want this information when they are faced with a crime but no suspects—any information whatever may advance their investigation.[32] Note that, if a comprehensive universal DNA database were available, these phenotypic deduction methods would be unnecessary.

DNA-based sex testing is widely used and generally reliable

The simplest and most common genetic distinction between human beings is their sex. Because men have a Y chromosome and women don't, it ought to be easy to tell whether a DNA sample is from a male or a female. A sex test would not only be useful in forensic and archaeological contexts, but also in prenatal testing in cases of a suspected sex-linked disorder.

A PCR-based assay that amplifies a Y-specific sequence might seem a simple solution, but is not a good test. If a PCR product is obtained, the sample is from a male, but if not, there are three possible explanations: the sample could be from a female, the DNA could be too degraded to yield a product, or the PCR could have failed for a technical reason. It is therefore important to have an assay that incorporates an internal control to show that the PCR is working properly. An independent pair of primers could be used to co-amplify a second locus, but this would not be ideal because absence of the Y-chromosomal product could then still be explained by a problem with the Y-specific primers. Ideally, the Y-specific and control sequences should be co-amplified by the same primer pair.

XY homology (or **gametology**), deriving from the common origin of the sex chromosomes as an ancestral autosomal pair and subsequent sequence exchanges (see **Appendix**), offers a natural solution to this problem. Many PCR primer pairs designed to amplify a Y-chromosomal sequence will also amplify an X-chromosomal sequence, and all that is necessary is to design an assay in which the X- and Y-specific products can be readily distinguished. The most widely used assay[53] amplifies part of intron 1 of the XY-homologous **amelogenin** gene (*AMELY/AMELX*; **Figure 18.5a**). A single pair of PCR primers produces a product of 112 bp from the Y chromosome and 106 bp from the X chromosome. A male will therefore yield both products, while a female will yield only the smaller. The 6-bp difference in size is easily resolved by capillary electrophoresis, and this assay is included in commercially available PCR multiplex kits that are widely used for DNA profiling (see Figure 18.3). Although this test is quite reliable and universally applied, there are rare instances where it will give the wrong answer (and, as discussed below, invade genetic privacy).

Sex reversal

46,XX males and 46,XY females have a sex-chromosomal constitution that is not concordant with their phenotypic sex. There are many underlying causes for these conditions. An example of XY femaleness is androgen insensitivity syndrome (AIS; OMIM 300068), an X-linked recessive disorder occurring in about 1 in 20,000 46,XY births, in which affected individuals have female external genitalia and breast development, no uterus, and undescended testes. In this syndrome the testis produces androgens, but a mutation in the androgen receptor gene means that the hormonal signal elicits no response in target tissues.

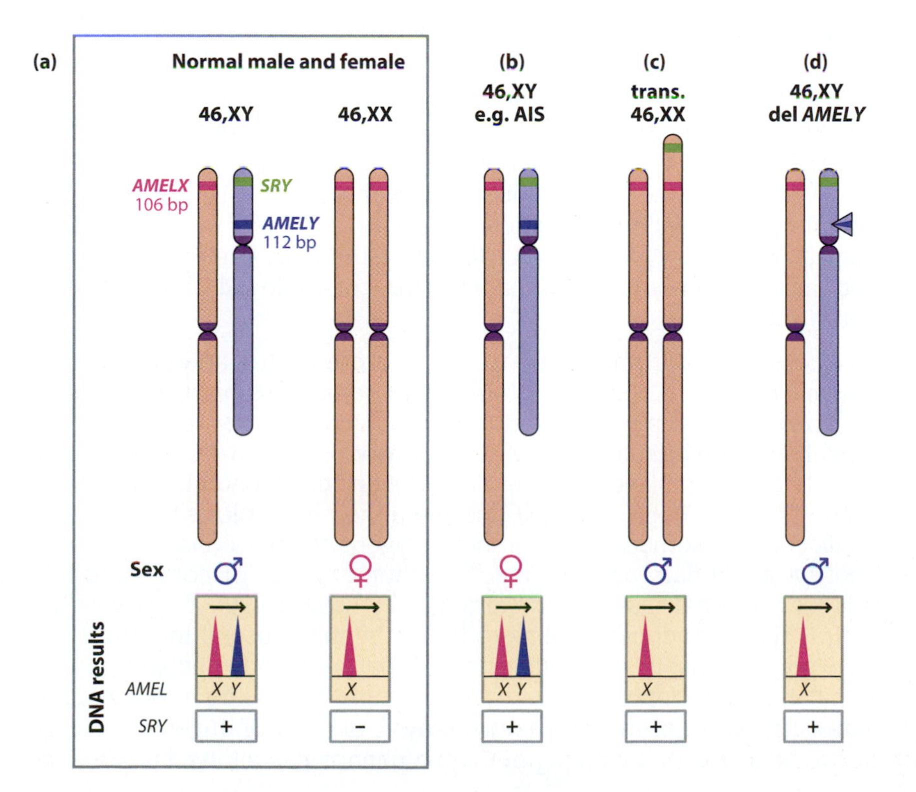

Figure 18.5: DNA-based sex testing. The amelogenin sex test distinguishes (a) normal males and females, but does not correctly sex (b) 46,XY females such as those with androgen insensitivity syndrome (AIS), (c) translocation 46,XX males, or (d) males carrying *AMELY* deletions. Testing of the presence of *SRY* is helpful in the latter two cases.

An amelogenin sex test here would suggest a male, though the phenotype is female (Figure 18.5b). XX maleness (OMIM 278850) has an overall frequency at birth of about 1 in 20,000, and is most commonly due to a translocation of a small portion of the Y chromosome, including the testis-determining gene *SRY*, to the X chromosome. Men carrying these translocation chromosomes are infertile because they do not carry Y-chromosomal long-arm genes necessary for spermatogenesis. Because the translocated portion of the Y chromosome does not include the *AMELY* locus (Figure 18.5c), the sex test suggests a female, though the phenotype is male. Other sex-reversal syndromes will also give contradictory results.

Deletions of the AMELY locus in normal males

Some males carry interstitial deletions of part of the short arm of the Y chromosome, apparently without phenotypic effect, but removing the *AMELY* locus.[49] Such males would be wrongly diagnosed as females using the standard sex test (Figure 18.5d). The overall incidence of *AMELY* deletions is low; for example, it was reported as 0.02% in ~29,000 Austrian males.[52] However, it may reach higher frequencies in some populations through drift: 5 of 270 Indian males (1.85%) showed such deletions.[54] To surmount some of these problems, a number of additions or alternatives to the basic amelogenin test have been proposed, including a PCR test for the presence of the *SRY* gene. While these may be useful in special cases, the amelogenin test is now so well established that it seems unlikely to be supplanted.

Some other phenotypic characteristics are predictable from DNA

As was discussed in Chapter 15, many human phenotypic characteristics (for example, stature, facial features, and pigmentation) have a strong genetic component. If we understood their genetic basis we might hope to be able to make reliable predictions about the phenotype of an individual from a DNA sample. However, there are two problems. First, as we have already seen, these **quantitative traits** are complex and **multigenic**, and our knowledge about the genes that underlie them is far from perfect. There is no guarantee that, even with a good understanding of these genes, an informative predictive test would emerge. Second, the environment has a part to play in many of these phenotypes. For instance, stature and morphology are influenced by diet, and pigmentation is influenced by climate, habits, and age, as well as by interventions via hair dye and novelty contact lenses.

Indirect information about phenotype can be sought by trying to predict population of origin (see following sub-section). If there is a strong prediction from a DNA profile that an individual is of African, or East Asian, or Northern European descent, such information is likely to influence our expectations of the appearance of the person who deposited the DNA sample. However, this indirect information may be misleading, particularly in **admixed** populations (Chapter 14).

Direct inference about phenotype would be more useful, and pigmentation is an obvious candidate. Skin color has proved rather less predictable than hair or eye color; the first pigmentation trait to be addressed by forensic scientists was red hair,[24] and its correlation with polymorphism in the melanocortin 1 receptor (*MC1R*) gene (**Section 15.3**). The frequencies of a number of *MC1R* variants in a group of people who self-described their hair color show that we can be confident that someone who is homozygous for the consensus sequence will not have red hair, and if someone is homozygous or compound heterozygous for a red-hair-associated variant, they have a >90% chance of having red hair. This test therefore seems to be a potentially useful investigative tool in some populations, if DNA data indicated that red-headed people (~5% of the UK population of European ancestry, for example) should be prioritized among the suspects. More recent systematic analysis of a set of 45 SNPs in 12 genes known to be associated with pigmentation phenotypes allowed prediction of a

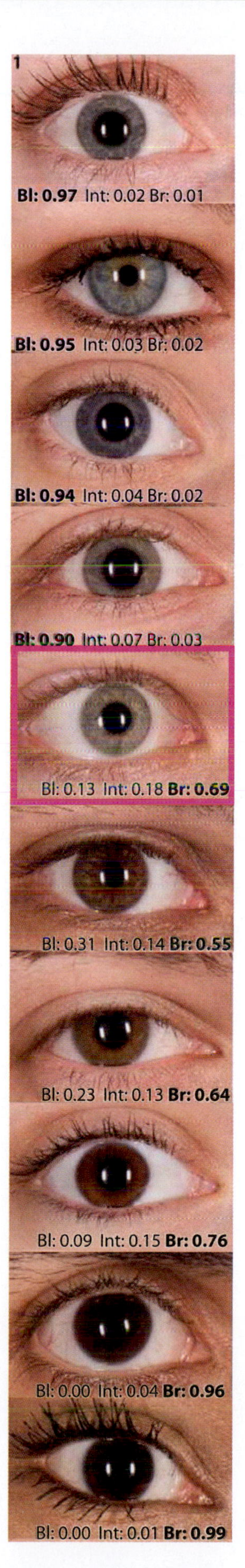

Figure 18.6: Prediction of eye color using DNA testing. Examples of eyes and their predicted colors based on testing with the six SNPs in the IrisPlex assay. Bl: blue; Int: intermediate; Br: brown. Values indicate individual probabilities for each eye color. One example here (in *red box*) is wrongly predicted. [From Walsh S et al. (2011) *Forensic Sci. Int. Genet.* 5, 170, with permission from Elsevier, in which colors of 37/40 eyes are predicted correctly and 2/40 tests are uninformative (equal probability of all three colors).]

much wider variety of hair colors;[7] for red, black, brown, and blond the predictive accuracy (where 1 indicates complete accuracy and 0.5 random prediction) was estimated respectively as 0.93, 0.87, 0.82, and 0.81. Hair color prediction based on this method has not yet been introduced into forensic casework. As well as hair color, the color of the iris is now also highly predictable from DNA evidence (**Figure 18.6**). The so-called IrisPlex assay types the six most informative SNPs, from six pigmentation genes (*HERC2*, *OCA2*, *SLC24A4*, *SLC45A2*, *TYR*, and *IRF4*), with a predictive accuracy of 0.93 and 0.91 for brown and blue eye color respectively,[57] and is validated for forensic use.

As was described in Box 15.1, many variants have been discovered that contribute to variation in stature. However, their effects are small, so their current predictive accuracy would be weak—180 loci explain only 10% of the variance in the trait.[38] A genomewide association study has also identified loci that contribute to normal variation in facial morphology among Europeans.[40] As new discoveries are made of genetic variants affecting physical traits, there is no doubt that the ingenuity of forensic scientists will be applied to them.

Reliability of predicting population of origin depends on what DNA variants are analyzed

The determination of the population of origin of the depositor of a crime-scene sample would be very useful information for law enforcement agencies. As has been discussed in **Section 10.2**, ~85% of genetic variation is found *within* human populations. However, significant variation still lies between them, and this might be useful in attempts to predict population of origin in a forensic setting. In practice, the potential benefit of prediction is assessed by considering the likely reduction in the number of suspects investigated, before the actual perpetrator is reached. A potential danger is that targeting one particular ethnic group in the absence of other evidence could lead to accusations of prejudice.

Prediction from forensic microsatellite multiplexes

The microsatellites used for individual identification were selected because of their high degree of polymorphism. This arises from their high mutation rates, and leads to very low interpopulation variance (F_{ST}; Box 5.2). They are therefore expected to be poorly suited to predicting the population of origin of a DNA sample. However, since a substantial investment of time and money has gone into establishing large databases, and since the systems are standardized, it is interesting to know how well these microsatellites perform. This has been assessed by genotyping the 15 microsatellites of the Identifiler® kit in the **CEPH-HGDP panel** of DNA samples.[44] Using the program STRUCTURE, classification of the samples into one of four major groups (African, European, East Asian, or American) works quite well (**Figure 18.7**), with an average error rate of ~10%, though achieving this result requires the exclusion of the Middle Eastern and South Asian samples, which cannot be distinguished from the Europeans. An online tool, PopAffiliator (cracs.fc.up.pt/popaffiliator), has been developed that allows users to estimate the population of origin of a DNA profile. Based on a machine-learning method and a database of 56,000 classified profiles, the algorithm classifies profiles into one of three major population groups (Eurasian, East Asian, or sub-Saharan African), with an accuracy of ~86%.

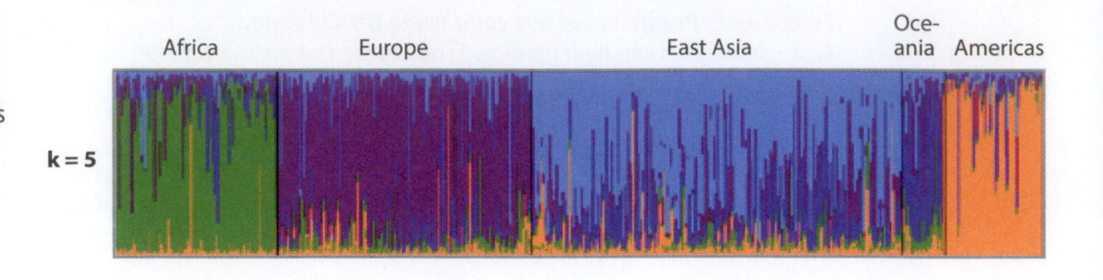

Figure 18.7: Population structure revealed by forensic microsatellites. CEPH-HGDP samples typed with the 15 CODIS and SGM+ autosomal microsatellites in the Identifiler® kit, and analyzed using the program STRUCTURE with $k = 5$. Four clusters are clearly visible, but the result is "noisy": some individuals (thin vertical lines) would not be predicted to fall within their regional population grouping; most individuals appear admixed (an artifact due to the type of variation analyzed here); and the Oceanian samples do not form a distinctive cluster. Note that Middle Eastern and South Asian samples have been excluded from this analysis, because they cannot be distinguished from Europeans, and if included would form part of a larger Eurasian cluster. [From Phillips C et al. (2011) *Forensic Sci. Int. Genet.* 5, 155. With permission from Elsevier.]

Prediction from other systems

While these methods may provide some information, clearly they would not be very useful discriminators among more closely related populations, or the real urban samples encountered in forensic work. Population genetic studies show that the use of genomewide SNP chips bearing a few hundred thousand SNPs can perform much better, for example in distinguishing among different European national populations[42] using principal component analysis (**Figure 18.8**). However, because of issues of DNA sample quality and quantity, these methods are not practicable in the forensic setting. Attempts have therefore been made to choose a smaller subset of particularly informative SNPs (**ancestry informative markers**, **AIMs**) that can be typed on forensic samples. One example is a 34-SNP multiplex that can be typed successfully on samples of <1 ng DNA.[45] The SNPs were selected to show very different allele frequencies between populations, based on previous information. This method gave a <1% error in classifying CEPH-HGDP samples (Box 10.3) as African, Asian, or European, but it is not clear how well it would perform for discriminating among continental subpopulations.

Some studies have used or recommended the Y chromosome[58] and mtDNA[13] to suggest population of origin. These loci show strong geographic differentiation (see **Appendix**) because their small effective population sizes lead to strong genetic drift. Mating practices may also contribute to interpopulation differences (see **Section 5.3**). However, since they each reflect only one out of many ancestors, they are vulnerable to mis-assignment; an example is the finding of a rare and typically African Y haplogroup associated with an English surname[37]—a Y-chromosomal test would have strongly (and wrongly) suggested that the bearer was African. At the population level, differently inherited segments of DNA can provide different indications: for example, a study of African-Americans[39] showed average levels of European admixture of 30% with ancestry-informative Y-SNPs, 8% with 24 autosomal AIM SNPs, and only 5% with mtDNA.

The problem of admixed populations

It is important to remember that the samples in collections such as the CEPH-HGDP panel, discussed above, were chosen to represent non-admixed indigenous populations. If forensic prediction methods were applied to urban samples from New York or London with more complex ancestries, they might not be so informative. Judging the effectiveness of prediction methods in practice will also be complicated by the non-anthropological way in which population of origin is defined by police forces; in the UK, for example, the "ethnic appearance" categories White European, Dark European, Afro-Caribbean, Asian, Oriental, and Arabian/Egyptian are used. Also it is not clear how a person would be classified who had parents from different ethnic groups.

A particularly controversial proposal was the so-called Human Provenance Project, launched in 2009. This UK Home Office-funded pilot exercise aimed to use DNA evidence (together with stable isotope analysis of tissue) to determine whether asylum-seekers attempting to enter the UK and claiming to be fleeing persecution in Somalia were actually of another nationality (for example, Kenyan); details of the proposal are sketchy, but there was concerted opposition

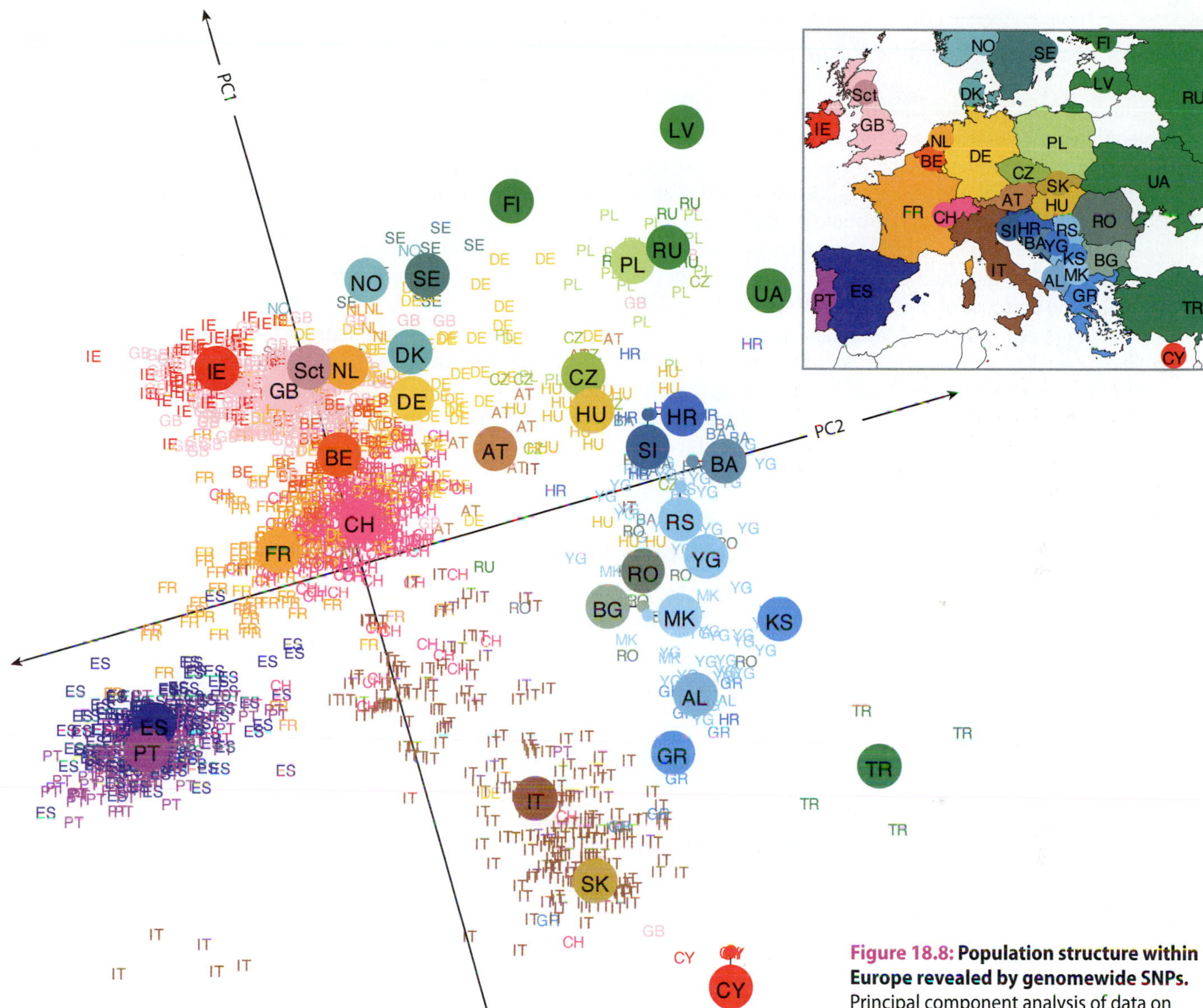

Figure 18.8: Population structure within Europe revealed by genomewide SNPs. Principal component analysis of data on 197,146 SNPs in 1387 Europeans. Small colored labels represent individuals and large colored points represent median PC1 and PC2 values for each country. The PC axes are rotated to emphasize the similarity to the geographic map of Europe. [From Novembre J et al. (2008) *Nature* 456, 98. With permission from Macmillan Publishers Ltd.]

from human geneticists and the proposal has now been dropped. The idea that DNA variants respect the borders of such recent constructs as nations is implausible.

18.3 DEDUCING FAMILY AND GENEALOGICAL RELATIONSHIPS

Typing any polymorphism in a mother–father–child trio can give information about the biological parentage of the child. As soon as DNA fingerprinting was developed, it was shown by family studies that the inheritance of the many minisatellite bands in a DNA fingerprint was **Mendelian**, and therefore that close family relationships could be verified to near-certainty by genetic analysis.[29] The landmark case was the verification that a Ghanaian boy, seeking to re-enter the UK, was indeed the son of the UK-resident woman claiming him as her child. This is a maternity test, which is unusual; generally (except in the rare cases of mix-ups in hospital nurseries), the mother is known and undisputed, and the question is about the father (**Box 18.2**). As with forensic applications, technology in **paternity testing** evolved from minisatellite-based methods to microsatellite profiling (**Figure 18.9**). Paternity testing is big business, with about 415,000 tests performed in 2008 in the USA, for example (www.aabb.org/sa/facilities/Documents/rtannrpt08.pdf).

Box 18.2: Nonpaternity in human populations

Commonly, if you ask a human geneticist what the per-generation nonpaternity rate is, they will give a quick answer of "about 10%." When you ask them how they know this, they are often not sure, and the figure has been described as an "urban myth." Sometimes the rate quoted is that found when typing polymorphisms in a family segregating a genetic disorder, and if so, this could represent an ascertainment bias, since the presence of the disorder in itself means that the family is not typical of the general population. Similarly, overall rates certainly cannot be estimated from nonpaternity casework (where the published rate varies between 14 and 50%) because when a test is carried out there is some prior reason for suspecting nonpaternity.[1]

Systematic studies of nonpaternity rates are few and far between. Early experience of cystic fibrosis screening[8] found seven nonpaternities out of 521 families (1.35%). One study in Switzerland[50] looked at 1607 children and their parents and found 11 exclusions, corresponding to a rate of <1%. However, it may be that rates vary between populations: that in a Mexican population[10] was estimated at ~12%. It is certainly true that there is variation in the degree of acceptability of nonpaternity in different cultures. For example, among the Himba, a semi-nomadic people of northwest Namibia, 17.6% of children born within marriage were not fathered by the husband.[51]

The probability of paternity can be estimated confidently

The technical and statistical issues surrounding paternity testing are similar to those in profiling for other purposes. A paternity case usually employs profiling evidence (E) to compare these two hypotheses:

H_1: the alleged father is indeed the father; and

H_0: the father is some unknown, unrelated man.

Genotypes are determined for child, alleged father, and mother, and the likelihood ratio X/Y is calculated, where X and Y are:

X = P(observing child's genotype given adults' genotypes and assuming H_1), which can be written $P(E|H_1)$

Y = P(observing child's genotype given adults' genotypes and assuming H_0), or $P(E|H_0)$

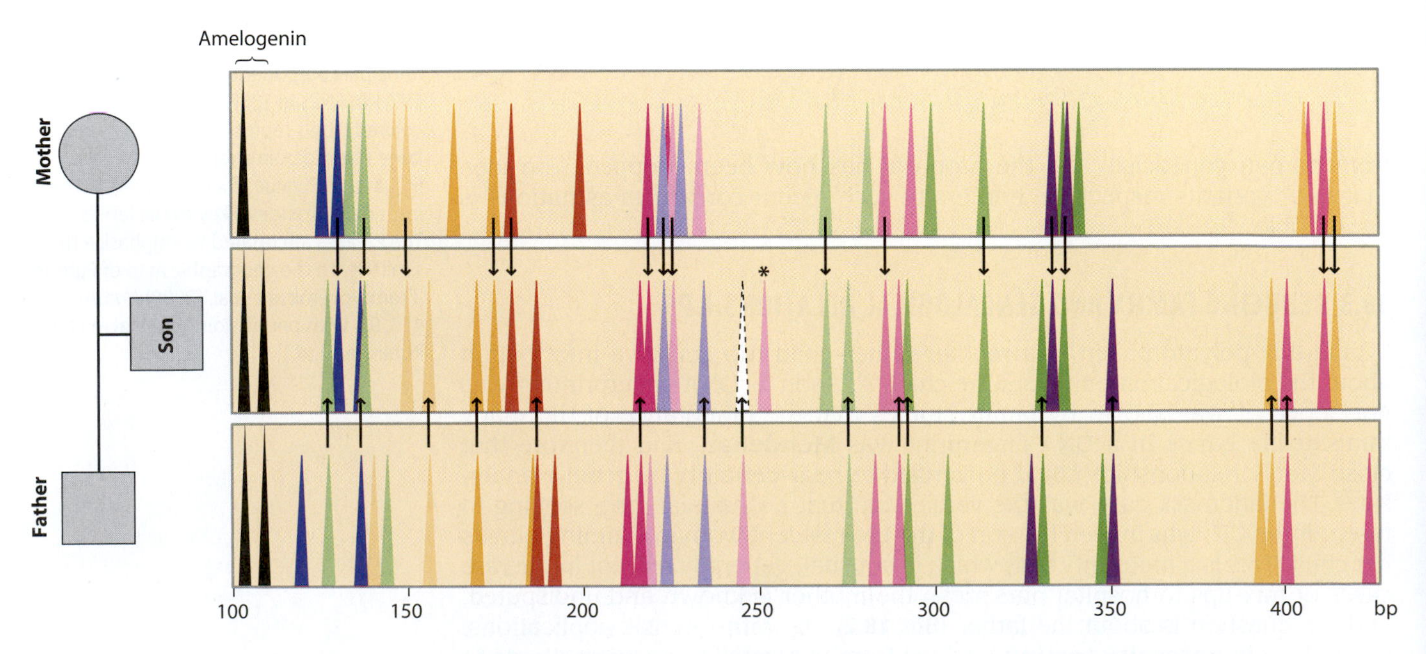

Figure 18.9: DNA-based paternity testing using a microsatellite multiplex.
Schematic electropherograms for a 15-locus microsatellite system are shown, with different colors representing different loci. The mother and father contribute one allele (peak) each to the son for each locus. One of the father's transmitted alleles has undergone a mutation (*asterisk*), and this must be taken into account when calculating the paternity index.

(a)

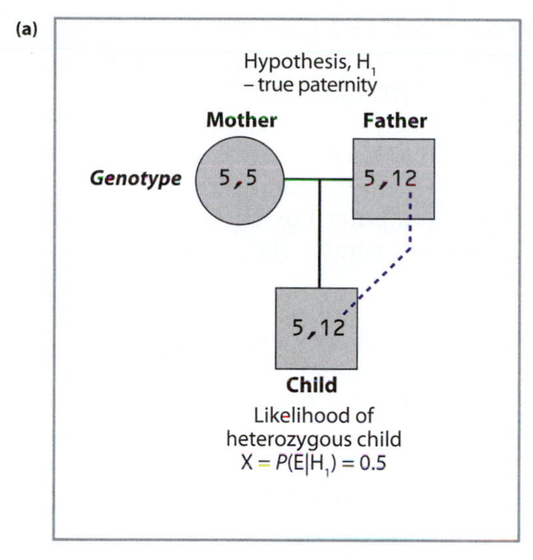

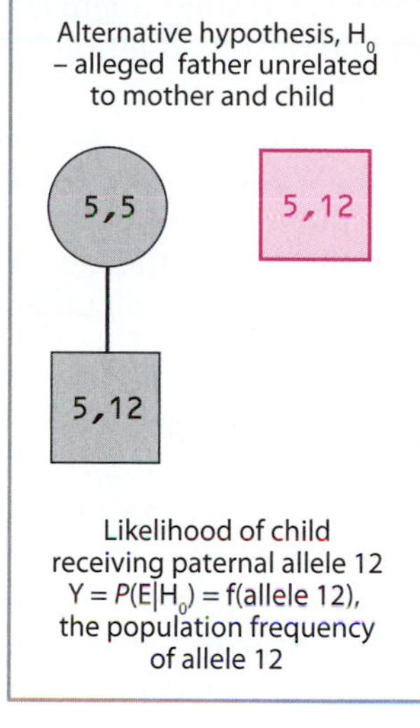

Likelihood ratio (paternity index):

$X/Y = P(E|H_1)/P(E|H_0)$

$= 1/[2f(\text{allele } 12)]$

Evidence in favor of true paternity strengthens as allele frequency becomes lower

(b)

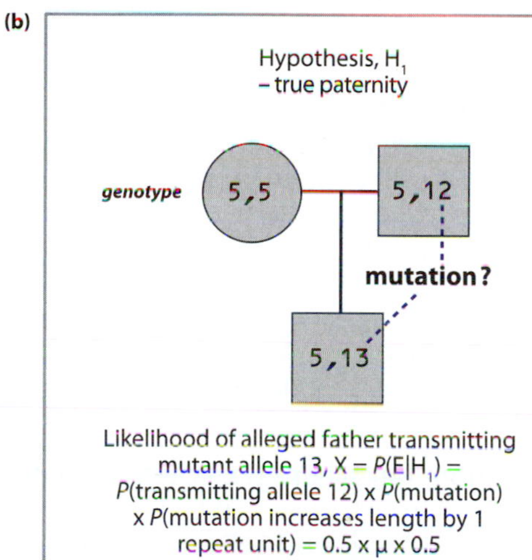

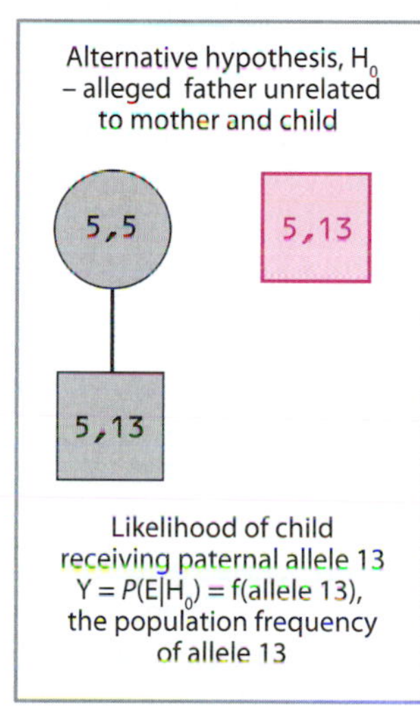

Likelihood ratio (paternity index):
$X/Y = P(E|H_1)/P(E|H_0)$

$= 1/[4f(\text{allele } 13)/\mu]$

Figure 18.10: Likelihoods in DNA-based paternity testing. Genotypes at a single autosomal microsatellite are shown, with numbers indicating repeat units. (a) Absence of mutation. Likelihoods for other allele combinations in the trio can be calculated. (b) Taking into account the possibility of single-step mutation in a paternal microsatellite allele. The assumption here is that a single-step increase or decrease is equally likely, and that other larger scale mutations can be neglected (as discussed in Section 3.4, this assumption is not very realistic). Formulae are available to take these rarer mutations into account. Note that average mutation rates, rather than allele-specific mutation rates are used.

X/Y, often known as the **paternity index**, is thus a measure of how much more readily the observed results are explained by a true relationship than by coincidence. A simple example using a single microsatellite is illustrated in **Figure 18.10a**, where the probability of true paternity increases with the rarity of a matching paternal allele. As long as the various loci tested are statistically independent, the paternity index for a set of loci is simply the product of the values for its constituents. Thus, when large numbers of microsatellites are used, probabilities of true paternity can become extremely high—of the order of a million-fold more likely than false paternity. Often, paternity casework requires a prior probability of true paternity to be stated, and by convention this is 0.5—the alleged father is equally likely to be the true father or not.

Because sexual relationships, even more than crimes, often occur among a fairly small group of people including family members, the possibility that the true father is actually a close relative of the suspected father (for example, a brother) needs to be considered. Disentangling the offspring of different possible incestuous relationships is even more complex, and computer algorithms are often used in these cases. Usually, an exclusion of a parent or child is more straightforward; however, there is a complicating factor here, too. A person who gives a sample of DNA to a forensic identification database leaves the same DNA at a crime scene. A parent who gives a DNA sample to a paternity-testing lab, however, will not necessarily have passed on exactly the same DNA to their child: there could have been a mutation (Figure 18.9 and Figure 18.10b). The mutation rates of the microsatellites used for forensic and paternity studies are low: paternity testing agencies report these mutations from casework (an amazing 18,000 observed mutations in almost 13 million allele transmissions), and for the 15 loci contained within the CODIS and SGM+ sets, rates vary between 0.01 and 0.22% per transmission (www.cstl.nist.gov/biotech/strbase/mutation.htm). What is more, the vast majority of confirmed mutations involve a single-step mutation rather than a multi-step change, and this property helps to distinguish them from mismatches due to nonpaternity. These factors can be taken into account when calculating the paternity index (Figure 18.10b).

Other aspects of kinship analysis

DNA-based kinship analysis is widely used in disaster victim identification, when conventional means are unreliable. In accidents and disasters, such as plane crashes or the 9/11 World Trade Center attacks, possible profiles of missing individuals can be reconstructed from relatives and reliably matched against profiles from human remains.[4] This is important in allowing the remains of victims to be returned to their families, and allowing death certificates to be issued. As relationships become more distant, however, the amount of shared DNA becomes less, and the degree of certainty also lessens. For a 10-locus profile, a parent contributes 10 alleles to a child; first cousins, on the other hand, share on average only 2.5 alleles by descent (and this varies substantially between cousin pairs), which is likely to provide no support for this relationship, given that individuals can often share some alleles by chance. Joint consideration of many different family members in calculating a likelihood ratio can improve success.[18] Familial searching of DNA databases, discussed in Section 18.1, is a variety of kinship analysis, in which the task is to discover, within a large database, those profiles most likely to belong to close relatives of a perpetrator. In practice, a kinship matching approach that takes into account the population frequency of matching alleles to construct a likelihood ratio, outperforms a simple allele-matching approach; the method is only practicable for first-degree relatives.[46]

For a male child whose paternity is in question, but where the potential father cannot be traced or is dead, Y-chromosomal microsatellites can be used (**Figure 18.11**). This is called **deficiency paternity testing**, and relies on matching the Y chromosome of the child with that of a potential grandfather or paternal uncle, for example. In the event of a match, the paternity is supported, with paternity index equal to the inverse of the frequency of the allele or haplotype (Figure 18.11a), but in the event of an exclusion, ambiguity remains, since it is impossible to determine in which generation the nonpaternity took place (see example in Figure 18.11b). Mutation at Y-specific microsatellites also needs to be taken into account, and the caveats stated above (**Section 18.1**) about population structure are again important. Maternal relationships within families can also be investigated using mtDNA. As has already been mentioned, victims of disasters can be identified, in effect, by maternity testing through mtDNA.

The Y chromosome and mtDNA are useful in genealogical studies

Because of their **uniparental inheritance**, Y chromosomes and mtDNA can provide a way to link individuals reliably across a large number of generations provided unbroken lines of patrilineal or matrilineal descent exist. Two examples illustrate this nicely: the identification of the remains of the Romanovs largely using mtDNA; and the investigation of the Thomas Jefferson paternity case using Y chromosomes.

The Thomas Jefferson paternity case

The Jefferson case is illustrated in **Figure 18.12**. The question to be addressed was whether Thomas Jefferson, third president of the USA, fathered any of the children of one of his slaves, Sally Hemings. Thomas's wife Martha predeceased him by 43 years after bearing him two daughters. For two of Sally's sons, Tom Woodson (claimed as a son of Sally in the oral history of his descendants) and Eston Hemings Jefferson, modern male-line descendants could be recruited.

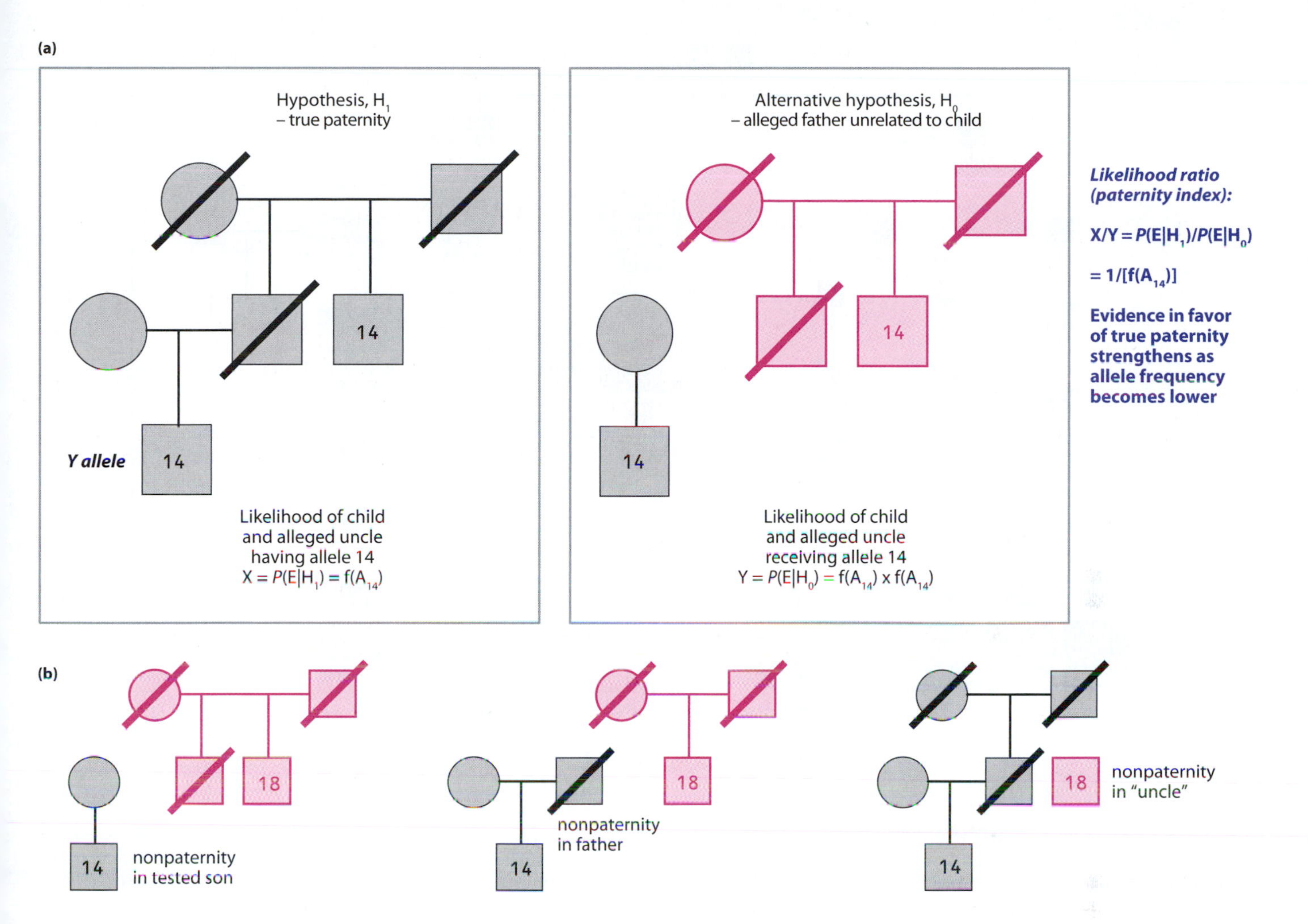

Figure 18.11: Likelihoods in deficiency paternity testing with a Y-chromosomal microsatellite.
(a) Evaluating the likelihood ratio (paternity index) in the case of a shared microsatellite allele between a son and alleged paternal uncle. (b) In the event of an exclusion that cannot easily be explained by a mutation (in this case, a four-step difference), there remains ambiguity about the nonpaternity event. Typing of further male-line relatives may resolve this.

Because Thomas did not have any legitimate sons of his own, modern male-line descendants of his paternal uncle, Field Jefferson, were used as a source of the Jefferson Y chromosome. As a defense against the observation that Sally's children resembled Thomas, his own legitimate descendants had claimed that the true father was one or both of Thomas's nephews Samuel and Peter Carr, the sons of his sister. To test this hypothesis, modern male-line descendants of the Carrs were also traced.

Using Y-specific **binary polymorphisms** (mostly SNPs), microsatellites, and a minisatellite, detailed Y-chromosomal haplotypes were constructed.[17] Four of the five Field Jefferson descendants shared the same Y-chromosomal haplotype (the fifth differed by one microsatellite repeat unit at one locus, most likely representing a mutation), and the Carr descendants shared a different haplotype. Therefore it could be asked which, if either, of these haplotypes Sally Hemings's descendants carried. The descendant of her last child, Eston, but not those of Tom Woodson, matched the Field Jefferson descendants, which was consistent with Thomas being Eston's father. The frequency of the Jefferson haplotype in the general population was difficult to estimate, but low: for example, the microsatellite haplotype was not observed in over 1200 other individuals analyzed. Subsequent analysis[34] showed the haplotype to belong to the generally rare Y-haplogroup T. Of course, any one of Thomas's contemporary patrilineal relatives, including his brother Randolph, could have been the true father, and it is therefore impossible to offer formal proof of paternity. However, historians have produced the important circumstantial evidence that Thomas and Sally were together at Monticello (Jefferson's home in Virginia) nine months before the birth of each of Sally's children.[23]

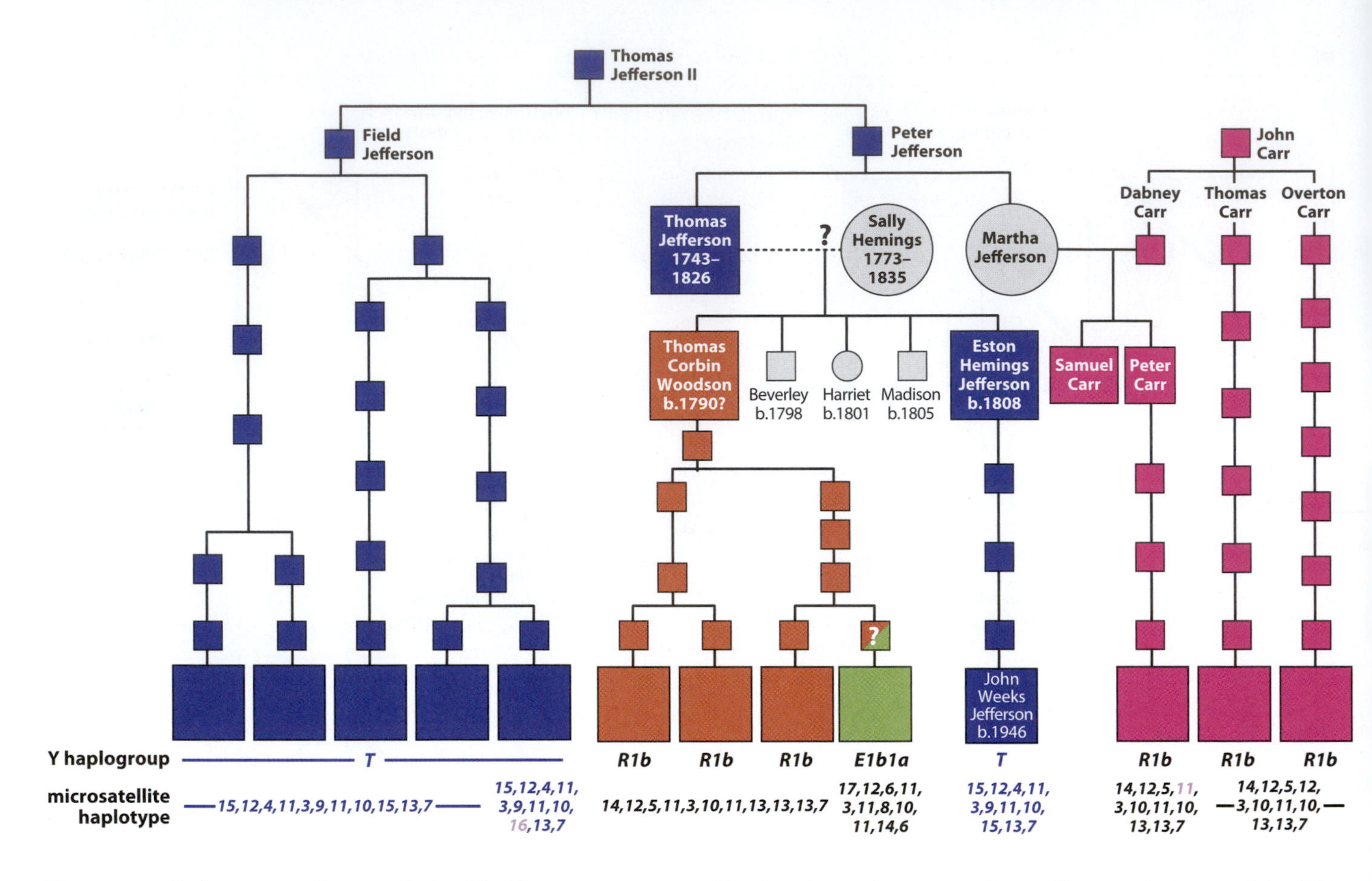

Figure 18.12: Y-chromosomal DNA analysis of the Thomas Jefferson paternity case.
Four of the Field Jefferson descendants and Eston Hemings Jefferson's descendant share an identical haplogroup-T Y chromosome, distinct from those in the Woodson and Carr descendants, and this is consistent with Thomas Jefferson's fathering Eston Hemings Jefferson, Sally Hemings's last child. Note that there is nonpaternity among the Woodson descendants, and a microsatellite mutation in one of the Field Jefferson descendants and the descendant of Peter Carr. Y haplogroup nomenclature has been updated according to Karafet TM et al. (2008) *Genome Res.* 18, 830 (see Appendix), and the microsatellite haplotype is the number of repeat units of each of 11 Y-specific loci. The MSY1 minisatellite was also typed (not shown). For details of polymorphisms, see Foster EA et al. (1998) *Nature* 396, 27.

DNA-based identification of the Romanovs

mtDNA analysis was famously applied[21, 27] to the question of establishing the identity of human remains thought to be those of Czar Nicholas II of Russia and his family (**Figure 18.13**), killed by the Bolsheviks in 1918. Nine skeletons were discovered in a communal grave in Ekaterinburg, Russia, in 1991. These had been tentatively identified as the remains of the Czar, the Czarina, three of their daughters, three servants, and the family's doctor. Sexing using the amelogenin test agreed with the skeletal analysis, and autosomal microsatellite profiling using five loci was consistent with the family relationship of five of the skeletons. Sequencing of mtDNA HVSI and **HVSII** showed that the Czarina and the three daughters had the same sequence, confirming the family relationship. A modern matrilineal relative of the Czarina, Prince Philip, the Duke of Edinburgh, had the same sequence also, supporting her identity. mtDNA analysis of the Czar's remains, however, produced a surprise: his sequence was identical to those of two living matrilineal relatives at all bases except one. Here, while the relatives possessed a T, the Czar had a mixture of T and C—a heteroplasmy. The remains of the Grand Duke of Russia Georgij Romanov, brother of the Czar, were later exhumed and he was also shown to carry the heteroplasmy.[27] This adds considerable weight to the conclusion that the Ekaterinburg remains included those of Czar Nicholas II, and also illustrates that mtDNA heteroplasmy can resolve within four generations to apparent homoplasmy.

Two children, Alexei and a daughter (presumed to be Anastasia), were missing from the grave, however, and there were suggestions that they survived. In 2007 a second grave was found nearby containing the burned bones of two individuals. These were tested using DNA methods adapted for very damaged material,[48] and it was possible to retrieve the entire mtDNA sequence, and also, from the male individual, a 17-locus Y microsatellite haplotype. Comparison of these (plus some autosomal profile evidence) with previously sampled skeletal material and living individuals, gave a picture consistent with the newly discovered remains being those of the two missing children.[11]

Y-chromosomal DNA has been used to trace modern diasporas

Genealogical studies such as those described above can be extended further back in time to encompass a lineage or set of lineages associated with a particular group. These studies are potentially useful in tracing the descendants of historical diasporas, such as that of the Jews, or that of people of African descent whose ancestors were taken as slaves to the Americas. Identification of a particular mtDNA or Y-chromosomal haplotype at high frequency in certain population groups can sometimes be taken as a "population marker," and commercial companies were quick to offer typing of these, to somehow legitimize a customer's membership of a group. It is important to remember that each of us has very many ancestors, and that this matrilineal or patrilineal tracing considers only a single ancestor out of many. Also, there are no ideal polymorphic variants that are present in all members of a population or group, and not present in others.

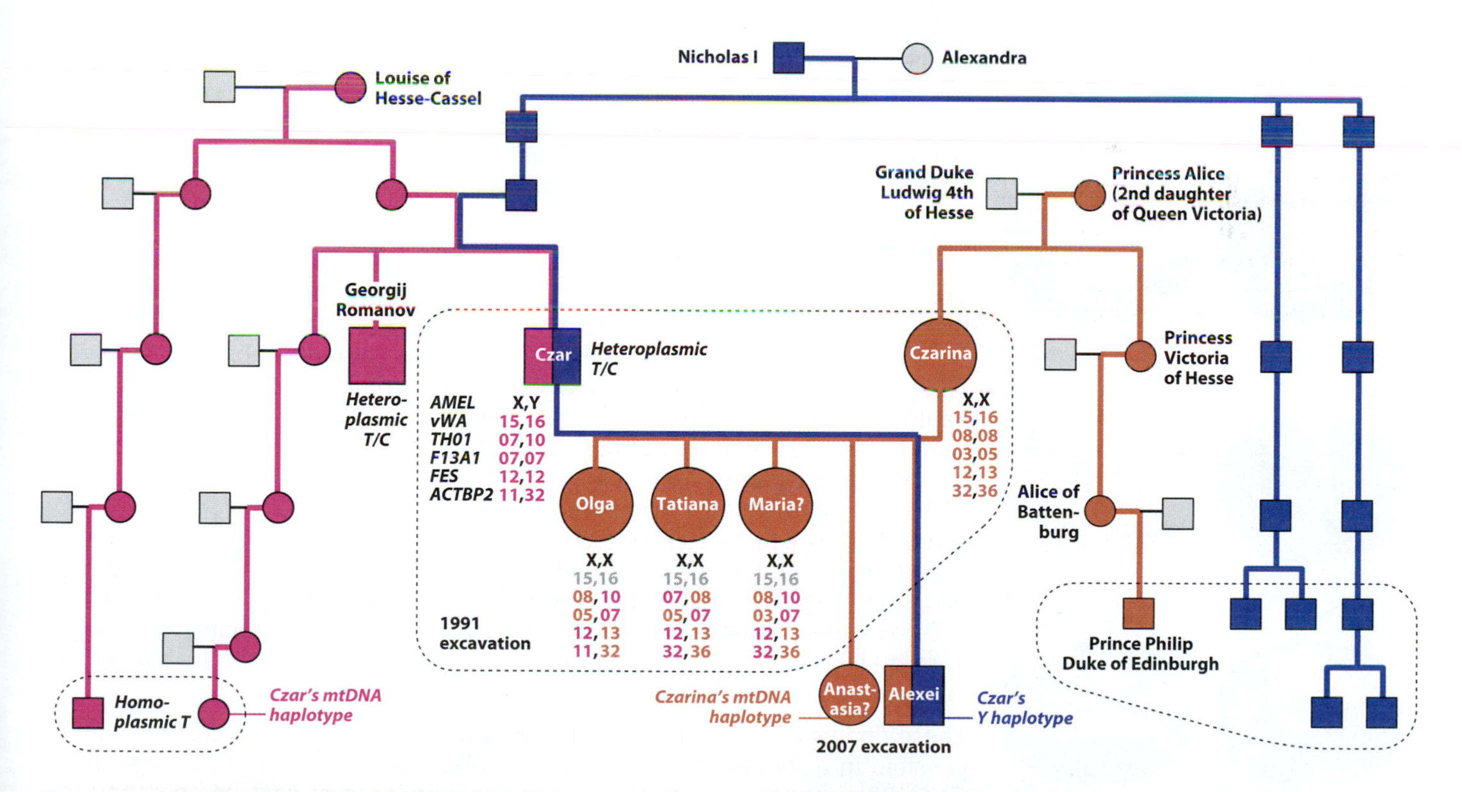

Figure 18.13: Confirmation of the identity of the Romanov remains by DNA analysis.
Autosomal microsatellite data confirm the family relationships of the five skeletons in the grave excavated in 1991, and exclude the other four adults as parents (not shown). Paternal alleles are shown in *magenta*, maternal in *orange*, and uninformative alleles in *gray*. Modern matrilineal relatives confirm the identity of the Czar and Czarina by mtDNA analysis. The Czarina's maternal lineage (haplogroup H) is shown by *orange* filled symbols, and that of the Czar (haplogroup T) in *magenta*. The Czar's exhumed brother, Georgij Romanov, shares with him a heteroplasmic position in mtDNA that is homoplasmic in two living relatives. The two skeletons found in the 2007 excavation were confirmed as the missing children using various analyses, including mtDNA analysis (*orange*), and the Romanov haplogroup R1b Y haplotype in Alexei (*blue*). Although Anastasia and Maria are both present among the complete set of remains, ambiguity remains as to which is which [Coble MD et al. (2009) *PloS One*, 4, e4838]. Tested living relatives are indicated by dotted ellipses. [Data from Gill P et al. (1994) *Nat. Genet.* 6, 130; Ivanov PL et al. (1996) *Nat. Genet.* 12, 417; and Rogaev EI et al. (2009) *Proc. Natl Acad. Sci. USA* 106, 5258.]

A particular six-locus Y-chromosomal microsatellite haplotype cluster within haplogroup J was shown to be at high frequencies in Jewish groups, and in members of the Cohanim priestly lineage in particular.[55] Priesthood status is patrilinearly inherited, and many men who claim Cohanim status have surname Cohen, or related names such as Kahn or Kane. The microsatellite diversity within the Cohen haplotype cluster was used to estimate a coalescence time for these Y chromosomes (see **Section 6.5**). The age reported, 2650 years (95% confidence interval, 2100–3250 years), seemed consistent with the Temple period in Jewish history, when the Cohanim were established following the exodus from Egypt. Subsequent work[25] has refined the lineage and its likely age: defined by the SNP P58 (rs34043621) within haplogroup J1 and a particular common haplotype for 17 Y microsatellites, it comprises 46% of self-defined Cohanim, and has an estimated TMRCA of 3190 ± 1090 years.

A study of the diversity of Y-chromosomal haplotypes in Asia revealed the previously unsuspected spread of a lineage that has a persuasive historical explanation.[60] A sublineage within haplogroup C*(xC3c), defined at high resolution with 16 microsatellites, was found in 16 populations throughout much of Asia, from the Pacific to the Caspian Sea, and made up ~8% of the Y chromosomes sampled from the region. The coalescence time of this **star-like** cluster of haplotypes was calculated at ~1000 years (95% confidence interval, 590–1300 years), and comparative diversity considerations suggested Mongolia as the source. The age, place of origin, and modern distribution of these chromosomes are consistent with their representing the Y-chromosomal lineage of Genghis Khan, his immediate male relatives, and their descendants. This fearsome emperor, who lived between circa 1162 and 1227, established a long-lasting male dynasty that ruled large parts of Asia for many generations. A Y chromosome that may have been carried by him is now carried by about 0.5% of the world's men, representing a spectacular founder effect that may have an impact upon allele frequencies of loci in other chromosomes. Further examples of such **socially selected** lineages have been found elsewhere in Asia (the Manchu lineage[59]), and in Ireland (the Uí Neíll lineage[41]).

Y-chromosomal haplotypes tend to correlate with patrilineal surnames

The patrilineal links indicated by shared Y-chromosomal haplotypes suggest that surnames, also usually patrilineal, might act as cultural markers for shared ancestry. This is clear enough in the Jefferson case, described above, but is it generally true? Do all men who share a patrilineal surname share a Y-chromosomal haplotype, descending from a founding male? A survey of the Y haplotypes of sets of unrelated men belonging to 40 British surnames[35] shows that, while the commonest surname, *Smith*, carries a variety of Y haplotypes resembling that of the general population, rarer names often show remarkably low Y haplotype diversity (**Figure 18.14**). Some, such as *Attenborough*, are almost fixed for a single haplotype. While these findings might suggest that rare names may each have had a single founder at the time of surname establishment (about 700 years ago in Britain), genetic drift in Y lineages is so marked that this conclusion would be unsafe—it could be that there were many founders, and all but one lineage has drifted to extinction.

These findings have a number of applications.[36] Surname prediction may be possible in a forensic setting, given suitable databases, and haplotype diversity within surnames can illuminate the parameters of past demography and

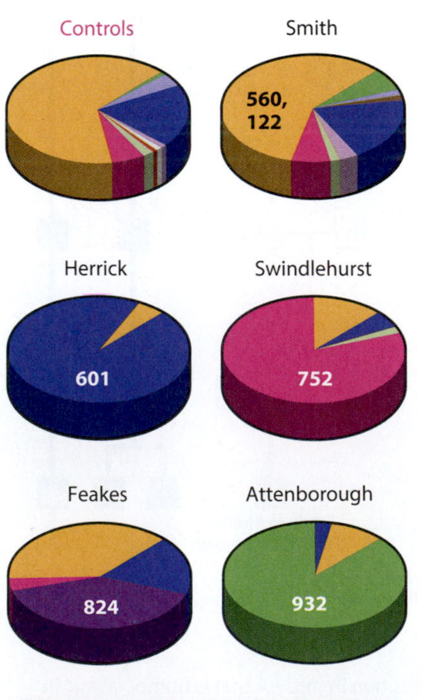

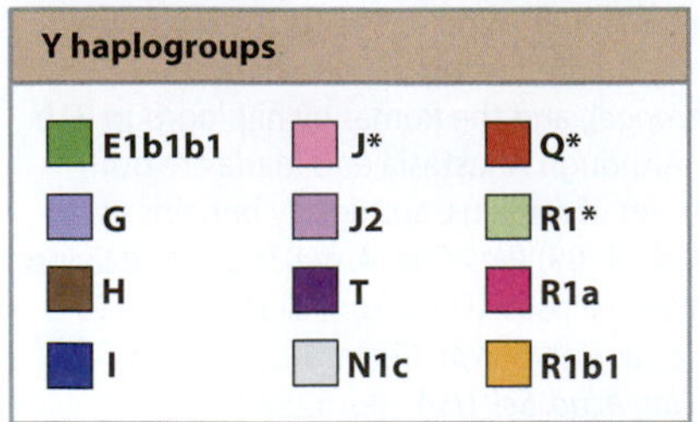

Figure 18.14: Y-chromosomal haplogroup diversity in British surnames. The pie charts illustrate the haplogroup frequencies found in a group of 110 men bearing different surnames (Controls), and in five groups of men each sharing a surname (sample sizes 20–58). Numbers of bearers (in England and Wales) of each surname are shown within the pie charts. *Smiths* resemble Controls, but the other, less frequent names show low haplogroup diversity reflecting a high degree of co-ancestry. [From King TE & Jobling MA (2009) *Mol. Biol. Evol.* 26, 1093. With permission from Oxford University Press.)]

migration. Sampling males based on the historical location of their surnames can provide a set of Y haplotypes that differs from current distributions,[5] and more closely reflects the patterns that existed before the extensive population movement that followed the Industrial Revolution. Most work to date on surname–Y-chromosome relationships has been done in the British Isles, and different rules may apply elsewhere in the world.

18.4 THE PERSONAL GENOMICS REVOLUTION

When an individual exploits the power of genomic analysis to learn about themselves, they are engaging in personal genomics. Since the first edition of this book was published in 2004, the haplotype structure of the genome has been investigated, stimulating technological development in methods of genomewide SNP analysis for medical genetic research into common, complex diseases—as a consequence one million SNPs can currently be analyzed for a few hundred dollars (**Section 4.5**). At the same time, remarkable technical innovations have allowed the sequencing of whole human genomes (**Section 4.4**), as we write at a cost of a few thousand dollars, and falling fast. These developments have made personal genomics possible, and companies have sprung up to satisfy consumer demand: with a credit card and an internet connection, anyone can now instigate a detailed investigation of their own genome, with all its promise and problems (**Opinion Box 17**).

The first personal genetic analysis involved the Y chromosome and mtDNA

The first forays of members of the public into genetic analysis began with investigations of the uniparentally inherited Y chromosome and mtDNA, driven by interest in ancestry and family history. Companies with names such as *Family Tree DNA*, *Oxford Ancestors*, *Relative Genetics*, *Ethnoancestry*, and *DNA Heritage* began by offering mtDNA HVSI sequencing and the typing of 10–25 Y-chromosomal microsatellites, but through competition and public demand have increased their resolution to whole mtDNA sequences, as many as 111 Y microsatellites, sets of Y-SNPs, and the sequencing of a ~200-kb segment of Y-chromosomal DNA for SNP discovery. The link between Y haplotypes and surnames has been exploited with great enthusiasm in "surname projects" (**genetic genealogy**), and these activities are feeding into academic research.[36] Public databases of Y haplotypes and surnames (for example, www.ysearch.org) contain information on over 100,000 people.

Personal genomewide SNP analysis is used for ancestry and health testing

Personal genetic testing using SNP chips (**Section 4.5**) designed for medical genetic research is currently offered by several companies, including *23andMe* and *deCODEme*. These companies provide information on ancestry to their clients, through comparing their genotypes with large databases of genomewide SNP variation in different populations, and in some cases seeking to discover relatives by identifying large tracts of shared genetic variation. However, these services are increasingly focused on health-related variation. Because the ≤1.2 million SNPs on these chips were chosen under a tag SNP design (**Section 4.5**), inferences about traits and phenotypes are in many cases indirect—based on linkage disequilibrium between a tag SNP and a causative variant that is either known but not itself present on the chip, or unknown. To address this, additional sets of SNPs including causative variants are often analyzed, and at the time of writing one company, *23andMe*, is piloting an exome sequencing service—the analysis by targeted next-generation sequencing of a ~50-Mb set of DNA sequences including most annotated protein-coding exons.

Personal genome sequencing provides the ultimate resolution

In principle, whole-genome sequencing should provide access to all the genetic variation that influences traits in an individual (aside from those in

Over the last five years genomics has been dominated by exponential advances in technology, as generating genetic data has become cheaper, faster, and easier at a rate that has shocked even the most optimistic of researchers. This pace of change promises to make genomic information ever more accessible, and not just to researchers and clinicians: through direct-to-consumer genetic testing and even do-it-yourself "genome hacking," it is now possible for anyone with sufficient motivation and disposable income to take a peek into their own DNA.

Such ready access to genomic information promises to transform medicine, by making it easier for clinicians to diagnose and predict both diseases and adverse drug responses. However, it will also make it easier for individuals to share their genetic information, either publicly or with specific commercial entities. That possibility has spurred a heated debate about the benefits and risks of genetic sharing.

Advocates of "public genomics" argue that genetic data sharing can be a public good, making it easier for researchers to find new genetic associations with disease. One non-profit venture, the Personal Genome Project (http://www.personalgenomes.org), seeks to create a fully open database of both genetic and medical information from over 100,000 volunteers to drive the discovery of new association between genetic variants and disease. On a smaller scale, the members of the Genomes Unzipped project (http://www.genomesunzipped.org), of which I am a founder, have publicly shared their results from a number of direct-to-consumer genetic tests to help foster informed discussion of the accuracy and utility of such tests.

Genetic sharing also has corporate advocates. One Silicon Valley-based company, *23andMe*, offers its customers their genotypes at a million sites of known variation for as little as $100. In return, the customers can volunteer both their genetic and phenotypic data (in the form of online survey answers) to the company's research arm, *23andWe*. They can also choose to share their data with other customers in a social network. The long-term goal of the company appears to be primarily the discovery of novel markers for diseases and drug responses that can be commercialized through partnerships with pharmaceutical and biotech companies.

Whether or not *23andMe*'s model will prove a commercial success remains to be seen, but there is little doubt that it has proved its worth as a mechanism for driving research: the company now boasts a database of over 100,000 genotyped individuals, at least 60,000 of whom have answered research surveys, and has published novel genetic associations for traits as diverse as hair curliness[15] and Parkinson disease.[12] Translating such discoveries into profit will prove challenging, as demonstrated by the 2009 bankruptcy of Iceland's *deCODE Genetics*, but the benefits of *23andMe*'s participant-centric approach to research have not been lost on academic researchers.

Critics are keen to point out that genetic sharing will sometimes bring unexpected consequences to participants. Some will be approached by distant relatives, while others will learn they are not in fact related to people they saw as biological kin. Some may discover, after their genomes have been shared across the Web, that they have serious disease-causing mutations they would rather have kept to themselves. Open databases of genetic information could be mined by investigators looking for matches with crime-scene DNA or with the genetic profiles of children in contested paternity cases. Those who publicly share their genetic data have also indirectly shared the genetic information of their relatives, so some of these consequences may be borne by their families as well.

In my view these risks are generally overblown, and could be further reduced by legislation that prevents discriminatory use of genetic information, such as the Genetic Information Nondiscrimination Act legislation passed by US Congress in 2008.[19] Researchers and companies should be clear with participants about the potential risks of sharing their information. However, if fully informed individuals choose to share their data, either out of altruism or for individual gain (such as access to services that help them interpret their genomes), they should not be prevented from doing so. Certainly researchers would gain much from access to large and fully public databases of genetic and health information like that proposed by the Personal Genome Project.

So, in an era where online sharing of personal information has become increasingly unremarkable, will genetic data continue to be viewed as fundamentally private? Only time will tell.

Daniel G. MacArthur *Department of Medicine, Massachusetts General Hospital, Boston, USA*

regions inaccessible to current technologies). J. Craig Venter, founder of Celera Genomics and driver of the commercially funded competitor of the publicly funded **Human Genome Project**, was the first identified person to have his own diploid genome analyzed, publishing the sequence and designating it (with characteristic modesty) HuRef. The Venter sequence used conventional (Sanger sequencing) technology, and for this reason attracted less attention than the subsequent publication of the next-generation genome sequence of another famous scientist, James D. Watson (**Table 18.1**).

In both cases their genome sequences contained many variants with likely functional consequences, but the vast majority of these were heterozygous and probably without noticeable effects. However, several variants were expected to influence phenotype—some of these are shown in Table 18.1, including examples that are discordant with the known phenotypes of Venter and Watson, and which may indicate sequencing errors or, more interestingly, our incomplete understanding of the penetrance of these alleles. Some variants are apparently straightforward and predictable in their effects, such as the *ABCC11* variant affecting the type of earwax (see Table 15.1). However, the contributions of variants to complex phenotypes such as behavior remain unclear—for example, Venter's *DRD4* (dopamine receptor D4) gene contains variants that both increase and decrease his probability of exhibiting **novelty-seeking** behavior. Watson has a family history of Alzheimer disease, and decided to redact the region containing the *APOE* gene, because certain alleles enhance risk of the disease. The redacted region spans ~2 Mb in order to prevent the inference of *APOE* genotypes from variants in linkage disequilibrium with it.[43]

George Church, another famous scientist who has had his own genome sequenced, has founded the Personal Genome Project (www.personalgenomes.org), which aims to enroll 100,000 public participants, and requires candidates to pass an entrance exam in genetics. Companies such as *Complete Genomics* offer direct-to-consumer genome sequencing at prices that are within the reach of a large number of people.

TABLE 18.1:
COMPARISON OF SOME FEATURES OF THE VENTER AND WATSON PERSONAL GENOMES

Individual sequenced	J. Craig Venter	James D. Watson
Reference	Levy S et al. (2007) *PLoS Biol.* 5, e254	Wheeler DA et al. (2008) *Nature* 452, 872
URL for data	huref.jcvi.org	jimwatsonsequence.cshl.edu
Technology	Capillary electrophoresis (Sanger)	454
Total estimated cost	~US$100 million	<US$1 million
Coverage	7.5×	6×
Discordance based on SNP chip data	0.12%	~1%
No. of reported nonsynonymous SNPs compared to ref. seq.	3882	3766
Redacted loci	none	*APOE* (Alzheimer disease; hyperlipidemia)
Examples of predicted phenotypes*	Alzheimer disease (*APOE*; *SORL1*; OMIM 104300) Hyperlipidemia (*APOE*; OMIM 107741) Wet earwax type (*ABCC11*; OMIM 117800) Lactase persistence (*LCT* rs4988235; OMIM 223100)—disagrees with Venter's phenotype Novelty seeking (*DRD4* rs180095; OMIM 601696) non-Novelty seeking (*DRD4* short VNTR alleles; OMIM 601696)	Duffy antigen absent (*FY*; OMIM 110700) Cockayne syndrome type B (*ERCC6*; OMIM 133540—disagrees with Watson's phenotype Usher syndrome 1b (*MYO7A*; OMIM 276900)—disagrees with Watson's phenotype Type 2 diabetes increased risk (*SLC2A2*; OMIM 125853) Lactase persistence (*LCT* rs4988235; OMIM 223100)

* Not all SNPs are annotated in both genomes.

Personal genomics offers both promise and problems

With the current state of knowledge, personal genomics reveals a mass of variation that we do not understand—to improve on this, many more genome sequences and much better information about phenotypes in both healthy and unhealthy people are needed.[28] There is, however, some unambiguous predictive information. While earwax type is more easily diagnosed with a cotton-wool bud than a genetic test, information about the presence of some disease-susceptibility alleles that have high penetrance will be useful, and most clearly so if something can be done to prevent the disease, or to detect and treat it early. Similarly, information about drug sensitivities may prove useful (**Section 17.5**), perhaps decades later, if the individual comes to need medical treatment; an example is *23andMe*'s test for variants affecting response to the anti-coagulant warfarin. When there is little to be done about an enhanced disease risk, as in the case of Alzheimer disease susceptibility, the knowledge may be a source of anxiety for a testee. For most traits, susceptibility alleles only explain a small proportion of the variance in a trait in population studies, and these may not be useful for predicting individual outcomes. Information on the disease risks of any individual has implications for their close relatives, with whom they share DNA, and yet these individuals are not usually asked to consent to the analysis—Watson did not ask his sons' permission to publish his genome sequence, for example. Is an individual always free to have their genome sequenced, even if they have an identical twin who might not want this to happen?

Return of results to participants in personal genomics is usually direct from the testing company, and does not involve a health professional (the company *Navigenics* is an exception). There is a danger here that participants will find the results meaningless, overwhelming, or unnecessarily worrying. The sequencing or genotyping process is not error-free, and our current understanding of variant consequences is far from perfect, as Venter and Watson discovered, so some predictions will be erroneous. Also, predictions based on linkage disequilibrium may not be similarly accurate in all populations, so if a testee's population of origin is an unusual one not tested in **genomewide** association studies, the risk may be wrongly estimated. There are also issues about data confidentiality—testees may want to keep disease susceptibilities private, but can they be sure that information about their genome will not be accidentally left on a train, released, sold, or stolen? Participants should be aware that confidentiality cannot be guaranteed.

Another name for personal genomics is **recreational genomics**, and this reflects the motivation of many participants. They do it out of curiosity, to learn about their personal evolutionary genetic history: their phenotypic quirks, their links to close and more distant relatives, including famous people (most of them currently geneticists), and less celebrated relatives like chimpanzees. They gain insight into their ancestors—from the last few generations, back to Neolithic farmers, the first humans, and even Neanderthals. The increasing public literacy about DNA that began with Y haplotypes and mtDNA is extending to the rest of the genome, aided by open-source genotyping analysis tools like **SNPedia** (www.snpedia.com) and **Promethease** (www.snpedia.com/index.php/Promethease) that anyone can use. Crowd-sourcing of self-funded research participants is revealing interesting associations between genetic variants and phenotypes that would be difficult to find in other ways.[15] Though we do not know where the coming tide of genomes will take us, we should be heartened by this "citizen science." Soon, perhaps, we will all be evolutionary geneticists.

SUMMARY

- There are a vast number of differences between any one copy of the human genome and another (except in identical twins), providing the potential for individual identification based on DNA.
- Forensic application of DNA-based identification is based on PCR multiplexes analyzing a set of autosomal microsatellites (DNA profile), and gives a very low probability that a profile chosen at random from the population will match that from a given individual.
- Very large forensic identification databases have been established containing DNA profiles of offenders and some non-offenders. These are powerful investigative tools.
- Y-chromosomal and mtDNA haplotypes are shared by members of a patriline or matriline, respectively, but have specialized uses in forensic analysis.
- Prediction of sex from a DNA sample is almost always reliable. Predictions of eye-color and hair-color phenotypes are becoming increasingly reliable.
- Prediction of population of origin may be of value if polymorphisms are carefully chosen and populations are not recently admixed, but has limited geographical resolution.
- DNA profiles can also be used for establishing family relationships among DNA samples, as in paternity testing. Comparisons must take into account the possibility of mutation, as well as the population frequencies of alleles.
- Y-chromosomal and mtDNA haplotypes can be used to follow lineages in genealogical cases, and to trace population diasporas in historical times. The association of Y haplotypes and surnames is significant, and has been exploited in a number of applications.
- The technological revolution in genotyping and sequencing has led to the era of personal genomics, in which individuals can have their genomes analyzed or sequenced, and may choose to release the data publicly. The predictive power of the information for health risks is currently weak, and there are ethical issues that need to be addressed.

QUESTIONS

Question 18-1: Use the allele frequencies given in Figure 18.4 to estimate the population frequency of the genotype: 7,14; 8,10; 14,19 respectively for the three microsatellites CSF1PO, D13S317, and vWA.

Question 18-2: Analysis of the three autosomal microsatellites CSF1PO, D13S317, and vWA respectively in a paternity case gave the following genotypes: child, 10,14; 11,12; 15,18; mother, 10,10; 9,11; 16,18; putative father, 12,14; 12,13; 14,15. Use the frequencies given in Figure 17.4 to estimate the paternity index in this case.

Question 18-3: A crime-scene sample gives the Identifiler® profile 13/14, 30/30, 10/11, 10/12, 15/17, 9/9, 11/12, 9/12, 19/23, 13/14, 17/18, 11/11, 14/15, 12/12, 22/24, for the loci D8S1179, D21S11, D7S820, CSF1PO, D3S1358, TH01, D13S317, D16S539, D2S1338, D19S433, vWA, TPOX, D18S51, D5S818, and FGA, respectively. Use sources mentioned in the text to predict the continent of origin of the sample donor.

Question 18-4: Three anonymous DNA samples have the following genotypes for eight SNPs typed by a personal genomics company:

Sample no.	1	2	3
rs4988235	G/G	G/G	A/G
rs2814778	T/T	C/C	T/T
rs1426654	G/G	G/G	A/A
rs17822931	T/T	C/C	T/C
rs77121243	T/T	T/A	T/T
rs3827760	C/C	T/T	T/T
rs7903146	C/C	C/T	T/T
rs1805009	G/G	G/G	C/C
rs2032597	no result	A/A	C/C

Use SNPedia, and Ensembl or OMIM, to predict the phenotypic consequences of these variants for each DNA donor; Chapters 15 and 16 may help. Also, visit the Websites of some personal genomics companies (see text for examples) to see what they say about these variants. What guess would you make about each DNA donor's continent of origin?

Question 18-5: Professor Pangloss is delighted to win a voucher allowing him to have his genome sequenced. His pessimistic sister Cassandra asks him not to go ahead. What does he do? Justify your answer.

REFERENCES

The references highlighted in purple are considered to be important (for this chapter) by the authors.

1. **Anderson KG** (2006) How well does paternity confidence match actual paternity? Evidence from worldwide nonpaternity rates. *Curr. Anthropol.* **47**, 513–520.
2. **Ballantyne KN, Goedbloed M, Fang R et al.** (2010) Mutability of Y-chromosomal microsatellites: rates, characteristics, molecular bases, and forensic implications. *Am. J. Hum. Genet.* **87**, 341–353.
3. **Betz A, Bassler G, Dietl G et al.** (2001) DYS STR analysis with epithelial cells in a rape case. *Forensic Sci. Int.* **118**, 126–130.
4. **Biesecker LG, Bailey-Wilson JE, Ballantyne J et al.** (2005) Epidemiology. DNA identifications after the 9/11 World Trade Center attack. *Science* **310**, 1122–1123.
5. **Bowden GR, Balaresque P, King TE et al.** (2008) Excavating past population structures by surname-based sampling: the genetic legacy of the Vikings in northwest England. *Mol. Biol. Evol.* **25**, 301–309.
6. **Brandstatter A, Niederstatter H, Pavlic M et al.** (2007) Generating population data for the EMPOP database – an overview of the mtDNA sequencing and data evaluation processes considering 273 Austrian control region sequences as example. *Forensic Sci. Int.* **166**, 164–175.
7. **Branicki W, Liu F, van Duijn K et al.** (2011) Model-based prediction of human hair color using DNA variants. *Hum. Genet.* **129**, 443–454.
8. **Brock DJH & Shrimpton AE** (1991) Non-paternity and prenatal genetic screening. *Lancet* **338**, 1151.
9. **Butler JM** (2012) Advanced Topics in Forensic DNA Typing: Methodology, 2nd ed. Elsevier.
10. **Cerda-Flores RM, Barton SA, Marty-Gonzalez LF et al.** (1999) Estimation of nonpaternity in the Mexican population of Nuevo Leon: a validation study with blood group markers. *Am. J. Phys. Anthropol.* **109**, 281–293.
11. **Coble MD, Loreille OM, Wadhams MJ, et al.** (2009) Mystery solved: the identification of the two missing Romanov children using DNA analysis. *PLoS One* **4**, e4838.
12. **Do CB, Tung JY, Dorfman E et al.** (2011) Web-based genome-wide association study identifies two novel loci and a substantial genetic component for Parkinson's Disease. *PLoS Genet.* **7**, e1002141.
13. **Egeland T, Bovelstad HM, Storvik GO & Salas A** (2004) Inferring the most likely geographical origin of mtDNA sequence profiles. *Ann. Hum. Genet.* **68**, 461–471.
14. **Egeland T, Fonnelop AE, Berg PR et al.** (2012) Complex mixtures: a critical examination of a paper by Homer et al. *Forensic Sci. Int. Genet.* **6**, 64–69.
15. **Eriksson N, Macpherson JM, Tung JY et al.** (2010) Web-based, participant-driven studies yield novel genetic associations for common traits. *PLoS Genet.* **6**, e1000993.
16. **Foreman LA & Evett IW** (2001) Statistical analyses to support forensic interpretation for a new ten-locus STR profiling system. *Int. J. Legal Med.* **114**, 147–155.
17. **Foster EA, Jobling MA, Taylor PG et al.** (1998) Jefferson fathered slave's last child. *Nature* **396**, 27–28.
18. **Ge J, Budowle B & Chakraborty R** (2010) DNA identification by pedigree likelihood ratio accommodating population substructure and mutations. *Investig. Genet.* **1**, 8.
19. Genetic Information Nondiscrimination Act of 2008, H.R. 493, 110th Congress, 2nd Sess. US Government (2008).
20. **Gershaw CJ, Schweighardt AJ, Rourke LC & Wallace MM** (2011) Forensic utilization of familial searches in DNA databases. *Forensic Sci. Int. Genet.* **5**, 16–20.
21. **Gill P, Ivanov PL, Kimpton C et al.** (1994) Identification of the remains of the Romanov family by DNA analysis. *Nat. Genet.* **6**, 130–135.
22. **Gill P, Jeffreys AJ & Werrett DJ** (1985) Forensic application of DNA "fingerprints." *Nature* **318**, 577–579.
23. **Gordon-Reed A** (1997) Thomas Jefferson and Sally Hemings: an American Controversy. University Press of Virginia.
24. **Grimes EA, Noake PJ, Dixon L & Urquhart A** (2001) Sequence polymorphism in the human melanocortin 1 receptor gene as an indicator of the red hair phenotype. *Forensic Sci. Int.* **122**, 124–129.
25. **Hammer MF, Behar DM, Karafet TM et al.** (2009) Extended Y chromosome haplotypes resolve multiple and unique lineages of the Jewish priesthood. *Hum. Genet.* **126**, 707–717.
26. **Homer N, Szelinger S, Redman M et al.** (2008) Resolving individuals contributing trace amounts of DNA to highly complex mixtures using high-density SNP genotyping microarrays. *PLoS Genet.* **4**, e1000167.
27. **Ivanov PL, Wadhams MJ, Roby RK et al.** (1996) Mitochondrial DNA sequence heteroplasmy in the Grand Duke of Russia Georgij Romanov establishes the authenticity of the remains of Tsar Nicholas II. *Nat. Genet.* **12**, 417–420.
28. **Janssens AC & van Duijn CM** (2010) An epidemiological perspective on the future of direct-to-consumer personal genome testing. *Investig. Genet.* **1**, 10.
29. **Jeffreys AJ, Brookfield JFY & Semeonoff R** (1985) Positive identification of an immigration test-case using human DNA fingerprints. *Nature* **317**, 818–819.
30. **Jeffreys AJ, Wilson V & Thein SL** (1985) Individual-specific "fingerprints" of human DNA. *Nature* **316**, 76–79.
31. **Jobling MA & Gill P** (2004) Encoded evidence: DNA in forensic analysis. *Nat. Rev. Genet.* **5**, 739–751.

32. **Kayser M & de Knijff P** (2011) Improving human forensics through advances in genetics, genomics and molecular biology. *Nat. Rev. Genet.* **12**, 179–192.

33. **Kayser M, Kittler R, Erler A et al.** (2004) A comprehensive survey of human Y-chromosomal microsatellites. *Am. J. Hum. Genet.* **74**, 1183–1197.

34. **King TE, Bowden GR, Balaresque PL et al.** (2007) Thomas Jefferson's Y chromosome belongs to a rare European lineage. *Am. J. Phys. Anthropol.* **132**, 584–589.

35. **King TE & Jobling MA** (2009) Founders, drift and infidelity: the relationship between Y chromosome diversity and patrilineal surnames. *Mol. Biol. Evol.* **26**, 1093–1102.

36. **King TE & Jobling MA** (2009) What's in a name? Y chromosomes, surnames and the genetic genealogy revolution. *Trends Genet.* **25**, 351–360.

37. **King TE, Parkin EJ, Swinfield G et al.** (2007) Africans in Yorkshire? The deepest-rooting clade of the Y phylogeny within an English genealogy. *Eur. J. Hum. Genet.* **15**, 288–293.

38. **Lango Allen H, Estrada K, Lettre G et al.** (2010) Hundreds of variants clustered in genomic loci and biological pathways affect human height. *Nature* **467**, 832–838.

39. **Lao O, Vallone PM, Coble MD et al.** (2010) Evaluating self-declared ancestry of U.S. Americans with autosomal, Y-chromosomal and mitochondrial DNA. *Hum. Mutat.* **31**, E1875–1893.

40. **Liu F, van der Lijn F, Schurmann C et al.** (2012) A genome-wide association study identifies five loci influencing facial morphology in Europeans. *PLoS Genet.* **8**, e1002932.

41. **Moore LT, McEvoy B, Cape E et al.** (2006) A Y-chromosome signature of hegemony in Gaelic Ireland. *Am. J. Hum. Genet.* **78**, 334–338.

42. **Novembre J, Johnson T, Bryc K et al.** (2008) Genes mirror geography within Europe. *Nature* **456**, 98–101.

43. **Nyholt DR, Yu CE & Visscher PM** (2009) On Jim Watson's APOE status: genetic information is hard to hide. *Eur. J. Hum. Genet.* **17**, 147–149.

44. **Phillips C, Fernandez-Formoso L, Garcia-Magarinos M et al.** (2011) Analysis of global variability in 15 established and 5 new European Standard Set (ESS) STRs using the CEPH human genome diversity panel. *Forensic Sci. Int. Genet.* **5**, 155–169.

45. **Phillips C, Salas A, Sanchez JJ et al.** (2007) Inferring ancestral origin using a single multiplex assay of ancestry-informative marker SNPs. *Forensic Sci. Int. Genet.* **1**, 273–280.

46. **Reid TM, Baird ML, Reid JP et al.** (2008) Use of sibling pairs to determine the familial searching efficiency of forensic databases. *Forensic Sci. Int. Genet.* **2**, 340–342.

47. **Roewer L** (2009) Y chromosome STR typing in crime casework. *Forensic Sci. Med. Pathol.* **5**, 77–84.

48. **Rogaev EI, Grigorenko AP, Moliaka YK et al.** (2009) Genomic identification in the historical case of the Nicholas II royal family. *Proc. Natl Acad. Sci. USA* **106**, 5258–5263.

49. **Santos FR, Pandya A & Tyler-Smith C** (1998) Reliability of DNA-based sex tests. *Nat. Genet.* **18**, 103.

50. **Sasse G, Muller H, Chakraborty R & Ott J** (1994) Estimating the frequency of non-paternity in Switzerland. *Hum. Hered.* **44**, 337–343.

51. **Scelza BA** (2011) Female choice and extra-pair paternity in a traditional human population. *Biol. Lett.* **7**, 889–891.

52. **Steinlechner M, Berger B, Niederstatter H & Parson W** (2002) Rare failures in the amelogenin sex test. *Int. J. Legal Med.* **116**, 117–120.

53. **Sullivan KM, Mannucci A, Kimpton CP & Gill P** (1993) A rapid and quantitative DNA sex test: fluorescence-based PCR analysis of X-Y homologous gene amelogenin. *Biotechniques* **15**, 636–638.

54. **Thangaraj K, Reddy AG & Singh L** (2002) Is the amelogenin gene reliable for gender identification in forensic casework and prenatal diagnosis? *Int. J. Legal Med.* **116**, 121–123.

55. **Thomas MG, Skorecki K, Ben-Ami H et al.** (1998) Origins of Old Testament priests. *Nature* **394**, 138–140.

56. **van Oorschot RA, Ballantyne KN & Mitchell RJ** (2010) Forensic trace DNA: a review. *Investig. Genet.* **1**, 14.

57. **Walsh S, Liu F, Ballantyne KN et al.** (2011) IrisPlex: a sensitive DNA tool for accurate prediction of blue and brown eye colour in the absence of ancestry information. *Forensic Sci. Int. Genet.* **5**, 170–180.

58. **Wetton JH, Tsang KW & Khan H** (2005) Inferring the population of origin of DNA evidence within the UK by allele-specific hybridization of Y-SNPs. *Forensic Sci. Int.* **152**, 45–53.

59. **Xue Y, Zerjal T, Bao W et al.** (2005) Recent spread of a Y-chromosomal lineage in northern China and Mongolia. *Am. J. Hum. Genet.* **77**, 1112–1116.

60. **Zerjal T, Xue Y, Bertorelle G et al.** (2003) The genetic legacy of the Mongols. *Am. J. Hum. Genet.* **72**, 717–721.

APPENDIX

This appendix summarizes information about two loci of outstanding importance for human evolutionary genetics: mitochondrial DNA and the Y chromosome.[12] This information is relevant to many chapters, particularly Chapters 9–14.

HAPLOGROUP NOMENCLATURE

Complex nomenclatures are used to describe Y-chromosomal and mtDNA haplogroups. This complexity is required to maintain some order in the face of changes to the phylogeny as additional variants and/or haplogroups are discovered. The nomenclatures are based on a cladistic appreciation of the underlying phylogeny that describes the relationships of the haplogroups. Major **clades** are identified by single capital letters (for example, haplogroup R), sublineages within these clades are given numerical suffixes (for example, haplogroup R1), and this can be continued using alternating lowercase letters and numbers until all lineages have been named (for example, R1b1b2). An alternative sometimes adopted for Y haplogroups is to use the name of the derived variant furthest from the root (R-M269 for the last example). As the number of variants and length of names increases, this alternative becomes increasingly attractive because of its brevity and stability: variant names do not change.

The set of chromosomes that share the derived state of a unique event polymorphism are by definition **monophyletic**. By contrast, a set of chromosomes defined by the sharing of the derived state of a deep-rooting variant and ancestral states at the more derived variants that define sublineages are potentially paraphyletic at the sublineage level. In other words, a new mutation may be found that is carried by some but not all these chromosomes. These sets of chromosomes are **paragroups** instead of haplogroups, and are highlighted by using a "*" (pronounced "star") suffix (for example, chromosomes belonging to haplogroup R but not R1 or R2 are called R*). Often not all of the variants defining known sublineages are typed when paragroups are identified and additional suffixes are added to reflect this. For example, if only variants defining the R1 but not the R2 sublineage are typed, the corresponding paragroup is named R*(xR1), where the "x" signifies "not."

Y-chromosomal phylogenies contain **multifurcations** that may be resolved into a set of bifurcations as additional variants are discovered, whereas many multifurcations in the mtDNA tree will remain unresolved as the existing phylogeny is already based on complete mtDNA genome sequence data. A new variant found to be derived in two lineages descending from a multifurcation defines a new clade which is named by the union of the names of the individual clades (for example, a clade that unites haplogroups D and E is known as DE, **Appendix Figure 1**).

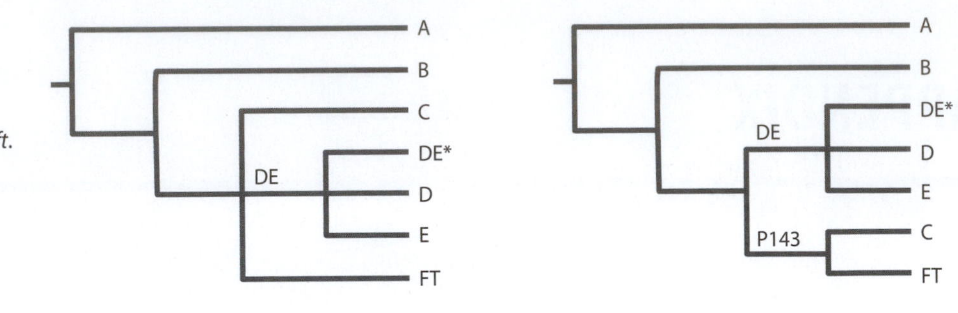

Appendix Figure 1: Haplogroups, paragroups, and resolving multifurcations.
In 2007, the most basal section of the Y haplogroup tree was drawn as on the *left*. A, B, C, D, E, DE, and FT are haplogroups, each defined by the derived state of one or more polymorphisms (and each could be subdivided further by additional polymorphisms). DE*, however, was not defined by the derived state of any polymorphism, but only by being derived for the DE polymorphisms and ancestral for the D and E polymorphisms. DE* is thus a paragroup. The tree also contains a trifurcation, where the branching order of the three branches C, DE, and FT was undefined. In 2008, a new polymorphism (P143) was reported which resolved this trifurcation (*right*). P143 showed that the first split was between DE on the one side and C plus FT on the other. The branching within haplogroup A has subsequently been refined further.

THE MITOCHONDRIAL GENOME

What are its origins?

The cytoplasmic location and energy-linked function of mitochondria (mt) derive from their origin as endosymbiotic prokaryotes (see **Section 2.8**). Each mitochondrion contains several identical mitochondrial genomes (mtDNA) that appear to be exclusively maternally inherited in humans (see Figure 2.19).

What genes are encoded within the mitochondrial genome?

The **endosymbiont** progenitor of modern mitochondria contained many more genes than are presently found in mtDNA. Most of these have been lost, with some being transferred to the nuclear genome. mtDNA retains 37 genes encoding proteins involved in energy production (oxidative phosphorylation) and mitochondrial protein synthesis (see Figure 2.21). These genes are tightly packed within the 16.5-kb circular genome, whose only substantial stretch of noncoding sequence contains a dedicated replication origin and is known as the "D-loop" ("Displacement-loop," because the two strands of DNA are held apart for part of the replication cycle and one is paired with a third strand), or the control region. The two strands of mtDNA differ in their base content, with the G-rich and C-rich strands known as the "heavy" and "light" strands, respectively.

What diseases are caused by mutations within mtDNA?

The maternal inheritance of mtDNA allows inherited disease-causing mutations to be mapped to this locus on the basis of the segregation of the disease through pedigrees. Most diseases affect organs with a high energy requirement, such as nerves and muscles. Pathogenic mutations include base substitutions, deletions, duplications, insertions, and inversions (see **Appendix Table 1** and http://www.mitomap.org/). Rearrangements are often flanked by short repeats of 2–13 bp, suggesting that mitochondria contain machinery for homology recognition and the splicing of DNA ends, despite the apparent lack of homologous recombination. Mitochondrial DNA mutations may also be acquired in somatic tissues, and may be involved in the aging process.

How has the study of mtDNA diversity developed?

The first full mitochondrial genome sequence was determined from human placenta in the lab of Fred Sanger in Cambridge, UK,[1] and subsequently became known as the Cambridge Reference Sequence (**CRS**); the nucleotide numbering of mtDNA sequences is based on a revised and corrected version of this, the **rCRS**,[2] although there are proposals to adopt a reconstructed ancestral sequence instead.[4] Diversity within mtDNA has been studied in approaching a million individuals by sequencing the most variable parts of the control region, the hypervariable segments I and II (HVSI and HVSII, or HVS1 and HVS2), often complemented by typing a number of informative SNPs from the coding region.

APPENDIX TABLE 1: EXAMPLES OF MITOCHONDRIAL DISEASES

Mutation	Disease
Base substitution in protein-coding gene: 11778 transition in NADH dehydrogenase 4 (*ND4*) gene	Leber's hereditary optic neuropathy (LHON; OMIM 535000)
Base substitution in tRNA gene: 3243 transition in tRNA Leucine 1 gene	mitochondrial encephalomyopathy, lactic acidosis, and stroke-like episodes (MELAS; OMIM 540000)
Deletion: 4977-bp deletion (8469 to13447) between 13-bp direct repeats	Kearns–Sayre syndrome (OMIM 530000)
Duplication: 266-bp duplication between 7-bp direct repeats in the control region	mitochondrial myopathy
Inversion: 7-bp inversion (3902–3908) in NADH dehydrogenase (*ND1*) gene	mitochondrial myopathy
Insertion: C inserted at 7472 in tRNA Serine 1 gene	progressive encephalopathy (PEM; OMIM 590080)

Tens of thousands of complete mtDNA sequences have also been determined, and this number is increasing rapidly.

How is information from the mtDNA variants in an individual combined?

Different parts of mtDNA accumulate variation at different rates, perhaps because of different DNA structures influencing mutation rates and different selective constraints influencing variant survival in the population (see **Section 3.3**). While the absence of mtDNA recombination (see Box 3.4) means that there is a single phylogeny for mtDNA lineages, the presence of parallel mutations and reversions at fast-mutating sites (as well as sequencing errors) makes it difficult to reconstruct this phylogeny correctly. The phylogeny drawn from complete sequences, however, is robust and an accepted haplogroup nomenclature (www.phylotree.org) is used in the literature. A simplified version of this phylogenetic tree is shown in **Appendix Figure 2**.

Why are all the deep-rooting clades called L?

The present nomenclature (described above) has been established piecemeal since the early days of mtDNA typing. Initial studies focused on Native American and Eurasian populations, and by the time large studies of Africa were initiated, few single-letter clades remained unassigned. The early African study which defined L1 and L2[5] assumed Asian origins of human mtDNA variation and misplaced the root of the phylogeny in an Asian branch, such that it appeared that most African lineages formed a monophyletic clade, but this is now known not to be the case.

Why is mtDNA so useful for exploring the human past?

- The high mutation rate makes it easy to detect diversity efficiently.
- The low effective population size leads to increased genetic drift, which generates geographic structure (for example, continent-specific lineages, see Appendix Figure 2).
- Exclusively maternal inheritance allows access to female-specific processes.
- The high copy number of mtDNA per cell facilitates the analysis of degraded contemporary samples and ancient DNA (see **Sections 4.10 and 18.1**).

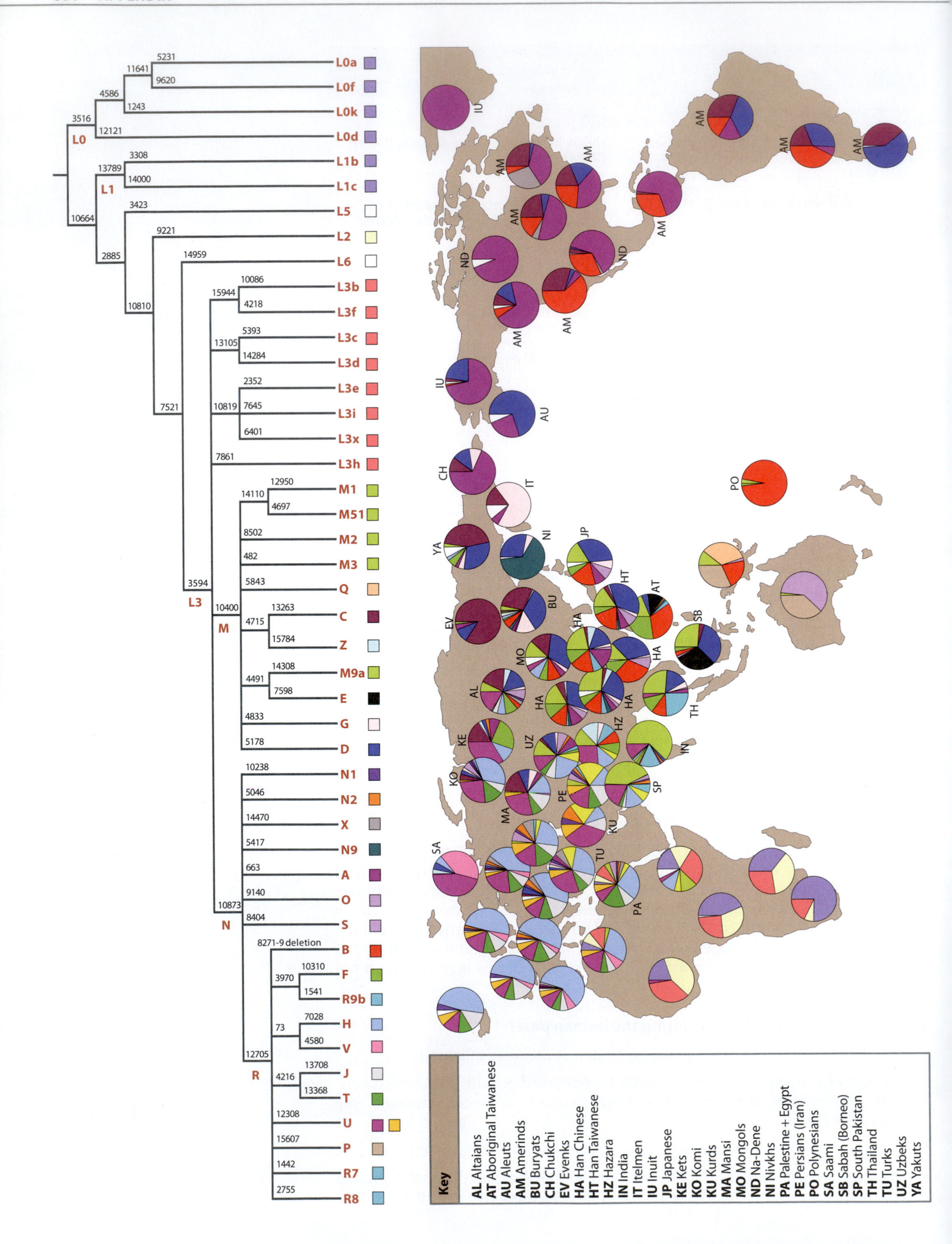

5231
11641
9620
4586
1243
3516
12121
L0
3308
13789
14000
L1
3423
10664
9221
2885
14959
10086
15944
4218
10810
5393
13105
14284
2352
7521
10819
7645
6401
7861
12950
14110
4697
8502
482
5843
3594
L3
10400
13263
4715
M
15784
14308
4491
7598
4833
5178
10238
5046
14470
5417
663
9140
10873
8404
N
8271-9 deletion
10310
3970
1541
7028
73
4580
12705
R
13708
4216
13368
12308
15607
1442
2755
L0a
L0f
L0k
L0d
L1b
L1c
L5
L2
L6
L3b
L3f
L3c
L3d
L3e
L3i
L3x
L3h
M1
M51
M2
M3
Q
C
Z
M9a
E
G
D
N1
N2
X
N9
A
O
S
B
F
R9b
H
V
J
T
U
P
R7
R8
IU
AM
ND
AU
CH
IT
PO
YA
NI
JP
HT
AT
EV
BU
HA
SB
MO
AL
TH
KE
UZ
HZ
IN
KO
PE
SP
MA
KU
SA
TU
PA
Key
AL Altaians
AT Aboriginal Taiwanese
AU Aleuts
AM Amerinds
BU Buryats
CH Chukchi
EV Evenks
HA Han Chinese
HT Han Taiwanese
HZ Hazara
IN India
IT Itelmen
IU Inuit
JP Japanese
KE Kets
KO Komi
KU Kurds
MA Mansi
MO Mongols
ND Na-Dene
NI Nivkhs
PA Palestine + Egypt
PE Persians (Iran)
PO Polynesians
SA Saami
SB Sabah (Borneo)
SP South Pakistan
TH Thailand
TU Turks
UZ Uzbeks
YA Yakuts

Appendix Figure 2: mtDNA phylogeny and haplogroup distribution. Left: simplified version of the mtDNA phylogenetic tree. The major clades L0, L1, L3, M, N, and R are indicated on the tree, together with numbers specifying the locations in the rCRS of key variants defining the branches. Haplogroup names are shown at the ends of the branches and are assigned a color, corresponding to the colors used on the map (*right-hand side*). Right: geographical distributions of mtDNA haplogroups. Each pie chart represents a population sample, with the haplogroup frequencies indicated. The phylogenetic resolution differs slightly between the tree and map, so some haplogroups share the same color, some are not represented on the map (*white*), and one is subdivided on the map. Map source: **http://www.scs.illinois.edu/~mcdonald/WorldHaplogroupsMaps.pdf**

What about possible selection pressures?

Many studies have investigated associations between mtDNA variants and non-disease phenotypes, which might reflect positive or negative selection. Two topics of particular interest have been adaptation to colder climates and resistance to sepsis. First, an excess of nonsynonymous mutations in mtDNA protein-coding genes in populations from North Eurasia and America was interpreted as evidence of cold adaptation.[8] However, other studies (for example, Kivisild et al.[7]) have found that the excess of nonsynonymous mutations is not restricted to populations living in cold climates, but rather appears to be a feature of mtDNA haplogroups that are relatively young in age. One possible explanation is that purifying selection has not had enough time to remove slightly deleterious mutations from the terminal branches of the phylogeny. Second, haplogroup H, which is the most frequent mtDNA clade in Europe, has been associated with an increased chance of long-term survival after sepsis compared with other haplogroups.[3] The functional basis for this association is uncertain; it could be that haplogroup H mitochondria have different electron transport rates or generate more reactive oxygen species which could reduce bacterial infection, thereby allowing haplogroup H to have been selected for during past disease epidemics.

THE Y CHROMOSOME

How has it evolved?

The human sex chromosomes are very different from each other (Figure 2.12), but they were once a pair of homologous autosomes. The process of divergence started when one of them acquired a male sex-determining function early in mammalian evolution, followed by a repression of recombination. The non-recombining portion of the Y has expanded over time and now comprises ~95% of the chromosome, with the **pseudoautosomal regions** confined to the ends of the chromosome (see **Section 2.8**).

What does the chromosome contain?

Given that women do not have a Y chromosome, it cannot contain genes important for survival that are not shared with the X chromosome. In fact, the Y chromosome is extremely gene-poor, with the 23 Mb of male-specific euchromatin coding for only 27 different proteins[11] (1.2/Mb) compared to the 1098 genes on the 155-Mb X chromosome[9] (7.1/Mb). Much of their common ancestral gene content has been lost from the Y chromosome, and it has been suggested that this decline presages an inevitable disappearance of the Y chromosome itself. However, the remaining genes, which include the sex-determining gene *SRY*, have been supplemented by more recently acquired genes with male-specific functions (for example, spermatogenesis), several of which have been amplified into multiple copies. Gene loss is due in part to the inability of the Y chromosome to eliminate mutant alleles by recombination with a non-mutant homolog, leading to the degeneration of many genes (Muller's ratchet).

In contrast to the lack of genes, the Y is enriched for many different types of repeats, including SINEs, **endogenous retroviruses**, and segmental duplications. The consequent susceptibility to rearrangements by **non-allelic homologous recombination** results in unusually high levels of structural polymorphism among humans (see **Section 3.6**) and structural divergence among great apes. The instability of the Y chromosome combined with the important role of many of its genes in spermatogenesis means that the commonest classes of pathogenic mutation on this chromosome are deletions that influence sperm production and can lead to male infertility[10] (see idiogram in Figure 2.12).

How similar are Y chromosomes within and between species?

The Y chromosome has unusual evolutionary characteristics compared to other nuclear loci: a higher mutation rate, higher *between species* sequence divergence, but lower *within species* sequence diversity. How can these seemingly discordant properties be reconciled? Higher mutation rates result from the exclusive passage of Y chromosomes through the male germ line (see Box 5.4), which is more mutagenic than the female germ line. This leads to greater Y-chromosomal sequence divergences between hominoid species than for other loci (see **Section 7.3**). However, there is lower diversity within a species because the Y chromosome is prone to genetic drift due to its smaller effective population size (see Table 5.2; **Section 5.3**). This results in a very recent common ancestor, and therefore less time to accrue diversity. The enhanced genetic drift more than outweighs the increased mutation rate. The high genetic drift also leads to large differences between populations (see **Section 5.5**), making the Y the most geographically informative single locus in the genome.

What molecular polymorphisms are found on the Y chromosome?

Since the discovery of the first molecular polymorphism in 1985, the number of Y-chromosome polymorphisms has accumulated at an ever-increasing rate. Hundreds of microsatellites have been identified from the reference sequence (see **Section 4.7**). Genome sequencing projects have led to the discovery of many thousands of SNPs,[13] although these have often not been validated or placed on the Y phylogeny. From the Y-chromosomal SNP mutation rate,[14] we can predict that each Y chromosome will on average carry a new SNP.

How should the polymorphic information from different variants be combined?

The molecular variants identified above have differing mutation rates; some are sufficiently slowly mutating that they can be considered unique in human evolution. These unique event variants are typically binary SNPs or indels and can easily be combined into haplotypes (see **Section 3.8**), known as haplogroups. The absence of recombination (see Box 3.4) means that these monophyletic haplogroups can be related by a single phylogeny (**Appendix Figure 3**) using the principle of **maximum parsimony** (see **Section 6.4**), and this phylogeny is the most detailed for any region of the genome. The global distribution of

Appendix Figure 3: Y-chromosomal phylogeny and haplogroup distribution.
Left: simplified version of the Y-chromosomal phylogenetic tree. The major clades are indicated on the tree, together with the names of key variants defining the branches. Haplogroup names are shown at the ends of the branches and are assigned a color, corresponding to the colors used on the map (*right-hand side*). Right: geographical distributions of Y-chromosomal haplogroups. Each pie chart represents a population sample, with the haplogroup frequencies indicated. The phylogenetic resolution differs slightly between the tree and map, so some haplogroups share the same color, and some are not represented on the map (*white*). Map source: **http://www.scs.illinois.edu/~mcdonald/WorldHaplogroupsMaps.pdf**

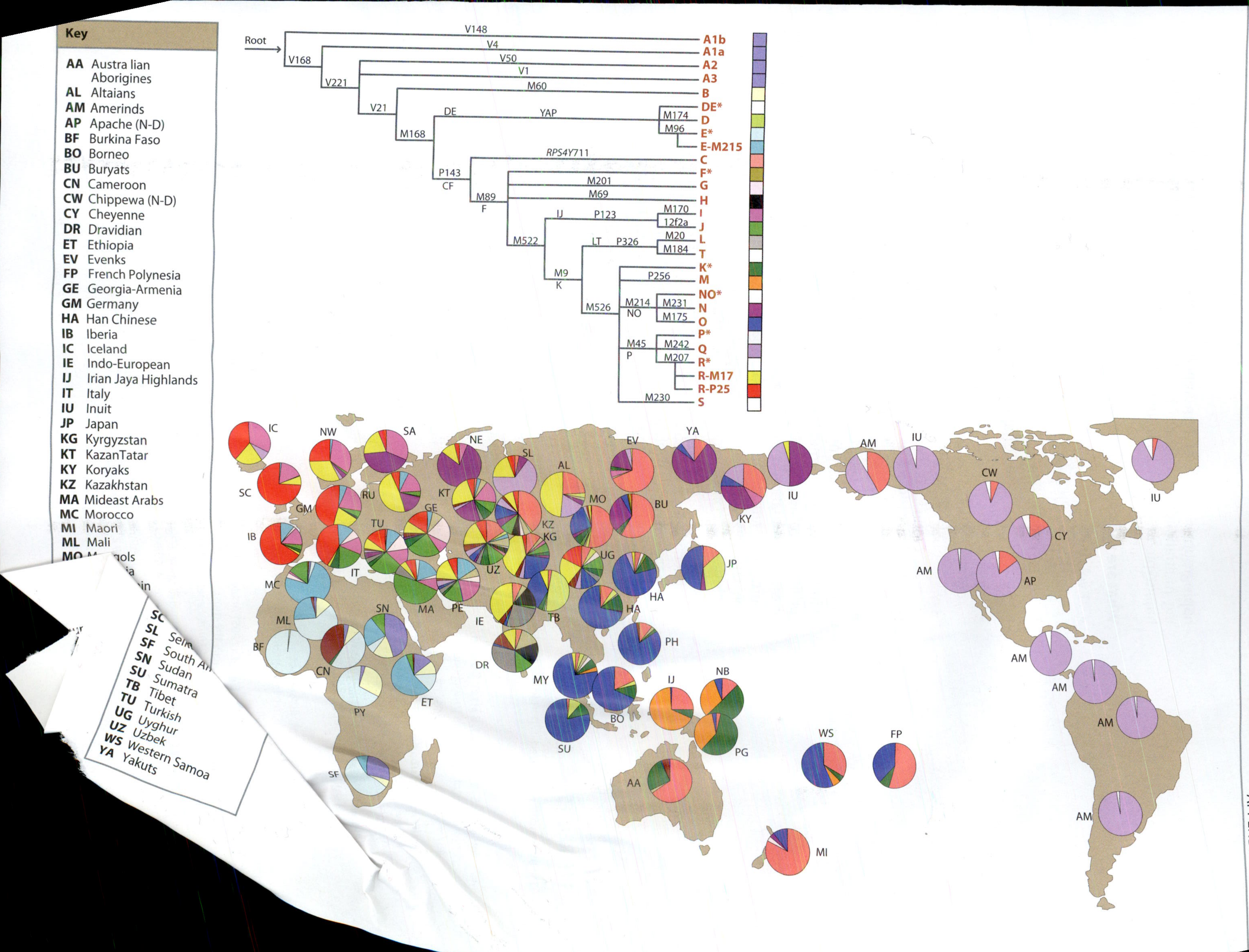
Key
AA Austra lian Aborigines
AL Altaians
AM Amerinds
AP Apache (N-D)
BF Burkina Faso
BO Borneo
BU Buryats
CN Cameroon
CW Chippewa (N-D)
CY Cheyenne
DR Dravidian
ET Ethiopia
EV Evenks
FP French Polynesia
GE Georgia-Armenia
GM Germany
HA Han Chinese
IB Iberia
IC Iceland
IE Indo-European
IJ Irian Jaya Highlands
IT Italy
IU Inuit
JP Japan
KG Kyrgyzstan
KT KazanTatar
KY Koryaks
KZ Kazakhstan
MA Mideast Arabs
MC Morocco
MI Maori
ML Mali
SN Sudan
SU Sumatra
TB Tibet
TU Turkish
UG Uyghur
UZ Uzbek
WS Western Samoa
YA Yakuts
Root
V148
V168
V4
V221
V50
V1
V21
M60
M168
DE
YAP
M174
M96
P143
CF
RPS4Y711
M89
F
M201
M69
IJ
P123
M170
12f2a
M522
LT
P326
M20
M184
M9
K
P256
M526
M214
NO
M231
M175
M45
P
M242
M207
M230
A1b
A1a
A2
A3
B
DE*
D
E*
E-M215
C
F*
G
H
I
J
L
T
K*
M
NO*
N
O
P*
Q
R*
R-M17
R-P25
S

haplogroups is also shown in Appendix Figure 3. Diversity of the rapidly mutating multiallelic variants can be examined within haplogroups to gain further haplotypic resolution, and some chronological insight.

What are the applications of studying Y-chromosomal diversity?

- Evolutionary studies (see Jobling and Tyler-Smith,[6] Chapters 9 to 14, and Appendix Figure 3)
- Genealogical investigations (see Chapter 18)
- Forensic work (see Chapter 18)
- Medical research, for example studies of male infertility, anomalies of sex determination, and deafness; associations with conditions such as HIV progression or coronary artery disease

Is there any evidence of selection on the Y chromosome?

Drawing inferences about the human past from Y-chromosomal diversity usually assumes that selection has not played a role in shaping present patterns of diversity. Is there any evidence to the contrary? Negative selection clearly removes Y chromosomes carrying severe defects in male sex determination or fertility from the population. Such chromosomes are thus likely to be too rare to affect population inferences, although variants leading to mild decreases in sperm production can in principle reach appreciable frequencies and a deletion known as "gr/gr" may provide an example of this.[10] The magnitude and evolutionary consequences of such weak negative selection remain unclear. No examples of Y-chromosome lineages with biological advantages have been identified, but social selection has been proposed as an explanation for the very rapid expansion of lineages associated with Genghis Khan, the Chinese emperor Nurhaci, and the Irish king Niall of the Nine Hostages (**Section 18.3**).

REFERENCES

The references highlighted in purple are considered to be important (for this chapter) by the authors.

1. **Anderson S, Bankier AT, Barrell BG et al.** (1981) Sequence and organization of the human mitochondrial genome. *Nature* **290**, 457–465.
2. **Andrews RM, Kubacka I, Chinnery PF et al.** (1999) Reanalysis and revision of the Cambridge reference sequence for human mitochondrial DNA. *Nat. Genet.* **23**, 147.

Baudouin SV, Saunders D, Tiangyou W et al. (2005) Mitochondrial DNA and survival after sepsis: a prospective study. *Lancet* **366**, [illegible]8–2121.

[illegible]M, van Oven M, Rosset S et al. (2012) A "Copernican" [illegible]ent of the human mitochondrial DNA tree from its root. *Am.* [illegible] **90**, 675–684.

5\. [illegible] A, Excoffier L et al. (1995) Analysis of mtDNA [illegible]populations reveals the most ancient of all human [illegible]logroups. *Am. J. Hum. Genet.* **57**, 133–149.

6\. Job[illegible] evol[illegible]

7\. Kivisi[illegible] evoluti[illegible] C (2003) The human Y chromosome: an [illegible]f age. *Nat. Rev. Genet.* **4**, 598–612.

8\. Mishma[illegible] shaped re[illegible] **100**, 171–[illegible] (2006) The role of selection in the [illegible] genomes. *Genetics* **172**, 373–387.

[illegible]l. (2003) Natural selection [illegible]ns. *Proc. Natl Acad. Sci. USA*

9. **Ross MT, Grafham DV, Coffey AJ et al.** (2005) The DNA sequence of the human X chromosome. *Nature* **434**, 325–337.
10. **Rozen SG, Marszalek JD, Irenze K et al.** (2012) *AZFc* deletions and spermatogenic failure: a population-based survey of 20,000 Y chromosomes. *Am. J. Hum. Genet.* **91**, 890–896.
11. **Skaletsky H, Kuroda-Kawaguchi T, Minx PJ et al.** (2003) The male-specific region of the human Y chromosome is a mosaic of discrete sequence classes. *Nature* **423**, 825–837.
12. **Underhill PA & Kivisild T** (2007) Use of Y chromosome and mitochondrial DNA population structure in tracing human migrations. *Annu. Rev. Genet.* **41**, 539–564.
13. **Wei W, Ayub Q, Chen Y et al.** (2013) A calibrated human Y-chromosomal phylogeny based on resequencing. *Genome Res.*, **23**, 388–395.
14. **Xue Y, Wang Q, Long Q et al.** (2009) Human Y chromosome base-substitution mutation rate measured by direct sequencing in a deep-rooting pedigree. *Curr. Biol.* **19**, 1453–1457.

GLOSSARY

α-thalassemia
A blood cell disorder resulting from a deficiency of the α-globin protein.

θ
See *population mutation parameter*.

μ
See *mutation rate*.

μg
Microgram.

π
See *nucleotide diversity*.

ρ
See *population recombination parameter*, or *rho statistic*.

1000 Genomes Project
An international collaborative project launched in 2008 that pioneered the application of *next-generation sequencing* to sequence the entire *genomes* and *exomes* of 2500 diverse human individuals.

^{14}C
Carbon-14.

17q21.31 microdeletion syndrome
A very rare developmental disorder caused by a *deletion* within *chromosome* 17.

3′ end
The end of a DNA or RNA molecule bearing a free hydroxyl (–OH) group (unattached to another nucleotide) on the 3′ carbon; of a *gene*, the end at which *transcription* terminates.

3′ UTR, 3′ untranslated region
The portion of an *mRNA* molecule downstream of the *stop codon*; includes the *polyadenylation signal*, and can include binding sites for *proteins* and *miRNAs*.

454 sequencing
A form of *next-generation sequencing* based on the *pyrosequencing* method, characterized by long (up to 1000 bp) read lengths but relatively low throughput and high cost per base.

5′ end
The end of a *DNA* or *RNA* molecule bearing a free hydroxyl (–OH) group (unattached to another *nucleotide*) on the 5′ carbon; of a *gene*, the end at which *transcription* starts.

5′ UTR, 5′ untranslated region
The portion of an *mRNA* upstream of the start codon; contains sequences needed for the initiation of *translation*.

A
See *adenine*.

abasic site
A site within damaged or *ancient DNA* where the *base* has been lost.

accessible genome
The proportion of the *reference sequence* (~85%) that can be analyzed readily by *next-generation sequencing* technologies; the rest is difficult to analyze because of complex and duplicated regions, or because the reference sequence assembly is poor.

acclimatized, acclimatization
Physiological mechanisms producing short-term and sometimes rapid reversible responses to environmental change.

acculturation
The spread of cultural practices by learning from neighboring groups, rather than by the spread of people themselves; a possible mode of spread of *agriculture* in Europe, for example.

aCGH
See *array-comparative genomic hybridization*.

Acheulean
One of the stone tool traditions of the *Lower Paleolithic*, including *bifaces* and cleavers; named after the French site of St. Acheul; also called *mode 2 technology*.

acrocentric
Of a *chromosome*: having the *centromere* close to one end.

adaptation
A character that has been favored by natural selection for its effectiveness in a particular role; an adaptation is not necessarily *adaptive*.

adaptive
Describing a character that functions currently to increase reproductive success; see *adaptation*.

adaptive immunity
The system of immune response that identifies a foreign agent, adapts to best recognize it, and remembers for future reference, and thus is able to mount a more effective response on future exposure to the same agent.

additive tree
A *phylogeny* with branch lengths proportional to the amount of evolutionary change.

adenine, A
A *purine* base found within DNA; pairs with *thymine*.

ADME (absorption, distribution, metabolism, and excretion) genes
Collective term for genes whose products influence drug efficacy, and variation in which can give a variable drug response.

admixture
The formation of a hybrid *population* through the mixing of two ancestral populations.

aDNA
See *ancient DNA*.

advantageous allele
An *allele* that increases the *fitness* of its carrier.

affected-sib-pair analysis
A *nonparametric* method of *linkage analysis* focusing on sib-pairs with a *complex disorder*.

AFLP
See *amplified fragment length polymorphism*.

agonist
A drug that triggers an action, for example, from a cell; contrasted with an antagonist, which blocks or nullifies an action.

agriculture
The cultivation of soil, production of crops, and raising of livestock. Can also refer more specifically to the intensive growing of crops in fields, with the more general term "farming" including the tending of animals.

AIDS, acquired immunodeficiency syndrome
A disease of the human immune system caused by *HIV*.

AIM
See *ancestry informative marker*.

Akaike's information criterion
A measure of the relative *goodness-of-fit* of a statistical model, which provides a means for model selection.

albinism
Complete or partial absence of pigment in the skin, hair, and eyes.

alkaloid
A nitrogen-rich organic compound generally from plants or fungi, often having a physiological effect on humans.

allele
One of two or more alternative forms of a *gene* or *DNA* sequence at a specific *chromosomal* location.

allele drop-in
The appearance of one or more extraneous *alleles* in a *DNA profile* due to contaminating DNA.

allele drop-out
The absence of one or more *alleles* from a *DNA profile* due to insufficient DNA quality or quantity.

allele frequency
The frequency of an *allele* within a *population*.

allele frequency spectrum
The spectrum of *allele frequencies* within a set of sequences; also *site frequency spectrum*.

allele heterogeneity
The number of different mutant *alleles*, usually of one *gene*, that cause a genetic disease.

allele-surfing
The spread of a particular *allele* due to its association with the edge of a wave of population expansion.

allelic architecture
The number and frequency of susceptibility *alleles* at a *locus*; also sometimes called *allelic spectrum*, or *genetic architecture*.

allelic ladder
A set of labeled *PCR* products corresponding to common *alleles* across the known size range of a set of *microsatellites*; used for the sizing of test fragments.

allelic spectrum
See *allelic architecture*.

Allen's rule
A rule stating that shorter appendages are favored in colder climates because they have relatively high ratios of mass to surface area, and therefore retain heat.

allozyme
A variant form of an *enzyme* characterized by particular *amino acid* sequence differences; distinguishable by *protein electrophoresis*, and one type of *classical polymorphism*.

alpha factor
The ratio of *mutations* that occur in the male *germ line* compared to those that occur in the female germ line; see *male-driven evolution*.

alphoid repeat
A repeat unit of ~170 bp found at the *centromeres* of chromosomes and organized into higher-order repeat structures a few kb in length.

alternative splicing
The utilization by the splicing machinery of alternative *exons* from within a *gene*, to produce different forms of *proteins* with different functions.

***Alu* element**
A member of the *SINE* class of interspersed repeated sequences. The element is about 290 bp long, and has a copy number of about 1 million. Named because it often contains a site for the *restriction enzyme AluI* (AGCT).

amelogenin
A gene with differentiable copies on the *X* and *Y chromosomes*, commonly used in a *PCR*-based *sex test*.

Amerind
A language superfamily proposed by Joseph Greenberg that includes all Native American languages not belonging to the *Eskimo-Aleut* and *Na-Dene* language families; regarded by many linguists as controversial.

AMH
See *anatomically modern humans*.

amino acid
One of 20 commonly used building blocks of *proteins*.

AMOVA
See *analysis of molecular variance*.

amplicon
A *DNA* sequence amplified in a specific *PCR* reaction.

amplified fragment length polymorphism, AFLP
Fragments amplified by *PCR* that are polymorphic in length due to variation in the position of *restriction* sites in the original source *DNA*.

analogous
Displaying similarity not due to common ancestry.

analysis of molecular variance, AMOVA
A method of estimating within- and among-population levels of *population* differentiation from molecular data.

anatomically modern humans, AMH
Homo sapiens showing modern head and body structure; contrasted with archaic forms of humans such as *Neanderthals*; anatomically modern humans do not necessarily show other modern characteristics such as modern behavior.

anatomical modernity
A package of physical traits differentiating modern humans from earlier forms of *Homo* such as Neanderthals, including a high and rounded forehead, chin, lack of brow ridges, *gracile* skeleton, and a distinct shape of the *clavicle*.

ancestral graph
A mathematical object similar to a *network* that can be used in *coalescent* analysis to model the *genealogy* of a *locus* under natural selection.

ancestral haplotype
The *haplotype* on which a new variant arose.

ancestral neutrality
The idea that complex disease susceptibility *alleles* were not disadvantageous in the past, and have become disadvantageous too recently for *purifying selection* to have had a significant effect on their frequencies.

ancestral recombination graph
A mathematical object similar to a *network* that can be used in *coalescent* analysis to model the *genealogy* of a *locus* that has undergone *recombination*.

ancestral state
Characteristic retained from ancestors; applies to both physical characteristics and DNA-based characters such as the allelic state of a human *SNP*, whose ancestral state can be deduced by examining its *orthologous* state in other species such as *chimpanzees*.

ancestry informative marker, AIM
A *polymorphism* having *alleles* that are significantly more common in one group of *populations* than in other groups. Useful in admixture studies.

ancient DNA, aDNA
DNA retrieved from degraded and sometimes very old biological material.

aneuploidy
Departure from the normal *diploid* number of 46 human *chromosomes*; for example, *trisomy 21*.

annealing
The stage of a *PCR* in which *oligonucleotide primers* are allowed to bind to their complementary sites in the single-stranded template; typically carried out at 50–60°C.

annotation
Of the *genome* sequence: addition of information about biological function, such as that a genomic segment codes for a *protein*, or acts as an *enhancer*.

anthropogenic
Of an environmental phenomenon: caused by human activity.

anthropozoonosis
Transfer of a human disease to other primates.

antibody
A *protein* produced by white blood cells in response to *antigen*; antigen–antibody reactions can be very specific.

anticodon
A specific set of three adjacent *nucleotides* in a *tRNA* molecule, which binds to the relevant *codon* in the *mRNA* through *base pairing* during the *translation* process.

antigen
A substance that elicits an immune response, through binding to an *antibody*.

Aotearoa
The Māori name for New Zealand.

approximate Bayesian computation, **ABC**
A method that calculates *summary statistics* from *coalescent* simulations, and compares these with the summary statistic of the observed data.

Arabian cradle
A model according to which the primary differentiation of non-African *populations* occurred in the Arabian Peninsula at the onset of the *out-of-Africa* dispersal.

arboreal locomotion
Type of movement of animals in trees.

archipelago
A cluster of closely positioned islands.

array
A closely spaced set of *DNA* or *protein* molecules immobilized on a substrate that allows the simultaneous investigation of many macromolecules; also *chip*.

array-comparative genomic hybridization, **array CGH**, **aCGH**
A genomewide method for measuring *copy-number variation*. Genomic DNAs from a test and a reference sample labeled with different fluorescent dyes are co-hybridized to an array of clone DNAs or *oligonucleotide probes*. The ratio of fluorescence indicates relative sequence copy number.

ascertainment bias
Distortions in a dataset caused by the way that *polymorphisms* or samples are collected.

ASD
Average squared distance.

association study
See *disease association study*.

assortative mating
Using phenotypic similarity to guide mate choice.

assortment
See *lineage assortment* or *independent assortment*.

Aurignacian
One of the earliest *Upper Paleolithic* cultures: defined by its tools and associated with *anatomically modern humans* in Europe. Named after the first site where it was discovered, Aurignac in the Pyrenees.

aurochs
Bos primigenius, the progenitor species to modern cattle; an unusual singular word with the plural aurochsen.

Australopithecine
Extinct African hominin that lived 2–4 MYA.

Austronesian
A large language family spoken throughout Island Southeast Asia and the Pacific.

autocorrelation
A correlation calculated between all items in a series.

autoradiography
The process of recording bands on a *Southern blot* that are hybridizing to radiolabeled *probes* by exposure to X-ray film.

autosome
One of the 22 *biparentally inherited* chromosomes, each present in two copies in males and females, and numbered 1 to 22.

autozygosity
Homozygosity in which the two *alleles* are *identical by descent*.

autozygosity mapping
Identification of genomic regions where both *homologs* have been inherited from a very recent ancestor and are therefore *identical by descent*, particularly in *consanguineous* families, to localize mutant *alleles* causing genetic disease.

azoospermia
The absence of sperm in the semen.

BAC
See *bacterial artificial chromosome*.

back mutation
A mutation from the derived state back to the ancestral state.

background selection
The reduction in diversity at linked polymorphic *loci* as a result of nearby *negative selection*.

bacterial artificial chromosome, **BAC**
A *DNA* vector in which inserts of up to 300 kb long can be propagated in bacterial cells.

baits
Oligonucleotides using in *sequence capture* experiments; also known as a *probes*.

balanced
Of a *translocation* or other *structural variant*: involving no change in sequence copy number.

balancing selection
A selective regime that favors more than one *allele* and thus prevents the *fixation* of any allele.

Bantu expansion
A complex series of population movements from western Africa into central and southern Africa, in which the Bantu languages, iron-working, and farming may have been co-dispersed.

bar coding
See *indexing*.

base
The informational part of a *nucleotide* molecule, the building block of *DNA*; the bases in DNA are *adenine* (A), *guanine* (G), *thymine* (T), and *cytosine* (C).

base analog
A compound other than a natural *base* that can be incorporated into DNA; analogs can act as *mutagens* if they possess different *base-pairing* properties to natural bases, leading to *base substitution*.

base composition
The proportion of the four bases in a given DNA sequence; often expressed simply as GC-content.

base-excision repair
Repair of a lesion in DNA on one strand of a helix, together with some surrounding bases, followed by resynthesis of

the gap using the undamaged *complementary* strand as a template.

base-modifying agent
A compound that modifies *bases* in DNA, often altering their *base-pairing* properties within DNA, and thus acting as a *mutagen*.

base pairing
The interaction of one base with another through *hydrogen-bond* formation: A pairs with T, and G with C. This interaction holds one strand of *DNA* together with its *complementary* strand.

base substitution
A *mutation* in which one *base* of *DNA* is exchanged for another.

Bayesian
Statistical methods based on Bayes' theorem that allow inferences to be drawn from both the data and any prior information.

behaviorally modern humans
Hominins possessing a combination of traits that demonstrate an ability to use complex symbolic thought and express cultural creativity.

Bergmann's rule
A rule stating that body size increases as climate becomes colder, because as mass increases, surface area does not increase proportionately.

Beringian land bridge, Beringia
Land that joins the North American and Eurasian continents across the Bering Straits at times of low sea level; named after the Danish explorer Vitus Bering.

biallelic
Of a *polymorphism*: possessing two *alleles* only (see *diallelic*, *binary polymorphism*).

biface
Teardrop-shaped stone tool worked on both sides and most of the margin, sometimes called a *handaxe*; biface is the preferred term because it does not imply that the tool was used as an axe.

bifurcation
The splitting of an ancestral lineage into two daughter lineages.

bimodal
Refers to a statistical distribution that has two modes.

binary polymorphism/marker
A *polymorphism* possessing two *alleles* only (see *biallelic*, *diallelic*).

bioinformatic
Of the computational analysis of biological data including sequence analysis and expression studies.

biological species concept
One of a number of species definitions, based on an ability to interbreed in the wild.

biopiracy
The misuse of the knowledge of indigenous people about animals and plants without their consent, often for profit.

biotin
Water-soluble vitamin that is widely used in biotechnology to conjugate molecules to physical substrates through its strong binding to the proteins avidin or *streptavidin*.

biparental inheritance
Inheritance from both parents; characteristic of *autosomes*.

bipedalism
Walking upright on two feet, like modern humans.

blades
Flakes, commonly made from flint or similar material, that are long and narrow.

BLAST, basic local alignment search tool
A bioinformatic tool for comparing a *DNA* or *amino acid* sequence against other entries in a database.

BLAT, BLAST-like alignment tool
A bioinformatic tool for aligning a *DNA* or *amino acid* sequence from an organism against that organism's *reference sequence*.

Bonferroni correction
In statistics, a method used to counteract the problem of *multiple testing*.

bonobo
The pygmy chimpanzee, *Pan paniscus*.

bootstrap
A method for assessing how well supported by the data are individual *clades* within a *phylogeny*.

bottleneck
The reduction of genetic diversity that results from a dramatic reduction in population size.

bp
Base pairs.

brachycephalic
"Broad headed," with a *cephalic index* over 80 (see *dolichocephalic*, *mesocephalic*).

brain size
A feature commonly deduced in fossils from the volume of the neurocranium, which in the *hominins* and apes is the upper part of the skull excluding the facial skeleton.

branch
Part of a *phylogenetic tree* connecting a pair of *nodes*; also *edge*.

broad-sense heritability
The proportion of genetic variance that can be attributed to additive, dominant, and *epistatic* genetic effects.

buccal cells
Cells taken from inside the cheek with a brush, swab, or mouthwash—a convenient source of small quantities of human genomic *DNA*.

build
A specific version of the human (or other species) genome sequence.

C
See *cytosine*.

Cambridge Reference Sequence, CRS
The original *reference sequence* of human *mtDNA*.

cAMP
Cyclic AMP.

candidate gene
A gene selected for possible involvement in a *phenotype* because of what is known about the physiological or biochemical basis of the phenotype, and the function of the gene product.

capillary DNA sequencing
Sanger sequencing carried out in a *capillary electrophoresis* apparatus.

capillary electrophoresis
Electrophoretic separation of *DNA* fragments, labeled with fluorescent dyes, in a fine capillary, with fragment detection by laser excitation; easily automatable, and used in *Sanger sequencing* and *microsatellite* typing, for example.

Cardial ware
A distinctive type of pottery associated with the spread of farming in the northern Mediterranean region; impressed with the serrated edge of the cardium (cockle) shell.

cardiomyopathy
Deterioration of the function of the myocardium (the heart muscle), usually leading to heart failure.

carotenoids
A group of pigments found in plants, including carotenes and xanthophylls.

carrier
A person who is *heterozygous* for a *recessive* allele. They carry the allele, but do not manifest the disease, although they can pass it on to a child if their partner has a mutant allele in the same gene.

cases and controls, **case-control**
A study design in which the subjects are divided into two groups by a single binary character.

caste system
Social hierarchy based on division of labor and power.

Caucasian
An old racial definition based on a skull from the Caucasus mountains, and including the peoples of western Europe. Commonly used to mean European, or of European descent.

cc
Cubic centimeters.

CCDS
See *consensus coding sequence*.

CDCV
See *common-disease common-variant*.

cDNA, **complementary DNA**
A *DNA* molecule made by *reverse transcription* of an *mRNA* molecule.

cell line
Cultured cells established from primary sampled tissues such as skin fibroblasts or white blood cells.

census population size
See *census size*.

census size, ***N***
The actual number of individuals in a *population* at a particular time.

centiMorgan, **cM**
A measure of *recombination* frequency: two loci separated by 1% recombination frequency are 1 centiMorgan (cM) apart.

centromere
The primary constriction of a *chromosome*, separating the short arm from the long arm, and the point at which *spindle fibers* attach to enable the chromosomes to move apart at cell division.

CEPH
Centre d'Etude du Polymorphisme Humain.

cephalic index, **CI**
The ratio of the breadth to the length of the skull multiplied by 100.

CEU
One of the population samples used in the *HapMap* project: 30 parent–child trios, Utah (US) residents of north and west European ancestry.

CF
See *cystic fibrosis*.

chain-termination sequencing
See *Sanger sequencing*.

chalcedony
A silica-based mineral that is composed of intergrowths of quartz and moganite.

Charcot–Marie–Tooth disease type 1A
A *dominant* genetic disease characterized by progressive motor and sensory neuropathy, and caused by *duplication* of the peripheral myelin protein 22 gene (*PMP22*) on chromosome 17.

Châtelperronian
One of the first *Upper Paleolithic* cultures, defined by the tools used and associated with *Neanderthals*.

CHB
One of the population samples used in the *HapMap* project: 45 unrelated individuals, Han Chinese from Beijing, China.

Chifumbaze complex
An archaeological culture involving the early iron-working communities of eastern and southern Africa, from about 2500–1000 YA.

childhood
Human-specific developmental stage of slow growth that occurs between the infant and juvenile stages.

chimpanzee
Common name for two living species of apes from the genus *Pan*.

chip
See *array*.

chromatin
The combination of *DNA* and *proteins* (mostly *histones*) making up *chromosomes*.

chromosomal assortment
Random distribution of maternal and paternal *chromosomes* during *gamete* formation.

chromosomal bar coding
Subregional *chromosome* painting by *probes* to produce a multicolored high-resolution pattern on chromosomes.

chromosomal segregation
The separation of *chromosomes* into daughter cells when a cell divides.

chromosome
The *DNA–protein* structure that contains part of the *nuclear genome*; *diploid* human cells have 46.

chromosome painting
The hybridization of labeled DNA from a single *chromosome* to a metaphase chromosome spread; useful in detecting chromosomal rearrangements within and between species.

CI
See *cephalic index*.

cis
A prefix of Latin origin, meaning "on the same side [as]" or "on this side [of]"; for example, a *cis*-acting factor or sequence exerts its influence on the same chromosome (as opposed to *trans*).

clade
An evolutionary branch.

cladistics
The science of reconstructing evolutionary relationships by identifying common ancestors through the sharing among *taxa* of *derived* characteristics, rather than the sharing of *ancestral* characteristics.

cladogram
An evolutionary tree that encapsulates the relative ancestries of different *taxa*.

classical polymorphism/marker
A *polymorphism* assayed by serological or protein *electrophoretic* means, used before *DNA* polymorphisms became available; for example, blood groups.

clavicle
The collarbone, a bone joining scapula and sternum.

click language
A language found in southern and east Africa (for example, the Khoisan languages) containing click consonants.

cline
A gradient of *allele* frequencies from one region to another, indicative of *migration* or *natural selection*.

Clovis point
A diagnostic stone tool fashioned by early settlers of the Americas.

cluster
In *next-generation sequencing*, a population of ~1000 identical *DNA* molecules that undergo sequencing within an *Illumina*™ device.

cluster analysis
A method for analyzing *population structure* that groups individuals into different clusters based on genetic similarity.

cM
See *centiMorgan*.

CNP
See *copy-number polymorphism*.

CNV
See *copy-number variation*.

coalesce
Join.

coalescence time
The time taken (going backward in time) for two or more lineages to *coalesce* into their ancestral lineage.

coalescent theory
A branch of population genetics that achieves considerable computational gains by considering the ancestry of lineages backward in time.

coalescent unit
In *coalescent* analysis, the unit in which time is measured.

CODIS
Combined DNA Index System.

co-dominant selection
A selective regime in which both *alleles* contribute to fitness, such that the fitness of the *heterozygote* is distinguishable from those of the two *homozygotes*.

codon
A series of three adjacent *bases* in *messenger RNA* (and, by extension, in DNA) that encodes a specific *amino acid* in a *protein*.

coefficient of kinship
A measure of relatedness between two individuals, representing the probability that two *alleles* sampled at random from each individual are identical. The value between a parent and its offspring is 0.25.

coefficient of relatedness
The proportion of a diploid *autosomal genome* that is shared by two individuals by common descent. The value between a parent and its offspring is 0.5.

cognate
In linguistics, a word in one language that shares its origin with a word in another.

colonization
The settlement by a group of previously uninhabited lands; this term can also be applied when the territory was previously occupied by another population for example, European colonization of the Americas.

combined likelihood ratio, CLR
A test that combines different aspects of the *site frequency spectrum* to identify *positive selection*.

commensal
A species that is not domesticated but has adopted a niche created by human activities.

common-disease common-variant (CDCV) hypothesis
The proposal, based largely on theoretical considerations, that genetic susceptibility to common diseases is expected to be due to common alleles at susceptibility loci.

comparative genomics
Genomewide comparisons of the DNA sequences of one species with another.

comparative phylogeography
Comparing the geographical distribution of lineages among a number of species to identify the evolutionary processes common to those species.

complementary
Of a strand of DNA that will base pair with another strand.

complex disorder
A disorder including genetic and nongenetic components, in which the mode of inheritance is not *Mendelian*.

composite of multiple signals, CMS
A test for *positive selection* that combines five different statistical tests.

compound heterozygote
An individual who is *heterozygous* for two different mutant *alleles*, one on each *homologous chromosome*, within a *gene* that causes a *recessive* genetic disease.

conchoidal fracture pattern
A fracture pattern showing smooth curved surfaces like the interior of a shell (conch).

consanguinity
Degree of inbreeding; consanguineous marriages involve closely related partners (for example, first cousins).

consensus coding sequence, CCDS
One of a core set of human and mouse *protein*-coding regions that are consistently annotated and of high quality.

conservative substitution
A *mutation* causing a *codon* to be replaced by another codon encoding an *amino acid* whose physicochemical properties are similar to those of the original amino acid. Often has no effect, or a comparatively mild effect.

constitutive
Always present (for example, of gene expression); not subject to change over time. Constitutively expressed genes are also referred to as *housekeeping* genes.

contig
A set of contiguous clone (for example, *BAC*) sequences, forming part of the *genome* sequence.

continental shelf
The submerged land of shallow gradient that lies between the sea shore and deep ocean bed.

control region
The segment of mtDNA containing the origin of replication and two *hypervariable segments*; equivalent to the *D-loop*.

convergent evolution
The evolution of similar forms by two lineages that do not share a common ancestor with that form. For example, the evolution of wings by birds and bats.

coprolite
Fossilized or semi-fossilized feces; a source for ancient DNA analysis.

copy-number polymorphism, CNP
A *copy-number variant* existing at a frequency of ≥1% in a population.

copy-number variation, CNV
Differences in the number of copies of a sequence (normally >1 kb in size) between *alleles*.

correlogram
A graphical display of how *autocorrelation* values vary depending on the distance between populations.

cosmid
A circular DNA vector in which inserts of typically ~40 kb long can be propagated in bacterial cells.

covalent bonds
Strong chemical bonds between atoms in a molecule requiring considerable energy to be broken.

coverage
In *next-generation sequencing* of DNA, the number of times a region of the targeted DNA is represented by a mapped sequence read. Typically represented as, for example, 4× coverage.

CpG
See *CpG dinucleotide*.

CpG dinucleotide
The sequence 5′–CG–3′ within a longer DNA molecule. The site of specific *DNA methylation*, important in the control of gene expression.

CpG island
A region of DNA, often associated with the 5′ region of a gene, in which unmethylated *CpG dinucleotides* are more plentiful than elsewhere in the genome.

craniodental
Pertaining to the *cranium* and teeth.

craniometry
The measurement of skull dimensions.

cranium
The bones of the skull except the *mandible* (lower jaw).

Creole language
A fully fledged hybrid language formed by the mixing of two parental languages.

cross-linking agent
A *mutagenic* compound that can create chemical cross-links between different parts of a DNA helix, or different helices.

crossing over, **or crossover**
The exchange of *DNA* between *chromosomes* at *meiosis*.

crowd epidemic disease
An infectious disease that requires a high host density if it is to be sustained in the population, for example, smallpox.

CRS
See *Cambridge Reference Sequence*.

C-terminus
The end of a *protein* bearing a free carboxy (–COOH) group, corresponding to the 3′-most translated region of a *gene*.

cultivation
The general planting and harvesting of plants.

cultural diffusion
See *acculturation*.

cultural drift
A process analogous to *genetic drift*, in which the frequencies of cultural traits in a population may fluctuate due to chance variation in which traits are transmitted; as with genetic drift, the effect is strongest in small populations.

cystic fibrosis, **CF**
An *autosomal recessive* disease due to mutations in the *CFTR gene*, in which the production of thick, sticky mucus leads to serious problems with the lungs and digestive system.

cytogenetic
Of the study of *chromosomes*, including their number, morphology, and evolution.

cytoplasm
The material within a cell excluding the *nucleus*.

cytoplasmic segregation
The unequal segregation of different *mitochondrial DNA* types within a cell into daughter cells at cell division.

cytosine, **C**
A *pyrimidine* base found within DNA; pairs with *guanine*.

D-loop (displacement loop)
The *control region* of *mtDNA*, so-called to reflect its unusual structure, in which one of the DNA strands is displaced by a partially replicated region.

DARC
Duffy antigen receptor for chemokines; see *Duffy blood group*.

dbSNP
An electronic database containing information about *SNPs* (www.ncbi.nlm.nih.gov/projects/SNP/).

deamination
A chemical change involving the loss of an amine group (–NH_2) from a molecule; the deamination of *cytosine* yields *uracil*, for example.

deficiency paternity testing
Testing of paternity when one parent is unavailable.

deleterious allele
An *allele* that decreases the fitness of an individual carrying it.

deletion
Loss of a segment of *DNA* from within a *chromosome*; often as opposed to *duplication*.

deme
See *subpopulation*.

demic diffusion
An advancing wave of a particular population, driven by *demographic* growth.

demographic history
A *population*'s size and distribution in previous generations.

demography
The study of the life-history of people; the characteristics of a *population*, such as size, age-structure, kinship-size etc.

denaturation
The separation of the strands of *DNA*, often by heating or treatment with alkali; in *PCR*, the first stage of a cycle, in which the temperature is typically 94°C.

denaturing high performance liquid chromatography, **DHPLC**
Method for *variant* detection in which *heteroduplexes* are detected by their altered retention time on chromatography columns under near-denaturing conditions.

dendrochronology
A means of dating based on the counting of annual growth rings in trees or timbers

Denisovan
Late Pleistocene *hominins* of unknown species whose finger bone and molar tooth that were discovered in Denisova Cave, Altai mountains, Siberia, have been studied for their DNA sequence.

de novo
Newly arising ("from new").

deoxyribonucleic acid, **DNA**
The informational macromolecule that encodes genetic information, composed of linked *deoxyribonucleotides*. DNA can be double or single stranded, with the strands linked by complementary *base pairing*.

deoxyribonucleotide
The monomeric molecular component of the polymer *DNA*; often abbreviated as *nucleotide*.

deoxyribose
The sugar part of a *deoxyribonucleotide*, the building block of *DNA*.

derived state
Of a DNA variant or phenotype: the recently evolved state, contrasted with the *ancestral state*.

developmental plasticity
The variation in developmental outcome as a result of environmental influences.

DHPLC
See *denaturing high performance liquid chromatography*.

diabetes mellitus
A condition in which the amount of glucose in the blood is elevated, causing long-term damage to eyes, kidneys, nerves, heart, and major arteries. Some forms of the disease have a genetic component. Often abbreviated to "diabetes."

diallelic
Of a polymorphism: possessing two *alleles* only (see *biallelic*, *binary marker*).

diaspora
A multidirectional dispersal from a homeland, often applied to Jewish populations.

dichromatic vision
Possession of two different types of color receptors in the retina of the eye.

dideoxynucleotide
A *nucleotide* analog lacking the hydroxyl (–OH) group on its 3′-carbon atom necessary for a phosphodiester bond to be made. When incorporated into a growing *DNA* chain, it terminates synthesis. Fluorescently labeled dideoxynucleotides are used in *Sanger sequencing*.

dideoxy sequencing
See *Sanger sequencing*.

differential lysis
In *forensic genetics*, a method to selectively enrich sperm cells in a mixture of different cell types.

diploid
Having two copies of the *genome*; for example, a cell or an organism.

direct repeats
Repeated sequences in the same orientation on a *chromosome*; *non-allelic homologous recombination* between such repeats can cause *deletions* and *duplications*.

disassortative mating
Using *phenotypic* disparity to guide mate choice.

disease association study
A study in which the frequencies of *alleles* in individuals with a disease are compared with those in normal control individuals in the hope of identifying a region of the *genome* associated with the disease.

disease heritage
The characteristic pattern of genetic diseases found within a population as a result of its evolutionary history.

distal
Toward the *telomere* of a *chromosome* arm (as opposed to *proximal*).

distance matrix
A table whose elements are distances between the categories (often *haplotypes* or populations) arrayed in rows and columns.

diversifying selection
A selective regime that favors greater diversity than that expected under *neutral* evolution.

dizygotic, DZ
Of twins: formed from two independent *zygotes*, and therefore no more closely related than any pair of sibs. Also called nonidentical, or fraternal twins (as opposed to *monozygotic, MZ*).

DMD
See *Duchenne muscular dystrophy*.

dN/dS ratio (also K_a/K_s ratio)
The ratio of *nonsynonymous* to *synonymous* changes in a DNA sequence.

DNA
See *deoxyribonucleic acid*.

DNA–DNA hybridization
See *hybridization*.

DNA fingerprint
A pattern of bands produced after *Southern blotting* and hybridization using *minisatellite* probes that is highly individual-specific and can be used in individual identification.

DNA methylation
Addition, by a specific methyltransferase *enzyme*, of a methyl ($-CH_3$) group to the 5-carbon of the *cytosine* ring of a *CpG* dinucleotide to give 5-methylcytosine; an *epigenetic* modification of DNA.

DNA polymerase
An *enzyme* responsible for the synthesis of a *DNA* strand using the *complementary* DNA strand as a template.

DNA profile
A pattern of peaks on an *electropherogram* produced after *PCR* amplification of highly variable *microsatellites* that is highly individual-specific and can be used in individual identification. Also refers to the set of digits representing the numbers of repeat units at each of the microsatellites.

DNA repair
The set of processes that maintain *DNA* integrity and minimize the number of *mutations*.

dolichocephalic
"Long headed," with a *cephalic index* below 75 (see *brachycephalic*, *mesocephalic*).

domestication
The selective breeding of a plant or animal species to make it more useful to humans.

domestication syndrome
A suite of changes in plants resulting from *domestication*.

dominant
Of an *allele* that shows its complete *phenotypic* effect when present in only one copy (*heterozygous*).

double-strand break, DSB
A lesion in *DNA* where both strands of the double helix are severed; an important *recombination* intermediate.

double-stranded DNA
DNA in its double-helical form, with two *complementary* polynucleotide strands attached to each other by *base pairing*.

drift
See *genetic drift*.

drifty gene hypothesis
The idea that, in ancestral populations, lipid-storage *genes* accumulated variation that was *neutral* and influenced mainly by *genetic drift* because the metabolism did not have to cope with excessive fat intake.

DSB
See *double-strand break*.

Duchenne muscular dystrophy, DMD
A serious *X-linked* recessive disease involving muscle weakness and respiratory failure.

Duffy blood group, Duffy antigen, FY
A red-blood-cell antigen encoded by the *DARC gene*, and named after the patient in which it was first discovered. Acts as the receptor for the malarial parasite *Plasmodium vivax*.

duplication
Copying of a segment of *DNA* within a *chromosome*; often as opposed to *deletion*.

dynamic mutation
A *mutation*, usually at a trinucleotide repeat locus, involving a large expansion of the repeat array; associated with trinucleotide repeat expansion disorders such as myotonic dystrophy and *Huntington's disease*.

DZ
See *dizygotic*.

ectopic
Of a *recombination* event, between non-allelic sequences ("in the wrong place").

edge
See *branch*.

effective mutation rate
The *mutation rate* that explains the accumulated diversity in a given population; usually applied to *microsatellites*, and tends to be considerably lower than the mutation rate observed in pedigree analysis.

effective population size, N_e
A quotient invented by Sewall Wright to compare the *genetic drift* experienced by different populations, and greatly affected by past small population sizes; contrasts with *census size*.

EHH
See *extended haplotype homozygosity*.

eigenvalue
In population-genetic *principal component analysis*, a value of an individual (or, less commonly, a *population*) on an *eigenvector*.

eigenvector
An individual principal component of a *principal component analysis*; used, for example, for analysis of *population structure*.

einkorn wheat
The first wheat to be successfully domesticated, *Triticum monococcum*.

electronic PCR, ePCR
Computational tools used to calculate theoretical *PCR* results using a given set of *primers* to amplify sequences from a sequenced genome; also known as *in silico* PCR.

electropherogram
A trace showing the intensity of fluorescence as a function of molecular weight; peaks in the trace at a particular wavelength (color) correspond to a specifically labeled molecule of a particular size. Seen as the output of many *capillary electrophoresis* devices in DNA sequencing, *SNP* typing, and *microsatellite* analysis.

electrophoresis
Separating macromolecules by using a voltage gradient; usually carried out using a gel or capillary matrix.

electrophoretic
See *electrophoresis*.

emmer wheat
The most important plant domesticate of the Neolithic, *Triticum dicoccum*.

emotional weeping
Crying in response to emotional state.

ENCODE project
A large-scale research consortium (the ENCyclopedia Of DNA Elements; genome.ucsc.edu/ENCODE/) that aims to find all the functional elements in the human *genome*.

endemic, endemicity *(adjective)*
Of a disease, constantly prevalent in a certain region; of a species, naturally found in a certain restricted region.

endocast
An internal cast, most relevantly of the interior of the *cranium*, thus revealing the structure of the exterior surface of the brain.

endogamy
The practice of marrying within a social group.

endogenous mutagens
Molecules produced within the body that can cause changes in DNA sequence (*mutations*).

endogenous retrovirus
Dispersed repetitive genomic element representing the remnant of a retroviral infection in our ancestors; often a substrate for recombination-mediated chromosomal rearrangements.

endonuclease
An enzyme that cleaves within a nucleic acid molecule (for example, DNA).

endophenotype
A readily definable quantitative parameter that underlies a disease *phenotype*.

endosymbiotic
Describing an organism living within the cell of another in symbiosis (an *endosymbiont*); the state of the progenitor of *mitochondria*.

enhancer
A DNA sequence that regulates gene expression from a distance.

enzyme
A *protein* that catalyzes a biological reaction; like any catalyst, it is not used up in the process.

ePCR
See *electronic PCR*.

epicanthal fold
The internal skin fold of the eyelid, seen in many Asian populations.

epigenetic
Inherited without involving a change to the DNA sequence itself; for example a pattern of DNA *methylation*, chemical modification of *histones*, or protein binding.

episodic selection
Selection that is not constant over time.

epistatic
Referring to interactions between genes influencing a complex trait.

Eskimo-Aleut
One of the three major families of Native American languages proposed by Joseph Greenberg, including the languages spoken by the Inuit.

estimator
A quantity calculated from the sample data to estimate an unknown parameter in the population.

ethidium bromide
An example of an *intercalating agent*, a flat molecule that can insert between the *base pairs* of a DNA double helix, often causing *frameshift* mutations. This chemical is very widely used in laboratories, since the DNA-ethidium bromide complex fluoresces under ultra-violet light.

etiology
The combination of factors that give rise to a disease.

euchromatin
The part of the genome containing transcriptionally active *DNA*, and which, unlike *heterochromatin*, is in a relatively extended conformation.

eugenicist
A practitioner, or follower, of eugenics, advocating the perceived improvement of human traits through the promotion of differential reproductive success.

eukaryotic
Describing an organism (for example, humans) whose cells possess membrane-bound *organelles*, particularly a *nucleus* that contains the genetic material (as opposed to *prokaryote*).

eustatic
Factors that alter the ratio of global water present in ice and liquid form.

evolutionary medicine
The application of evolutionary theory to the understanding of health and disease.

exclusion
In *forensic genetics*, the finding that a crime-scene sample has a different *DNA profile* to a suspect, or to another crime-scene sample, thus demonstrating that they are derived from different individuals; term also used in *paternity testing*.

exome
Collective term for the set of protein-coding *exons*, which can be analyzed via *sequence capture* and *next-generation sequencing*.

exon
One of the noncontiguous sections of a *gene* that together form the region present in the mature transcript; in many genes, exons comprise the coding region, but some genes have noncoding exons. See also *intron*.

exon skipping
Failure to retain a particular *exon* in a mature *mRNA* on *splicing*; can be caused by *mutation* of a *splice site*.

exonuclease
An *enzyme* that degrades a *DNA* molecule from one of its ends.

expected value, expectation
If x is a variable, then $E(x)$ (the expected value of x) describes the expectation of the value of x, averaged over a number of trials or experiments. It can be regarded as equivalent to the mean of x, except the mean would be calculated from data while $E(x)$ is calculated from all different possible values of x and their relative probabilities.

extended haplotype
homozygosity (EHH): A measure of the transmission of an extended *haplotype* without *recombination* - the probability that two randomly chosen *chromosomes* carrying a core haplotype of interest are *identical by descent* (as assayed by *homozygosity* at all *SNPs*) for the entire interval from the core region to a given distance.

extended haplotype test
A class of methods that detects recent *positive selection* by identifying extended *haplotypes* around selected *alleles*.

extension
The stage of a *PCR* in which *Taq* polymerase extends the new strands of *DNA* from the *oligonucleotide primers*; typically carried out at 72°C.

false negative
See *type 2 error*.

false positive
See *type 1 error*.

falsifiability
The possibility of saying with certainty that a hypothesis is not correct.

familial searching
In *forensic genetics*, screening a *forensic identification database* for *DNA profiles* sufficiently similar to a crime-scene profile to represent a close relative of its depositor.

favism
Hemolytic anemia occurring in *G6PD*-deficient individuals, and due to ingestion of vicine, an alkaloid glycoside present in broad (fava) beans, which increases oxidative stress.

Fay and Wu's *H*
A statistical test that distinguishes between a *DNA* sequence evolving neutrally and one evolving under *positive selection*.

fecundity
The biological capacity for reproduction.

Fertile Crescent
A crescent-shaped region in the Near East, encompassing the valleys of the Tigris and Euphrates rivers, and the delta and valley of the Nile. A key region for the early development of *agriculture*.

fertility
The actual numbers of offspring produced.

fiber FISH
Fluorescence in situ hybridization (*FISH*), in which fluorescently labeled probes are hybridized to stretched *chromatin* fibers, allowing a resolution of a few *kilobases*.

FISH
See *fluorescence in situ hybridization*.

fitness
The ability of an individual to survive and reproduce relative to the rest of the population.

fixation
The process by which one *allele* increases in a population until all other alleles go extinct and the locus becomes *monomorphic*.

fixation indices
Quotients devised by Sewall Wright to analyze the departure of *genotype* frequencies from *Hardy–Weinberg equilibrium*.

fixed
A locus that has undergone *fixation*.

flavin
A family of organic molecules derived from riboflavin (vitamin B_2).

flow-cell
The part of an Illumina™ *next-generation sequencing* device in which the sequencing takes place.

fluctuating selection
A mode of *selection* in which the selective pressure varies through time, for example, in infectious disease.

fluorescence *in situ* hybridization, FISH
Hybridization of a fluorescently labeled nucleic acid to a target *DNA* or *RNA*, usually immobilized on a microscope slide; a key technique in physical mapping of *chromosomes*.

folate
Vitamin B_9, required by the body for synthesis of DNA.

forensic genetics
The application of *genetics* for the resolution of legal cases.

forensic identification database
A large database of *DNA profiles* from individuals and crime scenes.

forensic population database
A database of *DNA profiles* from ~100–200 unrelated individuals from a particular population, used to estimate allele frequencies.

forward simulations
Successive *in silico* modeling of *population* and/or molecular processes forward in time to generate multiple replicates of artificial genetic diversity datasets.

founder analysis
A method for analyzing *haplotypes* (typically *mtDNA*), aiming to identify and date migrations into new territories. The method chooses founder sequence types in potential source populations and dates lineage clusters deriving from them in settlement zones.

founder effect
Reduced genetic diversity in a population founded by a small number of individuals.

fourfold degenerate
Describing a site within a *codon* at which any of the four possible *nucleotides* specifies the same *amino acid*.

frameshift mutation
A *mutation* in a coding region that causes the three-base *reading frame* to be shifted by adding or subtracting a number of bases that is not a multiple of three.

frequency-dependent selection
A selective regime under which lower frequency variants are favored, which prevents their elimination and promotes diversity.

frequentist
A class of statistical methods relying on a p value.

F_{ST}
A family of measures of population subdivision devised by Sewall Wright that can also be used as a *genetic distance* between populations.

FY
See *Duffy blood group*.

g
Gram.

G
See *guanine*.

G6PD
Glucose-6-phosphate dehydrogenase.

gain-of-function mutation
A mutation that causes a gene product to possess a new function, often leading to a dominantly inherited phenotype.

gamete
An egg or sperm; these cells are *haploid*.

gametic phase
The way in which the *alleles* of two linked loci are associated together in a *genotype*; also simply referred to as *phase*.

gametogenesis
Development and maturation of sex cells through meiosis.

gametolog
Similar sequences on the X and Y chromosomes, some of which share an origin in the *autosomal* pair from which the modern X and Y have evolved.

gamma distribution
A continuous probability distribution with two parameters: α (shape parameter) and β (rate parameter).

Gb
See *gigabase*.

G-banding
A staining method (employing the stain Giemsa) that reveals dark- and light-staining bands in *chromosomes*, used to distinguish chromosomes from each other and recognize individual chromosomes.

GC-content
The percentage of *base pairs* in a *genome* or sequence that are G–C, as opposed to A–T.

gene
Part of a *DNA* molecule that encodes a *protein* or functional *RNA* molecule.

gene–culture co-evolution
The interaction of genetic and cultural adaptations in human evolution.

gene conversion
A nonreciprocal exchange of sequence information between one *DNA* molecule and another; despite the name, not restricted to *genic* sequences.

gene diversity
A widely used measure of genetic diversity devised by Masatoshi Nei; refers to nongenic sequences, as well as *genes*.

gene dosage
The number of copies of a *gene* (or *DNA* sequence) present in a single cell. In humans, who are *diploid*, the most common value is two for *somatic cells*.

gene duplication
The generation of an identical copy of a segment of *DNA*, often located in tandem.

gene expression
Transcription of the information embedded in the DNA code of *genes* into *mRNA* and in the case of protein-coding genes its further *translation* into the amino acid code.

gene flow
The movement of *genes* resulting from the movement of people from one region to another, followed by successful reproduction and subsequent genetic contribution to the next generation.

gene genealogy
The ancestral relationships (genealogical history) among a number of sampled copies of a *DNA* segment; not restricted to *genes* themselves.

gene knockout
A targeted disruption of a particular *gene*, for example, in mice.

gene ontology
A *gene* classification system based on a catalog of each gene product's cellular component of expression, molecular function, and the biological process in which it is involved.

gene tree
A *phylogeny* relating the ancestry of sequences at the same locus in different species or individuals, where *bifurcations* relate molecular divergences, and not speciation or population events; not restricted to *genes* themselves.

general reversible model
A model of sequence evolution in which each *base substitution* has its own rate of *mutation*, identical in either direction, and *base composition* is taken into account.

generation time
The average time between the birth of a parent and the birth of their offspring.

generation time hypothesis
The suggestion that differences in *mutation* rates between different evolutionary lineages can be accounted for by the different number of *germ-line* replications per unit of time.

genetics
The study of inheritance.

genetic architecture
See *allelic architecture*.

genetic barrier
A geographical or cultural difference associated with elevated *allele* frequency change as a result of low *gene flow*.

genetic boundary
A geographical or cultural difference associated with elevated *allele* frequency change; may result from low *gene flow* or migration.

genetic code
The three-letter code that allows base sequences within *messenger RNA* molecules to be translated into specific *amino acids* within a *protein*.

genetic dating
A class of statistical methods used to estimate the time to the most recent common ancestor of a set of populations or molecules.

genetic differentiation
The process by which the genetic composition of two or more populations isolated from one another diverges over time.

genetic disease heritage
See *disease heritage*.

genetic distance
A measure of the evolutionary relatedness of two populations or two molecules OR
Distance on a genetic map, defined by the frequency of *recombination* events, and measured in *centiMorgans*.

genetic drift
The random fluctuation of *allele* frequencies in a finite population due to chance variations in the contribution of each individual to the next generation.

genetic genealogy
Use of *DNA*-based evidence to determine the genetic relationship between individuals.

genetic linkage
The co-inheritance of *alleles* at different *polymorphisms* on the same *chromosome* within a family.

genetic load
The burden on a population caused by the deaths required to eliminate deleterious *alleles*.

genetic map
An ordered sequence of polymorphic loci determined by analyzing *recombination* events. The distance between loci is determined by the probability of observing a recombination between them (as opposed to a *physical map*).

genetics
The study of inheritance and variation.

genic
Of, or relating to, a *gene*.

Genographic Project
A study, begun in 2005, that aims to map past human migration patterns by collecting and analyzing DNA samples from hundreds of thousands of people from around the world.

genome
The total content of genetic information in an organism.

genomewide association study, GWAS
A study of many common genomewide variants (usually *SNPs*) in different individuals to determine if any variant is associated with a trait.

genomewide scan
An attempt to locate a *gene* of interest by *linkage analysis* carried out using a large number of anonymous *polymorphisms* spread throughout the genome.

genomic disorder
A genetic disorder in which the underlying mutant *allele* is due to an underlying structural feature of the genome, such as *deletion* or *duplication* of a region through aberrant *recombination*.

genomics
The scientific discipline of mapping, sequencing, and analyzing *genomes*.

genotype
The combination of *allelic* states of a set of polymorphisms lying on a pair of *chromosomes*—the combination of two *haplotypes*.

genotypic relative risk, GRR
The increased chance that an individual with a particular *genotype* will develop a disease.

germ line
The cell lineage culminating in eggs and sperm, the cells responsible for passing genetic information from one generation to the next.

gigabase
Unit of a thousand million *bases*, or (usually) a thousand million *base pairs*, abbreviated Gb; one copy of the human *genome* contains about 3.2 Gb of DNA.

glacial period
Period when the average temperature was colder than at present and parts of the Earth were covered by ice sheets.

glacial refugia
Regions in which populations took refuge after their retreat from the harsh environments of extreme latitude during recent *glacial periods*.

glottochronology
Statistical methods for calculating linguistic divergence times from the proportion of word *cognates* shared between different languages.

glucosinolate
A class of toxic glucoside produced by most *Brassica* plants.

glume
A leaf-like structure that protects the seeds in an ear of wheat.

goodness-of-fit
A class of statistical methods used to assess competing models on the basis of their fit to empirical data.

gracile
Of body form and skeletal remains, lightly built (as opposed to *robust*).

grammatical
Pertaining to the phonology, rules of inflection, and sentence structure of a language.

great circle distance
The distance between two points on the surface of a sphere (for example, the world) calculated by using a trigonometric equation to take into account the curvature of the sphere.

group informed consent
Informed consent given not only by the participating individual, but also by family or wider group involved.

GRR
See *genotypic relative risk*.

guanine, G
A *purine* base found within *DNA*; pairs with *cytosine*.

GWAS
See *genomewide association study*.

hafted
Attached, for example of a stone point to an arrow or spear shaft.

half-life
For a radioactive isotope, the length of time after which there is a 50% chance that an atom will have undergone nuclear decay.

handaxe
See *biface*.

handedness
Unequal distribution of skills between the left and right hand.

haplogroup
Usually applied to a set of *mtDNA* or Y-chromosomal *haplotypes* that is defined by relatively slowly mutating *polymorphisms*, and that has more phylogenetic stability than other "haplotypes." For the Y chromosome, haplogroups are defined by *binary* polymorphisms such as *SNPs*, while haplotypes are usually defined by *microsatellites*.

haploid
Having one copy of the *genome*; can describe a cell (for example, a *gamete*), or an organism.

haploinsufficiency
A condition in which a *gene* product is only available from one *allele* of a *diploid* locus, and where this amount of product is not sufficient for normal function. One basis for a *dominantly* inherited phenotype.

haplosufficiency
A condition in which a gene product is only available from one *allele* of a *diploid* locus, but this amount of product is sufficient for normal function. This kind of allele has a *recessive* mode of inheritance.

haplotype
The combination of *allelic* states of a set of *polymorphisms* lying on the same *DNA* molecule, for example, a *chromosome*, or region of a chromosome.

haplotype block
The apparent haplotypic structure of recombining portions of the *genome* in which blocks of consecutive co-inherited *alleles* are separated by short boundary regions.

haplotypic background
See *ancestral haplotype*.

HapMap, The International HapMap Project
A large international project to characterize patterns of genetic variation and *linkage disequilibrium* in samples of individuals from geographically defined populations.

HAR
See *human accelerated region*.

hard sweep
A *selective sweep* in which a single *haplotype* increases in frequency due to recent *positive selection* on one *variant*.

Hardy–Weinberg equilibrium
The situation when the observed *genotype* frequencies within the population equate to the expected genotype frequencies calculated from known *allele* frequencies using the Hardy–Weinberg principle.

Hb^S
Hemoglobin S.

HD
See *Huntington's disease*.

Heinrich event
Brief spell of extreme cold during a *glacial period*.

hemizygous
Possessing only one *allele* as a result of normal *chromosome* constitution; for example, males are hemizygous for all *X chromosome*-specific loci.

hereditary hemorrhagic telangiectasia
A fully penetrant *dominant* disease that leads to recurrent and often severe nosebleeds and gastrointestinal bleeding.

heritability
The proportion of variance in a trait that is explained by genetic factors.

HERV
Human endogenous retrovirus.

heterochromatin
Highly condensed, transcriptionally inert segment of the *genome* often behaving abnormally in chromosomal banding; the part of the *genome* other than *euchromatin*.

heteroduplex
A double-stranded DNA molecule in which one strand contains a sequence difference with respect to the other.

heterogeneity
Of a genetic disorder, the phenomenon whereby *mutations* in more than one *gene* result in the same phenotype.

heteroplasmy
Possessing two or more different *mitochondrial DNA* sequences in the same cell, or individual (as opposed to *homoplasmy*).

heterozygosity
A measure of the diversity of a polymorphic locus; for a *diploid* locus, the average probability that the *alleles* carried by an individual are different from each other.

heterozygote
An individual who is *heterozygous*.

heterozygote advantage
A selective regime in which the *heterozygote* has greater fitness than either *homozygote*.

heterozygote disadvantage
A selective regime in which the *heterozygote* has lower fitness than either *homozygote*.

heterozygous
Carrying two different *alleles* at a particular *diploid* locus.

heuristic
Of a model or representative device, aiding understanding or explanation.

HGDP
See *Human Genome Diversity Project*.

HGDP-CEPH panel
A set of 1064 *lymphoblastoid cell lines* from 51 diverse human populations (see Box 10.2).

HGPS
See *Hutchinson-Gilford Progeria Syndrome*.

hierarchical population structure
See *population structure*.

high coverage
In *next-generation sequencing* of *DNA*, sequencing in which, on average, each *base* of the targeted DNA is represented by many (for example, ≥30) mapped *sequence reads*.

histocompatibility
The degree to which tissue from one individual will be tolerated by the immune system of another.

histone
One of a family of small, basic (positively charged), highly conserved *proteins* associated with *DNA* in the *chromosomes* of all cells except sperm (which instead use protamines).

hitchhiking
The increase in frequency of a neutral allele as a result of positive selection for a linked *allele*.

HIV
Human Immunodeficiency Virus, the cause of *AIDS*.

HKA test (Hudson–Kreitman–Aguadé test)
A test for *natural selection* in which within-species *polymorphism* and between-species divergence are compared at two or more *loci*.

HLA
The human leukocyte antigen system, a name conventionally used for the *major histocompatibility complex* (MHC) in humans.

Holliday junction
A four-stranded *DNA* structure occurring as an intermediate in *homologous recombination*.

Holocene
The last ~11 KY, with an unusually warm and stable climate.

hominid/hominin
A species (extinct or extant) more closely related to humans than to our closest living relatives, *chimpanzees* and *bonobos*.

homologous
Phenotypic or genotypic characters that share a common ancestor.

homologous recombination, HR
Recombination between sequences that share a high degree of sequence similarity, for example, allelic homologs, or non-allelic *paralogs*.

homoplasmy
Possessing only one *mitochondrial DNA* sequence type in a cell, or individual (as opposed to *heteroplasmy*).

homoplasy
The generation of the same state by an independent evolutionary path (*convergent evolution*).

homopolymeric
Describes a tract of adjacent identical *nucleotides* in *DNA*, for example, poly(A).

homozygosity
The state of being *homozygous*.

homozygote
An individual who is *homozygous* for a particular sequence variant.

homozygous
Carrying the same *allele* at each copy of a *diploid* locus.

horticulture
Non-intensive management of individual plants, rather than large populations of plants (*agriculture*).

hotspot
See *recombination hotspot*.

hotspot motif
A 13-bp degenerate sequence motif (CCNCCNTNNCCNC), which is over-represented in human *hotspots*, and estimated to be associated with the activity of about 40% of them; also referred to in this book as a *Myers motif* after its discoverer.

housekeeping
See *constitutive*.

HR
See *homologous recombination*.

human accelerated region, HAR
Region of the *genome* showing higher rate of evolution than other genome regions on average, specifically in the human lineage.

Human Genome Diversity Project, HGDP
A project aiming to study genetic diversity in many different human *populations*. It encountered difficulties in doing this,

but was successful in establishing the now widely used *CEPH*-HGDP cell-line panel.

Human Genome Project
An international collaborative project that accomplished the first sequencing of the human *genome*, beginning in 1987, and declared complete in 2003.

human leukocyte antigen
See *HLA*.

hunter-gatherer
A system of food supply based on the hunting of wild animals and gathering of wild plants.

Huntington's disease, HD
An *autosomal dominant* disease involving movement abnormalities, personality changes, and cognitive loss, caused by expansion of a *trinucleotide repeat* encoding a *polyglutamine* tract.

Hutchinson–Gilford Progeria Syndrome, HGPS
A serious *autosomal dominant* genetic disease with very early onset symptoms that resemble aging.

HVR
See *hypervariable region*.

HVS
See *hypervariable segment*.

hybridization
The process of annealing two *homologous DNA* molecules together.

hydantoin
Oxidized derivative of the *base cytosine* or *thymine*, blocking *DNA polymerases*—a problem in *ancient DNA* analysis.

hydrogen bond
Weak interatomic bond either within a molecule or between molecules; these are the bonds between *bases* in a *base pair* of *DNA*.

hyoid
Horseshoe-shaped lingual bone in the neck between chin and thyroid cartilage.

hyperglycemia
Elevated blood-sugar level.

hypervariable region, HVR
See *hypervariable segment*.

hypervariable segment, HVS
Part of the *mitochondrial DNA* molecule showing particularly high *DNA* sequence variability; two of them (HVSI, HVSII) lie within the *control region* of mtDNA; also known as *hypervariable regions, HVRs*.

hypoxia
A condition where the body, or part of the body, has inadequate oxygen supply.

IBD
See *isolation by distance*, or *identical by descent*.

Iberian peninsula
The region of southwestern Europe that comprises Spain and Portugal.

IBS
See *identical by state*.

ice age
See *glacial period*.

ice sheet
A layer of ice covering a tract of land.

identical by descent
See *identity by descent*.

identity by descent, IBD
Property of *alleles* in an individual or in two people that are identical because they were inherited from a common ancestor (as opposed to *identity by state*).

identity by state, IBS
Property of *alleles* in an individual or in two people that are identical because of coincidental mutational processes, and not because they were inherited from a common ancestor (*identity by descent*).

iHS
Integrated haplotype score, a kind of *extended haplotype test*.

Illumina™ sequencing
A high-throughput proprietary method of *next-generation sequencing*, relying on *DNA* synthesis and *chain termination*.

immunoreactivity
A method for assaying genetic diversity at the *protein* level by using the discriminating power of the immune system.

imprinted gene
A *gene* expressed in a parent-of-origin-specific manner. Imprinted genes can be either maternally or paternally suppressed.

imputation
Inference of the *allelic* states of *SNPs* that have not themselves been typed in a sample, by exploiting the *LD* structure of the *genome* and the availability of more extensively typed reference population datasets.

in silico
Within a computer (in silicon).

in utero
Within the uterus (during gestation).

in vitro
Within laboratory equipment (in glass).

in vivo
Within a living organism (in life).

inbreeding
Reproduction involving genetically closely related parental types.

inbreeding depression
The reduction in fitness observed in offspring resulting from *inbreeding* as a result of the elevated frequency of deleterious *recessive homozygous alleles*.

incomplete lineage sorting (ILS)
A difference between gene trees and species tree due to the short time gaps between multiple ancestral speciation events.

incidence
The proportion of people who develop a characteristic, such as a disease, over a particular period, for example a year or a lifetime.

inclusion
See *match*.

indel
A *variant* or *polymorphism* involving the *insertion* or *deletion* of *DNA* sequence.

independent assortment
The random segregation of each member of a *chromosome* pair to opposite poles of the cell during *meiosis*.

indexing
In *next-generation sequencing*, the use of a specific short (typically 6-bp) sequence in library preparation that allows a particular *sequence read* to be assigned to a particular individual within a *multiplexed* sequencing run.

individual identification
The use of highly variable *DNA* markers to distinguish one individual from many (or all) others.

induced pluripotent stem cells, iPS cells
Cells that have been reprogrammed to resemble stem cells, often by forced expression of transcription factors, or use of small molecules. iPS cells can be generated from adult cells such as fibroblasts, and subsequently differentiated into many cell types, allowing their study (or potential use) in medicine.

inferential methods
Statistical methods for making inferences about the processes that generated the data.

infinite alleles model
A *population genetics* model that assumes that each *mutation* generates a new *allele* not previously found within the population.

infinite sites model
A model of *nucleotide diversity* in which each *mutation* occurs at a different nucleotide; also *infinite alleles model*.

informative sites
A subset of *segregating sites* that are phylogenetically informative because neither allele is a *singleton*.

informed consent
Agreement to take part in an investigation that is given freely and based on an understanding of the work, including its risks and benefits.

insertion
A segment of DNA present in a chromosome that is absent from a reference or ancestral sequence.

intercalating agent
A *mutagenic* compound with a flat molecular shape, such as *ethidium bromide*, that can insert between *base pairs* in *DNA*, distorting its structure and often leading to *insertions* or *deletions*.

interfertile
Of two species, capable of mating and producing offspring, though these are not necessarily themselves of normal fertility.

interglacial
Warm geological interval between *glacial periods*; we are currently living in an interglacial.

International HapMap Project
See *HapMap*.

interpolation
The process of generating a value for a variable at a site intermediate between two sites at which the value for the variable is known empirically.

interstadial
Brief, relatively warm spell during a *glacial period*.

intron
A transcribed but noncoding section of a *gene* that separates the *exons*.

inversion
A *structural variant* in which a segment of *DNA* has the opposite orientation to that in the *reference sequence*.

ionizing radiation, **IR**
Energetic electromagnetic radiation capable of directly breaking the *DNA* backbone, or producing reactive ions that modify DNA, for example, gamma- and X-radiation.

iPS cells
See *induced pluripotent stem cells*.

IQ test
One of several standardized tests designed to assess intelligence.

island model
See *N-island model*.

isolated population
A population that has experienced little *gene flow* from surrounding populations and is therefore differentiated from them; also simply "isolate."

isolation by distance, **IBD**
The decline of population similarity with geographical distance as a result of mating distances being less than the range of the species.

isostatic
Factors influencing local sea levels as a result of vertical movements of the Earth's crust.

isotope ratio
A measure of relative abundances of different isotopes of the same chemical element in a particular sample that is commonly used, for example, in *radiocarbon dating*.

JC
See *Jukes–Cantor model*.

JPT
One of the *HapMap* population samples: 44 unrelated individuals, Japanese from Tokyo, Japan.

Jukes–Cantor model, **JC**
The simplest model of sequence evolution in which each *base substitution* has the same rate.

"Just So" story
An explanation for which there is little or no direct evidence; from Rudyard Kipling's tales, for example, "How the Leopard got his spots."

K_a/K_s ratio
See *dN/dS ratio*.

karyogram
The cytogenetic representation of a set of banded *chromosomes*.

karyotype
The complement of *chromosomes* in a cell or individual, revealed by *cytogenetic* analysis at certain stages of the cell cycle; for example, 46,XY for a normal male.

kb
See *kilobase*.

kg
Kilogram.

Khoe-San
The indigenous hunter-gatherer and pastoralist peoples of southern Africa, speaking languages belonging to the *Khoisan* family.

Khoisan
A language family of southern Africa, spoken by the *Khoe-San* people.

kilobase
A unit of a thousand *bases*, or (usually) a thousand *base pairs*, abbreviated kb.

kin-structured migration
The situation when migrating individuals are not a random sample from the source population but are members of the same family.

kinetochore
The *protein* structure at the *centromere* of a *chromosome* to which *spindle fibers* attach.

knuckle-walking
Walking on all fours, using the knuckles of the hands—like *chimpanzees*.

kriging
A method for interpolating a *surface* distribution of *gene* frequencies over a landscape.

kuru
An acquired neurodegenerative disease, caused by a *prion* and transmitted by eating the brains of affected individuals. Has been prominent among the Fore people of Papua New Guinea.

KY
Thousand years.

KYA
Thousand years ago.

L1 element
A member of the *LINE* class of interspersed repeated sequences. The full-length element is about 6.1 kb long, but many are truncated. The L1 copy number is about 0.5 million.

lactase
The *enzyme* that hydrolyzes the disaccharide *lactose* into glucose and galactose; full name is lactase phlorizin hydrolase.

lactase persistence
The persistence of intestinal *lactase* into adulthood, maintaining the ability to digest *lactose*, the major sugar of milk.

lactose
The major sugar in milk, a disaccharide that is digested by *lactase* into its constituent monosaccharides glucose and galactose.

lactose intolerance
The inability to digest *lactose* in adulthood; this is the ancestral condition, found in nearly all mammals.

language family
A group of languages sharing a common origin, such as Indo-European. The names, composition, and status of proposed language families are not always agreed among linguists.

language shift
The displacement of one language by another throughout an entire speech community.

Lapita
A cultural complex, denoted in part by a distinctive pottery style, that is associated with the first settlement of *Remote Oceania*.

last glacial maximum, **LGM**
18–22 KYA, a period in climate history when the land-based ice volume was at its maximum.

Later Stone Age
The period following the *Middle Stone Age* in Africa, defined by its cultural assemblages such as complex tools; the term *Upper Paleolithic* originated in Europe and is not used in Africa, but the two are broadly contemporary.

LBK
Linienbandkeramik; see *linear pottery*.

LCL
See *lymphoblastoid cell line*.

LCR
See *low-copy repeat*.

LD
See *linkage disequilibrium*.

leaky replacement model
Of human evolution: an intermediate model between the *out-of-Africa* and *multiregional* extremes, proposing replacement of archaic humans by African migrants with some *admixture*, for example with *Neanderthals* and *Denisovans*.

Levallois technique
A *Middle Paleolithic/Middle Stone Age* method of producing stone tools by first preparing a core and then removing the tool in a nearly finished state as one final flake (also *Mode 3* technology).

Levantine
Of the Levant, a geographic and cultural region consisting of the eastern Mediterranean seaboard between Anatolia and Egypt.

lexical
Relating to items of vocabulary.

lexicostatistics
The statistical study of vocabulary differences between related languages.

LGM
See *last glacial maximum*.

library
A large collection of *DNA* fragments carrying adapter sequences at their ends, amplified and ready for *next-generation sequencing*.

life history
Sequence and duration of developmental stages in an organism's lifetime.

ligation, **ligated**
The process of covalent joining of two DNA molecules by the formation of a phosphodiester bond using a DNA ligase enzyme.

likelihood
A statistical framework that considers which hypothesis out of a range of options best accounts for the observed outcome.

likelihood ratio test
A method for comparing the likelihood of different hypotheses.

LINE, **long interspersed nuclear element**
A dispersed long repeat sequence such as an *L1 element*.

lineage
A group of *taxa* sharing a common ancestor to the exclusion of other taxa.

lineage assortment
The process by which *polymorphisms* within a parental species are randomly distributed into daughter lineages during speciation. This can lead to discrepant *gene trees*.

lineage effects
Mutation rate differences among different evolutionary lineages.

linear pottery
A distinctive type of pottery associated with the early European Danubian farming culture of the *Neolithic*; also known as *Linienbandkeramik, LBK*.

linearity
The property of a summary statistic that describes how closely its relationship with another variable of interest (often time) approximates to a straight line.

lingua franca
A language used for communication among people whose first languages differ.

Linienbandkeramik, **LBK**
See *linear pottery*.

link
In a *phylogenetic network*, a line connecting two *nodes*.

linkage analysis
The mapping, often using pedigrees, of a *gene* or trait on the basis of its tendency to be co-inherited with polymorphic loci; see *LOD score*.

linkage disequilibrium, **LD**
Nonrandom association between *alleles* in a population due to their tendency to be co-inherited because of reduced *recombination* between them.

linked
Of *alleles* that are not independently inherited, such as those lying close together on a particular *chromosome*.

locus
The specific location of a *DNA* sequence on a *chromosome*; sometimes a unit of genetic inheritance, for example, segment of *autosomal* DNA between *recombination* breakpoints, or *mtDNA*; plural *loci*.

LOD score
The logarithm (in base 10) of the odds of *linkage*—the ratio of the likelihood that loci are linked to the likelihood that they are not linked.

long branch attraction
A phenomenon in *phylogenetics* where two or more long branches in a tree are misclassified as more closely related sister groups.

long PCR
A modified *PCR* protocol allowing long (up to 35 kb) fragments to be amplified.

long-term effective population size
Harmonic mean of *effective population size* measured at

different time-points, which approximately reflects the smallest bottleneck the *population* has gone through.

longitudinal study
In epidemiology, a research study that involves repeated observation or measurement of a cohort of individuals over a period of time, often decades—for example the Framingham Heart Study.

low-copy-number, **LCN**
In *forensic genetics*, *DNA profiling* where the amount of DNA is very low, and *PCR* cycle numbers need to be increased.

low-copy repeat, **LCR**
Repeated sequences present on a *chromosome* in two or a few copies, between which *non-allelic homologous recombination* can occur. Usually a few kilobases in size; see also *paralog*.

low coverage
In *next-generation sequencing* of DNA, sequencing in which, on average, each *base* of the targeted DNA is represented by only a few (typically 2–6) mapped *sequence reads*.

Lower Paleolithic, **LP**
The earliest part (and thus lower down in archaeological deposits) of the Paleolithic period or Old Stone Age. Its dates can vary for one place to another, but lie between ~2.6 MYA and ~200 KYA.

LP
See *Lower Paleolithic*.

LSA
See *Later Stone Age*.

lymphoblastoid cell line, **LCL**
An immortalized cell line that grows rapidly and indefinitely, established by transforming white blood cells using Epstein–Barr virus.

m
Meter.

macroevolution
The evolution of taxonomic groups above the species level.

MAF
See *minor allele frequency*.

major histocompatibility complex, **MHC**
The *locus* on human *chromosome* 6 that contains the major determinants of tissue compatibility between individuals; see *HLA*.

malaria
A mosquito-borne tropical and subtropical infectious disease caused by unicellular parasites of the genus *Plasmodium*. Symptoms typically include fever and headache, progressing in severe cases to coma or death.

MALDI
See *matrix-assisted laser desorption-ionization*.

male-driven evolution
The hypothesis that the majority of evolutionary *mutations* occur in the male *germ line* as a result of the greater number of cell replications than in the female germ line.

male-specific region of the Y, **MSY**
See *nonrecombining portion of the Y*.

Mammoth Steppe
A cold and dry, but productive, environment that existed during the last *glacial period* and supported abundant large animals, including mammoths.

mandible
Lower jaw bone.

Mantel test
A statistical test for correspondence between a number of distance matrices, for example, of geographical, genetic, and linguistic distances.

Māoris
Native Austronesian-speaking people of New Zealand.

marker
See *molecular marker*.

Markov chain Monte-Carlo, **MCMC**
A simulation method used in *Bayesian* calculations.

mass screen
In *forensic genetics*, the *DNA profiling* of a large number of volunteers within a geographical region in which a perpetrator is thought to live.

mass spectrometry, **MS**
An analytical method in which the mass/charge ratio of ions is measured, allowing their molecular weights to be deduced very accurately.

massively parallel sequencing
See *next-generation sequencing*.

mastodon
An extinct elephant-like mammal.

match
In *forensic genetics*, the finding that a crime-scene sample has the same *DNA profile* as a suspect, or as another crime-scene sample; term also used in *paternity testing*. The word *inclusion* is also used.

match probability
In *forensic genetics*, the chance that a *DNA profile* chosen at random from the population would match the profile found at the crime scene.

mate choice
The situation whereby one sex is of limited abundance and can exercise choice over their mates leading to sexual selection for attractive traits.

material culture
The physical remains of cultural processes.

matriline
All female-line relatives of a person (excluding the daughters of a man)—they will share an *mtDNA haplotype*.

matrilocality
A marital residence pattern whereby a husband moves to his wife's birthplace (as opposed to *patrilocality*).

matrix-assisted laser desorption-ionization, **MALDI**
A *mass spectrometry* technique allowing the analysis of large, nonvolatile, thermolabile intact molecules.

maximum likelihood, **ML**
A method for selecting the best hypothesis from a set of alternatives on the basis of which maximizes the likelihood of the outcome.

maximum parsimony, **MP**
A method for selecting the best evolutionary tree from a set of alternatives on the basis of which contains the fewest evolutionary changes.

Mb
See *megabase*.

MC1R
melanocortin-1 receptor.

McDonald–Kreitman test
A *codon*-based test for *natural selection* that compares the amount of *nonsynonymous* and *synonymous polymorphism* within a species to the analogous fixed differences between species.

MCMC
See *Markov chain Monte-Carlo*.

MDA
See *multiple displacement amplification*.

MDS
Multi-dimensional scaling.

median joining
A method for constructing *phylogenetic networks*, based on the limited introduction of likely ancestral sequences or *haplotypes* into a *minimum spanning network* of the observed sequences.

median network
A *phylogenetic network* displaying the relationships of both observed and unobserved *haplotypes*.

megabase
Unit of a million *bases*, or (usually) a million *base pairs*, abbreviated Mb.

megafauna
Large-bodied vertebrates, now mostly extinct except in Africa.

megalithic
Built of large stones (megaliths); including tombs, single standing stones, and alignments of stones. Characteristic of the *Neolithic* in the west of Europe.

meiosis
The series of events, involving two cell divisions, by which *diploid* cells produce *haploid gametes*.

meiotic drive
A process causing one type of *gamete* to be over- or under-represented in the gametes formed during *meiosis*, and hence in the next generation.

Melanesia
The group of Pacific islands that lies from New Guinea in the west to Fiji in the east. Now discredited as a useful unit of anthropological investigation as a result of the different settlement histories of the populations on these islands.

melanin
The most important pigment influencing skin and hair color.

melanocyte
The cell type in which *melanin* is synthesized; lies at the boundary between the dermis and epidermis.

melanosome
A vesicle within the *melanocyte* in which *melanin* is synthesized; transported into keratinocytes in the skin.

melting
The separation of the strands of *DNA*, often by heating or treatment with alkali; also called *denaturation*.

meltwater pulses
Rapid rises in sea levels that resulted from sudden ice-sheet thaws during the warming at the end of the last *ice age*.

membership coefficient
A property of an individual within a *cluster analysis*, reflecting the probability of belonging to a particular cluster.

menarche
In women, the age at which the first menstrual period occurs.

Mendelian
Of a pedigree pattern, *allele,* or disease, showing inheritance consistent with simple *recessive*, *dominant*, or sex-linked behavior.

Mesoamerica
Literally "middle America"—collective term for a group of cultures in Central America.

mesocephalic
Having a *cephalic index* between 75 and 80—neither long (*dolichocephalic*) nor broad headed (*brachycephalic*).

Mesolithic
Archaeological cultures falling between the *Paleolithic* and the *Neolithic*.

messenger RNA, **mRNA**
An intermediate *RNA* molecule transcribed from a *gene* that is used as the template for the production of a *protein* through *translation*.

meta-analysis
The analysis of analyses; the statistical analysis of a large collection of results from individual studies for the purpose of integrating the findings.

metabolic rate hypothesis
The suggestion that different rates of sequence change among evolutionary lineages result from differences in the metabolic rates of those lineages.

metacentric
Of a *chromosome*: having the *centromere* close to the middle.

metagenomic
Of a sample or analysis: comprising more than a single species' *genome*, for example, the human genome, but also genomes of other species such as bacteria and fungi. Applies to many *ancient DNA* analyses.

metaphase
A stage of cell division in which *chromosomes* are in a condensed state and aligned at the cell center prior to separation.

metaphase plate
A region in the center of the cell in which the *chromosomes* align after *replication* and prior to cell division in *mitosis*.

metaphase spread
The condensed *chromosomes* from a single cell at *metaphase* displayed on a slide.

metapopulation
A group of *populations* connected by *migration*.

methylation
See *DNA methylation*.

mg
Milligram.

MHC
See *major histocompatibility complex*.

microarray
A collection of microscopic *DNA* or *RNA* spots on a solid surface used for parallel processing of a large number of reactions, most commonly for the purpose of genotyping or gene expression analyses. Also known as a *chip*.

microbiome
The totality of microbes and their *genomes* in a particular environment, such as the human gut.

microblades
Small stone tools that are produced by chipping silica-rich stones.

microcephaly
A disorder in which the head is more than two standard deviations smaller than the average for age- and sex-matched individuals.

microevolution
The processes of evolutionary change operating within species, at the population level.

microhomology
Very limited (a few bp) sequence *homology* found at the breakpoints of some *double-strand break* repairs carried out by *nonhomologous end-joining*.

Micronesia
The group of small coral atolls in the Pacific that lie to the north of *Melanesia*.

microRNA, **miRNA**, **miR**
A short (22-nucleotide on average) regulatory *RNA* molecule that binds to complementary sequences in target *mRNAs*, usually resulting in repression of *translation*, or mRNA degradation and hence *gene* silencing.

microsatellite
A *DNA* sequence containing a number (usually ≤50) of tandemly repeated short (2–6 bp) sequences, such as $(GAT)_n$. Often polymorphic, and also known as a *short tandem repeat* (*STR*).

Middle Paleolithic, **MP**
The period between the *Lower Paleolithic* and *Upper Paleolithic* in Europe and Asia, defined by its artifacts such as flake tools. These were produced by both *Neanderthals* and modern humans.

Middle Stone Age, MSA
The period between the *Acheulean* and *Later Stone Age* in Africa, defined by artifacts such as flake tools, some but not all made by the *Levallois technique*.

migration
The process of movement of a population (or individual) from one inhabited area to another.

migration drift equilibrium
A stable level of *population* subdivision that is reached when *migration* acting to homogenize *subpopulations* balances the *genetic drift* that differentiates them.

minimum spanning network
A *phylogenetic network* displaying the relationships of observed *haplotypes*.

minisatellite
A DNA sequence containing a number (~10 to >1000) of tandemly repeated sequences, each unit typically 10–100 bp in length. Sometimes hypervariable, and useful in *DNA fingerprinting*.

minisatellite variant repeat-PCR, MVR-PCR
A *PCR*-based method for assaying the positions of repeat units within *minisatellite* arrays that have variant *DNA* sequences.

minisequencing
The determination of a *SNP*'s allelic state by incorporation of a single fluorescent *dideoxynucleotide* at the SNP site.

mini-STR
In *forensic genetics*, an assay at a *microsatellite* (*STR*) designed to minimize *PCR amplicon* size; for use with degraded DNA.

minor allele frequency, MAF
Of a *SNP*, the frequency at which the less common *allele* is found within a given population.

miR, miRNA
See *microRNA*.

mismatch distribution
The frequency distribution of pairwise differences between a set of *DNA* sequences or *haplotypes*.

mismatch repair
A process that replaces a mispaired *nucleotide* in a *DNA* duplex with the correct base.

missense mutation
See *nonsynonymous mutation*.

missing heritability
The difference between the *heritability* of a trait determined from genetic variation in families, and the amount of variance in the trait explained by genetic variants found, for example, in *genomewide association studies*.

mitochondrion (plural **mitochondria**)
A cellular *organelle* containing the molecular apparatus for several metabolic pathways, primarily concerned with energy generation; present in many (usually 1000s) of copies per cell, and contains its own circular *genome* (*mtDNA*). The organelle, and its genome, is maternally inherited.

mitosis
The series of events resulting in cell division; contrasted with *meiosis*.

ml
Milliliter.

ML
See *maximum likelihood*.

MNP
See *multinucleotide polymorphism*.

Mode 1 technology
See *Oldowan*.

Mode 2 technology
See *Acheulean*.

Mode 3
See *Levallois technique*.

Mode 4
Describing the tools of the *Upper Paleolithic* period.

model organism
A nonhuman species such as the mouse or zebrafish in which a characteristic of interest, such as a particular genetic variant, can be generated and its consequences investigated.

modifier gene
A *gene* that acts to modify the *phenotype* of a trait primarily caused by another gene.

molecular clock
The hypothesis that sequence evolution occurs at a sufficiently constant rate to allow divergence between two sequences to be accurately related to the time they split from a common ancestor.

molecular distance
A measure of evolutionary change between two *DNA* sequences or *haplotypes*.

molecular evolution
The study of changes to *DNA* through time.

molecular marker/polymorphism
A polymorphic *DNA* sequence deriving from a single *locus*, which can be used in *linkage analysis* or genetic diversity studies.

Mongoloid
One of the morphological types of human identified by early anthropologists, typified by populations of East Asia.

monogenic
Of a disease: due to mutant *alleles* in a single *gene*.

monophyletic
A group of lineages sharing a common ancestor to the exclusion of all others.

monozygotic, MZ
Of twins: formed from a single *zygote* that has split, and therefore genetically identical (as opposed to *dizygotic*, *DZ*).

Monte-Carlo methods
See *permutation test*.

Moore's law
A law formulated by Gordon E. Moore, describing a long-term trend whereby computing power doubles every two years.

morbidity
The disease state of an individual, or the incidence of illness in a population.

morphology
The characteristics of the form and structure of organisms.

most recent common ancestor, MRCA
See *time to most recent common ancestor, TMRCA*.

Mousterian
A Middle Paleolithic culture using tools made by the *Levallois technique*; mainly, but not exclusively, associated with *Neanderthals*.

Movius Line
A boundary running across Asia, roughly from the Caucasus Mountains to the Bay of Bengal, distinguishing areas to the south and west where *bifaces* were common from those to the north and east where they are rare; first proposed by the anthropologist Movius.

MP
See *maximum parsimony*, or *Middle Paleolithic*.

m_R
Estimator of admixture.

MRCA
See *most recent common ancestor*.

mRNA (messenger RNA)
The *RNA* product of *transcription* from a *gene*, and the template for *translation*, the production of a protein in the *ribosome*.

MS
See *mass spectrometry*.

MSA
See *Middle Stone Age*.

MSY, male-specific region of the Y chromosome
See *nonrecombining portion of the Y*.

mtDNA
The circular *genome* carried by the *mitochondrion*.

Muller's ratchet
The unidirectional accumulation of deleterious *mutations* by nonrecombining sequences as a result of the periodic elimination by *genetic drift* of the fittest *haplotype*.

multiallelic polymorphism
A *polymorphism* possessing more than two *alleles*, for example, a *microsatellite*.

multi-dimensional scaling
A type of multivariate analysis which allows multi-dimensional information to be displayed graphically (usually in two dimensions) with minimum loss of information.

multifactorial disorder
See *complex disorder*.

multifurcation
See *polytomy*.

multigenic
Involving the influence of many *genes*.

multinucleotide polymorphism, MNP
Polymorphism in which two or three adjacent *base pairs* are substituted with respect to a reference *genome*.

multiple displacement amplification, MDA
A non-*PCR*-based method for *whole-genome amplification*, employing degenerate primers and a special polymerase from a bacterial virus.

multiple testing problem
A statistical issue arising when the number of tests being carried out is so large that one or more of them is expected to reach a predetermined level of significance by chance.

multiplex
Of *PCR*, the simultaneous amplification of several loci using multiple *primer* pairs in a single reaction.

multiregional hypothesis/model
The proposal that modern human characteristics arose in many different parts of the world, possibly at many different times; contrasted with the *out-of-Africa* and *leaky replacement* models.

multivariate analyses
A class of statistical methods that extract information from multi-dimensional data.

mutagen
A chemical or physical cause of *mutation*.

mutation
Any change in a *DNA* sequence (usually with the exception of changes caused by *crossing over*).

mutation–drift balance
See *mutation–drift equilibrium*.

mutation–drift equilibrium
A stable level of genetic diversity reached when the rate at which new variants are introduced by *mutation* is balanced by their loss due to *genetic drift*.

mutation dropping
A procedure for placing *mutations* on the branches of a modeled *gene genealogy* in *coalescent* analysis.

mutation pressure
The decline in frequency of a *haplotype* as a result of *mutations* creating new haplotypes.

mutation rate
The rate (given as per generation, per year, or per thousand years) at which a particular *mutation* or mutational type occurs; given the Greek character μ (mu).

mutation–selection balance
A concept in population genetics where the rate of formation of new *alleles* by *mutation* is balanced by the removal of alleles by *purifying selection*.

mutational bias
A pattern of *mutation* that results in an unequal accumulation of certain *nucleotides*.

mutational target
A concept in disease genetics which refers to the number of *DNA bases* that could be potentially *mutated* to generate a new mutant *allele* that causes a certain disease.

MVR-PCR
See *minisatellite variant repeat-PCR*.

MY
Million years.

MYA
Million years ago.

Myers motif
See *hotspot motif*.

MZ
See *monozygotic*.

N_e
See *effective population size*.

***N*-island model**
A simple model of *population structure* in which two populations of identical size exchange alleles with equal probabilities.

Na-Dene
One of the three groups of Native American languages proposed by Joseph Greenberg.

NADP, nicotinamide adenine dinucleotide phosphate
See *NADPH*.

NADPH
The reduced form of *NADP*, a biochemical compound used in the synthesis of, for example, nucleic acids, and in the oxidation–reduction reactions involved in protecting against the toxicity of reactive oxygen species.

NAHR
See *non-allelic homologous recombination*.

narrow-sense heritability
Only that proportion of genetic variance that can be attributed to additive genetic effects.

natural selection
The differential contribution of individuals to the next generation on the basis of their ability to survive and reproduce.

NCBI
National Center for Biotechnology Information.

ncRNA
See *noncoding RNA*.

NDNAD
National DNA Database.

Neanderthal
A group of extinct humans who lived in Europe and West Asia between about 250 and 28 KYA. Also sometimes spelled "Neandertal."

Near Oceania
An area within the Pacific Ocean that encompasses islands first settled over 25 KYA.

nearly neutral theory
A population genetic hypothesis that suggests patterns of genetic diversity are best explained if *nonsynonymous* mutations are slightly deleterious.

nebulization
The use of compressed air to shear *DNA* into short fragments prior to *next-generation sequencing*.

negative epistasis
The *epistasis* interaction of *alleles* of different *genes*, having a negative effect on the trait of interest.

negative selection
A selective regime in which new *mutations* are almost exclusively deleterious and are removed from the population.

Negrito
Describing a group of dark-skinned *populations* of small stature that live in Oceania and Southeast Asia; for example, Andaman Islanders.

neighbor-joining, **NJ**
A fast method for phylogenetic reconstruction.

Neolithic
The New Stone Age; originally defined by the use of ground or polished stone implements and weapons; other characteristic features of the Neolithic are now taken to include the manufacture of pottery and the appearance of settlements.

neotenic ape
Theory according to which humans retain juvenile features of other apes.

neoteny
The retention of juvenile features in adults; a feature selected for in domesticated animals.

nephrotoxin
A toxin that particularly affects the kidney.

nested model
A concept in statistics where the simpler model can be obtained directly from the more complex model by constraining its parameters.

network
See *phylogenetic network*.

neutral
Of a *mutation* or *allele*, having no effect on selective fitness.

neutral allele
An *allele* that does not affect the fitness of the carrier.

neutral parameter
See *population mutation parameter*.

neutral theory
See *neutral theory of molecular evolution*.

neutral theory of molecular evolution
The theory that the majority of *mutations* do not influence the fitness of their carriers.

neutral variation
Genetic variation that has no effect on fitness, and hence unaffected by *natural selection*.

neutrality test
A statistical method for exploring whether observed genetic diversity is compatible with *neutral* evolutionary processes.

next-generation sequencing, **NGS**
A set of diverse sequencing technologies united by their development after *Sanger* or *capillary sequencing*, usually producing more data at lower cost; also known as *second-generation sequencing* and *massively parallel sequencing*.

ng
Nanogram.

NGS
See *next-generation sequencing*.

NJ
See *neighbor-joining*.

nm
Nanometer.

node
The position of a *taxon* within a *phylogeny*, at the end of an evolutionary *branch*.

nomenclature
A naming system used to classify diversity for facilitating communication.

non-allelic homologous recombination, **NAHR**
Recombination occurring between sequences that are very similar in sequence, but are not allelic (*paralogs*).

noncoding RNA, **ncRNA**
An *RNA* molecule that functions in its own right, without being translated into *protein*; includes molecules with key functions in *translation*, such as *transfer RNAs* and *ribosomal RNAs*, and regulation, such as *microRNAs*.

nonconservative mutation
A *mutation* causing a *codon* to be replaced by another codon encoding an *amino acid* whose physical properties are different from those of the original amino acid. Often has a serious effect on *protein* function.

nondisjunction
The failure of a *chromosome* pair to segregate correctly at *meiosis*: both chromosomes move to one pole of the cell, rather than one to each. Results in *aneuploidy*.

nonhomologous end-joining, **NHEJ**
A mechanism used to repair *double-strand breaks*, in which breakpoint repair either involves no sequence *homology*, or just a few *base pairs* of homology (*microhomology*).

nonparametric linkage analysis
A method of *linkage analysis* for complex diseases, where there is no model specifying how the disease is inherited, aiming to identify chromosomal regions more likely to be shared by family members with, than without, the disease.

nonrecombining loci
Segments of the *genome* in which no *recombination* events have occurred since the most recent common ancestor of all extant copies.

nonrecombining portion of the Y
The part of the *Y chromosome* that escapes from meiotic recombination; sequences in this region are truly male-specific (sometimes abbreviated to NRY, or NRPY). See also *MSY*.

nonsense-mediated decay
A *protein* quality-control mechanism, triggered by a *nonsense* variant, in which a defective mRNA transcript is degraded before it can be translated.

nonsense mutation
A *mutation* that occurs within a *codon* and changes it into a *stop codon*.

non-shattering
The *phenotype* of the *rachis* of domesticated plants, in which the seeds remain attached after ripening.

nonsynonymous mutation
A *base substitution* in a *gene* that results in an *amino acid* change.

NOR
Nucleolar organizer region.

novelty-seeking
A personality trait associated with exploratory activity in response to novel stimuli, and impulsiveness.

NRPY, **or NRY**
See *nonrecombining portion of the Y*.

N-terminus
The end of a *protein* bearing a free amino (–NH_2) group, corresponding to the 5′-most translated region of a *gene*.

nuclear genome
The *autosomes* and sex chromosomes, residing in the *nucleus*; excludes *mtDNA*.

nucleosome
A *DNA–protein* complex made up of an octamer of *histones*, around which *DNA* is coiled in *chromatin*.

nucleotide
The monomeric molecular component of the polymers *DNA* or *RNA*; sometimes used as a shorthand for *deoxyribonucleotide*, the specific building block of DNA.

nucleotide diversity
A measure of genetic diversity based on the probability of two randomly chosen sequences from within the population having different *bases* at the same *nucleotide* position; given the Greek character π (pi).

nucleotide excision repair
A versatile repair pathway, involved in the removal of a variety of bulky *DNA* lesions.

nucleus
The large body within the *cytoplasm* of a *eukaryotic* cell that contains the *chromosomes*.

null
Of an *allele*: undetectable in an assay, either because of *deletion* or a *variant* in a site (e.g. a *primer*-binding site) necessary for allele detection.

numt
An insertion of *mtDNA* sequence within the *nuclear genome* (nuclear mtDNA insertion).

nystagmus
Involuntary and rapid movement of the eyeball.

obsidian
A volcanic rock resembling glass that has often been used for making stone tools.

OCA
Oculocutaneous albinism.

odds ratio
A measure of the size of effect, often used in complex disease genetics to quantify the contribution of a single *allele* to a complex disease.

Oldowan
The oldest assemblage of archaeological remains recognized. *Lower Paleolithic*, dating back to 2.6 MYA, and named after the Olduvai Gorge in Tanzania, East Africa; also called *Mode 1 technology*.

oligonucleotide primers
Short (typically 18–24 base), single-stranded, chemically synthesized *DNA* molecules, usually of specific sequence and used in opposing pairs to prime synthesis of a specific DNA target in the *polymerase chain reaction* (PCR).

oligonucleotide probes
A collection of specific *oligonucleotides*, often immobilized on a solid support as an *array*, for detection of specific *DNA* sequences by *hybridization*.

onomastic
Evidence based on the study of names of people and places.

oocyte
Female gametocyte, immature egg cell.

oogenesis
Development of the egg cell.

oogonium
Primordial egg cell of a female fetus (plural oogonia).

open reading frame, **ORF**
A *DNA* sequence in which there are no *stop codons* in any of the three *reading frames* on either strand.

optically stimulated luminescence, **OSL**
A method of measuring ionizing radiation that can be used to estimate the time when a particular grain of a mineral was last exposed to the sunlight.

optimality criteria
A set of rules that allow a single best option to be chosen from a number of possibilities.

oral histories
The collection of stories about the past handed down by word of mouth between generations.

ORF
See *open reading frame*.

organelle
A discrete substructure within a cell, performing a specific function (for example, the *mitochondrion* or the *nucleus*).

ortholog
A sequence that is *orthologous* to another.

orthologous
Homologous sequences that have diverged since splitting from their common ancestor as a result of a speciation event.

OTU
Operational taxonomic unit.

OSL
See *optically stimulated luminescence*.

osteology
The scientific study of bones.

osteomalacia
An adult disease in which bones become softened, due to a deficiency of vitamin D (see *rickets*).

Ötzi
The Tyrolean Iceman, discovered in 1991 in a naturally mummified state in the ice of a glacier of the Ötztal Alps on the Austrian-Italian border, ~5300 years after his death.

out-of-Africa hypothesis
A model of human origins in which modern humans arose recently (~200 KYA) in Africa and migrated into the Old World, replacing existing archaic humans there without interbreeding.

outbreeding
Reproduction involving distantly related parents.

outbreeding depression
A reduction in fitness of offspring resulting from the mating of genetically distinct parents.

outgroup
The *taxon* from a group of taxa that is known to diverge earliest.

outlier analysis
A test for selection in which the value of a test statistic for a candidate gene is compared to empirical distribution of the statistic from the whole genome; an outlying observed value in the extremes of the distribution (e.g. the top or bottom 1% or 5%) can be regarded as statistically significant.

overdominant selection
A selective regime under which the *heterozygote* is the fittest genotype (see *heterozygote advantage*).

OWM
Old World Monkey.

paired-end sequencing
A common design in *next-generation sequencing* in which the two ends of each DNA molecule are sequenced, but the middle may not be.

pairwise
A comparison between two entities.

pairwise distance
The *genetic distance* between two *populations*, or two molecules.

pairwise sequentially Markovian coalescent analysis, **PSMC**
A *coalescent*-based analytical approach used to infer *demographic history* from a single *diploid genome*.

paleoclimatology
Study of climates in prehistory.

paleoecological
Relating to the study of climates in prehistory.

Paleoindians
Native American peoples prior to ~9 KYA.

Paleolithic
The Old Stone Age, defined by archaeological remains such as stone tools; subdivided according to time and geographical region.

paleontology
Study of fossils and prehistoric life.

paleopathology
Study of the signs of disease preserved in fossils.

palimpsest
Literally, a reused manuscript in which an original text has been over-written; often used as an analogy for the complex genetic record of past events.

panmixis (panmictic)
Random mating throughout the entire range of a population.

Papuan
The group of non-Austronesian languages spoken in and around the island of New Guinea, also used to refer to the inhabitants of Papua New Guinea in the eastern half of New Guinea.

PAR1
Pseudoautosomal region 1.

PAR2
Pseudoautosomal region 2.

paracentric
On one side of a *centromere*; for example, a paracentric *inversion* has both breakpoints on one chromosomal arm.

paragroup
In a phylogeny, a set of lineages defined by the sharing of the derived state of a deep-rooting variant and ancestral states at the more derived variants; a new variant may be found that is carried by some but not all these lineages.

parallelism
See *recurrent mutation*.

paralog
Highly similar non-allelic sequences resulting from a *duplication* event; includes *segmental duplications*.

paralog ratio test, PRT
A method that amplifies by *PCR* a sequence from a *copy-number variable* (CNV) region, and at the same time a sequence of a slightly different size from a non-CNV *paralogous* region as a control, allowing the ratio of the two to be determined.

paralogous sequence variant, PSV
A difference in *DNA* sequence between *paralogs*.

parameter
A numerical characteristic of a *population* that is typically unknown and therefore estimated by taking a sample from the population. A quantity used in a model (for example, mean) that can be calculated from data.

paraphyletic
A grouping that shares a common ancestor to the exclusion of many other lineages but does not include all descendants of that common ancestor.

parenchyma
The storage material of tubers and rhizomes.

p arm
The short ("*petit*") arm of a *chromosome*.

parsimony
The principle that the best explanation is that which requires the least number of causal factors.

pastoralism
The herding of animals; often a nomadic practice, as animals are followed from one region to another as conditions dictate.

paternal age effect
The increase in the rate of some kinds of *mutations* seen with increasing paternal age.

paternity index
In *paternity testing*, the likelihood ratio of the hypothesis that the alleged father is indeed the true father, to the alternative hypothesis that the father is some unknown, unrelated man; the higher its value, the more likely is true paternity.

paternity testing
Determining whether or not a particular man is the father of a child, using *DNA* analysis.

pathogenic
Relating to the causation of disease.

patriline
All male-line relatives of a man—they will share his Y chromosome *haplotype*.

patrilocality
The anthropologically defined practice whereby men tend to remain closer to their birthplace upon marriage than do women; opposite of *matrilocality*.

Patterson's *D*
A test statistic devised to analyze the relationship of ancient *genomes* to modern human genomes, but applicable to any four-way genome comparison.

PC
Principal component.

PCA
See *principal component analysis*.

PCR
See *polymerase chain reaction*.

PCR stutter
A phenomenon in the *PCR* amplification of *microsatellites* in which, as well as the true repeat array, the *polymerase* produces arrays that are one (or more) repeat units shorter or longer, due to replication slippage.

penetrance
The frequency with which a person carrying a particular *genotype* will manifest a disease.

peptide
A short *protein*, typically ten or fewer amino acids in length.

peptide bond
The bond between a pair of adjacent *amino acids* in a *protein*.

pericentric
Located close to the *centromere*, for example, a pericentric *inversion* has breakpoints near the centromere on both chromosomal arms.

perisylvian
A region of the cerebral cortex. The left perisylvian region encompasses Broca's and Wernicke's areas, associated with word formation and recognition respectively.

permutation test
A statistical test that assesses significance by randomizing the observed data many times, and comparing a test statistic calculated from these randomizations to the value of the test statistic calculated from the observed data.

personal genome/genomics
The genotyping or sequencing, and analysis, of the *genome* of an individual.

personalized medicine
Health care influenced by information, particularly genomic information, about an individual.

PFGE
See *pulsed-field gel electrophoresis*.

pg
Picogram.

phalanx
Here, a bone within the fingers or toes, notable as the source material for the *Denisovan* genome sequence.

pharmacogenetics
The study of *genetic* variation affecting therapeutic response to drugs.

pharmacogenomics
The study of *genetic* variation affecting therapeutic response to drugs at the whole-*genome* and *transcriptome* level.

phase
See *gametic phase*.

phenotype
The observable characteristics of a cell or organism.

phenylthiocarbamide, PTC
An organic compound that tastes either very bitter or neutral, depending on the *genotype* of the taster.

phoneme
The basic unit of sound in a language.

phylogenetic network
A graphical representation of evolutionary relationships that includes cycles or *reticulations*, and thus summarizes a collection of evolutionary trees.

phylogenetic reconstruction
See *phylogeny reconstruction*.

phylogenetics
The study of genetic diversity through the construction of evolutionary trees.

phylogenetic tree
See *phylogeny*.

phylogeny
A tree-like structure that represents evolutionary relationships among a set of *taxa*.

phylogeny reconstruction
The process of deducing the evolutionary tree underlying a set of data.

phylogeography
Analysis of the geographical distributions of different *clades* within a *phylogeny*.

physical anthropology
The study of humans through analysis of their physical, rather than social or cultural, characteristics.

physical map
In *genetics*, a map in which the relative positions of *polymorphisms* are defined by the physical distance in base pairs between them, rather than by *recombination* frequencies (a *genetic map*).

phytolith
Particles of silicon dioxide from leaves or fruits.

pidgin
A simplified new language that forms and changes at the contact zone of two or more languages.

pipeline
A set of procedures linked together to provide automation and efficiency, for example, a set of programs for the analysis of *next-generation sequencing* data.

plasmid
A circular *DNA* molecule maintained within a bacterial cell, and used as a *vector* for the cloning and propagation of specific DNA fragments.

pleiotropic, pleiotropism
A property of a gene that affects more than one *phenotype*.

Pleistocene
The geological epoch that lies between about 2 MYA and 10 KYA.

point estimate
Best single estimate of a *parameter* of interest, not including the uncertainty.

Poisson distribution
A statistical distribution that has the property that the mean equals the variance.

polarity
A difference between one end of a *DNA* or *RNA* molecule and another, provided by the asymmetry of the *sugar–phosphate backbone*.

poly(A) tail
A stretch of *adenine ribonucleotides* added after *transcription* to the 3′ end of most *mRNA* molecules.

polyadenylated
Of an *mRNA* molecule, carrying a tail of multiple adenosine *ribonucleotides* [*poly(A) tail*] at the 3′ end.

polyadenylation signal
A short sequence in a *pre-mRNA* (AAUAAA) that indicates that the molecule should be cleaved, and a *poly(A) tail* should be added.

polygenic adaptation
Positive selection acting on *standing variation* in multiple *variants* within different genes.

polyglutamine tract
A tract of adjacent glutamine *amino acids* in a *protein*, encoded by a CAG *tandem* repeat within the coding region of a *gene*; prone to pathogenic expansions, as in *Huntington's disease*.

polygyny
The practice of males taking multiple wives.

polymerase
An *enzyme* responsible for the synthesis of nucleic acids from their component *nucleotide* triphosphates; DNA polymerase replicates *DNA*, while RNA polymerase produces *RNA*.

polymerase chain reaction, PCR
The exponential amplification of a specific *DNA* sequence using specific *oligonucleotide primers*, a thermal cycling protocol, and a thermally stable DNA *polymerase*.

polymorphism
The existence of two or more variants (of *DNA* sequences, *proteins*, *chromosomes*, *phenotypes*) at significant frequencies in the population. For DNA: any sequence variant, or more properly any sequence variant at ≥1% frequency in a given population; see also *variant*.

Polynesia
A geographical area of the Pacific defined by a triangle with apices at *Aotearoa* (New Zealand), Hawaii, and *Rapanui* (Easter Island), containing islands inhabited by speakers of Polynesian languages.

Polynesian motif
A *haplotype* comprising four variants in *HVSI* of *mitochondrial DNA* found at highest frequencies in Polynesian populations.

polypeptide
A *protein*, *amino acids* joined by peptide bonds; sometimes implies a short protein.

polyphyletic
A grouping of evolutionary lineages that derive from many different ancestors and so do not share a common ancestor to the exclusion of any other lineages.

polyploidy
Having more than two copies (*diploidy*) of the *genome* per cell. Lethal in humans, but well tolerated in some plant species.

polytomy
An evolutionary split in which more than two daughter lineages derive from a single ancestor; also *multifurcation*.

population
A group of individuals that may be defined according to some shared characteristic which may be social or physical. Sometimes used in a theoretical sense to mean a group of individuals in which there is random mating.

population differentiation
The process by which *allele* frequencies in two or more populations diverge over time.

population genetics
The study of genetic diversity in *populations* and how it changes through time.

population mutation parameter (θ, or theta)
A fundamental parameter of *population genetics* that encapsulates the expected level of genetic diversity in a randomly mating, constant-sized population not subject to *selection* when an equilibrium is reached between *genetic drift* and *mutation*; also *neutral parameter*.

population recombination parameter (ρ, or rho)
A fundamental parameter of population genetics that encapsulates the expected level of *linkage disequilibrium* in a randomly mating, constant-sized population not subject to *selection* when an equilibrium is reached between *genetic drift* and *recombination*.

population stratification
See *population structure*.

population structure
The absence of random mating within a population, often taken to mean that the population can be more accurately represented as being a *metapopulation* comprising several *subpopulations*.

population subdivision
See *population structure*.

positional cloning
The isolation of a *gene* based on its position in the *genome*, as determined by *physical* and *genetic mapping* approaches.

positive selection
A selective regime that favors the increase in frequency and sometimes *fixation* of an *allele* that increases the *fitness* of its carrier.

post-cranial
Collective term for bones other than the *cranium* (skull excluding lower jaw).

posterior distribution; also **posterior probability distribution**
A concept in *Bayesian* statistics describing the distribution of probabilities of a certain event after considering the evidence gathered from an experiment.

pre-mRNA
The primary *RNA* product of *transcription*, before the *splicing* out of *introns* and other post-transcriptional modifications.

prepared core technology
A method of producing stone tools in which either a central core is shaped to fit a final purpose, or flakes are removed from the core for use in their own right, used after ~200 KYA.

prevalence
The proportion of people who have a characteristic such as a disease at any one time.

Primates
A mammalian order containing simians and prosimians, and including humans.

primer
See *oligonucleotide primer*.

primitive
Of a trait, present in the ancestors of a species or a group of species.

principal component analysis, PCA
A type of *multivariate analysis* that allows multi-dimensional information to be displayed graphically with minimum loss of information.

prion
A misfolded *protein* that can act as the infectious agent of a disease, such as *kuru*.

prior probability distribution
In *Bayesian* statistical inference, the distribution of probabilities of an outcome before the data are taken into account; often simplified to 'prior.'

probability of identity, PI
In *forensic genetics*, the chance of two individuals selected at random having identical *DNA profiles*.

probe
A specific *DNA* sequence, typically 200 bp to a few kilobases long, that can be labeled and used to detect specific target sequences in *hybridization*, for example, of a *Southern blot*; alternatively, a *nucleotide* sequence attached to a solid support as part of a DNA *chip*; alternatively, an *oligonucleotide* used in *sequence capture* experiments.

product rule
In *forensic genetics*, a principle followed in calculating *match probabilities*, where *genotype* frequencies at each *microsatellite* making up a *DNA profile* are multiplied together.

prokaryotic
Describing an organism (the Bacteria and Archaea) whose cells lack membrane-bound *organelles*, particularly a *nucleus* (as opposed to *eukaryote*).

Promethease
An open-source genotype analysis tool.

promoter
The regulatory region, located 5′ to a *gene*, containing sequences necessary for *transcription* to be initiated.

prosecutor's fallacy
A fallacy of statistical reasoning made in a court of law, which fails to compare the likelihood of observing the evidence given that the defendant is guilty, to the likelihood of observing the evidence given that someone else is guilty.

protein
A large biomolecule consisting of one or more chains of *amino acids*.

protein electrophoresis
The separation of *proteins*, including variants, by *electrophoresis*; used to identify some *classical polymorphisms*.

proteome
The entire set of *proteins* expressed by a *genome*, cell, tissue, or organism.

proto-Indo-European
The reconstructed language from which modern Indo-European languages are supposed to have developed.

proto-language
A reconstructed language ancestral to a group of extant languages (for example, *proto-Indo-European*).

proximal
Toward the *centromere* of a *chromosome* arm (as opposed to *distal*).

PRT
See *paralog ratio test*.

pruning
A process to remove reticulations in a phylogenetic network.

pseudoautosomal
Of a region of the *sex chromosomes*: displaying inheritance from both parents. These regions (PAR1, PAR2), at the tips of the *X* and *Y chromosomes*, are the only segments of the Y that undergo *crossing over*.

pseudogene
A nonfunctional *DNA* sequence that shows a high degree of similarity to a non-allelic *homologous gene*.

PSMC
See *pairwise sequentially Markovian coalescent analysis.*

PSV
See *paralogous sequence variant*, above.

psychometric
Relating to psychometry, the measurement of mental states or capacities.

publication bias
A bias in the scientific literature in which negative results tend not to be reported.

pulsed-field gel electrophoresis, PFGE
A method for separating large (up to *megabase*-scale) molecules of *DNA* by periodically alternating the direction of the electric field during *electrophoresis*.

purifying selection
See *negative selection*.

purine
A class of *base* containing two closed rings, including *adenine* and *guanine*.

pyrimidine
A class of *base* containing one closed ring, including *uracil*, *thymine*, and *cytosine*.

pyrosequencing
A *DNA sequencing* method that depends on detecting the release of pyrophosphate during synthesis; the basis of the 454 *next-generation sequencing* technology.

q arm
The long ("*queue*") arm of a *chromosome*.

qPCR
See *quantitative PCR*.

QTL
See *quantitative trait locus*.

quality score, Q
A measure of the quality of a base call in a sequencing reaction. Its scale is logarithmic, so that Q20 indicates a 1 in 100 chance of error, while Q30 indicates a 1 in 1000 chance.

quantitative genetics
A discipline used in the identification of *quantitative trait loci* contributing to complex traits. It aims to subdivide the total variance in a *phenotype* into its genetic and environmental variance components.

quantitative PCR, qPCR
A *PCR*-based method that monitors the accumulation of a specific PCR product in real time, measured against an independent control sequence. Can be used for *CNV* analysis.

quantitative trait
A trait (such as human height) showing continuous variation in a population, and influenced by variation in many *genes*.

quantitative trait locus, QTL
A polymorphic genetic locus identified through the statistical analysis of a continuously distributed trait (such as height). These traits are typically affected by more than one *gene*, as well as by the environment.

quorum-sensing
A system used by a bacterial population that modulates behavior based on the local population density of bacteria of the same, or different, species.

rachis
The "backbone" of an ear of seeds, in wheat, for example. Fragile in wild wheats when ripe, giving a *shattering* phenotype, but toughened in domesticated wheats to facilitate harvesting.

racism
The mistaken belief that humans are divided into distinct categories, "races," usually distinguished by a few *phenotypic* characteristics such as skin color; often associated with discriminatory behavior toward those considered to belong to other "races."

radiocarbon dating
A method of dating carbon-bearing materials on the basis of the decay of the carbon-14 isotope; suitable for ages up to ~50 KYA.

radioisotope
A form of an element with an unstable nucleus that decays, with the emission of radioactivity, for example, phosphorus-32.

random genetic drift
See *genetic drift*.

Rapanui
The Polynesian name for Easter Island, meaning "big Rapa."

rapidly mutating STRs, RMSTRs
In *forensic genetics*, *microsatellites* (STRs) with particularly high *mutation rates*, giving very diverse *DNA profiles*.

rCRS
See *revised Cambridge Reference Sequence*.

rDNA
Ribosomal DNA.

read
See *sequence read*.

read-depth
The number of *sequence reads* covering a position or region in a *genome*; equivalent to *coverage*.

read length
In DNA sequencing (and particularly *next-generation sequencing*), the number of successive *nucleotides* that can be reliably read in a single sequencing reaction.

reading frame
In *translation*, the means in which the continuous sequence of *mRNA* is read as a series of triplet *codons*.

recessive
An *allele* that will not manifest in a *phenotype* unless both alleles at a *diploid* locus are mutated (*homozygous*).

recombination
Exchange of *DNA* between members of a *chromosomal* pair, usually in *meiosis*.

recombination–drift equilibrium
The balance reached when the rate at which *genetic drift* removes *haplotypes* from the population is matched by the rate at which *recombination* generates new haplotypes.

recombination hotspot
A short (few kb) region of the *genome* in which *recombination* is significantly elevated over the genome average.

recreational genomics
See *personal genomics*.

recurrent mutation
A *mutation* that independently generates a derived state previously observed within the population.

reduced median network
A type of *phylogenetic network*.

redundant
Of the *genetic code*: the property that some *amino acids* are encoded by more than one *codon*.

reference sequence
One copy of the sequence of a species or strain, often the first to be determined, that is used as the standard. Reference sequences often contain gaps and errors, and are periodically revised, so it is important to know which build has been used in any study.

RefSNP
See *rs*.

relative rates test
A simple statistical test to examine whether or not all lineages are evolving at the same rate.

relative risk
The risk (of developing a disorder, for example) for the individual in question compared with the risk for a randomly chosen individual.

Remote Oceania
An area within the Pacific Ocean that encompasses islands first settled within the past 4000 years.

repeat-masking
The identification and annotation of sequences that are repeated within a *genome*, often so that they can be excluded from analysis.

replication
The copying of a double-stranded *DNA* molecule to yield two double-stranded daughter molecules; alternatively, the repeating of a study such as a *GWAS* in an independent sample.

replication origin
Part of a *DNA* molecule in which DNA *replication* begins.

replication slippage
Errors in *DNA replication* in which the DNA *polymerase* "slips" in a repeated tract, leading to gains or losses of repeat units. Thought to be a major mechanism of *microsatellite* mutation.

reporter gene
A *gene* used to test the ability of a linked upstream sequence element (for example, an *enhancer*) to influence its expression.

reproductive variance
The variation in number of offspring produced by different members of a group of individuals.

resequencing
Taking a specific known sequence from an existing source, such as a database, or an entire *genome*, and determining it in several different individuals. A way to discover sequence variation.

restriction endonuclease
See *restriction enzyme*.

restriction enzyme
An *endonuclease*, usually isolated from a bacterium, that cleaves double-stranded *DNA* at a specific short sequence, typically 4–8 bp in length; also *restriction endonuclease*.

restriction fragment length polymorphism, **RFLP**
A *polymorphism* identified (typically in *hybridization* analysis after *Southern blotting*) by differences in the lengths of restriction fragments. Can be due to polymorphism in the restriction sites themselves, or variation in the length of a sequence between the sites.

reticulation
A closed loop observed within a *phylogenetic network* indicating the potential existence of *homoplasy*.

retroposon
See *retrotransposon*.

retrotransposon
A mobile *DNA* element that inserts into a genomic location after *transcription* into *RNA* from an active genomic copy, then *reverse transcription* into DNA; examples are *Alu* and *L1* elements.

retrovirus
A virus with an *RNA* genome, and a *reverse transcriptase* function, allowing the genome to be copied into *DNA* prior to insertion into the chromosomes of a host cell; for example, *HIV*.

REV
General reversible model.

reverse transcriptase
An *enzyme* that uses *reverse transcription* to make a *DNA* copy (a *cDNA*) of an RNA molecule; used by *retroviruses* prior to host genome integration, and encoded by some *L1 elements*.

reverse transcription
Production of *DNA* from an *RNA* template by the *enzyme reverse transcriptase*.

reversion
See *back mutation*.

revised Cambridge Reference Sequence, **rCRS**
The revised version of the *reference sequence* of human *mtDNA*.

RFLP
See *restriction fragment length polymorphism*.

rho
See *population recombination parameter*.

rho statistic, **ρ**
An estimate used in genetic dating based on *phylogenetic network* analysis: the mean number of mutations (or time, based on the number of mutations) to the root of a set of *haplotypes* within the network.

ribonucleic acid, **RNA**
An informational macromolecule produced by *transcription* from a DNA template, composed of linked *ribonucleotides*. Usually single stranded.

ribonucleotide
The monomeric molecular component of the polymer *RNA*.

ribosomal RNA, **rRNA**
The *RNA* component of the *ribosome*; human ribosomes contain four rRNAs, and the *genes* encoding them are tandemly repeated in several clusters.

ribosome
A *cytoplasmic* multisubunit *protein–RNA* complex that performs the function of *translation* of *mRNA* into protein.

rickets
A childhood disease in which bones become softened, due to a deficiency of vitamin D.

risk ratio, **λ**; **also**, **relative risk**, **RR**
In epidemiology, the ratio of the risk of an event (such as a disease) occurring in one group (such as a group carrying a particular *allele*) compared to the risk of that same event occurring in another group. See also *odds ratio*.

RNA
See *ribonucleic acid*.

RNA polymerase
The *enzyme* that synthesizes *RNA* from a *DNA* template in the process of *transcription*.

robust
Of body form and skeletal remains, heavily built (as opposed to *gracile*).

Romanization
The replacement of many early European languages by languages related to Latin, due to the dominance of the Roman Empire.

rooted tree
A tree in which the *node* ancestral to all *taxa* is known.

rRNA
See *ribosomal RNA*.

rs, **RefSNP**
A number used to identify a particular *SNP*, for example, rs8179021.

R_{ST}
A statistic, analogous to F_{ST}, which assumes the *stepwise mutation model*, and is used to quantify *genetic distance*. Commonly used for *microsatellite* data.

Sahul
The land mass formerly formed at lower sea levels by the joining of the islands of Australia, New Guinea, and Tasmania.

Sanger sequencing
A method of *DNA* sequencing, invented by Fred Sanger,

involving the incorporation into a growing DNA strand of labeled *dideoxynucleotides*, which allow the detection of DNA fragments of different lengths, each one terminated at one of the four *bases*. Fragments are analyzed by *capillary electrophoresis*. Also known as *chain-termination sequencing*, *dideoxy sequencing*, and *capillary sequencing*.

Saqqaq
The name given to a 4500-year-old Greenlander from the Paleo-Eskimo Saqqaq culture whose *genome* was sequenced in 2010.

satellite
A large tandem-repeated *DNA* array spanning hundreds of *kilobases* to *megabases*, and composed of repeat units of a wide range of sizes that can display a higher-order structure. Some satellites (for example, at *centromeres*) are important functional components of *chromosomes*.

satellited chromosome
A chromosome carrying a cytogenetically visible *translocation* of material from the short arm of an *acrocentric chromosome* such as chromosome 15.

savanna/savannah
An ecosystem in which grassland predominates but trees may also be present.

seafloor topography
The geographical patterns of variation in sea depth.

second-generation sequencing
See *next-generation sequencing*.

secondary product revolution
Applications of domestic animals other than for food; for example, the use of hair and hides for clothing, and large animals for traction.

sedentism
Of a population: staying in one place (as opposed to nomadism or *pastoralism*).

segmental duplication
See *paralog*.

segregating sites
The *nucleotide* sites that are variable within a set of sequences.

segregation
See *chromosomal segregation*.

selection
See *natural selection*.

selection coefficient
A quotient used to compare the fitness of different *genotypes*.

selective sweep
A rapid increase in frequency of an advantageous *allele* and other *variants* linked to it.

sepsis
An over-reaction of the immune system, usually triggered by infection, and associated with organ dysfunction and a high mortality rate.

sequence alignment
The juxtaposition of a set of sequences such that *nucleotide* sites that derive from a common ancestor are aligned in a column.

sequence capture
The use of long (50–120-nt) *oligonucleotides* in solution to capture complementary single-stranded genomic *DNA* fragments for a specified region of the genome, which can then be used to construct a library for *next-generation sequencing*; also known as *target enrichment*.

sequence divergence
The number of changes that distinguish two or more *homologous DNA* or *protein* sequences.

sequence read
In *next-generation sequencing*, a short sequence derived from a sequenced sample that is usually mapped to a *reference sequence*; also simply *read*.

serial founder model
Of human expansion out of Africa: model in which a new population is established by a small number of individuals (founders), expands, and then the process is repeated, leading to a progressive loss of lineages.

serological
Pertaining to the study of blood serum.

sex-biased admixture
The generation of a hybrid population from two ancestral populations in which there are different contributions from males and females of either population.

sex chromosomes
The X and the Y *chromosomes*, the constitution of which differs between the sexes.

sex-reversal
Discordance of physical sex with *chromosomal* sex (for example, 46,XX maleness).

sex-specific admixture
The generation of a hybrid *population* from two ancestral populations in which the contribution from one of the ancestral populations comes entirely from a single sex.

sex test
A *PCR*-based test to determine the sex of a DNA donor, usually by amplifying part of the *amelogenin* gene on the *X* and *Y chromosomes*.

sexual dimorphism
Sex-specific differences in morphology.

sexual selection
A selective regime under which the characteristics that are selected for are those that enhance mate attractiveness or competitiveness.

SGM+
Second-generation multiplex, plus.

shattering
Release of individual seeds from the *rachis* of a plant (an ear of grass), for example, on ripening; characteristic of wild plants.

short tandem repeat, **STR**
See *microsatellite*.

shotgun sequencing
Analysis of a sequence by the generation and assembly of many short *sequence reads*; applies to *next-generation sequencing* methods.

shoveled incisors (also shovel-shaped incisors)
A particular tooth morphology of the upper incisors. The tongue-sides of the central and lateral incisors are shaped like a shovel, with ridges on their mesial and buccal edges.

sibling risk ratio, λ_S
The probability (relative to the population level of risk) that, if a certain individual has a particular disease, their sibling will also have that disease.

sickle-cell anemia
A severe *autosomal recessive* anemia with characteristic "sickling" of red blood cells, caused by *homozygosity* for the *Hb^S allele*.

silent-site substitution
See *synonymous substitution*.

SINE, **short interspersed nuclear element**
A dispersed short repeated sequence of high copy number, such as an *Alu element*.

single-copy
A DNA sequence regarded as existing in only one copy per *haploid genome*.

single-locus profiling
The use of *DNA* probes specific for one genomic locus (single-locas probes, SLPs), to derive a *DNA profile*.

single nucleotide polymorphism, **SNP**
A *variant* due to a *base substitution* or the *insertion* or *deletion* of a single base. Although the term *polymorphism* strictly implies ≥1% *minor allele frequency*, many variants described as SNPs have much lower frequencies than this; see also *single nucleotide variant*.

single nucleotide variant, **SNV**
A frequency-independent term for a variant involving a single *nucleotide* (*base substitution* or *indel*); see also *single nucleotide polymorphism*.

single-step mutation model, **SMM**
A simple model of *mutation* of *microsatellites*, in which each mutation increases or decreases an *allele* by a single repeat unit, with equal probability.

single-stranded DNA
A *DNA* molecule that comprises just a single polynucleotide chain.

singleton
An *allele* or *haplotype* that occurs only once within the population.

sister chromatids
The two copies of each *chromosome*, associated at their *centromeres* after DNA *replication* during *meiosis* or *mitosis*.

site frequency spectrum
The frequency distribution within the population of *variants* from a given sequence; also *allele frequency spectrum*.

SIV
Simian immunodeficiency virus.

skipping
See *exon skipping*.

SLP
See *single-locus profiling*.

SMM
See *single-step mutation model*.

SNP
See *single nucleotide polymorphism*.

SNP chip
See *chip*.

SNPedia
An open-source *genotype* analysis tool.

SNV
See *single nucleotide variant*.

social selection
An increased probability of production and survival of offspring due to social factors, for example, wealth or status.

soft sweep
A *selective sweep* in which *positive selection* acts on standing variation (one or more *variants* that have existed in the *population* for some time).

Solexa sequencing
An old term for *Illumina*™ *sequencing*.

soma, **somatic cells**
The cells of the body other than the *germ line*; adjective: somatic.

somatic mutations
Mutations taking place in the *somatic* cells, and not contributing to evolutionary change.

Southern blotting
The technique of detecting specific restriction fragments of *DNA*. Digested and size-separated DNA is transferred from an agarose gel to a filter by blotting in a transfer solution; fragments are detected subsequently by *hybridization* using a labeled *probe*.

spatial autocorrelation
A statistical method that examines how the *autocorrelation* in *allele* frequencies between two populations depends on the distance between them.

species tree
A *phylogeny* that relates the evolutionary relationships among a set of species, where *bifurcations* represent speciation events.

spermatogenesis
The developmental process, involving *meiosis*, by which sperm are produced.

sphenoid
A small bone at the base of the *cranium* whose development is thought to have a major role in determining cranial morphology.

spindle fibers
Fibers within the cell that attach to the *centromere* to move apart the replicated *chromosomes* at cell division.

splice site
A short sequence in a *gene* required for splicing of *pre-mRNA* into mature *mRNA*.

splicing
The process of removal of *intronic* sequences from a pre-*messenger RNA* molecule, and the linking together of *exonic* sequences to form a transcript containing a single unbroken *reading frame* for *translation*.

split decomposition
A method allowing the display of incompatible and ambiguous *links* in a *split network*.

split network
See *split decomposition*.

sporadic
Of a genetic disease, occurring in a patient without a family history of disease.

spread zone
A large geographical region in which one *language family* dominates.

SRY
The testis-determining *gene* on the Y chromosome—sex-determining region, Y.

standing variation
Variants that have existed in the *population* for some time.

star (*or* star-like) phylogeny
A special *topology* of a *phylogeny* in which each extant *taxon* is derived independently from the common ancestor of all taxa.

steppe
An environment with abundant grassland and few trees.

stepping-stone model
A model of population structure in which *gene flow* can only take place between neighboring *subpopulations*.

stepwise mutation model, **SMM**
A simple model of *microsatellite* evolution in which the length of the microsatellite varies by single units at a fixed rate independent of repeat length and with the same probability of expansion and contraction.

stochastic, **stochasticity**
The result of a random process, such that the outcome cannot be predicted precisely.

stop codon
A series of three adjacent *bases* in *messenger RNA* (and, by extension, in *DNA*) that instructs the protein *translation* machinery to stop translating—the end of the coding region of a *gene* (also termination codon).

stop mutation
See *nonsense mutation*.

STR
See *short tandem repeat*.

stratigraphy
Layers of sedimentary or volcanic rock or other deposits used in geology and archaeology for placing fossils or cultural material in relative time sequence.

streptavidin
A *protein* that binds tightly to *biotin*.

strontium isotope analysis
The strontium isotope composition (expressed as the $^{87}Sr/^{86}Sr$ ratio) of tooth enamel, reflecting local geology during childhood, and hence providing information about a skeleton's geographical origins.

structural variation
A variation in *genome* structure, usually defined as affecting more than 1 kb of *DNA* sequence. Includes *deletion*, *duplication*, *inversion*, and *copy-number variation*. Contrast with *single nucleotide variation*.

STRUCTURE
A software package for using multi-locus *genotype* data to investigate *population structure*.

structured coalescent
A variety of *coalescent* analysis that divides *alleles* into selected and non-selected alleles, and models each as a separate *gene genealogy*.

subpopulation
A randomly mating population that exchanges migrants with other populations to form a *metapopulation*.

sugar–phosphate backbone
The structural component of *DNA*, excluding the information-bearing *bases*.

summary statistic
A statistic that reduces complex data to a single value.

Sunda
The landmass formed at lower sea levels that encompassed many of the present-day islands of Southeast Asia.

supernumerary
Of a chromosome, an additional (often small) *chromosome* in excess of the normal 46; may have no phenotypic effect.

surface
A three-dimensional representation of how a single variable varies in two dimensions.

Swadesh word list
A list of 100 or 200 basic words, relatively resistant to word-borrowing, which are used to compare languages in *lexicostatistics*; named after US linguist Morris Swadesh.

synapomorphic
An *apomorphic* character state that is shared among at least two *taxa*.

synonymous substitution
A *base substitution* that replaces one *codon* with another that encodes the same *amino acid*.

synteny
The state of a set of genes being on the same *chromosome*.

synthetic map
A method for the geographical display of *principal component* information on *allele* frequencies from several *loci* simultaneously.

T
See *thymine*.

T cell
One of a group of lymphocyte types that play a central role in cell-mediated immunity. They carry T-cell receptors, and mature in the thymus.

tag SNP
A single *SNP* that alone defines most of the *haplotype* diversity of a *haplotype block*.

tagging
Of a *variant* (usually a *SNP*), in strong *linkage disequilibrium* with a disease *allele*.

Taiga
Cold coniferous forest covering large areas of northern Asia, Europe, and America.

Tajima's *D*
A test for *selection* that compares estimates of the *population mutation parameter* based on the number of *segregating sites* and *nucleotide diversity*.

tandem
Of a repeated sequence: indicates adjacent copies of repeat units, for example, *short tandem repeat*.

Taq
Thermus aquaticus.

target enrichment
See *sequence capture*.

tastant
Any substance capable of stimulating the sense of taste.

taxon
An evolutionary unit of investigation; plural, *taxa*.

Tay-Sachs disease
A severe autosomal recessive neurodegenerative disease.

TDT
See *transmission disequilibrium test*.

telomerase
A specialized *DNA polymerase*, which is active in the germ line and "regrows" the *telomere*, which loses repeat units from its end at every DNA *replication*.

telomere
A specialized *DNA–protein* structure at the tips of *chromosomes*, containing an array of short *tandem* hexanucleotide repeats (TTAGGG). Protects the chromosome from degradation and from fusing with other chromosomes.

template
A *DNA* molecule forming the substrate for the synthesis of another, in either cellular DNA *replication*, or *in vitro* DNA synthesis such as *PCR*.

teosinte
The wild progenitor of maize.

termination codon
See *stop codon*.

thalassemia
A group of mostly *autosomal recessive* anemias due to imbalances in the production of hemoglobin chains.

therapeutic index
Of a drug, the difference between the dose needed to achieve the desired effect and the toxic dose.

thermal cycler
A programmable heating block used to carry out *PCR*; also known as a PCR machine.

thermoluminescence, TL
Emission of light from some minerals on heating, due to the release of previously absorbed energy from electromagnetic radiation; used to date buried objects that have been heated in the past, including pottery.

theta, θ
See *population mutation parameter*.

third-generation sequencing
A new generation of diverse sequencing technologies that do not involve an initial *DNA* amplification step.

thrifty genotype hypothesis
A suggestion that limiting food resources in the past have favored *alleles* that promote efficient bodily storage of energy reserves; in conditions of plentiful nutrition, the *genotype* predisposes to non-insulin-dependent *diabetes mellitus*.

thrifty phenotype hypothesis
A suggestion that limiting maternal food resources cause physical and metabolic adaptations of the fetus that predispose to non-insulin-dependent *diabetes mellitus* in later life.

thymidine dimer
A chemically linked pair of adjacent *thymidine nucleotides* in DNA, often induced by UV radiation.

thymine, T
A *pyrimidine base* found within *DNA*; pairs with *adenine*.

time-of-flight, TOF
A detection system for ions in *mass spectrometry* that allows their mass to be calculated.

time to most recent common ancestor, TMRCA
The estimated time since a set of sequences in a genealogy most recently shared an ancestor (an *MRCA*).

TL
See *thermoluminescence*.

TMRCA
See *time to most recent common ancestor*.

tocopherol
A group of lipid-soluble molecules that is included within the term vitamin E.

TOF
See *time-of-flight*.

topology
The branching pattern of an evolutionary tree.

trans
A prefix of Latin origin, meaning "on the opposite side [as]" or "beyond"; a *trans*-acting factor or sequence exerts its influence from a different *chromosome* (as opposed to *cis*).

transcription
The production of an *RNA* molecule from a *gene* (*DNA*).

transcriptome
The combined variability of all transcripts produced by the *genome*, including the products of *alternative splicing* and many *noncoding RNA* species.

transfecting, transfection
The experimental introduction of *DNA* into cells.

transfer RNA, tRNA
An "adapter" molecule that can carry a specific *amino acid* and recognizes the appropriate *codon* in *messenger RNA*, thus allowing that amino acid to be incorporated into a growing *protein* molecule.

transformation
An *in vitro* process whereby cells taken from a primary tissue (for example, white blood cells) are "immortalized," so that they continue to divide indefinitely; alternatively, the uptake and incorporation of exogenous *DNA* by a cell.

transition
A *base substitution* in which a *pyrimidine base* (C or T) is exchanged for another pyrimidine, or a *purine* base (A or G) is exchanged for another purine; as opposed to *transversion*.

translation
The production of a *protein* from a *messenger RNA* molecule within the *ribosome*.

translocation
In genetics, the transfer of a segment of one *chromosome* into another; translocations can be reciprocal, or nonreciprocal.

transmission disequilibrium test, TDT
In its simplest form, a test for genetic association of an *allele* with a *phenotype* by testing for the over- or under-transmission of that allele from parents to children with or without that phenotype.

transnational isolate
An isolated population that has maintained its genetic distinctiveness despite being spread over a wide area.

transposition
The movement of a *DNA* sequence from one genomic location to another.

trans-species polymorphism
A *polymorphism* that has been maintained over a long period of time such that it is present in two species as a result of it being present in their common ancestor.

transversion
A *base substitution* in which a *pyrimidine base* (C or T) is exchanged for a *purine* base (A or G), or *vice versa*; as opposed to *transition*.

tree topology
The shape (branching pattern) of an evolutionary tree.

trichotomy
An internal node in a *phylogeny* from which three *taxa* descend.

trichromatic vision
Possession of three different types of color receptors in the retina of the eye.

trinucleotide repeat
A *microsatellite* composed of 3-bp repeat units; include pathogenic microsatellites prone to expansions, and causing diseases such as *Huntington's disease*.

trisomy 21
The state of having three copies of chromosome 21, which causes Down syndrome.

tRNA
See *transfer RNA*.

trypanosomiasis
A serious disease of cattle, caused by the trypanosome carried by the tsetse fly; also causes sleeping sickness in humans.

type 1 error
A false positive result; in *forensic genetics*, a false inclusion of a suspect.

type 2 error
A false negative result; in *forensic genetics*, a false exclusion of a suspect.

type specimen
The specimen used to delineate the defining characteristics of a type, for example, of a *hominid* species.

U
See *uracil*.

UEP
See *unique event polymorphism*.

ultraconserved element
A segment of *DNA* sequence that is much more highly conserved between the *genomes* of different, often distantly related species than the genome average, suggesting functional importance.

ultrametric tree
A *phylogeny* in which the summed branch lengths of every *taxon* to their common ancestor is equal.

Uluzzian
An early stage of the *Upper Paleolithic*.

unbiased estimator
A value that is corrected for sample size, for example, *heterozygosity*.

underdominant selection
A selective regime that favors either *homozygote* over the *heterozygote*. The heterozygote has the lowest fitness of all *genotypes*.

unequal crossing over
Recombination between repetitive *allelic* or non-allelic sequences that leads to *insertion* or *deletion* of repeats, or of *single-copy* sequences between repeats.

uniparental inheritance
Inheritance from one parent only; characteristic of *mtDNA* (from the mother) and the *Y chromosome* (from the father).

unique event polymorphism, UEP
A *polymorphism* representing a likely unique event in human history, such that all individuals sharing the *derived allele* share it by descent, rather than by state. All *Alu* insertion polymorphisms and many *SNPs* are examples of UEPs.

unlinked
Of sequences that are independently inherited, such as those lying on different *chromosomes*.

unrooted tree
A tree in which the *node* ancestral to all *taxa* is not known.

UP
See *Upper Paleolithic*.

UPGMA
Unweighted pair-group method with arithmetic mean.

Upper Paleolithic, **UP**
The latest part of the Paleolithic in Europe and Asia, defined by its cultural assemblages and roughly equivalent to the Later Stone Age in Africa; mainly associated with modern humans, but at least one Upper Paleolithic culture (*Châtelperronian*) is associated with *Neanderthals*.

uracil, **U**
A *pyrimidine base* found within *RNA*—analogous to *thymine* in DNA, it pairs with *adenine*.

UVR
Ultraviolet radiation.

variable number of tandem repeats polymorphism, **VNTR**
A *polymorphism* due to differing numbers of *tandemly* arranged repeat sequences; ranges from *satellites*, through *minisatellites*, to *microsatellites*. Sometimes used specifically to refer to minisatellites.

variable penetrance
Of a disease *allele*, conferring a range of probabilities of manifesting the disease.

variance
A statistic that describes how widely spread are a number of estimates of the same *parameter*, by comparing the difference between each estimate and the mean of all estimates.

variant
In our usage, any difference between *genome* copies with *allele* frequency $\leq 1\%$ (more frequent differences being *polymorphisms*); also used more loosely for any difference, regardless of frequency.

vector
An organism involved in the transmission of a pathogen between hosts, for example, the mosquito is a vector of malaria. Note that in molecular biology, the term refers to a *DNA* molecule such as a *plasmid* that is used to transfer *DNA* to host cells.

virtual heterozygosity
For a *haploid* system, the probability that two *haplotypes* drawn randomly from the population are different from each other; see *heterozygosity*.

virulence factor
A molecule produced by a pathogen that promotes infection, for example, by aiding colonization of the host, or evasion of its immune response.

VNTR
See *variable number of tandem repeats polymorphism*.

Wahlund effect
A deficiency of *heterozygotes* in a *metapopulation* due to *subpopulation* divergence.

walking distance
The distance between two points on the Earth that takes into account geographical barriers; as opposed to *great circle distance*.

Wallace line
The boundary delineated by Alfred Wallace that divides two regions with distinct fauna in Island Southeast Asia.

Wallacea
The biogeographically rich islands of Southeast Asia that lay between the *Sahul* or *Sunda* landmasses.

wave of advance
See *demic diffusion*.

WGA
See *whole-genome amplification*.

whole-genome amplification, **WGA**
Method for indiscriminate amplification of sequences from the entire *genome*; see also *multiple displacement amplification*.

wild-type
A non-mutant form of an *allele*, gene, cell, or organism.

Williams–Beuren syndrome
A rare neurodevelopmental disorder caused by a large *deletion* of the long arm of *chromosome* 7.

Wright–Fisher population model
A simple idealized population model that forms the basis for much population genetic analysis. The population has the properties of constant size, equal sex ratio, non-overlapping generations, and random mating and each individual has the same probability of contributing to the next generation.

X chromosome
One of the *sex chromosomes*, present in one copy in men, and two in women.

X-inactivation
The transcriptional inactivation of one of the two X chromosomes in the *somatic* tissues of female mammals.

X-linked
Of a *phenotype*, caused by an *allele* on the X chromosome, and therefore manifesting in males, but (in the case of *recessive* alleles) often not in females, who can be asymptomatic *carriers*.

xenobiotic
A chemical compound (for example, a drug) that is foreign to a living organism.

Y chromosome
One of the *sex chromosomes*, present only in men, and male-determining.

YA
Years ago.

YAC
See *yeast artificial chromosome*.

yeast artificial chromosome, **YAC**
A *DNA* vector in which inserts *megabases* in length can be propagated in yeast cells.

Younger Dryas event
A brief and rapid cooling during the general warming at the end of the *glacial period*; its end marks the beginning of the *Holocene*.

YRI
One of the *HapMap* population samples: 30 parent–child trios, Yorubans from Ibadan, Nigeria.

zebu
Indian cattle, *Bos indicus*.

zinc finger
A small protein motif containing multiple fingerlike protrusions that make tandem contacts with their target molecule. Some of these domains bind zinc, but many do not.

zoonosis
A disease of humans acquired from animals; pronounced "zoo-o-nosis."

zygote
The fertilized egg, from which an individual will develop.

INDEX

Note: abbreviations following page numbers are: B, box; F, figure; and T, table. Prefixes are ignored in the alphabetical sequence—thus α-Thalassemia will be found under the letter T.

C

D

E

F

H

I

J

K

T

U

V

W